# Lecture Notes in Computer Science 8392

*Commenced Publication in 1973*
Founding and Former Series Editors:
Gerhard Goos, Juris Hartmanis, and Jan van Leeuwen

### Editorial Board

David Hutchison, UK
Josef Kittler, UK
Alfred Kobsa, USA
John C. Mitchell, USA
Oscar Nierstrasz, Switzerland
Bernhard Steffen, Germany
Demetri Terzopoulos, USA
Gerhard Weikum, Germany

Takeo Kanade, USA
Jon M. Kleinberg, USA
Friedemann Mattern, Switzerland
Moni Naor, Israel
C. Pandu Rangan, India
Doug Tygar, USA

## Advanced Research in Computing and Software Science

Subline of Lectures Notes in Computer Science

### Subline Series Editors

Giorgio Ausiello, *University of Rome 'La Sapienza', Italy*
Vladimiro Sassone, *University of Southampton, UK*

### Subline Advisory Board

Susanne Albers, *University of Freiburg, Germany*
Benjamin C. Pierce, *University of Pennsylvania, USA*
Bernhard Steffen, *University of Dortmund, Germany*
Deng Xiaotie, *City University of Hong Kong*
Jeannette M. Wing, *Microsoft Research, Redmond, WA, USA*

Alberto Pardo   Alfredo Viola (Eds.)

# LATIN 2014: Theoretical Informatics

11th Latin American Symposium
Montevideo, Uruguay, March 31 – April 4, 2014
Proceedings

 Springer

Volume Editors

Alberto Pardo
Universidad de la República
Facultad de Ingeniería
Instituto de Computación
Julio Herrera y Reissig 565
11300 Montevideo, Uruguay
E-mail: pardo@fing.edu.uy

Alfredo Viola
Universidad de la República
Facultad de Ingeniería
Instituto de Computación
Julio Herrera y Reissig 565
11300 Montevideo, Uruguay
E-mail: viola@fing.edu.uy

ISSN 0302-9743                               e-ISSN 1611-3349
ISBN 978-3-642-54422-4                       e-ISBN 978-3-642-54423-1
DOI 10.1007/978-3-642-54423-1
Springer Heidelberg New York Dordrecht London

Library of Congress Control Number: 2014931658

LNCS Sublibrary: SL 1 – Theoretical Computer Science and General Issues

© Springer-Verlag Berlin Heidelberg 2014

This work is subject to copyright. All rights are reserved by the Publisher, whether the whole or part of the material is concerned, specifically the rights of translation, reprinting, reuse of illustrations, recitation, broadcasting, reproduction on microfilms or in any other physical way, and transmission or information storage and retrieval, electronic adaptation, computer software, or by similar or dissimilar methodology now known or hereafter developed. Exempted from this legal reservation are brief excerpts in connection with reviews or scholarly analysis or material supplied specifically for the purpose of being entered and executed on a computer system, for exclusive use by the purchaser of the work. Duplication of this publication or parts thereof is permitted only under the provisions of the Copyright Law of the Publisher's location, in ist current version, and permission for use must always be obtained from Springer. Permissions for use may be obtained through RightsLink at the Copyright Clearance Center. Violations are liable to prosecution under the respective Copyright Law.
The use of general descriptive names, registered names, trademarks, service marks, etc. in this publication does not imply, even in the absence of a specific statement, that such names are exempt from the relevant protective laws and regulations and therefore free for general use.
While the advice and information in this book are believed to be true and accurate at the date of publication, neither the authors nor the editors nor the publisher can accept any legal responsibility for any errors or omissions that may be made. The publisher makes no warranty, express or implied, with respect to the material contained herein.

*Typesetting:* Camera-ready by author, data conversion by Scientific Publishing Services, Chennai, India

Printed on acid-free paper

Springer is part of Springer Science+Business Media (www.springer.com)

# Preface

This volume contains the papers presented at the 11th Latin American Theoretical INformatics Symposium (LATIN 2014) held during March 31-April 4, 2014 in Montevideo, Uruguay. Previous editions of LATIN took place in São Paulo, Brazil (1992), Valparaíso, Chile (1995), Campinas, Brazil (1998), Punta del Este, Uruguay (2000), Cancún, México (2002), Buenos Aires, Argentina (2004), Valdivia, Chile (2006), Buzios, Brazil (2008), Oaxaca, México (2010) and Arequipa, Perú (2012).

The conference received 192 submissions from 42 countries. Each submission was reviewed by at least three Program Committee members, and carefully evaluated on quality, originality, and relevance to the conference. Overall, the Committee members wrote 588 reviews with the help of 254 external referees. Based on an extensive electronic discussion, the Committee selected 65 papers, leading to an acceptance rate of 34%. In addition to the accepted contributions, the symposium featured distinguished lectures by Ronitt Rubinfeld (Massachusetts Institute of Technology and Tel Aviv University), Robert Sedgewick (Princeton University), Gilles Barthe (IMDEA Software Institute), Gonzalo Navarro (Universidad de Chile), and J. Ian Munro (University of Waterloo).

The Imre Simon Test-of-Time Award started in 2012 and it is given to the authors of the LATIN paper deemed to be most influential among all those published at least ten years prior to the current edition of the conference. Papers published in the LATIN proceedings up to and including 2004 were eligible for the 2014 award. This year's winners were Graham Cormode and Sethu Muthu Muthukrishnan for their paper " An improved data stream summary: The count-min sketch and its applications", which appeared in LATIN 2004.

Many people helped to make LATIN 2014 possible. First, I would like to recognize the outstanding work of the members of the Program Committee. Their commitment contributed to a very detailed discussion on each of the submitted papers. The LATIN Steering Committee offered valuable advice and feedback; the conference benefitted immensely from their knowledge and experience. I would also like to recognize J. Ian Munro, Yoshiharu Kohayakawa and Michael Bender for their work in the Imre Simon Test-of-Time Award Committee.

Our industrial sponsors, Yahoo! Labs and Google provided much-needed funding. In particular, Yahoo! provided funds for the Imre Simon Award and Google for student grants. I thank Ricardo Baeza-Yates, Ravi Kumar and Prabhakar Raghavan for serving as contacts to those institutions.

The Centro Latinoamericano de Estudios en Informática (CLEI), the Comisión Sectorial de Investigaciones Científicas de la Universidad de la República (CSIC), the Programa de Desarrollo de las Ciencias Básicas (PEDECIBA) and the Agencia Nacional de Investigación e Innovación (ANII) also provided important seed funding. The Universidad ORT supported all the graphic design for the conference.

At the Universidad de la República, Alberto Pardo chaired the Local Arrangements Committee. His outstanding commitment in the most difficult moments of the organization was key to the success of LATIN. Guillermo Calderón administered the conference web site. The rest of the Local Arragements Committee, Javier Molina, Laura Molina and Alfonsina Pastori ably handled the innumerable logistical details that had to be resolved along the way. Finally, I thank my wife Graciela Pastori for the encouragement she offered during the year and a half that it took to make LATIN 2014 a reality.

January 2014                                               Alfredo Viola

# Organization

## Program Committee

| | |
|---|---|
| Ricardo Baeza-Yates | Yahoo! Labs, Spain |
| Jérémy Barbay | Universidad de Chile, Chile |
| Michael Bender | Stony Brook University, USA |
| Joan Boyar | University of Southern Denmark, Denmark |
| Vida Dujmovic | McGill University, Canada |
| Leah Epstein | University of Haifa, Israel |
| Cristina Fernandes | Universidade de São Paulo, Brazil |
| Maribel Fernandez | KCL London, England |
| Joachim von zur Gathen | University of Bonn, Germany |
| Gaston Gonnet | ETH Zurich, Switzerland |
| Marcos Kiwi | Universidad de Chile, Chile |
| Yoshiharu Kohayakawa | University of São Paulo, Brazil |
| Evangelos Kranakis | Carleton University, Canada |
| Ravi Kumar | Google, USA |
| Anna Lubiw | University of Waterloo, Canada |
| Conrado Martínez | Universitat Politècnica de Catalunya, Spain |
| Elvira Mayordomo | Universidad de Zaragoza, Spain |
| Marco Molinaro | Carnegie Mellon University, USA |
| Regina Motz | Universidad de la República, Uruguay |
| Lucia Moura | University of Ottawa, Canada |
| Daniel Panario | Carleton University, Canada |
| Sergio Rajsbaum | Universidad Nacional Autonoma de México, Mexico |
| Tamara Rezk | Inria, France |
| Andrea Richa | Arizona State University, USA |
| Jacques Sakarovitch | CNRS / ENST Paris, France |
| Nicolas Schabanel | CNRS - Université Paris Diderot (Paris 7), France |
| Rodrigo Silveira | Universitat Politècnica de Catalunya, Spain |
| Jose A. Soto | Universidad de Chile, Chile |
| Martin Strauss | University of Michigan, USA |
| Vilmar Trevisan | UFRGS, Brazil |
| Jorge Urrutia | Universidad Nacional Autonoma de México, Mexico |
| Tarmo Uustalu | Tallinn University of Technology, Estonia |
| Brigitte Vallée | CNRS/University of Caen, France |
| Alfredo Viola (Chair) | Universidad de la República, Uruguay |
| Santiago Zanella-Béguelin | Microsoft Research, England |

## Local Arrangements Committee

Guillermo Calderón  
Javier Molina  
Laura Molina  
Alfonsina Pastori  
Alberto Pardo (chair)

## Steering Committee

David Fernández-Baca     Iowa State University, USA  
Eduardo Sany Laber     PUC- Rio, Brazil  
Alejandro López-Ortiz     University of Waterloo, Canada  
Gonzalo Navarro     Universidad de Chile, Chile  
Marie-France Sagot     Inria Grenoble Rhône-Alpes and Université Claude Bernard (Lyon 1), France  
Yoshiko Wakabayashi     Universidade de São Paulo, Brazil

## Imre Simon Test-of-Time Award Committee

Michael Bender     Stony Brook University, USA  
Yoshiharu Kohayakawa     Universidade de São Paulo, Brazil  
J. Ian Munro (Chair)     University of Waterloo, Canada

## Sponsors

ANII (Agencia Nacional de Investigación e Innovación), Uruguay  
CLEI (Centro Latinoamericano de Estudios en Informática)  
CSIC (Comisión Sectorial de Investigación Científica, Universidad de la República), Uruguay  
Google, USA  
PEDECIBA Informática (Programa de Desarrollo de las Ciencias Básicas), Uruguay  
Universidad ORT, Uruguay  
Yahoo! Labs, Spain

## Additional Reviewers

Abdessalem, Talel  
Addario-Berry, Louigi  
Afshani, Peyman  
Akhavi, Ali  
Angelini, Patrizio  
Antoniadis, Antonios  
Ayala-Rincon, Mauricio  
Aziz, Haris  
Bacher, Axel  
Bampas, Evangelos  
Barba, Luis  
Barcelo, Pablo  
Bauer, Andrej  
Bazgan, Cristina  
Bernardi, Olivier  
Bodini, Olivier

Bonomo, Flavia
Bose, Prosenjit
Brandstadt, Andreas
Brewster, Rick
Brizuela, Carlos
Buchbinder, Niv
Buchin, Maike
Bulteau, Laurent
Buratti, Marco
Buriol, Luciana
Cai, Leizhen
Calinescu, Gruia
Camarão, Carlos
Campos, Victor
Castaneda, Armando
Castelli Aleardi, Luca
Chalermsook, Parinya
Chalopin, Jérémie
Chapelle, Mathieu
Chen, Yuxin
Chierichetti, Flavio
Christodoulakis, Manolis
Clément, Julien
Corteel, Sylvie
Costello, Kevin
Couillec, Yoann
Courcelle, Bruno
Csirmaz, Laszlo
Damian, Mirela
Dantas, Simone
Daudé, Hervé
David, Julien
de Carli Silva, Marcel
De La Clergerie, Eric
de Pina, José Coelho
de Rezende, Pedro J.
de Vries, Fer-Jan
Delgado, Jordi
Delporte-Gallet, Carole
Devismes, Stéphane
Dobrev, Stefan
Doerr, Benjamin
Dourado, Mitre
Drmota, Michael
Duchon, Philippe

Duffy, Chris
Duncan, Christian
Elizalde, Sergi
Eppstein, David
Esfandiari, Hossein
Fabrikant, Alex
Fagerberg, Rolf
Faliszewski, Piotr
Fauconnier, Hugues
Favrholdt, Lene Monrad
Feige, Uriel
Fertin, Guillaume
Fiala, Jiri
Find, Magnus
Flocchini, Paola
Fomin, Fedor
Fonseca, Guilherme
Fournier, Hervé
Fragoso Santos, Jose
Frati, Fabrizio
Ganapathi, Pramod
Gao, Jie
Gao, Shuhong
Gao, Zhicheng
Garg, Vijay
Gargano, Luisa
Gaspers, Serge
Georgiou, Konstantinos
Geremia, Ezequiel
Gittenberger, Bernhard
Green, Oded
Grossi, Roberto
Guha, Sudipto
Gutin, Gregory
Harutyunyan, Anna
Havet, Frederic
He, Meng
Hernandez, Cecilia
Hoppen, Carlos
Horak, Peter
Huang, Chien-Chung
Hwang, Hsien-Kuei
Hüffner, Falk
Ilcinkas, David
Iljazović, Zvonko

Im, Sungjin
Jansen, Bart
Jansen, Klaus
Jeż, Artur
Jimenez, Andrea
Josuat-Verges, Matthieu
Jungnickel, Dieter
Kanagal, Bhargav
Kiazyk, Stephen
King, James
Klostermeyer, Chip
Kniesburges, Sebastian
Kobourov, Stephen
Kononov, Alexander
Korman, Matias
Kosowski, Adrian
Kratochvil, Jan
Krivelevich, Michael
Krumke, Sven
Kuhn, Daniela
Kuznetsov, Petr
Labarre, Anthony
Lamb, Luis
Langerman, Stefan
Larsen, Kim S.
Lattanzi, Silvio
Lecroq, Thierry
Lee, Sang June
Lefmann, Hanno
Leme, Renato
Levin, Asaf
Lhote, Loick
Li, Minming
Loebenberger, Daniel
Lozano, Antoni
Lozin, Vadim
Lugosi, Gabor
Lumbroso, Jérémie
Löffler, Maarten
MacQuarrie, Fraser
Mahdian, Mohammad
Makowsky, Johann
Mandel, Arnaldo
Mansour, Toufik
Margalit, Oded

Markou, Euripides
Martin, Russell
Martinez-Moro, Edgar
Martins, Enide
Martín, Álvaro
McCauley, Samuel
Meer, Klaus
Milani, Alessia
Milanič, Martin
Molinero, Xavier
Morales Ponce, Oscar
Moseley, Benjamin
Mota, Guilherme O.
Moura, Arnaldo
Mucha, Marcin
Mueller, Moritz
Musicante, Martin
Nagarajan, Viswanath
Nantes, Daniele
Navarro, Gonzalo
Nesmachnow, Sergio
Nilsson, Bengt
Nüsken, Michael
Ollinger, Nicolas
Ott, Sebastian
Pacheco, Eduardo
Pagourtzis, Aris
Pajak, Dominik
Panagiotou, Konstantinos
Pathak, Vinayak
Paulusma, Daniel
Perez, Anthony
Perret, Ludovic
Pighizzini, Giovanni
Pilz, Alexander
Ponty, Yann
Popa, Alex
Pott, Alexander
Pruhs, Kirk
Pérez-Lantero, Pablo
Rad, Nader Jafari
Radke, Klaus
Raekow, Yona
Rahman, M. Sohel
Reyes, Nora

Richmond, Bruce
Rojas, Javiel
Saket, Rishi
Salinger, Alejandro
Salvy, Bruno
Sam, Sethserey
Sampaio, Rudini
Sato, Cristiane M.
Saumell, Maria
Saurabh, Saket
Sawada, Joe
Schaudt, Oliver
Schmid, Stefan
Schouery, Rafael
Schwartz, Roy
Seara, Carlos
Sereni, Jean-Sébastien
Serpette, Bernard
Shah, Rahul
Shirazipourazad, Shahrzad
Singer, Yaron
Sitchinava, Nodari
Soria, Michele
Sotelo, David
Stein, Maya
Stewart, Lorna
Stiller, Sebastian
Sviridenko, Maxim
Swenson, Krister

Tamir, Arie
Tannier, Eric
Telha, Claudio
Thraves, Christopher
Toft, Bjarne
Tomkins, Andrew
Tran, Huong
Travers, Corentin
Tsichlas, Kostas
Uchizawa, Kei
Umboh, Seeun
V. Silva, Pedro
van Leeuwen, Erik Jan
van Stee, Rob
Vassilvitskii, Sergei
Vee, Erik
Venkatasubramanian, Suresh
Verdonschot, Sander
Viera, Marcos
Vigneron, Antoine
Villard, Gilles
Wakabayashi, Yoshiko
Weber, Ken
Xia, Donglin
Yamamura, Akihiro
Yen, Hsu-Chun
Ziegler, Konstantin
Ziegler, Martin
Zito, Michele

# Abstracts

# Something for Almost Nothing: Advances in Sub-linear Time Algorithms

Ronitt Rubinfeld

CSAIL, MIT, Cambridge MA 02139
Blavatnik School of Computer Science, Tel Aviv University
ronitt@csail.mit.edu

**Abstract.** Linear-time algorithms have long been considered the gold standard of computational endciency. Indeed, it is hard to imagine doing better than that, since for a nontrivial problem, any algorithm must consider all of the input in order to make a decision. However, as extremely large data sets are pervasive, it is natural to wonder what one can do in sub-linear time. Over the past two decades, several surprising advances have been made on designing such algorithms. We will give a non-exhaustive survey of this emerging area, highlighting recent progress and directions for further research.

# Computer-Aided Cryptographic Proofs

Gilles Barthe

IMDEA Software Institute
gilles.barthe@imdea.org

EasyCrypt [6] is a computer-assisted framework for reasoning about the security of cryptographic constructions, using the methods and tools of provable security, and more specifically of the game-based techniques. The core of EasyCrypt is a relational program logic for a core probabilistic programming language with sequential composition, conditionals, loops, procedure calls, assignments and sampling from discrete distributions. The relational program logic is key to capture reductionist arguments that arise in cryptographic proofs. It is complemented by a (standard, non-relational) program logic that allows to reason about the probability of events in the execution of probabilistic programs; this program logic allows for instance to upper bound the probability of failure events, that are pervasive in game-based cryptographic proofs. In combination, these logics capture general reasoning principles in cryptography and have been used to verify the security of emblematic constructions, including the Full-Domain Hash signature [8], the Optimal Asymmetric Encryption Padding (OAEP) [7], hash function designs [3] and zero-knowledge protocols [5,1]. Yet, these logics can only capture instances of general principles, and lack mechanisms for stating and proving these general principles once and for all, and then for instantiating them as needed. To overcome this limitation, we have recently extended EasyCrypt with programming language mechanisms such as modules and type classes. Modules provide support for composition of cryptographic proofs, and for formalizing hybrid arguments, whereas type classes are convenient to model and reason about algebraic structures. Together, these extensions significantly expand the class of examples that can be addressed with EasyCrypt. For instance, we have used the latest version of EasyCrypt to verify the security of a class of authenticated key exchange protocols, and of a secure function evaluation protocol based on garbled circuits and oblivious transfer.

Our current work explores two complementary directions. On the one hand, we are extending the EasyCrypt infrastructure in order to derive security guarantees about implementations of cryptographic constructions. Indeed, practical attacks often target specific implementations and exploit some characteristics that are not considered in typical provable security proofs; as a consequence, several widely used implementations of provably secure schemes are vulnerable to attacks. In order to narrow the gap between provable security and implementations, we are extending EasyCrypt with support to reason about C-like implementations, and use the CompCert verified C compiler (http://compcert.inria.fr/) to carry the security guarantees down to executable implementations [2]. On the other hand, we are developing specialized formalisms to reason about the security of particular classes of constructions. For instance, we have recently developed the ZooCrypt framework [4], which supports automated analysis of chosen-plaintext and chosen ciphertext-security for public-key encryption schemes built from (partial-domain) one-way trapdoor permutations and random oracles. Using ZooCrypt, we have analyzed over a million (automatically generated) schemes, including many schemes from the literature. For chosen-plaintext security, ZooCrypt is able to report in nearly 99% of the cases a proof of security with a concrete security bound, or an attack. We are currently extending our approach to reason about encryption schemes based on Diffie-Hellmann groups and bilinear pairings, both in the random oracle and in the standard models.

More information about the project is available from the project web page

http://www.easycrypt.info

# References

1. Almeida, J.B., Barbosa, M., Bangerter, E., Barthe, G., Krenn, S., Zanella-Béguelin, S.: Full proof cryptography: verifiable compilation of efficient zero-knowledge protocols. In: 19th ACM Conference on Computer and Communications Security. ACM (2012)
2. Almeida, J.B., Barbosa, M., Barthe, G., Dupressoir, F.: Certified computer-aided cryptography: efficient provably secure machine code from high-level implementations. In: ACM Conference on Computer and Communications Security. ACM (2013)
3. Backes, M., Barthe, G., Berg, M., Grégoire, B., Skoruppa, M., Zanella-Béguelin, S.: Verified security of Merkle-Damgård. In: IEEE Computer Security Foundations. ACM (2012)
4. Barthe, G., Crespo, J.M., Grégoire, B., Kunz, C., Lakhnech, Y., Schmidt, B., Zanella-Béguelin, S.: Automated analysis and synthesis of padding-based encryption schemes. In: ACM Conference on Computer and Communications Security. ACM (2013)
5. Barthe, G., Grégoire, B., Hedin, D., Heraud, S., Zanella-Béguelin, S.: A Machine-Checked Formalization of Sigma-Protocols. In: IEEE Computer Security Foundations. ACM (2010)
6. Barthe, G., Grégoire, B., Heraud, S., Zanella-Béguelin, S.: Computer-aided security proofs for the working cryptographer. In: Rogaway, P. (ed.) CRYPTO 2011. LNCS, vol. 6841, pp. 71–90. Springer, Heidelberg (2011)
7. Barthe, G., Grégoire, B., Lakhnech, Y., Zanella-Béguelin, S.: Beyond Provable Security Verifiable IND-CCA Security of OAEP. In: Kiayias, A. (ed.) CT-RSA 2011. LNCS, vol. 6558, pp. 180–196. Springer, Heidelberg (2011)
8. Zanella-Béguelin, S., Barthe, G., Grégoire, B., Olmedo, F.: Formally certifying the security of digital signature schemes. In: IEEE Symposium on Security and Privacy. IEEE Computer Society (2009)

# "If You Can Specify It, You Can Analyze It"
# —The Lasting Legacy of Philippe Flajolet

Robert Sedgewick

Department of Computer Science, Princeton University
rs@cs.princeton.edu

**Abstract.** The "Flajolet School" of the analysis of algorithms and combinatorial structures is centered on an effective calculus, known as analytic combinatorics, for the development of mathematical models that are sufficiently accurate and precise that they can be validated through scientific experimentation. It is based on the generating function as the central object of study, first as a formal object that can translate a specification into mathematical equations, then as an analytic object whose properties as a function in the complex plane yield the desired quantitative results. Universal laws of sweeping generality can be proven within the framework, and easily applied. Standing on the shoulders of Cauchy, Polya, de Bruijn, Knuth, and many others, Philippe Flajolet and scores of collaborators developed this theory and demonstrated its effectiveness in a broad range of scientific applications. Flajolet's legacy is a vibrant field of research that holds the key not just to understanding the properties of algorithms and data structures, but also to understanding the properties of discrete structures that arise as models in all fields of science. This talk will survey Flajolet's story and its implications for future research.

"A man ... endowed with an an exuberance of imagination which puts it in his power to establish and populate a universe of his own creation".

# Encoding Data Structures

Gonzalo Navarro[*]

Department of Computer Science, University of Chile
gnavarro@dcc.uchile.cl

Classical data structures can be regarded as additional information that is stored on top of the raw data in order to speed up some kind of queries. Some examples are the suffix tree to support pattern matching in a text, the extra structures to support lowest common ancestor queries on a tree, or precomputed shortest path information on a graph.

Some data structures, however, can operate *without accessing the raw data*. These are called *encodings*. Encodings are relevant when they do not contain enough information to reproduce the raw data, but just what is necessary to answer the desired queries (otherwise, any data structure could be seen as an encoding, by storing a copy of the raw data inside the structure).

Encodings are interesting because they can occupy much less space than the raw data. In some cases the data itself is not interesting, only the answers to the queries on it, and thus we can simply discard the raw data and retain the encoding. In other cases, the data is used only sporadically and can be maintained in secondary storage, while the encoding is maintained in main memory, thus speeding up the most relevant queries.

When the raw data is available, any computable query on it can be answered with sufficient time. With encodings, instead, one faces a novel fundamental question: what is the *effective entropy* of the data with respect to a set of queries? That is, what is the minimum size of an encoding that can answer those queries without accessing the data? This question is related to Information Theory, but in a way inextricably associated to the data structure: the point is not how much information the data contains, but how much information is conveyed by the queries. In addition, as usual, there is the issue of how efficiently can be the queries answered depending on how much space is used.

In this talk I will survey some classical and new encodings, generally about preprocessing arrays $A[1, n]$ so as to answer queries on array intervals $[i, j]$ given at query time. I will start with the classical range minimum queries (which is the minimum value in $A[i, j]$?) which has a long history that culminated a few years ago in an asymptotically space-optimal encoding of $2n + o(n)$ bits answering queries in constant time. Then I will describe more recent (and partly open)

---

[*] Funded in part by Millennium Nucleus Information and Coordination in Networks ICM/FIC P10-024F, Chile.

problems such as finding the second minimum in $A[i,j]$, the $k$ smallest values in $A[i,j]$, the $k$th smallest value in $A[i,j]$, the elements that appear more than a fraction $\tau$ of the times in $A[i,j]$, etc. All these queries appear recurrently within other algorithmic problems, and they have also direct application in data mining.

# Succinct Data Structures ... Not Just for Graphs

J. Ian Munro

Cheriton School of Computer Science, University of Waterloo,
Waterloo, Ontario N2L 3G1, Canada
imunro@uwaterloo.ca

**Abstract.** Succinct data structures are data representations that use the (nearly) the information theoretic minimum space, for the combinatorial object they represent, while performing the necessary query operations in constant (or nearly constant) time. So, for example, we can represent a binary tree on $n$ nodes in $2n + o(n)$ bits, rather than the "obvious" $5n$ or so *words*, i.e. $5n \lg n$ bits. Such a difference in memory requirements can easily translate to major differences in runtime as a consequence of the level of memory in which most of the data resides. The field developed to a large extent because of applications in text indexing, so there has been a major emphasis on trees and a secondary emphasis on graphs in general; but in this talk we will draw attention to a much broader collection of combinatorial structures for which succinct structures have been developed. These will include sets, permutations, functions, partial orders and groups, and yes, a bit on graphs.

# Table of Contents

## Complexity 1

Conjugacy in Baumslag's Group, Generic Case Complexity,
and Division in Power Circuits .................................... 1
  *Volker Diekert, Alexei G. Myasnikov, and Armin Weiß*

Hierarchical Complexity of 2-Clique-Colouring Weakly Chordal Graphs
and Perfect Graphs Having Cliques of Size at Least 3 ................. 13
  *Helio B. Macêdo Filho, Raphael C.S. Machado, and
  Celina M.H. Figueiredo*

The Computational Complexity of the Game of Set and Its Theoretical
Applications ..................................................... 24
  *Michael Lampis and Valia Mitsou*

## Complexity 2

Independent and Hitting Sets of Rectangles Intersecting a Diagonal
Line ............................................................. 35
  *José R. Correa, Laurent Feuilloley, and José A. Soto*

Approximating Vector Scheduling: Almost Matching Upper and Lower
Bounds ........................................................... 47
  *Nikhil Bansal, Tjark Vredeveld, and Ruben van der Zwaan*

False-Name Manipulation in Weighted Voting Games Is Hard
for Probabilistic Polynomial Time ................................. 60
  *Anja Rey and Jörg Rothe*

A Natural Generalization of Bounded Tree-Width and Bounded
Clique-Width ..................................................... 72
  *Martin Fürer*

## Computational Geometry 1

Optimal Algorithms for Constrained 1-Center Problems ............... 84
  *Luis Barba, Prosenjit Bose, and Stefan Langerman*

A Randomized Incremental Approach for the Hausdorff Voronoi
Diagram of Non-crossing Clusters .................................. 96
  *Panagiotis Cheilaris, Elena Khramtcova, Stefan Langerman, and
  Evanthia Papadopoulou*

Upper Bounds on the Spanning Ratio of Constrained Theta-Graphs .... 108
  *Prosenjit Bose and André van Renssen*

Computing the $L_1$ Geodesic Diameter and Center of a Simple
Polygon in Linear Time.............................................. 120
  *Sang Won Bae, Matias Korman, Yoshio Okamoto, and Haitao Wang*

## Graph Drawing

The Planar Slope Number of Subcubic Graphs........................ 132
  *Emilio Di Giacomo, Giuseppe Liotta, and Fabrizio Montecchiani*

Smooth Orthogonal Drawings of Planar Graphs....................... 144
  *Muhammad Jawaherul Alam, Michael A. Bekos, Michael Kaufmann,
  Philipp Kindermann, Stephen G. Kobourov, and Alexander Wolff*

Drawing *HV*-Restricted Planar Graphs ............................. 156
  *Stephane Durocher, Stefan Felsner, Saeed Mehrabi, and
  Debajyoti Mondal*

Periodic Planar Straight-Frame Drawings with Polynomial
Resolution ........................................................ 168
  *Luca Castelli Aleardi, Éric Fusy, and Anatolii Kostrygin*

## Automata

A Characterization of Those Automata That Structurally Generate
Finite Groups ..................................................... 180
  *Ines Klimann and Matthieu Picantin*

Linear Grammars with One-Sided Contexts and Their Automaton
Representation .................................................... 190
  *Mikhail Barash and Alexander Okhotin*

## Computability

On the Computability of Relations on λ-Terms and Rice's Theorem -
The Case of the Expansion Problem for Explicit Substitutions ........ 202
  *Edward Hermann Haeusler and Mauricio Ayala-Rincón*

Computing in the Presence of Concurrent Solo Executions ............ 214
  *Maurice Herlihy, Sergio Rajsbaum, Michel Raynal, and
  Julien Stainer*

## Algorithms on Graphs

Combining All Pairs Shortest Paths and All Pairs Bottleneck Paths Problems .................................................. 226
*Tong-Wook Shinn and Tadao Takaoka*

(Total) Vector Domination for Graphs with Bounded Branchwidth ..... 238
*Toshimasa Ishii, Hirotaka Ono, and Yushi Uno*

Computing the Degeneracy of Large Graphs ......................... 250
*Martín Farach-Colton and Meng-Tsung Tsai*

## Computational Geometry 2

Approximation Algorithms for the Geometric Firefighter and Budget Fence Problems .................................................. 261
*Rolf Klein, Christos Levcopoulos, and Andrzej Lingas*

An Improved Data Stream Algorithm for Clustering ................. 273
*Sang-Sub Kim and Hee-Kap Ahn*

Approximation Algorithms for the Gromov Hyperbolicity of Discrete Metric Spaces ................................................... 285
*Ran Duan*

A (7/2)-Approximation Algorithm for Guarding Orthogonal Art Galleries with Sliding Cameras ................................... 294
*Stephane Durocher, Omrit Filtser, Robert Fraser, Ali D. Mehrabi, and Saeed Mehrabi*

## Algorithms

Helly-Type Theorems in Property Testing .......................... 306
*Sourav Chakraborty, Rameshwar Pratap, Sasanka Roy, and Shubhangi Saraf*

New Bounds for Online Packing LPs ................................ 318
*Matthias Englert, Nicolaos Matsakis, and Marcin Mucha*

Improved Minmax Regret 1-Center Algorithms for Cactus Networks with $c$ Cycles ................................................... 330
*Binay Bhattacharya, Tsunehiko Kameda, and Zhao Song*

Collision-Free Network Exploration ............................... 342
*Jurek Czyzowicz, Dariusz Dereniowski, Leszek Gąsieniec, Ralf Klasing, Adrian Kosowski, and Dominik Pająk*

## Random Structures

Powers of Hamilton Cycles in Pseudorandom Graphs .................. 355
  *Peter Allen, Julia Böttcher, Hiệp Hàn, Yoshiharu Kohayakawa, and Yury Person*

Local Update Algorithms for Random Graphs ....................... 367
  *Philippe Duchon and Romaric Duvignau*

Odd Graphs Are Prism-Hamiltonian and Have a Long Cycle .......... 379
  *Felipe De Campos Mesquita, Letícia Rodrigues Bueno, and Rodrigo De Alencar Hausen*

Relatively Bridge-Addable Classes of Graphs ....................... 391
  *Colin McDiarmid and Kerstin Weller*

## Complexity on Graphs 1

$O(n)$ Time Algorithms for Dominating Induced Matching Problems .... 399
  *Min Chih Lin, Michel J. Mizrahi, and Jayme L. Szwarcfiter*

Coloring Graph Powers: Graph Product Bounds and Hardness
of Approximation ................................................. 409
  *Parinya Chalermsook, Bundit Laekhanukit, and Danupon Nanongkai*

Convexity in Partial Cubes: The Hull Number ...................... 421
  *Marie Albenque and Kolja Knauer*

Connected Greedy Colourings ..................................... 433
  *Fabrício Benevides, Victor Campos, Mitre Dourado, Simon Griffiths, Robert Morris, Leonardo Sampaio, and Ana Silva*

## Analytic Combinatorics

On the Number of Prefix and Border Tables ....................... 442
  *Julien Clément and Laura Giambruno*

Probabilities of 2-Xor Functions .................................. 454
  *Élie de Panafieu, Danièle Gardy, Bernhard Gittenberger, and Markus Kuba*

Equivalence Classes of Random Boolean Trees and Application
to the Catalan Satisfiability Problem .............................. 466
  *Antoine Genitrini and Cécile Mailler*

## Analytic and Enumerative Combinatorics

The Flip Diameter of Rectangulations and Convex Subdivisions ........ 478
*Eyal Ackerman, Michelle M. Allen, Gill Barequet, Maarten Löffler, Joshua Mermelstein, Diane L. Souvaine, and Csaba D. Tóth*

Weighted Staircase Tableaux, Asymmetric Exclusion Process,
and Eulerian Type Recurrences ................................. 490
*Paweł Hitczenko and Svante Janson*

Counting and Generating Permutations Using Timed Languages ....... 502
*Nicolas Basset*

## Complexity on Graphs 2

Semantic Word Cloud Representations: Hardness and Approximation
Algorithms ................................................... 514
*Lukas Barth, Sara Irina Fabrikant, Stephen G. Kobourov, Anna Lubiw, Martin Nöllenburg, Yoshio Okamoto, Sergey Pupyrev, Claudio Squarcella, Torsten Ueckerdt, and Alexander Wolff*

The Complexity of Homomorphisms of Signed Graphs and Signed
Constraint Satisfaction ....................................... 526
*Florent Foucaud and Reza Naserasr*

Complexity of Coloring Graphs without Paths and Cycles ............. 538
*Pavol Hell and Shenwei Huang*

## Approximation Algorithms

Approximating Real-Time Scheduling on Identical Machines ........... 550
*Nikhil Bansal, Cyriel Rutten, Suzanne van der Ster, Tjark Vredeveld, and Ruben van der Zwaan*

Integrated Supply Chain Management via Randomized Rounding ...... 562
*Lehilton L.C. Pedrosa and Maxim Sviridenko*

The Online Connected Facility Location Problem .................... 574
*Mário César San Felice, David P. Williamson, and Orlando Lee*

Multiply Balanced $k$−Partitioning ................................ 586
*Amihood Amir, Jessica Ficler, Robert Krauthgamer, Liam Roditty, and Oren Sar Shalom*

On Some Recent Approximation Algorithms for MAX SAT ............ 598
*Matthias Poloczek, David P. Williamson, and Anke van Zuylen*

## Analysis of Algorithms

Packet Forwarding Algorithms in a Line Network ..................... 610
    *Antonios Antoniadis, Neal Barcelo, Daniel Cole, Kyle Fox,
Benjamin Moseley, Michael Nugent, and Kirk Pruhs*

Survivability of Swarms of Bouncing Robots ......................... 622
    *Jurek Czyzowicz, Stefan Dobrev, Evangelos Kranakis, and
Eduardo Pacheco*

Emergence of Wave Patterns on Kadanoff Sandpiles .................. 634
    *Kévin Perrot and Éric Rémila*

## Computational Algebra

A Divide and Conquer Method to Compute Binomial Ideals ........... 648
    *Deepanjan Kesh and Shashank K. Mehta*

How Fast Can We Multiply Large Integers on an Actual Computer? .... 660
    *Martin Fürer*

## Aplications to Bioinformatics

Sorting Permutations by Prefix and Suffix Versions of Reversals
and Transpositions ................................................ 671
    *Carla Negri Lintzmayer and Zanoni Dias*

Algorithmic and Hardness Results for the Colorful Components
Problems ......................................................... 683
    *Anna Adamaszek and Alexandru Popa*

## Budget Problems

On the Stability of Generalized Second Price Auctions with Budgets.... 695
    *Josep Díaz, Ioannis Giotis, Lefteris Kirousis,
Evangelos Markakis, and Maria Serna*

Approximation Algorithms for the Max-Buying Problem with Limited
Supply ........................................................... 707
    *Cristina G. Fernandes and Rafael C.S. Schouery*

Budget Feasible Mechanisms for Experimental Design ............... 719
    *Thibaut Horel, Stratis Ioannidis, and S. Muthukrishnan*

## Algorithms and Data Structures

LZ77-Based Self-indexing with Faster Pattern Matching............... 731
  *Travis Gagie, Paweł Gawrychowski, Juha Kärkkäinen,*
  *Yakov Nekrich, and Simon J. Puglisi*

Quad-K-d Trees ................................................... 743
  *Nikolett Bereczky, Amalia Duch, Krisztián Németh, and*
  *Salvador Roura*

Biased Predecessor Search......................................... 755
  *Prosenjit Bose, Rolf Fagerberg, John Howat, and Pat Morin*

**Author Index** .................................................. 765

# Conjugacy in Baumslag's Group, Generic Case Complexity, and Division in Power Circuits

Volker Diekert[1], Alexei G. Myasnikov[2], and Armin Weiß[1]

[1] FMI, Universität Stuttgart, Universitätsstr. 38, D-70569 Stuttgart, Germany
[2] Department of Mathematics, Stevens Institute of Technology, Hoboken, NJ, USA

**Abstract.** The conjugacy problem is the following question: given two words $x$, $y$ over generators of a fixed group $G$, decide whether $x$ and $y$ are conjugated, i.e., whether there exists some $z$ such that $zxz^{-1} = y$ in $G$. The conjugacy problem is more difficult than the word problem, in general. We investigate the conjugacy problem for two prominent groups: the Baumslag-Solitar group $\mathbf{BS}_{1,2}$ and the Baumslag(-Gersten) group $\mathbf{G}_{1,2}$. The conjugacy problem in $\mathbf{BS}_{1,2}$ is $\mathrm{TC}^0$-complete. To the best of our knowledge $\mathbf{BS}_{1,2}$ is the first natural infinite non-commutative group where such a precise and low complexity is shown. The Baumslag group $\mathbf{G}_{1,2}$ is an HNN-extension of $\mathbf{BS}_{1,2}$ and its conjugacy problem is decidable $\mathbf{G}_{1,2}$ by a result of Beese (2012). Here we show that conjugacy in $\mathbf{G}_{1,2}$ can be solved in polynomial time in a strongly generic setting. This means that essentially for all inputs conjugacy in $\mathbf{G}_{1,2}$ can be decided efficiently. In contrast, we show that under a plausible assumption the average case complexity of the same problem is non-elementary. Moreover, we provide a lower bound for the conjugacy problem in $\mathbf{G}_{1,2}$ by reducing the division problem in power circuits to the conjugacy problem in $\mathbf{G}_{1,2}$. The complexity of the division problem in power circuits is an open and interesting problem in integer arithmetic.

**Keywords:** Algorithmic group theory, power circuit, generic case complexity.

## 1 Introduction

More than 100 years ago Max Dehn introduced the word problem and the conjugacy problem as fundamental decision problems in group theory. Let $G$ be a finitely generated group. The input are two words $x$, $y$ written in generators. *Word problem*: Decide whether $x = y$ in $G$. *Conjugacy problem*: Decide whether $x \sim_G y$ in $G$, i.e., decide whether there exists $z$ such that $zxz^{-1} = y$ in $G$. In recent years, conjugacy played an important role in non-commutative cryptography, see e.g. [5,9,20]. These applications use that is is easy to create elements which are conjugated, but to check whether two given elements are conjugated might be difficult even if the word problem is easy. In fact, there are groups where the word problem is easy but the conjugacy problem is undecidable [16]. Frequently, in cryptographic applications the ambient group is fixed. The focus in this paper is on the conjugacy problem in $\mathbf{G}_{1,2}$. In 1969 Gilbert Baumslag defined the group $\mathbf{G}_{1,2}$ as an example of a one-relator group which enjoys certain remarkable properties. It was introduced as an infinite non-cyclic group all of whose finite quotients are cyclic [1]. In particular, it is not residually finite; but being one-relator it has a decidable word problem [15]. The group $\mathbf{G}_{1,2}$ is generated by two generators $a$ and $b$ subject to a single relation $bab^{-1}a = a^2bab^{-1}$. Another way to understand $\mathbf{G}_{1,2}$ is to view it as an HNN-extension of the even more prominent Baumslag-Solitar group $\mathbf{BS}_{1,2}$. The group $\mathbf{BS}_{1,2}$ is defined by a single relation $tat^{-1} = a^2$ where $a$ and $t$ are generators[1]. The complexity of the word problem and conjugacy problem in $\mathbf{BS}_{1,2}$ are very low; indeed, we show that they are $\mathsf{TC}^0$-complete. However, such a low complexity does not transfer to the complexity of the corresponding problems in HHN-extensions like $\mathbf{G}_{1,2}$. Gersten showed that the Dehn function of $\mathbf{G}_{1,2}$ is non-elementary [8]. Moreover, Magnus' break-down procedure [14] on $\mathbf{G}_{1,2}$ is non-elementary, too. This means that the time complexity for the standard algorithm to solve the word problem in $\mathbf{G}_{1,2}$ cannot be bounded by any fixed tower of exponentials. Therefore, for many years, $\mathbf{G}_{1,2}$ was the simplest candidate for a group with an extremely difficult word problem. However, Myasnikov, Ushakov, and Won showed in [18] that the word problem of the Baumslag group is solvable in polynomial time! In order to achieve a polynomial time bound they introduced a versatile data structure for integer arithmetic which they called *power circuit*. The data structure supports $+$, $-$, $\leq$, and $(x, y) \mapsto 2^x y$, a restricted version of multiplication which includes exponentiation $x \mapsto 2^x$. Thus, by iteration it is possible to represent huge values (involving the tower function) by very small circuits. Still, all operations above can be performed in polynomial time. On the other hand there are notoriously difficult arithmetical problems in power circuits, too. A very important one is division. The input are power circuits $C$ and $C'$ representing integers $m$ and $m'$; the question is whether $m$ divides $m'$. The problem is clearly decidable by converting $m$ and $m'$ into binary; but this procedure is non-elementary. So far, no idea for any better algorithm is known. It is plausible to assume that the problem "division in power circuits" has no elementary time complexity at all.

In the present paper we show a tight relation between the problems "division in power circuits" and conjugacy in $\mathbf{G}_{1,2}$. Our results concerning the Baumslag-Solitar group $\mathbf{BS}_{1,2}$, the Baumslag group $\mathbf{G}_{1,2}$, its generic case complexity, and division in power circuits are as follows.

- The conjugacy problem of $\mathbf{BS}_{1,2}$ is $\mathsf{TC}^0$-complete.
- There is a strongly generic polynomial time algorithm for the conjugacy problem in $\mathbf{G}_{1,2}$. This means, the difficult instances for the algorithm are exponentially sparse, and therefore, on random inputs, conjugacy can be solved efficiently.
- If "division in power circuits" is non-elementary in the worst case, then the conjugacy problem in $\mathbf{G}_{1,2}$ is non-elementary on the average.

Decidability of the conjugacy problem in $\mathbf{G}_{1,2}$ is not new, it was shown in [2][2] and decidability outside a so-called "black hole" follows already from [3]. Our work improves Beese's work leading to a polynomial time algorithm outside a proper subset of the "black hole" (and decidability everywhere). Thus, our result underlines that in special cases like $\mathbf{G}_{1,2}$ much better results than stated in [3] are possible. Let us also note that there are undecidable problems (hence no finite average case complexity is defined), like the halting problem for certain encodings of Turing machines, which have

---

[1] Adding a generator $b$ and a relation $bab^{-1} = t$ results in $\mathbf{G}_{1,2}$. Indeed, due to $bab^{-1} = t$, we can remove $t$ and we obtain exactly the presentation of $\mathbf{G}_{1,2}$ above.
[2] It is unknown whether the conjugacy problem in one-relator groups is decidable, in general.

generically linear time partial solutions. However, many of these examples depend on encodings and special purpose constructions. In our case we consider a natural problem where the average case complexity is defined, but the only known algorithm to solve it runs in non-elementary time on the average. Nevertheless, there is a polynomial $p$ (roughly of degree 4) such that the probability that the same algorithm requires more than $p(n)$ steps on random inputs converges exponentially fast to zero. The main technical difficulty in establishing a strongly generic polynomial time complexity is to show that a random walk of length $n$ in the Cayley graph of $\mathbf{G}_{1,2}$ ends with probability less than $(1-\varepsilon)^n$ in the subgroup $\mathbf{BS}_{1,2}$ for some $\varepsilon > 0$. Random walks in infinite graphs are widely studied in various areas, see e.g. [22] or the textbook [23].

In [7] we prove a more general statement about HNN extension of the form $G = \langle H, b \mid bab^{-1} = \varphi(a), a \in A \rangle$ with a finitely generated base group $H$ and $\Delta$ a finite symmetric set of generators for $G$. We show that the complement of $H$ (inside $\Delta^*$) is strongly generic if and only if $A \neq H \neq B$. With other words, the Schreier graph $\Gamma(G, H, \Delta)$ is non-amenable if and only if $A \neq H \neq B$. (For a definition of amenability and its equivalent characterizations see e.g. [4,12].) This applies to $\mathbf{G}_{1,2}$ because it is an HNN extension where $A \neq H \neq B$. Note that $\mathbf{BS}_{1,2}$ is an HNN extension with $A = H \neq B$, and indeed, the corresponding Schreier graph $\Gamma(G, H, \Delta)$ is amenable for the Baumslag-Solitar group $\mathbf{BS}_{1,2}$.

However, in the special case of $\mathbf{G}_{1,2}$ we can also apply a technique quite different from the that approach. In order to show our result about strongly generic polynomial time we define a "pairing" between random walks in the Cayley graph and Dyck words. We exhibit an $\varepsilon > 0$ such that for each Dyck word $w$ of length $2n$ the probability that a pairing with $w$ evaluates to 1 is bounded by $(1/4 - \varepsilon)^n$. The result follows since there are at most $4^n$ Dyck words.

## 1.1 Notation and Preliminaries

**Functions.** We use standard $\mathcal{O}$-notation for functions from $\mathbb{N}$ to non-negative reals $\mathbb{R}^{\geq 0}$. The *tower function* $\tau : \mathbb{N} \to \mathbb{N}$ is defined by $\tau(0) = 0$ and $\tau(i+1) = 2^{\tau(i)}$ for $i \geq 0$. It is primitive recursive. We say that a function $f : \mathbb{N} \to \mathbb{R}^{\geq 0}$ is *elementary*, if the growth of $f$ can be bounded by a fixed number of exponentials. It is called *non-elementary* if it is not elementary, but $f(n) \in \tau(\mathcal{O}(n))$.
**Circuit Complexity.** We deal with various complexity measures. On the lowest level we are interested in problems which can be decided by (uniform) $\mathsf{TC}^0$-circuits. These are circuits of polynomial size with constant depth where we allow Boolean gates and majority gates, which evaluate to 1 if and only if the majority of inputs is 1. For a precise definition and uniformity conditions we refer to the textbook [21].
**Time Complexity.** A uniform family of $\mathsf{TC}^0$-circuits computes a polynomial time computable function. We use a standard notion for worst-case and for average case complexity and random access machines (RAMs) as machine model. An algorithm $\mathcal{A}$ computes a function between domains $D$ and $D'$. In our applications $D$ comes always with a natural partition $D = \bigcup \{D^{(n)} \mid n \in \mathbb{N}\}$ where each $D^{(n)}$ is finite.
**Generic Case Complexity.** For many practical applications the "generic-case behavior" of an algorithm is more important than its average-case or worst-case behavior. We refer to [12,13] where the foundations of this theory were developed and to [17]

for applications in cryptography. The notion of *generic complexity* refers to partial algorithms which are defined on a (strongly) generic set $I \subseteq D$. Thus, they may refuse to give an answer outside $I$, but if they give an answer, the answer must always be correct. In our context it is enough to deal with totally defined algorithms and strongly generic sets. Thus, the answer is always computed and always correct, but the runtime is measured by a worst-case behavior over a strongly generic set $I \subseteq D$. Here a set $I$ is called *strongly generic*, if there exists an $\varepsilon > 0$ such that $|D^{(n)} \setminus I| / |D^{(n)}| \leq 2^{-\varepsilon n}$ for almost all $n \in \mathbb{N}$. This means the probability to find a random string outside $I$ converges exponentially fast to zero. Thus, if an algorithm $\mathcal{A}$ runs in polynomial time on a strongly generic set, then, for practical purposes, $\mathcal{A}$ behaves as a polynomial time worst-case algorithm. This is true although the average time complexity of $\mathcal{A}$ can be arbitrarily high.

**Group Theory.** We use standard notation and facts from group theory as found in the classical text book [14]. Groups $G$ are generated by some subset $S \subseteq G$. We let $\overline{S} = S^{-1}$ and we view $S \cup \overline{S}$ as an alphabet with involution; its elements are called *letters*. We have $\overline{\overline{a}} = a$ for letters and also for words by letting $\overline{a_1 \cdots a_n} = \overline{a_n} \cdots \overline{a_1}$ where $a_i \in S \cup \overline{S}$ are letters. Thus, if $g \in G$ is given by a word $w$, then $\overline{w} = g^{-1}$ in the group $G$. For a word $w$ we denote by $|w|$ its length. We say that $w$ is *reduced* if there is no factor $a\overline{a}$ for any letter. It is called *cyclically reduced* if $ww$ is reduced. For words (or group elements) we write $x \sim_G y$ to denote conjugacy, i.e., $x \sim_G y$ if and only if there exists some $z \in G$ such that $zx\overline{z} = y$ in $G$. We apply the standard (so called Magnus break-down) procedure for solving the word problem in HNN-extensions. Our calculations are fully explicit and accessible with basic knowledge in combinatorial group theory.

**Glossary.** $\mathsf{TC}^0$ circuit class. $x \sim_G y$ conjugacy in groups. $(\Gamma, \delta)$ power circuits. $\varepsilon(P)$, $\varepsilon(M)$ evaluation of nodes and markings. $\tau(n)$ tower function. Baumslag-Solitar group: $\mathbf{BS}_{1,2} = \langle a, t \mid tat^{-1} = a^2 \rangle$. Baumslag group: $\mathbf{G}_{1,2} = \langle a, b \mid bab^{-1}a = a^2ba^{-1}b^{-1} \rangle$. Subgroup relations $A = \langle a \rangle$, $T = \langle t \rangle \leq \mathbf{BS}_{1,2} = \mathbb{Z}[1/2] \rtimes \mathbb{Z} = H \leq \mathbf{G}_{1,2}$. Standard symmetric set of generators for $\mathbf{G}_{1,2}$ is $\Sigma = \{a, \overline{a}, b, \overline{b}\}^*$ and $\overline{z} = z^{-1}$ in groups.

**Proofs.** Missing proofs are in the forthcoming journal version and also in the paper on the arXiv server [7].

## 2 Power Circuits

In binary a number is represented as a sum $m = \sum_{i=0}^{k} b_i 2^i$ with $b_i \in \{0, 1\}$. Allowing $b_i \in \{-1, 0, 1\}$ we obtain a "compact representation" of integers, which may require less non-zero $b_i$s than the normal representation. The notion of power circuit is due to [19]. It generalizes compact representations and goes far beyond since it allows a compact representation of tower functions. Formally: a *power circuit* of size $n$ is given by a pair $(\Gamma, \delta)$. Here, $\Gamma$ is a set of $n$ vertices and $\delta$ is a mapping $\delta : \Gamma \times \Gamma \to \{-1, 0, +1\}$. The support of $\delta$ is the subset $\Delta \subseteq \Gamma \times \Gamma$ with $(P, Q) \in \Delta \iff \delta(P, Q) \neq 0$. Thus, $(\Gamma, \Delta)$ is a directed graph. Throughout we require that $(\Gamma, \Delta)$ is acyclic. In particular, $\delta(P, P) = 0$ for all vertices $P$. A *marking* is a mapping $M : \Gamma \to \{-1, 0, +1\}$. We can also think of a marking as a subset of $\Gamma$ where each element in $M$ has a sign (+ or −). If $M(P) = 0$ for all $P \in \Gamma$ then we simply write $M =$

$\emptyset$. Each node $P \in \Gamma$ is associated in a natural way with a successor marking $\Lambda_P$ : $\Gamma \to \{-1, 0, +1\}$, $Q \mapsto \delta(P, Q)$, consisting of the target nodes of outgoing arcs from $P$. We define the *evaluation* $\varepsilon(P)$ of a node ($\varepsilon(M)$ of a marking resp.) bottom-up in the directed acyclic graph by induction: $\varepsilon(\emptyset) = 0$, $\varepsilon(P) = 2^{\varepsilon(\Lambda_P)}$ for a node $P$, and $\varepsilon(M) = \sum_P M(P)\varepsilon(P)$ for a marking $M$. Note that leaves evaluate to 1, the evaluation of a marking is a real number, and the evaluation of a node $P$ is a positive real number. Thus, $\varepsilon(P)$ and $\varepsilon(M)$ are well-defined. We have $\varepsilon(\Lambda_P) = \log_2(\varepsilon(P))$, thus the successor marking plays the role of a logarithm. We are interested only in power circuits where all markings evaluate to integers; equivalently all nodes evaluate to some positive natural number in $2^{\mathbb{N}}$.

The *power circuit-representation* of an integer sequence $m_1, \ldots, m_k$ is given by a tuple $(\Gamma, \delta; M_1, \ldots, M_k)$ where $(\Gamma, \delta)$ is a power circuit and $M_1, \ldots, M_k$ are markings such that $\varepsilon(M_i) = m_i$. (Hence, a single power circuit can store several different numbers; a fact which has been crucial in the proof of Prop. 9, see [6].)

*Example 1.* We can represent every $n$-bit integer as a power circuit with $\mathcal{O}(n)$ vertices.

*Example 2.* A power circuit of size $n$ can realize $\tau(n)$ since a chain of $n$ nodes represents $\tau(n)$ as the evaluation of the last node.

**Proposition 3 ([18,6]).** *The following operations can be performed in quadratic time. Input a power circuit $(\Gamma, \delta)$ of size $n$ and two markings $M_1$ and $M_2$. Decide whether $(\Gamma, \delta)$ is indeed a power circuit, i.e., decide whether all markings evaluate to integers. If "yes": Decide whether $\varepsilon(M_1) \leq \varepsilon(M_2)$; and compute a new power circuit with markings $M$, $X$ and $U$ such that*

1. $\varepsilon(M) = \varepsilon(M_1) \pm \varepsilon(M_2)$.
2. $\varepsilon(M) = 2^{\varepsilon(M_1)} \cdot \varepsilon(M_2)$.
3. $\varepsilon(M_1) = 2^{\varepsilon(X)} \cdot \varepsilon(U)$ *and either* $U = \emptyset$ *or* $\varepsilon(U)$ *is odd.*

Let us mention that the complexity of the division problem in power circuits is open. The only known general algorithm transforms markings into binary numbers. This involves a non-elementary explosion.

## 3 Conjugacy in the Baumslag-Solitar Group $\mathrm{BS}_{1,2}$

The solution of the conjugacy problem in the Baumslag group $\mathbf{G}_{1,2}$ relies on the simpler solution for the Baumslag-Solitar group $\mathrm{BS}_{1,2}$. The aim of this section is to show that the conjugacy problem in $\mathrm{BS}_{1,2}$ is $\mathsf{TC}^0$-complete. The group $\mathrm{BS}_{1,2}$ is given by the presentation $\langle a, t \mid tat^{-1} = a^2 \rangle$. We have $ta = a^2 t$ and $at^{-1} = t^{-1}a^2$. This allows to represent all group elements by words of the form $t^{-p} a^r t^q$ with $p, q \in \mathbb{N}$ and $r \in \mathbb{Z}$. However, for $q \geq 0$, transforming $t^q a^r$ into this form leads to $a^s t^q$ with $s = 2^q r$, so the word $a^s t^q$ is exponentially longer than the word $t^q a^r$. We denote by $\mathbb{Z}[1/2] = \{p/2^q \in \mathbb{Q} \mid p, q \in \mathbb{Z}\}$ the ring of *dyadic fractions*. Multiplication by 2 is an automorphism of the underlying additive group and therefore we can define the semi-direct product $\mathbb{Z}[1/2] \rtimes \mathbb{Z}$ as follows. Elements are pairs $(r, m) \in \mathbb{Z}[1/2] \times \mathbb{Z}$. The multiplication in $\mathbb{Z}[1/2] \rtimes \mathbb{Z}$ is defined by

$$(r, m) \cdot (s, q) = (r + 2^m s, m + q).$$

Inverses can be computed by the formula $(r,m)^{-1} = (-r \cdot 2^{-m}, -m)$. It is straightforward to show that $a \mapsto (1,0)$ and $t \mapsto (0,1)$ defines an isomorphism between $\mathbf{BS}_{1,2}$ and $\mathbb{Z}[1/2] \rtimes \mathbb{Z}$. In the following we abbreviate $\mathbf{BS}_{1,2}$ ($= \mathbb{Z}[1/2] \rtimes \mathbb{Z}$) by $H$. There are several options to represent a group element $g \in H$. In a *unary representation* we write $g$ as a word over the alphabet with involution $\{a, \bar{a}, t, \bar{t}\}$. Another way is to write $g = (r,m)$ with $r \in \mathbb{Z}[1/2]$ and $m \in \mathbb{Z}$. In the following we use both notations interchangeably. The *binary representation* of $(r,m)$ consists of $r$ written in binary (as floating point number) and $m$ in unary. Let us write $(r,m)$ with $r = 2^k s$ and $k, s, m \in \mathbb{Z}$. We then have $(2^k s, m) = (0, k) \cdot (s, m - k)$ and the corresponding triple $[k, s, m - k] \in \mathbb{Z}^3$ is called the *triple-representation* of $(r,m)$; it is not unique. The *power circuit representation* of $g = [k, s, m - k]$ is given by a power circuit and markings $K, S, L$ such that $\varepsilon(K) = k$, $\varepsilon(S) = s$, and $\varepsilon(L) = m - k$. Note that if $g \in \{a, \bar{a}, t, \bar{t}\}^n$ satisfies $g = (r,m) \in H$, then $|r| \leq 2^n$ and $|m| \leq n$. Thus, a transformation from unary to binary notation is on the safe side.

**Proposition 4.** *Let $(r_1, m_1), \ldots, (r_n, m_n) \in \mathbb{Z}[1/2] \rtimes \mathbb{Z}$ given in binary representation for all $i$. Then there is a uniform construction of a $\mathsf{TC}^0$-circuit which calculates $(r,m) = (r_1, m_1) \cdots (r_n, m_n)$ in $\mathbb{Z}[1/2] \rtimes \mathbb{Z}$.*

The next proof uses a deep result of Hesse: integer division is in uniform $\mathsf{TC}^0$.

**Proposition 5.** *Let $f = (r,m), g = (s,q) \in \mathbb{Z}[1/2] \rtimes \mathbb{Z}$ be given in binary representation. Then there is a uniform construction of a $\mathsf{TC}^0$-circuit which decides $f \sim_H g$.*

*Proof.* Let $(r,m) \sim_H (s,q)$, i.e., there are $k \in \mathbb{Z}$, $x \in \mathbb{Z}[1/2]$ with $(x,k)(r,m) = (s,q)(x,k)$. In particular, $(r,m) \sim_H (s,q)$ if and only if $m = q$ and there are $k \in \mathbb{Z}$, $x \in \mathbb{Z}[1/2]$ such that
$$s = r \cdot 2^k - x \cdot (2^m - 1). \tag{1}$$
We have $(r,m) \sim_H (s,m)$ if and only if $(-r,-m) \sim_H (-s,-m)$ since $(-p,-m) \sim_H (-p2^{-m}, -m) = (p,m)^{-1}$ for all $p \in \mathbb{Z}[1/2]$. Therefore, without restriction $m \in \mathbb{N}$. Since a conjugation with $t^k$ maps $(r,m)$ to $(2^k r, m)$, we may assume that $r, s \in \mathbb{Z}$ and $m \in \mathbb{N}$. For $m = 0$ this means $(r,0) \sim_H (s,0)$ if and only if there is some $k \in \mathbb{Z}$ such that $s = r \cdot 2^k$. This can be decided in $\mathsf{TC}^0$. For $m = 1$ we can choose $x = r - s$ and the answer is "yes". For $m \geq 2$ we can multiply (1) by $2^\ell$ such that $x \cdot 2^\ell \in \mathbb{Z}$. We obtain $2^\ell \cdot (r \cdot 2^k - s) = 2^\ell x \cdot (2^m - 1)$, i.e., $2^\ell \cdot (r \cdot 2^k - s) \equiv 0 \bmod (2^m - 1)$. The number 2 is invertible modulo $2^m - 1$ and its order is $m$. Hence, actually for $m \geq 1$:
$$(r,m) \sim_H (s,m) \iff \exists k \in \mathbb{N} : 0 \leq k < m \wedge r \cdot 2^k - s \equiv 0 \bmod (2^m - 1). \tag{2}$$
It can be checked whether such a $k$ exists using Hesse's result for division [10,11]. □

**Theorem 6.** *The word problem as well as the conjugacy problem in $\mathbf{BS}_{1,2}$ is $\mathsf{TC}^0$-complete.*

*Remark 7.* Let us highlight that integer division can be reduced to the conjugacy problem in $\mathbf{BS}_{1,2}$. For $m \geq 1$ we obtain as a special case of (2) and a well-known fact from elementary number theory
$$(0,m) \sim_H (2^s - 1, m) \iff 2^m - 1 \mid 2^s - 1 \iff m \mid s. \tag{3}$$

If we allow a power circuit representation for integers, then this reduction from division to conjugacy can be computed in polynomial time. Hence, no elementary algorithm is known to solve the conjugacy problem in $\mathbf{BS}_{1,2}$ in power circuit representation, whereas the word problem remains solvable in cubic time by [6].

## 4 Conjugacy in the Baumslag Group $\mathbf{G}_{1,2}$

The Baumslag group $\mathbf{G}_{1,2}$ is an HNN-extension of the Baumslag-Solitar group $\mathbf{BS}_{1,2}$. We make this explicit. We let $\mathbf{BS}_{1,2}$ be our base group, generated by $a$ and $t$. Again, $\mathbf{BS}_{1,2}$ is abbreviated as $H$. The group $H$ contains infinite cyclic subgroups $A = \langle a \rangle$ and $T = \langle t \rangle$ with $A \cap T = \{1\}$. Let $b$ be a fresh letter which is added as a new generator together with the relation $bab^{-1} = t$. This defines the Baumslag group $\mathbf{G}_{1,2}$. It is generated by $a, t, b$ with defining relations $tat^{-1} = a^2$ and $bab^{-1} = t$. However, the generator $t$ is now redundant and we obtain $\mathbf{G}_{1,2}$ as a group generated by $a, b$ with a single defining relation $bab^{-1}a = a^2bab^{-1}$. We represent elements of $\mathbf{G}_{1,2}$ by $\beta$-factorizations. A *$\beta$-factorization* is written as a word $z = \gamma_0\beta_1\gamma_1\ldots\beta_k\gamma_k$ with $\beta_i \in \{b, \bar{b}\}$ and $\gamma_i \in \{a, \bar{a}, t, \bar{t}\}^*$. The number $k$ is called the $\beta$-*length* and is denoted as $|z|_\beta$. A *transposition* of a $\beta$-factorization $z = \gamma_0\beta_1\gamma_1\ldots\beta_k\gamma_k$ is given as $z' = \beta_i\gamma_i\ldots\beta_k\gamma_k\gamma_0\beta_1\gamma_1\ldots\beta_{i-1}\gamma_{i-1}$ for some $1 \leq i \leq k$. Clearly, $z \sim_{\mathbf{G}_{1,2}} z'$ in this case. Throughout we identify a power $c^{-\ell}$ with $\bar{c}^\ell$ for letters $c$ and $\ell \in \mathbb{N}$.

**Britton Reductions.** A *Britton reduction* considers some factor $\beta\gamma\bar{\beta}$ with $\gamma \in \{a, \bar{a}, t, \bar{t}\}^*$. There are two cases. First, if $\beta = b$ and $\gamma = a^\ell$ in $H$ for some $\ell \in \mathbb{Z}$ then the factor $b\gamma\bar{b}$ is replaced by $t^\ell$. Second, if $\beta = \bar{b}$ and $\gamma = t^\ell$ in $H$ for some $\ell \in \mathbb{Z}$ then the factor $\bar{b}\gamma b$ is replaced by $a^\ell$. At most $|z|_\beta$ Britton reduction are possible on a word $z$. Be aware! There can be a non-elementary blow-up in the exponents, see Ex. 8. If no Britton reduction is possible, then the word $x$ is called *Britton-reduced*. It is called *cyclically Britton-reduced* if $xx$ is Britton-reduced. Britton reductions are effective because we can check whether $\gamma = a^\ell$ (resp. $\gamma = t^\ell$) in $H$. Thus, on input $x \in \{a, \bar{a}, t, \bar{t}, b, \bar{b}\}^*$ we can effectively calculate a Britton-reduced word $\hat{x}$ with $x = \hat{x}$ in $\mathbf{G}_{1,2}$. The following assertions are standard facts for HNN-extensions, see [14]:

1. If $x$ is Britton-reduced then $x \in H$ if and only if $|x|_\beta = 0$.
2. If $x$ is Britton-reduced and $|x|_\beta = 0$ then $x = 1$ in $\mathbf{G}_{1,2}$ if and only if $x = 1$ in $H$.
3. Let $\beta_1\gamma_1\ldots\beta_k\gamma_k$ and $\beta'_1\gamma'_1\ldots\beta'_{k'}\gamma'_{k'}$ be $\beta$-factorizations of Britton-reduced words $x$ and $y$ such that $k \geq 2$ and $x = y$ in $\mathbf{G}_{1,2}$. Then we have $k = k'$ and $(\beta_1, \ldots, \beta_k) = (\beta'_1, \ldots, \beta'_{k'})$. Moreover, $\gamma'_1 \in \gamma_1 T$ if $\beta_2 = b$ and $\gamma'_1 \in \gamma_1 A$ if $\beta_2 = \bar{b}$.

*Example 8.* Define words $w_0 = t$ and $w_{n+1} = b\, w_n\, a\, \overline{w_n}\, \bar{b}$ for $n \geq 0$. Then we have $|w_n| = 2^{n+2} - 3$ but $w_n = t^{\tau(n)}$ in $\mathbf{G}_{1,2}$.

The *power circuit-representation* of a $\beta$-factorization $\gamma_0\beta_1\gamma_1\ldots\beta_k\gamma_k$ is the sequence $(\beta_1, \ldots, \beta_k)$ and a power circuit $(\Gamma, \delta)$ together with a sequence of markings $K_0, S_0, L_0, \ldots, K_k, S_k, L_k$ such that $[\varepsilon(K_i), \varepsilon(S_i), \varepsilon(L_i)] = [k_i, s_i, \ell_i]$ is the triple representation of $\gamma_i \in H$ for $1 \leq i \leq k$. It is known that the word problem of $\mathbf{G}_{1,2}$ is decidable in cubic time. Actually a more precise statement holds.

**Proposition 9** ([18,6]). *There is a cubic time algorithm which computes on input of a power circuit representation of $x = \gamma_0 \beta_1 \gamma_1 \ldots \beta_k \gamma_k$ a power circuit representation of a Britton-reduced word (resp. cyclically Britton-reduced word) $\widehat{x}$ such that $x = \widehat{x}$ in $\mathbf{G}_{1,2}$ (resp. $x \sim_{\mathbf{G}_{1,2}} \widehat{x}$). Moreover, the size for the power circuit representation of $\widehat{x}$ is linear in the size of the power circuit representation of $x$.*

*Remark 10.* A polynomial time algorithm for the result in Prop. 9 has been given first in [18], it has been estimated by $\mathcal{O}(n^7)$. This was lowered in [6] to cubic time.

**Theorem 11.** *The following computation can be performed in time $\mathcal{O}(n^4)$. Input: words $x, y \in \{a, \bar{a}, b, \bar{b}\}^*$. Decide whether $|\widehat{x}|_\beta > 0$ for a cyclically Britton-reduced form $\widehat{x}$ of $x$. If "yes", decide $x \sim_{\mathbf{G}_{1,2}} y$ and, in the positive case, compute a power circuit representation of some $z$ such that $zx\bar{z} = y$ in $\mathbf{G}_{1,2}$.*

*Proof.* Due to Prop. 9, we may assume that input words $x$ and $y$ are given as cyclically Britton-reduced words. In particular, $\widehat{x} = x$ and $|\widehat{x}|_\beta = n > 0$. Let us write $x = \gamma_0 b^{\varepsilon_1} \gamma_1 \ldots b^{\varepsilon_n} \gamma_n$ as its $\beta$-factorization where $\varepsilon_i = \pm 1$. If all $\varepsilon_i = +1$ then we replace $x$ and $y$ by $\bar{x}$ and $\bar{y}$. Hence, without restriction there exists some $\varepsilon_i = -1$. After a possible transposition we may assume that $x = b^{\varepsilon_1} \gamma_1 \ldots b^{\varepsilon_n} \gamma_n$ with $\varepsilon_1 = -1$. Since $y$ is cyclically Britton-reduced, too, Collins' Lemma ([14, Thm. IV.2.5]) tells us several things: If $x \sim_{\mathbf{G}_{1,2}} y$ then $|y|_\beta = n$ and after some transposition the $\beta$-factorization of $y$ can be written as $b^{\varepsilon_1} \gamma'_1 \ldots b^{\varepsilon_n} \gamma'_n$. Moreover, still by Collins' Lemma, we now have $x \sim_{\mathbf{G}_{1,2}} y \iff \exists k \in \mathbb{Z} : y = a^k x a^{-k}$ in $\mathbf{G}_{1,2}$. The key is that $k$ is unique and that we find an efficient way to calculate it,[3] see [7] for the calculations.

By Prop. 9, the tests $a^k x \bar{a}^k = y \in \mathbf{G}_{1,2}$ can be performed in cubic time. All other computations can be done in quadratic time by Prop. 4. Since all transpositions of the $\beta$-factorization for $y$ have to be considered this yields an $\mathcal{O}(n^4)$-algorithm. □

For the remainder of the section the situation is as follows: We have $x = (r, m) \in \mathbb{Z}[1/2] \rtimes \mathbb{Z}$ and $y = (s, q) \in \mathbb{Z}[1/2] \rtimes \mathbb{Z}$, both can be assumed to be in power circuit representation. We may assume $x \neq 1 \neq y$ in $\mathbf{G}_{1,2}$. After conjugation with some $t^k$ where $k$ is large enough we may assume that $r, m, s, q \in \mathbb{Z}$. If $m = 0$ then we replace $x$ by $bx\bar{b}$. Hence, $m \neq 0$ and, by symmetry, $q \neq 0$, too. By (2) and "division in power circuits", we are able to to test whether $(r, m) \sim_H (0, m)$ and $(s, q) \sim_H (0, q)$. Assume that one of the answers is "no". Say, $(r, m) \not\sim_H (0, m)$. Then there is no $h \in A \cup T \subseteq H$ such that $(r, m) \sim_H h$. Since then $\beta\gamma(r, m)\bar{\gamma}\bar{\beta}$ is Britton-reduced for all $\beta \in \{b, \bar{b}\}$, $\gamma \in \{a, \bar{a}, t, \bar{t}\}^*$ we obtain:

**Proposition 12.** *Let $r, m \in \mathbb{Z}$, $m \neq 0$. If $(r, m) \not\sim_H (0, m)$ then*

$$(r, m) \sim_{\mathbf{G}_{1,2}} (s, q) \iff (r, m) \sim_H (s, q).$$

By Prop. 12, we may assume $(r, m) \sim_H (0, m)$, $(s, q) \sim_H (0, q)$, and $(r, m) \not\sim_H (s, q)$. This involves perhaps non-elementary procedures. However, it remains to decide $(0, m) \sim_{\mathbf{G}_{1,2}} (0, q)$, only. The last test is polynomial time again, even for power circuits.

---
[3] Beese calculates in [2] this value $k$ and computes certain normal forms which are checked for equivalence. This leads to an exponential time algorithm.

**Proposition 13.** *Let $m, q \in \mathbb{Z}$. Then we have*

$$(0, m) \sim_{\mathbf{G}_{1,2}} (0, q) \iff (m, 0) \sim_H (q, 0) \iff \exists k \in \mathbb{Z} : m = 2^k q.$$

**Corollary 14.** *The following problem is decidable in at most non-elementary time. Input: Power circuit representations $x, y$ for elements of $\mathbf{G}_{1,2}$. Question: $x \sim_{\mathbf{G}_{1,2}} y$?*

**Corollary 15.** *If there is no elementary algorithm to solve the division problem in power circuits then the conjugacy problem in the Baumslag group $\mathbf{G}_{1,2}$ is non-elementary in the average case even for a unary representation of group elements.*

*Proof.* Assume that the conjugacy problem in the Baumslag group $\mathbf{G}_{1,2}$ is elementary on the average. We give an elementary algorithm to solve division in power circuits. Let $(\Gamma, \delta)$ be a power circuit of size $n$ with markings $M$ and $S$ such that $\varepsilon(M) = m$ and $\varepsilon(S) = s$. For each node in $P \in \Gamma$ it is easy to construct a word $w(P) \in \{a, \bar{a}, b, \bar{b}\}^*$ such that $t^{\varepsilon(P)} = w(P)$ in $\mathbf{G}_{1,2}$ and $|w(P)| \leq n^n$. Just follow the scheme from Ex. 8. Hence, in time $2^{\mathcal{O}(n \log n)}$ we can construct words $x$ and $y$ such that $x = (0, m)$ and $y = (2^s - 1, m)$ in $\mathbf{G}_{1,2}$. Now by Rem. 7 we have $m \mid s$ if and only if $x \sim_{\mathbf{G}_{1,2}} y$. The number of words of length $2^{\mathcal{O}(n \log n)}$ is at most $2^{2^{\mathcal{O}(n \log n)}}$. □

## 5 Generic Case Analysis

Let us define a preorder between functions from $\mathbb{N}$ to $\mathbb{R}^{\geq 0}$ as follows. We let $f \preceq g$ if there exist $k \in \mathbb{N}$ and $\varepsilon > 0$ such that for almost all $n$ we have

$$f(n) \leq n^k g(n) + 2^{-\varepsilon n}.$$

Moreover, we let $f \approx g$ if both, $f \preceq g$ and $g \preceq f$. We are mainly interested in functions $f \approx 0$. These functions form an ideal in the ring of functions which are bounded by polynomial growth. Moreover, if $f \approx 0$ then $g \approx 0$ for $g(n) \in f(\theta(n))$. The notion $f \approx g$ is therefore rather flexible and simplifies some formulae. We consider cyclically reduced words over $\Sigma = \{a, \bar{a}, b, \bar{b}\}$ of length $n$ with uniform distribution. This yields a function $p(n) = \Pr\left[\exists y : x \sim_{\mathbf{G}_{1,2}} y \land y \in H\right]$. We prove $p(n) \approx 0$. More precisely, we are interested in the following result.

**Theorem 16.** *There is a strongly generic algorithm that decides in time $\mathcal{O}(n^4)$ on cyclically reduced input words $x, y \in \{a, \bar{a}, b, \bar{b}\}^*$ with $|xy| \in \theta(n)$ whether $x \sim_{\mathbf{G}_{1,2}} y$.*

*Proof.* By Thm. 11, there is an algorithm deciding $x \sim_{\mathbf{G}_{1,2}} y$ which runs in time $\mathcal{O}(n^4)$ for inputs which cannot be conjugated to elements in $H$. Hence, we only have to bound the number of cyclically reduced words of length $m \in \theta(n)$ which can be conjugated to some element in $H$. For simplicity of notation we assume $m = n$. A reduced word in $\Sigma^n$ can be identified with a random walk without backtracking in the Cayley graph of $\mathbf{G}_{1,2}$ with generators $a$ and $b$. We encode reduced words over $\Sigma$ of length $n$ in a natural way as words in $\Omega = \Sigma \cdot \{1, 2, 3\}^{n-1}$. On $\Omega$ we choose a uniform probability (e.g., if the $i$-th letter is $b$ then the $i+1$-st letter is $a$, $\bar{a}$, or $b$ with equal probability $1/3$). Because we are interested in conjugacy, we compute the probability under the condition that $x \in \Omega$ is cyclically reduced. (Actually this does not change the results but makes the analysis smoother.) The probability that $x \in \Omega$ is cyclically reduced is at least $2/3$ for all $n$. Let $C \subseteq \Omega$ be the subset of cyclically reduced words. We show $\Pr\left[\exists y : x \sim_{\mathbf{G}_{1,2}} y \wedge y \in H \mid x \in C\right] \approx 0$. The question whether there exists some $y$ with $x \sim_{\mathbf{G}_{1,2}} y$ is answered by calculating Britton reductions for a transposition of $x$. The set $C$ is closed under transpositions and it is no restriction to assume that $|x|_\beta \geq 1$. Therefore, we can choose the transposition that $x' = vu$ where $x = uv$ such that the first letter of $x'$ is $\beta \in \{b, \bar{b}\}$. There are at most $n$ such transpositions. Hence,

$$\Pr\left[\exists y : x \sim_{\mathbf{G}_{1,2}} y \wedge y \in H \mid x \in C\right] \approx \Pr[x \in H \mid x \in C]$$
$$= \Pr[x \in H \wedge x \in C] \cdot \Pr[x \in C]^{-1} \leq \Pr[x \in H] \cdot \Pr[x \in C]^{-1} \leq \frac{3}{2}\Pr[x \in H].$$

It is therefore enough to prove $\Pr[x \in H] \approx 0$. We switch the probability space and we embed $\Omega$ into the space $\Sigma^*$ with a measure $\mu_0$ on $\Sigma^*$ which concentrates on $\Omega$, i.e., $\mu_0(\Omega) = 1$. Within $\Omega$ we still have a uniform distribution for $\mu_0$. In order to emphasize this change of view, we write $\Pr[\cdots] = \Pr_0[\cdots]$. We are now interested in words $x \in \{b, \bar{b}\} \cdot \Sigma^*$ which contain exactly $2m$ letters $\beta \in \{b, \bar{b}\}$ for $m \geq 1$. (The number $|x|_\beta$ must be even if $x \in H$.) Each such word can be written as a $\beta$-factorization of the form $x = \beta_1 \alpha_1 \ldots \beta_{2m} \alpha_{2m}$ where $\alpha_i = a^{e_i}$ with $e_i \in \mathbb{Z}$. This defines a new measure $\mu_m$ on $\Sigma^*$ which is defined as follows. We start a random walk without backtracking with either $b$ or $\bar{b}$ with equal probability. For the next letter there are always 3 possibilities, each is chosen with probability $1/3$. We continue as long as the random walk contains at most $2m$ letters from $\{b, \bar{b}\}$. This gives a corresponding probability on $\Sigma^*$ which is concentrated on those words with $|x|_\beta = 2m$. We denote the corresponding probability by $\Pr_m[\cdots]$. An easy calculation shows that $\Pr_0[x \in H] \approx \sum_{m=\lceil n/4 \rceil}^n \Pr_m[x \in H]$ (see [7]). Hence, it remains to show $\Pr_m[x \in H] \approx 0$.

From now on we work with the measure $\mu_n$ and the corresponding probability $\Pr_n[\cdots]$ for $n \geq 1$. Thus, we may assume that our probability space contains only those words $x$ which have $\beta$-factorizations of the form $x = \beta_1 \alpha_1 \ldots \beta_{2n} \alpha_{2n}$ with $\alpha_i \in a^{\mathbb{Z}}$.

**Lemma 17.** *We have $\Pr_n[x \in H] \leq (8/9)^n$.*

The proof of Lem. 17 is based on a "pairing" with Dyck words: Define a new alphabet $B = \{\lfloor, \rceil\}$ where $\lfloor$ is an opening left-bracket and $\rceil$ is the corresponding closing right-bracket. The set of Dyck words $D_n$ is the set of words in $B^{2n}$ with correct bracketing. The number of Dyck words is well-understood, we have $|D_n| = \frac{1}{n+1}\binom{2n}{n} \leq 4^n$. The connection between Dyck words and Britton reductions is as follows. Britton reductions are defined for words $\{a, \bar{a}, t, \bar{t}, b, \bar{b}\}^*$. Consider a $\beta$-factorization of the form $x = \beta_1 \alpha_1 \ldots \beta_{2n} \alpha_{2n}$ with $\alpha_i \in a^{\mathbb{Z}}$. If $x \in H$, then there exists a sequence of Britton reductions which transforms $x$ into $\widehat{x} \in \{a, \bar{a}, t, \bar{t}\}^*$. We call such a sequence a *successful Britton reduction*. Every successful Britton reduction defines in a natural way a Dyck word by assigning an opening bracket to position $i$ and a closing bracket to position $j$ if $\beta_i u \beta_j$ is replaced by a Britton reduction. Moreover, Britton reductions are confluent on $H$. In particular, this means that for $x \in H$ we can start a successful Britton reduction by replacing all factors $\beta_i a^e \beta_{i+1}$ with $\beta_i = b = \overline{\beta_{i+1}}$ and $e \in \mathbb{Z}$ by $t^e$ where $1 \leq i < 2n$. Thus, if such a successful Britton reduction is described by $d$, then we may assume that

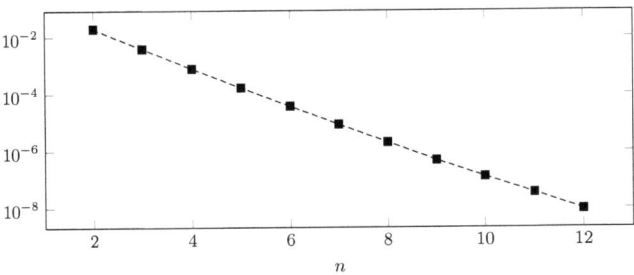

**Fig. 1.** Portion of reduced words $x \in H$ with $|x|_\beta = 2n$, sampling $11 \cdot 10^9$ words

$d_i d_{i+1} = \lfloor \rceil$ whenever $\beta_i a^e \beta_{i+1} = b a^e \bar{b}$. Vice versa, if $d_i d_{i+1} = \lfloor \rceil$, then we must have $\beta_i = b = \overline{\beta_{i+1}}$, otherwise $d$ is no description of any Britton reduction for $x$ at all. Note that for each $i$ with $d_i = \lfloor$ there is exactly one $j$ which matches $d_i$. The characterization of $j$ is that $d_{i+1} \cdots d_{j-1}$ is a Dyck word and $d_j = \rceil$. If $d$ describes a Britton reduction for $x$ and $(i, j)$ is a matching pair for $d$ then $\beta_i \overline{\beta_j} = \beta \overline{\beta}$ for some $\beta \in \{b, \bar{b}\}$. We therefore say that $x$ and $d$ *match* if the following two conditions are satisfied:

1. For all $1 \leq i < 2n$ we have $d_i d_{i+1} = \lfloor \rceil \iff \beta_i \beta_{i+1} = b \bar{b}$.
2. For all $1 \leq i < j \leq 2n$ where $d_i d_j = \lfloor \rceil$ is a matching pair we have $\beta_i \beta_j = \beta \overline{\beta}$.

We define $\langle x, d \rangle_\beta = 1$ if $x$ and $d$ match and $\langle x, d \rangle_\beta = 0$ otherwise. We refine this pairing by defining $\langle x, d \rangle = 1$ if $\langle x, d \rangle_\beta = 1$ and $d$ describes a successful Britton reduction proving $x \in H$. Otherwise we let $\langle x, d \rangle = 0$. Clearly,

$$\mathrm{Pr}_n[x \in H] \leq \sum_{d \in D_n} \mathrm{Pr}_n[\langle x, d \rangle = 1]. \tag{4}$$

Since $|D_n| \leq 4^n$, the proof of Lem. 17 reduces to show that for every $d \in D_n$ we have

$$\mathrm{Pr}_n[\langle x, d \rangle = 1] \leq (2/9)^n. \tag{5}$$

**Lemma 18.** *Let $d \in D_n$ be a Dyck word and $k = |\{i \mid d_i d_{i+1} = \lfloor \rceil\}|$. Then we have $\mathrm{Pr}_n[\langle x, d \rangle_\beta = 1] \leq (2/3)^{n-k}(2/9)^k$.*

**Lemma 19.** *Let $d \in D_n$ be a Dyck word and $k = |\{i \mid d_i d_{i+1} = \lfloor \rceil\}|$. Then we have*

$$\mathrm{Pr}_k[\langle x, d \rangle = 1 \mid \langle x, d \rangle_\beta = 1] \leq (5/16)^{n-k}.$$

Lem. 18 and Lem. 19 enable us to calculate $\mathrm{Pr}_n[\langle x, d \rangle = 1]$ as follows:

$$\mathrm{Pr}_n[\langle x, d \rangle = 1] = \mathrm{Pr}_k[\langle x, d \rangle = 1 \mid \langle x, d \rangle_\beta = 1] \cdot \mathrm{Pr}_n[\langle x, d \rangle_\beta = 1]$$
$$\leq (5/16)^{n-k} \cdot (2/3)^{n-k}(2/9)^k \leq (2/9)^n.$$

This shows (5) and therefore Lem. 17 which in turn implies Thm. 16. □

**Computer Experiments.** We have conducted computer experiments with a sample of $11 \cdot 10^9$ (i.e., 11 billion) random words $x \in \Sigma^*$ with $4 \leq |x|_\beta = 2n \leq 24$, see Fig. 1. Moreover, for $n = 14$ our random process did not find a single $x \in H$. The experiments confirm $\mathrm{Pr}_n[x \in H] \approx 0$. The initial values seem to suggest $\mathrm{Pr}_n[x \in H] \in \mathcal{O}(0.25^n)$. This is much better than the upper bound of Lem. 17, but our proof is using very rough estimations in (4) and (5). Hence, a difference is no surprise.

# References

1. Baumslag, G.: A non-cyclic one-relator group all of whose finite quotients are cyclic. J. Austr. Math. Soc. 10(3-4), 497–498 (1969)
2. Beese, J.: Das Konjugationsproblem in der Baumslag-Gersten-Gruppe. Diploma thesis, Fakultät Mathematik, Universität Stuttgart (2012) (in German)
3. Borovik, A.V., Myasnikov, A.G., Remeslennikov, V.N.: Generic Complexity of the Conjugacy Problem in HNN-Extensions and Algorithmic Stratification of Miller's Groups. IJAC 17, 963–997 (2007)
4. Ceccherini-Silberstein, T., Grigorchuk, R.I., de la Harpe, P.: Amenability and paradoxical decompositions for pseudogroups and discrete metric spaces. Tr. Mat. Inst. Steklova 224, 68–111 (1999)
5. Craven, M.J., Jimbo, H.C.: Evolutionary algorithm solution of the multiple conjugacy search problem in groups, and its applications to cryptography. Groups Complexity Cryptology 4, 135–165 (2012)
6. Diekert, V., Laun, J., Ushakov, A.: Efficient algorithms for highly compressed data: The word problem in Higman's group is in P. IJAC 22, 1–19 (2012)
7. Diekert, V., Miasnikov, A., Weiß, A.: Conjugacy in Baumslag's group, generic case complexity, and division in power circuits. CoRR, abs/1309.5314 (2013)
8. Gersten, S.M.: Isodiametric and isoperimetric inequalities in group extensions (1991)
9. Grigoriev, D., Shpilrain, V.: Authentication from matrix conjugation. Groups Complexity Cryptology 1, 199–205 (2009)
10. Hesse, W.: Division is in uniform $TC^0$. In: Orejas, F., Spirakis, P.G., van Leeuwen, J. (eds.) ICALP 2001. LNCS, vol. 2076, pp. 104–114. Springer, Heidelberg (2001)
11. Hesse, W., Allender, E., Barrington, D.A.M.: Uniform constant-depth threshold circuits for division and iterated multiplication. JCCS 65, 695–716 (2002)
12. Kapovich, I., Miasnikov, A.G., Schupp, P., Shpilrain, V.: Generic-case complexity, decision problems in group theory and random walks. J. Algebra 264, 665–694 (2003)
13. Kapovich, I., Myasnikov, A., Schupp, P., Shpilrain, V.: Average-case complexity and decision problems in group theory. Adv. Math. 190, 343–359 (2005)
14. Lyndon, R., Schupp, P.: Combinatorial Group Theory, 1st edn. (1977)
15. Magnus, W.: Das Identitätsproblem für Gruppen mit einer definierenden Relation. Math. Ann. 106, 295–307 (1932)
16. Miller III, C.F.: On group-theoretic decision problems and their classification. Annals of Mathematics Studies, vol. 68. Princeton University Press (1971)
17. Myasnikov, A., Shpilrain, V., Ushakov, A.: Group-based Cryptography. Advanced courses in mathematics. CRM Barcelona, Birkhäuser (2008)
18. Myasnikov, A.G., Ushakov, A., Won, D.W.: The Word Problem in the Baumslag group with a non-elementary Dehn function is polynomial time decidable. Journal of Algebra 345, 324–342 (2011)
19. Myasnikov, A.G., Ushakov, A., Won, D.W.: Power circuits, exponential algebra, and time complexity. IJAC 22, 51 pages (2012)
20. Shpilrain, V., Zapata, G.: Combinatorial group theory and public key cryptography. Appl. Algebra Eng. Comm. Comput. 17, 291–302 (2006)
21. Vollmer, H.: Introduction to Circuit Complexity. Springer, Berlin (1999)
22. Woess, W.: Random walks on infinite graphs and groups - a survey on selected topics. London Math. Soc. 26, 1–60 (1994)
23. Woess, W.: Random Walks on Infinite Graphs and Groups. Cambridge Univ. Press (2000)

# Hierarchical Complexity of 2-Clique-Colouring Weakly Chordal Graphs and Perfect Graphs Having Cliques of Size at Least 3*

Helio B. Macêdo Filho[1], Raphael C.S. Machado[2], and Celina M.H. Figueiredo[1]

[1] COPPE, Universidade Federal do Rio de Janeiro
[2] Inmetro — Instituto Nacional de Metrologia, Qualidade e Tecnologia

**Abstract.** A clique of a graph is a maximal set of vertices of size at least 2 that induces a complete graph. A $k$-clique-colouring of a graph is a colouring of the vertices with at most $k$ colours such that no clique is monochromatic. Défossez proved that the 2-clique-colouring of perfect graphs is a $\Sigma_2^P$-complete problem [J. Graph Theory 62 (2009) 139–156]. We strengthen this result by showing that it is still $\Sigma_2^P$-complete for weakly chordal graphs. We then determine a hierarchy of nested subclasses of weakly chordal graphs whereby each graph class is in a distinct complexity class, namely $\Sigma_2^P$-complete, $\mathcal{NP}$-complete, and $\mathcal{P}$. We solve an open problem posed by Kratochvíl and Tuza to determine the complexity of 2-clique-colouring of perfect graphs with all cliques having size at least 3 [J. Algorithms 45 (2002), 40–54], proving that it is a $\Sigma_2^P$-complete problem. We then determine a hierarchy of nested subclasses of perfect graphs with all cliques having size at least 3 whereby each graph class is in a distinct complexity class, namely $\Sigma_2^P$-complete, $\mathcal{NP}$-complete, and $\mathcal{P}$.

**Keywords:** $(\alpha, \beta)$-polar graphs, clique-colouring, hierarchical complexity, perfect graphs, weakly chordal graphs.

## 1 Introduction

Let $G = (V, E)$ be a simple graph with $n = |V|$ vertices and $m = |E|$ edges. A *clique* of $G$ is a maximal set of vertices of size at least 2 that induces a complete graph. A *$k$-clique-colouring* of a graph is a colouring of the vertices with at most $k$ colours such that no clique is monochromatic. Any undefined notation concerning complexity classes follows that of Marx [9].

A *cycle* is sequence of vertices starting and ending at the same vertex, with each two consecutive vertices in the sequence adjacent to each other in the graph. A *chord* of a cycle is an edge joining two nodes that are not consecutive in the cycle.

The *clique-number* $\omega(G)$ of a graph $G$ is the number of vertices of a clique with the largest possible size in $G$. A *perfect graph* is a graph in which every

---
* Partially supported by CNPq and FAPERJ. Fullpaper available at http://arxiv.org/abs/1312.2086

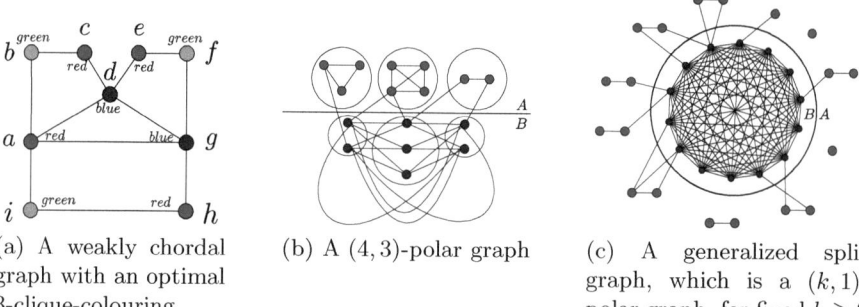

(a) A weakly chordal graph with an optimal 3-clique-colouring  (b) A (4,3)-polar graph  (c) A generalized split graph, which is a $(k,1)$-polar graph, for fixed $k \geq 2$

**Fig. 1.** Examples of $(\alpha, \beta)$-polar graphs

induced subgraph $H$ needs exactly $\omega(H)$ colours in its vertices such that no $K_2$ (not necessarily clique) is monochromatic. The celebrated *Strong Perfect Graph Theorem* of Chudnovsky et al. [3] says that a graph is perfect if neither it nor its complement contains a chordless cycle with an odd number of vertices greater than 4. A graph is *chordal* if it does not contain a chordless cycle with a number of vertices greater than 3, and a graph is *weakly chordal* if neither it nor its complement contains a chordless cycle with a number of vertices greater than 4.

Both clique-colouring and perfect graphs have attracted much attention due to a conjecture posed by Duffus et al. [5] that *perfect graphs are k-clique-colourable for some constant k*. This conjecture has not yet been proved. Following the chronological order, Kratochvíl and Tuza gave a framework to argue that 2-clique-colouring is $\mathcal{NP}$-hard and proved that 2-clique-colouring is $\mathcal{NP}$-complete for $K_4$-free perfect graphs [7]. Notice that $K_3$-free perfect graphs are bipartite graphs, which are clearly 2-clique-colourable. Moreover, 2-clique-colouring is in $\Sigma_2^P$, since it is co$\mathcal{NP}$ to check that a colouring of the vertices is a clique-colouring. A few years later, the 2-clique-colouring problem was proved to be a $\Sigma_2^P$-complete problem by Marx [9], a major breakthrough in the clique-colouring area. Défossez [4] proved later that 2-clique-colouring of perfect graphs remained a $\Sigma_2^P$-complete problem.

When restricted to chordal graphs, 2-clique-colouring is in $\mathcal{P}$, since all chordal graphs are 2-clique-colourable [10]. Notice that chordal graphs are a subclass of weakly chordal graphs, while perfect graphs are a superclass of weakly chordal graphs. In constrast to chordal graphs, not all weakly chordal graphs are 2-clique-colourable (see Fig. 1a).

We show that 2-clique-colouring of weakly chordal graphs is a $\Sigma_2^P$-complete problem, improving the proof of Défossez [4] that 2-clique-colouring is a $\Sigma_2^P$-complete problem for perfect graphs. As a remark, Défossez [4] constructed a graph which is not a weakly chordal graph as long as it has chordless cycles with even number of vertices greater than 5 as induced subgraphs. We determine a hierarchy of nested subclasses of weakly chordal graphs whereby each graph class is in a distinct complexity class, namely $\Sigma_2^P$-complete, $\mathcal{NP}$-complete, and $\mathcal{P}$.

A graph is $(\alpha, \beta)$-*polar* if there exists a partition of its vertex set into two sets $A$ and $B$ such that all connected components of the subgraph induced by $A$ and of the complementary subgraph induced by $B$ are complete graphs. Moreover, the order of each connected component of the subgraph induced by $A$ (resp. of the complementary subgraph induced by $B$) is upper bounded by $\alpha$ (resp. upper bounded by $\beta$) [2]. A *satellite* of an $(\alpha, \beta)$-polar graph is a connected component of the subgraph induced by $A$ (see Fig. 1b). In this work, we restrict ourselves to the $(\alpha, \beta)$-*polar* graphs with $\beta = 1$, so the subgraph induced by $B$ is complete and the order of each satellite is upper bounded by $\alpha$ (see Fig. 1c). Clearly, $(\alpha, 1)$-polar graphs are perfect, since they do not contain chordless cycles with an odd number of vertices greater than 4 nor their complements.

A *generalized split graph* is a graph $G$ such that $G$ or its complement is an $(\infty, 1)$-polar graph [11]. See Fig. 1c for an example of a generalized split graph, which is a $(2, 1)$-polar graph. The class of generalized split graphs plays an important role in the areas of perfect graphs and clique-colouring. This class was introduced by Prömel and Steger [11] to show that the strong perfect graph conjecture is at least asymptotically true by proving that almost all $C_5$-free graphs are generalized split graphs. Approximately 14 years later the strong perfect graph conjecture became the *Strong Perfect Graph Theorem* by Chudnovsky et al. [3]. Regarding clique-colouring, Bacsó et al. [1] proved that generalized split graphs are 3-clique-colourable and concluded that almost all perfect graphs are 3-clique-colourable [1]. This conclusion supports the conjecture due to Duffus et al. [5]. In fact, there is no example of a perfect graph where more than three colors would be necessary to clique-colour. Surprisingly, after more than 20 years, relatively little progress has been made on the conjecture.

The class of $(k, 1)$-polar graphs, for fixed $k \geq 3$, is incomparable to the class of weakly chordal graphs. Indeed, a chordless path with seven vertices $P_7$ and a complement of a chordless cycle with six vertices $\overline{C_6}$ are witnesses. Nevertheless, $(2, 1)$-polar graphs are a subclass of weakly chordal graphs, since they do not contain a chordless cycle with an even number of vertices greater than 5. We show that 2-clique-colouring of $(2, 1)$-polar graphs is a $\mathcal{NP}$-complete problem. Finally, the class of $(1, 1)$-polar graphs is precisely the class of split graphs. It is interesting to recall that 2-clique-colouring of $(1, 1)$-polar graphs is in $\mathcal{P}$, since $(1, 1)$-polar are a subclass of chordal graphs, which are 2-clique-colourable.

Giving continuity to our results, we investigate an open problem left by Kratochvíl and Tuza [7] to determine the complexity of 2-clique-colouring of perfect graphs with all cliques having size at least 3. Restricting the size of the cliques to be at least 3, we first show that 2-clique-colouring is still $\mathcal{NP}$-complete for $(3, 1)$-polar graphs, even if it is restricted to weakly chordal graphs with all cliques having size at least 3. Subsequently, we prove that the 2-clique-colouring of $(2, 1)$-polar graphs becomes polynomial when all cliques have size at least 3. Recall that the 2-clique-colouring of $(2, 1)$-polar graphs is $\mathcal{NP}$-complete when there are no restrictions on the size of the cliques.

We finish the paper answering the open problem of determining the complexity of 2-clique-colouring of perfect graphs with all cliques having size at least 3 [7],

**Table 1.** 2-clique-colouring complexity of perfect graphs and subclasses

| Class | | | | 2-clique-colouring complexity |
|---|---|---|---|---|
| - | Perfect | - | - | $\Sigma_2^P$-complete [4] |
| | | $K_4$-free | - | $\mathcal{NP}$-complete [7] |
| | | $K_3$-free (Bipartite) | - | $\mathcal{P}$ |
| | Weakly chordal | - | - | $\Sigma_2^P$-complete |
| | (3, 1)-polar | - | - | $\mathcal{NP}$-complete |
| | (2, 1)-polar | - | - | |
| | Chordal (includes Split) | - | - | $\mathcal{P}$ [10] |
| All cliques having size at least 3 | Perfect | - | - | $\Sigma_2^P$-complete |
| | Weakly chordal | - | - | |
| | | (3, 1)-polar | - | $\mathcal{NP}$-complete |
| | (2, 1)-polar | - | - | $\mathcal{P}$ |

by improving our proof that 2-clique-colouring is a $\Sigma_2^P$-complete problem for weakly chordal graphs. We replace each $K_2$ clique by a gadget with no clique of size 2, which forces distinct colours into two given vertices.

The paper is organized as follows. In Section 2, we show that 2-clique-colouring is still $\Sigma_2^P$-complete for weakly chordal graphs. We then determine a hierarchy of nested subclasses of weakly chordal graphs whereby each graph class is in a distinct complexity class, namely $\Sigma_2^P$-complete, $\mathcal{NP}$-complete, and $\mathcal{P}$. In Section 3, we determine the complexity of 2-clique-colouring of perfect graphs with all cliques having size at least 3, answering a question of Kratochvíl and Tuza [7]. We then determine a hierarchy of nested subclasses of perfect graphs with all cliques having size at least 3 whereby each graph class is in a distinct complexity class. We refer the reader to Table 1 for our results and related work about 2-clique-colouring complexity of perfect graphs.

## 2 Hierarchical Complexity of 2-Clique-Colouring of Weakly Chordal Graphs

Défossez proved that 2-clique-colouring of perfect graphs is a $\Sigma_2^P$-complete problem [4]. In this section, we strengthen this result by showing that it is still $\Sigma_2^P$-complete for weakly chordal graphs. We show a subclass of perfect graphs (resp. of weakly chordal graphs) in which 2-clique-colouring is neither a $\Sigma_2^P$-complete problem nor in $\mathcal{P}$, namely $(3,1)$-polar graphs (resp. $(2,1)$-polar graphs). Recall that 2-clique-colouring of $(1,1)$-polar graphs is in $\mathcal{P}$, since $(1,1)$-polar are a subclass of chordal graphs, thereby 2-clique-colourable. Notice that weakly chordal, $(2,1)$-polar, and $(1,1)$-polar (resp. perfect, $(3,1)$-polar, and $(1,1)$-polar) are nested classes of graphs.

Given a graph $G = (V, E)$ and adjacent vertices $a, g \in V$, we say that we add to $G$ a copy of an auxiliary graph $AK(a, g)$ of order 7 – depicted in Fig. 2a

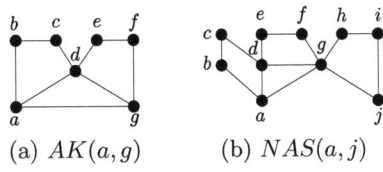

**Fig. 2.** Auxiliary graphs $AK(a,g)$ and $NAS(a,j)$

– if we change the definition of $G$ by doing the following: we first change the definition of $V$ by adding to it copies of the five vertices $b, \ldots, f$ of the auxiliary graph $AK(a,g)$; then we change the definition of $E$, adding to it copies of the eight edges $(u,v)$ of $AK(a,g)$. Similarly, given a graph $G = (V,E)$ and non-adjacent vertices $a, j \in V$, we say that we add to $G$ a copy of an auxiliary graph $NAS(a,j)$ of order 10 – depicted in Fig. 2b – if we change the definition of $G$ by doing the following: we first change the definition of $V$ by adding to it eight copies of the vertices $b, \ldots, i$ of the auxiliary graph $NAS(a,j)$; then we change the definition of $E$, adding to it copies of the thirteen edges $(u,v)$ of $NAS(a,j)$.

The auxiliary graph $AK(a,g)$ is constructed to force the same colour (in a 2-clique-colouring) to adjacent vertices $a$ and $g$, while the auxiliary graph $NAS(a,j)$ is constructed to force distinct colours (in a 2-clique-colouring) to non-adjacent vertices $a$ and $j$ (see Lemmas 1 and 2). Refer to the Appendix for the omitted proofs throughout the present work.

**Lemma 1.** *Let $G$ be a graph and $a, g$ be adjacent vertices in $G$. If we add to $G$ a copy of an auxiliary graph $AK(a,g)$, then in any 2-clique-colouring of the resulting graph, adjacent vertices $a$ and $g$ have the same colour.*

**Lemma 2.** *Let $G$ be a graph and $a, j$ be non-adjacent vertices in $G$. If we add to $G$ a copy of an auxiliary graph $NAS(a,j)$, then in any 2-clique-colouring of the resulting graph, non-adjacent vertices $a$ and $j$ have distinct colours.*

We improve the proof of Défossez [4], in order to determine the complexity of 2-clique-colouring for weakly chordal graphs. We prove that 2-clique-colouring weakly chordal graphs is $\Sigma_2^P$-complete by reducing the $\Sigma_2^P$-complete canonical problem QSAT2 to it. For a QSAT2 formula $\Psi = (X, Y, D)$, a weakly chordal graph $G$ is constructed such that graph $G$ is 2-clique-colourable if, and only if, there is a truth assignment of $X$, such that $\Psi$ is true for every truth assignment of $Y$.

**Theorem 1.** *The problem of 2-clique-colouring is $\Sigma_2^P$-complete for weakly chordal graphs.*

Now, our focus is on showing a subclass of weakly chordal graphs in which 2-clique-colouring is $\mathcal{NP}$-complete, namely (3, 1)-polar and (2, 1)-polar graphs.

Complements of bipartite graphs are a subclass of $(\infty, 1)$-polar graphs. Indeed, let $G = (V, E)$ be a complement of a bipartite graph, where $(A, B)$ is a partition of $V$ into two disjoint complete sets. Clearly, $G$ is a $(\infty, 1)$-polar graph.

Défossez [4] showed that it is co$\mathcal{NP}$-complete to check whether a 2-colouring of a complement of a bipartite graph is a 2-clique-colouring [4]. Hence, it is co$\mathcal{NP}$-hard to check if a colouring of the vertices of a $(\infty, 1)$-polar graph is a 2-clique-colouring. On the other hand, we show next that, if $k$ is fixed, listing all cliques of a $(k, 1)$-polar graph and checking if each clique is polychromatic can be done in polynomial-time, although the constant behind the big $O$ notation is impraticable. Theorem 2 shows that clique-colouring is in $\mathcal{NP}$ for $(k, 1)$-polar graphs, for fixed $k \geq 1$.

**Theorem 2.** *There exists an $O(n^2)$-time algorithm to check if a colouring of the vertices of a $(k, 1)$-polar graph, for a fixed $k \geq 1$, is a clique-colouring.*

We apply the ideas of the framework of Kratochvíl and Tuza [7] to determine the complexity of 2-clique-colouring of $(3, 1)$-polar graphs. We prove that 2-clique-colouring $(3, 1)$-polar graphs is $\mathcal{NP}$-complete by reducing the NAE-SAT problem to it. For a NAE-SAT formula $\Psi$, a $(3, 1)$-polar graph $G$ is constructed such that graph $G$ is 2-clique-colourable if, and only if, $\Psi$ is satisfiable. This is an intermediary step to achieve the complexity of 2-clique-colouring of $(2, 1)$-polar graphs, which are a subclass of weakly chordal graphs.

**Theorem 3.** *The problem of 2-clique-colouring is $\mathcal{NP}$-complete for $(3, 1)$-polar graphs.*

We use a reduction from 2-clique-colouring of $(3, 1)$-polar graphs to determine the complexity of 2-clique-colouring of $(2, 1)$-polar graphs. In what follows, we provide some notation to classify the structure of 2-clique-colouring of $(2, 1)$-polar graphs and of $(3, 1)$-polar graphs. We capture their similarities and make it feasible a reduction from 2-clique-colouring $(3, 1)$-polar graphs to 2-clique-colouring $(2, 1)$-polar graphs.

Let $G = (V, E)$ be a $(3, 1)$-polar graph. Let $K$ be a satellite of $G$. Consider the following four cases: $(\mathcal{K}_1)$ there exists a vertex of $K$ such that none of its neighbors is in partition $B$; $(\mathcal{K}_2)$ the complementary case of $\mathcal{K}_1$, where there exists a pair of vertices of $K$, such that the closed neighborhood of one vertex of the pair is contained in the closed neighborhood of the other vertex of the pair; $(\mathcal{K}_3)$ the complementary case of $\mathcal{K}_2$, where the intersection of the closed neighborhood of the vertices of $K$ is precisely $K$; and $(\mathcal{K}_4)$ the complementary case of $\mathcal{K}_3$. Clearly, any satellite $K$ is either in case $\mathcal{K}_1$, $\mathcal{K}_2$, $\mathcal{K}_3$, or $\mathcal{K}_4$. Refer to Fig. 3 for an example of each case of a satellite.

For a given $(3, 1)$-polar graph $G$, we proceed to obtain a $(2, 1)$-polar graph $G'$ that is 2-clique-colourable if, and only if, $G$ is 2-clique-colourable, as follows. For each satellite that is a triangle, if the triangle is in case $\mathcal{K}_4$, we replace it by an edge in which (i) both complete sets have the same neighboorhod contained in $B$ and (ii) the edge is also in case $\mathcal{K}_4$, otherwise we just delete triangle $K$. See Fig. 4 for examples. Such construction is done in polynomial-time. This algorithm and Theorem 2 imply the following theorem.

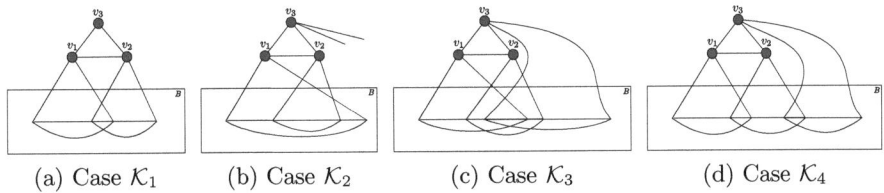

(a) Case $\mathcal{K}_1$   (b) Case $\mathcal{K}_2$   (c) Case $\mathcal{K}_3$   (d) Case $\mathcal{K}_4$

**Fig. 3.** A triangle satellite of an $(\alpha, \beta)$-polar graph

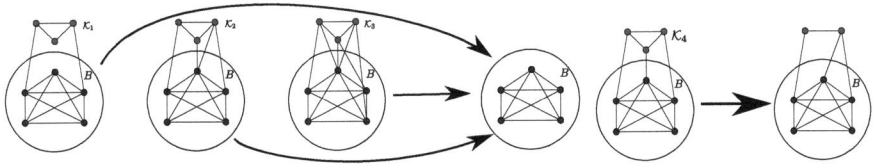

**Fig. 4.** An iteration to obtain a (2, 1)-polar graph $G'$, given a (3, 1)-polar graph $G$, such that $G$ is 2-clique-colourable if and only if $G'$ is 2-clique-colourable

**Theorem 4.** *The problem of 2-clique-colouring is $\mathcal{NP}$-complete for (2, 1)-polar graphs.*

As a remark, we noticed strong connections between hypergraph 2-colorability and 2-clique-colouring (2, 1)-polar graphs. Indeed, we have a simpler alternative proof showing that 2-clique-colouring (2, 1)-polar graphs is $\mathcal{NP}$-complete by a reduction from hypergraph 2-colouring. Please, refer to the Appendix for this alternative proof. In constrast to graphs, deciding if a given hypergraph is 2-colourable is an $\mathcal{NP}$-complete problem, even if all edges have cardinality at most 3 [8]. The reader may ask why we did not exploit only the alternative proof that is quite shorter than the original proof. The reason is related to the next section, where we show that even restricting the size of the cliques to be at least 3, the 2-clique-colouring of (3, 1)-polar graphs is still $\mathcal{NP}$-complete, while 2-clique-colouring of (2, 1)-polar graphs becomes a problem in $\mathcal{P}$.

## 3 Restricting the Size of the Cliques

Kratochvíl and Tuza [7] are interested in determining the complexity of 2-clique-colouring of perfect graphs with all cliques having size at least 3. We determine what happens with the complexity of 2-clique-colouring of (2, 1)-polar graphs, of (3, 1)-polar graphs, and of weakly chordal graphs, respectively, when all cliques are restricted to have size at least 3. The latter result address Kratochvíl and Tuza's question.

Given graph $G$ and $b_1, b_2, b_3 \in V(G)$, we say that we add to $G$ a copy of an auxiliary graph $BP(b_1, b_2, b_3)$ of order 6 – depicted in Fig. 5a – if we change the definition of $G$ by doing the following: we first change the definition of $V$ by adding to it copies of the vertices $a_1, a_2, a_3$ of the auxiliary graph $BP(b_1, b_2, b_3)$;

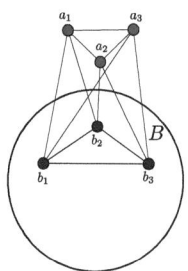
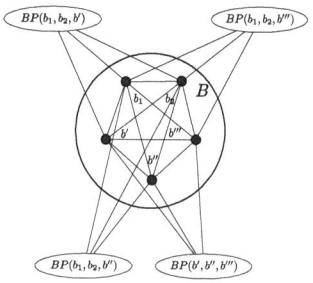

(a) Auxiliary graph $BP(b_1, b_2, b_3)$    (b) Auxiliary graph $BS(b_1, b_2)$

**Fig. 5.** Auxiliary graphs $BP(b_1, b_2, b_3)$ and $BS(b_1, b_2)$

second, we change the definition of $E$ by adding to it copies of the edges $(u, v)$ of $BP(b_1, b_2, b_3)$.

Similarly, given a graph $G$ and $b_1, b_2 \in V(G)$, we say that we add to $G$ a copy of an auxiliary graph $BS(b_1, b_2)$ of order 17 – depicted in Fig. 5b – if we change the definition of $G$ by doing the following: we first change the definition of $V$ by adding to it copies of the vertices $b'$, $b''$, $b'''$ of the auxiliary graph $BS(b_1, b_2)$; second, we change the definition of $E$ by adding to it edges so that $B(G) \cup \{b_1, b_2, b', b'', b'''\}$ is a complete set; finally, we add copies of the auxiliary graphs $BP(b_1, b_2, b')$, $BP(b_1, b_2, b'')$, $BP(b_1, b_2, b''')$, $BP(b', b'', b''')$.

**Lemma 3.** *Let $G$ be a weakly chordal graph (resp. (3, 1)-polar graph) and $b_1, b_2, b_3 \in V(G)$ (resp. $b_1, b_2, b_3 \in B(G)$). If we add to $G$ a copy of an auxiliary graph $BP(b_1, b_2, b_3)$, then the following assertions are true.*

- *The resulting graph $G'$ is weakly chordal (resp. (3, 1)-polar).*
- *If all cliques of $G$ have size at least 3, then all cliques of $G'$ have size at least 3.*
- *Any 2-clique-colouring of $G'$ assigns at least 2 colours to $b_1, b_2, b_3$.*
- *$G$ is 2-clique-colourable if $G'$ is 2-clique-colourable.*
- *$G'$ is 2-clique-colourable if there exists a 2-clique-colouring of $G$ that assigns at least 2 colours to $b_1, b_2, b_3$.*

**Lemma 4.** *Let $G$ be a weakly chordal graph (resp. (3, 1)-polar graph) and $b_1, b_2 \in V(G)$ (resp. $b_1, b_2 \in B(G)$). If we add to $G$ a copy of an auxiliary graph $BS(b_1, b_2)$, then the following assertions are true.*

- *The resulting graph $G'$ is weakly chordal (resp. (3, 1)-polar).*
- *If all cliques of $G$ have size at least 3, then all cliques of $G'$ have size at least 3.*
- *Any 2-clique-colouring of $G'$ assigns 2 colours to $b_1$ and $b_2$.*
- *$G$ is 2-clique-colourable if $G'$ is 2-clique-colourable.*
- *$G'$ is 2-clique-colourable if there exists a 2-clique-colouring of $G$ that assigns 2 colours to $b_1$ and $b_2$.*

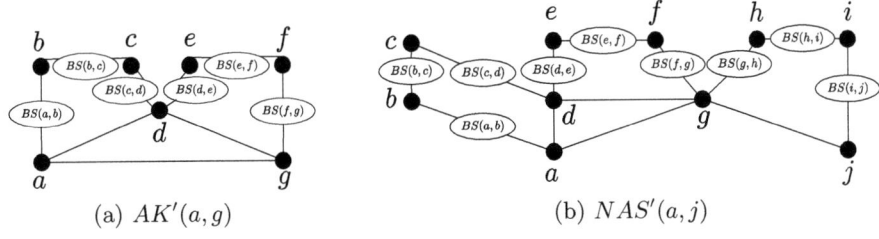

**Fig. 6.** Auxiliary graphs $AK'(a,g)$ and $NAS'(a,j)$

We strengthen the result that 2-clique-colouring of (3, 1)-polar graphs is $\mathcal{NP}$-complete, now even restricting all cliques to have size at least 3, which gives a subclass of weakly chordal graphs.

**Theorem 5.** *The problem of 2-clique-colouring is $\mathcal{NP}$-complete for (weakly chordal) (3, 1)-polar graphs with all cliques having size at least 3.*

On the other hand, we prove that 2-clique-colouring (2, 1)-polar graphs becomes polynomial when all cliques have size at least 3.

**Theorem 6.** *The problem of 2-clique-colouring is polynomial for (2, 1)-polar graphs with all cliques having size at least 3.*

In the proof that 2-clique-colouring weakly chordal graphs is a $\Sigma_2^P$-complete problem (Theorem 1), we constructed a weakly chordal graph with $K_2$ cliques to force distinct colours in their extremities (in a 2-clique-colouring). We can obtain a weakly chordal graph with no cliques of size 2 by adding copies of the auxiliary graph $BS(u, v)$, for every $K_2$ clique $\{u, v\}$. Auxiliary graphs $AK$ and $NAS$ become $AK'$ and $NAS'$, both depicted in Fig. 6. Finally, the weakly chordal graph constructed in Theorem 1 becomes a weakly chordal graph with no $K_2$ clique. Such construction is done in polynomial-time. Notice that, in the constructed graph of Theorem 1, every $K_2$ clique $\{u, v\}$ has 2 distinct colours in a clique-colouring. Hence, one can check with Lemmas 3 and 4 that the obtained graph is weakly chordal and it is 2-clique-colourable if, and only if, the constructed graph of Theorem 1 is 2-clique-colourable. This implies the following theorem.

**Theorem 7.** *The problem of 2-clique-colouring is $\Sigma_2^P$-complete for weakly chordal graphs with all cliques having size at least 3.*

As a direct consequence of Theorem 7, we have that 2-clique-colouring is $\Sigma_2^P$-complete for perfect graphs with all cliques having size at least 3.

**Corollary 1.** *The problem of 2-clique-colouring is $\Sigma_2^P$-complete for perfect graphs with all cliques having size at least 3.*

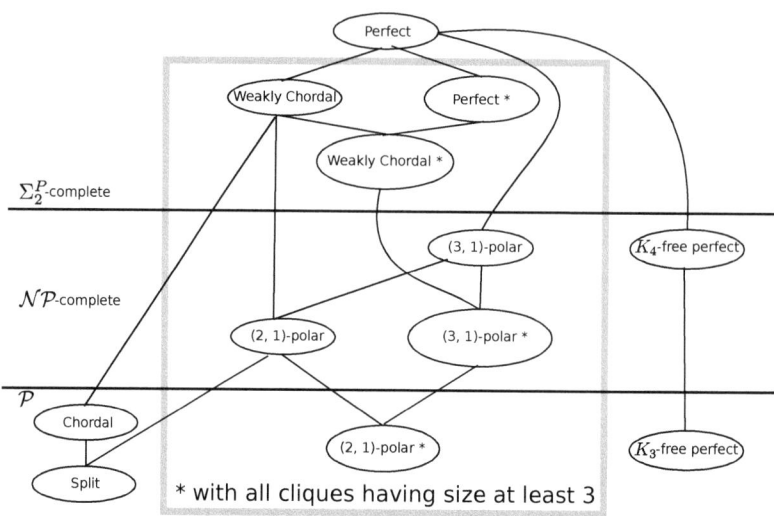

**Fig. 7.** 2-clique-colouring complexity of perfect graphs and subclasses

## 4  Final Considerations

Marx [9] proved complexity results for $k$-clique-colouring, for fixed $k \geq 2$, and related problems that lie in between two distinct complexity classes, namely $\Sigma_2^P$-complete and $\Pi_3^P$-complete. Marx approaches the complexity of clique-colouring by fixing the graph class and diversifying the problem. In the present work, our point of view is the opposite: we rather fix the (2-clique-colouring) problem and we classify the problem complexity according to the inputted graph class, which belongs to nested subclasses of weakly chordal graphs. We achieved complexities lying in between three distinct complexity classes, namely $\Sigma_2^P$-complete, $\mathcal{NP}$-complete and $\mathcal{P}$. Fig. 7 shows the relation of inclusion among the classes of graphs of Table 1. The 2-clique-colouring complexity for each class is highlighted.

Notice that the perfect graph subclasses for which the 2-clique-colouring problem is in $\mathcal{NP}$ mentioned so far in the present work satisfy that the number of cliques is polynomial. We remark that the complement of a matching has an exponential number of cliques and yet the 2-clique-colouring problem is in $\mathcal{NP}$, since no such graph is 2-clique-colourable. Now, notice that the perfect graph subclasses for which the 2-clique-colouring problem is in $\mathcal{P}$ mentioned so far in the present work satisfy that all graphs in the class are 2-clique-colourable. Macêdo Filho et al. [6] have proved that unichord-free graphs are 3-clique-colourable, but a unichord-free graph is 2-clique-colourable if and only if it is perfect. As a future work, we aim to find subclasses of perfect graphs where not all graphs are 2-clique-colourable and yet the 2-clique-colouring problem is in $\mathcal{P}$ when restricted to the class.

# References

1. Bacsó, G., Gravier, S., Gyárfás, A., Preissmann, M., Sebő, A.: Coloring the maximal cliques of graphs. SIAM J. Discrete Math. 17(3), 361–376 (2004)
2. Chernyak, Z.A., Chernyak, A.A.: About recognizing $(\alpha, \beta)$ classes of polar graphs. Discrete Math. 62(2), 133–138 (1986)
3. Chudnovsky, M., Robertson, N., Seymour, P., Thomas, R.: The strong perfect graph theorem. Ann. of Math. (2) 164(1), 51–229 (2006)
4. Défossez, D.: Complexity of clique-coloring odd-hole-free graphs. J. Graph Theory 62(2), 139–156 (2009)
5. Duffus, D., Sands, B., Sauer, N., Woodrow, R.E.: Two-colouring all two-element maximal antichains. J. Combin. Theory Ser. A 57(1), 109–116 (1991)
6. Macêdo Filho, H.B., Machado, R.C.S., Figueiredo, C.M.H.: Clique-colouring and biclique-colouring unichord-free graphs. In: Fernández-Baca, D. (ed.) LATIN 2012. LNCS, vol. 7256, pp. 530–541. Springer, Heidelberg (2012)
7. Kratochvíl, J., Tuza, Z.: On the complexity of bicoloring clique hypergraphs of graphs. J. Algorithms 45(1), 40–54 (2002)
8. Lovász, L.: Coverings and coloring of hypergraphs. In: Proc. Fourth Southeastern Conference on Combinatorics, Graph Theory, and Computing, pp. 3–12 (1973)
9. Marx, D.: Complexity of clique coloring and related problems. Theoret. Comput. Sci. 412(29), 3487–3500 (2011)
10. Poon, H.: Coloring Clique Hypergraphs. Master's thesis, West Virginia University (2000)
11. Prömel, H.J., Steger, A.: Almost all Berge graphs are perfect. Combin. Probab. Comput. 1(1), 53–79 (1992)

# The Computational Complexity of the Game of Set and Its Theoretical Applications

Michael Lampis[1,*] and Valia Mitsou[2,**]

[1] Research Institute for Mathematical Sciences (RIMS), Kyoto University
mlampis@kurims.kyoto-u.ac.jp
[2] CUNY Graduate Center
vmitsou@gc.cuny.edu

**Abstract.** The game of SET is a popular card game in which the objective is to form Sets using cards from a special deck. In this paper we study single- and multi-round variations of this game from the computational complexity point of view and establish interesting connections with other classical computational problems.

Specifically, we first show that a natural generalization of the problem of finding a single Set, parameterized by the size of the sought Set is W-hard; our reduction applies also to a natural parameterization of PERFECT MULTI-DIMENSIONAL MATCHING, a result which may be of independent interest. Second, we observe that a version of the game where one seeks to find the largest possible number of disjoint Sets from a given set of cards is a special case of 3-SET PACKING; we establish that this restriction remains NP-complete. Similarly, the version where one seeks to find the smallest number of disjoint Sets that overlap all possible Sets is shown to be NP-complete, through a close connection to the INDEPENDENT EDGE DOMINATING SET problem. Finally, we study a 2-player version of the game, for which we show a close connection to ARC KAYLES, as well as fixed-parameter tractability when parameterized by the number of rounds played.

## 1 Introduction

In this paper, we analyze the computational complexity of some variations of the game of SET and its interesting relations with other classical problems, like PERFECT MULTI-DIMENSIONAL MATCHING, SET PACKING, and INDEPENDENT EDGE DOMINATING SET.

The game of SET is a card game in which players seek to form Sets of cards from a special deck. Each card from this deck has a picture with 4 attributes (shape, color, number, shading), and each attribute can take one of 3 values (for example the shape can be oval, squiggle, or diamond, the color can be blue,

---

[*] Research supported by the Scientific Grant-in-Aid from Ministry of Education, Culture, Sports, Science and Technology of Japan.
[**] Part of this work was done while the second author was visiting RIMS, Kyoto University.

green, or purple, etc). To create a Set[1], the player needs to identify 3 cards in which, for each attribute independently, either all cards agree on the value, or they constitute a rainbow of all possible values. In a single round of the normal play, 12 cards are dealt and the players seek (simultaneously) a Set. The first player to find a Set wins the 3 cards constituting it. Then 3 new cards are dealt in the old ones' places and the game continues with the next round. For more information regarding the game and its rules as well as for other variations see the official website of the game http://www.setgame.com/set/index.html.

The game of SET has gained remarkable attention and popularity (especially among mathematicians) as well as many awards. The game has been the subject of both educational and technical research. A broad set of educational activities has been suggested, a collection of which can be found on the official website of SET. Furthermore, the game has been studied extensively from a more technical mathematical point of view, considering questions like "what is the maximum number of cards with $n$ attributes and 3 values that can be laid such that no Sets are formed" [5], or "for fixed $n$, how many non-isomorphic collections of $n$ cards are there" [4]. In [13], many other similar questions are posed. In addition to the game's popularity, one motivation for this intense study is that the problem has a very natural alternative mathematical formulation: if one describes the cards as four-dimensional vectors over the set $\{0, 1, 2\}$, then a Set is exactly a collection of three collinear points, that is, three points whose vectors add up to 0( mod 3). Nevertheless, the first and - to the best of our knowledge - only attempt to consider the game's computational complexity was made by Chaudhuri et al [2] in 2003, who showed that a generalization of the game is NP-complete. Our focus on this paper is to continue and refine this work by studying further aspects of the computational complexity of SET.

In order to study a game from the viewpoint of computational complexity theory, one needs to define a natural generalization of the game in question (as the original constant size game always has constant time and space complexity). In a round of SET, there are 3 parameters to consider: the number of laid cards during each round $m$, the number of attributes $n$ and the number of values $k$ (in the original game $m = 12$, $n = 4$ and $k = 3$). A subset of $k$ cards will be considered a Set if for all attributes, values either all agree or all differ. Of course these three parameters are not totally independent as the number of cards $m$ is upper-bounded by $k^n$. In any multi-round version of the game, an extra parameter $r$ being the number of rounds is added.

**Summary of Results.** We first talk about a single-round version of SET. This one-round version generalizes PERFECT MULTI-DIMENSIONAL MATCHING as was first observed in [2]. It is easy to see that the problem parameterized by the number of values $k$ is in XP (by the trivial algorithm that enumerates all size-$k$ sets of cards and checking whether any of them constitutes a Set). We prove that this parameterized version of the problem is W-hard. Our W-hardness proof applies

---

[1] The first letter of Set is capitalized to avoid a mix-up with the notion of mathematical set.

to PERFECT MULTI-DIMENSIONAL MATCHING as well, proving that PERFECT MULTI-DIMENSIONAL MATCHING parameterized by the size of the dimensions $k$ (while the number of dimensions $n$ is unbounded) is W[1]-hard. This result may be of independent interest, as this is a natural parameterization of a classic problem that has not been considered before. The only relevant parameterized result known about this problem is that MAXIMUM MULTI-DIMENSIONAL MATCHING parameterized by the size of the matching and the number of dimensions is FPT [3].

Next, we focus our attention to the case where the number of values is 3. As was suggested, there is a polynomial time algorithm to find whether there exists at least one Set, in other words to play just one round. The complexity remains polynomial even if we consider the question of enumerating all Sets. This generalizes the daily puzzles found either on the official website of SET or in the New York Times. In these puzzles we are given $m$ cards and need to find the maximum number of Sets assuming that we don't remove any cards from the table after finding a Set.

It becomes interesting to ask the same question for a multi-round game, where cards are gradually removed. This corresponds to the CO-OP version of the game, where players have to cooperate in order to find the maximum number of available Sets given that cards of found Sets are removed from the table. Another interesting variation is the one where we are looking for the minimum number of Sets that once picked destroy all existing Sets. Both problems can be seen as special cases of more general packing and covering problems. In the maximization version, one is looking for a maximum 3-SET PACKING, while in the minimization version one is looking for a minimum INDEPENDENT EDGE DOMINATING SET in a 3-uniform hypergraph. We show that both problems remain NP-Hard even on instances that correspond to the SET game. From the parameterized point of view, if one considers as the parameter the number of rounds $r$ to be played, a natural parameterization of the former problem asking whether there are at least $r$ mutually disjoint Sets is Fixed Parameter Tractable, following from the results of Chen et al. [3]. We establish that the natural parameterized version of the latter problem (find at most $r$ Sets to destroy all Sets) is also FPT, through a connection with the related INDEPENDENT EDGE DOMINATING SET problem on graphs.

Finally, we consider a two-player version of the $r$-round game, which can be seen as a restriction of the game ARC KAYLES in 3-uniform hypergraphs (where hyperedges should be valid Sets). The complexity of ARC KAYLES is currently unknown even on graphs and it has been a long-standing open question since the PSPACE-Completeness of its sibling problem NODE KAYLES was established in [11]. We prove that this multi-round 2-player version of SET is at least as hard as ARC KAYLES. Nevertheless, we prove that deciding whether the first player has a winning strategy in $r$ moves in 2-player SET is FPT parameterized by $r$. This implies the same result for ARC KAYLES on graphs.

The paper is divided as follows: In section 2 we present the W-hardness of the single-round version of SET. In section 3 we analyze the above-mentioned multi-round variations with $k = 3$. In section 4 we analyze the natural turn-based 2-player version. Last, in section 5 we give some conclusions and open problems.

## 2 W-hardness of $k$-Value 1-Set and Perfect Multi-dimensional Matching

In this section, we talk about a single-round generalization of the game of SET. We are dealt $m$ cards, each with $n$ attributes that can take one of $k$ values and we need to find a Set of size $k$ cards. This is the main problem considered by Chaudhuri et al. in [2]. Their main insight is that this problem can be seen as a hypergraph problem. Specifically, one may construct a hypergraph on $n \cdot k$ vertices, each representing an attribute-value pair. Now, cards can be represented as hyperedges, by including in each hyperedge the $k$ values that describe the corresponding card's attributes. It is not hard to see that a perfect matching in this $n$-partite hypergraph corresponds to a Set in the original instance. On the other hand, some Sets do not correspond to perfect matchings, because all cards may share the same value for some attributes. Nevertheless, Chaudhuri et al. have established that the two problems have the same complexity and finding a Set is essentially algorithmically equivalent to finding a perfect matching in this hypergraph.

Here we will exploit this connection between the two problems to analyze the complexity of finding a Set with respect to the three relevant parameters $m, n$, and $k$. If $k$ is unbounded, finding a Set was shown to be NP-hard in [2] even for just 3 attributes. If the cards have only 2 attributes, the game is in P. On the other hand, if $n$ is unbounded but the number of values $k$ is considered as a parameter the problem is trivially in XP. Here we will show that the trivial algorithm cannot be improved to an FPT algorithm, by proving that the problem is W[1]-hard. The first step in our reduction is to show that the relevant parameterization of PERFECT MULTI-DIMENSIONAL MATCHING is W[1]-hard, a result that may be of independent interest.

**Theorem 1.** PERFECT MULTI-DIMENSIONAL MATCHING *parameterized by the dimension size is W[1]-hard.*

*Proof.* We present a reduction from $k$-MULTICOLORED CLIQUE (proven to be W[1]-hard in [6]).

Given an instance of $k$-MULTICOLORED CLIQUE, in other words a $k$-partite graph $G(V, E)$ where each part has size $n$, we construct an instance of PERFECT MULTI-DIMENSIONAL MATCHING, a multigraph $G'(V', E')$ with $nk(k-1)$ dimensions where each dimension has $k + \binom{k}{2}$ different values, such that if $G$ has a clique of size $k$ then $G'$ has a multidimensional perfect matching.

For each ordered pair $(V_i, V_j)$ with $V_i, V_j, i \neq j$ being parts of $V$, we add $n$ dimensions which we group together in a group $(i; ij)$. Each of the $n$ dimensions in each group $(i; ij)$ of graph $G'$ corresponds to a different vertex in part $V_i$ of graph $G$. Each dimension will have $k + \binom{k}{2}$ different possible values, one value

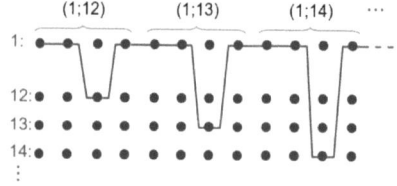
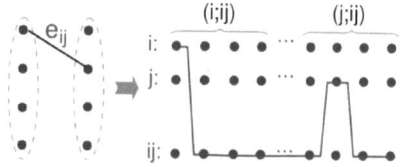

**Fig. 1.** The vertex-multiedge of $G'$ that corresponds to vertex $v_{13}$ of part $V_1$ in $G$

**Fig. 2.** The edge-multiedge of $G'$ that corresponds to the edge $e_{ij}$ of $G$

corresponding to each part $V_i$ and one value corresponding to each pair of parts $(V_i, V_j), i < j$.

Furthermore, for each vertex $v_{ij}$ in the original graph ($j^{th}$ vertex of part $V_i$) we create a multiedge as follows (see figure 1): it will contain the vertices labeled with $i$ for all dimensions but the $j^{th}$ dimension of each group $(i; ki)$, where $k \neq i$. For these dimensions we'll include the vertex labeled with $ki$. We call these *vertex-multiedges*.

Last, for each edge $e_{ij} \in E$ that connects the $a^{th}$ vertex of part $V_i$ with the $b^{th}$ vertex of part $V_j$ in the original graph, we create a multiedge as follows (see figure 2): we add all vertices labeled with $ij$ for all dimensions except for the $a^{th}$ dimension in the group $(i; ij)$ where we take the vertex with label $i$ and the $b^{th}$ dimension in group $(j; ij)$ that we take the vertex with label $j$. We call these *edge-multiedges*.

Notice that the above construction is polynomial in the size of the input and the parameter of $k$-MULTICOLORED CLIQUE. Also, the dimension size in the constructed instance of PERFECT MULTI-DIMENSIONAL MATCHING $k + \binom{k}{2}$ is quadratic in the parameter $k$ of $k$-MULTICOLORED CLIQUE.

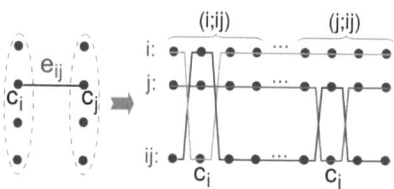

**Fig. 3.** Vertices of groups $(i; ij)$ and $(j; ij)$ that were not covered by the vertex-multiedges of $G'$ that correspond to vertices $v_{ic_i}$ or $v_{jc_j}$ of $G$ are covered by the edge-multiedge of $G'$ that corresponds to edge $e_{ij} = (v_{ic_i}, v_{jc_j})$ and vice versa

Now we prove that if $G$ has a clique of size $k$ then $G'$ has a perfect multidimensional matching and vice versa. Suppose that $G$ has a clique of size $k$. In other words, there should be a tuple $(v_{1c_1}, v_{2c_2}, \ldots v_{kc_k})$, with $v_{ic_i} \in V_i$, where all vertices in the tuple are connected with each other. We select in the matching the $k$ vertex-multiedges of $G'$ that correspond to the vertices in the clique of $G$

and the $\binom{k}{2}$ edge-multiedges of $G'$ that correspond to edges of $G$ that connect vertices in the clique. This selection is a perfect matching: each vertex-multiedge or edge-multiedge selects all vertices with labels that correspond to the vertex or edge that they represent, except for $k-1$ vertices for each vertex-multiedge and 2 vertices for each edge-multiedge as it is described above. Also, the edge-multiedge of $G'$ that corresponds to edge $e_{ij} = (v_{ic_i}, v_{jc_j})$ of $G$ covers those two vertices that the vertex-multiedges that correspond to $v_{ic_i}$ and $v_{jc_j}$ left uncovered, and vice versa (see figure 3).

On the other hand, if $G'$ has a perfect matching, then this matching contains exactly one vertex-multiedge and exactly one edge-multiedge of each value (otherwise there would be uncovered vertices or vertices covered twice by the matching). We select all vertices of $G$ that correspond to a vertex-multiedge in the matching. Now, all these vertices that we picked should be pairwise connected in $G$, because the edge-multiedges in the matching should be covering those vertices in $G'$ that the vertex-multiedges did not cover, which correspond to vertices in the clique. □

**Corollary 1.** *The game of Set parameterized by the number of values (or else the size of the Sets) is $W[1]$-hard.*

*Proof.* The "if" part of the above reduction also holds for the game of Set: if $G'$ has a multidimensional perfect matching it also has a Set. For the "only if" part, notice that if $G'$ has a Set then this Set is also a multidimensional perfect matching since no vertex-multiedge can pass through a value that belongs to another vertex-multiedge. □

## 3 Multi-round Variations of SET

In this and the next section we talk about multi-round variations of SET where the number of values (or in other words the size of the Sets) is 3. In this case, each card (vertex of the hypergraph) is described by a vector in $\mathbb{F}_3^n$. Note that, three cards form a Set if and only if their corresponding vectors add up to the all-0 vector. It is also easy to observe that every pair of cards can have up to one card that forms a Set with them. This property will prove useful later.

We will once again use a hypergraph formulation, though different from the one in the previous section. Specifically, we consider the 3-uniform hypergraph formed if we construct a vertex for each dealt card and a hyperedge (that is, a set of size 3) for each Set. It is clear that given a SET instance, one can in polynomial time construct this hypergraph.

We will first talk about a maximization variation: given a set of cards we ask the question whether there exist at least $r$ Sets that we can pick up before leaving no Sets on the table. We call this problem MAX 3-VALUE $r$-SET. Observe that this problem is a special case of 3-SET PACKING, which is a known NP-hard problem. We thus need to show that the problem remains NP-hard when restricted to instances realizable by SET cards. This is established in Theorem 2.

Then, we turn our attention to a minimization version: given a set of cards, is it possible by removing at most $r$ Sets ($3r$ cards) to eliminate all potential Sets? We call this problem MIN 3-VALUE $r$-SET. This problem is a special case of INDEPENDENT EDGE DOMINATING SET in 3-uniform hypergraphs. We show its NP-hardness even when restricted to hypergraphs realizable by SET cards. Then, we prove that the natural parameterized version of INDEPENDENT EDGE DOMINATING SET in 3-uniform hypergraphs with parameter $r$ is FPT, thus proving that the special case of a parameterization of this version of SET is also FPT.

### 3.1 NP-Hardness of the Maximization Version

**Theorem 2.** MAX 3-VALUE $r$-SET is NP-Hard.

*Proof.* We design a reduction from 3-SAT. Given a formula $\phi$ of 3-SAT we first create an equivalent formula $\phi'$ where each clause contains at most 3 literals and each variable appears exactly 3 times (two as positive and one as negative or two as negative and one as positive). Furthermore, any two clauses of $\phi'$ share at most one variable. A similar construction appears in [10]. Let $m$ be the number of clauses of $\phi'$ and $n$ the number of variables.

The main idea of the reduction is as follows: from formula $\phi'$ we create an instance of MAX 3-VALUE $r$-SET which consists of variable gadgets (one corresponding to each variable) and clause gadgets (one corresponding to each clause). The variable gadget of a variable $x$ contains five cards: three cards $x_1, x_2$ and $x_3$ for each appearance of $x$ in $\phi'$ ($x_1$ and $x_2$ corresponding to appearances with the same sign and $x_3$ to opposite), and two more cards: $x_{12}$ which forms a Set with $x_1$ and $x_2$, and $x_{123}$ which forms a Set with $x_3$ and $x_{12}$. Picking either Set is equivalent to making an assignment to $x$ (both Sets contain $x_{12}$, only one Set can be formed leaving either positive or negative appearances of $x$ unused). The cards $x_1, x_2, x_3$ will also appear in the clause gadgets and, intuitively, we will be able to select a Set from a clause gadget if and only if one of its $x_i$ vertices is free, corresponding to a true literal.

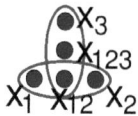
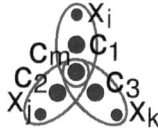

**Fig. 4.** The variable gadget  **Fig. 5.** The clause gadget

The clause gadget consists of four additional cards: one card per literal in the clause $c_1, c_2$, and $c_3$, and one additional card $c_m$ (for clauses of size 2 we do not introduce $c_3$). Furthermore, each card $x_{c_i}$ corresponding to the literal in the $i^{th}$ position of a clause $c$ forms a Set with cards $c_i$ and $c_m$. In order to be able to pick this Set (and satisfy c) $x_{c_i}$ should not have been picked during the assignment phase.

Observe that, if one sees the new instance as a 3-SET PACKING instance, it is not hard to establish that the instance has a solution of size $n + m$ if and only if $\phi'$ is satisfiable. The bonus point is that this instance is realizable with Set cards. Due to space limitations, card vectors and the rest of the proof is omitted. The complete proof can be found in the full version of this work [9]. □

### 3.2 Results on the Minimization Version

Next, we present yet another multi-round version of SET, MIN 3-VALUE $r$-SET.

We prove that MIN 3-VALUE $r$-SET is NP-hard via a simple reduction from INDEPENDENT EDGE DOMINATING SET (proven NP-hard in [7]).

**Theorem 3.** MIN *3*-VALUE $r$-SET *is NP-hard.*

The reduction appears in the full version.

Since the MIN 3-VALUE $r$-SET problem is hard, it makes sense to consider its naturally parameterized version: Given an arbitrary set of cards, do there exist $r$ Sets that overlap all other formed Sets? We show that a simple FTP algorithm can decide this question. As a matter of fact, the algorithm works on any 3-uniform hypergraph. Recall that the similar parameterization of the maximization problem is also known to be FPT, by relevant results on 3-SET PACKING [3].

**Theorem 4.** INDEPENDENT EDGE DOMINATING SET *in 3-uniform hypergraphs parameterized by the size of the edge dominating set is FPT.*

**Corollary 2.** MIN *3*-VALUE $r$-SET *parameterized by the number of Sets that will be picked is FPT.*

The proof of Theorem 4 can be found in the full version. Corollary 2 follows directly from Theorem 4.

## 4 A Two Player Game

In this section, we consider a natural two-player turn-based game that we call 2P 3-VALUE SET. Suppose that an arbitrary set of cards is on the table and two opposing players take turns playing. Each player may select three cards that form a Set and remove them from play. No additional cards are dealt. The game goes on until a player is unable to find a Set, in which case she loses.

We exploit the ideas developed for the single-player game MIN 3-VALUE $r$-SET. Although we will not completely settle the complexity of the two-player version, the result given in Theorem 3 implies directly that the two-player version of Set is at least as hard as ARC KAYLES.

ARC KAYLES is a two-player game played on an undirected graph. Two players take turns selecting edges from the graph, under the constraint that the edge they pick cannot share a common endpoint with any previously selected edges. The first player unable to move loses.

Though the complexity of the related version of the problem called NODE KAYLES was settled in the '70s by Schaefer [11], ARC KAYLES has been open ever since. It is not hard to see that, since the game in ARC KAYLES ends essentially when the two players have formed a minimal independent edge dominating set, we can say the following:

**Corollary 3.** 2P 3-VALUE SET *is at least as hard as* ARC KAYLES.

It will likely be hard to find a polynomial-time algorithm for ARC KAYLES, and therefore also for 2P 3-VALUE SET. A slightly more general version of ARC KAYLES is mentioned to be PSPACE-complete in [11]. The natural generalization of ARC KAYLES to hypergraphs with unbounded hyperedge size is PSPACE-hard by the complexity of poset games [8].

Let us consider a natural parameterization of 2P 3-VALUE SET. In this problem, the question is whether a winning outcome for the first player can be achieved within at most $r$ rounds (with $r$ being the parameter). Parameterized problems of this form have been considered in the past, beginning with [1], where it was established that the $r$-move parameterized version of NODE KAYLES is AW[*]-hard. 2P 3-VALUE SET (and thus ARC KAYLES too), as we show in Theorem 5, parameterized by the number of rounds turns out to be FPT.

**Theorem 5.** 2P 3-VALUE SET *parameterized by the number of allowed rounds $r$ is FPT.*

*Proof.* We will give a sketch of a kernelization argument. More details can be found in the full version of this work.

We observe that for the first player to have a strategy there must exist a small hitting set in the original hypergraph, which we can find in FPT time [12]. We then use the Set property that any two cards have a unique third to establish a "bounded degree property" in this graph. Specifically, any hyperedge of the original graph that uses a vertex $h_i$ of the hitting set may overlap with at most 2 hyperedges that use vertex $h_j$ of the hitting set, for $j \neq i$.

The idea now is to reduce the problem to a version of NODE KAYLES played on an $r$-partite graph, where parts are played in order. This is easily done by creating $r$ vertices (one in each part) to represent each hyperedge and connecting vertices from different parts that correspond to overlapping hyperedges. Using the above observations we can partition the vertices of the last set according to the hitting set vertex their hyperedges contain. If one of the parts formed has size $2r$ or more we now know that it's impossible to eliminate all its vertices without using the hitting set vertex. We can thus simplify it to a single vertex connected only to vertices corresponding to hyperedges that contain the same hitting set vertex.

This allows us to bound the size of the last partite set. We then proceed by induction: if two vertices in a partite set have the same neighbors in the next partite set, they are equivalent. Once again if we have too many we can reduce the graph. Thus, we can bound the size of the whole graph. □

The proof only uses the property of SET that every pair of cards has a unique third that forms a Set with them. Thus the game is FPT even when played on the more general class of 3-uniform hypergraphs having this property. Also, Corollary 4 follows directly from Theorems 3 and 5:

**Corollary 4.** *The natural parameterization of* ARC KAYLES *by the number of rounds played is FPT.*

## 5 Conclusions and Open Problems

In this paper we studied the computational complexity of the game of SET and presented some interesting connections with other well-studied problems, such as PERFECT MULTI-DIMENSIONAL MATCHING, INDEPENDENT EDGE DOMINATING SET and SET PACKING.

The one-round case of SET is now fairly well-understood. However there are quite a few interesting open problems one might consider in the multi-round case, especially the two-player version 2P 3-VALUE SET. It remains unknown whether this game is PSPACE-Complete. However, proving the hardness of ARC KAYLES on graphs would settle the complexity of this problem as well (which is an interesting open question on its own accord). Staying on ARC KAYLES, it might be interesting to show whether the game played on general 3-uniform hypergraphs is FPT. We remind the reader that our proof that 2P 3-VALUE SET is FPT is based on the property of SET that each pair of cards can have at most one third with which they all form a Set. That property is vital for the proof since it establishes that the line graph has essentially bounded degree. This is not true for a general 3-uniform hypergraph though.

## References

1. Abrahamson, K.R., Downey, R.G., Fellows, M.R.: Fixed-Parameter Tractability and Completeness IV: On Completeness for W[P] and PSPACE Analogues. Ann. Pure Appl. Logic 73(3), 235–276 (1995)
2. Chaudhuri, K., Godfrey, B., Ratajczak, D., Wee, H.: On the Complexity of the Game of Set (2003) (manuscript)
3. Chen, J., Feng, Q., Liu, Y., Lu, S., Wang, J.: Improved Deterministic Algorithms for Weighted Matching and Packing Problems. Theor. Comput. Sci. 412(23), 2503–2512 (2011)
4. Coleman, B., Hartshorn, K.: Game, Set, Math. Mathematics Magazine 85(2), 83–96 (2012)
5. Davis, B.L., Davis, Maclagan, D.: The Card Game Set (2003)
6. Fellows, M.R., Hermelin, D., Rosamond, F.A., Vialette, S.: On the Parameterized Complexity of Multiple-interval Graph Problems. Theor. Comput. Sci. 410(1), 53–61 (2009)

7. Garey, M.R., Johnson, D.S.: Computers and Intractability, vol. 174. Freeman, New York (1979)
8. Grier, D.: Deciding the Winner of an Arbitrary Finite Poset Game Is PSPACE-Complete. In: Fomin, F.V., Freivalds, R., Kwiatkowska, M., Peleg, D. (eds.) ICALP 2013, Part I. LNCS, vol. 7965, pp. 497–503. Springer, Heidelberg (2013)
9. Lampis, M., Mitsou, V.: The Computational Complexity of the Game of Set and its Theoretical Applications. arXiv preprint arXiv:1309.6504 (2013)
10. Papadimitriou, C.M.: Computational Complexity. Addison-Wesley, Reading (1994)
11. Schaefer, T.J.: On the Complexity of Some Two-Person Perfect-Information Games. J. Comput. Syst. Sci. 16(2), 185–225 (1978)
12. Wahlström, M.: Algorithms, measures and upper bounds for satisfiability and related problems. PhD thesis, Linköping (2007)
13. Zabrocki, M.: The joy of set (2001)

# Independent and Hitting Sets of Rectangles Intersecting a Diagonal Line

José R. Correa[1], Laurent Feuilloley[2], and José A. Soto[3]

[1] Department of Industrial Engineering, Universidad de Chile
correa@uchile.cl
[2] Department of Computer Science, ENS Cachan
lfeuillo@ens-cachan.fr
[3] Department of Mathematical Engineering and Center for Mathematical Modeling, Universidad de Chile
jsoto@dim.uchile.cl

**Abstract.** Finding a maximum independent set of a given family of axis-parallel rectangles is a basic problem in computational geometry and combinatorics. This problem has attracted significant attention since the sixties, when Wegner conjectured that the corresponding duality gap, i.e., the maximum possible ratio between the maximum independent set and the minimum hitting set, is bounded by a universal constant. In this paper we improve upon recent results of Chepoi and Felsner and prove that when the given family of rectangles is intersected by a diagonal, this ratio is between 2 and 4. For the upper bound we derive a simple combinatorial argument that first allows us to reprove results of Hixon, and Chepoi and Felsner and then we adapt this idea to obtain the improved bound in the diagonal intersecting case. From a computational complexity perspective, although for general rectangle families the problem is known to be NP-hard, we derive an $O(n^2)$-time algorithm for the maximum weight independent set when, in addition to intersecting a diagonal, the rectangles intersect below it. This improves and extends a classic result of Lubiw. As a consequence, we obtain a 2-approximation algorithm for the maximum weight independent set of rectangles intersecting a diagonal.

## 1 Introduction

Given a family of axis-parallel rectangles, two natural objects of study are the maximum number of rectangles that do not overlap and the minimum set of points stabbing every rectangle. These problems are known as maximum independent set MIS and minimum hitting set MHS respectively, and in the associated intersection graph they correspond to the maximum independent set and the minimum clique covering. We study these problems for restricted classes of rectangles, and focus on designing algorithms and on evaluating the *duality gap* $\delta_{\text{GAP}}$, i.e., the maximum ratio between these quantities. This term arises as MHS is the integral version of the dual of the natural linear programming relaxation of MIS.

From a computational complexity viewpoint, MIS and MHS of rectangles are strongly NP-hard [11,13], so attention has been put into approximation algorithms and polynomial time algorithms for special classes. The current best known approximation factor for MIS are $O(\log \log n)$ [3], and $O(\log n/\log \log n)$ for weighted MIS (WMIS) [4]. Very recently, Adamaszek and Wiese [1] designed a pseudo-polynomial time algorithm finding a $(1+\varepsilon)$-approximate solution for WMIS, but it is unknown whether there exist polynomial time constant factor approximation algorithms. A similar situation occurs for MHS: the current best approximation factor is $O(\log \log n)$ [2], while in general, the existence of a constant factor approximation is open. Polynomial time algorithms for these problems have been obtained for special classes. When all rectangles are intervals, the underlying intersection graph is an interval graph and even linear time algorithms are known for MIS, MHS and WMIS [12]. Moving beyond interval graphs, Lubiw [15] devised a cubic-time algorithm for computing a maximum weight independent family of point-intervals, which can be seen as families of rectangles having their upper-right corner along the same diagonal. More recently, Soto and Telha [17] considered the case where the upper-right and lower-left corners of all rectangles are two prescribed point sets of total size $m$. They designed an algorithm that computes both MIS and MHS in the time required to do $m$ by $m$ matrix multiplication, and showed that WMIS is NP-hard on this class. Finally, there are also known PTAS for special cases, including the results of Chan [4] for squares, and Mustafa and Ray [16] for unit height rectangles.

It is straightforward to observe that given a family of rectangles the size of a maximum independent set is at most that of a minimum hitting set. In particular, for interval graphs this inequality is actually an equality, and this still holds in the case studied by Soto and Telha [17], so that the duality gap is 1 for these classes. A natural question to ask is whether the duality gap for general families of rectangles is bounded. Indeed, already in the sixties Wegner [19] conjectured that the duality gap for arbitrary rectangles families equals 2, whereas Gyárfás and Lehel [9] proposed the weaker conjecture that this gap is bounded by a universal constant. Although these conjectures are still open, Károlyi and Tardos [14] proved that the gap is within $O(\log(\text{mis}))$, where mis is the size of a maximum independent set. For some special classes, the duality gap is indeed a constant. In particular, when all rectangles intersect a given diagonal line, Chepoi and Felsner [5] prove that the gap is between $3/2$ and $6$, and the upper bound has been further improved for more restricted classes [5,10].

### 1.1 Notation and Classes of Rectangle Families

Throughout this paper, $\mathcal{R}$ denotes a family of $n$ closed, axis-parallel rectangles in $\mathbb{R}^2$. A rectangle $r \in \mathcal{R}$ is defined by its lower-left corner $\ell^r$ and its upper-right corner $u^r$. For a point $v \in \mathbb{R}^2$ we let $v_x$ and $v_y$ be its $x$-coordinate and $y$-coordinate, respectively. Also, each rectangle $r \in \mathcal{R}$ is associated with a non-negative weight $w_r$. We also consider a monotone curve, given by a decreasing bijective real function, so that the boundary of each $r \in \mathcal{R}$ intersects the curve in at most 2 points. We use $a^r$ and $b^r$ to denote the higher and lower of these points respectively (which may coincide). We identify the rectangles in $\mathcal{R}$ with the set $[n] = \{1, \ldots, n\}$ so that $a_x^1 < a_x^2 < \cdots < a_x^n$. For any rectangle $i$, we define $f(i)$ as the rectangle $j$ (if it exists) following $i$ in the order of the $b$-points, that is, $b_x^i < b_x^j$ and no rectangle $k$ is such that $b_x^i < b_x^k < b_x^j$. For reference, see Figure 1.

A set of rectangles $\mathcal{Q} \subseteq \mathcal{R}$ is called independent if and only if no two rectangles in $\mathcal{Q}$ intersect. On the other hand, a set $H \subseteq \mathbb{R}^2$ of points is a hitting set of $\mathcal{R}$ if every rectangle $r \in \mathcal{R}$ contains at least one point in $H$. In this paper we consider the problem of finding an independent set of rectangles in $\mathcal{R}$ of maximum cardinality (MIS), and its weighted version (WMIS). We also consider the problem of finding a hitting set of $\mathcal{R}$ of minimum size (MHS). Let us denote by mis($\mathcal{R}$), wmis($\mathcal{R}$), mhs($\mathcal{R}$) the solutions to the above problems, respectively.

Since the solutions of the previous problems depend on properties of the intersection graph $\mathcal{I}(\mathcal{R}) = (\mathcal{R}, \{rr': r \cap r' \neq \emptyset\})$ of the family $\mathcal{R}$, we will assume that no two defining corners in $\{\ell^1, \ell^2, \ldots, \ell^n, u^1, u^2, \ldots, u^n\}$ have the same $x$-coordinates or $y$-coordinates (this is done without loss of generality by individually perturbing each rectangle). We will also assume that the curve mentioned in the first paragraph is the diagonal line $D$ given by the equation $y = -x$. This is assumed without loss of generality: by applying suitable piecewise linear transformations on both coordinates we can transform the rectangle family into one with the same intersection graph such that every rectangle intersects $D$. In what follows, call the closed halfplanes given by $y \geq -x$ and $y \leq -x$, the *halfplanes* of $D$. Note that both halfplanes intersect in $D$. The points in the bottom (resp. top) halfplane are said to be below (resp. above) the diagonal.

We study four special classes of rectangle families intersecting $D$.

**Definition 1** (Classes of rectangle families).

1. $\mathcal{R}$ is *diagonal-intersecting* if for all $r \in \mathcal{R}$, $r \cap D \neq \emptyset$.
2. $\mathcal{R}$ is *diagonal-splitting* if there is a side (upper, lower, left, right) such that $D$ intersects all $r \in R$ on that particular side.
3. $\mathcal{R}$ is *diagonal-corner-separated* if there is a halfplane of $D$ containing the same three corners of all $r \in \mathcal{R}$.
4. $\mathcal{R}$ is *diagonal-touching* if there is a corner (upper-right or lower-left) such that $D$ intersects all $r \in R$ exactly on that corner (in particular, either all the upper-right corners, or all the lower-left corners are in $D$.)

By rotating the plane, we can make the following assumptions: In the second class, we assume that the common side of intersection is the upper one; in the third class, that the upper-right corner is on the top halfplane of $D$ and the other three are in the bottom one; and in the last class, that the corner contained in $D$ is the upper-right one. Under these assumption, each type of rectangle family is more general than the next one. It is worth noting that in terms of their associated intersection graphs, the second and third classes coincide. Indeed, two rectangles of a diagonal-splitting rectangle family $\mathcal{R}$ intersect if and only if they have a point in common in the bottom halfplane of $D$. Therefore, we can replace each rectangle $r$ with the minimal possible one containing the region of $r$ that is below the diagonal, obtaining a diagonal-corner-separated family with the same intersection graph.

**Definition 2** (diagonal-lower-intersecting). A diagonal-intersecting family $\mathcal{R}$ is *diagonal-lower-intersecting* if whenever two rectangles in $\mathcal{R}$ intersect, they have a common point in the bottom halfplane of $D$.

The next lemma describes the relation between the graph classes associated to the families just defined. Its proof is deferred to the full version of the paper [7].

**Lemma 1.** Let $\mathcal{G}_{\text{int}} = \{\mathcal{I}(\mathcal{R}) \colon \mathcal{R} \text{ is diagonal-intersecting}\}$ be the class of intersection graphs arising from diagonal-intersecting families of rectangles. Let also $\mathcal{G}_{\text{low-int}}$, $\mathcal{G}_{\text{split}}$, $\mathcal{G}_{\text{c-sep}}$ and $\mathcal{G}_{\text{touch}}$ be the classes arising from diagonal-lower-intersecting, diagonal-splitting, diagonal-corner-separated, and diagonal-touching families of rectangles, respectively. Then

$$\mathcal{G}_{\text{touch}} \subsetneq \mathcal{G}_{\text{low-int}} = \mathcal{G}_{\text{split}} = \mathcal{G}_{\text{c-sep}} \subsetneq \mathcal{G}_{\text{int}}.$$

We observe that these classes have appeared in the literature under different names. For instance, Hixon [10] call the graphs in $\mathcal{G}_{\text{touch}}$ *hook graphs*, Soto and Thraves [18] call them AND(1) *graphs*, while those in $\mathcal{G}_{\text{int}}$ are called *separable rectangle graphs* by Chepoi and Felsner [5].

### 1.2 Our Results

Our main results, given in §2, are a quadratic-time algorithm to compute a wmis($\mathcal{R}$) when $\mathcal{R}$ is diagonal-lower-intersecting and a 2-approximation for the same problem when $\mathcal{R}$ is diagonal-intersecting. As far as we know, the former is the first polynomial time algorithm for WMIS on a natural class containing diagonal-touching rectangle families. Our algorithm improves upon previous work in the area. Specifically, for diagonal-touching rectangle families, the best known algorithm to solve WMIS is due to Lubiw [15], who designed a cubic-time algorithm for the problem in the context of *interval systems*. More precisely, a collection of *point-intervals* $Q = \{(p_i, I_i)\}_{i=1}^n$ is a family such that for all $i$, $p_i \in I_i$ and $I_i = [\text{left}(I_i), \text{right}(I_i)] \subseteq \mathbb{R}$ are a point and an interval, respectively. $Q$ is called *independent* if for $k \neq j$, $p_k \notin I_j$ or $p_j \notin I_k$. Given a finite collection $Q$ of weighted point-intervals, Lubiw designed a dynamic programming based algorithm to find a maximum weighted independent subfamily of $Q$. It is easy to see[1] that this problem is equivalent to that of finding wmis($\mathcal{R}$) for the diagonal-touching family $\mathcal{R} = \{r_i\}_{i=1}^n$ where $r_i$ is the rectangle with upper right corner $(p_i, -p_i)$ and lower left corner $(\text{left}(I_i), -\text{right}(I_i))$ and having the same weight as that of $(p_i, I_i)$. Lubiw's algorithm was recently rediscovered by Hixon [10].

As in Lubiw's, our algorithm is based on dynamic programming. However, rather than decomposing the instance into small triangles and computing the optimal solution for every possible triangle, our approach involves computing the optimal solutions for what we call a *harpoon*, which is defined for every pair of rectangles. We show that the amortized cost of computing the optimal solution for all harpoons is constant, leading to an overall quadratic time. Interestingly, it

---
[1] This equivalence has been noticed before [17].

is possible to show that our algorithm is an extension of the linear-time algorithm for maximum weighted independent set of intervals [12].

In §3 we give a short proof that the duality gap $\delta_{\text{GAP}}$, i.e., the maximum ratio mhs / mis, is always at most 2 for diagonal-touching families; we also show that $\delta_{\text{GAP}} \leq 3$ for diagonal-lower-intersecting families, and $\delta_{\text{GAP}} \leq 4$ for diagonal-intersecting families. These bounds yields simple 2, 3, and 4-approximation polynomial time algorithms for MHS on each class (they can also be used as approximation algorithms for MIS with the same guarantee, however, as discussed in the previous paragraph, we have an exact algorithm for WMIS on the two first classes, and a 2-approximation for the last one). The 4-approximation for MHS in diagonal-intersecting families is the best approximation known and improves upon the bound of 6 of Chepoi and Felsner [5], who also give a bound of 3 for diagonal-splitting families based on a different method. For diagonal-touching families, Hixon [10] independently showed that $\delta_{\text{GAP}} \leq 2$. To complement the previous results, we show that the duality gap for diagonal-lower-intersecting families is at least 2. We do this by exhibiting an infinite family of instances whose gap is arbitrarily close to 2. Similar instances were obtained, and communicated to us, by Cibulka et al. [6]. Note that this lower bound of 2 improves upon the 5/3 by Fon-Der-Flaass and Kostochka [8] which was the best known lower bound for the duality gap of general rectangle families.

In the full version of the paper [7], besides proving Lemma 1, we prove that computing a MIS on a diagonal-intersecting family is NP-complete. In light of our polynomial-time algorithm for diagonal-lower-intersecting families, the latter hardness result exhibits what is, in a way, a class at the boundary between polynomial-time solvability and NP-completeness. On the other hand, combining the results of Chalermsook and Chuzhoy [3] and Aronov et al. [2], we show that the duality gap is $O((\log \log \text{mis}(\mathcal{R}))^2)$ for a general family $\mathcal{R}$ of rectangles, improving on the logarithmic bound of Károlyi and Tardos [14].

## 2 Algorithms for WMIS

The idea behind Lubiw's algorithm [15] for WMIS on diagonal-touching families is to compute the optimal independent set $\text{OPT}_{ij}$ included in every possible triangle defined by the points $u^i$, $u^j$ (which are on $D$), and $(u_x^i, u_y^j)$ for two rectangles $i < j$. The principle exploited is that in $\text{OPT}_{ij}$ there exists one rectangle, say $i < k < j$, such that $\text{OPT}_{ij}$ equals the union of $\text{OPT}_{ik}$, the rectangle $k$, and $\text{OPT}_{kj}$. With this idea the overall complexity of the algorithm turns out to be cubic in $n$. We now present our algorithm, which works for the more general diagonal-lower-intersecting families, and that is based in a more elaborate idea involving what we call *harpoons*.

### 2.1 Algorithm for Diagonal-Lower-Intersecting Families

Let us first define some geometric objects that will be used in the algorithm. For any pair of rectangles $i < j$ we define $H_{i,j}$ and $H_{j,i}$, two shapes that we call

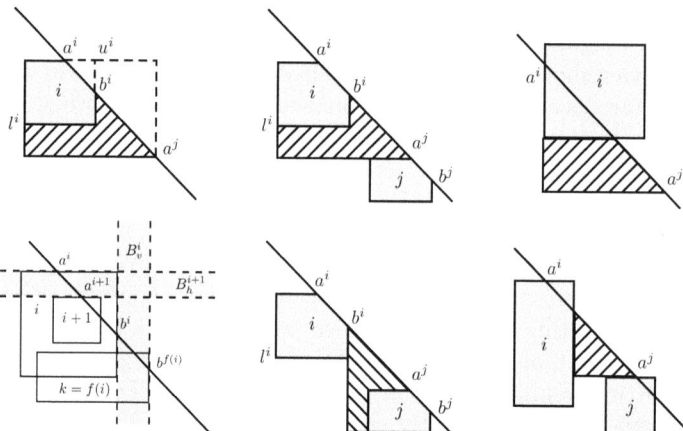

**Fig. 1.** On the left, the construction of a harpoon and the construction of the strips. On the middle, the harpoons $H_{ij}$ and $H_{ji}$, with $i < j$. On the right, other particular cases for the harpoon $H_{ij}$ with $i < j$ (the symmetric cases occur for $H_{ji}$).

harpoons. See Fig. 1. More precisely, the *horizontal harpoon* $H_{i,j}$ consists of the points below the diagonal $D$ obtained by subtracting rectangle $i$ from the closed box defined by the points $(\ell_x^i, a_y^i)$ and $a^j$. Similarly, the *vertical harpoon* $H_{j,i}$ are the points below $D$ obtained by subtracting $j$ from the box defined by the points $(b_x^j, \ell_y^j)$ and $b^i$. Also, for every rectangle $i$ with $i \geq 1$ (resp. such that $f(i)$ exists) we define $B_h^i$ (resp. $B_v^i$) as the open horizontal strip that goes through $a_{i-1}$ and $a_i$ (resp. as the open vertical strip that goes through $b_i$ and $b_{f(i)}$).

We say that a rectangle $r$ is contained in the set $H_{i,j}$ (and abusing notation, we write $r \in H_{i,j}$) if the region of $r$ below the diagonal is contained in $H_{i,j}$.

In our algorithm we will compute $S(i, j)$, the weight of the maximum independent set for the subset of rectangles contained in the harpoon $H_{i,j}$. We define two dummy rectangles 0 and $n + 1$, at the two ends of the diagonal such that the harpoons defined by these rectangles contain every other rectangle. As previously observed, two rectangles intersect in $\mathcal{R}$ if and only if they intersect below the diagonal. Therefore, wmis$(\mathcal{R}) = S(0, n + 1)$.

*Description of the algorithm:*

1. *Initialization.* In the execution of the algorithm we will need to know what rectangles have their lower-left corner in which strips. To compute this we do a preprocessing step. Define $\hat{B}_v^i$ and $\hat{B}_h^i$ as initially empty. For each rectangle $r \in \mathcal{R}$, check if $\ell^r$ is in $B_h^i$. If so, we add $r$ to the set $\hat{B}_h^i$. Similarly, if $\ell^r$ is in $B_v^i$, we add $r$ to the set $\hat{B}_v^i$.
2. *Main loop.* We compute the values $S(i,j)$ corresponding to the maximum-weight independent set of rectangles in $\mathcal{R}$ strictly contained in $H_{i,j}$. We do this by dynamic programming starting with the values $S(i,i) = 0$. Assume

that we have computed all $S(i,j)$ for all $i,j$ such that $|i-j| < \ell$. We now show how to compute these values when $|i-j| = \ell$.
 2.1 Set $S(i,j) = S(i,j-1)$ if $i < j$ and $S(i,j) = S(i,f(j))$ if $i > j$.
 2.2 Define $\hat{B}_{i,j}$ as $\hat{B}_h^j$ if $i < j$, or $\hat{B}_v^j$ if $i > j$.
 2.3 For each rectangle $k \in \hat{B}_{i,j}$ and strictly contained in harpoon $H_{i,j}$ do:
  2.3.1 Compute $m = w_k + \max\{S(i,k), S(k,i)\} + S(k,j)$.
  2.3.2 If $m > S(i,j)$, then $S(i,j) := m$.
3. *Output.* $S(0, n+1)$.

It is trivial to modify the algorithm to return not only wmis($\mathcal{R}$) but also the independent set of rectangles attaining that weight. We now establish the running time of our algorithm.

**Theorem 1.** *The previous algorithm runs in $O(n^2)$.*

*Proof.* The pre-processing stage needs linear time if the rectangles are already sorted, otherwise we require $O(n \log n)$ time. The time to compute $S(i,j)$ is $O(1 + |\hat{B}_{i,j}|)$ since checking if a rectangle is in a harpoon takes constant time. As the index of a rectangle is at most once in some $\hat{B}_h$ and at most once in some $\hat{B}_v$, the time to fill all the table $S(\cdot,\cdot)$ is:

$$\sum_{(i,j) \in [n]^2} O(1 + |\hat{B}_{i,j}|) = O(n^2).$$

The algorithm is then quadratic in the number of rectangles. □

In order to analyze the correctness of our algorithm we define a partial order over the rectangles in $\mathcal{R}$.

**Definition 3.** *The (strict) onion ordering $\prec$ in $\mathcal{R}$ is defined as*

$$i \prec j \iff \text{rectangles } i \text{ and } j \text{ are disjoint}, \ell_x^i < \ell_x^j, \text{ and } \ell_y^i < \ell_y^j.$$

It is immediate to see that $\prec$ is a strict partial ordering in $\mathcal{R}$. We say that $i$ is *dominated by* $j$ if $i \prec j$.

For any rectangle $k$ in a harpoon $H_{i,j}$, let $S_k(i,j)$ be the value of the maximum-weight independent set containing $k$ and rectangles in $H_{i,j}$ which are not dominated by $k$ in the onion ordering, and $\mathcal{S}_k(i,j)$ be the corresponding set of rectangles.

**Lemma 2.** *For any rectangle $k$ in $H_{i,j}$, the following relation holds:*

$$S_k(i,j) = w_k + \max\{S(i,k), S(k,i)\} + S(k,j).$$

*Proof.* Since $k \in H_{i,j}$, we have that $i, k$ and $j$ are mutually non-intersecting, and as indices, $\min(i,j) < k < \max(j,i)$. Assume that the harpoon is horizontal, i.e., $i < j$ (the proof for $i > j$ is analogous). In particular, we know that $a^i, b^i, a^k, b^k, a^j, b^j$ appear in that order on the diagonal. There are three cases for the positioning of the two rectangles $i$ and $k$. See Fig. 2.

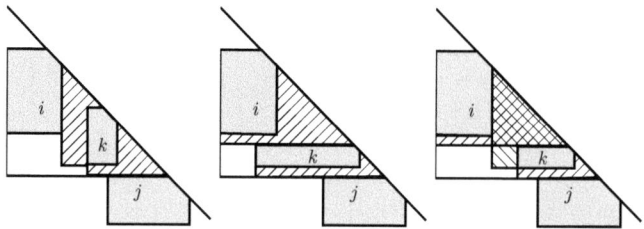

**Fig. 2.** The three cases for a rectangle in a horizontal harpoon

*First case*: $i$ and $k$ are separated by a vertical line, but not separated by a horizontal one. Noting that $H_{i,k} \subseteq H_{k,i}$, we conclude that all the rectangles of $\mathcal{S}_k(i,j) \setminus \{k\}$ are in $H_{k,i}$ or in $H_{k,j}$. Since $H_{k,i}$ and $H_{k,j}$ are disjoint, as shown on the first picture, we conclude the correctness of the formula.

*Second case*: $i$ and $k$ are separated by a horizontal line, but not by a vertical one. The proof follows almost exactly as in the first case.

*Third case*: $i$ and $k$ are separated by both a horizontal line and a vertical line. By geometric and minimality arguments, all the rectangles in $\mathcal{S}_k(i,j) \setminus \{k\}$ are in the union of the three harpoons $H_{i,k}$, $H_{k,i}$ and $H_{k,j}$ depicted. Finally, if there are two rectangles in $H_{i,k} \cup H_{k,i}$ then they must be in the same harpoon, so the formula holds. $\square$

**Theorem 2.** *Our algorithm returns a maximum weight independent set of $\mathcal{R}$.*

*Proof.* By induction. For the trivial harpoons $H_{i,i}$, the maximum independent set has weight 0, because this set is empty. The correctness of the theorem follows directly from the previous lemma and the next implications: For $i \neq j$,

$$i < j \Longrightarrow S(i,j) = \max\left\{S(i,j-1), \max_{k \in \hat{B}_h^j \cap H_{i,j}} S_k(i,j)\right\}.$$

$$j < i \Longrightarrow S(i,j) = \max\left\{S(i,f(j)), \max_{k \in \hat{B}_v^j \cap H_{i,j}} S_k(i,j)\right\}.$$

Indeed, assume that $i < j$ (the case $i > j$ is analogous). Let $\mathcal{S}$ be the MIS corresponding to $S(i,j)$, and let $m \in \mathcal{S}$ be minimal with respect to the onion ordering. If $m$ is in $H_{i,j-1}$ then $S(i,j) = S(i,j-1)$. Otherwise, $m$ is in $\hat{B}_h^j$ and since $\mathcal{S} \setminus \{m\}$ does not contain rectangles dominated by $m$, $S(i,j) = S_m(i,j)$. $\square$

## 2.2 An Approximation for Diagonal-Intersecting Families

We use the previous algorithm to get a 2-approximation for diagonal-intersecting rectangle families. This improves upon the 6-approximation (which is only for the unweighted case) of Chepoi and Felsner [5].

**Theorem 3.** *There exists a 2-approximation polynomial algorithm for* WMIS *on diagonal-intersecting rectangle families.*

*Proof.* Divide $\mathcal{R}$ into two subsets: the rectangle that intersect the diagonal on their upper side, and the ones that don't. It is easy to see that every rectangle in the second subset intersect the diagonal on its left side. Using symmetry, the left side case is equivalent to the upper side case. Therefore we can compute in polynomial time a WMIS in each subset. We output the heaviest one. Its weight is at least half of wmis($\mathcal{R}$). This algorithm gives a 2-approximation □

## 3 Duality Gap and Other Approximation Algorithms

In this section we explore the duality gap, that is, the largest possible ratio between mhs and mis, on some of the rectangle classes defined before.

**Theorem 4.** *The duality gap for diagonal-touching rectangle families is between 3/2 and 2. For diagonal-lower-intersecting families it is between 2 and 3, and for diagonal-intersecting families it is between 2 and 4.*

We will prove the upper bounds and the lower bounds separately.

*Proof of the upper bounds in Theorem 4.* Let $\mathcal{R}$ be a rectangle family in the plane, that can be in one of the three classes described on the theorem. In the case which $\mathcal{R}$ is diagonal-lower-intersecting we first replace each rectangle $r \in \mathcal{R}$ by the minimal one containing the region of $r$ that is below the diagonal. The modified family has the same intersection graph as before, but it is diagonal-corner-separated. In particular, the region of each rectangle that is above the diagonal is a triangle or a single point.

We use $\mathcal{R}_x$ and $\mathcal{R}_y$ to denote the projections of the rectangles in $\mathcal{R}$ on the $x$-axis and $y$-axis respectively. Both $\mathcal{R}_x$ and $\mathcal{R}_y$ can be regarded as intervals, and so we can compute in polynomial time the minimum hitting sets, $P_x$ and $P_y$, and the maximum independent sets, $\mathcal{I}_x$ and $\mathcal{I}_y$, of $\mathcal{R}_x$ and $\mathcal{R}_y$ respectively. Since interval graphs are perfect, $|P_x| = |\mathcal{I}_x|$ and $|P_y| = |\mathcal{I}_y|$.

Furthermore, since rectangles with disjoint projections over the $x$-axis (resp. over the $y$-axis) are disjoint, we also have

$$\text{mis}(\mathcal{R}) \geq \max\{|\mathcal{I}_x|, |\mathcal{I}_y|\} = \max\{|P_x|, |P_y|\}.$$

Observe that the collection $\mathcal{P} = P_x \times P_y \subset \mathbb{R}^2$ hits every rectangle of $\mathcal{R}$. From here we get the (trivial) bound mhs($\mathcal{R}$) $\leq |\mathcal{P}| \leq$ mis($\mathcal{R}$)$^2$ which holds for every rectangle family. When $\mathcal{R}$ is in one of the classes studied in this paper, we can improve the bound.

Let $\mathcal{P}^-$ and $\mathcal{P}^+$ be the sets of points in $\mathcal{P}$ that are below or above the diagonal, respectively. Consider the following subsets of $\mathcal{P}$:

$$\mathcal{F}^- = \{p \in \mathcal{P}^- : \nexists q \in \mathcal{P}^- \setminus \{p\}, p_x < q_x \text{ and } p_y < q_y\}.$$
$$\mathcal{F}^+ = \{p \in \mathcal{P}^+ : \nexists q \in \mathcal{P}^+ \setminus \{p\}, q_x < p_x \text{ and } q_y < p_y\}.$$
$$\mathcal{F}^* = \{p \in \mathcal{P}^+ : \nexists q \in \mathcal{P}^+ \setminus \{p\}, q_x \leq p_x \text{ and } q_y \leq p_y\}.$$

The set $\mathcal{F}^-$ (resp. $\mathcal{F}^+$) forms the closest "staircase" to the diagonal that is below (resp. above) it. The set $\mathcal{F}^*$ corresponds to the lower-left bending points of the staircase defined by $\mathcal{F}^+$. From here, it is easy to see that

$$\max\{|\mathcal{F}^-|, |\mathcal{F}^+|\} \leq |P_x| + |P_y| - 1 \leq 2\operatorname{mis}(\mathcal{R}) - 1.$$
$$|\mathcal{F}^*| \leq \max\{|P_x|, |P_y|\} \leq \operatorname{mis}(\mathcal{R}).$$

If $r \in \mathcal{R}$ is hit by a point of $\mathcal{P}^-$, let $p_1(r)$ be the point of $\mathcal{P}^- \cap r$ closest to the diagonal (in $\ell_1$-distance). Since $r$ intersects the diagonal, and the points of $\mathcal{P}$ form a grid, we conclude that $p_1(r) \in \mathcal{F}^-$. Similarly, if $r \in \mathcal{R}$ is hit by a point of $\mathcal{P}^+$, let $p_2(r)$ be the point of $\mathcal{P}^+ \cap r$ closest to the diagonal. Since $r$ intersects the diagonal, we conclude that $p_2(r) \in \mathcal{F}^+$. Furthermore, if the region of $r$ that is above the diagonal is a triangle, then $p_2(r) \in \mathcal{F}^*$.

If $\mathcal{R}$ is diagonal-touching, then every rectangle is hit by a point of $\mathcal{F}^-$, and so $\operatorname{mhs}(\mathcal{R}) \leq |\mathcal{F}^-| \leq 2\operatorname{mis}(\mathcal{R})-1$. If $\mathcal{R}$ is diagonal-lower-intersecting (and, after the modification discussed at the beginning of this proof, diagonal-corner-separated), then every rectangle is hit by a point of $\mathcal{F}^- \cup \mathcal{F}^*$, and so $\operatorname{mhs}(\mathcal{R}) \leq |\mathcal{F}^-|+|\mathcal{F}^*| \leq 3\operatorname{mis}(\mathcal{R})-1$. Finally, if $\mathcal{R}$ is diagonal-intersecting, then every rectangle is hit by a point of $\mathcal{F}^- \cup \mathcal{F}^+$, and so $\operatorname{mhs}(\mathcal{R}) \leq |\mathcal{F}^-|+|\mathcal{F}^+| \leq 4\operatorname{mis}(\mathcal{R})-2$. □

*Proof of the lower bounds of Theorem 4.* The lower bound of $3/2$ is achieved by any family $\mathcal{R}$ whose intersection graph $G$ is a 5-cycle. It is easy to see that $\mathcal{R}$ can be realized as a diagonal-touching family, that $\operatorname{mis}(\mathcal{R}) = 2$ and $\operatorname{mhs}(\mathcal{R}) = 3$, and so the claim holds.

The lower bound of 2 for diagonal-lower-intersecting and diagonal-intersecting families is asymptotically attained by a sequence of rectangle families $\{\mathcal{R}_k\}_{k \in \mathbb{Z}^+}$. We will describe the sequence in terms of infinite rectangles which intersect the diagonal, but it is easy to transform each $\mathcal{R}_k$ into a family of finite ones by considering a big bounding box.

For $i \in \mathbb{Z}^+$, define the $i$-th layer of the instance as $\mathcal{L}_i = \{U(i), D(i), L(i), R(i)\}$, and the $k$-th instance $\mathcal{R}_k = \bigcup_{i=1}^{k} \mathcal{L}_i$, where:

$$U(i) = [2i, 2i+1] \times [-(2i+\tfrac{1}{3}), +\infty), \quad D(i) = [2i+\tfrac{2}{3}, 2i+\tfrac{5}{3}] \times (-\infty, -2i],$$
$$L(i) = (-\infty, 2i+\tfrac{1}{3}] \times [-2i-1, -2i], \quad R(i) = [2i, \infty) \times [-(2i+\tfrac{5}{3}), -(2i+\tfrac{2}{3})].$$

Consider the instance $\mathcal{R}_k$ depicted in Figure 3 with $k$ layers of rectangles. $\mathcal{R}_k$ can be easily transformed into a diagonal-lower-intersecting family by "straightening" the staircase curve shown in the figure without changing its intersection graph. Let $I$ be a maximum independent set of rectangles in that instance. It is immediately clear that a minimum hitting set has size $2k$ since no point in the plane can hit more that two rectangles.

Let us prove that the size of a maximum independent set is at most $k + 2$, amounting to conclude that the ratio is arbitrarily close to 2. To this end, we let $i_D = \min\{i : D(i) \in I\}$ and $i_R = \min\{i : R(i) \in I\}$, and if no $D(i) \in I$ or no $R(i) \in I$, we let $i_D = k+1$ or $i_R = k+1$, respectively. When $i_D = i_R = k+1$, it is immediate that $|I| \leq k$. Assume then, without loss of generality, that $i_D < i_R$.

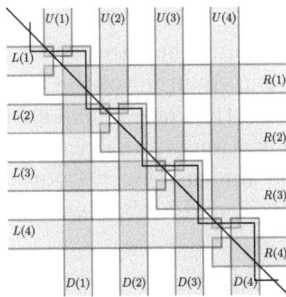

**Fig. 3.** The family $\mathcal{R}_4$. The diagonal line shows this family is diagonal-intersecting. The staircase line shows that it is actually lower-diagonal-intersecting.

Since for $i = 1, \ldots, i_D - 1$ the set $I$ neither contains rectangle $D(i)$ nor $R(i)$, we have that $I$ contains at most one rectangle on each of these layers. It follows that $\left|I \cap \bigcup_{i=1}^{i_D-1} \mathcal{L}_i\right| \leq i_D - 1$. Similarly, for $i = i_D + 1, \ldots, i_R - 1$ the set $I$ neither contains rectangle $L(i)$ nor $R(i)$, thus $\left|I \cap \bigcup_{i=i_D+1}^{i_R-1} \mathcal{L}_i\right| \leq i_R - i_D - 1$. Finally, we have that for $i = i_R + 1, \ldots, k$ the set $I$ neither contains rectangle $L(i)$ nor $U(i)$, and on layer $i_R$, $I$ contains at most 2 rectangles; thus $\left|I \cap \bigcup_{i=i_R}^{k} \mathcal{L}_i\right| \leq k - i_R + 2$. To conclude, note that $I$ may contain at most 2 rectangles of layer $i_D$, then

$$|I| = \sum_{i=1}^{k} |I \cap \mathcal{L}_i| \leq i_D - 1 + i_R - i_D - 1 + k - i_R + 2 + 2 = k + 2. \qquad \square$$

**Corollary 1.** *There is a simple 2-approximation polynomial time algorithm for MHS on diagonal-touching families, a 3-approximation for MHS on diagonal-lower-intersecting families, and a 4-approximation polynomial time algorithm for MHS on diagonal-intersecting families.*

*Proof.* The algorithm consists in computing and returning $\mathcal{F}^-$ for the first case, $\mathcal{F}^- \cup \mathcal{F}^*$ for the second one, and $\mathcal{F}^- \cup \mathcal{F}^+$ for the third one. $\qquad \square$

## 4 Discussion

To conclude the paper we mention open problems that are worth further investigation. First, note that the computational complexity of MHS is open for all classes of rectangle families considered in this paper. The complexity of recognizing the intersection graphs of different rectangles families is also open. It is known that the most general version of this problem, that is recognizing if a graph is the intersection graph of a family of rectangles, is NP-complete [20]. However, little is known for restricted classes. Finally, it would be interesting to determine the duality gap for the classes of rectangle families studied here.

**Acknowledgements.** We thank Pablo Pérez-Lantero, who got involved in this project, particularly in determining the hardness of MIS in the rectangle classes studied in this paper. Pablo is a coauthor of the full version. We also thank Vít Jelínek for allowing us to include the lower bound example in Figure 3, and Flavio Guíñez and Mauricio Soto for stimulating discussions. This work was partially supported by Núcleo Milenio Información y Coordinación en Redes ICM/FIC P10-024F and done while the second author was visiting Universidad de Chile.

# References

1. Adamaszek, A., Wiese, A.: Approximation Schemes for Maximum Weight Independent Set of Rectangles. In: FOCS 2013 (2013)
2. Aronov, B., Ezra, E., Sharir, M.: Small-size $\varepsilon$-nets for axis-parallel rectangles and boxes. SIAM J. Comp. 39, 3248–3282 (2010)
3. Chalermsook, P., Chuzhoy, J.: Maximum independent set of rectangles. In: SODA 2009 (2009)
4. Chan, T.M., Har-Peled, S.: Approximation algorithms for maximum independent set of pseudo-disks. In: SoCG 2009 (2009)
5. Chepoi, V., Felsner, S.: Approximating hitting sets of axis-parallel rectangles intersecting a monotone curve. Computational Geometry 46, 1036–1041 (2013)
6. Cibulka, J., Hladký, J., Kazda, A., Lidický, B., Ondráčková, E., Tancer, M., Jelínek, V.: Personal Communication (2011)
7. Correa, J.R., Feuilloley, L., Pérez-Lantero, P., Soto, J.A.: Independent and Hitting Sets of Rectangles Intersecting a Diagonal Line: Algorithms and Complexity. arXiv:1309.6659
8. Fon-Der-Flaass, D.G., Kostochka, A.V.: Covering boxes by points. Disc. Math. 120, 269–275 (1993)
9. Gyárfás, A., Lehel, J.: Covering and coloring problems for relatives of intervals. Disc. Math. 55, 167–180 (1985)
10. Hixon, T.S.: Hook graphs and more: Some contributions to geometric graph theory. Master's thesis, Technische Universitat Berlin (2013)
11. Fowler, R.J., Paterson, M.S., Tanimoto, S.L.: Optimal packing and covering in the plane are NP-complete. Inf. Process. Lett. 12, 133–137 (1981)
12. Hsiao, J.Y., Tang, C.Y., Chang, R.S.: An efficient algorithm for finding a maximum weight 2-independent set on interval graphs. Inf. Process. Lett. 43, 229–235 (1992)
13. Imai, H., Asano, T.: Finding the connected components and a maximum clique of an intersection graph of rectangles in the plane. J. of Algorithms 4, 310–323 (1983)
14. Károlyi, G., Tardos, G.: On point covers of multiple intervals and axis-parallel rectangles. Combinatorica 16, 213–222 (1996)
15. Lubiw, A.: A weighted min-max relation for intervals. J. Comb. Theory, Ser. B 53, 151–172 (1991)
16. Mustafa, N.H., Ray, S.: Improved results on geometric hitting set problems. Discrete Comput. Geom. 44, 883–895 (2010)
17. Soto, J.A., Telha, C.: Jump Number of Two-Directional Orthogonal Ray Graphs. In: Günlük, O., Woeginger, G.J. (eds.) IPCO 2011. LNCS, vol. 6655, pp. 389–403. Springer, Heidelberg (2011)
18. Soto, M., Thraves, C.: (c-)And graphs - more than intersection, more than geometric. arXiv:1306.1957 (2013) (submitted)
19. Wegner, G.: Über eine kombinatorisch-geometrische frage von hadwiger und debrunner. Israel J. of Mathematics 3, 187–198 (1965)
20. Yannakakis, M.: The complexity of the partial order dimension problem. SIAM J. Alg. Discr. Meth. 3, 351–358 (1982)

# Approximating Vector Scheduling: Almost Matching Upper and Lower Bounds*

Nikhil Bansal[1], Tjark Vredeveld[2], and Ruben van der Zwaan[1]

[1] Eindhoven University of Technology
{n.bansal,g.r.j.v.d.zwaan}@tue.nl
[2] Maastricht University
t.vredeveld@maastrichtuniversity.nl

**Abstract.** We consider the vector scheduling problem, a natural generalization of the classical makespan minimization problem to multiple resources. Here, we are given $n$ jobs represented as $d$-dimensional vectors in $[0,1]^d$ and $m$ identical machines, and the goal is to assign the jobs to machines such that the maximum *load* of each machine over all the coordinates is at most 1.

For fixed $d$, the problem admits an approximation scheme, and the best known running time is $n^{f(\varepsilon,d)}$ where $f(\varepsilon,d) = (1/\varepsilon)^{\tilde{O}(d)}$ ($\tilde{O}$ supresses polylogarithmic terms in $d$). In particular, the dependence on $d$ is doubly exponential.

In this paper we show that a double exponential dependence on $d$ is necessary, and give an improved algorithm with essentially optimum running time. Specifically, we show that:

- For any $\varepsilon < 1$, there is no $(1+\varepsilon)$-approximation with running time $\exp(o(\lfloor 1/\varepsilon \rfloor^{d/3}))$ unless the Exponential Time Hypothesis fails.
- No $(1+\varepsilon)$-approximation with running time $\exp(\lfloor 1/\varepsilon \rfloor^{o(d)})$ exists, unless $NP$ has subexponential time algorithms.
- Similar lower bounds also hold even if $\varepsilon m$ extra machines are allowed (i.e. with resource augmentation), for sufficiently small $\varepsilon > 0$.
- We complement these lower bounds with a $(1+\varepsilon)$-approximation that runs in time $\exp((1/\varepsilon)^{O(d \log \log d)}) + nd$. This gives the first efficient approximation scheme (EPTAS) for the problem.

## 1 Introduction

We consider the vector scheduling problem defined as follows. The input consists of a collection $J$ of $n$ jobs $\mathbf{p_1}, ..., \mathbf{p_n}$, viewed as $d$-dimensional vectors from $[0,1]^d$ and $m$ identical machines. The goal is to find an assignment of the jobs to the machines such that the load satisfies $\left\|\sum_{\mathbf{p} \in P_i} \mathbf{p}\right\|_\infty \leq 1$ for each machine $i \in [m]$, where $P_i$ is the set of jobs assigned to machine $i$. That is, the maximum load on any machine in any coordinate is most 1.

Vector scheduling is the natural multi-dimensional generalization of the classic multiprocessor scheduling problem (also known as makespan minimization or

---

* Supported by the NWO VIDI grant 639.022.211.

load balancing). In the latter problem, the goal is to assign $n$ jobs with arbitrary processing times to $m$ machines to minimize the maximum sum of processing times (load) over all the machines. For many applications however, the job may use different resources and the load of a job cannot be described by a single aggregate measure. For example, if jobs have both CPU and memory requirements, their processing requirement is best modeled as a two dimensional vector, where the value in each coordinate corresponds to each of the requirements. Clearly, an assignment of vectors is valid if and only if no machine is overloaded on any resource. Note that the assumption that the maximum load of a machine in any coordinate is 1 is without loss generality, as the different coordinates can be scaled independently.

In this paper we are concerned with approximation algorithms. We say that a polynomial-time algorithm is an $\alpha$-approximation (for $\alpha \geq 1$) if it finds an assignment with load at most $\alpha$, whenever there exists a feasible schedule with load at most 1.

### 1.1 Previous Work

Multiprocessor scheduling and the related bin-packing problem are one of the most fundamental problems in computer science with a long and rich history. We only describe the work on multiprocessor scheduling in the setting where the number of machines $m$ is a part of the input. It is well known that multiprocessor scheduling is strongly NP-hard [9].

The first polynomial time approximation scheme (PTAS), that is, a $(1+\varepsilon)$-approximation algorithm with polynomial running time for every fixed $\varepsilon > 0$, was obtained by Hochbaum and Shmoys [10]. The running time of their algorithm was $O(n^{O(1/\varepsilon^2)})$. Note that by the strong NP-Hardness of the problem one cannot hope to have a running time with polynomial dependence in $\varepsilon$ (i.e. an FPTAS), unless P=NP.

An efficient polynomial time approximation scheme (EPTAS), i.e. an algorithm with running time $f(\varepsilon)n^{O(1)}$, was implicit in [10] by replacing the dynamic program by an integer linear program (and using fast integer programming algorithms in fixed dimensions). A more general framework to obtain EPTASes for parallel machine scheduling was developed by Alon et al. [1] that runs in $f(1/\varepsilon) + O(n)$ time, where $f(\varepsilon)$ is a double exponential function in $1/\varepsilon$.

Recently, this running time was substantially improved by Jansen [13] to $O(2^{\tilde{O}(1/\varepsilon^2)} + n^{O(1)})$. His main idea is to use fast integer programming in fixed dimensions, together with an elegant result of Eisenbrand and Shmonin [5] about the existence of optimum integer solutions with small support. Most of these results also extend to the setting where the machine speeds differ, referred to as uniform machines (see e.g. [11,13]).

Fewer results are known for the case when the number of dimensions exceeds one. Chekuri and Khanna [4] gave the first polynomial-time approximation scheme for a fixed number of dimensions. They gave an algorithm with running time $n^{g(\varepsilon,d)}$, where $g$ is a singly exponential function of $d$. More precisely, $g(\varepsilon, d) = (1/\varepsilon)^{d \log \log d + o(d)}$ and hence the running time is $n^{(1/\varepsilon)^{\tilde{O}(d)}}$. This seems to be the currently best known running time for this problem. PTASes for several other generalizations are also known [2,6,7].

When $d$ is part of the input, Chekuri and Khanna [4] gave a polynomial time $O(\ln^2(d))$ approximation and proved that it is NP-hard to approximate the problem to within any constant factor. This approximation factor has been recently improved to $O(\log d)$ by Meyerson et al. [17]. The latter result even holds in the online setting.

### 1.2 Our Contribution

**Lower Bounds.** A natural question is whether there exists an approximation scheme for vector scheduling with running time that is singly exponential in $\varepsilon$ and $d$, e.g. $\exp(\text{poly}(\varepsilon, d))$. We rule out this possibility by showing the following strong lower bound.

**Theorem 1.** *For any $\varepsilon < 1$ with $1/\varepsilon \in \mathbb{N}$, there is a $d(\varepsilon)$ such that there is no $(1 + \varepsilon)$-approximation algorithm with running time $O\left(2^{o\left((1/\varepsilon)^{d/3}\right)}(nd)^{O(1)}\right)$ for vector scheduling in $d \geq d(\varepsilon)$ dimensions, unless the Exponential Time Hypothesis (ETH) fails.*

This follows from a relatively simple reduction from the 3-dimensional matching problem. The same reduction also implies the following hardness under a more standard complexity assumption.

**Theorem 2.** *For any $\varepsilon < 1$ with $1/\varepsilon \in \mathbb{N}$, there is a $d(\varepsilon)$ such that there is no $(1 + \varepsilon)$-approximation algorithm with running time $O\left(2^{(1/\varepsilon)^{o(d)}}(nd)^{O(1)}\right)$ for vector scheduling in $d \geq d(\varepsilon)$ dimensions, unless NP has subexponential time algorithms, i.e. $NP \subseteq \cap_{\varepsilon > 0} DTIME(2^{n^\varepsilon})$.*

One may wonder whether these lower bounds are robust or whether they crucially exploit the fact that no additional machines are allowed. It is instructive to consider the case of $d = 1$ (i.e. multiprocessor scheduling). Recall that no FPTAS is possible for the problem. However, if one allows some extra machines (say $\lceil \varepsilon m \rceil$ of them), then the running time dependence on $\varepsilon$ reduces dramatically and in particular, a FPTAS is possible. In fact, the known FPTASes for bin packing imply that even very few extra machines (poly-logarithmic in $m$) suffice [15,18], and in fact one does not even need to violate the capacity of any machine!

Somewhat surprisingly, we show that extra machines do not help for vector scheduling, provided that the desired approximation ratio is sufficiently small.

**Theorem 3.** *For any $\varepsilon < \varepsilon_0$ with $1/\varepsilon \in \mathbb{N}$, there is a $d(\varepsilon)$ such that there is no $(1 + \varepsilon)$-approximation algorithm with running time $O\left(2^{(1/\varepsilon)^{o(d)}}(nd)^{O(1)}\right)$ for vector scheduling in $d \geq d(\varepsilon)$ dimensions, even with $\lceil \varepsilon m \rceil$ extra machines, unless NP has subexponential time algorithms, i.e. $NP \subseteq \cap_{\varepsilon > 0} DTIME(2^{n^\varepsilon})$, where $\varepsilon_0 < 1$ is a universal constant. Assuming the ETH, no such algorithm can run in time $2^{\left(o(1/\varepsilon)^{d/6}\right)}$.*

**Upper Bounds.** To complement the lower bounds above, we show the following algorithmic result.

**Theorem 4.** *For any $\varepsilon > 0$ and $d \geq 1$, there is a deterministic $(1+\varepsilon)$-approximation algorithm for d-dimensional vector scheduling that runs in time $O\left(2^{(1/\varepsilon)^{O(d \log \log(d))}} + nd\right)$.*

**Techniques.** By the lower bounds above, the running time is essentially the best possible (modulo the $O(\log \log d)$ factor in the exponent), and the $nd$ term is simply the time required to read the input. Theorem 4 gives the first EPTAS for vector scheduling.

At a high level, the algorithm is similar to that of [13], and relies on integer programming in fixed dimensions and existence of optimum integer solutions with small support. However, there are some important differences between $d = 1$ and $d > 1$. In particular, for $d = 1$ the small jobs (with size $\leq \varepsilon$) do not cause any problems and can be filled in greedily later in the remaining spaces, after solving the problem for just big jobs. However, for $d \geq 2$, the big and small jobs (by small we mean jobs with value at most $\varepsilon$ in *every* dimension) interact in more complex says and must be considered together.

The following example highlights this difficulty. Consider the following instance with $m = 2$ and jobs with sizes: $\mathbf{p_1} = \left(\frac{1}{2}, 0\right), \mathbf{p_2} = \left(\frac{1}{2}, 0\right)$ and $\mathbf{p_i} = \left(\frac{\varepsilon}{2}, \varepsilon\right)$ for $3 \leq i \leq 2/\varepsilon$. Clearly, these jobs can be scheduled on two machines: place the first two jobs on separate machines, and split the small jobs evenly. However, if the first two (large) jobs are placed on the same machine, then the small jobs cannot be assigned feasibly anymore.[1]

Chekuri and Khanna [4] overcame this problem by 'guessing' for each machine the division between small and large jobs. This allows them to decouple the assignment of small and big vectors. However, as there are roughly $m^{(1/\varepsilon)^d}$ different divisions possible, with $\varepsilon$ precision, this is not useful to obtain an efficient polynomial time approximation scheme.

To get around this, we incorporate both large and small vectors in our integer linear program (ILP), but ensure that it has only few constraints by tracking only some coarse-grained information for the small jobs. We find an optimum solution to this ILP, which gives an integral assignment of large jobs but small jobs might be fractionally assigned. We then show how to assign the small jobs to machines, without overloading the machines. To do this, we first assign the jobs greedily guided by a potential function, which guarantees that the aggregate amount of overload on machines is small. This load is small enough that the jobs causing 'overloaded' can finally be redistributed in a round-robin manner. A naive implementation of the greedy assignment requires $O(mn)$ time (as for each job, we need to determine which machine causes the least increase in potential), so we also need some additional ideas to show how this can be done in linear time.

**Organization.** In Section 2 we state our notation and the hypotheses that our lower bounds are based on, and describe the relevant background on integer

---

[1] The two large jobs have total load $(1, 0)$. As the small jobs have total load $(1, 2)$, no matter how these are assigned to the two machines, one machine will have load at least $\min(\max(1+x, 2x), \max(1-x, 2(1-x)))$ which is $4/3$ (attained for $x = 1/3$).

programming. In Section 3 we prove our lower bounds for vector scheduling. We describe our algorithm in Section 4, however due to lack of space some ideas are only sketched and their detailed proofs are deferred to the full version of the paper.

## 2 Preliminaries

Let $[n]$ denote the set of positive integers 1 to $n$ i.e. $[n] := \{1, ..., n\}$. Let **1** be the all-ones vector. For two vectors $\mathbf{a}, \mathbf{b}$ we say that $\mathbf{a} \leq \mathbf{b}$ if $a_i \leq b_i, \forall i$. Throughout the paper the logarithm log is taken with base two and let $\exp(x)$ denote $2^x$. For a $d$-dimensional vector $\mathbf{v} = (v_1, ..., v_d)$, let $v_j$ denote its $j$-th coordinate. We say that a function $f(n)$ is sub-exponential if $f(n) \in O(2^{O(n^{o(1)})})$. Without loss of generality we assume that the number of machines is less than the number of jobs (otherwise assign one job per machine or conclude infeasibility).

Impagliazzo, Paturi and Zane formulated the Exponential Time Hypothesis which in combination with the sparsification lemma [3] can be stated as follows.

**Hypothesis 1 (Exponential Time Hypothesis (ETH) [12]).** *There is a positive real $s$ such that 3-CNF-Sat with $N$ variables and $M$ clauses cannot be solved in time $O(2^{sM}(N+M)^{O(1)})$.*

We will use the following well-known results for fast integer linear programs with few integer variables.

**Theorem 5 (Lenstra [16], Kannan [14], Frank and Tardos [8]).** *Consider a mixed-integer linear program $\min\{\mathbf{c}^T\mathbf{x} \mid A\mathbf{x} \geq \mathbf{b} \text{ and } \forall i \in \mathcal{I} : x_i \in \mathbb{Z}\}$ with $n$ variables and $m$ constraints, and where $\mathcal{I} \subseteq [n]$ denotes the set of indices of integer variables. Let $s$ denote the binary encoding length of the input. Then, there is an algorithm that finds a feasible solution or decides that there is no feasible solution in $O(n^{2.5n+o(n)} \cdot s)$ arithmetic operations.*

Relatively recently, based on an elegant pigeonhole argument, Eisenbrand and Shmonin [5] showed that every feasible integer linear program has an optimum solution with small support.

**Theorem 6 (Eisenbrand and Shmonin [5]).** *Let $\min\{\mathbf{c}^T\mathbf{y} \mid A\mathbf{y} = \mathbf{b}, \mathbf{y} \geq 0, \mathbf{y} \in \mathbb{Z}^n\}$ be an integer program, where $A \in \mathbb{Z}^{m \times n}$ and $\mathbf{c} \in \mathbb{Z}^n$. If this integer program has a finite optimum, then there exists an optimal solution $\mathbf{y}^* \in \mathbb{Z}^n_{\geq 0}$ with number of nonzero components at most $2(m+1)(\log(m+1)+s+2)$, where $s$ is the largest size (in binary representation) of any coefficient of $A$ and $\mathbf{c}$.*

## 3 Lower Bounds on the Running Time

We prove our lower bounds by a reduction from 3-dimensional matching (3-DM) to Vector Scheduling. In subsection 3.1 we describe the reduction, and in subsection 3.2 we prove that an approximate solution to the vector scheduling instance implies an exact solution for 3-DM and hence 3-CNF-Sat. The proof that resource augmentation does not help much, Theorem 3, is postponed to the full version.

Before we give our reduction, we first define the 3-dimensional matching problem. An instance of 3-DM consists of three disjoint sets $X$, $Y$, and $Z$, satisfying $|X| = |Y| = |Z| := q$, and a set $T \subset X \times Y \times Z$ of triples. The goal is to find a subset of triples $T' \subset T$ such that there are no each element of $X$, $Y$, and $Z$ occurs in exactly one triple of $T'$.

In [9] a reduction from 3-CNF-Sat to 3-DM is given, that transforms instances of 3-SAT with $N$ variables and $M$ clauses, into instances for 3-DM with $q = 6M$ and $|T| = 17M$. Therefore, the ETH (Theorem 1) implies there is no $O(2^{o(q)} q^{O(1)})$ time algorithm for 3-DM.

### 3.1 The Construction of a Vector Scheduling Instance from 3-DM

The main idea of the reduction is the following: We construct for each triple in $T$ a job (that we call a *triple-job*) and for each element in $X$, $Y$ or $Z$, as many jobs as the number of times this element occurs in the triples (we call such jobs *element-jobs*). For each element $i$, we designate exactly one of its jobs as the *real* element-job corresponding to $i$, and refer to the other element-jobs for $i$ as *dummy* jobs. The number of machines is equal to the number of triples. We will assign sizes to these jobs such that to obtain a schedule such that the maximum load in any coordinate is at most 1, we need to schedule each triple together with its corresponding three element-jobs, and moreover these element-jobs are either all *real* element-jobs or all dummy jobs.

Let $\varepsilon < 1$ be such that $1/\varepsilon$ is integer. Let $b = 1/\varepsilon - 1$ and let $\mathbf{b}$ denote the vector that has $b$ in every coordinate. By $\langle i \rangle$ we denote the $(b+1)$-ary encoding of the integer $i$ and by $\overline{\langle i \rangle}$ we denote its complement, that is, $\overline{\langle i \rangle} := \mathbf{b} - \langle i \rangle$. Let $\langle i \rangle_j$ denote the $j$-th digit from the right of $\langle i \rangle$. For ease of notation, we scale the jobs by a factor $b$. That is, all vectors are in $[0, b]^d$ and we ask the question whether we can schedule the jobs such that the maximum load in each coordinate is at most $b$. To make the proofs easier to read, we rename the elements in the sets $X$, $Y$ and $Z$: we assume that $X = Y = Z = \{1, ..., q\}$.

**The Formal Reduction.** Given an instance $(X, Y, Z; T)$ of 3-DM, let $n_X(i)$ denote the number of triples $(x, y, z)$ for which $x = i$; in a similar way, we define $n_Y(i)$ and $n_Z(i)$. For each element $i \in X$, we create $n_X(i)$ jobs, one *real* $X$-job $i$ and $n_X(i) - 1$ *dummy* $X$-jobs. In a similar way, we create $n_Y(j)$ $Y$-jobs for each element $j \in Y$ and $n_Z(k)$ $Z$-jobs for each element $k \in Z$. Finally, we have another $|T|$ triple-job, one for each triple $l \in T$. The number of machines is equal to $m := |T|$. Note that the total number of jobs is $\sum_{i \in X} n_X(i) + \sum_{j \in Y} n_Y(j) + \sum_{k \in Z} n_Z(k) + |T| = 4|T| = 4m$.

Recall that $|X| = |Y| = |Z| = q$, and let $\ell := \lceil \log_{(1/\varepsilon)} q \rceil$. We associate a vector to each of the jobs as follows. The number of dimensions of these vectors is $d := 7 + 3\ell$. The precise construction for the vectors for each of the jobs is described in Table 1.

In particular, the first four coordinates of a job indicate whether the job corresponds to an element in $X$, $Y$, $Z$ or to a triple in $T$. The following three coordinates encode for each $X$, $Y$, or $Z$-job whether it is a real job or a dummy

**Table 1.** Construction of the jobs from elements and triples of the 3-DM problem

| Job name | Values of the coordinates | | | | |
|---|---|---|---|---|---|
| | T/X/Y/Z | Real/dummy | Encoding of element(s) | | |
| real $X$-job $i$: | $0, b, 0, 0$ | $b, 0, 0$ | $\langle i \rangle_1, ..., \langle i \rangle_\ell$ | $0, ..., 0$ | $0, ..., 0$ |
| dummy $X$-job $i$: | $0, b, 0, 0$ | $0, b, 0$ | $\langle i \rangle_1, ..., \langle i \rangle_\ell$ | $0, ..., 0$ | $0, ..., 0$ |
| real $Y$-job $j$: | $0, 0, b, 0$ | $0, b, 0$ | $0, ..., 0$ | $\langle j \rangle_1, ..., \langle j \rangle_\ell$ | $0, ..., 0$ |
| dummy $Y$-job $j$: | $0, 0, b, 0$ | $0, 0, b$ | $0, ..., 0$ | $\langle j \rangle_1, ..., \langle j \rangle_\ell$ | $0, ..., 0$ |
| real $Z$-job $k$: | $0, 0, 0, b$ | $0, 0, b$ | $0, ..., 0$ | $0, ..., 0$ | $\langle k \rangle_1, ..., \langle k \rangle_\ell$ |
| dummy $Z$-job $k$: | $0, 0, 0, b$ | $b, 0, 0$ | $0, ..., 0$ | $0, ..., 0$ | $\langle k \rangle_1, ..., \langle k \rangle_\ell$ |
| triple $(i, j, k)$: | $b, 0, 0, 0$ | $0, 0, 0$ | $\overline{\langle i \rangle_1}, ..., \overline{\langle i \rangle_\ell}$ | $\overline{\langle j \rangle_1}, ..., \overline{\langle j \rangle_\ell}$ | $\overline{\langle k \rangle_1}, ..., \overline{\langle k \rangle_\ell}$ |

job. The last part of each job is reserved to encode the element to which the job corresponds.

### 3.2 Proof of the Reduction

We now show that the reduction has the desired properties.

**Lemma 1.** *(Completeness) If the 3-DM instance has a solution, then there assignment with load at most $b$.*

*Proof.* Consider the collection of disjoint triples that cover $X, Y$ and $Z$. For each job triple $(i, j, k)$, we place the corresponding triple-job, and the real element-jobs corresponding to $i$, $j$ and $k$ on a single machine. It is easily checked that every coordinate on every such machine has load at most $b$. For each of the remaining triples $(i, j, k)$, we place them on a machine with (any) of the dummy jobs for $i$, $j$ and $k$. It is easily verified that this is a feasible assignment. □

**Lemma 2.** *If the vector scheduling instance has a solution with load at most $(1 + \varepsilon)b$, then there is a solution to the corresponding 3-DM instance.*

*Proof.* Consider any solution with load at most $(1 + \varepsilon)b$. We begin with various properties of such a solution.

**Property 1.** *The load is exactly $b$ in each coordinate on each machine.*
*Proof.* The load of each machine is at most $(1 + \varepsilon)b = b + b/(b+1) < b + 1$. As all jobs have integer coordinates, the load of each machine is at most $b$.

Moreover, observe that the total amount of work in the $i$-th coordinate summed over all jobs is $mb$. In particular, this follows as $\sum_{i \in X} n_X(i) = \sum_{j \in Y} n_Y(j) = \sum_{k \in Z} n_Z(k) = |T| = m$. As all jobs are scheduled and the load is most $b$, it must be *exactly* $b$. □

*Property 2.* Each machine processes exactly one triple- one $X$-element-, one $Y$-element- and one $Z$-element-job.

*Proof.* This follows immediately from the values in the first four coordinates and the previous property. □

*Property 3.* Element-jobs on a machine are either all *real* or all *dummy*.

*Proof.* From Property 1 and the values in the fifth, sixth and seventh coordinate we see that the following three statements are simultaneously true:

1. There is exactly one real $X$-element or dummy $Z$-element (coordinate 5);
2. There is exactly one real $Y$-element or dummy $X$-element (coordinate 6);
3. There is exactly one real $Z$-element or dummy $Y$-element (coordinate 7).

The claim now follows, by combining with the fact there can be exactly one real or dummy element of each of the types $X,Y$ and $Z$ (this is ensured by coordinates 2,3 and 4). □

*Property 4.* If a machine has the triple-job $(i, j, k)$ and a (real or a dummy) element-job $a$, then $a$ is equal to $i$, $j$ or $k$, depending on whether $a$ is a $X$, $Y$ or $Z$-element.

*Proof.* We only consider the case that $a$ is an $X$-element; the other cases are similar. By Properties 1 and 2, we know that $\overline{\langle i \rangle} + \langle a \rangle = \mathbf{b}$. Therefore, $\langle a \rangle = \mathbf{b} - \overline{\langle i \rangle} = \mathbf{b} - (\mathbf{b} - \langle i \rangle) = \langle i \rangle$ and thus $a = i$. □

By the last property, if a machine processes three real element-jobs, then the corresponding three elements form a triple in the 3-DM instance. Let $T'$ consist of all triples corresponding to the triple-jobs that are scheduled together with real elements. Then, the triples in $T'$ have no overlap as there is only one real element-job corresponding to an element. Moreover, $T'$ covers all elements as all jobs and therefore also all real element-jobs need to be scheduled. □

Therefore we have the following lemma.

**Lemma 3.** *Given an instance of 3-Dimensional Matching with $|X| = |Y| = |Z| = q$, $T \subseteq X \times Y \times Z$, $b \in \mathbb{N}_+$, $b \geq 2$ and $\varepsilon = 1/(b-1)$. Then, there is a polynomial time reduction to an instance for Vector Scheduling with $4|T|$ vectors in dimension $d := 3 \left\lceil \log_{(1/\varepsilon)} q \right\rceil + 7$. Further, a $(1+\varepsilon)$-approximate solution to the Vector Scheduling instance defines a solution to the 3-DM problem.*

Thus, Lemma 3 in combination with the ETH and the reduction from 3-CNF-Sat to 3-dimensional matching yields Theorem 1. In similar vein we can derive Theorem 2: the proof is postponed to the full version.

*Proof (Theorem 1).* Suppose that there exists an $(1+\varepsilon)$-approximation for Vector Scheduling that runs in time $O\left(2^{o((1/\varepsilon)^{d/3})} n^{O(1)}\right)$. By Lemma 3 we get a $O\left(2^{o(q)} |T|^{O(1)}\right)$ time algorithm for 3-DM which in turn implies a $O\left(2^{o(M)} M^{O(1)}\right)$ time algorithm for 3-CNF-SAT which contradicts the ETH. □

## 4 Linear Time Approximation Algorithm

In this section we describe our deterministic linear time algorithm. Roughly, our algorithm works as follows. First, we preprocess the instance such that there are relatively few different types of large jobs at the cost of a small factor in the approximation guarantee. Second, we formulate and solve exactly a mixed-integer linear program that assigns large jobs integrally to machines and small jobs fractionally. We assign the small jobs integrally to machines in a greedy mannner guided by a potential function that tracks the aggregate "overload" on the machines, due to jobs that cause machines to have load exceeding $1+\varepsilon$. Then, we distribute these "overload" evenly over all machines ensuring the final loads of all machines is at most $1+\varepsilon$.

**Preprocessing.** The preprocessing steps are standard [4] and are summarized in Lemma 4. We defer the details to the full version.

**Lemma 4.** *Let $V$ be the original set of jobs and $W$ be the preprocessed set of jobs and $\varepsilon > 0$. Then for any $\mathbf{w} \in W$ and coordinate $i \in [d]$*

- *if $w_i \neq 0$ then $w_i/\|w\|_\infty \geq \varepsilon/d$;*
- *there exists a $k \in \mathbb{N}$ such that $w_i = \varepsilon^3/d^2 \cdot (1+\varepsilon)^k$;*

*and for any subset of jobs $V' \subset V$ such that $\sum_{\mathbf{v} \in V'} \mathbf{v} \leq 1$ with corresponding modified subset $W' \subseteq W$, we have $\sum_{\mathbf{w} \in W'} \mathbf{w} \leq \sum_{\mathbf{v} \in V'} \mathbf{v} \leq (1+\varepsilon)\sum_{\mathbf{w} \in W'} \mathbf{w} + \varepsilon$.*

### 4.1 The Mixed-Integer Linear Program

In this subsection we describe our mixed-integer linear program and also how to solve it faster. We distinguish between *small* and *big* jobs and treat them differently. A job $\mathbf{p}$ is *small* if $\|\mathbf{p}\|_\infty < \varepsilon^2/d$ and otherwise the vector is *big*.

Let $\mathcal{T}_{\text{big}}$ be the set of all *types* of big vectors, $\mathcal{T}_{\text{big}} := \{0, \varepsilon^3/d^2, (1+\varepsilon)\varepsilon^3/d^2, (1+\varepsilon)^2\varepsilon^3/d^2, ..., 1\}^d$. A big job $\mathbf{p}$ has type $\mathbf{t} \in \mathcal{T}_{\text{big}}$ if and only if $\mathbf{p} = \mathbf{t}$. Every big vector has a corresponding type, since the rounding procedure rounded these vectors to exactly these values.

We define the type of a small vector based on its relative size in each coordinate. Let $\mathcal{T}_{\text{small}}$ be the set, $\mathcal{T}_{\text{small}} := \{0, (1+\varepsilon)^{-\ell}, (1+\varepsilon)^{-\ell+1}, ..., (1+\varepsilon), 1\}^d$, where $\ell := \left\lceil \log_{(1+\varepsilon)}(d/\varepsilon) \right\rceil$. A small job $\mathbf{p}$ has type $\mathbf{t} = (t_1, \ldots, t_d) \in \mathcal{T}_{\text{small}}$ if and only if $\frac{(p)_j}{\|\mathbf{p}\|_\infty} = t_j$ for all coordinates $j \in [d]$. By Lemma 4 the biggest and smallest non-zero coordinates are at most a factor $d/\varepsilon$ apart, and hence $\frac{\mathbf{p}}{\|\mathbf{p}\|_\infty}$ is at least $\varepsilon/d$. Therefore, each small vector has exactly one type in $\mathcal{T}_{\text{small}}$. Thus, there are at most $T := \left\lceil 3\log_{(1+\varepsilon)}(d/\varepsilon) + 2 \right\rceil^d$ types of big and small jobs.

In our mixed-integer linear programming we have variables corresponding to configurations, which are a collection of big jobs together with available room for small jobs. We will call the (rounded) room for small jobs a *profile*. A profile is a vector from $\mathcal{F} := \{0, \varepsilon, (1+\varepsilon)^1\varepsilon, (1+\varepsilon)^2\varepsilon, ..., 1\}^d$. A configuration $C$ is a tuple $C = (B, \mathbf{f})$, where $B$ is a set of big jobs and $\mathbf{f}$ is a profile for small jobs such that together the big vectors and the profile fit on one machine while only exceeding the maximum load by a little i.e. $(\sum_{p \in B} p_j) + f_j \leq (1 + \varepsilon)$ for all coordinates $j$. As each big job has a coordinate of at least $\varepsilon/d$ there can be at most $d^2/\varepsilon$ big jobs on a machine or the load exceeds 1. As there at most $T$ types of big jobs, this implies that there are $N \leq T^{\lceil d^2/\varepsilon \rceil} \cdot T$ different configurations.

We now describe our mixed-integer linear program. Let $x_C$ denote the number of machines that have jobs assigned to them according to configuration $C$ and let $\mathcal{C}$ be the set of all configurations. Let $n(C, \mathbf{t})$ denote the number of big jobs of type $\mathbf{t}$ in configuration $C$, and let $n(\mathbf{t})$ denote the total number of big jobs of type $\mathbf{t}$ in the instance.

Let $y_{\mathbf{f},\mathbf{t}}$ be the *amount* with respect to the largest coordinate (i.e. sum of $\ell_\infty$ norm of the sizes), of small jobs of type $\mathbf{t}$ assigned to configurations with profile $\mathbf{f}$ for small jobs. Let $a(\mathbf{t}) := \sum_{\mathbf{p}:\mathbf{p} \text{ is of small type } \mathbf{t}} \|\mathbf{p}\|_\infty$ denote the total amount of small jobs of type $\mathbf{t}$ in the instance.

Consider the following program.

$$\min \sum_{C \in \mathcal{C}} x_C \qquad \text{(MILP)}$$

$$\text{s.t.} \sum_{C \in \mathcal{C}} x_C \cdot n(C, \mathbf{t}) \geq n(\mathbf{t}) \qquad \forall \mathbf{t} \in \mathcal{T}_{\text{big}} \qquad \text{(C1)}$$

$$\sum_{\mathbf{f} \in \mathcal{F}} y_{\mathbf{f},\mathbf{t}} \geq a(\mathbf{t}) \qquad \forall \mathbf{t} \in \mathcal{T}_{\text{small}} \qquad \text{(C2)}$$

$$\sum_{\mathbf{t} \in \mathcal{T}_{\text{small}}} y_{\mathbf{f},\mathbf{t}} \cdot t_i / \|\mathbf{t}\|_\infty \leq f_i \cdot \sum_{C:C=(B,\mathbf{f})} x_C \qquad \forall i \in [d], \mathbf{f} \in \mathcal{F} \qquad \text{(C3)}$$

$$\mathbf{x} \in \mathbb{Z}^{|\mathcal{C}|}$$

$$\mathbf{y}, \mathbf{x} \geq 0$$

The first constraint ensures that the big jobs are covered. The second constraint ensures that the small jobs are covered fractionally. The third constraint is not as obvious, and requires that for each profile $f$ of space for small jobs, the cumulative amount of small jobs of type $t$ that are assigned to profile $f$ must be no more than the total amount of profile $f$. It is easily seen that these are valid constraints for any feasible solution.

The following lemma follows directly by Theorems 5 and 6, and its proof is deferred to the full version.

**Lemma 5.** *An optimal solution to MILP can be found in time $O\left(\exp\left((1/\varepsilon)^{O(d \log \log d)}\right) \cdot \log(n)\right)$, where $s$ denotes the maximum length of the binary encoding of the mixed-integer linear program.*

### 4.2 Linear Time Algorithm

In this section we sketch how small jobs are assigned integrally to machines, using the solution to MILP. Our algorithm can be viewed as the derandomization of the following natural randomized algorithm.

**Randomized Algorithm.** Recall that $y_{\mathbf{f},\mathbf{t}}$ is the amount of small jobs of type $\mathbf{t}$ that are assigned to machines with profile $\mathbf{f}$. For each small job of type $\mathbf{t}$, let $\beta(\mathbf{f},\mathbf{t})$ denote the fraction of type $\mathbf{t}$ assigned to profile $\mathbf{f}$ : $\beta(\mathbf{f},\mathbf{t}) := y_{\mathbf{f},\mathbf{t}} / \sum_{\mathbf{g} \in \mathcal{F}} y_{\mathbf{g},\mathbf{t}}$.

For each small job $\mathbf{p}$ of type $\mathbf{t}$ do the following. Pick a profile $\mathbf{f}$ randomly with probability $\beta_{\mathbf{f},\mathbf{t}}$. Then we pick a machine uniformly at random among the ones with profile $\mathbf{f}$. We assign job $\mathbf{p}$ to this machine.

It is easily checked that the expected load on any machine is precisely the fractional load of the ILP solution. However, the randomness can cause some machines to be overloaded (i.e. have load more than $1 + \varepsilon$ is some coordinate). However, standard probabilistic tail bounds imply that the total aggregate load on overloaded machines is too not high (about $O(\text{poly}(\varepsilon/d) \cdot m)$. Thus for each overloaded machine, we can remove all the small jobs assigned to it, and redistribute these jobs in a round-robin manner over all machines.

**Deterministic Algorithm.** Recall that the ILP only gives an assignment of small job types to profiles, while we need an assignment of individual jobs to machines. We achieve this in three steps.

**(Step 1)** We first assign small job types to machines. To this end, we note that as there are few constraints involving $y$ variables, by standard polyhedral arguments there are few $y$ variables that are divided over many profiles. By greedily assigning small jobs to profiles as long as they fit w.r.t $y_{\mathbf{f},\mathbf{t}}$, only few small jobs remain which can then be evenly divided over the profiles only causing a $O(\varepsilon)$ increase in the loads.

**(Step 2)** By step 1 small jobs are integrally assigned to profiles and thus we can restrict our attention to machines using a fixed profile. The small jobs are now assigned to the machine that minimizes a potential function $\phi$. Let $\mathbf{p}_1, ..., \mathbf{p}_n$ be small jobs assigned to the same profile and $\lambda > 0$ some parameter, then $\phi^t := \sum_{k \in [d]} \sum_{i \in [m]} e^{\lambda(L_k^{t,i} - f_k^t)}$, where $L_k^{t,i}$ is the load of small jobs $\mathbf{p}_1, ..., \mathbf{p}_t$ assigned to machine $i$ on coordinate $k$ and let $f_k^t$ be the expected load of the first $t$ jobs in coordinate $k$ which is at most the free space reserved for small jobs by the solution for MILP.

The function $\phi$ is essentially a pessimistic estimator of the load exceeding the expectation summed over the machines and coordinates. Regardless how $\mathbf{p}_1, ..., \mathbf{p}_{t-1}$ are assigned, there is always an assignment for $\mathbf{p}_t$ that increases the potential function by at most a small multiplicative factor. By assigning at each step the job to the machine that minimizes the potential, we have an integral assignment and $\phi^n \leq md \exp(2\lambda^2 dp_{max})$, where $p_{max}$ is the maximum value over all small jobs and all coordinates.

**(Step 3)** In the above integral assignment, let $m(x)$ be the number of machines with load exceeding $1 + \varepsilon + x$ in any coordinate and let $L$ be the set of small jobs on machines with load at least $1 + \varepsilon$ in any coordinate. The bound on $\phi^n$ above implies that $m(x) \cdot e^{\lambda(\varepsilon+x)} \leq md \exp(2\lambda^2 dp_{max})$, which by setting $\lambda := 1/\varepsilon \log(d^3/\varepsilon^3)$ gives that $m(x) \leq \frac{O(md)}{(d^3/\varepsilon^3)^{1+x/\varepsilon}}$.

A direct calculation then yields that the total load of jobs on machines with load at least $1+\varepsilon$ is at most $O(\varepsilon m/d)$. By removing the small jobs from these machines and reassigning them in round-robin fashion over all machines, the loads are increased by at most $O(\varepsilon)$ on each machine.

**Implementation in Linear Time.** While step 1 and 3 can be implemented directly in linear time, step 2 is more complicated. For simplicity, let $\delta$ be the maximum value of any coordinate of any small jobs. The trivial implementation takes $O(md)$ time per job, by trying each machine for each job and evaluating the resulting value of $\phi$. To get around this, we do the following. Greedily glue small jobs of the same type together, as long as the resulting small job is still small. This results in jobs $B$ that have their coordinates in $\{0\} \cup [\varepsilon\delta/2d, \delta]$ and at most $|\mathcal{T}_{\text{small}}|$ jobs $S$ with coordinates in $[0, \delta/2]$. Then, round the coordinates of jobs from $B$ down to a multiple of $\varepsilon(\varepsilon\delta/2d)$, at the cost of a small increase in the loads. This ensures that during the assignment of jobs from $B$, the machines have only few distinct loads. This enables that for each job, the best machine can be found in $O(1)$ time and updating takes $O(d)$ time. A similar strategy for $S$ would fail, but notice that $|S|$ is relatively small. If $|S| \leq m$, then assign one job to each machine, increasing the loads by at most $O(\delta)$. If $|S| > m$ then the number of machines is at most $|\mathcal{T}_{\text{small}}|$ and the trivial algorithm takes $O(|\mathcal{T}_{\text{small}}|^2)$ time. This results in the claimed running time of Theorem 4.

## References

1. Alon, N., Azar, Y., Woeginger, G.J., Yadid, T.: Approximation schemes for scheduling on parallel machines. Journal of Scheduling 1, 55–66 (1998)
2. Bonifaci, V., Wiese, A.: Scheduling unrelated machines of few different types. CoRR, abs/1205.0974 (2012)
3. Calabro, C., Impagliazzo, R., Paturi, R.: A duality between clause width and clause density for SAT. In: IEEE Conference on Computational Complexity, pp. 252–260. IEEE Computer Society (2006)
4. Chekuri, C., Khanna, S.: On multidimensional packing problems. SIAM J. Comput. 33(4), 837–851 (2004)
5. Eisenbrand, F., Shmonin, G.: Carathéodory bounds for integer cones. Oper. Res. Lett. 34(5), 564–568 (2006)
6. Epstein, L., Tassa, T.: Vector assignment problems: a general framework. J. Algorithms 48(2), 360–384 (2003)
7. Epstein, L., Tassa, T.: Vector assignment schemes for asymmetric settings. Acta Inf. 42(6-7), 501–514 (2006)
8. Frank, A., Tardos, É.: An application of simultaneous diophantine approximation in combinatorial optimization. Combinatorica 7, 49–65 (1987)
9. Garey, M.R., Johnson, D.S.: Computers and Intractability: A Guide to the Theory of NP-Completeness. W. H. Freeman and Company (1979)
10. Hochbaum, D.S., Shmoys, D.B.: Using dual approximation algorithms for scheduling problems theoretical and practical results. J. ACM 34(1), 144–162 (1987)
11. Hochbaum, D.S., Shmoys, D.B.: A polynomial approximation scheme for scheduling on uniform processors: Using the dual approximation approach. SIAM J. Comput. 17(3), 539–551 (1988)
12. Impagliazzo, R., Paturi, R., Zane, F.: Which problems have strongly exponential complexity? J. Comput. Syst. Sci. 63(4), 512–530 (2001)

13. Jansen, K.: An eptas for scheduling jobs on uniform processors: Using an milp relaxation with a constant number of integral variables. SIAM J. Discrete Math. 24(2), 457–485 (2010)
14. Kannan, R.: Minkowskis convex body theorem and integer programming. Mathematics of Operations Research 12, 415–440 (1987)
15. Karmarkar, N., Karp, R.M.: An efficient approximation scheme for the onedimensional bin-packing problem. In: FOCS, pp. 312–320 (1982)
16. Lenstra, H.W.: Integer programming with a fixed number of variables. Mathematics of Operations Research 8(4), 538–548 (1983)
17. Meyerson, A., Roytman, A., Tagiku, B.: Online multidimensional load balancing. In: Raghavendra, P., Raskhodnikova, S., Jansen, K., Rolim, J.D.P. (eds.) APPROX/ RANDOM 2013. LNCS, vol. 8096, pp. 287–302. Springer, Heidelberg (2013)
18. Rothvoß, T.: Approximating bin packing within O(log OPT log log OPT) bins. In: FOCS (2013)

# False-Name Manipulation in Weighted Voting Games Is Hard for Probabilistic Polynomial Time

Anja Rey and Jörg Rothe

Institut für Informatik, Heinrich-Heine-Universität Düsseldorf, 40225 Düsseldorf, Germany

**Abstract.** False-name manipulation refers to the question of whether a player in a weighted voting game can increase her power by splitting into several players and distributing her weight among these false identities. Analogously to this splitting problem, the beneficial merging problem asks whether a coalition of players can increase their power in a weighted voting game by merging their weights. Aziz et al. [1] analyze the problem of whether merging or splitting players in weighted voting games is beneficial in terms of the Shapley–Shubik and the normalized Banzhaf index, and so do Rey and Rothe [20] for the probabilistic Banzhaf index. All these results provide merely NP-hardness lower bounds for these problems, leaving the question about their exact complexity open. For the Shapley–Shubik and the probabilistic Banzhaf index, we raise these lower bounds to hardness for PP, "probabilistic polynomial time," and provide matching upper bounds for beneficial merging and, whenever the new players' weights are given, also for beneficial splitting, thus resolving previous conjectures in the affirmative. It follows from our results that beneficial merging and splitting for these two power indices cannot be solved in NP, unless the polynomial hierarchy collapses, which is considered highly unlikely.

## 1 Introduction

Weighted voting games are an important class of succinctly representable, simple games. They can be used to model cooperation among players in scenarios where each player is assigned a weight, and a coalition of players wins if and only if their joint weight meets or exceeds a given quota. Typical real-world applications of weighted voting games include decision-making in legislative bodies (e.g., parliamentary voting) and shareholder voting (see the book by Chalkiadakis et al. [6] for further concrete applications and literature pointers). In particular, the algorithmic and complexity-theoretic properties of problems related to weighted voting have been studied in depth, see, e.g., the work of Elkind et al. [8,9], Bachrach et al. [4], Zuckerman et al. [26], and [6] for an overview.

Bachrach and Elkind [3] were the first to study false-name manipulation in weighted voting games: Is it possible for a player to increase her power by splitting into several players and distributing her weight among these false identities? Relatedly, is it possible for two or more players to increase their power in a weighted voting game by merging their weights? The most prominent measures of a player's power, or influence, in a weighted voting game are the Shapley–Shubik and Banzhaf power indices. Merging and extending the results of [3] and [2], Aziz et al. [1] in particular study the problem of whether merging or splitting players in weighted voting games is beneficial in terms of the Shapley–Shubik index [21,22] and the normalized Banzhaf index [5] (see Section 2 for formal definitions). Rey and Rothe [20] extend this study for the probabilistic Banzhaf index proposed by Dubey and Shapley [7]. All these results, however, provide merely NP-hardness lower bounds. Aziz et al. [1, Remark 13 on p. 72] note that "it is quite possible that our problems are not in NP" (and thus are not NP-complete). Faliszewski and Hemaspaandra [10] provide the best known upper bound for the beneficial merging problem for two players with respect to the Shapley–Shubik index: It is contained in the class PP, "probabilistic polynomial time," which is considered to be by far a larger class than NP, and they conjecture that this problem is PP-complete. Rey and Rothe [20] observe that the same arguments give a PP upper bound for beneficial merging also in terms of the probabilistic Banzhaf index, and they conjecture PP-completeness as well. They also note that the same arguments cannot be transferred immediately to the corresponding problem for the normalized Banzhaf index.

We resolve these conjectures in the affirmative by proving that beneficial merging and splitting (for given new weights) are PP-complete problems both for the Shapley–Shubik and the probabilistic Banzhaf index. Beneficial splitting in general (i.e., if the number of new false identities is given, but not their actual weights) belongs to $\text{NP}^{\text{PP}}$ and is PP-hard for the same two indices. Thus, none of these six problems can be in NP, unless the polynomial hierarchy collapses to its first level, which is considered highly unlikely.

## 2 Preliminaries

We will need the following concepts from cooperative game theory (see, e.g., the textbook by Chalkiadakis et al. [6]). A *coalitional game with transferable utilities*, $\mathscr{G} = (N, v)$, consists of a set $N = \{1, \ldots, n\}$ of *players* (or, synonymously, *agents*) and a *coalitional function* $v : \mathfrak{P}(N) \to \mathbb{R}$ with $v(\emptyset) = 0$, where $\mathfrak{P}(N)$ denotes the power set of $N$. $\mathscr{G}$ is said to be *monotonic* if $v(B) \leq v(C)$ whenever $B \subseteq C$ for coalitions $B, C \subseteq N$, and it is *simple* if it is monotonic and $v : \mathfrak{P}(N) \to \{0, 1\}$, that is, $v$ maps each coalition $C \subseteq N$ to a value that indicates whether $C$ wins (i.e., $v(C) = 1$) or loses (i.e., $v(C) = 0$), where we require that the grand coalition $N$ is always winning.

The *probabilistic Banzhaf power index* of a player $i \in N$ in a simple game $\mathscr{G}$ (see [7]) is defined by

$$\text{Banzhaf}(\mathscr{G}, i) = \frac{1}{2^{n-1}} \sum_{C \subseteq N \smallsetminus \{i\}} (v(C \cup \{i\}) - v(C)). \qquad (1)$$

Intuitively, this index measures the power of player $i$ in terms of the probability that $i$ turns a losing coalition $C \subseteq N \smallsetminus \{i\}$ into a winning coalition by joining it, and therefore is *pivotal* for the success of $C$.

For comparison, the *normalized Banzhaf index of $i$ in $\mathscr{G}$* defined by Banzhaf [5], who rediscovered a notion originally introduced by Penrose [18], is obtained by dividing the *raw Banzhaf index of $i$ in $\mathscr{G}$*, which is the term $\sum_{C \subseteq N \smallsetminus \{i\}} (v(C \cup \{i\}) - v(C))$ in (1), not by $2^{n-1}$, but by the sum of the raw Banzhaf indices of all players in $\mathscr{G}$; see [7,11,20] for a discussion of the differences between these two power indices.

Unlike the Banzhaf indices, the *Shapley–Shubik index of i in* $\mathcal{G}$ takes into account the order in which players enter coalitions and is defined by

$$\text{ShapleyShubik}(\mathcal{G},i) = \frac{1}{n!} \sum_{C \subseteq N \smallsetminus \{i\}} \|C\|! \cdot (n-1-\|C\|)! \cdot (v(C \cup \{i\}) - v(C)).$$

Since the number of coalitions is exponential in the number of players, specifying coalitional games by listing all values of their coalitional function would require exponential space. For algorithmic purposes, however, it is important that these games can be represented succinctly. Certain simple games can be represented compactly by weighted voting games. A *weighted voting game* (WVG) $\mathcal{G} = (w_1, \ldots, w_n; q)$ consists of nonnegative integer weights $w_i$, $1 \leq i \leq n$, and a quota $q$, where $w_i$ is the $i$th player's weight. For each coalition $C \subseteq N$, letting $w(C)$ denote $\sum_{i \in C} w_i$, $C$ wins if $w(C) \geq q$, and it loses otherwise. Requiring the quota to satisfy $0 < q \leq w(N)$ ensures that the empty coalition loses and the grand coalition wins. Weighted voting games have been intensely studied from a computational complexity point of view (see, e.g., [8,9,4,26] and [6, Chapter 4] for an overview).

Aziz et al. [1] introduce the merging and splitting operations for WVGs. We use the following notation. Given a WVG $\mathcal{G} = (w_1, \ldots, w_n; q)$ and a nonempty[1] coalition $S \subseteq \{1, \ldots, n\}$, let $\mathcal{G}_{\&S} = (w(S), w_{j_1}, \ldots, w_{j_{n-\|S\|}}; q)$ with $\{j_1, \ldots, j_{n-\|S\|}\} = N \smallsetminus S$ denote the new WVG in which the players in $S$ have been merged into one new player of weight $w(S)$.[2] Similarly, given a WVG $\mathcal{G} = (w_1, \ldots, w_n; q)$, a player $i$, and an integer $m \geq 2$, define the set of WVGs $\mathcal{G}_{i \div m} = (w_1, \ldots, w_{i-1}, w_{n+1}, \ldots, w_{n+m}, w_{i+1}, \ldots, w_n; q)$ in which $i$ with weight $w_i$ is split into $m$ new players $n+1, \ldots, n+m$ having the weights $w_{n+1}, \ldots, w_{n+m}$ such that $\sum_{j=1}^m w_{n+j} = w_i$. Note that there is a *set* of such WVGs $\mathcal{G}_{i \div m}$, since there might be several possibilities of distributing $i$'s weight $w_i$ to the new players $n+1, \ldots, n+m$ satisfying $\sum_{j=1}^m w_{n+j} = w_i$. If the new players' weights are also given beforehand, there is one unique new game, and splitting is the inverse function to merging. For a power index PI, the beneficial merging and splitting problems are defined as follows.

---
PI-BENEFICIALMERGE
---
**Given:** A WVG $\mathcal{G} = (w_1, \ldots, w_n; q)$ and a nonempty coalition $S \subseteq \{1, \ldots, n\}$.
**Question:** Is it true that $\text{PI}(\mathcal{G}_{\&S}, 1) > \sum_{i \in S} \text{PI}(\mathcal{G}, i)$?

---
PI-BENEFICIALSPLIT
---
**Given:** A WVG $\mathcal{G} = (w_1, \ldots, w_n; q)$, a player $i$, and an integer $m \geq 2$.
**Question:** Is it possible to split $i$ into $m$ new players $n+1, \ldots, n+m$ with weights $w_{n+1}, \ldots, w_{n+m}$ satisfying $\sum_{j=1}^m w_{n+j} = w_i$ such that in this new WVG $\mathcal{G}_{i \div m}$, it holds that $\sum_{j=1}^m \text{PI}(\mathcal{G}_{i \div m}, n+j) > \text{PI}(\mathcal{G}, i)$?

---

[1] We omit the empty coalition, since this would slightly change the idea of the problem.
[2] Note that the players' order doesn't matter when considering the normalized or probabilistic Banzhaf index.

The goal of this paper is to classify these problems in terms of their complexity for both the Shapley–Shubik and the probabilistic Banzhaf index. We assume that the reader is familiar with the basic complexity-theoretic concepts such as the complexity classes P and NP, the polynomial-time many-one reducibility, denoted by $\leq_m^p$, and the notions of hardness and completeness with respect to $\leq_m^p$ (see, e.g., the textbook by Papadimitriou [17]). Valiant [24] introduced #P as the class of functions that give the number of solutions of the instances of NP problems. For a decision problem $A \in$ NP, we denote this function by #$A$. For example, if SAT is the satisfiability problem from propositional logic, then #SAT denotes the function mapping any boolean formula $\varphi$ to the number of truth assignments satisfying $\varphi$. There are various notions of reducibility between functional problems in #P (see [10] for an overview, literature pointers, and discussion). Here, we need only the most restrictive one: We say *a function f parsimoniously reduces to a function g* if there exists a polynomial-time computable function $h$ such that for each input $x$, $f(x) = g(h(x))$. That is, for functional problems $f, g \in$ #P, a parsimonious reduction $h$ from $f$ to $g$ transfers each instance $x$ of $f$ into an instance $h(x)$ of $g$ such that $f(x)$ and $g(h(x))$ have the same number of solutions. We say that $g$ is #P-parsimonious-hard if every $f \in$ #P parsimoniously reduces to $g$. We say that $g$ is #P-parsimonious-complete if $g$ is in #P and #P-parsimonious-hard. It is known that, given a WVG $\mathscr{G}$ and a player $i$, computing the raw Banzhaf index is #P-parsimonious-complete [19], whereas computing the raw Shapley–Shubik index is not [10], although it is #P-hard in a weaker sense and, of course, is in #P as well.

Gill [12] introduced the class PP ("probabilistic polynomial time") that contains all decision problems $X$ for which there exist a function $f \in$ #P and a polynomial $p$ such that for all instances $x$, $x \in X$ if and only if $f(x) \geq 2^{p(|x|)-1}$. It is easy to see that NP $\subseteq$ PP; in fact, PP is considered to be by far a larger class than NP, due to Toda's theorem [23]: PP is at least as hard (in terms of polynomial-time Turing reductions) as any problem in the polynomial hierarchy (i.e., PH $\subseteq$ P$^{PP}$). NP$^{PP}$, the second level of Wagner's counting hierarchy [25], is the class of problems solvable by an NP machine with access to a PP oracle; Mundhenk et al. [16] identified NP$^{PP}$-complete problems related to finite-horizon Markov decision processes.

## 3 Beneficial Merging and Splitting Is PP-Hard

In this section we prove that beneficial merging and splitting is PP-hard, and we provide matching upper bounds for beneficial merging and splitting (for the latter, assuming that the new players' weights are given) both for the Shapley–Shubik and the probabilistic Banzhaf index. We start with the probabilistic Banzhaf index.

### 3.1 The Probabilistic Banzhaf Power Index

We will use the following result due to Faliszewski and Hemaspaandra [10, Lemma 2.3].

**Lemma 1 (Faliszewski and Hemaspaandra [10]).** *Let F be a #P-parsimonious-complete function. The problem* COMPARE-$F = \{(x,y) \mid F(x) > F(y)\}$ *is PP-complete.*

The well-known NP-complete problem SUBSETSUM (which is a special variant of the KNAPSACK problem) asks, given a sequence $(a_1, \ldots, a_n)$ of positive integers and a positive integer $q$, do there exist $x_1, \ldots, x_n \in \{0,1\}$ such that $\sum_{i=1}^{n} x_i a_i = q$? It is known that #SUBSETSUM is #P-parsimonious-complete (see, e.g., [13,17] for parsimonious reductions from #3-SAT via #EXACTCOVERBY3-SETS to #SUBSETSUM). Hence, by Lemma 1, we have the following.

**Corollary 1.** COMPARE-#SUBSETSUM *is* PP-*complete.*

Our goal is to provide a $\leq_m^p$-reduction from COMPARE-#SUBSETSUM to Banzhaf-BENEFICIALMERGE. However, to make this reduction work, it will be useful to consider two restricted variants of COMPARE-#SUBSETSUM, which we denote by COMPARE-#SUBSETSUM-R and COMPARE-#SUBSETSUM-RR, show their PP-hardness, and reduce COMPARE-#SUBSETSUM-RR to Banzhaf-BENEFICIALMERGE. This will be done in Lemmas 2 and 3 and in Theorem 1. In all restricted variants of COMPARE-#SUBSETSUM we may assume, without loss of generality, that the target value $q$ in a related SUBSETSUM instance $((a_1, \ldots, a_n), q)$ satisfies $1 \leq q \leq \alpha - 1$, where $\alpha = \sum_{i=1}^{n} a_i$.

---

**COMPARE-#SUBSETSUM-R**

**Given:** A sequence $A = (a_1, \ldots, a_n)$ of positive integers and two positive integers $q_1$ and $q_2$ with $1 \leq q_1, q_2 \leq \alpha - 1$, where $\alpha = \sum_{i=1}^{n} a_i$.

**Question:** Is the number of subsequences of $A$ summing up to $q_1$ greater than the number of subsequences of $A$ summing up to $q_2$, that is, does it hold that #SUBSETSUM$((a_1, \ldots, a_n), q_1) >$ #SUBSETSUM$((a_1, \ldots, a_n), q_2)$?

---

**Lemma 2.** COMPARE-#SUBSETSUM $\leq_m^p$ COMPARE-#SUBSETSUM-R.

*Proof.* Given an instance $(X,Y)$ of COMPARE-#SUBSETSUM, $X = ((x_1, \ldots, x_m), q_x)$ and $Y = ((y_1, \ldots, y_n), q_y)$, construct a COMPARE-#SUBSETSUM-R instance $(A, q_1, q_2)$ as follows. Let $\alpha = \sum_{i=1}^{m} x_i$ and define $A = (x_1, \ldots, x_m, 2\alpha y_1, \ldots, 2\alpha y_n)$, $q_1 = q_x$, and $q_2 = 2\alpha q_y$. This construction can obviously be achieved in polynomial time. It holds that integers from $A$ can only sum up to $q_x \leq \alpha - 1$ if they do not contain multiples of $2\alpha$, thus #SUBSETSUM$(A, q_1) = $ #SUBSETSUM$((x_1, \ldots, x_m), q_x)$. On the other hand, $q_2$ can only be obtained by multiples of $2\alpha$, since $\sum_{i=1}^{m} x_i = \alpha$ is too small. Thus, it holds that #SUBSETSUM$(A, q_2) = $ #SUBSETSUM$((y_1, \ldots, y_n), q_y)$. It follows that $(X,Y)$ is in COMPARE-#SUBSETSUM if and only if $(A, q_1, q_2)$ is in COMPARE-#SUBSETSUM-R. ☐

In order to perform the next step, we need to ensure that all integers in a COMPARE-#SUBSETSUM-R instance are divisible by 8. This can easily be achieved, by multiplying each integer in an instance $((a_1, \ldots, a_n), q_1, q_2)$ by 8, obtaining $((8a_1, \ldots, 8a_n), 8q_1, 8q_2)$ without changing the number of solutions for both related SUBSETSUM instances. Thus, from now on, without loss of generality, we assume that for a given COMPARE-#SUBSETSUM-R instance $((a_1, \ldots, a_n), q_1, q_2)$, it holds that $a_i, q_j \equiv 0 \bmod 8$ for $1 \leq i \leq n$ and $j \in \{1, 2\}$.

Now, we consider our even more restricted variant of this problem.

| COMPARE-#SUBSETSUM-RR |
|---|

**Given:** A sequence $A = (a_1, \ldots, a_n)$ of positive integers.
**Question:** Is the number of subsequences of $A$ summing up to $(\alpha/2) - 2$ greater than the number of subsequences of $A$ summing up to $(\alpha/2) - 1$, i.e., #SUBSETSUM$((a_1, \ldots, a_n), (\alpha/2) - 2) > $ #SUBSETSUM$((a_1, \ldots, a_n), (\alpha/2) - 1)$, where $\alpha = \sum_{i=1}^{n} a_i$?

**Lemma 3.** COMPARE-#SUBSETSUM-R $\leq_m^p$ COMPARE-#SUBSETSUM-RR.

*Proof.* Given an instance $(A, q_1, q_2)$ of COMPARE-#SUBSETSUM-R, where we assume that $A = (a_1, \ldots, a_n)$, $q_1$, and $q_2$ satisfy $a_i, q_j \equiv 0 \mod 8$ for $1 \leq i \leq n$ and $j \in \{1, 2\}$, we construct an instance $B$ of COMPARE-#SUBSETSUM-RR as follows. (This reduction is inspired by the standard reduction from SUBSETSUM to PARTITION due to Karp [14].)
Letting $\alpha = \sum_{i=1}^{n} a_i$, define

$$B = (a_1, \ldots, a_n, 2\alpha - q_1, 2\alpha + 1 - q_2, 2\alpha + 3 + q_1 + q_2, 3\alpha).$$

This instance can obviously be constructed in polynomial time. Observe that

$$T = \left(\sum_{i=1}^{n} a_i\right) + (2\alpha - q_1) + (2\alpha + 1 - q_2) + (2\alpha + 3 + q_1 + q_2) + 3\alpha = 10\alpha + 4,$$

and therefore, $(T/2) - 2 = 5\alpha$ and $(T/2) - 1 = 5\alpha + 1$. We show that $(A, q_1, q_2)$ is in COMPARE-#SUBSETSUM-R if and only if $B$ is in COMPARE-#SUBSETSUM-RR.

First, we examine which subsequences of $B$ sum up to $5\alpha$. Consider two cases.
**Case 1:** If $3\alpha$ is added, $2\alpha + 3 + q_1 + q_2$ cannot be added, as it would be too large. Also, $2\alpha + 1 - q_2$ cannot be added, leading to an odd sum. So, $2\alpha - q_1$ has to be added, as the remaining $\alpha$ are too small. Since $3\alpha + 2\alpha - q_1 = 5\alpha - q_1$, $5\alpha$ can be achieved by adding some $a_i$'s if and only if there exists a subset $A' \subseteq \{1, \ldots, n\}$ such that $\sum_{i \in A'} a_i = q_1$ (i.e., $A'$ is a solution of the SUBSETSUM instance $(A, q_1)$).
**Case 2:** If $3\alpha$ is not added, but $2\alpha + 3 + q_1 + q_2$, an even number can only be achieved by adding $2\alpha + 1 - q_2$, thus, $\alpha - 4 - q_1$ remain. $2\alpha - q_1$ is too large, while no subsequence of $A$ sums up to $\alpha - 4 - q_1$, because of the assumption of divisibility by 8. If neither $3\alpha$ nor $2\alpha + 3 + q_1 + q_2$ are added, the remaining $5\alpha + 1 - q_1 - q_2$ are too small.
Thus, the only possibility to obtain $5\alpha$ is to find a subsequence of $A$ adding up to $q_1$. Thus, #SUBSETSUM$(A, q_1) = $ #SUBSETSUM$(B, 5\alpha)$.
Second, for similar reasons, a sum of $5\alpha + 1$ can only be achieved by adding $3\alpha + (2\alpha + 1 - q_2)$ and a term $\sum_{i \in A'} a_i$, where $A'$ is a subset of $\{1, \ldots, n\}$ such that $\sum_{i \in A'} a_i = q_2$. Hence, #SUBSETSUM$(A, q_2) = $ #SUBSETSUM$(B, 5\alpha + 1)$.
Thus, the relation #SUBSETSUM$(A, q_1) > $ #SUBSETSUM$(A, q_2)$ holds if and only if #SUBSETSUM$(B, 5\alpha) > $ #SUBSETSUM$(B, 5\alpha + 1)$, which completes the proof. ❑

We now are ready to prove the main result of this section.

**Theorem 1.** *Banzhaf-*BENEFICIALMERGE *is PP-complete, even if only three players of equal weight merge.*

*Proof.* Membership of Banzhaf-BENEFICIALMERGE in PP has already been observed in [20, Theorem 3]. It follows from the fact that the raw Banzhaf index is in #P and that #P is closed under addition and multiplication by two,[3] and, furthermore, since comparing the values of two #P functions on two (possibly different) inputs reduces to a PP-complete problem. This technique (which was proposed by Faliszewski and Hemaspaandra [10] and applies their Lemma 2.10) works, since PP is closed under $\leq_m^p$-reducibility.

We show PP-hardness of Banzhaf-BENEFICIALMERGE by means of a $\leq_m^p$-reduction from COMPARE-#SUBSETSUM-RR, which is PP-hard by Corollary 1 via Lemmas 2 and 3. Our construction is inspired by the NP-hardness results by Aziz et al. [2] and Rey and Rothe [20].

Given an instance $A = (a_1, \ldots, a_n)$ of COMPARE-#SUBSETSUM-RR, construct the following instance for Banzhaf-BENEFICIALMERGE. Let $\alpha = \sum_{i=1}^n a_i$. Define the WVG

$$\mathcal{G} = (2a_1, \ldots, 2a_n, 1, 1, 1, 1; \alpha),$$

and let the merging coalition be $S = \{n+2, n+3, n+4\}$. Letting $N = \{1, \ldots, n\}$, it holds that

$$\text{Banzhaf}(\mathcal{G}, n+2) = \frac{1}{2^{n+3}} \left\| \left\{ C \subseteq \{1, \ldots, n+1, n+3, n+4\} \;\middle|\; \sum_{i \in C} w_i = \alpha - 1 \right\} \right\|$$

$$= \frac{1}{2^{n+3}} \left( \left\| \left\{ A' \subseteq N \;\middle|\; \sum_{i \in A'} 2a_i = \alpha - 1 \right\} \right\| + 3 \cdot \left\| \left\{ A' \subseteq N \;\middle|\; 1 + \sum_{i \in A'} 2a_i = \alpha - 1 \right\} \right\| \quad (2) \right.$$

$$\left. + 3 \cdot \left\| \left\{ A' \subseteq N \;\middle|\; 2 + \sum_{i \in A'} 2a_i = \alpha - 1 \right\} \right\| + \left\| \left\{ A' \subseteq N \;\middle|\; 3 + \sum_{i \in A'} 2a_i = \alpha - 1 \right\} \right\| \right) \quad (3)$$

$$= \frac{1}{2^{n+3}} \left( 3 \cdot \left\| \left\{ A' \subseteq N \;\middle|\; \sum_{i \in A'} 2a_i = \alpha - 2 \right\} \right\| + \left\| \left\{ A' \subseteq N \;\middle|\; \sum_{i \in A'} 2a_i = \alpha - 4 \right\} \right\| \right),$$

since the $2a_i$'s can only add up to an even number. The first of the four sets in (2) and (3) refers to those coalitions that do not contain any of the players $n+1$, $n+3$, and $n+4$; the second, third, and fourth set in (2) and (3) refers to those coalitions containing either one, two, or three of them, respectively. Since the players in $S$ have the same weight, players $n+3$ and $n+4$ have the same probabilistic Banzhaf index as player $n+2$.

---

[3] Again, note that this idea cannot be transferred straightforwardly to the normalized Banzhaf index, since in different games the indices have possibly different denominators, not only different by a factor of some power of two, as is the case for the probabilistic Banzhaf index.

Furthermore, the new game after merging is $\mathcal{G}_{\&\{n+2,n+3,n+4\}} = (3, 2a_1, \ldots 2a_n, 1; \alpha)$ and, similarly to above, the Banzhaf index of the first player is calculated as follows:

$\text{Banzhaf}(\mathcal{G}_{\&\{n+2,n+3,n+4\}}, 1)$

$= \dfrac{1}{2^{n+1}} \left\| \left\{ C \subseteq \{2, \ldots, n+2\} \mid \sum_{i \in C} w_i \in \{\alpha - 3, \alpha - 2, \alpha - 1\} \right\} \right\|$

$= \dfrac{1}{2^{n+1}} \left( \left\| \left\{ A' \subseteq N \mid \sum_{i \in A'} 2a_i \in \{\alpha - 3, \alpha - 2, \alpha - 1\} \right\} \right\| \right.$

$\left. + \left\| \left\{ A' \subseteq N \mid 1 + \sum_{i \in A'} 2a_i \in \{\alpha - 3, \alpha - 2, \alpha - 1\} \right\} \right\| \right)$

$= \dfrac{1}{2^{n+1}} \left( 2 \cdot \left\| \left\{ A' \subseteq N \mid \sum_{i \in A'} 2a_i = \alpha - 2 \right\} \right\| + \left\| \left\{ A' \subseteq N \mid \sum_{i \in A'} 2a_i = \alpha - 4 \right\} \right\| \right).$

Altogether, it holds that

$\text{Banzhaf}(\mathcal{G}_{\&\{n+2,n+3,n+4\}}, 1) - \sum_{i \in \{n+2, n+3, n+4\}} \text{Banzhaf}(\mathcal{G}, i)$

$= \dfrac{1}{2^{n+1}} \left( 2 \cdot \left\| \left\{ A' \subseteq N \mid \sum_{i \in A'} 2a_i = \alpha - 2 \right\} \right\| + \left\| \left\{ A' \subseteq N \mid \sum_{i \in A'} 2a_i = \alpha - 4 \right\} \right\| \right)$

$- \dfrac{3}{2^{n+3}} \left( 3 \cdot \left\| \left\{ A' \subseteq N \mid \sum_{i \in A'} 2a_i = \alpha - 2 \right\} \right\| + \left\| \left\{ A' \subseteq N \mid \sum_{i \in A'} 2a_i = \alpha - 4 \right\} \right\| \right)$

$= \left( \dfrac{1}{2^{n+1}} \cdot 2 - \dfrac{3}{2^{n+3}} \cdot 3 \right) \left\| \left\{ A' \subseteq N \mid \sum_{i \in A'} 2a_i = \alpha - 2 \right\} \right\|$

$+ \left( \dfrac{1}{2^{n+1}} - \dfrac{3}{2^{n+3}} \right) \left\| \left\{ A' \subseteq N \mid \sum_{i \in A'} 2a_i = \alpha - 4 \right\} \right\|$

$= -\dfrac{1}{2^{n+3}} \cdot \left\| \left\{ A' \subseteq N \mid \sum_{i \in A'} a_i = \dfrac{\alpha}{2} - 1 \right\} \right\| + \dfrac{1}{2^{n+3}} \cdot \left\| \left\{ A' \subseteq N \mid \sum_{i \in A'} a_i = \dfrac{\alpha}{2} - 2 \right\} \right\|,$

which is greater than zero if and only if $\left\| \{ A' \subseteq N \mid \sum_{i \in A'} a_i = (\alpha/2) - 2 \} \right\|$ is greater than $\left\| \{ A' \subseteq N \mid \sum_{i \in A'} a_i = (\alpha/2) - 1 \} \right\|$, which in turn is the case if and only if $A$ is in COMPARE-#SUBSETSUM-RR. ☐

It is known (see [20]) that both the beneficial merging problem for a coalition $S$ of size 2 and the beneficial splitting problem for $m = 2$ false identities can trivially be decided in polynomial time for the probabilistic Banzhaf index, since the sum of the power (in terms of this index) of two players is always equal to the power of the player that is obtained by merging them. Although it may seem as if this implied that merging or splitting were never beneficial regarding this index, this cannot be generalized to merging or splitting more than two players, by repeatedly applying the above result to pairs of players step by step. For example, as soon as two players merge, a third player's probabilistic Banzhaf index might have already changed in the new game, before merging her with another player in a subsequent step. Analogously to the proof of Theorem 1, it can be shown that the beneficial splitting problem for at least three false identities with given new weights is PP-complete.

On the other hand, a PP upper bound for the general beneficial splitting problem cannot be shown in any straightforward way. Here, we can only show membership in $NP^{PP}$, and we conjecture that this problem is even complete for this class.

**Theorem 2.** *Banzhaf-*BENEFICIALSPLIT *is PP-hard (even if the given player can only split into three players of equal weight) and belongs to* $NP^{PP}$.

*Proof.* With $m$ being part of the input, there are exponentially many possibilities to distribute the split player's weight to her false identities. Nondeterministically guessing such a distribution and then, for each distribution guessed, asking a PP oracle to check in polynomial time whether their combined Banzhaf power in the new game is greater than the original player's Banzhaf power in the original game, shows that Banzhaf-BENEFICIALSPLIT is in $NP^{PP}$.

In order to show PP-hardness for Banzhaf-3-BENEFICIALSPLIT, we use the same techniques as in Theorem 1, appropriately modified. ☐

## 3.2 The Shapley–Shubik Power Index

In order to prove PP-hardness for the merging and splitting problems with respect to the Shapley–Shubik index, we need to take a further step back.

EXACTCOVERBY3-SETS (X3C, for short) is another well-known NP-complete decision problem: Given a set $B$ of size $3k$ and a family $\mathscr{S}$ of subsets of $B$ that have size three each, does there exist a subfamily $\mathscr{S}'$ of $\mathscr{S}$ such that $B$ is exactly covered by $\mathscr{S}'$?

**Theorem 3.** *ShapleyShubik-*BENEFICIALMERGE *is PP-complete, even if only two players of equal weight merge.*

*Proof.* The PP upper bound, which has already been observed for two players in [10], can be shown analogously to the proof of Theorem 1.

For proving the lower bound, observe that the size of a coalition a player is pivotal for is crucial for determining the player's Shapley–Shubik index. Pursuing the techniques of Faliszewski and Hemaspaandra [10], we examine the problem COMPARE-#X3C, which is PP-complete by Lemma 1. We will apply the following useful properties of X3C instances shown as Lemma 2.7 in [10]: Every X3C instance $(B', \mathscr{S}')$ can be transformed into an X3C instance $(B, \mathscr{S})$, where $\|B\| = 3k$ and $\|\mathscr{S}\| = n$, that satisfies $k/n = 2/3$ without changing the number of solutions, i.e., #X3C$(B, \mathscr{S}) = $ #X3C$(B', \mathscr{S}')$.

Now, by the properties of the standard reduction from X3C to SUBSETSUM (which in particular preserves the number of solutions, i.e., #X3C parsimoniously reduces to #SUBSETSUM, as well as the "input size" $n$ and the "solution size" $k$), we can assume that in a given COMPARE-#SUBSETSUM instance each subsequence summing up to the given integer $q$ is of size $2n/3$. Following the track of the reductions from COMPARE-#SUBSETSUM via COMPARE-#SUBSETSUM-R to COMPARE-#SUBSETSUM-RR in Lemmas 2 and 3, a solution $A' \subseteq \{1,\ldots,n\}$ to a given instance $A = (a_1,\ldots,a_n)$ of the latter problem ($A'$ satisfying either $\sum_{i \in A'} a_i = (\alpha/2) - 2$ or $\sum_{i \in A'} a_i = (\alpha/2) - 1$, where $\alpha = \sum_{i=1}^n a_i$) can be assumed to satisfy $\|A'\| = k = (n+2)/3$. Under this assumption, we show PP-hardness of ShapleyShubik-BENEFICIALMERGE via a reduction from COMPARE-#SUBSETSUM-RR. Given such an instance, we construct the WVG $\mathscr{G} = (a_1,\ldots,a_n,1,1;\alpha/2)$ and consider coalition $S = \{n+1,n+2\}$. Let $N = \{1,\ldots,n\}$ and define $X = \text{\#SUBSETSUM}(A,(\alpha/2)-1)$ and $Y = \text{\#SUBSETSUM}(A,(\alpha/2)-2)$. Then,

ShapleyShubik$(\mathscr{G}, n+1) = $ ShapleyShubik$(\mathscr{G}, n+2)$

$$= \frac{1}{(n+2)!}\left(\left(\sum_{\substack{C \subseteq N \text{ such that} \\ \sum_{i \in C} a_i = (\alpha/2)-1}} \|C\|!(n+1-\|C\|)!\right) + \left(\sum_{\substack{C \subseteq N \text{ such that} \\ \sum_{i \in C} a_i = (\alpha/2)-2}} (\|C\|+1)!(n-\|C\|)!\right)\right)$$

$$= \frac{1}{(n+2)!}(X \cdot k!(n+1-k)! + Y \cdot (k+1)!(n-k)!).$$

Merging the players in $S$, we obtain $\mathscr{G}_{\&S} = (2,a_1,\ldots,a_n;\alpha/2)$. The Shapley–Shubik index of the new player in $\mathscr{G}_{\&S}$ is

$$\text{ShapleyShubik}(\mathscr{G}_{\&S},1) = \frac{1}{(n+1)!} \sum_{\substack{C \subseteq N \text{ such that} \\ \sum_{i \in C} a_i \in \{(\alpha/2)-1,(\alpha/2)-2\}}} \|C\|!(n-\|C\|)!$$

$$= \frac{1}{(n+1)!}(X+Y) \cdot k!(n-k)!.$$

All in all,

ShapleyShubik$(\mathscr{G}_{\&S},1) - ($ShapleyShubik$(\mathscr{G},n+1) + $ShapleyShubik$(\mathscr{G},n+2))$

$$= \frac{(X+Y) \cdot k!(n-k)!}{(n+1)!} - \frac{2(X \cdot k!(n+1-k)! + Y \cdot (k+1)!(n-k)!)}{(n+2)!}$$

$$= \frac{k!(n-k)!}{(n+2)!}(n-2k)(-X+Y). \tag{4}$$

Since we assumed that $k = (n+2)/3$ and we can also assume that $n > 4$ (because we added four integers in the construction in the proof of Lemma 3), it holds that $n - 2k = (n-4)/3 > 0$. Thus the term (4) is greater than zero if and only if $Y$ is greater than $X$, which is true if and only if $A$ is in COMPARE-#SUBSETSUM-RR. ❑

Analogously to the probabilistic Banzhaf index, we can show that also for the Shapley–Shubik index it is PP-complete to decide if splitting a player into players with given weights is beneficial. For the more general case where the number of false identities but no actual weights are given, we can raise the previously known lower bound to PP-hardness. However, the upper bound of PP cannot be transferred straightforwardly.

**Theorem 4.** ShapleyShubik-BENEFICIALSPLIT *is PP-hard (even if the given player can only split into two players of equal weight) and belongs to* $\text{NP}^{\text{PP}}$.

*Proof.* The upper bound of $\text{NP}^{\text{PP}}$ holds due to analogous arguments as in the proof of Theorem 2. Also, PP-hardness can be shown analogously to the proof of Theorem 2, appropriately modified to use the arguments from the proof of Theorem 3 instead of those from the proof of Theorem 1. ❑

## 4 Conclusions and Open Questions

Resolving previous conjectures in the affirmative, we have pinpointed the precise complexity of the beneficial merging problem in weighted voting games for the Shapley–Shubik and the probabilistic Banzhaf index by showing that it is PP-complete. We have obtained the same result for beneficial splitting (a.k.a. false-name manipulation) whenever the new players' weights are given. For a given number of false identities, but unknown weights, we raised the known lower bound from NP-hardness to PP-hardness and showed that it is contained in $\text{NP}^{\text{PP}}$. For this problem, it remains open whether it can be shown to be complete for $\text{NP}^{\text{PP}}$, a huge complexity class that by Toda's theorem [23] contains the entire polynomial hierarchy. $\text{NP}^{\text{PP}}$ is an interesting class, but somewhat sparse in natural complete problems. The only (natural) $\text{NP}^{\text{PP}}$-completeness results we are aware of are due to Littman et al. [15], who analyze a variant of the satisfiability problem and questions related to probabilistic planning, and due to Mundhenk et al. [16], who study problems related to finite-horizon Markov decision processes.

Another interesting open question is whether our results can be transferred also to the beneficial merging and splitting problems for the normalized Banzhaf index. Finally, it would be interesting to know to which classes of simple games, other than weighted voting games, our results can be extended.

**Acknowledgments.** We thank the anonymous reviewers for their helpful comments. This work has been supported in part by DFG grant RO-1202/14-1.

## References

1. Aziz, H., Bachrach, Y., Elkind, E., Paterson, M.: False-name manipulations in weighted voting games. Journal of Artificial Intelligence Research 40, 57–93 (2011)
2. Aziz, H., Paterson, M.: False name manipulations in weighted voting games: Splitting, merging and annexation. In: Proceedings of the 8th International Joint Conference on Autonomous Agents and Multiagent Systems, pp. 409–416. IFAAMAS (May 2009)
3. Bachrach, Y., Elkind, E.: Divide and conquer: False-name manipulations in weighted voting games. In: Proceedings of the 7th International Joint Conference on Autonomous Agents and Multiagent Systems, pp. 975–982. IFAAMAS (May 2008)
4. Bachrach, Y., Elkind, E., Meir, R., Pasechnik, D., Zuckerman, M., Rothe, J., Rosenschein, J.S.: The cost of stability in coalitional games. In: Mavronicolas, M., Papadopoulou, V.G. (eds.) SAGT 2009. LNCS, vol. 5814, pp. 122–134. Springer, Heidelberg (2009)

5. Banzhaf III, J.: Weighted voting doesn't work: A mathematical analysis. Rutgers Law Review 19, 317–343 (1965)
6. Chalkiadakis, G., Elkind, E., Wooldridge, M.: Computational Aspects of Cooperative Game Theory. Synthesis Lectures on Artificial Intelligence and Machine Learning. Morgan and Claypool Publishers (2011)
7. Dubey, P., Shapley, L.: Mathematical properties of the Banzhaf power index. Mathematics of Operations Research 4(2), 99–131 (1979)
8. Elkind, E., Chalkiadakis, G., Jennings, N.: Coalition structures in weighted voting games. In: Proceedings of the 18th European Conference on Artificial Intelligence, pp. 393–397. IOS Press (July 2008)
9. Elkind, E., Goldberg, L., Goldberg, P., Wooldridge, M.: On the computational complexity of weighted voting games. Annals of Mathematics and Artificial Intelligence 56(2), 109–131 (2009)
10. Faliszewski, P., Hemaspaandra, L.: The complexity of power-index comparison. Theoretical Computer Science 410(1), 101–107 (2009)
11. Felsenthal, D., Machover, M.: Voting power measurement: A story of misreinvention. Social Choice and Welfare 25(2), 485–506 (2005)
12. Gill, J.: Computational complexity of probabilistic Turing machines. SIAM Journal on Computing 6(4), 675–695 (1977)
13. Hunt, H., Marathe, M., Radhakrishnan, V., Stearns, R.: The complexity of counting problems. SIAM Journal on Computing 27(4), 1142–1167 (1998)
14. Karp, R.: Reducibility among combinatorial problems. In: Miller, R., Thatcher, J. (eds.) Complexity of Computer Computations, pp. 85–103. Plenum Press (1972)
15. Littman, M., Goldsmith, J., Mundhenk, M.: The computational complexity of probabilistic planning. Journal of Artificial Intelligence Research 9(1), 1–36 (1998)
16. Mundhenk, M., Goldsmith, J., Lusena, C., Allender, E.: Complexity results for finite-horizon Markov decision process problems. Journal of the ACM 47(4), 681–720 (2000)
17. Papadimitriou, C.: Computational Complexity, 2nd edn. Addison-Wesley (1995)
18. Penrose, L.: The elementary statistics of majority voting. Journal of the Royal Statistical Society 109(1), 53–57 (1946)
19. Prasad, K., Kelly, J.: NP-completeness of some problems concerning voting games. International Journal of Game Theory 19(1), 1–9 (1990)
20. Rey, A., Rothe, J.: Complexity of merging and splitting for the probabilistic Banzhaf power index in weighted voting games. In: Proceedings of the 19th European Conference on Artificial Intelligence, pp. 1021–1022. IOS Press (August 2010)
21. Shapley, L.: A value for $n$-person games. In: Kuhn, H., Tucker, A. (eds.) Contributions to the Theory of Games. Annals of Mathematics Studies 40, vol. II. Princeton University Press (1953)
22. Shapley, L., Shubik, M.: A method of evaluating the distribution of power in a committee system. The American Political Science Review 48(3), 787–792 (1954)
23. Toda, S.: PP is as hard as the polynomial-time hierarchy. SIAM Journal on Computing 20(5), 865–877 (1991)
24. Valiant, L.: The complexity of computing the permanent. Theoretical Computer Science 8(2), 189–201 (1979)
25. Wagner, K.: The complexity of combinatorial problems with succinct input representations. Acta Informatica 23, 325–356 (1986)
26. Zuckerman, M., Faliszewski, P., Bachrach, Y., Elkind, E.: Manipulating the quota in weighted voting games. In: Proceedings of the 23rd AAAI Conference on Artificial Intelligence, pp. 215–220. AAAI Press (July 2008)

# A Natural Generalization of Bounded Tree-Width and Bounded Clique-Width

Martin Fürer[*]

Department of Computer Science and Engineering
Pennsylvania State University
University Park, PA 16802, USA
furer@cse.psu.edu
http://cse.psu.edu/~furer

**Abstract.** We investigate a new width parameter, the fusion-width of a graph. It is a natural generalization of the tree-width, yet strong enough that not only graphs of bounded tree-width, but also graphs of bounded clique-width, trivially have bounded fusion-width. In particular, there is no exponential growth between tree-width and fusion-width, as is the case between tree-width and clique-width. The new parameter gives a good intuition about the relationship between tree-width and clique-width.

**Keywords:** tree-width, clique-width, fusion-width, FPT, XP.

## 1 Introduction

Tree-width is a very natural concept. In an intuitive direct way, it measures how similar a graph is to a tree. Many graph problems are not only easy for trees, but also for other tree-like graphs. Indeed there is a huge number of efficient algorithms for graphs of bounded tree-width.

While graphs of bounded tree-width are sparse, there are some dense graphs, like the complete graph $K_n$ or the complete bipartite graph $K_{nn}$ for which most computational problems have a trivial solution. Graphs of bounded clique-width are intended to cover classes of graphs for which many problems have efficient solutions, even though they contain many dense graphs.

Not unlike tree-width, the concept of clique-width [11] is based on a type of graph decompositions [7] chosen to allow fast algorithms for large classes of graphs. As the name clique-width indicates, this measure is designed to ensure that complete graphs have a very small width. But neither does clique-width measure a closeness to a clique in a natural way, nor is there an intuitive width (as in tree-width) involved in the definition of this concept. Even though the definition of clique-width is fairly simple, it is harder to obtain an intuition for the classes of graphs with small clique-width.

Bounded clique-width is an extension of bounded tree-width in the sense that every class of bounded tree-width is also a class of bounded clique-width [11,6].

---

[*] Research supported in part by NSF Grant CCF-0964655 and CCF-1320814.

But the worst case clique-width of graphs of tree-width $k$ has been upper [11,6] and lower [6] bounded by an exponential in $k$. The fact that this containment result is a difficult theorem suggests that the extension from tree-width to clique-width might not be very natural.

Equivalent to clique-width up to a factor of 2 is the notion of *k-NLC (node label controlled) graphs*. Partial $k$-trees have been shown to be $(2^{k+1} - 1)$-NLC trees [21]. $k$-NLC trees are a sub-class of the $k$-NLC graphs.

In contrast to clique-width (and the width measure produced by NLC trees), we propose a natural generalization of tree-width, which simultaneously generalizes clique-width. Furthermore, containment in the new class is obtained basically without increasing the parameter in both cases. We call the new measure *fusion-width*. We initially choose the name multi-tree-width to emphasize that it is a very natural extension of tree-width, by which it is motivated (even though it is much more powerful). We follow the strong suggestion of two referees to name it differently. The main difference is that while tree-width deals with single vertices or pairs of vertices at a time, fusion-width deals with multiple vertices (with the same label) or pairs of sets of vertices at a time.

We show that the clique-width grows at most exponentially in the fusion-width, implying that all classes of graphs with bounded fusion-width also have bounded clique-width. Furthermore, for some classes of graphs, there is really such an exponential growth.

Other width parameters generalizing both tree-width and clique-width without blowing up the parameter are rank-width [18] and boolean width [5]. The rank-width has the additional nice property of being computable in FPT [15]. For an overview of width parameters, see [16]. There are infinite classes of graphs where the clique-width is exponentially bigger than the boolean width.

The fusion-width has the additional property of being easy to work with and of being the most natural generalization of tree-width and clique-width. This is an essential strength of the new notion of fusion-width.

## 2 Definitions

For the definitions of *FPT (fixed parameter tractable)*, *XP (fixed parameter polynomial time)*, and *tree decomposition*, see e.g. [12].

**Definition 1.** *The* tree-width $tw(G)$ *[19] of a graph $G$ is the smallest integer $k$, such that $G$ has a tree decomposition with largest bag size $k + 1$.*

It is NP-complete to decide whether the tree-width of a graph is at most $k$ (if $k$ is part of the input) [1]. For every fixed $k$, there is a linear time algorithm deciding whether the tree-width is at most $k$, and if that is the case, producing a corresponding tree decomposition [2]. For arbitrary $k$, this task can still be approximated. A tree decomposition of width $O(k \log n)$ [4] and even $5k$ [3] can be found in polynomial time.

Closely related to tree-width is the notion of *branch-width* [20].

**Definition 2.** *A k-expression is an expression formed from the atoms $i(v)$, the two unary operations $\eta_{i,j}$ and $\rho_{i\to j}$, and one binary operation $\oplus$ as follows.*

- *$i(v)$ creates a vertex $v$ with label $i$, where $i$ is from the set $\{1,\ldots,k\}$.*
- *$\eta_{i,j}$ creates an edge between every vertex with label $i$ and every vertex with label $j$ for $i \neq j$.*
- *$\rho_{i\to j}$ changes all labels $i$ to $j$.*
- *$\oplus$ does a disjoint union of the generated labeled graphs.*

*Finally, the generated graph is obtained by deleting the labels.*

**Definition 3.** *The* clique-width $cw(G)$ *of a graph is the smallest $k$ such that the graph can be defined by a $k$-expression [7,11].*

Computing the clique-width is NP-hard [13]. Thus, one usually assumes that a graph is given together with a $k$-expression. For constant $k$, the clique-width can be approximated by a constant factor in polynomial time [18], in fact, this factor can be made smaller than 3 [17].

## 3 The Fusion-Width

We define a new width measure fw$(G)$ (fusion-width of $G$) with the properties

$$\text{fw}(G) \leq \text{tw}(G) + 2 \text{ and } \text{fw}(G) \leq \text{cw}(G).$$

We want these containments to be obvious. Still, we would like all tasks known to be solvable in polynomial time for graphs of bounded clique-width (and therefore of bounded tree-width) to be solvable in polynomial time also for graphs of bounded fusion-width.

These inequalities immediately imply that the class of graphs of bounded clique-width is contained in the class of graphs of bounded fusion-width. The definition of fusion-width is obtained as a simple extension of the definition of clique-width by a new operation $\theta_i$, merging all vertices with label $i$.

**Definition 4.** *A k-fusion-tree expression is an expression formed from the atoms $i\langle m\rangle$, the three unary operations $\eta_{i,j}$, $\rho_{i\to j}$, and $\theta_i$, and one binary operation $\oplus$ as follows.*

- *$i\langle m\rangle$ creates a graph consisting of $m$ isolated vertices labeled $i$, where $i$ is from the set $\{1,\ldots,k\}$.*
- *$\eta_{i,j}$ creates an edge between every vertex with label $i$ and every vertex with label $j$ for $i \neq j$.*
- *$\rho_{i\to j}$ changes all labels $i$ to $j$.*
- *$\theta_i$ merges all vertices labeled $i$ into one vertex. The new vertex is labeled $i$ and is adjacent to every vertex not labeled $i$ to which some vertex labeled $i$ was adjacent before the operation.*
- *$\oplus$ does a disjoint union.*

*Finally, the generated graph is obtained by deleting all the labels.*

It might be more natural not to require $i \neq j$ for $\eta_{i,j}$. This is insignificant, but would have the nice effect of giving any clique a fusion-width of 1 instead of 2. Thus the simplest graphs in this measure would just be the collections of disjoint cliques. Nevertheless, we stick to the traditional $\eta_{i,j}$ operation.

**Definition 5.** *The fusion-width $fw(G)$ of a graph $G$ is the smallest $k$ such that there is a $k$-fusion-tree expression generating $G$.*

The operation $i\langle m \rangle$ is introduced for convenience and to emphasize the possibility to create huge collections of identical vertices and huge bipartite graphs (together with $\eta_{i,j}$). We always want to emphasize the difference between fusion-width and tree-width here. Otherwise, except for the $\theta_i$ operation, we have just the clique-width operations, and the slightly more efficient version of vertex creation.

Compared to clique-width, the definition of fusion-width contains the completely new operation $\theta_i$. It is introduced to directly mimic the tree-width construction. An immediate consequence is that bounded tree-width graphs have also bounded fusion-width, without a difficult proof and without an exponential blow-up in the width parameter. This is in sharp contrast to the relationship between bonded tree-width and bounded clique-width.

Thanks to a previous referee, we know that Courcelle and Makowsky [8] have already defined the parameter $fw(G)$ when they showed that labelled graphs of bounded clique-width are closed their fusion operator. They call the parameter $cwd'(G)$, viewing it as an alternative notion of clique-width (justified by bounded clique-width being equivalent to bounded $cwd'(G)$). They don't propose $cwd'(G)$ to be used as a new width measure nor do they relate $cwd'(G)$ to tree-width.

**Theorem 1.** *Graphs with clique-width $k$ have fusion-width at most $k$. Furthermore, the number of operations does not increase from the associated $k$-expression to the associated $k$-fusion-tree expression.*

*Proof.* This is trivial, because all the operations of $k$-expressions are also operations of the $k$-fusion-tree expressions (with the obvious variation of replacing $i(v)$ by $i\langle 1 \rangle$). □

The width parameter increases by at most 2 from tree-width to fusion-width. This tiny increase has two causes. An increase of 1 is just due to the somewhat artificial push-down by 1 in the definition of tree-width. We don't use it for fusion-width, because we want to align the measure with clique-width. The other increase by 1 is needed to have an extra label for the vertices that have already received all their incident edges.

**Theorem 2.** *Graphs of tree-width $k$ have fusion-width at most $k+2$.*

*Proof.* We start with a tree decomposition with bag size $k+1$, and transform it into a $k+2$-fusion-tree expression in a bottom-up way. One special label is reserved for vertices that have already been handled, i.e., all their incident edges have been produced. Here we refer to the $k+1$ other labels as regular labels.

In each bag, we use different labels for different vertices. Thus, when handling a bag, it is trivial to introduce all edges present in that bag by the corresponding $\eta_{i,j}$ operations.

Choosing such a labeling is easy to do top down. We select an arbitrary node as the root of the decomposition tree, and assign distinct labels to the vertices in its bag. Before we assign labels to vertices in the bag of a node $j$ not appearing in the bag of the parent node, we assume that every vertex appearing in the bag of $j$ as well as in the bag of its parent node of the decomposition tree has already received its label. If there are still vertices in the bag of node $j$ without a label, then there are enough unused regular labels for these vertices, because we have $k+1$ regular labels and at most $k+1$ vertices in the bag of $j$.

The only slightly tricky part of the fusion-tree expression is the handling of the fact that the same vertices occur in the bags of both children of a node in the tree decomposition. This is handled by using distinct vertices with the same label. When handling a node in the decomposition tree, such that more than one of its children contain the same vertex $v$ in their bags, a merge operation $\theta_i$ is issued with $i$ being the common label of all occurrences of $v$ in the subtrees. □

Naturally, the following corollary is an immediate consequence.

**Corollary 1.** *If a problem can be solved in time $T(n, k)$ for graphs with $n$ vertices and fusion-width $k$, then it can be solved in time $T(n, k+2)$ for graphs with $n$ vertices and tree-width $k$.*

This corollary should be compared with the corresponding important result for clique-width.

If a problem can be solved in time $T(n, k)$ for graphs with $n$ vertices and clique-width $k$, then it can be solved in time $T(n, 3 \cdot 2^{k-1})$ for graphs with $n$ vertices and tree-width $k$ [6].

Instead of using this result, one would rather apply the corollary with its much stronger conclusion, provided that the premises are comparable.

We believe that whenever there is a nice argument that a natural problem can be solved for graphs of bounded clique-width, then we are able to nicely handle the operations $\eta_{i,j}$, $\rho_{i \to j}$, and $\oplus$. Usually, we would then also have a nice argument that the problem could be solved for graphs of bounded tree-width, and we could nicely handle the operation $\theta_i$. In such a situation, we would be able to handle all the fusion-width operations nicely, and therefore would also have an elegant algorithm whose running time should not be too bad as a function of the fusion-width.

The allowance of merging vertices with the $\theta_i$ operation might cause two concerns. First, it is more powerful than necessary for our results. It would be sufficient to restrict it to sets of vertices of size 2. Nevertheless, we opted for the more flexible notion, because it does not cause any problems. A second concern looks more important. As the construction of graphs allows them to grow and shrink, it is reasonable to ask whether there are graphs of bounded fusion-width requiring super-polynomial size $k$-fusion-tree expressions. This is not the case

**Proposition 1.** *Every graph of fusion-width $k$ has a $k$-fusion-tree expression of size $O(|V| + |E|)$.*

*Proof.* Whenever some vertices are merged due to a $\theta_i$ operation, it is possible that some edges are merged too. Assume, there is a vertex $v$ that has been used to create some edge set $E_v$ with some $\eta_{i,j}$ operations. Further assume that when $v$ is merged with some set of vertices, every edge of $E_v$ is merged with at least one other edge. Then we obtain the same graph by omitting the creation of vertex $v$. In other words, every vertex ever created is either useless, or it is an isolated vertex in the resulting graph $G = (V, E)$, or it is responsible for at least one edge. Now, assume no useless vertices (that have no effect and disappear in a later merge operation) are ever created. Then the number of vertices ever created is at most $|V|+|E|$. Furthermore, it is obvious, that without unnecessary label change operations $\rho_{i \to j}$, the graph $G$ has a $k$-fusion-tree expression of size $O(|V| + |E|)$. □

## 4 Illustration with the Independent Set Polynomial

Naturally, we know that finding a maximum independent set is possible in polynomial time for graphs of bounded clique-width [9]. In fact the far reaching meta-theorem of Courcelle et al. [9] shows that this result is not just valid for the maximum independent set problem, but for every problem expressible in monadic second order logic with quantification only over sets of vertices (not edges). Furthermore, the resulting algorithm shows the problem to be in FPT with the clique-width as the parameter.

Here, we look at a more difficult problem. Instead of just finding the size of a maximum independent set for graphs of bounded clique-width, we count the number of independent sets of all sizes. We present a fixed parameter polynomial time algorithm for this counting problem. We refer to [10,14] for more discussions of the fixed parameter tractability of counting problems.

Let $[k] = \{1,\ldots,k\}$ be the set of vertex labels. We define the $[k]$-labeled independent set polynomial of a $[k]$-labeled graph $G$ by

$$P(x, x_1, \ldots, x_k) = \sum_{i=1}^{n} \sum_{(n_1,\ldots,n_k) \in \{0,1\}^k} a_{i;n_1,\ldots,n_k} \, x^i \prod_{j=1}^{k} x_j^{n_j}$$

where $n_j \in \{0,1\}$ and $a_{i;n_1,\ldots,n_k}$ is the number of independent sets of size $i$ in $G$ which contain some vertices with label $j$ if and only if $n_j = 1$.

We define the independent set polynomial of a graph $G$ by

$$I(x) = \sum_{i=1}^{n} a_i x^i$$

where $a_i$ is the number of independent sets of size $i$ in $G$.

Then the independent set polynomial $I(x)$ can immediately be expressed by the $[k]$-labeled independent set polynomial $P(x, x_1, \ldots, x_k)$.

**Proposition 2.** *The independent set polynomial $I(x)$ of a $[k]$-labeled graph $G$ is*

$$I(x) = \sum_{(n_1,\ldots,n_k)\in\{0,1\}^k} P(x,1,\ldots,1) = \sum_{i=1}^{n} \sum_{(n_1,\ldots,n_k)\in\{0,1\}^k} a_{i,n_1,\ldots,n_k} x^i.$$

**Theorem 3.** *Given a graph $G$ with $n$ vertices and bounded fusion-width $k$, and a polynomial size $k$-fusion-tree expression generating $G$, the independent set polynomial $I(x)$ of $G$ can be computed in FPT, i.e., in time $f(k)n^{O(1)}$ for some function $f$.*

*Proof.* We have to show how to compute the $[k]$-labeled independent set polynomial $P(x, x_1, \ldots, x_k)$ of a $[k]$-labeled graph $G$. We compute it recursively bottom-up for the given $k$-fusion-tree expression. For the edgeless graph with $m$ vertices and label $i$ generated by $i\langle m \rangle$, the $[k]$-labeled independent set polynomial is

$$x + \sum_{j=1}^{m} \binom{m}{j} x^j x_i.$$

This polynomial is computable in time polynomial in $n$, because w.l.o.g, we can assume $m \leq n$. Otherwise, some set of vertices would be constructed together (by the $i\langle m \rangle$ operation) and destroyed together (with the same merge operation $\theta_i$). Such a redundancy can easily be removed in a preprocessing phase.

In the following, assume for some $[k]$-labeled graph $H$, the $[k]$-labeled independent set polynomial is $\tilde{P}(x, x_1, \ldots, x_k)$.

Let $G$ be obtained from $H$ by the operation $\eta_{i,j}$. Then the $[k]$-labeled independent set polynomial $P(x, x_1, \ldots, x_k)$ of $G$ is obtained from the $[k]$-labeled independent set polynomial $\tilde{P}(x, x_1, \ldots, x_k)$ of $H$ by deleting all monomials that are multiples of $x_i x_j$. These monomials count sets that are no longer independent after inserting all the edges between vertices labeled $i$ and $j$.

W.l.o.g., we can assume that before a merge operation $\theta_i$ is done, there are only 2 vertices labeled $i$. This assumption is allowed for two reasons.

1. If later a $\theta_i$ operation is done, then every $i\langle m \rangle$ operation can be replaced by an $i\langle 1 \rangle$ operation without changing the graphs obtained after the $\theta_i$ operation. Creating many equivalent vertices (with the same neighbors) and merging them later has the same effect as creating just one vertex.
2. Every $\theta_i$ operation can be replaced by a collection of $\theta_i$ operations done immediately after a disjoint union $\oplus$ or a relabeling operation $\rho_{i \to j}$ has created a graph with two vertices labeled $i$.

We describe the $\theta_i$ operation not in isolation, but only immediately after a disjoint union $\oplus$ or a relabeling operation $\rho_{i \to j}$.

We now describe how to obtain the polynomial $P(x, x_1, \ldots, x_k)$ of $G$ from the polynomials $P_r(x, x_1, \ldots, x_k)$ of $H_r$ ($r = 1, 2$), where $G$ is obtained from $H_1$ and $H_2$ by the operation $\theta_{i_1} \ldots \theta_{i_\ell}(H_1 \oplus H_2)$. For ease of notation, assume that $\{x_1, \ldots, x_\ell\} = \{x_{i_1}, \ldots, x_{i_\ell}\}$.

- Form the product $P_1(x, x_1, \ldots, x_k) \cdot P_2(x, x_1, \ldots, x_k)$.
- Delete all monomials where some $x_j$ with $1 \leq j \leq \ell$ appears with an exponent of 1.
  If $u$ and $v$ merge into vertex $w$, then we either want both $u$ and $v$ in the independent set (to account for an independent set containing $w$), or neither (to account for an independent set not containing $w$).
- Replace $x_j^2$ by $x_j/x$ for $1 \leq j \leq \ell$.
  Division by $x$ compensates the double count of a vertex in the independent set (counting $u$ and $v$ for $w$).
- Replace $x_j^2$ by $x_j$ for $\ell + 1 \leq j \leq k$.
  If label $j$ is not merged, then $x_j$ just indicates whether there are any vertices labeled $j$ in the independent set.

The case of a simple disjoint union ($G = H_1 \oplus H_2$) is the special case $\ell = 0$ of the just described situation. Here, we just compute the product $P_1(x, x_1, \ldots, x_k) \cdot P_2(x, x_1, \ldots, x_k)$ and delete all monomials $x_j$ (for $1 \leq j \leq k$) appearing with an exponent 1 to obtain $P_r(x, x_1, \ldots, x_k)$.

We now consider the relabeling operation $\rho_{i \to j}$. First assume, there will be no succeeding $\theta_j$ operation. Let $G$ be obtained from $H$ by the operation $\rho_{i \to j}$. Then the $[k]$-labeled independent set polynomial $P(x, x_1, \ldots, x_k)$ of $G$ is obtained from $\tilde{P}(x, x_1, \ldots, x_k)$ by substituting $x_j$ for $x_i$ and then replacing $x_j^2$ by $x_j$.

If $G$ is obtained from $H$ by the operation $\theta_j \rho_{i \to j}$, then we assume that in $H$ there is just one vertex labeled $i$ and one vertex labeled $j$. In this case, we proceed as follows, with the same reasoning as for the disjoint union $\oplus$ operation.

- In the given polynomial $\tilde{P}(x, x_1, \ldots, x_k)$ of $H$, substitute $x_j$ for $x_i$.
- Delete all monomials where $x_j$ appears with an exponent of 1.
- Replace $x_j^2$ by $x_j$.

To prove the polynomial time claim, it is important to notice that all polynomials have at most $k+1$ variables, and all monomials are linear in each of their variables. Thus there are at most $2^{k+1}$ monomials. The number of arithmetic operations is $O(k^2 n)$, as every efficient $k$-fusion-tree expression has at most $O(k^2)$ unary operations in a row. Thus for $k$ a constant, the time is at most $O(k^2 2^{2k} n)$ if a trivial polynomial multiplication algorithm is used. With fast polynomial multiplication, based on fast Fourier transform, the time goes down to $O(k^3 2^k n)$. As the total number of independent sets is at most $2^n$, it is sufficient to do the computations with numbers of length $O(n)$. Thus, each arithmetic operation requires even with school multiplication only quadratic time. □

Note that we do not claim that this was a difficult theorem. To the contrary, the point was to illustrate that a fast algorithm for a typical problem like computing the independent set polynomial restricted to bounded clique-width can be extended to a fast algorithm for this problem for graphs with the same fusion-width, i.e., for a much larger class of graphs.

## 5 Relations between Tree-Width, Clique-Width and Fusion-Width

We have $\text{fw}(G) \leq \text{tw} G + 2$ by Theorem 2. The following inequality is trivial, as $k$-fusion-tree expressions are strictly more powerful than $k$-expressions.

**Proposition 3.** *[8] $\text{fw}(G) \leq \text{cw}(G)$.*

The following main result immediately implies that the graphs of bounded clique-width are exactly the graphs of bounded fusion-width. In fact this implication has already been shown by Courcelle and Makowsky [8], as they prove the existence of a function $f$ with $\text{cw}(G) \leq f(\text{fw}(G))$. We still present our direct proof, because we get a much stronger result, and also because the logic framework of [8] might not be so widely accessible. Many of the proof ideas are from the corresponding Theorem of Corneil and Rotics [6], relating tree-width to clique-width.

**Theorem 4.** *Graphs with fusion-width $\text{fw}(G) = k$ have clique-width $\text{cw}(G)$ at most $k2^k$.*

*Proof.* We assume, we are given a $k$-fusion-tree expression $E$ describing a graph $G$, and we want to construct a $k2^k$-expression describing the same graph $G$.

We have to focus on the operation $\theta_i$ merging all vertices labeled $i$ into one vertex. This is the only operation that has to be eliminated, because it is allowed in $k$-fusion-tree expressions determining the fusion-width, but not in $k$-expressions determining the clique-width.

We view the parse trees $T$ of the $k$-fusion-tree expression. We want to transform it into a parse tree $T'$ of a $k$-expression. The main idea is that if the vertices of some label $i$ are involved in a merge operation, we focus on the highest location $\ell$ in $T$ where such a merge occurs involving label $\hat{i}$, where $\hat{i}$ is either $i$ or a label $i$ has been changed to.

At the corresponding location $\ell'$ in $T'$, we create the single vertex $v$ to which the vertices labeled $\hat{i}$ have been merged by $\theta_{\hat{i}}$ using the operation $\hat{i}\langle 1 \rangle$. This means that all the vertices $v_1, \ldots v_p$ which are finally merged into $v$ are not available further down in the tree $T'$. Therefore all operations in $T$ involving the labels of $v_1, \ldots v_p$ have to be delayed until the vertex creation operation at $\ell'$.

Let $L$ be the set of labels used in $T$. The new labels in $T'$ are from $L \times \mathcal{P}(L)$, where $\mathcal{P}$ is the powerset of $L$. This way, every new label can retain its own (old) identity and in addition remember all the other old labels to which its vertices should actually be adjacent according to the edge constructing operations $\eta_{i,j}$ issued in the subtree of the current node in $T$.

Whenever a label changes due to a renaming operation $\rho_{i \to j}$, in $T$, then every occurrence of $i$ (in the first or second component) of a label in $T'$ is changed to $j$.

We say that a label $i = i_1$ is subject to a merge operation as label $\hat{i} = i_{q+1}$, if there is a sequence of label change operations $\rho_{i_p \to i_{p+1}}$ ($p = 1, 2, \ldots, q$, $q \geq 0$) (i.e., there might be no label changes and $\hat{i} = i$), such that after these changes, label $i_{q+1}$ is involved in a merge operation.

Now, any $\eta_{i,j}$ operation in $T$ is handled as follows.

- If neither label $i$ nor $j$ are subject to a merge operation, then in the corresponding location (involving several nodes) of $T'$, $\eta_{i',j'}$ is issued for all labels $i'$ with first component $i$ and labels $j'$ with first component $j$.
- If label $i$ is subject to a merge operation as label $\hat{i}$, and $j$ is not subject to a merge operation, then the label $i$ is added to the second component of $j$. This way nodes labeled $j$ "remember" to create an edge to label $\hat{i}$ later. This edge is created immediately after the last merge of label $\hat{i}$.
- If label $i$ is subject to a last merge operation as label $\hat{i}$ after $j$ has been subject to its last merge operation, then the label $i$ is added to the second component of $j$. This way nodes labeled $j$ "remember" to create an edge to label $\hat{i}$ later. This edge is created immediately after the last merge of label $\hat{i}$. □

**Corollary 2.** *A class of graphs is of bounded fusion-width if and only if it is a class of bounded clique-width.*

*Proof.* This follows immediately from Proposition 3 and Theorem 4. □

**Corollary 3.** *If a problem can be solved in time $O(f(k)n^c)$ for graphs of fusion-width at most $k$, then it can be solved in time $O(f(k+2)n^c)$ for graphs of tree-width at most $k$.*

This immediate corollary compares favorably with the following fact. If a problem can be solved in time $O(f(k)n^c)$ for graphs of clique-width at most $k$, then it can be solved in time $O(2^{f(k)}n^c)$ for graphs of tree-width at most $k$.

This statement cannot be much improved, as clique-width can be exponentially larger than fusion-width.

**Theorem 5.** *[6] For any $k$, there is a graph $G$ with $\mathrm{tw}(G) = k$ and $\mathrm{cw}(G) \geq 2^{\lfloor k/2 \rfloor - 1}$.*

**Corollary 4.** *For any $k$, there is a graph $G$ with fusion-width $\mathrm{fw}(G) = k$ and clique-width $\mathrm{cw}(G) \geq 2^{\lfloor k/2 \rfloor - 2}$.*

*Proof.* This follows from Theorem 2 and Theorem 5 by picking a graph with the properties of Theorem 5 and noticing that by Theorem 2 its fusion-width is at most $k + 2$. Then $2^{\lfloor (k-2)/2 \rfloor - 1} = 2^{\lfloor k/2 \rfloor - 2}$ produces the result. □

Corollary 4 shows that our example of the independent set polynomial proves the fusion-width to be a powerful notion. There are graphs with fusion-width $\mathrm{fw}(G) = k$ and clique-width $\mathrm{cw}(G) \geq 2^{\lfloor k/4 \rfloor - 2}$. If for such a graph, we have a $k$-fusion-tree expression, then we can compute its independent set polynomial in time $O(k^3 2^k n)$ by the method of Theorem 3. Using just the fact that the clique-width $\mathrm{cw}(G) \geq 2^{\lfloor k/4 \rfloor - 2}$, we would be unlikely to find a better algorithm based on clique-width. Thus we would only have an an algorithm that is doubly exponential in $k$ for computing the independent set polynomial of these graphs.

We believe that the independent set polynomial is not an isolated instance. It has merely been used to illustrate the convenience and power of the fusiion-with parameter. Many other examples could be used instead.

## 6 Conclusion

We have introduced a new width measure, the fusion-width. Its purpose is twofold. It provides a tool for handling generally difficult problems for a large class of graphs. It also sheds a light on the essence of the generalization from bounded tree-width to bounded clique width. It is the ability at each stage of the construction not only to add edges between a limited number of vertices, but to add complete bipartite graphs between a limited number of sets of vertices.

## 7 Open Problems

What is the complexity of determining the fusion-width of a graph? Is it in XP (fixed parameter polynomial time) or even in FPT (fixed parameter tractable)? How well can fusion-width be approximated?

Find interesting classes of graphs with a large clique-width and a small fusion-width.

What is the relationship between fusion-width, rank-width, and boolean width?

**Acknowlegement.** The help of previous anonymous reviewers has improved this paper significantly.

## References

1. Arnborg, S., Corneil, D.G., Proskurowski, A.: Complexity of finding embeddings in a $k$-tree. SIAM Journal of Alg. and Discrete Methods 8, 277–284 (1987)
2. Bodlaender, H.L.: A linear-time algorithm for finding tree-decompositions of small treewidth. SIAM J. Comput. 25(6), 1305–1317 (1996)
3. Bodlaender, H.L., Drange, P.G., Dregi, M.S., Fomin, F.V., Lokshtanov, D., Pilipczuk, M.: An $O(c^k n)$ 5-approximation algorithm for treewidth. In: Proc. 54th FOCS 2013, pp. 499–508. IEEE (2013)
4. Bodlaender, H.L., Gilbert, J.R., Hafsteinsson, H., Kloks, T.: Approximating treewidth, pathwidth, frontsize, and shortest elimination tree. J. Algorithms 18(2), 238–255 (1995)
5. Bui-Xuan, B.M., Telle, J.A., Vatshelle, M.: Boolean-width of graphs. Theor. Comput. Sci. 412(39), 5187–5204 (2011)
6. Corneil, D.G., Rotics, U.: On the relationship between clique-width and treewidth. SIAM J. Comput. 34(4), 825–847 (2005)
7. Courcelle, B., Engelfriet, J., Rozenberg, G.: Handle-rewriting hypergraph grammars. J. Comput. Syst. Sci. 46(2), 218–270 (1993)
8. Courcelle, B., Makowsky, J.A.: Fusion in relational structures and the verification of monadic second-order properties. Mathematical Structures in Computer Science 12(2), 203–235 (2002)
9. Courcelle, B., Makowsky, J.A., Rotics, U.: Linear time solvable optimization problems on graphs of bounded clique-width. Theory Comput. Syst. 33(2), 125–150 (2000)
10. Courcelle, B., Makowsky, J.A., Rotics, U.: On the fixed parameter complexity of graph enumeration problems definable in monadic second-order logic. Discrete Applied Mathematics 108(1-2), 23–52 (2001)

11. Courcelle, B., Olariu, S.: Upper bounds to the clique width of graphs. Discrete Applied Mathematics 101(1-3), 77–114 (2000)
12. Downey, R., Fellows, M.R.: Parameterized complexity. Monographs in computer science. Springer, New York (1999), Downey, R.G., Fellows, M.R. New Zealand authors. Includes bibliographical references (p. [489]-516) and index
13. Fellows, M.R., Rosamond, F.A., Rotics, U., Szeider, S.: Clique-width minimization is np-hard. In: Kleinberg, J.M. (ed.) STOC, pp. 354–362. ACM (2006)
14. Fischer, E., Makowsky, J.A., Ravve, E.V.: Counting truth assignments of formulas of bounded tree-width or clique-width. Discrete Applied Mathematics 156(4), 511–529 (2008)
15. Hliněný, P., Oum, S.: Finding branch-decompositions and rank-decompositions. In: Arge, L., Hoffmann, M., Welzl, E. (eds.) ESA 2007. LNCS, vol. 4698, pp. 163–174. Springer, Heidelberg (2007)
16. Hlineny, P., Oum, S., Seese, D., Gottlob, G.: Width parameters beyond tree-width and their applications. Comput. J. 51(3), 326–362 (2008)
17. Oum, S.: Approximating rank-width and clique-width quickly. ACM Trans. Algorithms 5(1), 10:1–10:20 (2008), http://doi.acm.org/10.1145/1435375.1435385
18. Oum, S., Seymour, P.D.: Approximating clique-width and branch-width. J. Comb. Theory, Ser. B 96(4), 514–528 (2006)
19. Robertson, N., Seymour, P.D.: Graph minors. III. Planar tree-width. J. Comb. Theory, Ser. B 36(1), 49–64 (1984)
20. Robertson, N., Seymour, P.D.: Graph minors. X. Obstructions to tree-decomposition. J. Comb. Theory, Ser. B 52(2), 153–190 (1991)
21. Wanke, E.: k-NLC graphs and polynomial algorithms. Discrete Applied Mathematics 54(2-3), 251–266 (1994)

# Optimal Algorithms for Constrained 1-Center Problems[*]

Luis Barba[1,2], Prosenjit Bose[1], and Stefan Langerman[2,**]

[1] Carleton University, Ottawa, Canada
jit@scs.carleton.ca
[2] Université Libre de Bruxelles, Brussels, Belgium
{lbarbafl,slanger}@ulb.ac.be

**Abstract.** We address the following problem: Given two subsets $\Gamma$ and $\Phi$ of the plane, find the minimum enclosing circle of $\Gamma$ whose center is constrained to lie on $\Phi$. We first study the case when $\Gamma$ is a set of $n$ points and $\Phi$ is either a set of points, a set of segments (lines) or a simple polygon. We propose several algorithms, the first solves the problem when $\Phi$ is a set of $m$ segments (or $m$ points) in expected $\Theta((n+m)\log \omega)$ time, where $\omega = \min\{n, m\}$. Surprisingly, when $\Phi$ is a simple $m$-gon, we can improve the expected running time to $\Theta(m + n \log n)$. Moreover, if $\Gamma$ is the set of vertices of a convex $n$-gon and $\Phi$ is a simple $m$-gon, we can solve the problem in expected $\Theta(n+m)$ time. We provide matching lower bounds in the algebraic computation tree model for all the algorithms presented in this paper. While proving these results, we obtained a $\Omega(n \log m)$ lower bound for the following problem: Given two sets $A$ and $B$ in $\mathbb{R}$ of sizes $m$ and $n$, respectively, decide if $A$ is a subset of $B$.

**Keywords:** minimum enclosing circle, facility location problems.

## 1 Introduction

Let $P$ be a set of $n$ points in the plane. The minimum enclosing circle problem, originally posed by Sylvester in 1857 [17], asks to identify the center and radius of the minimum enclosing circle of $P$. For ease of notation we say that every circle containing $P$ is a $P$-circle. Several independent solutions were proposed to solve the problem in $O(n \log n)$ time [10,15,16]. Megiddo [14] settled the complexity of this problem and presented a $\Theta(n)$-time algorithm using prune and search.

Finding the minimum $P$-circle is also known as the 1-center facility location problem: Given the position of a set of clients (represented by $P$), compute the optimal location for a facility such that the maximum distance between a client and its closest facility is minimized. The aforementioned algorithms provide solutions to this problem. However, in most situations the location of the facility is constrained by external factors such as the geography and features of the terrain. Therefore, the study of constrained versions of the 1-center problem is of importance and has received great attention from the research community [3,4,5,8].

---

[*] Research supported in part by NSERC.
[**] Directeur de recherches du F.R.S.-FNRS.

Megiddo [14] proposed a linear time algorithm to find the minimum $P$-circle whose center is constrained to lie on a given line. Extending these ideas, Hurtado et al. [8] presented a $\Theta(n+m)$-time algorithm to find the minimum $P$-circle whose center is constrained to lie inside a convex $m$-gon. Bose and Toussaint [4] generalized this result by restricting the center of the $P$-circle to lie inside a simple $m$-gon $Q$. They proposed an $O((n+m)\log(n+m)+k)$-time algorithm, where $k$ is the number of intersections of $Q$ with the farthest-point Voronoi diagram of $P$. The dependency on $k$ was later removed from the running time [5].

Bose et al. [3] addressed the query version of the problem and proposed an $O(n \log n)$-time preprocessing on $P$, that allows them to find the minimum $P$-circle with center on a given query line in $O(\log n)$ time. Using this result, they showed how to compute the minimum $P$-circle, whose center is constrained to lie on a set of $m$ segments, in $O((n+m)\log n)$ time. However, when $m = O(1)$, the problem can be solved in $O(n)$ time by using Megiddo's algorithm [14] a constant number of times. Moreover, when $n = O(1)$, the problem can be solved in $O(m)$ time by finding the farthest point of $P$ from every given segment. Therefore, one would expect an algorithm that behaves like the algorithm presented in [3] when $m = O(n)$ but that converges to a linear running time as the difference between $n$ and $m$ increases (either to $O(n)$ or to $O(m)$). In this paper we show that such an algorithm exists and prove its optimality. When constraining the center on a simple $m$-gon however, the order of the vertices along its boundary allows us to further speed up our algorithm, provided that $m$ is larger than $n$.

Let $M$ be a set of $m$ points, let $S$ be a set of $m$ segments and let $Q$ be a simple polygon on $m$ vertices. We say that a $P$-circle $C$ has its center on $M$, on $S$ or on $Q$ if the center of $C$ is either a point of $M$, lies on a segment of $S$ or belongs to $Q$, respectively. The $(P, M)$-*optimization problem* asks to find the minimum $P$-circle with center on $M$. Given a radius $r$, the $(P, M)_r$-*decision problem* asks if there is a $P$-circle of radius $r$ with center on $M$. Analogous problems exist for $S$ and $Q$. In Section 2, we show a $\Theta((n+m)\log \omega)$-time algorithm for the $(P,S)_r$-decision problem where $\omega = \min\{n, m\}$. In Section 3 we transform it to solve the $(P, S)$-optimization problem with the same running time. In Section 4, we show a matching lower bound in the algebraic computation tree model provided that $n \leq m$ and the restriction is on a set of points, lines, segments or even on a simple polygon. When $m > n$ however, we only prove a matching lower bound when the center is restricted to be on a set of points, segments or lines, yet the lower bound breaks down when the restriction is on a simple polygon. Indeed, given a simple $m$-gon $Q$, we show an $\Theta(m + n \log n)$-time algorithm for the $(P, Q)$-optimization problem. To put this in perspective, note that whenever $m = \Omega(n \log n)$, the algorithm runs in $\Theta(m)$ time. Since the bottleneck of this algorithm is the computation of the farthest-point Voronoi diagram, if we assume that $P$ is the set of vertices of a given convex $n$-gon we can reduce the running time to $\Theta(m + n)$ [1,11]. Finally, we show a matching lower bound for these algorithms, thereby solving the problem for all ranges of $n$ and $m$ and all possible restrictions on points, lines, segments and simple polygons.

As a side note, while proving these lower bounds, we stumbled upon the following problem: Given two sets $A$ and $B$ of $\mathbb{R}$ of sizes $m$ and $n$, respectively, is $A \subseteq B$? While a lower bound of $\Omega(n \log n)$ is known in the case where $n = m$ [2], no lower bounds were known when $m$ and $n$ differ. Using a method of Yao [18] and the topology of affine subspace families, we were able to prove an $\Omega(n \log m)$ lower bound, even when $A$ is restricted to be a sorted set of $m$ real numbers.

Although the main techniques used in this paper have been around for a while [7,12], they are put together in a different way in this paper, showing the potential of several tools that were not specifically designed for this purpose. Furthermore, these results provide significant improvements over previous algorithms when $n$ and $m$ differ widely as it is the case in most applications. Due to space constraints, in this extended abstract we provide only proof sketches. The full version of this paper is included as an appendix.

**Preliminaries.** Given a subset $X$ of the plane, the interior and convex hull of $X$ are denoted by $\text{INT}(X)$ and $\text{CH}(X)$, respectively. A point $x$ is *enclosed* by a circle $C$ if $x \in \text{CH}(C)$; otherwise we say that $x$ is *excluded* by $C$. An $X$-circle is a circle that encloses every point of $X$.

Given a point $x \in \mathbb{R}^2$, let $\bigcirc_r(x)$ be the circle with radius $r$ and center on $x$. Let $P$ be a set of $n$ points in $\mathbb{R}^2$. Given $W \subseteq P$, let $\Lambda_r(W) = \bigcap_{p \in W} \text{CH}(\bigcirc_r(p))$, i.e., the intersection of every disk of radius $r$ with center on a point of $W$. Notice that $\Lambda_r(W)$ is a convex set whose boundary is composed of circular arcs each with the same curvature. A point $p \in W$ *contributes* to $\Lambda_r(W)$ if there is an arc of the circle $\bigcirc_r(p)$ on the boundary of $\Lambda_r(W)$. We refer to this arc as the *contribution of $p$* to $\Lambda_r(W)$. As the curvature of all circles is the same, a point contributes with at most one arc to the boundary of $\Lambda_r(W)$.

Given two subsets $X$ and $Y$ of the plane, let $B_X(Y)$ be the minimum $X$-circle with center on $Y$ and let $b_X(Y)$ denote its center. If $X = P$, we let $\rho(Y)$ denote the radius of $B_P(Y)$, i.e., $\rho(Y)$ is the radius of the minimum $P$-circle with center on $Y$. Let $C_P$ be the minimum $P$-circle, $c_P$ be its center and let $r_P$ be its radius.

**Observation 1.** *Given a point $x \in \mathbb{R}^2$ and a real number $r \geq r_P$, $\rho(x) \leq r$ if and only if $x \in \Lambda_r(P)$.*

## 2 Solving the Decision Problem on a Set of Segments

Let $S$ be a set of $m$ segments and let $r > r_P$. In this section we present an $O((n+m)\log \omega)$ time algorithm to solve the $(P,S)_r$-decision problem for the given radius $r$, where $\omega = \min\{n,m\}$. Notice that by Observation 1, if we could compute $\Lambda_r(P)$, we could decide if there is a $P$-circle of radius $r$ with center on $S$ by checking if there is a segment of $S$ that intersects $\Lambda_r(P)$. However, we cannot compute $\Lambda_r(P)$ explicitly as this requires $\Omega(n \log n)$ time. Thus, we approximate it using $\varepsilon$-nets and use it to split both $S$ and $P$ into a constant number of subsets each representing a subproblem of smaller size. Using divide and conquer we determine if there is an intersection between $S$ and $\Lambda_r(P)$ by solving the decision problem recursively for each of the subproblems. The algorithm runs in $O(\min\{\log n, \log m\})$ phases and on each of them we spend $O(n+m)$ time.

**The Algorithm.** Initially, compute the minimum $P$-circle $C_P$, its center $c_P$ and its radius $r_P$ in $O(n)$ time [14]. In $O(m)$ time we can verify if $c_P$ lies on a segment of $S$. If it does, then $C_P$ is the minimum $P$-circle with center on $S$. Otherwise, as we assume from now on, the radius of $B_P(S)$ is greater than $r_P$.

Consider a family of convex sets $\mathcal{G}$ defined as follows. A set $G \in \mathcal{G}$ is the intersection of $A_1 \cap \ldots \cap A_6$, where each $A_i$ is either the interior of a circle or an open half-plane supported by a straight line ($A_i$ may be equal to $A_j$ for some $i \neq j$). Given a family $\mathcal{Y}$ of geometric objects in the plane (segments, lines or points), we define a set of ranges on $\mathcal{Y}$ as follows. For each $G \in \mathcal{G}$, let $G_{\mathcal{Y}} = \{y \in \mathcal{Y} : G \cap y \neq \emptyset\}$ and let $\mathcal{G}_{\mathcal{Y}} = \{G_{\mathcal{Y}} : G \in \mathcal{G}\}$ be the family of subsets of $\mathcal{Y}$ induced by $\mathcal{G}$.

Fix a constant $0 < \varepsilon \leq 1$ and consider the range space defined by $S$ and $\mathcal{G}_S$. As the VC-dimension of this range space is finite, we can compute an $\varepsilon$-net $N_S$ of $(S, \mathcal{G}_S)$ of size $O(1)$ in $O(m)$ time [13], i.e., any convex set $G \in \mathcal{G}$ that intersects more than $\varepsilon m$ segments of $S$ must intersect at least one segment of $N_S$.

For each segment $s$ of $N_S$, compute $B_P(s)$ in $O(n)$ time [14] and mark three points of $P$ that uniquely define this circle by lying on its boundary. Let $r_{min}$ be the radius of the minimum circle among the computed $P$-circles. If $r_{min} \leq r$, then there is a positive answer to the $(P, S)_r$-decision problem and the decision algorithm finishes. Otherwise, let $P^0 \subset P$ be the set of marked points and note that $|P^0| \leq 3|N_S| = O(1)$. By the minimality of $r_{min}$, any point in the interior of $\Lambda_{r_{min}}(P^0)$ is at distance at least $r_{min}$ from the segments of $N_S$. That is, the interior of $\Lambda_{r_{min}}(P^0)$ intersects no segment of $N_S$. As $r_{min} > r$, we know that $\Lambda_r(P^0) \subset \Lambda_{r_{min}}(P^0)$ and hence $\Lambda_r(P^0)$ intersects no segment of $N_S$.

We refine this intersection using another $\varepsilon$-net. Let $\mathcal{C} = \{\bigcirc_r(p) : p \in P\}$ be the set of circles of radius $r$ centered at the points of $P$. Compute an $\varepsilon$-net $N_P$ of the range space $(\mathcal{C}, \mathcal{G}_\mathcal{C})$ in $O(n)$ time [13]. That is, if a convex set $G \in \mathcal{G}$ intersects more than $\varepsilon n$ circles of $\mathcal{C}$, then $G$ intersects at least one circle of $N_P$.

Let $P^1 = \{p \in P : \bigcirc_r(p) \in N_P\}$ i.e., $P^1$ is the subset of $P$ defining $N_P$ where $|P^1| = O(1)$. Notice that for every $p \in P^1$, $\Lambda_r(P^1)$ is enclosed by $\bigcirc_r(p)$, i.e. the circle $\bigcirc_r(p)$ does not intersect the open set $\Lambda_r(P^1)$. Let $P^+ = P^0 \cup P^1$, as $\Lambda_r(P^+)$ is contained in both $\Lambda_r(P^0)$ and $\Lambda_r(P^1)$, we observe the following.

**Lemma 1.** *No segment of $N_S$ and no circle of $N_P$ intersects $int(\Lambda_r(P^+))$.*

We assume the existence of a set of points $\Pi \supset P^+$ of constant size such that $\Pi$ inherits the properties of $P^+$. This set and its properties will be described later.

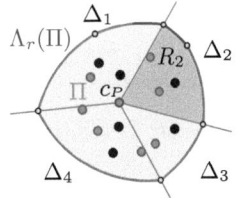

Because $r > r_P$, $c_P$ is enclosed by $\bigcirc_r(p)$ for every $p \in P$ and hence, $c_P$ lies in the interior of $\Lambda_r(\Pi)$. Therefore, we can consider $k = O(1)$ rays with apex at $c_P$ that pass through some of the vertices along the boundary of $\Lambda_r(\Pi)$. These rays split the plane into $k$ cones $\mathcal{D} = \{\Delta_1, \ldots, \Delta_k\}$.

For every $1 \leq i \leq k$, let $R_i = \Delta_i \cap \Lambda_r(\Pi)$ be a "slice" of $\Lambda_r(\Pi)$; see Fig. 1. By constructing

**Fig. 1.** The set $\Pi \subset P$ is shown in red. The vertex set of $\Lambda_r(\Pi)$ is used to split the plane into cones $\Delta_1, \ldots, \Delta_4$ by shooting rays from $c_P$. The "slice" $R_2$ is the portion of $\Lambda_r(\Pi)$ inside $\Delta_2$.

these cones such that each contains at most four vertices of $\Lambda_r(\Pi)$, we guarantee that each element in $\mathcal{R} = \{R_1, \ldots, R_k\}$ is a convex region of the family $\mathcal{G}$ used to define the $\varepsilon$-nets. Because $P^+ \subset \Pi$, we know that $\Lambda_r(\Pi) \subset \Lambda_r(P^+)$. Therefore, by Lemma 1 the interior of each region $R_i$ intersects no segment of $N_S$ and no circle of $N_P$. Because $N_S$ and $N_P$ are both $\varepsilon$-nets, we obtain the following.

**Lemma 2.** *For each $1 \leq i \leq k$, at most $\varepsilon m$ segments of $S$ intersect the region $R_i$ and at most $\varepsilon n$ circles of $\mathcal{C}$ intersect $R_i$.*

Due to space constraints, we omit a full description of $\Pi$. However, we provide a summary of its properties and a sketch of its construction.

**Lemma 3.** *We can construct $\Pi \supset P^+$ and the partition cones $\mathcal{D} = \{\Delta_1, \ldots, \Delta_k\}$ in $O(n)$ time such that: (1) for every $s \in S$, if $s$ intersects $\Lambda_r(\Pi)$, then $s \cap \Lambda_r(\Pi)$ is contained in exactly one cone of $\mathcal{D}$, and (2) for any point $p \in P$, if $p$ contributes to $\Lambda_r(P)$, then its contribution is contained in exactly one cone of $\mathcal{D}$.*

*Proof sketch.* For a given direction, we can shoot a ray from $c_P$ in that direction and compute the first circle from $\mathcal{C}$ that this ray intersects. Thus, this circle defines an actual arc of the boundary of $\Lambda_r(P)$ in the direction of the ray. Moreover, we can compute its neighboring arc along this boundary, i.e., we can find an actual vertex of $\Lambda_r(P)$ and the points of $P$ that define it. By doing this for a constant number of directions, given by the vertices of $\Lambda_r(P^+)$, we obtain a constant size subset $Y$ of the vertices of $\Lambda_r(P)$. Using this vertices to shoot the rays from $c_P$, we construct $\mathcal{D}$ and ensure that a point $p \in P$ will contribute to $\Lambda_r(P)$ inside only one of these cones. Moreover, the segments of $S$ cannot cross the segment connecting a vertex of $Y$ with $c_P$. Therefore, using convexity arguments we show that if a segment $s \in S$ intersects $\Lambda_r(\Pi)$, then it can do it in at most one cone of $\mathcal{D}$. □

The idea is to use divide and conquer using Lemma 2. That is, we split both $P$ and $S$ into $k$ subsets according to their intersection with the elements of $\mathcal{R}$, where each pair represents a subproblem. Finally, we prove that the $(P, S)_r$-decision problem has a positive answer if and only if some subproblem has a positive answer.

**Lemma 4.** *We can compute sets $S_1, \ldots, S_k \subset S$ in $O(m)$ time such that $|S_i| < \varepsilon m$ and $\sum_{i=1}^{k} |S_i| \leq m$. Moreover, $S_i$ contains all segments of $S$ that intersect $\Lambda_r(\Pi)$ inside $R_i$.*

*Proof sketch.* Let $S_i$ be the set of segments of $S$ that intersect $R_i$. The construction of $S_1, \ldots, S_k$ can be performed in $O(m)$ time since the size of $R_i$ is constant. By Lemma 3, a segment of $S$ belongs to at most one set of the partition and hence, $\sum_{i=1}^{k} m_i \leq m$. Moreover, for any $1 \leq i \leq k$ at most $\varepsilon m$ segments of $S$ intersect $R_i$ by Lemma 2. Consequently, $|S_i| < \varepsilon m$. □

**Lemma 5.** *We can compute sets $P_1, \ldots P_k \subset P$ in $O(n)$ time such that $|P_i| < \varepsilon n$, $\sum_{i=1}^{k} |P_i| \leq n$ and $\Lambda_r(P) = \Lambda_r(P_1 \cup \ldots \cup P_k)$. Moreover, if a point $p \in P_i$ contributes to $\Lambda_r(P)$, then this contribution intersects $R_i$.*

*Proof sketch.* Let $P_i = \{p \in P : \bigcirc_r(p) \text{ intersects } R_i\}$. This partition of $P$ can be computed in $O(n)$ time as the size of $R_i$ is constant. For each point $p \in P$ that contributes to $\Lambda_r(P)$, $\bigcirc_r(p)$ has to intersect $\Lambda_r(\Pi)$. Hence, the contribution of $p$ to $\Lambda_r(P)$ intersects at least one region of $\mathcal{R}$. By Lemma 3, a point of $P$ contributes to $\Lambda_r(\Pi)$ inside only one cone of $\mathcal{D}$, i.e., a point of $P$ belongs to at most one of the computed sets. By Lemma 2, at most $\varepsilon n$ circles of $\mathcal{C}$ intersect $R_i$, i.e., $|P_i| < \varepsilon n$. □

**Theorem 2.** *The $(P, S)_r$-decision problem has positive answer if and only if there is a $P_i$-circle of radius $r$ with center on $S_i$ for some $1 \le i \le k$.*

*Proof sketch.* Let $C$ be a $P$-circle of radius $r$ with center $c$ lying on a segment $s \in S$. Since the cones of $\mathcal{D}$ partition the plane, $c$ belongs to some cone $\Delta_i$ for some $1 \le i \le k$, i.e., $s$ intersects the cone $\Delta_i$. By Observation 1, $s$ intersects $\Lambda_r(P) \subset \Lambda_r(\Pi)$. Consequently, by Lemma 4, $s$ belongs to $S_i$. Assume that $c$ lies on the arc being the contribution of some point $p \in P$. Because $c$ is in $\Delta_i$ and on the boundary of $\Lambda_r(P) \subset \Lambda_r(\Pi)$, $\bigcirc_r(p)$ intersects $\Lambda_r(\Pi) \cap \Delta_i = R_i$. Thus, by Lemma 5 $p$ belongs to $P_i$, i.e., $C$ is a $P_i$-circle of radius $r$ with center on $S_i$. The other implication is similar and can be found in the full version of the paper. □

By Lemmas 4 and 5, in $O(n + m)$ time we can either give a positive answer to the decision algorithm, or compute sets $P_1, \ldots, P_k$ and $S_1, \ldots, S_k$ in order to define $k$ decision subproblems each stated as follows: Decide if there is a $P_i$-circle of radius $r$ with center on $S_i$. Because Theorem 2 allows us to solve each subproblem independently, we proceed until we find a positive answer on some branch of the recursion, or until either $P_i$ or $S_i$ reaches $O(1)$ size and can be solved in linear time. Since $|S_i| < \varepsilon m$ and $|P_i| < \varepsilon n$ by Lemmas 4 and 5, the number of recursion steps needed is $O(\min\{\log n, \log m\})$. Furthermore, by Lemmas 4 and 5, the size of all subproblems at the $i$-th level of the recursion is bounded above by $n + m$.

**Lemma 6.** *Given sets $P$ of $n$ points and $S$ of $m$ segments and $r > 0$, the $(P, S)_r$-decision problem can be solved in $O((n + m) \log \omega)$ time, where $\omega = \min\{n, m\}$.*

## 3 Converting Decision to Optimization

In the previous section, we showed an algorithm for the $(P, S)_r$-decision problem. However, our main objective is to solve its optimization version. To do that, we use the technique presented by Chan [6]. This technique requires an efficient algorithm to partition the problem into smaller subproblems, where the global solution is the minimum among the subproblems solutions. By presenting an $O(n + m)$-time partition algorithm, we obtain a randomized algorithm for the $(P, S)$-optimization problem having an expected running time of $O((n + m) \log \omega)$, where $\omega = \min\{n, m\}$. As the partition of the plane into cones used in the previous section has no correlation with the structure of $\Lambda_r(P)$ as $r$ changes, the partition of $P$ used in this section requires a different approach. However, the partition of $S$ is very similar.

**Lemma 7.** *We can compute sets $P'_1, \ldots, P'_h \subset P$ and $S'_1, \ldots, S'_h \subset S$ in $O(n + m)$ time such that $|P'_i| < \varepsilon n$, $|S'_i| < \varepsilon m$ and $B_P(S)$ is the circle of minimum radius among the elements in the set $\{B_{P'_1}(S'_1), \ldots, B_{P'_h}(S'_h)\}$.*

*Proof sketch.* Given any subset of $\mathbb{R}^2$, it can be *embedded* into $\mathbb{R}^3$ by identifying $\mathbb{R}^2$ with the plane $Z_0 = \{(x, y, z) \in \mathbb{R}^3 : z = 0\}$. As a first step embed $P$ into $\mathbb{R}^3$. Given a point $p \in P$, let $\gamma_p$ be the boundary of the 3-dimensional cone, lying above $p$, with apex on $p$ and 45° slope with respect to the plane $Z_0$.

Consider an $O(1)$ size sample $P^+$ of $P$ whose properties will be specified later and let $\Gamma = \{\gamma_p : p \in P^+\}$. Construct the farthest-point Voronoi diagram of $P^+$ and triangulate it. Then, compute $\Lambda_r(P^+)$ and pseudo-triangulate it by joining $c_P$ with every vertex on its boundary. By choosing $P^+$ carefully, we can guarantee that at most $\varepsilon m$ segments of $S$ intersect each of the pseudo-triangles.

Let $\mathcal{T}$ be the geometric graph obtained as the union of the triangulation of the farthest-point Voronoi diagram of $P^+$ and the pseudo-triangulation of $\Lambda_r(P^+)$. Then, we embed $\mathcal{T}$ in the plane $Z_0$. Since $|P^+| = O(1)$, the size of $\mathcal{T}$ is also constant. Recall that the furthest-point Voronoi diagram of $P^+$ is the upper envelope $\mathcal{U}$ of $\Gamma$ when projected onto the plane $Z_0$. That is, a point $x \in Z_0$ is farther from $p$ if and only if $\gamma_p$ is the last cone intersected by a ray shooting upwards, orthogonally to the plane $Z_0$, from $x$. Let $\mathcal{U}^+$ be the set of points lying strictly above $\mathcal{U}$.

Consider the vertical lifting of $\mathcal{T}$, which is simply the union of the vertical lines passing through the points on every edge of this triangulation. This vertical lifting partitions $\mathbb{R}^3$ into $O(1)$ solid prisms each defined by the intersection of at most three vertical halfspaces or cylinders. Finally, intersect each of these prisms with $\mathcal{U}^+$ to obtain a family of convex regions $\mathcal{Y} = \{Y_1, \ldots, Y_h\}$ for some $h \in O(1)$. Since $\mathcal{U}^+$ intersects no cone of $\Gamma$, no region of $\mathcal{Y}$ intersects the boundary of a cone in $\Gamma$. By choosing $P^+$ carefully, we can guarantee that for each $1 \leq i \leq h$, at most $\varepsilon n$ cones of $\Gamma$ intersect $Y_i$. Moreover, we can also guarantee that the vertical lifting of at most $\varepsilon m$ segments of $S$ intersect each $Y_i$. For every $1 \leq i \leq h$, let $P'_i = \{p \in P : \gamma_p \cap Y_i \neq \emptyset\}$ and note that $P'_i$ can be computed in $O(n)$ time. Let $S'_i = \{s \in S : \text{the vertical lifting of } s \text{ intersects } Y_i\}$ and note that $S'_i$ can be computed in $O(m)$ time. Moreover, we have that $|P'_i| < \varepsilon n$ and $|S'_i| < \varepsilon m$.

Recall that $b_P(S)$ is the center of the minimum $P$-circle $B_P(S)$ with center on $S$. Let $s^* \in S$ be the segment where $b_P(S)$ lies and let $p \in P$ be a point on the boundary of $B_P(S)$. We claim that $s^*$ and $p$ belong to the same subproblem, i.e., belong to $S'_j$ and $P'_j$, respectively, for some $1 \leq j \leq k$. Notice that if this claim is true, then all the points of $P$ through which $B_P(S)$ passes belong to $P'_j$. That is, $B_{P'_j}(S'_j)$ and $B_P(S)$ are defined as the circles with center on $s^*$ passing through the same set of points, i.e., $B_{P'_j}(S'_j) = B_P(S)$. Thus, by computing $B_{P'_i}(S'_i)$ for each $1 \leq i \leq k$, the minimum $P$-circle with center on $S$ can be obtained by choosing the minimum among $B_{P'_1}(S'_1), \ldots, B_{P'_h}(S'_h)$.

We proceed to prove that $s^*$ and $p$ belong to the same subproblem. Since $B_P(S)$ contains every point in $P^+$, $b_P(S)$ lies in $\Lambda_r(P^+)$. Therefore, $b_P(S)$ lies inside the projection of $Y_j$ for some $1 \leq j \leq h$. i.e, $s^* \in S'_j$. Consider the ray $\sigma$ shooting upwards (perpendicular to $Z_0$) from $b_P(S)$. Since $p$ lies on the boundary

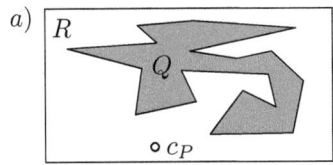 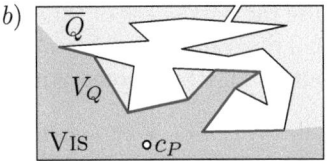

**Fig. 2.** a) A simple polygon $Q$ and a rectangle $R$ enclosing both $c_P$ and $Q$. By removing $Q$ from $R$ we obtain a polygon with one hole. b) The simple polygon $\overline{Q}$ obtained by connecting the hole with the boundary of $R$. In red, the visible chain $V_Q$ of $\overline{Q}$ from $c_P$.

of $B_P(S)$, $p$ is farther away from $b_P(S)$ than any other point of $P$. That is, $\gamma_p$ is the last cone intersected by $\sigma$. Therefore, $\gamma_p$ intersects $Y_j$ and consequently $p \in P'_j$. □

By Lemmas 6 and 7, we can use Chan's technique [6] to obtain the following.

**Theorem 3.** *Given a set $P$ of $n$ points and a set $S$ of $m$ segments in the plane, the $(P, S)$-optimization problem can be solved in expected $O((n + m) \log \omega)$ time where $\omega = \min\{n, m\}$.*

When constraining to a simple $m$-gon, the sequence of points along its boundary allows us to improve upon Theorem 3 provided that $m \geq n$.

**Theorem 4.** *Given a set $P$ of $n$ points and a simple polygon $Q$ of $m$ vertices, the $(P, Q)$-optimization problem can be solved in expected $O(m + n \log n)$ time.*

*Proof sketch.* If $c_P$ lie inside $Q$, then $C_P$ is the solution to our problem. Therefore, we assume that $c_P$ lies outside $Q$. In this case, we allow ourselves to compute the farthest-point Voronoi diagram of $P$ explicitly in $O(n \log n)$ time. Using this structure, we can compute $\Lambda_r(P)$ in $O(n)$ time for any given value of $r > r_P$ which is key to the speed up of the $(P, Q)_r$-decision algorithm.

In $O(m)$ time, compute a rectangle $R$ sufficiently large to enclose $Q$ and $c_P$ in its interior. Let $\overline{Q} = R - \text{INT}(Q)$ which is a polygon with one hole. This polygon can be turned into a simple polygon by adding a thin corridor connecting the hole with the exterior in such a way that no point on this corridor is visible from $c_P$. In this way, $c_P$ lies in the interior of $\overline{Q}$; see Fig. 3 for an illustration.

Compute the visibility polygon VIS of $\overline{Q}$ from $c_P$ in $O(m)$ time using the algorithm from Joe and Simpson [9]. Finally, let $V_Q$ be the polygonal chain obtained by removing the edges of the boundary of VIS that have an endpoint lying on the boundary of the rectangle $R$ (there may be none). Because the boundary of $\Lambda_r(P)$ is a Jordan curve, $\Lambda_r(P)$ intersects $Q$ if and only if $\Lambda_r(P)$ intersects $V_Q$.

A polygonal chain is *star-shaped* if there exists a set of points called its *kernel* such that every point on this chain is visible from every point in its kernel. Note that $V_Q$ is a star-shaped polygonal chain with $c_P$ in its kernel. Thus, since $\Lambda_r(P)$ can be computed in $O(n)$ time from the Voronoi diagram of $P$, we can decide if $V_Q$ intersect $\Lambda_r(P)$ in $O(n + m)$ time. Hence, we can solve the $(P, Q)_r$-decision problem in linear time. By considering the set of segments along the boundary of $Q$, we can use Lemma 7 to construct $O(1)$ subproblems such that the solution to the $(P, Q)$-optimization problem is the minimum among the subproblems solutions. Consequently, we can use Chan's technique [6] to obtain our result. □

Because the bottleneck of this algorithm is the construction of the farthest-point Voronoi diagram, whenever $P$ is the set of vertices of a convex polygon, we can compute its farthest-point Voronoi diagram in linear time [11,1].

**Corollary 1.** *Let $N$ be a convex n-gon and let $Q$ be a simple m-gon. The minimum enclosing circle of $N$, whose center is constrained to lie on $Q$, can be found in expected $\Theta(m+n)$ time.*

## 4 Lower Bounds

We prove lower bounds for the decision problems: We show inputs where the decision problem is equivalent to answering a membership query in a set with "many" disjoint components. We then use Ben-Or's Theorem [2] to obtain lower bounds for any decision algorithm that solves this membership problem.

**Lemma 8.** *Let $P$ be a set of $n$ points and a let $M$ be a set of $m$ points (m segments or m lines). Given a radius $r$, the $(P,M)_r$-decision problem has a lower bound of $\Omega(m \log n)$ in the algebraic computation tree model.*

*Proof sketch.* We construct a set of points $P$ such that for any point set $M$, with certain constraints, the $(P,M)_r$-decision problem has a lower bound of $\Omega(m \log n)$.

Let $r > 0$ and let $r'$ be a number such that $0 < r' < r$. Let $P$ be the set of vertices of a regular n-gon circumscribed on a circle of radius $r'$. Because $r > r'$, $\Lambda_r(P)$ is a non-empty convex region whose boundary is composed of circular arcs. Notice that by Observation 1, the decision algorithm has an affirmative answer if and only if there is a point of $M$ lying in $\Lambda_r(P)$. Let $C$ be the circumcircle of the vertices of $\Lambda_r(P)$. Partition this circle into $\varphi_1 = C \cap \Lambda_r(P)$ and $\varphi_0 = C - \varphi_1$. Because $\varphi_1$ consists of exactly $n$ points being the vertices of $\Lambda_r(P)$, $\varphi_0$ consists of $n$ disconnected open arcs all lying outside of $\Lambda_r(P)$; see Fig. 3(a). Moreover, a point on $C$ supports a $P$-circle of radius $r$ if and only if it lies on $\varphi_1$.

Consider the restriction of the decision problem where $M$ is constrained to lie on $C$. Notice that any lower bound for this restricted problem is also a lower bound for the general decision problem. Because an input on $m$ points for this restricted problem can be seen as a point in $\mathbb{R}^{2m}$, its input space defines a subspace $C^m = C \times \ldots \times C$ of $\mathbb{R}^{2m}$. Moreover, this set of points can be split into two regions, the "yes" and the "no" region (with respect to the decision problem), where the "no" region is equal to $\varphi_0^m$. That is, a point $(x_1, y_1, x_2, y_2, \ldots, x_m, y_m)$ lies in the "no" region $\varphi_0^m$ if for every index $1 \leq j \leq m$, the point $(x_j, y_j)$ lies inside $\varphi_0 \subset C$.

Because $\varphi_0$ contains $n$ disjoint components, $\varphi_0^m$ contains $O(n^m)$ disjoint components being the product-space of $m$ copies of $\varphi_0$. Recall that the $(P,M)_r$-decision problem is equivalent to answering if the input, seen as a point in $\mathbb{R}^{2m}$, lies in the "no" region. Therefore, by Ben-Or's Theorem [2] we obtain a lower bound of $\Omega(m \log n)$ for every decision algorithm in the algebraic computation tree model. □

**Lemma 9.** *Let $P$ be a set of $n$ points and a let $Q$ be a simple polygon on $m$ vertices such that $m \geq n$. Given a radius $r$, the $(P,Q)_r$-decision problem has a lower bound of $\Omega(m + n \log n)$ in the algebraic computation tree model.*

*Proof sketch.* In this proof, we construct a simple $m$-gon $Q$ such that for any input $P$ on $n$ points, the $(P,Q)_r$-decision problem has a lower bound of $\Omega(m+n\log n)$.

Let $r > 0$ and let $N = \{p_1,\ldots,p_n\}$ be the set of vertices of a regular $n$-gon whose circumcircle $C$ has radius smaller than $r$ and center on $c$. Let $\varepsilon > 0$ and let $r_\varepsilon = r + \varepsilon$. Because $r_\varepsilon$ is greater than the radius of $C$, $\Lambda_{r_\varepsilon}(N)$ is non-empty. Consider the middle points of every arc along $\Lambda_{r_\varepsilon}(N)$ and label them so that $m_i$ is the middle point on the arc opposite to $p_i$. Let $C'$ be any circle with center on $c$ and radius greater than $r_\varepsilon$. For every $1 \leq i \leq n$, let $q_i$ be the intersection point of $C'$ with the ray shooting from $c$ that passes through $p_i$. Let $Q'$ be a star-shaped polygon with vertex set $\{m_1,\ldots,m_n\} \cup \{q_1,\ldots,q_n\}$ where edges connect consecutive vertices in the radial order around $c$; see Fig. 3(b).

Let $R$ be a sufficiently large rectangle to enclose $Q'$ and let $Q = R \setminus \mathrm{INT}(Q')$. Remove the star-shaped hole of $Q$ by connecting the boundary of $R$ with an edge of $Q$ using a small corridor; see Fig. 3(c) for an illustration.

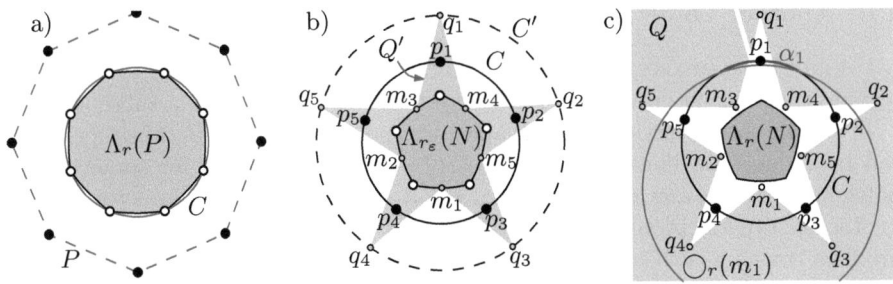

**Fig. 3.** a) The construction presented in Lemma 8. b) The construction of the star-shaped polygon $Q'$ used in Lemma 9. c) The polygon $Q$ constructed in Lemma 9 being disjoint from $\Lambda_r(P)$. The arc $\alpha_1$ (in red) is the arc of $C$ excluded by $\bigcirc_{r_\varepsilon}(m_1)$.

Consider the restriction of the decision problem where every point of $P$ is constrained to lie on $C$. Note that any lower bound for this restricted problem is also a lower bound for the general problem. Recall that every input on $n$ points constrained to lie on $C$ can be indistinctly seen as a point in $C^n \subset \mathbb{R}^{2n}$ and vice versa. Let $\gamma_0$ be a subset of $C^n$ such that $(x_1, y_1, \ldots, x_n, y_n) \in \gamma_0$ if and only if the decision problem on $Q$ with input $\{(x_1, y_1), \ldots, (x_n, y_n)\}$ and radius $r$ has a negative answer. Let $\gamma_1 = C^n - \gamma_0$. Because $r < r_\varepsilon$, $\Lambda_r(N)$ doesn't intersect $Q$, i.e., $N \in \gamma_0$. Note that every point of $N$ lies inside $\bigcirc_r(m_i)$ except for $p_i$, i.e., there is a portion of $C$ excluded by $\bigcirc_r(m_i)$. For every $1 \leq i \leq n$, let $\alpha_i$ be the arc of $C$ excluded by the circle $\bigcirc_r(m_i)$ where $p_i$ lies on $\alpha_i$.

By letting $\varepsilon$ sufficiently small, $\alpha_i$ is disjoint from $\alpha_j$ for any $i \neq j$. Moreover, every point lying on $C \setminus \alpha_i$ is enclosed by $\bigcirc_r(m_i)$. Therefore, if an input $P$ on $n$ points has no point lying on $\alpha_i$ for some $1 \leq i \leq n$, then $\bigcirc_r(m_i)$ is a $P$-circle of radius $r$, i.e., $P$ belongs to $\gamma_1$. Notice that every permutation of the same input of $n$ points induces a different point in $\mathbb{R}^{2n}$. That is, a set of $n$ points in the plane can be represented by $n!$ different points in $\mathbb{R}^{2n}$. Recall that each arc $\alpha_i$ has exactly one point of $N$ (say $p_i$) on it. Because $Q$ is disjoint from $\Lambda_r(N)$, every one of the $n!$ points representing $N$ in $\mathbb{R}^{2n}$ lies in $\gamma_0$. In the full version of the paper, we show that if $P_0$ and $P_1$ are two points in $C^n$ representing different permutations of $P$, then they are in disjoint connected components of $\gamma_0$. The idea is that any continuous transformation from $P_0$ to $P_1$ will reach a state in which an arc $\alpha_i$ is empty of input points, meaning that the it belongs to $\gamma_1$. Thus, as there are $n!$ permutations of $N$ and each of them belongs to a different connected component in $\gamma_0$, $\gamma_0$ contains at least $n!$ disjoint connected regions. By Ben-Or's Theorem [2] we obtain a lower bound of $\Omega(n \log n)$ for the restricted $(P, M)_r$-decision problem in the algebraic computation tree model. To obtain this lower bound, the $m$-gon $Q$ needs to have at least $n$ vertices, i.e., $m \geq n$.

Finally, notice that any decision algorithm has also a lower bound of $\Omega(m)$ since every vertex of $Q$ has to be considered. Otherwise, an adversary could perturb a vertex so that the solution switches to from a negative to a positive answer without affecting the execution of the algorithm. □

## 4.1 Another Lower Bound When Constraining to Sets of Points

Let $A$ and $B$ be two sets of $m$ and $n$ numbers in $[0, 1]$ such that $m \leq n$ and $A$ is sorted in increasing order. The *A-B-subset problem* asks if $A$ is a subset of $B$.

In the extended version of this paper, we show that any $A$-$B$-subset problem can be reduced in linear time to a $(P, Q)_r$-decision problem for some simple polygon $Q$. Hence, any lower bound for the $A$-$B$-subset problem is a lower bound for the $(P, Q)_r$-decision problem. In fact, the lower bound for the $A$-$B$-subset problem considers arbitrary sets of real numbers. Furthermore, the lower bound holds when $A$ is given as a fixed sorted set prior to the design of the algorithm. Note that a set of $n$ numbers can be represented by a point in $\mathbb{R}^n$ and vice versa.

**Lemma 10.** *Let $n, m$ be two integers such that $n \geq m$. For any $A \in \mathbb{R}^m$ such that $A$ is given in sorted order, there is a lower bound of $\Omega(n \log m)$ in the algebraic computation tree model for the $A$-$B$-subset problem given any $B \in \mathbb{R}^n$.*

*Proof sketch.* Let $A = \{a_1, \ldots, a_m\}$ be a sorted set of $m$ real numbers and think of it as a point in $\mathbb{R}^m$. Let $\gamma_1$ be the subspace of $\mathbb{R}^n$ containing all points representing a set $B$ such that $A \subseteq B$, i.e., the "yes" region.

An *A-constraint* is an equation of the form $(x_i = a_j)$ for some $1 \leq i \leq n$, $1 \leq j \leq m$. Two A-constraints $(x_i = a_j)$ and $(x_h = a_k)$ are *compatible* if $i \neq h$ ($j$ may be equal to $k$). A point $X \in \mathbb{R}^n$ *satisfies* an A-constraint $(x_i = a_j)$ if its $i$-th coordinate is equal to $a_j$. A set $\varphi$ of pairwise compatible A-constraints is *complete* if it contains exactly one A-constraint of the form $(x_j = a_i)$ for every $a_i \in A$, i.e., it contains exactly $m$ A-constraints, one for each element of $A$. Given a complete set $\varphi$ of A-constraints, let $K_\varphi = \{X \in \mathbb{R}^n : X \text{ satisfies every A-constraint in } \varphi\}$. Notice that $dim(K_\varphi) = n - m$ and $codim(K_\varphi) = m$. Moreover, if a point $B$ belongs to $K_\varphi$, then $A \subseteq B$, i.e., every point in $K_\varphi$ belongs to $\gamma_1$. Additionally, if $A \subseteq B$, then $B$ belongs to some $K_{\varphi'}$ for some complete set $\varphi'$ of A-constraints. Therefore, if we let $\mathscr{A} = \{K_\varphi : \varphi \text{ is a complete set of A-constraints}\}$, then $\gamma_1 = \cup \mathscr{A}$. In the full version of this paper, we study the topological structure of $\cup \mathscr{A}$ and obtain an $\Omega(n \log m)$ lower bound for the membership problem in $\gamma_1$.

We consider the poset induced by the intersection semilattice of $\mathscr{A}$ ordered by the reverse inclusion. We then consider the Möebius function on the elements of this poset and use result (16) of [18] to obtain our lower bound. □

**Corollary 2.** *Given a set $P$ of $n$ points and a simple polygon $Q$ on $m$ vertices (or a set of $m$ segments or a set of $m$ points), the $(P,Q)_r$-decision problem has a lower bound of $\Omega(n \log m)$ in the algebraic computation tree model.*

# References

1. Aggarwal, A., Guibas, L., Saxe, J., Shor, P.: A linear time algorithm for computing the Voronoi diagram of a convex polygon. In: Proceedings of STOC, pp. 39–45. ACM, New York (1987)
2. Ben-Or, M.: Lower bounds for algebraic computation trees. In: Proceedings of STOC, pp. 80–86. ACM, New York (1983)
3. Bose, P., Langerman, S., Roy, S.: Smallest enclosing circle centered on a query line segment. In: Proceedings of CCCG, pp. 167–170 (2008)
4. Bose, P., Toussaint, G.: Computing the constrained Euclidean, geodesic and link centre of a simple polygon with applications. In: Proceedings of CGI, pp. 102–111 (1996)
5. Bose, P., Wang, Q.: Facility location constrained to a polygonal domain. In: Rajsbaum, S. (ed.) LATIN 2002. LNCS, vol. 2286, pp. 153–164. Springer, Heidelberg (2002)
6. Chan, T.M.: Geometric applications of a randomized optimization technique. Discrete and Computational Geometry 22, 547–567 (1999)
7. Chazelle, B., Edelsbrunner, H., Guibas, L., Sharir, M.: Diameter, width, closest line pair, and parametric searching. DCG 10, 183–196 (1993)
8. Hurtado, F., Sacristan, V., Toussaint, G.: Some constrained minimax and maximin location problems. Studies in Locational Analysis 15, 17–35 (2000)
9. Joe, B., Simpson, R.B.: Corrections to Lee's visibility polygon algorithm. BIT Numerical Mathematics 27, 458–473 (1987)
10. Lee, D.T.: Farthest neighbor Voronoi diagrams and applications. Report 80-11-FC-04, Dept. Elect. Engrg. Comput. Sci. (1980)
11. Lee, D.T.: On finding the convex hull of a simple polygon. International Journal of Parallel Programming 12(2), 87–98 (1983)
12. Matousek, J.: Computing the center of planar point sets. Discrete and Computational Geometry 6, 221 (1991)
13. Matoušek, J.: Construction of epsilon nets. In: Proceedings of SCG, pp. 1–10. ACM, New York (1989)
14. Megiddo, N.: Linear-time algorithms for linear programming in $\mathbb{R}^3$ and related problems. SIAM J. Comput. 12(4), 759–776 (1983)
15. Preparata, F.: Minimum spanning circle. In: Preparata, F.P. (ed.) Steps in Computational Geometry. University of Illinois (1977)
16. Shamos, M., Hoey, D.: Closest-point problems. In: Proceedings of FOCS, pp. 151–162. IEEE Computer Society, Washington, DC (1975)
17. Sylvester, J.J.: A Question in the Geometry of Situation. Quarterly Journal of Pure and Applied Mathematics 1 (1857)
18. Yao, A.C.-C.: Decision tree complexity and Betti numbers. In: Proceedings of STOC, pp. 615–624. ACM, New York (1994)

# A Randomized Incremental Approach for the Hausdorff Voronoi Diagram of Non-crossing Clusters*

Panagiotis Cheilaris[1], Elena Khramtcova[1],
Stefan Langerman[2], and Evanthia Papadopoulou[1]

[1] Faculty of Informatics, Università della Svizzera italiana, Lugano, Switzerland
[2] Départment d'Informatique, Université Libre de Bruxelles, Brussels, Belgium

**Abstract.** In the Hausdorff Voronoi diagram of a set of *point-clusters* in the plane, the distance between a point $t$ and a cluster $P$ is measured as the maximum distance between $t$ and any point in $P$, and the diagram reveals the nearest cluster to $t$. This diagram finds direct applications in VLSI computer-aided design. In this paper, we consider "non-crossing" clusters, for which the combinatorial complexity of the diagram is linear in the total number $n$ of points on the convex hulls of all clusters. We present a randomized incremental construction, based on point-location, to compute the diagram in expected $O(n \log^2 n)$ time and expected $O(n)$ space, which considerably improves previous results. Our technique efficiently handles non-standard characteristics of generalized Voronoi diagrams, such as sites of non-constant complexity, sites that are not enclosed in their Voronoi regions, and empty Voronoi regions.

## 1 Introduction

Given a set $S$ of sites in some space, the *Voronoi region* of a site $s \in S$ is the geometric locus of points in the given space that are closer to $s$ than to any other site. In the classic Voronoi diagram, each site is a point and closeness is measured according to the Euclidean distance. In this work, we consider the *Hausdorff Voronoi diagram*. The containing space is $\mathbb{R}^2$, each site is a cluster of points (i.e., a set of points), and closeness of a point $t \in \mathbb{R}^2$ to a cluster $P$ is measured by the *farthest distance* $d_f(t, P) = \max_{p \in P} d(t, p)$, where $d(\cdot, \cdot)$ is the Euclidean distance between two points. The farthest distance $d_f(t, P)$ equals the *Hausdorff distance* between $t$ and cluster $P$, hence the name of the diagram.

Our motivation for investigating the Hausdorff Voronoi diagram comes from VLSI circuit design, where this diagram can be used to efficiently estimate the *critical area* of a VLSI layout for various types of open faults [20,21].

### 1.1 Previous Work

Let $k$ be the number of clusters in the input family, and $n$ be the total number of points on the convex hulls of all clusters. We denote by $\operatorname{conv} P$ the convex

---
* Supported in part by the Swiss National Science Foundation project 20GG21-134355, under the auspices of the ESF EUROCORES program EuroGIGA/VORONOI.

hull of cluster $P$ and by CH($P$) the sequence of points of $P$ on the boundary of the convex hull in counterclockwise order.

**Definition 1.** *Two clusters $P$ and $Q$ are called* non-crossing *if the convex hull of $P \cup Q$ admits at most two supporting segments with one endpoint in $P$ and one endpoint in $Q$. See Fig. 1.*

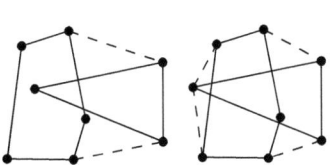

**Fig. 1.** Non-crossing and crossing clusters with supporting segments (dashed lines)

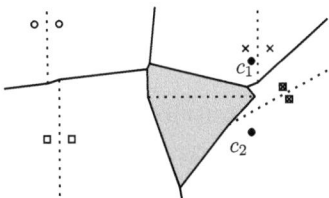

**Fig. 2.** HVD of five 2-point clusters; region of $C = \{c_1, c_2\}$ (gray)

The combinatorial complexity (size) of the Hausdorff Voronoi diagram is $O(n + m)$, where $m$ is the number of supporting segments reflecting crossings between all pairs of crossing clusters, and this is tight [22]. In the worst case, $m$ is $\Theta(n^2)$. If all clusters are *non-crossing* ($m = 0$) the diagram has linear size. There are *plane sweep* [20] and *divide and conquer* [22] algorithms for constructing the Hausdorff Voronoi diagram of arbitrary clusters. Both algorithms have a $K \log n$ term in their time complexity, where $K$ is a parameter reflecting the number of pairs of clusters such that one is contained in a specially defined enclosing circle of the other, for example, the minimum enclosing circle [22]. However, $K$ can be $\omega(n)$ (superlinear) even in the case of non-crossing clusters. The Hausdorff Voronoi diagram is equivalent to an upper envelope of a family of lower envelopes of an arrangement of hyperplanes in $\mathbb{R}^3$ (each envelope corresponds to a cluster) [13]. Edelsbrunner *et al.* [13] give a construction algorithm of $O(n^2)$ time complexity. [1] Although the time complexity is optimal in the worst case, it remains quadratic even for non-crossing clusters for which the size of the diagram is linear. A more recent parallel algorithm [10] constructs the Hausdorff Voronoi diagram of non-crossing clusters in $O(p^{-1} n \log^4 n)$ time with $p$ processors, which implies a divide and conquer sequential algorithm of time complexity $O(n \log^4 n)$ and space complexity $O(n \log^2 n)$.

The Hausdorff Voronoi diagram of a family of non-crossing clusters is an instance of abstract Voronoi diagrams [16]. Using the randomized incremental framework of Klein et al. [17], it can be computed in expected $O(bn \log n)$ time, where $b$ is the time it takes to construct the bisector between two clusters [1]. If there are clusters of linear size, then $b$ can be $\Theta(n)$. The framework was successfully applied to compute the Voronoi diagram of disjoint polygons [18] in $O(k \log n)$ time, where $k$ is the number of the sites, and $n$ is their total size.

---

[1] The reported $O(n^2 \alpha(n))$ time complexity (where $\alpha(n)$ is the inverse Ackermann function) improves to $O(n^2)$ due to the $O(n^2)$ bound on the size of the diagram.

It is not easy, however, to apply a similar approach to the Hausdorff Voronoi diagram because of a fundamental difference between the farthest and the nearest distance from a point to a convex polygon [12].

The Hausdorff Voronoi diagram is a min-max diagram, where every point $t$ in the plane lies in the region of the *closest* cluster with respect to the *farthest* distance. The "dual" max-min diagram is the *farthest color* Voronoi diagram [2, 14]. For disjoint simple polygons, this is the *farthest polygon* Voronoi diagram [8].

### 1.2  Our Contribution

In this paper we give a randomized incremental algorithm to compute the Hausdorff Voronoi diagram of a family of $k$ non-crossing clusters, based on point location. Clusters are inserted in random order one by one, while the diagram computed so far is maintained in a dynamic data structure, where generalized point location queries can be answered efficiently. To insert a cluster, a representative point in the new Voronoi region of this cluster is first identified and located, and then the new region is traced while the data structure is updated, e.g., [6, 11, 15].

In case of the Hausdorff Voronoi diagram, a major technical challenge is to quickly identify a representative point that lies in the new Voronoi region. This is difficult because: (a) the region of the new cluster might not contain any of its points, (b) the region of the new cluster might be empty, and (c) sites have non-constant size and thus the computation of a bisector or answering an *in-circle test* require non-constant time. Furthermore, the addition of a new cluster may make an existing region empty.

The dynamic data structure that we use is a variant of the *Voronoi hierarchy* [15], which in turn is inspired by the Delaunay hierarchy [11], and which we augment with the ability to efficiently handle the difficulties listed above. We also exploit a technique by Aronov et al. [4] to efficiently query the static farthest Voronoi diagram of a cluster. The expected running time of our algorithm is $O(n \log n \log k)$ and the expected space complexity is $O(n)$. The augmentation of the Voronoi hierarchy introduced in this paper may be of interest for incremental constructions of other non-standard types of generalized Voronoi diagrams. Our algorithm can also be implemented in *deterministic* $O(n)$ space and $O(n \log^2 n (\log \log n)^2)$ expected running time, using the dynamic point location data structure by Baumgarten et al. [5], while applying a simplified type of *parametric search* similarly to Cheong et al. [8].

Due to the lack of space some proofs and technical details are omitted in this version of the paper. Please refer to the online full version [7].

## 2  Preliminaries

Throughout this paper, we consider a family $F = \{C_1, \ldots, C_k\}$ of non-crossing clusters of points. We assume that no two clusters have a common point, and no four points lie on the same circle.

For a point $c \in C$, the *farthest Voronoi region* of $c$ is $\mathrm{freg}_C(c) = \{p \mid \forall c' \in C \setminus \{c\} \colon d(p,c) > d(p,c')\}$. The farthest Voronoi diagram of $C$ is denoted as $\mathrm{FVD}(C)$ and its graph structure as $\mathcal{T}(C)$. If $|C| > 1$, $\mathcal{T}(C)$ is a tree defined as $\mathbb{R}^2 \setminus \bigcup_{c \in C} \mathrm{freg}_C(c)$, and $\mathcal{T}(C) = c$, if $C = \{c\}$. A point at infinity along an arbitrary unbounded edge of $\mathcal{T}(C)$ is treated as the root of $\mathcal{T}(C)$, denoted as $\mathrm{root}(C)$.

For a cluster $C \in F$, the *Hausdorff Voronoi region* of $C$ is

$$\mathrm{hreg}_F(C) = \{p \mid \forall C' \in F \setminus \{C\} \colon d_\mathrm{f}(p,C) < d_\mathrm{f}(p,C')\}.$$

For a point $c \in C$, $\mathrm{hreg}_F(c) = \mathrm{hreg}_F(C) \cap \mathrm{freg}_C(c)$. The closure of $\mathrm{freg}_C(c)$, $\mathrm{hreg}_F(C)$, and $\mathrm{hreg}_F(c)$ is denoted by $\overline{\mathrm{freg}}_C(c)$, $\overline{\mathrm{hreg}}_F(C)$, and $\overline{\mathrm{hreg}}_F(c)$, respectively. When there is no ambiguity on the set under consideration, we omit the subscript from the above notation. The partitioning of the plane into non-empty Hausdorff Voronoi regions, together with their bounding edges and vertices, is called the *Hausdorff Voronoi diagram* of $F$, and it is denoted as $\mathrm{HVD}(F)$. Below we review some useful definitions and properties of the Hausdorff Voronoi diagram, which appeared in previous work [22].

The Hausdorff Voronoi diagram is *monotone*, that is, a region of the diagram can only shrink with the insertion of a new cluster. The structure of the Hausdorff Voronoi region of a point $c \in C$ is shown in Fig. 3. Its boundary consists of two chains: (1) the *farthest boundary* that belongs to $\mathcal{T}(C)$ and is internal to $\mathrm{hreg}(C)$, $(\mathrm{bd}\,\mathrm{hreg}(c) \cap \mathrm{bd}\,\mathrm{freg}(c))$; (2) the *Hausdorff boundary* $(\mathrm{bd}\,\mathrm{hreg}(c) \cap \mathrm{bd}\,\mathrm{hreg}(C))$. Neither chain can be empty, if $\mathrm{hreg}(C) \neq \emptyset$ and $|C| > 1$. There are three types of vertices on the boundary of $\mathrm{hreg}(c)$: (1) Standard Voronoi vertices that are equidistant from $C$ and two other clusters, referred in this paper as *pure* vertices. Pure vertices appear on the Hausdorff boundary of $\mathrm{hreg}(c)$. (2) *Mixed* vertices that are equidistant to three points of two clusters ($C$ and another cluster). The mixed vertices which are equidistant to two points of $C$ and one point of another cluster are called *C-mixed* vertices; there are exactly two of them on the boundary of $\mathrm{hreg}(c)$ and they delimit both the farthest boundary of $c$ and the Hausdorff boundary of $c$. (3) Vertices of $\mathcal{T}(C)$ on the farthest boundary of $c$.

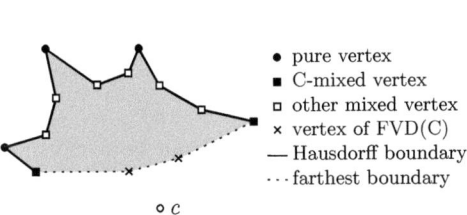

**Fig. 3.** Features of the Hausdorff Voronoi region of a point $c \in C$

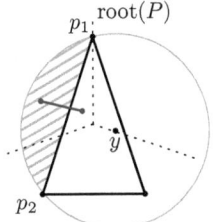

**Fig. 4.** The 2-point cluster $Q$ (red), forward limiting w.r.t. the 3-point cluster $P$ (black) with $P$-circle $\mathcal{K}_y$; portion $\mathcal{K}_y^f$ (shaded)

A line-segment $\overline{c_1c_2}$ is a *chord* of cluster $C$ if $c_1, c_2 \in \mathrm{CH}(C)$ and $c_1 \neq c_2$. In Fig. 4, $\overline{p_1p_2}$ is a chord of cluster $P$.

**Definition 2 ($C$-circle $\mathcal{K}_y$; $\mathcal{K}_y^f$, $\mathcal{K}_y^r$ [22]).** *Let $uv$ be an edge of $\mathcal{T}(C)$ bisecting a chord $\overline{c_1c_2}$ of $C \in F$. A circle centered at $y \in uv$ of radius $d(y, c_1) = d_f(y, C)$ is called the $C$-circle of $y$ and is denoted as $\mathcal{K}_y$. The chord $\overline{c_1c_2}$ partitions $\mathcal{K}_y$ in two parts: $\mathcal{K}_y^f$ and $\mathcal{K}_y^r$, where $\mathcal{K}_y^f$ is the part that encloses the two points of $C$ that define $\mathrm{root}(C)$. In case $y$ and $\mathrm{root}(C)$ are on the same edge of $\mathcal{T}(C)$, $\mathcal{K}_y^f$ is the portion of $\mathcal{K}_y$ that is enclosed in the halfplane bounded by $c_1c_2$ which does not contain $\mathrm{root}(C)$.*

**Definition 3 (Rear/forward limiting cluster [22]).** *A cluster $P \in F \setminus \{C\}$ is rear limiting with respect to $C$, if there is a $C$-circle $\mathcal{K}_y$ such that $P$ is enclosed in $\mathcal{K}_y^r \cup \mathrm{conv}\, C$. Similarly, $P$ is forward limiting with respect to $C$, if there is a $C$-circle $\mathcal{K}_y$ such that $P$ is enclosed in $\mathcal{K}_y^f \cup \mathrm{conv}\, C$. See Fig. 4.*

The following properties can be directly derived from Lemma 2 and Properties 2, 3 [22].

**Properties [22]**

1. If $\mathrm{hreg}(C) \neq \emptyset$, then $\mathrm{hreg}(C) \cap \mathcal{T}(C)$ consists of exactly one non-empty connected component.
2. Consider a point $v$ of $\mathcal{T}(P)$, such that $v \notin \mathrm{hreg}(P)$. Let $Q$ be a cluster, which is closer to $v$ than $P$. Then, only one of the subtrees of $\mathcal{T}(P)$ rooted at $v$, might contain points which are closer to $P$ than to $Q$.
3. Let $uv$ be an edge of $\mathcal{T}(P)$. If both $u$ and $v$ are closer to $Q$ than to $P$ then $\mathrm{hreg}_F(P)$ cannot intersect $uv$.
4. Region $\mathrm{hreg}_F(P) = \emptyset$, if and only if there is a cluster $Q \subset \mathrm{conv}\, P$, or there is a pair of clusters $\{Q, R\}$ such that $Q$ is rear limiting and $R$ is forward limiting with the same $P$-circle. Pair $\{Q, R\}$ is called a *killing pair* for $P$.

## 3  A Randomized Incremental Algorithm

Let $C_1, \ldots, C_k$ be a random permutation of the clusters in family $F$, and let $F_i = \{C_1, \ldots, C_i\}$ for $1 \leq i \leq k$. The algorithm iteratively constructs $\mathrm{HVD}(F_1), \ldots, \mathrm{HVD}(F_k) = \mathrm{HVD}(F)$. The cluster $C_i$ is inserted in $\mathrm{HVD}(F_{i-1})$ as follows:

1. Identify a point $t$ that is closer to $C_i$ than to any cluster in $F_{i-1}$ (i.e., $t \in \mathrm{hreg}_{F_i}(C_i)$) or determine that no such point exists (i.e., $\mathrm{hreg}_{F_i}(C_i) = \emptyset$).
2. If $t$ exists, grow $\mathrm{hreg}_{F_i}(C_i)$ starting from $t$ and update $\mathrm{HVD}(F_{i-1})$ to derive $\mathrm{HVD}(F_i)$; otherwise, $\mathrm{HVD}(F_i) = \mathrm{HVD}(F_{i-1})$.

The main challenge is to perform Step 1 efficiently. Step 2 can be performed in linear time [22]. Throughout this section, we skip the subscript $F_i$ and let $\mathrm{hreg}(C_i)$ stand for $\mathrm{hreg}_{F_i}(C_i)$.

To identify a representative point $t$ in $\mathrm{hreg}(C_i)$ (Step 1) it is enough to search along $\mathcal{T}(C_i)$, by Property 1. However, $\mathrm{hreg}(C_i) \cap \mathcal{T}(C_i)$ might not contain a vertex of $\mathcal{T}(C_i)$, see e.g., the gray region in Fig. 2. In this case, $\mathrm{hreg}(C_i)$ is either empty, or intersects exactly one edge of $\mathcal{T}(C_i)$, which is called a *candidate edge*.

**Definition 4.** Let $uv$ be an edge of $\mathcal{T}(C_i)$. Let $Q^u, Q^v \in F_{i-1}$ be the clusters closest to $u$ and $v$ respectively. We call $uv$ a candidate edge, if $Q^u \neq Q^v$ and $uv$ satisfies the following predicate:
$\text{cand}(uv) = d_f(u, Q^u) < d_f(u, C_i) < d_f(u, Q^v) \wedge d_f(v, Q^v) < d_f(v, C_i) < d_f(v, Q^u)$.

By Properties 2 and 3 we derive the following.

**Lemma 1.** Suppose $\text{hreg}(C_i) \cap \mathcal{T}(C_i)$ does not contain any vertex of $\mathcal{T}(C_i)$. Then at most one edge $uv$ of $\mathcal{T}(C_i)$ can be a candidate edge, in which case $\text{hreg}(C_i) \cap \mathcal{T}(C_i) \subset uv$. Otherwise $\text{hreg}(C_i) = \emptyset$.

A high-level description of Step 1 is as follows: We traverse $\mathcal{T}(C_i)$ starting at $\text{root}(C_i)$, checking its vertices and pruning if possible appropriate subtrees according to Property 3. In this process we either determine $t$ as a vertex of $\mathcal{T}(C_i)$, or we determine a candidate edge $uv$, or $\text{hreg}(C_i) = \emptyset$. Pseudocode is given as Procedure 1 below, which should be run with $u = \text{root}(C_i)$.

In more detail, to check if a vertex $w$ suits as $t$, determine the cluster $Q^w \in F_{i-1}$, which is nearest to $w$, by point location in $\text{HVD}(F_{i-1})$. If $d_f(w, C_i) < d_f(w, Q^w)$, then $t = w$. To compute $d_f(w, P)$ for a cluster $P$, do point location in $\text{FVD}(P)$. If Procedure 1 identifies a candidate edge, the representative point $t$ is determined by performing *parametric point location* along the candidate edge in $\text{HVD}(F_{i-1})$ (see Def. 5).

---

**Procedure 1.** Tracing the subtree of $\mathcal{T}(C_i)$ rooted at $u$ (within Step 1)

---
**Require:** $d_f(u, C_i) > d_f(u, Q^u)$.
  Let $v$ and $w$ be children of $u$.
  Locate $v$ and $w$ in $\text{HVD}(F_{i-1})$ to obtain $Q^v$ and $Q^w$ respectively.
  **if** $d_f(v, C_i) < d_f(v, Q^v)$ **or** $d_f(w, C_i) < d_f(w, Q^w)$ **then return** $v$ or $w$ respectively.
  **if** either $uv$ or $uw$ is a candidate edge **then**
      **return** the $uv$ or $uw$ respectively.
  **if** $d_f(v, C_i) < d_f(v, Q^u)$ **then**          ▷ Otherwise, prune the subtree of $w$
      Set $u = w$ and recurse.
  **if** $d_f(w, C_i) < d_f(w, Q^u)$ **then**          ▷ Otherwise, prune the subtree of $v$
      Set $u = v$ and recurse.

---

**Definition 5 (Parametric point location).** *Given* $\text{HVD}(F_{i-1})$ *and a candidate edge* $uv \subset \mathcal{T}(C_i)$ *determine the cluster* $P_j \in F_{i-1}$ *and the point* $t \in uv$ *such that* $d_f(t, C_i) = d_f(t, P_j) = \min_{P \in F_{i-1}} d_f(t, P)$. *If such point $p$ does not exist, return* nil.

Parametric point location in the Hausdorff Voronoi diagram is performed using the data structure that stores the diagram. Its performance determines the time complexity of our algorithm. In Sections 4 and 5, we describe the data structures and the algorithms used to answer the necessary queries.

## 4  Separator Decomposition

In this section we describe a data structure to efficiently perform point location and answer so-called *segment queries* in a *tree-type* of planar subdivision such as a farthest Voronoi diagram.

It is well-known [19] that any tree with $h$ vertices has a vertex called *centroid*, removal of which decomposes the tree into subtrees of at most $h/2$ vertices each. The centroid can be found in $O(h)$ time [19]. Thus, the farthest Voronoi diagram of a cluster $P$ can be organized as a balanced tree, whose nodes correspond to vertices of the diagram. This representation is called the *separator decomposition*, it is denoted as $\mathrm{SD}(P)$, and can be built as follows:

- Find a centroid $c$ of $\mathcal{T}(P)$. Create a node for $c$ and assign it as the root node.
- Remove $c$ from $\mathcal{T}(P)$. Recursively build the trees for the remaining three connected components, and link them as subtrees of the root.

Point location in $\mathrm{SD}(P)$ for a query point $q$, is performed as follows. Starting from the root of $\mathrm{SD}(P)$, perform a constant-time test of the query point $q$ against a node of $\mathrm{SD}(P)$, to decide in which of the node's subtrees to continue. When a leaf of $\mathrm{SD}(P)$ is reached, choose $p$ among the owners of the three regions that are adjacent to the corresponding vertex of $\mathrm{FVD}(P)$. The test of $q$ against a node $\alpha$ of $\mathrm{SD}(P)$ is due to Aronov et al. [4]. In more detail, let the node $\alpha$ correspond to a vertex $w$ of $\mathrm{FVD}(P)$. Let the points $p_1, p_2, p_3 \in P$ be the owners of the three regions of $\mathrm{FVD}(P)$, incident to $w$. Consider the rays $r_i, i = 1, 2, 3$, with origin at $w$ and direction $\overrightarrow{p_i w}$ respectively. Each ray $r_i$ lies entirely inside $\mathrm{freg}_P(p_i)$, and thus $r_1, r_2$ and $r_3$ subdivide the plane into three sectors with exactly one connected component of $\mathcal{T}(P) \setminus \{w\}$ in each sector. Choose the sector that contains $q$, and pick the corresponding subtree of $\alpha$.

A *segment query* in a farthest Voronoi diagram is defined as follows. Let $C, P \in F$. Given $\mathrm{FVD}(P)$ and a segment $uv \subset \mathcal{T}(C)$ such that $d_\mathrm{f}(u, C) < d_\mathrm{f}(u, P)$ and $d_\mathrm{f}(v, C) > d_\mathrm{f}(v, P)$, find the point $x \in uv$ that is equidistant from both $C$ and $P$ ($d_\mathrm{f}(x, C) = d_\mathrm{f}(x, P)$).

If $\mathrm{FVD}(P)$ is represented as a separator decomposition, a segment query can be performed efficiently similarly to a point location query with the difference that we test a segment against a node of $\mathrm{SD}(P)$. In particular, consider a node of $\mathrm{SD}(P)$ corresponding to a vertex $w$ of $\mathrm{FVD}(P)$. Let rays $r_i, i = 1, 2, 3$, be defined as above. Consider the (at most two) intersection points of $uv$ with the rays $r_i$. If any of these points is equidistant to $C$ and $P$, return it. Otherwise, since $P$ and $C$ are non-crossing, there is exactly one subsegment $u'v' \subset uv$ such that $d_\mathrm{f}(u', C) < d_\mathrm{f}(u', P)$ and $d_\mathrm{f}(v', C) > d_\mathrm{f}(v', P)$, where $u'$, $v'$ can be any of $u$, $v$, or the intersection points. The subsegment $u'v'$ can be computed in constant time, together with one of the three sectors where $u'v'$ is contained. If we reached a leaf of $\mathrm{SD}(P)$, we are left with a single edge $e$ of $\mathcal{T}(P)$. Suppose $e$ bisects the chord $\overline{p_1 p_2}$ of $P$, and the current $u'v'$ bisects the chord $\overline{c_1 c_2}$ of $C$. Then, return as point $x$ the center of the circle passing through $p, c_1, c_2$, where $p$ is the point among $p_1, p_2$ farthest from $x$.

**Lemma 2.** *The separator decomposition* $\mathrm{SD}(P)$ *of a cluster* $P \in F$ *can be built in* $O(n_p \log n_p)$ *time, where* $n_p$ *is the number of vertices of* $\mathrm{FVD}(P)$. *Both the point location and the segment query in* $\mathrm{SD}(P)$ *require* $O(\log n_p)$ *time.*

## 5 Voronoi Hierarchy for the Hausdorff Voronoi Diagram

Consider a set $S$ of $k$ sites. The *Voronoi hierarchy* of $S$ is a sequence of *levels* $S = S^{(0)} \supseteq \ldots \supseteq S^{(h)}$. For $\ell \in \{1, \ldots, h\}$, level $S^{(\ell)}$ is a random sample of $S^{(\ell-1)}$ according to a Bernoulli distribution with parameter $\beta \in (0, 1)$. For each level $S^{(\ell)}$ the data structure stores the Voronoi diagram of $S^{(\ell)}$. The Voronoi hierarchy is inspired by the Delaunay hierarchy given by Devillers [11].

In the Hausdorff Voronoi diagram sites are clusters of non-constant size each. We first adapt some known properties of the hierarchy to be valid in such an environment. Then, we consider several enhancements of the hierarchy to handle efficiently the Hausdorff Voronoi diagram and its queries, such as point location through walks, dynamic updates, including the handling of empty Voronoi regions, and parametric point location along a segment.

**Lemma 3.** *Let the underlying Voronoi diagram have size* $O(n)$, *where* $n$ *is the total size of the sites. Then for any set* $S$ *of* $k$ *sites of total size* $n$, *the Voronoi hierarchy of* $S$ *has* $O(n)$ *expected size and* $O(\log k)$ *expected number of levels.*

To perform *point location* in the Voronoi hierarchy for a query point $q$, we start at level $h$, and for each level $\ell$, we determine the site $s^\ell \in S^{(\ell)}$ that is closest to $q$, by performing a *walk*. Each step of the walk moves from a site $s \in S^{(\ell)}$ to a neighbor of $s$, such that the distance to $q$ is reduced. A walk at level $\ell-1$ starts from $s^\ell$. The answer to the query is $s^0$.

**Lemma 4.** *Let* $s_0^\ell, \ldots, s_r^\ell = s^\ell$ *be the sequence of sites visited at level* $\ell$ *during the point location of a query point* $q$. *Assuming that* $d_\mathrm{f}(q, s_i^\ell) < d(q, s_{i-1}^\ell)$, *for* $i \in \{1, \ldots, r\}$, *and either* $s^{\ell+1} = s_0^\ell$, *or* $d_\mathrm{f}(q, s_0^\ell) < d_\mathrm{f}(q, s^{\ell+1})$, *the expectation of the length* $r$ *of the walk at level* $\ell$ *is constant.*

In the original Voronoi hierarchy for a set of disjoint convex objects [15], one step of the walk to determine the correct neighboring site consists of a binary search among the neighbors of the site. For a Hausdorff Voronoi diagram, however, there is no natural ordering for the set of neighbors of a site. In addition, the subset of points in a cluster that contribute to the diagram reduces over time.

*A single step of the walk for the Hausdorff Voronoi diagram.* Consider point location in the Voronoi hierarchy for a family $\mathcal{F}$ of non-crossing clusters and a query point $q$. Let $C \in \mathcal{F}^{(\ell)}$ be the current cluster being considered at level $\ell$. We need to determine a cluster $Q$ at level $\ell$ whose region neighbors the region of $C$ and whose distance from $q$ gets reduced. Let $\hat{C} \subset C$ denote the set of all *active* points $c \in C$ that contribute a face to $\mathrm{hreg}_{\mathcal{F}^{(\ell)}}(C)$ at the current level $\ell$ ($\mathrm{hreg}_{\mathcal{F}^{(\ell)}}(c) \neq \emptyset$). Let $\mathrm{hreg}_{\mathcal{F}}^{(\ell)}(\cdot)$ denote $\mathrm{hreg}_{\mathcal{F}^{(\ell)}}(\cdot)$.

The cluster $Q$ is determined as follows. Let $c \in \hat{C}$ be the active point that is farthest from query point $q$ ($q \in \overline{\text{freg}}_{\hat{C}}(c)$). To determine point $c$ it is enough to draw the tangents from $q$ to $\text{CH}(\hat{C})$. Let $v_1, \ldots, v_j$ be the pure vertices in $\text{hreg}_{\mathcal{F}}^{(\ell)}(c)$ (see Fig. 5) in counterclockwise order, and let $Q^0, \ldots, Q^j, Q^{j+1}$ be their respective adjacent clusters. The rays $\overrightarrow{cv_1}, \ldots, \overrightarrow{cv_j}$ partition $\overline{\text{freg}}_{\hat{C}}(c)$ into $j+1$ unbounded regions. The walk moves from $C$ to $Q^i$ such that the ray $\overrightarrow{cq}$ immediately follows $\overrightarrow{cv_i}$ or immediately precedes $\overrightarrow{cv_{i+1}}$. For example, in Fig. 5, $c \in \hat{C}^{(\ell)}$ is the farthest active point from $q$ ($d_f(q, \hat{C}^{(\ell)}) = d(q,c)$). Region $\text{freg}_{\hat{C}}(c)$ is shown gray and its boundary is drawn bold. The walk moves from $C$ to $Q = Q^2$. We organize $\hat{C}$ as a sorted list of its points and for each point $c \in \hat{C}$ we maintain a sorted list of all Voronoi vertices adjacent to $\text{hreg}_{\mathcal{F}}^{(\ell)}(c)$. It can be shown that $d_f(q, \hat{Q}) \leq d_f(q, \hat{C})$, thus, the above procedure is correct. Note that $d_f(q, Q)$ may be greater than $d_f(q, C)$ because $d_f(q, \hat{C})$ may be different from $d_f(q, C)$ if $q \notin \text{hreg}(C)$.

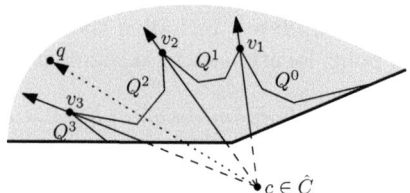

**Fig. 5.** The step of a walk from the cluster $C$

*Parametric point location in the Voronoi hierarchy.* We are given $\text{HVD}(F_{i-1})$, stored as a Voronoi hierarchy, and the candidate edge $uv \in \mathcal{T}(C_i)$. For each level $\ell$ of the Voronoi hierarchy, starting from the last level $h$, we search for the cluster $Q^\ell \in F_{i-1}^{(\ell)}$ and a point $u^\ell \in uv$ such that $u^\ell \in \text{hreg}_{F_{i-1}^{(\ell)}}(Q^\ell)$ and $d_f(u^\ell, C_i) = d_f(u^\ell, Q^\ell)$. If at some level there is no such point, return *nil*. Else return the cluster $C^0$ and the point $u^0$ determined at level 0.

In more detail, suppose that $u^{\ell+1}$ and $Q^{\ell+1}$ have been computed, for some $\ell \in \{0, \ldots, h-1\}$. To compute $u^\ell$ and $Q^\ell$, we determine a sequence $u^{\ell+1} = a_0, a_1, \ldots, a_r = u^\ell$ of points on $uv$. Let $Q^{a_j}$ be the cluster in $F_{i-1}^{(\ell)}$ nearest to $a_j$. It is determined by a walk at level $\ell$ starting with $Q^{a_{j-1}}$. Then point $a_{j+1}$ is the point on $uv$, equidistant from $C_i$ and $Q^{a_j}$ ($d_f(a_{j+1}, C_i) = d_f(a_{j+1}, Q^{a_j})$). If $a_j$ is equidistant from $C_i$ and $Q^{a_j}$, we are done at level $\ell$; continue to level $\ell - 1$ with $u^\ell = a_j$ and $Q^\ell = Q^{a_j}$. Else, if $d_f(v, Q^{a_j}) > d_f(v, C_i)$, perform a segment query to determine $a_{j+1}$. Otherwise, report that a point $t$ does not exist.

**Lemma 5.** *The expected number of visits of clusters at level $\ell$ during the parametric point location is $O(1)$.*

*Handling the empty regions.* After inserting $C_i$ at level 0 of the hierarchy for $\text{HVD}(F_{i-1})$, we insert $C_i$ into the series of higher levels. When $C_i$ is inserted at a given level, however, a region of another cluster may become empty.

We call a cluster $P$ *critical at level* $\ell$ if $\mathrm{hreg}_{F_{i-1}}^{(\ell-1)}(P) \neq \emptyset$, $\mathrm{hreg}_{F_i}^{(\ell-1)}(P) = \emptyset$, and $\mathrm{hreg}_{F_i}^{(\ell)}(P) \neq \emptyset$. Such a cluster $P$ becomes an obstacle to correct point location in the Voronoi hierarchy for $\mathrm{HVD}(F_i)$. Indeed, if a query point lies in $\mathrm{hreg}_{F_i}^{(\ell)}(P)$, we do not know where to continue the point location at level $\ell - 1$.

To fix the problem, $P$ can be deleted from all levels, but this is computationally expensive. Instead, we link $P$ to at most two other clusters $Q, R \in F_i^{(\ell-1)}$, such that every point $q \in \mathbb{R}^2$ is closer to either $Q$ or to $R$ than to $P$. Property 4 guarantees that such a cluster or two clusters exist (a cluster contained in $\mathrm{conv}\, P$ or a killing pair for $P$, respectively).

We now describe how to find a killing pair for $P$. While inserting $C_i$ at level $\ell - 1$, we store all (deleted) $P$-mixed vertices of $\mathrm{hreg}_{F_{i-1}}^{(\ell-1)}(P)$ in a list $V$. At level $\ell$, for each $P$-mixed vertex $v$ of $\mathrm{hreg}_{F_i}^{(\ell)}(P)$, we check if $v$ is closer to $C_i$ or to $P$. If $d_\mathrm{f}(v, C_i) \geq d_\mathrm{f}(v, P)$, let $c$ be the point in $C_i$ for which $d_\mathrm{f}(v, C_i) = d(v, c)$. Note that $c \notin \mathrm{conv}\, P$, which will be useful. The linking is performed as follows:

- If all $P$-mixed vertices of $\mathrm{hreg}_{F_{i-1}}^{(\ell)}(P)$ are closer to $C_i$ than to $P$, link only to $C_i$. (This happens only if $C_i \notin F^{(\ell)}$.)
- Else find the cluster $K \in F^{(\ell-1)}$ such that $\{K, C_i\}$ is a killing pair for $P$. If $C_i \in F^{(\ell)}$, link only to cluster $K$. Otherwise, link to both $K$ and $C_i$.

What remains is to determine cluster $K$. To this aim, we use list $V$ and point $c$. Each vertex $u \in V$ is equidistant from two points $p_1, p_2 \in P$, and one point $q \in Q$, for some $Q \in F^{(\ell-1)}$. We simply check whether $c$ and $q$ are on different sides of the chord $\overline{p_1 p_2}$. If they are, then we set $K = Q$ and we stop. By Property 4, $\{K, C_i\}$ is a killing pair for $P$, and thus the linking is correct.

We summarize the result on the Voronoi hierarchy in the following theorem, which is easily derived from Lemmas 3 to 5 and the discussion in Section 5.

**Theorem 1.** *The Voronoi hierarchy for the Hausdorff Voronoi diagram of a family of $k$ clusters of total complexity $n$ has expected size $O(n)$. Both the point location query and the parametric point location take expected $O(\log n \log k)$ time. Insertion of a cluster takes amortized $O((N/k)\log n)$ time, where $N$ is the total number of update operations in all levels during the insertion of all $k$ clusters.*

## 6 Complexity Analysis

The running time of our algorithm depends on the number of update operations (insertions and deletions) during the construction of the diagram. Using the Clarkson-Shor technique [9], we prove that the expectation of this number is linear, when clusters are inserted in random order. Note that in contrast to the standard probabilistic argument, our proof does not assume sites (clusters) to have constant size.

**Theorem 2.** *The expected number of update operations is $O(n)$.*

Theorem 2 can be easily extended to all levels of the Voronoi hierarchy. The total time for the construction of the separator decomposition for all clusters is $O(n \log n)$ (see Lemma 2). For each cluster $C \in F$, we perform $O(|C|)$ point location queries and at most one parametric point location in Voronoi hierarchy. By Lemma 2 and Theorems 1 and 2, we conclude.

**Theorem 3.** *The Hausdorff Voronoi diagram of non-crossing clusters can be constructed in $O(n \log n \log k)$ expected time and $O(n)$ expected space.*

Deterministic $O(n)$ space could be achieved by using a dynamic point location data structure for a planar subdivision [3,5]. On this data structure, the parametric point location can be performed as a simplified form of the parametric search, as described by Cheong et al. [8]. The time complexity of such a query is $t_q^2$, where $t_q$ is the time complexity of point location in the chosen data structure. In particular, the data structure by Baumgarten et al. [5] has $t_q \in O(\log n \log \log n)$, which leads to the construction of the Hausdorff Voronoi diagram with expected running time $O(n \log^2 n (\log \log n)^2)$ and deterministic space $O(n)$.

## 7 Discussion and Open Problems

We have provided improved complexity algorithms for constructing the Hausdorff Voronoi diagram of a family of non-crossing point clusters based on randomized incremental construction and point location. There is still a gap in the complexity of constructing the Hausdorff Voronoi diagram between our $O(n \log^2 n)$ expected time algorithm and the well-known $\Omega(n \log n)$ time lower bound. An open problem is to close or reduce this gap. It is interesting that in the $L_\infty$ metric, a simple $O(n \log n)$-time $O(n)$-space algorithm, based on plane sweep, is known [23]. We are currently considering the application of randomized incremental construction through conflict and history graphs. In future research we plan to consider families of arbitrary point clusters that may be crossing. In this case, the size of the diagram can vary from linear to quadratic, and therefore, an output-sensitive algorithm is most desirable. Another direction for research is to study the problem for clusters of segments, clusters of convex polygons or other shapes, rather than clusters of points.

## References

1. Abellanas, M., Hernandez, G., Klein, R., Neumann-Lara, V., Urrutia, J.: A combinatorial property of convex sets. Discrete Comput. Geom. 17(3), 307–318 (1997)
2. Abellanas, M., Hurtado, F., Icking, C., Klein, R., Langetepe, E., Ma, L., Palop, B., Sacristán, V.: The farthest color Voronoi diagram and related problems. In: 17th Eur. Workshop on Comput. Geom. (EWCG), pp. 113–116 (2001)
3. Arge, L., Brodal, G.S., Georgiadis, L.: Improved dynamic planar point location. In: 47th Ann. IEEE Symp. Found. Comput. Sci. (FOCS), pp. 305–314 (2006)

4. Aronov, B., Bose, P., Demaine, E.D., Gudmundsson, J., Iacono, J., Langerman, S., Smid, M.: Data structures for halfplane proximity queries and incremental Voronoi diagrams. In: Correa, J.R., Hevia, A., Kiwi, M. (eds.) LATIN 2006. LNCS, vol. 3887, pp. 80–92. Springer, Heidelberg (2006)
5. Baumgarten, H., Jung, H., Mehlhorn, K.: Dynamic point location in general subdivisions. J. Algorithm 17(3), 342–380 (1994)
6. Boissonnat, J.-D., Wormser, C., Yvinec, M.: Curved Voronoi diagrams. In: Boissonnat, J.-D., Teillaud, M. (eds.) Effective Computational Geometry for Curves and Surfaces, pp. 67–116. Springer, Heidelberg (2006)
7. Cheilaris, P., Khramtcova, E., Langerman, S., Papadopoulou, E.: A randomized incremental approach for the Hausdorff Voronoi diagram of non-crossing clusters. CoRR abs/1312.3904 (2013)
8. Cheong, O., Everett, H., Glisse, M., Gudmundsson, J., Hornus, S., Lazard, S., Lee, M., Na, H.S.: Farthest-polygon Voronoi diagrams. Comput. Geom. 44(4), 234–247 (2011)
9. Clarkson, K., Shor, P.: Applications of random sampling in computational geometry, II. Discrete Comput. Geom. 4, 387–421 (1989)
10. Dehne, F., Maheshwari, A., Taylor, R.: A coarse grained parallel algorithm for Hausdorff Voronoi diagrams. In: 35th Int. Conf. on Parallel Processing (ICPP), pp. 497–504 (2006)
11. Devillers, O.: The Delaunay Hierarchy. Int. J. Found. Comput. S. 13, 163–180 (2002)
12. Edelsbrunner, H.: Computing the extreme distances between two convex polygons. J. Algorithm 6(2), 213–224 (1985)
13. Edelsbrunner, H., Guibas, L.J., Sharir, M.: The upper envelope of piecewise linear functions: algorithms and applications. Discrete Comput. Geom. 4, 311–336 (1989)
14. Huttenlocher, D.P., Kedem, K., Sharir, M.: The upper envelope of Voronoi surfaces and its applications. Discrete Comput. Geom. 9, 267–291 (1993)
15. Karavelas, M., Yvinec, M.: The Voronoi diagram of convex objects in the plane. Technical report RR-5023, INRIA (2003)
16. Klein, R.: Concrete and Abstract Voronoi Diagrams. LNCS, vol. 400. Springer, Heidelberg (1989)
17. Klein, R., Mehlhorn, K., Meiser, S.: Randomized incremental construction of abstract Voronoi diagrams. Comput. Geom. 3(3), 157–184 (1993)
18. McAllister, M., Kirkpatrick, D., Snoeyink, J.: A compact piecewise-linear Voronoi diagram for convex sites in the plane. Discrete Comput. Geom. 15(1), 73–105 (1996)
19. Megiddo, N., Tamir, A., Zemel, E., Chandrasekaran, R.: An $O(n \log^2 n)$ algorithm for the $k$th longest path in a tree with applications to location problems. SIAM J. Comput. 10(2), 328–337 (1981)
20. Papadopoulou, E.: The Hausdorff Voronoi diagram of point clusters in the plane. Algorithmica 40(2), 63–82 (2004)
21. Papadopoulou, E.: Net-aware critical area extraction for opens in VLSI circuits via higher-order Voronoi diagrams. IEEE T. Comput. Aid D. 30(5), 704–716 (2011)
22. Papadopoulou, E., Lee, D.T.: The Hausdorff Voronoi diagram of polygonal objects: a divide and conquer approach. Int. J. Comput. Geom. Ap. 14(6), 421–452 (2004)
23. Papadopoulou, E., Xu, J.: The $L_\infty$ Hausdorff Voronoi diagram revisited. In: 8th Int. Symp. on Voronoi Diagr. in Sci. and Eng. (ISVD), pp. 67–74 (2011)

# Upper Bounds on the Spanning Ratio of Constrained Theta-Graphs*

Prosenjit Bose and André van Renssen

School of Computer Science, Carleton University, Ottawa, Canada
jit@scs.carleton.ca, andre@cg.scs.carleton.ca

**Abstract.** We present tight upper and lower bounds on the spanning ratio of a large family of constrained $\theta$-graphs. We show that constrained $\theta$-graphs with $4k + 2$ ($k \geq 1$ and integer) cones have a tight spanning ratio of $1 + 2\sin(\theta/2)$, where $\theta$ is $2\pi/(4k + 2)$. We also present improved upper bounds on the spanning ratio of the other families of constrained $\theta$-graphs.

## 1 Introduction

A geometric graph $G$ is a graph whose vertices are points in the plane and whose edges are line segments between pairs of points. Every edge is weighted by the Euclidean distance between its endpoints. The distance between two vertices $u$ and $v$ in $G$, denoted by $d_G(u, v)$, is defined as the sum of the weights of the edges along the shortest path between $u$ and $v$ in $G$. A subgraph $H$ of $G$ is a $t$-spanner of $G$ (for $t \geq 1$) if for each pair of vertices $u$ and $v$, $d_H(u, v) \leq t \cdot d_G(u, v)$. The smallest value $t$ for which $H$ is a $t$-spanner is the *spanning ratio* or *stretch factor*. The graph $G$ is referred to as the *underlying graph* of $H$. The spanning properties of various geometric graphs have been studied extensively in the literature (see [4,10] for a comprehensive overview of the topic). We look at a specific type of geometric spanner: $\theta$-graphs.

Introduced independently by Clarkson [7] and Keil [9], $\theta$-graphs partition the plane around each vertex into $m$ disjoint cones, each having aperture $\theta = 2\pi/m$. The $\theta_m$-graph is constructed by, for each cone of each vertex $u$, connecting $u$ to the vertex $v$ whose projection along the bisector of the cone is closest. Ruppert and Seidel [11] showed that the spanning ratio of these graphs is at most $1/(1 - 2\sin(\theta/2))$, when $\theta < \pi/3$, i.e. there are at least seven cones. Recent results include a tight spanning ratio of $1 + 2\sin(\theta/2)$ for $\theta$-graphs with $4k + 2$ cones [1], where $k \geq 1$ and integer, and improved upper bounds for the other three families of $\theta$-graphs [6].

Most of the research, however, has focused on constructing spanners where the underlying graph is the complete Euclidean geometric graph. We study this problem in a more general setting with the introduction of line segment *constraints*. Specifically, let $P$ be a set of points in the plane and let $S$ be a set

---

* Research supported in part by NSERC and Carleton University's President's 2010 Doctoral Fellowship.

of line segments between two vertices in $P$, called *constraints*. The set of constraints is planar, i.e. no two constraints intersect properly. Two vertices $u$ and $v$ can see each other if and only if either the line segment $uv$ does not properly intersect any constraint or $uv$ is itself a constraint. If two vertices $u$ and $v$ can see each other, the line segment $uv$ is a *visibility edge*. The *visibility graph* of $P$ with respect to a set of constraints $S$, denoted $Vis(P, S)$, has $P$ as vertex set and all visibility edges as edge set. In other words, it is the complete graph on $P$ minus all edges that properly intersect one or more constraints in $S$.

This setting has been studied extensively within the context of motion planning amid obstacles. Clarkson [7] was one of the first to study this problem and showed how to construct a linear-sized $(1+\epsilon)$-spanner of $Vis(P, S)$. Subsequently, Das [8] showed how to construct a spanner of $Vis(P, S)$ with constant spanning ratio and constant degree. The Constrained Delaunay Triangulation was shown to be a 2.42-spanner of $Vis(P, S)$ [3]. Recently, it was also shown that the constrained $\theta_6$-graph is a 2-spanner of $Vis(P, S)$ [2]. In this paper, we generalize the recent results on unconstrained $\theta$-graphs to the constrained setting. There are two main obstacles that differentiate this work from previous results. First, the main difficulty with the constrained setting is that induction cannot be applied directly, as the destination need not be visible from the vertex closest to the source (see Figure 5, where $w$ is not visible from $v_0$, the vertex closest to $u$). Second, when the graph does not have $4k + 2$ cones, the cones do not line up as nicely as in [2], making it more difficult to apply induction.

In this paper, we overcome these two difficulties and show that constrained $\theta$-graphs with $4k+2$ cones have a spanning ratio of at most $1+2\sin(\theta/2)$, where $\theta$ is $2\pi/(4k + 2)$. Since the lower bounds of the unconstrained $\theta$-graphs carry over to the constrained setting, this shows that this spanning ratio is tight. We also show that constrained $\theta$-graphs with $4k + 4$ cones have a spanning ratio of at most $1+2\sin(\theta/2)/(\cos(\theta/2) - \sin(\theta/2))$, where $\theta$ is $2\pi/(4k+4)$. Finally, we show that constrained $\theta$-graphs with $4k+3$ or $4k+5$ cones have a spanning ratio of at most $\cos(\theta/4)/(\cos(\theta/2) - \sin(3\theta/4))$, where $\theta$ is $2\pi/(4k+3)$ or $2\pi/(4k+5)$.

## 2 Preliminaries

We define a *cone* $C$ to be the region in the plane between two rays originating from a vertex referred to as the apex of the cone. When constructing a (constrained) $\theta_{(4k+x)}$-graph, for each vertex $u$ consider the rays originating from $u$ with the angle between consecutive rays being $\theta = 2\pi/(4k + x)$, where $k \geq 1$ and integer and $x \in \{2, 3, 4, 5\}$. Each pair of consecutive rays defines a cone. The cones are oriented such that the bisector of some cone coincides with the vertical halfline through $u$ that lies above $u$. Let this cone be $C_0$ of $u$ and number the cones in clockwise order around $u$. The cones around the other vertices have the same orientation as the ones around $u$. We write $C_i^u$ to indicate the $i$-th cone of a vertex $u$. For ease of exposition, we only consider point sets in general position: no two points lie on a line parallel to one of the rays that define the cones, no two points lie on a line perpendicular to the bisector of a cone, and no three points are collinear.

Let vertex $u$ be an endpoint of a constraint $c$ and let the other endpoint $v$ lie in cone $C_i^u$. The lines through all such constraints $c$ split $C_i^u$ into several *subcones*. We use $C_{i,j}^u$ to denote the $j$-th subcone of $C_i^u$. When a constraint $c = (u,v)$ splits a cone of $u$ into two subcones, we define $v$ to lie in both of these subcones. We consider a cone that is not split to be a single subcone.

We now introduce the constrained $\theta_{(4k+x)}$-graph: for each subcone $C_{i,j}$ of each vertex $u$, add an edge from $u$ to the closest vertex in that subcone that can see $u$, where distance is measured along the bisector of the original cone (*not the subcone*). More formally, we add an edge between two vertices $u$ and $v$ if $v$ can see $u$, $v \in C_{i,j}^u$, and for all points $w \in C_{i,j}^u$ that can see $u$, $|uv'| \le |uw'|$, where $v'$ and $w'$ denote the projection of $v$ and $w$ on the bisector of $C_i^u$ and $|xy|$ denotes the length of the line segment between two points $x$ and $y$. Note that our assumption of general position implies that each vertex adds at most one edge for each of its subcones.

Given a vertex $w$ in the cone $C_i$ of vertex $u$, we define the *canonical triangle* $T_{uw}$ to be the triangle defined by the borders of $C_i^u$ and the line through $w$ perpendicular to the bisector of $C_i^u$. Note that subcones do not define canonical triangles. We use $m$ to denote the midpoint of the side of $T_{uw}$ opposing $u$ and $\alpha$ to denote the unsigned angle between $uw$ and $um$ (see Figure 1). Note that for any pair of vertices $u$ and $w$, there exist two canonical triangles: $T_{uw}$ and $T_{wu}$. We say that a region is *empty* if it does not contain any vertex of $P$.

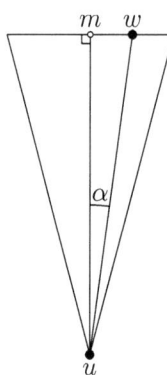

**Fig. 1.** The canonical triangle $T_{uw}$

## 3 Some Useful Lemmas

In this section, we list a number of lemmas that are used when bounding the spanning ratio of the various graphs. Note that these lemmas are not new, as they are already used in [2,6], though some are expanded to work for all four families of constrained $\theta$-graphs. We start with a nice property of visibility graphs from [2].

**Lemma 1.** *Let $u$, $v$, and $w$ be three arbitrary points in the plane such that $uw$ and $vw$ are visibility edges and $w$ is not the endpoint of a constraint intersecting the interior of triangle $uvw$. Then there exists a convex chain of visibility edges from $u$ to $v$ in triangle $uvw$, such that the polygon defined by $uw$, $wv$ and the convex chain is empty and does not contain any constraints.*

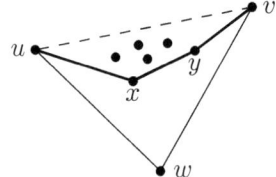

**Fig. 2.** The convex chain between vertices $u$ and $v$, where thick lines are visibility edges

Next, we use two lemmas from [6] to bound the length of certain line segments. Note that Lemma 2 is extended such that it also holds for the constrained $\theta_{(4k+2)}$-graph. We use $\angle xyz$ to denote the smaller angle between line segments $xy$ and $yz$.

**Lemma 2.** *Let $u$, $v$ and $w$ be three vertices in the $\theta_{(4k+x)}$-graph, $x \in \{2,3,4,5\}$, such that $w \in C_0^u$ and $v \in T_{uw}$, to the left of $uw$. Let $a$ be the intersection of the side of $T_{uw}$ opposite $u$ and the left boundary of $C_0^v$. Let $C_i^v$ denote the cone of $v$ that contains $w$ and let $c$ and $d$ be the upper and lower corner of $T_{vw}$. If $1 \leq i \leq k-1$, or $i = k$ and $|cw| \leq |dw|$, then $\max\{|vc| + |cw|, |vd| + |dw|\} \leq |va| + |aw|$ and $\max\{|cw|, |dw|\} \leq |aw|$.*

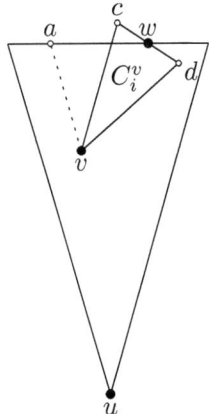 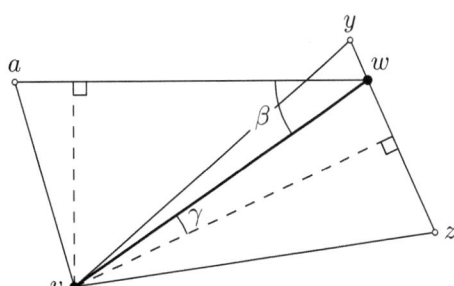

**Fig. 3.** The situation where we apply Lemma 2

**Fig. 4.** The situation where we apply Lemma 3

**Lemma 3.** *Let $u$, $v$ and $w$ be three vertices in the $\theta_{(4k+x)}$-graph, $x \in \{2,3,4,5\}$, such that $w \in C_0^u$, $v \in T_{uw}$ to the left of $uw$, and $w \notin C_0^v$. Let $a$ be the intersection of the side of $T_{uw}$ opposite $u$ and the line through $v$ parallel to the left boundary of $T_{uw}$. Let $y$ and $z$ be the corners of $T_{vw}$ opposite to $v$. Let $\beta = \angle awv$ and let $\gamma$ be the unsigned angle between $vw$ and the bisector of $T_{vw}$. Let $\mathbf{c}$ be a positive constant. If $\mathbf{c} \geq \frac{\cos\gamma - \sin\beta}{\cos(\frac{\theta}{2} - \beta) - \sin(\frac{\theta}{2} + \gamma)}$, then $|vp| + \mathbf{c} \cdot |pw| \leq |va| + \mathbf{c} \cdot |aw|$, where $p$ is $y$ if $|yw| \geq |zw|$ and $z$ if $|yw| < |zw|$.*

## 4 Constrained $\theta_{(4k+2)}$-Graph

In this section we prove that the constrained $\theta_{(4k+2)}$-graph has spanning ratio at most $1 + 2 \cdot \sin(\theta/2)$. Since this is also a lower bound [1], this proves that this spanning ratio is tight.

**Theorem 1.** *Let $u$ and $w$ be two vertices in the plane such that $u$ can see $w$. Let $m$ be the midpoint of the side of $T_{uw}$ opposing $u$ and let $\alpha$ be the unsigned angle* between $uw$ and $um$. There exists a path connecting $u$ and $w$ in the constrained $\theta_{(4k+2)}$-graph of length at most

$$\left(\left(\frac{1+\sin\left(\frac{\theta}{2}\right)}{\cos\left(\frac{\theta}{2}\right)}\right)\cdot\cos\alpha+\sin\alpha\right)\cdot|uw|.$$

*Proof.* We assume without loss of generality that $w \in C_0^u$. We prove the theorem by induction on the area of $T_{uw}$. Formally, we perform induction on the rank, when ordered by area, of the triangles $T_{xy}$ for all pairs of vertices $x$ and $y$ that can see each other. Let $a$ and $b$ be the upper left and right corner of $T_{uw}$, and let $A$ and $B$ be the triangles $uaw$ and $ubw$ (see Figure 5).

Our inductive hypothesis is the following, where $\delta(u, w)$ denotes the length of the shortest path from $u$ to $w$ in the constrained $\theta_{(4k+2)}$-graph:

- If $A$ is empty, then $\delta(u, w) \leq |ub| + |bw|$.
- If $B$ is empty, then $\delta(u, w) \leq |ua| + |aw|$.
- If neither $A$ nor $B$ is empty, then $\delta(u, w) \leq \max\{|ua| + |aw|, |ub| + |bw|\}$.

We first show that this induction hypothesis implies the theorem: $|um| = |uw| \cdot \cos\alpha$, $|mw| = |uw| \cdot \sin\alpha$, $|am| = |bm| = |uw| \cdot \cos\alpha \cdot \tan(\theta/2)$, and $|ua| = |ub| = |uw| \cdot \cos\alpha / \cos(\theta/2)$. Thus the induction hypothesis gives that $\delta(u, w)$ is at most $|uw| \cdot (((1+\sin(\theta/2))/\cos(\theta/2)) \cdot \cos\alpha + \sin\alpha)$.

**Base Case:** $T_{uw}$ has rank 1. Since the triangle is a smallest triangle, $w$ is the closest vertex to $u$ in that cone. Hence the edge $(u, w)$ is part of the constrained $\theta_{(4k+2)}$-graph, and $\delta(u, w) = |uw|$. From the triangle inequality, we have $|uw| \leq \min\{|ua| + |aw|, |ub| + |bw|\}$, so the induction hypothesis holds.

**Induction Step:** We assume that the induction hypothesis holds for all pairs of vertices that can see each other and have a canonical triangle whose area is smaller than the area of $T_{uw}$.

If $(u, w)$ is an edge in the constrained $\theta_{(4k+2)}$-graph, the induction hypothesis follows by the same argument as in the base case. If there is no edge between $u$ and $w$, let $v_0$ be the vertex closest to $u$ in the subcone of $u$ that contains $w$, and let $a_0$ and $b_0$ be the upper left and right corner of $T_{uv_0}$ (see Figure 5). By definition, $\delta(u, w) \leq |uv_0| + \delta(v_0, w)$, and by the triangle inequality, $|uv_0| \leq \min\{|ua_0| + |a_0v_0|, |ub_0| + |b_0v_0|\}$. We assume without loss of generality that $v_0$ lies to the left of $uw$, which means that $A$ is not empty.

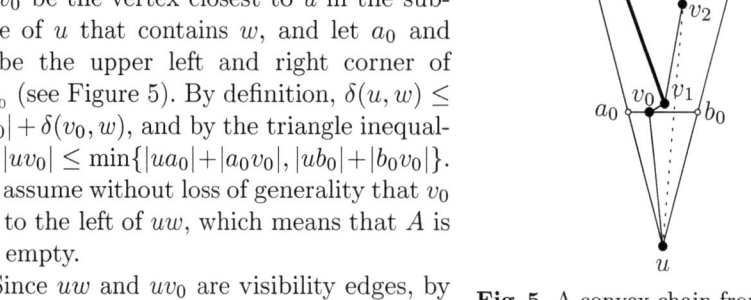

Fig. 5. A convex chain from $v_0$ to $w$

Since $uw$ and $uv_0$ are visibility edges, by applying Lemma 1 to triangle $v_0uw$, a convex chain $v_0, ..., v_l = w$ of visibility edges connecting $v_0$ and $w$ exists (see Figure 5). Note that, since $v_0$ is the closest visible vertex to $u$, every vertex along the convex chain lies above the horizontal line through $v_0$.

We now look at two consecutive vertices $v_{j-1}$ and $v_j$ along the convex chain. There are four types of configurations (see Figure 6): (i) $v_j \in C_k^{v_{j-1}}$, (ii) $v_j \in C_i^{v_{j-1}}$ where $1 \leq i < k$, (iii) $v_j \in C_0^{v_{j-1}}$ and $v_j$ lies to the right of or has the same $x$-coordinate as $v_{j-1}$, (iv) $v_j \in C_0^{v_{j-1}}$ and $v_j$ lies to the left of $v_{j-1}$. By convexity, the direction of $\overrightarrow{v_j v_{j+1}}$ is rotating counterclockwise for increasing $j$. Thus, these configurations occur in the order Type (i), Type (ii), Type (iii), Type (iv) along the convex chain from $v_0$ to $w$. We bound $\delta(v_{j-1}, v_j)$ as follows:

**Type (i):** If $v_j \in C_k^{v_{j-1}}$, let $a_j$ and $b_j$ be the upper and lower left corner of $T_{v_j v_{j-1}}$ and let $B_j = v_{j-1} b_j v_j$. Note that since $v_j \in C_k^{v_{j-1}}$, $a_j$ is also the intersection of the left boundary of $C_0^{v_{j-1}}$ and the horizontal line through $v_j$. Triangle $B_j$ lies between the convex chain and $uw$, so it must be empty. Since $v_j$ can see $v_{j-1}$ and $T_{v_j v_{j-1}}$ has smaller area than $T_{uw}$, the induction hypothesis gives that $\delta(v_{j-1}, v_j)$ is at most $|v_{j-1} a_j| + |a_j v_j|$.

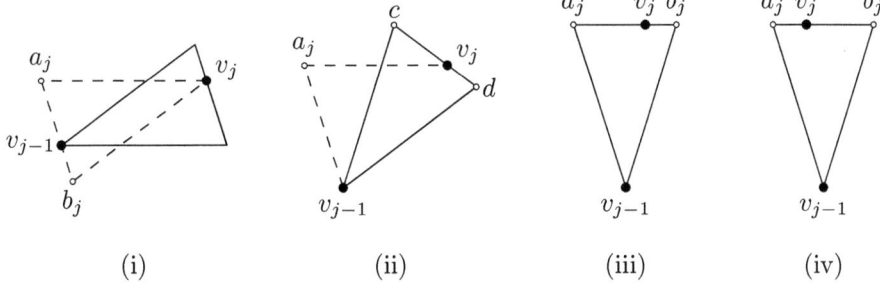

(i)  (ii)  (iii)  (iv)

**Fig. 6.** The four types of configurations

**Type (ii):** If $v_j \in C_i^{v_{j-1}}$ where $1 \leq i < k$, let $c$ and $d$ be the upper and lower right corner of $T_{v_{j-1} v_j}$. Let $a_j$ be the intersection of the left boundary of $C_0^{v_{j-1}}$ and the horizontal line through $v_j$. Since $v_j$ can see $v_{j-1}$ and $T_{v_{j-1} v_j}$ has smaller area than $T_{uw}$, the induction hypothesis gives that $\delta(v_{j-1}, v_j)$ is at most $\max\{|v_{j-1} c| + |c v_j|, |v_{j-1} d| + |d v_j|\}$. Since $v_j \in C_i^{v_{j-1}}$ where $1 \leq i < k$, we can apply Lemma 2 (where $v$, $w$, and $a$ from Lemma 2 are $v_{j-1}$, $v_j$, and $a_j$), which gives us that $\max\{|v_{j-1} c| + |c v_j|, |v_{j-1} d| + |d v_j|\} \leq |v_{j-1} a_j| + |a_j v_j|$.

**Type (iii):** If $v_j \in C_0^{v_{j-1}}$ and $v_j$ lies to the right of or has the same $x$-coordinate as $v_{j-1}$, let $a_j$ and $b_j$ be the left and right corner of $T_{v_{j-1} v_j}$ and let $A_j = v_{j-1} a_j v_j$ and $B_j = v_{j-1} b_j v_j$. Since $v_j$ can see $v_{j-1}$ and $T_{v_{j-1} v_j}$ has smaller area than $T_{uw}$, we can apply the induction hypothesis. Regardless of whether $A_j$ and $B_j$ are empty or not, $\delta(v_{j-1}, v_j)$ is at most $\max\{|v_{j-1} a_j| + |a_j v_j|, |v_{j-1} b_j| + |b_j v_j|\}$. Since $v_j$ lies to the right of or has the same $x$-coordinate as $v_{j-1}$, we know that $|v_{j-1} a_j| + |a_j v_j| \geq |v_{j-1} b_j| + |b_j v_j|$, so $\delta(v_{j-1}, v_j)$ is at most $|v_{j-1} a_j| + |a_j v_j|$.

**Type (iv):** If $v_j \in C_0^{v_{j-1}}$ and $v_j$ lies to the left of $v_{j-1}$, let $a_j$ and $b_j$ be the left and right corner of $T_{v_{j-1} v_j}$ and let $A_j = v_{j-1} a_j v_j$ and $B_j = v_{j-1} b_j v_j$. Since $v_j$ can see $v_{j-1}$ and $T_{v_{j-1} v_j}$ has smaller area than $T_{uw}$, we can apply the induction hypothesis. Thus, if $B_j$ is empty, $\delta(v_{j-1}, v_j)$ is at most $|v_{j-1} a_j| + |a_j v_j|$ and if $B_j$ is not empty, $\delta(v_{j-1}, v_j)$ is at most $|v_{j-1} b_j| + |b_j v_j|$.

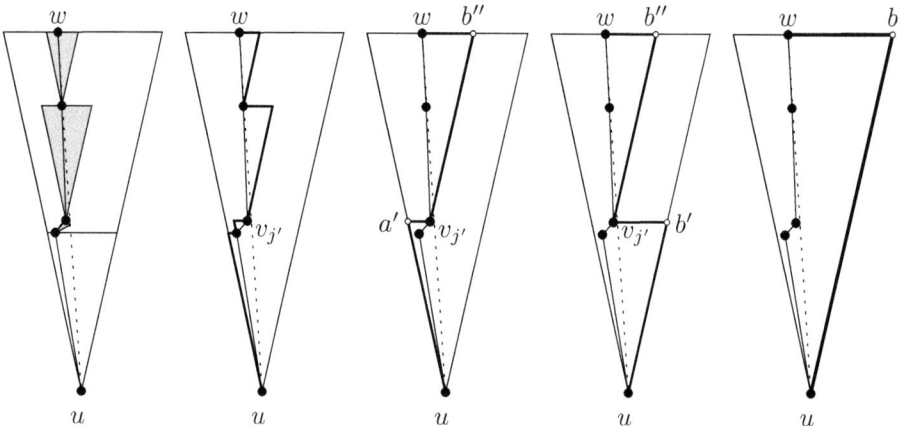

**Fig. 7.** Visualization of the paths (thick lines) in the inequalities of case (c)

To complete the proof, we consider three cases: (a) $\angle awu \leq \pi/2$, (b) $\angle awu > \pi/2$ and $B$ is empty, (c) $\angle awu > \pi/2$ and $B$ is not empty.

**Case (a):** If $\angle awu \leq \pi/2$, the convex chain cannot contain any Type (iv) configurations: for Type (iv) configurations to occur, $v_j$ needs to lie to the left of $v_{j-1}$. However, by construction, $v_j$ lies on or to the right of the line through $v_{j-1}$ and $w$. Hence, since $\angle awv_{j-1} < \angle awu \leq \pi/2$, $v_j$ lies to the right of or has the same $x$-coordinate as $v_{j-1}$. We can now bound $\delta(u,w)$ by using these bounds: $\delta(u,w) \leq |uv_0| + \sum_{j=1}^{l} \delta(v_{j-1}, v_j) \leq |ua_0| + |a_0 v_0| + \sum_{j=1}^{l}(|v_{j-1} a_j| + |a_j v_j|) = |ua| + |aw|$.

**Case (b):** If $\angle awu > \pi/2$ and $B$ is empty, the convex chain can contain Type (iv) configurations. However, since $B$ is empty and the area between the convex chain and $uw$ is empty (by Lemma 1), all $B_j$ are also empty. Using the computed bounds on the lengths of the paths between the points along the convex chain, we can bound $\delta(u,w)$ as in the previous case.

**Case (c):** If $\angle awu > \pi/2$ and $B$ is not empty, the convex chain can contain Type (iv) configurations and since $B$ is not empty, the triangles $B_j$ need not be empty. Recall that $v_0$ lies in $A$, hence neither $A$ nor $B$ are empty. Therefore, it suffices to prove that $\delta(u,w) \leq \max\{|ua| + |aw|, |ub| + |bw|\} = |ub| + |bw|$. Let $T_{v_{j'} v_{j'+1}}$ be the first Type (iv) configuration along the convex chain (if it has any), let $a'$ and $b'$ be the upper left and right corner of $T_{uv_{j'}}$, and let $b''$ be the upper right corner of $T_{v_{j'} w}$. We now have that $\delta(u,w) \leq |uv_0| + \sum_{j=1}^{l} \delta(v_{j-1}, v_j) \leq |ua'| + |a'v_{j'}| + |v_{j'} b''| + |b'' w| \leq |ub| + |bw|$ (see Figure 7). □

Since $((1 + \sin(\theta/2))/\cos(\theta/2)) \cdot \cos\alpha + \sin\alpha$ is increasing for $\alpha \in [0, \theta/2]$, for $\theta \leq \pi/3$, it is maximized when $\alpha = \theta/2$, and we obtain the following corollary:

**Corollary 1.** *The constrained $\theta_{(4k+2)}$-graph is a $\left(1 + 2 \cdot \sin\left(\frac{\theta}{2}\right)\right)$-spanner of $Vis(P, S)$.*

## 5 Generic Framework for the Spanning Proof

Next, we modify the spanning proof from the previous section and provide a generic framework for the spanning proof for the other three families of $\theta$-graphs. After providing this framework, we fill in the blanks for the individual families.

**Theorem 2.** *Let $u$ and $w$ be two vertices in the plane such that $u$ can see $w$. Let $m$ be the midpoint of the side of $T_{uw}$ opposing $u$ and let $\alpha$ be the unsigned angle between $uw$ and $um$. There exists a path connecting $u$ and $w$ in the constrained $\theta_{(4k+x)}$-graph of length at most*

$$\left( \frac{\cos \alpha}{\cos\left(\frac{\theta}{2}\right)} + \left( \cos\alpha \cdot \tan\left(\frac{\theta}{2}\right) + \sin\alpha \right) \cdot c \right) \cdot |uw|,$$

*where $c \geq 1$ is a constant that depends on $x \in \{3, 4, 5\}$. For the constrained $\theta_{(4k+4)}$-graph, $c$ equals $1/(\cos(\theta/2) - \sin(\theta/2))$ and for the constrained $\theta_{(4k+3)}$-graph and $\theta_{(4k+5)}$-graph, $c$ equals $\cos(\theta/4)/(\cos(\theta/2) - \sin(3\theta/4))$.*

*Proof.* We prove the theorem by induction on the area of $T_{uw}$. Formally, we perform induction on the rank, when ordered by area, of the triangles $T_{xy}$ for all pairs of vertices $x$ and $y$ that can see each other. We assume without loss of generality that $w \in C_0^u$. Let $a$ and $b$ be the upper left and right corner of $T_{uw}$ (see Figure 5).

Our inductive hypothesis is the following, where $\delta(u, w)$ denotes the length of the shortest path from $u$ to $w$ in the constrained $\theta_{(4k+x)}$-graph: $\delta(u, w) \leq \max\{|ua| + |aw| \cdot c, |ub| + |bw| \cdot c\}$.

We first show that this induction hypothesis implies the theorem. Basic trigonometry gives us the following equalities: $|um| = |uw| \cdot \cos\alpha$, $|mw| = |uw| \cdot \sin\alpha$, $|am| = |bm| = |uw| \cdot \cos\alpha \cdot \tan(\theta/2)$, and $|ua| = |ub| = |uw| \cdot \cos\alpha/\cos(\theta/2)$. Thus the induction hypothesis gives that $\delta(u, w)$ is at most $|uw| \cdot (\cos\alpha/\cos(\theta/2) + (\cos\alpha \cdot \tan(\theta/2) + \sin\alpha) \cdot c)$.

**Base Case:** $T_{uw}$ has rank 1. Since the triangle is a smallest triangle, $w$ is the closest vertex to $u$ in that cone. Hence the edge $(u, w)$ is part of the constrained $\theta_{(4k+x)}$-graph, and $\delta(u, w) = |uw|$. From the triangle inequality and the fact that $c \geq 1$, we have $|uw| \leq \min\{|ua| + |aw| \cdot c, |ub| + |bw| \cdot c\}$, so the induction hypothesis holds.

**Induction Step:** We assume that the induction hypothesis holds for all pairs of vertices that can see each other and have a canonical triangle whose area is smaller than the area of $T_{uw}$.

If $(u, w)$ is an edge in the constrained $\theta_{(4k+x)}$-graph, the induction hypothesis follows by the same argument as in the base case. If there is no edge between $u$ and $w$, let $v_0$ be the vertex closest to $u$ in the subcone of $u$ that contains $w$, and let $a_0$ and $b_0$ be the upper left and right corner of $T_{uv_0}$ (see Figure 5). By definition, $\delta(u, w) \leq |uv_0| + \delta(v_0, w)$, and by the triangle inequality, $|uv_0| \leq \min\{|ua_0| + |a_0v_0|, |ub_0| + |b_0v_0|\}$. We assume without loss of generality that $v_0$ lies to the left of $uw$.

Since $uw$ and $uv_0$ are visibility edges, by applying Lemma 1 to triangle $v_0 uw$, a convex chain $v_0, ..., v_l = w$ of visibility edges connecting $v_0$ and $w$ exists (see Figure 5). Note that, since $v_0$ is the closest visible vertex to $u$, every vertex along the convex chain lies above the horizontal line through $v_0$.

We now look at two consecutive vertices $v_{j-1}$ and $v_j$ along the convex chain. When $v_j \notin C_0^{v_{j-1}}$, let $c$ and $d$ be the upper and lower right corner of $T_{v_{j-1} v_j}$. We distinguish four types of configurations: (i) $v_j \in C_i^{v_{j-1}}$ where $i > k$, or $i = k$ and $|cw| > |dw|$, (ii) $v_j \in C_i^{v_{j-1}}$ where $1 \leq i \leq k-1$, or $i = k$ and $|cw| \leq |dw|$, (iii) $v_j \in C_0^{v_{j-1}}$ and $v_j$ lies to the right of or has the same $x$-coordinate as $v_{j-1}$, (iv) $v_j \in C_0^{v_{j-1}}$ and $v_j$ lies to the left of $v_{j-1}$. By convexity, the direction of $\overrightarrow{v_j v_{j+1}}$ is rotating counterclockwise for increasing $j$. Thus, these configurations occur in the order Type (i), Type (ii), Type (iii), Type (iv) along the convex chain from $v_0$ to $w$. We bound $\delta(v_{j-1}, v_j)$ as follows:

**Type (i):** $v_j \in C_i^{v_{j-1}}$ where $i > k$, or $i = k$ and $|cw| > |dw|$. Since $v_j$ can see $v_{j-1}$ and $T_{v_j v_{j-1}}$ has smaller area than $T_{uw}$, the induction hypothesis gives that $\delta(v_{j-1}, v_j)$ is at most $\max\{|v_{j-1} c| + |cv_j| \cdot \mathbf{c}, |v_{j-1} d| + |dv_j| \cdot \mathbf{c}\}$.

Let $a_j$ be the intersection of the left boundary of $C_0^{v_{j-1}}$ and the horizontal line through $v_j$. We aim to show that $\max\{|v_{j-1} c| + |cv_j| \cdot \mathbf{c}, |v_{j-1} d| + |dv_j| \cdot \mathbf{c}\} \leq |v_{j-1} a_j| + |a_j v_j| \cdot \mathbf{c}$. We use Lemma 3 to do this. However, since the precise application of this lemma depends on the family of $\theta$-graphs and determines the value of $\mathbf{c}$, this case is discussed in the spanning proofs of the three families.

**Type (ii):** $v_j \in C_i^{v_{j-1}}$ where $1 \leq i \leq k-1$, or $i = k$ and $|cw| \leq |dw|$. Since $v_j$ can see $v_{j-1}$ and $T_{v_j v_{j-1}}$ has smaller area than $T_{uw}$, the induction hypothesis gives that $\delta(v_{j-1}, v_j)$ is at most $\max\{|v_{j-1} c| + |cv_j| \cdot \mathbf{c}, |v_{j-1} d| + |dv_j| \cdot \mathbf{c}\}$.

Let $a_j$ be the intersection of the left boundary of $C_0^{v_{j-1}}$ and the horizontal line through $v_j$. Since $v_j \in C_i^{v_{j-1}}$ where $1 \leq i \leq k-1$, or $i = k$ and $|cw| \leq |dw|$, we can apply Lemma 2 in this case (where $v$, $w$, and $a$ from Lemma 2 are $v_{j-1}$, $v_j$, and $a_j$) and we get that $\max\{|v_{j-1} c| + |cv_j|, |v_{j-1} d| + |dv_j|\} \leq |v_{j-1} a_j| + |a_j v_j|$ and $\max\{|cv_j|, |dv_j|\} \leq |a_j v_j|$. Since $\mathbf{c} \geq 1$, this implies that $\max\{|v_{j-1} c| + |cv_j| \cdot \mathbf{c}, |v_{j-1} d| + |dv_j| \cdot \mathbf{c}\} \leq |v_{j-1} a_j| + |a_j v_j| \cdot \mathbf{c}$.

**Type (iii):** If $v_j \in C_0^{v_{j-1}}$ and $v_j$ lies to the right of or has the same $x$-coordinate as $v_{j-1}$, let $a_j$ and $b_j$ be the left and right corner of $T_{v_{j-1} v_j}$. Since $v_j$ can see $v_{j-1}$ and $T_{v_{j-1} v_j}$ has smaller area than $T_{uw}$, we can apply the induction hypothesis. Thus, since $v_j$ lies to the right of or has the same $x$-coordinate as $v_{j-1}$, $\delta(v_{j-1}, v_j)$ is at most $|v_{j-1} a_j| + |a_j v_j| \cdot \mathbf{c}$.

**Type (iv):** If $v_j \in C_0^{v_{j-1}}$ and $v_j$ lies to the left of $v_{j-1}$, let $a_j$ and $b_j$ be the left and right corner of $T_{v_{j-1} v_j}$. Since $v_j$ can see $v_{j-1}$ and $T_{v_{j-1} v_j}$ has smaller area than $T_{uw}$, we can apply the induction hypothesis. Thus, since $v_j$ lies to the left of $v_{j-1}$, $\delta(v_{j-1}, v_j)$ is at most $|v_{j-1} b_j| + |b_j v_j| \cdot \mathbf{c}$.

To complete the proof, we consider two cases: (a) $\angle awu \leq \frac{\pi}{2}$, (b) $\angle awu > \frac{\pi}{2}$.

**Case (a):** We need to prove that $\delta(u, w) \leq \max\{|ua| + |aw|, |ub| + |bw|\} = |ua| + |aw|$. We first show that the convex chain cannot contain any Type (iv) configurations: for Type (iv) configurations to occur, $v_j$ needs to lie to the left of $v_{j-1}$. However, by construction, $v_j$ lies on or to the right of the line through $v_{j-1}$ and $w$. Hence, since $\angle awv_{j-1} < \angle awu \leq \pi/2$, $v_j$ lies to the right of $v_{j-1}$. We can now bound $\delta(u,w)$ by using these bounds: $\delta(u,w) \leq |uv_0| + \sum_{j=1}^{l} \delta(v_{j-1}, v_j) \leq |ua_0| + |a_0v_0| + \sum_{j=1}^{l}(|v_{j-1}a_j| + |a_jv_j| \cdot c) \leq |ua| + |aw| \cdot c$.

**Case (b):** If $\angle awu > \pi/2$, the convex chain can contain Type (iv) configurations. We need to prove that $\delta(u,w) \leq \max\{|ua|+|aw|, |ub|+|bw|\} = |ub|+|bw|$. Let $T_{v_{j'}v_{j'+1}}$ be the first Type (iv) configuration along the convex chain (if it has any), let $a'$ and $b'$ be the upper left and right corner of $T_{uv_{j'}}$, and let $b''$ be the upper right corner of $T_{v_{j'}w}$. We now have that $\delta(u,w) \leq |uv_0|+\sum_{j=1}^{l}\delta(v_{j-1},v_j) \leq |ua'| + |a'v_{j'}| \cdot c + |v_{j'}b''| + |b''w| \cdot c \leq |ub| + |bw| \cdot c$ (see Figure 7). □

## 6 The Constrained $\theta_{(4k+4)}$-Graph

In this section we complete the proof of Theorem 2 for the constrained $\theta_{(4k+4)}$-graph.

**Theorem 3.** *Let $u$ and $w$ be two vertices in the plane such that $u$ can see $w$. Let $m$ be the midpoint of the side of $T_{uw}$ opposite $u$ and let $\alpha$ be the unsigned angle between $uw$ and $um$. There exists a path connecting $u$ and $w$ in the constrained $\theta_{(4k+4)}$-graph of length at most*

$$\left(\frac{\cos\alpha}{\cos\left(\frac{\theta}{2}\right)} + \frac{\cos\alpha \cdot \tan\left(\frac{\theta}{2}\right) + \sin\alpha}{\cos\left(\frac{\theta}{2}\right) - \sin\left(\frac{\theta}{2}\right)}\right) \cdot |uw|.$$

*Proof.* We apply Theorem 2 using $c = 1/(\cos(\theta/2) - \sin(\theta/2))$. The assumptions made in Theorem 2 still apply. It remains to show that for the Type (i) configurations, we have that $\max\{|v_{j-1}c|+|cv_j|\cdot c, |v_{j-1}d|+|dv_j|\cdot c\} \leq |v_{j-1}a_j|+|a_jv_j|\cdot c$, where $c$ and $d$ are the upper and lower right corner of $T_{v_{j-1}v_j}$ and $a_j$ is the intersection of the left boundary of $C_0^{v_{j-1}}$ and the horizontal line through $v_j$.

We distinguish two cases: (a) $v_j \in C_k^{v_{j-1}}$ and $|cw| > |dw|$, (b) $v_j \in C_{k+1}^{v_{j-1}}$. Let $\beta$ be $\angle a_jv_jv_{j-1}$ and let $\gamma$ be the angle between $v_jv_{j-1}$ and the bisector of $T_{v_{j-1}v_j}$.

*Case (a):* When $v_j \in C_k^{v_{j-1}}$ and $|cw| > |dw|$, the induction hypothesis for $T_{v_{j-1}v_j}$ gives $\delta(v_{j-1}, v_j) \leq |v_{j-1}c| + |cv_j| \cdot c$. We note that $\gamma = \theta - \beta$. Hence Lemma 3 gives that the inequality holds when $c \geq (\cos(\theta-\beta) - \sin\beta)/(\cos(\theta/2-\beta) - \sin(3\theta/2-\beta))$. As this function is decreasing in $\beta$ for $\theta/2 \leq \beta \leq \theta$, it is maximized when $\beta$ equals $\theta/2$. Hence $c$ needs to be at least $(\cos(\theta/2) - \sin(\theta/2))/(1-\sin\theta)$, which can be rewritten to $1/(\cos(\theta/2) - \sin(\theta/2))$.

*Case (b):* When $v_j \in C_{k+1}^{v_{j-1}}$, $v_j$ lies above the bisector of $T_{v_{j-1}v_j}$ and the induction hypothesis for $T_{v_{j-1}v_j}$ gives $\delta(v_{j-1},v_j) \leq |v_jd| + |dv_{j-1}| \cdot c$. We note that $\gamma = \beta$. Hence Lemma 3 gives that the inequality holds when $c \geq (\cos\beta - \sin\beta)/(\cos(\theta/2 - \beta) - \sin(\theta/2 + \beta))$. As this function is decreasing in $\beta$ for $0 \leq \beta \leq \theta/2$, it is maximized when $\beta$ equals 0. Hence $c$ needs to be at least $1/(\cos(\theta/2) - \sin(\theta/2))$. □

Since $\cos\alpha/\cos(\theta/2) + (\cos\alpha \cdot \tan(\theta/2) + \sin\alpha)/(\cos(\theta/2) - \sin(\theta/2))$ is increasing for $\alpha \in [0, \theta/2]$, for $\theta \leq \pi/4$, it is maximized when $\alpha = \theta/2$, and we obtain the following corollary:

**Corollary 2.** *The constrained $\theta_{(4k+4)}$-graph is a $\left(1 + \frac{2 \cdot \sin\left(\frac{\theta}{2}\right)}{\cos\left(\frac{\theta}{2}\right) - \sin\left(\frac{\theta}{2}\right)}\right)$-spanner of $Vis(P, S)$.*

## 7  The Constrained $\theta_{(4k+3)}$-Graph and $\theta_{(4k+5)}$-Graph

In this section we complete the proof of Theorem 2 for the constrained $\theta_{(4k+3)}$-graph and $\theta_{(4k+5)}$-graph.

**Theorem 4.** *Let $u$ and $w$ be two vertices in the plane such that $u$ can see $w$. Let $m$ be the midpoint of the side of $T_{uw}$ opposite $u$ and let $\alpha$ be the unsigned angle between $uw$ and $um$. There exists a path connecting $u$ and $w$ in the constrained $\theta_{(4k+3)}$-graph of length at most*

$$\left(\frac{\cos \alpha}{\cos\left(\frac{\theta}{2}\right)} + \frac{(\cos \alpha \cdot \tan\left(\frac{\theta}{2}\right) + \sin \alpha) \cdot \cos\left(\frac{\theta}{4}\right)}{\cos\left(\frac{\theta}{2}\right) - \sin\left(\frac{3\theta}{4}\right)}\right) \cdot |uw|.$$

*Proof.* We apply Theorem 2 using $\mathbf{c} = \cos(\theta/4)/(\cos(\theta/2) - \sin(3\theta/4))$. The assumptions made in Theorem 2 still apply. It remains to show that for the Type (i) configurations, we have that $\max\{|v_{j-1}c| + |cv_j| \cdot \mathbf{c}, |v_{j-1}d| + |dv_j| \cdot \mathbf{c}\} \le |v_{j-1}a_j| + |a_j v_j| \cdot \mathbf{c}$, where $c$ and $d$ are the upper and lower right corner of $T_{v_{j-1}v_j}$ and $a_j$ is the intersection of the left boundary of $C_0^{v_{j-1}}$ and the horizontal line through $v_j$.

We distinguish two cases: (a) $v_j \in C_k^{v_{j-1}}$ and $|cw| > |dw|$, (b) $v_j \in C_{k+1}^{v_{j-1}}$. Let $\beta$ be $\angle a_j v_j v_{j-1}$ and let $\gamma$ be the angle between $v_j v_{j-1}$ and the bisector of $T_{v_{j-1}v_j}$.

*Case (a):* When $v_j \in C_k^{v_{j-1}}$ and $|cw| > |dw|$, the induction hypothesis for $T_{v_{j-1}v_j}$ gives $\delta(v_{j-1}, v_j) \le |v_{j-1}c| + |cv_j| \cdot \mathbf{c}$. We note that $\gamma = 3\theta/4 - \beta$. Hence Lemma 3 gives that the inequality holds when $\mathbf{c} \ge (\cos(3\theta/4 - \beta) - \sin \beta)/(\cos(\theta/2 - \beta) - \sin(5\theta/4 - \beta))$. As this function is decreasing in $\beta$ for $\theta/4 \le \beta \le 3\theta/4$, it is maximized when $\beta$ equals $\theta/4$. Hence $\mathbf{c}$ needs to be at least $(\cos(\theta/2) - \sin(\theta/4))/(\cos(\theta/4) - \sin \theta)$, which is equal to $\cos(\theta/4)/(\cos(\theta/2) - \sin(3\theta/4))$.

*Case (b):* When $v_j \in C_{k+1}^{v_{j-1}}$, $v_j$ lies above the bisector of $T_{v_{j-1}v_j}$ and the induction hypothesis for $T_{v_{j-1}v_j}$ gives $\delta(v_{j-1}, v_j) \le |v_j d| + |dv_{j-1}| \cdot \mathbf{c}$. We note that $\gamma = \theta/4 + \beta$. Hence Lemma 3 gives that the inequality holds when $\mathbf{c} \ge (\cos(\theta/4 + \beta) - \sin \beta)/(\cos(\theta/2 - \beta) - \sin(3\theta/4 + \beta))$, which is equal to $\cos(\theta/4)/(\cos(\theta/2) - \sin(3\theta/4))$. □

**Theorem 5.** *Let $u$ and $w$ be two vertices in the plane such that $u$ can see $w$. Let $m$ be the midpoint of the side of $T_{uw}$ opposite $u$ and let $\alpha$ be the unsigned angle between $uw$ and $um$. There exists a path connecting $u$ and $w$ in the constrained $\theta_{(4k+5)}$-graph of length at most*

$$\left(\frac{\cos \alpha}{\cos\left(\frac{\theta}{2}\right)} + \frac{(\cos \alpha \cdot \tan\left(\frac{\theta}{2}\right) + \sin \alpha) \cdot \cos\left(\frac{\theta}{4}\right)}{\cos\left(\frac{\theta}{2}\right) - \sin\left(\frac{3\theta}{4}\right)}\right) \cdot |uw|.$$

Due to space constraints the proof of this theorem can be found in [5].

When looking at two vertices $u$ and $w$ in the constrained $\theta_{(4k+3)}$-graph and $\theta_{(4k+5)}$-graph, we notice that when the angle between $uw$ and the bisector of $T_{uw}$ is $\alpha$, the angle between $wu$ and the bisector of $T_{wu}$ is $\theta/2 - \alpha$. Hence the worst case spanning ratio becomes the minimum of the spanning ratio when looking at $T_{uw}$ and the spanning ratio when looking at $T_{wu}$.

**Theorem 6.** *The constrained $\theta_{(4k+3)}$-graph and $\theta_{(4k+5)}$-graph are $\frac{\cos\left(\frac{\theta}{4}\right)}{\cos\left(\frac{\theta}{2}\right)-\sin\left(\frac{3\theta}{4}\right)}$-spanners of $Vis(P,S)$.*

*Proof.* The spanning ratio of the constrained $\theta_{(4k+3)}$-graph and $\theta_{(4k+5)}$-graph is at most:

$$\min\left\{\begin{array}{c}\frac{\cos\alpha}{\cos\left(\frac{\theta}{2}\right)} + \frac{\left(\cos\alpha\cdot\tan\left(\frac{\theta}{2}\right)+\sin\alpha\right)\cdot\cos\left(\frac{\theta}{4}\right)}{\cos\left(\frac{\theta}{2}\right)-\sin\left(\frac{3\theta}{4}\right)}, \\ \frac{\cos\left(\frac{\theta}{2}-\alpha\right)}{\cos\left(\frac{\theta}{2}\right)} + \frac{\left(\cos\left(\frac{\theta}{2}-\alpha\right)\cdot\tan\left(\frac{\theta}{2}\right)+\sin\left(\frac{\theta}{2}-\alpha\right)\right)\cdot\cos\left(\frac{\theta}{4}\right)}{\cos\left(\frac{\theta}{2}\right)-\sin\left(\frac{3\theta}{4}\right)}\end{array}\right\}$$

Since $\cos\alpha/\cos(\theta/2) + (\cos\alpha\cdot\tan(\theta/2)+\sin\alpha)\cdot c$ is increasing for $\alpha \in [0,\theta/2]$, for $\theta \leq 2\pi/7$, the minimum of these two functions is maximized when the two functions are equal, i.e. when $\alpha = \theta/4$. Thus the constrained $\theta_{(4k+3)}$-graph and $\theta_{(4k+5)}$-graph has spanning ratio at most:

$$\frac{\cos\left(\frac{\theta}{4}\right)}{\cos\left(\frac{\theta}{2}\right)} + \frac{\left(\cos\left(\frac{\theta}{4}\right)\cdot\tan\left(\frac{\theta}{2}\right)+\sin\left(\frac{\theta}{4}\right)\right)\cdot\cos\left(\frac{\theta}{4}\right)}{\cos\left(\frac{\theta}{2}\right)-\sin\left(\frac{3\theta}{4}\right)} = \frac{\cos\left(\frac{\theta}{4}\right)\cdot\cos\left(\frac{\theta}{2}\right)}{\cos\left(\frac{\theta}{2}\right)\cdot\left(\cos\left(\frac{\theta}{2}\right)-\sin\left(\frac{3\theta}{4}\right)\right)}$$

□

## References

1. Bose, P., De Carufel, J.-L., Morin, P., van Renssen, A., Verdonschot, S.: Optimal bounds on theta-graphs: More is not always better. In: CCCG, pp. 305–310 (2012)
2. Bose, P., Fagerberg, R., van Renssen, A., Verdonschot, S.: On plane constrained bounded-degree spanners. In: Fernández-Baca, D. (ed.) LATIN 2012. LNCS, vol. 7256, pp. 85–96. Springer, Heidelberg (2012)
3. Bose, P., Keil, J.M.: On the stretch factor of the constrained Delaunay triangulation. In: ISVD, pp. 25–31 (2006)
4. Bose, P., Smid, M.: On plane geometric spanners: A survey and open problems. In: CGTA (2011) (accepted)
5. Bose, P., van Renssen, A.: Upper bounds on the spanning ratio of constrained theta-graphs. CoRR, abs/1401.2127 (2014)
6. Bose, P., van Renssen, A., Verdonschot, S.: On the spanning ratio of theta-graphs. In: Dehne, F., Solis-Oba, R., Sack, J.-R. (eds.) WADS 2013. LNCS, vol. 8037, pp. 182–194. Springer, Heidelberg (2013)
7. Clarkson, K.: Approximation algorithms for shortest path motion planning. In: STOC, pp. 56–65 (1987)
8. Das, G.: The visibility graph contains a bounded-degree spanner. In: CCCG, pp. 70–75 (1997)
9. Keil, J.: Approximating the complete Euclidean graph. In: Karlsson, R., Lingas, A. (eds.) SWAT 1988. LNCS, vol. 318, pp. 208–213. Springer, Heidelberg (1988)
10. Narasimhan, G., Smid, M.: Geometric Spanner Networks. Cambridge University Press (2007)
11. Ruppert, J., Seidel, R.: Approximating the $d$-dimensional complete Euclidean graph. In: CCCG, pp. 207–210 (1991)

# Computing the $L_1$ Geodesic Diameter and Center of a Simple Polygon in Linear Time*

Sang Won Bae[1], Matias Korman[2,3], Yoshio Okamoto[4], and Haitao Wang[5]

[1] Kyonggi University, Suwon, South Korea
swbae@kgu.ac.kr
[2] National Institute of Informatics, Tokyo, Japan
korman@nii.ac.jp
[3] JST, ERATO, Kawarabayashi Large Graph Project
[4] The University of Electro-Communications, Tokyo, Japan
okamotoy@uec.ac.jp
[5] Utah State University, Logan, USA
haitao.wang@usu.edu

**Abstract.** In this paper, we show that the $L_1$ geodesic diameter and center of a simple polygon can be computed in linear time. For the purpose, we focus on revealing basic geometric properties of the $L_1$ geodesic balls, that is, the metric balls with respect to the $L_1$ geodesic distance. More specifically, in this paper we show that any family of $L_1$ geodesic balls in any simple polygon has Helly number two, and the $L_1$ geodesic center consists of midpoints of shortest paths between diametral pairs. These properties are crucial for our linear-time algorithms, and do not hold for the Euclidean case.

## 1 Introduction

Let $P$ be a simple polygon with $n$ vertices in the plane. The *diameter* and *radius* of $P$ with respect to a certain metric $d$ are the most natural and important among several common measures of $P$. The diameter with respect to $d$ is defined to be the maximum distance over all pairs of points in $P$, that is, $\max_{p,q \in P} d(p,q)$, while the radius is defined to be the min-max value $\min_{p \in P} \max_{q \in P} d(p,q)$. Here, the polygon $P$ is considered as a closed and bounded space and thus the diameter and radius of $P$ with respect to $d$ are well defined. A pair of points in $P$ realizing the diameter is called a *diametral pair* and the *center* is defined to be the set of points $c \in P$ such that $\max_{q \in P} d(c,q)$ is equal to the radius.

One of the most natural metrics on a simple polygon $P$ is induced by the length of the Euclidean shortest paths that stay within $P$, namely, the *(Euclidean)*

---

* S.W. Bae was supported by Basic Science Research Program through the National Research Foundation of Korea (NRF) funded by the Ministry of Science, ICT & Future Planning (2013R1A1A1A05006927). Y. Okamoto was supported by Grant-in-Aid for Scientific Research from Ministry of Education, Science and Culture, Japan, and Japan Society for the Promotion of Science (JSPS). H. Wang was supported in part by NSF under Grant CCF-1317143.

*geodesic distance*. The problem of computing the diameter and center of a simple polygon with respect to the geodesic distance has been intensively studied in computational geometry since the early 1980s. The diameter problem was first studied by Chazelle [6], where a $O(n^2)$-time algorithm was given. The running time was afterwards improved to $O(n \log n)$ by Suri [20]. Finally, Hershberger and Suri [10] presented a linear-time algorithm based on a fast matrix search technique. Recently, Bae et al. [3] considered the diameter problem for polygons with holes.

The first algorithm for finding the Euclidean geodesic center was given by Asano and Toussaint [2]. Their algorithm runs in $O(n^4 \log n)$-time, and was afterwards reduced to $O(n \log n)$ by Pollack, Sharir, and Rote [16]. Since then, it has been a longstanding open problem whether the geodesic center can be computed in linear time (as also mentioned later by Mitchell [13]).

Another popular metric with a bit different flavor is the *link distance*, which measures the smallest possible number of links (or turns) of piecewise linear paths. The currently best algorithms that compute the link diameter or center run in $O(n \log n)$ time [7,12,19]. The *rectilinear link distance* measures the minimum number of links when feasible paths in $P$ are constrained to be rectilinear. It is known that the problem with respect to the rectilinear link distance can be solved in linear time by Nilsson and Schuierer [14,15].

In order to tackle the open problem of computing the Euclidean geodesic center, we investigate another natural metric: the $L_1$ metric. To the best of our knowledge, only a special case where the input polygon is rectilinear has been considered in the literature. This result is by Schuierer [17], where he showed how to compute the $L_1$ geodesic diameter and center of a simple rectilinear polygon in time.

This paper aims to provide a clear and complete exposition on the diameter and center of general simple polygons with respect to the $L_1$ geodesic distance. We first focus on revealing basic geometric properties of the geodesic balls (that is, the metric balls with respect to the $L_1$ geodesic distance). Among other results, we show that any family of $L_1$ geodesic balls has Helly number two (see Theorem 1). This is a critical property that does not hold for the Euclidean geodesic distance, and thus we identify that the main difficulty of the open problem lies there.

We then show that the method of Hershberger and Suri [10] for computing the Euclidean diameter extends to $L_1$ metrics, and that the running time is preserved. However, the algorithms for computing the Euclidean center do not easily extend to rectilinear metrics. Indeed, even though the approach of Pollack et al. [16] can be adapted for the $L_1$ metric, the running time will increase to $O(n \log n)$. On the other hand, the algorithm of Schuierer [17] for the rectilinear simple polygons heavily exploits properties derived from rectilinearity. Thus, its extension to general simple polygons is not straightforward either.

In this paper we use a different approach: using the previously mentioned Helly-type theorem, we show that the $L_1$ geodesic center coincides with the intersection of a finite number of geodesic balls. Afterwards we show how to

**Table 1.** Summary of currently best results on computing the diameter and center of a simple polygon $P$ with respect to various metrics on $P$

|  | Metric | Restriction on $P$ | Diameter | | Center | |
|---|---|---|---|---|---|---|
| Geodesic | Euclidean | simple | $O(n)$ | [10] | $O(n \log n)$ | [16] |
|  | $L_1$ | rect. simple | $O(n)$ | [17] | $O(n)$ | [17] |
|  |  | simple | $O(n)$ | [Thm. 3] | $O(n)$ | [Thm. 4] |
| Link | regular | simple | $O(n \log n)$ | [19] | $O(n \log n)$ | [7,12] |
|  | rectilinear | rect. simple | $O(n)$ | [14] | $O(n)$ | [15] |

compute their intersection in linear time. Table 1 summarizes the currently best results on computing the diameter and center of a simple polygon with respect to the most common metrics, including our new results.

Due to page limit, several proofs are omitted. They can be found in the extended version of this paper [4].

## 2 Preliminaries

For any subset $A \subset \mathbb{R}^2$, we denote by $\partial A$ and $\text{int} A$ the boundary and the interior of $A$, respectively. For $p, q \in \mathbb{R}^2$, denote by $\overline{pq}$ the line segment with endpoints $p$ and $q$. For any path $\pi$ in $\mathbb{R}^2$, let $|\pi|$ be the length of $\pi$ under the $L_1$ metric, or simply the $L_1$ *length*. Note that $|\overline{pq}|$ equals the $L_1$ distance between $p$ and $q$.

The following is a basic observation on the $L_1$ length of paths in $\mathbb{R}^2$. A path is called *monotone* if any vertical or horizontal line intersects it in at most one connected component.

**Fact 1.** *For any monotone path $\pi$ between $p, q \in \mathbb{R}^2$, it holds that $|\pi| = |\overline{pq}|$.*

Let $P$ be a simple polygon with $n$ vertices. We regard $P$ as a compact set in $\mathbb{R}^2$, so its boundary $\partial P$ is contained in $P$. An $L_1$ *shortest path* between $p$ and $q$ is a path joining $p$ and $q$ that lies in $P$ and minimizes its $L_1$ length. The $L_1$ *geodesic distance* $d(p, q)$ is the $L_1$ length of an $L_1$ shortest path between $p$ and $q$. We are interested in two quantities: the $L_1$ *geodesic diameter* $\text{diam}(P)$ and *radius* $\text{rad}(P)$ of $P$, defined to be $\text{diam}(P) := \max_{p,q \in P} d(p, q)$ and $\text{rad}(P) := \min_{p \in P} \max_{q \in P} d(p, q)$. Any pair of points $p, q \in P$ such that $d(p, q) = \text{diam}(P)$ is called a *diametral pair*. The $L_1$ *geodesic center* $\text{cen}(P)$ of $P$ is $\text{cen}(P) := \{c \in P \mid \max_{q \in P} d(c, q) = \text{rad}(P)\}$.

Analogously, a path lying in $P$ minimizing its *Euclidean* length is called the *Euclidean shortest path*. It is well known that there is always a unique Euclidean shortest path between any two points in a simple polygon [8]. We let $\pi_2(p, q)$ be the unique Euclidean shortest path from $p \in P$ to $q \in P$. The following states a crucial relation between Euclidean and $L_1$ shortest paths in a simple polygon.

**Fact 2 (Hershberger and Snoeyink [9]).** *For any two points $p, q \in P$, the Euclidean shortest path $\pi_2(p, q)$ is also an $L_1$ shortest path between $p$ and $q$.*

Notice that this does not imply the coincidence between the Euclidean and the $L_1$ geodesic diameters or centers, as the lengths of paths are measured differently.

Nonetheless, Fact 2 enables us to exploit several structures for Euclidean shortest paths such as the shortest path map.

A *shortest path map* for a source point $s \in P$ is a subdivision of $P$ into regions according to the combinatorial structure of shortest paths from $s$. For the Euclidean shortest paths, Guibas et al. [8] showed that the shortest path map $SPM(s)$ can be computed in $O(n)$ time. Once we have $SPM(s)$, the Euclidean geodesic distance from $s$ to any query point $q \in P$ can be answered in $O(\log n)$ time, and the actual path $\pi_2(s,q)$ in additional time proportional to the complexity of $\pi_2(s,q)$. Fact 2 implies that the map $SPM(s)$ also plays a role as a shortest path map for the $L_1$ geodesic distance so that a query $q \in P$ can be processed in the same time bound to evaluate the $L_1$ geodesic distance $d(s,q)$ or to obtain the shortest path $\pi_2(s,q)$.

Throughout the paper, unless otherwise stated, $P$ refers to a simple polygon, a shortest path and the geodesic distance always refer to an $L_1$ shortest path and the $L_1$ geodesic distance $d$, and the geodesic diameter/center is always assumed to be with respect to the $L_1$ geodesic distance $d$.

## 3 The $L_1$ Geodesic Balls

Geodesic balls (or geodesic disks) are metric balls under the geodesic distance $d$. More precisely, the $L_1$ *(closed) geodesic ball* centered at point $s \in P$ with radius $r \in \mathbb{R}$, denoted by $\mathbf{B}_s(r)$, is the set of points $x \in P$ such that $d(s,x) \le r$. Note that if $r < 0$, it holds that $\mathbf{B}_s(r) = \emptyset$. In this section, we reveal several geometric properties of the geodesic balls $\mathbf{B}_s(r)$, which build a basis for our further discussion.

### 3.1 P-convex Sets

A set $A \subseteq P$ is *P-convex* if for any $p, q \in A$, the Euclidean shortest path $\pi_2(p, q)$ is a subset of $A$. The $P$-convex sets are also known as the *geodesically convex* sets in the literature [20]. Pollack et al. [16] achieved their $O(n \log n)$-time algorithm computing the Euclidean geodesic center based on the $P$-convexity of Euclidean geodesic balls. A set $A$ is *path-connected* if an only if, for any $x, y \in A$ there exists a path $\pi$ connecting them such that $\pi \subseteq A$. With this definition we can introduce an equivalent condition of $P$-convexity.

**Lemma 1.** *For any subset $A \subseteq P$ of $P$, the following are equivalent.*

*(i) $A$ is $P$-convex.*
*(ii) $A$ is path-connected and for any line segment $\ell \subset P$, $A \cap \ell$ is connected.*

We are interested in the boundary of a $P$-convex set. Let $A \subseteq P$ be a $P$-convex set. Consider any convex subset $Q \subseteq P$. Since $\pi_2(p,q) = \overline{pq}$ for any $p, q \in Q$, the intersection $A \cap Q$ is also a convex set due to the $P$-convexity of $A$. Based on this observation, we show the following lemma.

**Lemma 2.** *Let $A \subseteq P$ be a closed $P$-convex set. Then, any connected component $C$ of $\partial A \cap \mathrm{int} P$ is a convex curve, being either open or closed.*

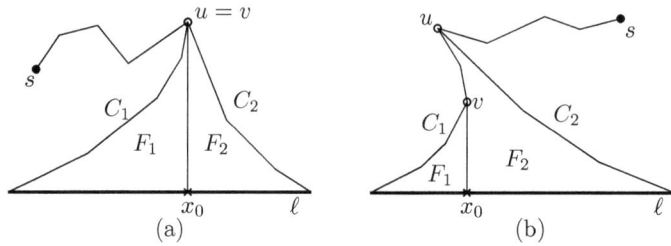

**Fig. 1.** Illustration of the proof of Lemma 3: (a) When both chains $C_1$ and $C_2$ of the funnel are monotone. (b) When $C_1$ is not monotone.

Note that if a connected component $C$ of $\partial A \cap \text{int} P$ is not a closed curve, then $C$ is an open curve excluding its endpoints, which lie on $\partial P$. This implies that the curve $C$ divides $P$ into two connected components such that $\text{int} A$ lies on one side of $C$, regardless of whether $C$ is open or closed.

### 3.2 Geometric Properties of $L_1$ Geodesic Balls

In the following, we show several geometric properties of geodesic balls $\mathbf{B}_s(r)$ which follow from the $P$-convexity of $\mathbf{B}_s(r)$. Note that, to the best of our knowledge, most of these properties of $\mathbf{B}_s(r)$ have not been discussed before in the literature.

We start with a simple observation. By Fact 2, $\pi_2(s,p)$ is an $L_1$ shortest path from $s$ to $p \in P$. Since $\pi_2(s,p)$ makes turns only at vertices of $P$, the ball is equal to the union of some $L_1$ balls centered at the vertices of $P$. More precisely, $\mathbf{B}_s(r) = \bigcup_{v \in V \cup \{s\}} \mathbf{B}_v(r - d(s,v))$, where $V$ denotes the set of vertices of $P$. This immediately implies the following observation.

**Observation 1.** *For any $s \in P$ and $r > 0$, the geodesic ball $\mathbf{B}_s(r)$ is a simple polygon in $P$ and each side of $\mathbf{B}_s(r)$ either lies on $\partial P$ or has slope $1$ or $-1$.*

**Lemma 3.** *Given a point $s \in P$ and a horizontal or vertical line segment $\ell \subset P$, the function $f(x) = d(s,x)$ over $x \in \ell$ is convex.*

*Proof.* Without loss of generality, we assume that $\ell$ is horizontal. The case where $\ell$ is vertical can be handled in a symmetric way. Consider the union of all Euclidean shortest paths $\pi_2(s,x)$ from $s$ to $x$ over $x \in \ell$, which forms a *funnel* $F$ with apex $u$ and base $\ell$ plus $\pi_2(s,u)$. The funnel $F$ consists of two concave chains $C_1$ and $C_2$ through vertices of $P$ and the endpoints of $\ell$ so that $C_1$ connects the apex $u$ and the left endpoint of $\ell$ and $C_2$ connects $u$ and the right endpoint of $\ell$. See Fig. 1. Note that the apex $u$ is also a vertex of $P$ unless $u = s$.

Each of the two concave chains $C_1$ and $C_2$ is either monotone or not. Recall that a path is called monotone if and only if any vertical or horizontal line intersects it at most once. Observe that at least one of them must be monotone, since the apex $u$ must see a point of $\ell$. Without loss of generality, we assume that $C_2$ is monotone in either way. Let $v$ be a vertex of $F$ defined as follows: $v = u$ if

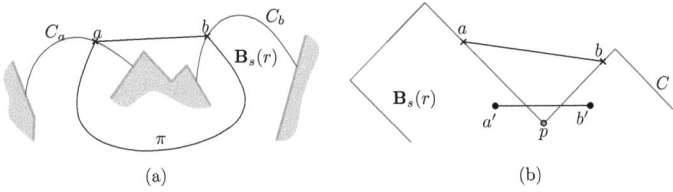

**Fig. 2.** Illustration of the proof of Lemma 4. The region shaded by gray depicts $\mathbb{R}^2 \setminus P$.

both chains are monotone; if $C_1$ is not monotone, then $v$ is the rightmost vertex of $C_1$ so that $v$ cuts $C_1$ into two monotone concave chains. Let $x_0 \in \ell$ be the perpendicular foot point of $v$ on $\ell$.

We claim that $x_0$ minimizes $f(x) = d(s,x)$ over $x \in \ell$ and moreover that $f(x) = f(x_0) + |\overline{xx_0}|$. This implies the lemma. First observe that $f(x_0) = |\pi_2(s,u)| + |\pi_2(u,v)| + |\overline{vx_0}| = |\pi_2(s,u)| + |\overline{ux_0}|$ since $\pi_2(s,u)$ is an $L_1$ shortest path by Fact 2 and the path $\pi_2(u,v) \cup \overline{vx_0}$ is monotone, whose length is equal to that of $\overline{ux_0}$ by Fact 1. Now, consider the partition of $F$ into $F_1$ and $F_2$ cut by segment $\overline{vx_0}$, such that $F_2$ contains $C_2$ and $F_1$ contains the subchain of $C_1$ after $v$. Also, let $\ell_1$ and $\ell_2$ be the corresponding partition of $\ell$ with $\ell_i \subset F_i$ for $i = 1, 2$. By Fact 1, for any point $x \in \ell_1$, there exists a monotone path from $v$ to $x$, so $d(v,x) = |\overline{vx}|$; for any $x \in \ell_2$, there exists a monotone path from $u$ to $x$, so $d(u,x) = |\overline{ux}|$. Since $\pi_2(s,x)$ for $x \in \ell_1$ always passes through $v$, $f(x) = |\pi_2(s,v)| + d(v,x) = |\pi_2(s,v)| + |\overline{vx}|$. Similarly, for $x \in \ell_2$, we have $f(x) = |\pi_2(s,u)| + |\overline{ux}|$. Moreover, since $\pi_2(u,v)$ is also monotone, we have $f(x) = |\pi_2(s,v)| + |\overline{vx}|$ for any $x \in \ell$. Since $\ell$ is horizontal, $|\overline{vx}| = |\overline{vx_0}| + |\overline{x_0 x}|$ holds. Therefore, we have $f(x) = f(x_0) + |\overline{xx_0}|$. This proves our claim. □

We are ready to prove the $P$-convexity of any $L_1$ geodesic ball.

**Lemma 4.** *For any point $s \in P$ and any real $r \in \mathbb{R}$, the $L_1$ geodesic ball $\mathbf{B}_s(r)$ is $P$-convex.*

*Proof.* The case where $r \leq 0$ is trivial, so assume $r > 0$. Suppose that $\mathbf{B}_s(r)$ is not $P$-convex. Since $\mathbf{B}_s(r)$ is a simple polygon (Observation 1), any line segment in $P$ intersects $\mathbf{B}_s(r)$ in finitely many connected components. Thus, by Lemma 1, there exists a line segment $\ell \subset P$ such that $\ell$ crosses $\partial \mathbf{B}_s(r) \cap \text{int} P$ exactly twice. Let $a, b \in \ell$ be the two intersection points such that $\overline{ab} \setminus \{a, b\}$ lies in $P \setminus \mathbf{B}_s(r)$.

We then observe that $a$ and $b$ belong to a common connected component $C$ of $\partial \mathbf{B}_s(r) \cap \text{int} P$. Suppose for a contradiction that $a$ and $b$ belong to different components $C_a$ and $C_b$, respectively. See Fig. 2(a). Since $\mathbf{B}_s(r)$ is path-connected and closed, there exists a path $\pi$ between $a$ and $b$ such that $\pi \subset \mathbf{B}_s(r)$. Consider the simple closed curve $L := \overline{ab} \cup \pi$. Since $C_a \neq C_b$ and $a \in C_a$, $L$ separates the two endpoints of $C_a$, that is, an endpoint of $C_a$ lies in the region bounded by $L$. However, this is impossible since $P$ is simple, a contradiction. Hence, both $a$ and $b$ lie in a common connected component $C$ of $\partial \mathbf{B}_s(r) \cap \text{int} P$.

By Observation 1, $C$ is a polygonal curve consisting of line segments with slope 1 or $-1$. Since $a, b \in C$ and $\overline{ab}$ is not contained in $\mathbf{B}_s(r)$, $C$ has a reflex corner $p$ incident to two line segments whose slopes are 1 and $-1$, respectively.

See Fig. 2(b) for an illustration. Then, we can find a horizontal or vertical line segment $\overline{a'b'}$ sufficiently close to $p$ such that $a', b' \in \mathbf{B}_s(r)$ and $\mathbf{B}_s(r) \cap \overline{a'b'}$ consists of two connected components. Take any point $x \in \overline{a'b'} \setminus \mathbf{B}_s(r)$. Since $a', b' \in \mathbf{B}_s(r)$ but $x \notin \mathbf{B}_s(r)$, we have a strict inequality $d(s, x) > r \geq d(s, a')$ and $d(s, x) > r \geq d(s, b')$, a contradiction to Lemma 3. □

The $P$-convexity of the geodesic balls, together with Lemma 2 and Observation 1, immediately implies the following corollary.

**Corollary 1.** *For $s \in P$ and $r > 0$, each connected component $C$ of $\partial \mathbf{B}_s(r) \cap \mathrm{int} P$ is a convex polygonal curve consisting of line segments of slope $1$ or $-1$.*

The following corollary can also be easily derived from Lemma 4.

**Corollary 2.** *For any point $s \in P$ and any $r > 0$, the geodesic ball $\mathbf{B}_s(r)$ intersects any line segment in $P$ in a connected subset.*

A real-valued function $f$ is called *quasiconvex* if its sublevel set $\{x \mid f(x) \leq a\}$ for any $a \in \mathbb{R}$ is convex. Corollary 2 implies the following.

**Corollary 3.** *Given a point $s \in P$ and a line segment $\ell \subset P$, the function $f(x) = d(s, x)$ over $x \in \ell$ is quasiconvex.*

Indeed, the geodesic distance function $d(s, x)$ over $x \in \ell$ is not only quasiconvex but convex; this can be shown by a more careful geometric analysis. Nonetheless, the quasiconvexity will be sufficient for our overall purpose.

### 3.3 Helly-Type Theorem for Geodesic Balls

Here, we discuss the intersection of a family of $L_1$ geodesic balls, and show that the $L_1$ geodesic balls have Helly number two. More precisely, we claim the following theorem.

**Theorem 1.** *Let $\mathcal{B}$ be a family of closed $L_1$ geodesic balls. If the intersection of every two members of $\mathcal{B}$ is nonempty, then $\bigcap \{B \mid B \in \mathcal{B}\} \neq \emptyset$.*

In the following, we prove Theorem 1. For the purpose, we make use of a Helly-type theorem on simple polygons proven by Breen [5].

**Theorem 2 (Breen [5]).** *Let $\mathcal{P}$ be a family of simple polygons in the plane. If every three (not necessarily distinct) members of $\mathcal{P}$ have a simply connected union and every two members of $\mathcal{P}$ have a nonempty intersection, then $\bigcap \{P \mid P \in \mathcal{P}\} \neq \emptyset$.*

Thus, we are done by showing that the union of two or three balls is simply connected, provided that any two of them have a nonempty intersection. This can be done based on the above discussion on the geodesic balls with Lemma 2 and Corollary 1.

**Lemma 5.** *Let $B_1, B_2, B_3$ be any three closed $L_1$ geodesic balls such that every two of them have a nonempty intersection. Then, the union $B_1 \cup B_2 \cup B_3$ is simply connected.*

*Proof.* By the assumption, the union $B_1 \cup B_2 \cup B_3$ is obviously connected. Assume to the contrary that the union $B_1 \cup B_2 \cup B_3$ has a hole $H$, that is, the boundary of the union has more than one connected component and one of them is $\partial H$. The hole $H$ is also a simple polygon whose boundary $\partial H$ consists of portions of $\partial B_i \cap \text{int} P$ and $\partial P$ for $i = 1, 2, 3$. We consider the connected components $C_1, C_2, \ldots, C_m$ of $\partial B_i \cap \text{int} P$ that appear on $\partial H$.

First, we observe that $m \leq 3$. Otherwise, if $m > 3$, then there are two components $C_1, C_2 \subset \partial B_1 \cap \text{int} P$ after reordering the indices, without loss of generality. Since $H \subseteq P$ by the simplicity of $P$, there exists a path $\pi$ between two points $p_1 \in C_1$ and $p_2 \in C_2$ such that $\pi \setminus \{p_1, p_2\} \subset P \setminus B_1$. Lemma 2, however, implies that $C_1$ partitions $P$ into two components and $B_1$ lies in one side of $C_1$, which implies the nonexistence of such a path $\pi$, a contradiction.

The above argument also implies that only a single component of $\partial B_i \cap \text{int} P$ appears on $\partial H$ for each $i = 1, 2, 3$. Let $C_i$ be the component of $\partial B_i \cap \text{int} P$ that appears on $\partial H$, if exists. By Corollary 1, each $C_i$ is a convex polygonal curve, consisting of line segments of slope 1 or $-1$. This implies that each $C_i$ appears on $\partial H$ in a connected set. Also, we have $m \geq 2$ since a single convex chain cannot make such a hole $H$.

Next, we claim that $\partial H$ does not contain any portion of $\partial P$, or equivalently $\partial H \subset \text{int} P$. Suppose to the contrary that $\partial H \cap \partial P \neq \emptyset$. Let $C \subset \partial H \cap \partial P$ be a connected portion. Without loss of generality, we assume that $C$ is adjacent to $C_1$ and $C_2$ so that an endpoint of $C_1$ and an endpoint of $C_2$ lie on $C$. Since $B_1 \cap B_2 \neq \emptyset$, we observe that $C_1 \cap C_2 \neq \emptyset$. This shows that $H$ is not a hole of the union $B_1 \cup B_2 \cup B_3$ by the simplicity of $P$, wherever the third chain $C_3$ is located.

Therefore, $H$ is bounded only by $C_1, C_2, C_3$ that are polygonal convex chains formed by line segments of slope 1 or $-1$ as observed above. A geometric analysis concludes that $H$ must be degenerate to a line segment. Since the balls $B_i$ are closed sets, this is impossible, a contradiction. □

## 4 The $L_1$ Geodesic Diameter

In this section, we show that the $L_1$ geodesic diameter of $P$, $\text{diam}(P)$, and a diametral pair can be computed in linear time by extending the approach of Suri [18] and Hershberger and Suri [10] to the $L_1$ case. For any point $s \in P$, let $\phi(s)$ be the maximum geodesic distance from $s$ to any other point in $P$, that is, $\phi(s) = \max_{q \in P} d(s, q)$. A point $q \in P$ such that $d(s, q) = \phi(s)$ is called a *farthest neighbor of $s$*. Obviously, $\text{diam}(P) = \max_{s \in P} \phi(s)$ and $\text{rad}(P) = \min_{s \in P} \phi(s)$. The following lemma is a key observation for our purpose.

**Lemma 6.** *For any $s \in P$, all farthest neighbors of $s$ lie on the boundary $\partial P$ of $P$, and at least one of them is a vertex of $P$.*

**Corollary 4.** *There exist two vertices $v_1$ and $v_2$ of $P$ such that $d(v_1, v_2) = \text{diam}(P)$, that is, $(v_1, v_2)$ is a diametral pair.*

Thus, the problem of computing diam($P$) is solved by finding the farthest vertex-pair. Let $v_1, \ldots, v_n$ be the vertices of $P$ ordered counterclockwise along $\partial P$. Let $v_a$ and $v_b$ be vertices of $P$ such that $v_a$ is a farthest neighbor of $v_1$ and $v_b$ is a farthest neighbor of $v_a$. The existence of $v_a$ and $v_b$ is guaranteed by Lemma 6. We assume that $a < b$; otherwise, we take the mirror image of $P$ for the following discussion. The three vertices $v_1, v_a, v_b$ divide $\partial P$ into three chains: $U_1 = (v_2, \ldots, v_{a-1})$, $U_2 = (v_{a+1}, \ldots, v_{b-1})$, and $U_3 = (v_{b+1}, \ldots, v_n)$. Let $W_1, W_2, W_3$ be the chains complimentary to $U_1, U_2, U_3$, respectively, that is, $W_1 = (v_a, \ldots, v_n, v_1)$, $W_2 = (v_b, \ldots, v_n, v_1, \ldots, v_a)$, and $W_3 = (v_1, \ldots, v_b)$. We then observe the following, which we prove with Lemma 6.

**Lemma 7.** *For any $i = 1, 2, 3$ and $u \in U_i$, there is a vertex $w \in W_i$ that is a farthest neighbor of $u$.*

Lemma 7 implies that computing a farthest vertex from every vertex of $P$ can be done by handling three pairs $(U_i, W_i)$ of two disjoint chains that partition the vertices of $P$. Note that an analogy of Lemma 7 with respect to the Euclidean geodesic distance was first observed by Suri [20, Lemma 8], and used for computing the Euclidean geodesic diameter [10, 20].

This motivates the *restricted farthest neighbor* problem: Given two disjoint chains of vertices of $P$, $U = (u_1, \ldots, u_p)$ and $W = (w_1, \ldots, w_m)$ that together partition the vertices of $P$, where the vertices $u_1, \ldots, u_p, w_1, \ldots, w_m$ are ordered counterclockwise and $p + m = n$, find a farthest vertex on $W$ from each $u \in U$. With respect to the Euclidean geodesic distance, Suri [20] presented an $O(n \log n)$-time algorithm for the problem, and later Hershberger and Suri [10] improved it to $O(n)$ time based on the matrix searching technique by Aggarwal et al. [1]. In the following, we show with Fact 2 that the method of Hershberger and Suri [10] can be applied to solve the problem with respect to the $L_1$ geodesic distance $d$.

**Lemma 8.** *Let $U$ and $W$ be two disjoint chains of vertices of $P$ that together partition the vertices of $P$. One can compute in $O(n)$ time a farthest vertex over $w \in W$ for every $u \in U$ with respect to the $L_1$ geodesic distance $d$.*

We are now ready to conclude this section with a linear-time algorithm. We first find $v_a$ and $v_b$ such that $v_a$ is a farthest neighbor of $v_1$ and $v_b$ is a farthest neighbor of $v_a$. This can be done in $O(n)$ time by computing the shortest path maps $SPM(v_1)$ and then $SPM(v_a)$ due to Guibas et al. [8] and Fact 2. We then have the three chains $U_1, U_2, U_3$ and their compliments $W_1, W_2, W_3$. Next, we apply Lemma 8 to solve the three instances $(U_i, W_i)$ for $i = 1, 2, 3$ of the restricted farthest neighbor problem, resulting in a farthest neighbor of each vertex of $P$ by Lemma 7. Corollary 4 guarantees that the maximum over the $n$ pairs of vertices is a diametral pair of $P$. All the effort in the above algorithm is bounded by $O(n)$ time. We finally conclude the main result of this section.

**Theorem 3.** *The $L_1$ geodesic diameter of a simple polygon with $n$ vertices, along with a pair of vertices that is diametral, can be computed in $O(n)$ time.*

## 5  The $L_1$ Geodesic Center

In this section, we study the $L_1$ geodesic radius $\mathrm{rad}(P)$ and center $\mathrm{cen}(P)$ of a simple polygon $P$, and present a simple algorithm that computes the center $\mathrm{cen}(P)$ in linear time.

Consider the geodesic balls $\mathbf{B}_p(r)$ centered at all points $p \in P$ with radius $r$, and imagine their intersection as $r$ grows continuously. By definition, the the first nonempty intersection happens when $r = \mathrm{rad}(P)$. Equivalently, by Theorem 1, $r = \mathrm{rad}(P)$ is the smallest radius such that $\mathbf{B}_p(r) \cap \mathbf{B}_q(r) \neq \emptyset$ for any $p, q \in P$. From the triangular inequality we have $\mathrm{rad}(P) \geq \mathrm{diam}(P)/2$

**Lemma 9.** *For any simple polygon $P$, it holds that* $\mathrm{rad}(P) = \mathrm{diam}(P)/2$.

We note that Schuierer [17] claimed Lemma 9 but no proof was given. In this paper, we instead provide proofs based on the Helly-type theorem for the $L_1$ geodesic balls (Theorem 1). It is worth mentioning that Theorem 2 was also used to prove a similar relation between the diameter and center with respect to the rectilinear link distance [11]. To explicitly compute the center $\mathrm{cen}(P)$, we first need a technical lemma.

**Lemma 10.** *Let $a, b \in P$ be any two points with $\overline{ab} \subset P$. Then, for any $r > 0$, it holds that $\mathbf{B}_a(r) \cap \mathbf{B}_b(r) = \bigcap_{s \in \overline{ab}} \mathbf{B}_s(r)$.*

Lemma 10 together with Lemma 6 implies that $\mathrm{cen}(P) = \bigcap_{v \in V} \mathbf{B}_v(\mathrm{rad}(P))$, where $V$ denotes the set of vertices of $P$. Moreover, this observation together with Lemma 9 implies that the $L_1$ geodesic center $\mathrm{cen}(P)$ is a line segment.

### 5.1  Computing the Center in Linear Time

Now, we describe our algorithm for computing $\mathrm{cen}(P)$ in linear time. Compute the diameter $\mathrm{diam}(P)$ and a diametral pair of vertices $(v_1, v_2)$ in $O(n)$ time by Theorem 3. Then, we know that $\mathrm{rad}(P) = \mathrm{diam}(P)/2$ by Lemma 9. Compute the intersection of two geodesic balls $\mathbf{B}_{v_1}(\mathrm{rad}(P))$ and $\mathbf{B}_{v_2}(\mathrm{rad}(P))$, which is a line segment of slope 1 or $-1$. Extend the line segment obtained above to a diagonal $\ell = \overline{ab}$, where $a, b \in \partial P$. The above two steps can be performed in linear time. In particular, for the first step, $\mathbf{B}_{v_1}(\mathrm{rad}(P))$ and $\mathbf{B}_{v_2}(\mathrm{rad}(P))$ can be found by computing the shortest path maps $SPM(v_1)$ and $SPM(v_2)$ and traversing the cells of the maps, and computing their intersection is done by a local search at the midpoint of $\pi_2(v_1, v_2)$ since it is guaranteed that $\mathbf{B}_{v_1}(\mathrm{rad}(P)) \cap \mathbf{B}_{v_2}(\mathrm{rad}(P))$ is a line segment by Corollary 1.

Since $\mathrm{cen}(P) = \bigcap_{v \in V} \mathbf{B}_v(\mathrm{rad}(P))$ and $\mathbf{B}_{v_1}(\mathrm{rad}(P)) \cap \mathbf{B}_{v_2}(\mathrm{rad}(P)) \subseteq \ell$, we conclude that $\mathrm{cen}(P) \subseteq \ell$. The last task is thus to identify $\mathrm{cen}(P)$ from $\ell$. Here, we present a simple method based on further geometric observations. Recall that for any $s \in P$ and any line segment $l \subset P$, the geodesic distance function $d(s, x)$ over $x \in l$ is quasiconvex as stated in Corollary 3. A more careful analysis based on Fact 1 gives us the following.

**Lemma 11.** *Given a point $s \in P$ and a line segment $\overline{ab} \subset P$ with slope 1 or $-1$, let $f(x) = d(s,x)$ be the geodesic distance from $s$ to $x$ over $x \in \overline{ab}$. Then, there are two points $x_1, x_2 \in \overline{ab}$ with $|\overline{ax_1}| \leq |\overline{ax_2}|$ such that we have*

$$f(x) = \begin{cases} d(s,a) - |\overline{ax}| & \text{if } x \in \overline{ax_1} \\ d(s,x_1) = d(s,x_2) & \text{if } x \in \overline{x_1 x_2} \\ d(s,b) - |\overline{bx}| & \text{if } x \in \overline{x_2 b}. \end{cases}$$

*In particular, the function $f$ attains its minimum at any point $x \in \overline{x_1 x_2}$.*

For any vertex $v \in V$ of $P$, let $\ell_v \subseteq \ell$ be the intersection $\mathbf{B}_v(\mathrm{rad}(P)) \cap \ell$. Since $\mathrm{cen}(P) = \bigcap_{v \in V} \mathbf{B}_v(\mathrm{rad}(P))$ and $\mathrm{cen}(P) \subseteq \ell$, it holds that $\mathrm{cen}(P) = \bigcap_{v \in V} \ell_v$.

**Lemma 12.** *For any vertex $v$ of $P$, $\ell_v$ can be computed in $O(1)$ time, provided that $d(v,a)$ and $d(v,b)$ have been evaluated.*

Thus, our last task can be completed as follows: Compute the two shortest path maps $SPM(a)$ and $SPM(b)$ with sources $a$ and $b$, respectively, by running the algorithm by Guibas et al. [8]. This evaluates $d(v,a)$ and $d(v,b)$ for all vertices $v$ of $P$ in linear time. Next, we compute $\ell_v$ for all vertices $v$ of $P$ by Lemma 12 and find their common intersection, which finally identifies $\mathrm{cen}(P)$. All the effort to obtain $\mathrm{cen}(P)$ is bounded by $O(n)$.

**Theorem 4.** *The $L_1$ geodesic radius and center of a simple polygon with $n$ vertices can be computed in $O(n)$ time.*

## 6 Concluding Remarks

In this paper, we presented a comprehensive study on the $L_1$ geodesic diameter and center of simple polygons, resulting in optimal linear-time algorithms. Our approach relies on observations about the $L_1$ geodesic balls, in particular, the $P$-convexity (Lemma 4) and the Helly-type theorem (Theorem 1). These are shown to be key tools to show structural properties of the diameter and center.

One would be interested in extending this framework to polygons with holes, namely, *polygonal domains*. However, it is not difficult to see that only few of the observations we made extend for general polygonal domains. First and foremost, an $L_1$ (also, Euclidean) geodesic ball may not be $P$-convex when $P$ has a hole. In addition, the Helly number of $L_1$ geodesic balls in a polygonal domain is strictly larger than two: one can easily construct three balls around a hole such that every two of them intersect but the three have no common point. Also, Lemma 6 (the existence of a farthest neighbor that is a vertex) does not always hold in polygonal domains. Bae et al. [3] have exhibited several examples of polygonal domains in which a farthest neighbor with respect to the Euclidean geodesic distance is a unique point in the interior. This construction can be easily extended to the $L_1$ geodesic distance.

# References

1. Aggarwal, A., Klawe, M., Moran, S., Shor, P., Wilbur, R.: Geometric applications of a matrix-searching algorithm. Algorithmica 2, 195–208 (1987)
2. Asano, T., Toussaint, G.: Computing the geodesic center of a simple polygon. Technical Report SOCS-85.32. McGill University (1985)
3. Bae, S.W., Korman, M., Okamoto, Y.: The geodesic diameter of polygonal domains. Discrete Comput. Geom. 50(2), 306–329 (2013)
4. Bae, S.W., Korman, M., Okamoto, Y., Wang, H.: Computing the $L_1$ geodesic diameter and center of a simple polygon in linear time. ArXiv e-prints (2013), arXiv:1312.3711
5. Breen, M.: A Helly-type theorem for simple polygons. Geometriae Dedicata 60, 283–288 (1996)
6. Chazelle, B.: A theorem on polygon cutting with applications. In: Proc. 23rd Annu. Sympos. Found. Comput. Sci. (FOCS 1982), pp. 339–349 (1982)
7. Djidjev, H., Lingas, A., Sack, J.R.: An $O(n \log n)$ algorithm for computing the link center of a simple polygon. Discrete Comput. Geom. 8, 131–152 (1992)
8. Guibas, L.J., Hershberger, J., Leven, D., Sharir, M., Tarjan, R.: Linear time algorithms for visibility and shortest path problems inside triangulated simple polygons. Algorithmica 2, 209–233 (1987)
9. Hershberger, J., Snoeyink, J.: Computing minimum length paths of a given homotopy class. Comput. Geom.: Theory and Appl. 4(2), 63–97 (1994)
10. Hershberger, J., Suri, S.: Matrix searching with the shortest path metric. SIAM J. Comput. 26(6), 1612–1634 (1997)
11. Katz, M.J., Morgenstern, G.: Settling the bound on the rectilinear link radius of a simple rectilinear polygon. Inform. Proc. Lett. 111, 103–106 (2011)
12. Ke, Y.: An efficient algorithm for link-distance problems. In: Proc. 5th Annu. Sympos. Comput. Geom. (SoCG 1989), pp. 69–78 (1989)
13. Mitchell, J.S.B.: Shortest paths and networks. In: Handbook of Discrete and Computational Geometry, ch. 27, 2nd edn., pp. 607–641. CRC Press, Inc. (2004)
14. Nilsson, B.J., Schuierer, S.: Computing the rectilinear link diameter of a polygon. In: Bieri, H., Noltemeier, H. (eds.) CG-WS 1991. LNCS, vol. 553, pp. 203–215. Springer, Heidelberg (1991)
15. Nilsson, B.J., Schuierer, S.: An optimal algorithm for the rectilinear link center of a rectilinear polygon. Comput. Geom.: Theory and Appl. 6, 169–194 (1996)
16. Pollack, R., Sharir, M., Rote, G.: Computing the geodesic center of a simple polygon. Discrete Comput. Geom. 4(6), 611–626 (1989)
17. Schuierer, S.: Computing the $L_1$-diameter and center of a simple rectilinear polygon. In: Proc. Int. Conf. on Computing and Information (ICCI 1994), pp. 214–229 (1994)
18. Suri, S.: The all-geodesic-furthest neighbors problem for simple polygons. In: Proc. 3rd Annu. Sympos. Comput. Geom. (SoCG 1987), pp. 64–75 (1987)
19. Suri, S.: Minimum Link Paths in Polygons and Related Problems. Ph.D. thesis. Johns Hopkins Univ. (1987)
20. Suri, S.: Computing geodesic furthest neighbors in simple polygons. J. Comput. Syst. Sci. 39(2), 220–235 (1989)

# The Planar Slope Number of Subcubic Graphs*

Emilio Di Giacomo, Giuseppe Liotta, and Fabrizio Montecchiani

Dip. di Ingegneria, Università degli Studi di Perugia
{emilio.digiacomo,giuseppe.liotta,
fabrizio.montecchiani}@unipg.it

**Abstract.** A subcubic planar graph is a planar graph whose vertices have degree at most three. We show that the subcubic planar graphs with at least five vertices have planar slope number at most four, which is worst case optimal. This answers an open question by Jelínek et al. [6]. As a corollary, we prove that the subcubic planar graphs with at least five vertices have angular resolution $\pi/4$, which solves an open problem by Kant [7] and by Formann et al. [4].

## 1 Introduction

A *straight-line drawing* of a graph $G$ is a representation of $G$ where the vertices are drawn as distinct points in the plane and the edges are drawn as line segments connecting the two corresponding end-points and not passing through any other point representing a vertex. Minimizing the number of slopes used in a straight-line drawing is a desirable aesthetic requirement and an interesting theoretical problem which has received considerable attention since its first definition by Wade and Chu [17].

The *slope number* of a graph $G$ is defined as the minimum number of distinct slopes required by any straight-line drawing of $G$ [17]. Let $\Delta$ be the maximum degree of a graph $G$ and let $m$ be the number of edges of $G$, then the slope number of $G$ is at least $\frac{\Delta}{2}$ and at most $m$, as no more than two edges incident to the same vertex can have the same slope and at most one slope per edge can be used. It has been shown that there exist graphs with $\Delta \geq 5$ whose slope number is unbounded [1,16], while the slope number of graphs with $\Delta = 4$ is still unknown.

The above results consider drawings that may contain edge crossings. A *planar straight-line drawing* is a straight-line drawing that contains no edge crossings. The *planar slope number* of a planar graph $G$ is defined as the minimum number of distinct slopes required by any planar straight-line drawing of $G$. Keszegh, Pach and Pálvölgyi [9] proved that the planar slope number of a planar graph $G$ is bounded by a function which is $O(2^{O(\Delta)})$; besides this upper bound, in the same paper a lower bound of $3\Delta - 6$, for $\Delta \geq 3$ is also proved [9]. The gap between these two bounds is large and the upper bound is probably far from being optimal, as pointed out by the authors. Jelínek et al. [6] study the *plane slope number* of plane partial 3-trees, i.e., planar partial 3-tree with a fixed combinatorial embedding. The plane slope number of an embedded planar graph $G$ is the minimum number of distinct slopes required by any straight-line drawing of $G$ that preserves the given embedding. Clearly the planar slope

---
* Research supported in part by the MIUR project AMANDA "Algorithms for MAssive and Networked DAta", prot. 2012C4E3KT\_001.

A. Pardo and A. Viola (Eds.): LATIN 2014, LNCS 8392, pp. 132–143, 2014.
© Springer-Verlag Berlin Heidelberg 2014

number is bounded by the plane slope number. Jelínek et al. [6] proved that the plane slope number of any plane partial 3-tree with maximum degree $\Delta$ is at most $O(\Delta^5)$. Knauer, Micek and Walczak [12] focus on a subclass of planar partial 3-trees, showing that the (outer)planar slope number of outerplanar graphs is at most $\Delta - 1$, for $\Delta \geq 4$, and this bound is tight. Very recently, Lenhart et al. [13] proved $O(\Delta)$ upper bounds on the planar and plane slope numbers of partial 2-trees.

Special interest has been devoted to the slope number of (sub)cubic graphs, i.e., graphs having vertex degree (at most) 3. Keszegh et al [10] proved that the slope number of cubic graphs is five. This result has been improved by Mukkamala and Szegedy [15] who proved that the slope number of simply connected cubic graphs is four. Finally, Mukkamala and Pálvölgyi showed that the four basic slopes $\{0, \frac{\pi}{4}, \frac{\pi}{2}, \frac{3\pi}{4}\}$ suffice for every cubic graph [14]. Concerning the planar slope number, Kant [7] and independently Dujmović et al. [3] proved that cubic 3-connected planar graphs have planar slope number three except for three edges on the outerface. Jelínek et al. [5] showed that subcubic series-parallel graphs have planar slope number three, which is worst-case optimal. Jelínek et al. also asked to prove an upper bound on the planar slope number of subcubic planar graphs analogous to those in [11,14,15]. An answer to this question is given by the following theorem which is the main contribution of our paper.

**Theorem 1.** *Let $G$ be a subcubic planar graph with $n \geq 5$ vertices. The planar slope number of $G$ is at most four and this bound is tight.*

Note that for $n \leq 4$ four slopes are not sufficient in general. Namley, it is known that six slopes are necessary ad sufficient for $K_4$. On the other hand, each subcubic planar graph with $n \leq 4$ vertices is a subgraph of $K_4$ and therefore 6 slopes are sufficient. The proof of Theorem 1 is based on an algorithm that computes a planar straight-line drawing of a subcubic planar graph on a grid of polynomial area using only slopes in the set $\{0, \frac{\pi}{4}, \frac{\pi}{2}, \frac{3\pi}{4}\}$. A byproduct of our proof technique is therefore the following.

**Corollary 1.** *Every subcubic planar graph with $n \geq 5$ vertices has a straight-line planar drawing whose angular resolution is $\frac{\pi}{4}$ and whose area is $O(n^2)$.*

About Corollary 1, we recall that Formann et al. [4] initiate the study of straight-line planar drawings with good angular resolution. Among the many questions that stimulated further research, Formann et al. ask whether every subcubic planar graph has a planar straight-line drawing such that the smallest angle is a constant independent of the size of the graph. An answer to this fundamental question has been already given in a paper by Kant [7], who claims that every subcubic planar graph with $n \geq 6$ has a planar straight-line drawing with all angles at least $\frac{\pi}{3}$ except for four angles which are at least $\frac{\pi}{6}$. This claim is correct if restricted to 3-connected subcubic planar graphs, but unfortunately incorrect in the general case as observed by Dujmović et al. [3], who provided as a counter-example a family of connected subcubic planar graphs requiring a linear number of angles less than $\frac{\pi}{3}$. In [7], Kant also asks the following question: Does every subcubic planar graph admit a straight-line planar drawing such that the smallest angle is at least $\frac{\pi}{4}$? Corollary 1 answers in the affirmative both the question by Formann et al. and the one by Kant. We also remark that an angular resolution of $\frac{\pi}{4}$ is worst-case optimal for subcubic planar graphs with at least five vertices.

The remainder of this paper is organized as follows. Preliminaries are introduced in Section 2. Section 3 studies 2-connected subcubic planar graphs. The 1-connected and 3-connected graphs are briefly addressed in Section 4. Open problems are listed in Section 5. For reasons of space many technical details are omitted.

## 2 Preliminaries

We call the slopes $\{0, \frac{\pi}{4}, \frac{\pi}{2}, \frac{3\pi}{4}\}$ the *four canonical slopes*. Let $\Gamma$ be a straight-line grid drawing of $G$, let $v$ be a vertex of $G$ and let $s$ be one of the four canonical slopes, we denote by $l_s^v$ the line with slope $s$ and passing through the point of $\Gamma$ representing $v$.

A graph is $k$-*connected* if removing at most $k-1$ vertices cannot make the graph disconnected. A graph that is $k$-connected and not $(k+1)$-connected, for $k \geq 1$, is called *simply $k$-connected graph*. A 1-connected graph is also called a connected graph.

Let $G$ be a simply 2-connected graph. A *separation pair* is a pair of vertices whose removal disconnects $G$. A *split pair* is either a separation pair or a pair of adjacent vertices. A *split component* of a split pair $\{u, v\}$ is either an edge $(u, v)$ or a maximal subgraph $G_{uv} \subset G$ such that $\{u, v\}$ is not a split pair of $G_{uv}$. Vertices $\{u, v\}$ are the poles of $G_{uv}$. The *SPQR-tree* $T$ of $G$ with respect to an edge $e$ is a rooted tree that describes a recursive decomposition of $G$ induced by its split pairs. In what follows, we call *nodes* the vertices of $T$, to distinguish them from the vertices of $G$. The nodes of $T$ are of four types $S,P,Q,$ or $R$. Each node $\mu$ of $T$ has an associated 2-connected multigraph called the *skeleton of $\mu$*. At each step, given the current split component $G^*$, its split pair $\{s, t\}$, and a node $\nu$ in $T$, the node $\mu$ of the tree corresponding to $G^*$ is introduced and attached to its parent vertex $\nu$, while the decomposition possibly recurs on some split component of $G^*$. At the beginning of the decomposition the parent of $\mu$ is a $Q$-node corresponding to $e = (u, v)$, $G^* = G \setminus e$, and $\{s, t\} = \{u, v\}$.

**Base Case:** $G^*$ consists of a single edge between $s$ and $t$. Then, $\mu$ is a $Q$-node whose skeleton is $G^*$ itself plus the reference edge between $s$ and $t$.

**Parallel Case:** The split pair $\{s, t\}$ has $G_1, \ldots, G_k$ ($k \geq 2$) split components. Then, $\mu$ is a $P$-node whose skeleton is a set of $k+1$ parallel edges between $s$ and $t$, one for each split component $G_i$ plus the reference edge between $s$ and $t$. The decomposition recurs on $G_1, \ldots, G_k$ with $\mu$ as parent node.

**Series Case:** $G^*$ is not 2-connected and it has at least one cut vertex (a vertex whose removal disconnects $G^*$). Then, $\mu$ is an $S$-node whose skeleton is defined as follows. Let $v_1, \ldots, v_{k-1}$, where $k \geq 2$, be the cut vertices of $G^*$. The skeleton of $\mu$ is a path $e_1, \ldots, e_k$, where $e_i = (v_{i-1}, v_i)$, $v_0 = s$ and $v_k = t$, plus the reference edge between $s$ and $t$ which makes the path a cycle. The decomposition recurs on the split components corresponding to each $e_1, \ldots, e_k$ with $\mu$ as parent node.

**Rigid Case:** None of the other cases is applicable. A split pair $\{s', t'\}$ is maximal with respect to $\{s, t\}$, if for every other split pair $\{s^*, t^*\}$, there is a split component that includes the vertices $s', t', s, t$. Let $\{s_1, t_1\}, \ldots, \{s_k, t_k\}$ be the maximal split pairs of $G^*$ with respect to $\{s, t\}$ ($k \geq 1$), and, for $i = 1, \ldots, k$, let $G_i$ be the union of all the split components of $\{s_i, t_i\}$. Then $\mu$ is an $R$-node whose skeleton is obtained from $G^*$ by replacing each component $G_i$ with an edge between $s_i$ and $t_i$, plus the reference edge between $s$ and $t$. The decomposition recurs on each $G_i$ with $\mu$ as parent node.

If we consider the $SPQR$-tree of a graph $G$ as an unrooted tree, we get the same tree no matter what edge of the graph was chosen as the reference edge [2]. Therefore, choosing a different edge as the reference edge is equivalent to root $T$ at the $Q$-node corresponding to the new reference edge.

## 3 Simply 2-Connected Subcubic Planar Graphs

In this section we describe how to draw simply 2-connected graphs using the four basic slopes. We exploit their $SPQR$-tree. With respect to series and parallel components, our technique is similar to the one by Jelínek et al. [5] for two terminal series-parallel graphs. For the case of rigid components we use a technique based on a decomposition of 3-connected planar graphs called *canonical ordering* [8]. The resulting drawings have $O(n^2)$ area. We remark that the area requirement is not addressed in [5].

Let $G = (V, E)$ be a 3-connected plane graph, i.e., a 3-connected planar graph with a prescribed planar embedding. Let $\delta = \{V_1, \ldots, V_K\}$ be an ordered partition of $V$, that is, $V_1 \cup \cdots \cup V_K = V$ and $V_i \cap V_j = \emptyset$ for $i \neq j$. Let $G_i$ be the subgraph of $G$ induced by $V_1 \cup \cdots \cup V_i$ and denote by $C_i$ the outerface of $G_i$. The partition $\delta$ is a *canonical ordering* of $G$ if:

- $V_1 = \{v_1, v_2\}$, where $v_1$ and $v_2$ lie on the outerface of $G$ and $(v_1, v_2) \in E$.
- $V_K = \{v_n\}$, where $v_n$ lies on the outerface of $G$, $(v_1, v_n) \in E$, and $v_n \neq v_2$.
- Each $C_i$ ($i > 1$) is a cycle containing $(v_1, v_2)$.
- Each $G_i$ is 2-connected and internally 3-connected.
- For each $i = \{2, \ldots, K-1\}$, one of the following conditions holds:
  (1) $V_i$ is a *singleton* $v_i$ which belongs to $C_i$ and has at least one neighbor in $G \setminus G_i$.
  (2) $V_i$ is a *chain* $\{v_i^1, \ldots, v_i^l\}$, each $v_i^j$ has at least one neighbor in $G \setminus G_i$, and both $v_i^1$ and $v_i^l$ have one neighbor on $C_{i-1}$ and are the only two neighbors of $V_i$ in $G_{i-1}$.

Observe that, if the graph $G$ is a subcubic planar graph and $V_i$ is a singleton, then $v_i$ has exactly two neighbors in $C_i$ (otherwise $G_i$ would not be 2-connected) and therefore exactly one neighbor in $G \setminus G_i$. Similarly, if $V_i$ is a chain, then all its vertices will have exactly one neighbor in $G \setminus G_i$, since they already have two neighbors in $G_i$. Therefore, for each $V_i$, $i = 2, \ldots, K-1$, there are exactly two vertices in $G_{i-1}$ which are adjacent one to $v_i^1$ and one to $v_i^l$ if $V_i$ is a chain, or both to $v_i$ if $V_i$ is a singleton. We denote them as the *leftmost* and the *rightmost predecessor* of $V_i$, respectively. Kant [8] proved that every 3-connected plane graph has a canonical ordering. The technique described by Kant to compute a canonical ordering is such that one can arbitraily choose two adjacent vertices $u$ and $w$ on the outerface so that $u = v_1$ and $w = v_2$ in the computed canonical ordering. We now prove a lemma that will be used to draw the $R$-nodes of simply 2-connected subcubic planar graphs.

**Lemma 1.** *Let $G$ be an $n$-vertex subcubic planar graph containing two vertices of degree 2 such that connecting these two vertices makes $G$ 3-connected and keeps it planar. $G$ admits a planar straight-line grid drawing with the four canonical slopes in area at most $(2n-6) \times (n-3)$.*

*Proof sketch:* Let $u$ and $w$ be the two vertices of $G$ with degree 2. Add a dummy edge between $u$ and $w$ so that $G$ is planar and 3-connected. Then, construct the planar embedding of $G$ such that the dummy edge $(u, w)$ belongs to the boundary of the outerface. Following such an embedding, compute the canonical ordering $\delta = \{V_1, \ldots, V_K\}$ of $G$, in such a way that $u = v_1$ and $w = v_2$. We denote by $G_i$ the subgraph of $G$ induced by the vertices $V_1 \cup V_2 \cup \cdots \cup V_i$, and by $n_i$ the number of vertices of $G_i$. Every degree 2 vertex of $G_i$ different from $v_1$ and $v_2$ is called an *attaching vertex*. Since $G_i$ is internally 3-connected, each attaching vertex belongs to $C_i$. Two attaching vertices in $G_i$ are *consecutive* if there is no attaching vertex between them when walking clockwise along $C_i$. We describe a drawing algorithm, called QUASI3CONNDRAWER, that inductively constructs a drawing of $G$ by adding a set $V_i$ per time. The base case of the induction is the construction of the drawing of the graph $G_2$. We denote by $\Gamma_i$ the drawing after the addition of $V_i$ ($i = 2, 3, \ldots, K$), i.e., the drawing of $G_i$. We assume that each drawing $\Gamma_i$ ($i = 2, \ldots, K$) satisfies the following invariants: **I1.** $\Gamma_i$ is a planar straight-line drawing using the four canonical slopes. **I2.** For every pair of consecutive attaching vertices $u$ and $w$, with $u$ preceding $w$ when walking clockwise along $C_i$, the path from $u$ to $w$ in $C_i$ consists of $n_1 \geq 0$ edges drawn with slope $\frac{3\pi}{4}$, one edge drawn with slope 0, and $n_2 \geq 0$ edges drawn with slope $\frac{\pi}{4}$, in this order. **I3.** For every pair of consecutive attaching vertices $u$ and $w$, with $u$ preceding $w$ when walking clockwise along $C_i$, there exists a set $E_i(u, w)$ of horizontal edges, whose removal disconnects the drawing into two subdrawings, one containing $u$ and one containing $w$.

The drawing $\Gamma_2$ of $G_2$ is computed as follows. The vertices $v_1$ and $v_2$ are drawn at points $(0, 0)$ and $(|V_2| + 1, 0)$, respectively. Let $v_2^i$, for $i = 1, \ldots, |V_2|$, be the vertices of $V_2$; vertex $v_2^i$ is placed at point $(i, 0)$. Invariants **I1**, **I2**, and **I3** trivially hold for $\Gamma_2$.

We describe now how to add the set of vertices $V_i$, for $i = 3, 4, \ldots, K - 1$. The addition of $V_K = \{v_n\}$ requires to be handled in a slightly different way. In order to place the vertices of $V_i$ at integer coordinates, we define an operation that expands the drawing $\Gamma_{i-1}$. The expansion operation applies to a drawing $\Gamma_{i-1}$ and to two consecutive attaching vertices $u$ and $w$, and it takes an integer $N$ as a parameter, which represents the amount of the enlargement required. We denote the operation as $\exp(\Gamma_{i-1}, u, w, N)$. Let $E_{i-1}(u, w)$ be the set of edges defined as in Invariant **I3** and let $\Gamma'_{i-1}$ and $\Gamma''_{i-1}$ be the two subdrawings obtained by removing $E_{i-1}(u, w)$, with $\Gamma''_{i-1}$ to the right of $\Gamma'_{i-1}$. The expansion operation increases the $x$-coordinates of all the vertices of $\Gamma''_{i-1}$ by $N$ units. In other words, the subdrawing $\Gamma''_{i-1}$ is shifted to the right and the edges of $E_i(u, w)$ become $N$ units longer. Notice that, the expansion operation increases the horizontal distance between $u$ and $w$ by $N$ units. It is easy to see that the drawing produced by the expansion operation still satisfies Invariants **I1**, **I2**, and **I3**.

We are now ready to describe how to add $V_i$ to $\Gamma_{i-1}$. Let $p$ and $q$ be the leftmost and rightmost predecessors of $V_i$, respectively and let $r$ be the intersection point of $l^p_{\pi/4}$ and $l^q_{3\pi/4}$. Notice that, $p$ and $q$ are two consecutive attaching vertices and therefore, by Invariant **I2**, point $r$ is above $p$ and $q$ (see Fig. 1(a)).

Suppose first that $V_i$ is a singleton. If $r$ has integer coordinates, then $v_i$ is placed at $r$. Otherwise we execute an expansion operation $exp(\Gamma_{i-1}, p, q, 1)$ that increases the horizontal distance between $p$ and $q$ by one unit so that $r$ has integer coordinates.

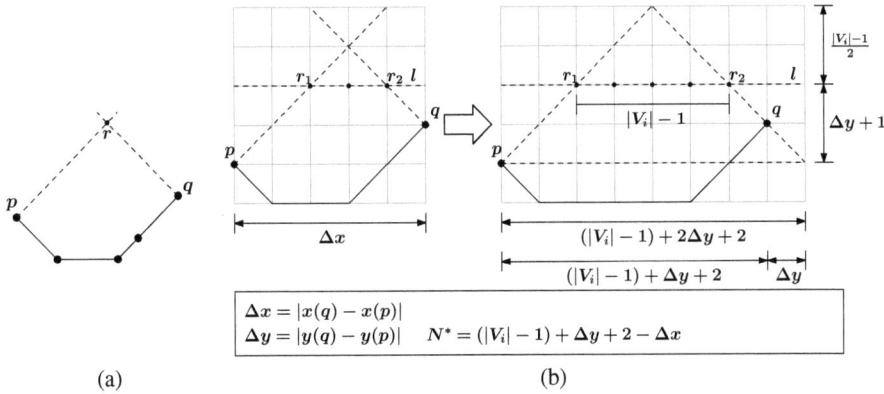

**Fig. 1.** Lemma 1: (a) The intersection point of $l^p_{\pi/4}$ and $l^q_{3\pi/4}$ is above $p$ and $q$. (b) An example of the expansion required to add $V_i$.

Suppose now that $V_i$ is a chain. Let $l$ be the horizontal straight line $y = y_l$ with $y_l = \max\{y(p), y(q)\} + 1$, let $r_1$ be the intersection point between $l^p_{\pi/4}$ and $l$, and let $r_2$ be the intersection point between $l^q_{3\pi/4}$ and $l$. Let $n_p$ the number of grid points along $l$ between $r_1$ and $r_2$, including $r_1$ and $r_2$. Notice that $r_1$ could be to the right of $r_2$; in this case however $x(r_1) = x(r_2) + 1$ and we set $n_p = -1$. If $n_p$ is at least $|V_i|$, then we have enough points to place the vertices of $V_i$ on $l$. Otherwise we execute an expansion operation $exp(\Gamma_{i-1}, p, q, N^*)$ with $N^* = (|V_i| - 1) + \Delta y + 2 - \Delta x$, where $\Delta x = |x(q) - x(p)|$ and $\Delta y = |y(q) - y(p)|$. After the expansion, the number of grid points along $l$ between $r_1$ and $r_2$ (including $r_1$ and $r_2$) is $|V_i|$ (see Fig. 1(b)). We place $v_i^1$ at point $r_1$, $v_i^{|V_i|}$ at point $r_2$, and $v_i^j$ ($j = 2, 3, \ldots, |V_i| - 1$) at point $(x(v_i^{j-1}) + 1, y_l)$.

It is immediate to see that the drawing is a straight-line drawing and that it uses the four basic slopes. About planarity, observe that when adding $V_i$ ($i = 3, \ldots, K$), we draw a planar subgraph (either a path or a single vertex) in the outerface of $\Gamma_{i-1}$ and attach it to two vertices of the boundary of this face; the resulting drawing $\Gamma_i$ is clearly planar. Thus Invariant **I1** holds. The proofs of Invariant **I2** and **I3** are omitted.

We conclude the description of the drawing technique by explaining how to add the set $V_K = \{v_n\}$. Let $a = v_1$, $b$ and $c$, be the three vertices adjacent to $v_n$, ordered by increasing $x$-coordinates. Place $v_n$ at the intersection point between $l^b_{\pi/2}$ and $l^c_{3\pi/4}$ (notice that this point has integer coordinates). Then, move $v_1$ to the intersection point between $l^{v_n}_{\pi/4}$ and $l^{v_2}_0$. Observe that, before the addition of $v_n$, $v_1$ is adjacent to a single vertex, namely the first vertex of $V_2$, with a horizontally drawn edge. Therefore $v_1$ can be moved along the horizontal line $l^{v_2}_0$ and placed at the intersection point of $l^{v_1}_0$ with $l^{v_n}_{\pi/4}$ (also this point has integer coordinates). It is easy to see that Invariant **I1** holds for $\Gamma_K$. Namely, the drawing is clearly a straight-line drawing using the four basic slopes. Since we add a single vertex on the outerface of $\Gamma_{K-1}$ and this vertex is connected only to vertices in the boundary of this face, the resulting drawing is clearly planar. Invariant **I2** and **I3** do not apply since $\Gamma_K$ does not have attaching vertices.

We prove now the bound on the area of the final drawing $\Gamma = \Gamma_K$. Denote by $w(\Gamma_i)$ the width of $\Gamma_i$ ($i = 2, 3, \ldots, K$). We have $w(\Gamma_2) = |V_1| + |V_2| - 1$. Each time we add a set $V_i$ ($i = 3, 4, \ldots, K-1$) we possibly perform an expansion operation that enlarges the width of the previous drawing by at most $|V_i|$ units (the enlargement is $|V_i|-1-n_p$, and since $n_p \geq -1$ it is at most $|V_i|$). It follows that $w(\Gamma_{K-1}) = \sum_{i=1}^{k-1} |V_i|-1 = n-2$. As explained, the placement of vertex $v_n$ does not increase the width of the drawing; however, in order to connect $v_n$ to $v_1$, the latter needs to be moved thus enlarging the width of the drawing. In the worst case (which happens when the second vertex $b$ adjacent to $v_n$ is the leftmost of $\Gamma_{K-1} \setminus \{v_1\}$) $v_1$ is moved by $n-4$ units, which results in $w(\Gamma) \leq 2n-6$. It is easy to see that $\Gamma$ is completely contained inside an isosceles right triangle with the hypothenuse as the base, thus the height of $\Gamma$ is at most $n-3$. □

Before giving the main result of this section, we shall introduce some additional notation and prove a few useful properties.

Let $G$ be a simply 2-connected planar graph and let $T$ be its $SPQR$-tree with respect to any reference edge $e$. Let $\mu$ be a node of $T$. The *pertinent graph* of $\mu$ is the subgraph of $G$ whose $SPQR$-tree (with respect to the reference edge of $\mu$) is the subtree of $T$ rooted at $\mu$. The *virtual edge* of $\mu$ is the edge in the skeleton of the parent of $\mu$ in $T$ that represents the pertinent graph of $\mu$. Thus, for every internal node $\mu$ of $T$ (i.e., which is not a $Q$-node), each edge in its skeleton is the virtual edge of one of its children. In what follows, we call $S$-*edge* a virtual edge of an $S$-node, $P$-*edge* a virtual edge of a $P$-node, $R$-*edge* a virtual edge of an $R$-node, and $Q$-*edge* an edge of $G$.

Let $\mu$ be a node of $T$, we associate with $\mu$ another graph, called the *frame of $\mu$*, as follows. If $\mu$ is a $P$-node, we replace each virtual edge of the skeleton of $\mu$ with the frame of the node it represents. Also, we remove the reference edge. If $\mu$ is not a $P$-node, we only remove the reference edge. Observe that, every vertex in a frame represents exactly one vertex in $G$.

Let $G$ be a simply 2-connected subcubic planar graph and let $T$ be its $SPQR$-tree with respect to any reference edge $e$. Since $G$ has maximum degree 3, the following properties hold.

*Property 1.* The frame of an $S$-node of $T$ whose parent is not the root of $T$ is a path $e_1, \ldots, e_{k_\mu}$, where $k_\mu \geq 2$, and $e_1, e_{k_\mu}$ are two $Q$-edges.

*Property 2.* The frame of a $P$-node of $T$ is composed of either two $S$-edges or of one $S$-edge and one $Q$-edge.

*Property 3.* The frame of an $R$-node of $T$ can contain only $S$-edges and $Q$-edges.

*Property 4.* $T$ contains at least one $S$-node.

**Lemma 2.** *Let $G$ be an $n$-vertex simply 2-connected subcubic planar graph. $G$ admits a planar straight-line grid drawing with the four canonical slopes in area $O(n) \times O(n)$.*

*Proof sketch:* Let $T$ be an $SPQR$-tree of $G$ with respect to an arbitrary reference edge $e$. By Property 4, $T$ contains at least one $S$-node $\nu$. Let $\rho$ be a $Q$-node of $T$, which is a child of $\nu$. By Property 1, $\rho$ always exists. Change the root of $T$ to $\rho$. We describe an algorithm that draws $G$ through a bottom-up visit of $T$, handling $\rho$ and $\nu$ as special case. For each node $\mu$ of $T$, first draw its frame, then replace each virtual edge with the (already computed) drawing of the pertinent graph represented by the edge itself.

Let $n_\mu$ be the number of vertices of $G_\mu$, the drawing $\Gamma_\mu$ of $G_\mu$ respects the following three invariants: **I1.** $\Gamma_\mu$ is a planar straight-line drawing using the four canonical slopes. **I2.** $\Gamma_\mu$ is contained in an isosceles right triangle whose base is the hypotenuse with the poles as endpoints. **I3.** The area of $\Gamma_\mu$ is $O(n_\mu) \times O(n_\mu)$.

For the sake of simplicity, when we refer to the drawing of a virtual edge $e$, we mean the drawing of the pertinent graph associated with $e$. The end-vertices of a virtual edge coincide with the poles of the associated pertinent graph. Thanks to invariant **I2**, in order to replace a virtual edge $e$ with its drawing $\Gamma_e$ it is sufficient to identify the base of the triangle $\tau_e$ containing $\Gamma_e$ with the segment representing $e$, provided that this segment length is greater than or equal to the width of $\Gamma_e$. Denote by $l_e$ the straight line containing the segment representing $e$ and by $H_1$ and $H_2$ the two half-planes defined by $l_e$. We can make this replacement in two ways depending on whether $\tau_e$ belongs to $H_1$ or $H_2$. In order to guarantee Invariant **I2**, if $e$ is drawn horizontally, we place $\tau_e$ so that it is above $l_e$. If $e$ is drawn with slope $\frac{\pi}{4}$ or $\frac{3\pi}{4}$, we place $\tau_e$ so that it is below $l_e$. If $e$ is drawn vertically, we place $\tau_e$ so that it is to the left of $l_e$.

Furthermore, observe that if $e$ is drawn as a segment with slope $\frac{\pi}{4}$ or $\frac{3\pi}{4}$, in order to replace it with $\Gamma_e$, $\Gamma_e$ has to be rotated. If we rotate by an angle $h\frac{\pi}{4}$ (with $h$ integer) a drawing that respects invariant **I1**, the rotated drawing still respects this invariant. Nevertheless, if $h$ is odd the rotated drawing is not an integer grid drawing anymore. To cope with this issue, if $h$ is odd, before rotating $\Gamma_e$ we scale it up by a factor $\sqrt{2}$. The resulting drawing is again a grid drawing requiring twice the initial area.

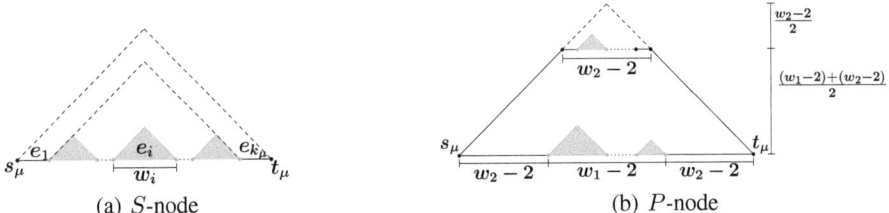

**Fig. 2.** Lemma 2: Construction for (a) $S$-nodes and (b) $P$-nodes

Let $\mu$ be the current node of $T$ to be drawn and let $s_\mu$ and $t_\mu$ be its poles. If $\mu$ is a $Q$-node, draw $s_\mu$ and $t_\mu$ at points $(0,0)$ and $(1,0)$, respectively. It is immediate to see that all the invariants hold for $\Gamma_\mu$.

**Series Case.** If $\mu$ is an $S$-node, recall that, by Property 1, its frame is a path $e_1, \ldots, e_{k_\mu}$, such that $e_1$ and $e_{k_\mu}$ are $Q$-edges. Let $w_i$ be the width of the drawing of $e_i$, for any $1 \leq i \leq k_\mu$. Denote by $v_0 = s_\mu, v_1, \ldots, v_{k_\mu} = t_\mu$ the vertices of the frame of $\mu$ in the order they appear along the path. We draw the frame of $\mu$ as follows (see also Fig. 2(a)). Vertex $v_0$ is drawn at point $(0,0)$, and $v_i$, for every $i = 1, \ldots, k_\mu$, is drawn at point $(x(v_{i-1}) + w_i, 0)$. Once the frame is drawn, each virtual edge is replaced with its drawing. It is easy to see that all invariants hold for $\Gamma_\mu$. Furthermore, we observe that, not only $\Gamma_\mu$ is contained in an isosceles right triangle $\tau$, but the subdrawing induced by the subgraph $G_\mu \setminus \{s_\mu, t_\mu\}$ is also contained in an isosceles right triangle $\tau'$ such that $\tau'$ is contained into $\tau$ but for one shared side.

**Parallel Case.** If $\mu$ is a $P$-node, then, by Property 2, its frame is composed of two $S$-edges, or of one $S$-edge and one $Q$-edge. Let $e_1$ and $e_2$ be these two virtual edges and let $w_1$ and $w_2$ be the width of their drawings. In the first case, we combine the drawings of $e_1$ and $e_2$ as follows (see also Fig. 2(b)). Denote by $\tau_1$ and $\tau_2$ the two triangles that, by Invariant **I2**, contain the drawing of $e_1$ and the drawing of $e_2$, respectively. Place the drawing of $e_1$ and $e_2$ so that the base of $\tau_2$ is on the first grid line above $\tau_1$ and the midpoints of the basis of $\tau_1$ and $\tau_2$ have the same $x$-coordinate. Let $s_i$ be the vertex adjacent to $s_\mu$ in the drawing of $e_i$ ($i=1,2$) and let $t_i$ be the vertex adjacent to $t_\mu$ in the drawing of $e_i$ ($i=1,2$). Place $s_\mu$ at the intersection point between $l^{s_2}_{\pi/4}$ and $l^{s_1}_0$, and $t_\mu$ at the intersection point between $l^{t_2}_{3\pi/4}$ and $l^{t_1}_0$. Invariant **I1** and **I2** clearly hold. About **I3**, observe that the base of the triangle containing $\Gamma_\mu$ has length less than $w_1 + w_2$.

**Rigid Case.** If $\mu$ is an $R$-node, its two poles have degree two in the frame of $\mu$, and they are not connected, otherwise $\mu$ would be a $P$-node. Hence, the frame of $\mu$ can be drawn using the algorithm QUASI3CONNDRAWER. Notice that if we draw the frame of $\mu$ with the algorithm QUASI3CONNDRAWER the edges of the resulting drawing could be too short to be replaced by the drawings of their pertinent graphs. Moreover, there might not be enough space inside the faces to accommodate the drawings of the virtual edges in the boundary of the face. Hence we assign to each virtual edge $(u,v)$ a weight $w(u,v)$ equal to the width of its drawing, and modify the algorithm QUASI3CONNDRAWER so that the edges are sufficiently long and the faces have sufficient space.

We denote by $\delta = \{V_1, \ldots, V_K\}$ the canonical ordering computed by our algorithm and by $G_i$ the subgraph induced by $V_1 \cup V_2 \cup \cdots \cup V_i$ ($i = 1, 2, \ldots, K$). The drawing of $G_2$ is computed as follows. Recall that $G_2$ is a path starting at $v_1$ passing through all the vertices of $V_2$ and ending at $v_2$. Let $u_1 = v_1, u_2, \ldots, u_{n_2} = v_2$ be the vertices of $G_2$ in the order they appear when walking along $G_2$ from $v_1$ to $v_2$. Vertex $u_1$ is placed at $(0,0)$ and each vertex $u_j$ is placed on line $y=0$ at distance $w(u_{j-1}, u_j)$ from $u_{j-1}$.

Now, consider $V_i = \{v_i^1, \ldots, v_i^{|V_i|}\}$, for $i = 3, \ldots, K-1$ and let $p$ and $q$ be the leftmost and rightmost predecessors of $V_i$, respectively. Let $u_0 = p$, $u_j = v_i^j$ ($j = 1, 2, \ldots, |V_i|$), and $u_{|V_i|+1} = q$ and denote by $e_j$ the edge $(u_j, u_{j+1})$ for $j = 0, 1, \ldots, |V_i|$. Using the algorithm QUASI3CONNDRAWER we would perform an expansion operation $exp(\Gamma_{i-1}, p, q, N^*)$ for a value of $N^*$ sufficient to guarantee that the horizontal distance between $u_{j-1}$ and $u_j$ ($j = 1, 2, \ldots, |V_i|+1$) is equal to one. In the current case instead we need to expand $\Gamma_{i-1}$ so that the edges are sufficiently long and the faces have sufficient space. To this aim we choose a value of $N^*$ equal to:
$\sum_{j=1}^{|V_i|-1} w(e_j) + 2\max\{w(e_0), w(e_{|V_i|})\} + 2\max\{\Delta y + 1 - d, \frac{w(e)}{2} - 1\} - \Delta y - \Delta x$
where $\Delta x = |x(q) - x(p)|$, $\Delta y = |y(q) - y(p)|$, $e$ is the horizontally drawn edge in the path from $p$ to $q$ along $C_{i-1}$ (this edge exists by invariant **I2** of Lemma 1), and $d$ is the vertical distance between the line containing $e$ and the lowest among $p$ and $q$. As shown in Fig. 3, this choice guarantees that the face $f_i$ created by the addition of $V_i$ can accommodate the drawings of the virtual edges that need to be placed inside $f_i$.

Consider now the addition of $V_K$, let $a = v_1$, $b$ and $c$ be the neighbors of $v_n$, with $x(a) \leq x(b) \leq x(c)$. Also in this case we need to enlarge $\Gamma_{K-1}$ so that the edges are sufficiently long and the faces have sufficient space. We perform the expansion operation $exp(\Gamma_{K-1}, b', c', N^*)$, where $b'$ and $c'$ are the end-vertices of the horizontal edge

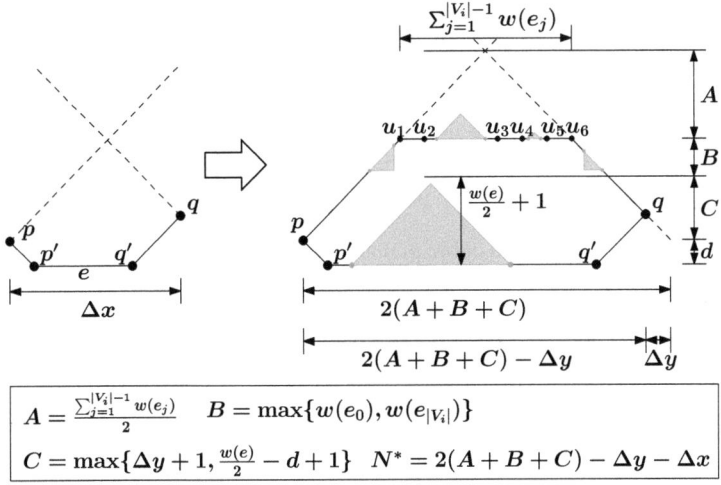

**Fig. 3.** Lemma 2: Enlargement for the addition of $V_i$ to draw an $R$-node

in the path from $b$ to $c$ along $C_{K-1}$, and $N^*$ is given by: $N^* = \max\{N_1^*, N_2^*\}$, with $N_1^* = w(v_n, c) + \frac{w(b',c')}{2} + 1 - d$ and $N_2^* = \frac{w(v_n, b)}{2} + w(a, v_n) + w(a, a')$; in the previous formulas $a'$ is the vertex adjacent to $a$ in $\Gamma_{K-1}$, and $d$ is the vertical distance between the line containing $(b', c')$ and the lowest among $b$ and $c$. This choice guarantees that the two faces created by the addition of $V_K$ can accommodate the drawings of the virtual edges that need to be placed inside them.

Once the drawing of the frame is computed, replace each virtual edge with its drawing. Observe that, some virtual edge $e$ has been stretched due to the expansion operations. If $e$ is a $Q$-edge, then there are no problems when replacing $e$ with its drawing. If $e$ is an $S$-edge, then it is enough to sufficiently stretch the first $Q$-edge of its drawing. It is easy to see that Invariants **I1** and **I2** hold. About Invariant **I3**, let $N_i$ be the number of vertices of the subgraph of $G_\mu$ obtained from $G_i$ by replacing each virtual edge with its pertinent graph. It can be proved, by induction on $i$, that $\Gamma_i$ has area $O(N_i) \times O(N_i)$, which implies that $G_\mu = G_K$ has area $O(n_\mu) \times O(n_\mu)$. The root $\rho$ of $T$ and its child $\nu$ must be handled differently from the other nodes of $T$. Details are omitted. □

## 4 Simply Connected and 3-Connected Subcubic Planar Graphs

In this section we briefly sketch the ideas to draw simply connected and 3-connected subcubic planar graphs. A *cut vertex* is a vertex whose removal disconnects $G$, while a *bridge* is an edge whose removal disconnects $G$. In a subcubic graph $G$, the absence of cut vertices implies the absence of bridges and vice versa. Therefore a simply connected subcubic planar graph $G$ consists of a set of components connected by bridges. Each component is either a simply 2-connected graph or a single vertex. The idea is to compute a drawing of each 2-connected component according to Lemma 2 and then to combine these drawings into a drawing of $G$. We can prove the following lemma.

**Lemma 3.** *Let $G$ be an $n$-vertex simply connected subcubic planar graph. $G$ admits a planar straight-line grid drawing with the four canonical slopes in area $O(n) \times O(n)$.*

All vertices of a 3-connected subcubic planar graph have degree 3, i.e., the graph is in fact cubic. If the outerface of the graph has degree 3 and cannot be changed, 6 slopes are always required [3]. For this reason, in our construction we always choose a "big" face as the outerface. The following property can be easily proved.

*Property 5.* Every 3-connected cubic planar graph has a face with degree less than 6.

Based on Property 5 we present three different algorithms depending on whether $G$ has a face of degree 3, 4 or 5. All algorithms are based on a common idea: We remove some vertices/edges from the outerface of $G$ (and add some dummy edges to it if necessary) so that $G$ can be drawn by algorithm QUASI3CONNDRAWER. We then suitably reinsert the removed vertices and edges using only the four canonical slopes.

**Lemma 4.** *Let $G$ be a 3-connected cubic planar graph with $n \geq 6$ vertices and a face $F$ of degree 3. $G$ admits a planar straight-line grid drawing with the four canonical slopes in area at most $(n-4) \times (n-3)$.*

*Proof.* Consider the planar embedding of $G$ with $F$ as the outerface. Denote by $a, b, c$ the three vertices of $F$, in clockwise order. $F$ is not adjacent to another face $F'$ of degree 3. By contradiction, let $F'$ be a face of degree 3 sharing one edge with $F$, say $(a, b)$. Denote by $d$ the vertex of $F'$ which is not in $F$. Since $G$ is 3-connected, there exists a path $\Pi$ from $d$ to $c$ (not passing through $a$ and $b$). Note that, $\Pi$ cannot consist of a single edge, otherwise $n=4$. Hence, $\Pi$ is formed by at least three vertices and $\{d, c\}$ is a separation pair, a contradiction. Thus, $F$ is adjacent to faces of degree at least 4. Compute a canonical ordering of $G$, and assume, without loss of generality, that $a = v_1$, $c = v_2$ and $b = v_n$. Construct a drawing of $G \setminus (v_1, v_2)$ using the algorithm QUASI3CONNDRAWER. We now explain how to modify the drawing in order to add edge $(v_1, v_2)$. To this aim we move vertices $v_1, v_2$, and $v_n$ as follows. Let $u_2$ and $u_{k-1}$ be the two extreme vertices of $V_2$. By recalling algorithm QUASI3CONNDRAWER, one can prove that the drawing of $G \setminus \{v_1, v_2, v_n\}$ is contained in a isosceles right triangle whose vertices are $\{v_3, v'_3, v_{n-1}\}$. We move $v_n$ to the point $(x(v_{n-1}), y(v_{n-1}) + 1)$, $v_1$ to the intersection point of $l^{v_n}_{3\pi/4}$ and $l^{u_2}_{\pi/2}$, and $v_2$ to the intersection point of $l^{v_n}_{\pi/4}$ and $l^{u_{k-1}}_{\pi/2}$. It is easy to see that the drawing of $G \setminus \{v_1, v_2, v_n\}$ has area at most $(n-4) \times (n-4)/2$, and therefore the final drawing has area at most $(n-4) \times (n-3)$. □

The cases when $G$ has a face of degree 4 or 5 require slightly larger area, the details are omitted. The following lemma summarizes all the cases.

**Lemma 5.** *Let $G$ be a 3-connected cubic planar graph with $n \geq 6$ vertices. $G$ admits a planar straight-line grid drawing with the four canonical slopes in area at most $(2n - 14) \times (2n - 14)$.*

Lemmas 2, 3 and 5 imply Theorem 1, while Corollary 1 immediately follows. The bound of Theorem 1 is tight because there exist subcubic planar graphs that cannot be drawn with less than four slopes.

## 5 Open Problems

Every planar graph with degree $\Delta$ have planar slope number $O(2^{O(\Delta)})$ [9]. Is there a better upper bound when $\Delta = 4$? It has been recently proved that those planar graphs whose $SPQR$-tree does not have $R$-nodes have $O(\Delta)$ slope number [13]. Since an $R$-node corresponds to a 3-connected planar subgraph (minus an edge), we ask whether the 3-connected planar graphs have a planar slope number polynomial in $\Delta$.

## References

1. Barát, J., Matousek, J., Wood, D.R.: Bounded-degree graphs have arbitrarily large geometric thickness. Electr. J. Comb. 13(1) (2006)
2. Di Battista, G., Tamassia, R.: On-line planarity testing. SIAM J. Comput. 25(5), 956–997 (1996)
3. Dujmović, V., Eppstein, D., Suderman, M., Wood, D.R.: Drawings of planar graphs with few slopes and segments. Comput. Geom. 38(3), 194–212 (2007)
4. Formann, M., Hagerup, T., Haralambides, J., Kaufmann, M., Leighton, F.T., Symvonis, A., Welzl, E., Woeginger, G.J.: Drawing graphs in the plane with high resolution. SIAM J. Comput. 22(5), 1035–1052 (1993)
5. Jelínek, V., Jelínková, E., Kratochvíl, J., Lidický, B., Tesař, M., Vyskočil, T.: The planar slope number of planar partial 3-trees of bounded degree. In: Eppstein, D., Gansner, E.R. (eds.) GD 2009. LNCS, vol. 5849, pp. 304–315. Springer, Heidelberg (2010)
6. Jelínek, V., Jelínková, E., Kratochvíl, J., Lidický, B., Tesar, M., Vyskocil, T.: The planar slope number of planar partial 3-trees of bounded degree. Graphs and Combinatorics 29(4), 981–1005 (2013)
7. Kant, G.: Hexagonal grid drawings. In: Mayr, E.W. (ed.) WG 1992. LNCS, vol. 657, pp. 263–276. Springer, Heidelberg (1993)
8. Kant, G.: Drawing planar graphs using the canonical ordering. Algorithmica 16(1), 4–32 (1996)
9. Keszegh, B., Pach, J., Pálvölgyi, D.: Drawing planar graphs of bounded degree with few slopes. SIAM J. Discrete Math. 27(2), 1171–1183 (2013)
10. Keszegh, B., Pach, J., Pálvölgyi, D., Tóth, G.: Drawing cubic graphs with at most five slopes. Comput. Geom. 40(2), 138–147 (2008)
11. Keszegh, B., Pach, J., Pálvölgyi, D., Tóth, G.: Cubic graphs have bounded slope parameter. J. Graph Algorithms Appl. 14(1), 5–17 (2010)
12. Knauer, K., Micek, P., Walczak, B.: Outerplanar graph drawings with few slopes. In: Gudmundsson, J., Mestre, J., Viglas, T. (eds.) COCOON 2012. LNCS, vol. 7434, pp. 323–334. Springer, Heidelberg (2012)
13. Lenhart, W., Liotta, G., Mondal, D., Nishat, R.: Planar and plane slope number of partial 2-trees. In: Wismath, S., Wolff, A. (eds.) GD 2013. LNCS, vol. 8242, pp. 412–423. Springer, Heidelberg (2013)
14. Mukkamala, P., Pálvölgyi, D.: Drawing cubic graphs with the four basic slopes. In: van Kreveld, M., Speckmann, B. (eds.) GD 2011. LNCS, vol. 7034, pp. 254–265. Springer, Heidelberg (2011)
15. Mukkamala, P., Szegedy, M.: Geometric representation of cubic graphs with four directions. Comput. Geom. 42(9), 842–851 (2009)
16. Pach, J., Pálvölgyi, D.: Bounded-degree graphs can have arbitrarily large slope numbers. Electr. J. Comb. 13(1) (2006)
17. Wade, G.A., Chu, J.-H.: Drawability of complete graphs using a minimal slope set. The Computer Journal 37(2), 139–142 (1994)

# Smooth Orthogonal Drawings of Planar Graphs*

Muhammad Jawaherul Alam[1], Michael A. Bekos[2], Michael Kaufmann[2],
Philipp Kindermann[3], Stephen G. Kobourov[1], and Alexander Wolff[3]

[1] Department of Computer Science, University of Arizona, USA
{mjalam,kobourov}@cs.arizona.edu
[2] Wilhelm-Schickhard-Institut für Informatik, Universität Tübingen, Germany
{bekos,mk}@informatik.uni-tuebingen.de
[3] Lehrstuhl für Informatik I, Universität Würzburg, Germany
http://www1.informatik.uni-wuerzburg.de/en/staff

**Abstract.** In *smooth orthogonal layouts* of planar graphs, every edge is an alternating sequence of axis-aligned segments and circular arcs with common axis-aligned tangents. In this paper, we study the problem of finding smooth orthogonal layouts of low *edge complexity*, that is, with few segments per edge. We say that a graph has *smooth complexity k*—for short, an $SC_k$-layout—if it admits a smooth orthogonal drawing of edge complexity at most $k$.

Our main result is that every 4-planar graph has an $SC_2$-layout. While our drawings may have super-polynomial area, we show that for 3-planar graphs, cubic area suffices. We also show that any biconnected 4-outerplane graph has an $SC_1$-layout. On the negative side, we demonstrate an infinite family of biconnected 4-planar graphs that require exponential area for an $SC_1$-layout. Finally, we present an infinite family of biconnected 4-planar graphs that do not admit an $SC_1$-layout.

## 1 Introduction

In the visualization of technical networks such as the structure of VLSI chips [8] or UML diagrams [10] there is a strong tendency to draw edges as rectilinear paths. The problem of laying out networks in such a way is called *orthogonal graph drawing* and has been studied extensively. For drawings of (planar) graphs to be readable, special care is needed to keep the number of bends small. In a seminal work, Tamassia [11] showed that one can efficiently minimize the total number of bends in orthogonal layouts of *embedded 4-planar graphs*, that is, planar graphs of maximum degree 4 whose combinatorial embedding (the cyclic order of the edges around each vertex) is given. In contrast to this, minimizing the number of bends over all embeddings of a 4-planar graph is NP-hard [6].

---

* Research of M.J. Alam and S.G. Kobourov is supported in part by NSF grants CCF-1115971 and DEB 1053573. The work of M.A. Bekos is implemented within the framework of the Action "Supporting Postdoctoral Researchers" of the Operational Program "Education and Lifelong Learning" (Action's Beneficiary: General Secretariat for Research and Technology), and is co-financed by the European Social Fund (ESF) and the Greek State. M. Kaufmann as well as Ph. Kindermann and A. Wolff acknowledge support by the ESF EuroGIGA project GraDR (DFG grants Ka 812/16-1 and Wo 758/5-1, respectively).

In a so far unrelated line of research, circular-arc drawings of graphs have become a popular matter of research in the last few years. Inspired by American artist Mark Lombardi (1951–2000), Duncan et al. [4] introduced and studied *Lombardi drawings*, which are circular-arc drawings with the additional requirement of *perfect angular resolution*, that is, for each vertex, all pairs of consecutive edges form the same angle. Among others, Duncan et al. treat drawings of $d$-regular graphs where all vertices have to lie on one circle. They show that under this restriction, for some subclasses, Lombardi drawings can be constructed efficiently, whereas for the others, the problem is NP-hard. They also show [5] that ordered trees can always be Lombardi drawn in polynomial area, whereas straight-line drawings with perfect resolution may need exponential area.

Very recently, Bekos et al. [2] introduced the *smooth orthogonal* graph layout problem that combines the two worlds; the rigidity and clarity of orthogonal layouts with the artistic style and aesthetic appeal of Lombardi drawings. Formally, a smooth orthogonal drawing of a graph is a drawing on the plane where (i) each vertex is drawn as a point; (ii) edges leave and enter vertices horizontally or vertically, (iii) each edge is drawn as an alternating sequence of axis-aligned line segments and circular-arc segments such that consecutive segments have a common *horizontal* or *vertical* tangent at their intersection point. In the case of (4-) planar graphs, it is additionally required that (iv) there are no edge-crossings. Note that, by construction, (smooth) orthogonal drawings of 4-planar graphs have angular resolution within a factor of two of optimal.

Figure 1 shows a real-world example: a smooth orthogonal drawing of an Austrian regional bus and train map. Extending our model, the map has (multi-) edges that enter vertices diagonally (as in *Grünau im Almtal Postamt*; bottom right).

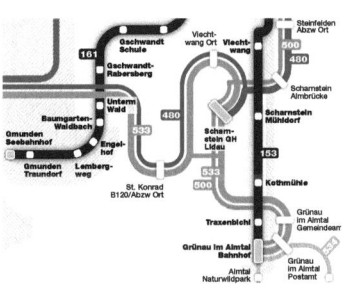

**Fig. 1.** Clipping of the public transport map *Gmunden – Vöcklabruck – Salzkammergut*, Austria [12]

For usability, it is important to keep the visual complexity of such drawings low. In a (smooth) orthogonal drawing, the *complexity* of an edge is the number of segments it consists of, that is, the number of inflection points plus one. Then, a natural optimization goal is to minimize, for a given (embedded) planar graph, the *edge complexity* of a drawing, which is defined as the maximum complexity over all edges. We say that a graph has *orthogonal complexity* $k$ if it admits an orthogonal drawing of edge complexity at most $k$, for short, an $OC_k$-*layout*. Accordingly, we say that a graph has *smooth complexity* $k$ if it admits a smooth orthogonal drawing of edge complexity at most $k$, for short, an $SC_k$-*layout*. We seek for drawings of 4-planar graphs with low smooth complexity.

*Our Contribution.* Known results and our contributions to smooth orthogonal drawings are shown in Table 1. The main result of our paper is that any 4-planar graph admits an $SC_2$-layout (see Sections 2 and 3). Our upper bound of 2 for the smooth complexity of 4-planar graphs improves the previously known bound of 3 and matches the corresponding lower bound [2]. In contrast to the known algorithm for $SC_3$-layout [2], which is based on an algorithm for $OC_3$-layout of Biedl and Kant [3], we use an algorithm of Liu at al. [9] for $OC_3$-layout, which avoids *S-shaped edges* (see Figure 2b, top). Such edges

**Table 1.** Comparison of our results to the results of Bekos et al. [2]

| graph class | our contribution | | | Bekos et al. [2] |
|---|---|---|---|---|
| | complexity | area | reference | |
| biconnected 4-planar | $SC_2$ | super-poly | Theorem 1 | $SC_3$ |
| 4-planar | $SC_2$ | super-poly | Theorem 2 | |
| 3-planar | $SC_2$ | $\lfloor n^2/4 \rfloor \times \lfloor n/2 \rfloor$ | Theorem 3 | |
| biconnected 4-outerplane | $SC_1$ | exponential | Theorem 4 | |
| triconnected 3-planar | | | | $SC_1$ |
| Hamiltonian 3-planar | | | | $SC_1$ |
| 4-planar, poly-area | $\not\supseteq SC_1$ | | full version [1] | |
| $OC_3$, octahedron | | | | $\not\subseteq SC_1$ |
| $OC_2$ | $\not\subseteq SC_1$ | | full version [1] | |

are undesirable since they force their endpoints to lie on a line of slope $\pm 1$ in a smooth orthogonal layout (see Figure 2b, bottom). Our construction requires super-polynomial area. Hence, we have made no effort in proving a concrete bound.

Further, we prove that every biconnected 4-outerplane graph admits an $SC_1$-layout (see Section 4), expanding the class of graphs with $SC_1$-layout from triconnected or Hamiltonian 3-planar graphs [2]. Note that in our result, the outerplane embedding can be prescribed, while in the other results the algorithms need the freedom to choose an appropriate embedding.

We complement our positive results by two negative results; see the last three lines of Table 1. Due to lack of space, the detailed proofs are given in [1]. Many problems remain open: does polynomial area suffice for $SC_2$-layouts of 4-planar graphs? Do larger graph classes admit $SC_1$? What's the computational complexity of deciding $SC_1$?

## 2 Smooth Layouts for Biconnected 4-Planar Graphs

In this section, we prove that any biconnected 4-planar graph admits an $SC_2$-layout. Given a biconnected 4-planar graph, we first compute an $OC_3$-layout, using an algorithm of Liu et al. [9]. Then we turn the result of their algorithm into an $SC_2$-layout.

Liu et al. choose two vertices $s$ and $t$ and compute an $st$-ordering of the input graph. An $st$-ordering is an ordering $(s = 1, 2, \ldots, n = t)$ of the vertices such that every $j$ $(2 < j < n-1)$ has neighbors $i$ and $k$ with $i < j < k$. Then they go through all vertices as prescribed by the $st$-ordering, placing vertex $i$ in row $i$. Calling an edge of which exactly one end-vertex is already drawn an *open edge*, they maintain the following invariant:

($I_1$) In each iteration, every open edge is associated with a *column* (a vertical grid line).

An algorithm of Biedl and Kant [3] yields an $OC_3$-layout similar to that of Liu et al. However, Liu et al. additionally show how to modify their algorithm such that it produces $OC_3$-layouts without the undesirable S-shapes; see Fig. 2b (top).

In their modified algorithm, Liu et al. search for paths in the drawing that consist only of S-shapes; every vertex lies on at most one such path. They place all vertices on

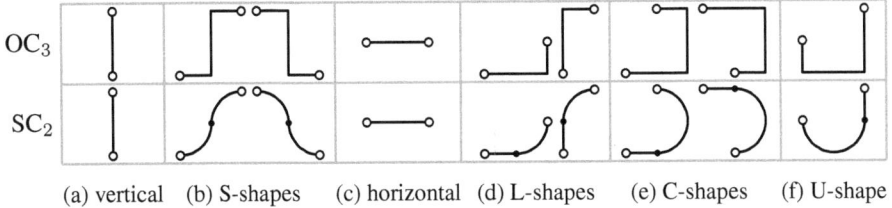

**Fig. 2.** Converting shapes from the $OC_3$-layout to $SC_2$

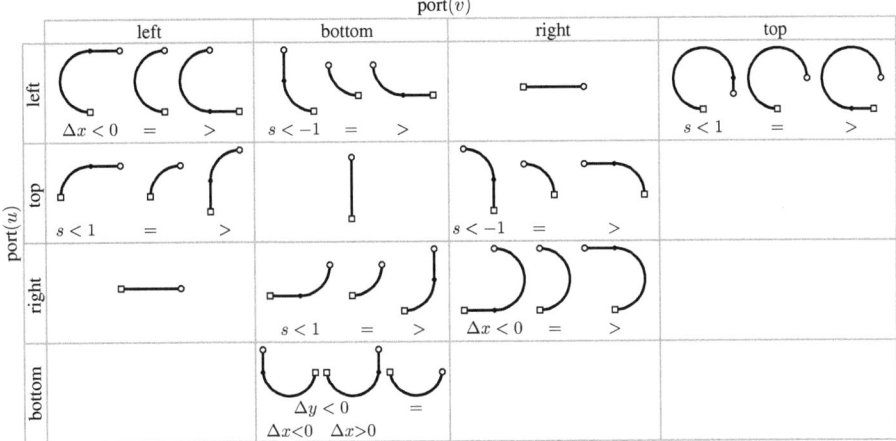

**Fig. 3.** Cases for drawing the edge $(u, v)$ based on the port assignment. In each case, $u$ is the lower of the two vertices $(y(u) < y(v))$. As shorthand, we use $\Delta x = x(u) - x(v)$, $\Delta y = y(u) - y(v)$, and $s = \text{slope}(u, v) = \Delta x / \Delta y$.

such a path in the same row, without changing their column. This essentially converts all S-shapes into horizontal edges. Now every edge (except $(1, 2)$ and $(1, n)$) is drawn as a vertical segment, horizontal segment, L-shape, or C-shape; see Fig. 2. The edge $(1, 2)$ is drawn as a U-shape and the edge $(1, n)$, if it exists, is either drawn as a C-shape or (only in the case of the octahedron) as a three-bend edge that uses the left port of vertex 1 and the top port of vertex $n$.

We convert the output of the algorithm of Liu et al. from $OC_3$ to $SC_2$. The coordinates of the vertices and the port assignment of their drawing define a (non-planar) $SC_2$-layout using the conversion table in Fig 3. In order to avoid crossings, we carefully determine new vertex positions scanning the drawing of Liu et al. from bottom to top.

We now introduce our main tool for the conversion: a *cut*, for us, is a $y$-monotone curve consisting of horizontal, vertical, and circular segments that divides the current drawing into a left and a right part, and only intersects horizontal segments and semicircles of the drawing. In the following, we describe how one can find such a cut from any starting point at the top of the drawing; see Fig. 4. (In spite of the fact that we define the cut going from top to bottom, "to its left" will, as usually, mean "with smaller $x$-coordinate".)

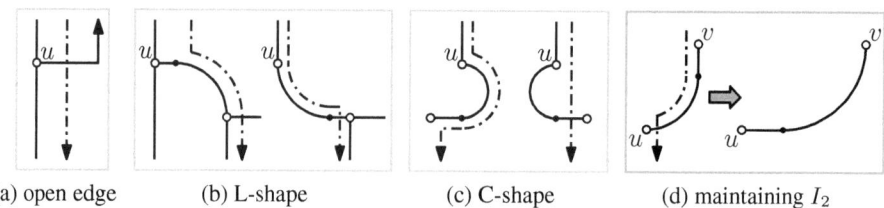

(a) open edge  (b) L-shape  (c) C-shape  (d) maintaining $I_2$

**Fig. 4.** Finding a cut

When such a cut encounters a vertex $u$ to its right with an outgoing edge associated with its left port, then the cut continues by passing through the segment incident to $u$. On the other hand, if the port has an incoming L-shaped or C-shaped edge, the cut just follows the edge. The case when the cut encounters a vertex to its left is handled symmetrically.

Let $v$ be a vertex incident to two incoming C-shapes $(u, v)$ and $(w, v)$. If $y(w) \leq y(u)$ we call the C-shape $(u, v)$ *protected* by $(w, v)$; otherwise, we call it *unprotected*. In order to ensure that a cut passes only through horizontal segments and that our final drawing is planar, our algorithm will maintain the following new invariants:

($I_2$) An L-shape never contains a vertical segment (as in Fig. 2d right); it always contains a horizontal segment (as in Fig. 2d left) or a single quarter-circle.
($I_3$) An unprotected C-shape never contains a horizontal segment incident to its top vertex (as in Fig. 2e right); it always contains a horizontal segment incident to its bottom vertex (as in Fig. 2e left) or no straight-line segment.
($I_4$) The subgraph induced by the vertices that have already been drawn has the same embedding as in the drawing of Liu et al.

Below, we treat L- and C-shapes of complexity 1 as if they had a horizontal segment of length 0 incident to their bottom vertex. Note that we always cut around protected C-shapes, so we will never end up in their interior. Now we are ready to state the main theorem of this section by presenting our algorithm for $SC_2$-layouts.

**Theorem 1.** *Every biconnected 4-planar graph admits an $SC_2$-layout.*

*Proof.* In the drawing $\Gamma$ of Liu et al., vertices are arranged in rows. Let $V_1, \ldots, V_r$ be the partition of the vertex set $V$ in rows $1, \ldots, r$. Following Liu et al., the vertices in each such set induce a path in $G$. We place vertices in the order $V_1, \ldots, V_r$. In this process, we maintain a planar drawing $\Gamma'$ and the invariants $I_1$ to $I_4$. As Liu et al., we place the vertices on the integer grid. We deal with the special edges $(1, 2)$ and $(1, n)$ at the end, leaving their ports, that is, the bottom and left port of vertex 1 and the top port of vertex $n$, open.

For invariant $I_1$, we associate each open edge with the column on which the algorithm of Liu et al. places it. If their algorithm draws the first segment of the open edge horizontally (from the source vertex to the column), we use the same segment for our drawing. We use the same ports for the edges as their algorithm. Thus, our drawing keeps the embedding of Liu et al., maintaining invariant $I_4$.

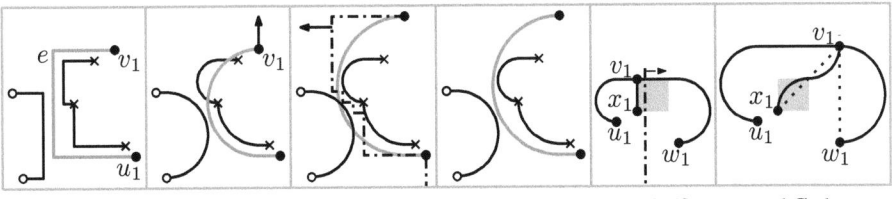

(a) $OC_3$-layout  (b) move $v_1$ up  (c) find a cut  (d) $SC_2$-layout  (e) & (f) protected C-shape

**Fig. 5.** Handling C-shapes

Assume that we have placed $V_1, \ldots, V_{i-1}$ and that the vertices in $V_i$ are $v_1, \ldots, v_c$ in left-to-right order (the case $v_1 = v_c$ is possible; this is the only case in which a vertex can have incoming L- or C-shapes at both its left and right port). Vertex $v_j$ ($1 \leq j \leq c$) is placed in the column with which the edge entering the bottom port of $v_j$ is associated. If the left port of $v_1$ is used by an incoming L- or C-shape $e = (u_1, v_1)$, we place $v_1$ (and the other vertices in $V_i$) on a row high enough so that a smooth drawing of $e$ does not create any crossings with edges lying on the right side of $e$ in $\Gamma$; see Fig. 5b.

In order to make sure that the new drawing of $e$ does not create crossings with edges on the left side of $e$ in $\Gamma$, we need to "push" those edges to the left of $e$. We do this by computing a cut that starts from $v_1$, separates the vertices and edges that lie on the left side of $e$ in $\Gamma$ from those on the right side, passes $u_1$ slightly to the left, and continues downwards as described above; see Fig. 5c. Since, by invariant $I_4$, our drawing so far is planar and each edge is drawn in a $y$-monotone fashion, we can find a cut, too, that is $y$-monotone. We move everything on the left side of the cut further left such that $e$ has no more crossings. Note that the cut intersects only horizontal edge segments. These will simply become longer by the move.

Let $\Delta x_i = x(v_i) - x(u_i)$ and $\Delta y_i = y(v_i) - y(u_i)$ for $i = 1, \ldots, c$. It is possible that the drawing of $e$ violates invariant $I_3$—if $u_1$ lies to the left of $v_1$. We consider two cases. First, assume that the edge $(u_1, v_1)$ is the only incoming C-shape at $v_1$. Note that if $c > 1$ this is always the case. In this case, we simply define a cut that starts slightly to the right of $v_1$, follows $e$, intersects $e$ slightly to the left of $u_1$, and continues downwards. Then we move everything on the left side of the cut by $\Delta x_1 + 1$ units to the left. Next, assume that $c = 1$ and there is another C-shape $(w_1, v_1)$ entering the right port of $v_1$; see Fig. 5e. We assume w.l.o.g. that $y(w_1) \leq y(u_1)$. Let $(x_1, v_1)$ be the edge incident to the bottom port of $v_1$. In this case, we first find a cut that starts slightly to the right of $v_1$, follows $(x_1, v_1)$, passes $x_1$ slightly to the right, and continues downwards. Then we move everything on the right side of the cut by $y(v_1) - y(x_1)$ units to the right. Thus, there is an empty square to the right of $(x_1, v_1)$ of side length $y(v_1) - y(x_1)$. Now we place $v_1$ at the intersection of the slope-1 diagonal through $x_1$ and the vertical line through $w_1$. Due to this placement, we can draw $(x_1, v_1)$ using two quarter-circles with a common horizontal tangent in the top right corner of the empty square; see Fig. 5f. Note that the edge $(u_1, v_1)$ is protected by $(w_1, v_1)$, so it can have a horizontal segment incident to $v_1$. This establishes $I_3$.

It is also possible that the drawing of $e$ violates invariant $I_2$—if slope$(u_1, v_1) > 1$. In this case we define a cut that starts slightly to the left of $v_1$, intersects $e$ and continues downwards. Then we move everything on the left side of the cut by $\Delta y_1$ units to the left. This establishes $I_2$.

We treat $v_c$, the rightmost vertex in the current row, symmetrically to $v_1$.

For the case that $v_1$ does not have incoming C-shapes at both its left and right port, we still have to treat the edges entering vertices $v_1, \ldots, v_c$ from below. Note that these edges can only be vertical or L-shaped. Vertical edges can be drawn without violating the invariants. However, invariant $I_2$ may be violated if an edge $e_i = (u_i, v_i)$ entering the bottom port of vertex $v_i$ is L-shaped; see Fig. 4d. Assume that $x(u_i) < x(v_i)$. In this case we find a cut that starts slightly to the left of $v_i$, follows $e_i$, intersects $e_i$ slightly to the right of $u_i$, and continues downwards. Then we move everything on the left side of the cut by $\Delta y_i$ units to the left. This establishes $I_2$. We handle the case $x(u_i) > x(v_i)$ symmetrically.

We thus place the vertices row by row and draw the incoming edges for the newly placed vertices, copying the embedding of the current subgraph from $\Gamma$. This completes the drawing of $G - \{(1,2), (1,n)\}$. Note that vertex 1 has no incoming edge and vertex 2 has only one incoming edge, that is, $(1,2)$. Thus, the bottom port of both vertices is still unused. We draw the edge $(1,2)$ as a U-shape. Finally, we finish the layout by drawing the edge $(1, n)$. By construction, the left port of vertex 1 is still unused. Note that vertex $n$ has no outgoing edges, so the top port of $n$ is still free. Hence, we can draw the edge $(1, n)$ as a horizontal or vertical segment followed by a three-quarter-circle. To avoid crossings, we may have to move vertex $n$ upwards. This way, we will get a horizontal segment at vertex 1, and the three-quarter-circle will completely lie outside of the rest of the drawing. This completes the proof of Theorem 1. □

## 3 Smooth Layouts for Arbitrary 4-Planar Graphs

In this section, we describe how to create $SC_2$-layouts for arbitrary 4-planar graphs. To achieve this, we decompose the graph into biconnected components, embed them separately and then connect them. For the connection it is important that one of the connector vertices lies on the outer face of its component. Within each component, the connector vertices have degree at most 3; if they have degree 2, we must make sure that their incident edges don't use opposite ports. Following Biedl and Kant [3], we say that a degree-2 vertex $v$ is drawn *with right angle* if the edges incident to $v$ use two neighboring ports.

**Lemma 1.** *Any biconnected 4-planar graph admits an $SC_2$-layout such that all degree-2 vertices are drawn with right angle.*

*Proof.* Let $v$ be a degree-2 vertex. We now show how to adjust the algorithm of Section 2 such that $v$ is drawn with right angle. By construction, the top and the bottom ports of $v$ are used. Let $(u, v)$ be the edge entering $v$ from below (we allow $v = 1$ and $u = 2$). We modify the algorithm such that $(u, v)$ uses the left or right rather than the bottom port of $v$. We consider three cases; $(u, v)$ is either L-shaped, U-shaped, or vertical. These cases are handled when $v$ is inserted into the smooth orthogonal drawing.

First, we assume that $(u, v)$ is L-shaped; see Fig. 6a. Then, we can simply move $v$ to the same row as $u$, making the edge horizontal.

(a) an L-shape becomes a horizontal edge

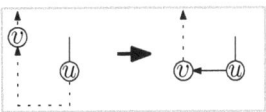
(b) a U-shape becomes a horizontal edge

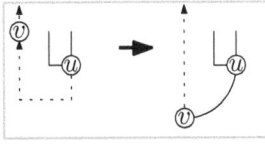
(c) a U-shape becomes an L-shape

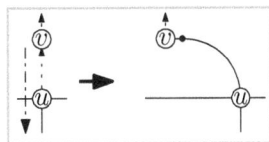
(d) a vertical edge becomes an L-shape

**Fig. 6.** Modification of the placement of degree-2 vertices

Now, we assume that $(u, v)$ is U-shaped; see Fig. 6b, 6c. Then $u = 1$ and $v = 2$ or vice versa. If both have degree 2, we move the higher vertex to the row of the lower vertex (if necessary) and replace the U-shaped edge by a horizontal edge. Otherwise we move the vertex with degree-2, say $v$, downwards to row $y(u) - \Delta x$ such that we can replace the U-shape by an L-shape.

Otherwise, $(u, v)$ is vertical; see Fig. 6d. Then, we compute a cut that starts slightly below $v$, follows $(u, v)$ downwards, passing $u$ to its left. We move all vertices (including $u$, but not $v$) that lie on the right side of this cut (by at least $\Delta y$) to the right. Then we can draw $(u, v)$ as an L-shape that uses the right port of $v$.

Observe that, in each of the three cases, we redraw all affected edges with $SC_2$. Hence, the modified algorithm still yields an $SC_2$-layout. At the same time, all degree-2 vertices are drawn with right angle as desired. □

Now we describe how to connect the biconnected components. Recall that a *bridge* is an edge whose removal disconnects a graph $G$. We call the two endpoints of a bridge *bridge heads*. A *cut vertex* is a vertex whose removal disconnects the graph, but is not a bridge head.

**Theorem 2.** *Any 4-planar graph admits an $SC_2$-layout.*

*Proof.* Let $G_0$ be some biconnected component of $G$, and let $v_1, \ldots, v_k$ be the cut vertices and bridge heads of $G$ in $G_0$. For $i = 1, \ldots, k$, if $v_i$ is a bridge head, let $v'_i$ be the other head of the bridge, otherwise let $v'_i = v_i$. Let $G_i$ be the subgraph of $G$ containing $v'_i$ and the connected components of $G - v'_i$ not containing $G_0$. Following Lemma 1, $G_0$ can be drawn such that all degree-2 vertices are drawn with right angles.

The algorithm of Section 2 that we modified in the proof of Lemma 1 places the last vertex ($n$) at the top of the drawing and thus on the outer face. When drawing $G_i$, we choose $v'_i$ as this vertex. By induction, $G_i$ can be drawn such that all degree-2 vertices are drawn with right angles.

In order to connect $G_i$ to $G_0$, we make $G_0$ large enough to fit $G_i$ into the face that contains the free ports of $v_i$. We may have to rotate $G_i$ by a multiple of 90° to achieve the following. If $v_i$ is a cut vertex, we make sure that $v'_i$ uses the ports of $v_i$ that are free in $G_0$. Then we identify $v_i$ and $v'_i$. Otherwise we make sure that a free port of $v_i$ and a free port of $v'_i$ are opposite. Then we draw the bridge $(v_i, v'_i)$ horizontally or vertically. This completes our proof.

For an example run of our algorithm, see the full version [1]. For graphs of maximum degree 3, we can make our drawings more compact. This is due to the fact that we can avoid C-shaped edges (and hence cuts) completely. In the presence of L-shapes only, it suffices to stretch the orthogonal drawing by a factor of $n$.

**Theorem 3.** *Every biconnected 3-planar graph with $n$ vertices admits an $SC_2$-layout using area $\lfloor n^2/4 \rfloor \times \lfloor n/2 \rfloor$.*

*Proof.* It is known that every biconnected 3-planar graph except $K_4$ has an $OC_2$-layout using area $\lfloor n/2 \rfloor \times \lfloor n/2 \rfloor$ from Kant [7]. Now we use the same global stretching as Bekos et al. when they showed that every $OC_2$-layout can be transformed into an $SC_2$-layout [2, Thm. 2]: we stretch the drawing horizontally by the height of the drawing, that is, by a factor of $\lfloor n/2 \rfloor$. This makes sure that we can replace every bend by a quarter circle without introducing crossings. Figure 7 shows an $SC_1$-layout of $K_4$; completing our proof. □

**Fig. 7.** $SC_1$-layout of $K_4$

## 4  $SC_1$-Layouts of Biconnected 4-Outerplane Graphs

In this section, we consider *4-outerplane* graphs, that is, 4-outerplanar graphs with an outerplanar embedding. We prove that any biconnected 4-outerplane graph admits an $SC_1$-layout. To do so, we first prove the result for a subclass of 4-outerplane graphs, which we call $(2, 3)$-restricted outerplane graphs; then we generalize. Recall that the *weak dual* of a plane graph is the subgraph of the dual graph whose vertices correspond to the bounded faces of the primal graph.

**Definition 1.** *A 4-outerplane graph is called $(2, 3)$-restricted if it contains a pair of consecutive vertices on the outer face, $x$ and $y$, such that $\deg(x) = 2$ and $\deg(y) \leq 3$.*

**Lemma 2.** *Any biconnected $(2, 3)$-restricted 4-outerplane graph admits an $SC_1$-layout.*

*Proof.* Let $x$ and $y$ be two vertices, consecutive on the outer face of the given graph $G$ such that $\deg(x) = 2$ and $\deg(y) \leq 3$. Let also $T$ be the weak dual tree of $G$ rooted at the node, say $v^*$, of $T$ corresponding to the bounded face, say $f^*$, containing both $x$ and $y$. We construct the $SC_1$-layout $\Gamma$ for $G$ by traversing $T$, starting with $v^*$. When we traverse a node of $T$, we draw the corresponding face of $G$ with $SC_1$.

Consider the case when we have constructed a drawing $\Gamma(H)$ for a connected subgraph $H$ of $G$ and we want to add a new face $f$ to $\Gamma(H)$. For each vertex $u$ of $H$, let $p_u = (x(u), y(u))$ denote the point at which $u$ is drawn in $\Gamma(H)$. The *remaining degree* of $u$ is the number of vertices adjacent to $u$ in $G - H$. Since we construct $\Gamma(H)$ face by face, the remaining degree of each vertex in $H$ is at most two. The *free ports* of $u$ are the ones that are not occupied by an edge of $H$ in $\Gamma(H)$. During the construction of $\Gamma$, we maintain the following four invariants:

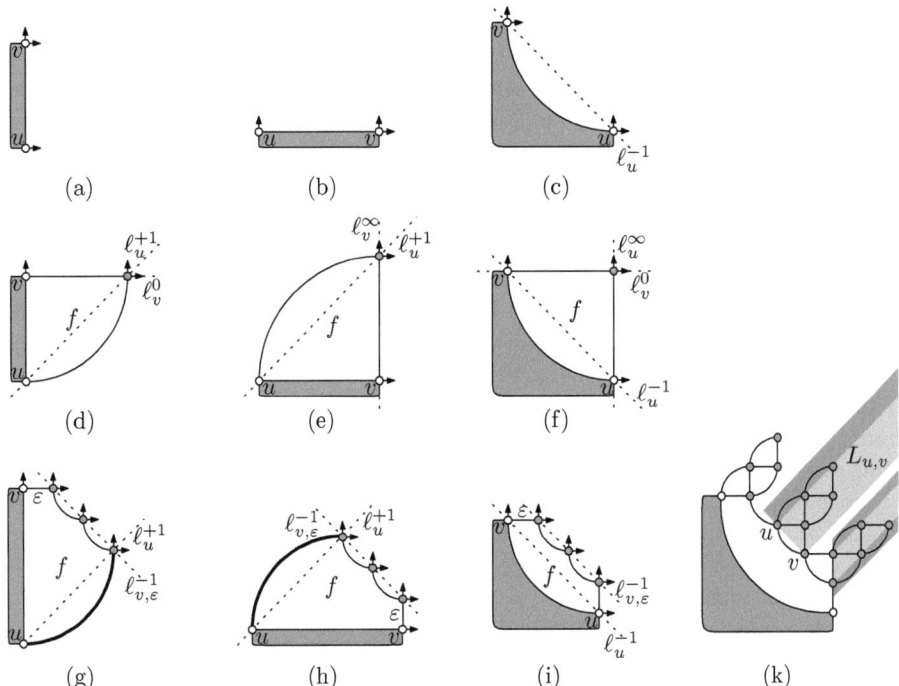

**Fig. 8.** (a)-(i) Different cases that arise when drawing face $f$ of $G$. (k) A sample drawing.

($J_1$) $\Gamma(H)$ is an $SC_1$-layout that preserves the planar embedding of $G$, and each edge is drawn either as an axis-parallel line segment or as a quarter-circle in $\Gamma(H)$. (Note that we do not use semi- and 3/4-circles.)
($J_2$) For each vertex $u$ of $H$, the free ports of $u$ in $\Gamma(H)$ are consecutive around $u$, and they point to the outer face of $\Gamma(H)$.
($J_3$) Vertices with remaining degree exactly 2 are incident to an edge drawn as a quarter-circle.
($J_4$) If an edge $(u,v)$ is drawn as an axis-parallel segment, then at least one of $u$ and $v$ has remaining degree at most 1. If $(u,v)$ is vertical and $y(u) < y(v)$, then $u$ has remaining degree at most 1 and the free port of $u$ in $\Gamma(H)$ is horizontal; see Figs. 8a, 8d and 8g. Symmetrically, if $(u,v)$ is horizontal and $x(u) < x(v)$, then $u$ has remaining degree at most 1 and the free port of $u$ in $\Gamma(H)$ is vertical; see Figs. 8b, 8e and 8h.

We now show how we add the drawing of the new face $f$ to $\Gamma(H)$. Since $G$ is biconnected and outerplanar, and due to the order in which we process the faces of $G$, $f$ has exactly two vertices, say $u$ and $v$, which have already been drawn (as $p_u$ and $p_v$). The two vertices are adjacent. Depending on how the edge $(u,v)$ is drawn in $\Gamma(H)$, we draw the remaining vertices and edges of $f$.

Let $k \geq 3$ be the number of vertices on the boundary of $f$. The slopes of the line segment $\overline{p_u p_v}$ is in $\{-1, 0, +1, \infty\}$, where $\infty$ means that $\overline{p_u p_v}$ is vertical. For $s \in \{-1, 0, +1, \infty\}$, we denote by $\ell_u^s$ the line with slope $s$ through $p_u$. Similarly, we denote by $\ell_{u,\varepsilon}^s$ the line with slope $s$ through the point $(x(u) + \varepsilon, y(u))$, for some $\varepsilon > 0$. Figs. 8d–8f show the drawing of $f$ for $k = 3$, and Figs. 8g–8i for any $k \geq 4$.

Note that the lengths of the line segments and the radii of the quarter-circles that form $f$ are equal (except for the radii of the bold-drawn quarter-circles of Figs. 8g and 8h which are determined by the remaining edges of $f$). Hence, the lengths of the line segments and the radii of the quarter-circles that form any face that is descendant of face $f$ in $T$ are smaller than or equal to the lengths of the line segments and the radii of the quarter-circles that form $f$. Our construction ensures that all vertices of the subgraph of $G$ induced by the subtree of $T$ rooted at $f$ lie in the interior or on the boundary of the diagonal semi-strip $L_{uv}$ delimited by $\ell_u^{+1}$, $\ell_v^{+1}$, and $\overline{p_u p_v}$ (see Fig. 8k). The only edges of this subgraph that are drawn in the complement of $L_{uv}$ (and are potentially involved in crossings) are incident to two vertices that both lie on the boundary of $L_{uv}$. In this particular case, however, the degree restriction implies that $L_{uv}$ is surrounded from above and/or below by two empty diagonal semi-strips of at least half the width of semi-strip $L_{uv}$, which is enough to ensure planarity for two reasons. First, any face that is descendant of face $f$ in $T$ is formed by line segments and quarter-circles of radius that are at most as big as the corresponding ones of face $f$. Second, due to the degree restrictions, if two neighboring children of $f$ are triangles, the left one cannot have a right child and vice versa.

Let us summarize. Figures 8d–8i show that the drawing of $f$ ensures that invariants $(J_1)$–$(J_4)$ of our algorithm are satisfied for $H \cup \{f\}$. We begin by drawing the root face $f^*$. Since $G$ is $(2, 3)$-restricted, $f^*$ has two vertices $x$ and $y$ consecutive on the outer face with $\deg(x) = 2$ and $\deg(y) \leq 3$. We draw edge $(x, y)$ as a vertical line segment. Then the remaining degrees of $x$ and $y$ are 1 and 2, respectively, which satisfies the invariants for face $f^*$. Hence, we complete the drawing of $f^*$ as in Fig. 8d or 8g. Traversing $T$ in pre-order, we complete the drawing of $G$. □

Next, we show how to deal with general biconnected 4-outerplane graphs. Suppose $G$ is not $(2, 3)$-restricted. As the following lemma asserts, we can always construct a biconnected $(2, 3)$-restricted 4-outerplane graph by deleting a vertex of degree 2 from $G$.

**Lemma 3.** *Let $G = (V, E)$ be a biconnected 4-outerplane graph that is not $(2, 3)$-restricted. Then $G$ has a degree-2 vertex whose removal yields a $(2, 3)$-restricted biconnected 4-outerplane graph.*

*Proof.* The proof is by induction on the number of vertices. The base case is a maximal biconnected outerplane graph on six vertices, which is the only non-$(2, 3)$-restricted graph with six or less vertices. It is easy to see that in this case the removal of any degree-2 vertex yields a biconnected $(2, 3)$-restricted 4-outerplane graph. Now assume that the hypothesis holds for any biconnected 4-outerplane graph with $k \geq 6$ vertices. Let $G_{k+1}$ be a biconnected 4-outerplane graph on $k + 1$ vertices, which is not $(2, 3)$-restricted. Let $\mathcal{F}$ be a face of $G_{k+1}$ that is a leaf in its weak dual. Then $\mathcal{F}$ contains only one internal edge and exactly two external edges since, if it contained more than two external edges, $G_{k+1}$ would be $(2, 3)$-restricted. Therefore, $\mathcal{F}$ consists of three vertices, say $a$, $b$ and $c$, consecutive on the outer face and $\deg(a) = \deg(c) = 4$, since otherwise $G_{k+1}$ would be $(2, 3)$-restricted. By removing $b$, we obtain a new graph, say $G_k$, on $k$ vertices. If $a$ or $c$ is incident to a degree-2 vertex in $G_k$, then $G_k$ is $(2, 3)$-restricted.

Otherwise, by our induction hypothesis, $G_k$ has a degree-2 vertex whose removal yields a $(2, 3)$-restricted outerplanar graph. Since this vertex is neither adjacent to $a$ nor $c$, the removal of this vertex makes $G_{k+1}$, too, $(2, 3)$-restricted. □

**Theorem 4.** *Any biconnected 4-outerplane graph admits an $SC_1$-layout.*

*Proof.* If the given graph $G$ is $(2, 3)$-restricted, then the result follows from Lemma 2. Thus, assume that $G$ is not $(2, 3)$-restricted. Then, $G$ contains a degree-2 vertex, say $b$, whose removal yields a biconnected $(2, 3)$-restricted 4-outerplane graph, say $G'$. Hence, we can apply the algorithm of Lemma 2 to $G'$ and obtain an outerplanar $SC_1$-layout $\Gamma(G')$ of $G'$. Since this algorithm always maintains consecutive free ports for each vertex and the neighbors of $b$ are on the outer face of $\Gamma(G')$, we insert insert $b$ and its two incident edges to obtain an $SC_1$-layout $\Gamma(G)$ of $G$ as follows. Let $a$ and $c$ be the neighbors of $b$ and assume w.l.o.g. that $c$ is drawn above $a$. If edge $(a, c)$ is drawn as a quarter-circle, then a 3/4-circle arc from $p_c$ to $p_b$ and a quarter-circle from $p_b$ to $p_a$ suffice. Otherwise, line segment $\overline{p_a p_b}$ and a quarter-circle from $p_b$ to $p_c$ do the job. □

# References

1. Alam, M.J., Bekos, M.A., Kaufmann, M., Kindermann, P., Kobourov, S.G., Wolff, A.: Smooth orthogonal drawings of planar graphs. Arxiv report arxiv.org/abs/1312.3538 (2013)
2. Bekos, M.A., Kaufmann, M., Kobourov, S.G., Symvonis, A.: Smooth orthogonal layouts. In: Didimo, W., Patrignani, M. (eds.) GD 2012. LNCS, vol. 7704, pp. 150–161. Springer, Heidelberg (2013)
3. Biedl, T., Kant, G.: A better heuristic for orthogonal graph drawings. Comput. Geom. Theory Appl. 9(3), 159–180 (1998)
4. Duncan, C.A., Eppstein, D., Goodrich, M.T., Kobourov, S.G., Nöllenburg, M.: Lombardi drawings of graphs. J. Graph Algorithms Appl. 16(1), 85–108 (2012), http://dx.doi.org/10.7155/jgaa.00251
5. Duncan, C.A., Eppstein, D., Goodrich, M.T., Kobourov, S.G., Nöllenburg, M.: Drawing trees with perfect angular resolution and polynomial area. Discrete Comput. Geom. 49(2), 157–182 (2013)
6. Garg, A., Tamassia, R.: On the computational complexity of upward and rectilinear planarity testing. SIAM J. Comput. 31(2), 601–625 (2001)
7. Kant, G.: Drawing planar graphs using the canonical ordering. Algorithmica 16(1), 4–32 (1996)
8. Leiserson, C.E.: Area-efficient graph layouts (for VLSI). In: Proc. 21st Annu. IEEE Symp. Foundat. Comput. Sci. (FOCS 1980), pp. 270–281 (1980)
9. Liu, Y., Morgana, A., Simeone, B.: A linear algorithm for 2-bend embeddings of planar graphs in the two-dimensional grid. Discrete Appl. Math. 81(1-3), 69–91 (1998)
10. Seemann, J.: Extending the Sugiyama algorithm for drawing UML class diagrams: Towards automatic layout of object-oriented software diagrams. In: Di Battista, G. (ed.) GD 1997. LNCS, vol. 1353, pp. 415–424. Springer, Heidelberg (1997)
11. Tamassia, R.: On embedding a graph in the grid with the minimum number of bends. SIAM J. Comput. 16(3), 421–444 (1987)
12. Waldherr, H.: Network Gmunden–Vöcklabruck–Salzkammergut of the Publ. Transp. Assoc. OÖVG, Austria (November 2012), http://www.ooevv.at/uploads/media/OOE2_Salzkammergut_V17_END.pdf (accessed September 10, 2013)

# Drawing $HV$-Restricted Planar Graphs

Stephane Durocher[1,*], Stefan Felsner[2,**],
Saeed Mehrabi[1,***], and Debajyoti Mondal[1,***]

[1] Department of Computer Science, University of Manitoba, Canada
[2] Institut für Mathematik, Technische Universität Berlin, Germany
{durocher,mehrabi,jyoti}@cs.umanitoba.ca, felsner@math.tu-berlin.de

**Abstract.** A strict orthogonal drawing of a graph $G = (V, E)$ in $\mathbb{R}^2$ is a drawing of $G$ such that each vertex is mapped to a distinct point and each edge is mapped to a horizontal or vertical line segment. A graph $G$ is $HV$-restricted if each of its edges is assigned a horizontal or vertical orientation. A strict orthogonal drawing of an $HV$-restricted graph $G$ is good if it is planar and respects the edge orientations of $G$. In this paper we give a polynomial-time algorithm to check whether a given $HV$-restricted plane graph (i.e., a planar graph with a fixed combinatorial embedding) admits a good orthogonal drawing preserving the input embedding, which settles an open question posed by Mañuch, Patterson, Poon and Thachuk (GD 2010). We then examine $HV$-restricted planar graphs (i.e., when the embedding is not fixed). Here we completely characterize the 2-connected maximum-degree-three $HV$-restricted outerplanar graphs that admit good orthogonal drawings.

## 1 Introduction

An *orthogonal drawing* $\Gamma$ of an undirected graph $G = (V, E)$ in $\mathbb{R}^2$ is a drawing of $G$ in the plane, where each vertex of $G$ is mapped to a distinct point and each edge of $G$ is mapped to an orthogonal polyline. $\Gamma$ is called *planar* if no two edges in $\Gamma$ cross, however, two edges can meet at their common endpoints. Otherwise, the drawing is a *non-planar orthogonal drawing*. Orthogonal drawings have been extensively studied over the last two decades [1,3,8,14,16] because of its applications in many practical fields such as VLSI floor-planning, circuit schematics, and entity relationship diagrams.

An orthogonal drawing is *strict* if every edge in the drawing is represented by a single vertical and horizontal line segment. In 1987, Tamassia [14] gave a polynomial-time algorithm to decide whether a plane graph (i.e., when the embedding is fixed) admits a strict orthogonal drawing preserving the input

---

[*] Work of the author is supported in part by the Natural Sciences and Engineering Research Council of Canada (NSERC).
[**] Work of the author is supported in part by DFG grant FE-340/7-2 and ESF EuroGIGA project GraDR.
[***] Work of the author is supported in part by a University of Manitoba Graduate Fellowship.

embedding. Later, Garg and Tamassia [6] proved that deciding strict orthogonal drawability is NP-hard for planar graphs (i.e., when the embedding is not fixed). However, polynomial time algorithms have been developed for some well-known subclasses of planar graphs. For example, Di Battista et al. [1] showed that the problem is polynomial-time solvable for series-parallel graphs and maximum-degree-three planar graphs. Nomura et al. [13] showed that every maximum-degree-three outerplanar graph admits a planar strict orthogonal drawing if and only if it contains no cycle of three vertices.

Many variants of strict orthogonal drawings impose constraints on how the edges of the input graph have to be drawn. One of these variants describes the input graph $G$ as an *LRDU-restricted graph* that associates each vertex-edge incidence of $G$ with an orientation, i.e., left(L), right(R), up(U), or down(D), and asks to find an orthogonal drawing of $G$ that respects the prescribed orientations. Another variant considers *HV-restricted graphs*, where the orientation of an edge is either horizontal(H), or vertical(V). By a *good orthogonal drawing* we denote a planar strict orthogonal drawing that preserves the input edge orientations.

In this paper we only examine strict orthogonal drawings of $HV$-restricted plane and planar graphs, and hence from now on we omit the term 'strict'.

**$HV$-Restricted Plane Graphs.** In 1985, Vijayan and Wigderson [15] gave an algorithm that can decide in linear time whether an $LRDU$-restricted plane graph admits a good orthogonal drawing, but takes $O(n^2)$ time to construct such a drawing when it exists. Later, Hoffmann and Kriegel [7] gave a linear-time construction. The task of characterizing $HV$-restricted plane graphs is more involved. The difficulty arises from the exponential number of choices for drawing $HV$-restricted paths, where the drawing of an $LRDU$-restricted path is unique, as illustrated in Figures 1(a)–(c). Recently, Maňuch et al. [10] examined several results on the non-planar orthogonal drawings of $LRDU$- and $HV$-restricted graphs. They proved that non-planar orthogonal drawability maintaining edge orientations can be decided in polynomial-time for $LRDU$-restricted graphs, but is NP-hard for $HV$-restricted graphs. An interesting open question in this context, as posed by Maňuch et al. [10], is to determine the complexity of deciding good orthogonal drawability of $HV$-restricted plane graphs. In Section 2 we settle this question by giving a polynomial-time algorithm to recognize $HV$-restricted plane graphs. Here we assume that a planar embedding of the input graph is given, and our algorithm decides whether there exists a solution that respects the input embedding.

**$HV$-Restricted Planar Graphs.** A problem analogous to drawing $LRDU$-restricted graphs in $\mathbb{R}^2$ has been well studied in $\mathbb{R}^3$, but polynomial-time algorithms are known only for cycles [4] and theta graphs [5]. The exponential number of possible orthogonal embeddings in $\mathbb{R}^3$ makes the problem very difficult. Similarly, we find the problem of characterizing $HV$-restricted planar graphs that admit good orthogonal drawings in $\mathbb{R}^2$ nontrivial even for outerplanar graphs, where the difficulty arises from the exponential number of choices for plane embeddings of the input graph.

To further illustrate the challenge, here we prove that the $HV$-restricted outerplanar graph of Figure 1(d) does not admit a good orthogonal drawing. Suppose for a contradiction that $\Gamma$ is a good orthogonal drawing of $G$, and consider the drawing of the face $F = (a, b, ..., f)$ in $\Gamma$. Since the edges $(a, b)$ and $(c, d)$ are horizontally oriented and $(a, f)$ is vertically oriented, either $(a, b)$ lies above $(e, f)$, or $(e, f)$ lies above $(a, b)$ in $\Gamma$. If $(a, b)$ lies above $(e, f)$ as in Figure 1(e), then the drawing of cycle $a, b, i, j$ would create an edge crossing (irrespective of whether it lies inside or outside of $F$). Similarly, if $(e, f)$ lies above $(a, b)$ as in Figure 1(f), then the drawing of cycle $e, f, h, g$ would create an edge crossing. Drawing both of these cycles without crossing would imply a unique drawing of $F$, as shown in Figure 1(g). However, in this case we cannot draw the cycle $c, d, k, l$ without edge crossings. In Section 3 we characterize 2-connected maximum-degree-three outerplanar graphs that admit good orthogonal drawings. Our proof is constructive, i.e., given an $HV$-restricted 2-connected maximum-degree-three outerplanar graph $G$, in polynomial time we can decide whether $G$ admits a good orthogonal drawing, and find such a drawing if it exists. Note that the construction can choose any feasible embedding (i.e., the embedding is not fixed), and the output is not necessarily outerplanar.

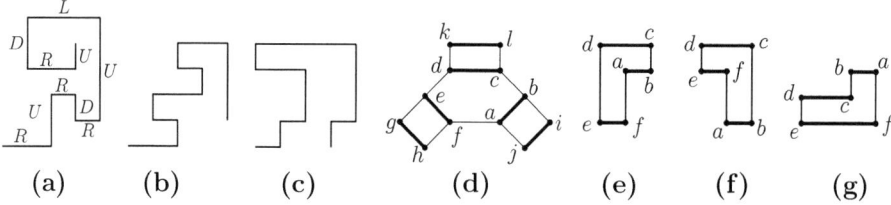

**Fig. 1.** (a) Drawing of an $LRDU$-restricted path. (b)–(c) Two different drawings of an $HV$-restricted path. (d) An $HV$-restricted outerplanar graph $G$ with maximum degree three, where the horizontal and vertical orientations are shown in black and gray, respectively. (e)–(g) Drawing of the face $F$.

## 2 Drawing $HV$-Restricted Plane Graphs

In this section we give a polynomial-time algorithm that checks whether a given $HV$-restricted plane graph admits a good orthogonal drawing that preserves the input embedding. If the answer is affirmative, the algorithm certifies its answer by constructing a good orthogonal drawing.

We will first identify some necessary conditions and later show that they are also sufficient for the existence of the good drawing. The first condition is that every vertex has most two incident edges with label $H$ and at most two with label $V$, and if the degree is four, the labels alternate. This condition is easily checked and from now on we assume it to be satisfied by the input.

Assume that a good drawing exists and consider a face $f$ in the drawing. The face is represented by a polygon, hence, if $f$ has $k$ corners, then the sum of all interior angles of $f$ must be $(k-2)\pi$ (the outer face makes an exception, here the angles sum to $(k+2)\pi$). Since $f$ is an orthogonal polygon, the angle contributed by each corner is a multiple of $\pi/2$. From the given edge orientations we can infer the angle of some corners precisely: if a corner has two incident edges with the same label, then it contributes an angle of $\pi$, and if a corner corresponds to a vertex of degree one, it contributes $2\pi$. The interesting corners are those where the incident edges have different labels, these corners contribute either $\pi/2$ or $3\pi/2$. Dual to the angle condition for faces we also have the obvious condition for vertices: around each vertex the sum of angles is $2\pi$.

Associate a variable $x_c$ with each corner $c$ of the plane graph. The above conditions can all be written as linear equations in these variables. This yields a linear system $Ax = b$ and the unified necessary condition that the system has a solution $\bar{x}$ where each component $\bar{x}_c$ is in $\{1, 2, 3, 4\}$. Such a solution is called a *global admissible angle assignment*. Similar quests for global angle assignments have been studied in rectangular drawing problems, where Miura et al. [11] reduced the problem to perfect matching, and in the context of orthogonal drawing with bends, where Tamassia [14] modeled an angle assignment problem with minimum-cost maximum-flow.

Instead of directly using the linear system stated above, we use the fact that the value of some variables $x_c$ is prescribed by the input. The value for the remaining variables and hence a global admissible angle assignment can be determined using a maximum-flow problem.

To construct the flow network start with the angle graph $A(G)$ of the plane graph $G$. The vertex set is $V_{A(G)} = V_G \cup F_G$, i.e., the vertices of $A(G)$ are the vertices and faces of $G$ or stated in just another way: the vertices of $A(G)$ are the vertices of $G$ together with the vertices of the dual $G^*$. The edges of $A(G)$ correspond to the corners of $G$: if $v \in V_G$ and $f \in V_F$ are incident at a corner $c$ then there is an edge $e_c = (v, f)$ in $E_{A(G)}$.

Next step is to remove an edge $e_c = (v, f)$ from $A(G)$ when the value of the variable $x_c$ is prescribed by the input, i.e., in the following situations:

(a) If the two edges of a corner have the same orientation and the edges are distinct, then the corner is assigned a $\pi$ angle, i.e., $x_c = 2$.
(b) If the vertex corresponding to a corner is of degree one, then the corner is assigned a $2\pi$ angle, i.e., $x_c = 4$.
(c) If the two edges of a corner have different orientations and the vertex is of degree three or more, then the corner is assigned a $\pi/2$ angle, i.e., $x_c = 1$.

Let $A^\star(G)$ be the graph after removing all these edges. Since $A(G)$ is a plane graph the same is true for $A^\star(G)$. Figure 2(a) shows an example of a graph $G$ together with the network $A^\star(G)$.

Since we want to use a fast maximum-flow algorithm, we describe the flow-problem using a planar flow network with multiple sources and sinks. It only remains to decide for some vertices of degree two in $G$ which of its corners is of size $\pi/2$ and which is of size $3\pi/2$. We model a $\pi/2$ corner with a flow of one unit entering the corresponding vertex.

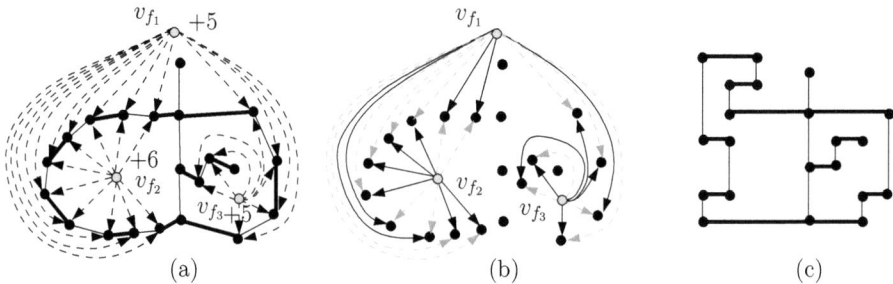

**Fig. 2.** (a) An $HV$-restricted plane graph $G$ (induced by solid edges), and its corresponding flow network $A^\star(G)$ (induced by dotted edges). The edges with horizontal (respectively, vertical) orientations in $G$ are bold (respectively, thin). (b) A feasible flow in $A^\star(G)$, where each solid edge correspond to one unit of flow. (c) A corresponding orthogonal drawing of $G$.

An original vertex $v \in V_G$ is incident to an edge in $A^\star(G)$ if and only if $v$ is a vertex of degree two in $G$. With these vertices we assign a demand of 1. The capacities of all the edges are also restricted to 1. Finally, we have to set the excess of all $f \in F_G$. We know the total angle sum of $f$ and the angles that have been assigned in the reduction step from $A(G)$ to $A^\star(G)$. Since all the remaining angles are of size $\pi/2$ or $3\pi/2$, we can compute how many of size $3\pi/2$ are needed, this number $z_f$ is the excess of $f$. (Note that if the computation yields a $z_f$ that is not an integer, then $G$ does not admit a good orthogonal realization). Similarly, we can also compute the number $z'_f$ of $\pi/2$ angles that we need. For example, for the face $f_2$ in Figure 2(a), we consider $3z_{f_2} + z'_{f_2} = 18$ and $z_{f_2} + z'_{f_2} = 10$, which solves to $(z'_{f_2}, z_{f_2}) = (4, 6)$. Since all edges $e_c \in E_{A^\star(G)}$ connect a source $f$ to a sink $v$, we may think of them as directed edges $f \to v$. Figure 2(b) illustrates a maximum flow for the flow-network of Figure 2(a).

We claim that a flow satisfying all the constraints (demand/excess/capacity) exists if and only if $G$ admits a good orthogonal drawing preserving the input embedding. If a flow $y \in \{0,1\}^{E_{A^\star(G)}}$ exists, then we get a solution vector for the linear system by defining $x_c = 3 - 2y_c$ for all $e_c \in E_{A^\star(G)}$. Together with the variables defined by conditions (a) – (c) we obtain a global admissible angle assignment which by definition satisfies:

1. The sum of angles around each vertex $v$ in $G$ is $2\pi$.
2. For every edge $(u, v)$ in $G$, the angle assignment at the corners of $u$ and $v$ is consistent with respect to the two faces that are incident to $(u, v)$.
3. The total assigned angle of every face $f$ is the angle sum required for polygons with that many corners. All angles are multiples of $\pi/2$, i.e., the induced representation is orthogonal.

These conditions on an angle assignment are sufficient to construct a plane orthogonal representation that respects the input embedding [14]. In fact the orthogonal drawing can be computed in linear time. Figure 2(c) shows an orthogonal representation corresponding to the flow of Figure 2(b).

For the converse, if $G$ admits a good orthogonal drawing $\Gamma$ respecting the input embedding, then the angles at the degree two vertices readily imply a flow in the network satisfying the constraints. We thus obtain the following theorem.

**Theorem 1.** *Given an HV-restricted plane graph $G$ with $n$ vertices, one can check in $T(n)$ time whether $G$ admits a good orthogonal drawing preserving the input embedding, and construct such a drawing if it exists. Here $T(n)$ is the time to find maximum flows in multiple-source multiple-sink directed planar graphs.*

Since the maximum flow problem for a multiple-source and multiple-sink directed planar graph can be solved in $O(n \log^3 n)$-time [2], one can check whether a given $HV$-restricted plane graph that admits a good orthogonal drawing preserving the input embedding in $O(n \log^3 n)$ time. Note that we precisely know the production or demand of each node in the flow network, and hence we are actually finding a feasible flow. There are faster algorithms in such cases, e.g., Klein et al. [9] gave an algorithm to find a feasible integral flow in $O(n \log^2 n)$-time. Later, Mozes and Wulff-Nilsen [12] improved the running time to $O(n \log^2 n / \log \log n)$.

## 3 Drawing 2-Connected Outerplanar Graphs with $\Delta = 3$

In this section we give a polynomial-time algorithm to determine whether an arbitrary 2-connected $HV$-restricted outerplanar graph with maximum degree three admits a good orthogonal drawing, and construct such a drawing if it exists. Note that the good orthogonal drawing we produce is not necessarily an outerplanar embedding. We first introduce some notation.

Let $G$ be an $HV$-restricted planar graph. By a *segment* of $G$, we denote a maximal path in $G$ such that all the edges on that path have the same orientation. A graph is *outerplanar* if it admits a planar drawing with all its vertices on the outer face. Let $G$ be a 2-connected $HV$-restricted embedded outerplanar graph with $\Delta = 3$, where $\Delta$ is the maximum degree of $G$. Let $e$ be an edge of $G$. Then by $\lambda_e$ we denote the orientation of $e$ in $G$. Let $F$ be an inner face of $G$. Note that $G$ is an embedded graph. Thus any edge of $G$ is an *inner edge* if it does not lie on the boundary of the outer face of $G$, and all the remaining edges of $G$ are the *outer edges*. An inner edge $e$ of $G$ on the boundary of $F$ is called *critical* if the two edges preceding and following $e$ have the same orientation that is different from $\lambda_e$. For example, in Figure 1(d), the edge $(a,b)$ is a critical edge of the inner face $F = (a, b, ..., f)$. An edge $e$ is *h-critical* (respectively, *v-critical*) if it is a critical edge and $\lambda_e = H$ (respectively, $\lambda_e = V$). For some inner face $F$ in $G$, let $E_v(F)$ and $E_h(F)$ be the number of distinct edges of $F$ with vertical and horizontal orientations, respectively. By $C_v(F)$ and $C_h(F)$ we denote the number of $v$-critical and $h$-critical edges of $F$.

Let $pqrs$ be a rectangle, and let $a$ and $b$ be two points in the proper interior of $qr$ and $rs$, respectively, as shown in Figures 3(a) and (b). Construct a rectangle $sbcd$, where $c$ and $d$ lie outside of the rectangle $pqrs$. Then the region consisting of the rectangles $pqrs$ and $sbcd$ is called a *flag*. A flag includes all the segments on its boundary except the segment $aq$. The rectangles $pqrs$ and $sbcd$ are called the *banner* and *post*, respectively. The segments $ar$ and $br$ are called the *borders* of the flag.

### 3.1 Necessary and Sufficient Conditions

Throughout this section, $G$ denotes an arbitrary 2-connected $HV$-restricted embedded outerplanar graph with $\Delta = 3$; see Figure 3(c) for an example. We now prove the following theorem, which is the main result of this section.

**Theorem 2.** *Let $G$ be a 2-connected $HV$-restricted embedded outerplanar graph with maximum degree three. Then $G$ admits a good planar orthogonal drawing if and only if the following three conditions hold.*

($C_1$) *For every inner face $f$, the sequence of orientations of the edges in clockwise order contains $HVHV$ as a subsequence.*
($C_2$) *For every inner face $f$, if $C_v(f) = E_v(f)$, then $C_v(f)$ is even. Similarly, if $C_h(f) = E_h(f)$, then $C_h(f)$ is even.*
($C_3$) *Every vertex of $G$ has at most two edges of the same orientation.*

### 3.2 Necessity

We first show that Conditions ($C_1$)–($C_3$) are necessary for $G$ to admit a good planar orthogonal drawing. We use the following two lemmas.

**Lemma 1.** *Let $\Gamma$ be a good orthogonal drawing of $G$, and let $(b,c)$ be an inner edge of some face $f = (a, b, c, d, \ldots, a)$. Figure 3(d) illustrates an example. Since $(b,c)$ is an inner edge, there is another face $f' = (b, x, \ldots, y, c, b)$ that does not contain any edge of $f$ except $(b, c)$. Let $H^+$ and $H^-$ be the two half-planes determined by the straight line through $(b, c)$. If $(b, c)$ is a critical edge in $f$, then either both $(a, b)$ and $(c, d)$ lie in $H^+$, or both lie in $H^-$.*

*Proof.* Without loss of generality assume that $\lambda_{bc} = H$. Since $(b,c)$ is a critical edge, $\lambda_{bc} \neq \lambda_{ab}$ and $\lambda_{ab} = \lambda_{cd}$. If $(a,b)$ and $(c,d)$ lie in $H^+$ and $H^-$, respectively, then one of $x$ and $y$ must lie interior to $f$ and the other must lie exterior to $f$. Therefore, the path $b, x, ..., y, c$ must create an edge crossing with $f$, which contradicts that $\Gamma$ is a good orthogonal drawing. □

Let $x(v)$ and $y(v)$ denote the $x$- and $y$-coordinates of a vertex $v$. We now use Lemma 1 to prove the following.

**Lemma 2.** *Let $\Gamma$ be good orthogonal drawing of $G$. Let $f$ be an inner face in $\Gamma$, and let $(a, b)$ and $(c, d)$ be two edges on $f$ (without loss of generality assume that $(a, b)$ is above $(c, d)$), where $\lambda_{ab} = \lambda_{cd} = H$, $x(a) > x(b)$ and $x(d) > x(c)$. Let $P = (a, b, ..., c, d)$ be a path on the boundary of $f$ in anticlockwise order, e.g., see the path $P_l$ in Figure 3(e). If all the vertically oriented edges of $P$ are critical, then the number of such critical edges on $P$ must be odd. This property holds symmetrically for $P = (b, a, ..., d, c)$.*

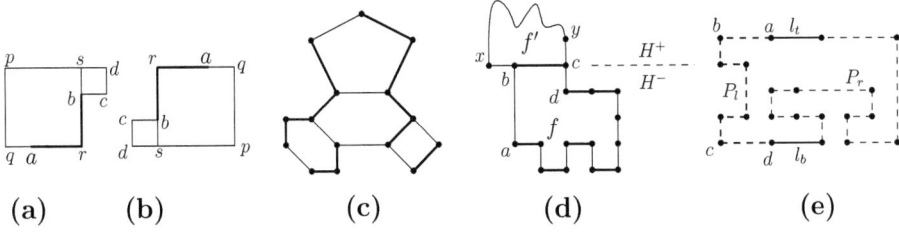

**Fig. 3.** (a)–(b) Two flags, where the borders are shown in bold. (c) An outerplanar graph $G$ with $\Delta = 3$. (d) Illustration for Lemma 1. (e) Illustration for $P_l$ and $P_r$, where $P_l$ contains three $v$-critical edges and $P_r$ contains five $v$-critical edges.

*Proof.* Consider a traversal of the edges of $P$ starting at $a$. Let $e$ be a $v$-critical edge on $P$, and let $e'$ and $e''$ be the edges preceding and following $e$, respectively. By Lemma 1, $e'$ and $e''$ must lie on the same side of $e$ in $\Gamma$. Therefore, if we traverse $e'$ from left to right, then we have to traverse $e''$ from right to left, and vice versa. In other words, every $v$-critical edge reverses the direction of traversal. Since we traverse $(a, b)$ and $(c, d)$ from opposite directions and all the vertically oriented edges of $P$ are critical, we need an odd number of $v$-critical edges on $P$ to complete the traversal. □

We are now ready to prove the necessity part of Theorem 2.

If $(C_1)$ does not hold for some $f$, then the face $f$ does not admit a planar orthogonal drawing. Because, drawing $f$ would require the sum of the interior angles of the corresponding polygon to be at least $2\pi$.

If $(C_2)$ does not hold, then without loss of generality assume that for some $f$, $C_v(f) = E_v(f)$ and $C_v(f)$ is odd. Let $\Gamma_f$ be a drawing of $f$ such that $l_t$ and $l_b$ are topmost and bottommost horizontal edges in $\Gamma_f$. Then we can find two disjoint paths $P_l$ and $P_r$ by traversing $f$ anticlockwise and clockwise from $l_t$ to $l_b$, respectively, as shown in Figure 3(e). Since $C_v(f)$ is odd, either $P_l$ or $P_r$ must contain an even number of $v$-critical edges, which contradicts Lemma 2.

If $(C_3)$ does not hold at some vertex $v$, then the drawing of its incident edges would contain edge overlapping.

### 3.3 Sufficiency

To prove the sufficiency we assume that $G$ satisfies $(C_1)$–$(C_3)$, and then construct a good orthogonal drawing of $G$. The idea is to first draw an arbitrary inner face $f$ of $G$, and then the other faces of $G$ by a depth first search on the faces of $G$ starting at $f$.

Let $f = (v_1, v_2, \ldots, v_r, \ldots, v_s, \ldots, v_t(= v_1))$ be the vertices of $f$ in clockwise order. Let $P = (v_r, \ldots, v_s, \ldots, v_t)$ be a maximal path on $f$ such that all the edges on path $P_v = (v_r, \ldots, v_s)$ (respectively, $P_h = (v_s, \ldots, v_t)$) have vertical (respectively, horizontal) orientation. The maximality of $P$ ensures that $\lambda_{v_1 v_2} = V$ and $\lambda_{v_{r-1} v_r} = H$. An example of such a path $P$ in the face of Figure 4(a) is $a(=v_r), b, c(=v_s), d, e(=v_t)$. Observe that $\lambda_{v_1 v_2} = \lambda_{eg} = V$ and $\lambda_{v_{r-1} v_r} = \lambda_{ia} = H$. We now have the following lemma.

**Lemma 3.** *Given an inner face $f$ of $G$ that satisfies conditions $(C_1)$–$(C_3)$, and a drawing of two consecutive segments $P_h$ and $P_v$ of $f$. One can find a good orthogonal drawing $\Gamma_f$ of $f$ that satisfies the following properties.*

- *Lemma 1 holds for every critical edge $e$ in $\Gamma_f$, i.e., the two edges preceding and following $e$ lie in the same side of $e$.*
- *$\Gamma_f$ is contained in a flag $F$ with borders $P_h$ and $P_v$.*
- *If $P_h$ is a critical edge, then the post of $F$ (if exists) is incident to $P_v$. Similarly, if $P_v$ is a critical edge, then the post of $\Gamma_f$ (if exists) is incident to $P_h$. (Note that since $\Delta = 3$, both $P_h$ and $P_v$ cannot be critical).*

*Proof.* Due to space constraints, here we only sketch the steps of the proof.

We first prove that if $f$ satisfies Conditions $(C_1)$–$(C_3)$, then $f$ admits a good orthogonal drawing such that Lemma 1 holds for every critical edge of $f$. Our proof is constructive. We construct two drawings $\Gamma_{f_1}$ and $\Gamma_{f_2}$ of $f$, and prove that one of these two drawings satisfies the lemma. Since $f$ satisfies $(C_1)$, $P$ must contain at least three vertices. We first draw the path $P$ maintaining edge orientations. Let the drawing be $\Gamma_P$. We next draw $P' = (v_1, v_2, \ldots, v_r)$ in two different ways that give the drawings $\Gamma_{f_1}$ and $\Gamma_{f_2}$, as follows.

**Construction of $\Gamma_{f_1}$.** We construct $\Gamma_{f_1}$ in three steps. At Step 1, we draw $P'$ starting at $v_1$ such that every $v$-critical edge $e$ of $P'$ satisfies Lemma 1. However, the position of $v_r$ in the drawing of $P'$ may not coincide with its position in $\Gamma_P$. Let the resulting drawing of $P'$ be $\Gamma_{P'}$. At Step 2, we modify $\Gamma_{P'}$ such that Lemma 1 holds for every $h$-critical edge, except possibly $(v_{r-1}, v_r)$. While modifying $\Gamma_{P'}$, we ensure that the $v$-critical edges still satisfy Lemma 1. Therefore, after Step 2, the resulting drawing $\Gamma'_{P'}$ has all its critical edges, except possibly $(v_{r-1}, v_r)$, satisfying Lemma 1. At Step 3, we modify the drawing such that the positions of $v_r$ in $\Gamma'_{P'}$ and $\Gamma_P$ coincide. Thus after Step 3, we obtain a drawing $\Gamma_{f_1}$ of $f$ that respects all the edge orientations, furthermore, all the critical edges, except possibly $(v_{r-1}, v_r)$, satisfy Lemma 1.

**Construction of $\Gamma_{f_2}$.** To construct $\Gamma_{f_2}$, we start drawing $P'$ at $v_r$ of $\Gamma_P$, and then the construction is symmetric, i.e., here we treat the horizontal (respectively, vertical) orientations as the vertical (respectively, horizontal) orientations.

**Either $\Gamma_{f_1}$ or $\Gamma_{f_2}$ satisfies Lemma 3.** We first prove that one of $\Gamma_{f_1}$ and $\Gamma_{f_2}$ is a good orthogonal drawing and Lemma 1 holds for each of its critical edge. The idea of the proof is as follows. We first prove that both $\Gamma_{f_1}$ and $\Gamma_{f_2}$ are good. We next prove that if Lemma 1 does not hold for the critical edges in $\Gamma_{f_1}$, then $P'$ cannot contain any $v$-critical edge and $P_h$ cannot be an $h$-critical edge. We show that in such a scenario, Lemma 1 must hold for every critical edge in $\Gamma_{f_2}$. As a byproduct of our construction, we obtain the remaining two properties of $\Gamma_f$, i.e., $\Gamma_f$ is contained in a flag $F$ with borders $P_h$ and $P_v$, and if $P_h$ (respectively, $P_v$) is a critical edge, then the post of $F$ (if exists) is incident to $P_v$ (respectively, $P_h$). □

We are now ready to describe the drawing of $G$. We first construct the drawing $\Gamma_f$ for some inter face $f$ of $G$. We then draw the other inner faces of $G$ by a depth first search on the faces of $G$, such that after adding a new inner face, the resulting drawing remains

($P_1$) a good orthogonal drawing, and
($P_2$) each critical edge respects Lemma 1.

Let $\Gamma_k$ be a drawing of the set of inner faces $f_1(=f), f_2, \ldots, f_k$ that we have already constructed. Let $f_{k+1}$ be an inner face of $G$ that has not been drawn yet, but has an edge $(b,c)$ in common with some face $f_j$, where $1 \le j \le k$. Without loss of generality assume that $\lambda_{bc} = V$ in $\Gamma_k$. Furthermore, since $G$ is outerplanar, $f_{k+1}$ cannot have any edge other than $(b,c)$ in common with $f_j$. Let $l_v$ be a segment of $f_{k+1}$ that contains $(b,c)$, and let $l_h$ be another segment of $f_{k+1}$ incident to $l_v$. We now construct $\Gamma_{k+1}$ considering the following cases.

**Case 1 (None of $b$ and $c$ is an end vertex of $l_v$):** In this case none of the end vertices of the path formed by $l_v$ and $l_h$ belongs to $\Gamma_k$. Since $G$ satisfies Condition ($C_3$), the edges of $f_j$ that are incident to $b$ and $c$ must be horizontal, i.e., $(b,c)$ must be a $v$-critical edge of $f_j$. Since $\Gamma_k$ is a good orthogonal drawing, there is enough space to create a flag $F$ with borders $l_v$ and $l_h$ such that the banner and post of $F$ do not create any edge crossing. Figure 4(c) illustrates such an example. By Lemma 3, we can draw $f_{k+1}$ inside $F$ maintaining Properties ($P_1$) and ($P_2$). Thus the resulting drawing $\Gamma_{k+1}$ satisfy ($P_1$)–($P_2$).

**Case 2 (Exactly one of $b$ and $c$ is an end vertex of $l_v$):** If $b$ (respectively, $c$) is an end vertex of $l_v$, then we choose $l_h$ such that it contains $b$ (respectively, $c$). Therefore, none of the end vertices of the path formed by $l_v$ and $l_h$ belongs to $\Gamma_k$. Figure 4(d) illustrates such an example. Similar to Case 1, we now draw $\Gamma_{k+1}$ satisfying ($P_1$)–($P_2$).

**Case 3 (Both $b$ and $c$ are end vertices of $l_v$):** Observe that in this case $l_v = (b,c)$. Let $a,b,c,d$ be a path of $f_{k+1}$. Since $l_v = (b,c)$ is a maximal set of edges with vertical orientation, we have $\lambda_{ab} = \lambda_{bc} = H$. Thus $l_v = (b,c)$ is a $v$-critical edge of $f_{k+1}$. We now create a flag $F$ with borders $l_v$ and $l_h$ such that the post

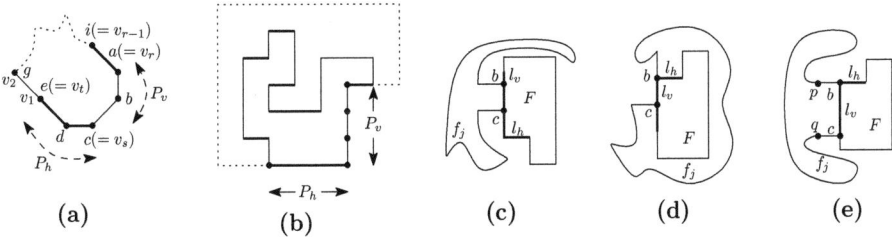

**Fig. 4.** (a) An inner face of $G$. (b) Illustration for $\Gamma_f$. (c)–(e) Illustration for the construction of $\Gamma_{k+1}$.

of the flag is incident to $l_h$. Note that since $l_v$ is critical, by Lemma 3, we do not require a flag with its post incident to $l_v$. We now can draw $f_{k+1}$ inside $F$ maintaining $(P_2)$ and $(P_3)$. Figure 4(e) illustrates such an example. It may initially appear from the figure that drawing of $f_{k+1}$ inside $F$ may overlap the boundary of $f_j$, i.e., consider the Figure 4(e) with $\lambda_{cq} = V$. However, by definition of a flag, $F$ does not contain the part of its boundary that overlaps $f_j$, and hence drawing $f_{k+1}$ would not create any edge overlapping.

## 4 Conclusion

In Section 2 we have developed a polynomial-time algorithm to decide good orthogonal drawability of $HV$-restricted plane graphs. An interesting open question in this context, as Maňuch et al. [10] asked, is to determine the complexity of deciding good orthogonal drawability for $HV$-restricted planar graphs.

**Problem 1.** What is the time complexity of deciding whether an arbitrary $HV$-restricted planar graph admits a planar orthogonal drawing preserving the given edge orientations?

In Section 3 we have characterized $HV$-restricted 2-connected maximum-degree-three outerplanar graphs that admit good orthogonal drawings. If we relax the 2-connected constraint, then our characterization no longer holds. For example, the $HV$-restricted outerplanar graph $G$ of Figure 5(b) satisfies Conditions $(C_1)$-$(C_3)$ of Theorem 2, but does not admit any good orthogonal drawing.

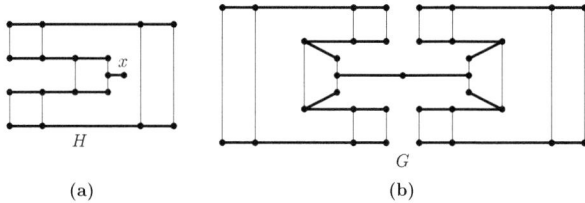

**Fig. 5.** Illustration for the graphs (a) $H$ and (b) $G$

Observe that $G$ is constructed from two copies of the graph $H$ of Figure 5(a), where the vertices with label $x$ are identified. Since in any good orthogonal drawing of $H$ the vertex $x$ lies in some inner face, any orthogonal drawing of $G$ preserving edge orientations must contain edge crossing. Hence a natural open question is to extend our result for arbitrary outerplanar graphs.

**Problem 2.** Characterize the class of $HV$-restricted outerplanar graphs that admit planar orthogonal drawings preserving the given edge orientations.

# References

1. Battista, G.D., Liotta, G., Vargiu, F.: Spirality and optimal orthogonal drawings. SIAM Journal on Computing 27, 1764–1811 (1998)
2. Borradaile, G., Klein, P.N., Mozes, S., Nussbaum, Y., Wulff-Nilsen, C.: Multiple-source multiple-sink maximum flow in directed planar graphs in near-linear time. In: Ostrovsky, R. (ed.) IEEE 52nd Annual Symposium on Foundations of Computer Science (FOCS), pp. 170–179. IEEE (2011)
3. Cornelsen, S., Karrenbauer, A.: Accelerated bend minimization. Journal of Graph Algorithms and Applications 16(3), 635–650 (2012)
4. Di Battista, G., Kim, E., Liotta, G., Lubiw, A., Whitesides, S.: The shape of orthogonal cycles in three dimensions. Discrete & Computational Geometry 47(3), 461–491 (2012)
5. Di Giacomo, E., Liotta, G., Patrignani, M.: Orthogonal 3D shapes of theta graphs. In: Goodrich, M.T., Kobourov, S.G. (eds.) GD 2002. LNCS, vol. 2528, pp. 142–149. Springer, Heidelberg (2002)
6. Garg, A., Tamassia, R.: On the computational complexity of upward and rectilinear planarity testing. SIAM Journal on Computing 31(2), 601–625 (2001)
7. Hoffmann, F., Kriegel, K.: Embedding rectilinear graphs in linear time. Information Processing Letters 29(2), 75–79 (1988)
8. Kant, G.: Drawing planar graphs using the canonical ordering. Algorithmica 16, 4–32 (1996)
9. Klein, P.N., Mozes, S., Weimann, O.: Shortest paths in directed planar graphs with negative lengths: A linear-space $O(n \log^2 n)$-time algorithm. ACM Transactions on Algorithms 6(2), 236–245 (2010)
10. Maňuch, J., Patterson, M., Poon, S.-H., Thachuk, C.: Complexity of finding non-planar rectilinear drawings of graphs. In: Brandes, U., Cornelsen, S. (eds.) GD 2010. LNCS, vol. 6502, pp. 305–316. Springer, Heidelberg (2011)
11. Miura, K., Haga, H., Nishizeki, T.: Inner rectangular drawings of plane graphs. International Journal of Computational Geometry and Applications 16(2-3), 249–270 (2006)
12. Mozes, S., Wulff-Nilsen, C.: Shortest paths in planar graphs with real lengths in $O(n \log^2 n/\log \log n)$-time. In: de Berg, M., Meyer, U. (eds.) ESA 2010, Part II. LNCS, vol. 6347, pp. 206–217. Springer, Heidelberg (2010)
13. Nomura, K., Tayu, S., Ueno, S.: On the orthogonal drawing of outerplanar graphs. IEICE Transactions on Fundamentals of Electronics, Communications and Computer Sciences E88-A(6), 1583–1588 (2005)
14. Tamassia, R.: On embedding a graph in the grid with the minimum number of bends. SIAM Journal on Computing 16(3), 421–444 (1987)
15. Vijayan, G., Wigderson, A.: Rectilinear graphs and their embeddings. SIAM Journal on Computing 14(2), 355–372 (1985)
16. Zhou, X., Nishizeki, T.: Orthogonal drawings of series-parallel graphs with minimum bends. SIAM Journal on Discrete Mathematics 22(4), 1570–1604 (2008)

# Periodic Planar Straight-Frame Drawings with Polynomial Resolution

Luca Castelli Aleardi, Éric Fusy, and Anatolii Kostrygin

LIX - École Polytechnique
{amturing,fusy}@lix.polytechnique.fr, anatoly.kostrygin@polytechnique.org

**Abstract.** We present a new algorithm to compute periodic (planar) straight-line drawings of toroidal graphs. Our algorithm is the first to achieve two important aesthetic criteria: the drawing fits in a straight rectangular frame, and the grid area is polynomial, precisely the grid size is $O(n^4 \times n^4)$. This solves one of the main open problems in a recent paper by Duncan et al. [3].

## 1 Introduction

The main goal of graph drawing algorithms is to compute a drawing which is easily readable. One basic problem consists in mapping the vertices and edges of a graph onto a region in the plane or a portion of a 3D surface. Most of the time edges are represented as smooth curves (very often as straight-line segments), and the drawing is required to be *crossing-free*. Sometimes the drawing is asked to satisfy some further aesthetic criteria, in order to obtain a pleasing and readable result. For example, one could seek for *good vertex resolution* for ensuring that vertices are not too close to one another. In the planar case, an elegant solution to this problem is provided by the *barycentric embedding* by Tutte [12]. Tutte showed how to compute vertex positions by solving a system of linear equations: the method applies to a 3-connected planar graph and the resulting drawing is guaranteed to be crossing-free, also allowing to fix the positions of outer vertices (which are mapped to the vertices of a given convex polygon). The solution can be also reformulated in terms of a system of springs converging to an equilibrium position, and has inspired a huge number of force-directed embedding algorithms. A drawback of Tutte's method is that one cannot achieve a good vertex resolution, since matrix computations lead to vertex coordinates of exponential size. A solution to this problem are the so-called *straight-line grid drawings* [4,8,11]: the graph is embedded on a regular grid whose area is typically polynomial with respect to the size of the graph. A further advantage of this approach is that the algorithm performs essentially arithmetic computations on integers of bounded magnitude (no roundings are needed). For the planar case, many classes of algorithms have been proposed to solve this task, achieving very good vertex resolution (quadratic area) and time complexity (running in linear time).

*Statement of the main result.* A *quasi-triangulation* is a graph (topologically) embedded in the plane such that all inner faces are triangles and the outer

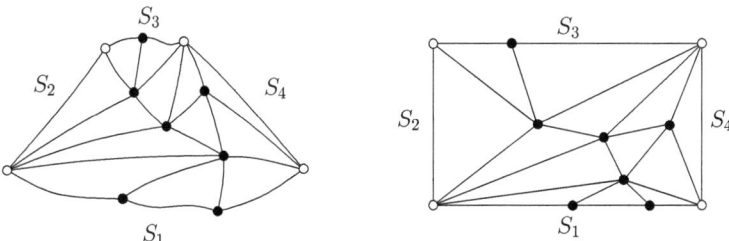

**Fig. 1.** Left: a 4-scheme triangulation $G$ (corners are white). Right: a straight-frame drawing of $G$.

face contour is a simple cycle. The vertices and edges are called *outer* or *inner* whether they are incident to the outer face or not. Define a *k-scheme triangulation* as a quasi-triangulation with $k$ marked outer vertices, called *corners*, such that each path of the outer face contour between two consecutive corners is chordless. Call $S_1, \ldots, S_k$ these $k$ outer paths (in clockwise order around the outer face), and denote by $|S_i|$ the length of $S_i$.

In this article we focus on the case $k = 4$. Given a 4-scheme triangulation $G$, a *straight-frame drawing* of $G$ is a planar straight-line drawing of $G$ such that the outer face contour is an axis-aligned rectangle whose corners are the corners of $G$, with $S_1$ and $S_3$ as horizontal sides and $S_2$ and $S_4$ as vertical sides, see Fig. 1. In a recent article, Duncan et al. [3] prove the following:

**Theorem 1 (Duncan et al. [3]).** *Each 4-scheme triangulation with $n$ vertices admits a straight-frame drawing on a (regular) grid of size $O(n^2 \times n)$.*

A motivation for such drawings is to draw toroidal graphs [9,10]. Indeed, it is shown in [3] (more general results in higher genus are shown in [1]) that any triangulation $\tilde{G}$ on the torus can be cut along a subgraph (cut-graph) such that the resulting unfolded graph $G$ is naturally a 4-scheme triangulation, with the additional property that $|S_1| = |S_3|$ and $|S_2| = |S_4|$. A 4-scheme triangulation $G$ satisfying $|S_1| = |S_3|$ and $|S_2| = |S_4|$ is called *balanced*, and a straight-frame drawing of $G$ is called *periodic* if the abscissas of vertices of the same rank along $S_1$ and $S_3$ (ordered from left to right) coincide, and the ordinates of vertices of the same rank along $S_2$ and $S_4$ (ordered from bottom to top) coincide. As shown in Fig. 2, when $G$ arises from a toroidal triangulation $\tilde{G}$, this is exactly the condition to satisfy so that the drawing lifts to a periodic representation of $G$ in the plane (i.e., to a drawing of $G$ on the flat torus). The authors of [3,1] cite it as an important open problem to compute periodic straight-frame drawings with polynomial grid size[1]. Our main result is the following.

---

[1] Tutte's spring embedding algorithm gives a solution, but with exponential grid size; the two recent articles [2,5] yield periodic drawings for toroidal triangulations of respective grid areas $O(n^{5/2})$ and $O(n^4)$, but these drawings do not fit inside a straight frame, so that when looking at an elementary cell, there is the aesthetic disadvantage of having edges crossing the boundary of the cell, as in Fig. 2(b).

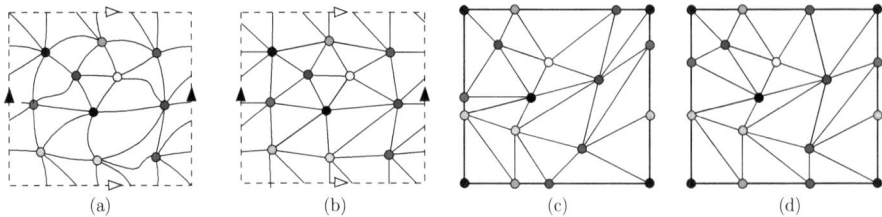

**Fig. 2.** (a) A toroidal triangulation $G$; (b) a periodic straight-line drawing of $G$ which is not straight-frame; (c) a non-periodic straight-frame drawing of $G$ (where $G$ is planarly unfolded using a cut-graph); (d) a periodic straight-frame drawing of $G$.

**Theorem 2.** *Each balanced 4-scheme-triangulation admits a periodic straight-frame drawing on a (regular) grid of size $O(n^4 \times n^4)$. The drawing can be computed in linear time.*

Our algorithm makes use of a new decomposition strategy for 4-scheme triangulations, we cut the triangulation into two components $A, B$ along a certain path "close to" $S_3$, the "lower part" $B$ is drawn using the algorithm of Duncan et al. [3] specially adapted for our purpose (to make the ordinates of matching vertices in the lower parts of $S_2$ and $S_4$ coincide), and then we draw the "upper part" $A$ with a suitably fixed outer frame (this is the most technical part, which requires a further decomposition of $A$ into several pieces). At first we will show that we can have grid-size $O(n^5 \times n^5)$, then we will argue in Section 5.2 that the grid-size can be improved to $O(n^4 \times n^4)$.

## 2 The Duncan et al. Algorithm, Adapted

### 2.1 Description of the Algorithm in [3]

From now on, a 4-scheme triangulation is shortly called a 4ST. Define a *half-4-scheme triangulation*, shortly written H4ST, as a graph $H$ embedded in the plane satisfying all conditions of a 4ST, except that the paths $S_2$ and $S_4$ are allowed to be empty, and the path $S_3$ is allowed to have chords and to meet $S_1$; $S_1$ is called the *bottom-path* of $H$. Given a 4ST $G$, an important ingredient in [3] is the *river lemma*, which guarantees the existence of a so-called *river*, that is, a path $P$ in the dual graph of $G$, such that the first edge of $P$ crosses $S_1$, the last edge of $P$ crosses $S_3$, and the two components $H, H'$ of $G$ separated by $P$ are H4ST (with $S_2$ as bottom path of the left component and $S_4$ as bottom-path of the right component), see Fig. 3(a).

A straight-line drawing of a H4ST $H$ is called *admissible* if $S_1$ is horizontal, $S_2$ and $S_4$ are vertical, and all edges in $S_3$ have slope in $\{-1, 0, +1\}$. By an extension of the shift algorithm introduced by de Fraysseix, Pach and Pollack [4] (which treats the case where $S_2$ and $S_4$ are empty), Duncan et al. show that it is possible to obtain an admissible straight-line drawing of $H$ on a grid of

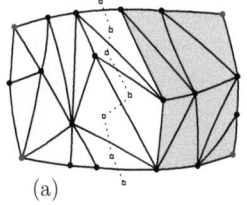 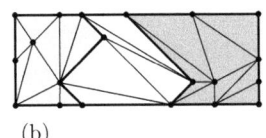 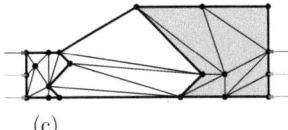

**Fig. 3.** (a) a 4ST $G$; the river between $S_1$ and $S_3$ decomposes $G$ into two H4ST $H, H'$ (shaded). (b) The straight-frame drawing of $G$ as given in [3], which results from the two drawings of $H, H'$ put together; (c) our modified version (with an oblique edge on $S_3$), where the ordinates of the vertices on $S_2$ coincide with the ordinates of the corresponding vertices on $S_4$.

size $O(n \times n^2)$, more precisely a grid of size $O(n \times (d+1)n)$, where $n$ is the number of vertices of $H$ and $d$ is the graph-distance between $S_1$ and $S_3$. To obtain a straight-frame drawing of a 4ST $G$, decompose $G$ along a river between $S_1$ and $S_3$ into two H4ST components $H$ and $H'$, draw $H$ and $H'$ using the shift algorithm (with $S_2$ as the bottom-path of $H$ and $S_4$ as the bottom-path of $H'$), and then shift the left boundary of the component whose drawing has the smaller width so that the widths of the drawings of $H$ and $H'$ coincide. Then rotate the drawing of $H$ by $\pi/2$ clockwise, rotate the drawing of $H'$ by $\pi/2$ counterclockwise, and put the two drawing in front of each other, leaving enough horizontal space between them so that the edges connecting $H$ to $H'$ have slope smaller than 1 in absolute value. Since the edges on the boundaries of (the rotated copies of) $H$ and $H'$ are either vertical or of slope in $\{-1, +1\}$, the edges between $H$ and $H'$ do not introduce crossings, so the resulting drawing of $G$ is planar, see Fig. 3(b). Overall the grid-size is $O(n^2 \times n)$, more precisely $O(n(d+1) \times n)$, with $d$ the graph-distance between $S_2$ and $S_4$.

### 2.2 Our Modified Version of the Algorithm

A first simple adaptation we do is to do all (abscissa) shift operations by 2 instead of doing them by 1. Given a H4ST $G$, let $p$ be the number of edges of $S_1$. For a vector $I$ of $p$ even integers, $I$ is called the *initial interspace vector*, consider the shift algorithm for $G$ starting with $S_1$ drawn as a horizontal line with interspaces given by $I$. Let $F(I) = (f_1, \ldots, f_p)$ be the vector representing the interspaces between consecutive vertices on $S_1$ at the end of the algorithm; $F(I)$ is called the *final interspace vector*. For our purpose, the advantage of doing the shift operations by 2 is to guarantee that the components of $F(I)$ are even.

*Property 1.* Let $F_0 = F(I_0)$ be the final interspace vector when using $I_0 = (2, 2, \ldots, 2)$ as initial interspace vector on $S_1$. Then for any vector $F$ of $p$ even integers such that $F \geq F_0$ (component-wise), it is possible to re-execute the shift-algorithm so as to have $F$ as final interspace vector.

*Proof.* In a similar way as in [2], one can adopt a reformulation of the shift algorithm in terms of vertical strips insertions (of width 2 here) and edge stretch. With this reformulation it is easy to see that $F(I) - I$ is an invariant (is the same for any vector $I$ of $p$ even integers). □

We can now give another strategy (suited in view of showing Theorem 2) for drawing a 4ST $G$. The drawing we obtain is not straight-frame, but is straight-frame except for an oblique edge along $S_3$, and is such that, with $m = \min(|S_2|, |S_4|)$, the $m$ first components of the interspace vectors along $S_2$ and $S_4$ (ordered from bottom to top) coincide. At first, similarly as in [3] we cut $G$ along a river (between $S_1$ and $S_3$) into two components $H$ and $H'$. Then we draw independently $H$ and $H'$ using the shift algorithm (with width 2 strip insertions). Let $F_0$ ($F_0'$) be the final interspace vector of $H$ (resp. of $H'$), starting with initial interspace vector $(2, \ldots, 2)$. For $1 \leq i \leq m$ let $u_i$ be the maximum of the $i$th components of $F_0$ and $F_0'$. By Property 1, one can redraw $H$ and $H'$ so that the $m$ first components of the final interspace vectors of $H$ and of $H'$ are $(u_1, \ldots, u_m)$. In addition one can check that the widths of both drawings are at most $8n$, and the two widths differ by at most $4n$. Similarly as in [3] we rotate $H$ (resp. $H'$) by $\pi/2$ clockwise (resp. counterclockwise) and place the drawings in front of each other, leaving horizontal space $8n$ between them, enough to draw the edges between $H$ and $H'$ crossing-free. We obtain (see Fig. 3(c)):

**Lemma 1.** *For any 4ST $G$ with $n$ vertices, and $m = \min(|S_2|, |S_4|)$, there is a straight-line drawing of $G$, where $S_2$ and $S_4$ are vertical with their interspace vectors equal at the $m$ first components, $S_1$ is horizontal, and $S_3$ is horizontal except for an oblique edge of slope in $[-1/2, 1/2]$. The grid size is $O(n^2 \times n)$, more precisely $O(n(d+1) \times n)$ with $d$ the graph-distance between $S_2$ and $S_4$, and the drawing can be computed in linear time.*

## 3 A New Binary Decomposition for 4ST

We now introduce a new way to decompose a 4ST $G$ into two components $A, B$, in such a way that proving Theorem 2 will reduce to drawing $B$ using Lemma 1, and then drawing $A$ using a certain fixed outer frame. A path $P$ is said to be *just below* $S_3$ if $P$ connects a vertex of $S_2$ to a vertex of $S_4$, all non-extremal vertices of $P$ avoid $S_2 \cup S_3 \cup S_4$, and each vertex on $P$ (including the extremities) has at least one neighbour on $S_3$.

**Lemma 2.** *Each 4ST $G$ has a chordless path just below $S_3$.*

*Proof.* A path $P$ is said to be *below* $S_3$ if it satisfies the same conditions as "just below", but dropping the condition that each vertex of $P$ must have a neighbour on $S_3$. Let $E$ be the non-empty (since it contains $S_1$) set of paths below $S_3$. For $P, P' \in E$, write $P \leq P'$ if no edge of $P$ is above $P'$. It is easy to see that $E$ admits a unique maximal element $P_0$, and the fact that $G$ is triangulated ensures that all vertices of $P_0$ have at least one neighbour on $S_3$. Then one can extract a chordless subpath with same extremities out of $P_0$. □

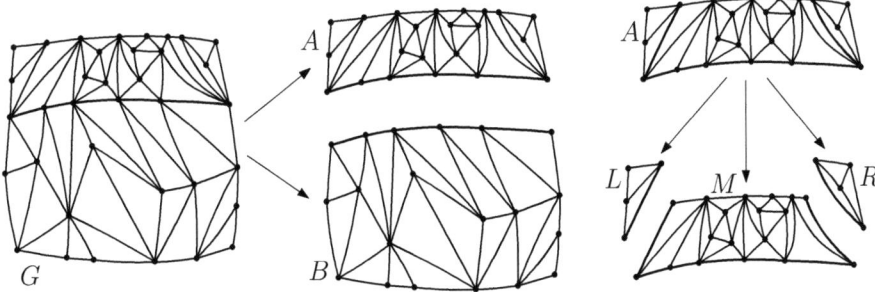

**Fig. 4.** Left: a 4ST $G$, where we distinguish a chordless path $P$ just below $S_3$ (shown bolder); middle: cutting along $P$ yields two components $A, B$; right: the band-graph $A$ is further decomposed into 3 pieces $L, M, R$

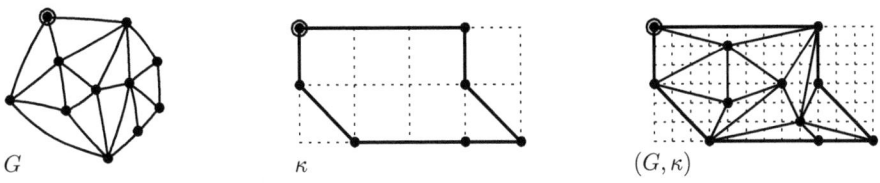

**Fig. 5.** Left: a quasi-triangulation $G$; middle: a fixed frame $\kappa$ for $G$ on a grid of size $4 \times 2$; right: a drawing of $G$ fitting in $\kappa$, using refinement factor 3 for the grid (hence the factor 3 is suitable for $(G, \kappa)$)

Let $P$ be a chordless path just below $S_3$. Cutting $G$ along $P$ yields two 4ST denoted $A$ and $B$, with $B$ below $P$ and $A$ above $P$. We draw $B$ using Lemma 1. Then we have to draw $A$ —which is called the *band-graph*— in such a way that the drawing obtained by pasting the drawing of $B$ with the drawing of $A$ yields a periodic drawing of $G$. To state the drawing result for $A$, we introduce the notion of *fixed frame*. Given a quasi-triangulation $G$, with $C$ its outer cycle, a fixed frame $\kappa$ for $G$ is a crossing-free drawing of $C$ on a regular grid $w \times h$ ($w$ and $h$ are the width and height of the fixed frame). For $\gamma \geq 1$ the grid is *refined by factor* $\gamma$ by replacing each unit cell by a $\gamma \times \gamma$ regular grid. We say that the factor $\gamma$ is *suitable* for the pair $(G, \kappa)$ if, after refining the grid of $\kappa$ by factor $\gamma$, $G$ admits a straight-line drawing with $\kappa$ as the outer face contour of the drawing, see Fig. 5 for an example. We will prove the following result in Section 4:

**Lemma 3.** *For each $K \geq 1$ and any fixed frame $\kappa$ for $A$ ($A$ is the band-graph of $G$, which has $n$ vertices) of the form shown in the left drawing of Figure 6, there is $\gamma_0$ in $O(n \cdot \max(n, K))$ such that any even factor $\gamma \geq \gamma_0$ is suitable for $(A, \kappa)$.*

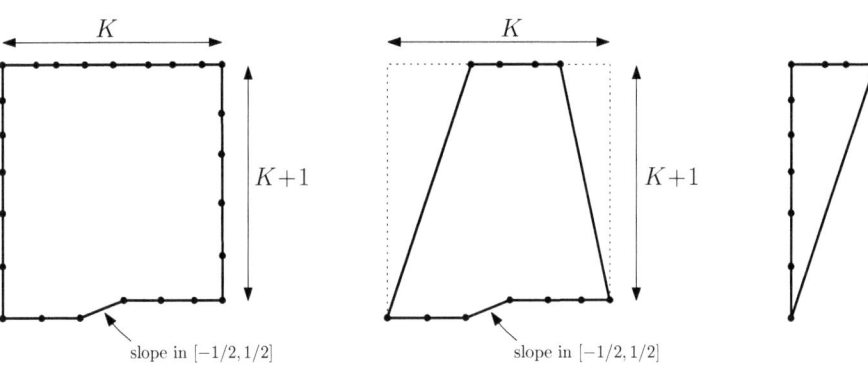

**Fig. 6.** Left: frame for the band-graph $A$, middle: frame for the middle piece $M$ of $A$, right: frame for the left piece $L$ of $A$

We claim that Lemma 1 together with Lemma 3 imply Theorem 2 (at the moment with grid-size $O(n^5 \times n^5)$). Indeed, once $B$ is drawn using Lemma 1, one easily designs a fixed frame of the form of Fig. 6 and such that pasting a drawing of $A$ in this frame with the drawing of $B$ yields a periodic drawing of $G$ (note that the lower part of $\kappa$ is determined by the property of fitting with the boundary-path $S_3(B)$, and the upper part of $\kappa$ is determined by the property of fitting with the boundary-path $S_1(B)$ in order to get the periodicity property). Note that the parameter $K$ equals the width of the drawing of $B$, which is $O(n^2)$. Since the refinement factor for $A$ is in $O(n \cdot \max(n, K))$, the grid-size of the resulting drawing of $G$ is $O(K^2 n \times K^2 n)$, which is $O(n^5 \times n^5)$.

*Remark 1.* For convenience and to keep it short, we have written the proof in worst-case style, but there is room for improvements in favorable instances, in particular to reduce the height (which is $K + 1$ in the worst case, as in Figure 6) of the frames for the band-graph.

## 4 Proof of Lemma 3

At first we state a useful lemma that easily follows from Property 1:

**Lemma 4.** *Let $G$ be a H4ST with $n$ vertices, with $p$ the length of the bottom-boundary. Let $V$ be a vector of $p$ positive integers. Then, for any even $\gamma \geq 4n$, there is an admissible drawing of $G$ whose (bottom-)interspace vector is $\gamma \cdot V$.*

*Proof.* With initial interspace vector $(2, 2, \ldots, 2)$, the final interspace vector is an even vector $F_0$ whose components are bounded by $4n$. Hence $\gamma \cdot V$ dominates $F_0$ for any $\gamma \geq 4n$, so the result follows from Property 1. □

Let $A$ be the band-graph of a 4ST $G$ with $n$ vertices, i.e., $A$ is obtained as the upper component after cutting along a chordless path $P$ just below $S_3(G)$. Let $a$ be the extremity of $P$ on $S_2$ and $b$ the extremity of $P$ on $S_4$. Let $u_a$ be the leftmost neighbour of $a$ along $S_3$ and let $u_b$ be the rightmost neighbour of $b$ along $S_3$. Then cutting $A$ along the edges $\{a, u_a\}$ and $\{b, u_b\}$ yields three pieces: a left-piece $L$, a middle piece $M$, and a right-piece $R$ (see the right drawing of Fig. 4). Note that $L$ (and similarly $R$) is either empty (if $u_a$ is the top-left corner of $G$) or otherwise is naturally a 3-scheme triangulation (shortly called a 3ST), whose three corners are $a$, $u_a$ and the topleft corner of $G$. And $M$ is naturally a 4ST such that $|S_2(M)| = |S_4(M)| = 1$. If $\kappa$ is a fixed frame for $A$ as in Lemma 3 (left drawing of Fig. 6), let $\kappa_M$ be the fixed frame for $M$ inherited from $\kappa$ (i.e., drawing the chords $\{a, u_a\}$, $\{b, u_b\}$, and deleting what is top-left of $\{a, u_a\}$ and top-right of $\{b, u_b\}$), of the form shown in the middle drawing of Fig. 6. We have:

**Lemma 5.** *Any even factor $\gamma \geq 4n$ is suitable for $(M, \kappa_M)$.*

*Proof.* If we decompose $M$ into two components using a river (between $S_2$ and $S_4$), the fact that $|S_2| = |S_4| = 1$ ensures that the two resulting H4ST $H$ and $H'$ have no left nor right vertical boundary: such H4ST are called *flat*. We draw the upper component $H'$ so as to respect the interspaces of the upper boundary of $\kappa$; according to Lemma 4 this is possible for any even refinement factor $\gamma \geq 4n$. Next we draw $H$, which is a bit more difficult due to the presence of the oblique edge at the bottom-boundary. We use the property that, since every vertex on $H$ is adjacent to a vertex on $S_3(G)$, then $H$ is even more constrained: there is a flat H4ST "attached to" each edge of the bottom-path of $H$. Denote by $G_e$ the flat H4ST at the oblique edge $e$. By an easy modification of the shift drawing algorithm in [4], for any even refinement factor $\gamma \geq 4n$, one can draw $G_e$ so that $e$ fits with the oblique edge of $\kappa$, and as usual with the slopes of the upper boundary of $G_e$ in $\{-1, 0, +1\}$. Finally we have to draw the flat H4ST $H_\ell$ to the left of $e$ (and similarly the flat H4ST $H_r$ to the right of $e$) so as to respect the positions of corresponding vertices of $\kappa$. By Lemma 4, this is possible for any even refinement factor $\gamma \geq 4n$. Finally note that all the H4ST considered are flat (they have no lateral path); and because of the shape of the frame (see Fig. 6), all edges connecting $H$ to $H'$ have slope greater than 1 in absolute value. Hence these edges can be added crossing-free. □

It remains to draw the left piece $L$ (and similarly the right piece $R$) with a fixed frame as shown in the right drawing of Fig. 6.

**Lemma 6.** *Let $\kappa$ be a fixed frame of the form in the right drawing of Fig. 6, and let $T$ be a 3ST (with the 3 side lengths compatible with $\kappa$). Then, for some $\gamma_0$ in $O(n \cdot \max(n, K))$, any even factor $\gamma \geq \gamma_0$ is suitable for $(T, \kappa)$.*

*Proof.* The generic situation is shown in the left drawing of Fig. 7 (where for convenience, the right drawing of Fig. 6 has undergone an horizontal mirror). A crucial role is played by the most downleft chord $e$; there is a flat H4ST $H_r$ on the right of $e$ aligned along the horizontal side, and similarly there is a flat H4ST $H_a$ above $e$ aligned along the vertical side. By Lemma 4 we can draw these two H4ST so as to respect the interspaces of the outer frame, at the price of a refinement factor $O(n)$. To draw the chords in a crossing-free way, we have

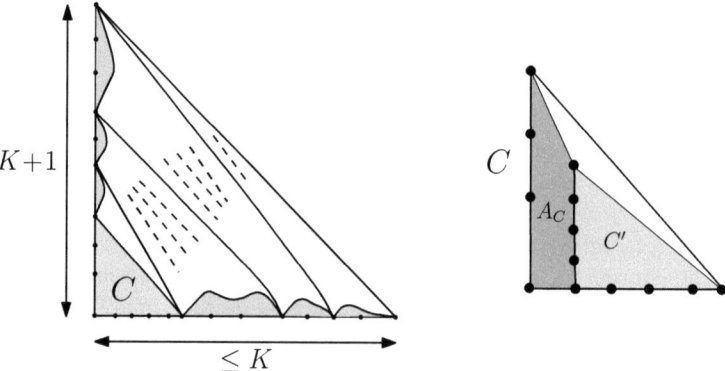

**Fig. 7.** Left: generic situation for a 3ST with one side (hypotenuse) reduced to an edge. Right: decomposition of $C$.

to do a further operation: refine by factor $K+2$ and divide the ordinates of $H_r$ by $K+2$ and the abscissas of $H_a$ by $K+2$. The effect is to make the slopes of the upper contour of $H_r$ strictly smaller than $1/(K+1)$ in absolute value (and similarly for $H_a$). We can now draw the chords in a crossing-free way (indeed, since the chords connect two points in a $(K+1) \times (K+1)$ grid they have slope not smaller than $1/(K+1)$). We now have to draw the piece $C$ downleft of $e$, and assume without loss of generality that $e$ has slope smaller than $-1$. Note that (except for the trivial case where $C$ has just one inner face) we get a 4ST $\tilde{C}$ when deleting the topright (hypotenuse) edge. Let $P$ be a chordless path "just to the right" of the left vertical boundary path of $\tilde{C}$. Denote by $A_C$ (band-graph) the part of $\tilde{C}$ to the left of $P$ and by $C'$ the part of $\tilde{C}$ to the right of $P$. Using Lemma 4 we draw $C'$ (with $P$ as left vertical boundary) so that its bottom path fits with the interspaces prescribed by $\kappa$. Next we draw the band-graph $A_C$ (note that the drawing of $C'$, together with the prescription of $\kappa$, completely fix the outer frame of $A_C$). An important point is that, in the 3-piece decomposition of $A_C$ (as shown in the right drawing of Fig. 6), the left-piece $L$ and the right-piece $R$ are trivial, due to the absence of chords in $C$. This means that we do not have to recurse, and can easily conclude by Lemma 5 (in fact a simpler version of Lemma 5, without bothering about an oblique edge). □

## 5 Finishing the Proof of Theorem 2

### 5.1 Remaining Cases

In order to carry out the decomposition strategy based on the path just below $S_3$ we have assumed that there is no chord between $S_1$ and $S_3$. Let $G$ be a 4ST. If there is a chord between $S_1$ and $S_3$ but there is no chord between $S_2$ and $S_4$, then we can just rotate $G$ by $\pi/2$ (so as to exchange $S_1, S_3$ with $S_2, S_4$). If there is also a chord between $S_2$ and $S_4$, then it is easy to see that there must be a chord incident to a corner of $G$. So we just have to find a drawing strategy for any 4ST $G$ that has a chord incident to a corner, say without loss of generality that there is a chord $e$ that connects the bottom-left corner to a vertex on $S_3$. If we cut along $e$ we obtain two components $L, R$, where $L$ is a 3ST and $R$ is a 4ST (possibly a 3ST if $e$ goes to the top-right corner). Then draw $R$ using Lemma 1 (with $e$ as oblique edge); since $|S_2(R)| = 1$, the drawing of $R$ has grid-size $O(n \times n)$. If we want a periodic straight-frame drawing of $G$, the drawing of $R$ completely fixes the outer frame for $L$ (hence the outer frame of $L$ is of size $O(n \times n)$). And using Lemma 6, we can refine the grid by factor $O(n^2)$ to fit in the fixed frame. We conclude that, when there is a chord incident to a corner of $G$, then $G$ has a periodic straight-frame drawing on a grid of size $O(n^3 \times n^3)$.

## 5.2 Getting Grid-Size $O(n^4 \times n^4)$

We now argue that, in the case where there is no chord incident to a corner of $G$, then $G$ actually admits a periodic straight-frame drawing of grid-size $O(n^4 \times n^4)$. For $G$ a 4ST with $n$ vertices, let $d_h$ be the graph-distance between $S_1$ and $S_3$, and let $d_v$ be the graph-distance between $S_2$ and $S_4$. Then it easily follows from the Menger vertex-disjoint theorem and the fact that $G$ is innerly triangulated that $d_h \times d_v \leq n$, hence $\min(d_h, d_v) \leq n^{1/2}$ (it follows from this observation that, up to possibly rotating $G$ by $\pi/2$ to ensure that $d_h \leq d_v$, the Duncan et al. algorithm gives a grid-size $O(n \times n^{3/2})$). We have seen in Section 5.1 that, if there is no chord incident to a corner, then either there is no chord between $S_1$ and $S_3$ or no chord between $S_2$ and $S_4$. Let us assume witout loss of generality that there is no chord between $S_1$ and $S_3$.

If there is also no chord between $S_2$ and $S_4$, consider a chordless path $P$ just below $S_3$, and a chordless path $P'$ "just to the left" of $S_4$. Let $G'$ be the 4ST obtained by deleting the band above $P$ and the band to the right of $P'$. Let $d'_h$ be the graph-distance between $S_1(G')$ and $S_3(G')$ and let $d'_v$ be the graph-distance between $S_2(G')$ and $S_4(G')$. By the argument just above, either $d'_h$ or $d'_v$ is bounded by $n^{1/2}$. If $d'_v \leq n^{1/2}$ we do the binary decomposition using $P$, yielding two components $A, B$ (with $B$ a 4ST and $A$ a band-graph). Recall that the grid-size of the resulting drawing of $G$ is $O(nK^2 \times nK^2)$, where $K$ is $O(nd)$, with $d$ the graph-distance in $B$ between $S_2(B)$ and $S_4(B)$. In addition, since $G'$ is obtained from $B$ by removing a band-graph on the right side (recall that every vertex on the left side of the band-graph is adjacent to a vertex on the right side of the band-graph), then $d \leq d'_h + 1$. Hence $d$ is $O(n^{1/2})$, so that $K$ is $O(n^{3/2})$, hence the grid-size is $O(n^4 \times n^4)$. Finally, if there is a chord between $S_2$ and $S_4$, then this chord can not be above $P$ (otherwise the chord would be incident to the top-left corner or to the top-right corner). Hence $d = 1$, so that $K$ is $O(n)$, which guarantees a grid-size $O(n^3 \times n^3)$ for the drawing of $G$.

# 6 Application to Spherical Drawings

*Computing geodesic spherical drawings* In this section we consider the problem of drawing a planar triangulation on the sphere so that faces are mapped to non-overlapping spherical triangles. This problem is closely related to the *surface parameterization* problem, which has several applications (such as texture mapping, morphing and remeshing) and has attracted a great attention in the computer graphics and geometric modeling communities [6,7]. In the spherical case, the goal is to define a bijective correspondence between the surface of a sphere (the parameter domain) and a surface mesh. A *geodesic spherical drawing* of a graph $G$ is a drawing such that vertices are mapped to distinct points on the unit sphere $\mathcal{S}^2$, and edges are drawn as non-crossing minor arcs of great circles (geodesics on $\mathcal{S}^2$). Given a graph of size $n$, we say that a spherical drawing has *polynomial vertex resolution* whether the geodesic length of the shortest edge is bounded by $\Omega(\frac{1}{n^c})$ (for some constant $c$).

As stated by the result below, we are able to compute spherical drawings guaranteeing linear time performance and polynomial vertex resolution (these requirements could not be achieved by prior works). The idea is quite simple and consists in constructing a convex mesh representation $\mathcal{M}$ of the input graph $G$ (refer to Fig. 8). First compute a special partition of the faces of $G$: each sub-graph is drawn in the plane using previous results (boundary vertices have to preserve inter-path distances). Then we glue all drawings together in order to obtain a polyhedron $\mathcal{M}$ contained in the interior of unit sphere. Finally perform a central projection (from the origin) of the vertices of $\mathcal{M}$ on $\mathcal{S}^2$: this bijectively maps edges to geodesic arcs on $\mathcal{S}^2$.

**Theorem 3.** *Given a planar triangulation $G$ with $n$ vertices, we can compute in linear time a geodesic spherical drawing of $G$, having resolution $\Omega(\frac{1}{n})$.*

A simple way to achieve this result is to pick a vertex $v^N$ of bounded degree $d$ (Euler's relations ensures the existence of a vertex of degree at most 5). Assuming $v^N$ has degree at least 4, we can construct a pyramid with rectangular base having $v^N$ as apex, and whose base corners are four vertices among the neighbors of $v^N$ (refer to Fig. 8). This leads to a partition of the faces of $G$ into five pieces: four triangulations and one 4-scheme triangulation (the base). Each piece, the lateral and bottom faces of the pyramid, admits a planar grid drawing of size $O(n) \times O(n)$: just apply the Duncan et. al algorithm to the base, recalling that there are a constant number of boundary vertices (as $v^N$ has bounded degree). If $v^N$ has degree 3 then construct a triangular pyramid in a similar way: this time it remains only to draw the four faces of the pyramid using the shift algorithm.

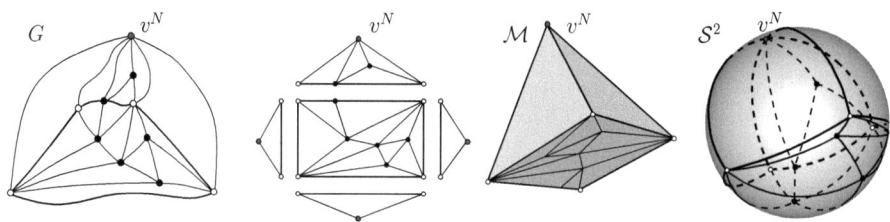

**Fig. 8.** These pictures illustrate the computation of a geodesic spherical drawing

We can also use a similar strategy with other polyhedral shapes, in order to obtain a better distribution of the vertices on the sphere. For example, picking two vertices of bounded degree (and at distance at least 3) it is possible to obtain a prism representation of $G$: lateral faces are then drawn using Property 1.

**Acknowledgments.** We would like to thank the anonymous referees for their helpful comments. This work is supported by the ANR grant EGOS 12 JS02 002 01.

# References

1. Chambers, E., Eppstein, D., Goodrich, M., Loffler, M.: Drawing Graphs in the Plane with a Prescribed Outer Face and Polynomial Area. JGAA 16(2), 243–259 (2012)
2. Castelli Aleardi, L., Devillers, O., Fusy, É.: Canonical Ordering for Triangulations on the Cylinder, with Applications to Periodic Straight-Line Drawings. In: Didimo, W., Patrignani, M. (eds.) GD 2012. LNCS, vol. 7704, pp. 376–387. Springer, Heidelberg (2013)
3. Duncan, C., Goodrich, M., Kobourov, S.: Planar drawings of higher-genus graphs. Journal of Graph Algorithms and Applications 15, 13–32 (2011)
4. de Fraysseix, H., Pach, J., Pollack, R.: How to draw a planar graph on a grid. Combinatorica 10(1), 41–51 (1990)
5. Gonçalves, D., Lévêque, B.: Toroidal maps: Schnyder woods, orthogonal surfaces and straight-line representation. arXiv:1202.0911 (2012)
6. Gotsman, C., Gu, X., Sheffer, A.: Fundamentals of spherical parameterization for 3D meshes. ACM Trans. on Graphics 22(3), 358–363 (2003)
7. Grimm, C.: Parameterization using Manifolds. Int. J. of Shape Modeling 10(1), 51–82 (2004)
8. Kant, G.: Drawing planar graphs using the canonical ordering. Algorithmica 16(1), 4–32 (1996)
9. Kocay, W., Neilson, D., Szypowski, R.: Drawing graphs on the torus. Ars Combinatoria 59, 259–277 (2001)
10. Mohar, B.: Straight-line representations of maps on the torus and other flat surfaces. Discrete Mathematics 15, 173–181 (1996)
11. Schnyder, W.: Embedding planar graphs on the grid. In: SODA, pp. 138–148 (1990)
12. Tutte, W.: How to draw a graph. Proc. of London Math. Soc. 13, 734–767 (1963)

# A Characterization of Those Automata That Structurally Generate Finite Groups

Ines Klimann* and Matthieu Picantin*

Univ Paris Diderot, Sorbonne Paris Cité, LIAFA,
UMR 7089 CNRS, F-75013 Paris, France
{klimann,picantin}@liafa.univ-paris-diderot.fr

**Abstract.** Antonenko and Russyev independently have shown that any Mealy automaton with no cycle with exit—that is, where every cycle in the underlying directed graph is a sink component—generates a finite (semi)group, regardless of the choice of the production functions. Antonenko has proved that this constitutes a characterization in the non-invertible case and asked for the invertible case, which is proved in this paper.

**Keywords:** automaton groups, Mealy automata, finiteness problem.

## 1 Introduction

The class of automata (semi)groups contains multiple interesting and complicated (semi)groups with sometimes unusual features [5]. For instance, the article [12] constructs simple Mealy automata generating infinite torsion groups and so contributes to the Burnside problem, and, the article [4] produces Mealy automata generating the first examples of (semi)groups with intermediate growth and so answers the Milnor problem. Over the years, important results have started revealing their full potential.

In the last decades, the classical decision problems have been investigated for such (semi)groups. The word problem is solvable using standard minimization techniques, while the conjugacy problem is undecidable for automata groups [25]. Of special interest for our concern here, the finiteness problem was proved to be undecidable for automata semigroups [11] and remains open for automata groups despite several positive and promising results [1, 2, 6, 7, 15–18, 22, 23].

The family of automata with no cycle with exit was investigated by Antonenko and by Russyev independently. Focused on the invertible case, Russyev stated in [20] that any invertible Mealy automata with no cycle with exit generates a finite group. Meanwhile, Antonenko showed in [2] (see also [3]) the same result in the non-invertible case and proved the following maximality result: for any automaton with at least one cycle with exit, it is possible to choose (highly

---

* Both authors are partially supported by the french *Agence Nationale pour la Recherche*, through the Project MealyM ANR-JCJC-12-JS02-012-01.

non-invertible) production functions such that the semigroup generated by the induced Mealy automaton is infinite.

In this paper, we fill the visible gap by extending the aforesaid maximality result to the invertible case: for any automaton with at least one cycle with exit, it is possible to choose invertible production functions such that the group generated by the induced Mealy automaton is infinite.

The proof of this new result makes use of original arguments for the current framework, whose common idea is to put a special emphasis on the *dual automaton*, obtained by exchanging the roles of the stateset and the alphabet. Thereby it continues to validate the general strategy first suggested in the paper [1], then followed and continuously developed in [15, 16].

The new maximality result provides a precious milestone in the ongoing work by De Felice and Nicaud (see [9] for a first paper) who propose to design random generators for finite groups based on those invertible Mealy automata with no cycle with exit. Their aim is to simulate interesting distributions that might offer a wide diversity of different finite groups by trying to avoid the classical concentration phenomenon around a typical object, namely symmetric or alternating groups [10, 14], which is significant in already studied distributions. Once implemented, such generators would be very useful to test the performance and robustness of algorithms from computational group theory. They would also be of great use when trying to check a conjecture, by testing it on various random inputs, since exhaustive tests are impossible due to a combinatorial explosion.

The structure of the paper is the following. Basic notions on Mealy automata and automaton (semi)groups are presented in Sect. 2. In Sect. 3, we introduce new tools and prove the main result.

## 2 Mealy Automata

This section contains material for the proofs: first classical definitions and then considerations already made in [15] to maintain the paper self-contained.

### 2.1 Automaton Groups and Semigroups

If one forgets initial and final states, a *(finite, deterministic, and complete) automaton* $\mathcal{A}$ is a triple

$$(A, \Sigma, \delta = (\delta_i : A \to A)_{i \in \Sigma}) ,$$

where the *stateset* $A$ and the *alphabet* $\Sigma$ are non-empty finite sets, and where the $\delta_i$ are functions called *transition functions*.

The transitions of such an automaton are

$$x \xrightarrow{i} \delta_i(x) .$$

An automaton is *reversible* if all its transition functions are permutations of the stateset. Note that in this case each state has exactly one incoming transition labelled by each letter.

A *Mealy automaton* is a quadruple

$$\left(A, \Sigma, \delta = (\delta_i : A \to A)_{i \in \Sigma}, \rho = (\rho_x : \Sigma \to \Sigma)_{x \in A}\right),$$

such that both $(A, \Sigma, \delta)$ and $(\Sigma, A, \rho)$ are automata. In other terms, a Mealy automaton is a letter-to-letter transducer with the same input and output alphabet. If $\mathcal{A} = (A, \Sigma, \delta)$ is an automaton, a finite sequence of functions $\rho = (\rho_x : \Sigma \to \Sigma)_{x \in A}$ is called a *coloring* for $\mathcal{A}$: we denote by $(\mathcal{A}, \rho)$ the Mealy automaton $(A, \Sigma, \delta, \rho)$ and we say that $\mathcal{A}$ is *colored by* $\rho$ (according to the nomenclature in [8]). The graphical representation of a Mealy automaton is standard, see Fig. 1.

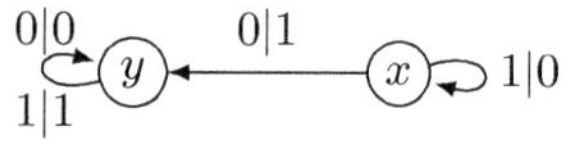

**Fig. 1.** An example of a Mealy automaton: the so-called adding machine

The transitions of a Mealy automaton are

$$x \xrightarrow{i|\rho_x(i)} \delta_i(x) \ .$$

A Mealy automaton $\mathcal{M} = (A, \Sigma, \delta, \rho)$ is *reversible* if the automaton $(A, \Sigma, \delta)$ is reversible and *invertible* if the functions $\rho_x$ are permutations of the alphabet. In this latter case, its *inverse* is the Mealy automaton $\mathcal{M}^{-1}$ with stateset $A^{-1} = \{x^{-1}, x \in A\}$ and set of transitions

$$x^{-1} \xrightarrow{j|i} y^{-1} \in \mathcal{M}^{-1} \quad \Longleftrightarrow \quad x \xrightarrow{i|j} y \in \mathcal{M}.$$

A Mealy automaton $\mathcal{M}$ is *bireversible* if both $\mathcal{M}$ and $\mathcal{M}^{-1}$ are invertible and reversible.

In a Mealy automaton $\mathcal{M} = (A, \Sigma, \delta, \rho)$, the sets $A$ and $\Sigma$ play dual roles. So we may consider the *dual (Mealy) automaton* defined by

$$\mathfrak{d}(\mathcal{M}) = (\Sigma, A, \rho, \delta) \ ,$$

see an example on Fig. 2. Obviously, a Mealy automaton is reversible if and only if its dual is invertible.

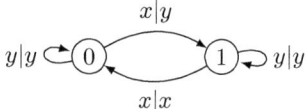

**Fig. 2.** The dual automaton of the Mealy automaton of Fig. 1

Let $\mathcal{M} = (A, \Sigma, \delta, \rho)$ be a Mealy automaton. We view $\mathcal{M}$ as an automaton with an input and an output tape, thus defining mappings from input words over $\Sigma$ to output words over $\Sigma$. Formally, for $x \in A$, the map $\rho_x : \Sigma^* \to \Sigma^*$, extending $\rho_x : \Sigma \to \Sigma$, is defined by:

$$\forall i \in \Sigma, \ \forall \mathbf{s} \in \Sigma^*, \qquad \rho_x(i\mathbf{s}) = \rho_x(i)\rho_{\delta_i(x)}(\mathbf{s}) \ .$$

By convention, the image of the empty word is itself. The mapping $\rho_x$ is length-preserving and prefix-preserving. We say that $\rho_x$ is the *production function* associated with $x$ in $\mathcal{M}$ or, more briefly, if there is no ambiguity, the *production function* of $x$. For $\mathbf{u} = x_1 \cdots x_n \in A^n$ with $n > 0$, we set $\rho_{\mathbf{u}} : \Sigma^* \to \Sigma^*$, $\rho_{\mathbf{u}} = \rho_{x_n} \circ \cdots \circ \rho_{x_1}$.

Denote dually by $\delta_i : A^* \to A^*, i \in \Sigma$, the production functions associated with the dual automaton $\mathfrak{d}(\mathcal{M})$. For $\mathbf{s} = i_1 \cdots i_n \in \Sigma^n$ with $n > 0$, we set $\delta_{\mathbf{s}} : A^* \to A^*$, $\delta_{\mathbf{s}} = \delta_{i_n} \circ \cdots \circ \delta_{i_1}$.

The semigroup of mappings from $\Sigma^*$ to $\Sigma^*$ generated by $\rho_x, x \in A$, is called the *semigroup generated by* $\mathcal{M}$ and is denoted by $\langle \mathcal{M} \rangle_+$. When $\mathcal{M}$ is invertible, its production functions are permutations on words of the same length and thus we may consider the group of mappings from $\Sigma^*$ to $\Sigma^*$ generated by $\rho_x, x \in A$; it is called the *group generated by* $\mathcal{M}$ and is denoted by $\langle \mathcal{M} \rangle$.

The automaton of Fig. 1 generates the semigroup $\mathbb{N}$ and the group $\mathbb{Z}$. The orbit of the word $0^n$ under the action of $\rho_x$ is of size $2^n$: it acts like a binary addition until $1^n$ (considering the most significant bit on the right). In fact the Mealy automaton of Fig. 1 is called the *adding machine* [13].

Remind some known facts on finiteness of the automaton (semi)group:

(F1) To prune a Mealy automaton by deleting its states which are not reachable from a cycle (see the precise definition of a cycle in Subsection 3.1) does not change the finiteness or infiniteness of the generated (semi)group [2].
(F2) An invertible Mealy automaton generates a finite group if and only if it generates a finite semigroup [1, 24].
(F3) A Mealy automaton generates a finite semigroup if and only if so does its dual [1, 19, 21].
(F4) An invertible-reversible but not bireversible Mealy automaton generates an infinite group [1].

Whenever the alphabet is unary, the generated group is trivial and there is nothing to say. **Throughout this paper, the alphabet has at least two elements.**

### 2.2 On the Powers of a Mealy Automaton and Its Connected Components

Let $\mathcal{M} = (A, \Sigma, \delta, \rho)$ be a Mealy automaton.

Considering the underlying graph of $\mathcal{M}$, it makes sense to look at its connected components. If $\mathcal{M}$ is reversible, its connected components are always strongly connected (its transition functions are permutations of a finite set).

A convenient and natural operation is to raise $\mathcal{M}$ to the power $n$, for some $n > 0$: its *n-th power* is the Mealy automaton

$$\mathcal{M}^n = \left( A^n, \Sigma, (\delta_i : A^n \to A^n)_{i \in \Sigma}, (\rho_{\mathbf{u}} : \Sigma \to \Sigma)_{\mathbf{u} \in A^n} \right) .$$

If $\mathcal{M}$ is reversible, so is each of its powers.

If $\mathcal{M}$ is reversible, we can be more precise on the behavior of the connected components of its powers. As highlighted in [15], they have a very peculiar form: if $\mathcal{C}$ is a connected component of $\mathcal{M}^n$ for some $n$ and $\mathbf{u}$ is a state of $\mathcal{C}$, we obtain a connected component of $\mathcal{M}^{n+1}$ by choosing a state $x \in A$ and building the connected component of $\mathbf{u}x$, denoted by $\mathcal{D}$. For any state $\mathbf{v}$ of $\mathcal{C}$, there exists a state of $\mathcal{D}$ prefixed with $\mathbf{v}$:

$$\exists \mathbf{s} \in \Sigma^* \mid \delta_{\mathbf{s}}(\mathbf{u}) = \mathbf{v} \quad \text{and} \quad \delta_{\mathbf{s}}(\mathbf{u}x) = \mathbf{v}\delta_{\rho_{\mathbf{u}}(\mathbf{s})}(x) .$$

Furthermore, if $\mathbf{u}y$ is a state of $\mathcal{D}$, for some state $y \in A$ different from $x$, then $\delta_{\mathbf{s}}(\mathbf{u}x)$ and $\delta_{\mathbf{s}}(\mathbf{u}y)$ are two different states of $\mathcal{D}$ prefixed with $\mathbf{v}$, because of the reversibility of $\mathcal{M}^{n+1}$: the transition function $\delta_{\rho_{\mathbf{u}}(\mathbf{s})}$ is a permutation.

Hence $\mathcal{D}$ can be seen as consisting of several full copies of $\mathcal{C}$ and $\#\mathcal{C}$ divides $\#\mathcal{D}$. They have the same size if and only if, once fixed some state $\mathbf{u}$ of $\mathcal{C}$, for any different states $x, y \in A$, $\mathbf{u}x$ and $\mathbf{u}y$ cannot both belong to $\mathcal{D}$.

If all of those connected components of $\mathcal{M}^{n+1}$ built from $\mathcal{C}$ have the same size as $\mathcal{C}$, we say that $\mathcal{C}$ *splits up totally*. If all the connected components of an automaton split up totally, we say that the automaton *splits up totally*.

## 3 A Maximal Family for Groups

Antonenko and Russyev both investigated a family of Mealy automata such that the finiteness of the generated group (Russyev [20]) or semigroup (Antonenko [2]) is inherent to the structure of the automaton, regardless of its production functions. In fact, though they use different definitions and names, they study the same family: automata where every cycle is a sink component. Antonenko has proved that this family is maximal in the non-invertible case: if an automaton admits a cycle which is not a sink component, it can be colored to generate an infinite semigroup.

To prove his result, Antonenko analyzes different cases and, in each situation, exhibits an element of infinite order in the semigroup. In this section we prove the maximality of the former family for groups, using completely different techniques. We adopt and adapt Russyev's nomenclature.

### 3.1 How to Exit from a Cycle?

Let $\mathcal{A} = (A, \Sigma, \delta)$ be an automaton. A *cycle* of length $n \in \mathbb{N}$ in the automaton $\mathcal{A}$ is a sequence of transitions of $\mathcal{A}$

$$x_1 \xrightarrow{i_1} x_2, \quad \ldots, \quad x_{n-1} \xrightarrow{i_{n-1}} x_n, \quad x_n \xrightarrow{i_n} x_1$$

where $x_1, \ldots, x_n$ are pairwise different states in $A$ and $i_1, \ldots, i_n$ are some letters of $\Sigma$.

The *label* of this cycle *from the state* $x_k$ is the word $i_k \cdots i_n i_1 \cdots i_{k-1}$.

This cycle is *with external exit* if there exist $k$ with $1 \leq k \leq n$ and $i \in \Sigma$ satisfying $\delta_i(x_k) \notin \{x_1, \ldots, x_n\}$. It is *with internal exit* if there exist $k$ with $1 \leq k \leq n$ and $i \in \Sigma$ satisfying $\delta_i(x_k) \in \{x_1, \ldots, x_n\}$ and $\delta_i(x_k) \neq \delta_{i_k}(x_k)$. We could say that a cycle is *with exit* without specifying the nature of the exit. In all other cases, this cycle is *without exit*. Examples are given in Fig. 3.

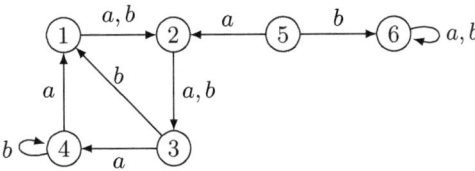

**Fig. 3.** The cycle $1 \xrightarrow{a} 2 \xrightarrow{a} 3 \xrightarrow{b} 1$ is with external exit; the cycle $1 \xrightarrow{a} 2 \xrightarrow{a} 3 \xrightarrow{a} 4 \xrightarrow{a} 1$ is with internal exit; and the cycle $6 \xrightarrow{a} 6$ is without exit

Note that the existence of a cycle with internal exit induces the existence of a (possibly shorter) cycle with external exit. For example in Fig. 3, the cycle $1 \xrightarrow{a} 2 \xrightarrow{a} 3 \xrightarrow{a} 4 \xrightarrow{a} 1$ has two internal exits: $4 \xrightarrow{b} 4$ and $3 \xrightarrow{b} 1$; the first one leads to the cycle $4 \xrightarrow{b} 4$ with external exit, while the second one leads to the cycle $1 \xrightarrow{a} 2 \xrightarrow{a} 3 \xrightarrow{b} 1$ with external exit.

**Proposition 1 ([2, 20]).** *Whenever an automaton $\mathcal{A}$ admits no cycle with exit, whatever choice is made for the coloring $\rho$, the colored automaton $(\mathcal{A}, \rho)$ generates a finite (semi)group.*

### 3.2 A Pumping Lemma for the Reversible Two-State Automata

It is proved in [15, Lemma 9] that in the case of a reversible Mealy automaton $\mathcal{M}$ with exactly two states, if some power of $\mathcal{M}$ splits up totally, then all the later powers of $\mathcal{M}$ split up totally. We can deduce the following result which can be seen as a pumping lemma: if the generated semigroup is infinite, sufficiently long paths can be considered in the dual automaton to turn indefinitely in a cycle. More formally:

**Lemma 2 (Pumping Lemma).** *Let $\mathcal{M}$ be a reversible Mealy automaton with two states $\{x, y\}$. The automaton $\mathcal{M}$ generates an infinite semigroup if and only if, for any integer $N \in \mathbb{N}$, there exists a word $\mathbf{u} \in \{x, y\}^*$ of length at least $N$ such that the states $\mathbf{u}x$ and $\mathbf{u}y$ belong to the same connected component of $\mathcal{M}^{|\mathbf{u}|+1}$.*

## 3.3 The Family of Automata with no Cycle with Exit Is Maximal for Groups

We prove here that any automaton which admits a cycle with exit can be colored in order to generate an infinite group. We analyze several simple cases in Lemmas 3, 4, and 5 which contribute to prove the general case of Theorem 6.

**Lemma 3.** *Any automaton over a binary alphabet with a cycle with external exit can be colored to generate an infinite group.*

*Proof.* Let $\mathcal{A}$ be an automaton over a binary alphabet $\{0, 1\}$ with a cycle $\mathcal{C}$ with external exit as shown in Fig. 4.

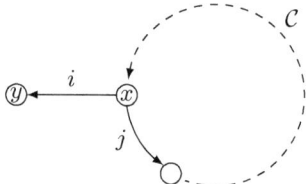

**Fig. 4.** The cycle $\mathcal{C}$ is with external exit: $y = \delta_i(x)$, $y \notin \mathcal{C}$

Take the following permutations on the alphabet : $\rho_y$ permutes the letters of the alphabet and $\rho_z$ stabilizes the alphabet for any other state $z$ (in particular for any state of $\mathcal{C}$).

Let $\mathbf{s} \in \{0, 1\}^+$ be the label of $\mathcal{C}$ from $x$. For any $n \in \mathbb{N}$, the words $\mathbf{s}^n i0$ and $\mathbf{s}^n i1$ belong to a same connected component of $\mathfrak{d}(\mathcal{A}, \rho)^{|\mathbf{s}|+2}$: $\rho_x(\mathbf{s}^n i0) = \mathbf{s}^n i1$. The Mealy automaton $\mathfrak{d}(\mathcal{A}, \rho)$ is reversible and has two states, so we can apply the Pumping Lemma and conclude on the infiniteness of $\langle\!\langle (\mathcal{A}, \rho) \rangle\!\rangle$ by (**F2**) and (**F3**). □

**Lemma 4 (River of no return Lemma).** *Let $\mathcal{A}$ be an automaton and $\mathcal{C}$ a cycle of $\mathcal{A}$. If $\mathcal{C}$ admits an external exit to some state and is not reachable from this state, then $\mathcal{A}$ can be colored to generate an infinite group.*

*Proof.* The idea of this proof is to mimic the adding machine (see Fig. 1). Again, Fig. 4 illustrates the situation: $\mathcal{C}$ admits an external exit to the state $y$, the additional hypothesis being that $\mathcal{C}$ is not reachable from $y$ (and the alphabet is not supposed binary any longer).

Denote the label of $\mathcal{C}$ from $x$ by $\mathbf{s} = j\mathbf{t}$ with $j \in \Sigma$ and $\mathbf{t} \in \Sigma^*$. We choose the following production functions on the alphabet: $\rho_x$ is the transposition of $i$ and $j$ and $\rho_z$ is the identity for any other state $z$.

As for the adding machine, the orbit of $(j\mathbf{t})^n$ under the action of $\rho_x$ has size $2^n$. Therefore the element $x$ is of infinite order and so is the group $\langle\!\langle (\mathcal{A}, \rho) \rangle\!\rangle$. □

**Lemma 5.** *Any reversible automaton with a cycle with exit can be colored to generate an infinite group.*

*Proof.* Let $\mathcal{A} = (A, \Sigma, \delta)$ be a reversible automaton with a cycle with exit.

As $\mathcal{A}$ is reversible, it admits some states $x, y, z$ with $x \neq y$ such that there exist a transition from $x$ to $z$ and a transition from $y$ to $z$. We can choose the permutations $\rho_x$ and $\rho_y$ such that these transitions have the same output and take identity for all the other permutations.

The colored automaton $(\mathcal{A}, \rho)$ is invertible and reversible but not bireversible. Hence it generates an infinite group by (**F4**). □

The next theorem is the main result of this paper.

**Theorem 6.** *Any automaton with a cycle with exit can be colored into an invertible Mealy automaton generating an infinite group.*

*Proof.* Let $\mathcal{A} = (A, \Sigma, \delta)$ be an automaton with a cycle with exit.

By (**F1**), we can suppose, without loss of generality, that $\mathcal{A}$ is pruned. If there exists a transition not belonging to a cycle, as the starting state of this transition is reachable from a cycle, Lemma 4 applies and we are done.

We can assume now that any transition belongs to (at least) one cycle.

If $\mathcal{A}$ is reversible, it can be colored to generate an infinite group by Lemma 5.

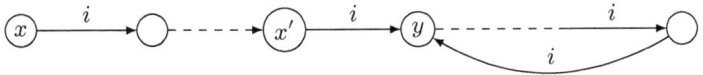

**Fig. 5.** Path from $x$ with all transitions labelled by $i$

We can suppose now that $\mathcal{A}$ is not reversible: there exist a state $x$ and a letter $i$ such that $x$ has no incoming transition labelled by $i$. Consider the path starting at $x$ with all transitions labelled by $i$ as shown in Fig. 5. This path loops on some state $y \neq x$. Denote by $\mathcal{C}$ the resulting cycle. We have $y = \delta_{i^{n+k\#C}}(x)$ for some minimal $n > 0$ and for all $k \geq 0$. Now let $x'$ denote the state $\delta_{i^{n-1}}(x)$: we have $x' \notin \mathcal{C}$.

The transition $x' \xrightarrow{i} y$ belongs to some cycle by hypothesis and this cycle is not $\mathcal{C}$ by construction. Therefore $\mathcal{C}$ admits an external exit $y' \xrightarrow{j} y''$ (with $j \neq i$). Hence $x'$ is reachable from $y$ and so from $\mathcal{C}$, by hypothesis, but does not belong to $\mathcal{C}$, by construction. The automaton $\mathcal{B} = (A, \{i, j\}, (\delta_i, \delta_j))$ contains the cycle $\mathcal{C}$ and the transition $y' \xrightarrow{j} y''$. So $\mathcal{B}$ can be colored to generate an infinite group by Lemma 3, say with $\rho = (\rho_z : \{i, j\} \to \{i, j\})_{z \in A}$. This group is a quotient of any group obtained by completing each $\rho_z$ from $\{i, j\}$ into $\Sigma$, and we can conclude. □

## 4 Conclusion

Until the finiteness problem for automaton groups has not been proved to be decidable—and even more if it is eventually proved to be undecidable, random generation of finite groups *via* Mealy automata is possible only for some automata known to generate finite groups. The family of Mealy automaton with no cycle with exit is the largest family generating finite groups, whatever choice is made for the production functions.

**Acknowledgments.** The authors would like to thank Jean Mairesse who has detected a serious gap in a previous version of this paper.

## References

1. Akhavi, A., Klimann, I., Lombardy, S., Mairesse, J., Picantin, M.: On the finiteness problem for automaton (semi)groups. Int. J. Algebra Comput. 22(6), 26p. (2012)
2. Antonenko, A.S.: On transition functions of Mealy automata of finite growth. Matematychni Studii 29(1), 3–17 (2008)
3. Antonenko, A.S., Berkovich, E.L.: Groups and semigroups defined by some classes of Mealy automata. Acta Cybernetica 18(1), 23–46 (2007)
4. Bartholdi, L., Reznykov, I., Sushchanskiĭ, V.: The smallest Mealy automaton of intermediate growth. J. Algebra 295(2), 387–414 (2006)
5. Bartholdi, L., Silva, P.: Groups defined by automata (2010), arXiv:cs.FL/1012.1531
6. Bondarenko, I., Bondarenko, N., Sidki, S., Zapata, F.: On the conjugacy problem for finite-state automorphisms of regular rooted trees (with an appendix by Raphaël M. Jungers). Groups Geom. Dyn. 7(2), 323–355 (2013)
7. Cain, A.: Automaton semigroups. Theor. Comput. Sci. 410(47-49), 5022–5038 (2009)
8. D'Angeli, D., Rodaro, E.: Groups and Semigroups Defined by Colorings of Synchronizing Automata (2013), arXiv:math.GR/1310.5242
9. De Felice, S., Nicaud, C.: Random generation of deterministic acyclic automata using the recursive method. In: Bulatov, A.A., Shur, A.M. (eds.) CSR 2013. LNCS, vol. 7913, pp. 88–99. Springer, Heidelberg (2013)
10. Dixon, J.D.: The probability of generating the symmetric group. Math. Z. 110, 199–205 (1969)
11. Gillibert, P.: The finiteness problem for automaton semigroups is undecidable. arXiv:cs.FL/1304.2295 (2013)
12. Grigorchuk, R.: On Burnside's problem on periodic groups. Funktsional. Anal. i Prilozhen. 14(1), 53–54 (1980)
13. Grigorchuk, R., Nekrashevich, V., Sushchanskiĭ, V.: Automata, dynamical systems, and groups. Tr. Mat. Inst. Steklova 231, 134–214 (2000)
14. Jaikin-Zapirain, A., Pyber, L.: Random generation of finite and profinite groups and group enumeration. Ann. of Math. (2) 173(2), 769–814 (2011)
15. Klimann, I.: The finiteness of a group generated by a 2-letter invertible-reversible Mealy automaton is decidable. In: Proc. 30th STACS. LIPIcs, vol. 20, pp. 502–513 (2013)
16. Klimann, I., Mairesse, J., Picantin, M.: Implementing computations in automaton (semi)groups. In: Moreira, N., Reis, R. (eds.) CIAA 2012. LNCS, vol. 7381, pp. 240–252. Springer, Heidelberg (2012)
17. Maltcev, V.: Cayley automaton semigroups. Int. J. Algebra Comput. 19(1), 79–95 (2009)
18. Mintz, A.: On the Cayley semigroup of a finite aperiodic semigroup. Int. J. Algebra Comput. 19(6), 723–746 (2009)

19. Nekrashevych, V.: Self-similar groups. Mathematical Surveys and Monographs, vol. 117. American Mathematical Society, Providence (2005)
20. Russyev, A.: Finite groups as groups of automata with no cycles with exit. Algebra and Discrete Mathematics 9(1), 86–102 (2010)
21. Savchuk, D., Vorobets, Y.: Automata generating free products of groups of order 2. J. Algebra 336(1), 53–66 (2011)
22. Sidki, S.: Automorphisms of one-rooted trees: growth, circuit structure, and acyclicity. J. Math. Sci. 100(1), 1925–1943 (2000); Algebra, 12
23. Silva, P., Steinberg, B.: On a class of automata groups generalizing lamplighter groups. Int. J. Algebra Comput. 15(5-6), 1213–1234 (2005)
24. Steinberg, B., Vorobets, M., Vorobets, Y.: Automata over a binary alphabet generating free groups of even rank. Int. J. Algebra Comput. 21(1-2), 329–354 (2011)
25. Šuniḱ, Z., Ventura, E.: The conjugacy problem in automaton groups is not solvable. Journal of Algebra 364, 148–154 (2012)

# Linear Grammars with One-Sided Contexts and Their Automaton Representation

Mikhail Barash[1,2] and Alexander Okhotin[2]

[1] Department of Mathematics and Statistics, University of Turku,
Turku FI-20014, Finland, {mikhail.barash,alexander.okhotin}@utu.fi
[2] Turku Centre for Computer Science, Turku FI-20520, Finland

**Abstract.** The paper considers a family of formal grammars that extends linear context-free grammars with an operator for referring to the left context of a substring being defined, as well as with a conjunction operation (as in linear conjunctive grammars). These grammars are proved to be computationally equivalent to an extension of one-way real-time cellular automata with an extra data channel. The main result is the undecidability of the emptiness problem for grammars restricted to a one-symbol alphabet, which is proved by simulating a Turing machine by a cellular automaton with feedback. The same construction proves the $\Sigma_2^0$-completeness of the finiteness problem for these grammars.

## 1 Introduction

The idea of defining context-free rules applicable only in certain contexts dates back to the early work of Chomsky. However, the mathematical model improvised by Chomsky, which he named a "context-sensitive grammar", turned out to be too powerful for its intended application, as it could simulate a space-bounded Turing machine. Recently, the authors [3] made a fresh attempt on implementing the same idea. Instead of the string-rewriting approach from the late 1950s, which never quite worked out for this task, the authors relied upon the modern understanding of formal grammars as a first-order logic over positions in a string, discovered by Rounds [16]. This led to a family of grammars that allows such rules as $A \to BC \mathbin{\&} \triangleleft D$, which asserts that all strings representable as a concatenation $BC$ and preceded by a left context of the form $D$ have the property $A$. The semantics of such grammars are defined through logical deduction of items of the form *"a substring $v$ written in left context $u$ has a property $A$"* [3], and the resulting formal model inherits some of the key properties of formal grammars, including parse trees, an extension of the Chomsky normal form [3,4], a form of recursive descent parsing [2] and a variant of the Cocke–Kasami–Younger parsing algorithm that works in time $O\bigl(\frac{n^3}{\log n}\bigr)$ [14].

This paper aims to investigate the *linear subclass* of grammars with one-sided contexts, where linearity is understood in the sense of Chomsky and Schützenberger, that is, as a restriction to concatenate nonterminal symbols only to terminal strings. An intermediate family of *linear conjunctive grammars*, which allows using the conjunction operation, but no context specifications, was earlier studied by the second author [11,12]. Those grammars were found to be computationally equivalent to *one-way real-time cellular automata* [6,17], also known under a proper name of *trellis automata* [5,7].

This paper sets off by developing an analogous automaton representation for linear grammars with one-sided contexts. The proposed *trellis automata with feedback*, defined in Section 3, augment the original cellular automaton model by an extra communication channel, which adds exactly the same power as context specifications do in grammars. This representation implies the closure of this language family under complementation, which, using grammars alone, would require a complicated construction.

The main contribution of the paper is a method for simulating a Turing machine by a trellis automaton with feedback processing an input string over a one-symbol alphabet. This method subsequently allows uniform undecidability proofs for linear grammars with contexts, which parallels the recent results for conjunctive grammars due to Jeż [8] and Jeż and Okhotin [9,10], but is based upon an entirely different underlying construction.

The new construction developed in this paper begins in Section 4 with a simple example of a 3-state trellis automaton with feedback, which recognizes the language $\{\, a^{2^k-2} \mid k \geqslant 2 \,\}$. To compare, ordinary trellis automata over a one-symbol alphabet recognize only regular languages [5]. The next Section 5 presents a simulation of a Turing machine by a trellis automaton with feedback, so that the latter automaton, given an input $a^n$, simulates $O(n)$ first steps of the Turing machine's computation on an empty input, and accordingly can accept or reject the input $a^n$ depending on the current state of the Turing machine.

This construction is used in the last Section 6 to prove the undecidability of the emptiness problem for linear grammars with one-sided contexts over a one-symbol alphabet. The finiteness problem for these grammars is proved to be complete for the second level of the arithmetical hierarchy.

## 2 Grammars with One-Sided Contexts

Grammars with contexts were introduced by the authors [3,4] as a model capable of defining context-free rules applicable only in contexts of a certain form.

**Definition 1 ([3]).** *A grammar with left contexts is a quadruple $G = (\Sigma, N, R, S)$, where*

- *$\Sigma$ is the alphabet of the language being defined;*
- *$N$ is a finite set of auxiliary symbols ("nonterminal symbols" in Chomsky's terminology), which denote the properties of strings defined in the grammar;*
- *$R$ is a finite set of grammar rules, each of the form*

$$A \to \alpha_1 \,\&\, \ldots \,\&\, \alpha_k \,\&\, \triangleleft\beta_1 \,\&\, \ldots \,\&\, \triangleleft\beta_m \,\&\, \trianglelefteq\gamma_1 \,\&\, \ldots \,\&\, \trianglelefteq\gamma_n, \qquad (1)$$

  *with $A \in N$, $k \geqslant 1$, $m, n \geqslant 0$ and $\alpha_i, \beta_i, \gamma_i \in (\Sigma \cup N)^*$;*
- *$S \in N$ represents syntactically well-formed sentences of the language.*

Every rule (1) is comprised of *conjuncts* of three kinds. Each conjunct $\alpha_i$ specifies the form of the substring being defined, a conjunct $\triangleleft\beta_i$ describes the form of its left context, while a conjunct $\triangleleft\gamma_i$ refers to the form of the left context concatenated with the current substring. To be precise, let $w \in \Sigma^*$ be the whole string being defined, and consider defining its substring $v$ by a rule (1), where $w = uvx$ for $u, v, x \in \Sigma^*$. Then, each conjunct $\alpha_i$ describes the form of $v$, each *left context operator* $\triangleleft\beta_i$ describes the form of $u$, and each *extended left context operator* $\triangleleft\gamma_i$, describes the form of $uv$. The conjunction means that all these conditions must hold at the same time.

If no context specifications are used in the grammar, that is, if $m = n = 0$ in each rule (1), then this is a *conjunctive grammar* [11,13]. If, furthermore, only one conjunct is allowed in each rule ($k = 1$), this is an ordinary context-free grammar. A grammar is called *linear*, if every conjunct refers to at most one nonterminal symbol, that is, $\alpha_1, \ldots, \alpha_k, \beta_1, \ldots, \beta_m, \gamma_1, \ldots, \gamma_n \in \Sigma^* N \Sigma^* \cup \Sigma^*$.

The language generated by a grammar with left contexts is defined by deduction of elementary statements of the form "a substring $v \in \Sigma^*$ in the left context $u \in \Sigma^*$ has the property $X \in \Sigma \cup N$", denoted by $X(u\langle v \rangle)$. A full definition applicable to every grammar with left contexts is presented in the authors' previous paper [3]; this paper gives a definition specialized for linear grammars.

**Definition 2.** *Let $G = (\Sigma, N, R, S)$ be a linear grammar with left contexts, and consider deduction of items of the form $X(u\langle v \rangle)$, with $u, v \in \Sigma^*$ and $X \in N$. Each rule $A \to w$ defines an axiom scheme $\vdash_G A(x\langle w \rangle)$, for all $x \in \Sigma^*$. Each rule of the form $A \to x_1 B_1 y_1 \& \ldots \& x_k B_k y_k \& \triangleleft x'_1 D_1 y'_1 \& \ldots \& \triangleleft x'_m D_m y'_m \& \triangleleft x''_1 E_1 y''_1 \& \ldots \& \triangleleft x''_n E_n y''_n$ defines the following scheme for deduction rules for all $u, v \in \Sigma^*$:*

$$\{B_i(ux_i\langle v_i\rangle)\}_{1\leqslant i \leqslant k}, \{D_i(x'_i\langle u_i\rangle)\}_{1\leqslant i \leqslant m}, \{E_i(x''_i\langle w_i\rangle)\}_{1\leqslant i \leqslant n} \vdash_G A(u\langle v \rangle),$$

*where $x_i v_i y_i = v$, $x'_i u_i y'_i = u$ and $x''_i w_i y''_i = uv$. Then the language defined by a nonterminal symbol $A$ is $L_G(A) = \{\, u\langle v\rangle \mid u, v \in \Sigma^*,\ \vdash_G A(u\langle v\rangle) \,\}$. The language defined by the grammar $G$ is the set of all strings with an empty left context defined by $S$, that is, $L(G) = \{\, w \mid w \in \Sigma^*,\ \vdash_G S(\varepsilon\langle w \rangle) \,\}$.*

This definition is illustrated in the grammar below.

*Example 1.* The following grammar defines the singleton language $\{abac\}$:

$$S \to aBc$$
$$B \to bA \,\&\, \triangleleft A$$
$$A \to a$$

The string *abac* is generated as follows:

|  |  |
|---|---|
| $\vdash A(\varepsilon\langle a \rangle)$ | $(A \to a)$ |
| $\vdash A(ab\langle a \rangle)$ | $(A \to a)$ |
| $A(ab\langle a \rangle), A(\varepsilon\langle a \rangle) \vdash B(a\langle ba \rangle)$ | $(B \to bA \,\&\, \triangleleft A)$ |
| $B(a\langle ba \rangle) \vdash S(\varepsilon\langle abac \rangle)$ | $(S \to aBc)$ |

The next example defines a language that is known to have no linear conjunctive grammar [19].

*Example 2 (Törmä [18]).* The following linear grammar with contexts defines the language $\{\, a^n b^{in} \mid i, n \geqslant 1\,\}$:

$$S \to aSb \mid B \,\&\, \triangleleft S \mid \varepsilon$$
$$B \to bB \mid \varepsilon$$

The rule $S \to B \,\&\, \triangleleft S$ appends as many symbols $b$ as there are $as$ in the beginning of the string.

Every grammar with contexts can be transformed to a certain normal form [3,4,14], which extends the Chomsky normal form for ordinary context-free grammars. This extension allows multiple conjuncts of the form $BC$ and context specifications $\triangleleft D$, that is, every rule in a normal form grammar is either of the form $A \to a \,\&\, \triangleleft D_1 \,\&\, \ldots \,\&\, \triangleleft D_m$ or $A \to B_1 C_1 \,\&\, \ldots \,\&\, B_k C_k$. A similar normal form can be established for the linear subclass of grammars.

**Theorem 1.** *For every linear grammar with left contexts, there exists another linear grammar with left contexts that defines the same language and has all rules of the form*

$$A \to bB_1 \,\&\, \ldots \,\&\, bB_\ell \,\&\, C_1 c \,\&\, \ldots \,\&\, C_k c \tag{2a}$$
$$A \to a \,\&\, \triangleleft D_1 \,\&\, \ldots \,\&\, \triangleleft D_m, \tag{2b}$$

*where $A, B_i, C_i, D_i \in N$, $a, b, c \in \Sigma$, $\ell + k \geqslant 1$ and $m \geqslant 0$.*

The transformation is carried out along the same lines as in the general case. The first step is elimination of *null conjuncts*, that is, any rules of the form $A \to \varepsilon \,\&\, \ldots$. This is followed by elimination of *null contexts* $\triangleleft \varepsilon$, and on *unit conjuncts*, as in the rule $A \to B \,\&\, \ldots$. The final step is elimination of *extended left contexts* $\triangleleft E$, which are all expressed through proper left contexts $\triangleleft D$ [14]. Each step applies to linear grammars with contexts and preserves their linearity.

## 3 Automaton Representation

Linear conjunctive grammars are known to be computationally equivalent to one of the simplest types of cellular automata: the *one-way real-time cellular automata*, also known under the proper name of *trellis automata*. This section presents a generalization of trellis automata, which similarly corresponds to linear grammars with one-sided contexts.

An ordinary trellis automaton processes an input string of length $n \geqslant 1$ using a uniform array of $\frac{n(n+1)}{2}$ nodes, as presented in Figure 1(left). Each node computes a value from a fixed finite set $Q$. The nodes in the bottom row obtain their values directly from the input symbols using a function $I \colon \Sigma \to Q$. The rest of the nodes compute the function $\delta \colon Q \times Q \to Q$ of the values in their predecessors. The string is accepted if and only if the value computed by the topmost node belongs to the set of accepting states $F \subseteq Q$.

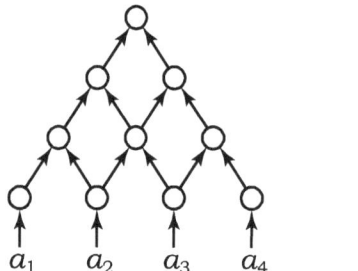 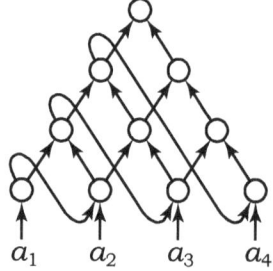

**Fig. 1.** Trellis automata (left) and trellis automata with feedback (right)

**Theorem A (Okhotin [12]).** *A language $L \subseteq \Sigma^+$ is defined by a linear conjunctive grammar if and only if $L$ is recognized by a trellis automaton.*

In terms of cellular automata, every horizontal row of states in Figure 1(left) represents an automaton's configuration at a certain moment of time. An alternative motivation developed in the literature on trellis automata [5,6,7] is to consider the entire grid as a digital circuit with uniform structure of connections. In order to obtain a similar representation of linear grammars with left contexts, the trellis automaton model is extended with another type of connections, illustrated in Figure 1(right).

**Definition 3.** *A trellis automaton with feedback is a sextuple $M = (\Sigma, Q, I, J, \delta, F)$, in which:*

- *$\Sigma$ is the input alphabet,*
- *$Q$ is a finite non-empty set of states,*
- *$I \colon \Sigma \to Q$ is a function that sets the initial state for the first symbol,*
- *$J \colon Q \times \Sigma \to Q$ sets the initial state for every subsequent symbol, using the state computed on the preceding substring as a feedback,*
- *$\delta \colon Q \times Q \to Q$ is the transition function, and*
- *$F \subseteq Q$ is the set of accepting states.*

*The behaviour of the automaton is described by a function $\Delta \colon \Sigma^* \times \Sigma^+ \to Q$, which defines the state $\Delta(u\langle v\rangle)$ computed on each string with a context $u\langle v\rangle$ by*

$$\Delta(\varepsilon\langle a\rangle) = I(a),$$
$$\Delta(w\langle a\rangle) = J\bigl(\Delta(\varepsilon\langle w\rangle), a\bigr),$$
$$\Delta(u\langle bvc\rangle) = \delta\bigl(\Delta(u\langle bv\rangle), \Delta(ub\langle vc\rangle)\bigr).$$

*The language recognized by the automaton is $L(M) = \{\, w \in \Sigma^+ \mid \Delta(\varepsilon\langle w\rangle) \in F\,\}$.*

**Theorem 2.** *A language $L \subseteq \Sigma^+$ is defined by a linear grammar with left contexts if and only if $L$ is recognized by a trellis automaton with feedback.*

The proof is by effective constructions in both directions.

**Lemma 1.** *Let $G = (\Sigma, N, R, S)$ be a linear grammar with left contexts, in which every rule is of the forms (2a)–(2b), and define a trellis automaton with feedback $M = (\Sigma, Q, I, J, \delta, F)$ by setting $Q = \Sigma \times 2^N \times \Sigma$,*

$$I(a) = (a, \{A \mid A \to a \in R\}, a)$$
$$J((b, X, c), a) = (a, \{A \mid \exists \text{ rule (2b) with } D_1, \ldots, D_m \in X\}, a)$$
$$\delta((b, X, c'), (b', Y, c)) = (b, \{A \mid \exists \text{ rule (2a) with } B_i \in X \text{ and } C_i \in Y\}, c)$$
$$F = \{(b, X, c) \mid S \in X\}.$$

*For every string with context $u\langle v \rangle$, let $b$ be the first symbol of $v$, let $c$ be the last symbol of $v$, and let $Z = \{A \mid u\langle v \rangle \in L_G(A)\}$. Then $\Delta(u\langle v \rangle) = (b, Z, c)$.*
*In particular, $L(M) = \{w \mid \varepsilon\langle w \rangle \in L_G(S)\} = L(G)$.*

**Lemma 2.** *Let $M = (\Sigma, Q, I, J, \delta, F)$ be a trellis automaton with feedback and define the grammar with left contexts $G = (\Sigma, N, R, S)$, where $N = \{A_q \mid q \in Q\} \cup \{S\}$, and the set $R$ contains the following rules:*

$$A_{I(a)} \to a \ \& \ \triangleleft \varepsilon \qquad (a \in \Sigma)$$
$$A_{J(q,a)} \to a \ \& \ \triangleleft A_q \qquad (q \in Q, a \in \Sigma)$$
$$A_{\delta(p,q)} \to b A_q \ \& \ A_p c \qquad (p, q \in Q, b, c \in \Sigma)$$
$$S \to A_q \qquad (q \in F)$$

*Then, for every string with context $u\langle v \rangle$, $\Delta(u\langle v \rangle) = r$ if and only if $u\langle v \rangle \in L_G(A_r)$. In particular, $L(G) = \{w \mid \Delta(\varepsilon\langle w \rangle) \in F\} = L(M)$.*

This automaton representation is useful for establishing some basic properties of linear grammars with contexts, which would be more difficult to obtain using grammars alone. For instance, one can prove their closure under complementation by taking a trellis automaton with feedback and inverting its set of accepting states. Another property is the closure of the family under concatenating a linear conjunctive language from the right; thus, in particular, the language used by Terrier [17] to show that linear conjunctive languages are not closed under concatenation, can be defined by a linear grammar with contexts.

## 4 Defining a Non-regular Unary Language

Ordinary context-free grammars over a unary alphabet $\Sigma = \{a\}$ define only regular languages. Unary linear conjunctive languages are also regular, because a trellis automaton operates on an input $a^n$ as a deterministic finite automaton [5]. The non-triviality of unary conjunctive grammars was discovered by Jeż [8], who constructed a grammar for the language $\{a^{4^k} \mid k \geqslant 0\}$ using iterated conjunction and concatenation of languages.

This paper introduces a new method for constructing formal grammars for non-regular languages over a unary alphabet, which makes use of a left context operator, but does not rely upon non-linear concatenation. The simplest case of the new method is demonstrated by the following automaton, which can be transformed to a grammar by Lemma 2.

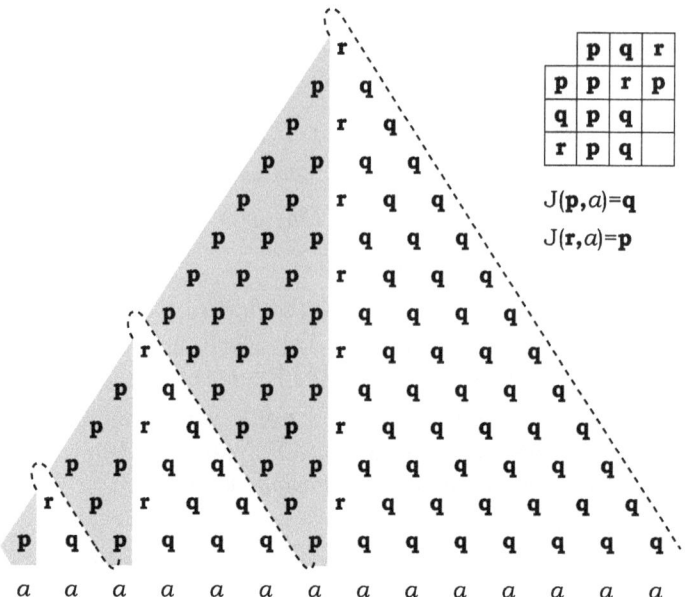

**Fig. 2.** How the automaton in Example 3 recognizes $\{\,a^{2^k-2} \mid k \geqslant 2\,\}$

*Example 3.* Consider a trellis automaton with feedback $M = (\Sigma, Q, I, J, \delta, F)$ over the alphabet $\Sigma = \{a\}$ and with the set of states $Q = \{p, q, r\}$, where $I(a) = p$ is the initial state, the feedback function gives states $J(p, a) = q$ and $J(r, a) = p$, and the transition function is defined by $\delta(s, p) = p$ for all $s \in Q$, $\delta(q, q) = \delta(r, q) = q$, $\delta(p, q) = r$ and $\delta(p, r) = p$. The only accepting state is $r$. Then $M$ recognizes the language $\{\,a^{2^k-2} \mid k \geqslant 2\,\}$.

The computation of this automaton is illustrated in Figure 2. The state computed on each one-symbol substring $a^\ell \langle a \rangle$ is determined by the state computed on $\varepsilon \langle a^\ell \rangle$ according to the function $J$. Most of the time, $\Delta(\varepsilon \langle a^\ell \rangle) = p$ and hence $\Delta(a^\ell \langle a \rangle) = q$, and the latter continues into a triangle of states $q$. Once for every power of two, the automaton computes the state $r$ on $\varepsilon \langle a^{2^k-2} \rangle$, which sends a signal through the feedback channel, so that $J$ sets $\Delta(a^{2^k-2} \langle a \rangle) = p$. This in turn produces the triangle of states $p$ and the next column of states $r$.

It is now known that linear grammars with contexts over a one-symbol alphabet are non-trivial. How far does their expressive power go? For conjunctive grammars (which allow non-linear concatenation, but no context specifications), Jeż and Okhotin [9,10] developed a method for manipulating base-$k$ notation of the length of a string in a grammar, which allowed representing the following language: for every trellis automaton $M$ over an alphabet $\{0, 1, \ldots, k-1\}$, there is a conjunctive grammar generating $L_M = \{\,a^\ell \mid$ the base-$k$ notation of $\ell$ is in $L(M)\,\}$ [9]. This led to the following undecidability method: given a Turing machine $T$, one first constructs a trellis automaton $M$ for the language $\mathrm{VALC}(T) \subseteq \Sigma^*$ of computation histories of $T$; then, assuming that the symbols in $\Sigma$ are digits in some base-$k$ notation, one can define the unary version of VALC($T$) by a conjunctive grammar.

Linear grammars with contexts are an entirely different model, and the automaton in Example 3 has nothing in common with the basic unary conjunctive grammar discovered by Jeż [8], in spite of defining almost the same language. The new model seems to be unsuited for manipulating base-$k$ digits, and the authors took another route to undecidability results, which is explained below.

## 5 Simulating a Turing Machine

The overall idea is to augment the automaton in Example 3 to calculate some additional data, so that its computation on a unary string simulates any fixed Turing machine running on the empty input. Each individual cell $\Delta(a^k\langle a^\ell\rangle)$ computed by the automaton should hold some information about the computation of the Turing machine, such as the contents of a certain tape square at a certain time. Then the automaton can accept its input $a^n$ depending on the state of the computation of the Turing machine at time $f(n)$.

Consider the computation in Figure 2, which is split into regions by vertical $r$-columns. The bottom line of states $q$ in each region shall hold the tape contents of the Turing machine. The new automaton should simulate several steps of the Turing machine, and then transfer its resulting tape contents to the top diagonal border of this region. The transfer of each letter is achieved by sending a signal to the right, reflecting it off the vertical $r$-column, so that it arrives at the appropriate cell in the top border. From there, the tape contents shall be moved to the bottom line of the next region through the feedback data channel. Because of the reflection, the tape symbols arrive at the next region *in the reverse order*.

In order to simulate a Turing machine using this method, it is useful to assume a machine of the following special kind. This machine operates on an initially blank two-way infinite tape, and proceeds by making left-to-right and right-to-left sweeps over this tape, travelling a longer distance at every sweep. At the first sweep, the machine makes one step to the left, then, at the second sweep, it makes 3 steps to the right, then 7 steps to the left, 15 steps to the right, etc. In order to simplify the notation, assume that the machine always travels *from right to left* and flips the tape after completing each sweep.

**Definition 4.** *A sweeping Turing machine is a quintuple* $T = (\Gamma, \mathcal{Q}, q_0, \nabla, \mathcal{F})$, *where*

- $\Gamma$ *is a finite tape alphabet containing a blank symbol* $\square \in \Gamma$,
- $\mathcal{Q}$ *is a finite set of states,*
- $q_0 \in \mathcal{Q}$ *is the initial state and* $\mathcal{F} \subseteq \mathcal{Q}$ *is the set of accepting states,*
- $\nabla \colon \mathcal{Q} \times \Gamma \to \mathcal{Q} \times \Gamma$ *is a transition function, and*
- $\mathcal{F}$ *is a finite set of flickering states.*

*A configuration of* $T$ *is a string of the form* $[\![k]\!]uqav$, *where* $k \geqslant 1$ *is the number of the sweep, and* $uqav$ *with* $u, v \in \Gamma^*$, $a \in \Gamma$ *and* $q \in \mathcal{Q}$ *represents the tape contents* $uav$ *with the head scanning the symbol* $a$ *in the state* $q$.

*The initial configuration of the machine is $[\![1]\!]\square q_0 \square$. Each k-th sweep deals with a tape with $2^k$ symbols, and consists of $2^k - 1$ steps of the following form:*

$$[\![k]\!]ubqcv \vdash_T [\![k]\!]uq'bc'v \qquad (\nabla(q,c) = (q',c')).$$

*Once the machine reaches the last symbol, it flips the tape, appends $2^k$ blank symbols and proceeds with the next sweep:*

$$[\![k]\!]qcw \vdash_T [\![k+1]\!]\square^{2^k} w^R qc$$

*A sweeping Turing machine never halts; at the end of each sweep, it may flicker by entering a state from $\mathcal{F}$. Define the set of numbers accepted by $T$ as $S(T) = \{ k \mid [\![1]\!]\square q_0 \square \vdash_T^* [\![k]\!]q_\mathrm{f} cw \text{ for } q_\mathrm{f} \in \mathcal{F} \}$.*

A sweeping Turing machine is simulated by the following trellis automaton with feedback over a one-symbol alphabet.

**Construction 1.** *Let $T = (\Gamma, \mathcal{Q}, q_0, \nabla, \mathcal{F})$ be a sweeping Turing machine. Construct a trellis automaton with feedback $M = (\{a\}, Q, I, J, \delta, F)$ as follows. Its set of states is $Q = \{{}^Z\mathbf{p}_y^x \mid x, y \in \Gamma \cup \mathcal{Q}\Gamma, Z \in \{\circ, \bullet\}\} \cup \{{}^Z\mathbf{q}^x \mid x \in \Gamma \cup \mathcal{Q}\Gamma, Z \in \{\circ, \bullet\}\} \cup \{\mathbf{r}\}$. Each superscript $x$ represents a tape symbol at the current position, which is augmented with a state, if the head is in this position. Each subscript $y$ similarly contains a symbol and possibly a state, representing the contents of some other tape square, which is being sent as a signal to the left. A bullet marker "$\bullet$" marks the beginning of the tape, whereas each state ${}^Z\mathbf{p}_y^x$ or ${}^Z\mathbf{q}^x$ with $Z = \circ$ shall be denoted by $\mathbf{p}_y^x$ and $\mathbf{q}^x$, respectively.*

*Let $I(a) = \mathbf{p}_{\square q_0}^\square$, $J(\mathbf{r}, a) = \mathbf{p}_\square^\square$, and $J({}^Z\mathbf{p}_y^x, a) = {}^Z\mathbf{q}^y$. For all $x, y, x', y' \in \Gamma \cup \mathcal{Q}\Gamma$ and $Z, Z' \in \{\circ, \bullet\}$, the following transitions are defined:*

$$\delta\left({}^Z\mathbf{q}^x,\ {}^{Z'}\mathbf{q}^{x'}\right) = {}^Z\mathbf{q}^x \qquad \text{(propagation; } x, x' \in \Gamma,$$
$$\text{and } x \in \mathcal{Q}\Gamma \text{ with } Z = \bullet)$$
$$\delta\left({}^Z\mathbf{q}^x,\ {}^{Z'}\mathbf{p}_{y'}^{x'}\right) = {}^{Z'}\mathbf{p}_{y'}^x \qquad \text{(propagation)}$$
$$\delta\left(\mathbf{p}_y^x,\ {}^{Z'}\mathbf{q}^{x'}\right) = \mathbf{r} \qquad \text{(r-column)}$$
$$\delta\left({}^Z\mathbf{p}_y^x,\ \mathbf{r}\right) = \mathbf{p}_x^\square \qquad \text{(reflection)}$$
$$\delta\left({}^Z\mathbf{p}_y^x,\ {}^{Z'}\mathbf{p}_{y'}^{x'}\right) = {}^{Z'}\mathbf{p}_{y'}^x \qquad \text{(propagation)}$$
$$\delta\left(\mathbf{r},\ {}^{Z'}\mathbf{p}_{y'}^{x'}\right) = {}^\bullet\mathbf{p}_{y'}^{x'} \qquad \text{(new region in top diagonal)}$$
$$\delta\left(\mathbf{r},\ {}^{Z'}\mathbf{q}^{x'}\right) = \mathbf{q}^\square \qquad \text{(first q-column after r-column)}$$

*A transition $\nabla(q, c) = (q', c')$ of the Turing machine is simulated as follows:*

$$\delta\left(\mathbf{q}^{cq},\ \mathbf{q}^y\right) = \mathbf{q}^{c'} \qquad \text{(rewriting the symbol; } y \in \Gamma)$$
$$\delta\left({}^Z\mathbf{q}^x,\ \mathbf{q}^{cq}\right) = {}^Z\mathbf{q}^{xq'} \qquad \text{(moving the head; } x \in \Gamma)$$

*The set of accepting states is $F = \{ \mathbf{p}_{cq_\mathrm{f}}^\square \mid c \in \Gamma,\ q_\mathrm{f} \in \mathcal{F} \}$.*

The first thing to note about this construction is that if all attributes attached to the letters $p, q, r$ are discarded, then the resulting automaton is exactly the one from Example 3. This ensures the overall partition of the computation into regions illustrated in Figure 2.

Each region after second $r$-column corresponds to a sweep of the Turing machine. The bottom row of states contains the machine's configuration in the beginning of the sweep, where each state $\mathbf{q}^x$ holds the symbol in one square of the tape. The leftmost cell is marked by a bullet ($^\bullet\mathbf{q}^x$). The cell in the middle of the bottom row ($\mathbf{q}^{xq}$) corresponds to the rightmost square of the tape, which contains the state of the machine. The cells in the right half of the bottom row contain the state $\mathbf{q}^\square$. Each of the several rows above holds the tape contents after another step of computation. After $2^k - 1$ steps of simulation the head reaches the leftmost square, which marks the end of the current sweep.

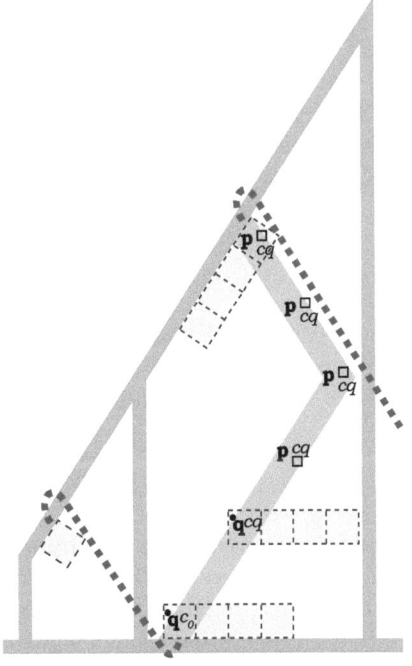

Then, each tape symbol is propagated by a signal to the right using the states $\mathbf{p}_y^x$. Every such state holds two symbols: $x$ is carried to the right, to be reflected off the right border, and $y$ is a leftbound symbol that has already been reflected. As a result, the top diagonal border is filled with the states of the form $\mathbf{p}_y^x$, and their subscripts $y$ form the resulting contents of the tape, reversed. These symbols are sent to the next region by the function $J$.

With this simulation running, the last state $q \in \mathcal{Q}$ reached by the Turing machine upon completing each $k$-th sweep shall always end up in a predefined position exactly in the middle of the top diagonal border. It will be $\Delta(\varepsilon \langle a^{2^{k+2}+2^{k+1}-2} \rangle) = \mathbf{p}_{cq}^\square$, and the trellis automaton with feedback accepts this string if and only if $q \in \mathcal{F}$.

The following theorem states the correctness of the construction.

**Theorem 3.** *Let $T = (\Gamma, \mathcal{Q}, q_0, \nabla, \mathcal{F})$ be a sweeping Turing machine and let $M = (\{a\}, Q, I, J, \delta, F)$ be a trellis automaton with feedback obtained in Construction 1. Then $L(M) = \{\, a^{2^{k+2}+2^{k+1}-2} \mid k \in S(T) \,\}$.*

## 6 Implications

The simulation of Turing machines by a trellis automaton with feedback over a one-symbol alphabet is useful for proving undecidability of basic decision problems for these automata. Due to Theorem 2, the same undecidability results equally hold for linear grammars with contexts.

The first decision problem is testing whether the language recognized by an automaton (or defined by a grammar) is empty. The undecidability of the *emptiness problem* follows from Theorem 3. To be precise, the problem is complete for the complements of the r.e. sets.

**Theorem 4.** *The emptiness problem for linear grammars with left contexts over a one-symbol alphabet is $\Pi_1^0$-complete. It remains in $\Pi_1^0$ for any alphabets.*

*Proof.* The non-emptiness problem is clearly recursively enumerable, because one can simulate a trellis automaton with feedback on all inputs, accepting if it ever accepts. If the automaton accepts no strings, the algorithm does not halt.

The $\Pi_1^0$-hardness is proved by reduction from the Turing machine halting problem. Given a machine $T$ and an input $w$, construct a sweeping Turing machine $T_w$, which first prints $w$ on the tape (over $1 + \log |w|$ sweeps, using around $|w|$ states), and then proceeds by simulating $T$, using one sweep for each step of $T$. If the simulated machine $T$ ever halts, then $T_w$ changes into a special state $q_\mathrm{f}$ and continues moving its head until the end of the current sweep.

Construct a trellis automaton with feedback $M$ simulating the machine $T_w$ according to Theorem 3, and define its set of accepting states as $F = \{\,\mathbf{p}_{cq_\mathrm{f}}^{\square} \mid c \in \Sigma\,\}$. Then, by the theorem, $M$ accepts some string $a^\ell$ if and only if $T_w$ ever enters the state $q_\mathrm{f}$, which is in turn equivalent to $T$'s halting on $w$. □

The second slightly more difficult undecidability result asserts that testing the finiteness of a language generated by a given grammar is complete for the second level of the arithmetical hierarchy.

**Theorem 5.** *The finiteness problem for linear grammars with left contexts over a one-symbol alphabet is $\Sigma_2^0$-complete. It remains $\Sigma_2^0$-complete for any alphabet.*

*Proof (a sketch).* Reduction from the finiteness problem for a Turing machine, which is $\Sigma_2^0$-complete, see Rogers [15, §14.8]. Given a Turing machine $T$, construct a sweeping Turing machine $T'$, which simulates $T$ running on all inputs, with each simulation using a segment of the tape. Initially, $T'$ sets up to simulate $T$ running on $\varepsilon$, and then it regularly begins new simulations. Every time one of the simulated instances of $T$ accepts, the constructed machine "flickers" by entering an accepting state in the end of one of its sweeps. Construct a trellis automaton with feedback $M$ corresponding to this machine. Then $L(M)$ is finite if and only if $L(T)$ is finite. □

## 7 Conclusion

At the first glance, linear grammars with contexts seem like a strange model. However, they are motivated by the venerable idea of a rule applicable in a context, which is worth being investigated. Also, trellis automata with feedback at the first glance seem like a far-fetched extension of cellular automata. Its motivation comes from the understanding of a trellis automaton as a circuit with uniform connections [5], to which one can add a new type of connections. Both models are particularly interesting for being equivalent.

A suggested topic for future research is to investigate the main ideas in the literature on trellis automata [5,6,7,17] and see whether they can be extended to trellis automata with feedback, and hence to linear grammars with contexts.

## References

1. Aizikowitz, T., Kaminski, M.: $LR(0)$ conjunctive grammars and deterministic synchronized alternating pushdown automata. In: Kulikov, A., Vereshchagin, N. (eds.) CSR 2011. LNCS, vol. 6651, pp. 345–358. Springer, Heidelberg (2011)
2. Barash, M.: Recursive descent parsing for grammars with contexts. In: SOFSEM 2013 Student Research Forum (2013)
3. Barash, M., Okhotin, A.: Defining contexts in context-free grammars. In: Dediu, A.-H., Martín-Vide, C. (eds.) LATA 2012. LNCS, vol. 7183, pp. 106–118. Springer, Heidelberg (2012)
4. Barash, M., Okhotin, A.: An extension of context-free grammars with one-sided context specifications (September 2013) (submitted)
5. Čulík II, K., Gruska, J., Salomaa, A.: Systolic trellis automata. International Journal of Computer Mathematics 15, 195–212, 16, 3–22 (1984)
6. Dyer, C.: One-way bounded cellular automata. Information and Control 44, 261–281 (1980)
7. Ibarra, O.H., Kim, S.M.: Characterizations and computational complexity of systolic trellis automata. Theoretical Computer Science 29, 123–153 (1984)
8. Jeż, A.: Conjunctive grammars can generate non-regular unary languages. International Journal of Foundations of Computer Science 19(3), 597–615 (2008)
9. Jeż, A., Okhotin, A.: Conjunctive grammars over a unary alphabet: undecidability and unbounded growth. Theory of Computing Systems 46(1), 27–58 (2010)
10. Jeż, A., Okhotin, A.: Complexity of equations over sets of natural numbers. Theory of Computing Systems 48(2), 319–342 (2011)
11. Okhotin, A.: Conjunctive grammars. Journal of Automata, Languages and Combinatorics 6(4), 519–535 (2001)
12. Okhotin, A.: On the equivalence of linear conjunctive grammars to trellis automata. RAIRO Informatique Théorique et Applications 38(1), 69–88 (2004)
13. Okhotin, A.: Conjunctive and Boolean grammars: the true general case of the context-free grammars. Computer Science Review 9, 27–59 (2013)
14. Okhotin, A.: Improved normal form for grammars with one-sided contexts. In: Jurgensen, H., Reis, R. (eds.) DCFS 2013. LNCS, vol. 8031, pp. 205–216. Springer, Heidelberg (2013)
15. Rogers Jr., H.: Theory of Recursive Functions and Effective Computability (1967)
16. Rounds, W.C.: LFP: A logic for linguistic descriptions and an analysis of its complexity. Computational Linguistics 14(4), 1–9 (1988)
17. Terrier, V.: On real-time one-way cellular array. Theoretical Computer Science 141(1-2), 331–335 (1995)
18. Törmä, I.: Personal communication (February 2013)
19. Yu, S.: A property of real-time trellis automata. Discrete Applied Mathematics 15(1), 117–119 (1986)

# On the Computability of Relations on λ-Terms and Rice's Theorem - The Case of the Expansion Problem for Explicit Substitutions

Edward Hermann Haeusler[1] and Mauricio Ayala-Rincón[2]

[1] Departamento de Informática, PUC-Rio, Brasil
hermann@inf.puc-rio.br

[2] Departamentos de Computação e Matemática, Universidade de Brasília, Brasil
ayala@unb.br

**Abstract.** Explicit substitutions calculi are versions of the λ-calculus having a concretely defined operation of substitution. An Explicit substitutions calculus, $\lambda_\xi$, extends the language $\Lambda$, of the λ-calculus including operations and rewriting rules that explicitly implement the implicit substitution involved in β-reduction in $\Lambda$. $\Lambda_\xi$, that is the language of $\lambda_\xi$, might have terms without any computational meaning, i.e., that do not arise from pure lambda terms in $\Lambda$. Thus, it is relevant to answer whether for a given $t \in \Lambda_\xi$, there exists $s \in \Lambda$ such that $s \to^*_{\lambda_\xi} t$, i.e., whether there exists a pure λ-term reducing in the extended calculus to the given term. This is known as the *expansion problem* and was proved to be undecidable for a few explicit substitutions calculi by using *Scott*'s theorem. In this note we prove the undecidability of the expansion problem for the $\lambda\sigma$ calculus by using a version of *Rice*'s theorem. This method is more straightforward and general than the one based on Scott's theorem.

**Keywords:** Explicit Substitution, Lambda-Calculus, Rice's and Scott's Theorem.

## 1 Introduction

Explicit substitutions calculi (e.g. [1], [10]) are extensions of the λ-calculus in which the operation of substitution involved in β-contraction, that is

$$(\lambda_x.M)\ N \to_\beta M\{x \leftarrow N\},$$

where "$M\{x \leftarrow N\}$" denotes the term obtained from $M$ by simultaneously replacing all occurrences of $x$ by $N$, is made explicit.

These calculi can be classified either by their properties as in [10] or in those based on a representation of variables as names, and those based on nameless representation of variables following De Bruijn indices [6].

The problem of expansion of explicit substitutions calculi is of genuine computational interest and has been answered to be undecidable for a few calculi such as $\lambda_s$ [9] and $\lambda_v$ [3] in [2] and then, using the same technique, for $\lambda\sigma$ [1] in [5]. Essentially, for an explicit substitutions calculus $\lambda_\xi$, the technique applied in these works is based on

- the application of Scott's theorem, which states that any non trivial subset of $\Lambda$ that is closed under $\beta$-conversion is non-recursive, and
- the encoding in the language of $\Lambda_\xi$ of the problem of whether two terms in $\Lambda$ have a common expansion term in $\Lambda$, which is also undecidable.

This technique can be applied both for calculi with variables as names or as indices, since these representations of the $\lambda$-calculus are well-known to be isomorphic. Here we show how a variant of Rice's theorem can be applied to prove undecidability of the expansion problem for the $\lambda\sigma$-calculus [1]. This technique is more general and straightforwardly applicable to other calculi of explicit substitutions, since it avoids the second step of encoding of an undecidable problem of the pure $\lambda$-calculus inside the extended language of the explicit substitutions calculi used in [2].

## 2 Preliminaries

### 2.1 De Bruijn Notation and Explicit Substitutions

Avoiding names in the $\lambda$-calculus is an effective way of clarifying the meaning of $\lambda$-terms and, for the unification process, of eliminating redundant renaming. De Bruijn developed in [6] a notation where names of bound variables are replaced by indices which relate these bound variables to their corresponding abstractors.

It is clear that the correspondence between an occurrence of a bound variable and its associated abstractor operator is uniquely determined by its *depth*, that is the number of abstractors between them. Hence, $\lambda$-terms can be written in a term algebra over the natural numbers $\mathbb{N}$, representing depth's indices, the application operator (_ _) and a sole abstractor operator $\lambda\_$; i.e., $\mathcal{T}(\{(\_\ \_), \lambda\_\} \cup \mathbb{N})$. For instance, $\lambda_x.(x\ \lambda_y.(x\ y))$ is written in De Bruijn notation as $\lambda(\underline{1}\ \lambda(\underline{2}\ \underline{1}))$.

In De Bruijn's notation, indexing the occurrences of free variables is given by a *referential* according to a fixed enumeration of the set of variables $\mathcal{V}$, say $x, y, z, \ldots$, and prefixing all $\lambda$-terms with $\ldots\lambda_z.\lambda_y.\lambda_x.\_$. For instance in this referential, $\lambda_u.(x\ \lambda_v.((x\ (u\ z))\ v))$ is written in De Bruijn notation as $\lambda(\underline{2}\ \lambda((\underline{3}\ (\underline{2}\ \underline{5}))\ \underline{1}))$. Now we can define the $\lambda$-calculus in De Bruijn notation.

**Definition 1 ($\lambda$-calculus in De Bruijn notation).** *The set $\Lambda_{dB}$ of $\lambda$-terms in De Bruijn notation is built inductively as:* $\Lambda_{dB} ::= \underline{n} \mid (\Lambda_{dB}\ \Lambda_{dB}) \mid \lambda\Lambda_{dB}$, *where $\underline{n}$ is the $n^{th}$ De Bruijn index for $n \in \mathbb{N} \setminus \{0\}$.*

We write De Bruijn indices as underlined naturals to distinguish them from naturals and scripts. Since in this paper all considered calculi of explicit substitutions are built over the language of $\Lambda_{dB}$, we will use $\Lambda$ to denote $\Lambda_{dB}$.

The attempt to define $\beta$-reduction in De Bruijn notation as $(\lambda a\ b) \to \{\underline{1}/b\}a$, where $\{\underline{1}/b\}a$ is the first-order substitution of the index 1 in $a$ with $b$ fails because:

- when eliminating the leading abstractor all indices associated with free variable occurrences in $a$ should be decremented;
- when propagating inside the body of the abstraction $a$, the substitution $\{\underline{1}/b\}$ through $\lambda$'s, the indices of the substitution (initially $\underline{1}$) and of the free variables in $b$ should be incremented.

Hence, in order to define $\beta$-reduction we need new operators for detecting and updating (i.e., incrementing and decrementing) free variables.

**Definition 2 ($n$-lift).** Let $a \in \Lambda_{dB}$. The $i$-lift of $a$, denoted $a^{+i}$ is defined inductively as follows:

i) $(a_1\ a_2)^{+i} = (a_1^{+i}\ a_2^{+i})$  

ii) $(\lambda a_1)^{+i} = \lambda a_1^{+(i+1)}$

iii) $\underline{n}^{+i} = \begin{cases} \underline{n+1}, & \text{if } n > i \\ \underline{n}, & \text{if } n \leq i \end{cases}$ for $n \in \mathbb{N}$.

The **lift** of a term $a$ is its 0-lift and is denoted briefly as $a^+$.

**Definition 3 (Substitution).** The application of the **substitution** with $b$ at the depth $n-1, n \in \mathbb{N} \setminus \{0\}$, denoted $\{\underline{n}/b\}a$, on a term $a$ in $\Lambda_{dB}$ is defined inductively as follows:

i) $\{\underline{n}/b\}(a_1\ a_2) = (\{\underline{n}/b\}a_1\ \{\underline{n}/b\}a_2)$  

ii) $\{\underline{n}/b\}\lambda a_1 = \lambda\{\underline{n+1}/b^+\}a_1$

iii) $\{\underline{n}/b\}\underline{m} = \begin{cases} \underline{m-1}, & \text{if } m > n \\ b, & \text{if } m = n \\ \underline{m}, & \text{if } m < n \end{cases}$ if $m \in \mathbb{N}$.

**Definition 4 ($\beta$-reduction).** The $\beta$-**reduction** in the $\lambda$-calculus with De Bruijn indices is defined as $(\lambda a\ b) \to_\beta \{\underline{1}/b\}a$.

It is well-known that the $\lambda$-calculus is confluent. This property holds also for the rewriting system defined over $\Lambda_{dB}$ by the $\beta$-reduction rule. We stress here that since both representations of the $\lambda$-calculus are isomorphic, instead of notation $\Lambda_{dB}$, the notation $\Lambda$ will be used.

Undecidability of the relation $=_\beta$, that is the equivalence closure of $\to_\beta$, is a direct consequence of Scott's theorem.

We will use standard rewriting notations, that is for a (rewriting) relation $R$ the super script $*$ is used to denote its reflexive transitive closure, $R^*$; $=_R$ denotes its equivalence closure; $\mathcal{R}$ its inverse and, by a matter of elegance, instead $\mathcal{R}^*, {}^*\mathcal{R}$ denotes the reflexive transitive closure of its inverse. This notation is very natural when relations are denoted by symbols such as $\to$ and $\rhd$; thus, their inverses are denoted as $\leftarrow$ and $\lhd$. The symbol $\circ$ denotes composition of relations.

## 2.2 The $\lambda\sigma$-calculus and the Expansion Problem

We state the traditional definition of the *Expansion Problem*

**Definition 5 (The Expansion Problem).** *Consider a calculus $\lambda_\xi$ that extends the pure $\lambda$-calculus by means of implementing its implicit substitution over* the language $\Lambda_\xi \supset \Lambda$. The expansion problem for this calculus is the problem of answering whether for a given $t \in \Lambda_\xi$, there exists $s \in \Lambda$ such that $s \to^*_{\lambda_\xi} t$.

The $\lambda\sigma$-calculus works on two-sorted terms: *(proper) terms* (over which $a, b, \ldots$ range), and *substitutions* (over which $s, t, \ldots$ range).

**Definition 6 (The $\lambda\sigma$-calculus).** *The $\lambda\sigma$-calculus is defined as the calculus of the rewriting system $\lambda\sigma$ given in Table 1 where*

$$\text{TERMS} \quad a ::= \underline{1} \mid (a\ a) \mid \lambda a \mid a[s] \quad \text{and} \quad \text{SUBS} \quad s ::= id \mid \uparrow \mid a.s \mid s \circ s$$

*The set of $\lambda\sigma$-terms built in this form will be denoted as $\Lambda_\sigma$ (one can discriminate between substitutions and proper terms by using superscripts $t$ and $s$: $\Lambda_\sigma^t$ and $\Lambda_\sigma^s$). Observe that $\Lambda \subset \Lambda_\sigma$. Use of the symbol $\circ$ for composition of substitutions and of relations will easily be discriminated according to the context.*

**Table 1.** The set of rewriting rules of the $\lambda\sigma$-calculus

| | | |
|---|---|---|
| (Beta) | $(\lambda a\ b)$ | $\longrightarrow a[b \cdot id]$ |
| (Id) | $a[id]$ | $\longrightarrow a$ |
| (VarCons) | $\underline{1}[a \cdot s]$ | $\longrightarrow a$ |
| (App) | $(a\ b)[s]$ | $\longrightarrow (a[s])\,(b[s])$ |
| (Abs) | $(\lambda a)[s]$ | $\longrightarrow \lambda a[\underline{1} \cdot (s \circ \uparrow)]$ |
| (Clos) | $(a[s])[t]$ | $\longrightarrow a[s \circ t]$ |
| (IdL) | $id \circ s$ | $\longrightarrow s$ |
| (IdR) | $s \circ id$ | $\longrightarrow s$ |
| (ShiftCons) | $\uparrow \circ (a \cdot s)$ | $\longrightarrow s$ |
| (Map) | $(a \cdot s) \circ t$ | $\longrightarrow a[t] \cdot (s \circ t)$ |
| (Ass) | $(s \circ t) \circ u$ | $\longrightarrow s \circ (t \circ u)$ |
| (VarShift) | $\underline{1} \cdot \uparrow$ | $\longrightarrow id$ |
| (SCons) | $\underline{1}[s] \cdot (\uparrow \circ s)$ | $\longrightarrow s$ |

For every substitution $s$ we define the *iteration of the composition of $s$* inductively as $s^0 = id$ and $s^{n+1} = s \circ s^n$, where $id$ is the substitution identity. Note that the only De Bruijn index used is $\underline{1}$, each other index $\underline{n}$ is encoded as $\underline{1}[\uparrow^{n-1}]$. Also, notice that by rule $(Id)$, $\underline{1}[\uparrow^{1-1}] \to_{Id} \underline{1}$.

The equational theory associated with the rewriting system $\lambda\sigma$ defines a congruence denoted $=_{\lambda\sigma}$ (i.e., the theory built from the equivalence closure of the $\lambda\sigma$ rewriting system). The congruence obtained by dropping the rule *Beta*, that is the *generation* rule of this calculus, is denoted as $=_\sigma$. When one restricts reduction to only these rules, one will use expressions such as $\sigma$-reduction, $\sigma$-normal form, $\sigma$-calculus, etc, with the obvious meaning. In general, for an explicit substitutions calculus $\lambda_\xi$, one has a generation rule which starts the simulations of $\beta$-reductions by application of the other rules in the $\xi$-calculus.

The rewriting system $\lambda\sigma$ is known to simulate $\beta$-reduction and also to be confluent [1]. The former means that for terms $a, b \in \Lambda$, whenever $a \to_\beta b$, $a \to^*_{\lambda\sigma} b$; the latter, that for terms $c, a, b \in \Lambda_\sigma$, if $c \to^*_{\lambda\sigma} a$ and $c \to^*_{\lambda\sigma} b$, then there exists $d \in \Lambda_\sigma$ such that $a \to^*_{\lambda\sigma} d$ and $b \to^*_{\lambda\sigma} d$, or in other words, that $^*\!\leftarrow \circ \to^* \subseteq \to^* \circ \,^*\!\leftarrow$.

Given $a, b \in \Lambda$, $a =_\beta b$ if and only if $a =_{\lambda\sigma} b$.

The relations $\to^*_{\lambda\sigma}$ and $=_{\lambda\sigma}$ are undecidable. This is a consequence of the fact that $\lambda\sigma$ consistently simulates $\beta$-reduction; thus, if $\lambda\sigma$ derivability and/or conversion were decidable, correspondingly $\beta$ derivability and/or conversion would be also decidable.

Given $a, b \in \Lambda$, the question: Does $c \in \Lambda$ exist such that $c \to^*_{\lambda\sigma} a$ and $c \to^*_{\lambda\sigma} b$? is undecidable too.

Also, given $a, b \in \Lambda_\sigma$, the question whether there exists $c \in \Lambda$ such that $c \to^*_{\lambda\sigma} a$ and $c \to^*_{\lambda\sigma} b$, is undecidable.

Given $a, b \in \Lambda$, there exists $c' \in \Lambda_\sigma$ such that $c' \to^*_{\lambda\sigma} a$ and $c' \to^*_{\lambda\sigma} b$ if, and only if, there exists $c \in \Lambda$ such that $c \to^*_\beta a$ and $c \to^*_\beta b$.

Analogously to Scott's theorem for $\Lambda$, one has that any non trivial subset of $\Lambda_\sigma$ that is closed under $=_{\lambda\sigma}$ is non recursive.

The approach in [2] to prove undecidability of the expansion problem for $\Lambda_\sigma$ consists in encoding inside the problem of expansion an undecidable problem in $\Lambda$. For doing this one considers the following context in $\Lambda_\sigma$:

$$(\lambda(\underline{1}[(\underline{1} \cdot ((\Box \cdot id) \circ \uparrow))] \; \underline{1}[(\underline{1} \cdot ((\Box \cdot id) \circ \uparrow))]))$$

Replacing the holes in this context by the same term, say $c$, expansion with the rule *(App)* is possible obtaining $(\lambda((\underline{1} \; \underline{1})[\underline{1} \cdot ((c \cdot id) \circ \uparrow)])])$.

Then, by expansion with the rule *(Abs)* one obtains the term $(\lambda(\underline{1} \; \underline{1}))[c \cdot id]$ and, finally by expansion with rule *(Beta)*, the term $(\lambda\lambda(\underline{1} \; \underline{1}) \; c)$.

Thus, the undecidable question whether given $a, b \in \Lambda$, there exists a $c \in \Lambda$ into which $a$ and $b$ expand, that is a $a \;^*\!\leftarrow_\beta c \to^*_\beta b$, is equivalent to the question whether the term at the top of Fig. 1, which illustrates the sketch of this standard proof, expands to a pure lambda term.

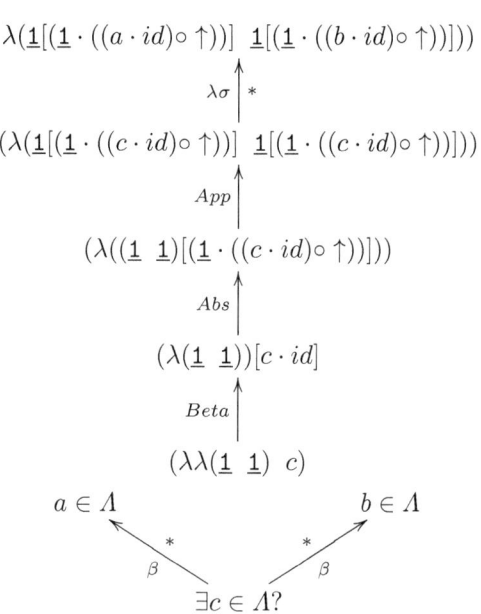

Fig. 1. Standard proof sketch

This technique, as mentioned in the introduction, requires both the application of Scott's theorem and considerations about the structure of terms in $\Lambda_\sigma$. Here we show that the analysis can be generalized and simplified for explicit substitutions calculi $\lambda_\xi$ that are consistent extensions of the $\lambda$-calculus. Indeed, this is a consequence of the fact that consistent extensions of $\lambda$ are non recursive or in other words that $\lambda$ is essentially undecidable. In fact, suppose that $\lambda_\xi$ extends $\lambda$ and is consistent, then the set $\{M : M =_{\lambda_\xi} \boldsymbol{I}\}$, where $\boldsymbol{I}$ denotes the $\boldsymbol{I}$ combinator, is non empty and different of $\Lambda_\xi$ because $\lambda_\xi$ is consistent. Thus, by Scott's theorem this set is not recursive and consequently neither $\lambda_\xi$.

For the case of $\lambda\sigma$ as well as for other explicit substitutions calculi, the analysis follows since they are consistent extensions of $\lambda$.

## 2.3 Basic Abstract Recursion Theory Terminology

In what follows we consider computable functions as a class of functions from $\mathbb{N}$ to $\mathbb{N}$. We assume the standard notions of primitive recursive functions, partially recursive functions and recursive functions (a partially recursive function that is total). We also assume a rather abstract view of an algorithm as its code and the code is simply a natural number. Every natural number is a code for at least one partially recursive function. We could use the notation $f_i$, $i \in \mathbb{N}$, to denote the partial recursive function that $i$ codes. In a more abstract setting, Rogers [13] suggests axioms for an indexed class $\Phi = (\varphi_i)_{i \in \mathbb{N}}$ of partial functions from $\mathbb{N}$ to $\mathbb{N}$ be considered as algorithms (codes) and their corresponding denotations. In this way, $\varphi_i$ is the function "computed" by code $i$. Rogers proved that such indexed classes are isomorphic, providing a stronger (more abstract) evidence to Turing-Church thesis. In this brief note we focus on Rice's theorem, but we would like to notice that in [11] the reader can find a quite pedagogical presentation of Rogers' theorem. In the sequel we present the formal definition of this abstract view of a class of Turing-computable functions, namely, Abstract Family of Algorithms (**AFA**), as it is in [11].

We will use $\uparrow$ to denote that the (partial) function $\varphi_m(i)$ is undefined in $i$ and $\downarrow$ to denote that it is defined.

**Definition 7 (Abstract Family of Algorithms).** *A collection $\Phi = (\varphi_i)_{i \in \mathbb{N}}$ of partial functions from $\mathbb{N}$ into $\mathbb{N}$ is such that:*

1. *$(\varphi_i)_{i \in \mathbb{N}}$ contains all partial recursive functions.*
2. *There is $u \in \mathbb{N}$, such that, for every $i \in \mathbb{N}$ $\varphi_u(\langle i, x \rangle) = \varphi_i(x)$, where $\langle m, n \rangle$ is the code for pairs of naturals $m, n$.*
3. *There is a primitive recursive function $c$, such that $\varphi_{c(i,j)} = \varphi_j \circ \varphi_i$.*

The following definition is the abstract version of a computational step in an algorithm. In the case of $\lambda$-calculus or an explicit substitution calculus it might be either the abstraction of reduction or expansion ($\beta$-reduction, for example) between $\lambda$ or $\lambda_\xi$-terms.

**Definition 8 (Pre-step-computable relation).** *Let $\Phi$ be an **AFA** and $\triangleright$ a binary relation on $\mathbb{N}$. We say that $\triangleright$ is pre-$\Phi$-step-computable, iff, there is a primitive recursive function $f$, such that, for each $i, j \in \mathbb{N}$:*

1. if $i \triangleright j$ then there is a finite set $I_o$ of natural numbers, such that, for every $k \in I_o$, $f(\langle i, k \rangle) = j$, $i \neq j$, and;
2. for all $n \in \mathbb{N}$, if $\varphi_j(n) \downarrow$ then $\varphi_j(n) = \varphi_i(n)$, and;
3. for every $k' \in I_o$ and $j' = f(\langle i, k' \rangle)$, $i \triangleright j'$, and;
4. it is required that for every $i, j, i', j' \in \mathbb{N}$, $c(i, j) \triangleright c(i', j')$, if and only if, $i \triangleright i'$ and $j = j'$, or, $i = i'$ and $j \triangleright j'$.

We denote by $\triangleright^+$ the transitive closure of $\triangleright$.

**Definition 9.** *Let $\Phi$ be an **AFA** and $\triangleright$ a pre-$\Phi$-step-computable binary relation. We say that $\triangleright$ is a $\Phi$-step-computable relation, if and only if, for every $i \in \mathbb{N}$, if $\varphi_i$ is a total function[1] then, $i \downarrow = \{j : i \triangleright^+ j\}$ is finite.*

We have the following lemma.

**Lemma 1.** *Let $\Phi$ be an **AFA** and $\triangleright$ a $\Phi$-step-computable binary relation. Let $i, j \in \mathbb{N}$, such that, $\varphi_i = \varphi_j$, and $((i \downarrow) - (j \downarrow))$ is a finite set. In this case $i \triangleright^+ j$ holds.*

**Definition 10 (Saturated by a relation).** *Let $\mathcal{C}$ be a set of natural numbers and $\triangleright$ a relation on $\mathbb{N}$. $\mathcal{C}$ is saturated by $\triangleright$, iff, for all $i, j \in \mathbb{N}$, if $i \triangleright j$ then, if $i \in \mathcal{C}$ then $j \in \mathcal{C}$, and; if $j \in \mathcal{C}$ then $i \in \mathcal{C}$.*

## 3 Abstract Rice's Theorem

The statement of Rice's theorem using **AFA** as stated in [11] is as below. The original statement can be found in [12] and the abstract version we are dealing with is according [13].

**Theorem 1 (Rice).** *Let $\mathcal{C} = \{i : \Psi(\varphi_i)\}$, where $\Psi(x)$ is a property on partial functions from $\mathbb{N}$ to $\mathbb{N}$. $\mathcal{C}$ is recursive, iff, $\mathcal{C} = \emptyset$ or $\mathcal{C} = \mathbb{N}$.*

We state and prove a slightly modified version of Rice's theorem, using step-computable relations instead of satisfaction by a "semantical" property. It is the key result used in proving the undecidability of expansion.

**Theorem 2 (Modified Rice).** *Let $\Phi$ be an **AFA** and $\triangleright$ be a $\Phi$-step-computable relation on $\mathbb{N}$ and $\mathcal{C} \subset \mathbb{N}$ be a set saturated by $\triangleright$. $\mathcal{C}$ is recursive, iff, $\mathcal{C} = \mathbb{N}$ or $\mathcal{C} = \emptyset$.*

*Proof.* If $\mathcal{C} \neq \emptyset$ and $\mathcal{C} \neq \mathbb{N}$, there is $k \in \mathcal{C}$ and wlog we can consider $i_\perp \notin \mathcal{C}$, where $\varphi_{i_\perp} = \emptyset \neq \varphi_k$, with $\emptyset$ being the undefined function[2]. Let $g$ be the recursive function defined by

$$g(m) = \varphi_m(m); k$$

---
[1] For every $n \in \mathbb{N}$, $\varphi_i(n) \downarrow$.
[2] $\emptyset(x) \uparrow$, for each, $x \in \mathbb{N}$.

The right-hand side can be written as $\varphi_k \circ \pi_2 \circ \langle \varphi_m(m), id \rangle$ in notation of partial recursive functions. If $\varphi_p = \pi_2$, $\varphi_I = id$ then, the formal code of this functional notation in terms of $\Phi$ is : $c(c(\langle s(u, \langle m, m \rangle), I \rangle, p), k)$, where $s$ is the s-1-1 instance of the well-known s-m-n function, i.e., $\varphi_{s(i,j)}(x) = \varphi_i(\langle j, x \rangle)$. We have two cases to consider:

- $\varphi_m(m) \uparrow$, and hence $\varphi_{g(m)} = \emptyset = \varphi_{i_\perp}$, or;
- $\varphi_m(m) \downarrow$, and in this case it is a total function, in fact a constant total function, an hence, because of item 4 of definition 8, and the fact that there are only finitely many $n$, such that, $s(u, s(u, \langle m, m \rangle)) \triangleright n$, and lemma 1, we can conclude that $g(m) \triangleright^+ k$. So $g(m) \in \mathcal{C}$.

Thus, if $\varphi_m(m) \uparrow$ then $g(m) \notin \mathcal{C}$ and, if $\varphi_m(m) \downarrow$ then $g(m) \in \mathcal{C}$. Since it is undecidable whether $\varphi_m(m) \uparrow$ or not, then $\mathcal{C}$ cannot be recursive.

Note that the above theorem is a rather abstract version of Scott's theorem. If $\Phi$ is $\Lambda$, the $\beta$-reduction, $\rightarrow_\beta$, is a $\Phi$-step-computable relation and a set closed by $\rightarrow_\beta$ is a saturated set. The abstract version of the expansion property within a computational formalism (**AFA**) is undecidable and this is a corollary of the above theorem.

It is interesting to point out that, when theorem 2 is compared to its concrete version (Scott's theorem), we can see that some natural numbers are the normal terms (values) while other are algorithms. For any $j, i \in \mathbb{N}$, $\varphi_j(i)$ represents the algorithm/value $t_j t_i \in \Lambda$, where $t_j$ and $t_i$ are represented by $j$ and $i$ respectively.

Other completely different application of the above theorem concerns the calculus $\lambda_\mathbf{x}$ ([14] [4]). The reduction in this calculus is a step-computable relation too. In particular, there is a reduction $(\lambda z.M)N \rightarrow M\langle z := N \rangle$, and corresponding rules for reducing the closure $M\langle z := N \rangle$. In this calculus the set $Clo$ of terms $t$ such that $t \rightarrow^* t'$, $t_1 \langle z := t_2 \rangle$ occurs in $t'$, is closed by the step-computable relation $\rightarrow$. $Clo$ includes every term of $\lambda_\mathbf{x}$, since for each closure $t_1 \langle z := t_2 \rangle$ there is the term $(\lambda z.t_1)t_2$. This case contradicts the theorem. $Clo$ being the set of all terms is a trivial subset of the set of terms, and is hence recursive. Our observation amounts to the fact that the expansion problem (see next section) of this calculus is decidable.

An important observation concerns the seamless use of the finitely branching nature of the step-computable relation in the proof of theorem 2. We can see that the following proposition ensures that the condition on saturated sets could be weakened to a closure condition. Let $\mathcal{C}$ be a set and $Sat(\mathcal{C})$ the least saturated set containing $\mathcal{C}$. We have the following proposition. In this proposition we use the construct $ConcurrentRun(S)$ that runs concurrently (time sharing) the finite list of algorithms $S$ on the same input data. The output of this construct is the output of the first member in $S$ that stops yielding a result.

**Proposition 1.** *Let $\Phi$ be an **AFA** and $\triangleright$ be a $\Phi$-step-computable relation on $\mathbb{N}$. Let $\mathcal{C} \subset \mathbb{N}$ be a set closed by $\triangleright$, i.e., for every $i \in \mathcal{C}$, if $i \triangleright j$ then $j \in \mathcal{C}$. Then, $\mathcal{C}$ is recursive, iff, the set $Sat(\mathcal{C})$ is recursive.*

*Proof.* We first observe that as $\rhd$ is $\Phi$-step-computable, then for every term $i \in \mathcal{C}$ the set $J_i = \{j : j \rhd i \text{ and } j \notin \mathcal{C}\}$ is finite. As $\Phi$ is an **AFA**, it is computationally equivalent to your favorite Turing-complete computational model. Thus, there is an index $m$ equivalent to $ConcurrentRun(J_i)$. Since every member of $J_i$ computes the same partial function that $i$, any reduction to the halting problem using $i$ corresponds to an equivalent reduction using $m$. This implies that, see the proof of theorem 2, $\mathcal{C}$ is recursive iff $Sat(\mathcal{C})$ is recursive.

**Corollary 1.** *Let $\varphi$ be an **AFA** and $\rhd$ a $\Phi$-step-computational relation. Let $\lhd$ be the inverse of $\rhd$. For a given $j$, the question whether there exists $k \in \mathbb{N}$ such that $k \lhd j$ is undecidable.*

*Proof.* Directly from proposition 1 and theorem 2. In fact this is a joint corollary of both results.

It is worth noting that as **AFA** is concerned with the semantics of a computational model, there is no need to mention consistency as it is usual in Scott's theorem on $\Lambda$.

The case when we consider a relation extending the $\beta$-reduction, as it is the case of explicit substitutions is considered in the next section.

## 4 $\lambda$-terms and Undecidable Properties

The modified Rice's theorem statement is used to prove that some relations between algorithms, in an **AFA** $\Phi = (\varphi_i)_{i \in \mathbb{N}}$ and extensions of these algorithms in a family $\Psi = (\psi_i)_{i \in \mathbb{N}}$, based on respective step-computable relations $\rhd_\Psi$ and $\rhd_\Phi$, cannot be recursive.

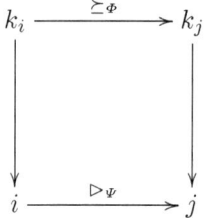

Fig. 2. $\Psi$ is an extension of $\Phi$: not sharing codes

There are two main ways of abstracting the concept of a family of partial functions $\Psi = (\psi_i)_{i \in \mathbb{N}}$ extending an **AFA** $\Phi = (\varphi_i)_{i \in \mathbb{N}}$. As it is the case of an explicit substitutions calculus $\lambda_\xi$ and $\lambda$, as mentioned in section 2.2. The first way is to consider the extension as an integral one, so to say, do not point out the additional mechanism in $\Psi$ regarding $\Phi$. The second way of defining such a concept is to indicate which these additional formal mechanism is. For example, in the case of $\lambda_\xi$, the substitution calculus $\xi$ and its corresponding reductions are these additional mechanisms.

**Remark 1.** When defining another **AFA** as extension, without indicating anything as a key additional mechanism, the extension is equivalent to the **AFA** it extends, and hence, every undecidable property holding in $\Phi$ also holds in $\Psi$ and vice-versa. Thus, defining the extension $\Psi$ as an **AFA** is not informative, since the undecidable relations are already known from $\Phi$.

As we have seen above a more informative way of abstracting the computational relation between $\lambda$-terms is to consider the extension language defined by two computable relations.

**Definition 11 (Finitely-step-computable).** *Consider $\Psi$ as a collection of partial functions from $\mathbb{N}$ into $\mathbb{N}$. A finitely-step-computable relation $\succeq\ \subseteq\ \mathbb{N}\times\mathbb{N}$ is such that, for each $i,j \in \mathbb{N}$, there is a finite $I_o \subseteq \mathbb{N}$ and a primitive recursive function $f$, such that, $i \succeq j$, iff, there is $k \in I_o$ and $f(\langle i,k\rangle) = j$ and for all $n \in \mathbb{N}$, if $\varphi_j(n) \downarrow$ then $\varphi_j(n) = \varphi_i(n)$.*

In the sequel we write sometimes $\triangleright_\Phi$ and $\succeq_\Psi$, in order to explicitly indicate to which class of functions the relation is regarded to.

**Definition 12 (Extension).** *Let $\Phi$ be an **AFA** and $\Psi$ be a collection of partial functions from $\mathbb{N}$ into $\mathbb{N}$. Consider $\triangleright$ a $\Phi$-step-computable relation and $\succeq$ a finitely-step-computable relation associated to $\Psi$. We say that $\Psi$ extends $\Phi$, iff, for all $i,j \in \mathbb{N}$ there are $k_i, k_j \in \mathbb{N}$, such that, if $i \triangleright j$ then $k_i \succeq k_j$, and $\varphi_i = \psi_{k_i}$.*

The expansion problem is better examined when we consider $\Psi$ sharing codes (algorithmic codes) with $\Phi$, i.e, in the definition above, $k_i = i$ and $k_j = j$. In this case, we have the expansion problem stated as: For every $j \in \mathbb{N}$, there is $i \in \mathbb{N}$, such that, $i \succeq j$ and $\psi_i = \psi_j$ and $\varphi_i = \psi_i$. In this case we say that $i$ is the expansion of $j$ in $\Psi$.

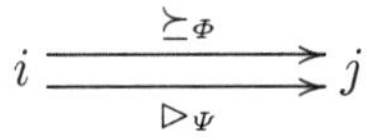

**Fig. 3.** $\Psi$ is an extension of $\Phi$: sharing codes

In Fig. 2 a diagram depicting the general extension is shown, where the codes are not shared between the formalism, and, in Fig. 3, a simpler version of sharing of codes is shown. The latter is the one we use for the sake of clarity.

**Theorem 3 (Expansion).** *Let $\Phi$ be an **AFA** and $\Psi$ one of its extensions. The question whether for every $j \in \mathbb{N}$ there exist $i \in \mathbb{N}$, such that, $i \succeq j$, is undecidable.*

**Lemma 2.** *Let $\Phi$ be an **AFA** and $\Psi$ be one of its extensions, then $\triangleright_\Phi\ \subseteq\ \succeq_\Psi$.*

*Proof.* This is a direct consequence of definition 12.

**Lemma 3.** *Let $\Phi$ be an **AFA** and $\Psi$ one of its extensions. Let $\mathcal{C}$ be a $\succeq_\Psi$ saturated set. $\mathcal{C}$ is recursive, iff, either $\mathcal{C} = \mathbb{N}$ or $\mathcal{C} = \emptyset$.*

*Proof.* This proof is slightly similar to the proof of Rice's theorem (theorem 2). We will repeat all steps here, in order to call reader's attention to the central, and sensible, difference between this case and 2. If $\mathcal{C} \neq \emptyset$ and $\mathcal{C} \neq \mathbb{N}$, there is $k \in \mathcal{C}$ and wlog we can consider $i_\emptyset \notin \mathcal{C}$, where $\varphi_{i_\emptyset} = \emptyset \neq \varphi_k$. **Pay attention to this main difference**: $k$ can be considered as not being a code in $\Phi$. This means that $\varphi_k \neq \psi_k$, for if it was a code in $\Phi$ and as $\Psi$ is an extension, then $\varphi_k = \psi_k$. Remember that we are considering extensions sharing codes. Anyway, $k$ is not a code in $\Phi$, for if there is no such $k$, then $\mathcal{C}$ is formed only by codes of $\Phi$, being recursive and non-trivial this is a contradiction. Here an application of lemma 2 is used to show that $\mathcal{C}$, in this case, is saturated by $\triangleright$ too.

Now that we know that $\varphi_k \neq \psi_k$, we keep tracing the proof of theorem 2. Let $g$ be the (partial recursive) function defined by

$$g(m) = \varphi_m(m); k$$

The formal code of the right-hand side of this definition is obtained as before. Consider $n$, such that $\varphi_n = \varphi_{g(m)}$, then $n \succeq j$, iff, either:

- $\varphi_j = \emptyset = \varphi_{i_o}$ and $\varphi_m(m) \uparrow$, or;
- $\varphi_m(m) \downarrow$ and $k \succeq j$ and $n \succeq k$. In this case $n \in \mathcal{C}$, since $\mathcal{C}$ is saturated by $\succeq$ and $k \in \mathcal{C}$.

Thus, on the one hand, if $\varphi_m(m) \uparrow$ then $j \notin \mathcal{C}$ and $n \notin \mathcal{C}$ either. On the other hand, if $\varphi_m(m) \downarrow$ then $n \in \mathcal{C}$. Since it is undecidable whether $\varphi_m(m) \uparrow$ or not, then $\mathcal{C}$ cannot be recursive.

Using the lemma above (lemma 3) we can verify that the undecidability of expansion holds. Now we can prove theorem 3.

*Proof.* of theorem 3. Let $j \in \mathbb{N}$ be an arbitrary natural number, thought of as a code in $\Psi$. Form the set $\mathcal{C} = \{i \;::\; i \succeq j\}$. Since $\mathcal{C}$ is non-trivial, it cannot be recursive by lemma 3. Thus the expansion property is undecidable.

We can also use lemma 3 to prove the undecidability of other properties on extensions of the **AFA**. As a matter of illustration we prove the undecidability of joined expansion.

**Definition 13 (Join-expansion).** *Consider an* **AFA** $\Phi$ *and one of its extensions* $\Psi$. *Let* $\succeq$ *be finitely-step-computable relation on* $\mathbb{N} \times \mathbb{N}$. *Let* $i$ *and* $j$, *such that*, $\varphi_i \neq \psi_i$ *and* $\varphi_j \neq \psi_j$. *This means that* $i$ *and* $j$ *are extension codes. Define* $i \Downarrow j$, *iff, there is* $k$ *such that* $k \succeq i$ *and* $k \succeq j$, *such that,* $\varphi_k = \psi_k$. *This last fact means that* $k$ *is a code in* $\Phi$.

**Proposition 2 (Undecidability of $\Downarrow$).** *It is undecidable whether* $i \Downarrow j$, *for any* $i, j \in \mathbb{N}$.

*Proof.* Any set $S$ saturated by $\succeq$ is such that, $i \Downarrow j$, iff, either $i, j \in S$ or $i, j \notin S$. Since, by lemma 3 there is no non-trivial and recursive saturated set $S$, the relation $\Downarrow$ cannot be recursive.

## 5 Conclusion

In this note we showed how to obtain abstract definitions and proofs of well-known notions and properties related to $\lambda$-calculus and explicit substitutions communities. Starting from the abstract setting to computability, initiated by Hartley Rogers, we defined abstract notions of reducibility ($\beta$-reduction) and of extension calculus and proved that the expansion and join-expansion properties are undecidable. Abstract versions of Rice's theorem, theorem 2 and lemma 3 were used as key auxiliary results.

It is important to emphasize that Rice's theorem in this abstract setting is strongly related to Scott's theorem. However, from what was shown here, there is no need of additional encoding to prove expansion undecidability as it is discussed at the introduction. This points out that Rice's theorem is broader as a tool to prove undecidability properties than Scott's theorem. Although, both theorems, at the statement level, are equivalent.

Further considerations from our investigations include the application of Rice's theorem to prove undecidability of other properties of computational interest such as those related to higher-order unification (e.g., [8,7]) in more sophisticated λ-calculus extensions.

## References

1. Abadi, M., Cardelli, L., Curien, P.-L., Lévy, J.-J.: Explicit Substitutions. J. of Functional Programming 1(4), 375–416 (1991)
2. Arbiser, A.: The Expansion Problem in Lambda Calculi with Explicit Substitution. J. Log. Comput. 18(6), 849–883 (2008)
3. Benaissa, Z.-E.-A., Briaud, D., Lescanne, P., Rouyer-Degli, J.: $\lambda v$, a Calculus of Explicit Substitutions which Preserves Strong Normalization. J. of Functional Programming 6(5), 699–722 (1996)
4. Bloo, R., Rose, K.H.: Preservation of Strong Normalisation in Named Lambda Calculi with Explicit Substitution and Garbage Collection. In: CSN-95: Computer Science in the Netherlands, pp. 62–72 (1995)
5. da Silva, F.H.: Expansibilidade em Cálculos de Substituições Explícitas. Master's thesis, Graduate Program in Informatics, Universidade de Brasília (December 2012) (in Portuguese)
6. de Bruijn, N.G.: Lambda-Calculus Notation with Nameless Dummies, a Tool for Automatic Formula Manipulation, with Application to the Church-Rosser Theorem. Indag. Mat. 34(5), 381–392 (1972)
7. de Moura, F.L.C., Ayala-Rincón, M., Kamareddine, F.: Higher-Order Unification: A structural relation between Huet's method and the one based on explicit substitutions. J. Applied Logic 6(1), 72–108 (2008)
8. Dowek, G., Hardin, T., Kirchner, C.: Higher-order Unification via Explicit Substitutions. Information and Computation 157(1/2), 183–235 (2000)
9. Kamareddine, F., Ríos, A.: A λ-calculus à la de Bruijn with Explicit Substitutions. In: Swierstra, S.D. (ed.) PLILP 1995. LNCS, vol. 982, pp. 45–62. Springer, Heidelberg (1995)
10. Kesner, D.: The Theory of Calculi with Explicit Substitutions Revisited. In: Duparc, J., Henzinger, T.A. (eds.) CSL 2007. LNCS, vol. 4646, pp. 238–252. Springer, Heidelberg (2007)
11. Machtey, M., Young, P.: An introduction to the general theory of algorithms. Theory of computation series. Elsevier North-Holland, New York (1978)
12. Rice, H.G.: Classes of Recursively Enumerable Sets and Their Decision Problems. Trans. Amer. Math. Soc. 74, 358–366 (1953)
13. Rogers Jr., H.: Theory of recursive functions and effective computability. MIT Press, Cambridge (1987)
14. Rose, K.H.: Explicit Cyclic Substitutions. In: Rusinowitch, M., Remy, J.-L. (eds.) CTRS 1992. LNCS, vol. 656, pp. 36–50. Springer, Heidelberg (1993)

# Computing in the Presence of Concurrent Solo Executions

Maurice Herlihy[1], Sergio Rajsbaum[2], Michel Raynal[3,4], and Julien Stainer[4]

[1] Brown University, Providence (RI), USA
[2] Instituto de Mathematicas, UNAM, D.F. 04510, Mexico
[3] Institut Universitaire de France
[4] IRISA Université de Rennes 1, INRIA, Campus de Beaulieu, 35042 Rennes Cedex, France

**Abstract.** In a *wait-free* model any number of processes may crash. A process runs *solo* when it computes its local output without receiving any information from other processes, either because they crashed or they are too slow. While in wait-free shared-memory models at most one process may run solo in an execution, any number of processes may have to run solo in an asynchronous wait-free message-passing model.

This paper is on the computability power of models in which several processes may concurrently run solo. It first introduces a family of round-based wait-free models, called the $d$-*solo* models, $1 \leq d \leq n$, where up to $d$ processes may run solo. The paper gives then a characterization of the colorless tasks that can be solved in each $d$-solo model. It also introduces the $(d, \epsilon)$-*solo approximate agreement* task, which generalizes $\epsilon$-approximate agreement, and proves that $(d, \epsilon)$-solo approximate agreement can be solved in the $d$-solo model, but cannot be solved in the $(d+1)$-solo model. The paper studies also the relation linking $d$-set agreement and $(d, \epsilon)$-solo approximate agreement in asynchronous wait-free message-passing systems.

These results establish for the first time a hierarchy of wait-free models that, while weaker than the basic read/write model, are nevertheless strong enough to solve non-trivial tasks.

## 1 Introduction

*Distributed computability.* The computability power of a distributed model depends on its communication, timing, and failure assumptions. A basic result is the impossibility to solve consensus in an asynchronous read/write [16] or message-passing [8] system even if only one process may crash. When looking at the communication medium and assuming asynchronous processes prone to crash failures, a read/write system and a message-passing system have the same computability power if and only if less than half of the processes may crash [1]. If a majority of the processes may crash, the message passing model is weaker than the shared memory model because partitions can occur.

The power of a distributed model has been studied in detail with respect to *tasks*, which are the distributed equivalent of functions in sequential computing. Each process gets only one part of the input, and after communicating with the others, decides on an output value, such that collectively, the various local outputs produced by the processes respect the task specification, which is defined from the local inputs of the processes. This paper concentrates on the class of *colorless tasks* (e.g., [3,12]), where the specification is in terms of possible inputs and outputs, but without referring to which process gets which input or produces which output. Among the previously studied notable tasks, many are colorless, such as consensus [8], set agreement [5], approximate agreement [6], loop agreement [13] while some are not, like renaming [2].

*Wait-freedom and solo execution.* This paper considers *wait-free* distributed crash-prone asynchronous models. Wait-free has two (complementary) meanings. First, it means that the model allows up to $n-1$ processes to crash, where $n$ is total number of processes. Its other meaning expresses a liveness condition, namely it requires that every non-faulty process progresses and eventually decides (i.e., computes a result) whatever the behavior of the other processes [11].

In a wait-free model where processes must satisfy the wait-freedom liveness condition, a process has to make progress even in the extreme cases where all other processes have crashed, or are too slow, and consequently be forced to decide without knowing their input values. Hence, for each process, there are executions where this process perceives itself as being the only process participating in the computation.

More generally, we say that a process executes *solo* if it computes its local output without knowing the input values of the other processes.

*Two extreme wait-free models: shared memory and message passing.* In a model where processes communicate by reading and writing shared registers, at most one process can run solo in any execution. This is because, when a process runs solo, it writes and reads from the shared memory, and eventually writes its decision. Any other process that starts running, will be able to read the history left by the solo process in the memory.

When considering message-passing communication, all processes may have to run solo concurrently in the extreme case, where messages are arbitrarily delayed, and each process perceives the other processes as having crashed. Only tasks that can be solved without communication can be computed in this model.

*Investigating the computability power of intermediary models.* The aim of the paper is to study the computability power of asynchronous models in which processes may run solo in the same execution. More precisely, assuming that up to $d$ processes may run solo, the paper addresses the following questions:

– How to define a computation model in which up to $d$ processes may run solo?
– Which tasks can be computed in such a model?

The aim is to study these questions in a clean theoretical framework, and (for the first time) investigate models weaker than the basic wait-free read/write model. However, we hope that our results are relevant to other intermediate models, such as distributed models over fixed or wireless networks.

To simplify the technical development, following [4], the paper develops a theoretical round-based framework, iterated model (IIS) that has been proved useful in many other papers. Processes execute an infinite sequence of asynchronous rounds and communicate through specific objects called *immediate snapshot* objects. Such objects are high-level read/write objects such that a new object instance is associated with each round and, when it executes a round $r$, a process can access only the object associated with round $r$. A main interest of the IIS model is that, from a task computability point of view, it has the same power as the read/write wait-free model [4]. Also, the topology of the IIS model is easier to analyze, establishing a good foundation to analyze task solvability in various distributed computing models.

*Contributions.* The following contributions answer previous questions:

- The definition of a family of *d-solo models,* each parametrized with an integer $d$, $1 \leq d \leq n$. The 1-solo model corresponds to the IIS model (which is equivalent to the read/write wait-free model [4]), while the $n$-solo model corresponds to the round-based wait-free message-passing model.
- A characterization of the set of colorless tasks that can be solved in the $d$-solo model, $1 \leq d \leq n$. Via a new form of complex subdivisions, this characterization connects topology with *colorless* algorithms.
- Any *d-solo* model with $d \geq 2$, is weaker than the read/write wait-free model, yet there are natural, non-trivial tasks that can be solved in the *d-solo* model. One of these tasks, called $(d, \epsilon)$-*solo approximate agreement* (in short $(d, \epsilon)$-SAA) is such that $(d, \epsilon)$-SAA can be solved in the $d$-solo model, for any $\epsilon > 0$, but not in the $(d + 1)$-solo model. Hence, more tasks can be solved in the $d$-solo model than in the $(d + 1)$-solo model, for $1 \leq d < n$, which establishes a hierarchy of solo models.
- Finally, the $d$-solo model is related to $d$-set agreement. This relation shows that, for $d < n$, $d$-set agreement is strong enough to solve $(d, \epsilon)$-solo approximate agreement but is too weak to solve $(d-1, \epsilon)$-solo approximate agreement in the wait-free message-passing model. This provides us with a better insight on a bound on the "maximal partitioning" allowed to solve $(d - 1, \epsilon)$-solo approximate agreement in the wait-free message-passing model.

The $(d, \epsilon)$-solo approximate agreement task is a generalization of approximate agreement [6]. The input of each process consists of a point in the Euclidean space $\mathbb{R}^N$ ($N \geq d$). The *validity* property states that each process $p_i$ has to decide a point which is in the convex hull of all the input points. The *agreement* property states that at most $d$ processes may decide any point in the convex hull of the input points (let $CH$ be the convex hull defined by these at most $d$ points), while the other processes have to decide values whose distance to $CH$ is at most $\epsilon$. Actually, the convex hull of solo processes is an "attractor" for the set of decided values.

When $d = 1$, validity and agreement imply that the Euclidean distance between any pair of points decided by the processes has to be upper bounded by a predefined constant. Thus, $(1, \epsilon)$-solo approximate agreement problem in $\mathbb{R}^m$ is essentially the problem that has been recently considered in the context of $t$ Byzantine failures and asynchronous message-passing systems [17,20], where it is shown that it can be solved iff $n > t(m + 2)$.

The colorless tasks that are solvable in the wait-free iterated immediate snapshot (IIS) model have been characterized in [12]. Due to the simulations in [4,10], this characterization holds for the usual read/write wait-free model. Section 4 extends the characterization of [12] to the $d$-solo model, $1 \leq d \leq n$. Our characterization in terms of colorless algorithms permits the use of standard subdivisions, instead of chromatic subdivisions used in previous papers. We believe colorless algorithms are interesting in themselves, and indeed, for $d = 1$, if a colorless task is solvable, it is solvable by a colorless algorithm. For $d > 1$ we defer the proof that colorless algorithms and general algorithms can solve a very similar class of tasks.

One of the central results of topology is the Simplicial Approximation Theorem [18], which establishes what is a "discrete version" of a continuous map. This theorem is also central for the wait-free characterization theorem of [15] and its $t$-resilient extension (e.g., [12]). However, this theorem cannot be used in a $d$-solo model, $d > 1$, because it is no longer the case that the diameter of the simplexes in a subdivision is reduced. Not even the Relative Simplicial Approximation Theorem [21] can be directly used.

Finally, it is important to notice that our $d$-solo model addresses different issues than the $d$-concurrency model of [9], where it is shown that with $d$-set agreement any number of processes can emulate $d$ state machines of which at least one remains highly available. While $d$-concurrency is used to reduce the concurrency degree to at most $d$ processes that are always allowed to cooperate, $d$-solo allows up to $d$ processes to run independently (i.e., without any cooperation).

*Roadmap.* The paper is composed of 6 sections. Section 2 introduces base definitions, the communication objects, and the $d$-solo model. Section 3 investigates colorless tasks in the $d$-solo model, while Section 4 focuses on what can be computed in the presence of concurrent solo executions. Then, Section 5 defines the $(d, \epsilon)$-solo approximate agreement problem, shows that it can be solved in the $d$-solo model and cannot in the $(d+1)$-solo model, thereby defining a strict hierarchy of distributed computing models. Section 6 concludes the paper. Due to page limitation, topology notions, all proofs, additional technical developments, and relations between $d$-set agreement and $(d, \epsilon)$-solo approximate agreement in wait-free message-passing systems are given in [14].

## 2  Tasks, Processes, Communication Object, and Iterated Model

*Tasks.* A task is a one-shot distributed computing problem specified in terms of an input/output relation $\Delta$. Each process starts with a private input value and must eventually compute a private output value. The task specifies the possible initial configurations. An initial configuration $I$ specifies the input value of each process. Similarly, the output values produced by the processes in an execution represents an output configuration $O$.

A task $(\mathcal{I}, \mathcal{O}, \Delta)$ is defined by a set of input configurations $\mathcal{I}$, a set of possible output configurations $\mathcal{O}$, and a relation $\Delta$ which specifies which output configurations $O \in \mathcal{O}$ are correct for each input $I \in \mathcal{I}$. A more formal description appears in Section 3.1 and in previous papers such as in [15].

*Processes.* The system model is made up of $n$ asynchronous (deterministic) sequential processes, $p_1, \ldots, p_n$, which proceed in asynchronous rounds [19]. The index $i$ of process $p_i$ is sometimes used to denote $p_i$. Up to $n - 1$ processes may crash. Once a process crashes, it never recovers. We say the model is *wait-free*.

*Rounds and communication objects.* A communication object $CO[r]$ is associated with each round $r$ and this object is the only means for the processes to communicate during round $r$. The rounds are *communication-closed* [7] which means that, when a process executes a round, it can communicate with other processes only through the object associated with this round.

More precisely, $CO[r]$ is a one-shot object (i.e., each process accesses it only once) which provides the processes with a single operation denoted communicate$(i, v)$, where $v$ is the value that the invoking process $p_i$ wants to communicate to the other processes during round $r$. Such an invocation returns to $p_i$ a set of pairs (process identity, value) deposited into $CO[r]$ by other processes during round $r$.

*Iterated model.* Each process $p_i$ executes the algorithm skeleton described in Figure 1, where the local computation parts are related to the particular task that is solved. The local variable $r_i$ is the local round number, $\ell s_i$ contains $p_i$'s local state, while $view_i$ contains all the pairs $(j, \ell s_j)$ communicated to $p_i$ during the current round. The local transition function $\delta_i()$ defines the new local state of $p_i$ according to its previous local state and the pairs $(j, \ell s_j)$ it has obtained from $CO^d[r]$ (the parameter $d$ is explained below in Section 2.1). To solve a task, it is necessary to instantiate accordingly $\delta_i()$, the predicate decision() and the function dec_val(): decision() allows $p_i$ to decide, while dec_val() allows it to compute the decided value. As we are interested in computability and not efficiency, we assume a *full information* algorithm, i.e., at the end of each round $r_i$, $\ell s_i$ contains the value of $view_i$, and $\delta_i$ can be task independent. However, we will see in Section 3 that in some cases, tasks can be solved without communicating all a process knows.

(01) $r_i \leftarrow 0$; $\ell s_i \leftarrow$ initial local state;
(02) **loop forever** $r_i \leftarrow r_i + 1$; $view_i \leftarrow CO^d[r_i].\text{communicate}(i, \ell s_i)$;
(03) $\quad\quad\quad \ell s_i \leftarrow \delta_i(\ell s_i, view_i)$; **if** decision($\ell s_i$) **then** dec_val($\ell s_i$) **end if**
(04) **end loop**.

**Fig. 1.** Generic iterated model

### 2.1 Communication Object

The communication objects $CO^d[1]$, $CO^d[2]$, etc., of an execution are parametrized by a solo-dimension $d$, $1 \leq d \leq n$. As previously indicated, an object $CO^d[r]$ contains a set of pairs, one per process. Each pair $(i, v)$ is such that $i$ is a process index and $v$ the value communicated by $p_i$, and $CO^d[r]$ contains at most one pair per process.

*Definition.* The behavior of every object $CO^d$ is defined as follows. Considering an execution during which each of the $n$ processes $\{p_1, \ldots, p_n\}$ accesses the object (at most once) using its local state $\ell s_i$ as input, one can represent this execution by an *ordered partition*, i.e., a tuple of non-empty sets $(P_1, \ldots, P_z)$ such that (1) for any distinct $i, j \in \{1, \ldots, z\}$: $P_i \cap P_j = \emptyset$, and (2) $\bigcup_{i=1}^{z} P_i = \{p_1, \ldots, p_n\}$. From an operational view, the ordered partition $(P_1, \ldots, P_z)$ describes the sequence of concurrent accesses to the object $CO^d$.

The behavior of $CO^d$ is defined from a $d$-*ordered partition*, where a $d$-ordered partition is an ordered partition $(\pi_1, \ldots, \pi_{z'})$ such that $0 \leq |\pi_1| \leq d$ (the size of the first set of the partition can be 0 and cannot exceed $d$). More precisely, the $d$-ordered partition $(\pi_1, \ldots, \pi_{z'})$ associated with $CO^d$ is:

- If $|P_1| > d$: $(\pi_1, \ldots, \pi_{z'}) = (\emptyset, P_1, \ldots, P_z)$, and
- If $|P_1| \leq d$: $(\pi_1, \ldots, \pi_{z'}) \in \{(\emptyset, P_1, \ldots, P_z), (P_1, \ldots, P_z)\}$.

$(\pi_1, \ldots, \pi_{z'}) = (P_1, \ldots, P_z)$ captures the cases where, initially, $d$ (or less) processes execute solo. In the other cases we have $(\pi_1, \ldots, \pi_{z'}) = (\emptyset, P_1, \ldots, P_z)$, because initially either too many processes execute concurrently (first item), or, while no more than $d$ processes execute concurrently, none of them executes solo.

The values $view_i$, $1 \leq i \leq n$, obtained by the processes when the behavior of $CO^d$ is represented by the *d-ordered partition* $(\pi_1, \ldots, \pi_{z'})$ are defined as follows:
$$(i \in \pi_1) \Rightarrow (view_i = \{(i, \ell s_i)\}), \text{ and}$$
$$(x > 1 \land i \in \pi_x) \Rightarrow (view_i = \{(j, \ell s_j) : j \in \pi_y \land y \leq x\}).$$

This means that the view of each process $p_i$ belonging to $\pi_1$ (where $0 \leq |\pi_1| \leq d$) contains only its own contribution, namely the pair $(i, ls_i)$. Differently, the view of a process $p_i$ in $\pi_x$, where $x > 1$, contains all the pairs $(j, \ell s_j)$ deposited in $CO^d$ by the processes $p_j$ of the sets $\pi_y$ such that $y \leq x$. Thus, each process of $\pi_1$ appears as executing solo, while each other process of a set $p_x$, $x \neq 1$, sees the contributions provided (a) by all the processes $p_i$ belonging to the "previous" sets $\pi_y$ ($y < x$), and (b) by all the processes from its "concurrency" set $\pi_x$. (The immediate snapshot object described in [3] implements $CO^d$ for $d = 1$.) Examples of communication objects are presented in [14].

**Object properties.** Given an object $CO^d$, the next properties follows from its definition (See examples of $CO^d$ objects in the Appendix).
- Solo execution upper bound. $0 \leq |\{i \text{ such that } |view_i| = 1\}| \leq d$.
- Self-inclusion. $\forall i : (i, -) \in view_i$.
- Containment. $\forall i, j : \big((|view_i| \leq |view_j|) \land |view_j| > 1)\big) \Rightarrow (view_i \subseteq view_j)$.

## 2.2 A Spectrum of Solo Models

It follows from their definition that $CO^d$ is stronger (more constraining) than $CO^{d+1}$ in the sense that the subdivided complex of $CO^d$ is included the one of $C^{d+1}$. Intuitively, this means that $CO^d$ includes "more synchrony" than $CO^{d+1}$.

*The d-solo model.* The generic framework described in Figure 1 instantiated with $CO^d$ objects is called the *d-solo model*. It is denoted $\mathcal{ACS}^d_{n,n-1}$(ASC stands for Asynchronous Concurrent Solo) where the first subscript denotes the total number of processes, while the second subscript denotes the upper bound on the number of processes allowed to crash.

*Hierarchy of d-solo models.* Let $A \succeq_T B$ mean that any task that can be solved in the model $B$ can be solved in the model $A$, and $A \simeq_T B \stackrel{def}{=} (A \succeq_T B) \land (B \succeq_T A)$.

Let $\mathcal{ARW}_{n,n-1}$ denote the base wait-free (asynchronous) read/write model. It follows from the fact that (for task solvability) the IIS model and $\mathcal{ARW}_{n,n-1}$ have the same computability power [4], and IIS is nothing more than $\mathcal{ACS}^1_{n,n-1}$, that we have $\mathcal{ARW}_{n,n-1} \simeq_T \mathcal{ACS}^1_{n,n-1}$.

Let $\mathcal{AMP}_{n,n-1}$ denote the classical (non-iterated) message-passing system where up to $(n-1)$ processes may crash. As all processes except one may crash and communication is asynchronous (hence messages can be arbitrarily delayed), the tasks that can be solved in $\mathcal{AMP}_{n,n-1}$ are the tasks that can be wait-free solved without communication. But, this set of tasks is exactly the set of tasks that can be solved in $\mathcal{ACS}^n_{n,n-1}$. Hence, $\mathcal{ACS}^n_{n,n-1} \simeq_T \mathcal{AMP}_{n,n-1}$.

It follows from the definition of the communication objects $CO^d$ and $CO^{d+1}$ that any task solvable in $\mathcal{ACS}^{d+1}_{n,n-1}$ is solvable in $\mathcal{ACS}^d_{n,n-1}$. We have consequently the following hierarchy of models: $\mathcal{ARW}_{n,n-1} \simeq_T \mathcal{ACS}^1_{n,n-1} \succeq_T \cdots \succeq_T \mathcal{ACS}^d_{n,n-1} \succeq_T \cdots \succeq_T \mathcal{ACS}^n_{n,n-1} \simeq_T \mathcal{AMP}_{n,n-1}$. We will see in Section 5 that $A \succeq_T B$ can be replaced by $A \succ_T B$ (all the tasks solvable in $B$ are solvable in $A$, and there is one task solvable in $A$ and not in $B$).

## 3 Colorless Tasks and the $d$-Solo Model

This section focuses on colorless tasks that can be solved in the $d$-solo model. After having defined colorless tasks it shows that, for these tasks, one can use a restricted form of the algorithm in Figure 1. It then, introduces the notions of a $(d, R)$-subdivision task and a $(d, R)$-agreement task. (More topology notions are given in [14].)

### 3.1 Colorless Tasks

A colorless task is a special kind of task where the processes cannot use their ids during the computation. This implies that the task specification is not in terms of ids. A colorless task specifies which sets of values are valid input configurations, and which are valid output decisions, but not which value is assigned to which process. Thus, a process may adopt the input value or the output value of another process.

Formally, a *colorless task* is a triple $(\mathcal{I}^*, \mathcal{O}^*, \Delta^*)$, where $\mathcal{I}^*$ is a colorless *input complex*, $\mathcal{O}^*$ is a colorless *output complex*, and $\Delta^* : \mathcal{I}^* \to 2^{\mathcal{O}^*}$ is a *carrier map*. A colorless complex is a family of sets, over some basic set of values, such that if a set is in the complex, then all its subsets are also in the complex. A set in the complex is called a *simplex*. Simplexes of size 1, are called *vertices*, and of size 2, *edges*. Indeed, a graph is a 1-dimensional complex. In the case of a colorless complex, a vertex is just a value, either an input or an output value, while in a colored complex, a vertex is a pair of values, one is a process id, and the other is an input our output value. If $\sigma$ is an input simplex in $\mathcal{I}^*$, the carrier map $\Delta^*(\sigma)$ is a subcomplex of $\mathcal{O}^*$ satisfying *monotonicity*:
$$\forall \sigma, \sigma' \in \mathcal{I}^* \ : \ \Delta^*(\sigma \cap \sigma') \subseteq \Delta^*(\sigma) \cap \Delta^*(\sigma').$$

Operationally, the meaning of a colorless task is the following. If $\sigma \in \mathcal{I}^*$, then the processes can start an execution with input values from $\sigma$; different processes may propose the same vertex or different vertices from $\sigma$. Processes eventually decide (not necessarily distinct) vertices that belong to the same output simplex $\tau \in \mathcal{O}^*$, such that $\tau \in \Delta^*(\sigma)$. If the system consists of $n$ processes, then the processes can start with at most $n$ different input values, and hence, processes will never start on a simplex $\sigma$ of $\mathcal{I}$ of dimension greater than $n - 1$ (the dimension of $\sigma$ is $|\sigma| - 1$). Thus, for $n$ processes, only the simplexes of $\mathcal{I}$ of dimension $\leq n - 1$ are relevant, i.e., the $n - 1$ *skeleton* of $\mathcal{I}$, denoted $\text{Skel}^{n-1}\mathcal{I}$. For example, in a system of two processes, $n = 2$, only the 1-skeleton of $\mathcal{I}$ is of interest, which is the graph consisting of the vertices and 1-simplices of $\mathcal{I}$.

### 3.2 Colorless Algorithms

A *colorless algorithm* is an algorithm in the form of Figure 1, but where the local computation made by $\delta_i$ in line (3) is very restricted. Although a colorless algorithm is not as powerful as an algorithm with no restrictions, it simplifies that exposition, and in the full version we show that they can solve a similar class of colorless tasks.

Informally, in a colorless algorithm processes behave in an anonymous way: processes consider the shared memory as if it is a set. (A colorless complex is denoted with a $*$ superscript, as in $\mathcal{K}^*$). In each round, a process deposits its input in the set, and gets back a view of the contents of the set. If two processes deposit the same value in the set, only one copy is stored. When a process gets back a set of values, there is no information of which process deposited which value. A process "forgets" which is its own value in the set. The set of values that a process receives at the end of a round, becomes its input to the next round.

Formally, in an execution, the initial local state of a process $p_i$ is a vertex $v_i$ of $\mathcal{I}^*$, and is assigned in line 1 to $\ell s_i$. Furthermore, the set of all initial states $v_i$ (not necessarily distinct) is a simplex $\sigma$ of $\mathcal{I}^*$. We may write, $\sigma = \{\ell s_1[0], \ldots, \ell s_n[0]\}$, where $\ell s_i[0]$ denotes the initial value of $\ell s_i$. Notice that $|\sigma|$ may be less than $n$ because different processes may start with the same input value.

The local transition $\delta_i$ eliminates process ids. Namely, during any round $r$ and for any process $p_i$, if we denote by $\ell s_i[r]$ the value of $\ell s_i$ at the end of round $r$, in line 2 of the algorithm, $view_i$ is assigned the value returned by $CO^d[r]$.communicate$(i, \ell s_i[r-1])$, and this value is a set of pairs $\{(i_1, \ell s_{i_1}[r-1]), \ldots, (i_k, \ell s_{i_k}[r-1])\}$ that includes ids $i_1, \ldots, i_k$, but when the function $\delta_i$ is applied to this set it returns a set $\sigma_i^r = \{\ell s_{i_1}[r-1], \ldots, \ell s_{i_k}[r-1]\}$. We assume every process executes the same number of rounds, $R \geq 0$, and in the last round, produces an output value dec_val$(\ell s_i)$ (all processes use the same function dec_val).

For an $R$ round colorless algorithm in the $d$-dimensional model, the *algorithm complex* is defined as follows. For each input simplex $\sigma \in \mathcal{I}^*$, the subcomplex $\mathcal{P}^*(\sigma)$ represents the executions $r$ where all processes start with inputs from $\sigma$ (at least one process starts with each of the vertices in $\sigma$). Moreover, in the algorithm complex for the $d$-dimensional model we do not want to include the $(d-1)$-dimensional model, so we consider only runs where the processes that in a round see more than one process, they see at least $d+1$ processes. The complex $\mathcal{P}^*(\sigma)$ contains a top dimensional simplex $\tau = \{\ell s_i\}$ for each such $R$ round execution of the algorithm starting in $\sigma$, where the vertices $\ell s_i$ of $\tau$ are the values of $\ell s_i[r]$ at the end of this execution, for each process $p_i$ (without repetitions, as the simplex is a set). The complex $\mathcal{P}^*$ is the union of $\mathcal{P}^*(\sigma)$ over all $\sigma \in \mathcal{I}^*$. It is easy to prove that $\mathcal{P}^*(\cdot)$ is a strict carrier map from $\mathcal{I}^*$ to the algorithm complex $\mathcal{P}^*$.

We will explain the significance of the next lemma later on, when we discuss subdivisions.

**Lemma 1.** *Consider a 1-round colorless algorithm and an input simplex $\sigma \in \mathcal{I}^*$. The simplexes of $\mathcal{P}^*(\sigma)$ are of the form $\tau = \{\tau_1, \ldots, \tau_z\}$, where each $\tau_i \subseteq \sigma$, and there is an $l, 0 \leq l \leq d$ such that (1) for all $i, 0 \leq i \leq l, |\tau_i| = 1$, so $\cup_{0 \leq i \leq l} \tau_i$ is a face $\sigma'$ of $\sigma$, (2) for all $j, l < j \leq z, \sigma' \subsetneq \tau_j$, and (3) for all $j, l < j \leq z - 1, \tau_j \subsetneq \tau_{j+1}$.*

If $\mathcal{P}^*(\cdot)$ is a carrier map from $\mathcal{I}^*$ to the algorithm complex $\mathcal{P}^*$, and dec_val is a simplicial map from $\mathcal{P}^*$ to $\mathcal{O}^*$, we say that dec_val *is carried by* $\Delta^*$ if for each $\sigma \in \mathcal{I}^*$ and each $\tau \in \mathcal{P}^*(\sigma)$, the simplex dec_val$(\tau)$ belongs to $\Delta^*(\sigma)$.

**Lemma 2.** *If the colorless task $(\mathcal{I}^*, \mathcal{O}^*, \Delta^*)$ is solvable by a colorless algorithm then there exists an algorithm complex $\mathcal{P}^*$, and a simplicial map dec_val from $\mathcal{P}^*$ to $\mathcal{O}^*$ that is carried by $\Delta^*$.*

### 3.3 $(d, R)$-Subdivision and $(d, R)$-Agreement Tasks

*The $(d, R)$-subdivision task.* Which is the simplest task a colorless algorithm can solve in the $d$-dimensional model? It is the task solved when each process executes $R$ rounds, then stops, and its decision function is the identity! Namely, dec_val$(\ell s_i) = \ell s_i$ i.e. a process decides the set of values $\ell s_i[R]$ it retrieves from the communication object during the $R^{th}$ round. Given any input complex $\mathcal{I}^*$ and any integer $R \geq 0$, we call this task the $(d, R)$-*subdivision task* over $\mathcal{I}^*$. The output complex $\mathcal{O}^*$ of this task is of course equal to the algorithm complex $\mathcal{P}^*$, with the simplicial map dec_val being the identity. For the carrier map, $\Delta^*(\sigma)$ includes all simplexes $\tau$ that correspond to executions starting in $\sigma$, i.e., $\Delta^*(\sigma) = \mathcal{P}^*(\sigma)$. In particular, for $R = 0$, $\mathcal{I}^* = \mathcal{O}^*$, and $\Delta^*$ is the identity carrier map, which sends a simplex $\sigma$ to the complex consisting of $\sigma$ and all its faces (which we often denote by $\sigma$, abusing notation).

By definition, the $(d, R)$-subdivision task over $\mathcal{I}^*$ is solvable in the $d$-dimensional model, and moreover, by a colorless algorithm. In fact, it is the basic building block to solve every other colorless task, as shown in Theorem 1. We will justify the name "subdivision task" when we see how to specify the task without resorting to executions of some model in Section 3.4.

*The $(d, R)$-agreement task.* When the vertices of $\mathcal{I}^*$ are points in Euclidean space, the $(d, R)$-subdivision task can be used directly to solve a task that we call $(d, R)$-*agreement task* over $\mathcal{I}^*$, which is defined combinatorially in Section 5. In the $(d, R)$-subdivision task, processes propose sets of values in each round. We can encode such a set of values as its barycenter $b$, and then the process can directly propose $b$. We shall see in Section 5, that, although both tasks are essentially the same, when we work with barycenters processes compute output values within $\epsilon$ of each other (except for at most $d$ processes that may run solo), and we can make $\epsilon$ as small as we want, by choosing a large enough value of $R$.

Operationally, the $(d, R)$-agreement task over $\mathcal{I}^*$ is defined as follows. Processes execute $R$ rounds of a colorless algorithm in the $d$-dimensional model. In each round $r$, each process $p_i$ computes its value $\ell s_i[r]$ that will be the input to the next round, in line 3 of the algorithm, by taking the barycenter of the values that it gets back from the object in line 2. The barycenter computed in round $R$ is the output of of the process.

### 3.4 The Structure of Colorless Algorithms

The structure of a colorless complex is explained in terms of *subdivisions* (due to page limitation, more developments can be found in [14]). Examples of subdivisions of a simplex are illustrated on the figure that follows at the right of the page.

Perhaps the simplest subdivision is the *stellar* subdivision. Given a complex (abusively denoted $\sigma^m$) consisting of an $m$-simplex $\sigma^m = \{s_0, \ldots, s_m\}$ and all its faces, the complex $\mathrm{Stel}(\sigma^m, b)$ is constructed by taking a *cone* with apex $b$ over the boundary complex $\partial \sigma^m$.

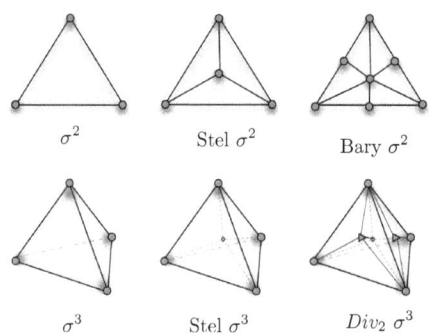

The *barycentric* subdivision, Bary $\sigma^m$, is perhaps the most widely used in topology. A simplex $\tau$ is in Bary $\sigma^m$ if and only if there exists a sequence $\sigma_0 \subset \ldots \subset \sigma_z$ of faces of $\sigma^m$, and the set of vertices of $\tau$ is the set of the barycenters of the these faces, denoted $\hat{\sigma}_i, 0 \leq i \leq z$.

For the $d$-solo models, we need to define a family of subdivisions that goes from the stellar to the barycentric subdivision. The *$d$-dimensional subdivision* of a complex $\mathcal{K}$

denoted $\text{Div}_d \, \mathcal{K}$, is the barycentric subdivision of $\mathcal{K}$ relative to $\text{Skel}^{d-1}\mathcal{K}$. Intuitively, we do not subdivide $\text{Skel}^{d-1}\mathcal{K}$ because we consider executions where up to $d$ processes run solo, they get their own view in an invocation of a $CO^d$ object. See the construction of Figure 2 and topology notions in Appendix. As usual, the *R-iterated d-dimensional subdivision*, $\text{Div}_d^R \, \mathcal{K}$, is obtained by repeating the subdivision process $R$ times.

```
(01)  Div_d Skel^{d-1} σ^m ← Skel^{d-1} σ^m; % each vertex is labeled by its name
(02)  for k from d to m do % Construct Div_d Skel^k σ^m %
(03)      for each simplex σ^k in σ^m do
(04)          insert a vertex b in the barycenter of σ^k;
              % this barycenter is labeled with the set of vertices of σ_k
(05)          construct the cone with apex at b over Div_d ∂σ^k;
              % over the already subdivided boundary of σ^k %
(06)          add the cone to Div_d Skel^k σ^m
(07)      end for loop
(08)  end for loop.
```

**Fig. 2.** Constructing the subdivision $\text{Div}_d \, \sigma^m$ of a simplex $\sigma^m$ for the $d$-solo model

The next lemma follows from the fact that the construction of $\text{Div}_d$ in Figure 2 corresponds exactly to the description given in Lemma 1, and the fact in the system there are $n$ processes, so they can start with at most $n$ different input values (so only the input simplexes in $\mathcal{I}^*$ of dimension at most $n-1$ are relevant).

**Lemma 3.** *If $\mathcal{P}_R^*$ is the R-round algorithm complex of a colorless algorithm in the d-solo model with input complex $\mathcal{I}^*$, then $\mathcal{P}_R^*$ is an R-iterated, d-dimensional subdivision of the $n-1$ skeleton of $\mathcal{I}^*$.*

Returning to the $(d, R)$-subdivision task, we can now justify its name, simply by recalling that its output complex is equal to the algorithm complex:

**Lemma 4.** *The $(d, R)$-subdivision task over $\mathcal{I}^*$ for n processes is a triple $(\mathcal{I}^*, \mathcal{O}^*, \Delta^*)$, where $\mathcal{O}^*$ is the R-iterated, d-dimensional subdivision of the $n-1$ skeleton of $\mathcal{I}^*$, and $\Delta^*$ is equal to the corresponding subdivision carrier map.*

## 4 What Can Be Computed in the Presence of Solo Executions?

This section presents a characterization of the colorless tasks that can be solved in each one of the $d$-solo models. Consider an $r$ round colorless algorithm that solves the colorless task $(\mathcal{I}^*, \mathcal{O}^*, \Delta^*)$. At the end of the $r$-th round, processes have to decide an output value, by executing dec_val$(\ell s_i)$ in line 3. The result of dec_val$(\ell s_i)$ is a vertex in $\mathcal{O}^*$, and different processes may decide different vertices as long as they belong to the same simplex of $\mathcal{O}^*$. This means that dec_val is a simplicial map from $\mathcal{P}_r^*$ to $\mathcal{O}^*$. Moreover, dec_val is carried by $\Delta^*$, in the sense that for $\sigma \in \mathcal{I}^*$: dec_val$(\mathcal{P}_r^*(\sigma)) \subseteq \Delta^*(\sigma)$, which means that for any input simplex $\sigma$, any $r$ round execution ends in a simplex $\tau$ of $\mathcal{P}_r^*$, and the decision that the processes make in $\tau$, form an output simplex dec_val$(\tau)$ of $\mathcal{O}^*$. This output simplex dec_val$(\tau)$ must be in $\Delta^*(\sigma)$, to satisfy the task's specification.

**Theorem 1.** *The colorless task $\mathcal{T}^* = (\mathcal{I}^*, \mathcal{O}^*, \Delta^*)$ is solvable with $n$ processes in the $d$-solo model by a colorless algorithm if and only if there is an $R \geq 0$ and a simplicial map $\phi : Div_d^R\, Skel^{n-1} \mathcal{I}^* \to \mathcal{O}^*$ carried by $\Delta^*$.*

## 5  $(d, \epsilon)$-Solo Approx. Agreement and Strict Hierarchy of Models

We now study the properties of the $(d, R)$-agreement task of Section 3.3 in terms of a precision parameter $\epsilon$, showing that this task can be solved in the $d$-solo model while it cannot be solved in the $(d+1)$-solo model.

Let $\epsilon$ be a positive real. The $(d, \epsilon)$-solo approximate agreement problem (in short $(d, \epsilon)$-SAA) is a generalization of the $\epsilon$-*approximate agreement* problem [6]. The $(1, \epsilon)$-solo approximate agreement instance implies $2\epsilon$-approximate agreement. Assuming the input of each process is a point of the $d$-dimensional Euclidean space $\mathbb{R}^d$, $(d, \epsilon)$-solo approximate agreement is defined by the following properties. (This definition is discussed and compared to other definitions in [14].)

– Validity. Any output lies within the convex hull of the inputs.
– Agreement. There is a set of processes $S$, $1 \leq |S| \leq d$, such that any process $p_i$ that is not in $S$ decides a value $o_i$ (point) such that the Euclidean distance between $o_i$ and $CH$ is at most $\epsilon$, where $CH$ is the convex hull of the points decided by the processes in $S$.
– Termination. If a process $p_i$ does not crash, it decides a value.

It follows from this definition that up to $d$ processes are allowed to decide any set of points within the convex hull (as an example each of them may decide the point it proposes). These processes define the set $S$, and intuitively, the values they decide are collectively "represented" by their convex hull $CH$. Finally, the values decided by the other processes are constrained by the values decided by the processes in $S$.

The next theorem shows that, from a task solvability point of view, the $d$-solo model is stronger than the $(d+1)$-solo model.

**Theorem 2.** *If the domain of the possible input values (a) is bounded and (b) contains a regular simplex of dimension $d$ whose edge length is strictly greater than $2\epsilon d\sqrt{\frac{2d}{d+1}}$, then the $(d, \epsilon)$-solo approximate agreement problem is solvable in $\mathcal{ACS}^d_{n,n-1}$ but not in $\mathcal{ACS}^{d+1}_{n,n-1}$.*

## 6  Conclusion

A process executes solo when its computes its local result without knowing the input values of the other participating processes. This paper addressed round-based asynchronous wait-free executions in which up to $d$ processes may execute solo in each round. Among several contributions, the paper presented a strict hierarchy of wait-free iterated models, called $d$-solo models, and a topology-based characterization of the colorless tasks which can be solved in such $d$-solo models, $1 \leq d \leq n$. The paper also introduced a colorless task, denoted $(d, \epsilon)$-solo approximate agreement (a generalization of the classic approximate agreement task), which can be solved in the $d$-solo model and cannot be solved in the $(d+1)$-solo model.

**Acknowledgments.** This work has been partially supported by the French ANR project DISPLEXITY devoted to computability and complexity in distributed computing), and a UNAM-PAPIIT grant.

# References

1. Attiya, H., Bar-Noy, A., Dolev, D.: Sharing memory robustly in message passing systems. Journal of the ACM 42(1), 121–132 (1995)
2. Attiya, H., Bar-Noy, A., Dolev, D., Peleg, D., Reischuk, R.: Renaming in an asynchronous environment. Journal of the ACM 37(3), 524–548 (1990)
3. Borowsky, E., Gafni, E.: Immediate atomic snapshots and fast renaming. In: Proc. 12th ACM Symposium on Principles of Distributed Computing (PODC 1993), pp. 41–51 (1993)
4. Borowsky, E., Gafni, E.: A simple algorithmically reasoned characterization of wait-free computations. In: Proc. 16th ACM PODC 1997, pp. 189–198 (1997)
5. Chaudhuri, S.: More choices allow more faults: set consensus problems in totally asynchronous systems. Information and Computation 105, 132–158 (1993)
6. Dolev, D., Lynch, N.A., Pinter, S.S., Stark, E.W., Weihl, W.E.: Reaching approximate agreement in the presence of faults. Journal of the ACM 33(3), 499–516 (1986)
7. Elrad, T.E., Francez, N.: Decomposition of distributed programs into communication-closed layers. Science of Computer Programming 2(3), 155–173 (1982)
8. Fischer, M.J., Lynch, N.A., Paterson, M.S.: Impossibility of distributed consensus with one faulty process. Journal of the ACM 32(2), 374–382 (1985)
9. Gafni, E., Guerraoui, R.: Generalized universality. In: Katoen, J.-P., König, B. (eds.) CONCUR 2011. LNCS, vol. 6901, pp. 17–27. Springer, Heidelberg (2011)
10. Gafni, E., Rajsbaum, S.: Distributed programming with tasks. In: Lu, C., Masuzawa, T., Mosbah, M. (eds.) OPODIS 2010. LNCS, vol. 6490, pp. 205–218. Springer, Heidelberg (2010)
11. Herlihy, M.P.: Wait-free synchronization. ACM TOPLAS 13(1), 124–149 (1991)
12. Herlihy, M.P., Rajsbaum, S.: The topology of shared-memory adversaries. In: Proc. 29th ACM Symp. on Principles of Distr. Computing (PODC 2010), pp. 105–113. ACM Press (2010)
13. Herlihy, M.P., Rajsbaum, S.: A classification of wait-free loop agreement tasks. Theoretical Computer Science 291(1), 55–77 (2003)
14. Herlihy, M.P., Rajsbaum, S., Raynal, M., Stainer, J.: Computing in the Presence of Concurrent Solo Executions. Tech. Report 2004, IRISA (France), 19 pages (2013)
15. Herlihy, M.P., Shavit, N.: The topological structure of asynchronous computability. Journal of the ACM 46(6), 858–923 (1999)
16. Loui, M.C., Abu-Amara, H.H.: Memory requirements for agreement among unreliable asynchronous processes. Adv. in Comp. Research, vol. 4, pp. 163–183. JAI Press (1987)
17. Mendes, H., Herlihy, M.P.: Multidimensional approximate agreement in Byzantine asynchronous systems. In: 45th ACM Symp. on the Theory of Comp. (STOC 2013), pp. 391–400 (2013)
18. Munkres, J.R.: Elements of algebraic topology, 547 pages. Addison-Wesley (1984)
19. Raynal, M.: Concurrent programming: algorithms, principles, and foundations, 515 pages. Springer (2013) ISBN 978-3-642-32027-9
20. Vaidya, N.H., Garg, V.K.: Byzantine vector consensus in complete graphs. In: Proc. 32nd ACM Symposium on Principles of Distributed Computing (PODC 2013). ACM Press (2013)
21. Zeeman, E.C.: Relative simplicial approximation. Proc. Mathematical Proceedings of the Cambridge Philosophical Society 60, 39–43 (1964)

# Combining All Pairs Shortest Paths and All Pairs Bottleneck Paths Problems*

Tong-Wook Shinn and Tadao Takaoka

Department of Computer Science and Software Engineering
University of Canterbury
Christchurch, New Zealand

**Abstract.** We introduce a new problem that combines the well known All Pairs Shortest Paths (APSP) problem and the All Pairs Bottleneck Paths (APBP) problem to compute the shortest paths for all pairs of vertices for all possible flow amounts. We call this new problem the All Pairs Shortest Paths for All Flows (APSP-AF) problem. We firstly solve the APSP-AF problem on directed graphs with unit edge costs and real edge capacities in $\tilde{O}(\sqrt{t}n^{(\omega+9)/4}) = \tilde{O}(\sqrt{t}n^{2.843})$ time, where $n$ is the number of vertices, $t$ is the number of distinct edge capacities (flow amounts) and $O(n^\omega) < O(n^{2.373})$ is the time taken to multiply two $n$-by-$n$ matrices over a ring. Secondly we extend the problem to graphs with positive integer edge costs and present an algorithm with $\tilde{O}(\sqrt{t}c^{(\omega+5)/4}n^{(\omega+9)/4}) = \tilde{O}(\sqrt{t}c^{1.843}n^{2.843})$ worst case time complexity, where $c$ is the upper bound on edge costs.

## 1 Introduction

Finding the shortest paths between pairs of vertices in a graph is one of the most extensively studied problems in algorithms research. The shortest paths problem is often categorized into the Single Source Shortest Paths (SSSP) problem, which is to compute the shortest paths between one source vertex to all other vertices in the graph, and the All Pairs Shortest Paths (APSP) problem, which is to compute the shortest paths between all possible pairs of vertices on the graph.

Arguably the most famous algorithm for the APSP problem is Floyd's algorithm that runs in $O(n^3)$ time [5]. There have been many attempts at providing sub-cubic time bounds for solving the APSP problem on dense graphs with real edge costs [6,3,12,16,8,2,9], all achieving time improvements by logarithmic factors. The current best time bound is $O(n^3 \log \log n / \log^2 n)$ by Han and Takaoka [9]. If the graph has integer edge costs, faster matrix multiplication over a ring [10] can be utilized to achieve deeply sub-cubic time bounds. Alon, Galil and Margalit achieved $O(n^{(3+\omega)/2})$ time bound for solving the APSP problem on directed unweighted graphs, where $O(n^\omega)$ is the time bound on multiplying two $n$-by-$n$ matrices over a ring [1]. This time complexity translates to $O(n^{2.687})$

---

* This research was supported by the EU/NZ Joint Project, Optimization and its Applications in Learning and Industry (OptALI).

with $\omega < 2.373$ [14]. The best time bound for this problem is currently $O(n^{2.530})$ by Zwick [15], thanks to Le Gall's recent achievement in rectangular matrix multiplication [7].

Another well studied problem in graph theory is finding the maximum bottleneck between pairs of vertices. The bottleneck of a path is the minimum capacity of all edges on the path. The bottleneck for the pair of vertices $(u, v)$ is the maximum bottleneck of all paths from $u$ to $v$. The problem of finding the paths that give the maximum bottlenecks for all pairs of vertices is formally known as the All Pairs Bottleneck Paths (APBP) problem. Vassilevska, Williams and Yuster achieved $O(n^{2+\omega/3}) = O(n^{2.791})$ time bound for solving the APBP problem on graphs with real edge capacities [13], and this has subsequently been improved to $O(n^{(\omega+3)/2}) = O(n^{2.687})$ by Duan and Pettie [4].

Let us consider the shortest path that gives us the bottleneck value of $b$ from vertex $u$ to vertex $v$. In other words, we can push flows of amounts up to $b$ from $u$ to $v$ using this path. If the flow demand from $u$ to $v$ is less than $b$, however, there may be a shorter path. This information is useful if we wish to minimize the path cost (distance) for varying flow demands. Thus we combine the two well known APSP and APBP problems and compute the shortest paths for all pairs for all possible flow demands. We call this new problem the All Pairs Shortest Paths for All Flows (APSP-AF) problem. Note that this is different from the All Pairs Bottleneck Shortest Paths (APBSP) problem [13], which is to compute the bottlenecks of the shortest paths for all pairs. There are obvious practical applications for the APSP-AF problem in any form of network analysis, such as computer networks, transportation and logistics, etc.

In this paper we present two algorithms for solving the APSP-AF problem on directed graphs with positive integer edge costs and real edge capacities. Firstly we present an algorithm to solve the problem on graphs with unit edge costs in $O(\sqrt{t}n^{(\omega+9)/4}) = O(\sqrt{t}n^{2.843})$ time, where $t$ is the number of distinct edge capacities. We then extend this algorithm to solve the problem on graphs with positive integer edge costs of at most $c$ in $O(\sqrt{t}c^{(\omega+5)/4}n^{(\omega+9)/4}) = O(\sqrt{t}c^{1.843}n^{2.843})$ time, which is reduced to the complexity of the first algorithm when $c = 1$.

## 2 Preliminaries

Let $G = (V, E)$ be a directed graph with non-negative integer edge costs and real edge capacities. Let $n = |V|$ and $m = |E|$. Vertices (or nodes) are given by integers such that $\{1, 2, 3, ..., n\} \in V$. Let $(i, j)$ denote the edge from vertex $i$ to vertex $j$. Let $cost(i, j)$ denote the cost and $cap(i, j)$ denote the capacity of the edge $(i, j)$. Let $t$ be the number of distinct $cap(i, j)$, and let $c$ be the upper bound on $cost(i, j)$. We define path *length* as the number of edges on the path, irrespective of their costs or capacities. We define path *cost* or *distance* as the sum of all edge costs on the path.

We represent $G$ in a series of matrices. Let $R^\ell = \{r_{ij}^\ell\}$ be the reachability matrix, for $0 < \ell < n$, where $r_{ij}^\ell = 1$ if $j$ is reachable from $i$ via some path of length up to $\ell$ and $r_{ij}^\ell = 0$ otherwise. $r_{ii}^\ell = 1$ for all $\ell$. $r_{ij}^1 = 1$ if an edge exists from $i$ to $j$, and 0 otherwise. $R^1$ is called the adjacency matrix of $G$. Let $C^\ell = \{c_{ij}^\ell\}$ be the capacity matrix, where $c_{ij}^\ell$ represents the maximum possible capacity (or bottleneck) from $i$ to $j$ via any paths of lengths up to $\ell$. $c_{ii}^\ell = \infty$ for all $\ell$. $c_{ij}^1 = cap(i,j)$ if there is an edge from $i$ to $j$, and 0 otherwise. Let $D^\ell = \{d_{ij}^\ell\}$ be the distance matrix, where $d_{ij}^\ell$ represents the shortest possible distance from $i$ to $j$ via any paths of lengths up to $\ell$. $d_{ii}^\ell = 0$ for all $\ell$. $d_{ij}^1 = cost(i,j)$ if there is an edge from $i$ to $j$, and $\infty$ otherwise.

Let $X * Y$ denote the $(min, +)$-product and $X \star Y$ denote the $(max, min)$-product of the two matrices $X$ and $Y$, where:

$$X * Y = \min_{k=1}^{n}\{x_{ik} + y_{kj}\} \quad X \star Y = \max_{k=1}^{n}\{\min\{x_{ik}, y_{kj}\}\}$$

Clearly the $(min, +)$-product is applicable to the distance matrix whereas the $(max, min)$-product is applicable to the capacity matrix.

## 3 Review of the Algorithm by Alon, Galil and Margalit

Our algorithm for solving the APSP-AF problem is largely based on the algorithm given by Alon et al. [1]. Therefore we provide a review of this algorithm using the same set of terminologies as an earlier review of the same algorithm by Takaoka [11]. The algorithm under review computes the All Pairs Shortest Distances (APSD) on directed graphs with unit edge costs. In summary this algorithm achieves sub-cubic time bound by utilizing faster matrix multiplication over a ring to perform Boolean matrix multiplication, and also using the novel idea of *Bridging Sets*.

Algorithm 1 consists of two phases. We refer to the first part of the algorithm as the *acceleration phase*, and the second part of the algorithm as the *cruising phase*. The acceleration phase repeatedly performs Boolean matrix multiplication with the adjacency matrix to compute APSD for all pairs with distances up to $\ell = r$, where $r$ is a constant such that $1 < r < n$. Clearly this only works on graphs with unit edge costs where the path length and the path cost are equivalent. The algorithm then switches to the cruising phase where the ordinary multiplication method is used with the help of bridging sets, $S_i$, where $S_i$ is a set of "via" vertices for all rows $i$ of the distance matrix $D$. That is, when computing $d_{ik}^\ell + d_{kj}^\ell$ for the $(min, +)$-product, we inspect only the set of vertices in $S_i$ for $k$ rather than inspecting all $O(n)$ elements. Alon et al. have shown that with path lengths equal to $r$, the size of the bridging set $S_i$ for each row $i$ is bounded by $O(n/r)$ [1]. Hence we start the cruising phase with $|S_i| = O(n/r)$ for each row $i$.

The acceleration phase takes $O(rn^\omega)$ time, and the cruising phase performs repeated squaring of the distance matrix in $O(n^2 \cdot \frac{n}{r})$ time. Alon et al. chose to increase the path length by a factor of $\frac{3}{2}$ in each iteration of the cruising phase. This factor of $\frac{3}{2}$ is somewhat arbitrary, as any factor greater than 1 and less than 2 can be used. Because the size of the bridging set decreases by a constant factor in each iteration, we end up with a geometric series if we add up

**Algorithm 1.** Algorithm by Alon, Galil and Margalit

/* Acceleration Phase*/
**for** $\ell = 2$ to $r$ **do**
    $R^\ell \leftarrow R^{\ell-1} \times R^1$ /* Boolean matrix multiplication */
    **for** $i \leftarrow 1$ to $n$; $j \leftarrow 1$ to $n$ **do**
        **if** $r^\ell_{ij} = 1$ and $d^{\ell-1}_{ij} = \infty$ **then** $d^\ell_{ij} \leftarrow \ell$
        **if** $d^{\ell-1}_{ij} < \ell$ **then** $d^\ell_{ij} \leftarrow d^{\ell-1}_{ij}$

/* Cruising Phase*/
**while** $\ell < n$ **do**
    $\ell' \leftarrow \lceil \frac{3\ell}{2} \rceil$
    **for** $i \leftarrow 1$ to $n$ **do**
        Scan $i^{th}$ row of $D^\ell$ with $j$ and find the smallest set of equal $d^\ell_{ij}$ such that
        $\lceil \ell/2 \rceil \leq d^\ell_{ij} \leq \ell$ and let the set of corresponding $j$ be $S_i$
    **for** $i \leftarrow 1$ to $n$; $j \leftarrow 1$ to $n$ **do**
        $m_{ij} \leftarrow \min_{k \in S_i} \{d^\ell_{ik} + d^\ell_{kj}\}$ /* Squaring $D^\ell$ with $S_i$ */
        **if** $d^\ell_{ij} \leq \ell$ **then**
            $d^{\ell'}_{ij} \leftarrow d^\ell_{ij}$
        **else if** $m_{ij} \leq \ell'$ **then**
            $d^{\ell'}_{ij} \leftarrow m_{ij}$
$\ell \leftarrow \ell'$

the time complexities of each iteration, and hence the first squaring dominates the time complexity. The total time complexity of $O(n^{(3+\omega)/2}) = O(n^{2.687})$ of this algorithm comes from balancing the time complexities of the two phases to retrieve the best value for $r$, that is, setting $rn^\omega = n^2 \cdot \frac{n}{r}$ then solving for $r$.

## 4 APSP-AF on Graphs with Unit Edge Costs

We first consider solving the All Pairs Shortest Distances for All Flows (APSD-AF) problem on directed graphs with unit edge costs, that is, computing only the shortest distances rather than actual path. Path lengths and path distances are used interchangeably in this section. To re-iterate the APSD-AF problem, for each pair of vertices $(i, j)$ for each possible flow amount, we want to compute the shortest distance. Thus our aim here is to obtain a set of $(d, f)$ pairs for all pairs of vertices, where $f$ is the maximum flow amount that can be pushed through the shortest path whose length (distance) is $d$. We refer to the distinct capacity values as *maximal flows*. i.e. there are $t$ maximal flows. Assume that the maximal flows are sorted in increasing order. If we wish to push $f$ such that $f_1 < f < f_2$ for consecutive maximal flows $f_1$ and $f_2$, then clearly $f$ is represented by $f_2$.

Let $U$ be a matrix such that $u_{ij}$ is a set of $(d, f)$ pairs as described above. Let both $(d, f)$ and $(d', f')$ be in $u_{ij}$ such that $d < d'$. We keep $(d', f')$ iff $f < f'$. In other words, a longer path is only useful to us if it can accommodate a greater flow. If $d = d'$, we keep the pair that provides the bigger flow. Since there can

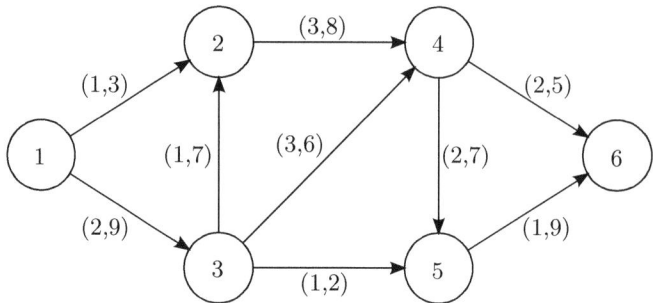

**Fig. 1.** An example graph with $n = 6$, $m = 9$, $t = 7$ and $c = 3$. Numbers in the parenthesis beside each edge shows the edge cost and capacity, respectively.

only be $n - 1$ different values of $d$, each $u_{ij}$ has at most $n - 1$ pairs of $(d, f)$. We assume the pairs are sorted in ascending order of $d$. We make an interesting observation here that once all $(d, f)$ pairs for all $u_{ij}$ are computed (i.e. the APSP-AF problem is solved), the first pairs for all $u_{ij}$ is the solution to the APBSP problem, and the last pairs for all $u_{ij}$ is the solution to the APBP problem.

*Example 1.* If the graph in Figure 1 had unit edge costs instead of the varying integer edge costs, solving APSD-AF on the graph would result in three $(d, f)$ pairs from vertex 1 to vertex 6, that is, $u_{1,6} = \{(3,5), (4,6), (5,7)\}$.

We now introduce Algorithm 2 to solve the APSD-AF problem on directed graphs with unit edge costs. Let $P^f$ be the approximate distance matrix for shortest paths that can accommodate flows up to $f$. In the acceleration phase, we compute the maximum bottleneck values for all possible path lengths up to $r$ for all pairs, where $r$ is a constant such that $1 < r < n$. Then from the results gathered in the acceleration phase, we prepare a series of distance matrices, $P^f$, one for each maximal flow value $f$, and move onto the cruising phase where we compute the shortest distances for all pairs for all flows by repeatedly squaring each $P^f$.

**Lemma 1.** *Algorithm 2 correctly solves APSD-AF on directed graphs with unit edge costs.*

*Proof.* In the acceleration phase, instead of performing Boolean matrix multiplication as in Algorithm 1, we compute the $(max, min)$-product with the capacity matrices $C^1$ and $C^{\ell-1}$. After each matrix multiplication, if a path of greater capacity has been found for the vertex pair $(i, j)$, we append the pair $(\ell, c_{ij}^\ell)$ to $u_{ij}$ since we have found a longer path that can accommodate a greater flow. Thus after the $r^{th}$ iteration of the acceleration phase, all relevant $(d, f)$ pairs for all $u_{ij}$ are found such that $d \leq r$.

After the acceleration phase we initialize the approximate distance matrices $P^f$ from $U$, one matrix for each maximal flow $f$, in preparation for the cruising

**Algorithm 2.** Solve APSD-AF on graphs with unit edge costs

/* Initialization for acceleration phase */
**for** $i \leftarrow 1$ to $n$; $j \leftarrow 1$ to $n$ **do**
    $u_{ij} \leftarrow \phi$ /* $\phi$ is empty */

/* Acceleration phase */
**for** $\ell \leftarrow 2$ to $r$ **do**
    $C^\ell \leftarrow C^{\ell-1} \star C^1$ /* $(max, min)$ matrix multiplication */
    **for** $i \leftarrow 1$ to $n$; $j \leftarrow 1$ to $n$; $i \neq j$ **do**
        **if** $c_{ij}^\ell > c_{ij}^{\ell-1}$ **then**
            Append $(\ell, c_{ij}^\ell)$ to $u_{ij}$

/* Initialization for cruising phase */
$P^f \leftarrow I$ for all maximal flows $f$ /* $I$ has 0 diagonal elements and $\infty$ for others */
**for** $i \leftarrow 1$ to $n$; $j \leftarrow 1$ to $n$; $i \neq j$ **do**
    Let $u_{ij} = \{(d_1, f_1), (d_2, f_2), ..., (d_s, f_s)\}$ for some $s$ /* We skip empty $u_{ij}$ */
    $k \leftarrow 1$ /* $k$ iterates from 1 to $s$ */
    **for all** maximal flows $f$ in increasing order **do**
        **if** $f > f_k$ **then**
            $k \leftarrow k + 1$ /* The next $d_k$ value is needed */
        **if** $k > s$ **then**
            break /* We proceed to the next $u_{ij}$ */
        $p_{ij}^f \leftarrow d_k$

/* Cruising phase */
**for all** maximal flows $f$ **do**
    Perform cruising phase of Algorithm 1 on $P^f$

/* Finalization */
**for** $i \leftarrow 1$ to $n$; $j \leftarrow 1$ to $n$; $i \neq j$ **do**
    **for all** maximal flows $f$ in increasing order **do**
        $d \leftarrow p_{ij}^f$
        Let the last pair of $u_{ij}$ be $x = (d', f')$ /* If $u_{ij}$ is empty, $x = \phi$ */
        **if** $x = \phi$ or ($f > f'$ and $d < \infty$) **then**
            **if** $d = d'$ /* This condition is false if $x = \phi$ */ **then**
                Replace $x$ with $(d, f)$
            **else**
                Append $(d, f)$ to $u_{ij}$

phase. Note that if the $(d, f)$ pair for a given flow value $f$ does not exist in $u_{ij}$, we take the next pair $(d', f')$ in $u_{ij}$ (if one exists) and let $p_{ij}^f = d'$.

At this stage, if $p_{ij}^f < \infty$, $p_{ij}^f$ is already the length of the shortest path from $i$ to $j$ that can push flow $f$. Thus the actual aim of the cruising phase of this algorithm is to compute the shortest distance for all other elements in $P^f$ such that $p_{ij}^f = \infty$ at the start of the cruising phase. Note that unless $G$ is strongly connected, some elements of $P^f$ will remain at $\infty$ until the end of the algorithm. The aim of the cruising phase is achieved by repeatedly squaring each $P^f$ with the help of the bridging set, as proven in [1].

Retrieving sets of $(d, f)$ pairs after the cruising phase from each resulting $P^f$ is simply a reverse process of the initialization for the cruising phase, and thus our search for all sets of $(d, f)$ pairs for all $(i, j)$ is complete after finalization. □

**Lemma 2.** *Algorithm 2 runs in $O(\sqrt{t}n^{(\omega+9)/4}) = O(\sqrt{t}n^{2.843})$ worst case time.*

*Proof.* For the acceleration phase we use the the current best known algorithm to compute the $(max, min)$-product in each iteration, which gives us the time bound of $O(rn^{(3+\omega)/2})$ [4]. The time complexity for the cruising phase is $O(tn^3/r)$ since there are a total of $t$ maximal flows, each taking $O(n^3/r)$ time to finish the computation of APSD. The time bound for the initialization for the cruising phase and the finalization is $O(tn^2)$, which is absorbed by $O(tn^3/r)$ since $n/r > 1$. We balance the time complexities of the acceleration phase and the cruising phase by setting $r = \sqrt{t}n^{(3-\omega)/4}$, and this gives us the total worst case time complexity of $O(\sqrt{t}n^{(\omega+9)/4})$. □

If $t = \Omega(n^{(\omega+1)/2})$, the value we choose for $r$ may exceed $n$. In such a case, we simply stay in the acceleration phase until $r = n-1$. Thus a more accurate worst case time complexity of Algorithm 2 is actually $O(\min \{n^{(5+\omega)/2}, \sqrt{t}n^{(\omega+9)/4}\})$. For all subsequent time complexities, for simplicity, we only show the time bound for actually going into the cruising phase.

A straightforward method of solving the APSD-AF problem is to repeatedly compute APSD for each maximal flow value $f$ using only edges that have capacities greater than or equal to $f$. This method is equivalent to starting the cruising phase at $r = 1$, giving us the time complexity of $O(tn^{2.530})$ if we use Zwick's algorithm to solve APSD for each maximal flow value [15]. For $t > n^{0.626}$, Algorithm 2 is faster. Note that a simple decremental algorithm where edges are removed in the reverse order of capacities while repeatedly solving APSD cannot be used to solve the APSD-AF problem because edges with larger capacities may later be required to provide shorter paths for a smaller maximal flow values.

**Theorem 1.** *There exists an algorithm that can solve APSP-AF on directed graphs with unit edge costs in $\tilde{O}(\sqrt{t}n^{(\omega+9)/4})$ worst case time.*

*Proof.* As noted earlier there can be $O(n)$ $(d, f)$ pairs for each vertex pair $(i, j)$. Since the lengths of each path can be $O(n)$, explicitly listing all paths takes $O(n^4)$ time. We get around this by modifying Algorithm 2 to extend the $(d, f)$ pair to the $(d, f, s)$ triplet, where $s$ is the *successor* node, such that retrieving the actual path from $(d, f, s)$ can be performed by simply following the successor nodes. In the acceleration phase witnesses for the $(max, min)$-product can be retrieved with an extra polylog factor [4], and the successor nodes can be computed from the witnesses in each iteration in $O(n^2)$ time [15]. In the cruising phase retrieving the witnesses, and hence the successor nodes, is a simple exercise since ordinary matrix multiplication is performed. Therefore extending $(d, f)$ to $(d, f, s)$ only takes an additional polylog factor.

The explicit path for a given flow demand from $i$ to $j$ can be generated in time linear to the path length as follows. Firstly we perform binary search for the triplet $(d, f, s)$ in $u_{ij}$ with $f$ as the key to find the minimum distance $d$ such that $f$ is greater than or equal to the given flow requirement. We then traverse the successor nodes $s$ one by one, using $d$ to look up each subsequent successor node in $O(1)$ time. □

## 5  APSP-AF on Graphs with Integer Edge Costs

We now consider solving the APSD-AF problem on directed graphs with integer edge costs and real edge capacities, where the edge cost is bounded by $c$. Note that with integer edge costs we need to make a clear distinction between path lengths and distances. One approach for solving this problem is to use the method described in [1] to replace $G$ with an expanded graph $G'$ such that all edges in $G'$ have unit edge costs, then applying the algorithm on $G'$ to solve the problem on $G$. $G'$ is created by attaching a chain of $c-1$ artificial vertices to each real vertex such that the artificial edges linking the artificial vertices in each chain have unit edge costs and capacities of $\infty$. We then replace each real edge $(i, j)$ with an artificial edge with unit edge cost and capacity of $cap(i, j)$ by choosing one of the artificial vertices of $i$ (or $i$ itself) as the source vertex and the real vertex $j$ as the destination, such that there exists a path from $i$ to $j$ with length equal to $cost(i, j)$. See Figure 2 for an illustration of how a graph is expanded. The expanded graph $G'$ has $O(cn)$ vertices, and we can clearly solve APSD-AF on $G$ by solving APSD-AF on $G'$ in $O(\sqrt{t}(cn)^{(9+\omega)/4})$ time.

*Example 2.* Solving APSD-AF on the graph in Figure 1 results in a total of five $(d, f)$ pairs from vertex 1 to 6, that is, $u_{(1)(6)} = \{(4, 2), (6, 3), (7, 5), (8, 6), (9, 7)\}$.

We can do better, however, with the key observation that only the acceleration phase of Algorithm 2 is restricted to graphs with unit edge costs. In other words, we can complete the acceleration phase on the expanded graph $G'$, gather the intermediate results, and then finish off the remaining computation after contracting the graph back to $G$. We need care here, as the path lengths in $G'$ are actually equivalent to the path costs in $G$, and the bridging sets in the cruising phase are determined from the path lengths rather than the path costs. Therefore we need to make substantial changes to Algorithm 2 to keep track of both the path lengths and the path costs of $G$ in the acceleration phase, as well as modifying the cruising phase to correctly use the path lengths of $G$ in determining the bridging sets.

Firstly we extend the pair $(d, f)$ to the triplet $(h, d, f)$, where $h$ is the path length in $G$, $d$ is the path cost in $G$ (i.e. the path length in $G'$) and $f$ is the maximal flow. We introduce $U' = \{u'_{ij}\}$ where $u'_{ij}$ is a set of triplets $(h, d, f)$ for all pairs of vertices in $G'$. We omit the superscript $\ell$ that denotes the path length in the following matrix definitions. Let $C' = \{c'_{ij}\}$ be the capacity matrix of $G'$ and let $W = \{w_{ij}\}$ be the witness matrix for the $(max, min)$-product. Let $Q^f = \{q^f_{ij}\}$ such that $q^f_{ij}$ is the length of the path that gives the path cost

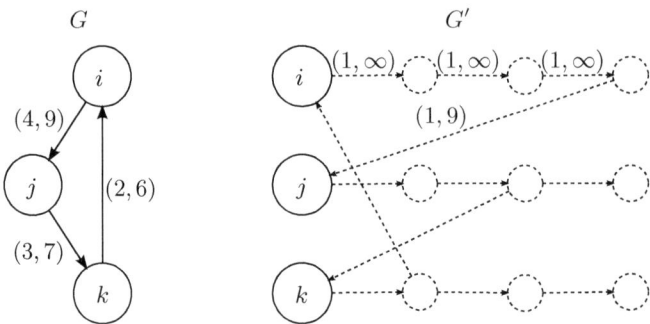

**Fig. 2.** Expanding $G$ to $G'$ with $c = 4$

(distance) of $p_{ij}^f$, where $P^f = \{p_{ij}^f\}$ is the distance matrix as defined in Section 4. That is, $p_{ij}^f$ is the minimum path cost (distance) of all paths from $i$ to $j$ that can push flow of amount $f$. Note that $h$ in the triplet $(h, d, f)$ is no longer required once all $Q^f$ are initialized before the start of the cruising phase.

**Lemma 3.** *Algorithm 3 correctly solves APSD-AF on directed graphs with non-negative integer edge costs.*

*Proof.* We start by creating $G'$ from $G$ then proceed to the acceleration phase. We only need to show that the actual path length $h$ in the triplet $(h, d, f)$ is correctly determined, since we have already discussed the $(d, f)$ pairs in Section 4. What is effectively happening in the acceleration phase of Algorithm 3 is that the path length information is carried from one real vertex to the next real vertex by the artificial vertices in between. Since we are multiplying by $C'^{(1)}$ in each iteration, the witness $w_{ij}$ will always be the vertex that comes straight before the destination vertex $j$ on the path from $i$ to $j$. That is, it is not possible for any vertices (real or artificial) to exist between $k = w_{ij}$ and $j$. Therefore we retrieve the last $(h, d, f)$ triplet from $u'_{ik}$ and increment the given $h$ iff $j$ is a real vertex. Thus the correct path length in $G$ is given by $h$ at the end of the acceleration phase since we are not counting the artificial vertices in the path.

The changes made to the cruising phase are to ensure that the bridging sets $S_i$ is correctly determined from the path lengths rather than the path costs. Note that in Algorithm 2 this distinction was unnecessary because we were only considering graphs with unit edge costs. Clearly the correctness of the crusing phase remains intact by keeping $q_{ij}^{f,\ell}$ updated alongside $p_{ij}^{f,\ell}$. □

**Lemma 4.** *Algorithm 3 runs in $\tilde{O}(\sqrt{t}c^{(\omega+5)/4}n^{(\omega+9)/4}) = \tilde{O}(\sqrt{t}c^{1.843}n^{2.843})$ worst case time.*

*Proof.* The time complexity of the acceleration phase is $\tilde{O}(r(cn)^{(3+\omega)/2})$ since there are $O(cn)$ vertices in $G'$. After the $r^{th}$ iteration in the acceleration phase, we have computed the bottleneck for all paths of lengths up to $r$, but this is path lengths in the expanded graph $G'$, and not $G$. We divide $r$ by $c$ to retrieve the

**Algorithm 3.** Solve APSD-AF on directed graphs with non-negative integer edge costs

/* Initialization for acceleration phase */
Create $G'$ from $G$ /* $G$ is expanded to $G'$ */
for $i \leftarrow 1$ to $cn$; $j \leftarrow 1$ to $cn$ do
  $u'_{ij} \leftarrow \phi$

/* Acceleration phase */
for $\ell \leftarrow 2$ to $r$ do
  $C'^{(\ell)} \leftarrow C'^{(\ell-1)} \star C'^1$ /* Witnesses given as $W = \{w_{ij}\}$ */
  for $i \leftarrow 1$ to $cn$; $j \leftarrow 1$ to $cn$; $i \neq j$ do
    if $c'^{(\ell)}_{ij} > c'^{(\ell-1)}_{ij}$ then
      Let $k = w_{ij}$, and $(h, d, f) \leftarrow$ last triplet in $u'_{ik}$ /* If empty, $h = 0$ */
      if $j \in G$ /* If $j$ is a real vertex */ then
        Append $(h+1, \ell, c'^{(\ell)}_{ij})$ to $u'_{ij}$
      else
        Append $(h, \ell, c'^{(\ell)}_{ij})$ to $u'_{ij}$

/* Initialization for cruising phase, $\ell = r$ */
$U \leftarrow$ rows and columns in $U'$ for real vertices /* $G'$ is contracted back to $G$ */
$P^{f,\ell}, Q^{f,\ell} \leftarrow I$ for all maximal flows $f$
for $i \leftarrow 1$ to $n$; $j \leftarrow 1$ to $n$; $i \neq j$ do
  Let $u_{ij} = \{(h_1, d_1, f_1), ..., (h_s, d_s, f_s)\}$ for some $s$ /* Skip empty $u_{ij}$ */
  $k \leftarrow 1$ /* $k$ iterates from 1 to $s$ */
  for all maximal flows $f$ in increasing order do
    if $f > f_k$ then $k \leftarrow k + 1$ /* The next $d_k$ value is needed */
    if $k > s$ then break /* We proceed to the next $u_{ij}$ */
    $p^{f,\ell}_{ij} \leftarrow d_k$; $q^{f,\ell}_{ij} \leftarrow h_k$

/* Cruising phase */
for all maximal flows $f$ do
  while $\ell < n$ do
    $\ell' \leftarrow \lceil \frac{3\ell}{2} \rceil$
    for $i \leftarrow 1$ to $n$ do
      Scan $i^{th}$ row of $Q^{f,\ell}$ with $j$ to find the smallest set of equal $q^{f,\ell}_{ij}$ such that $\lceil \ell/2 \rceil \leq q^{f,\ell}_{ij} \leq \ell$ and let the set of corresponding $j$ be $S_i$
    for $i \leftarrow 1$ to $n$; $j \leftarrow 1$ to $n$ do
      $m_{ij} \leftarrow \min_{k \in S_i} \{p^{f,\ell}_{ik} + p^{f,\ell}_{kj}\}$
      $k \leftarrow$ the vertex that gives above $m_{ij}$ such that $p^{f,\ell}_{ik} + p^{f,\ell}_{kj}$ is minimum
      if $m_{ij} < p^{f,\ell}_{ij}$ then
        $p^{f,\ell'}_{ij} \leftarrow m_{ij}$; $q^{f,\ell'}_{ij} \leftarrow q^{f,\ell}_{ik} + q^{f,\ell}_{kj}$
      else
        $p^{f,\ell'}_{ij} \leftarrow p^{f,\ell}_{ij}$; $q^{f,\ell'}_{ij} \leftarrow q^{f,\ell}_{ij}$
    $\ell \leftarrow \ell'$

/* Finalization - same as Algorithm 2 */

lower bound on the path lengths in the original graph $G$ after the acceleration phase. Therefore the time complexity of the cruising phase is $O(tcn^3/r)$. Both the time complexities for initialization for cruising phase and finalization are again absorbed by the time complexity of the cruising phase. We balance the time complexities of the acceleration phase and the cruising phase by setting $r = \sqrt{t}c^{(-1-\omega)/4}n^{(3-\omega)/4}$, which gives us the total worst case time complexity of $\tilde{O}(\sqrt{t}c^{(\omega+5)/4}n^{(\omega+9)/4})$. □

**Theorem 2.** *There exists an algorithm to solve the APSP-AF problem on directed graphs with positive integer edge costs in $\tilde{O}(\sqrt{t}c^{(\omega+5)/4}n^{(\omega+9)/4})$ worst case time complexity.*

*Proof.* Clearly we can take a similar approach to the method described in the proof of Theorem 1. We can still use the path cost (distance) to look up each successor node in $O(1)$ time. We note that the witnesses in the acceleration phase can be artificial vertices, but the corresponding real vertices can be retrieved in $O(1)$ time simply by storing this information when $G$ is expanded to $G'$. □

If $c = 1$, $\tilde{O}(\sqrt{t}c^{(\omega+5)/4}n^{(\omega+9)/4})$ becomes $\tilde{O}(\sqrt{t}n^{(\omega+9)/4})$, hence we have successfully generalized the APSP-AF problem from graphs with unit edge costs to graphs with integer edge costs. To compare with the straightforward method of repeatedly solving the APSP problem for each maximal flow value using Zwick's algorithm, we use the formula $\omega(1, r, 1) = 2 + (\omega - 2)(r - \alpha)/(1 - \alpha)$ where $O(n^{\omega(1,r,1)})$ is the time taken to multiply an $n$-by-$n^r$ matrix with an $n^r$-by-$n$ matrix, and $\alpha$ is a constant such that multiplying an $n$-by-$n^\alpha$ matrix with an $n^\alpha$-by-$n$ remains within the $\tilde{O}(n^2)$ time bound. We let $\omega = 2.376$ and $\alpha = 0.294$ in this comparison [15]. The time complexity of the straightforward method becomes $\tilde{O}(tn^{2+\mu})$, where $c = n^x$ such that the equation $\omega(1, \mu, 1) = 1 + 2\mu - x$ is satisfied. Clearly for relatively smaller values for $c$ and larger values for $t$, Algorithm 3 is faster.

Finally we make a note that the idea of expanding the graph to $G'$ for the acceleration phase then contracting it back to $G$ for the cruising phase can retrospectively be applied to Algorithm 1 to give a sharper time bound than $O((cn)^{(3+\omega)/2})$, which is the time bound given by Alon et al. in their original paper [1]. The time bound of $O((cn)^{(3+\omega)/2})$ is sub-cubic for $c < n^{0.117}$. Using our new approach of contracting the graph after the acceleration phase, the time bound can be improved to $O(c^{(1+\omega)/2}n^{(3+\omega)/2})$, which is sub-cubic for $c < n^{0.186}$. For solving the same problem as Algorithm 1, however, other algorithms are already known that remain sub-cubic for larger values of $c$ [11,15].

## 6 Concluding Remarks

The key achievements of this paper are: 1) the introduction of a new problem that clearly has numerous practical applications in network analysis involving both path costs and capacities, 2) non-trivial extension of Algorithm 1 to solve the new problem that is more complex than the APSP problem, and 3) a better method to utilize the artificial graph for integer edge costs resulting in an improved time bound for not only our new algorithm, but also an existing algorithm for solving the APSP problem.

Solving the new APSP-AF problem on other types of graphs (e.g. undirected, real edge costs, etc) as well as finding efficient algorithms for the single source version of the problem remain on the agenda for future research.

## References

1. Alon, N., Galil, Z., Margalit, O.: On the Exponent of the All Pairs Shortest Path Problem. In: Proc. 32nd FOCS, pp. 569–575 (1991)
2. Chan, T.: More algorithms for all-pairs shortest paths in weighted graphs. In: Proc. 39th STOC, pp. 590–598 (2007)
3. Dobosiewicz, W.: A more efficient algorithm for the min-plus multiplication. International Journal of Computer Mathematics 32, 49–60 (1990)
4. Duan, R., Pettie, S.: Fast Algorithms for (max,min)-matrix multiplication and bottleneck shortest paths. In: Proc. 19th SODA, pp. 384–391 (2009)
5. Floyd, R.: Algorithm 97: Shortest Path. Communications of the ACM 5, 345 (1962)
6. Fredman, M.: New bounds on the complexity of the shortest path problem. SIAM Journal on Computing 5, 83–89 (1976)
7. Le Gall, F.: Faster Algorithms for Rectangular Matrix Multiplication. In: Proc. 53rd FOCS, pp. 514–523 (2012)
8. Han, Y.: An $O(n^3(\log\log n/\log n)^{5/4})$ time algorithm for all pairs shortest paths. In: Azar, Y., Erlebach, T. (eds.) ESA 2006. LNCS, vol. 4168, pp. 411–417. Springer, Heidelberg (2006)
9. Han, Y., Takaoka, T.: An $O(n^3 \log\log n/\log^2 n)$ Time Algorithm for All Pairs Shortest Paths. In: Fomin, F.V., Kaski, P. (eds.) SWAT 2012. LNCS, vol. SWAT, pp. 131–141. Springer, Heidelberg (2012)
10. Schönhage, A., Strassen, V.: Schnelle Multiplikation Großer Zahlen. Computing 7, 281–292 (1971)
11. Takaoka, T.: Sub-cubic Cost Algorithms for the All Pairs Shortest Path Problem. Algorithmica 20, 309–318 (1995)
12. Takaoka, T.: A faster algorithm for the all-pairs shortest path problem and its application. In: Chwa, K.-Y., Munro, J.I. (eds.) COCOON 2004. LNCS, vol. 3106, pp. 278–289. Springer, Heidelberg (2004)
13. Vassilevska, V., Williams, R., Yuster, R.: All Pairs Bottleneck Paths and Max-Min Matrix Products in Truly Subcubic Time. Journal of Theory of Computing 5, 173–189 (2009)
14. Williams, V.: Breaking the Coppersmith-Winograd barrier. In: Proc. 44th STOC (2012)
15. Zwick, U.: All Pairs Shortest Paths using Bridging Sets and Rectangular Matrix Multiplication. Journal of the ACM 49, 289–317 (2002)
16. Zwick, U.: A Slightly Improved Sub-Cubic Algorithm for the All Pairs Shortest Paths Problem with Real Edge Lengths. Algorithmica 46, 278–289 (2006)

# (Total) Vector Domination for Graphs with Bounded Branchwidth

Toshimasa Ishii[1], Hirotaka Ono[2], and Yushi Uno[3]

[1] Graduate School of Economics and Business Administration, Hokkaido University,
Sapporo 060-0809, Japan
[2] Department of Economic Engineering, Faculty of Economics, Kyushu University,
Fukuoka 812-8581, Japan
[3] Department of Mathematics and Information Sciences, Graduate School of Science,
Osaka Prefecture University, Sakai 599-8531, Japan

**Abstract.** Given a graph $G = (V, E)$ of order $n$ and an $n$-dimensional non-negative vector $d = (d(1), d(2), \ldots, d(n))$, called demand vector, the vector domination (resp., total vector domination) is the problem of finding a minimum $S \subseteq V$ such that every vertex $v$ in $V \setminus S$ (resp., in $V$) has at least $d(v)$ neighbors in $S$. The (total) vector domination is a generalization of many dominating set type problems, e.g., the dominating set problem, the $k$-tuple dominating set problem (this $k$ is different from the solution size), and so on, and its approximability and inapproximability have been studied under this general framework. In this paper, we show that a (total) vector domination of graphs with bounded branchwidth can be solved in polynomial time. This implies that the problem is polynomially solvable also for graphs with bounded treewidth. Consequently, the (total) vector domination problem for a planar graph is subexponential fixed-parameter tractable with respect to $k$, where $k$ is the size of solution.

## 1 Introduction

Given a graph $G = (V, E)$ of order $n$ and an $n$-dimensional non-negative vector $d = (d(1), d(2), \ldots, d(n))$, called *demand vector*, the *vector domination* (resp., *total vector domination*) is the problem of finding a minimum $S \subseteq V$ such that every vertex $v$ in $V \setminus S$ (resp., in $V$) has at least $d(v)$ neighbors in $S$. These problems were introduced by [21], and they contain many existing problems, such as the minimum dominating set and the $k$-tuple dominating set problem (this $k$ is different from the solution size) [22,23], and so on. Indeed, by setting $d = (1, \ldots, 1)$, the vector domination becomes the minimum dominating set forms, and by setting $d = (k, \ldots, k)$, the total vector dominating set becomes the $k$-tuple dominating set. If in the definition of total vector domination, we replace open neighborhoods with closed ones, we get the *multiple domination*. In this paper, we sometimes refer to these problems just as *domination problems*. Table 1 of [9] summarizes how related problems are represented in the scheme of domination problems. Many variants of the basic concepts of domination and their applications have appeared in [23,24].

Since the vector or multiple domination includes the setting of the ordinary dominating set problem, it is obviously NP-hard, and further it is NP-hard to approximate within ($c \log n$)-factor, where $c$ is a positive constant, e.g., 0.2267 [1,26]. As for the approximability, since the domination problems are special cases of a set-cover type integer programming problem, it is known that the polynomial-time greedy algorithm achieves an $O(\log n)$-approximation factor [15]; it is already optimal in terms of order. We can see further analyses of the approximability and inapproximability in [8,9].

In this paper, we focus on another aspect of designing algorithms for domination problems, that is, the polynomial-time solvability of the domination problems for graphs of bounded treewidth or branchwidth. In [3], it is shown that the vector domination problem is W[1]-hard with respect to treewidth. This result and Courcelle's meta-theorem about MSOL [11] imply that the vector domination is unlikely expressible in MSOL; it is not obvious to obtain a polynomial time algorithm.

In this paper, we present a polynomial-time algorithm for the domination problems of graphs with bounded branchwidth. The branchwidth is a measure of the "global connectivity" of a graph, and is known to be a counterpart of treewidth. It is known that $\max\{bw(G), 2\} \leq tw(G) + 1 \leq \max\{3bw(G)/2, 2\}$, where $bw(G)$ and $tw(G)$ denote the branchwidth and treewidth of graph $G$, respectively [28]. Due to the linear relation of these two measures, polynomial-time solvability of a problem for graphs with bounded treewidth implies polynomial-time solvability of a problem for graphs with bounded branchwidth, and vice versa. Hence, our results imply that the domination problems (i.e., vector domination, total vector domination and multiple domination) can be solved in polynomial time for graphs with bounded treewidth; the polynomial-time solvability for all the problems (except the dominating set problem) in Table 1 of [9] is newly shown. Also, they answer the question by [8,9] about the complexity status of the domination problems of graphs with bounded treewidth.

Furthermore, by using the polynomial-time algorithms for graphs of bounded treewidth, we can show that these problems for a planar graph are subexponential fixed-parameter tractable with respect to the size of the solution $k$, that is, there is an algorithm whose running time is $2^{O(\sqrt{k} \log k)} n^{O(1)}$. To our best knowledge, these are the first fixed-parameter algorithms for the total vector domination and multiple domination, whereas the vector domination for planar graphs has been shown to be FPT [27]. For the latter case, our algorithm greatly improves the running time.

Note that the polynomial-time solvability of the vector domination problem for graphs of bounded treewidth has been independently shown very recently [7]. They considered a further generalization of the vector domination problem, and gave a polynomial-time algorithm for graphs of bounded clique-width. Since $cw(G) \leq 3 \cdot 2^{tw(G)-1}$ holds where $cw(G)$ denotes the clique-width of graph $G$ ([10]), their polynomial-time algorithm implies the polynomial-time solvability of the vector domination problem for graphs of bounded treewidth and bounded branchwidth.

### 1.1 Related Work

For graphs with bounded treewidth (or branchwidth), the ordinary domination problems can be solved in polynomial time. As for the fixed-parameter tractability, it is known that even the ordinary dominating set problem is W[2]-complete with respect to solution size $k$; it is unlikely to be fixed-parameter tractable [17]. In contrast, it can be solved in $O(2^{11.98\sqrt{k}}k + n^3)$ time for planar graphs, that is, it is subexponential fixed-parameter tractable [16]. The subexponent part comes from the inequality $bw(G) \leq 12\sqrt{k} + 9$, where $k$ is the size of a dominating set of $G$. Behind the inequality, there is a unified property of parameters, called *bidimensionality* [14]. Namely, the subexponential fixed-parameter algorithm of the dominating set for planar graphs (more precisely, $H$-minor-free graphs [13]) is based on the bidimensionality.

A maximization version of the ordinary dominating set is also considered. *Partial Dominating Set* is the problem of maximizing the number of vertices to be dominated by using a given number $k$ of vertices. In [2], it was shown that partial dominating set problem is FPT with respect to $k$ for $H$-minor-free graphs. Later, [18] gives a subexponential FPT with respect to $k$ for apex-minor-free graphs, also a superclass of planar graphs. Although partial dominating set is an example of problems to which the bidimensionality theory cannot be applied, they develop a technique to reduce an input graph so that its treewidth becomes $O(\sqrt{k})$.

For the vector domination, a polynomial-time algorithm for graphs of bounded treewidth has been proposed very recently [7], as mentioned before. In [27], it is shown that the vector domination for $\rho$-degenerated graphs can be solved in $k^{O(\rho k^2)} n^{O(1)}$ time, if $d(v) > 0$ holds for $\forall v \in V$ (positive constraint). Since any planar graph is 5-degenerated, the vector domination for planar graphs is fixed-parameter tractable with respect to solution size, under the positive constraint. Furthermore, the case where $d(v)$ could be 0 for some $v$ can be easily reduced to the positive case by using the transformation discussed in [3], while increasing the degeneracy by at most 1. It follows that the vector domination for planar graphs is FPT with respect to solution size $k$. However, for the total vector domination and multiple domination, neither polynomial time algorithm for graphs of bounded treewidth nor fixed-parameter algorithm for planar graphs has been known.

Other than these, several generalized versions of the dominating set problem are also studied. $(k, r)$-center problem is the one that asks the existence of set $S$ of $k$ vertices satisfying that for every vertex $v \in V$ there exists a vertex $u \in S$ such that the distance between $u$ and $v$ is at most $r$; $(k, 1)$-center corresponds to the ordinary dominating set. The $(k, r)$-center for planar graphs is shown to be fixed-parameter tractable with respect to $k$ and $r$ [12]. For $\sigma, \rho \subseteq \{0, 1, 2, \ldots\}$ and a positive integer $k$, $\exists[\sigma, \rho]$-dominating set is the problem that asks the existence of set $S$ of $k$ vertices satisfying that $|N(v) \cap S| \in \sigma$ holds for $\forall v \in S$ and $|N(v) \cap S| \in \rho$ for $\forall v \in V \setminus S$, where $N(v)$ denotes the open neighborhood of $v$. If $\sigma = \{0, 1, \ldots\}$ and $\rho = \{1, 2, \ldots\}$, $\exists[\sigma, \rho]$-dominating set is the ordinary dominating set problem, and if $\sigma = \{0\}$ and $\rho = \{0, 1, 2, \ldots\}$, it is the independent set. In [6], the parameterized complexity of $\exists[\sigma, \rho]$-dominating set with respect to treewidth is also considered.

### 1.2 Our Results

Our results are summarized as follows:

- We present a polynomial-time algorithm for the vector domination of graph $G = (V, E)$ with bounded branchwidth. The running time is roughly $O(n^{6bw(G)+2})$.
- We present polynomial-time algorithms for the total vector domination and multiple domination of graph $G$ with bounded branchwidth. The running time is roughly $O(2^{9bw(G)/2} n^{6bw(G)+2})$.

- Let $G$ be a planar graph. Then, we can check in $O(n^3 + \min\{k, d^*\}^{40\sqrt{k}+34}n)$ time whether $G$ has a vector dominating set with cardinality at most $k$ or not, where $d^* = \max\{d(v) \mid v \in V\}$.
- Let $G$ be a planar graph. Then, we can check in $O(n^3 + 2^{30\sqrt{k}+51/2} \min\{k, d^*\}^{40\sqrt{k}+34}n)$ time whether $G$ has a total vector dominating set and a multiple dominating set with cardinality at most $k$ or not.

It should be noted that it is actually possible to design directly polynomial time algorithms for graphs with bounded treewidth, but they are slower than the ones for graphs with bounded branchwidth; from this reason, we design branch decomposition-based algorithms.

As far as the authors know, the second and fourth results give the first polynomial time algorithms and the first fixed-parameter algorithm for the total vector domination and multiple domination of graphs with bounded branchwidth (or treewidth) and planar graphs, respectively. As for the vector domination, we give an $O(n^{6bw(G)+2})$-time algorithm, whose running time is $O(n^{6(tw(G)+1)+2})$ in terms of the treewidth, whereas the recent paper [7] gives an $O(cw(G)|\sigma|(n+1)^{5cw(G)})$-time algorithm, where $|\sigma|$ is the encoding length of $k$-expression used in the algorithm, and is bounded by a polynomial in the input size for fixed $k$. Since $cw(G) \leq 3 \cdot 2^{tw(G)-1}$ holds, this is an $O(2^{tw(G)}|\sigma|(n+1)^{7.5 \cdot 2^{tw(G)}})$-time algorithm.

Also, the third result shows that the vector domination of planar graphs is subexponential FPT with respect to $k$, and it greatly improves the running time of existing $k^{O(k^2)}n^{O(1)}$-time algorithm ([27]). It was shown in [5] that for the ordinary dominating set problem (equivalently, the vector domination (or multiple domination) with $d = (1, 1, \ldots, 1)$) in planar graphs, there is no $2^{o(\sqrt{k})}n^{O(1)}$-time algorithm unless the Exponential Time Hypothesis (i.e., the assumption that there is no $2^{o(n)}$-time algorithm for $n$-variable 3SAT [25]) fails. Hence, in this sense, our algorithm in third result (or the fourth results for the multiple domination) is optimal if $d^*$ is a constant.

The third and fourth results give subexponential fixed-parameter algorithms of the domination problems for planar graphs. It should be noted that the domination problems themselves do not have the bidimensionality, mentioned in the previous subsection, due to the existence of the vertices with demand 0. Instead, by reducing irrelevant vertices, we obtain a similar inequality about the branchwidth and the solution size of the domination problems, which leads to the subexponential fixed-parameter algorithms.

The remainder of the paper is organized as follows. In Section 2, we introduce some basic notations and then explain the branch decomposition. Section 3 is the main part of the paper, and presents our dynamic programming based algorithms for the considered problems. Section 4 explains how we extend the algorithms of Section 3 to fixed-parameter algorithms for planar graphs.

## 2 Preliminaries

A graph $G$ is an ordered pair of its vertex set $V(G)$ and edge set $E(G)$ and is denoted by $G = (V(G), E(G))$. Let $n = |V(G)|$ and $m = |E(G)|$. We assume throughout this paper that all graphs are undirected, and simple, unless otherwise stated. Therefore, an edge $e \in E(G)$ is an unordered pair of vertices $u$ and $v$, and we often denote it by $e = (u, v)$. Two vertices $u$ and $v$ are *adjacent* if $(u, v) \in E(G)$. For a graph $G$, the (*open*) *neighborhood* of a vertex $v \in V(G)$ is the set $N_G(v) = \{u \in V(G) \mid (u, v) \in E(G)\}$, and the *closed neighborhood* of $v$ is the set $N_G[v] = N_G(v) \cup \{v\}$.

For a graph $G = (V, E)$, let $d = (d(v) \mid v \in V)$ be an $n$-dimensional non-negative vector. Then, we call a set $S \subseteq V$ of vertices a *$d$-vector dominating set* (resp., *$d$-total vector dominating set*) if $|N_G(v) \cap S| \geq d(v)$ holds for every vertex $v \in V \setminus S$ (resp., $v \in V$). We call a set $S \subseteq V$ of vertices a *$d$-multiple dominating set* if $|N_G[v] \cap S| \geq d(v)$ holds for every vertex $v \in V$. We may drop $d$ in these notations if there are no confusions.

**Branch Decomposition.** A *branch decomposition* of a graph $G = (V, E)$ is defined as a pair $(T = (V_T, E_T), \tau)$ such that (a) $T$ is a tree with $|E|$ leaves in which every non-leaf node has degree 3, and (b) $\tau$ is a bijection from $E$ to the set of leaves of $T$. Throughout the paper, we shall use the term *node* to denote an element in $V_T$ for distinguishing it from an element in $V$.

For an edge $f$ in $T$, let $T_f$ and $T \setminus T_f$ be two trees obtained from $T$ by removing $f$, and $E_f$ and $E \setminus E_f$ be two sets of edges in $E$ such that $e \in E_f$ if and only if $\tau(e)$ is included in $T_f$. The *order function* $w : E(T) \to 2^V$ is defined as follows: for an edge $f$ in $T$, a vertex $v \in V$ belongs to $w(f)$ if and only if there exist an edge in $E_f$ and an edge in $E \setminus E_f$ which are both incident to $v$. The *width* of a branch decomposition $(T, \tau)$ is $\max\{|w(f)| \mid f \in E_T\}$, and the *branchwidth* of $G$, denoted by $bw(G)$, is the minimum width over all branch decompositions of $G$.

In general, computing the branchwidth of a given graph is NP-hard [30]. On the other hand, Bodlaender and Thilikos [4] gave a linear time algorithm which checks whether the branchwidth of a given graph is at most $k$ or not, and if so, outputs a branch decomposition of minimum width, for any fixed $k$. Also, as shown in the following lemma, it is known that for planar graphs, it can be done in polynomial time for any given $k$, where a graph is called *planar* if it can be drawn in the plane without generating a crossing by two edges.

**Lemma 1.** *Let $G$ be a planar graph.*
*(i) ([30]) It can be checked in $O(n^2)$ time whether $bw(G) \leq k$ or not for a given integer $k$.*
*(ii) ([20]) A branch decomposition of $G$ with width $bw(G)$ can be constructed in $O(n^3)$ time.* □

Here, we introduce the following basic properties about branch decompositions, which will be utilized in the subsequent sections (the proof is omitted).

**Lemma 2.** *Let $(T, \tau)$ be a branch decomposition of $G$.*
*(i) For tree $T$, let $x$ be a non-leaf node and $f_i = (x, x_i)$, $i = 1, 2, 3$, be an edge incident to $x$ (note that the degree of $x$ is three). Then, $w(f_i) \setminus (w(f_j) \cup w(f_k)) = \emptyset$ for every $\{i, j, k\} = \{1, 2, 3\}$. Hence, $w(f_i) \subseteq w(f_j) \cup w(f_k)$.*
*(ii) Let $f$ be an edge of $T$, $V_1$ be the set of all end-vertices of edges in $E_f$, and $V_2$ be the set of all end-vertices of edges in $E \setminus E_f$. Then, $(V_1 \setminus w(f)) \cap (V_2 \setminus w(f)) = \emptyset$ holds. Also, there is no edge in $E$ connecting a vertex in $V_1 \setminus w(f)$ and a vertex in $V_2 \setminus w(f)$.*

## 3 Domination Problems in Graphs of Bounded Branchwidth

In this section, we propose dynamic programming algorithms for the vector domination problem, the total vector domination problem, and the multiple domination problem, by utilizing a branch decomposition of a given graph. The techniques are based on the one developed by Fomin and Thilikos for solving the dominating set problem with bounded branchwidth [19]. Throughout this section, for a given graph $G = (V, E)$, the demand of each vertex $v \in V$ is denoted by $d(v)$, and let $d^* = \max\{d(v) \mid v \in V\}$.

Here, we show the following theorem for the vector domination problem.

**Theorem 1.** *If a branch decomposition of $G$ with width $b$ is given, a minimum vector dominating set in $G$ can be found in $O((d^* + 2)^b \{(d^* + 1)^2 + 1\}^{b/2} m)$ time.*

Due to the assumption of the above theorem, we need to consider how we obtain a branch decomposition of $G$ for the completeness of an algorithm of the vector domination problem. For a branch decomposition, there exists an $O(2^{b \lg 27} n^2)$-time algorithm that given a graph $G$ and an integer $b$, reports $bw(G) \geq b$, or outputs a branch decomposition of $G$ with width at most $3b$ [29,13]. Thus, the time to find a branch decomposition with width at most $3bw(G)$ is $O(\log bw(G) 2^{bw(G) \lg 27} n^2)$ (smaller than the time complexity below), and we have the following corollary.

**Corollary 1.** *A minimum vector dominating set in $G$ can be found in $O((d^*+2)^{3bw(G)} \{(d^* +1)^2 + 1\}^{3bw(G)/2} n^2)$ time.* □

Below, for proving this theorem, we will give a dynamic programming algorithm for finding a minimum vector dominating set in $G$ in $O((d^* + 2)^b \{(d^* + 1)^2 + 1\}^{b/2} m)$ time, based on a branch decomposition of $G$.

Let $(T', \tau)$ be a branch decomposition of $G = (V, E)$ with width $b$, and $w' : E(T') \to 2^V$ be the corresponding order function. Let $T$ be the tree from $T'$ by inserting two nodes $r_1$ and $r_2$, deleting one arbitrarily chosen edge $(x_1, x_2) \in E(T')$, adding three new edges $(r_1, r_2), (x_1, r_2)$, and $(x_2, r_2)$; namely, $T = (V(T') \cup \{r_1, r_2\}, E(T') \cup \{(r_1, r_2), (x_1, r_2), (x_2, r_2)\} \setminus \{(x_1, x_2)\})$. Here, we regard $T$ as a rooted tree with root $r_1$. Let $w(f) = w'(f)$ for every $f \in E(T) \cap E(T')$, $w(x_1, r_2) = w(x_2, r_2) = w'(x_1, x_2)$, and $w(r_1, r_2) = \emptyset$.

Let $f = (y_1, y_2) \in E$ be an edge in $T$ such that $y_1$ is the parent of $y_2$. Let $T(y_2)$ be the subtree of $T$ rooted at $y_2$, $E_f = \{e \in E \mid \tau(e) \in V(T(y_2))\}$, and $G_f$ be the subgraph of $G$ induced by $E_f$. Note that $w(f) \subseteq V(G_f)$ holds, since each vertex in $w(f)$ is an end-vertex of some edge in $E_f$ by definition of the order function $w$. In the following, each vertex $v \in w(f)$ will be assigned one of the following $d(v) + 2$ colors $\{\top, 0, 1, 2, \ldots, d(v)\}$. The meaning of the color of a vertex $v$ is as follows: for a vertex set (possibly, a vector dominating set) $D$,

- $\top$ means that $v \in D$.
- $i \in \{0, 1, \ldots, d(v)\}$ means that $v \notin D$ and $|N_{G_f}(v) \cap D| \geq d(v) - i$.

Notice that a vertex colored by $i > 0$ may need to be dominated by some vertices in $V \setminus V(G_f)$ for the feasibility. Given a coloring $c \in \{\top, 0, 1, 2, \ldots, d^*\}^{|w(f)|}$, let $D_f(c) \subseteq V(G_f)$ be a vertex set with the minimum cardinality satisfying the following (1)–(3), where $c(v)$ denotes the color assigned to a vertex $v \in V$:

$$c(v) = \top \text{ if and only if } v \in D_f(c) \cap w(f). \tag{1}$$
$$\text{If } c(v) = i, \text{ then } v \in w(f) \setminus D_f(c) \text{ and } |N_{G_f}(v) \cap D_f(c)| \geq d(v) - i. \tag{2}$$
$$|N_{G_f}(v) \cap D_f(c)| \geq d(v) \text{ holds for every vertex } v \in V(G_f) \setminus (w(f) \cup D_f(c)). \tag{3}$$

Intuitively, $D_f(c)$ is a minimum vector dominating set in $G_f$ under the assumption that the color for every vertex in $w(f)$ is restricted to $c$. Note that a vertex in $w(f)$ is allowed not to meet its demand in $G_f$, because it can be dominated by some vertices in $V \setminus V(G_f)$. Also note that every vertex in $V(G_f) \setminus w(f)$ is not adjacent to any vertex in $V \setminus V(G_f)$ by Lemma 2(ii), and it needs to be dominated by vertices only in $V(G_f)$ for the feasibility. We define $A_f(c)$ as $A_f(c) = |D_f(c)|$ if $D_f(c)$ exists and $A_f(c) = \infty$ otherwise.

Our dynamic programming algorithm proceeds bottom-up in $T$, while computing $A_f(c)$ for all $c \in \{\top, 0, 1, 2, \ldots, d^*\}^{|w(f)|}$ for each edge $f$ in $T$. We remark that since $w(r_1, r_2) = \emptyset$ and $G_{(r_1,r_2)} = G$ for the root edge $(r_1, r_2)$, the only coloring $c$ in $A_{(r_1,r_2)}(c)$ is the empty coloring and $A_{(r_1,r_2)}(c)$ is the cardinality of a minimum vector dominating set. The algorithm consists of two types of procedures: one is for leaf edges and the other is for non-leaf edges, where a *leaf edge* denotes an edge incident to a leaf of $T$.

**Procedure for Leaf Edges:** In the first step of the algorithm, we compute $A_f(c)$ for each edge $f$ incident to a leaf of $T$. Then, for all colorings $c \in \{\top, 0, 1, 2, \ldots, d^*\}^{|w(f)|}$, let $A_f(c)$ be the number of vertices colored by $\top$ if $D_f(c)$ exists and $G_f$ and $c$ satisfy (1) – (3), and $A_f(c) = \infty$ otherwise.

Let $f$ be a leaf edge incident to a leaf node $x$ in $T$ and $e = (v_1, v_2)$ be the edge in $G$ with $\tau(e) = x$. Then, notice that we have $w(f) = \{v_i\}$ if the degree of $v_j$ is 1 for $\{i, j\} = \{1, 2\}$, and $w(f) = \{v_1, v_2\}$ otherwise, and that $V(G_f) = \{v_1, v_2\}$. Hence, for a fixed $c$, we can check in $O(1)$ time if (1) – (3) hold. This step takes $O((d^* + 2)^2)$ time.

**Procedure for Non-Leaf Edges:** After the above initialization step, we visit non-leaf edges of $T$ from leaves to the root of $T$. Let $f = (y_1, y_2)$ be a non-leaf edge of $T$ such that $y_1$ is the parent of $y_2$, $y_3$ and $y_4$ are the children of $y_2$, and $f_1 = (y_2, y_3)$ and $f_2 = (y_2, y_4)$. Now we have already obtained $A_{f_j}(c')$ for all $c' \in \{\top, 0, 1, 2, \ldots, d^*\}^{|w(f_j)|}$, $j = 1, 2$. By Lemma 2(i), we have $w(f) \subseteq w(f_1) \cup w(f_2)$, $w(f_1) \subseteq w(f_2) \cup w(f)$, and $w(f_2) \subseteq w(f) \cup w(f_1)$; let $X_1 = w(f) \setminus w(f_2)$, $X_2 = w(f) \setminus w(f_1)$, $X_3 = w(f) \cap w(f_1) \cap w(f_2)$, and $X_4 = w(f_1) \setminus w(f) \ (= w(f_2) \setminus w(f))$.

We say that a coloring $c \in \{\top, 0, 1, 2, \ldots, d^*\}^{|w(f)|}$ of $w(f)$ is *formed* from a coloring $c_1$ of $w(f_1)$ and a coloring $c_2$ of $w(f_2)$ if the following (P1)–(P5) hold.

(P1) For every $v \in X_1 \cup X_2 \cup X_3$ with $c(v) = \top$,

 (a) For every $v \in X_1 \cup X_3$, $c_1(v) = \top$ if and only if $c(v) = \top$.
 (b) For every $v \in X_2 \cup X_3$, $c_2(v) = \top$ if and only if $c(v) = \top$.

(P2) For every $v \in X_4$, $c_1(v) = \top$ if and only if $c_2(v) = \top$.

(P3) For every $v \in X_j \setminus D_{c_1,c_2}$ where $\{j, j'\} = \{1, 2\}$ and $D_{c_1,c_2} = \{v \in X_1 \cup X_2 \cup X_3 \cup X_4 \mid c_1(v) = \top \text{ or } c_2(v) = \top\}$,

If $c(v) = i$, then $c_j(v) = \min\{d(v), i + |D_{c_1,c_2} \cap N_{G_f}(v) \cap X_{j'}|\}$.
(Intuitively, if $v \in X_j \setminus D_{c_1,c_2}$ needs to be dominated by at least $d(v) - i$ vertices in $G_f$, then at least $\max\{0, d(v) - i - |D_{c_1,c_2} \cap N_{G_f}(v) \cap X_{j'}|\}$ vertices from $V(G_{f_j})$ are necessary.)

(P4) For every $v \in X_3 \setminus D_{c_1,c_2}$,

If $c(v) = i$, then $c_1(v) = \min\{d(v), i + |D_{c_1,c_2} \cap N_{G_f}(v) \cap X_2| + i_1\}$ and $c_2(v) = \min\{d(v), i + |D_{c_1,c_2} \cap N_{G_f}(v) \cap X_1| + i_2\}$ for some non-negative integers $i_1, i_2$ with $i_1 + i_2 = \max\{0, d(v) - i - |D_{c_1,c_2} \cap N_{G_f}(v)|\}$.
(Intuitively, if $v \in X_3 \setminus D_{c_1,c_2}$ needs to be dominated by at least $d(v) - i$ vertices in $G_f$, then at least $\max\{0, d(v) - i - |D_{c_1,c_2} \cap N_{G_f}(v)|\}$ vertices from $(V(G_{f_1}) \setminus w(f_1)) \cup (V(G_{f_2}) \setminus w(f_2))$ are necessary for dominating $v$. If $i_1$ (resp., $i_2$) vertices among those vertices belong to $V(G_{f_2}) \setminus w(f_2)$ (resp., $V(G_{f_1}) \setminus w(f_1)$), then at least $\max\{0, d(v) - i - |D_{c_1,c_2} \cap N_{G_f}(v) \cap X_{j'}| - i_j\}$ vertices from $V(G_{f_j})$ are necessary for $\{j, j'\} = \{1, 2\}$.)

(P5) For every $v \in X_4 \setminus D_{c_1,c_2}$,

$c_1(v) = \min\{d(v), |D_{c_1,c_2} \cap N_{G_f}(v) \cap X_2| + i_1\}$ and $c_2(v) = \min\{d(v), |D_{c_1,c_2} \cap N_{G_f}(v) \cap X_1| + i_2\}$ for some non-negative integers $i_1, i_2$ with $i_1 + i_2 = \max\{0, d(v) - |D_{c_1,c_2} \cap N_{G_f}(v)|\}$. (This case can be treated in a similar way to (P4).)

The following two lemmas show that there exist a coloring $c_1$ of $w(f_1)$ and a coloring $c_2$ of $w(f_2)$ forming $c$ such that $D_{f_1}(c_1) \cup D_{f_2}(c_2)$ satisfies (1)–(3) and $|D_{f_1}(c_1) \cup D_{f_2}(c_2)| = A_f(c)$. Namely, we have $A_f(c) = \min\{A_{f_1}(c_1) + A_{f_2}(c_2) - |D_{c_1,c_2} \cap (X_3 \cup X_4)| \mid c_1, c_2$ forms $c\}$ (the proofs of these lemmas are omitted due to space limitation).

**Lemma 3.** *Let $c \in \{\top, 0, 1, 2, \ldots, d^*\}^{|w(f)|}$ be a coloring of $w(f)$. If a coloring $c_1$ of $w(f_1)$ and a coloring $c_2$ of $w(f_2)$ forms $c$, then $D_{f_1}(c_1) \cup D_{f_2}(c_2)$ satisfies (1)–(3) for $f$.*

**Lemma 4.** *Let $c \in \{\top, 0, 1, 2, \ldots, d^*\}^{|w(f)|}$ be a coloring of $w(f)$. There exist a coloring $c_1$ of $w(f_1)$ and a coloring $c_2$ of $w(f_2)$ forming $c$ such that $|D_{f_1}(c_1) \cup D_{f_2}(c_2)| \leq A_f(c)$.*

Thus, for all colorings $c \in \{\top, 0, 1, 2, \ldots, d^*\}^{|w(f)|}$, we can compute $A_f(c)$ from the information of $f_1$ and $f_2$. By repeating these procedure bottom-up in $T$, we can find a minimum vector dominating set in $G$.

Here, for a fixed $c$, we analyze the time complexity for computing $A_f(c)$. Let $D_c = \{v \in w(f) \mid c(v) = \top\}$, $x_j = |X_j|$ for $j = 1, 2, 3, 4$, $z_3 = |X_3 \setminus D_c|$. Under the assumption that $X_4$ is colored by a fixed coloring $c_4$, the number of pairs of a coloring $c_1$ of $w(f_1)$ and a coloring $c_2$ of $w(f_2)$ forming $c$ is at most $(d^* + 1)^{z_3}(d^* + 1)^{z_4}$ where $z_4$ denotes the number of vertices in $X_4$ not colored by $\top$ in $c_4$, since the number of pairs $(i_1, i_2)$ in (P4) or (P5) is at most $d^* + 1$ for each vertex in $X_3 \setminus D_c$ or each vertex in $X_4$ not colored by $\top$.

Hence, for an edge $f$, the number of pairs forming $c$ is at most $(d^* + 2)^{x_1+x_2} \sum_{z_3=0}^{x_3} \binom{x_3}{z_3}$ $(d^* + 1)^{z_3} \sum_{z_4=0}^{x_4} \binom{x_4}{z_4}(d^* + 1)^{z_4}(d^* + 1)^{z_3}(d^* + 1)^{z_4} = (d^* + 2)^{x_1+x_2}\{(d^* + 1)^2 + 1\}^{x_3+x_4}$ in total. Now we have $x_1 + x_2 + x_3 \leq b$, $x_1 + x_3 + x_4 \leq b$, and $x_2 + x_3 + x_4 \leq b$ (recall that $b$ is the width of $(T', \tau)$). By considering a linear programming problem which maximizes $(x_1 + x_2) \log(d^* + 2) + (x_3 + x_4) \log\{(d^* + 1)^2 + 1\}$ subject to these inequalities, we can observe that $(d^* + 2)^{x_1+x_2}\{(d^* + 1)^2 + 1\}^{x_3+x_4}$ attains the maximum when $x_1 = x_2 = x_4 = b/2$ and $x_3 = 0$. Thus, it takes in total $O((d^* + 2)^b\{(d^* + 1)^2 + 1\}^{b/2})$ time to compute $A_f(c)$ for all colorings $c$ of $w(f)$.

Since $|E(T)| = O(m)$ and the initialization step takes $O((d^* + 2)^2 m)$ time in total, we can obtain $A_{(r_1, r_2)}(c)$ in $O(((d^* + 2)^b \{(d^* + 1)^2 + 1\}^{b/2} m)$ time.

Summarizing the arguments given so far, we have shown Theorem 1. For the total vector domination and the multiple domination, we obtain the following theorems, though the proofs are omitted.

**Theorem 2.** *(i) If a branch decomposition of $G$ with width $b$ is given, a minimum total vector dominating set in $G$ can be found in $O(2^{3b/2}(d^* + 1)^{2b} m)$ time.*
*(ii) If a branch decomposition of $G$ with width $b$ is given, a minimum multiple dominating set in $G$ can be found in $O(2^{3b/2}(d^* + 1)^{2b} m)$ time.* □

## 4 Subexponential Fixed Parameter Algorithm for Planar Graphs

We consider the problem of checking whether a given graph $G$ has a $d$-vector dominating set with cardinality at most $k$. As mentioned in Subsection 1.1, if $G$ is $\rho$-degenerated, then the problem can be solved in $k^{O(\rho k^2)} n^{O(1)}$ time. Since a planar graph is 5-degenerated, it follows that the problem with a planar graph can be solved in $k^{O(k^2)} n^{O(1)}$ time. In this section, we give a subexponential fixed-parameter algorithm, parameterized by $k$, for a planar graph; namely, we will show the following theorem.

**Theorem 3.** *If $G$ is a planar graph, then we can check in $O(n^3 + (\min\{d^*, k\} + 2)^{b^*} \{(\min\{k, d^*\} + 1)^2 + 1\}^{b^*/2} n)$ time whether $G$ has a $d$-vector dominating set with cardinality at most $k$ or not, where $b^* = \min\{12\sqrt{k + z} + 9, 20\sqrt{k} + 17\}$ and $z = |\{v \in V \mid d(v) = 0\}|$.*

This time complexity is roughly $O(n^3 + 2^{O(\sqrt{k} \log k)} n)$, which is subexponential with respect to $k$; this improves the running time of the previous fixed-parameter algorithm.

Let $V_0 = \{v \in V \mid d(v) = 0\}$ and $z = |V_0|$. In [19, Lemma 2.2], it was shown that if a planar graph $G'$ has an ordinary dominating set (i.e., a $(1, 1, \ldots, 1)$-vector dominating set) with cardinality at most $k$, then $bw(G') \leq 12\sqrt{k} + 9$. This bound is based on the *bidimensionality* [14], and was used to design the subexponential fixed-parameter algorithm with respect to $k$ for the ordinary dominating set problem. In the case of our domination problems, however, it is difficult to say that they have the bidimensionality, due to the existence of $V_0$ vertices. Instead, we give a similar bound on the branchwidth not w.r.t $k$ but w.r.t $k + z$ as follows: For any (total, multiple) $d$-vector dominating set $D$ of $G$ ($|D| \leq k$), $D \cup V_0$ is an ordinary dominating set of $G$, and this yields $bw(G) \leq 12\sqrt{k + z} + 9$.

Actually, it is also possible to exclude $z$ from the parameters, though the coefficient of the exponent becomes larger. To this end, we use the notion of $(k, 2)$-center. Recall that a $(k, r)$-center of $G'$ is a set $W$ of vertices of $G'$ with size $k$ such that any vertex in $G'$ is within distance $r$ from a vertex of $W$. For a $(k, r)$-center, a similar bound on the branchwidth is known: if a planar graph $G'$ has a $(k, r)$-center, then $bw(G') \leq 4(2r + 1)\sqrt{k} + 8r + 1$ ([12, Theorem 3.2]). Here, we use this bound. We can assume that for $v \in V_0$, $N_G(v) \not\subseteq V_0$ holds, because $v \in V_0$ satisfying $N_G(v) \subseteq V_0$ is never selected as a member of any optimal solution; it is *irrelevant*, and we can remove it. That is, every vertex in $V_0$ has at least one neighbor from $V \setminus V_0$. Then, for any (total, multiple)

$d$-vector dominating set $D$ of $G$ ($|D| \leq k$), $D$ is a $(k, 2)$-center of $G$. This is because any vertex in $V \setminus V_0$ is adjacent to a vertex in $D$ and any vertex in $V_0$ is adjacent to a vertex in $V \setminus V_0$. Thus, we have $bw(G) \leq 20\sqrt{k} + 17$.

In summary, we have the following lemma.

**Lemma 5.** *Assume that $G$ is a planar graph without irrelevant vertices, i.e., $N_G(v) \not\subseteq V_0$ holds for each $v \in V_0$. Then, if $G$ has a (total, multiple) vector dominating set with cardinality at most $k$, then we have $bw(G) \leq \min\{12\sqrt{k+z} + 9, 20\sqrt{k} + 17\}$.* □

Combining this lemma with the algorithm in Section 3, we can check whether a given graph has a vector dominating set with cardinality at most $k$ according to the following steps 1 and 2:

Step 1: Let $b^* = \min\{12\sqrt{k+z} + 9, 20\sqrt{k} + 17\}$. Check whether the branchwidth of $G$ is at most $b^*$. If so, then go to Step 2, and otherwise halt after outputting 'NO'.

Step 2: Construct a branch decomposition with width at most $b^*$, and apply the dynamic programming algorithm in Section 3 to find a minimum vector dominating set for $G$.

By Lemma 1, Theorem 1, and the fact that any planar graph $G'$ satisfies $|E(G')| = O(|V(G')|)$, it follows that the running time of this procedure is $O(n^3 + (d^* + 2)^{b^*}\{(d^* + 1)^2 + 1\}^{b^*/2} n)$. Hence, in the case of $d^* \leq k$, Theorem 3 has been proved.

The case of $d^* > k$ can be reduced to the case of $d^* \leq k$ by the following standard kernelization method, which proves Theorem 3. Assume that $d^* > k$. Let $V_{\max}(d)$ be the set of vertices $v$ with $d(v) = d^*$. For the feasibility, we need to select each vertex $v \in V_{\max}(d)$ as a member in a vector dominating set. Hence, if $|V_{\max}(d)| > k$, then it turns out that $G$ has no vector dominating set with cardinality at most $k$. Assume that $|V_{\max}(d)| \leq k$. Then, it is not difficult to see that we can reduce an instance $I(G, d, k)$ with $G$, $d$, and $k$ to an instance $I(G', d', k')$ such that $G' = G \setminus V_{\max}(d)$ (i.e., $G'$ is the graph obtained from $G$ by deleting $V_{\max}(d)$), $d'(v) = \max\{0, d(v) - |N_G(v) \cap V_{\max}(d)|\}$ for all vertices $v \in V(G')$, and $k' = \max\{0, k - |V_{\max}(d)|\}$. Based on this observation, we can reduce $I(G, d, k)$ to an instance $I(G'', d'', k'')$ with $\max\{d''(v) \mid v \in V(G'')\} \leq k'' \leq k$ or output 'YES' or 'NO' in the following manner:

(a) After setting $G' := G$, $d' := d$, and $k' := k$, repeat the procedures (b1)–(b3) while $k' < d'^*(= \max\{d'(v) \mid v \in V(G')\})$.

(b1) If $k' < |V_{\max}(d')|$, then halt after outputting 'NO.'
(b2) If $k' \geq |V_{\max}(d')|$ and $V(G') = V_{\max}(d')$, then halt after outputting 'YES.'
(b3) Otherwise after setting $G'' := G' \setminus V_{\max}(d')$, $d''(v) := \max\{0, d'(v) - |N_{G'}(v) \cap V_{\max}(d')|\}$ for each $v \in V(G'')$, and $k'' := \max\{0, k' - |V_{\max}(d')|\}$, redefine $G''$, $d''$, and $k''$ as $G'$, $d'$, and $k'$, respectively.

Next, we consider the total vector domination problem and the multiple domination problem. For these problems, since all vertices $v \in V$ need to be dominated by $d(v)$ vertices, the condition that $d^* \leq k$ is necessary for the feasibility. Similarly, we have the following theorem by Theorem 2.

**Theorem 4.** *Assume that a given graph $G$ is planar, and let $b^* = \min\{12\sqrt{k+z} + 9, 20\sqrt{k} + 17\}$.*

(i) *We can check in* $O(n^3 + 2^{3b^*/2}(\min\{d^*, k\} + 2)^{2b^*} n)$ *time whether G has a total vector dominating set with cardinality at most k or not.*

(ii) *We can check in* $O(n^3 + 2^{3b^*/2}(\min\{d^*, k\} + 2)^{2b^*} n)$ *time whether G has a multiple dominating set with cardinality at most k or not.* □

Before concluding this section, we mention that the above results can be extended to *apex-minor-free graphs*, a superclass of planar graphs. An *apex graph* is a graph with a vertex $v$ such that the removal of $v$ leaves a planar graph. A graph $G$ has a graph $H$ as a *minor* if a graph isomorphic to $H$ can be obtained from $G$ by a sequence of deleting vertices, deleting edges, or contracting edges. A graph class is apex-minor-free if it does not contain any graph which has some fixed apex graph as a minor. For apex-minor-free graphs, the following lemma is known.

**Lemma 6.** *([18, Lemma 2]) Let G be an apex-minor-free graph. If G has a $(k, r)$-center, then the treewidth of G is $O(r\sqrt{k})$.*

From this lemma, the linear relation of treewidth and branchwidth, and the $2^{O(bw(G))}n^2$-time algorithm for for computing a branch decomposition with width $O(bw(G))$ (mentioned after Theorem 1), we obtain the following corollary.

**Corollary 2.** *We can check in $2^{O(\sqrt{k}\log k)} n^{O(1)}$ time whether an apex-minor-free graph G has a (total, multiple) vector dominating set with cardinality at most k or not.*

# References

1. Alon, N., Moshkovitz, D., Safra, S.: Algorithmic construction of sets for *k*-restrictions. ACM Transactions on Algorithms TALG 2, 153–177 (2006)
2. Amini, O., Fomin, F.V., Saurabh, S.: Implicit branching and parameterized partial cover problems. Journal of Computer and System Sciences 77, 1159–1171 (2011)
3. Betzler, N., Bredereck, R., Niedermeier, R., Uhlmann, J.: On bounded-degree vertex deletion parameterized by treewidth. Discrete Applied Mathematics 160, 53–60 (2012)
4. Bodlaender, H.L., Thilikos, D.M.: Constructive linear time algorithms for branchwidth, in Automata, Languages and Programming. In: Degano, P., Gorrieri, R., Marchetti-Spaccamela, A. (eds.) ICALP 1997. LNCS, vol. 1256, pp. 627–637. Springer, Heidelberg (1997)
5. Cai, L., Juedes, D.: On the existence of subexponential parameterized algorithms. Journal of Computer and System Sciences 67, 789–807 (2003)
6. Chapelle, M.: Parameterized complexity of generalized domination problems on bounded tree-width graphs. arXiv preprint arXiv:1004.2642 (2010)
7. Cicalese, F., Cordasco, G., Gargano, L., Milanič, M., Vaccaro, U.: Latency-bounded target set selection in social networks. In: Bonizzoni, P., Brattka, V., Löwe, B. (eds.) CiE 2013. LNCS, vol. 7921, pp. 65–77. Springer, Heidelberg (2013)
8. Cicalese, F., Milanič, M., Vaccaro, U.: Hardness, approximability, and exact algorithms for vector domination and total vector domination in graphs. In: Owe, O., Steffen, M., Telle, J.A. (eds.) FCT 2011. LNCS, vol. 6914, pp. 288–297. Springer, Heidelberg (2011)
9. Cicalese, F., Milanic, M., Vaccaro, U.: On the approximability and exact algorithms for vector domination and related problems in graphs. Discrete Applied Mathematics 161, 750–767 (2013)
10. Corneil, D.G., Rotics, U.: On the relationship between clique-width and treewidth. SIAM Journal on Computing 34, 825–847 (2005)

11. Courcelle, B.: The monadic second-order logic of graphs. I. recognizable sets of finite graphs. Information and Computation 85, 12–75 (1990)
12. Demaine, E.D., Fomin, F.V., Hajiaghayi, M., Thilikos, D.M.: Fixed-parameter algorithms for $(k, r)$-center in planar graphs and map graphs. ACM Transactions on Algorithms (TALG) 1, 33–47 (2005)
13. Demaine, E.D., Fomin, F.V., Hajiaghayi, M., Thilikos, D.M.: Subexponential parameterized algorithms on bounded-genus graphs and H-minor-free graphs. Journal of the ACM (JACM) 52, 866–893 (2005)
14. Demaine, E.D., Hajiaghayi, M.: The bidimensionality theory and its algorithmic applications. The Computer Journal 51, 292–302 (2008)
15. Dobson, G.: Worst-case analysis of greedy heuristics for integer programming with nonnegative data. Mathematics of Operations Research 7, 515–531 (1982)
16. Dorn, F.: Dynamic programming and fast matrix multiplication. In: Azar, Y., Erlebach, T. (eds.) ESA 2006. LNCS, vol. 4168, pp. 280–291. Springer, Heidelberg (2006)
17. Downey, R.G., Fellows, M.R.: Fixed-parameter tractability and completeness, Cornell University, Mathematical Sciences Institute (1992)
18. Fomin, F.V., Lokshtanov, D., Raman, V., Saurabh, S.: Subexponential algorithms for partial cover problems. Information Processing Letters 111, 814–818 (2011)
19. Fomin, F.V., Thilikos, D.M.: Dominating sets in planar graphs: branch-width and exponential speed-up. SIAM Journal on Computing 36, 281–309 (2006)
20. Gu, Q.-P., Tamaki, H.: Optimal branch-decomposition of planar graphs in $O(n^3)$ time. ACM Transactions on Algorithms (TALG) 4, 30 (2008)
21. Harant, J., Pruchnewski, A., Voigt, M.: On dominating sets and independent sets of graphs. Combinatorics, Probability and Computing 8, 547–553 (1999)
22. Harary, F., Haynes, T.W.: Double domination in graphs. Ars Combinatoria 55, 201–214 (2000)
23. Haynes, T.W., Hedetniemi, S.T., Slater, P.J.: Domination in graphs: advanced topics, vol. 40. Marcel Dekker (1998)
24. Haynes, T.W., Hedetniemi, S.T., Slater, P.J.: Fundamentals of domination in graphs. Marcel Dekker (1998)
25. Impagliazzo, R., Paturi, R., Zane, F.: Which problems have strongly exponential complexity? Journal of Computer and System Sciences 63, 512–530 (2001)
26. Lund, C., Yannakakis, M.: On the hardness of approximating minimization problems. Journal of the ACM (JACM) 41, 960–981 (1994)
27. Raman, V., Saurabh, S., Srihari, S.: Parameterized algorithms for generalized domination. In: Yang, B., Du, D.-Z., Wang, C.A. (eds.) COCOA 2008. LNCS, vol. 5165, pp. 116–126. Springer, Heidelberg (2008)
28. Robertson, N., Seymour, P.D.: Graph minors. X. obstructions to tree-decomposition. Journal of Combinatorial Theory, Series B 52, 153–190 (1991)
29. Robertson, N., Seymour, P.D.: Graph minors. XIII. the disjoint paths problem, Journal of Combinatorial Theory, Series B 63, 65–110 (1995)
30. Seymour, P.D., Thomas, R.: Call routing and the ratcatcher. Combinatorica 14, 217–241 (1994)

# Computing the Degeneracy of Large Graphs*

Martín Farach-Colton and Meng-Tsung Tsai

Rutgers University, New Brunswick NJ 08901, USA
{farach,mtsung.tsai}@cs.rutgers.edu

**Abstract.** Any ordering of the nodes of an $n$-node, $m$-edge simple undirected graph $G$ defines an acyclic orientation of the edges in which each edge is oriented from the earlier node in the ordering to the later. The *degeneracy* on an ordering is the maximum outdegree it induces, and the *degeneracy* of a graph is smallest degeneracy of any node ordering. Small-degeneracy orderings have many applications.

We give an algorithm for generating an ordering whose degeneracy approximates the minimum possible, that is, it approximates the degeneracy of the graph. Although the optimal ordering itself can be computed in $\mathcal{O}(m)$ time and $\mathcal{O}(m)$ space, such algorithms are infeasible for large graphs. Our approximation algorithm is semi-streaming: it uses less space, can achieve a constant approximation ratio, and accesses the graph in logarithmic read-only passes.

## 1 Introduction

Any ordering of the nodes of an $n$-node, $m$-edge simple undirected graph $G$ defines an acyclic orientation of the edges in which each edge is oriented from the earlier node in the ordering to the later. The *degeneracy* of an ordering is the maximum outdegree it induces. The *degeneracy*, $d(G)$, of $G$ is the smallest degeneracy of any ordering[1], and an ordering whose degeneracy is $d(G)$ is called a *degenerate ordering*. An ordering is $d$-*degenerate* if it has degeneracy at most $d$.

Degenerate orderings have many uses. Given a degenerate ordering, one can: decompose a graph into at most twice to the minimum number of disjoint forests [2,4]; decompose a graph into at most six times to the minimum number of disjoint planar graphs [4,11]; speed up the counting of the number of short paths or cycles [2], for example, counting the exact number of 3-cycles in $\mathcal{O}(md(G))$ time; find a component of density at least half the maximum density of any subgraph, i.e. a 1/2-approximation [7]; identify a dominating set of cardinality at most $\mathcal{O}(d^2(G))$ times the cardinality of minimum dominating set [19] and some variations of dominating set [12], e.g. $k$-dominating set; etc. Although most of

---

* Work supported by NSF Grants IIS-1247750 and CCF-1114930.
[1] The degeneracy of a graph was originally defined to be the maximum of minimum degree among all subgraphs [2,4,5,7,14,22,26]. The definition here is a slight modification of the *coloring number* [4,5,14] of a graph, a dual definition of degeneracy. The coloring number of a graph was shown to be one larger than the degeneracy [4,5,14], and our definition yields the same value as the original definition of degeneracy.

these problems can be solved exactly in polynomial time and space $\mathcal{O}(n)$ to $\mathcal{O}(m)$ [7,15,16,18], the approximation algorithms based on degenerate orderings are faster, use less space or yield better approximation factors for large graphs. For example, such orderings yield a better approximation algorithm for decomposing a graph into minimum number of planar subgraphs than other algorithms using $O(n)$ space [17,21]. Although all of the results listed originally relied on (optimally) degenerate orderings, we show that orderings that are nearly degenerate, that is, orderings whose degeneracy approximates rather than matching the graph degeneracy, also yield good approximation algorithms.

Known algorithms to compute a low-degeneracy ordering do not scale to graphs that are larger than memory. Several models of computation have been proposed for computing on large graphs, such as the restrictive *semi-streaming* [13, 23–25, 27] model that allows only $\mathcal{O}(n\operatorname{polylog} n)$ space and sequential read-only passes through the graph, the *W-stream* [24,27] model which is similar but also allows the algorithm $\mathcal{O}(m)$-size read-write space on disk, the *Stream-Sort* [24,25,27] model, in which sorting the graph needs only one pass, and the *Stream-with-annotations* [6] model that as *semi-streaming* but assume a powerful helper can be queried for a small number of annotations. When read-write space is restricted to $\mathcal{O}(n\operatorname{polylog} n)$ space and accessing the graph is restricted to a constant or logarithmic number of sequential passes on entire the graph, some graph problems, e.g. connectivity or minimum spanning tree, have known optimal solutions. Other graph problems, e.g. counting the number of 3-cycles and maximum matching, can be approximated [1,3]. Some fundamental problems, such as breath-first search, depth-first search, topological sorting, and directed connectivity, are believed to be difficult [24,25].

All known algorithms for computing degenerate orders have a structure that is similar to topological sort. Therefore, we seek to approximate the graph degeneracy. We use the *semi-streaming* model; that is, $\mathcal{O}(n\operatorname{polylog} n)$ space and constant/logarithm sequentially read-only passes on the entire graph are allowed, which is the most restricted model among the mentioned three. Our goal is to minimize the number of passes on the disk while finding a node ordering of low degeneracy, in particular one whose degeneracy is a good approximation of the degeneracy of the graph.

A simple semi-streaming algorithm can find a $\sqrt{n}d(G)$-degenerate ordering of nodes in one pass, by sorting the node by (undirected) degree [26]. We improve the approximation factor to a constant at the cost of a logarithmic number of passes while maintaining $n$ working space. Our algorithm can be made to use space that is less than $n$, and as our space usage decreases, our approximation factor degrades. Our algorithm outputs a sequence of some subset of nodes in each pass. At the end of all passes, the concatenation of all sequences is the desired ordering.

**Theorem 1.** *Given a simple undirected $n$-node $m$-edge graph $G$, an $\alpha d(G)$-degenerate ordering of nodes of $G$, i.e. an $\alpha$-approximation, can be computed in $\mathcal{O}((m+n)\mathcal{P})$ time, using $s(n)$ space and $\mathcal{P} = \mathcal{O}(\log_{1+\varepsilon/2} n/s(n) + \log_{\alpha/2} s(n))$*

sequential passes on the entire graph, where $\alpha = (2+\varepsilon)n/s(n)$ for any $\varepsilon > 0$ and $1 \leq s(n) \leq n$.

Note that $\alpha$ is inversely related to space $s(n)$. For example, if $s(n) = n/10$, then $\alpha = 20(1+\varepsilon)$, and if $s(n) = n/\log n$, let $\alpha = 2\log n(1+\varepsilon)$, for any $\varepsilon > 0$. In addition, it is possible to perform fewer than $\log n$ passes: our algorithm requires $\mathcal{O}(\log\log n/\varepsilon + \log n/\log\log n)$ passes, for small $\varepsilon > 0$, when $s(n) = \mathcal{O}(n/\operatorname{polylog} n)$.

**Organizations.** In Section 2, we revisit some properties of graph degeneracy and give a sketch of our algorithms. We propose the space-efficient approximation algorithms in Section 3 and analyze their complexities. Last, in Section 4, we discuss how the found ordering be applied to applications.

## 2 Preliminaries

We begin by reviewing some known results about degeneracy. A greedy algorithm finds $d(G)$ and a corresponding ordering of nodes in $\mathcal{O}(m)$ time using $\mathcal{O}(m)$ space [5, 22]. The greedy algorithm is based on the following two observations: $d(G)$ is at least the minimum degree, $\delta(G)$, of $G$, because the first node in any ordering has outdegree equal to its (undirected) degree; $d(G) \geq d(H)$ for any subgraph $H \subseteq G$, since one can apply the optimal orientation for $G$ to subgraph $H$. Hence, the algorithm can greedily pick node $v$ of minimum degree as the first node in the ordering and reduce the graph $G$ to a subgraph $G \setminus v$. This greedy step will not increase the maximum out-degree in the resulting ordering because

$$\max\{d(v), d(G\setminus v)\} \leq d(G).$$

This greedy algorithm needs to update the degree of nodes next to $v$ in order to find the node of minimum degree in $G \setminus v$, which requires a full graph scan in the semi-streaming model, or $\mathcal{O}(n)$ passes in total. To reduce the number of passes on the graph, one can find a subset of vertices $W$ whose degrees are within the range $[\delta(G), c\delta(G)]$ for any $c \geq 1$. Removing $W$ in a round yields a greedy algorithm that approximates $d(G)$ by a factor of $c$, since

$$\hat{d}(G) \leq \max\{c\delta(G), d(G\setminus W)\} \leq cd(G).$$

Although this algorithm finds sets of nodes, rather than single nodes, in each pass, the number of passes remains $\Theta(n)$ in the worst case, as illustrated in Figure 1.

To further reduce the number of passes to log, let the *density* of a $G$ be $m/n$, and let $d^*(G)$ be the maximum density among all subgraphs of $G$. It is known that $d^*(G) \leq d(G) \leq 2d^*(G)$ [14]. The first inequality is true by the pigeon-hole principle: if $m$ edges are assigned as out-edges to $n$ nodes, then some node will get at least $m/n$ edges, and this is true of all subgraphs as well. The second inequality is true because every $n$-node $m$-edge graph $G$ has a node of degree

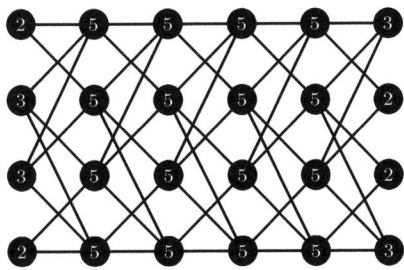

**Fig. 1.** In the above graph $G$, if one removes nodes of degree within $[\delta(G), 2\delta(G)]$ in each round, then the removal of all nodes requires three rounds. Each round removes the extreme column at either end, leaving a structure with "internal" nodes of degree 5 and end columns of degree 2/3/3/2. This graph can be extended to an arbitrary length by concatenating multiple instances of $G$. This longer graph shares the property that only the first and last column are removed in each round. Therefore, greedily removing nodes of degree with $[\delta(G), 2\delta(G)]$ requires $\Theta(n)$ rounds.

no more than $2m/n$; hence, one can always remove a vertex of degree at most $2d^*(G)$, or, equivalently, $d(G) \leq 2d^*(G)$.

If one iteratively removes a subset of vertices whose degrees are no more than $cm/n$, where $c = 2 + \varepsilon$ for any $\varepsilon > 0$, then the approximated degeneracy $\hat{d}(G)$ is at most $cd^*(G)$, i.e., we achieve a $c$-approximation of both $d^*(G)$ and $d(G)$. After the removal of vertices, the number of surviving vertices is at most $2n/c$ because the sum of degree is at most $2m$, which means that there are at most $2n/c$ nodes of degree more than $cm/n$. Thus, the number of passes is logarithmic, specifically, $\mathcal{O}(\log n/(\log c - 1))$, yielding a tradeoff between the approximation factor and the number of passes.

The space can be further reduced based on the ideas of counting sketches used in streaming algorithms [8–10]. That is, we use $s(n) < n$ space to count the degree of each node. Since $s(n)$ is not sufficient to count the degree of $n$ nodes individually, a counter is possibly shared among some nodes. Indeed, the counter is the sum of degrees of nodes assigned to this counter. Therefore, the counter is an overestimate of degree. If the overestimate is bounded, then we get a bounded approximation. We explore these ideas more fully below.

## 3 Algorithms

We present two algorithms: one that uses $n$ space and achieves a constant approximation factor and one that uses less than $n$ space and yields a smooth tradeoff between the space used and the approximation achieved. These algorithms share some high-level structure, which we present first.

The proposed algorithms require multiple passes on graph $G(V, E)$. Let $V_1$ be $V$. In the $i^{\text{th}}$ pass, the algorithms identify a subset $V_{i+1} \subseteq V_i$. Let $n_i = |V_i|$ and $m_i$ be the number of edges in the subgraph of $G$ induced by $V_i$. Step $i$ outputs the nodes in $V_i \setminus V_{i+1}$ in any arbitrary order. The algorithms terminate when there are no more nodes to output, that is, when $n_{i+1} = 0$. The desired ordering of nodes is the concatenation of the output of all phases.

So far, this algorithm follows the same outline as the other greedy algorithms. The difference will be in which nodes we output in each phase. A node becomes a *candidate* node in phase $i$ if $d_i(u) \le \alpha m_i/n_i$ for some $\alpha = 2 + \varepsilon$, $\varepsilon > 0$, where $d_i(u)$ is the degree of $u$ in the subgraph of $G$ induced by $V_i$. Once a node is a candidate, it remains a candidate. Both of our algorithms output only candidate nodes, and they both output candidates as soon as they detect that a node becomes a candidate. The difference between our algorithms is that if we use linear space, we can detect that a node is a candidate as soon as it becomes one, so that we output nodes more aggressively. We achieve a smaller number of phases and a better approximation ratio. If we have sublinear space, it will be more difficult to detect candidacy.

In the first algorithm, we will go further and guarantee to output all nodes $u$ with $d_i(u) \le \alpha m_i/n_i$, that is, we will detect candidate nodes as soon as they become candidates. In that case,

$$n_{i+1} \le (2m_i)/(\alpha m_i/n_i) = 2n_i/\alpha.$$

This implies that the algorithm terminates after a logarithmic number of phases. When we use sublinear space, we will not be able to guarantee that all low-degree nodes are output, so we will need to be more careful in order to guarantee a logarithmic number of phases.

**Using Space of Size** $n$**.** We start with an easy version such that the memory has size of $n^2$, which is used for $n$ counters, $d_i(u)$ for $u \in V$, where $d_i(u)$ is reused as $d_{i+1}(u)$ for all $i$. Each $d_i(u)$ keeps the information whether node $u$ is output before the $i^{th}$ pass, in which case it is set to $-1$. If the node has not be output yet, $d_i(u)$ is used to keep track of the degree of $u$ in the subgraph of $G$ induced by $V_i$. The value of $d_i(u)$ for all $u$ can be updated in one pass through the graph, as follows. If $d_i(u)$ is negative, do nothing; otherwise, reset $d_i(u)$ to 0. During the scan, process the edge, one by one. When processing edge $(u,v)$, if $d_i(u) \ge 0$ and $d_i(v) \ge 0$, then it means edge $(u,v)$ is still the subgraph induced by $V_i$; increment $d_i(u)$ and $d_i(v)$ by one.

Having $d_i(u)$ for $u \in V_i$, check each $u$ to see if $d_i(u) \ge \alpha m_i/n_i$. Output any such node $u$ as it is identified, and set $d_i(u) = -1$.

**Lemma 1.** *An $\alpha d(G)$-degenerate ordering of nodes, for $\alpha = 2+\varepsilon, \varepsilon > 0$, can be generated by a semi-streaming algorithm using $n$ space, $\mathcal{P} = \mathcal{O}(\log n/\log(\alpha/2))$ passes, and $\mathcal{O}((m+n)\mathcal{P})$ time.*

*Proof.* Consider an ordering produced by the algorithm. If node $u$ is output in the $i^{th}$ pass, then $u$ belongs to $V_i \setminus V_{i+1}$ and has outdegree at most $\alpha m_i/n_i$. The $n$ counters allows the algorithm to output any such candidate node during

---

[2] Any algorithm will require a constant amount of memory, so when we report space usage of $n$, we mean in addition to the $O(1)$ overhead for running the algorithm.

the phase that it becomes a candidate. Therefore, $n_{i+1} \leq 2n_i/\alpha$. The number of passes is therefore $\mathcal{O}(\log n / \log(\alpha/2))$.

As for the approximation factor, since each output node $u$ in the $i^{\text{th}}$ pass has degree $d_i(u)$, then the eventual outdegree of $u$ is at most

$$d_i(u) \leq \alpha m_i/n_i \leq \alpha d^*(G) \leq \alpha d(G),$$

yielding an $\alpha d(G)$-degenerate ordering of nodes. □

**Using Space of Size $s(n) \leq n$.** The small-space algorithm proceeds in two sections. At first, there is not enough space to count the degree of every one. As some nodes are output, the number of remaining nodes $n_i$ drops. When $n_i$ drops to no more than $s(n)/2$, then we can use a hash table to count the degrees and proceed as that using space of size $n$. Therefore, we focus on the first part of the algorithm.

If $n_i > s(n)/2$, then the $s(n)$ space is not sufficient to keep track of which nodes are still active and to keep track of $d_i(u)$ individually for each $u \in V_i$. Therefore, the algorithm switches over when $n_i \leq s(n)/2$. To count degrees with fewer counter, as with counting sketches [8–10], we map $n_i$ nodes to $s(n)$ space by a hash function $h$, and $\hat{d}_i(h(u))$ is used to count the degree of node $u$, although other nodes $v$ may share the same counter if $h(v) = h(u)$. Formally,

$$\hat{d}_i(h(u)) = \sum_{v \in V_i, h(v)=h(u)} d_i(v).$$

Note that by $d_i(u)$ and $\hat{d}_i(h(u))$ we denote the real degree of node $u$ in the subgraph induced by $V_i$ and its estimate, respectively. Clearly, $\hat{d}_i(h(u)) \geq d_i(u)$. We use a simple, deterministic hash function,

$$h(u) = u \bmod s(n).$$

The small-space procedure mimics the $n$-counter algorithm. First, recall that when a counter is negative, it means that a node has been output. Since several nodes can share the same counter, if we output one node $u$ that maps to a counter, we will output all nodes that map to that counter, and we will set that counter $\hat{d}(h(u)) = -1$, as before.

**Lemma 2.** *An $\alpha d(G)$-degenerate ordering of nodes, for $\alpha = (2+\varepsilon)n/s(n), \varepsilon > 0$, can be generated by a semi-streaming algorithm using $s(n)$ space, $\mathcal{P} = \mathcal{O}(\log n / \log(\alpha s(n)/(2n)))$ passes, and $\mathcal{O}((m + s(n))\mathcal{P})$ time.*

*Proof.* Recall that $n_1(=n), n_2, \ldots, n_{\mathcal{P}+1}$ is the number of nodes remaining after each pass. In the $i^{\text{th}}$ pass, the expectation of estimators $\hat{d}_i(x)$ for $x \in \{1, 2, \ldots, s(n)\}$ is $\mathrm{E}[\hat{d}_i(x)] = 2m_i/s(n)$. By Markov's inequality, we have

$$p_1 \equiv \Pr\left[\hat{d}_1(x) > \frac{\alpha s(n)}{2n_1}\mathrm{E}[\hat{d}_1(x)]\right] < \frac{2n_1}{\alpha s(n)} < 1.$$

The number of estimators that have value more than $(\alpha s(n))/(2n_1)\mathrm{E}[\hat{d}_1(x)]$ is at most

$$s_2 < s(n)\frac{2n_1}{\alpha s(n)} < s(n).$$

Then $V_2$, the identified subset of $V_1$, has cardinality at most $n_2 = s_2\lceil n/s(n)\rceil$ or precisely $s_2 n/s(n)$, assuming that $n$ is a multiple of $s(n)$. The assumption can be handled by adding some isolated nodes to increase $n$ to a multiple of $s(n)$ and ignoring the nodes of index more than the original $n$ as they are output. Similarly, due to $n_2 < n_1 = n$,

$$p_2 \equiv \Pr\left[\hat{d}_2(x) > \frac{\alpha s(n)}{2n_2}\mathrm{E}[\hat{d}_2(x)]\right] < \frac{2n_2}{\alpha s(n)} < 1.$$

The number of estimators that have value more than $(\alpha s(n))/(2n_2)\mathrm{E}[\hat{d}_2(x)]$ is

$$s_3 = s(n)p_2 < s(n)\frac{2}{\alpha s(n)}n_2 = s(n)\frac{2}{\alpha s(n)}s_2\frac{n}{s(n)} < s_2.$$

Therefore, $\mathcal{P} = \mathcal{O}(\log n/\log(\alpha s(n)/(2n)))$ because the above derivation is

$$s_{i+1} < \frac{2n}{\alpha s(n)}s_i < s_i$$

in general. The outdegree of nodes output in the $i^{\text{th}}$ pass is bounded by

$$\hat{d}_i(h(u)) \leq \frac{\alpha s(n)}{2n_i}\mathrm{E}[\hat{d}_i(x)] \leq \frac{\alpha s(n)}{2n_i}\frac{2m_i}{s(n)} \leq \alpha d^*(G) \leq \alpha d(G). \qquad \square$$

We compare the proposed algorithms in Table 1. The approximation factor $\alpha$ and the required number of passes $\mathcal{P}$ of the proposed sublinear-space algorithm are those of the linear-space algorithm multiplied by a factor of $n/s(n)$, respectively. There is therefore a tradeoff between $\alpha$ and $\mathcal{P}$ on the one hand and $s(n)$ on the other. We can combine the two algorithms as follows. When the space is not sufficient to accommodate all nodes, we use the sublinear-space algorithm. Once the remaining nodes are at most $s(n)/2$, we use the linear-space algorithm, which converges faster, changing the base of the log from $\alpha s(n)/(2n)$ (a constant close to 1) to $\alpha/2$ (a constant if $s(n) = \Theta(n)$ or logarithmic if $s(n) = \Theta(n)(n/\operatorname{polylog}))$. Hence, the required number of passes is bounded by

$$\mathcal{O}(\log_{\alpha s(n)/(2n)} n/s(n) + \log_{\alpha/2} s(n))$$
$$= \mathcal{O}(\log(n/s(n))/\log(1+\varepsilon/2) + \log s(n)/\log((1+\varepsilon/2)n/s(n))).$$

It is remarkable that, for small $\varepsilon$, only $\mathcal{O}(\log\log n/\varepsilon + \log n/\log\log n)$ passes are needed to achieve an approximation factor of $\mathcal{O}(\log^k n)$ when $s(n) = n/\log^k n$.

## 4 Applications

Here we show how to use a small degeneracy ordering to approximate some problems in streaming models.

**Table 1.** Comparison of proposed algorithms. Approximations are for small $\varepsilon$.

| space | approximation factor ($\alpha$) | # of passes ($\mathcal{P}$) for any $\varepsilon$ | $\mathcal{P}$ for small $\varepsilon$ |
|---|---|---|---|
| $n$ | $2+\varepsilon$ | $\log n / \log((2+\varepsilon)/2)$ | $\approx 2\log n/\varepsilon$ |
| $s(n)$ | $(2+\varepsilon)n/s(n)$ | $\log n/(\log(\alpha s(n)/(2n)))$ | $\approx 2\log n/\varepsilon$ |

***Forest/planar subgraph decomposition.*** Let $a(G)$, the *arboricity* of $G$, be the minimum number of disjoint forests into which graph $G$ can be decomposed. Let $\theta(G)$, the *thickness* of $G$, be the minimum of disjoint planar graph into which graph $G$ can be decomposed. The relationships among degeneracy, arboricity and thickness are illustrated in Figure 2.

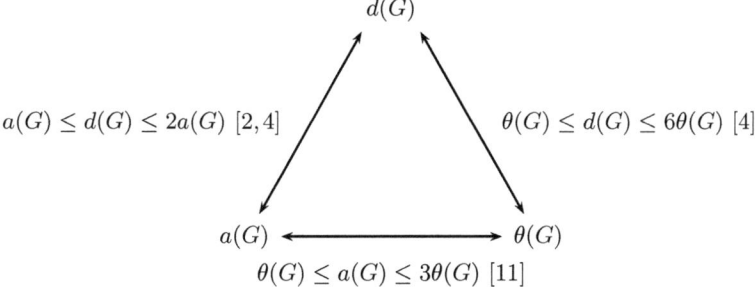

**Fig. 2.** The relationships among $d(G), a(G)$ and $\theta(G)$

One can use a $(2+\varepsilon)d(G)$-degenerate ordering to approximate the optimal decomposition of a graph into disjoint forests/planar subgraphs, as follows. For each $u$, assign a distinct color to each edge connecting a neighbor that appears later in the ordering. Since the ordering is $(2+\varepsilon)d(G)$-degenerate, $(4+\varepsilon)d(G)$ are required, i.e., a 4-approximation. Monochrome edges never form a cycle, that is, they form a forest. Since a forest is also a planar graph, forest-decomposition can also be applied to planar-decomposition, yielding a $(12+\varepsilon)$-approximation.

Given a small-degeneracy ordering, subgraph decomposition can therefore be approximated in the Sort-stream model, using the space $s(n)$, as follows. Associate with each edge the rank of its incident nodes in the ordering. This can be done in $n/s(n)$ sequential writable passes. Then, sort the stream by the smaller rank of incident nodes in one pass, as is assumed to be allowable in the Sort-stream model. Perform one more sequential scan to assign a distinct color to each edge that match in the smaller rank of incident nodes. Last, sort the stream by the edge color.

***Sever-Client Load Balancing.*** For this application, we need to generalize the result on ordinary graph $G$ to a simple $r$-hypergraph $\mathcal{G}$. In a hyergraph, each hyperedge is a subset of nodes. An $r$-hypergraph is a hypergraph where each edge has at least two nodes (no self loops) and at most $r$. In this case, $d^*(G) \leq d(G) \leq rd^*(G)$ because, on one hand, $n$ nodes are assigned by $m$

edges and there exists one node has been assigned no fewer than $m/n$ edges by the Pigeon-hole principle; on the other hand, there exists one node with degree no more than $rm/n$ in each $r$-hypergraph, which, by the optimal greedy construction of degeneracy, shows the right-hand inequality.

**Corollary 1.** *Given a simple undirected n-node m-edge r-hypergraph $\mathcal{G}$, using space $s(n)$, where $1 \leq s(n) \leq n$, an $\alpha d(\mathcal{G})$-degenerate ordering of nodes of $\mathcal{G}$, i.e. an $\alpha$-approximation, can be computed in $\mathcal{O}((m+n)\mathcal{P})$ time, using $s(n)$ space and $\mathcal{P} = \mathcal{O}(\log_{1+\varepsilon/r}(n/s(n)) + \log_{\alpha/r} s(n))$ sequential passes on the entire graph, where $\alpha = (r+\varepsilon)n/s(n)$ for any $\varepsilon > 0$.*

Consider the server-client load balancing problem [20]. A sever-client model is a bipartite graph such that all edges are incident on one server node and one client node. The task is to assign each client to a server while minimizing the number clients assigned to a single server.

Suppose that the degree of each client is at most $r$, and consider the following induced $r$-hypergraph: the nodes are the server nodes and each hyperedge in the induced hypergraph is the set of servers adjacent to a client node. Then an $(r+\varepsilon)d(\mathcal{G})$-degenerate ordering of the hypergraph is a $(r+\varepsilon)$-approximation of the server-client load balancing problem, where the approximated assignment of each client to a server is then obtained by the degenerate ordering.

An $(r+\varepsilon)d(G)$-degenerate ordering induces an assignment of each client to a server. As mentioned above, the induced assignment minimizes the heaviest server load to within a factor of $(r+\varepsilon)$ of optimal. In addition, based on the algorithm introduced in [20], a good approximation algorithm can be used to speedup their exact computation. In [20], they prove that, given a sub-optimal assignment, there exists an alternating path connecting the server of maximum load to another server. The better the initial solution, the fewer alternating-path rewrites that are needed, thus improving the time complexity of finding an optimal solution.

***Finding an Approximated Dominating Set.*** In [19], it was shown how to identify a dominating set based on a $d$-degenerate ordering, for any $d$. Let the neighbor nodes appear earlier (later) in the ordering be earlier (later) neighbors. The algorithm initializes an empty dominating set $\hat{D}$ and adds some nodes to this set greedily as follows. Traverse the nodes in the ordering from the first to the last. When processing node $u$, if $u$ is dominated by $\hat{D}$, do nothing; otherwise, some nodes need to be added to $\hat{D}$. The choice of added nodes is made depending on whether $u$ has a later neighbor. If so, add all later neighbors to $\hat{D}$; otherwise, add $u$ to $\hat{D}$. The set $\hat{D}$ so constructed is a dominating set. The greedy process requires the space $\mathcal{O}(n)$ and one sequential pass on the adjacency list, each list sorted by lower rank, can be handled as in the graph decomposition.

In [19], it is shown that $|\hat{D}| \leq \mathcal{O}(d^2)|D|$, where $D$ is the smallest dominating set. More applications that apply small degeneracy ordering to some variations of dominating set, e.g. $k$-dominating set, are introduced in [12].

# References

1. Ahn, K.J., Guha, S.: Linear programming in the semi-streaming model with application to the maximum matching problem. In: Aceto, L., Henzinger, M., Sgall, J. (eds.) ICALP 2011, Part II. LNCS, vol. 6756, pp. 526–538. Springer, Heidelberg (2011)
2. Alon, N., Yuster, R., Zwick, U.: Finding and counting given length cycles. Algorithmica 17(3), 209–223 (1997)
3. Bar-Yossef, Z., Kumar, R., Sivakumar, D.: Reductions in streaming algorithms, with an application to counting triangles in graphs. In: Proceedings of the Thirteenth Annual ACM-SIAM Symposium on Discrete Algorithms, SODA 2002, pp. 623–632. Society for Industrial and Applied Mathematics (2002)
4. Bollobás, B.: Extremal graph theory. Academic Press (1978)
5. Bollobás, B.: The evolution of sparse graphs. In: Graph Theory and Combinatorics, Proc. Cambridge Combinatorial Conf., pp. 35–57. Academic Press (1984)
6. Chakrabarti, A., Cormode, G., McGregor, A.: Annotations in data streams. In: Albers, S., Marchetti-Spaccamela, A., Matias, Y., Nikoletseas, S., Thomas, W. (eds.) ICALP 2009, Part I. LNCS, vol. 5555, pp. 222–234. Springer, Heidelberg (2009)
7. Charikar, M.: Greedy approximation algorithms for finding dense components in a graph. In: Jansen, K., Khuller, S. (eds.) APPROX 2000. LNCS, vol. 1913, pp. 84–95. Springer, Heidelberg (2000)
8. Charikar, M., Chen, K., Farach-Colton, M.: Finding frequent items in data streams. In: Widmayer, P., Triguero, F., Morales, R., Hennessy, M., Eidenbenz, S., Conejo, R. (eds.) ICALP 2002. LNCS, vol. 2380, pp. 693–703. Springer, Heidelberg (2002)
9. Cormode, G., Hadjieleftheriou, M.: Finding frequent items in data streams. Proc. VLDB Endow. 1(2), 1530–1541 (2008)
10. Cormode, G., Muthukrishnan, S.: An improved data stream summary: the count-min sketch and its applications. J. Algorithms 55(1), 58–75 (2005)
11. Dean, A.M., Hutchinson, J.P., Scheinerman, E.R.: On the thickness and arboricity of a graph. Journal of Combinatorial Theory, Series B 52(1), 147–151 (1991)
12. Dvořák, Z.: Constant-factor approximation of the domination number in sparse graphs. European Journal of Combinatorics 34(5), 833–840 (2013)
13. Feigenbaum, J., Kannan, S., McGregor, A., Suri, S., Zhang, J.: On graph problems in a semi-streaming model. In: Díaz, J., Karhumäki, J., Lepistö, A., Sannella, D. (eds.) ICALP 2004. LNCS, vol. 3142, pp. 531–543. Springer, Heidelberg (2004)
14. Frank, A., Gyarfas, A.: How to orient the edges of a graph. In: Combinatorics Volume I, Proc. of the Fifth Hungarian Colloquium on Combinatorics, vol. I, pp. 353–364 (1976)
15. Gabow, H., Westermann, H.: Forests, frames, and games: algorithms for matroid sums and applications. In: Proceedings of the Twentieth Annual ACM Symposium on Theory of Computing, STOC 1988 pp. 407–421. ACM (1988)
16. Goldberg, A.V.: Finding a maximum density subgraph. Tech. rep. (1984)
17. Kawano, S., Yamazaki, K.: Worst case analysis of a greedy algorithm for graph thickness. Information Processing Letters 85(6), 333–337 (2003)
18. Kowalik, L.: Approximation scheme for lowest outdegree orientation and graph density measures. In: Asano, T. (ed.) ISAAC 2006. LNCS, vol. 4288, pp. 557–566. Springer, Heidelberg (2006)
19. Lenzen, C., Wattenhofer, R.: Minimum dominating set approximation in graphs of bounded arboricity. In: Lynch, N.A., Shvartsman, A.A. (eds.) DISC 2010. LNCS, vol. 6343, pp. 510–524. Springer, Heidelberg (2010)

20. Liu, P., Wang, D.W., Wu, J.J.: Efficient parallel i/o scheduling in the presence of data duplication. In: International Conference on Parallel Processing, pp. 231–238 (2003)
21. Mansfield, A.: Determining the thickness of graphs is NP-hard. Math. Proc. Cambridge Philos. Soc. 93, 9–23 (1983)
22. Matula, D.W., Beck, L.L.: Smallest-last ordering and clustering and graph coloring algorithms. J. ACM 30(3), 417–427 (1983)
23. Muthukrishnan, S.: Data streams: Algorithms and applications. Tech. rep. (2003)
24. O'Connell, T.C.: A survey of graph algorithms under extended streaming models of computation. In: Fundamental Problems in Computing, pp. 455–476. Springer, Netherlands (2009)
25. Ruhl, J.M.: Efficient Algorithms for New Computational Models. Ph.D. thesis, Massachusetts Institute of Technology (September 2003)
26. Schank, T., Wagner, D.: Finding, counting and listing all triangles in large graphs, an experimental study. In: Nikoletseas, S.E. (ed.) WEA 2005. LNCS, vol. 3503, pp. 606–609. Springer, Heidelberg (2005)
27. Zhang, J.: A survey on streaming algorithms for massive graphs. In: Managing and Mining Graph Data, Advances in Database Systems, vol. 40, pp. 393–420. Springer, US (2010)

# Approximation Algorithms for the Geometric Firefighter and Budget Fence Problems

Rolf Klein[1], Christos Levcopoulos[2], and Andrzej Lingas[2]

[1] Universität Bonn, Institut für Informatik I, D-53117 Bonn
rklein@uni-bonn.de
[2] Department of Computer Science, Lund University, 22100 Lund
{Christos.Levcopoulos,Andrzej.Lingas}@cs.lth.se

**Abstract.** Let $R$ denote a connected region inside a simple polygon, $P$. By building 1-dimensional barriers in $P \setminus R$, we want to separate from $R$ part(s) of $P$ of maximum area. In this paper we introduce two versions of this problem. In the *budget fence* version the region $R$ is static, and there is an upper bound on the total length of barriers we may build. In the basic *geometric firefighter* version we assume that $R$ represents a fire that is spreading over $P$ at constant speed (varying speed can also be handled). Building a barrier takes time proportional to its length, and each barrier must be completed before the fire arrives. In this paper we are assuming that barriers are chosen from a given set $B$ that satisfies a certain linearity condition. For example, this condition is satisfied for barrier curves in general position, if any two barriers cross at most once.

Even for simple cases (e. g., $P$ a convex polygon and $B$ the set of all diagonals), both problems are shown to be NP-hard. Our main result is an efficient $\approx 11.65$ approximation algorithm for the firefighter problem. Since this algorithm solves a much more general problem—a hybrid of scheduling and maximum coverage—it may find wider application. We also provide a polynomial-time approximation scheme for the budget fence problem, for the case where barriers chosen from $B$ must not cross.

## 1 Introduction

The *firefighter problem in graphs* has recently received significant attention [2,7,11,13]. It models a situation where a fire, infection, computer virus, *etc.*, spreads through a network, and the goal is to save as many network nodes as possible by a suitable placement of firefighters.

At the beginning a fire breaks out at the source vertex of the input graph. At each subsequent time step a bounded number of firefighters (just one in the standard version) may be placed at vertices that are not already on fire, to defend them. Once defended, a vertex will never catch fire. After the firefighters have been placed, the fire spreads from each burning vertex to all its undefended neighbors. The process ends when the fire can no longer spread. All vertices which are not on fire are considered to be *saved*. The objective is to determine a placement of firefighters that maximizes the number of vertices saved.

This graph firefighter problem is NP-hard already for trees [11,12], and hard to approximate within $n^\alpha$, for any $\alpha < 1$, in polynomial-time in the general case [7]. Only trees are known to admit polynomial-time constant-factor ($e/(e-1) \approx 1.5819$) approximation algorithms [7].

In this paper we propose a natural *geometric firefighter problem*. Instead of a graph, we have a polygonal region $P$ with a distinguished point $R$ where a fire starts spreading through $P$ at a given constant speed. Instead of placing firefighters, we can build 1-dimensional barriers also at a given constant speed in the area still free of fire, one at a time. Thus, building a barrier takes time proportional to its length. A barrier must be built continuously and each barrier point must be completed before the fire reaches it. The goal is to maximize the area of $P$ that is separated from the fire by the barriers. (Our results can be modified to apply to other variants of the problem. For example, the fire can spread at various speeds, and the time it takes to build various barriers can also vary.)

We also consider a simpler version termed the *budget fence problem*. For a polygonal region $P$, a contaminated subregion $R$, and a fence budget $l$, we want to separate a maximum area of $P$ from $R$ by drawing barriers within $P \setminus R$ of total length not exceeding $l$. In this static case no time constraints need to be observed.

Both problems have several variants depending on the type of polygonal region and the set $B$ of barriers allowed. In either case we require that $B$ fulfills the following *linearity condition*. If a point $p$ in $P$ can be separated from $R$ by a subset of barriers of $B$, then there is a single barrier that already separates $p$ from $R$; see Figure 1 for illustrations.

### 1.1 Our Contributions

In Theorem 2 we show the NP-hardness of the geometric firefighter problem for convex polygons and diagonals as barriers. Theorem 3 states that the problem is also NP-hard for star-shaped polygons and unrestricted sets of barriers. Its proof is in the full version of the paper. These hardness results carry over to the budget fence problem.

Our main result is a constant ($\approx 11.65$) approximation greedy algorithm for the firefighter problem in simple polygons that runs in time polynomial in the size of the barrier set $B$; see Theorems 4 and 5. This algorithm solves in fact a more general problem which is a hybrid of scheduling and maximum coverage. We are given a finite set of non-splittable jobs with release, duration, and completion time demands. Each job covers some part of a universe. The objective is to feasibly schedule a subset of jobs so that the profit from the total part of universe covered is maximized. Since this hybrid problem is a generalization of the maximum coverage problem, it cannot be approximated within a factor smaller than $e/(e-1) \approx 1.5819$ in polynomial time unless $P = NP$ [10].

In Theorem 6 we address the budget fence problem in a simple polygon. Whereas in the firefighter problem the barriers can be thought of as trenches that may cross each other, we require that those fence segments the algorithm selects from the given set $B$ must have pairwise disjoint interiors. We present a polynomial-time approximation scheme (PTAS) based on dynamic programming. It also yields a low-constant approximation for the geometric firefighter problem with disjoint barriers in simple polygons, provided the geodesic distances of the given barriers from the fire source do not differ too much; see Corollary 1.

## 1.2 Related Results

Several different generalizations and variants of the graph firefighter problem have been studied in the literature [2,7,11,13]. Recently, this problem has also been studied in the context of random geometric graphs, whose nodes correspond to random points according to a random distribution; in this setting, two points are connected by an edge iff the distance between them is shorter than a prespecified constant [5].

There are other papers beside Khuller et al. [17] which study generalizations of the maximum coverage problem, e.g., Cohen and Katzir [9]. Some of them are also partly related to scheduling, since they consider picking one element or set at a time, e.g., in Bansal et al. [3]. However, these generalizations are still very different from our hybrid problem, and therefore none of the techniques used in those papers seem applicable to our problem. Other papers relating set cover to scheduling also deal with very different problems. See, for example, Bansal and Pruhs [4], Hassin and Levin [16] and Ghaderi et al. [15]. Following the general approach for submodular function maximization, as described in Chekuri et al. [8], it may be possible to design an algorithm for our hybrid problem, although perhaps with a much worse approximation ratio.

To the best of our knowledge, the budget fence problem has not been studied except for very special cases, e.g., for rectilinear strips [1]. A related problem which has been studied is to select, among a given set of curves, a minimum number of curves which separates two sets of points [6].

## 2 Barriers and Linearity

By a *barrier* in a polygonal region we mean a curve of constant algebraic degree that does not intersect itself and either has both endpoints on the polygon's boundary—a cut—, or its endpoints coincide so that it forms a loop. We shall assume that barriers are in general position; endpoints of cuts may coincide, but otherwise intersections are proper crossings. Barriers must be built (drawn) from one endpoint to the other at constant speed, a single one at a time, but there is no travel cost between different barrier locations. In the firefighter problem, no point $b$ on a barrier can be built after the expanding fire has reached $b$.

As part of the problem definition, a set $B$ of allowed barriers is specified. We assume that $B$ satisfies the following *linearity condition*. For a subset $A$ of barriers let saved($A$) denote the part of $P$ that is separated from $R$ by the barriers of $A$ (i.e., complement of the closed cell containing region $R$, in the arrangement of all barriers of $A$ inside $P$). Set $B$ is called *linear* if for each subset $A \subseteq B$

$$\text{saved}(A) \subseteq \bigcup_{a \in A} \text{saved}(a) \qquad (*)$$

holds; Figure 1 shows two examples. Intuitively, this means that sets of barriers cannot "cooperate" to save some piece which cannot be saved by any one of the individual barriers in the set.

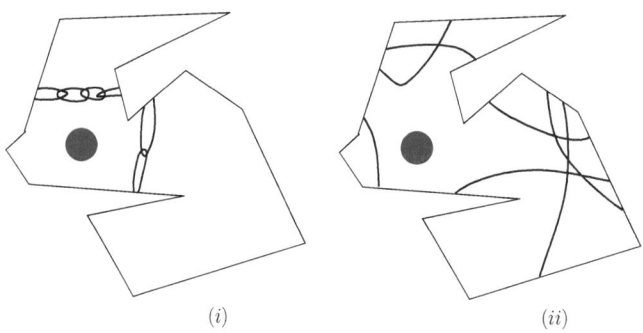

**Fig. 1.** (i) A non-linear barrier set. Each barrier itself separates only a miniscule area from the fire, but together they fence off large parts of $P$. (ii) A linear barrier set; any two barriers cross at most once.

To test a barrier set $B$ for linearity is not as complicated as this definition would suggest. For this purpose, we use the following lemma, whose proof is given in the full version.

**Lemma 1.** *For each barrier set $B$ in a simple polygon the following holds.*

1. *If $B$ contains only cuts and no loops then each point that can be saved by a set of barriers can even be saved by (at most) two of them. Consequently, condition $(*)$ needs only be checked for all subsets $A$ of size 2.*
2. *For a general set $B$ of size $m$, linearity can be tested in time polynomial in $m$, provided any two barriers cross only a constant number of times.*
3. *If any two barriers in $B$ cross at most once then $B$ is linear; Figure 1 (ii) is an example.*

## 3 NP-Hardness

A natural approach to show NP-hardness of the geometric firefighter problem is a reduction from the graph firefighter problem, which is known to be NP-hard even for trees of maximum outdegree three [11]. This approach yields the following theorem whose proof is given in the full version.

**Theorem 1.** *The geometric firefighter problem is NP-hard for simple polygons, when the barriers can be chosen from a given set of diagonals of identical length.*

If we allow for a large range of the lengths of barriers then we obtain NP-hardness for the geometric firefighter problem with diagonal barriers even for convex polygons.

**Theorem 2.** *The geometric firefighter problem is NP-hard even for convex polygons, if the possible barriers are restricted to all possible diagonals.*

*Proof.* We reduce from the subset-sum problem [14]. Let $\{a_1, a_2, ..., a_k\}$ be the set of positive integers, $s := \sum_{i=1}^{k} a_i$, and let $t$ be the desired target sum in the subset-sum problem we want to solve by using the firefighter algorithm for convex polygons.

The convex polygon $P$ is constructed as follows. It has $3k$ vertices. Of these vertices, $2k$ lie on a circle $C$ of radius $r$ (to be determined later) centered at the fire source. Every third vertex $v_i$ of the convex polygon lies in the interior of this circle $C$. Vertex $v_i$ can be cut off by a diagonal of length $a_i$; see Figure 2. The

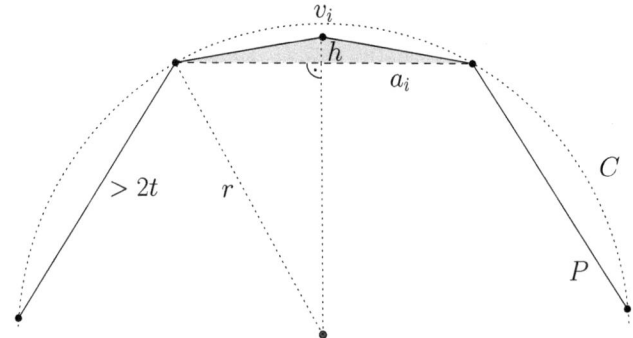

**Fig. 2.** Constructing firefighter instances to show NP-hardness

resulting triangle is of height $h$, independent of index $i$, so that its area equals $h \cdot a_i / 2$. Two consecutive vertices of polygon $P$ situated on $C$ are at distance greater than $2t$. (The height $h$ is the same for all triangles. The precise location of the vertex $v_i$ is not essential, as long as it lies on or inside the circle $C$.)

Let us assume that barriers can be built at speed one. Defining the fire's speed, $v$, by $v(t + 0.5) = r$ ensures that the firefighter can build barriers of total length at most $t + 0.5$ before the fire reaches the circle and the whole process terminates. Hence, no diagonal of length $> 2t$ can be built. On the other hand, if we make radius $r$ large enough to satisfy

$$\left(\frac{a_i}{2}\right)^2 < \left(1 - \frac{t^2}{(t+0.5)^2}\right) r^2,$$

then $vt < \sqrt{r^2 - (a_i/2)^2}$ holds, meaning that the fire has not reached the $i$th triangle at time $t$. Therefore, finding the optimal solution for the firefighting problem is equivalent to finding the subset of integers from the subset-sum problem whose sum is as close to $t$ as possible, without exceeding $t$.

We observe that an additive error less than $1/k^2$ in the lengths of the diagonals would be tolerable in this construction. Thus, vertices can be described by rationals of length polynomial in the bit length of the input. □

The proof of Theorem 2 would not work if the firefighter were allowed to build barriers freely anywhere in the polygon. But it turns out that the complexity of the problem does not decrease in this more liberal setting. The following result is shown in the full version of this paper.

**Theorem 3.** *The geometric firefighter problem and the budget fence problem are NP-hard for star-shaped polygons even if there is no restriction as to where the barriers can be built, and curved barriers are allowed.*

## 4 An Approximation Algorithm for the Geometric Firefighter Problem

In this section we present an efficient greedy algorithm for solving the geometric firefighter problem for a finite barrier set $B$, that achieves a constant approximation factor.

Our algorithm works in a more general setting, related to job scheduling and maximum coverage. Each barrier $b \in B$ of the firefighter problem can be considered a job. It has a duration (the time needed to build it) and a completion time (the last point in time where the fire permits the construction of $b$ to be completed). Each job is assumed to cover a subset of some finite universe $U$. The elements of $U$ carry profits, and a job's profit equals the sum of profits of all elements it covers.

The goal is to compute a feasible job schedule whose total profit (i.e., the sum of all profits of jobs scheduled) is maximized. Since this problem generalizes the max-coverage problem, it inherits its inapproximability results. Thus, no approximation factor smaller than $\approx 1.5819$ can be guaranteed by any polynomial time algorithm, unless $P = NP$ [10].

We will now present an approximation algorithm GlobalGreedy that runs in low polynomial time in the size of $B$ and guarantees an approximation factor of $\approx 11.65$. After proving this result, we comment more precisely on how this can be applied to the geometric firefighter problem.

GlobalGreedy maintains a feasible time schedule $L$ of jobs, with precise start and completion times for each of the jobs in $L$. (We may think of the schedule as a list of jobs, sorted according to their scheduled start time, although a more advanced data structure can be employed in order to perform searches and changes of the list more efficiently.) The algorithm starts with the empty schedule $L$. It considers each input job $J$ exactly once for possible insertion into the schedule $L$. If $J$ is rejected, it will never be considered again for inclusion. If $J$ is scheduled and inserted into the schedule $L$, its scheduled starting time will never change. However, it may still happen that $J$ is later deleted from $L$, in order to make it possible for some other job to be scheduled (partly) during the scheduled time for $J$. Once $J$ is deleted, it will never again be considered for insertion into the schedule.

Since GlobalGreedy inspects each job only once for possible insertion, the order in which the jobs are considered is crucial. To define this order, and for easier reference in the subsequent proof, we assign colors to all elements of universe $U$. These colors may change during the process. In the beginning, all elements are colored red. Each time GlobalGreedy inserts a job $J$ into the schedule $L$ the following happens: all red elements covered by $J$ are irrevocably associated to $J$, and change color to green. We will call these elements the *property elements* of $J$, and denote by the *property profit* of $J$ the total profits of all the property elements of $J$.

Finally, if some job $J$ is later deleted from the schedule $L$, in order to make place in the schedule for some other job, then the property elements of $J$ change color from green to grey during the deletion of $J$.

GlobalGreedy starts with the empty schedule $L$, and colors all elements of the universe red. Then it runs the following while-loop, one iteration for every input job $J$, possibly altering the schedule $L$, until there are no more jobs left unconsidered. Finally it outputs the schedule $L$. The order in which the jobs are considered is defined inside the while loop. A parameter $\mu \in (0,1)$ specifies when jobs will be deleted from the schedule $L$ in order to accommodate job $J$ currently under consideration.

**while** there is still any unconsidered job **do**
Consider an unconsidered job $J$ which maximizes the ratio of the total profit of the red elements it covers, divided by its duration. Insert $J$ into schedule $L$ if and only if this can be done respecting the deadline for $J$ and without any re-schedulings of other jobs, except for possibly deleting consecutive jobs in $L$ whose property profits are altogether not greater than $\mu$ times the total profit of all red elements covered by $J$. (In case there are several options for when to schedule $J$ satisfying this condition, choose one of them arbitrarily.) If $J$ is inserted, change to green the color of all red elements covered by $J$. For each job $J'$ possibly deleted from schedule $L$, change from green to grey the property elements of $J'$.
**end-while**

**Theorem 4.** *The above algorithm GlobalGreedy runs in polynomial time and achieves an approximation factor of $\approx 11.65$ if parameter $\mu$ is set to $\sqrt{2} - 1 \approx 0.41$.*

*Proof.* We will need the following lemma. For ease of reference, let us denote by $\Pi(\text{green})$ the total profit of all green elements in $U$, at a given time, and similarly for the other colors.

**Lemma 2.** *At each time, we have*

$$\Pi(grey) \leq \frac{\mu}{1-\mu}\Pi(green).$$

*Proof.* By induction on the number of job insertions with deletions. Before the first job is deleted from the schedule $L$, there are no grey elements, and the lemma holds. Suppose that, upon inserting job $J$, jobs with a total green profit of $z$ are deleted from $L$. All this green profit turns grey. But, by definition of GlobalGreedy, job $J$ wins at least $z/\mu$ new green profit, so that

$$\frac{\mu}{1-\mu}\Pi(\text{green})' - \Pi(\text{grey})' \geq \frac{\mu}{1-\mu}(\Pi(\text{green}) - z + \frac{z}{\mu}) - (\Pi(\text{grey}) + z)$$

$$= \frac{\mu}{1-\mu}\Pi(\text{green}) - \Pi(\text{grey}) \geq 0.$$

□

Let us run algorithm GlobalGreedy. All jobs in the final schedule are called green, and those jobs who were inserted into $L$ and later removed, grey.

Now let us consider a schedule OPT achieving maximum profit. We change to blue the color of all elements still red that are covered by OPT (noting that no green or grey elements become blue). Such a blue element is assigned, as a blue property element, to the first job $J$ in OPT that covers it. Job $J$ cannot be green or grey, because then it would no longer cover red elements. Consequently, if we name such a job $J$ blue, the three color classes are pairwise disjoint.

In order to prove Theorem 4 we are using a paying scheme, where each green or grey job $J$ pays money to any blue job $J'$ performed at least partially during the same time as $J$ (note that every job ever inserted into $L$ is assigned a unique execution time that will never be altered by GlobalGreedy). The paying scheme is specified as follows.

*The Paying Scheme.*

Case 1: The execution time of $J'$ is totally included within the execution time of $J$. In this case, $J$ gives to $J'$ money equal to the property profit of $J$, times the ratio of the duration of $J'$ divided by the duration of $J$.

Case 2: In all other cases where the execution intervals of $J$ and $J'$ overlap, $J$ gives to $J'$ money equal to $1/\mu$ times the property profit of $J$.

**Lemma 3.** *A green or grey job $J$ pays at most $1 + 2/\mu$ times its property profit to blue jobs.*

*Proof.* $J$ pays $1/\mu$ times its property profit to at most two blue job whose execution intervals include the start or end time of $J$. In addition, $J$ pays money to blue jobs whose scheduled times are totally included within the execution interval of $J$, in linear proportion to their respective duration; together, these payments do not exceed the property profit of $J$. □

The following lemma shows that each blue job gets well-payed by this scheme.

**Lemma 4.** *By the above paying scheme, each blue job receives an amount of money not smaller than its property profit.*

*Proof.* We define the *efficiency ratio* of any colored job $J$ to be the ratio of its property profit $\pi$ divided by the duration $\delta$ of $J$.

Let $J'$ be any blue job. Let us study the iteration of the while-loop when GlobalGreedy considered $J'$ for possible inclusion into the schedule $L$. Let $J_1, J_2, ..., J_k$ be the jobs in schedule $L$ during that step, whose execution times (partially) overlap with the execution time of $J'$. Since $J'$ became a blue job, it was rejected by GlobalGreedy. Thus, the total property profits of $J_1, J_2, ..., J_k$ must have been be at least $\mu$ times the red property profit of $J'$, which is no less than $\mu$ times its current blue property profit because all blue property profits of $J'$ were red during that iteration.

Moreover, by the order according to which GlobalGreedy considers jobs for possible insertion into $L$, it follows that the efficiency ratio of each one of the jobs $J_1, J_2, ..., J_k$ must be at least as large as the efficiency ratio of $J'$.

Let us first handle the case where $k = 1$. If the execution time of $J'$ is totally included within the execution time of $J_1$, job $J'$ receives from $J_1$

$$\pi(J_1) \frac{\delta(J')}{\delta(J_1)} \geq \pi(J') \frac{\delta(J')}{\delta(J')} = \pi(J'),$$

according to Case 1 of the paying scheme. In all other configurations, $J_1$ gives to $J'$ the amount of

$$\frac{1}{\mu} \pi(J_1) \geq \frac{1}{\mu} (\mu \pi(J')) = \pi(J')$$

by Case 2, and we are done. If $k > 1$, then the same arguments apply to $J_1, J_2, ..., J_k$. □

By Lemma 3 and Lemma 4,

$$\Pi(\text{blue}) \leq \left(1 + \frac{2}{\mu}\right) (\Pi(\text{green}) + \Pi(\text{grey})).$$

Now Lemma 2 allows us to bound all profits by green profits, and we obtain for the profits of OPT and GlobalGreedy

$$|\text{OPT}| \leq \Pi(\text{green}) + \Pi(\text{grey}) + \Pi(\text{blue})$$
$$\leq 2 \frac{\mu + 1}{\mu(1 - \mu)} \Pi(\text{green}) = 2 \frac{\mu + 1}{\mu(1 - \mu)} |\text{GlobalGreedy}|.$$

The factor is minimized for $\mu = \sqrt{2} - 1$ to the value $6 + 4\sqrt{2} \approx 11.657$. This completes the proof of Theorem 4. □

The same proof works if the jobs to be scheduled have release times, in addition to duration and completion times. Also, their durations might depend on their actual start times, as long as there are only a polynomial number of changes. We summarize our main result as follows, observing that $3/2 - \sqrt{2}$ is the inverse of $6 + 4\sqrt{2}$.

**Theorem 5.** *Let $U$ be a set of elements, each associated with a real profit, and let $\mathcal{J}$ be a set of jobs, where each job $J$ in $\mathcal{J}$ covers a given subset of universe $U$, and is given a release time, a completion deadline and a duration. Algorithm GlobalGreedy runs in polynomial time and constructs a feasible schedule whose total profit is at least $3/2 - \sqrt{2} \approx 0.086$ times the maximum possible profit.*

In the geometric firefighter problem, universe $U$ can be defined as follows. We take the arrangement of all barriers of $B$ inside polygon $P$, pick a representative point $u_c$ from each cell $c$, and let the profit of $u_c$ be the area of $c$; observe that all points in $c$ share their fates with $u_c$, with respect to the fire. Universe $U$ equals the set of the points $u_c$, and each barrier $b$ covers those points $u_c$ whose cells are separated by $b$ from the fire's starting point. We observe that $U$ is of polynomial size because any two barriers cross at most a constant number of times, due to their bounded algebraic degrees. Also, by the linearity of $B$, each cell that can be saved from the fire at all, can be saved by a single barrier; see Section 2. Thus, the above theorem yields an $\approx 0.086$ approximation to the geometric firefighter problem (on a Real RAM that can compute square roots in constant time).

## 5 A PTAS for the Budget Fence Problem and a Special Case of the Firefighter Problem

Recall that an instance of the budget fence problem consists of a simple polygon $P$ of $n$ edges, a "contaminated" connected region $R$ contained within $P$ (it can be degenerated to a point), the available total fence length $l$, and an allowed set $B$ of barriers, none of them intersecting the interior of $R$. The objective is to fence off from $R$ the largest possible area of polygon $P$ by barriers from $B$ of total length not exceeding $l$. While $B$ may contain candidates that cross each other, we want to use only barriers whose interiors are pairwise disjoint.

In the absence of time constraints, this problem seems of a more combinatorial nature than the firefighter problem. First, let us assume that $B$ consists of diagonals of $P$. Then the following standard dynamic programming and discretization approach yields a close approximation.

For each pair $a$, $b$ of vertices of $P$, and a number parameter $s$, we consider the problem $Q(a, b, s)$ of finding a shortest simple path saving at least an area of size $s$, from $a$ to $b$ in clockwise direction; the path can contain diagonals from $B$ and edges of $P$ between $a$ and $b$, but only the total length of diagonals counts.

Let $max$ be the maximum area that can be saved by a single diagonal. Clearly, no path can save more area than $max \times n$. We may assume the parameter $s$ to belong to $S = \{i \times max \times n/p(n) \mid 0 \leq i \leq p(n) \land i \in Z^+\}$, where $p(n)$ is an appropriate polynomial. Then, we can solve $Q(a, b, s)$ by considering solutions to all pairs $Q(a, c, s_1)$ and $Q(c, b, s_2)$, where $c$ is a vertex between $a$ and $b$ in clockwise direction on the perimeter of $P$, $s_1, s_2 \in S$, and $s_1 + s_2 \geq s$. We pick the pair that minimizes the length of the path from $a$ to $b$ through $c$ and compare it with the direct diagonal connection between $a$ and $b$ if this diagonal exists in $B$. We obtain an approximate solution of the budget fence problem for $P$ by picking the largest $s \in S$ such that the minimum length path solving $Q(a, a, s)$ for some vertex $a$ of $P$ has length less than or equal to $l$.

Any feasible solution includes at most $\lceil n/2 \rceil$ diagonals. By the definitions of $S$ and our dynamic programming scheme, the area saved by each of them can be underestimated by at most $\frac{n \times max}{p(n)}$. Hence, since the optimal solution saves at least $max$, the approximation factor of our dynamic programming method is $\frac{1}{1 - n^2/p(n)}$. It follows that it is sufficient to set $p(n)$ to $cn^2$, for a sufficiently large constant $c$, in order to obtain a $(1 + \epsilon)$ approximation.

Note that this method can be immediately adapted to work for any finite linear barrier set $B$ that does not contain loops. Thus, we have the following result.

**Theorem 6.** *There is a PTAS for the budget fence problem on a simple polygon with a finite barrier set $B$ that contains no loops.*

Remarkably, a good approximation to the budget fence problem can also yield a good approximation to the firefighter problem, under the assumptions from above ($B$ contains no loops, all barriers chosen must have pairwise disjoint interiors). In fact, we obtain the following corollary from Theorem 6. Its proof is given in the full version of this paper.

**Corollary 1.** *Let $min_B$ and $max_B$ be, respectively, the minimum and maximum geodesic distance of a barrier in $B$ from the fire source in $P$. For any $\delta, \epsilon > 0$, the firefighter problem in $P$ can be approximated within $2 \lceil \log_{1+\delta} \frac{max_B}{min_B} \rceil (1+\delta)(1+\epsilon)$ in polynomial time.*

## 6  Generalizations and Refinements

In this paper we have introduced a geometric version of the firefighter problem, and the closely related budget fence problem. There is a number of generalizations and interesting questions deserving further research.

For example, Algorithm GlobalGreedy could as well be applied to a situation where some parts of the polygonal domain $P$ are more important than others. Moreover, by adjusting the deadlines of the jobs, we can handle the cases when the speed of the fire and/or of building the barriers may vary.

The proof of a constant approximation ratio for GlobalGreedy would still work even if the barrier considered in the next iteration of the while-loop has only approximately largest efficiency ratio. The approximation constant would increase somewhat, but it would still be a constant. This observation may enable faster computation of the next candidate barrier to be considered, and improve the overall time performance of the algorithm.

Because of space considerations, more refinements and generalizations are only included in the full version of the paper.

# References

1. Altshuler, Y., Bruckstein, A.M.: On Short Cuts or Fencing in Rectangular Strips. arXiv:1911.5920v1[cs.CG] (November 26, 2010)
2. Anshelevich, E., Chakrabarty, D., Hate, A., Swamy, C.: Approximability of the Firefighter Problem: Computing Cuts over Time. Algorithmica 62(1-2), 520–536 (2012), preliminary version in proc. ISAAC 2009
3. Bansal, N., Gupta, A., Krishnaswamy, R.: A Constant Factor Approximation Algorithm for Generalized Min-Sum Set Cover. In: Proc. SODA, pp. 1539–1545 (2010)
4. Bansal, N., Pruhs, K.: The Geometry of Scheduling. In: FOCS 2010, pp. 407–414 (2010)
5. Barghi, A., Winkler, P.: Firefighting on a random geometric graph. Random Structures & Algorithms, doi:10.1002/rsa.20511 (first published online: June 27, 2013)
6. Cabello, S., Giannopoulos, P.: The Complexity of Separating Points in the Plane. In: Proc. 29th ACM Symposium on Computational Geometry, pp. 379–386 (2013)
7. Cai, L., Verbin, E., Yang, L.: Firefighting on Trees $(1 - 1/e)$–Approximation, Fixed Parameter Tractability and a Subexponential Algorithm. In: Hong, S.-H., Nagamochi, H., Fukunaga, T. (eds.) ISAAC 2008. LNCS, vol. 5369, pp. 258–269. Springer, Heidelberg (2008)
8. Chekuri, C., Vondrák, R., Zenklusen, R.: Submodular Function Maximization via the Multilinear Relaxation and Contention Resolution Schemes. Prel. version in STOC 2011, 783–792 (2011), Revised version in `http://arxiv.org/pdf/1105.4593v3.pdf` (July 30, 2012)
9. Cohen, R., Katzir, L.: The Generalized Maximum Coverage Problem. Information Processing Letters 108, 15–22 (2008)
10. Feige, U.: A threshold of ln n for approximating set cover. J. ACM 45(4), 634–652 (1998)
11. Finbow, S., King, A., MacGillivray, G., Rizzi, R.: The firefighter problem for graphs of maximum degree three. Discrete Mathematics 307(16), 2094–2105 (2007)
12. Finbow, S., MacGillivray, G.: The Firefighter Problem: a survey of results, directions and questions. Australasian J. Comb. 43, 57–78 (2009)
13. Floderus, P., Lingas, A., Persson, M.: Towards more efficient infection and fire fighting. Int. J. Found. Comput. Sci. 24(1), 3–14 (2013), preliminary version in proc. CATS 2011
14. Garey, M.R., Johnson, D.S.: Computers and Intractability. A Guide to the Theory of NP-completeness. W.H. Freeman and Company, New York (1979)
15. Ghaderi, R., Esnaashari, M., Meybodi, M.R.: An Adaptive Scheduling Algorithm for Set Cover Problem in Wireless Sensor Networks: A Cellular Learning Automata Approach. International Journal of Machine Learning and Computing 2(5) (October 2012)
16. Hassin, R., Levin, A.: An Approximation Algorithm for the Minimum Latency Set Cover Problem. In: Brodal, G.S., Leonardi, S. (eds.) ESA 2005. LNCS, vol. 3669, pp. 726–733. Springer, Heidelberg (2005)
17. Khuller, S., Moss, A., Naor, J.S.: The budgeted maximum coverage problem. Information Processing Letters 70, 39–45 (1999)

# An Improved Data Stream Algorithm for Clustering*

Sang-Sub Kim and Hee-Kap Ahn

Department of Computer Science and Engineering, POSTECH, Pohang,
Republic of Korea
{helmet1981,heekap}@postech.ac.kr

**Abstract.** We present a single-pass, $(1.8+\varepsilon)$-factor, $O(1/\varepsilon)$-space data stream algorithm for the Euclidean 2-center problem for any fixed $d \geqslant 1$. This is an improvement on the approximation factor over the $(2+\varepsilon)$-factor and $O(1/\varepsilon)$-space algorithms of Ahn et al. [3] and Guha [8]. It can also be considered as an improvement on the space over the $(1+\varepsilon)$-factor and $O(1/\varepsilon^d)$-space algorithm of Zarrabi-Zadeh [11], while sacrificing the approximation factor a little bit. To our best knowledge, this is the first breakthrough with an approximation factor below 2 using $O(1/\varepsilon)$ space for any fixed $d$. Our algorithm also extends to the $k$-center problem with $k > 2$.

## 1 Introduction

Clustering is a fundamental problem arising from many applications such as data mining [7,9], image processing [13], and astrophysics [4,6]. In a standard Euclidean $k$-*clustering problem*, we are given a set $P$ of points in $d$-dimensional space $\mathbb{R}^d$ and are to find $k$ points $c_1, c_2, \ldots, c_k$ of $\mathbb{R}^d$ such that

$$\max_{p \in P} \{ \min_{1 \leqslant i \leqslant k} |pc_i| \} \text{ is minimized,}$$

where $|pq|$ denotes the Euclidean distance between any two points $p$ and $q$ in the space. This problem is also known as the Euclidean $k$-center problem.

In the *data streaming model*, input data is given as a sequence of items that are allowed to be examined in only a few passes. A typical constraint of data streaming algorithms is that they have limited amount (typically smaller than the input size) of memory available, and therefore it is important to design an algorithm whose space complexity does not depend on the size of input.

In this paper we consider the Euclidean $k$-center problem for streaming points in $\mathbb{R}^d$, where each point arrives one by one in a stream and is allowed to be examined only once [2], and a small amount of information can be stored in a device.

---

* This work was supported by the National Research Foundation of Korea(NRF) grant funded by the Korea government(MSIP) (No. 2011-0030044).

*Previous work.* There has been work on computing $k$ centers for small $k$. In a fixed dimension, Hershberger and Suri [10] gave an algorithm for maintaining extreme points in a number of different directions. This algorithm can be adapted to a $(1+\varepsilon)$-approximation algorithm using $O(1/\varepsilon^{(d-1)/2})$ space for the Euclidean 1-center problem. Recently, Agarwal and Sharathkumar [1] gave a $(1+\sqrt{3})/2 + \varepsilon \approx 1.3661$-approximation algorithm using $O((1/\varepsilon^3)\log(1/\varepsilon))$ space. They also showed that any single-pass stream algorithm using space polynomially bounded in $d$ cannot approximate the optimal 1-center within factor $(1+\sqrt{2})/2 > 1.207$. Later, their approximation factor was improved to 1.22 by Chan and Pathak [5].

For $k=2$, Zarrabi-Zadeh presented an approximation algorithm that uses $O(1/\varepsilon^d)$ space and returns a solution with factor $(1+\varepsilon)$ [11]. Guha's algorithm uses $O((1/\varepsilon)\log(1/\varepsilon))$ space and returns an approximate solution with factor $(2+\varepsilon)$ [8]. Very recently, Ahn et al. gave a $(2+\varepsilon)$-factor and $O(1/\varepsilon)$-space algorithm [3].

There has also been work on computing $k$ centers using the minimum space in any arbitrary dimension. Zarrabi-Zadeh and Chan [14] gave a 1.5-approximation algorithm for the Euclidean 1-center problem using only one center and one radius. Poon and Zhu [12] gave a 5.708-approximation algorithm for the Euclidean 2-center problem using two centers and one radius.

*Our results.* We present a single-pass data stream algorithm that uses $O(1/\varepsilon)$ space and returns two centers with $(1.8+\varepsilon)$-approximation to the 2-center problem in any fixed $d \geqslant 1$. Each update takes $O(1/\varepsilon)$ time.

This is an improvement on the approximation factor of $(2+\varepsilon)$ by Ahn et al. [3] and Guha [8]. To our best knowledge, this is the first breakthrough with an approximation factor below 2 using $O(1/\varepsilon)$ space for any fixed $d$. It can also be considered as an improvement on the space over the $(1+\varepsilon)$-factor and $O(1/\varepsilon^d)$-space algorithm of Zarrabi-Zadeh [11], while sacrificing the approximation factor a little bit.

Our algorithm also extends to the $k$-center problem with $k > 2$. For any fixed dimension $d \geqslant 1$, our algorithm uses $O(2^k(k+3)!/\varepsilon)$ space and returns $k$ centers that guarantees a $(1.8+\varepsilon)$-approximation. Each update takes $O(2^k(k+2)!/\varepsilon)$ time.

## 2 Preliminaries

Let $\mathcal{X} = \langle p_1, p_2, \ldots, p_n \rangle$ be a sequence of $n$ points in $d$-dimensional Euclidean space. We denote by $\mathcal{X}_j = \langle p_1, \ldots, p_j \rangle$ the sequence of the first $j$ points of $\mathcal{X}$.

Let $B(c,r)$ denote a ball of radius $r$ centered at $c$, and let $r(B)$ and $c(B)$ denote the radius and the center of a ball $B$, respectively.

We denote by $pq$ the straight line segment connecting any two points any two points $p$ and $q$ in $\mathbb{R}^d$, and by $|pq|$ the length of $pq$. For a compact set $A$, we denote by $\partial A$ the boundary of $A$.

**Lemma 1.** *Let $B_0$ be a unit ball in $\mathbb{R}^d$. Any line segment of length at least 1.2 contained in $B_0$ intersects $B(c(B_0), 0.8)$.*

*Proof.* When $d = 2$, we know from some basic geometry that any line segment of length at least 1.2 intersects $B(c(B_0), 0.8)$. See Fig. 1 (a). For $d > 2$, we can show the lemma by choosing the 2-dimensional plane passing through $c(B_0)$ and the line segment. □

## 3  The 2-Center Problem

In this section we design an approximation for the 2-center problem and prove the following.

**Theorem 1.** *For any fixed $d \geqslant 1$, our single-pass data stream algorithm uses $O(1/\varepsilon)$ space and returns two centers that guarantees a $(1.8 + \varepsilon)$-approximation for the Euclidean 2-center problem. Each update takes $O(1/\varepsilon)$ time.*

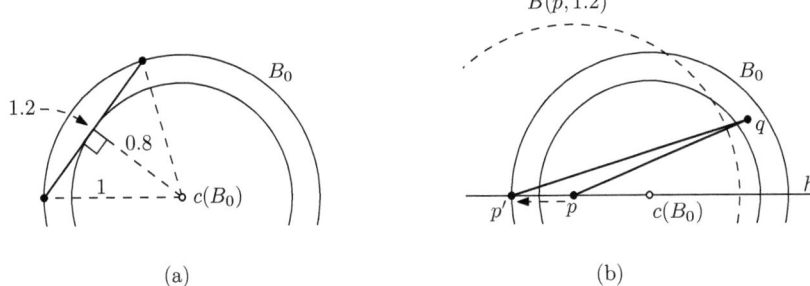

**Fig. 1.** (a) Proof of Lemma 1. (b) Proof of Lemma 2.

Let $B_1^*$ and $B_2^*$ denote an optimal pair of two congruent balls for $\mathcal{X}$. We let $r^* = r(B_1^*) = r(B_2^*)$ and $c_i^* = c(B_i^*)$ for $i = 1, 2$. By $\delta^*$ we denote the distance between $B_1^*$ and $B_2^*$, that is, $\delta^* := \max\{0, |c_1^* c_2^*| - 2r^*\}$.

In Section 3.1, we give a description of our algorithm for a given $r' > 0$ for the case $\delta^* \leqslant 2r^*$. Then we show that for $r'$ with $1.2r^* \leqslant r' < (1.2 + 2\varepsilon/3)r^*$, the algorithm returns a solution with $(1.8 + \varepsilon)$-approximation. We explain how we get such an $r'$ and present a full description of our algorithm for the case $\delta^* \leqslant 2r^*$ in Section 3.2. Then we consider the case $\delta^* > 2r^*$ in Section 3.3.

When the solution of a case encloses the points of $\mathcal{X}$ inserted so far, the solution is said to be *feasible* for them. Our algorithm maintains all feasible solutions, and returns the best one when all points of $\mathcal{X}$ are processed.

### 3.1  The Case $\delta^* \leqslant 2r^*$

Without loss of generality, we assume $p_1 \in B_1^*$. Before explaining the main idea, we need the following technical lemma.

**Lemma 2.** *Let $B_0$ be a unit ball in $\mathbb{R}^d$. For any line segment $pq$ of length at least 1.2 contained in $B_0$, every point $x$ of $pq$ at distance at least 0.6 from both endpoints is contained in $B(c(B_0), 0.8)$.*

*Proof.* Let $pq$ denote a line segment of length at least 1.2 contained $B_0$ in the plane. Without loss of generality, we assume that $p$ lies on the horizontal line $h$ through $c(B_0)$ and to the left of $c(B_0)$, and the other endpoint $q$ lies to the right of $p$ and above $h$ as illustrated in Fig. 1(b).

Then $pq \setminus B(c(B_0), 0.8)$ consists of at most two segments, one incident to $p$ and one incident to $q$. Imagine that we translate $p$ to the left along $h$ until it hits the boundary of $B_0$. Then the length of $pq$ and the length of each segment of $pq \setminus B(c(B_0), 0.8)$ does not get decreased during the translation of $p$. Let $p'$ denote the translated point of $p$ at $h \cap \partial B_0$. From Lemma 1, we know that both segments of $p'q \setminus B(c(B_0), 0.8)$ have length at most 0.6. Since $|pq| \geqslant 1.2$, we can conclude that there is a point $x$ on $pq$ at distance 0.6 from each endpoint of $pq$ such that $x \in B(c(B_0), 0.8)$.

For higher dimensions, we can show the lemma by choosing the 2-dimensional plane passing through $c(B_0)$ and a line segment of length at least 1.2 contained in $B_0$. □

Let us describe our algorithm. Our algorithm is given a fixed value $r'$ and maintains at most three *candidate solutions*, where each candidate solution consists of a ball or a pair of balls with larger radius $r'$ or $3r'/2$. Once any of candidate solutions does not contain every input points arrived so far, the algorithm sets it infeasible. Once any of candidate solutions becomes infeasible, the algorithm simply abandons it. Once all points of $\mathcal{X}$ are processed, the algorithm returns the feasible solution with smallest larger radius if there is any feasible solution.

1. Initially, there is no candidate solution.
2. When $p_1$ is inserted, we create one candidate solution $B(p_1, r')$.
3. Let $p_2^o$ be the first point of $\mathcal{X}$ that arrives after $p_1$ and does not lie in the current candidate solution, that is, $p_2^o \notin B(p_1, r')$.
4. If $p_2^o$ arrives, we replace the current candidate solution with two candidate solutions, one corresponding to the case of $p_2^o \in B_1^*$ and one corresponding to the case of $p_2^o \in B_2^*$.
5. For each case, let $p_3^o$ be the first point of $\mathcal{X}$ that arrives after $p_2^o$ and does not lie in the corresponding candidate solution. If $p_3^o$ arrives, we replace the corresponding candidate solution with two new candidate solutions, one corresponding to the subcase of $p_3^o \in B_1^*$ and one corresponding to the subcase of $p_3^o \in B_2^*$.
6. For each subcase, let $p_4^o$ be the first point of $\mathcal{X}$ that arrives after $p_3^o$ and does not lie in the corresponding candidate solution. Again, if $p_4^o$ arrives, we replace the corresponding candidate solution with a new candidate solution.
7. If there are more points arriving after $p_4^o$ and lying outside of a current candidate solution, we abandon the solution.
8. Among all feasible solutions, our algorithm returns the one with smallest larger radius as the final solution.

From now on, we assume that we are given $r'$ with $1.2r^* \leq r' < (1.2 + 2\varepsilon/3)r^*$. By the previous two lemmas, we get the following lemma.

**Lemma 3.** *If $p_2^o \in B_1^*$, then we have $B_1^* \cup B(p_1, r') \subset B(c_{12}, 3r'/2)$, where $c_{12} = \partial B(p_1, r'/2) \cap p_1 p_2^o$.*

*Proof.* We first show that $B_1^* \subset B(c_{12}, 3r'/2)$. For any point $x \in B_1^*$, $|xc_{12}| \leq |xc_1^*| + |c_1^* c_{12}|$. See Fig. 2(a). By Lemma 2, we have $c_1^* \in B(c_1^*, 0.8r^*)$ and $|c_1^* c_{12}| \leq 0.8r^*$. Therefore, $|xc_{12}| \leq |xc_1^*| + |c_1^* c_{12}| \leq r^* + 0.8r^* = 1.8r^* \leq 3r'/2$.

Next we show that $B(p_1, r') \subset B(c_{12}, 3r'/2)$. For any point $y \in B(p_1, r')$, $|yc_{12}| \leq |yp_1| + |p_1 c_{12}| \leq r' + r'/2 = 3r'/2$. □

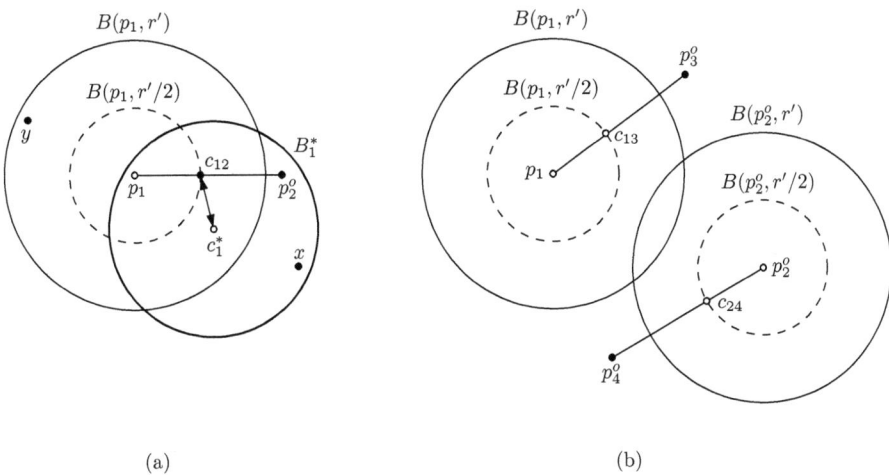

**Fig. 2.** (a) Proof of Lemma 3. (b) An illustration of Case 3b: $p_2^o \in B_2^*$ and $p_3^o \in B_1^*$.

Table 1 summarizes all possible subcases and the candidate solutions that our algorithm maintains for $1.2r^* \leq r' < (1.2 + 2\varepsilon/3)r^*$.

**Case 1: There is no $p_2^o$.** This implies that all points of $\mathcal{X}$ lie in $B(p_1, r')$. We simply return $B(p_1, r')$ and a dummy ball as the solution of the 2-center problem.

**Case 2a: $p_2^o \in B_1^*$ and there is no $p_3^o$.** We maintain one ball $B(p_1, r')$ until $p_2^o$ is inserted. Then we replace the ball with $B(c_{12}, 3r'/2)$, where $c_{12} = p_1 p_2^o \cap \partial B(p_1, r'/2)$. We have $B_1^* \cup B(p_1, r') \subset B(c_{12}, 3r'/2)$ by Lemma 3. We return $B(c_{12}, 3r'/2)$ and a dummy ball as the solution.

**Case 2b: $p_2^o \in B_1^*$, $p_3^o \in B_2^*$, and there is no $p_4^o$.** The points of $\mathcal{X}$ that are not in $B(c_{12}, 3r'/2)$ are inserted after $p_2^o$, and they are all in $B_2^*$. Let $p_3^o$ be the first such point. When $p_3^o$ is inserted, we create another ball $B(p_3^o, r')$. Since there is no $p_4^o$, every point of $\mathcal{X}$ lies in $B(c_{12}, 3r'/2)$ or $B(p_3^o, r')$, and we return them as the solution.

**Table 1.** Nine possible subcases for $r'$ with $1.2r^* \leqslant r' < (1.2 + 2\varepsilon/3)r^*$

|  | $p_2^o$ | $p_3^o$ | $p_4^o$ | Candidate solution |
|---|---|---|---|---|
| Case 1 | no $p_2^o$ | | | $B(p_1, r')$ and a dummy ball |
| Case 2a | $B_1^*$ | no $p_3^o$ | | $B(c_{12}, 3r'/2)$ and a dummy ball |
| Case 2b | $B_1^*$ | $B_2^*$ | no $p_4^o$ | $B(c_{12}, 3r'/2)$ and $B(p_3^o, r')$ |
| Case 2c | $B_1^*$ | $B_2^*$ | $B_2^*$ | $B(c_{12}, 3r'/2)$ and $B(c_{34}, 3r'/2)$ |
| Case 3a | $B_2^*$ | no $p_3^o$ | | $B(p_1, r')$ and $B(p_2^o, r')$ |
| Case 3b | $B_2^*$ | $B_1^*$ | no $p_4^o$ | $B(c_{13}, 3r'/2)$ and $B(p_2^o, r')$ |
| Case 3c | $B_2^*$ | $B_1^*$ | $B_2^*$ | $B(c_{13}, 3r'/2)$ and $B(c_{24}, 3r'/2)$ |
| Case 3d | $B_2^*$ | $B_2^*$ | no $p_4^o$ | $B(p_1, r')$ and $B(c_{23}, 3r'/2)$ |
| Case 3e | $B_2^*$ | $B_2^*$ | $B_1^*$ | $B(c_{14}, 3r'/2)$ and $B(c_{23}, 3r'/2)$ |

**Case 2c:** $p_2^o \in B_1^*$ and $p_3^o, p_4^o \in B_2^*$. Let $p_4^o$ be the first point of $\mathcal{X}$ that does not lie in $B(c_{12}, 3r'/2) \cup B(p_3^o, r')$. When $p_4^o$ is inserted, we replace $B(p_3^o, r')$ with $B(c_{34}, 3r'/2)$, where $c_{34} = p_3^o p_4^o \cap \partial B(p_3^o, r'/2)$. Again by Lemma 3, every point of $\mathcal{X} \setminus B(c_{12}, 3r'/2)$ lies in $B(c_{34}, 3r'/2)$. We return $B(c_{12}, 3r'/2)$ and $B(c_{34}, 3r'/2)$ as the solution.

The only remaining case is that $p_2^o \in B_2^*$. There are five subcases.

**Case 3a:** $p_2^o \in B_2^*$ **and there is no** $p_3^o$. Since there is no $p_3^o$, every point of $\mathcal{X}$ lies in $B(p_1, r')$ or $B(p_2^o, r')$. We simply return the two balls as the solution.

**Case 3b:** $p_2^o \in B_2^*$, $p_3^o \in B_1^*$, **and there is no** $p_4^o$. Let $p_3^o$ be the first point of $\mathcal{X}$ that does not lie in $B(p_1, r') \cup B(p_2^o, r')$. When $p_3^o$ is inserted, we replace $B(p_1, r')$ with $B(c_{13}, 3r'/2)$, where $c_{13} = p_1 p_3^o \cap \partial B(p_1, r'/2)$. We return $B(c_{13}, 3r'/2)$ and $B(p_2^o, r')$ as the solution.

**Case 3c:** $p_2^o \in B_2^*$, $p_3^o \in B_1^*$ and $p_4^o \in B_2^*$. Let $p_4^o$ denote the first point of $\mathcal{X}$ that is not in $B(c_{13}, 3r'/2) \cup B(p_2^o, r')$. When $p_4^o$ is inserted, we replace $B(p_2^o, r')$ with $B(c_{24}, 3r'/2)$, where $c_{24} = p_2^o p_4^o \cap \partial B(p_2^o, r'/2)$. Fig. 2(b) illustrates this case. Lemma 3 implies that every point of $\mathcal{X} \setminus B(c_{13}, 3r'/2)$ lies in $B_2^*$ and that $B_2^* \subset B(c_{24}, 3r'/2)$. We return $B(c_{13}, 3r'/2)$ and $B(c_{24}, 3r'/2)/$ as the solution.

**Case 3d:** $p_2^o, p_3^o \in B_2^*$ **and there is no** $p_4^o$. This case can be handled in exactly the same way for Case 3b, except that the roles of $B(p_1, r')$ and $B(p_2^o, r')$ are interchanged for the subsequence of $\mathcal{X}$ from $p_3^o$. When $p_3^o$ is inserted, we replace $B(p_2^o, r')$ with $B(c_{23}, 3r'/2)$, where $c_{23} = p_2^o p_3^o \cap \partial B(p_2^o, r'/2)$. Since every point of $\mathcal{X}$ lies in $B(p_1, r')$ or $B(c_{23}, 3r'/2)$, we return them as the solution.

**Case 3e:** $p_2^o, p_3^o \in B_2^*$ **and** $p_4^o \in B_1^*$. Let $p_4^o$ denote the first point of $\mathcal{X}$ that is not in $B(p_1, r') \cup B(c_{23}, 3r'/2)$. Since $B_2^* \subset B(c_{23}, 3r'/2)$, $p_4^o \in B_1^*$. When $p_4^o$ is inserted, we replace $B(p_1, r')$ with $B(c_{14}, 3r'/2)$, where $c_{14} = p_1 p_4^o \cap \partial B(p_1, r'/2)$. By Lemma 3, every point of $\mathcal{X} \setminus B(c_{23}, 3r'/2)$ lies in $B_1^*$ and $B_1^* \subset B(c_{14}, 3r'/2)$. We return $B(c_{23}, 3r'/2)$ and $B(c_{14}, 3r'/2)$ as the solution.

**Lemma 4.** *For $\delta^* \leqslant 2r^*$ and $r'$ with $1.2r^* \leqslant r' < (1.2 + 2\varepsilon/3)r^*$, our algorithm uses $O(1)$ space and returns two centers that guarantees $(1.8+\varepsilon)$-approximation. Our algorithm spends $O(1)$ update time for each point of $\mathcal{X}$.*

*Proof.* Since our algorithm considers all possible input cases of streaming points, there is at least one feasible solution. Since every feasible solution has its larger radius at most $3r'/2$, the final solution has larger radius at most $3r'/2 \leqslant (1.8 + \varepsilon)r^*$.

For space complexity, our algorithm maintains at most two balls in each case, and therefore it uses $O(1)$ space. Whenever the next point is inserted, the algorithm updates the solution for each subcase in $O(1)$ time. Therefore, the algorithm spends $O(1)$ update time for each point of $\mathcal{X}$. □

### 3.2 Finding $r'$

In this section, we explain how we get $r'$ such that $1.2r^* \leqslant r' < (1.2 + 2\varepsilon/3)r^*$. The basic procedure works as follows: Our algorithm maintains $m = \lceil 18/\varepsilon \rceil$ candidate lengths. See Fig. 3(b). Let $\mathcal{L}_j = \{i \cdot \ell_j \mid \text{for } i = 1, \ldots m\}$ denote the set of such $m$ lengths for $\mathcal{X}_j$, where $\ell_j$ denotes a certain nonnegative value for $\mathcal{X}_j$. For each candidate length $\ell$, we assume that $r' = \ell$ and run the algorithm in Section 3.1. This is similar to LAYERPARTITION procedure of the algorithm by Ahn et al. [3].

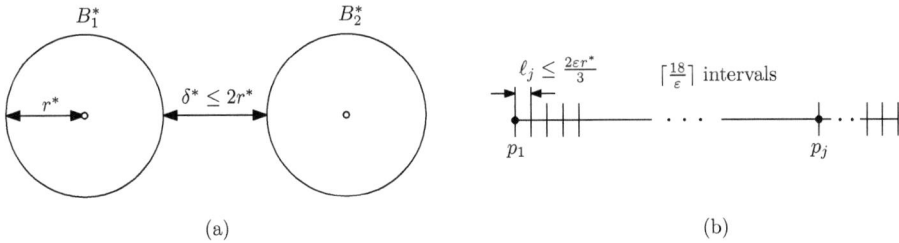

**Fig. 3.** (a) For $\delta^* \leqslant 2r^*$, we have $|p_i p_j| \leqslant 6r^*$ for any $1 \leqslant i, j \leqslant n$. (b) Our algorithm maintains $\lceil 18/\varepsilon \rceil$ candidate lengths for $r'$.

More precisely, we maintain the $m$ candidates as follows. The algorithm starts with two input points $p_1$ and $p_2$, and sets $\ell_2 := |p_1 p_2|/m$. Assume that we have processed the points of $\mathcal{X}_{j-1}$ and computed $\mathcal{L}_{j-1}$. For the next point $p_j$, if $|p_1 p_j| \leqslant m \cdot \ell_{j-1}$, we let $\mathcal{L}_j := \mathcal{L}_{j-1}$. Otherwise, we compute $\mathcal{L}_j$ from $\mathcal{L}_{j-1}$ as follows. Let $x$ be the integer satisfying $2^{x-1} \cdot m \cdot \ell_{j-1} < |p_1 p_j| \leqslant 2^x \cdot m \cdot \ell_{j-1}$. We let $\ell_j := 2^x \ell_{j-1}$.

If the new candidate $r' = i \cdot \ell_j$ is at most $m \cdot \ell_{j-1}$, it always coincides with an old candidate, say $y \cdot \ell_{j-1}$ for some $y \in \{1, \ldots m\}$, and we take the solutions of $y \cdot \ell_{j-1}$ for points of $\mathcal{X}_{j-1}$ as the initial solutions and update them for the insertion of $p_j$. Otherwise, there are two cases: either $r' = i \cdot \ell_j > |p_1 p_j|$ or

$r' = i \cdot \ell_j \leq |p_1 p_j|$. For the former case, all points of $\mathcal{X}_j$ lie in $B(p_1, r')$, and therefore it corresponds to Case 1 of the algorithm in Section 3.1. For the latter case, $p_j$ is the only point of $\mathcal{X}_j$ that does not lie in $B(p_1, r')$, and therefore it corresponds to Case 2a or Case 3a. The algorithm computes solutions for both cases.

**Lemma 5.** *For $\delta^* \leq 2r^*$, our algorithm uses $O(1/\varepsilon)$ space and returns two centers that guarantees $(1.8 + \varepsilon)$-approximation. Our algorithm spends $O(1/\varepsilon)$ update time for each point of $\mathcal{X}$.*

*Proof.* When we are done with processing all points of $\mathcal{X}$, there are $m$ candidate lengths $i \cdot \ell_n \in \mathcal{L}_n$ for $i = 1, \ldots, m$. If $m \cdot \ell_n < 1.2r^*$, then Case 1 of the algorithm in Section 3.1 returns a feasible solution, $B(p_1, r')$ and a dummy ball, for $r' := m \cdot \ell_n$. Otherwise, we show that there is an $i \cdot \ell_n \in \mathcal{L}_n$ such that $1.2r^* \leq i \cdot \ell_n < (1.2 + 2\varepsilon/3)r^*$. Let $p_t$ be the last point that changed the set of candidates. Then $|p_1 p_t| \leq m \cdot \ell_n \leq 2|p_1 p_t| \leq 12r^*$ (See Fig. 3(a)). Therefore there is a candidate $i \cdot \ell_n$ satisfying our assumption.

Whenever the next point is inserted, the algorithm updates the set of $O(1/\varepsilon)$ candidate lengths, if needed, and updates the solutions for each candidate length in $O(1)$ time. Therefore, the algorithm spends $O(1/\varepsilon)$ update time for each point of $\mathcal{X}$. □

### 3.3 The Case $\delta^* > 2r^*$

For $\delta^* > 2r^*$, the points of $\mathcal{X}$ are well separated. We simply use MERGEEXPAND procedure [3] together with Chan's minimum enclosing ball algorithm [14] for streaming points.

**Lemma 6.** *For $\delta^* > 2r^*$, MERGEEXPAND guarantees an optimal partition to the 2-center problem.*

MERGEEXPAND maintains three enclosing balls, one for each optimal partition of points inserted so far, and one for all points inserted so far. We simply use Chan's minimum enclosing ball algorithm for each enclosing ball which guarantees 1.5-approximation.

**Lemma 7.** *For $\delta^* > 2r^*$, our algorithm uses $O(1)$ space and return two centers that guarantees $1.5$-approximation. Each update takes $O(1)$ time.*

## 4 Extension to the $k$-Center problem

In this section we design an approximation for the $k$-center problem. Let $f(p)$ denote a point in $\mathcal{X}$ that is farthest from $p$. We consider two cases: $|p_1 f(p_1)| \leq 4kr^*$ and $|p_1 f(p_1)| > 4kr^*$.

In Section 4.1, we give a description of our algorithm for a given $r' > 0$ for the case $|p_1 f(p_1)| \leq 4kr^*$. Then we show for $r'$ with $1.2r^* \leq r' < (1.2 + 2\varepsilon/3)r^*$, the algorithm returns a solution satisfying $(1.8+\varepsilon)$-approximation. We explain how we get $r'$ and present a full description of our algorithm for the case $|p_1 f(p_1)| \leqslant 4kr^*$ in Section 4.2. Then we consider the case $|p_1 f(p_1)| > 4kr^*$ in Section 4.3. Our algorithm maintains all feasible solutions, and returns the best one when all points of $\mathcal{X}$ are processed.

### 4.1 The Case $|p_1 f(p_1)| \leqslant 4kr^*$

We extend the idea in Section 3.1 to the $k$-center problem. Let SELECTBALL$(B, p)$ denote a procedure that takes a ball $B$ and a point $p$, and returns $B(c, 3r(B)/2)$, where $c = c(B)p \cap \partial B(c(B), r(B)/2)$.

Our algorithm $k$-Centers maintains a set $\mathcal{S}$ of pairs $(\mathcal{B}^s, \mathcal{B}^c)$ of sets of balls. We call $\mathcal{B}^s$ a set of *selected* balls and $\mathcal{B}^c$ a set of *candidate* balls.

**Algorithm** $k$-*Centers*
**Input:** A sequence $\mathcal{X}$ of $n$ points, an integer $k > 2$, and a radius $r'$.
**Output:** A feasible solution consisting of $k$ centers with radius at most $3r'/2$ if exists. An empty set otherwise.
1.   $\mathcal{S}_1 \leftarrow \{(\emptyset, \{B(p_1, r')\})\}$
2.   **for** $i \leftarrow 2$ **to** $n$
3.     **for** each pair $(\mathcal{B}^s, \mathcal{B}^c) \in \mathcal{S}_{i-1}$
4.       **if** $p_i$ lies in one of balls in $\mathcal{B}^s \cup \mathcal{B}^c$
5.         **then** $\mathcal{S}_i \leftarrow \mathcal{S}_i \cup \{(\mathcal{B}^s, \mathcal{B}^c)\}$
6.     **for** each pair $(\mathcal{B}^s, \mathcal{B}^c) \in \mathcal{S}_{i-1} \setminus \mathcal{S}_i$
7.       **if** $|\mathcal{B}^s| + |\mathcal{B}^c| < k$
8.         **then** $\mathcal{S}_i \leftarrow \mathcal{S}_i \cup \{(\mathcal{B}^s, \mathcal{B}^c \cup \{B(p_i, r')\})\}$
9.       **for** each candidate ball $B \in \mathcal{B}^c$
10.        **if** $p_i \in$ SELECTBALL$(B, p_i)$
11.          **then** $\mathcal{S}_i \leftarrow \mathcal{S}_i \cup \{(\mathcal{B}^s \cup \{\text{SELECTBALL}(B, p_i)\}, \mathcal{B}^c \setminus \{B\})\}$
12.  **return** $\mathcal{B}^s \cup \mathcal{B}^c$ of a solution in $\mathcal{S}_n$

A proof of the follwoing lemma will appear in the full version of this paper.

**Lemma 8.** *For $|p_1 f(p_1)| \leqslant 4kr^*$ and $r'$ with $1.2r^* \leqslant r' < (1.2 + 2\varepsilon/3)r^*$, algorithm $k$-Centers uses $O((2k)!/((k-1)!2^k))$ space and returns $k$ centers that guarantees $(1.8+\varepsilon)$-approximation to the $k$-center problem. Our algorithm spends $O((2k)!/((k-1)!2^k))$ update time for each point of $\mathcal{X}$.*

### 4.2 Finding $r'$

In this section, we explain how we get $r'$ such that $1.2r^* \leq r' < (1.2 + 2\varepsilon/3)r^*$. The basic procedure works as follows: We maintain $m = \lceil 12k/\varepsilon \rceil$ candidate lengths and update them together with their solutions, as we did in Section 3.2. For each candidate length $\ell$, we assume that $r' = \ell$ and run $k$-*Centers* algorithm in Section 4.1. This is almost the same to the idea of LAYERPARTITION by Ahn et al. [3], except we use $k$-*Centers* as subroutine. By Lemma 5 and Lemma 8, we have the following corollary.

**Corollary 1.** *For $|p_1 f(p_1)| \leq 4kr^*$, our algorithm uses $O((2k)!/(\varepsilon(k-1)!2^k))$ space and returns $k$ centers that guarantees $(1.8 + \varepsilon)$-approximation. Our algorithm spends $O((2k)!/(\varepsilon(k-1)!2^k))$ update time for each point of $\mathcal{X}$.*

### 4.3 The Case $|p_1 f(p_1)| > 4kr^*$

Let $\mathcal{B} = \{B_1, B_2, \ldots, B_k\}$ denote the set of $k$ concentric balls, centered at $p_1$, satisfying following conditions. (1) $B_1 \subset B_2 \subset B_3 \subset \ldots \subset B_k$. (2) $\mathcal{X} \cap B_k \neq \mathcal{X}$. (3) $r(B_j) - r(B_{j-1}) > 2r^*$ for $j = 1, 2, \ldots, k$, where $B_0 = B(p_1, 0)$.

**Lemma 9.** *For $|p_1 f(p_1)| > 4kr^*$, there is some ball $B_j \in \mathcal{B}$ such that $B_j$ does not intersect any ball in $\mathcal{B}^*$ and partitions $\mathcal{B}^*$ into two nonempty subsets along its boundary.*

*Proof.* Because of Condition (2) of $\mathcal{B}$, $b_1$ has radius greater than $2r^*$ and $b_1$ contains every optimal ball in $\mathcal{B}^*$ that contains $p_1$. Note that there is at least one such an optimal ball in $\mathcal{B}^*$.

Again by Condition (2), no ball in $\mathcal{B}^*$ can intersect more than one layer boundary. By the pigeonhole principle, there must be at least one layer $B_j$ for $j = 1, \ldots, k$ whose boundary does not intersect any optimal ball in $\mathcal{B}^*$ and partitions $\mathcal{B}^*$ into two nonempty subsets. □

We maintain $m = 2k$ concentric balls centered at $p_1$ and choose $k$ balls among them satisfying the conditions of $\mathcal{B}$ and one more ball for update. The boundary of each ball $B_j$ divides $\mathbb{R}^d$ into two subspaces, $B_j$ and $\mathbb{R}^d \setminus B_j$. For each subspace of $B_j$, we maintain $t$-center solutions for $t = 1, 2, \ldots, k-1$. It is easy to see that there are integers $\ell$ with $1 \leq \ell \leq k$ and $c$ with $1 \leq c \leq k-1$ such that $B_\ell$ contains $c$ optimal balls and does not intersect $k - c$ optimal balls. Details will be given in the full version of this paper.

Now we analyze the space and update time complexity for our algorithm to the $k$-center problem.

**Theorem 2.** *For any fixed $d \geq 1$, our single-pass data stream algorithm uses $O(2^k(k+3)!/\varepsilon)$ space and returns $k$ centers that guarantees a $(1.8 + \varepsilon)$-approximation for the Euclidean $k$-center problem. Each update takes $O(2^k(k+2)!/\varepsilon)$ time.*

*Proof.* Let $L(k)$ denote the size of the data structure for a $k$-center, let $M(k)$ denote the size of the data structure for a $k$-center for the case $|p_1 f(p_1)| > 4kr^*$, and let $N(k)$ denote the size of the data structure for a $k$-center for the case $|p_1 f(p_1)| \leq 4kr^*$. We have

$$L(k) = N(k) + M(k), \text{ with base cases } L(1) = L(2) = O(1/\varepsilon).$$
$$M(k) = 2(k+1)\big(L(k-1) + L(k-2) + \cdots + L(2) + L(1)\big)$$
$$N(k) = O((2k)!/(\varepsilon 2^k(k-1)!)) \leq O((2^k(k+1)!)/\varepsilon)$$

By letting $\mathcal{L}(k) = \sum_{i=1}^{k} L(k)$, we have

$$\begin{aligned}
L(k) &= N(k) + 2(k+1)\mathcal{L}(k-1) \\
&= N(k) + 2(k+1)\big(2k(\mathcal{L}(k-2)) + N(k-1) + \mathcal{L}(k-2)\big) \\
&= N(k) + 2(k+1)N(k-1) + 2(k+1)(2k+1)\mathcal{L}(k-2) \\
&= N(k) + 2(k+1)N(k-1) \\
&\quad + 2(k+1)(2k+1)\big(2(k-1)\mathcal{L}(k-3) + N(k-2) + \mathcal{L}(k-3)\big) \\
&= N(k) + 2(k+1)N(k-1) + 2(k+1)(2k+1)N(k-2) \\
&\quad 2(k+1)(2k+1)(2(k-1)+1)\mathcal{L}(k-3)
\end{aligned}$$

Therefore, we can rearrange the equation as follows.

$$\begin{aligned}
L(k) &= N(k) + \sum_{i=1}^{k-1}\left(2(k+1)\cdot N(k-i)\prod_{j=0}^{i-2}(2(k-j)+1)\right) \\
&\leqslant N(k) + \sum_{i=1}^{k-1}\left(2(k+1)\cdot N(k-i)\cdot\frac{2^{i-1}(k+1)!}{(k-i+2)!}\right) \\
&\leqslant N(k) + \sum_{i=1}^{k-1}\left(\frac{2^{i}(k+2)!}{(k-i+2)!}\cdot N(k-i)\right) \\
&\leqslant O\Big(\frac{2^{k}(k+1)!}{\varepsilon}\Big) + \sum_{i=1}^{k-1}\left(\frac{2^{i}(k+2)!}{(k-i+2)!}\cdot O\Big(\frac{2^{k-i}(k-i+1)!}{\varepsilon}\Big)\right) \\
&\leqslant O\Big(\frac{2^{k}(k+1)!}{\varepsilon}\Big) + \sum_{i=1}^{k-1} O\Big(\frac{2^{k}(k+2)!}{\varepsilon}\Big) \leqslant O\Big(\frac{2^{k}(k+3)!}{\varepsilon}\Big)
\end{aligned}$$

We reuse $L(k)$, $M(k)$, and $N(k)$ denote the update time for each related algorithm. Each ball $B \in \mathcal{B}$ maintains two sets of data structures: one set consisting of data structures for points lying inside $b$ and another set consisting of data structures for points lying outside of $b$. When a new point $p$ arrives, only one of two sets is updated depending on whether $p \in b$ or not. Therefore we have a bit difference $M(k)$ and this makes the different result.

$$\begin{aligned}
M(k) &= (k+1)\big(L(k-1) + L(k-2) + \cdots + L(2) + L(1)\big) \\
L(k) &= N(k) + (k+1)\mathcal{L}(k-1) = \cdots \\
&= N(k) + \sum_{i=1}^{k-1}\left((k+1)\cdot N(k-i)\prod_{j=0}^{i-2}((k-j)+1)\right) \\
&\leqslant O\Big(\frac{2^{k}(k+1)!}{\varepsilon}\Big) + \sum_{i=1}^{k-1} O\Big(\frac{2^{k-i}(k+2)!}{\varepsilon}\Big) \leqslant O\Big(\frac{2^{k}(k+2)!}{\varepsilon}\Big)
\end{aligned}$$

□

# References

1. Agarwal, P.K., Sharathkumar, R.: Streaming algorithms for extent problems in high dimensions. In: Proc. of the 21st ACM-SIAM Sympos. Discrete Algorithms, pp. 1481–1489 (2010)
2. Aggarwal, C.C.: Data streams: models and algorithms. Springer (2007)
3. Ahn, H.-K., Kim, H.-S., Kim, S.-S., Son, W.: Computing $k$-center over streaming data for small $k$. In: Proc. of the 23rd Int. Sympos. Algorithms and Computation, pp. 54–63 (2012)
4. Bonnell, I., Bate, M., Vine, S.: The hierarchical formation of a stellar cluster. Monthly Notices of the Royal Astronomical Society 343(2), 413–418 (2003)
5. Chan, T.M., Pathak, V.: Streaming and dynamic algorithms for minimum enclosing balls in high dimensions. In: Proc. of the 12th Int. Conf. on Algorithms and Data Structures, pp. 195–206 (2011)
6. Clarke, C., Bonnell, I., Hillenbrand, L.: The formation of stellar clusters. In: Mannings, V., Boss, A., Russell, S. (eds.) Protostars and Planets IV, pp. 151–177. University of Arizona Press, Tucson (2000)
7. Fayyad, U., Piatetsky-Shapiro, G., Smyth, P.: From data mining to knowledge discovery in databases. AI Magazine 17, 37–54 (1996)
8. Guha, S.: Tight results for clustering and summarizing data streams. In: Proc. of the 12th Int. Conf. on Database Theory, pp. 268–275. ACM (2009)
9. Han, J., Kamber, M.: Data mining: concepts and techniques. Morgan Kaufmann (2006)
10. Hershberger, J., Suri, S.: Adaptive sampling for geometric problems over data streams. Computational Geometry 39(3), 191–208 (2008)
11. Zarrabi-Zadeh, H.: Core-preserving algorithms. In: Proc. of 20th Canadian Conf. on Computational Geometry, pp. 159–162 (2008)
12. Poon, C.K., Zhu, B.: Streaming with minimum space: An algorithm for covering by two congruent balls. In: Lin, G. (ed.) COCOA 2012. LNCS, vol. 7402, pp. 269–280. Springer, Heidelberg (2012)
13. Sonka, M., Hlavac, V., Boyle, R.: Image processing, analysis, and machine vision, 3rd edn. Thomson Learning (2007)
14. Zarrabi-Zadeh, H., Chan, T.: A simple streaming algorithm for minimum enclosing balls. In: Proc. of 18th Canadian Conf. on Computational Geometry, pp. 139–142 (2006)

# Approximation Algorithms for the Gromov Hyperbolicity of Discrete Metric Spaces

Ran Duan[*]

Max-Planck-Institut für Informatik, Saarbrücken, Germany
duanran@mpi-inf.mpg.de

**Abstract.** This paper discusses new approximation algorithms for computing the Gromov hyperbolicity of an $n$-point discrete metric space. We give a $(1+\epsilon)$-approximation algorithm with running time $\tilde{O}(\epsilon^{-1}n^{1+\omega})$, where $O(n^\omega) = O(n^{2.373})$ is the time complexity of matrix multiplications. Here an $\alpha$-approximation $\delta'$ means $\delta' \leq \delta^* \leq \alpha\delta'$ for the Gromov hyperbolicity $\delta^*$. We also give a $(2+\epsilon)$-approximation algorithm with running time $\tilde{O}(\epsilon^{-1}n^\omega)$. These are faster than the previous $O(n^{(5+\omega)/2})$-time algorithm for the exact solution and the $O(n^{(3+\omega)/2})$-time algorithm for a 2-approximation [Fournier, Ismail and Vigneron 2012], which directly perform (max, min)-product of matrices.

## 1 Introduction

The Gromov hyperbolicity of a metric space is an important concept in metric geometry [1,9]. In the definition by the *four-point condition*, a metric space $(M, d)$ is $\delta$-hyperbolic if for any $x, y, z, r \in M$, the two largest of the distance sums $d(x,y)+d(z,r)$, $d(x,z)+d(y,r)$, and $d(x,r)+d(y,z)$ differ by at most $2\delta$. The Gromov hyperbolicity $\delta^*$ of $(M,d)$ is the smallest such $\delta$. Thus, the trivial algorithm takes $O(n^4)$ time when the space is discrete and $|M| = n$. Recently, Fournier, Ismail and Vigneron [7] gave a straightforward algorithm with running time $O(n^{(5+\omega)/2}) = O(n^{3.687})$, which utilizes the algorithm for the (max, min)-product of matrices with running time $O(n^{(3+\omega)/2})$ [6]. They also gave a 2-approximation $O(n^{(3+\omega)/2})$-time algorithm and a $2\log_2 n$-approximation $O(n^2)$-time algorithm, which used the tree-metric embedding by Gromov [9]. Here an $\alpha$-approximation $\delta'$ means $\delta' \leq \delta^* \leq \alpha\delta'$, and $O(n^\omega)$ is the time needed for multiplying two $n \times n$ matrices. $\omega$ had been less than 2.376 for a long time [5] and now $\omega$ is proved to be less than 2.373 [11].

Many papers studied the Gromov hyperbolicity of particular types of graphs, and the properties on graphs with different hyperbolicities. [8,2,3,10]

*Our results.* In this paper, we give two approximation algorithms for computing the Gromov hyperbolicity of an $n$-point discrete metric space. The first one is a $(1+\epsilon)$-approximation algorithm of running time $\tilde{O}(\epsilon^{-1}n^{1+\omega}) = \tilde{O}(\epsilon^{-1}n^{3.373})$. As in [7,6], we need to compute a (max, min)-product of the Gromov product

---

[*] The author is supported by an Alexander von Humboldt Postdoctoral Fellowship.

matrix for every base $r$ to obtain the exact Gromov hyperbolicity $\delta^*$, and the (max, min)-matrix multiplication takes $O(n^{(3+\omega)/2})$ time. However, if we know that all the elements of the matrix of Gromov product are within a constant (or polylogarithmic) factor of $\delta^*$, then the $(1+\epsilon)$-approximate (max, min)-product can be computed in $\tilde{O}(\epsilon^{-1}n^\omega)$ time by a simpler approach. (Here we will need to find a 2-approximation of $\delta^*$ first, which takes $O(n^{(3+\omega)/2})$ time as in [7].) When some elements $(i,j), (i,k)$ in a line of the matrix differ by a large number times the approximation of $\delta^*$, we can bound $(j,k)$ to be in a smaller range since otherwise $\delta^*$ will be much larger than the approximation. After sorting the elements in such a line, we can partition the matrix into several parts, some of which are within a small range, so we can deal with them directly. For other parts, recursively run the procedure. We can see the running time is still $\tilde{O}(\epsilon^{-1}n^\omega)$, so the total running time is $\tilde{O}(\epsilon^{-1}n^{1+\omega})$ for all bases.

The second algorithm finds a $(2+\epsilon)$-approximation of $\delta^*$ with running time $\tilde{O}(\epsilon^{-1}n^\omega) = \tilde{O}(\epsilon^{-1}n^{2.373})$. As in the previous algorithm, we can first find a $O(\log n)$-approximation of $\delta^*$ with running time only $O(n^2)$ [7], then run the recursive procedure above on an arbitrarily selected base with this approximation. By [1], it will be a $(2+\epsilon)$-approximation. Note that if at the beginning we are given an undirected graph without the distance metric $d$, we can compute the $(1+\epsilon)$-approximation of the distance matrix in $\tilde{O}(\epsilon^{-1}n^\omega)$ time by [12].

## 2 Definitions and Basic Algorithms

### 2.1 Definitions of Gromov Hyperbolic Spaces

A metric space $(M,d)$ is $\delta$-hyperbolic ($\delta \geq 0$) if it satisfies the *four point condition*: For any $x,y,z,r \in M$, the two largest of the distance sums $d(x,y)+d(z,r)$, $d(x,z)+d(y,r)$, and $d(x,r)+d(y,z)$ differ by at most $2\delta$. The *Gromov hyperbolicity* $\delta^*$ of $(M,d)$ is the smallest $\delta^* \geq 0$ such that $(M,d)$ is $\delta^*$-hyperbolic.

There is another equivalent definition of Gromov hyperbolicity. For $x,y,r \in M$, the *Gromov product* of $x,y$ at $r$ is defined as:

$$(x|y)_r = \frac{1}{2}(d(x,r) + d(y,r) - d(x,y)). \tag{1}$$

The point $r$ is called the base point. The metric space $(M,d)$ is $\delta$-hyperbolic if, for any $x,y,z,r \in M$,

$$(x|z)_r \geq \min\{(x|y)_r, (y|z)_r\} - \delta. \tag{2}$$

So the Gromov hyperbolicity $\delta^*$ can be found as:

$$\delta^* = \max_{x,y,z,r}\{\min\{(x|y)_r, (y|z)_r\} - (x|z)_r\}. \tag{3}$$

Also define $\delta_r$ for a fixed base point $r$ to be:

$$\delta_r = \max_{x,y,z}\{\min\{(x|y)_r, (y|z)_r\} - (x|z)_r\}. \tag{4}$$

## 2.2 Basic Algorithms

An exact algorithm and a 2-approximation algorithm are given in [7], using the (max, min)-product for matrices. The (max, min)-product of two real matrices $A, B$ is defined as:

$$(A \otimes B)[i, j] = \max_k \min\{A[i, k], B[k, j]\} \quad (5)$$

Duan and Pettie [6] gave an algorithm for computing the (max, min)-product of two $n \times n$ matrices in $O(n^{(3+\omega)/2})$ time.

For a fixed base point $r$, we construct the $n \times n$ matrix $A_r$ in which $A_r[x, y] = (x|y)_r$, then we can see $\delta_r$ is the largest element of $A_r \otimes A_r - A_r$, so it can be computed in $O(n^{(3+\omega)/2})$ time. Since $\delta_r$ for any base point $r$ is a 2-approximation of $\delta^*$ [1], we have a 2-approximation algorithm in time $O(n^{(3+\omega)/2})$. Also note that since $(M, d)$ is a metric space, every $A_r$ is symmetric.

By computing $A_r \otimes A_r - A_r$ for every $r \in M$, we can find the largest element $\delta^*$. This takes $O(n^{(5+\omega)/2}) = O(n^{3.686})$ time.

## 3 Approximation Algorithms

In this section except in Theorem 2, we assume that we have already computed a 2-approximation $\delta'$ in $O(n^{(3+\omega)/2})$ time by the method of [7], that is, $\delta' \leq \delta^* \leq 2\delta'$. So for any base point $r$, we can guarantee that all the elements in $A_r \otimes A_r - A_r$ are at most $2\delta'$. We need to compute a $(1 + \epsilon)$-approximation of $A_r \otimes A_r - A_r$ for every $r$.

Note that although we have faster algorithms for rectangular matrix multiplications [4], we only need to divide rectangular matrices into square matrices in this paper. So when we multiply an $n^a \times n^a$ matrix with an $n^a \times n^b$ matrix, or an $n^a \times n^b$ matrix with an $n^b \times n^a$ matrix, where $b > a$, the running time is $O(n^{a(\omega-1)+b})$.

In Section 3.1, we will discuss the case when all the elements in the matrices are within a small number of multiples of $\delta'$. In Section 3.2, we provide the algorithm for the general case which uses the algorithm of Section 3.1 as a subroutine.

### 3.1 A Scaling Algorithm

Given an $n^a \times n^b$ matrix $A$ and an $n^b \times n^c$ matrix $B$, if all the elements in $A$ and $B$ are within a range of size $K \cdot \delta'$ ($K$ is a small number), that is, there exists $p$ such that $p \leq A[i, j] \leq p + K\delta'$ and $p \leq B[i', j'] \leq p + K\delta'$ for every $i, j, i', j'$, then we can scale $A, B$ to $A', B'$:

$$A'[i, j] = \lfloor (A[i, j] - p)/(\epsilon \cdot \delta') \rfloor \quad (6)$$
$$B'[i', j'] = \lfloor (B[i', j'] - p)/(\epsilon \cdot \delta') \rfloor \quad (7)$$

We can see the elements in $A'$ and $B'$ are all integers between $[0, \epsilon^{-1}K]$. For such matrices of small integers, we have the following lemma:

**Lemma 1.** *Given an $n^a \times n^b$ matrix $A'$ and an $n^b \times n^c$ matrix $B'$, in which all elements are integers between $[0, N]$, we can compute the (max, min)-product $C' = A' \otimes B'$ in $O(Nn^{\omega(a,b,c)})$ time, where $\omega(a,b,c)$ is the exponential of the time complexity for computing the Boolean matrix multiplication of an $n^a \times n^b$ matrix and an $n^b \times n^c$ matrix.*

*Proof.* For every integer $q = N, \cdots, 0$, we find the matrices $A_q$ and $B_q$:

$$A_q[i,j] = \begin{cases} 1 \text{ if } A'[i,j] \geq q; \\ 0 \text{ Otherwise.} \end{cases}$$

$$B_q[i,j] = \begin{cases} 1 \text{ if } B'[i,j] \geq q; \\ 0 \text{ Otherwise.} \end{cases}$$

Compute $A_q \cdot B_q$ for all $q = N, \cdots, 0$. If $(A_q \cdot B_q)[i,j] = 1$, then there exists $k$ such that $\min\{A'[i,k], B'[k,j]\} \geq q$, so $C'[i,j] \geq q$. So, for every $i,j$, we pick the largest $q$ such that $(A_q \cdot B_q)[i,j] = 1$, which means $C'[i,j] = q$. Thus, to compute $C'$, we need to compute $N+1$ matrix multiplications.

**Fact 1.** *For real numbers $a_1, a_2, \cdots, a_n$, we have:*

$$\min\{\lfloor a_1 \rfloor, \lfloor a_2 \rfloor, \cdots, \lfloor a_n \rfloor\} = \lfloor \min\{a_1, a_2, \cdots, a_n\} \rfloor \tag{8}$$
$$\max\{\lfloor a_1 \rfloor, \lfloor a_2 \rfloor, \cdots, \lfloor a_n \rfloor\} = \lfloor \max\{a_1, a_2, \cdots, a_n\} \rfloor \tag{9}$$

**Lemma 2.** *We can compute an approximation $C$ of $C^* = A \otimes B$ in which $C^*[i,j] - \epsilon \cdot \delta' < C[i,j] \leq C^*[i,j]$ for all $i,j$ in $O(\epsilon^{-1} K \cdot n^{\omega(a,b,c)})$ time.*

*Proof.* We first compute $C' = A' \otimes B'$ by Lemma 1. From Fact 1, $C'[i,j] = \lfloor (C^*[i,j] - p)/(\epsilon \cdot \delta') \rfloor$, where $C^*$ is the (max, min)-product of $A$ and $B$. Let $C$ be an $n^a \times n^c$ matrix, in which $C[i,j] = (\epsilon \cdot \delta')C'[i,j] + p$, then we have:

$$C^*[i,j] - \epsilon \cdot \delta' < C[i,j] \leq C^*[i,j] \tag{10}$$

This approximation $C$ can be computed in $O(\epsilon^{-1} K \cdot n^{\omega(a,b,c)})$ time by Lemma 1.

### 3.2 The Main Algorithm

This section will discuss the main procedure for computing an approximate (max, min)-product of two $n \times n$ matrices with additive error $(-\epsilon\delta', 0]$, since we need a $(1+\epsilon)$-approximation of $A_r \otimes A_r - A_r$ for every base $r$.

The recursive procedure Max-Min for finding the approximate (max, min)-product of $A \otimes A$ is described as follows: (We will describe the whole algorithm first, then prove the lemmas used in it.)

- The input is an $n \times n$ symmetric matrix A, and the output is an $n \times n$ matrix C.
- Sort all the elements of every row of $A$;

- Select a row of $A$ in which the maximum and the minimum elements differ by more than $8\log_2 n \cdot \delta'$. If there is no such a row, which means all rows and columns are within a range of $8\log_2 n \cdot \delta'$, so all elements in $A$ are within a range of $16\log_2 n \cdot \delta'$, then find the approximate (max, min) product $C$ of $A$ and $A$ by the procedure in Section 3.1 in $O(\epsilon^{-1}\log n \cdot n^\omega)$ time;
- Sort all the columns of $A$ by the elements of the row selected at previous step in increasing order, also rearrange all the rows in the same order so that $A$ is still symmetric, then $A[i,1] \leq A[i,2] \leq \cdots \leq A[i,n]$ and $A[i,n] - A[i,1] > 8\log_2 n \cdot \delta'$, if the row in step 2 is now row $i$;
- In the list $\{A[i,1], A[i,2], A[i,4], \cdots, A[i,2^{\lfloor \log_2 n-1 \rfloor}], A[i,\lfloor \frac{n}{2} \rfloor + 1]\}$, find the first two consecutive elements $A[i,j], A[i,k]$ in which $A[i,k] - A[i,j] > 4\delta'$, so $k \leq 2j, k-1 \leq n/2$ and $A[i,j] - A[i,1] \leq 4\log_2 n \cdot \delta'$. If there is no such $j,k$, we can see $A[i,\lfloor \frac{n}{2} \rfloor + 1] - A[i,1] \leq 4\log_2 n \cdot \delta'$, then we start from the end to get a list: $\{A[i,n], A[i,n-1], A[i,n-3], \cdots, A[i,n+1-2^{\lfloor \log_2 n-1 \rfloor}], A[i,n-\lfloor \frac{n}{2} \rfloor]\}$. Find the first two consecutive elements $A[i,j], A[i,k]$ in which $A[i,j] - A[i,k] > 4\delta'$, which must exist since otherwise $A[i,n] - A[i,1] \leq 8\log_2 n \cdot \delta'$. Since these two cases starting from beginning and end of line $i$ are symmetric, w.l.o.g., we only consider the first case in the following discussion, that is, $A[i,k] - A[i,j] > 4\delta'$, $k \leq 2j, k-1 \leq n/2$ and $A[i,j] - A[i,1] \leq 4\log_2 n \cdot \delta'$.
- Based on the $j$ and $k$ of the previous step, we can partition the symmetric matrix $A$ into:

$$A = \begin{pmatrix} J & L^T & P^T \\ L & K & Q^T \\ P & Q & S \end{pmatrix}, \quad (11)$$

where each line and column is partitioned into 3 parts: $\{1, \cdots, j\}, \{j+1, \cdots, k-1\}, \{k, \cdots, n\}$. By Lemma 5, the following statements hold:
  - All elements in $A$ are at least $A[i,1] - 2\delta'$;
  - Elements in $P$ are at most $A[i,j] + 2\delta'$, so elements in $P$ are within a range of $O(\log n \cdot \delta')$;
  - Elements in $S$ are larger than $A[i,j] + 2\delta'$.
- Recursively call this approximate Max-Min procedure for $(k-1) \times (k-1)$ symmetric matrix $\begin{pmatrix} J & L^T \\ L & K \end{pmatrix}$, and for $(n-j) \times (n-j)$ symmetric matrix $\begin{pmatrix} K & Q^T \\ Q & S \end{pmatrix}$. (Note that there is no problem with matrix overlapping since only maximum operations are needed when we put them together.)
- To compute $A \otimes A$, we also need to compute the following: $P^T \otimes P$, $P^T \otimes Q$, $J \otimes P^T$, $Q \otimes L$, $P^T \otimes S$, $L \otimes P^T$, $P \otimes P^T$.
  - For $\begin{pmatrix} J \\ L \end{pmatrix} \otimes P^T$, since the elements in $P^T$ are all at most $A[i,j] + 2\delta'$, we can replace the elements larger than $A[i,j] + 2\delta'$ in $J$ and $L$ by $A[i,j] + 2\delta'$. Then run the algorithm in Section 3.1 since all the elements after modification are within a range of $O(\log n \cdot \delta')$. Since $k \leq 2j, n-k > j$, computing a $(k-1) \times j$ and a $j \times (n-k+1)$ matrix multiplication takes $O(j^{\omega-1}(n-k))$ time. The running time is thus $O(\epsilon^{-1}\log n \cdot j^{\omega-1}(n-k))$ by Lemma 2.

- Similarly, we can compute $P^T \otimes (P, Q)$ in $O(\epsilon^{-1} \log n \cdot j^{\omega-1}(n-k))$ time.
- Compute $P^T \otimes S$ is easier since all elements in $S$ are larger than all elements in $P$. This only takes $O(j(n-k))$ time.
- The elements in $L^T \otimes Q^T$ are all at most $A[i,j] + 4\delta'$. This is because all the elements in $A \otimes A - A$ are at most $2\delta'$ by Lemma 4, and $L^T \otimes Q^T$ will be minus $P^T$ in $A \otimes A - A$. So we can replace all elements larger than $A[i,j] + 4\delta'$ in $L^T$ and $Q^T$ by $A[i,j] + 4\delta'$, and compute $L^T \otimes Q^T$ by the scaling procedure in Section 3.1. The running time is also $O(\epsilon^{-1} \log n \cdot j^{\omega-1}(n-k))$.
- We do not need to compute $P \otimes P^T$ since its position in $A \otimes A$ is at $S$, and elements of $S$ are all larger than $P \otimes P^T$. Set $P \times P^T = O$, where $O$ is an $(n-k+1) \times (n-k+1)$ zero matrix.

**Lemma 3.** *The matrix $C$ returned by this procedure satisfies the following for every $i, j$:*

$$(A \otimes A)[i,j] - \epsilon \delta' < C[i,j] \leq (A \otimes A)[i,j], \tag{12}$$

*when $(A \otimes A)[i,j] \geq A[i,j]$.*

*Proof.* If the result $C$ is obtained directly from the procedure in Section 3.1, it is trivial.

Otherwise $C$ is obtained both from recursive calls and from the procedure in Section 3.1 for rectangular matrices. We only need to check if $\min\{A[x,y], A[y,z]\}$ for every $y$ is taken into account for every $C[x,z]$. From the partition, $A \otimes A$ equals: (Here $(A_1 \oplus A_2)[i,j] = \max\{A_1[i,j], A_2[i,j]\}$).

$$\begin{pmatrix} B_{11} & B_{12} & B_{13} \\ B_{12}^T & B_{22} & B_{23} \\ B_{13}^T & B_{23}^T & B_{33} \end{pmatrix} \tag{13}$$

where

$$B_{11} = J \otimes J \oplus L^T \otimes L \oplus P^T \otimes P \tag{14}$$
$$B_{12} = J \otimes L^T \oplus L^T \otimes K \oplus P^T \otimes Q \tag{15}$$
$$B_{13} = J \otimes P^T \oplus L^T \otimes Q^T \oplus P^T \otimes S \tag{16}$$
$$B_{22} = L \otimes L^T \oplus K \otimes K \oplus Q^T \otimes Q \tag{17}$$
$$B_{23} = L \otimes P^T \oplus K \otimes Q^T \oplus Q^T \otimes S \tag{18}$$
$$B_{33} = P \otimes P^T \oplus Q \otimes Q^T \oplus S \otimes S \tag{19}$$

and,

$$\begin{pmatrix} J & L^T \\ L & K \end{pmatrix} \otimes \begin{pmatrix} J & L^T \\ L & K \end{pmatrix} = \begin{pmatrix} J \otimes J \oplus L^T \otimes L & J \otimes L^T \oplus L^T \otimes K \\ L \otimes J \oplus K \otimes L & L \otimes L^T \oplus K \otimes K \end{pmatrix} \tag{20}$$

$$\begin{pmatrix} K & Q^T \\ Q & S \end{pmatrix} \otimes \begin{pmatrix} K & Q^T \\ Q & S \end{pmatrix} = \begin{pmatrix} K \otimes K \oplus Q^T \otimes Q & K \otimes Q^T \oplus Q^T \otimes S \\ Q \otimes K \oplus S \otimes Q & Q \otimes Q^T \oplus S \otimes S \end{pmatrix} \tag{21}$$

We can check that every product of sub-matrices are considered except $P \otimes P^T$. By Lemma 4, $P \otimes P^T$ is at the position of $S$, but elements of $S$ are all larger than

$P \otimes P^T$, not satisfying $(A \otimes A)[i,j] \geq A[i,j]$. Thus, $P \otimes P^T$ is not considered in the algorithm.

In the rest of this section we prove the statements in the procedure and analyze the running time.

**Lemma 4.** *In every recursive calls of the Max-Min procedure with matrix $A'$, $A' = A[u \cdots v, u \cdots v]$ for some $u < v$. All the elements in $A' \otimes A' - A'$ are at most $2\delta'$.*

*Proof.* It is trivial for the original $A$ since $\delta'$ is a 2-approximation. For $A'$ in the recursive calls, we can see it must be $A[u \cdots v, u \cdots v]$, that is, the diagonal of $A'$ is contained in the diagonal of $A$. If $A[x,y], A[y,z] \in A'$, $A[x,z]$ is also in $A'$, and the position of $A' \otimes A'$ in $A \otimes A$ will also at the position $[u \cdots v, u \cdots v]$. So the statements hold.

**Lemma 5.** *When partitioning the matrix into 9 parts by partitioning each line into 3 parts $\{1, \cdots, j\}, \{j+1, \cdots, k-1\}, \{k, \cdots, n\}$ as in the algorithm, the following statements hold:*

1. *All elements in $A$ are at least $A[i,1] - 2\delta'$;*
2. *Elements in $P$ are at most $A[i,j] + 2\delta'$;*
3. *Elements in $S$ are larger than $A[i,j] + 2\delta'$.*

*Proof.* By Lemma 4, $A$ in this procedure is a matrix of Gromov product for some subsets of $M$ at some base $r$. Suppose there is an element $A[p,q] < A[i,1] - 2\delta'$, so $(p|q)_r < A[i,1] - 2\delta' \leq A[i,p] - 2\delta' = (i|p)_r - 2\delta'$ and $(p|q)_r < (i|q)_r - 2\delta'$ similarly. So $\min\{(i|p)_r, (i|q)_r\} - (p|q)_r > 2\delta'$, contradicting that $\delta'$ is a 2-approximation. Thus (1) is proved.

If $A[p,q] \in P$, then $p \geq k$ and $q \leq j$, so $A[i,p] - A[i,q] > 4\delta'$. If $A[p,q]$ is also larger than $A[i,j] + 2\delta' \geq A[i,q] + 2\delta'$, then $\min\{A[i,p], A[p,q]\} - A[i,q] > 2\delta'$, coming to a contradiction.

If $A[p,q] \in S$, $p,q \geq k$, $A[i,p]$ and $A[i,q]$ are both larger than $A[i,j] + 4\delta'$. So $A[p,q]$ must be larger than $A[i,j] + 2\delta'$ since otherwise $\min\{A[i,p], A[i,q]\} - A[p,q] > 2\delta'$.

**Lemma 6.** *The procedure Max-Min runs in $O(\epsilon^{-1} \log n \cdot n^\omega)$ time.*

*Proof.* Denote the running time of Max-Min with an $n \times n$ matrix by $\Gamma(n)$, then: ($K_1$ and $K_2$ are constants replacing $O(\cdot)$.)

$$\Gamma(n) = \begin{cases} K_1 \epsilon^{-1} \log n \cdot n^\omega & \text{If directly computed by Section 3.1;} \\ \Gamma(k-1) + \Gamma(n-j) + K_2 \epsilon^{-1} \log n \cdot j^{\omega-1}(n-k) & \\ & \text{Here } k \leq 2j \text{ and } k-1 < n/2. \end{cases}$$

Assume $\Gamma(n) = K_3 \epsilon^{-1} \log n \cdot n^\omega$ where $K_3 > K_1, K_2$. If the result is directly computed by the scaling algorithm in Section 3.1, it is trivial. Otherwise assume it is proved $\Gamma(k-1) = K_3 \epsilon^{-1} \log n(k-1)^\omega$ and $\Gamma(n-j) = K_3 \epsilon^{-1} \log n(n-j)^\omega$.

We denote the number of elements of a matrix $B$ by $|B|$, then:

$$\left|\begin{pmatrix} J & L^T \\ L & K \end{pmatrix}\right| + \left|\begin{pmatrix} K & Q^T \\ Q & S \end{pmatrix}\right| + 2|P| - |K| = |A| \tag{22}$$

Since $k - j - 1 < j$ and $k - j - 1 < n/2 < n - k + 1$, so $|P| > |K|$. so we have:

$$(k-1)^2 + (n-j)^2 + j(n-k+1) < n^2 \tag{23}$$

Since $\omega \geq 2$, we have:

$$(k-1)^\omega + (n-j)^\omega + j^{\omega-1}(n-k) < n^\omega \tag{24}$$

and,

$$\Gamma(k-1) + \Gamma(n-j) + K_2\epsilon^{-1}\log n \cdot j^{\omega-1}(n-k) < K_3\epsilon^{-1}\log n \cdot n^\omega = \Gamma(n) \tag{25}$$

Thus, $\Gamma(n) = O(\epsilon^{-1}\log n \cdot n^\omega)$ is proved.

The main results are summarized in the following two theorems.

**Theorem 1.** *A $(1+\epsilon)$-approximation of the Gromov hyperbolicity $\delta^*$ of a discrete metric space $(M, d)$ can be computed in $\tilde{O}(\epsilon^{-1}n^{1+\omega})$ time.*

*Proof.* We can compute a 2-approximation $\delta'$ in $O(n^{(3+\omega)/2})$ time by the method of [7]. Then run the Max-Min procedure with $A_r$ for every base $r \in M$, so we will get an approximate of $C_r = A_r \otimes A_r$ in which every elements $(A_r \otimes A_r)[i,j] - \epsilon\delta' < C_r[i,j] \leq (A_r \otimes A_r)[i,j]$, when $(A_r \otimes A_r)[i,j] \geq A_r[i,j]$, by Lemma 3. Pick the largest element $\delta''$ of all $C_r - A_r$ for every $r \in M$, we have: $\delta^* - \epsilon\delta' < \delta'' \leq \delta^*$ (since $\delta^* \geq 0$). Note that $\delta' < \delta^*$, we can obtain a $(1+\epsilon)$-algorithm with running time $\tilde{O}(\epsilon^{-1}n^{1+\omega})$ since we need to run the algorithm for every base $r$.

**Theorem 2.** *A $(2+\epsilon)$-approximation of the Gromov hyperbolicity $\delta^*$ of a discrete metric space can be computed in $\tilde{O}(\epsilon^{-1}n^\omega)$ time.*

*Proof.* They provide a $2\log_2 n$-approximation algorithm in $O(n^2)$ time [7]. When we get a $2\log_2 n$-approximation $\delta'$, run the Max-Min procedure for an arbitrarily chosen base $r$, in which the estimation $\delta'$ is replaced by $\delta'' = \log_2 n \cdot \delta'$ and the approximate factor is $\epsilon'' = \frac{\epsilon}{\log_2 n}$. Since $\delta^* \leq 2\log_2 n \cdot \delta'$, $\delta^* \leq 2\delta''$, so all the statements in the Max-Min procedure hold. Also $\delta' \leq \delta^*$, so $\delta'' \leq \log_2 n \cdot \delta^*$, and $\epsilon''\delta'' \leq \epsilon\delta^*$. Finally we get a solution $\delta'_r$ for the base $r$: (if $\delta_r \geq 0$)

$$\delta_r - \epsilon''\delta'' < \delta'_r \leq \delta_r \tag{26}$$

$$\frac{1}{2}\delta^* - \epsilon\delta^* \leq \delta_r - \epsilon\delta^* < \delta'_r \leq \delta_r \leq \delta^* \tag{27}$$

Since $\delta_r$ for any base $r$ is a 2-approximation of $\delta^*$ [1], so $\delta_r \geq 0$, and a $(2+\epsilon)$-approximation of the Gromov hyperbolicity $\delta^*$ can be obtained in $\tilde{O}(\epsilon^{-1}n^\omega)$ time.

# References

1. Bonk, M., Schramm, O.: Embeddings of Gromov hyperbolic spaces. Geom. Funct. Anal. 10, 266–306 (2000)
2. Chen, W., Fang, W., Hu, G., Mahoney, M.W.: On the hyperbolicity of small-world and tree-like random graphs. In: Chao, K.-M., Hsu, T.-s., Lee, D.-T. (eds.) ISAAC 2012. LNCS, vol. 7676, pp. 278–288. Springer, Heidelberg (2012)
3. Chepoi, V., Dragan, F.F., Estellon, B., Habib, M., Vaxès, Y., Xiang, Y.: Additive spanners and distance and routing labeling schemes for hyperbolic graphs. Algorithmica 62(3-4), 713–732 (2012)
4. Coppersmith, D.: Rectangular matrix multiplication revisited. J. Complex. 13(1), 42–49 (1997)
5. Coppersmith, D., Winograd, T.: Matrix multiplication via arithmetic progressions. In: Proc. 19th ACM Symp. on the Theory of Computing (STOC), pp. 1–6 (1987)
6. Duan, R., Pettie, S.: Fast algorithms for (max, min)-matrix multiplication and bottleneck shortest paths. In: SODA 2009: Proceedings of the twentieth Annual ACM-SIAM Symposium on Discrete Algorithms, pp. 384–391. Society for Industrial and Applied Mathematics, Philadelphia (2009)
7. Fournier, H., Ismail, A., Vigneron, A.: Computing the Gromov hyperbolicity of a discrete metric space. CoRR, abs/1210.3323 (2012)
8. Gavoille, C., Ly, O.: Distance labeling in hyperbolic graphs. In: Deng, X., Du, D.-Z. (eds.) ISAAC 2005. LNCS, vol. 3827, pp. 1071–1079. Springer, Heidelberg (2005)
9. Gromov, M.: Hyperbolic groups. In: Gersten, S. (ed.) Essays in Group Theory. Mathematical Sciences Research Institute Publications, vol. 8, pp. 75–263. Springer, New York (1987)
10. Krioukov, D.V., Papadopoulos, F., Kitsak, M., Vahdat, A., Bogu, M.: Hyperbolic geometry of complex networks. CoRR, abs/1006.5169 (2010)
11. Williams, V.V.: Multiplying matrices faster than Coppersmith-Winograd. In: Proceedings of the 44th Symposium on Theory of Computing, STOC 2012, pp. 887–898. ACM, New York (2012)
12. Zwick, U.: All pairs shortest paths using bridging sets and rectangular matrix multiplication. J. ACM 49(3), 289–317 (2002)

# A (7/2)-Approximation Algorithm for Guarding Orthogonal Art Galleries with Sliding Cameras

Stephane Durocher[1,*], Omrit Filtser[2], Robert Fraser[1],
Ali D. Mehrabi[3,**], and Saeed Mehrabi[1,***]

[1] Department of Computer Science, University of Manitoba, Canada
[2] Department of Computer Science, Ben-Gurion University of the Negev, Israel
[3] Department of Mathematics and Computer Science,
Eindhoven University of Technology, The Netherlands
{durocher,fraser,mehrabi}@cs.umanitoba.ca, omritna@cs.bgu.ac.il,
amehrabi@win.tue.nl

**Abstract.** Consider a sliding camera that travels back and forth along an orthogonal line segment $s$ inside an orthogonal polygon $P$ with $n$ vertices. The camera can see a point $p$ inside $P$ if and only if there exists a line segment containing $p$ that crosses $s$ at a right angle and is completely contained in $P$. In the minimum sliding cameras (MSC) problem, the objective is to guard $P$ with the minimum number of sliding cameras. In this paper, we give an $O(n^{5/2})$-time (7/2)-approximation algorithm to the MSC problem on any simple orthogonal polygon with $n$ vertices, answering a question posed by Katz and Morgenstern (2011). To the best of our knowledge, this is the first constant-factor approximation algorithm for this problem.

## 1 Introduction

In the classical art gallery problem, we are given a polygon and the objective is to cover the polygon with the union of visibility regions of a set of point guards while minimizing the number of guards. The problem was introduced by Klee in 1973 [13]. Two years later, Chvátal [2] showed that $\lfloor n/3 \rfloor$ point guards are always sufficient and sometimes necessary to guard the polygon. The orthogonal art gallery problem was first studied by Kahn et al. [4] who proved that $\lfloor n/4 \rfloor$ guards are always sufficient and sometimes necessary to guard the interior of a simple orthogonal polygon. Lee and Lin [9] showed that the problem of guarding a simple polygon using the minimum number of guards is NP-hard. Moreover, the problem was also shown to be NP-hard for orthogonal polygons [14]. Since then, the problem and its many variants have been studied extensively for different

---

* Work of the author is supported in part by the Natural Sciences and Engineering Research Council of Canada (NSERC).
** Work of the author is supported by the Netherlands' Organization for Scientific Research (NWO).
*** Work of the author is supported in part by a University of Manitoba Duff Roblin Fellowship.

types of polygons (e.g., orthogonal polygons [14] and polyominoes [1]), different types of guards (e.g., points and line segments) and different visibility types. See the surveys by O'Rourke [13] or Urrutia [15] for a detailed history of the art gallery problem.

Recently, Katz and Morgenstern [6] introduced a variant of the art gallery problem in which *sliding cameras* are used to guard an orthogonal polygon. Let $P$ be an orthogonal polygon with $n$ vertices. A sliding camera travels back and forth along an orthogonal line segment $s$ inside $P$. The camera can see a point $p \in P$ if the there is a point $q \in s$ such that $\overline{pq}$ is a line segment normal to $s$ that is completely inside $P$. In the minimum sliding cameras (**MSC**) problem, the objective is to guard $P$ using a the minimum number of sliding cameras.

In this paper, we give an $O(n^{5/2})$-time (7/2)-approximation algorithm to the minimum sliding cameras (MSC) problem on any simple orthogonal polygon. To do this, we introduce the *minimum guarded sliding cameras* (**MGSC**) *problem*. In the MGSC problem, the objective is to guard $P$ using a set of minimum cardinality of guarded sliding cameras. A sliding camera $s$ is guarded by a sliding camera $s'$ if every point on $s$ is seen by some point on $s'$. Note that $s$ and $s'$ could be perpendicular, in which case $s'$ and $s$ mutually guard each other if and only if they intersect. If $s$ and $s'$ mutually guard each other and have the same orientation (e.g., both are horizontal), then the visibility region of $s$ is identical to that of $s'$. Consequently, when minimizing the number of sliding cameras in the MGSC problem, it suffices to consider solutions in which each horizontal sliding camera is guarded by a vertical sliding camera and vice-versa. We first establish a connection between the MGSC problem and a related guarding problem on grids.

A grid $D$ is a connected union of vertical and horizontal line segments; each maximal line segment in the grid is called a *grid segment*. A *point guard* $x$ in grid $D$ is a point that sees a point $y$ in the grid if the line segment $xy \subset D$. Moreover, a *sliding camera* $p \in D$ is a point guard that moves along a grid segment $s \in D$. The camera $p$ can see a point $q$ on the grid if and only if there exists a point $p' \in s$ such that the line segment $\overline{p'q} \subset D$; that is, point $q$ is seen by camera $p$ if either $q$ is located on $s$ or $q$ belongs to a grid segment that intersects $s$. Note that sliding cameras are called *mobile guards* in grid guarding problems [7,8]. A *guarded* set of point guards and a guarded set of sliding cameras on grids are defined analogously to a guarded set of sliding cameras in polygons. A simple grid is defined as follows:

**Definition 1 (Kosowski et al. [8]).** *A grid is simple if (i) the endpoints of all of its segments lie on the outer face of the planar subdivision induced by the grid, and (ii) there exists an $\epsilon > 0$ such that every grid segment can be extended by $\epsilon$ in both directions such that its new endpoints are still on the outer face.*

Throughout the paper, we denote a simple orthogonal polygon by $P$; note that the polygon $P$ is a closed region. The rest of the paper is organized as follows. Section 2 presents related work. In Section 3, we give our (7/2)-approximation algorithm to the MSC problem and we conclude the paper in Section 4.

## 2 Related Work and Definitions

The minimum sliding cameras problem was introduced by Katz and Morgenstern [6]. They first considered a restricted version of the problem in which only vertical cameras are allowed; by reducing the problem to the minimum clique cover problem on chordal graphs, they solved the problem exactly in polynomial time. For the generalized case, where both vertical and horizontal cameras are allowed, they gave a 2-approximation algorithm for the MSC problem under the assumption that the polygon $P$ is $x$-monotone. Durocher and Mehrabi [3] showed that the MSC problem is NP-hard when the polygon $P$ is allowed to have holes. They also gave an exact algorithm that solves in polynomial time a variant of the MSC problem in which the objective is to minimize the sum of the lengths of line segments along which cameras travel.

The guard problem on grids was first formulated by Ntafos [12]. He proved that a set of (stationary) point guards of minimum cardinality covering a grid of $n$ grid segments has $nm$ guards, where $m$ is the size of the maximum matching in the intersection graph of the grid that can be found in $O(n^{5/2})$ time. Malafijeski and Zylinski [10] showed that the problem of finding a minimum-cardinality set of guarded point guards for a grid is NP-hard. Katz et al. [5] showed that the problem of finding a minimum number of sliding cameras covering a grid is NP-hard. Moreover, Kosowski et al. [7] proved that the problem of finding the minimum number of guarded sliding cameras covering a grid (we call this problem the **MMGG** problem) is NP-hard. Due to these hardness results, Kosowski et al. [8] studied the MMGG problem on some restricted classes of grids. In particular, they show the following result on simple grids:

**Theorem 1 (Kosowski et al. [8].).** *There exists an $O(n^2)$-time algorithm for solving the MMGG problem on simple grids, where $n$ is the number of grid segments.*

Throughout the paper, we denote optimal solutions for the MSC problem and the MGSC problem on $P$ by $OPT_P$ and $OPT_{GP}$, respectively. We denote the set of reflex vertices of $P$ by $V(P)$ and let $H_u$ and $V_u$ be the maximum-length horizontal and vertical line segments, respectively, inside $P$ through a vertex $u \in V(P)$. Let $L(P) = \{H_u \mid u \in V(P)\} \cup \{V_u \mid u \in V(P)\}$. Let $L$ and $L'$ be two orthogonal line segments (with respect to $P$) inside $P$; the *visibility region* of $L$ is the union of the points in $P$ that are seen by the sliding camera that travels along $L$. Moreover, we say that $L$ *dominates* $L'$ if the visibility region of $L'$ is a subset of that of $L$.

## 3 A (7/2)-Approximation Algorithm for the MSC Problem

In this section, we present an $O(n^{5/2})$-time (7/2)-approximation algorithm for the MSC problem.

## 3.1 Relating the MGSC and MMGG Problems

Consider an optimal solution $X$ for the MSC problem and let $X'$ be the multiset of line segments obtained by duplicating every line segment in $X$ (i.e., each line segment of $X$ occurs twice in $X'$). We observe that $X'$ is a feasible solution for the MGSC problem and, therefore, we have the following observation.

**Observation 1.** *An optimal solution for the MGSC problem on $P$ is a 2-approximation to an optimal solution for the MSC problem on $P$.*

We first consider how to apply a solution for the MGSC problem to the MSC problem. For the MGSC problem, the idea is to reduce the MGSC problem to the MMGG problem. Given any simple orthogonal polygon $P$, we construct a grid $G_P$ associated with $P$ as follows: initially, let $G_P$ be the set of all line segments in $L(P)$. Now, for any pair of reflex vertices $u$ and $v$ where $H_u$ dominates $H_v$ (resp., $V_u$ dominates $V_v$) in $P$, we remove $H_v$ (resp., $V_v$) from $G_P$; if two segments mutually dominate each other, remove one of the two arbitrarily. Let $T_G$ be the set of remaining grid segments in $G_P$. Observe that $G_P$ can be constructed in $O(n^2)$ time, where $n$ is the number of vertices of $P$. We first show the following result:

**Lemma 1.** *Grid $G_P$ is a simple and connected grid.*

*Proof.* It is straightforward from the construction of $G_P$ that both endpoints of each grid segment in $T_G$ lie on the boundary of $P$; this means that the endpoints of every grid segment in $T_G$ lie on the outer face of $G_P$ and, therefore, $G_P$ is simple. To show that $G_P$ is connected, we first observe that the grid induced by the line segments in $L(P)$ is connected. We now need to show that the grid remains connected after removing the set of grid segments that are dominated by other grid segments. Let $s \in L(P)$ be a grid segment that is removed from $L(P)$ (i.e., $s \notin T_G$). It is straightforward to see that the set of grid segments that are intersected by $s$ are also intersected by $s' \in T_G$, where $s'$ is the grid segment that dominates $s$. Therefore, grid $G_P$ is connected. □

The objective is to solve the MMGG problem on $G_P$ exactly and to use the solution $S$, the set of guarded grid segments computed, as the solution to the MGSC problem. However, $S$ is not always a feasible solution to the MGSC problem since some regions in $P$ might remain unguarded; see Figure 1 for an example. In the following, we characterize the regions of $P$ that may remain unguarded by the line segments in $S$; we call these the *critical regions* of $P$.

Consider $S$ and choose any unguarded point $p$ inside $P$. Let $R_p$ be a maximal axis-aligned rectangle contained in $P$ that covers $p$ and is also not guarded by the line segments in $S$. We observe that (i) some line segments in $T_G$ can guard $R_p$, and that (ii) no such line segments are in $S$ since $R_p$ is unguarded. Consider the maximal regions in $P$ that lie immediately above, below, left, and right of $R_p$; any sliding camera that sees any part of $R_p$ must intersect one of these regions. See Figure 2(a) for an example; note that the hatched region cannot contain any line segment in $S$ since $R_p$ is unguarded. Moreover, the hatched region must contain

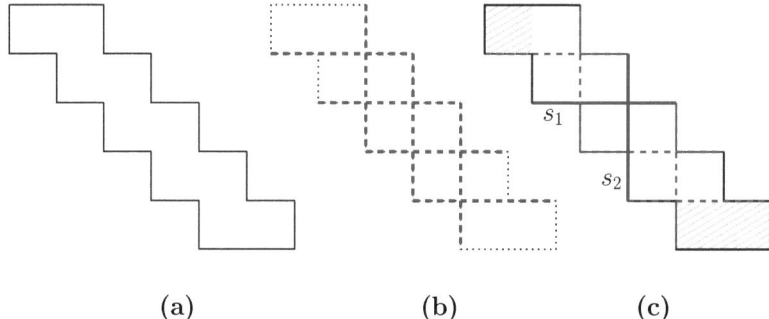

**Fig. 1.** (a) A simple orthogonal polygon $P$. (b) Grid $G_P$ with set $T_G$ of grid segments shown in red. (c) The set $S = \{s_1, s_2\}$ (represented by solid red line segments) is an optimal solution for the MMGG problem on $G_P$, but $s_1$ and $s_2$ cannot guard $P$ entirely; in particular, the hatched regions of $P$ are not guarded.

at least one line segment of $T_G$ in both horizontal and vertical directions and without loss of generality we can assume that the length of these line segments is maximal (i.e., both endpoints are on the boundary of $P$).

The rectangle $R_p$ defines a partition of $P$ into three parts: the *vertical slab* through $R_p$, (i.e., the slab whose sides are aligned with the vertical sides of $R_p$), the subpolygon of $P$ to the left of the vertical slab and the subpolygon of $P$ to the right of the vertical slab. Similarly, another partition of $P$ can be obtained by considering the *horizontal slab* through $R_p$; see Figure 2(b) for an example. We know that the union of the visibility regions of the line segments in $S$ is a connected subregion of the plane. Therefore, the set $S$ can only be found on one side of each of the partitions of $P$ and so $S$ must be in one *corner* of the partitioned polygon. Without loss of generality, assume that $S$ is on the bottom left corner of the partitioned polygon (see Figure 2(b)).

Let $\overline{S} \subseteq T_G$ be the set of line segments that can see $R_p$. Note that $\overline{S}$ is non-empty. This is because $\overline{S}$ is a subset of $T_G$ and the line segments in $\overline{S}$ can see $R_P$; but $R_P$ is a subregion of the polygon $P$. Since line segments in $T_G$ guard the polygon entirely, there has to be at least one line segment in $\overline{S}$. Therefore, polygon $P$ is partitioned into three subpolygons (see Figure 2(c)): the lower-left corner that is the location of $S$ denoted by $P_S$, the lower-right and upper-left corners that correspond to line segments in $\overline{S}$ denoted by $P_{\overline{S}}$, and the upper-right corner of the polygon that is unguarded denoted by $P_U$. Each line segment in $\overline{S}$ intersects at least one line segment in $S$ since the line segments in $\overline{S}$ are not in $S$ and $S$ is feasible solution for the MMGG problem (see Figure 2(c)).

**Lemma 2.** *No line segment in $T_G$ that is orthogonal to a line segment in $\overline{S}$ can intersect $P_U$.*

*Proof.* To derive a contradiction, suppose without loss of generality that there is one such vertical line segment $s$ intersecting $P_U$ (as shown in Figure 2(c)). Since $S$ is connected and is a feasible solution for the MMGG problem, there

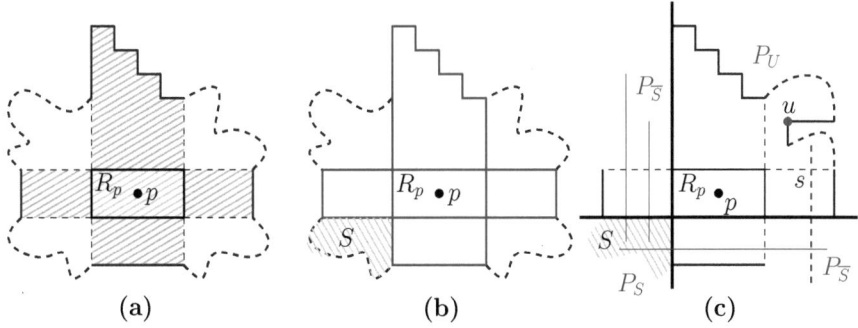

**Fig. 2.** (a) Point $p$ inside the polygon $P$ with rectangle $R_p$ hatched in gold. The purple hatched region indicates the subregion of $P$ covered by growing $R_p$ orthogonally towards the boundaries of $P$; the boundary of $P$ is shown in blue. (b) The horizontal rectangle and the vertical histogram shown in red indicate, respectively, the horizontal and vertical slabs of the partitions induced by rectangle $R_p$. The hatched region of $P$ on the bottom left corner of the partition indicates the location of the set $S$. (c) The partition of $P$ into three subpolygons $P_S$, $P_{\overline{S}}$ and $P_U$. The line segments in $\overline{S}$ (i.e., the set of line segments that can see $R_p$) are shown in red; observe that each line segment in $\overline{S}$ intersects at least one line segment in $S$. The line segment $s$ illustrates Lemma 2 and vertex $u$ is in support of Lemma 3.

must be a line segment in $S$ that guards $s$. So, $S$ must contain a line segment in $P_{\overline{S}}$, which is a contradiction. □

By Lemma 2, we conclude that $R_p$ is guarded by the line segments in $\overline{S}$, but not by any line segment in $S$, and, furthermore, $S$ is restricted to a corner as described above (i.e., the subpolygon $P_S$). We now show that each of the regions of $P$ that are not guarded by the line segments in $S$ must be a staircase with the reflex vertices oriented towards $S$.

To derive a contradiction, suppose that there exists a reflex vertex $u$ (as shown in Figure 2(c)) in the unguarded subpolygon $P_U$ of $P$. However, no line segment in $T_G$ can intersect $P_U$ by Lemma 2. This contradicts the existence of $u$. Therefore, any unguarded region of $P$ by $S$ must be bounded by the line segments in $L(P)$[1] on adjacent horizontal and vertical sides, and by a staircase of $P$ on the other sides; we call these regions the *critical regions* of $P$, and denote $R_C$ to be the set of critical regions of $P$. We now have the following lemma.

**Lemma 3.** *Every point of $P$ that is not inside a critical region of $P$ is visible to at least one line segment in $S$. Moreover, each critical region of $P$ is a staircase.*

Let $OPT_{GG}$ denote an optimal solution for the MMGG problem on $G_P$. We first prove that $|OPT_{GG}| \leq |OPT_{GP}|$.

---

[1] If the bounding line segment is not in $T_G$ then a dominating line segment must be in $T_G$. See Figure 4 for an example.

**Lemma 4.** *For any feasible solution $M$ for the MGSC problem on $P$, there exists a feasible solution $S'$ for the MMGG problem on $G_P$ such that $|S'| \leq |M|$.*

*Proof.* Let $M$ be a feasible solution to the MGSC problem on $P$; that is, $M$ is a guarded set of orthogonal line segments inside $P$ that collectively guard $P$. We construct a feasible solution $S'$ for the MMGG problem such that $|S'| \leq |M|$. To compute $S'$, for each horizontal line segment $s \in M$ (resp., vertical line segment $s \in M$), move $s$ up or down (resp., to the left or to the right) until it is collinear with a line segment $s \in L(P)$. If $s \in T_G$, then add $s$ to $S'$; otherwise, add $s'$ to $S'$, where $s' \in L(P)$ is the line segment that dominates $s$. Note that there exists at least one such line segment $s'$ because otherwise the line segment $s$ would have not been removed from $L(P)$. It is straightforward to see that the union of visibility regions of line segments in $M$ is a subset of the visibility regions of line segments in $S'$. Since the camera travelling along each line segment in $M$ is seen by at least one other camera (see the definition of a guarded set of sliding cameras) and the grid $G_P$ is entirely contained in $P$, we conclude that $S'$ is a feasible solution for the MMGG problem on $G_P$. The inequality $|S'| \leq |M|$ follows from the fact that each line segment in $M$ corresponds to at most one line segment in $S'$. This completes the proof of the lemma. □

Next we need to find a set of minimum cardinality of orthogonal line segments inside $P$ that collectively guard the critical regions of $P$.

### 3.2 Guarding Critical Regions: A (3/2)-Approximation Algorithm

In this section, we give an approximation algorithm for the problem of guarding the critical regions of $P$. The algorithm relies on reducing the problem to the minimum edge cover problem in graphs. The minimum edge cover problem in graphs is solvable in $O(n^{5/2})$ time, where $n$ is the number of graph vertices [11]. Recall $R_C$, the set of critical regions of $P$. We first need the following result:

**Lemma 5.** *Every critical region in $R_C$ is guarded entirely by some line segment in $L(P)$.*

*Proof.* Observe that if $P$ is a rectangle, then the MSC problem is trivial to solve. Suppose that $P$ is not a rectangle and so it has at least one reflex vertex. Furthermore, suppose that some regions of $P$ are not guarded by $S$ (the set of segments returned by solving the MMGG problem on $G_P$), i.e., the set $R_C$ of critical regions of $P$ is non-empty. Let $R \in R_C$ be a critical region of $P$. The lemma is implied by the fact that there exists at least one reflex vertex on the boundary of $R$; this is because $P$ is not a rectangle and the set of line segments in $S$ do not guard $R$. It is now straightforward to see that one of the orthogonal line segments in $L(P)$ that passes through either the lowest or the highest reflex vertex of $R$ can see the critical region $R$ entirely. □

We construct a graph $H_P$ associated with $P$ as follows: for each critical region $R \in R_C$, we add a vertex $v_R$ to $H_P$. Two vertices $v_R$ and $v_{R'}$ are adjacent in $H_P$

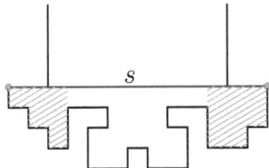

**Fig. 3.** Any orthogonal line segment inside $P$ can guard at most two critical regions of $P$ entirely

if and only if there exists an orthogonal line segment inside $P$ that can guard both critical regions $R$ and $R'$ entirely. Finally, we add a self-loop edge for every isolated vertex of $H_P$.

**Lemma 6.** *Any orthogonal line segment inside $P$ can guard at most two critical regions of $P$ entirely.*

*Proof.* Let $s$ be an orthogonal line segment inside $P$. Observe that since each critical region of $R$ is a staircase, line segment $s$ must hit the boundary of $R$ in order to guard $R$ entirely. That is, the only way for $s$ to guard $R$ entirely is that at least one of its endpoints lies on the boundary of $P$, covering one entire edge of $R$; see Figure 3 for an example. Therefore, $s$ can guard at most two critical regions of $P$. □

**Lemma 7.** *The problem of guarding the critical regions of $P$ using only those line segments that may individually guard a critical region reduces to the minimum edge cover problem on $H_P$.*

*Proof.* We prove that (i) for any solution $S$ to the minimum edge cover problem on $H_P$, there exists a solution $S'$ for guarding the critical regions of $P$ such that $|S'| = |S|$, and that (ii) for any solution $S'$ to the problem of guarding the critical regions of $P$, there exists a solution $S$ to the minimum edge cover problem on $H_P$ such that $|S| = |S'|$.

**Part 1.** Choose any edge cover $S$ of $H_P$. We construct a solution $S'$ for guarding the critical regions of $P$ as follows. For each edge $e = (v_R, v_{R'}) \in S$ let $s_e$ be the line segment in $P$ that can see both critical regions $R$ and $R'$ of $P$; we add $s_e$ to $S'$. It is straightforward to see that the line segments in $S'$ collectively guard all critical regions of $P$.

**Part 2.** Choose any solution $S'$ for guarding the critical regions of $P$. We now construct a solution $S$ for the minimum edge cover problem on $H_P$. By Lemma 6, we know that every line segment in $S'$ can see at most two critical regions of $P$. First, for each line segment in $S'$ that can see exactly one critical region $R$ of $P$, we add the self-loop edge of $H_P$ that corresponds to $R$ in $S$. Next, for each line segment $s \in S'$ that can see two critical regions of $P$, we add to $S'$ the edge in $H_P$ that corresponds to $s$. Since any line segment in $S'$ can see at most two critical regions of $P$, we conclude that every vertex of $H_P$ is incident to at least one edge in $S$ and, therefore, $S$ is a feasible solution for the minimum edge cover problem on $H_P$. □

In general, it is possible for the solution $S'$ to be non-optimal. Only those edges which may individually guard a critical region were considered, while an optimal guarding solution may use two line segments to collectively guard a critical region, as shown in Figure 4. By Lemma 5, $S'$ requires at most one edge for each critical region and, therefore, the number of guards returned by our algorithm is at most equal to the number of critical regions. If an optimal solution uses two segments to collectively guard one critical region, these two edges suffice to guard three critical regions, while our solution uses three segments to guard the same three critical regions. This results in an approximation factor of $3/2$ in the number of segments used to guard the set of critical regions.

We now examine the running time of the algorithm. Let $n$ denote the number of vertices of $P$. To compute the critical regions of $P$, we first compute the set of staircases of $P$ and, for each of them, we check to see whether they are guarded by the set of line segments in $S$. Each critical region of $P$ can be found in $O(n)$ time and so the set of unguarded critical regions of $P$ is easily computed in $O(n^2)$ time. Moreover, the graph $H_P$ can be constructed in $O(n^2)$ time by checking whether there is an edge between every pair of vertices of the graph. Therefore, by Lemma 7 and the fact that the minimum edge cover problem is solvable on a graph with $n$ vertices in $O(n^{5/2})$ time, we have the following lemma.

**Lemma 8.** *There exists a $(3/2)$-approximation algorithm for solving the problem of guarding the critical regions of a simple orthogonal polygon $P$ with $n$ vertices in $O(n^{5/2})$ time.*

Given any simple orthogonal polygon $P$, we find a set of sliding cameras that guards $P$ by first solving the instance $G_P$ of the MMGG problem determined by $P$. This may leave a set of critical regions within $P$ that remain unguarded. We then add a second set of sliding cameras to guard these critical regions. Recall the set $S$, an optimal solution to the MMGG problem on $G_P$, which can be found in $O(n^2)$ time, where $n$ is the number of vertices of $P$. By Lemma 8, we approximate the problem of guarding the critical regions of $P$ in $O(n^{5/2})$ time; let $S_C$ be the solution returned by the algorithm. Since the union of the critical regions of $P$ is a subset of $P$, any feasible solution to the MSC problem also guards the critical regions of $P$. Therefore, $|S_C| \leq (3/2) \cdot |OPT_P|$. Moreover, by Lemma 4 we know $|OPT_{GG}| \leq |OPT_{GP}|$ and since $|OPT_{GP}| \leq 2 \cdot |OPT_P|$ we have that $|OPT_{GG}| \leq 2 \cdot |OPT_P|$. Therefore, by combining $S$ and $S_C$ we obtain a feasible solution to the MSC problem whose cardinality is at most $7/2$ times $|OPT_P|$; that is, $|S \cup S_C| \leq (7/2) \cdot |OPT_P|$. This gives the main result:

**Theorem 2.** *There exists an $O(n^{5/2})$-time $(7/2)$-approximation algorithm for the MSC problem on any simple orthogonal polygon $P$ with $n$ vertices.*

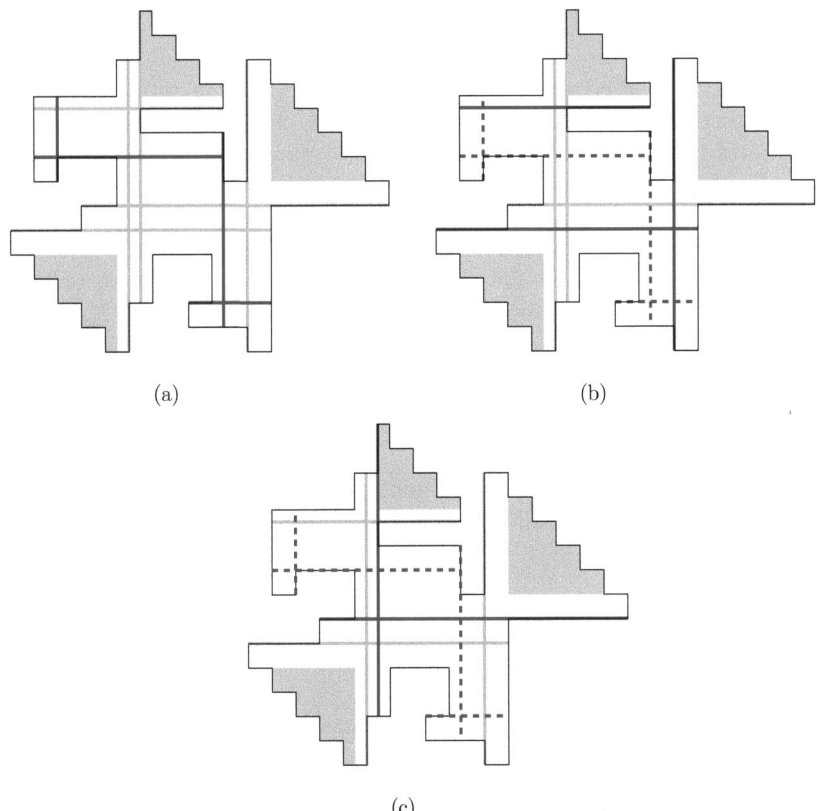

**Fig. 4.** An example of a polygon for which two line segments may collectively guard a critical region. (a) The polygon has unguarded critical regions after finding an optimal MMGG solution. The thick red line segments indicate guarding lines, light gray line segments are unused, and the shaded regions are unguarded. (b) A possible solution (solid red line segments) for guarding the critical regions using the edge guarding approach. (c) An optimal critical region guarding solution for the polygon.

As a consequence of our main result, we note that $|S_C| \leq (3/2) \cdot |OPT_P| \leq (3/2) \cdot |OPT_{GP}|$ and again by Lemma 4 we know $|OPT_{GG}| \leq |OPT_{GP}|$. Therefore, $|S \cup S_C| \leq (5/2) \cdot |OPT_{GP}|$. To show that the set $S \cup S_C$ is a feasible solution for the MGSC problem, we first observe that every line segment in $S$ is guarded by at least one other line segment in $S$. Moreover, every grid segment that is not in $S$ is guarded by some line segment in $S$ because $S$ is a feasible solution for the MMGG problem. This means that every line segment in $S_C$ is guarded by at least one line segment in $S$. Therefore, we have the following result.

**Corollary 1.** *Given a simple orthogonal polygon $P$ with $n$ vertices, there exists an $O(n^{5/2})$-time $(5/2)$-approximation algorithm for the MGSC problem on $P$.*

## 4 Conclusion

In this paper, we studied a variant of the art gallery problem, introduced by Katz and Morgenstern [6], where sliding cameras are used to guard an orthogonal polygon and the objective is to guard the polygon with minimum number of sliding cameras. We gave an $O(n^{5/2})$-time (7/2)-approximation algorithm to this problem by deriving a connection between a guarded variant of this problem (i.e., the MGSC problem) and the problem of guarding simple grids with sliding cameras. The complexity of the problem remains open for simple orthogonal polygons. Giving an $\alpha$-approximation algorithm, for any $\alpha < 7/2$, is another direction for future work. Finally, studying the MGSC problem (the complexity of the problem or improved approximation results) might be of independent interest.

**Acknowledgements.** The authors thank Mark de Berg and Matya Katz for insightful discussions of the sliding cameras problem.

## References

1. Biedl, T.C., Irfan, M.T., Iwerks, J., Kim, J., Mitchell, J.S.B.: The art gallery theorem for polyominoes. Disc. & Comp. Geom. 48(3), 711–720 (2012)
2. Chvátal, V.: A combinatorial theorem in plane geometry. J. Comb. Theory, Ser. B 18, 39–41 (1975)
3. Durocher, S., Mehrabi, S.: Guarding orthogonal art galleries using sliding cameras: Algorithmic and hardness results. In: Chatterjee, K., Sgall, J. (eds.) MFCS 2013. LNCS, vol. 8087, pp. 314–324. Springer, Heidelberg (2013)
4. Kahn, J., Klawe, M.M., Kleitman, D.J.: Traditional galleries require fewer watchmen. SIAM J. on Algebraic Disc. Methods 4(2), 194–206 (1983)
5. Katz, M.J., Mitchell, J.S.B., Nir, Y.: Orthogonal segment stabbing. Comp. Geom. 30(2), 197–205 (2005)
6. Katz, M.J., Morgenstern, G.: Guarding orthogonal art galleries with sliding cameras. Int. J. of Comp. Geom. & App. 21(2), 241–250 (2011)
7. Kosowski, A., Malafiejski, M., Zylinski, P.: Weakly cooperative mobile guards in grids. In: Proc. JCDCG, pp. 83–84 (2004)
8. Kosowski, A., Małafiejski, M., Żyliński, P.: An efficient algorithm for mobile guarded guards in simple grids. In: Gavrilova, M.L., Gervasi, O., Kumar, V., Tan, C.J.K., Taniar, D., Laganá, A., Mun, Y., Choo, H. (eds.) ICCSA 2006, Part I. LNCS, vol. 3980, pp. 141–150. Springer, Heidelberg (2006)
9. Lee, D.T., Lin, A.K.: Computational complexity of art gallery problems. IEEE Trans. on Info. Theory 32(2), 276–282 (1986)
10. Małafiejski, M., Żyliński, P.: Weakly cooperative guards in grids. In: Gervasi, O., Gavrilova, M.L., Kumar, V., Laganá, A., Lee, H.P., Mun, Y., Taniar, D., Tan, C.J.K. (eds.) ICCSA 2005, Part I. LNCS, vol. 3480, pp. 647–656. Springer, Heidelberg (2005)
11. Micali, S., Vazirani, V.V.: An $O(\sqrt{|v|}|E|)$ algorithm for finding maximum matching in general graphs. In: Proc. FOCS, pp. 17–27 (1980)

12. Ntafos, S.C.: On gallery watchmen in grids. Info. Process. Lett. 23(2), 99–102 (1986)
13. O'Rourke, J.: Art gallery theorems and algorithms. Oxford University Press, Inc., New York (1987)
14. Schuchardt, D., Hecker, H.: Two NP-hard art-gallery problems for ortho-polygons. Math. Logic Quarterly 41(2), 261–267 (1995)
15. Urrutia, J.: Art gallery and illumination problems. In: Handbook of Comp. Geom., pp. 973–1027. North-Holland (2000)

# Helly-Type Theorems in Property Testing

Sourav Chakraborty[1], Rameshwar Pratap[1], Sasanka Roy[1], and Shubhangi Saraf[2]

[1] Chennai Mathematical Institute,
Chennai, India
{sourav,rameshwar,sasanka}@cmi.ac.in
[2] Department of Mathematics and Department of Computer Science,
Rutgers University
shubhangi.saraf@rutgers.edu

**Abstract.** Helly's theorem is a fundamental result in discrete geometry, describing the ways in which convex sets intersect with each other. If $S$ is a set of $n$ points in $\mathbb{R}^d$, we say that $S$ is $(k,G)$-clusterable if it can be partitioned into $k$ clusters (subsets) such that each cluster can be contained in a translated copy of a geometric object $G$. In this paper, as an application of Helly's theorem, by taking a constant size sample from $S$, we present a testing algorithm for $(k,G)$-clustering, *i.e.*, to distinguish between two cases: when $S$ is $(k,G)$-clusterable, and when it is $\epsilon$-far from being $(k,G)$-clusterable. A set $S$ is $\epsilon$-far ($0 < \epsilon \leq 1$) from being $(k,G)$-clusterable if at least $\epsilon n$ points need to be removed from $S$ to make it $(k,G)$-clusterable. We solve this problem for $k=1$ and when $G$ is a symmetric convex object. For $k > 1$, we solve a *weaker* version of this problem. Finally, as an application of our testing result, in clustering with outliers, we show that one can find the *approximate* clusters by querying a constant size sample, with high probability.

## 1 Introduction

Given a set of $n$ points in $\mathbb{R}^d$, deciding whether all the points can be contained in a unit radius ball is a well known problem in Computational Geometry. Of course, the goal is to solve this problem as quickly as possible. In order to solve this problem exactly, one has to look at all the $n$ points in the worst case scenario. But if $n$ is too large, an algorithm with linear running time may not be fast enough. Thus, one may be interested in "solving" the above problem by taking a very small size sample and outputting the "right answer" with high probability. In this paper, we consider the *promise* version of this problem. More precisely, for the given *proximity parameter* $\epsilon$ (where $0 < \epsilon \leq 1$), our goal is to distinguish between the following two cases:

- all the points can be contained in a unit radius ball,
- no unit radius ball can contain more than $(1-\epsilon)$ fraction of points.

The above *promise* problem falls in the realm of property testing (see [10], [9] and [17]). In property testing, the goal is to look at a very small fraction of the input and decide whether the input satisfies the property or is "far" from satisfying it. Property testing algorithms for computational geometric problems have been studied earlier in

[6], [5] and [1]. In this paper, we study the above problem in property testing setting and give a simple algorithm to solve it. The algorithm queries only a constant number of points (where the constant depends on the dimension $d$ and $\epsilon$, but is independent of $n$) and correctly distinguishes between the two cases mentioned above with probability at least 2/3. While the algorithm is very simple, the proof of correctness is a little involved, for which we use Helly's theorem. Helly's theorem ([11]) states that if a family of convex sets in $\mathbb{R}^d$ has a non-empty intersection for every $d + 1$ sets, then the whole family has a non-empty intersection. In fact, since Helly's theorem also works for symmetric convex bodies, we can solve the above problem for any symmetric convex body instead of just a unit radius ball. Thus, we have

**Theorem 1.** *Let $A$ be a symmetric convex body. If $S$ is a set of $n$ points in $\mathbb{R}^d$ as input with the proximity parameter $\epsilon$ (where $0 < \epsilon \leq 1$), then there is an algorithm $\mathcal{A}$ that randomly samples $O(\frac{d}{\epsilon^{d+1}})$ many points and*

- *$\mathcal{A}$ accepts, if all the points in $S$ can be contained in a translated copy of $A$,*
- *$\mathcal{A}$ rejects with probability $\geq 2/3$, if any translated copy of $A$ can contain at most $(1 - \epsilon)n$ points.*

*The running time of $\mathcal{A}$ is $O(\frac{d}{\epsilon^{d+1}})$.*

One would like to generalize the above problem for more than one object, *i.e.*, given $k$ translated copies of object $B$, the goal is to distinguish between the following two cases with high probability:

- all $n$ points can be contained in $k$ translated copies of $B$,
- at least $\epsilon$ fraction of points cannot be contained in any $k$ translated copies of $B$.

We would like to conjecture that a similar algorithm, as stated in Theorem 1, would also work for the generalized $k$ object problem. Unfortunately, Helly's theorem does not hold for the $k$ object setting, but we would like to conjecture that a version of the Helly-type theorem does hold for this setting. Assuming the above conjecture, we can obtain a similar algorithm for the $k$ object setting. We can also unconditionally solve a *weaker* version of the $k$ object problem.

**Connection to Clustering:** We can also view this problem in the context of clustering. Clustering ([15],[12], [2]) is a common problem that arises in the analysis of large data sets. In a typical clustering problem, we have a set of $n$ input points in $d$ dimensional space and our goal is to partition the points into $k$ clusters. There are two ways to define the cluster size (cost):

- the maximum pairwise distance between an arbitrary pair of points in the cluster,
- twice the maximum distance between a point and a chosen centroid.

The first one is called as $k$-center clustering for diameter cost and the second one is called as $k$-center clustering for radius cost. In the $k$-center problem, our goal is to minimize the maximum of these distances. Computing $k$-center clustering is NP-hard: even for 2 clusters in general Euclidean space (of dimension $d$); and also for general number of $k$ clusters even on a plane.

In this paper, we assume that the cluster can be of symmetric convex shape also. Given a set $S$ of $n$ points and a symmetric convex body $A$ in $\mathbb{R}^d$, we say that the set of points is $(k, A)$-clusterable if all the points can be contained in $k$ translated copies of $A$. In the *promise* version of the problem, for a given proximity parameter $\epsilon$ (where $0 < \epsilon \leq 1$), our goal is to distinguish between the cases when $S$ is $(k, A)$-clusterable and when it is $\epsilon$-far from being $(k, A)$-clusterable. We say that $S$ is $\epsilon$-far from being $(k, A)$-clusterable if at least $\epsilon n$ points need to be removed from $S$ in order to make it $(k, A)$-clusterable.

We solve the above problem for $k = 1$ with constant number of queries. For $k > 1$, we solve a *weaker* version of the problem. In order to solve the *promise* version of the problem, we have designed a randomized algorithm which is generally called as *tester*.

Our algorithms can also be used to find an *approximately good* clustering. In clustering with outliers (anomalies), when we have the ability to ignore some points as outliers, we present a randomized algorithm that takes a constant size sample from input and outputs radii and centers of the clusters. The benefit of our algorithm is that we construct an *approximate* representation of such clustering in time which is independent of the input size.

The most interesting part of our result is that we initiate application of Helly-type theorem in property testing in order to solve the clustering problem.

### 1.1 Other Related Work

Alon *et al.* [1] presented testing algorithm for $(k, b)$-*clustering*. A set of points is said to be $(k, b)$-*clusterable* if it can be partitioned into $k$ *clusters*, where radius (or diameter) of every cluster is at most $b$. Section 5 of [1] presents a testing algorithm for radius cost under the $L_2$ metric. The analysis of this algorithm can be easily generalized to any metric under which each cluster is determined by a *simple* convex set (a convex set in $\mathbb{R}^d$ is called *simple* if its VC-dimension is $O(d)$).

For testing 1-center clustering, our result and the result from [1] give constant query testing algorithms. Although the two results have incomparable query complexity (in terms of number of queries depending on $\epsilon$), for testing $k$-center clustering, we give a *weaker* query complexity algorithm which works for fixed $k$ and $d$, and for $\epsilon \in (\epsilon', 1]$ where $\epsilon' = \epsilon'(k, t)$ (where $t$ is a constant which depends on the shape of the geometric object). Alon *et al.* used the sophisticated VC-dimension technique while we have used Helly-type results.

### 1.2 Organization of the Paper

In Section 2, we introduce the notations, definitions and state Helly and *Helly-type* theorems that are used in this paper. In Section 3, we design the *tester* for $(1, A)$-cluster testing for a given symmetric convex body $A$. In Section 4, we design the *tester* for $(k, G)$-cluster testing for a given geometric object $G$. In Section 5, as an application of results from Sections 3 and 4, we present an algorithm to find *approximate* clusters with outliers.

## 2 Preliminaries

### 2.1 Definitions

**n-piercing:** *A family of sets is called n-pierceable if there exists a set S of n points such that each member of the family has a non-empty intersection with S.*

**Homotheticity:** *Let A and B be two geometric bodies in $\mathbb{R}^d$. A is homothetic to B if there exist $v \in \mathbb{R}^d$ and $\lambda > 0$ such that $A = v + \lambda B$ (where $\lambda$ is called scaling factor of B). In particular, when $\lambda = 1$, A is said to be a **translated copy** of B.*

**Symmetric Convex Body:** *A convex body A is called symmetric if it is centrally symmetric with respect to the origin, i.e., a point $v \in \mathbb{R}^d$ lies in A if and only if its reflection through the origin $-v$ also lies in A. In other words, for every pair of points $v_1, v_2 \in \mathbb{R}^d$, if $v_1 \in v_2 + A$, then $v_2 \in v_1 + A$ and vice versa. Circles, ellipses, n-gons (for even n) with parallel opposite sides are examples of symmetric convex bodies.*

### 2.2 Property Testing

In property testing, the goal is to query a very small fraction of the input and decide whether the input satisfies a certain predetermined property or is "far" from satisfying it. Let $x = \{0, 1\}^n$ be a given input string. Then, a property testing algorithm, with query complexity $q(|x|)$ and proximity parameter $\epsilon$ for a decision problem $L$, is a randomized algorithm that makes at most $q(|x|)$ queries to $x$ and distinguishes between the following two cases:

- if $x$ is in $L$, then the algorithm *Accepts* $x$ with probability at least $\frac{2}{3}$,
- if x is $\epsilon$-far from $L$, then the algorithm *Rejects* $x$ with probability at least $\frac{2}{3}$.

Here, "$x$ is $\epsilon$-far from $L$" means that the Hamming distance between $x$ and any string in $L$ is at least $\epsilon|x|$. A property testing algorithm is said to have *one-sided error* if it satisfies the stronger condition that the accepting probability for instances $x \in L$ is 1 instead of $\frac{2}{3}$.

### 2.3 Helly's and Fractional Helly's Theorem

In 1913, Eduard Helly proved the following theorem:

**Theorem 2.** *(Helly's Theorem [11]) Given a finite family of convex sets $C_1, C_2, ..., C_n$ in $\mathbb{R}^d$ (where $n \geq d+1$) such that if intersection of every $d+1$ of these sets is non-empty, then the whole collection has a non-empty intersection.*

Katchalski and Liu proved the following result which can be viewed as a fractional version of the Helly's Theorem.

**Theorem 3.** *(Fractional Helly's Theorem [16]) For every $\alpha$ (where $0 < \alpha \leq 1$), there exists $\beta = \beta(d, \alpha)$ with the following property. Let $C_1, C_2, ..., C_n$ be convex sets in $\mathbb{R}^d$ (where $n \geq d + 1$) and if at least $\alpha \binom{n}{d+1}$ of the collection of subfamilies of size $d + 1$ has a non-empty intersection, then there exists a point contained in at least $\beta n$ sets.*

Independently, Kalai [8] and Eckhoff [13] proved that $\beta(d, \alpha) = 1 - (1 - \alpha)^{\frac{1}{(d+1)}}$.

### 2.4 *Helly-type* Theorem for More Than One Piercing in Convex Bodies

Helly's theorem on intersections of convex sets focuses on 1-pierceable families. Danzer et al. [7] investigated the following Helly-type problem : If $d$ and $m$ are positive integers, what is the least $h = h(d,m)$ such that a family of boxes (with parallel edges) in $\mathbb{R}^d$ is $m$-pierceable if each of its $h$-membered subfamilies is $m$-pierceable? Following is the main result of their paper:

**Theorem 4.**  1. $h(d,1) = 2$ *for all d (where $d \geq 1$)*;
2. $h(1,m) = m + 1$ *for all m;*
3. $h(d,2) = \begin{cases} 3d \text{ for odd } d; \\ 3d - 1 \text{ for even } d; \end{cases}$
4. $h(2,3) = 16$;
5. $h(d,m) = \infty$ *for $d \geq 2, n \geq 3$ and $(d,m) \neq (2,3)$.*

Katchalski et al. proved a result for families of homothetic triangles in a plane ([14]). This result is similar to the intersection property of axis parallel boxes in $\mathbb{R}^d$, studied by Danzer et al. This result can also be considered as a Helly-type theorem for more than one piercing of convex bodies. Theorem 5, below, presents the main result of their paper.

**Theorem 5.** *Let $\mathcal{T}$ be a family of homothetic triangles in a plane. If any nine of them can be pierced by two points, then all the members of $\mathcal{T}$ can be pierced by two points.*

## 3 Robust Helly for One Piercing of Symmetric Convex Body

Helly's theorem is a fundamental result in discrete geometry, describing the ways in which convex sets intersect with each other. In our case, we will focus on those subset of convex sets whose intersection properties behave *symmetric* in certain ways. Observation 6 explains this in detail. In order to design the *tester* for $(1, A)$-cluster testing problem, we will crucially use this observation, Helly's and fractional Helly's theorem.

**Observation 6.** *Let $A$ be a symmetric convex body in $\mathbb{R}^d$ containing $n$ points, then $n$ translated copies of $A$ centered at these $n$ points have a common intersection. Moreover, a translated copy of $A$ centered at a point in the common intersection contains all these $n$ points.*

**Lemma 7.** *Given a set $S$ of $n$ points in $\mathbb{R}^d$, if every $d + 1$ (where $d + 1 \leq n$) of them are contained in (a translated copy of) a symmetric convex body $A$, then all the $n$ points are contained in (a translated copy of) $A$.*

**Proof:** Consider a set $\mathcal{B}$ of translated copies of $A$ centered at points in $S$. Since every $d + 1$ of the given points are contained in (a translated copy of) $A$, by Observation 6, every $d + 1$ elements in $\mathcal{B}$ has a non-empty intersection. By Helly's theorem, all elements in $\mathcal{B}$ have a non-empty intersection. Let $q$ be a point from this intersection. Then $q$ belongs to every element in $\mathcal{B}$ and hence, by Observation 6, all the centers of the elements in $\mathcal{B}$, i.e., all the $n$ points in $S$, are contained in (a translated copy of) $A$ centered at $q$. □

**Lemma 8.** *Let $S$ be a set of $n$ points in $\mathbb{R}^d$ (where $n \geq d+1$). If at least $\epsilon n$ (where $0 < \epsilon \leq 1$) points cannot be contained in any translated copy of a symmetric convex body $A$, then at least $\epsilon^{d+1}$ fraction of all the $d+1$ size subsets of $S$ (number of such subsets is $\binom{n}{d+1}$) cannot be contained in any translated copy of $A$.*

**Proof:** Consider a set $\mathcal{B}$ of translated copies of $A$ centered at points in $S$. Now, by fractional Helly's theorem, for every $\alpha$ (where $0 < \alpha \leq 1$), there exists $\beta = \beta(d, \alpha)$ such that if at least an $\alpha$ fraction of $\binom{n}{d+1}$ subsets (of size $d+1$) in $\mathcal{B}$ has a non-empty intersection, then there exists a point (say $p$) which is contained in at least $\beta$ fraction of elements of $\mathcal{B}$.

Consider a translated copy of $A$ centered at $p$. By Observation 6, for every $\alpha$ (where $0 < \alpha \leq 1$), there exists $\beta = \beta(d, \alpha)$ such that if at least an $\alpha$ fraction of $\binom{n}{d+1}$ subsets (of size $d+1$) in $S$ are contained in $A$, then at least $\beta n$ points are contained in $A$.

Thus, if at least $(1-\beta)n$ points cannot be contained in $A$, then at least $1-\alpha$ fraction of $\binom{n}{d+1}$ subsets (of size $d+1$) in $S$ cannot be contained in $A$. (Contrapositive of the above statement.)

Since $\beta = 1 - (1-\alpha)^{\frac{1}{(d+1)}}$ ([8], [13]), choosing $1 - \beta$ as $\epsilon$ makes $1 - \alpha$ equal to $\epsilon^{d+1}$, which are the required values of the parameters. □

**Theorem 9.** *Consider a set of $n$ points in $\mathbb{R}^d$ ($n \geq d+1$) located such that at least $\epsilon n$ (where $0 < \epsilon \leq 1$) points cannot be contained in any translated copy of a symmetric convex body $A$. If we randomly sample $\frac{1}{\epsilon^{d+1}} \ln \frac{1}{\delta}$ (where $0 < \delta \leq 1$) many sets of $d+1$ points, then there exists a set in the sample which cannot be contained in any translated copy of $A$, with probability at least $1 - \delta$.*

**Proof:** By Lemma 8, if at least $\epsilon n$ points cannot be contained in (any translated copy of) $A$, then at least $\epsilon^{d+1}$ fraction of $\binom{n}{d+1}$ sets (of size $d+1$) cannot be contained in (any translated copy of) $A$. A set of $d+1$ points cannot be contained in $A$ with probability $\epsilon^{d+1}$. Hence, the probability that it can be contained in $A$ is $1 - \epsilon^{d+1}$. Thus, the probability that all the sampled sets are contained in $A$ is $\leq (1 - \epsilon^{d+1})^{\frac{1}{\epsilon^{d+1}} \ln \frac{1}{\delta}} \leq e^{-\ln \frac{1}{\delta}} = \delta$. □

Algorithm 1 is a randomized algorithm, *tester*, for $(1, A)$-cluster testing problem.

---

**Data:** A set $S$ of $n$ points in $\mathbb{R}^d$ (input is given as black-box), $0 < \delta, \epsilon \leq 1$.
**Result:** Returns a set of $d+1$ points, if it exists, which cannot be contained in $A$ or accepts (*i.e.*, all the points can be contained in $A$).

1 **repeat**
2     select a set (say $W$) of $d+1$ points uniformly at random from $S$
3     **if** $W$ *cannot be contained in* $A$ **then**
4         return $W$ as witness
5     **end**
6 **until** $\frac{1}{\epsilon^{d+1}} \ln \frac{1}{\delta}$ *many times*;
7 **if** *no witness found* **then**
8     return /* *all the points can be contained in $A$* */
9 **end**

**Algorithm 1.** $(1, A)$-cluster testing in a symmetric convex body $A$

---

This algorithm has a one sided error, i.e., if all the points can be contained in a symmetric convex body $A$ then it accepts the input, else it outputs a witness with probability at least $1 - \delta$. Correctness of the algorithm follows from Theorem 9. Thus, in the problem of testing $(1, A)$-clustering for a symmetric convex body $A$, the sample size is independent of the input size and hence the property is *testable*. Moreover, the *tester* works for all the possible values of $\epsilon$ (for $0 < \epsilon \leq 1$).

## 4 Robust Helly for More Than One Piercing of Convex Bodies

### 4.1 Helly-type Results for More Than One Piercing of Convex Bodies

The following lemma says that a *"Helly-type"* result is not true for circles even for 2-piercing. The result can be easily generalized for higher dimensions also. (The proof of the following lemma was suggested by Prof. Jeff Kahn in a private communication.)

**Lemma 10.** *Consider a set of $n$ circles in a plane. For any constant $w$ (where $w < n$), the condition that every $w$ circles are pierced at two points is not sufficient to ensure that all the circles in the set are pierced at two points.*

We present a proof of the above lemma in the full version of this paper [3].

Using arguments similar to the proof of above lemma, it is easy to prove that a *"Helly-type"* result for more than one piercing is also not true for a set of translated ellipsoids. Katchalski *et al.* [14] and Danzer *et al.* [7] proved a *"Helly-type"* result for more than one piercing of triangles and boxes, respectively. According to [14], a *"Helly-type"* result for more than one piercing is not true for centrally symmetric hexagon (with parallel opposite edges). Similar type of result is true for triangles and pentagons (with pair of parallel edges) which are not symmetric convex bodies. Thus, among symmetric convex bodies (spheres, ellipsoids and $n$-gons (for $n \leq 6$)), a *"Helly-type"* result for more than one piercing is possible only for parallelograms. We have following observation regarding the same (we present a proof in the full version of this paper [3]):

**Observation 11.** *Let $S$ be a set of $n$ points in $\mathbb{R}^d$. If every set of $h$ points (for finite possible values of $h$, see Theorem 4) in $S$ is contained in $m$ (where $m > 0$) translated parallelograms, then all the $n$ points are contained in $m$ translated parallelograms.*

### 4.2 Fractional Helly for More Than One Piercing of Convex Bodies

We now design a *weaker* version of *tester* for $(k, G)$-clustering (where $G$ is a bounded geometric object and $k > 1$). The tester works for some particular value of $\epsilon \in (\epsilon'(k, t), 1]$, where $t$ is some constant that depends on the shape of geometric object.

We state the following conjecture for more than one piercing of convex bodies.

**Conjecture 12.** *For every $\alpha$ (where $0 < \alpha \leq 1$), there exists $\beta = \beta(\alpha, k, d)$ with the following property. Let $C_1, C_2, .., C_n$ be convex sets in $\mathbb{R}^d$, $n \geq k(d+1)$, such that at least $\alpha \cdot \binom{n}{k(d+1)}$ of the collection of subfamilies of size $k(d+1)$ are pierced at $k$ points, then at least $\beta n$ sets are pierced at $k$ points. Also, $\beta$ approaches 1 as $\alpha$ approaches 1.*

**Lemma 13.** *If Conjecture 12 is true, then we have the following: Consider a set of $n$ points in $\mathbb{R}^d$ (where $n \geq k(d+1)$). If at least $\epsilon n$ (where $0 < \epsilon \leq 1$) points cannot be contained in any $k$ translated copies of symmetric convex body $A$, then at least $\gamma(\beta(\epsilon, k, d))$ fraction of $\binom{n}{k(d+1)}$ sets cannot be contained in any $k$ translated copies of $A$.*

**Proof:** Proof of this lemma is similar to the proof of Lemma 8. □

In the above lemma, $\gamma$ is an appropriately chosen function to compute the value of $1 - \alpha$, i.e., the fraction of $\binom{n}{k(d+1)}$ sets which cannot be contained in $k$ translated copies of $A$.

Now, we prove a *weaker* version of Conjecture 12. We show that for bounded geometric objects, a weaker version of fractional Helly for more than one piercing is true. We use *greedy* approach to prove the same. We prove it for some $\epsilon \in (\epsilon', 1]$, where $\epsilon' = \epsilon'(k, t)$ (where $t$ is a constant that depends on the shape of the geometric object). The result is true only for constant $k$ and $d$.

**Lemma 14.** *Consider $k$ translated copies of a geometric object $G$ and a set of $n$ points in $\mathbb{R}^d$ (for constant $k$ and $d$). Then there exist $\epsilon' = \epsilon'(k, t)$ (where $\epsilon'(k, t) = 1 - \frac{1}{2(t+1)(k+1)}$, $t$ is a constant that depends on the shape of the geometric object) such that for all $\epsilon \in (\epsilon', 1]$, if at least $\epsilon n$ points cannot be contained in any $k$ translated copies of $G$, then there exist at least $\Omega(n^{k+1})$ many witnesses of $k+1$ points which cannot be contained in any $k$ translated copies of $G$.*

**Proof:** We say a geometric object $G$ is *best* if it encloses the maximum number of points from the given set of $n$ points. Now, we start with such a best object. Let us say the best object contains at least $c_0(1 - \epsilon)n$ points (where $0 < c_0 \leq 1$). Now draw an object, $L_G$, concentric and homothetic with respect to $G$, having a scaling factor of $2 + \varepsilon$ (for $0 < \varepsilon \ll 1$, see the definition of Homotheticity in Subsection 2.1 where $v = 0$ and $\lambda = 2 + \varepsilon$). The annulus obtained by two concentric objects $G$ and $L_G$ can be filled with constant many (say $t\,(= \kappa^d - 1)$, [1] we present a proof of this in the full version of this paper [3]) translated copies of $G$. Since we started with the best object, the annulus contains at most $tc_0(1 - \epsilon)n$ points. Hence, the number of points which are outside $L_G$ is at least $\epsilon n - tc_0(1 - \epsilon)n = \epsilon_1 n$, where $\epsilon_1 = \epsilon - tc_0(1 - \epsilon)$. We throw away all the points in the annulus. Now, we are left with best object that containing at least $c_0(1 - \epsilon)n$ points and the remaining space containing at least $\epsilon_1 n$ points.

Now, we repeat the above process on $\epsilon_1 n$ points and would keep on repeating it until every point is either deleted or contained in some translated copies of $G$. Thus, total number of points that we have deleted from annuli is at most $t\Sigma_{i \geq 0} c_i(1 - \epsilon_i)n$ and total number of points that are inside translated copies of $G$ is at least $\Sigma_{i \geq 0} c_i(1 - \epsilon_i)n$ (where $\epsilon_0 = \epsilon$).

By construction, the total number of points inside translated copies of $G$ and the points that have been deleted from annuli is at least $n$. Thus,

---

[1] $\kappa$ is (ceiling of) the ratio of side length of the smallest $d$-cube circumscribing $L_G$ to that of the largest $d$-cube (homothetic w.r.t. smallest $d$-cube circumscribing $L_G$) inscribing $G$.

$$\Sigma_{i\geq 0}c_i(1-\epsilon_i)n + t\Sigma_{i\geq 0}c_i(1-\epsilon_i)n \geq n \quad (\text{where } \epsilon_0 = \epsilon).$$

$$\Sigma_{i\geq 0}c_i(1-\epsilon_i)n \geq \frac{n}{t+1}.$$

Let $G_i$ denotes the $i$-th geometric object and $|G_i|$ denotes the number of points contained in it. Thus,

$$\Sigma_{i\geq 0}|G_i| \geq \frac{n}{t+1}.$$

By assumption, $k$ translated copies of $G$ can contain at most $(1-\epsilon)n$ points. Thus, $|G_i| \leq (1-\epsilon)n$. Since $\epsilon > 1 - \frac{1}{2(t+1)(k+1)}$,

$$|G_i| < \frac{n}{2(t+1)(k+1)}.$$

Now, our goal is to make $k+1$ buckets, $S_1, S_2, .., S_{k+1}$, from $G_i$'s such that each bucket contains at least $\frac{n}{2(t+1)(k+1)}$ points and at most $\frac{n}{(t+1)(k+1)}$ points. We construct these buckets by adding points from $G_i$'s until its size become at least $\frac{n}{2(t+1)(k+1)}$. Since each $|G_i| < \frac{n}{2(t+1)(k+1)}$ and $\Sigma_{i\geq 0}|G_i| \geq \frac{n}{t+1}$, this construction is possible. Thus, for a particular bucket $S_i$,

$$\frac{n}{2(t+1)(k+1)} \leq |S_i| \leq \frac{n}{(t+1)(k+1)}.$$

Now, choosing one point from each of the $(k+1)$ buckets gives a set of $k+1$ points as a witness, which cannot be contained in $k$ translated copies of $G$. Thus, there are at least $\left(\frac{1}{2(t+1)(k+1)}\right)^{k+1} n^{k+1} \ (= \Omega(n^{k+1}))$ many witnesses. □

**Theorem 15.** *Consider $k$ translated copies of a geometric object $G$ and a set of $n$ points in $\mathbb{R}^d$ (for constant $k$ and $d$). Then there exist $\epsilon' = \epsilon'(k,t)$ (where $t$ is a constant that depends on the shape of the geometric object) such that for all $\epsilon \in (\epsilon', 1]$, at least $\epsilon n$ points cannot be contained in any $k$ translated copies of $G$. Now, if we randomly sample $\frac{1}{c}\ln\frac{1}{\delta}$ (where $0 < \delta \leq 1$ and $cn^{k+1}$ is the number of witnesses, see Lemma 14) many sets of size $k+1$, then there exists a set in the sample which cannot be contained in any $k$ translated copies of $G$, with probability at least $1-\delta$.*

We present a proof of the above theorem in the full version of this paper [3].

Similar to *tester* for $(1, A)$-cluster testing problem, we present a *tester* (Algorithm 2) for problem $(k, G)$-cluster testing. If all the points can be contained in $k$ translated copies of $G$ then algorithm accepts the input, else it outputs a witness with probability at least $1 - \delta$. Correctness of the algorithm follows from Theorem 15. Thus, similar to testing $(1, A)$-clustering, this property is also *testable*. But, the *tester* only works for constant $k$ and $d$ and for $\epsilon \in (\epsilon', 1]$ (see Lemma 14).

**Data:** A set $S$ of $n$ points in $\mathbb{R}^d$ (input is given as black-box), $0 < \delta \le 1$ and $\epsilon \in (\epsilon', 1]$.
**Result:** Returns a set of $k+1$ points, if it exists, which cannot be contained in $k$ translated copies of $G$, or accepts (*i.e.*, all the points can be contained in it).

1 **repeat**
2     select a set (say $W$) of $k+1$ points uniformly at random from $S$
3     **if** $W$ *cannot be contained in $k$ translated copies of $G$* **then**
4         return $W$ as witness
5     **end**
6 **until** $\frac{1}{c} \ln \frac{1}{\delta}$ *many times*;
7 **if** *no witness found* **then**
8     return /* *all the points can be contained in $k$ translated copies of $G$* */
9 **end**

**Algorithm 2.** $(k, G)$-cluster testing in geometric objects

## 5 Application in Clustering with Outliers

While considering the clustering problem, we mostly assume that data is perfectly clusterable. But a few random points (outliers, noise) could be added in the data by an adversary. For example, in the $k$-center clustering, if an adversary adds a point in the data which is very far from the original set of well clustered points, then in the optimum solution that point becomes center of its own cluster and the remaining points are forced to clustered with $(k-1)$ centers only. Also, it is even difficult to locate when a point becomes an outlier. For example: consider a set of points where we need to find its optimal $k$-center clustering. Take a point from that set and keep moving it far from the remaining set. Now, it is very difficult to locate correctly at which place that point becomes center of its own cluster and the remaining points are left with $(k-1)$-center clusters.

In this work, we consider clustering with outliers by ignoring some fraction of points. Thus, in the case when points are perfectly clusterable, ignoring some fraction of points does not affect the result too much, and the case when outliers are present, the algorithm has the ability to ignore them while computing the final clusters. It may seems that the ability to ignore some fraction of points makes the problem easier, but on the contrary it does not. Because it has not only to decide which point to include in the cluster but also to decide which point to include first. There may be two extreme approaches to solve this problem: 1) Decide which points are outliers and run the clustering algorithm; 2) Do not ignore any points, and after getting final clusters decide which ones are outliers. Unfortunately, neither of these two approaches works well. The first one scales poorly because there are exponentially many choices, and the second one may significantly change the final outcome when outliers are indeed present. This motivates the study of clustering with outliers (see[4]).

Theorem 9 has an application to 1-center clustering with outliers. More precisely, for $0 < \epsilon, \delta \leq 1$, when we have the ability to ignore at least $\epsilon n$ points as outliers, we present a randomized algorithm which takes a constant size sample from input and correctly output the radius and center of the *approximate* cluster with probability at least $1 - \delta$.

---
**Data**: A set $S$ of $n$ points in $\mathbb{R}^d$ (input is given as black-box), $0 < \epsilon, \delta \leq 1$.
**Result**: Report center and radius of cluster which contain all but at most $\epsilon n$ points.
1 Uniformly and independently, select $m = \frac{d+1}{\epsilon^{d+1}} \ln \frac{1}{\delta}$ points from $S$.
2 Compute minimum enclosing ball containing all the sample points and report its center and radius.

---
**Algorithm 3.** 1-center clustering with outliers

**Theorem 16.** *Given a set of $n$ points in $\mathbb{R}^d$ and $0 < \epsilon, \delta \leq 1$, Algorithm 3 correctly outputs, with probability at least $1 - \delta$, a ball containing all but at most $\epsilon n$ points in constant time by querying a constant size sample (constant depending on $d$ and $\epsilon$). Moreover, if $r_{outlier}$ is the smallest ball containing all but at most $\epsilon n$ points and $r_{min}$ is the smallest ball containing all the points, then Algorithm 3 outputs the radius $r$ such that $r_{outlier} \leq r \leq r_{min}$.*

We present a proof of the above theorem in the full version of this paper [3].

The problem of clustering with outliers can be generalized for $k$-center clustering. If Conjecture 12 is true, then it has an application to $k$-center clustering with outliers. For given $0 < \epsilon, \delta \leq 1$, ignoring at least $\epsilon n$ points as outliers, we present a randomized algorithm which takes a constant size sample from the input and correctly output the radii and $k$ centers of the *approximate* clusters with probability at least $1 - \delta$.

---
**Data**: A set $S$ of $n$ points in $\mathbb{R}^d$ (input is given as black-box), $0 < \epsilon, \delta \leq 1$.
**Result**: Reports $k$ centers and radii of clusters which contains all but at most $\epsilon n$ points.
1 Uniformly and independently, select $m = \frac{k(d+1)}{\gamma(\beta(\epsilon,k,d))} \ln \frac{1}{\delta}$ points from $S$.
2 Compute $k$ minimum enclosing balls containing all the sample points and report their centers and radii.

---
**Algorithm 4.** $k$-center clustering with outliers

**Theorem 17.** *Consider a set of $n$ points in $\mathbb{R}^d$. If Conjecture 12 is true and $0 < \epsilon, \delta \leq 1$, then with probability at least $1 - \delta$, Algorithm 4 output $k$ balls containing all but at most $\epsilon n$ points in constant time by querying a constant size sample (constant depending on $k$, $d$ and $\epsilon$). Moreover, for $1 \leq i \leq k$, if $r^{(i)}_{outlier}$ is the radius of the optimal $i$-th cluster by ignoring at most $\epsilon n$ points as outliers and $r^{(i)}_{min}$ is the radius of the optimal $i$-th cluster when all points are present, then Algorithm 4 outputs the radius $r^{(i)}$ such that $r^{(i)}_{outlier} \leq r^{(i)} \leq r^{(i)}_{min}$.*

We present a proof of the above theorem in the full version of this paper [3].

## 6 Conclusion and Open Problems

In this paper, we initiated an application of the Helly (and *Helly-type*) theorem in property testing. For $(1, A)$-cluster testing in a symmetric convex body $A$, we showed that testing can be done with constant number of queries and hence proved that the property is *testable*. Alon et al. [1] also solved a similar problem with constant number of queries, using combination of sophisticated arguments in geometric and probabilistic analysis. For 1-center clustering, our result had an incomparable query complexity in relation (in terms of number of queries depending on $\epsilon$) with the result of Alon et al. We stated a conjecture related to fractional *Helly-type* theorem for more than one piercing of convex bodies. Using a greedy approach, we proved a weaker version of the conjecture which we used for testing $(k, G)$-clustering. We also gave a characterization of the type of symmetric convex body for which Helly-type result for more that one piercing would be true. Finally, as an application of testing result in clustering with outliers, we showed that one can find, with high probability, the *approximate* clusters by querying a constant size sample.

## References

1. Alon, N., Dar, S., Parnas, M., Ron, D.: Testing of clustering. SIAM J. Discrete Math. 16(3), 393–417 (2003)
2. Anderberg, M.R.: Cluster Analysis for Applications. Academic Press (1973)
3. Chakraborty, S., Pratap, R., Roy, S., Saraf, S.: Helly-type theorems in property testing. CoRR, abs/1307.8268 (2013)
4. Charikar, M., Khuller, S., Mount, D.M., Narasimhan, G.: Algorithms for facility location problems with outliers, pp. 642–651 (2001)
5. Czumaj, A., Sohler, C.: Property testing with geometric queries. In: Meyer auf der Heide, F. (ed.) ESA 2001. LNCS, vol. 2161, pp. 266–277. Springer, Heidelberg (2001)
6. Czumaj, A., Sohler, C., Ziegler, M.: Property testing in computational geometry. In: Paterson, M. (ed.) ESA 2000. LNCS, vol. 1879, pp. 155–166. Springer, Heidelberg (2000)
7. Danzer, L., Branko, B.G.: Intersection properties of boxes in $\mathbb{R}^d$. Combinatorica 2(3), 237–246 (1982)
8. Eckhoff, J.: An upper bound theorem for families of convex sets. Geom. Dediata 19(75), 217–227 (1985)
9. Goldreich, O.: Combinatorial property testing (a survey). Electronic Colloquium on Computational Complexity (ECCC) 4(56) (1997)
10. Goldreich, O., Goldwasser, S., Ron, D.: Property testing and its connection to learning and approximation. J. ACM 45(4), 653–750 (1998)
11. Helly, E.: Über Mengen konvexer Köper mit gemeinschaftlichen Punkten (germen). Jahresber. Deutsch.Math. Verein (32), 175–176 (1923)
12. Jain, A.K., Dubes, R.C.: Algorithms for Clustering. Prentice-Hall (1988)
13. Kalai, G.: Intersection patterns of convex sets. Israel J. Math. (48), 161–174 (1984)
14. Katchalski, M., Nashtir, D.: On a conjecture of danzer and grunbaum. Proc. A.M.S (124), 3213–3218 (1996)
15. Kaufman, L., Rousseeuw, P.J.: Finding Groups in Data: An Introduction to Cluster Analysis. John Wiley (1990)
16. Katchalski, M., Liu, A.: A problem of geometry in $\mathbb{R}^n$. Proc. A.M.S (75), 284–288 (1979)
17. Ron, D.: Property testing: A learning theory perspective. Foundations and Trends in Machine Learning 1(3), 307–402 (2008)

# New Bounds for Online Packing LPs*

Matthias Englert[1], Nicolaos Matsakis[1], and Marcin Mucha[2]

[1] DIMAP and Dept. of Computer Science, University of Warwick
{M.Englert,N.Matsakis}@warwick.ac.uk
[2] Institute of Informatics, University of Warsaw
mucha@mimuw.edu.pl

**Abstract.** Solving linear programs online has been an active area of research in recent years and was used with great success to develop new online algorithms for a variety of problems. We study the setting introduced by Ochel et al. as an abstraction of lifetime optimization of wireless sensor networks.

In this setting, the online algorithm is given a packing LP and has to monotonically increase LP variables in order to maximize the objective function. However, at any point in time, the adversary only provides an $\alpha$-approximation of the remaining slack for each constraint. This is designed to model scenarios in which only estimates of remaining capacities (e.g. of batteries) are known, and they get more and more accurate as the remaining capacities approach 0.

Ochel et al. (ICALP'12) gave a $\Theta(\ln \alpha/\alpha)$-competitive online algorithm for this online packing LP problem and showed an upper bound on the competitive ratio of any online algorithm, even randomized, of $O(1/\sqrt{\alpha})$. We significantly improve the upper bound and show that any deterministic online algorithm for LPs with $d$ variables is at most $O(d^2 \alpha^{1/d}/\alpha)$-competitive. For randomized online algorithms we show an upper bound of $O(m^2 \alpha^{1/m}/\alpha)$ for LPs with $m^{m! \ln \alpha}$ variables. For LPs with sufficiently many variables, these bounds are $O(\ln^2 \alpha/\alpha)$, nearly matching the known lower bound.

On the other hand, we also show that the known lower bound can be significantly improved if the number of variables in the LP is small. Specifically, we give a deterministic $\Theta(1/\sqrt{\alpha})$-competitive online algorithm for packing LPs with two variables. This is tight, since the previously known upper bound of $O(1/\sqrt{\alpha})$ still holds for 2-dimensional LPs.

## 1 Introduction

In recent years, there has been great interest in methods for solving linear programs online, mainly to facilitate the development of new online algorithms with

---

* The first and second author are supported by the Centre for Discrete Mathematics and its Applications (DIMAP), University of Warwick, EPSRC award EP/D063191/1. The third author is supported by NCN grant N N206 567940.

improved competitive guarantees. Buchbinder and Naor [3] give a general online primal-dual approach to (approximately) solve the following type of packing (and the dual covering version) linear program online.

The underlying packing LP is of the form

$$\max \quad b_1 x_1 + \ldots + b_d x_d$$

$$\text{subject to} \quad A \begin{pmatrix} x_1 \\ \vdots \\ x_d \end{pmatrix} \leq \begin{pmatrix} c_1 \\ \vdots \\ c_m \end{pmatrix}$$

$$x_1, \ldots, x_d \geq 0 \, ,$$

where $A$ is matrix with non-negative entries, and all $b_i$ and $c_i$ are positive.

In the online version, an online algorithm plays against an adversary which only reveals entries of $b$ and $A$ online. More specifically, the online algorithm is only allowed to increase variables, not decrease them. The algorithm also must guarantee that all LP constraints are satisfied. The vector $c$ is given to the algorithm upfront, but $b$ and $A$ are initially hidden. The adversary then successively reveals all coefficients of a variable $x_j$ at a time of its choosing, i.e., in round $j$, $b_j$ and, for all $i$, $A_{ij}$ are revealed to the online algorithm. The goal is to maximize the objective function $b^T x$.

The primal-dual technique by Buchbinder and Naor and their extensions have been applied with great success to develop new improved online algorithms for a variety of online problems, among them the $k$-server problem [2] and generalized caching [1].

Ochel, Radke, and Vöcking [5] introduce a related model in which $b$ and $A$ are initially given to the online algorithm and instead $c$ is only gradually revealed. At each point in time, the adversary reveals a vector $\ell^t$, which can be seen as the current right hand side values of the LP, and the online algorithm responds by increasing variables $x_i$. The algorithm may never decrease a variable and has to ensure that the constraint $Ax \leq \ell^t$ is satisfied. This is not possible without imposing further restrictions on $\ell^t$. Therefore we require that

1. The revealed $\ell^t$ are (component wise) lower bounds on $c$, i.e., $\ell^t \leq c$.
2. If $x$ is the current online solution, the next revealed vector $\ell^t$ has to satisfy $(c - Ax) \leq \alpha(\ell^t - Ax)$.

In other words, the remaining slacks of constraints, if $\ell^t$ is taken as the right hand side, are an $\alpha$-approximation of the remaining slacks with respect to the *true* right hand side $c$. We call $\ell^t - Ax$ the *revealed remaining slacks* and $c - Ax$ the *true remaining slacks* and study the performance of online algorithms in dependence of the problem parameter $\alpha$.

Ochel et al. [5] give the problem of lifetime optimization in wireless sensor networks as a motivating application (see, e.g., [4]). There, the right hand sides of the constraints correspond to battery lifetimes of sensors. We only know the lower bounds on the remaining lifetimes, but the true values are always within a fixed factor of the revealed values. Given this information, we need to choose among a set of broadcasting scenarios in a way that maximizes the number of broadcasts performed before empty batteries prevent any further broadcasts.

### 1.1 Our Results

Ochel et al. [5] give a $\Theta(\ln \alpha/\alpha)$-competitive online algorithm for their online packing LP problem and show an upper bound on the competitive ratio of any online algorithm, even randomized, of $O(1/\sqrt{\alpha})$.

We significantly improve the upper bound. For LPs involving $d$ or more variables we show an upper bound of $O(d^2 \alpha^{1/d}/\alpha)$ on the competitive ratio of any deterministic algorithm. At the cost of increasing the number of variables, we obtain a similar bound on the competitive ratio of any randomized algorithm against an oblivious adversary. With $m^{\lceil m! \ln \alpha \rceil}$ or more variables we construct an upper bound of $O(m^2 \alpha^{1/m}/\alpha)$.

For $d = \Omega(\ln \alpha)$ in the case of deterministic algorithms and $d = \Omega(\alpha^{(\ln \alpha)})$ in the case of randomized algorithms, this results in an upper bound of $O(\ln^2 \alpha/\alpha)$, which nearly matches the known lower bound by Ochel et al.

However, we also demonstrate that the achievable competitive ratio crucially depends on the number of variables $d$ in the LP. We give a simple $\Theta(1/\sqrt{\alpha})$-competitive deterministic online algorithm that beats the general upper bound of $O(\ln^2 \alpha/\alpha)$ for LPs only involving two variables. This is tight, since the previously known general upper bound of $O(1/\sqrt{\alpha})$ still holds for 2-dimensional LPs.

The paper is organized as follows. In Section 2 we give the upper-bound on the competitive ratio achievable by deterministic algorithms. The techniques developed in the process are then extended in Section 3 to handle randomized algorithms. In Section 4 we describe and analyze an $O(1/\sqrt{\alpha})$-competitive algorithm for 2-dimensional LPs. We end with conclusions and open problems in Section 5.

## 2 Deterministic Upper Bound

In this section, we describe our upper bound construction and prove the following theorem.

**Theorem 1.** *The competitive ratio of any deterministic online algorithm is at most $O(d^2 \alpha^{1/d}/\alpha)$, for LPs involving $d$ or more variables.*

Note that the bound in the claim above is minimized for $d = \Theta(\ln \alpha)$. Using this value, gives us the following corollary.

**Corollary 1.** *The competitive ratio of any deterministic online algorithm is at most $O(\ln^2 \alpha/\alpha)$.*

We proceed with the construction of the adversary to prove Theorem 1. The construction will use exactly $d$ variables. The theorem follows since any additional variables $x_{d+1}, x_{d+2}, \ldots$ can be made irrelevant by adding constraints of the form $x_{d+1} \leq 0, x_{d+2} \leq 0, \ldots$

The basic idea behind our construction is to present an LP to the online algorithm that is completely symmetric. Once the online algorithm has increased some variable $x_i$ so much that the revealed remaining slacks of some constraints become small, the adversary decides that these are exactly the constraints for which the revealed right hand sides were already quite close to the true right hand sides. As a consequence, $x_i$ cannot be increased much further in the future and the online algorithm is, more or less, left with a similarly constructed input for the remaining $d - 1$ variables.

The initial linear program presented to the online algorithm is

$$\max \sum_{i=0}^{d-1} x_i$$

$$\forall \text{ permutations } \pi : \sum_{i=0}^{d-1} \alpha^{i/d} \cdot x_{\pi(i)} \leq \alpha$$

$$x_0, x_1, \ldots, x_{d-1} \geq 0 \ .$$

The adversary maintains a set of *active* constraints and a set of *active* variables. Initially all $d!$ constraints and all $d$ variables are active.

The adversary proceeds in $d$ rounds numbered $d-1, d-2, \ldots, 0$. Round $r$ ends in the first step in which there exists an active constraint with a revealed remaining slack of at most $\alpha^{r/d}$. At the end of a round, the adversary does the following:

- Determine the index $k_r$ of an active variable of maximum value among the active variables.
- Increase the right hand side of all active constraints that do not correspond to permutations with $\pi(r) = k_r$ to $\alpha \cdot \alpha^{r/d}$. Note that this is possible, i.e., this does not violate the condition that revealed remaining slacks always have to be an $\alpha$-approximation of the true remaining slacks, since before this point in time, all constraints have a revealed remaining slack of at least $\alpha^{r/d}$.
- Remove all these constraints from the set of active constraints.
- Remove the variable with index $k_r$ from the set of active variables.

*Remark 1.* It might happen that the online algorithm ends its execution before all rounds are completed. In this case, the adversary still executes the steps above (without waiting for slacks of constraints to trigger the next round).

We start our proof with two easy observations. At the end of round $r$, we remove all constraints with $\pi(r) \neq k_r$ from the set of active constraints. Since round numbers are decreasing, we have the first observation.

**Observation 2.** *A permutation $\pi$ corresponds to a constraint active in round $r$ iff $\pi(s) = k_s$ for all $s > r$.*

Let $x_i^r$ be the value of the variable $x_i$ at the end of round $r$, for $i, r \in \{0, \ldots, d-1\}$. Since for non-negative numbers $a_1 \geq a_2 \geq \ldots$ and non-negative numbers $b_i$, $\sum_i a_i \cdot b_{\pi(i)}$ is maximized if $b_{\pi(1)} \geq b_{\pi(2)} \geq \ldots$ we get the second observation.

**Observation 3.** Let $r = 0, \ldots, d-1$ be a round. Also, let $\pi$ be a permutation corresponding to a constraint active in round $r$ and such that

$$x^r_{\pi(r)} \geq x^r_{\pi(r-1)} \geq \ldots \geq x^r_{\pi(0)}.$$

Note that such a permutation exists due to Observation 2. Then, the constraint corresponding to $\pi$ has the smallest revealed remaining slack among all permutations active in round $r$.

For any $r = 0, \ldots, d-1$, let $\pi_r$ be the permutation corresponding to the constraint that causes round $r$ to end. Due to Observation 3,

$$x^r_{\pi_r(r)} \geq x^r_{\pi_r(r-1)} \geq \ldots \geq x^r_{\pi_r(0)}$$

and therefore, we may assume without loss of generality that $\pi_r(r) = k_r$.

**Lemma 1.** *For any round $r < d-1$ we have*

$$x^r_{\pi_r(i)} \geq x^{r+1}_{\pi_{r+1}(i)}$$

*for any $i = 0, \ldots, d-1$*

*Proof.* Due to Observation 2, we have $\pi_r(i) = \pi_{r+1}(i) = k_i$ for $i > r+1$ since the constraints corresponding to $\pi_r$ and $\pi_{r+1}$ are both active in rounds $d-1, \ldots, r+1$. Since $\pi_r(i)$ is also active in round $r$ and $\pi_{r+1}(r+1) = k_{r+1}$, we also have $\pi_r(i) = \pi_{r+1}(i) = k_i$ for $i = r+1$. Variables can only increase, this implies for $i \geq r+1$, $x^r_{\pi_{r+1}(i)} = x^r_{k_i} \geq x^{r+1}_{k_i} = x^{r+1}_{\pi_{r+1}(i)}$.

It remains to prove the lemma for $i \leq r$. Again because $\pi_r(i) = \pi_{r+1}(i) = k_i$ for $i > r+1$, the sequence $\pi_{r+1}(r+1), \pi_{r+1}(r), \ldots, \pi_{r+1}(0)$ is a permutation of the sequence $\pi_r(r+1), \pi_r(r), \ldots, \pi_r(0)$. Therefore, the sets of variables $\{x_{\pi_{r+1}(r+1)}, \ldots, x_{\pi_{r+1}(0)}\}$ and $\{x_{\pi_r(r+1)}, \ldots, x_{\pi_r(0)}\}$ are identical. Hence, since variables can only increase, the sequence

$$x^r_{\pi_r(r+1)} \geq x^r_{\pi_r(r)} \geq \ldots \geq x^r_{\pi_r(0)}$$

is obtained from

$$x^{r+1}_{\pi_{r+1}(r+1)} \geq x^{r+1}_{\pi_{r+1}(r)} \geq \ldots \geq x^{r+1}_{\pi_{r+1}(0)}$$

by increasing the values of some variables and rearranging the sequence so that it is sorted. The claim follows. □

We can now show the key lemma.

**Lemma 2.** *For any $r \in \{0, \ldots, d-1\}$ and $i \leq r$, we have $x^r_{\pi_r(i)} \leq (d-r)\alpha^{1/d}$.*

*Proof.* We use downward induction on $r$. The claim is clear for $r = d-1$, since during round $d-1$ all constraints still have right hand sides equal to $\alpha$ and for any variable $x_i$ there is a constraint that contains this variable with a coefficient of $\alpha^{1-1/d}$.

Consider now any round $r < d - 1$. We have

$$\sum_{i=0}^{d-1} \alpha^{i/d} \cdot x^r_{\pi_r(i)} = \sum_{i=0}^{d-1} \alpha^{i/d} \cdot x^{r+1}_{\pi_{r+1}(i)} + \sum_{i=0}^{d-1} \alpha^{i/d} \cdot \left( x^r_{\pi_r(i)} - x^{r+1}_{\pi_{r+1}(i)} \right).$$

The first sum is lower bounded by $\alpha - \alpha^{(r+1)/d}$ by the definition of $\pi_{r+1}$. Moreover, by Lemma 1 we have $x^r_{\pi_r(i)} \geq x^{r+1}_{\pi_{r+1}(i)}$ for any $i = 0, \ldots, d-1$. Therefore

$$\sum_{i=0}^{d-1} \alpha^{i/d} \cdot x^r_{\pi_r(i)} \geq \alpha - \alpha^{(r+1)/d} + \sum_{i=0}^{d-1} \alpha^{i/d} \cdot \left( x^r_{\pi_r(i)} - x^{r+1}_{\pi_{r+1}(i)} \right).$$

with all the terms in the right sum being non-negative. Since the constraint corresponding to $\pi_r$ is active and feasible in round $r$, the left hand side is upper-bounded by $\alpha$. By ignoring all terms except the one corresponding to $i = r$ in the sum of the right hand side we obtain

$$x^r_{\pi_r(r)} \leq x^{r+1}_{\pi_{r+1}(r)} + \frac{\alpha^{(r+1)/d}}{\alpha^{r/d}} = x^{r+1}_{\pi_{r+1}(r)} + \alpha^{1/d} \leq (d-r)\alpha^{1/d},$$

where the last inequality follows from induction.

The claim for $x^r_{\pi_r(i)}$ with $i < r$ follows as well, since $x^r_{\pi_r(0)} \leq \cdots \leq x^r_{\pi_r(r)} \leq (d-r)\alpha^{1/d}$. □

**Lemma 3.** *After the online algorithm terminates we have $x_i = O(d\alpha^{1/d})$ for all $i = 0, \ldots, d-1$.*

*Proof.* Let $r$ be the last round that is fully performed. Consider any permutation $\pi$ corresponding to a constraint that is active when the algorithm ends. From Observation 2 we know that these are exactly the constraints satisfying $\pi(t) = k_t$ for all $t \geq r$. Choose one such $\pi$ so that it also satisfies

$$x_{\pi(r-1)} \geq x_{\pi(r-2)} \geq \cdots \geq x_{\pi(0)},$$

where $x_i$ is the final value of a variable.

We will first prove the claim for variables that are not active when the algorithm ends. Any such variable has index $k_t = \pi_t(t) = \pi(t)$ for some $t \geq r$. We have

$$\sum_{i=0}^{d-1} \alpha^{i/d} \cdot x^t_{\pi_t(i)} \geq \alpha - \alpha^{t/d},$$

and similarly to the reasoning in the the proof of Lemma 2 we can argue that

$$x_{k_t} \leq x^t_{k_t} + \frac{\alpha^{t/d}}{\alpha^{t/d}} \leq (d-t)\alpha^{1/d} + 1.$$

Consider now the variables that are still active when the algorithm ends, i.e. $x_{\pi(i)}$ for $i < r$. It is enough to prove the claim for $x_{\pi(r-1)}$, since it is the largest one. We have

$$\sum_{i=0}^{d-1} \alpha^{i/d} \cdot x^r_{\pi_r(i)} \geq \alpha - \alpha^{r/d},$$

and again using the reasoning from the proof of Lemma 2 we get that

$$x_{\pi(r-1)} \leq x^r_{\pi_r(r-1)} + \frac{\alpha^{r/d}}{\alpha^{(r-1)/d}} \leq (d-r)\alpha^{1/d} + \alpha^{1/d} = (d-r+1)\alpha^{1/d}. \qquad \square$$

It remains to show a lower bound on the profit an optimal offline strategy can achieve.

**Lemma 4.** *An optimal offline algorithm can obtain a profit of $\alpha$.*

*Proof.* An offline algorithm can set $x_{k_0}$ to $\alpha$ and all other variables to 0. This solution is feasible.

For any $i \geq 1$, consider any of the $(d-1)!$ constraints in which $x_{k_0}$ has a coefficient of $\alpha^{i/d}$. In other words, a constraint corresponding to a permutation with $\pi(i) = k_0$. Such a constraint becomes inactive by the end of round $i$ the latest, since $\pi(i) = k_0 \neq k_i$. Due to the adversary's strategy this means that, once the constraint becomes inactive, the right hand side increases to at least $\alpha \cdot \alpha^{i/d}$ and therefore the constraint is satisfied.

Constraints in which the coefficient of $x_{k_0}$ is 1 are clearly satisfied as well, since the right hand side of all constraints is at least $\alpha$. $\qquad \square$

By combining Lemma 3 and Lemma 4 we obtain Theorem 1.

## 3 Randomized Upper Bound

The upper bound on the competitive ratio of randomized online algorithms against oblivious adversaries is based on the construction from the previous section and uses a technique that is, at least implicitly, also used by Ochel et al. [5]. Recall that each round ends when the slack of at least one active constraint drops below a certain threshold value. The adversary then identifies the offending (i.e. largest) active variable and renders all constraints, except for those in which this variable appears with a certain coefficient, irrelevant. This is done by increasing the right hand side of these constraints sufficiently. The offending variable becomes inactive and cannot be further increased by much anymore.

To obtain our upper bound on randomized algorithms, we use the standard approach based on Yao's min-max principle; instead of proving bounds for randomized algorithms, we construct a distribution on the inputs that (in expectation) foils any deterministic algorithm.

*Technical issues.* Our previous problem description involves a constant interaction between the online algorithm and an adversary; each increase of LP variables by the algorithm is followed by an update of the right hand sides of the LP constraints by the adversary.

In order to define oblivious adversaries, we need to remove this interaction and allow the adversary to specify its complete behavior upfront. For this, the input consists, as before, of a packing LP whose constraints are given by $Ax \leq c$. Additionally, for each constraint $i$, an adversary specifies a monotonically increasing function $\ell_i(\lambda_i)$ of the left hand side of the constraint $\lambda_i := (Ax)_i$. This function models the right hand side of the constraint in dependence of the current value of the left hand side. The function has to satisfy $\ell_i(\lambda_i) = c_i$ for $\lambda_i \geq c_i$ and $c_i - \lambda_i \leq \alpha(\ell_i(\lambda_i) - \lambda_i)$ for $\lambda_i < c_i$.

If $\lambda_i$ is the current value of the left hand side of the $i$-th constraint, the online algorithm does know all values of $\ell_i(z)$ for $z \leq \lambda_i$ but does not know any values for $z > \lambda_i$.

Note that we only need basic threshold functions to construct the deterministic upper bound from the previous section. Define functions $f_j(\lambda)$, for $j \in [0, d-1]$, as

$$f_j(\lambda) := \begin{cases} \alpha & \lambda < \alpha - \alpha^{j/d} \\ \alpha^{1+j/d} & \lambda \geq \alpha - \alpha^{j/d} \end{cases}.$$

The LP is the same as the one in the previous section. The monotonically increasing function assigned to a constraint that corresponds to permutation $\pi$, is $f_r$ if $r$ is the largest integer such that $\pi(r) \neq k_r$. (Recall that $k_r$ is the index of the largest active variable at the end of round $r$.) If no such integer exist, $f_0$ is assigned to the constraint.

This way we can achieve the same upper bound as in the previous section, but with an explicit, predefined behavior of the adversary. Two things need to be noted here. First of all, in order to actually foil an online algorithm with a fixed adversary one needs to know the $k_r$ values. Therefore, we might need a different fixed adversary for different algorithms. Second, even if we do know the $k_r$ values, the resulting adversary is not identical to the one from the previous section. Adversaries from the previous section always increase the right-hand sides of constraints as soon as they know what their final values should be. The adversaries defined here delay the increase (almost) as long as possible. However, it can be easily verified that the bound of Theorem 1 still holds, as long as the adversary is built with the correct $k_r$ values.

*Parallel adversaries.* In the construction of the previous section, the order in which the variables become inactive defines a permutation $\pi_o$ of $\{0, \ldots, d-1\}$. In other words, if the variable with index $k_r$ becomes inactive in round $r$ we have $\pi_o(r) = k_r$.

Consider an adversary that acts in exactly the same way as before, but it guesses the permutation $\pi_o$ and proceeds as if the variables would actually become inactive in this order. If the adversary guesses correctly, the construction works as intended and the algorithm can only obtain a value of $O(d^2 \alpha^{1/d})$, while the optimum value is $\alpha$. If the adversary chooses the permutation uniformly at random, the success probability is $1/(d!)$, i.e., with this probability we achieve the same upper bound as in the previous section. However, if the adversary guesses incorrectly the algorithm can perform better than $O(d^2 \alpha^{1/d})$, perhaps even obtain a value of $\alpha$, while the optimal value remains $\alpha$. We are not going to attempt to analyze the average performance of an algorithm in this setting. Instead, we will increase the success probability for the adversary from $1/(d!)$ to about $1 - 1/\alpha$.

The idea is to have $K$ adversaries with random $\pi_o$ sequences working in parallel and require the algorithm to beat all of them, for some sufficiently large $K$. This is done as follows. Suppose we want to have $K$ $d$-dimensional adversaries. We construct a packing program with $d^K$ variables $\{x_{i_1,\ldots,i_K} | 0 \leq i_1, \ldots, i_K \leq d-1\}$. The objective function is the sum of all variables.

For the $k$-th adversary, we add $d!$ constraints to the LP. The constraints have the same form as the ones in the previous section but instead of variables $x_j$ the variables are

$$x_j^k := \sum_{(i_1,\ldots,i_K):i_k=j} x_{i_1,\ldots,i_K} .$$

For each adversary $k$, a permutation $\pi^k$ of $\{0,\ldots,d-1\}$ is chosen independently and uniformly at random and the functions $f_r$ are randomly assigned to constraints based on this permutation as described earlier.

Note that, for any adversary $k$, the objective function is

$$\sum_{0 \leq i_1,\ldots,i_K \leq d-1} x_{i_1,\ldots,i_K} = \sum_{j=0}^{d-1} x_j^k .$$

Therefore the objective function is also equal to $\min_k \sum_{j=0}^{d-1} x_j^k$.

We now allow the online algorithm to increase $x_j^k$ directly instead of increasing the underlying $x_{i_1,\ldots,i_K}$ variables. Note that, in reality, the algorithm cannot increase the $x_j^k$ variables completely independently of each other, since increasing one of the underlying $x_{i_1,\ldots,i_K}$ variables always affects multiple $x_j^k$ variables. However, allowing the algorithm to directly and individually increase $x_j^k$ variables can only give the online algorithm more power.

This completes the construction of an input that combines $K$ adversaries from the previous section, each of them independently guessing a random order in which variables becomes inactive. The profit of the online algorithm against one of the adversaries is bounded by $O(d^2 \alpha^{1/d})$ with probability $1/(d!)$ (namely if the adversary guessed the correct permutation) and by $\alpha$ otherwise.

The total profit of the online algorithm is bounded by the minimum profit the algorithm achieves against any of the $K$ adversaries. By choosing $K = \lceil d! \ln \alpha \rceil$ we get that the expected overall profit of the online algorithm is bounded by

$$\left(1 - \frac{1}{d!}\right)^K \alpha + \left(1 - \left(1 - \frac{1}{d!}\right)^K\right) O(d^2 \alpha^{1/d}) = O(d^2 \alpha^{1/d}) .$$

The optimal profit is always $\alpha$. To see this set $x_{\pi_1(0),\ldots,\pi_K(0)} = \alpha$ and all other variables to 0. Consider the constraints that belong to the $k$-th adversary. Then $x_j^k$ is equal to $\alpha$ for $j = \pi^k(0)$ and 0 otherwise. This is exactly the feasible solution from Lemma 4, applied to the constraints of the $k$-th adversary. Altogether, this gives the desired theorem.

**Theorem 4.** *The competitive ratio of any randomized online algorithm against an oblivious adversary is at most $O(m^2 \alpha^{1/m}/\alpha)$ for LPs involving $m^{\lceil m! \ln \alpha \rceil}$ variables.*

## 4 Tight Lower Bound for Two Dimensions

In this section we give a simple deterministic $\Theta(1/\sqrt{\alpha})$-competitive algorithm for packing linear programs involving two variables $x_1$ and $x_2$.

For convenience, in a first step, the algorithm normalizes variables such that the objective function is $x_1 + x_2$ instead of the more general form $b_1 x_1 + b_2 x_2$, with positive $b_1$ and $b_2$. For this we divide all entries $a_{ij}$ of the constraint matrix by $b_j$. This does not change the profit of an optimal solution. In fact, an increase of $x_i$ by $\varepsilon$ in the normalized LP exactly corresponds to an increase of $x_i$ by $\varepsilon/b_i$ in the old LP. Both of these increases would increase the objective function of the respective LP by the same amount, namely $\varepsilon$.

Now, let $x^*(z) = (x_1^*(z), x_2^*(z))$ denote an optimal solution of the linear program where the capacities, that is, the right hand sides of constraints, are given by the vector $z$. Let $x = (x_1, x_2)$ be the current online solution.

At time $t$, our algorithm ALG, which uses a parameter $\gamma$, does the following:

1. If any constraint is tight, STOP.
2. Else if $x_1^*(\ell^t) > x_1 \gamma$, increase $x_1$ infinitesimally.
3. Else if $x_2^*(\ell^t) > x_2 \gamma$, increase $x_2$ infinitesimally.
4. Else, STOP.

**Theorem 5.** *For $\gamma = 1 + 1/\sqrt{\alpha}$, the optimum profit is at most $(\sqrt{\alpha} + 1)$ times the profit obtained by ALG.*

*Proof.* Let $x' = (x_1', x_2')$ indicate the point where $ALG$ stops. Since $ALG$ will either stop due to Step 1 or Step 4, we distinguish these two cases:

**ALG Stops Due to Step 1.** There is a tight constraint $k$. Let us write this constraint as $a_1 x_1 + a_2 x_2 \leq \ell_k^t$. Since $(c - Ax) \leq \alpha(\ell^t - Ax)$, this implies $a_1 x_1' + a_2 x_2' = c_k$.

Assume, without loss of generality, that $a_1 \geq a_2$. Then, the optimum profit can be at most $c_k/a_2$.

For every point in time $t$, $x^*(\ell^t)$ has to satisfy $a_1 x_1^*(\ell^t) + a_2 x_2^*(\ell^t) \leq \ell_k^t \leq c_k$. Therefore, $x_1^*(\ell^t) \leq c_k/a_1$. Additionally, since ALG increases variable $x_1$ only if $x_1 \gamma < x_1^*(\ell^t) \leq c_k/a_1$, it holds $x_1' \leq c_k/(a_1 \gamma)$. But since $a_1 x_1' + a_2 x_2' = c_k$, we obtain $a_2 x_2' = c_k - a_1 x_1' \geq (1 - 1/\gamma)c_k$.

Therefore, the profit of ALG is at least $(1 - 1/\gamma)c_k/a_2$. Since the optimum profit is at most $c_k/a_2$ and with $\gamma = 1 + 1/\sqrt{\alpha}$, the profit of ALG is at least a $1/(\sqrt{\alpha} + 1)$ fraction of the optimum profit.

**ALG Stops Due to Step 4.** In this case, by the choice of the stopping condition, $x_1^*(\ell^t) \leq x_1' \gamma$ and $x_2^*(\ell^t) \leq x_2' \gamma$. Adding the two inequalities, we get $\|x'\gamma\|_1 \geq \|x^*(\ell^t)\|_1$.

Since $(c - Ax) \leq \alpha(\ell^t - Ax)$, we have $\ell^t \geq ((\alpha - 1)Ax + c)/\alpha$. Hence,

$$\|x'\gamma\|_1 \geq \|x^*(\ell^t)\|_1$$
$$\geq \left\|x^*\left(\frac{(\alpha-1)Ax'+c}{\alpha}\right)\right\|_1$$
$$\geq \left(1 - \frac{1}{\alpha}\right)\|x^*(Ax')\|_1 + \frac{\|x^*(c)\|_1}{\alpha}$$
$$\geq \left(1 - \frac{1}{\alpha}\right)\|x'\|_1 + \frac{\|x^*(c)\|_1}{\alpha}.$$

The second inequality follows from the fact that $\ell^t \geq ((\alpha - 1)Ax + c)/\alpha$; therefore the polytope defined by setting the capacity vector to $((\alpha-1)Ax+c)/\alpha$ is enclosed by the polytope defined by setting the capacity vector to $\ell^t$. The last inequality follows from the fact that point $x'$ is a feasible solution if the capacities are set to $Ax'$.

Solving for $x'$ gives $\|x'\|_1 \geq \|x^*(c)\|_1/(\alpha\gamma - \alpha + 1) = \|x^*(c)\|_1/(\sqrt{\alpha} + 1)$, showing again that the optimum profit cannot be greater than $(\sqrt{\alpha} + 1)$ times the profit of ALG. □

## 5 Conclusions

Although we significantly narrow the gap between lower and upper bound for this online LP problem the obvious open question whether it is possible to beat the competitive ratio of $\Omega(\ln^2 \alpha/\alpha)$ in general, either with a randomized or even with a deterministic algorithm, remains.

Our new bounds also suggest that it is interesting to study the influence of the number of variables $d$ in the LP on the achievable competitive ratio. We would be interested in bounds that are tight for any fixed number of variables, not just when the number of variables is very large.

This is further emphasized by the fact that, if $d$ is very large, there seems little difference in the power of randomized and deterministic online algorithms. However, taking $d$ into account and considering the particular interesting case of moderately large values of $d$, randomization has the potential to greatly improve performance. It is, for instance, easy to see that the algorithm that picks one of the $d$ variables uniformly at random and increases that variable until a constraint becomes tight is $1/d$-competitive. Take for example $d = 2$, where this is a significant improvement over the $\Theta(1/\sqrt{\alpha})$ competitive ratio deterministic algorithms can achieve. In fact, this trivial randomized algorithm beats the general deterministic upper bound as long as, say, $d \leq \sqrt[3]{\alpha}/2$.

Another interesting question is whether our upper bound construction can be realized by some natural combinatorial packing problem.

# References

1. Adamaszek, A., Czumaj, A., Englert, M., Räcke, H.: An O(logk)-competitive algorithm for generalized caching. In: Proceedings of the 23rd ACM-SIAM Symposium on Discrete Algorithms (SODA), pp. 1681–1689 (2012)
2. Bansal, N., Buchbinder, N., Madry, A., Naor, J.: A polylogarithmic-competitive algorithm for the k-server problem. In: Proceedings of the 52nd IEEE Symposium on Foundations of Computer Science (FOCS), pp. 267–276 (2011)
3. Buchbinder, N., Naor, J.: Online primal-dual algorithms for covering and packing. Mathematics of Operations Research 34(2), 270–286 (2009)
4. Calinescu, G., Kapoor, S., Olshevsky, A., Zelikovsky, A.: Network lifetime and power assignment in ad hoc wireless networks. In: Di Battista, G., Zwick, U. (eds.) ESA 2003. LNCS, vol. 2832, pp. 114–126. Springer, Heidelberg (2003)
5. Ochel, M., Radke, K., Vöcking, B.: Online packing with gradually improving capacity estimations and applications to network lifetime maximization. In: Czumaj, A., Mehlhorn, K., Pitts, A., Wattenhofer, R. (eds.) ICALP 2012, Part II. LNCS, vol. 7392, pp. 648–659. Springer, Heidelberg (2012)

# Improved Minmax Regret 1-Center Algorithms for Cactus Networks with $c$ Cycles

Binay Bhattacharya[1], Tsunehiko Kameda[1], and Zhao Song[2]

[1] School of Computing Science, Simon Fraser University, Canada
{binay,tiko}@sfu.ca
[2] Department of Computer Science, The University of Texas at Austin, USA
zhaos@utexas.edu

**Abstract.** In a facility location problem, if the vertex weights are uncertain one may look for a "robust" solution that minimizes "regret." We present an $O(n \log n)$ (resp. $O(cn \log n)$) time algorithm for a tree (resp. $c$-cycle cactus), where $n$ is the number of vertices and $c$ is a constant. Our tree algorithm presents an improvement over the previously known algorithms that run in $O(n \log^2 n)$ time. There is no previously published result tailored specifically for a cactus network. The best algorithm for a general network takes $O(mn \log n)$ time, where $m$ is the number of edges.

## 1 Introduction

Deciding where to locate facilities to minimize the communication or transportation costs is known as the *facility location problem*. For a recent review of this subject, the reader is referred to [14]. The cost of a vertex is formulated as the distance from the nearest facility weighted by the weight of the vertex. In the "classical" *p-center* problem, the objective is to find $p$ facility locations such that the maximum cost over all vertices is minimized. This problem has attracted much research interest since the publication of the seminal paper on this topic by Hakimi [13]. It can be applied to the locating of fire stations, distribution centers, etc. Megiddo [16] computed the 1-center of a tree network with non-negative vertex weights in $O(n)$ time, where $n$ is the number of vertices. Megiddo and Tamir also studied this problem [17].

In the *minmax regret* version of this problem, there is uncertainty in the weights of the vertices and/or edge lengths, and only their ranges are known [11,15]. A particular *realization* (assignment of a weight to each vertex) is called a *scenario*. Intuitively, the planner of a facility proposes a location $x$. Then the adversary finds a scenario that makes $x$ bad (costly), i.e., its "regret" is "deep." The purpose of the planner is to make this cost as small as possible, no matter which scenario the adversary comes up with. For a general graph with uncertain edge lengths, the minmax-regret 1-center problem was shown to be strongly NP-hard [1]. For a tree, Averbakh and Berman [3] solved the problem in $O(n^6)$ time, and Burkard and Dollani [8] improved it to $O(n^3 \log n)$. For a general graph with fixed edge lengths, Averbakh and Berman [2] solved it in $O(mn^2 \log n)$ time, where $m$ is the number of edges. For a tree with fixed edge lengths, they solved the problem in $O(n^2)$ time [2,3]. For a tree with uncertain edge lengths and uniform vertex weights, they presented an $O(n^2 \log n)$ time algorithm [3], which was later improved to $O(n \log n)$ by Burkard and Dollani [8]. More recently, Yu et al. [19] made further improvements, coming up with more efficient algorithms when the edge lengths are fixed and the vertex weights are positive. The time complexities of their algorithms for a general network and a tree network are $O(mn \log n)$ and $O(n \log^2 n)$, respectively. We simplified the tree algorithm in [7], although the time complexity still remained $O(n \log^2 n)$. A network that is more general than a tree and is often considered is a *cactus* network. It is characterized by the property that no edge belongs to more than one cycle. Under the assumption that the edge lengths are fixed, the table below summarizes the best previously known time complexities and our new results presented in this paper, where $c$ ($\geq 1$) is a constant, and all results are valid if the minimum weight of at least one vertex is non-negative.

| Network type | Best known | This paper |
|---|---|---|
| Path | $O(n)$ [7] | |
| Tree | $O(n \log^2 n)$ [19] | $O(n \log n)$ (Sec. 3) |
| Cycle | $O(n \log n)$ [7] | |
| $c$-cycle cactus | | $O(cn \log n)$ (Sec. 4) |
| General network | $O(mn \log n)$ [19] | |

The improved results are achieved by efficiently answering various query questions for the tree/cactus networks in a dynamic setting where an arbitrary vertex of the network is allowed to increase its weight. This paper is organized as follows. In Sec. 2, we define basic terms and review some relevant facts. Sec. 3 discusses the tree network, and present an $O(n \log n)$ time algorithm. Sec. 4 shows that if a cactus has $c$ cycles, our algorithm runs in $O(cn \log n)$ time. Sec. 5 concludes the paper.

## 2 Preliminaries

Let $G = (V, E)$ denote a network with vertex set $V = \{v_1, v_2, \ldots, v_n\}$. Each vertex $v_i \in V$ is associated with an interval of integer weights $W(v_i) = [\underline{w}_i, \overline{w}_i]$, where $0 \leq \underline{w}_i \leq \overline{w}_i$, and each edge $e \in E$ is associated with a non-negative *length*. We assume that the distances between a point on an edge and its end vertices are prorated fractions of the edge length. We often use $G$ to denote the set of all points on the edges and vertices of $G$. For any pair of points $p, q \in G$, the shortest distance between them is denoted by $d(p, q)$. Let $\mathcal{S}$ denote the Cartesian product of all $W(v_i)$, $v_i \in V$: $\mathcal{S} \triangleq \prod_{v_i \in V}[\underline{w}_i, \overline{w}_i]$. Each element $s \in \mathcal{S}$ is called a *scenario*. Let $w_i^s$ be the weight of $v_i$ under $s$. The *cost* of a point $x \in G$ *with respect to* $v_i \in V$ under scenario $s$ is $d(v_i, x)w_i^s$, and the *cost* of $x$ under $s$ is defined by

$$F^s(x) \triangleq \max_{v_i \in V} d(v_i, x) w_i^s. \quad (1)$$

The point $x$ that minimizes $F^s(x)$ is called a (*classical*) *1-center* under $s$, and is denoted by $c(s)$. A vertex $v$ that maximizes (1) is called a *critical* vertex for

$x$ [3]. Throughout this paper the term *center* refers to this weighted 1-center. We define the *regret* of $x$ under $s$ by [15]

$$R^s(x) \triangleq F^s(x) - F^s(c(s)), \tag{2}$$

and the *maximum regret* of $x$ by

$$R^*(x) \triangleq \max_{s \in \mathcal{S}} R^s(x). \tag{3}$$

The scenario that maximizes $R^s(x)$ for a given $x \in G$ is called the *worst case scenario* for $x$. We seek location $x^* \in G$, called the *minmax regret 1-center*, that minimizes $R^*(x)$. A scenario $s_i$ *dominates* another scenario $s_j$ at point $x$ if $R^{s_i}(x) \geq R^{s_j}(x)$. If a scenario is dominated by no other scenario at any point in an interval, then it is said to be *non-dominated* in that interval.

The *base scenario*, $s_0$, is defined by $w_i^{s_0} = \underline{w}_i$ for all $i$. For $i = 1, 2, \ldots, n$, let us define the *single-max scenario* $s_i$ by $w_i = \overline{w}_i$ and $w_j = \underline{w}_j$ for all $j \neq i$, and let $\mathcal{S}^*$ denote the set of all single-max scenarios. Averbakh and Berman proved the following theorem for the trees, but their proof is valid for the general network.

**Theorem 1.** [3] *For any point $x$ in $G$, there is a worst case scenario for $x$ that is a single-max scenario $s_j$ such that vertex $v_j$ is a critical vertex for $x$.* □

Theorem 1 implies

$$R^*(x) = \max_{v_j \in V} \{d(v_j, x)\overline{w}_j - F^{s_j}(c(s_j))\}. \tag{4}$$

Averbakh and Berman [2] convert $G = (V, E)$ to its *auxiliary network* $G' = (V', E')$, where $V \subset V'$ and $E \subset E'$, in such a way that the minmax regret 1-center problem on $G$ becomes the classical 1-center problem on $G'$.[1] Assume that $F^s(c(s))$ for all $s \in \mathcal{S}^*$ are available, and let $M$ be a sufficiently large, positive integer. They append an edge $(v_i, v_i') \in E'$ of length $\{M - F^{s_i}(c(s_i))\}/\overline{w}_i$ to every vertex $v_i \in V$, where $v_i' \in V'$ is a new vertex of degree 1, called the *dummy vertex* corresponding to $v_i$. The vertices of $G'$ that belong to $V$ have weights equal to zero, and the edges in $E$ retain the original lengths. For each vertex $v_i' \in V' \backslash V$ its weight is set to $\overline{w}_i$. Clearly, $G'$ is a weighted cactus network. The construction of $G'$ requires $c(s_j)$ for all $v_j$, but it is easy to prove

**Lemma 1.** *If a scenario $s_j$ is dominated at all points, then $c(s_j)$ can be ignored in constructing $G'$.* □

**Theorem 2.** [2] *The minmax regret 1-center problem on a network $G$ can be solved by computing the classical weighted 1-center problem on $G'$.* □

It is known that the classical weighted 1-center problem on a cactus network can be computed in $O(n \log n)$ time [4]. After computing the 1-centers under all single-max scenarios, we can invoke Theorem 2. Therefore, we concentrate on finding those 1-centers. For $v_i \in V$ and $x \in G$, we introduce *cost functions* of $v_i$, $\overline{f}_i(x) = d(x, v_i)\overline{w}_i$ and $\underline{f}_i(x) = d(x, v_i)\underline{w}_i$. If $v_j$ is critical for $c(s_j)$, and $\overline{f}_j(c(s_j)) = \underline{f}_k(c(s_j))$ holds, then we call vertex $v_k$ a *counterbalance* for $v_j$.

---

[1] They introduced this operation for trees, but it works for a general network.

**Lemma 2.** *If $v_j$ is critical under $s_j$ for a center $c(s_j)$[2] then there is a center $c(s_j)$ that lies on a path between $v_j$ and $c(s_0)$.*

*Proof.* Let $v_k$ be a counterbalance for $v_j$, so that $d(v_k, c(s_j))\underline{w}_k = d(v_j, c(s_j))\overline{w}_j$ holds. The assertion of the lemma is obvious if $c(s_0)$ lies on the path between $v_k$ and $v_j$. So assume that it doesn't. Let $v$ be the vertex where the path from $c(s_0)$ to $v_j$ and that from $c(s_0)$ to $v_k$ diverge $c(s_j)$ is on the path from $v$ to $v_k$. Then we have $d(v_k, c(s_j)) < d(v_k, c(s_0))$, hence

$$d(v_k, c(s_j))\underline{w}_k < d(v_k, c(s_0))\underline{w}_k, \tag{5}$$

which implies that the cost of $c(s_j)$ (LHS of (5)) is less than that of $c(s_0)$, a contradiction. We assumed above that at least one edge on the path between $c(s_j)$ and $v$ has a positive length. Otherwise, $c(s_j)$ can be set to $v$, and it is on a path between $c(s_0)$ and $v_j$. □

However, it is easy to show that the counterbalance $v_k$ for $v_j$ may not be on any path between $v_j$ and $c(s_0)$.

**Proposition 1.** *The center $c(s_j)$ under $s_j$ is given by*

$$c(s_j) = \operatorname{argmin}_x \{\max\{\overline{f}_j(x), F^{s_0}(x)\}\}. \tag{6}$$

□

Let $H$ be a subgraph of graph $G$, and generalize the definition in (1) to

$$F^s(x, H) \triangleq \max_{v_i \in V \cap H} d(v_i, x) w_i^s. \tag{7}$$

The point $x$ that minimizes $F^s(x, H)$ is called a *restricted 1-center* relative to $H$ under $s$, and is denoted by $c_H(s)$. The difference

$$R^s(x, H) \triangleq F^s(x, H) - F^s(c_H(s), H) \tag{8}$$

is called the *restricted regret* of $x$ relative to $H$ under $s$. Here, $x$ need not belong to $H$.

## 3 Tree Network

Assume that a given tree network $T = (V, E)$ is balanced and binary, having height $O(\log n)$, and its classical 1-center $c(s_0)$ under the base scenario $s_0$ is at the root $r$.[3] If $T$ is not binary, we can introduce $O(n)$ vertices of 0 weight and $O(n)$ edges of 0 length to convert it into a binary tree [18]. We also assume that $T$ is balanced. If not, *spine tree decomposition* [5,6,7], can convert it into a structure that has properties of a balanced binary tree in linear time. Let $T(v)$

---

[2] A center may not be unique.
[3] If $c(s_0)$ is on an edge, we insert vertex $r$ of weight 0 there.

denote the subtree of $T$ rooted at vertex $v$. For each vertex $v_i \in V$, we introduce the *upper envelope* for the cost functions of the vertices in $T(v_i)$ under $s_0$ by

$$E_{T(v_i)}(x) \triangleq F^{s_0}(x, T(v_i)), x \notin T(v_i). \tag{9}$$

It is a monotone function, and consists of one or more linear segments, so that we can consider it as a sequence of bending points, starting at $v_i$. Let $\pi(v_i, v_k)$ denote the (shortest) path between two vertices $v_i, v_k \in T$. In the simple example in Fig. 1(a), several cost functions, such as $E_{T(v_2)}(x) = \underline{f}_2(x)$, are shown for $x \in \pi(v_1, v_j)$. We compute $E_{T(v_i)}(x)$, for all $v_i \in V$, bottom up. At a leaf vertex

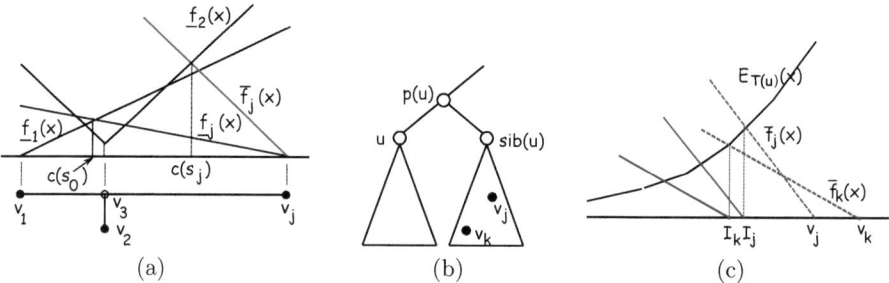

**Fig. 1.** (a) Center $c(s_j)$ under single-max scenario $s_j$; (b) $v_j, v_k \in T(sib(u))$; (c) $s_k$ is dominated by $s_j$

$v_i$ of $T$, the upper envelope is $\underline{f}_i(x) = d(x, v_i)\underline{w}_i$. At a higher level vertex $v_i$ of $T$ with two child vertices, $v_k$ and $v_l$, the linear segments of $E_{T(v_i)}(x)$ can be computed from those of $E_{T(v_k)}(x)$, $E_{T(v_l)}(x)$, and $\underline{f}_i(x) = d(x, v_i)\underline{w}_i$ in linear time. Clearly, the number of linear segments in all the upper envelopes at each level[4] of $T$ is $O(n)$. Since $T$ has height $O(\log n)$, we have

**Lemma 3.** [7] *The upper envelopes at all the vertices in a balanced tree can be computed in $O(n \log n)$ time. The space needed for them is also $O(n \log n)$.* □

We now look for center $c(s_j)$ under the single-max scenario $s_j$, using (6). Note that $F^{s_0}(x)$ can be replaced by the maximum of the upper envelopes $E_{T(v_i)}(x)$ for all maximal subtrees $T(v_i)$ such that $v_j \notin T(v_i)$, and $\underline{f}_k(x)$ for all $v_k$ such that $v_k \in \pi(v_j, r)$. We compute the intersection of $\overline{f}_j(x)$ with each of these $O(\log n)$ functions, and determine the most costly one among them. Proposition 1 implies that the corresponding point gives $c(s_j)$, provided its cost is not less than the cost of $c(s_0)$. This straightforward approach takes $O(\log^2 n)$ per scenario $s_j$ [7]. To reduce the required time, we now try to find $c(s_j)$ for all non-dominated scenarios $s_j$ together, by iterating on subtrees rather than on $v_j$.

Consider any vertex $u$, and its sibling $sib(u)$, as shown in Fig. 1(b). We need to compute the intersection of $\overline{f}_j(x)$ and $E_{T(u)}(x)$ for all $v_j \in T(sib(u))$. For a

---
[4] Level $l$ consists of all the vertices at distance $l$ from the root.

given $v_j$, if we iterate the above computation for every vertex $u \in V \setminus \{r\}$ such that $v_j \in T(sib(u))$, then we will have examined the intersections of $\overline{f}_j(x)$ with $\underline{f}_i(x)$ for all $i \neq j$ such that $v_i \notin \pi(v_j, r)$. We can now extract the intersections involving the cost function $\overline{f}_j(x)$ of a particular $v_j$. Under the scenario in which only $r$ takes the maximum weight, $c(s_0)$ is the 1-center. Fig. 1(c) shows how this can be done efficiently. The intersection of $\overline{f}_j(x)$ (resp. $\overline{f}_k(x)$) and $E_{T(u)}(x)$ on $\pi(v_j, u)$ (resp. $\pi(v_k, u)$) is named $I_j$ (resp. $I_k$). Note that we have $I_j = c_{T(u)}(s_j)$ (resp. $I_k = c_{T(u)}(s_k)$) and it is a candidate for $c(s_j)$ (resp. $c(s_k)$) provided $I_j \in \pi(v_j, p(u))$ (resp. $I_k \in \pi(v_k, p(u))$), where $p(u)$ denotes the parent of vertex $u$. See Lemma 2. Assume that $\overline{w}_j > \overline{w}_k$ holds and $v_k$ is farther from $sib(u)$ than $v_j$. In Fig. 1(c), the restricted regrets $R^{s_j}(x, T(u))(= F^{s_j}(x, T(u)) - F^{s_j}(c_{T(u)}(s_j), T(u)))$ and $R^{s_k}(x, T(u))$ are shown by solid lines. Thus we have $R^{s_j}(x, T(u)) > R^{s_k}(x, T(u))$, which implies that $s_k$ is dominated by $s_j$. If we process $v_k$ after $v_j$, this domination is detected, enabling us to ignore $v_k$. In the algorithm given below, we take advantage of this fact by constructing a list of vertices that are ordered by the distance from $u$. For each vertex $v_j$, we collect all the candidates for $c(s_j)$, i.e., $c_{T(u) \cup \{v_j\}}(s_j)$ for all vertices $u \in V$, and pick the most costly one as $c(s_j)$. This can be done by executing Algorithm Balanced_Tree$(T; u)$ below for each $u \in V \setminus \{r\}$.

**Algorithm** Balanced_Tree$(T; u)$

1. Let $L$ be a list of the vertices of $T(sib(u))$ ordered from the nearest vertex to the farthest from $sib(u)$. Initialize the "previous intersection" to be at $(x, height) = (p(u), 0)$.
2. For each vertex $v_j \in L$, in the order in $L$, carry out steps 3 to 5.
3. Find if the intersection of $\overline{f}_j(x)$ and $E_{T(u)}(x)$ is to the upper right of the "previous intersection" (in the context of Fig. 1(c)), by evaluating $\overline{f}_j(x)$ at position $x$ of the "previous intersection." It is the case if the value is larger than the *height* of the "previous intersection." If not, skip Steps 4 and 5.
4. Calculate the intersection point as follows: for each segment of $E_{T(u)}(x)$ above the the *height* of the "previous intersection," test if it intersects $\overline{f}_j(x)$, and if so, determine the intersection. Set the "previous intersection" to this intersection.
5. Record it as a candidate for $c(s_j)$, provided it lies on $\pi(v_j)$ and its cost is more than that of $c(s_0)$. □

During preprocessing, we construct ordered list $L$ for the vertices of each subtree $T(sib(u))$. This can be done bottom up in $T$ for all $sib(u)$, i.e., for all subtrees. If the test of Step 3 fails, then this intersection cannot be a candidate for $c(s_j)$. The condition in Step 5 is justified by Lemma 2. After executing Algorithm Balanced_Tree$(T; u)$ for every $u \in V \setminus \{r\}$, for each non-dominated single-max scenario $s_j$, the true $c(s_j)$ is the maximum among all the values of $x$ recorded for $s_j$ in Step 5.

**Lemma 4.** *The 1-centers $c(s_j)$ for all non-dominated scenarios $s_j$ of a balanced binary tree can be computed in $O(n \log n)$ time.*

*Proof.* Algorithm Balanced_Tree runs in time linear in $|T(u)| + |T(sib(u))|$, where $|T(u)|$ denotes the number of vertices in $T(u)$. This is because Step 3 takes constant time, and there is no back-tracking on the $O(|T(u)|)$ linear segments in $E_{T(u)}(x)$. Note that

$$\sum_{u \in V} |T(u)| + |T(sib(u))| < 2 \sum_{v \in V} |T(v)| = O(n \log n). \tag{10}$$

□

If the given tree is not balanced and binary, Lemma 4 is still valid, if we use *spine tree decomposition* [5,6,7]. Thus, Theorem 2 and Lemma 4 imply

**Theorem 3.** *The minmax regret 1-center of a tree network can be computed in $O(n \log n)$ time.* □

## 4 Cactus Network with Constant Number of Cycles

### 4.1 Unicyclic Network

A *unicyclic* network, $G = (V, E)$, contains just one cycle $C$ with circumference $l_C$. For $a, b \in C$, the clockwise section of $C$ from $a$ to $b$ is denoted by $C(a, b)$ and its length denoted by $d(a, b)$. If $d(a, b) = l_C/2$, we say that $a$ (resp. $b$) is the *antipode* [12] of $b$ (resp. $a$), denoted by $a = \alpha(b)$ (resp. $b = \alpha(a)$). We can assume without loss of generality that the degree of each cycle vertex is at most 3. Otherwise, we can insert dummy vertices of weight 0 and dummy edges of length 0. A subgraph that hangs from vertex $u \in C$, excluding $u$ and the edge to $u$, is called a *graft*,[5] and is denoted by $\Gamma(u)$. See Fig. 2(a). We use $\Gamma^c(u)$ to denote the complement of $\Gamma(u)$, obtained by removing $\Gamma(u)$ and the edge connecting $\Gamma(u)$ and $u$ from $G$.

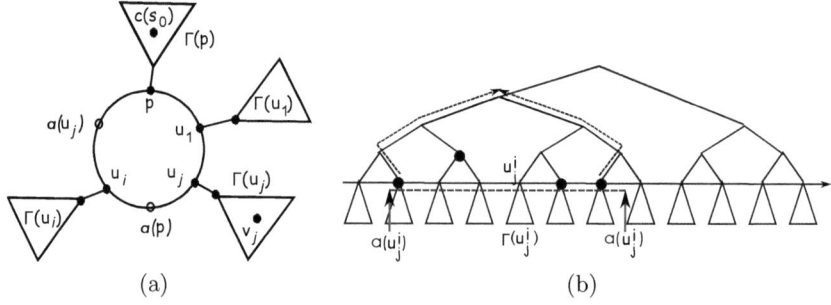

**Fig. 2.** (a) Cycle $C$; (b) Cycle envelope tree $\mathcal{T}_C$

---

[5] The definition of graft in [9] includes $u$.

We first find a 1-center $c(s_0)$ of $G$, using the $O(n \log n)$ time algorithm of Ben-Moshe et al. [4]. If $c(s_0) \notin C$, then let $p \in C$ be the vertex such that either $\Gamma(p)$ or the edge connecting $\Gamma(p)$ and $p$ (including $p$) contains $c(s_0)$. We call $\Gamma(p)$ the *parent graft* of $C$. If $c(s_0) \in C$ and it is not at a vertex, then we create a dummy vertex $p$ with 0 weight at $c(s_0)$. We now construct a tree $T$ with $r = c(s_0)$ as its root, by cutting cycle $C$ open by removing the edge on which the antipode $\alpha(p)$ lies.[6] See Fig. 2(a). Let $u_0 (=p), u_1, u_2, \ldots, u_{g-1}$ be the cycle vertices, clockwise along $C$. By our degree assumption, for each $k$, $u_k$ is connected to at most one graft. We assume that $T$ is a balanced binary tree, hence its height is $O(\log n)$. (If not, algorithmic properties and the tools developed here for the balanced trees can easily be extended, using *spine tree decomposition* [5,6,7].) As in Sec. 3, for each vertex $u \in T$, we compute the upper envelope $E_{T(u)}(x)$ under $s_0$ in $O(n \log n)$ time.

Our goal is to find $c(s_j)$ for each vertex $v_j \in V$ such that $s_j$ is not dominated. (See Lemma 1.) Let $u_j \in C$ be the cycle vertex such that $v_j$ is either in graft $\Gamma(u_j)$ that hangs from the cycle vertex $u_j$ or $v_j = u_j$.[7] See Fig. 2(a). If $v_j \in \Gamma(p)$, then we can use our tree algorithm in Sec. 3 to find $c(s_j)$. (Lemma 2.) Therefore, we assume from now on that $u_j \ne p$. The following procedure identifies the part of $G$ where $c(s_j)$ is located.

**Procedure In/Out_Test$(C; v_j)$**

1. If $F^{s_0}(p, \Gamma(p)) > F^{s_j}(p, \Gamma^c(p))$, then $c(s_j)$ lies in $\Gamma(p)$ or on the edge connecting $\Gamma(p)$ and $p$.
2. If $F^{s_j}(u_j, \Gamma(u_j)) > F^{s_0}(u_j, \Gamma^c(u_j))$, then $c(s_j)$ lies in $\Gamma(u_j)$ or on the edge connecting $\Gamma(u_j)$ and $u_j$.
3. In all other cases, $c(s_j)$ lies on $C$. □

To carry out the above tests, we need to compute different costs. Among them, it is easy to compute $F^{s_0}(p, \Gamma(p))$ once for all in $O(n \log n)$ time, since $\Gamma(p)$ is a tree. We can also easily compute $F^{s_j}(p, \Gamma^c(p)) = \max\{E_{T(p)}(p), \overline{f}_j(p)\}$ and $F^{s_j}(u_j, \Gamma(u_j)) = \max\{E_{T(u_j)}(u_j), \overline{f}_j(u_j)\}$, using the upper envelopes $\{E_{T(u)}(x) \mid u \in V\}$ that we have available.

It is more difficult to compute $F^{s_0}(u_i, \Gamma^c(u_i))$ for $0 \le i \le g-1$. We first construct the *cycle envelope tree*, denoted by $\tau_C$. Lay out the two periods of cycle vertices $u_0 (=p), \ldots, u_{g-1}$, together with their associated grafts horizontally on a line of length $2l_C$. See Fig. 2(b). We now build $\tau_C$ as a load-balanced[8] tree on top of the cycle vertices, using $|\Gamma(u_i)| + 1$ as the load of $u_i$, where $|\Gamma(u_i)|$ denotes the number of vertices in graft $\Gamma(u_i)$, and "+1" accounts for $u_i$. Note that, given a cycle vertex $u_j$, we can always find its antipode $\alpha(u_j)$ to the left and right of an occurrence of $u_j$ at the leaf level (where the cycle vertices lie) of $\tau_C$, as indicated by the dashed line from $\alpha^l(u_j)$ to $\alpha^r(u_j)$ in Fig. 2(b).

---

[6] If there is a vertex at $\alpha(u_i)$, then remove one of the two cycle edges incident to it.
[7] Strictly speaking, we should use $u_{\phi(j)}$ for some mapping $\phi(j)$, but we abuse notation for the sake of brevity.
[8] It is commonly called *weight-balanced*. We use "load" to avoid confusion with the vertex weight.

For each node $u$ of $\tau_C$, we now define two upper envelopes, $E_u^{cw}(x)$ and $E_u^{ccw}(x)$, reflecting the costs of the vertices in the subtree rooted at $u$, i.e., $\tau_C(u)$. Function $E_u^{cw}(x)$ (resp. $E_u^{ccw}(x)$) is the upper envelope for the costs of the vertices (cycle vertices and those in grafts) in $\tau_C(u)$ at point $x$ on the clockwise (resp. counterclockwise) side of the cycle vertices in $\tau_C(u)$. To be more precise, $E_u^{cw}(x)$ and $E_u^{ccw}(x)$ are valid for $x \in C$, provided $\tau_C(u)$ spans no more than a half period of $C$ and the distance from $x$ to the farthest cycle vertex in $\tau_C(u)$ is no more than $l_C/2$. They are also valid for $x \in \Gamma(u_j)$, where $u_j \notin \tau_C(u)$, provided $\tau_C(u)$ spans no more than a half period of $C$ and the distance from $u_j$ to the farthest vertex in $\tau_C(u)$ is no more than $l_C/2$. These functions can be represented by a sequence of bending points. We used a similar tree for a cycle in [7], and the proof given there carries over to prove

**Lemma 5.** *We can compute $E_u^{cw}(x)$ and $E_u^{ccw}(x)$ for all nodes $u$ of $\tau_C$ in $O(n \log n)$ time and $O(n \log n)$ space, where $E_u^{cw}(x)$ and $E_u^{ccw}(x)$ are given as upper envelopes of $O(\log n)$ monotone functions.* □

**Lemma 6.** $F^{so}(u_j, \Gamma^c(u_j))$ *in Step 2 of Procedure* In/Out_Test$(C; v_j)$ *can be computed in $O(\log n)$ time.*

*Proof.* Let $r_i$ be the node of $\tau_C$ where the upward paths from $\alpha^l(u_i)$ and $\alpha^r(u_i)$ meet. Then path $\pi(u_i, r_i)$ and the two paths from $\alpha^l(u_i)$ and $\alpha^r(u_i)$ to $r_i$ all have length $O(\log n)$. Thus there are $O(\log n)$ upper envelopes associated with the maximal subtrees hanging from these three paths. (The roots of these subtrees are shown as black circles in Fig. 2(b).) We thus can compute $F^{so}(u_j, \Gamma^c(u_j))$ in $O(\log n)$ time. □

From the discussion in the paragraph following Procedure In/Out_Test$(C; v_j)$ and Lemma 6, it follows that

**Lemma 7.** *Procedure* In/Out_Test$(C; v_j)$ *runs in $O(\log n)$ time.* □

Our general approach is to first apply Unicycle_In/Out_Test$(C; v_j)$ to each query vertex $v_j \in V$. If the condition of Step 1 holds, then we use our tree algorithm from Sec. 3. Let us first discuss the case where the tests in both Steps 1 and 2 fail, in which case we look for $c(s_j)$ on $C$. Let $E_C(x)$, $x \in C$, be the upper envelope for one period (of length $l_C$) of $C$. It is clear that $c(s_j)$ is the lowest point of the upper envelope $\max\{\overline{f}_j(x), E_C(x)\}$. As in our cycle algorithm in [7], we construct a kind of *binary heap*, $H_C$, in $O(n)$ time as follows. Its leaves are the bending points of $E_C(x)$, ordered clockwise (and laid out from left to right), and repeated twice, and their values are the heights of the bending points. The value of each non-leaf node $u$ is the minimum of the values associated with its two child nodes. Thus it is the minimum among the values associated with all leaves of the subtree $H_C(u)$ rooted at $u$.

**Lemma 8.** *The lowest point in the upper envelope $\max\{\overline{f}_j(x), E_C(x)\}$ can be computed in $O(n \log n)$ time total for all $v_j \in V$.*

*Proof.* If $\overline{f}_j(x)$ lies totally above (resp. below) $E_C(x)$ for all $x \in C$, then $u_j$ (resp. the lowest point of $E_C(x)$) is what is being sought. So assume that $\overline{f}_j(x)$ and $E_C(x)$ cross each other. We now compute the intersection point $a$ (resp. $b$) of $\overline{f}_j(x)$ and $E_C(x)$ that is closest to $v_j$ on the left (resp. right) side of $v_j$. This can be done using Chazelle et al.'s ray-shooting algorithm [10] in $O(\log n)$ time for each $v_j$. Since $\overline{f}_j(x)$ is V-shaped with the bottom of the V at $u_j$, the lowest point in the upper envelope of $\overline{f}_j(x)$ and $E_C(x)$ is the lowest point in $E_C(x)$ between these two intersection points.

We then identify the bending point $a'$ (resp. $b'$) in $E_C(x)$ that is closest to $a$ (resp. $b$) on its right (resp. left) side. It is easy to see that the lowest point of the upper envelope of $\overline{f}_j(x)$ and $E_C(x)$ can be found by looking for the minimum value among the values stored at $a'$ and $b'$ plus those stored at the right (resp. left) child nodes of the path from the leaf $a'$ (resp. $b'$) to the root of the heap $H_C$. Since there are $O(\log n)$ such nodes, it takes $O(\log n)$ time to find the minimum. Repeating the above for all $v_j$ takes $O(n \log n)$ time. □

For $i = 0, 1, \ldots, g-1$, let $V_i$ denote the set of vertices that comprise $\Gamma(u_i)$ plus $u_i$. The following algorithm computes the centers in $\{c(s_j) \mid s_j \in \mathcal{S}^*\}$ that are not dominated.

**Algorithm.** Unicycle($G(V, E)$):
1. Compute $\{E_{\Gamma(u_i)}(x) \mid u_i \in C\}$, $\{E_{T_C(u)}^{cw}(x), E_{T_C(u)}^{ccw}(x) \mid u \in \tau_C\}$ and $E_C(x)$, and construct the heap $H_C$.
2. In the order $i = 0, 1, \ldots, g-1$, for each $v_j \in V_i$, carry out Steps 3–6.
3. Perform Procedure Unicycle_In/Out_Test($C; v_j$).
4. In Case 1, invoke Balanced_Tree($T; u$) for relevant $u$.
5. In Case 3, find the lowest point in the upper envelope of $\overline{f}_j(x)$ and $E_C(x)$.
6. In Case 2, find the intersection of $\overline{f}_j(x)$ and each of the following, and pick the intersection that lies on $\pi(v_j, u_j)$ and has the highest cost.
   (a) $F^{s_0}(x, \Gamma^c(u_j))$.
   (b) $\underline{f}_k(x)$, where $v_k \in \pi(v_j, u_j)$.
   (c) $\overline{E}_u(x)$, for each $u$ that is a child of a vertex on $\pi(v_j, u_j)$. □

**Lemma 9.** *Algorithm* Unicycle($G(V, E)$) *correctly computes* $\{c(s_j) \mid s_j \in \mathcal{S}^* \land s \text{ is non-dominated}\}$ *and runs in* $O(n \log n)$ *time, if $T$ is a balanced binary tree.*

*Proof.* It is obvious that the three cases of Procedure In/Out_Test exhaust all possibilities. We have already discussed Steps 4 and 5. In Step 6 (Case 2), the counterbalance for $v_j$ can be either in $\Gamma^c(u_j)$ or on $\pi(v_j, u_j)$ or in a subtree hanging from $\pi(v_j, u_j)$. They are dealt with in Substeps 6(a), 6(b), and 6(c), respectively.

Let us compute the running time. By Lemma 7, Step 3 takes $O(n \log n)$ time for all $v_j$. Lemma 4 implies that Step 4 takes $O(n \log n)$ time even if it is invoked for all $u \in V$. Step 5 takes $O(n \log n)$ time for all $v_j \notin C$ by Lemma 8. Both Substeps 6(b) and 6(c) can be solved by deleting $\Gamma^c(u_j)$ and applying our tree algorithm in $O(n \log n)$ time for all $v_j$'s. See Lemma 4. As for Substep 6(a), $F^{s_0}(x, \Gamma^c(u_j))$ for $x \in \Gamma(u_j)$ and $x$ on the edge connecting $\Gamma(u_j)$ and $u_j$ can be extracted from $\tau_C$ as the upper envelope of a set of $O(\log n)$ functions. We can use Chazelle et al.'s algorithm [10] to find the lowest intersection between each of them and $\overline{f}_j(x)$ in $O(\log n)$ time. We then take their maximum. Given $u_i \in C$, this computation takes $O(|\Gamma(u_i)|\log n)$ time for all vertices in $\Gamma(u_i)$, where $|\Gamma(u_i)|$ denotes the number of vertices in $\Gamma(u_i)$. The total time for all vertices is thus $O(n \log n)$. □

As commented before, if $T$ is not a balanced binary tree, we can use spine tree decomposition [5,6,7]. From Theorem 2 and Lemma 9, we have

**Theorem 4.** *The minmax regret 1-center of a unicyclic network can be computed in $O(n \log n)$ time.* □

### 4.2 Cactus Network with $c$ Cycles

If there are $c$ cycles, then we essentially repeat the above operations $c$ times. Let us denote the $c$ cycles by $C_1, C_2, \ldots, C_c$. We first find the 1-center $c(s_0)$ of $G$ in $O(n \log n)$ time [4]. For cycle $C_i$, let $p_i$ be the vertex of $C_i$ closest to $c(s_0)$. Cut $C_i$ open by removing the edge on which $\alpha(p_i)$ lies and call the resulting tree $T$. We can essentially use Algorithm Unicycle($G(V, E)$) for each cycle $C_i$, except that we additionally need to compute the upper envelope for $\Gamma(p_i)$ for each $p_i$, which takes $O(n \log n)$ time.

**Theorem 5.** *If the number of cycles in a cactus is bounded by a constant $c\ (\geq 1)$, then centers $\{c(s_j) \mid s_j \in \mathcal{S}^* \wedge s\ \text{is non-dominated}\}$ of a cactus network can be computed in $O(cn \log n)$ time. Thus its minmax regret 1-center can also be computed in $O(cn \log n)$ time.* □

## 5 Conclusion

We have presented an $O(n \log n)$ time algorithm for finding the minmax regret 1-center in a tree. Our algorithm for a $c$–cycle cactus network runs in $O(cn \log n)$ time. One of the questions we would like to answer is whether our data structures designed for the queries could be modified to handle more general situations such as reducing the weight of a vertex, or the insertion/deletion of a vertex.

## References

1. Averbakh, I.: Complexity of robust single-facility location problems on networks with uncertain lengths of edges. Disc. Appl. Math. 127, 505–522 (2003)
2. Averbakh, I., Berman, O.: Minimax regret $p$-center location on a network with demand uncertainty. Location Science 5, 247–254 (1997)
3. Averbakh, I., Berman, O.: Algorithms for the robust 1-center problem on a tree. European Journal of Operational Research 123(2), 292–302 (2000)
4. Ben-Moshe, B., Bhattacharya, B., Shi, Q., Tamir, A.: Efficient algorithms for center problems in cactus networks. Theoretical Compter Science 378(3), 237–252 (2007)

5. Benkoczi, R.: Cardinality constrained facility location problems in trees. Ph.D. thesis, School of Computing Science, Simon Fraser University, Canada (2004)
6. Benkoczi, R., Bhattacharya, B., Chrobak, M., Larmore, L.L., Rytter, W.: Faster algorithms for $k$-medians in trees. In: Rovan, B., Vojtáš, P. (eds.) MFCS 2003. LNCS, vol. 2747, pp. 218–227. Springer, Heidelberg (2003)
7. Bhattacharya, B., Kameda, T., Song, Z.: Minmax regret 1-center on a path/cycle/tree. In: Proc. 6th Int'l Conf. on Advanced Engineering Computing and Applications in Sciences (ADVCOMP), pp. 108–113 (2012), http://www.thinkmind.org/index.php?view=article&articleid=advcomp_2012_5_20_20093
8. Burkard, R., Dollani, H.: A note on the robust 1-center problem on trees. Annals of Operational Research 110, 69–82 (2002)
9. Burkard, R., Krarup, J.: A linear algorithm for the pos/neg-weighted 1-median problem on a cactus. Computing 60, 193–215 (1998)
10. Chazelle, B., Guibas, L.J.: Fractional cascading: II. Applications. Algorithmica 1, 163–191 (1986)
11. Chen, B., Lin, C.S.: Minmax-regret robust 1-median location on a tree. Networks 31, 93–103 (1998)
12. Goldman, A.: Optimal center location in simple networks. Transportation Science 5, 212–221 (1971)
13. Hakimi, S.: Optimum locations of switching centers and the absolute centers and medians of a graph. Operations Research 12, 450–459 (1964)
14. Hale, T.S., Moberg, C.R.: Location science research: A review. Annals of Operations Research 123, 21–35 (2003)
15. Kouvelis, P., Vairaktarakis, G., Yu, G.: Robust 1-median location on a tree in the presence of demand and transportation cost uncertainty. Tech. Rep. Working Paper 93/94-3-4, Department of Management Science, The University of Texas, Austin (1993)
16. Megiddo, N.: Linear-time algorithms for linear-programming in $R^3$ and related problems. SIAM J. Computing 12, 759–776 (1983)
17. Megiddo, N., Tamir, A.: New results on the complexity of $p$-center problems. SIAM J. Computing 12, 751–758 (1983)
18. Tamir, A.: An $O(pn^2)$ algorithm for the $p$-median and the related problems in tree graphs. Operations Research Letters 19, 59–64 (1996)
19. Yu, H.I., Lin, T.C., Wang, B.F.: Improved algorithms for the minmax-regret 1-center and 1-median problem. ACM Transactions on Algorithms 4(3), 1–1 (2008)

# Collision-Free Network Exploration*

Jurek Czyzowicz[1], Dariusz Dereniowski[2], Leszek Gasieniec[3], Ralf Klasing[4],
Adrian Kosowski[4], and Dominik Pająk[4]

[1] Université du Québec en Outaouais, Canada
[2] Gdańsk University of Technology, Poland
[3] University of Liverpool, UK
[4] LaBRI, CNRS — Université de Bordeaux — Inria, France

**Abstract.** A set of mobile agents is placed at different nodes of a $n$-node network. The agents synchronously move along the network edges in a *collision-free* way, i.e., in no round may two agents occupy the same node. In each round, an agent may choose to stay at its currently occupied node or to move to one of its neighbors. An agent has no knowledge of the number and initial positions of other agents. We are looking for the shortest possible time required to complete the collision-free *network exploration*, i.e., to reach a configuration in which each agent is guaranteed to have visited all network nodes and has returned to its starting location.

We first consider the scenario when each mobile agent knows the map of the network, as well as its own initial position. We establish a connection between the number of rounds required for collision-free exploration and the degree of the minimum-degree spanning tree of the graph. We provide tight (up to a constant factor) lower and upper bounds on the collision-free exploration time in general graphs, and the exact value of this parameter for trees. For our second scenario, in which the network is unknown to the agents, we propose collision-free exploration strategies running in $O(n^2)$ rounds for tree networks and in $O(n^5 \log n)$ rounds for general networks.

## 1 Introduction

The graph searching problem is a task of central importance in many contexts, including network maintenance, terrain patrolling, and robotics. Its different aspects have been thoroughly investigated, cf. [8]. The *rendezvous search* problem has been often presented as a game with two mobile players walking within the *search space* and having the common goal of arriving at the same time at the same location (see [4]). On the other hand, the *exploration problem* consists in examining all elements of the search space by a mobile agent (e.g. visiting all graph nodes or traversing all its edges), e.g., in order to find a hidden target (see [1,13]).

In this paper we propose a new graph searching problem in which each of a set of mobile agents must explore a given undirected graph, in such a way that two agents may never visit the same node of the graph at the same time. This property of the model,

---

* Research partially supported by ANR project DISPLEXITY and by NCN under contract DEC-2011/02/A/ST6/00201. Dariusz Dereniowski has been partially supported by a scholarship for outstanding young researchers founded by the Polish Ministry of Science and Higher Education. The full text of the paper is available at: http://hal.inria.fr/hal-00736276.

which we call *collision avoidance*, is motivated by the fact that the processes executed by mobile agents (software agents or physical robots) sometimes require exclusive access to network resources. Our problem may have practical applications. For example, mobile software agents may need exclusive access to a node's resources when updating its data. Robots (or nano-robots) distributing interacting chemical or pharmacological agents within a battlefield or a human body must avoid to be simultaneously present at a small distance apart. Individuals, one of which is highly infectious or socially conflicting should avoid a meeting. According to our knowledge, this problem has not been studied in the past, although a question related to its offline version has been given some attention in the context of routing (cf. [3]).

In our considerations, time is divided into synchronous rounds. Initially, each agent is placed at a different node and in each round it may choose to move to a neighboring node or to stay motionless. The agents are independent in the sense that they cannot communicate and none of them knows the number of other agents, their initial placement in the graph, and is unaware of the current location of the other agents. The agents move independently, and each of them executes the same algorithm. The effectiveness of the algorithm is measured in terms of the *collision-free exploration time*, i.e., the number of rounds until all potentially existing agents are certain to have completed the exploration and returned to their initial location. Details of our model are discussed at the end of this section.

**Our Results.** We consider two scenarios, differing in the amount of global information about the network topology which is available to each agent. Our results are summarized in Table 1.

For the first scenario, considered in Section 2, we assume that a map of the network is *a priori* known to the agents. We show that a collision-free exploration strategy exists for any graph, and provide efficient solutions for trees and general graphs. We start by considering the case of trees, proposing a strategy which involves the simultaneous activation of agents located at the endpoints forming a matching in some optimal edge-coloring of the tree. This strategy is shown to yield optimal exploration time. We then extend this approach from the case of trees to the case of general graphs, by requiring that the agents perform exploration using only the edges of a well-chosen spanning tree of the graph. Somewhat surprisingly, it turns out that this approach is asymptotically the best possible, i.e., within a constant factor of the optimum. To prove the corresponding lower bound on the collision-free exploration time in graphs, we establish a tight connection between our problem and the fractional relaxation of the LP formulation of the minimum-degree spanning tree problem.

In the second scenario, discussed in Section 3, we deal with synchronous agents possessing only local knowledge about the graph to explore. In particular, no knowledge of the size of the graph is assumed. We suppose that each agent executes a local, distributed algorithm, in every round making a decision based on the information concerning the currently occupied node and the identifiers of the neighboring nodes. For this scenario, we show that a collision-free exploration is always feasible in finite time and we give algorithms for trees and general graphs. Our collision-free exploration strategies are of length $O(n^2)$ for trees and $O(n^5 \log n)$ for arbitrary graphs, and make use of the application of universal exploration sequences.

**Table 1.** The time of optimal collision-free graph exploration. $\Delta(G)$ denotes the maximum degree of a node in graph $G$, and $\Delta^*(G) = \Delta(T)$, where $T$ is a minimum-degree spanning tree of $G$.

| Scenario | Tree | General graph |
|---|---|---|
| With complete map: | $n\Delta(G)$<br>Thm. 1 | $\Theta(n\Delta^*(G))$<br>Thm. 2 |
| With local knowledge: | $O(n^2)$<br>Thm. 3 | $O(n^5 \log n)$<br>Thm. 4 |

Throughout the paper, we assume that the strategies for collision-free exploration are required to return the agent to their initial location. This assumption allows us to see our strategies as an analogue of the classical Traveling Salesman Problem with mutually-exclusive salesmen on an unweighted graph, and also allows the agents to engage in perpetual (periodic) exploration of the graph. After minor modification of the proofs, all the results presented in Table 1 also hold up to constant factors for the variant of the problem in which agents may end exploration at an arbitrary node of the graph.

**Related Work.** The offline setting of our question is related to the following problem (cf. [3]), which was studied in the context of routing. Each vertex of a given graph is initially occupied by a "pebble", which has to be moved to a destination, so that the destinations of different pebbles are different. In every synchronous round a set of edges is selected and the pebbles at each edge endpoints are interchanged. [3] attempts to minimize the number of rounds so that all pebbles reach their destination, giving lower and upper bounds for different classes of graphs. The routing model of [3] inherently implies the usage of matchings - the technique that we choose to apply in some results of our paper. The $3n$ upper bound for trees given in [3] was improved to $\frac{1}{2}n + O(\log n)$ in [15]. [12] and [14] independently extended this model to allow more than one pebble per origin and destination node.

For the (classical) graph exploration problem with local knowledge, a lot of attention has been given to exploration in *anonymous networks*, in which the agent, when located at a node, has to decide on its next move based only on its own local memory state, the local port ordering at the node, and the port by which it entered the current node. It has been shown in [9] that an agent must be equipped with at least $n$ states (i.e., $\Omega(\log n)$ bits of memory) to be able to explore all anonymous graphs with $n$ nodes. On the positive side, unknown anonymous graphs can be deterministically explored by following so called universal traversal/exploration sequences. These exist for any number of nodes, and have polynomial length [2]. The exploration time obtained using such an approach is $O(n^5 \log n)$, i.e., a factor of about $n^2$ greater than the (expected) cover time of a corresponding random walk.

The problem of graph exploration without collisions was also studied in the case when two agents also collide when traversing one edge in opposite directions. In [5] the authors study the maximal number of agents that can explore graph without collisions in a synchronous setting. The asynchronous Look-Compute-Move model is considered in [6] where the authors study the maximal and minimal number of agents that are necessary and sufficient to solve the problem for a ring. In both these papers it is assumed that each agent can observe (or compute) the positions of the other agents.

**Model and Definitions.** We assume that the nodes of each $n$-node network have unique identifiers in $\{1, \ldots, n\}$. The identifier of a node $v$ is denoted by $id(v)$. Several agents are initially located at pairwise different nodes of the network. The initial position of each agent $a$ is denoted by $home(a)$. Each agent is unaware of the number and initial positions of the other agents, and all agents are given the same algorithm that determines their behavior in the subsequent rounds.

Each agent can perceive the identifier $id(v)$ of the currently occupied node $v$ and can perceive the identifiers of all neighbors of $v$. Moreover, the agent can distinguish the edges incident to $v$ according to the identifiers of the nodes located at the endpoints of the edges. The latter assumption is necessary to properly perform the navigation in a node labeled network.

The agents are synchronous and hence the time is divided into rounds of equal duration. Each round is divided into two stages. In the first stage each agent $a$ makes a decision (by executing its algorithm) that determines its behavior in the second stage of the round. The decision can be three-fold: it may decide to stay in this particular round at the currently occupied node, to move from the currently occupied node to one of its neighbors, or decide that its exploration is completed. In the second stage of the round, all agents simultaneously perform the action corresponding to their decision. If, as a result, two agents located at some adjacent nodes $u$ and $v$ decide to move from $u$ to $v$ and from $v$ to $u$, respectively, then they traverse the same edge in this round, but remain unaware of this event, i.e., the two agents do not communicate and do not perceive each other. We require that the algorithm given to the agents ensures the following:
 – at the end of each round no two agents are present on the same node of the network,
 – by the end of some round $t \geq 0$, all the agents have decided that the exploration is completed,
 – each agent has visited each node of the network in one of the rounds $1, \ldots, t$,
 – each agent $a$ is present at $home(a)$ at the end of round $t$.

Note that, in this setting, the execution of the agent's algorithm (and thus the behavior of the agent) only depends on the input to the algorithm and on the identifiers of the nodes visited by the agent. Thus, in particular, an agent is unable to ever discover the initial or current position of any other agent or the number of agents in the network.

With respect to additional information available to the agents, we study two scenarios in this work: either the agents have no prior knowledge of network topology and no knowledge of global parameters, or the complete map of the network is given to all agents. In the latter case the map consists of node identifiers, but provides no information on the locations of other agents. Note that if, together with a complete map of the network, all agents receive as an input information on the initial positions of all agents, then our exploration problem becomes similar to the off-line routing problems considered e.g. in [3,14,15].

Let $G = (V(G), E(G))$ be any network. For any node $v$ of $G$ let $N_G(v)$ be the set of neighbors of $v$ in $G$. We use the symbol $\Delta(G)$ to denote the *degree* of $G$, defined as $\Delta(G) = \max\{|N_G(v)| : v \in V(G)\}$. ($|N_G(v)|$ is called the *degree* of $v$.) Given a set of edges $X \subseteq E(G)$, define $G[X]$ to be the network with nodes in $V(G)$ and edges in $X$,

$G[X] = (V(G), X)$. Note that $G[X]$ is not necessarily connected. A connected network $H$ such that $V(H) \subseteq V(G)$ and $E(H) \subseteq E(G)$ is called a *connected component* of $G$ if there exists no connected network $H'$ such that $V(H) \subseteq V(H') \subseteq V(G)$ and $E(H) \subseteq E(H') \subseteq E(G)$ and $H \neq H'$.

Any sequence $\mathcal{R} = (v_0, v_1, \ldots, v_l)$ of nodes of a network $G$ is called a *route in $G$* if $v_i = v_{i-1}$ or $\{v_i, v_{i-1}\}$ is an edge of $G$ for each $i = 1, \ldots, l$. We say that $l$ is the *length* of $\mathcal{R}$ and we write $\mathcal{R}_i = v_i$ for each $i = 0, \ldots, l$. The route $\mathcal{R}$ covers $G$ if for each node $v$ of $G$ there exists $i \in \{0, \ldots, l\}$ such that $v = \mathcal{R}_i$. The route $\mathcal{R}$ is *closed* if $\mathcal{R}_0 = \mathcal{R}_l$, where $l$ is the length of $\mathcal{R}$. Let $a$ be an agent. We say that the route $\mathcal{R}$ of length $l$ is a *route of $a$* if: (i) $\mathcal{R}_0 = home(a)$ and $a$ is present at $\mathcal{R}_i$ at the end of round $i$, $i = 1, \ldots, l$, and (ii) $a$ does not move in any round $r > l$.

We say that a route $\mathcal{R}$ of length $l$ is an *exploration strategy for $a$* if (i) $\mathcal{R}$ is a route of $a$, (ii) $\mathcal{R}$ is closed, (iii) $\mathcal{R}$ covers $G$. Two routes $\mathcal{R}$ and $\mathcal{R}'$ of length $l$ are *collision-free* if $\mathcal{R}_i \neq \mathcal{R}'_i$ for each $i = 0, \ldots, l$. Let $\mathcal{A} = \{a_1, \ldots, a_k\}$, $1 \leq k \leq n$, be the set of agents that are initially located at the nodes of $G$. Let $\mathcal{R}(a)$ be the exploration strategy for each agent $a \in \mathcal{A}$. We say that $\mathcal{R}(a_1), \ldots, \mathcal{R}(a_k)$ are *collision-free* if $\mathcal{R}(a_i)$ and $\mathcal{R}(a_j)$ are collision-free for each $i, j \in \{1, \ldots, k\}, i \neq j$. Let $t$ be the minimum integer such that for each set of agents placed arbitrarily on the nodes of $G$ there exist collision-free exploration strategies, each of length at most $t$, for the agents. Then, $t$ is called the *collision-free exploration time of $G$*.

## 2 Network Exploration with a Map

In this section we consider the problem of collision-free exploration in the case when each agent is given a complete map of the network to be explored. We start by discussing the simpler case of tree networks in Subsection 2.1, and in Subsection 2.2 we generalize our approach from trees to arbitrary networks, showing its asymptotic optimality by proving a corresponding lower bound.

### 2.1 Tree Exploration with a Map

We start with some notation and two preliminary lemmas that are the main tool in the analysis of an algorithm given in this section.

Given a tree network $T$, we say that a function $c \colon E(T) \to \{1, \ldots, d\}$ is a *$d$-edge-coloring* of $T$ if $c(e) \neq c(e')$ for any two adjacent edges in $T$.

Let $d$ be an integer, let $c$ be a $d$-edge-coloring of $G$, and let $v$ be any node of $G$. Define $\mathcal{T}(v, d, c) = (v_0, v_1, v_2, \ldots)$ to be an infinite route in $G$ starting at $v$ such that:
(i) if $c(\{v_{i-1}, u\}) \neq 1 + (i-1) \bmod d$ for each neighbor $u$ of $v_{i-1}$ in $G$, then $v_i = v_{i-1}$,
(ii) if $c(\{v_{i-1}, u\}) = 1 + (i-1) \bmod d$ for some neighbor $u$ of $v_{i-1}$, then $v_i = u$.
Then, define $\mathcal{T}^l(v, d, c)$, $l \geq 0$, to be the prefix of $\mathcal{T}(v, d, c)$ of length $l$, and $\mathcal{T}^l_i(v, d, c)$ to be $v_i$ for each $i = 0, \ldots, l$.

We now give two preliminary lemmas in which we prove that if $u$ and $v$ are two distinct nodes of $T$, then the routes $\mathcal{T}^{dn}(u, d, c)$ and $\mathcal{T}^{dn}(v, d, c)$ are collision-free, and each of them is closed and covers the tree network.

**Lemma 1.** *Let $T$ be a tree network. If $c$ is a $d$-edge-coloring of $T$, then for any two distinct nodes $u$ and $v$ of $T$ the routes $\mathcal{T}^l(u,d,c)$ and $\mathcal{T}^l(v,d,c)$ are collision-free for each $l \geq 0$.*

**Lemma 2.** *Let $T$ be a tree network and let $d$ be an integer. If $c$ is a $d$-edge-coloring of $T$ and $v$ is a node of $T$, then the route $\mathcal{T}^{dn}(v,d,c)$ is closed and covers $T$.*

It remains to observe that the considered routes can be implemented as exploration strategies. Indeed, each agent $a$ is able to construct some $d$-edge-coloring $c$ of $T$ (the same for all agents, e.g., lexicographically first with respect to some chosen ordering of all colorings) with $d = \Delta(T)$, and hence it is able to 'follow' $\mathcal{T}^{n\Delta(T)}(home(a), \Delta(T), c)$. We formulate this strategy in the form of the algorithm below.

---

**Algorithm** Tree-Exploration($T$)
 **Input:** A node-labeled tree network $T$.
**begin**
  Let $v$ be the initial position of the executing agent.
  Compute the lexicographically first $\Delta(T)$-edge-coloring $c$ of $T$
  **for** each round $r \leftarrow 1$ **to** $n\Delta(T)$ **do**
    **if** there exists an edge $\{v,u\}$ such that $c(\{v,u\}) = 1 + (r-1) \bmod \Delta(T)$
    **then** move from $v$ to $u$ in round $r$, set $v \leftarrow u$.
    **else** stay at $v$ in round $r$.
**end** Tree-Exploration

---

For an agent $a$ following Algorithm Tree-Exploration, its route is of length $n\Delta(T)$, and given as $\mathcal{R}^{n\Delta(T)}(a) = \mathcal{T}^{n\Delta(T)}(home(a), \Delta(T), c)$, where $c$ is the $\Delta(T)$-edge-coloring computed in the Algorithm. Consequently, taking into account Lemmas 1 and 2, we have the following.

**Proposition 1.** *Let $T$ be a tree network and let $a_1, \ldots, a_k$, $1 \leq k \leq n$, be the agents initially located at pairwise different nodes of $T$. Suppose that the agent $a_i$ uses Algorithm Tree-Exploration to compute its route $\mathcal{R}^{n\Delta(T)}(a_i)$, for each $i = 1, \ldots, k$. Then, $\mathcal{R}^{n\Delta(T)}(a_1), \ldots, \mathcal{R}^{n\Delta(T)}(a_k)$ are exploration strategies, and are collision-free.* □

It turns out that there exist no shorter collision-free exploration strategies than those constructed with Algorithm Tree-Exploration.

**Theorem 1.** *The collision-free exploration time of any $n$-node tree network $T$ is precisely equal to $n\Delta(T)$.*

*Proof.* The upper bound follows from Proposition 1. Now, we prove the lower bound, i.e., that the collision-free exploration time of $T$ is at least $n\Delta(T)$. Let $u$ be a fixed node of degree $\Delta(T)$ in $T$. First assume that there are $n$ agents in $T$. We say that an agent $a$ is *active* in round $r$ if $a$ goes from $v$ to $u$ in round $r$ for some $v \in N_T(u)$. In each round at most one agent is active. For each agent $a$ there exist at least $\Delta(T)$ rounds in which $a$ is active, because the route of $a$ needs to be closed and $T$ is a tree. Since there are

$n$ agents in total, we obtain that there are at least $n\Delta(T)$ rounds in which an agent is active. This proves that there exists an agent $a$ that is active in round $n\Delta(T)$, and hence its exploration strategy is of length at least $n\Delta(T)$. Finally, observe that $a$ constructs the same route regardless of the number of agents present in the network. This is due to the fact that $T$ and $id(home(a))$ is the entire input to the algorithm that $a$ executes. □

We finish this section by remarking on the complexity of Algorithm Tree-Exploration. For any tree network $T$ on $n$ nodes, there exists a $\Delta(T)$-edge-coloring of $T$ and it can be computed in $O(n)$-time. Consequently, the total time of an agent's local computations when running Algorithm Tree-Exploration is $O(n\Delta(T))$.

### 2.2 General Network Exploration with a Map

We say that $T$ is a *spanning tree* of $G$ if $T$ is a tree such that $V(T) = V(G)$ and $E(T) \subseteq E(G)$. Then, $T$ is a *minimum degree spanning tree* of $G$ if $T$ is a spanning tree of $G$ and the degree of $T$ is minimum over the degrees of all spanning trees of $G$. Define $\Delta^*(G) = \Delta(T)$, where $T$ is a minimum degree spanning tree of $G$. We propose the following solution to the collision-free exploration problem.

---
**Algorithm** Network-Exploration$(G)$
 **Input:** A node-labeled network $G$.
**begin**
   Compute the lexicographically first minimum-degree spanning tree $T^*$ of $G$.
   Call Algorithm Tree-Exploration$(T^*)$.
**end** Network-Exploration

---

**Proposition 2.** *Let $G$ be a network and let $a_1, \ldots, a_k$, $1 \le k \le n$, be the agents initially located at pairwise different nodes of $G$. Suppose that the agent $a_i$ uses Algorithm* Network-Exploration *to compute its route $\mathcal{R}(a)$, $i = 1, \ldots, k$. Then, $\mathcal{R}(a_1), \ldots, \mathcal{R}(a_k)$ are collision-free exploration strategies of length $n\Delta^*(G)$.* □

Now, the following theorem implies that our result is asymptotically tight, i.e., it implies that Algorithm Network-Exploration constructs exploration strategies whose length is within a constant factor from the optimum.

**Theorem 2.** *The collision-free exploration time of any network $G$ is $\Theta(n\Delta^*(G))$.*

*Proof.* The fact that the collision-free exploration time of $G$ is $O(n\Delta^*(G))$ follows from Proposition 2.

Now, we prove the lower bound of $\Omega(n\Delta^*(G))$. Observe that if $\Delta^*(G) \le 3$, then the theorem follows, because each exploration strategy must be of length $\Omega(n)$. To finish the proof, suppose that there exist exploration strategies for the agents, such that the length of each exploration strategy is at most $n(\Delta^*(G) - 3)/2$.

For each node $v$ of $G$ let $E_v = \{\{v, u\} : u \in N_G(v)\}$. Consider the following linear program (LP) with variables $(x_e : e \in E(G))$, which satisfies the following set of constraints [7,11]:

$$\sum_{e \in E(G)} x_e = n - 1 \tag{1}$$

$$\sum_{e \in E(G[S])} x_e \leq |S| - 1, \quad \text{for each } S \subseteq V(G) \tag{2}$$

$$\sum_{e \in E_v} x_e \leq t, \quad \text{for each } v \in V(G) \tag{3}$$

$$0 \leq x_e \leq 1, \tag{4}$$

where $t$ is an integer and $n$ is the number of nodes of $G$. Any solution to the above problem is called a *fractional spanning tree* of *degree t* of $G$. Informally speaking, if $(x_e \colon e \in E(G))$, is a solution to (1)-(4), then $x_e$ is the 'fraction' of the edge $e$ that is included in the resulting fractional spanning tree. Note that any integer solution, i.e. the one in which $x_e \in \{0, 1\}$ for each $e \in E(G)$, is a spanning tree of degree at most $t$ of $G$.

Suppose that $n$ agents $a_1, \ldots, a_n$ are present in the network $G$. Let $\mathcal{R}(a_1), \ldots, \mathcal{R}(a_n)$ be some collision-free exploration strategies for the agents. Suppose that the length of each exploration strategy is at most $nt/2$. Based on these exploration strategies, we now construct a solution to the LP in (1)-(4). For each $i = 1, \ldots, n$, let $T_i$ be any spanning tree of $G$ such that if $e \in E(T_i)$, then there exists a round $r$ such that $e = \{\mathcal{R}_{r-1}(a_i), \mathcal{R}_r(a_i)\}$ (in other words, $a_i$ traverses $e$ in some round). Such a $T_i$ exists, because $\mathcal{R}(a_i)$ covers $G$, $i = 1, \ldots, n$. Define:

$$f_i(e) = \begin{cases} 1/n, & \text{if } e \in E(T_i) \\ 0, & \text{if } e \notin E(T_i) \end{cases} \quad \text{and} \quad x_e = \sum_{i=1}^{n} f_i(e) \text{ for each } e \in E(G). \tag{5}$$

Now, we prove that $x_e$'s defined in (5) form a solution to the LP in (1)-(4).

First note that $\sum_{e \in E(G)} f_i(e) = (n-1)/n$ for each $i = 1, \ldots, n$, because $f_i$ assigns $1/n$ to exactly $n - 1$ edges of $G$, which follows from the fact that $T_i$ is a spanning tree of $G$, $i = 1, \ldots, n$. Thus, (1) holds.

Now, let $S \subseteq V(G)$ be selected arbitrarily. For each $i = 1, \ldots, n$, $|E(T_i) \cap E(G[S])| = |E(T_i[S])| \leq |S| - 1$, because $T_i[S]$ is, by definition, a collection of node-disjoint trees on set $S$. Hence, (2) follows.

Let $v$ be any node of $G$ and let $X = E_v \cap (E(T_1) \cup \cdots \cup E(T_n))$. For each $r$ there exist at most two edges in $X$ traversed by an agent in round $r$. Hence,

$$\sum_{e \in X} \sum_{i=1}^{n} f_i(e) \leq \frac{nt}{2} \cdot \frac{2}{n} = t.$$

Note that if $e \in E_v \setminus X$, then $\sum_{i=1}^{n} x_e = 0$. This proves that (3) holds.

Finally, (4) follows directly from (5).

We have proved that the existence of exploration strategies of length $nt/2$ implies the existence of a solution to (1)-(4). Moreover, we have the following.

*Claim* ([11]). *If there exists a solution to (1)-(4), then there exists an integer solution to (1),(2),(4) with the additional constraint*

$$\sum_{e \in E_v} x_e \leq t+2 \text{ for each } v \in V(G)$$

*which replaces (3).*

We remark that such an integer solution defines a spanning tree of $G$, given by the set of edges $\{e \in E(G) \colon x_e = 1\}$.

In view of the definition of $x_e$'s in (5), it follows that if there exist exploration strategies of length at most $nt/2$ for the $n$ agents, then there exists a spanning tree $T^*$ of $G$, and the degree of $T^*$ is at most $t+2$. By assumption, there exist in $G$ exploration strategies of length at most $n(\Delta^*(G)-3)/2$, hence, putting $t = \Delta^*(G)-3$, it follows that $G$ has a spanning tree of degree at most $\Delta^*(G)-1$, a contradiction with the definition of $\Delta^*(G)$. □

We finish this section with a complexity remark. Finding a minimum-degree spanning tree is in general an NP-hard problem. We can, however, modify the approach to obtain an exploration strategy of length $n(\Delta^*(G)+1)$ that can be computed efficiently. We make use of a $O(mn\alpha(m,n)\log n)$-time algorithm that for a given $G$ finds its spanning tree $T$ of degree $\Delta(T) \leq \Delta^*(G)+1$, where $m$ and $n$ are, respectively, the number of edges and nodes of $G$, and $\alpha$ is the inverse Ackermann function [10]. By using the tree $T$ in Algorithm Network-Exploration instead of $T^*$ we obtain an exploration strategy of length $n(\Delta^*(G)+1)$ for agent $a$, and this strategy is computed in time $O(mn\alpha(m,n)\log n)$. On the other hand, computing the precise value of collision-free exploration time is a hard problem.

**Proposition 3.** *The problem of deciding, for a given network $G$ and integer $l$, whether the collision-free exploration time of $G$ is at most $l$, is* NP-*complete.*

## 3 Local Network Exploration

In this section we consider the problem of collision-free exploration in the setting when the agents do not receive any information about the network in which they operate. Recall that we assume, that each node $v \in V$ is equipped with a unique identifier $id(v) \in \{1, 2, \ldots, n\}$, and each agent located at $v$ is only aware of the identifier $id(v)$ and the identifiers of the neighbors of $v$ at the endpoints of respective edges incident to $v$. In Section 3.1 we consider tree networks, and in Section 3.2 we show how any network can be explored.

Let $G$ be any network. For the purposes of this section we introduce an edge-labeling function $id'$ defined as

$$id'(\{u,v\}) = id(u) + id(v) \text{ for each } \{u,v\} \in E(G). \quad (6)$$

We recall without proof the following essential property of function $id'$.

**Lemma 3.** *Let $G$ be any $n$-node network. Then, $id'$ is a $2n$-edge-coloring of $G$.* □

## 3.1 Local Exploration of Tree Networks

In this section we provide an algorithm which defines collision-free routes of agents, and is guaranteed to perform exploration if the explored network is a tree. For any integer $b \geq 2$ define the following sequence of integers $U(b) = (1, \ldots, 2b, \ldots, 1, \ldots, 2b)$, where $1, \ldots, 2b$ is repeated $b$ times, and let $U_i(b)$, $i \in \{1, \ldots, 2b^2\}$, be its $i$-th element.

Define *phase $p$*, as the sequence of rounds $(1 + \sum_{j=1}^{p-1} |U(2^j)|, \ldots, \sum_{j=1}^{p} |U(2^j)|)$ and denote by $\ell(p) = |U(2^p)|$ the number of rounds of phase $p$. Note that

$$\ell(p) = 2^{2p+1} \text{ for each } p \geq 1, \tag{7}$$

and that phase $p$ consists of the rounds in which the behavior of any agent $a$ is determined in the $p$-th iteration of the 'while' loop of its execution of Local-Tree-Exploration, whenever $p$ does not exceed the total number of iterations executed.

Recall that an agent upon vising a node $v$ receives a list of identifiers of all neighbors of $v$. Thus the agent is aware if it visited all nodes during a phase.

---

**Algorithm** Local-Tree-Exploration
**begin**
    Let $v$ be the initial position of the executing agent.
    $b \leftarrow 2$
    $r \leftarrow 0$
    **while** not all nodes have been visited so far **do** {start a new phase}
        **for** $s \leftarrow 1$ **to** $|U(b)|$ in round $r + s$ **do**
            **if** there exists an edge $\{v, u\}$ such that $id(u) \leq b$
            and $id(v) \leq b$ and $id'(\{v, u\}) = U_s(b)$
            **then** move from $v$ to $u$ {in round $r + s$}; set $v \leftarrow u$.
            **else** stay at $v$. {in round $r + s$}
        **end for**
        $r \leftarrow r + |U(b)|$
        $b \leftarrow 2b$
    **end while**
    Backtrack all previous moves, i.e., $a$ moves from $v$ to $u$ in round $r + i$ if and only if $a$ moved from $u$ to $v$ in round $r - i + 1$ for each $i = 1, \ldots, r$.
**end** Local-Tree-Exploration

---

Denote by $\mathcal{R}(a, p)$ the route of an agent $a$ restricted to its moves in phase $p$, $p \geq 1$.

We denote by $T_p$ the subgraph of $T$ induced by all edges $e$ whose endpoints have identifiers at most $2^p$, $T_p = T[\{\{u, v\} \in E(T) : id(u) \leq 2^p \wedge id(v) \leq 2^p\}]$.

Finally, define $\ell = 2 \sum_{p=1}^{\lceil \log_2 n \rceil} \ell(p)$.

We now prove that each agent $a$ moves in phase $p$ 'inside' the connected component $T'$ of $T_p$ that contains the vertex occupied by $a$ at the beginning of phase $p$.

**Lemma 4.** *Let $p \geq 1$ be an integer, let $T$ be a tree network and let $a$ be an agent. Let $v$ be the vertex occupied by $a$ at the beginning of phase $p$. Then, $\mathcal{R}(a, p)$ is a route in the connected component of $T_p$ that contains $v$, and $\mathcal{R}(a, p) = \mathcal{T}^{\ell(p)}(v, 2^{p+1}, id')$.*

Note that the length of the route $\mathcal{R}(a, p)$ of $a$ in phase $p$ is bounded by $\ell(p)$, hence is, in general, 'unrelated' to the number of nodes of $T'$. For this reason, $T'$ need not be completely explored. However, by the definition of $T_p$, we have that $T_p = T$ (and $T' = T$) if and only if $p \geq \lceil \log_2 n \rceil$. We use this observation to show that all agents perform backtracking and stop after exactly the same phase $p = \lceil \log_2 n \rceil$, and that in this phase each of them visits all nodes of $T$.

**Lemma 5.** *Let $T$ be a n-node tree network. For each agent a the number of iterations of the 'while' loop of Algorithm* Local-Tree-Exploration *executed by a equals $\lceil \log_2 n \rceil$. Moreover, $\mathcal{R}(a, \lceil \log_2 n \rceil)$ covers $T$.*

We now argue that the agents will never meet while moving during any given phase $p$.

**Lemma 6.** *Let $a$ and $a'$ be any two agents, let $T$ be a tree network, and let $p \geq 1$ be an integer. The routes $\mathcal{R}(a,p)$ and $\mathcal{R}(a',p)$ are collision-free.*

**Theorem 3.** *Let $T$ be a tree network and let $a_1, \ldots, a_k$, $1 \leq k \leq n$, be the agents initially located at pairwise different nodes of $T$. Suppose that the agent $a_i$ uses Algorithm* Local-Tree-Exploration *to compute its route $\mathcal{R}^\ell(a_i)$, for each $i = 1, \ldots, k$. Then, $\mathcal{R}^\ell(a_1), \ldots, \mathcal{R}^\ell(a_k)$ are collision-free exploration strategies of length $O(n^2)$.*

### 3.2 Local Exploration of General Networks

For the purposes of analysis, we introduce some auxiliary notation concerning the so-called *anonymous graph model*. In this model nodes are anonymous, and each edge has two port numbers assigned, each to one of its endpoints, in such a way that the ports at edges incident to any node form a set of consecutive integers, starting from 1. An agent located at a node $v$ can only perform its next move based on the local port numbers.

Before continuing, we provide several comments and informal intuitions concerning this model. First note that a collision-free exploration is, in general, impossible in arbitrary anonymous port-labeled networks. (This is the case, for example, for two agents located initially in symmetric, and thus indistinguishable, positions at the endpoints of a 3-node path.) However, we will overcome this difficulty by designing an auxiliary port-labeled network $A(G)$ based on the node-labeled network $G$, that has the property that each edge has identical port numbers at both of its endpoints, and in such a case the collision-free exploration will be guaranteed to exist. The behavior of an agent can be seen as navigating in our node-labeled network $G$ by navigating in the underlying 'virtual' port-labeled network $A(G)$. In particular, the function $id'$ defined in (6) provides both port numbers for each edge. Hence, each agent, while present at any node $v$ can compute the port number of the edges incident to $v$. Then, the agent 'simulates' its next move in the port-labeled network and, based on that, performs the move in the node-labeled network.

As a tool for our analysis we use the theory of *universal sequences* (formal definitions are provided below) that has been developed for regular port-labeled networks. Such a universal sequence, once computed by all agents, is then used to find a collision-free exploration strategy in the port-labeled network. In view of our earlier comment, the latter results in the collision-free exploration strategy in the node-labeled network.

We say that a network is *d-regular* if all nodes of the network have degrees equal to $d$. Given a port-labeled network $A$ and a node $v$ of $A$, we say that an agent $a$ initially located at $v$ *follows* a sequence of integers $U = (x_1, \ldots, x_l)$, with $1 \leq x_i \leq d$ for $i = 1, \ldots, l$, if for each $i = 1, \ldots, l$, in round $i$ the agent $a$ performs a move along the edge with port number $x_i$ at its current node. By a slight extension of notation, we allow a port-labeled network to have self-loops (with exactly one port number assigned to the loop); a traversal of the self-loop is assumed not to change the location of the agent.

We say that a sequence $U$ of integers is $(n,d)$-*universal* if for each node $v$ of each regular $n$-node network $A$ of degree $d$, an agent initially placed at $v$ visits each node of $A$ by following $U$. Aleliunas et al. [2] have shown non-constructively that for each $n > 0$ and $d > 0$, there exists a $(n,d)$-universal sequence of length $O(d^2 n^3 \log n)$ for networks with self-loops. Note that a $(n,d)$-universal sequence can be computed (rather inefficiently) by examining all sequences of the considered length and for each such candidate sequence one can generate all $n$-node port-labeled regular networks of degree $d$. Once a sequence $U$ and a network $A$ are selected, it can be tested if following $U$ from each node of $A$ results in visiting all nodes of $A$.

Given a node-labeled network $G$, we define the corresponding port-labeled network $A(G)$ so that there exists a bijection $\varphi \colon V(G) \to V(A(G))$ such that $\{u,v\} \in E(G)$ if and only if $\{\varphi(u), \varphi(v)\} \in E(A(G))$, and for each $\{u,v\} \in E(G)$ the port numbers at both endpoints of edge $\{\varphi(u), \varphi(v)\} \in E(A(G))$ are equal to $id'(\{u,v\})$. Since, according to Lemma 3, $id'$ is an edge-coloring of $G$, no two edges of $A(G)$ sharing a node have the same port number at this node. Then, for each node $u \in V(G)$ we add $2n - |N_G(u)|$ loops at $\varphi(u)$ in $A(G)$. As a result, the degree of each node of $A(G)$ is $2n$, and the length of the universal sequences constructed following [2], which we will use when exploring $A(G)$, will not exceed $O(n^5 \log n)$. In what follows, we will identify exploration of $G$ with exploration of $A(G)$.

The algorithm will use a similar concept of exploring a growing subgraphs in consequtive phases as in algorithm Local-Tree-Exploration. The difference is that during a phase instead of a sequence $U(b)$, we will use here some universal exploration sequence.

**Theorem 4.** *There exists an algorithm that allows any set of agents located initially at distinct nodes of any network $G$, and having no information about $G$, to compute collision-free exploration strategies of length $O(n^5 \log n)$.*

## References

1. Albers, S., Henzinger, M.R.: Exploring unknown environments. SIAM J. Comput. 29(4), 1164–1188 (2000)
2. Aleliunas, R., Karp, R.M., Lipton, R.J., Lovasz, L., Rackoff, C.: Random walks, universal traversal sequences, and the complexity of maze problems. In: Proceedings of the 20th Annual Symposium on Foundations of Computer Science, FOCS 1979, pp. 218–223. IEEE Computer Society, Washington, DC (1979)
3. Alon, N., Chung, F.R.K., Graham, R.L.: Routing permutations on graphs via matchings. In: STOC, pp. 583–591 (1993); Also SIAM J. Discrete Math. 7(3), 513–530 (1994)
4. Alpern, S., Gal, S.: Theory of Search Games and Rendezvous. Kluwer Acad. Publ. (2003)
5. Baldoni, R., Bonnet, F., Milani, A., Raynal, M.: Anonymous graph exploration without collision by mobile robots. Inf. Process. Lett. 109(2), 98–103 (2008)

6. Blin, L., Milani, A., Potop-Butucaru, M., Tixeuil, S.: Exclusive perpetual ring exploration without chirality. In: Lynch, N.A., Shvartsman, A.A. (eds.) DISC 2010. LNCS, vol. 6343, pp. 312–327. Springer, Heidelberg (2010)
7. Edmonds, J.: Matroids and the greedy algorithm. Math. Programming 1, 127–136 (1971)
8. Fomin, F.V., Thilikos, D.M.: An annotated bibliography on guaranteed graph searching. Theor. Comput. Sci. 399(3), 236–245 (2008)
9. Fraigniaud, P., Ilcinkas, D., Peer, G., Pelc, A., Peleg, D.: Graph exploration by a finite automaton. Theoretical Computer Science 345(2-3), 331–344 (2005)
10. Fürer, M., Raghavachari, B.: Approximating the minimum-degree steiner tree to within one of optimal. J. Algorithms 17(3), 409–423 (1994)
11. Goemans, M.X.: Minimum bounded degree spanning trees. In: FOCS, pp. 273–282 (2006)
12. Krizanc, D., Zhang, L.: Many-to-one packed routing via matchings. In: COCOON, pp. 11–17 (1997)
13. Panaite, P., Pelc, A.: Exploring unknown undirected graphs. J. Algorithms 33(2), 281–295 (1999)
14. Pantziou, G.E., Roberts, A., Symvonis, A.: Many-to-many routings on trees via matchings. Theor. Comput. Sci. 185(2), 347–377 (1997)
15. Zhang, L.: Optimal bounds for matching routing on trees. In: SODA, pp. 445–453 (1997)

# Powers of Hamilton Cycles in Pseudorandom Graphs

Peter Allen[1,*], Julia Böttcher[1,**], Hiệp Hàn[2,***], Yoshiharu Kohayakawa[2,†], and Yury Person[3,‡,§]

[1] Department of Mathematics, London School of Economics, Houghton Street, London WC2A 2AE, U.K.
{p.d.allen,j.boettcher}@lse.ac.uk
[2] Instituto de Matemática e Estatística, Universidade de São Paulo, Rua do Matão 1010, 05508–090 São Paulo, Brazil
{yoshi,hh}@ime.usp.br
[3] Goethe-Universität, Institute of Mathematics, Robert-Mayer-Str. 10, 60325 Frankfurt, Germany
person@math.uni-frankfurt.de

**Abstract.** We study the appearance of powers of Hamilton cycles in pseudorandom graphs, using the following comparatively weak pseudorandomness notion. A graph $G$ is $(\varepsilon, p, k, \ell)$-pseudorandom if for all disjoint $X, Y \subseteq V(G)$ with $|X| \geq \varepsilon p^k n$ and $|Y| \geq \varepsilon p^\ell n$ we have $e(X,Y) = (1 \pm \varepsilon)p|X||Y|$. We prove that for all $\beta > 0$ there is an $\varepsilon > 0$ such that an $(\varepsilon, p, 1, 2)$-pseudorandom graph on $n$ vertices with minimum degree at least $\beta pn$ contains the square of a Hamilton cycle. In particular, this implies that $(n,d,\lambda)$-graphs with $\lambda \ll d^{5/2}n^{-3/2}$ contain the square of a Hamilton cycle, and thus a triangle factor if $n$ is a multiple of 3. This improves on a result of Krivelevich, Sudakov and Szabó [*Triangle factors in sparse pseudo-random graphs*, Combinatorica **24** (2004), no. 3, 403–426]. We also obtain results for higher powers of Hamilton cycles and establish corresponding counting versions. Our proofs are constructive, and yield deterministic polynomial time algorithms.

## 1 Introduction and Results

The appearance of certain graphs $H$ as subgraphs is a dominant topic in the study of random graphs. In the random graph model $G(n,p)$ this question turned

---

[*] Partially supported by FAPESP (Proc. 2010/09555-7)
[**] Partially supported by FAPESP (Proc. 2009/17831-7)
[***] Supported by FAPESP (Proc. 2010/16526-3)
[†] Partially supported by CNPq (308509/2007-2, 477203/2012-4), FAPESP (2013/03447-6, 2013/07699-0) and the NSF (DMS-1102086).
[‡] Partially supported by GIF grant no. I-889-182.6/2005.
[§] The cooperation of the authors was supported by a joint CAPES–DAAD project (415/ppp-probral/po/D08/11629, Proj. no. 333/09). The authors are grateful to NUMEC/USP, Núcleo de Modelagem Estocástica e Complexidade of the University of São Paulo, and Project MaCLinC/USP, for supporting this research.

out to be comparatively easy for graphs $H$ of constant size, but much harder for graphs $H$ on $n$ vertices, i.e., *spanning* subgraphs. Early results were however obtained in the case when $H$ is a Hamilton cycle, for which this question is by now very well understood [7,18,19,20,26].

When we turn to other spanning subgraphs $H$ rather little was known for a long time, until a remarkably general result by Riordan [27] established good estimates for a big variety of spanning graphs $H$. In particular his result determines the threshold for the appearance of a spanning hypercube, and the threshold for the appearance of a spanning square lattice, as well as of the $k$th-power of a Hamilton cycle for $k > 2$. Here the *$k$th power* of $H$ is obtained from $H$ by adding all edges between distinct vertices of distance at most $k$ in $H$. For the square of a Hamilton cycle the corresponding approximate threshold was only obtained recently by Kühn and Osthus [25].

Observe that the $k$th power of a Hamilton cycle contains $\lfloor n/(k+1) \rfloor$ vertex disjoint copies of $K_{k+1}$, a so-called *spanning $K_{k+1}$-factor*. It came as another breakthrough in the area and solved a long-standing problem when Johansson, Kahn and Vu [16] established the threshold for spanning $K_{k+1}$-factors in $G(n,p)$ (or more generally of certain $F$-factors).

## 1.1 Pseudorandom Graphs

Thomason [28] asked whether it is possible to single out some deterministic properties enjoyed by $G(n,p)$ with high probability which imply a similarly rich collection of structural results. He thus initiated the study of pseudorandom graphs and suggested a deterministic property similar to the following notion of jumbledness. An $n$-vertex graph $G$ is $(p,\beta)$-*jumbled* if

$$\left| e(A,B) - p|A||B| \right| \leq \beta \sqrt{|A||B|} \tag{1}$$

for all disjoint $A, B \subseteq V(G)$. The random graph $G(n,p)$ is with high probability $(p,\beta)$-jumbled with $\beta = O(\sqrt{pn})$, so this definition is justified. Moreover, this pseudorandomness notion indeed implies a rich structure (see, e.g., [9,10,12]). However, for spanning subgraphs of general jumbled graphs (with a suitable minimum degree condition) not much is known.

One special class of jumbled graphs, which has been studied extensively, is the class of $(n,d,\lambda)$-graphs. Its definition relies on spectral properties. For a graph $G$ with eigenvalues $\lambda_1 \geq \lambda_2 \geq \cdots \geq \lambda_n$ of the adjacency matrix of $G$, we call $\lambda(G) := \max\{|\lambda_2|, |\lambda_n|\}$ the *second eigenvalue* of $G$. An $(n,d,\lambda)$-*graph* is a $d$-regular graph on $n$ vertices with $\lambda(G) \leq \lambda$. The connection between $(n,d,\lambda)$-graphs and jumbled graphs is established by the well-known *expander mixing lemma* (see, e.g., [6]), which states that if $G$ is an $(n,d,\lambda)$-graph, then

$$\left| e(A,B) - \tfrac{d}{n}|A||B| \right| \leq \lambda(G)\sqrt{|A||B|} \tag{2}$$

for all disjoint subsets $A, B \subseteq V(G)$. Hence $G$ is $\left(\tfrac{d}{n}, \lambda(G)\right)$-jumbled.

One main advantage of $(n,d,\lambda)$-graphs are the powerful tools from spectral graph theory which can be used for their study. Thanks to these tools various results concerning spanning subgraphs of $(n, d, \lambda)$-graphs $G$ were obtained. It turns out that already an almost trivial eigenvalue gap guarantees a spanning matching: If $\lambda \leq d - 2$ and $n$ is even, then $G$ has a perfect matching [23]. Moreover, if $\lambda \leq d(\log \log n)^2/(1000 \log n \log \log \log n)$ then $G$ has a Hamilton cycle [22]. The only other embedding result for spanning subgraphs of $(n, d, \lambda)$-graphs that we are aware of concerns triangle factors. Krivelevich, Sudakov and Szabó [24] proved that an $(n, d, \lambda)$-graph $G$ with $3|n$ and $\lambda = o(d^3/n^2 \log n)$ contains a spanning triangle factor.

It is instructive to compare this last result with corresponding lower bound constructions. Krivelevich, Sudakov and Szabó also remarked that by using a blow-up of a construction of Alon [3] one can obtain for each $d' = d'(n')$ with $\Omega((n')^{2/3}) \leq d' \leq n'$ an $(n, d, \lambda)$-graph with $n = \Theta(n')$, $d = \Theta(d')$ and $\lambda = \Theta(d^2/n)$ which is triangle-free and thus contains no spanning triangle factor. They conjectured that in fact $(n, d, \lambda)$-graphs are so symmetric that the upper bound on $\lambda$ they proved for triangle factors can be improved, possibly all the way down to this lower bound. In this paper we bring the upper bound closer to the conjectured lower bound and establish more generally an embedding result for $k$th powers of Hamilton cycles (see Corollary 4).

## 1.2 Our Results

The pseudorandomness notion we shall work with in this paper is weaker than that of $(n, d, \lambda)$-graphs, and in fact even weaker than jumbledness.

**Definition 1.** *Let $\varepsilon > 0$ and let $k$, $\ell$ with $k \leq \ell$ be positive integers. For given $p = p(n)$ we call an n-vertex graph $G$ $(\varepsilon, p, k, \ell)$-pseudorandom if*

$$\bigl|e(X,Y) - p|X||Y|\bigr| \leq \varepsilon p|X||Y| \qquad (3)$$

*for any disjoint subsets $X$, $Y \subseteq V(G)$ with $|X| \geq \varepsilon p^k n$ and $|Y| \geq \varepsilon p^\ell n$.*

It is easy to check that a graph which is $(p, \varepsilon^2 p^s n)$-jumbled is $(\varepsilon, p, k, \ell)$-pseudorandom for all $k$ and $\ell$ with $k + \ell = 2s - 2$, but the jumbledness condition imposes tighter control on the edge density between (for example) linear sized subsets. An easy application of Chernoff's inequality and the union bound show that $G(n,p)$ is $(\varepsilon, p, k, \ell)$-pseudorandom with high probability if $p \gg (n^{-1} \log n)^{1/(\max\{k,\ell\}+1)}$, while $G(n,p)$ only gets $(p, \varepsilon^2 p^{(k+\ell+2)/2} n)$-jumbled if $p \gg n^{-1/(k+\ell+1)}$. Our major motivation for using this weaker pseudorandomness condition is that it is all we require.

Our main result states that sufficiently pseudorandom graphs which also satisfy a mild minimum degree condition contain spanning powers of Hamilton cycles.

**Theorem 2.** *For every $k \geq 2$ and $\beta > 0$ there is an $\varepsilon > 0$ such that for any $p = p(n)$ with $0 < p < 1$ the following holds. Let $G$ be a graph on $n$ vertices with minimum degree $\delta(G) \geq \beta p n$.*

(a) If $G$ is $(\varepsilon, p, 1, 2)$-pseudorandom then $G$ contains a square of a Hamilton cycle.
(b) If $G$ is $(\varepsilon, p, k-1, 2k-1)$-pseudorandom and $(\varepsilon, p, k, k+1)$-pseudorandom then $G$ contains a $k$th power of a Hamilton cycle.

We remark that our proof of Theorem 2 also yields a deterministic polynomial time algorithm for finding a copy of the $k$th power of the Hamilton cycle. The proof technique (see Section 2.2 for an overview) is partly inspired by the methods used in [2] (which have similarities to those of Kühn and Osthus [25]).

It is immediate from the discussion above that our theorem implies the following result for jumbled graphs.

**Corollary 3 (Powers of Hamilton cycles in jumbled graphs).** *For every $k \geq 2$ and $\beta > 0$ there is an $\varepsilon > 0$ such that for any $p = p(n)$ with $0 < p < 1$ the following holds. Let $G$ be a graph on $n$ vertices with minimum degree $\delta(G) \geq \beta pn$.*
(a) *If $G$ is $(p, \varepsilon p^{5/2} n)$-jumbled then $G$ contains a square of a Hamilton cycle.*
(b) *If $G$ is $(p, \varepsilon p^{3k/2} n)$-jumbled then $G$ contains a $k$th power of a Hamilton cycle.*

As a consequence we also obtain a corresponding corollary for $(n, d, \lambda)$-graphs.

**Corollary 4 (Powers of Hamilton cycles in $(n, d, \lambda)$-graphs).** *For all $k \geq 2$ there is $\varepsilon > 0$ such that for every $(n, d, \lambda)$-graph $G$,*
(a) *if $\lambda \leq \varepsilon d^{5/2} n^{-3/2}$ then $G$ contains a square of a Hamilton cycle,*
(b) *if $\lambda \leq \varepsilon d^{3k/2} n^{1-3k/2}$ then $G$ contains a $k$th power of a Hamilton cycle.*

In particular, under these conditions $G$ contains a spanning triangle factor and a $K_{k+1}$-factor, respectively. Thus we improve on the result of Krivelevich, Sudakov and Szabó [24] for triangle factors and extend it to $K_{k+1}$-factors.

As remarked above even for $k = 2$ our upper bound for $\lambda$ does not match the known lower bound. For $k > 2$ the situation gets even more complicated since 'good' lower bounds for the appearance of $K_{k+1}$ (let alone $k$th powers of Hamilton cycles) in $(n, d, \lambda)$-graphs are not available. The best we can do is to observe that $G(n, p)$ with $(\ln n/n)^{1/(k-\varepsilon)} \ll p \ll n^{-1/k}$ almost surely has no $k$th power of a Hamilton cycle, and that such a graph for any fixed $\varepsilon > 0$ is almost surely $(\varepsilon, p, k-1-\varepsilon, k-1-\varepsilon)$-pseudorandom.

## 1.3 Counting

Closely related to the question of the appearance of a certain subgraph in random or pseudorandom graphs is the question of how many copies of this subgraph are actually present. Janson [15] and Cooper and Frieze [13] studied this problem for Hamilton cycles in $G(n, p)$. Motivated by these results Krivelevich [21] recently turned to counting Hamilton cycles in sparse $(n, d, \lambda)$-graphs $G$. He showed that for every $\varepsilon > 0$ and sufficiently large $n$, if $\lambda \leq d/(\log n)^{1+\varepsilon}$ and $\log \lambda \ll \log d - \log n / \log d$ then $G$ contains $n!(d/n)^n (1+o(1))^n$ Hamilton cycles. This count is close to the expected number of labeled Hamilton cycles in $G(n,p)$ with $p = d/n$, which is $n!(d/n)^n$.

Krivelevich remarked that jumbled graphs may have isolated vertices and thus no Hamilton cycles at all. The same applies to our notion of pseudorandomness. If however, as in our main result, we combine this pseudorandomness with a minimum degree condition to avoid this obstacle, we do obtain a corresponding result concerning the number of Hamilton cycle powers in such graphs. Again, we obtain a count close to $p^{kn}n!$, which is the expected number of labeled copies of the $k$th power of a Hamilton cycle in $G_{n,p}$. Note that (unlike Krivelevich) we do not provide a corresponding upper bound.

**Theorem 5.** *For every $k \geq 2$, $\beta, \nu > 0$ there is a constant $c > 0$, such that for every $\varepsilon = \varepsilon(n) \leq c/\log^2 n$ and $p = p(n)$ with $0 < p < 1$ the following holds. Let $G$ be a graph on $n$ vertices with minimum degree $\delta(G) \geq \beta pn$. Suppose that $G$ is $(\varepsilon, p, 1, 2)$-pseudorandom if $k = 2$, and $(\varepsilon, p, k-1, 2k-1)$-pseudorandom and $(\varepsilon, p, k, k+1)$-pseudorandom if $k > 2$. Then $G$ contains at least $(1-\nu)^n p^{kn} n!$ copies of the $k$th power of a Hamilton cycle.*

With some minor modifications, this result follows from our proof of Theorem 2. For the sake of clarity, we sketch these modifications after discussing the proof of Theorem 2.

### 1.4 Organisation

The remainder of this extended abstract is organised as follows. In Section 2 we give some basic definitions, outline our proof strategy and provide the main lemmas precisely (without proof). We sketch how to modify the proof of Theorem 2 to get Theorem 5 in Section 3, and close with some remarks and open problems in Section 4.

## 2 Main Lemmas and an Outline of the Proof Theorem 2

### 2.1 Notation

An $s$-tuple $(u_1, \ldots, u_s)$ of vertices is an ordered set of vertices. We often denote tuples by bold symbols, and occasionally also omit the brackets and write $\boldsymbol{u} = u_1, \ldots, u_s$.

Given a graph $H$, the graph $H^k$, called the $k$th *power* of $H$, is the graph on $V(H)$ where two distinct vertices $u$ and $v$ are adjacent if and only if their distance in $H$ is at most $k$.

For simplicity we also call the $k$th power of a path a $k$-*path*, and the $k$th power of a cycle a $k$-*cycle*. We will usually specify $k$-paths and $k$-cycles by giving the (cyclic) ordering of the vertices in the form of a vertex tuple. We say that the start $s$-tuple of a $k$-path $P = (u_1, \ldots, u_\ell)$ is $(u_s, \ldots, u_1)$, and the end $s$-tuple is $(u_{\ell-s+1}, \ldots, u_\ell)$ (the vertices $u_{s+1}, \ldots, u_{\ell-s}$ are said to be internal). In these definitions, we shall often have $s = k$.

For a given graph $G$ let $N_X(x)$ be the set of neighbours of $x$ in $X \subseteq V(G)$. For an $\ell$-tuple $\boldsymbol{x}_\ell = (x_1, \ldots, x_\ell)$ of vertices let $N_X(x_1, \ldots, x_\ell)$ denote the common neighbourhood of $x_1, \ldots, x_\ell$ in $X$, and let $\deg_X(x_1, \ldots, x_\ell) = |N_X(x_1, \ldots, x_\ell)|$.

We say that $\boldsymbol{x}_\ell$ is $(\varrho, p)$-*connected* to a vertex set $X$ if $x_1, \ldots, x_\ell$ forms a clique in $G$ and

$$\deg_X(x_i, \ldots, x_\ell) \geq \varrho \left(\frac{p}{2}\right)^{\ell-i+1} |X|. \tag{4}$$

for every $i \in [\ell] = \{1, \ldots, \ell\}$. To motivate this definition, note that the bound in (4) corresponds to the expected number of common neighbours of $(x_i, \ldots, x_\ell)$ in $X$ in the random graph $G(n, p)$, up to a constant factor.

A vertex set $Y \subseteq X$ *witnesses* that $\boldsymbol{x}_\ell$ is $(\varrho, p)$-connected to $X$ if for every $i \in [\ell]$ we have $|Y \cap N_X(x_i, \ldots, x_\ell)| \geq \varrho \left(\frac{p}{2}\right)^{\ell-i+1} |X|$.

*Remark 6.* Since the sets $N_X(x_1, \ldots, x_\ell), N_X(x_2, \ldots, x_\ell), \ldots, N_X(x_\ell)$ are nested we have that if $\boldsymbol{x}_\ell$ is $(\varrho, p)$-connected to $X$, then there is a set $Y \subseteq X$ with $|Y| = \varrho \frac{p}{2} |X|$ vertices which witnesses this connectedness.

*Remark 7.* If $0 < p \leq 1/2$ and $\varepsilon < 1/8$, and the $n$-vertex graph $G$ is $(\varepsilon, p, k, \ell)$-pseudorandom, then $G$ has a vertex $y$ of degree at most $3n/4$. Moreover, letting $X = V(G) \setminus (\{y\} \cup N(y))$ and $Y = \{y\}$ we see that (3) does not hold. It follows that $1 < \varepsilon p^\ell n$, or equivalently $p^\ell n > \varepsilon^{-1}$. A similar statement holds if $1/2 \leq p < 1$, taking $X = N(y)$. Thus assuming the $n$-vertex graph $G$ to be $(\varepsilon, p, k, \ell)$-pseudorandom for any $0 < p < 1$ implicitly means we assume $p^\ell n > \varepsilon^{-1}$.

## 2.2 Outline of the Proof

Suppose that $G$ is an $(\varepsilon, p, k-1, k)$-pseudorandom graph on $n$ vertices. One crucial observation, which forms the starting point of our proof, is that it is relatively easy to find an almost spanning $k$-path in $G$. Indeed, it is not hard to check (see the Extension lemma, Lemma 8) that $G$ contains copies of $K_k$ and that typically such a $K_k$-copy is well-connected to the rest of the graph in the following sense. There are many vertices which extend this $K_k$-copy to a $k$-path on $k+1$ vertices. Iterating this argument we can greedily build a $k$-path $P'$ covering most of $G$. Let $L$ be the set of leftover vertices.

The true challenge is to incorporate the few remaining vertices into $P'$ and to close $P'$ into a $k$-cycle. To tackle the second of these tasks we will establish the Connection lemma (Lemma 11), which asserts that any two pairs of $k$-cliques in $G$ which are sufficiently well-connected to a set $U$ of vertices can be connected by a short $k$-path with interior vertices in $U$. At this point, if $k > 2$, we shall need to require that $G$ be $(\varepsilon, p, k-1, 2k-1)$-pseudorandom.

For the first task, we make use of the *reservoir method* developed in [2] (see also [25] for a similar method). In essence, the fundamental idea of this method is to ensure that $P'$ contains a sufficiently big proportion of vertices which are free to be taken out of $P'$ and used otherwise. More precisely, we shall construct (see the Reservoir lemma, Lemma 9) a path $P$ with the *reservoir property*: There is a subset $R$ of $V(P)$, called the *reservoir*, such that for any $W \subseteq R$ there is a $k$-path in $G$ whose vertex set is $V(P) \setminus W$ and whose ends are the same as those of $P$. We also call $P$ a *reservoir path*. We then use the greedy method outlined above to extend $P$ to an almost spanning $k$-path $P'$. For this step, if $k > 2$, we shall need to require that $G$ be $(\varepsilon, p, k, k+1)$-pseudorandom.

With the reservoir property we are now in good shape to incorporate the leftover vertices $L$ into $P'$ (and then close the path into a cycle): We show, using the Covering lemma (Lemma 10), that we can find a $k$-path $P''$ in $L \cup R$ covering all vertices of $L$ and using only a small fraction of $R$ (this is possible because $R$ is much bigger than $L$). Finally we connect both ends of $P'$ and $P''$ using some of the remaining vertices of $R$ with the help of the Connection lemma (again, this is possible because many vertices of $R$ remain).

Now the only problem is that some vertices of $R$ may be used twice, in $P'$ and in $P''$ or the connections. But this is where the reservoir property comes into play. This property asserts that there is a $k$-path $\widetilde{P}$ which uses all vertices of $P'$ except these vertices. Finally $\widetilde{P}$ and $P''$ together with the connections form the desired spanning $k$-cycle.

### 2.3 Main Lemmas

The proof of Theorem 2 relies on four main lemmas, the Extension lemma, the Reservoir lemma, the Covering lemma and the Connection lemma, which we will state and explain in the following (in this extended abstract, the proofs are omitted).

Our first lemma, the Extension lemma, states that in a sufficiently pseudorandom graph all well-connected $k$-tuples have a common neighbour which together with the last $k-1$ vertices of this $k$-tuple form again a well-connected $k$-tuple.

**Lemma 8 (Extension lemma).** *Given $k \geq 2$ and $\delta > 0$ there is an $\varepsilon > 0$ such that for all $0 < p < 1$, all $(\varepsilon, p, k-1, k)$-pseudorandom graphs $G$ on $n$ vertices, and all disjoint vertex sets $L$ and $R$ with $|L|, |R| \geq \delta n$ the following holds.*

*Let $x = (x_1, \ldots, x_k)$ be a $k$-tuple which is $(\frac{1}{8}, p)$-connected to both $L$ and $R$. Then there is a vertex $x_{k+1}$ of $L \cap N(x_1, \ldots, x_k)$ such that $(x_2, \ldots, x_{k+1})$ is $(\frac{1}{6}, p)$-connected to both $L$ and $R$.*

We stress that in this lemma we require and obtain well-connectedness to two sets $L$ and $R$. This will enable us in the proof of Theorem 2 to extend a $k$-path alternatively using vertices of the leftover set $L$ or the reservoir set $R$.

We remark moreover that the assumed $(\frac{1}{8}, p)$-connectedness is weaker than the $(\frac{1}{6}, p)$-connectedness in the conclusion. This is useful when we repeatedly apply the Extension lemma. It is possible to prove such a statement because the factor $\frac{1}{2}$ in the definition of connectedness allows for some leeway.

Our second lemma allows us to construct the reservoir path $P$ described in the outline, given a suitable reservoir $R$ (see properties $(a)$ and $(d)$). In addition, this lemma guarantees well-connectedness of the ends of this path to the reservoir and to the remaining vertices in the graph (see properties $(b)$ and $(c)$). This is necessary so that we can extend the reservoir path and later connect it to the path covering the leftover vertices $L$ using $R$.

**Lemma 9 (Reservoir lemma).** *Given $k \geq 2$, $0 < \delta < 1/4$ and $0 < \beta < 1/2$ there exists an $\varepsilon > 0$ such that the following holds.*

*Let $0 < p < 1$ and let $G = (V, E)$ be an n-vertex graph. Suppose that $G$ is $(\varepsilon, p, 1, 2)$-pseudorandom if $k = 2$, and $(\varepsilon, p, k-1, 2k-1)$-pseudorandom and $(\varepsilon, p, k, k+1)$-pseudorandom if $k > 2$. Let $R \subseteq V$ satisfy $\delta^2 n/(200k) \leq |R| \leq \delta n/(200k)$ and $\deg_{V \setminus R}(v) \geq \beta p n/2$ for all $v \in R$. Then there is a $k$-path $P$ in $G$ with the following properties.*

(a) *$R \subseteq V(P)$, $|V(P)| \leq 50k|R|$ and all vertices from $R$ being internal for $P$.*
(b) *The start and end $k$-tuples of $P$ are $(\frac{1}{8}, p)$-connected to $V \setminus V(P)$.*
(c) *The start and end $k$-tuples of $P$ are $(\frac{1}{2}, p)$-connected to $R$.*
(d) *For any $W \subseteq R$, there is a $k$-path with the vertex set $V(P) \setminus W$ whose start and end $k$-tuples are identical to those of $P$.*

Our third lemma enables us to cover the leftover vertices $L$ with a $k$-path (see property (a)). This lemma allows us in addition to specify a set $S$ to which the start and end tuples of this path have to maintain well-connectedness (see property (b)). When we cover the leftover vertices in the proof of the main theorem, $S$ will be a big proportion of $R$ and we will use the well-connectedness to connect the path covering $L$ and the extended reservoir path.

We remark that the requirements and conclusions of Lemma 9 and Lemma 10 overlap substantially.

**Lemma 10 (Covering lemma).** *Given $k \geq 2$, $0 < \delta < 1/4$ and $0 < \beta < 1/2$ there exists an $\varepsilon > 0$ such that the following holds.*

*Let $0 < p < 1$ and let $G = (V, E)$ be an n-vertex graph. Suppose that $G$ is $(\varepsilon, p, 1, 2)$-pseudorandom if $k = 2$, and $(\varepsilon, p, k-1, 2k-1)$-pseudorandom and $(\varepsilon, p, k, k+1)$-pseudorandom if $k > 2$. Let $L$ and $S$ be disjoint subsets of $V(G)$ with $|L| \leq \delta n/(200k)$ and $|S| \geq \delta n$ such that $\deg_S(v) \geq \beta \delta p n/2$ for all $v \in L$. Then there is a $k$-path $P$ contained in $L \cup S$ with the following properties.*

(a) *$L \subseteq V(P)$ and $|V(P)| \leq 50k|L|$.*
(b) *The start and end $k$-tuples of $P$ are in $S$ and are $(\frac{1}{8}, p)$-connected to $S \setminus V(P)$.*

Our fourth and final main lemma allows us to connect two $k$-tuples with a short $k$-path.

**Lemma 11 (Connection lemma).** *For all $k \geq 2$ and $\delta > 0$ there is an $\varepsilon > 0$ such that the following holds.*

*Let $0 < p < 1$ and let $G$ be an $n$ vertex graph. Suppose that $G$ is $(\varepsilon, p, 1, 2)$-pseudorandom if $k = 2$, and $(\varepsilon, p, k-1, 2k-1)$-pseudorandom if $k > 2$. Let $U \subseteq V(G)$ be a vertex set of size $|U| \geq \delta n$. If $\boldsymbol{x}$ and $\boldsymbol{y}$ are two disjoint $k$-tuples which are $(\delta, p)$-connected to $U$, then there exists a $k$-path $P$ with ends $\boldsymbol{x}$ and $\boldsymbol{y}$ of length at most $7k$ such that $V(P) \subseteq U \cup V(\boldsymbol{x}) \cup V(\boldsymbol{y})$.*

We remark that in the proof of Theorem 2 it is not especially important that the connecting $k$-path guaranteed by this lemma is of constant length. However, Lemma 11 is also used in the proof of Lemma 9, and in this proof we need that the connecting $k$-paths are of length independent of $n$.

## 3 Enumerating Powers of Hamilton Cycles

To prove Theorem 5 we would ideally like to show that we can construct the $k$th power of a Hamilton cycle vertex by vertex, and that when we have $t$ vertices remaining uncovered, we have at least $(1-\nu)p^k t$ choices for the next vertex; then the theorem would follow immediately. However, we obviously do not construct $k$th powers of Hamilton cycles in this way: we have very little control over choice in constructing the reservoir paths and connecting paths. Moreover for the promised number $(1-\nu)^n p^{kn} n!$ of Hamilton cycle powers even the Extension lemma, Lemma 8, does not provide the desired number of choices in the greedy portion of the construction where we do choose one vertex at a time. (We remark though that the proof of this lemma, together with the rest of our proof does immediately provide us with $c^n p^{(1-\nu)kn} \big((1-\nu)n\big)!$ Hamilton cycle powers for some absolute constant $c > 0$.)

Thus we have to upgrade the Extension lemma in two ways. Firstly, we have to modify it to give us more choices in each step (after a few initial steps). Secondly, it turns out that to obtain the desired number of Hamilton cycle powers we have to apply the Extension lemma for longer, that is, the leftover set will in the end only contain $\mathcal{O}\big(n/(\log n)^2\big)$ vertices. Thus we have to change the Extension lemma to deal with this different situation. This comes at the cost of slightly tightening the pseudorandomness requirement.

In the lemma below we will guarantee that for an end $k$-tuple $\boldsymbol{x}$ of a $k$-path there are $\big(1 - \tfrac{\nu}{2k}\big)\deg_L(\boldsymbol{x})$ valid extensions, where $L$ is the current set of leftover vertices. One may argue that this will provide us with the right number of Hamilton cycle powers if we can guarantee in addition that $\deg_L(\boldsymbol{x}) \geq \big((1-\tfrac{\nu}{2k})p\big)^k|L|$. Recall however that we will want to use this lemma after constructing the reservoir path with the help of Lemma 9, which guarantees $(\tfrac{1}{8}, p)$-connectedness to $L$, a property which only gives a weaker lower bound on $\deg_L(\boldsymbol{x})$ than desired. In order to overcome this shortcoming we will in the first few applications of the Counting version of the Extension lemma transform this $(\tfrac{1}{8}, p)$-connectedness to a stronger property which gives the desired bound. Conditions $(ii)$ and $(iii)$, and conclusions $(b)$ and $(c)$ take care of this.

**Lemma 12 (Counting version of the Extension lemma).** *Given $k \geq 2$ and $\nu > 0$, if $C = 2^{k+23}k^4/\nu$ then the following holds. Let $0 < p < 1$ and $G$ be an $\big(1/(C \log n)^2, p, k-1, k\big)$-pseudorandom graph on $n$ vertices. Let $L$ and $R$ be disjoint vertex sets with $|L|, |R| \geq n/(200k \log n)^2$. Suppose that there is $0 \leq j \leq k$ such that $\boldsymbol{x} = (x_1, \ldots, x_k)$ satisfies*

(i) *$\boldsymbol{x}$ is $\big(\tfrac{1}{8}, p\big)$-connected to $R$,*

(ii) *$\deg_L(x_i, \ldots, x_k) \geq \tfrac{1}{8}\big(\tfrac{p}{2}\big)^{k-i+1}|L|$ for each $1 \leq i \leq j$,*

(iii) $\deg_L(x_i, \ldots, x_k) \geq \left((1 - \frac{\nu}{2k})p\right)^{k-i+1}|L|$ for each $j < i \leq k$.

Then at least $\left(1 - \frac{\nu}{2k}\right)\deg_L(x_1, \ldots, x_k)$ vertices $x_{k+1} \in N_L(x_1, \ldots, x_k)$ satisfy that

(a) $(x_2, \ldots, x_{k+1})$ is $(\frac{1}{6}, p)$-connected to $R$,

(b) $\deg_L(x_{i+1}, \ldots, x_{k+1}) \geq \frac{1}{8}\left(\frac{p}{2}\right)^{k-i+1}|L|$ for each $1 \leq i \leq j-1$,

(c) $\deg_L(x_{i+1}, \ldots, x_{k+1}) \geq \left((1 - \frac{\nu}{2k})p\right)^{k-i+1}|L|$ for each $j-1 < i \leq k$.

## 4 Concluding Remarks

**Hamilton Cycles.** For Hamilton cycles, a simple modification of our arguments for squared Hamilton cycles yields that $(\varepsilon, p, 0, 1)$-pseudorandom graphs with minimum degree $\beta pn$ are Hamiltonian for sufficiently small $\varepsilon = \varepsilon(\beta)$. This bound is essentially best possible (for our notion of pseudorandomness) since the disjoint union of $G_{n-pn,p}$ and $K_{pn}$ is easily seen to be asymptotically almost surely $(\varepsilon, p, 0, 1-\varepsilon)$-pseudorandom and have minimum degree at least $pn/2$.

**Improving the Pseudorandomness Requirements.** It would be interesting to obtain stronger results on the pseudorandomness required to find $k$th powers of Hamilton cycles. We believe that a generalisation of our result for the $k = 2$ case is true.

*Conjecture 13.* For all $k \geq 2$ the pseudorandomness requirement in Theorem 2 can be replaced by $(\varepsilon, p, k-1, k)$-pseudorandomness.

As remarked in the introduction even in the $k = 2$ case we do not know whether Theorem 2 is sharp. It would also be very interesting (albeit probably very hard) to find better lower bound examples than those mentioned in the introduction.

In the evolution of random graphs, triangles, spanning triangle factors and squares of Hamilton cycles appear at different times: In $G(n, p)$ the threshold for triangles is $p = n^{-1}$, but only at $p = \Theta(n^{-2/3}(\log n)^{1/3})$ each vertex of $G(n, p)$ is contained in a triangle with high probability, which is also the threshold for the appearance of a spanning triangle factor [16]. Squares of Hamilton cycles on the other hand are with high probability not present in $G(n, p)$ for $p \leq n^{-1/2}$, and Kühn and Osthus [25] recently showed that for $p \geq n^{-1/2+\varepsilon}$ they are present. Our Theorem 2 is also applicable to random graphs, however, the range is worse: $p \gg (\ln n/n)^{1/3}$ for squares of Hamilton cycles and $p \gg (\ln n/n)^{1/(2k)}$ for general $k$th powers of Hamilton cycles, in the light of Riordan's result [27] which implies the optimal bound $p \gg n^{-1/k}$ for $k \geq 3$.

Pseudorandom graphs behave differently. For $(n, d, \lambda)$-graphs it is known that as soon as these graphs are forced to have one triangle, every vertex is contained in a triangle. This motivated Krivelevich, Sudakov and Szabó [24] to conjecture that indeed these graphs already contain a spanning triangle factor. We do not know whether triangle factors and squares of Hamilton cycles require different levels of pseudorandomness.

*Question 14.* Do spanning triangle factors and squares of Hamilton cycles appear for the same pseudorandomness requirements (up to constant factors)?

**Universality.** For random graphs, the study of when almost every $G(n,p)$ contains as subgraphs all $n$-vertex graphs or all $(1-\varepsilon)n$-vertex graphs with maximum degree bounded by a constant $\Delta$ was initiated in [5]. In this case $G(n,p)$ is also called *universal* for these graphs. The authors of [5] showed that $G(n,p)$ contains all graphs on $(1-\varepsilon)n$ vertices with maximum degree at most $\Delta$ if $p \geq Cn^{-1/\Delta}\log^{1/\Delta} n$. In [14] this result was extended to such subgraphs on $n$ vertices. Recently, Conlon [11] announced that for the first of these two results he can lower the probability to $p = n^{-\varepsilon - 1/\Delta}$ for some (small) $\varepsilon = \varepsilon(\Delta) > 0$. The best known lower bound results from the fact that $p = \Omega(n^{-2/(\Delta+1)})$ is necessary for $G(n,p)$ to contain a spanning $K_{\Delta+1}$-factor.

For pseudorandom graphs we were only recently able to establish universality results of this type, which follow from our work on a Blow-up lemma for pseudorandom graphs (see below). We can prove that $(p, cp^{\frac{3}{2}\Delta + \frac{1}{2}}n)$-jumbled graphs on $n$ vertices with minimum degree $\beta pn$ are universal for spanning graphs with maximum degree $\Delta$ [1]. We believe that these conditions are not optimal.

*Question 15.* Which pseudorandomness conditions (plus minimum degree conditions) imply universality for spanning graphs of maximum degree $\Delta$?

It is worth noting that Alon and Capalbo [4] explicitly constructed almost optimally sparse universal graphs for spanning graphs with maximum degree $\Delta$. These graphs have some pseudorandomness properties, but they also contain cliques of order $\log^2 n$, which random graphs of the same density certainly do not.

**Blow-up Lemmas.** For dense graphs the Blow-up lemma [17] is a powerful tool for embedding spanning graphs with bounded maximum degree (versions of this lemma for certain graphs with unbounded maximum degree have recently been developed in [8]). Already Krivelevich, Sudakov and Szabó [24] remark that their result on triangle factors in sparse pseudorandom graphs can be viewed as a first step towards the development of a Blow-up lemma for (subgraphs of) sparse pseudorandom graphs.

We see the results presented here as a further step in this direction. In fact, in recent work [1], we establish a blow-up lemma for spanning graphs with bounded maximum degree in sparse pseudorandom graphs. However, quite naturally, the pseudorandomness requirements for this more general result are more restrictive than those used here.

# References

1. Allen, P., Böttcher, J., Hàn, H., Kohayakawa, Y., Person, Y.: Blow-up lemmas for sparse graphs (in preparation)
2. Allen, P., Böttcher, J., Kohayakawa, Y., Person, Y.: Tight Hamilton cycles in random hypergraphs. Random Structures Algorithms (to appear), doi: 10.1002/rsa.20519

3. Alon, N.: Explicit Ramsey graphs and orthonormal labelings. Electronic Journal of Combinatorics 1, Research paper 12, 8pp (1994)
4. Alon, N., Capalbo, M.: Sparse universal graphs for bounded-degree graphs. Random Structures Algorithms 31(2), 123–133 (2007)
5. Alon, N., Capalbo, M., Kohayakawa, Y., Rödl, V., Ruciński, A., Szemerédi, E.: Universality and tolerance (extended abstract). In: Proc. 41 IEEE FOCS, pp. 14–21. IEEE (2000)
6. Alon, N., Spencer, J.H.: The probabilistic method, vol. 57. Wiley Interscience (2000)
7. Bollobás, B.: The evolution of sparse graphs. In: Graph theory and combinatorics (Cambridge, 1983), pp. 35–57. Academic Press, London (1984)
8. Böttcher, J., Kohayakawa, Y., Taraz, A., Würfl, A.: An extension of the blow-up lemma to arrangeable graphs, arXiv:1305.2059
9. Chung, F.R.K., Graham, R.L., Wilson, R.M.: Quasi-random graphs. Combinatorica 9(4), 345–362 (1989)
10. Chung, F., Graham, R.: Sparse quasi-random graphs. Combinatorica 22(2), 217–244 (2002)
11. Conlon, D.: Talk at RSA 2013 (2013)
12. Conlon, D., Fox, J., Zhao, Y.: Extremal results in sparse pseudorandom graphs. arXiv:1204.6645
13. Cooper, C., Frieze, A.M.: On the number of Hamilton cycles in a random graph. J. Graph Theory 13(6), 719–735 (1989)
14. Dellamonica Jr., D., Kohayakawa, Y., Rödl, V., Ruciński, A.: An improved upper bound on the density of universal random graphs. In: Fernández-Baca, D. (ed.) LATIN 2012. LNCS, vol. 7256, pp. 231–242. Springer, Heidelberg (2012)
15. Janson, S.: The numbers of spanning trees, Hamilton cycles and perfect matchings in a random graph. Combin. Probab. Comput. 3(1), 97–126 (1994)
16. Johansson, A., Kahn, J., Vu, V.: Factors in random graphs. Random Structures Algorithms 33(1), 1–28 (2008)
17. Komlós, J., Sárközy, G.N., Szemerédi, E.: Blow-up lemma. Combinatorica 17(1), 109–123 (1997)
18. Komlós, J., Szemerédi, E.: Limit distribution for the existence of Hamiltonian cycles in a random graph. Discrete Math. 43(1), 55–63 (1983)
19. Korshunov, A.D.: Solution of a problem of Erdős and Renyi on Hamiltonian cycles in nonoriented graphs. Sov. Math., Dokl. 17, 760–764 (1976)
20. Korshunov, A.D.: Solution of a problem of P. Erdős and A. Renyi on Hamiltonian cycles in undirected graphs. Metody Diskretn. Anal. 31, 17–56 (1977)
21. Krivelevich, M.: On the number of Hamilton cycles in pseudo-random graphs. Electron. J. Combin. 19(1), Paper 25, 14pp (2012)
22. Krivelevich, M., Sudakov, B.: Sparse pseudo-random graphs are Hamiltonian. J. Graph Theory 42(1), 17–33 (2003)
23. Krivelevich, M., Sudakov, B.: Pseudo-random graphs, More sets, graphs and numbers. Bolyai Soc. Math. Stud. 15, 199–262 (2006)
24. Krivelevich, M., Sudakov, B., Szabó, T.: Triangle factors in sparse pseudo-random graphs. Combinatorica 24(3), 403–426 (2004)
25. Kühn, D., Osthus, D.: On Pósa's conjecture for random graphs. SIAM J. Discrete Math. 26(3), 1440–1457 (2012)
26. Pósa, L.: Hamiltonian circuits in random graphs. Discrete Mathematics 14(4), 359–364 (1976)
27. Riordan, O.: Spanning subgraphs of random graphs. Combin. Probab. Comput. 9(2), 125–148 (2000)
28. Thomason, A.: Pseudo-random graphs. Random Graphs 85, 307–331 (1987)

# Local Update Algorithms for Random Graphs

Philippe Duchon and Romaric Duvignau

Univ. Bordeaux, LaBRI, UMR 5800, F-33400 Talence, France
CNRS, LaBRI, UMR 5800, F-33400 Talence, France
{philippe.duchon,romaric.duvignau}@labri.fr

**Abstract.** We study the problem of maintaining a given distribution of random graphs under an arbitrary sequence of vertex insertions and deletions. Since our goal is to model the evolution of dynamic logical networks, we work in a local model where we do not have direct access to the list of all vertices. Instead, we assume access to a global primitive that returns a random vertex, chosen uniformly from the whole vertex set. In this preliminary work, we focus on a simple model of uniform directed random graphs where all vertices have a fixed outdegree. We describe and analyze several algorithms for the maintenance task; the most elaborate of our algorithms are asymptotically optimal.

**Keywords:** random graphs, dynamic graphs, logical network maintenance, randomness preservation.

## 1 Introduction

In decentralized networks and in particular in peer-to-peer networks[1] (P2P), the structure of the network is maintained locally by the nodes following a predefined network protocol. In most modern implementations (e.g., Gnutella [1], GNUnet [2], Freenet [3]), the structure of the network is highly dependent on the sequence of updates (node insertions or deletions), and as a consequence a malicious sequence of updates may result in a disconnected or badly structured network. Moreover, designing and analyzing the update algorithms defined by such network protocols is made harder by this dependence on the update sequence. In particular, analysis of the network evolution is typically performed under nicely behaving update schemes: new arrivals follow a Poisson process and living times follow an exponential distribution (used in the analysis of [4] and [5]), or some similar dynamicity hypothesis (as in [6] for bounding the mixing time of a Markov chain modeling the P2P network of [7]).

In order to avoid this dependence and to make the analysis of the network evolution somewhat easier, we propose local update algorithms that maintain *exactly* some probability distribution of random graphs, so that the distribution of the graph resulting from an arbitrary update sequence does not depend on the sequence itself, but only on some small parameter (ideally, only the number of nodes in the final graph).

---

[1] Dynamic decentralized networks for sharing data or computing resources.

This goal is very similar to what is achieved by the update algorithms for some randomized data structures such as randomized binary search trees [8,9] or randomized skip lists [10]. An immediate benefit of using such an exact maintenance protocol lies in the analysis: only one probability distribution per network size has to be studied, and the analysis does not depend on some probabilistic model for the update sequence.

In this paper, we give a precise, general definition of what such a distribution preserving protocol should be, and illustrate the notion for a specific model of random graphs: uniform $k$-out graphs (simple digraphs with out-degree $k$); these graphs have good network properties associated with the uniform distribution. We describe and analyze several insertion and deletion algorithms that only work locally in the graph - global knowledge of the whole network is not assumed.

The problem of locally updating $k$-out graphs can be reformulated as follows: given a random uniform $k$-out graph $G$ with $n$ vertices, find a randomized procedure in order to insert (resp. delete) a vertex such that the resulting graph $G'$ is a random uniform $k$-out graph with $n+1$ (resp. $n-1$) vertices. While we do not assume global knowledge of the whole network, our algorithms are assumed to know the *size* of the graph.

Our algorithms make extensive use of random sampling primitives, and need the ability to pick a uniform random vertex. Since we do not want to assume centralized knowledge of the whole graph, we restrict this ability to the use of a special primitive `RandomVertex()`, RV for short; we also use other random generation functions whose output distributions are known in advance and do not depend on the graph.

This RV function is somewhat implicitly assumed in many decentralized protocols in the literature when considering that a new node has to know some *friend* node in order to insert properly in the network, and is often referred to as an *external mechanism*. It is usually not explicitly required that such a friend node be uniformly distributed over the network, which is a strong assumption, but this hypothesis avoids any centralization of the network on a particular subset of the nodes. Similar mechanisms are used in the literature, e.g., hashing a name allows a node to contact a random node in Chord [11]; the protocol in [5] assumes a centralized server that caches a $D$-uniform subset of the nodes ($D$ being a predefined constant); the random walks of [7] allow a new node to pick some uniformly distributed nodes during its insertion to build an almost random regular graph; the tokens used in the protocol of [12] also play the role of sampling uniform nodes in order to preserve the connectivity of the network over time.

Such a procedure is often very costly as it consumes network bandwidth or momentarily breaks the decentralization (in particular in the server case). Hence we shall essentially measure the cost of an update algorithm by the number of times it needs to call the RV primitive during its execution. Our most effective algorithms are optimal in this regard: algorithm INS2 asymptotically uses $k$ calls to RV, and algorithm DEL3 asymptotically uses $o(1)$ such calls (in expectation, for both algorithms).

The expected complexity of the update algorithms, when disregarding the cost of RV, should also be small; ideally it should not depend too much on the network size. This last criterion can be thought of as a safeguard in order to prevent the algorithm from saving calls to RV by exploring the entire graph (which, in the model we consider, is connected with asymptotic probability 1). The algorithms presented here have constant expected time.

The rest of this paper is organized as follows. In the next section, we define some notation, and give definitions for what we call distribution-preserving insertion and deletion algorithms; we also define precisely the random graph model we work with. Section 3 contains our major contributions; we describe and analyze our various distribution-preserving algorithms for insertion and deletion in random $k$-out graphs. This is followed by a conclusion, where we discuss some directions for further research. Due to space constraints, we omit all proofs that our algorithms are distribution-preserving.

## 2 Notation and Models

### 2.1 Notation

All graphs we consider in this paper are simple, loopless directed graphs. For any such graph $G$ and any vertex $u$, we note $N_G^+(u)$ for the outgoing neighborhood of vertex $u$, i.e., the set of vertices $v$ such that $(u,v)$ is an edge of the graph. The *closed* neighborhood, obtained by adding the vertex itself, is noted $N_G^+[u]$. Similarly, we note $N_G^-(u)$ for the incoming neighborhood of $u$: the set of vertices $v$ such that $(v,u)$ is an edge. We write $E|_{V'}$ for the edge set $E$ restricted to the subset $V'$ of the vertex set.

Throughout the paper, $n$ stands for the size of the vertex set $V$.

When $X$ is some random variable and $\rho$ is some probability distribution, we write $X \sim \rho$ to mean that $X$ follows the distribution $\rho$. To ease the notation and the reading, we take the liberty of noting the addition and removal of a singleton in an additive fashion, so that $S + x$ stands for $S \cup \{x\}$ and $S - x$ for $S \setminus \{x\}$.

### 2.2 Distribution Preserving Algorithms

We shall assume that all possible nodes of our graphs belong to some countable set $\Omega$ that we call our underlying *universe*. Such a set might be thought of as the space of IP addresses for computers over the Internet, $\mathbb{N}^2$ for an approximation of some 2D geometric space, and so on. For any finite subset $V$ of $\Omega$, we note $\mathcal{G}_V$ the set of all simple digraphs with vertex set $V$ and $\mathcal{G} = (\mathcal{G}_V)_{V \subset \Omega}$.

We now define what we consider to be valid update algorithms with regards to some family of probability distributions.

**Definition 1.** *Let $\mu = (\mu_V)_{V \subset \Omega}$ be a family of probability distributions such that each $\mu_V$ has support $\mathrm{Supp}(\mu_V) \subseteq \mathcal{G}_V$. A randomized algorithm $\mathcal{A}$ is a $\mu$-preserving insertion algorithm (resp. deletion algorithm) if, for any $V$ and any $u \in \Omega \setminus V$ (resp. $u \in V$), if $G \sim \mu_V$ then we have $\mathcal{A}(G,u) \sim \mu_{V+u}$ (resp. $\mathcal{A}(G,u) \sim \mu_{V-u}$).*

*Remark 1.* An equivalent formulation of the definition would be: $\mathcal{A}$ is a valid insertion algorithm (resp. deletion algorithm) if for any finite subset $V$ of $\Omega$, for any vertex $u$ in $\Omega \setminus V$ (resp. in $V$), and for any graph $g'$ in $\mathcal{G}_{V+u}$ (resp. $\mathcal{G}_{V-u}$), we have:
$$\sum_{g \in \mathcal{G}_V} \mu_V(g) \cdot \mathbb{P}(\mathcal{A}(g, u) = g') = \mu_{V'}(g')$$
with $V' = V + u$ (resp. $V - u$).

It is important to note, in the above definition, that although the input graph is assumed to be random, the vertex to be deleted[2] is not; the equations should hold for *any* deterministic choice of vertex. This is important because we want our results to be valid if this choice of vertices to be inserted or deleted is given to an adversary; still, because this choice of vertex is assumed to be deterministic, it cannot depend on the graph: one cannot, for instance, choose to delete one of the vertices with the highest indegree.

Explicit calculations show that a stronger model in which an adversary would be allowed to "see" the current graph, and then arbitrarily choose a vertex to be deleted, would only accept as a solution an algorithm that built a new graph independent of the previous one - essentially defeating our goal of maintaining dynamic graphs. The algorithms we present in this paper, working in a weaker adversary model, are significantly more efficient.

### 2.3 Random $k$-out Graphs

We now introduce the simple model of directed graphs for which we will describe distribution preserving algorithms.

**Definition 2.** *A $k$-out graph is a simple directed graph with all vertices of out-degree exactly $k$.*

We define $\mathcal{G}^k = (\mathcal{G}^k_V)_{V \subset \Omega, |V| \geq k+1}$, where $\mathcal{G}^k_V$ stands for the set of all $k$-out graphs having $V$ as vertex set, as well as $\nu^k = (\nu^k_V)_{V \subset \Omega, |V| \geq k+1}$, where $\nu^k_V$ is the uniform distribution over $\mathcal{G}^k_V$. Clearly, $\mathcal{G}^k_V$ is empty if $V$ has fewer than $k+1$ elements; accordingly, the behavior of our deletion algorithms is undefined if they are applied to a graph of size exactly $k + 1$.

*Remark 2.* A random graph $G = (V, E) \sim \nu^k_V$ iff for all $v \in V$, $N^+_G(v)$ is uniformly distributed over the $k$-subsets of $V - v$ and the $N^+_G(v)$ are mutually independent. Notice that, for all $u, v \in V$, $N \subset V - v$ with $|N| = k$, $\mathbb{P}(N^+_G(v) = N) = 1/\binom{n-1}{k}$; also, $\mathbb{P}((u, v) \in E) = k/(n-1)$, and for any family $(u_i, v_i)_{1 \leq i \leq \ell}$ of such potential pairs, the corresponding events are mutually independent if and only if all $u_i$ are distinct. Consequently, in a $\nu^k_V$-distributed graph with $n$ vertices, the indegree of any one vertex follows the binomial distribution with parameters $n - 1$ and $k/(n-1)$.

---

[2] Or inserted; but this does not matter with the distribution considered here.

## 2.4 From Centralized to Decentralized Algorithms

We describe all our algorithms as centralized, sequential algorithms. It should be noted that they have some "local" features. The deletion algorithms only need to examine the preexisting graph $G$ up to distance 2 (measured in the underlying undirected graph) from the vertex $u$ to be removed (we need to examine the outgoing edges from predecessors of $u$). Also, the algorithms end up only removing edges that were incident to $u$, and adding new edges to predecessors of $u$. Similarly, our insertion algorithms only need to examine the direct neighborhoods of those vertices returned by calls to RV, or, in the case of algorithm INS2, neighborhoods of vertices along short paths from vertices obtained through RV.

In light of this, it should be clear that our algorithms could easily be transposed into a decentralized, message-passing model where each vertex corresponds to a process that has access to its predecessors and successors. In this, we assume that knowing the "identity" of a vertex is sufficient to allow direct communication with the corresponding process – and that we are somehow given access to a RV primitive. The question of running such algorithms in an unreliable network, or concurrently, is way beyond the scope of our work.

# 3 Distribution Preserving Algorithms for $k$-out Graphs

## 3.1 Basic Random Samplers

We give a precise description in this short section of the various random samplers used by our algorithms. We make a clear distinction between those that do not use RV (called *internal*) and those that do (*external*).

**Internal Procedures.** Since we are concerned about minimizing the number of calls to RV, we allow our algorithms to use an alternate source of randomness. The following standard probability distributions can be used by our algorithms:

- **Bernoulli**($p$) returns 1 with probability $p$ and 0 with probability $1-p$.
- **Binomial**($n, p$) returns $0 \leq j \leq n$ with probability $\binom{n}{j} p^j (1-p)^{n-j}$.
- **Uniform**($S$) returns each element $x \in S$ with probability $1/|S|$.
- **Permute**($S$) returns the elements of the set $S$ in a random uniform order.

We also need once a subroutine RandomVector, not making use of RV, which, when given a random ordered $\ell$-subset $(U_1, \ldots, U_\ell)$ of a set $S$ and $m = |S|$, outputs $\ell$ independent uniform elements $V_1, \ldots, V_\ell$ of $S$, implemented as follows:

$X \leftarrow 0$
for $1 \leq i \leq \ell$ do
   if **Bernoulli**($X/m$) then
      $V_i \leftarrow$ **Uniform**($\{U_1, ..., U_X\}$)
   else
      $X \leftarrow X + 1$
      $V_i \leftarrow U_X$

**External Procedures.** Practically speaking, we would often like to sample an element from $V \setminus S$, where $S$ is typically a *small* and *known* set. Whenever this is needed, we write RVAvoiding$(S)$ – shortened RVA when $S$ is deducible from the context – which is assumed to return a vertex from $V \setminus S$ instead of $V$. This is implemented using RV by calling the primitive until an element outside of $S$ is sampled. The number of calls needed is geometrically distributed with expectation $n/(n - |S|)$, which is close to 1 when $|S| \ll n$.

In the algorithms, we will often only count $\mathcal{R}$, the number of calls to RVA, since if the set to avoid is bounded by some constant then the total expected number of times RV is called is always $\mathbb{E}(\mathcal{R})(1 + \mathcal{O}(1/n))$.

We extend the framework with another function RandomSubset(r) that returns a uniform random $r$-subset of $V$. It is implemented using RVA, starting with $W_0 = \emptyset$ and building iteratively a set $W_i$ of size $i$ avoiding $W_{i-1}$. Its asymptotic expected cost is $r$.

### 3.2 Insertion Algorithms

**Insertion with Cost 2k.** We first introduce a natural insertion algorithm. When inserting a new vertex $u$, we sample a uniform $k$-subset of $V$ for its outgoing neighborhood, and a random variable $L \sim$ Binomial$(n, k/n)$. Finally, a uniform $L$-subset of $V$ is chosen as its incoming neighborhood. We then choose randomly, for each selected predecessor, one of its outgoing edges and *redirect* it towards $u$. While simple, this algorithm preserves the uniform distribution over $k$-out graphs.

---

**Algorithm 1.** INS1

**Input:** a digraph $G = (V, E)$, a vertex $u \notin V$
**Output:** a digraph $G'$
1: $V' \leftarrow V + u;\ n \leftarrow |V|$
2: $S \leftarrow$ **RandomSubset**$(k)$
3: $E' \leftarrow E \cup \{(u, v) \mid v \in S\}$
4: $L \leftarrow$ **Binomial**$(n, k/n)$
5: $W \leftarrow$ **RandomSubset**$(L)$
6: **for all** $v \in W$ **do**
7: $\quad X \leftarrow$ **Uniform**$(N_G^+(v))$
8: $\quad E' \leftarrow E' - (v, X) + (v, u)$
9: $G' \leftarrow (V', E')$

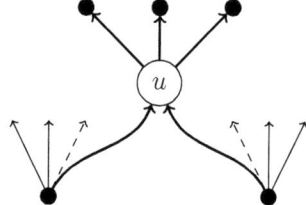

**Fig. 1.** An execution of INS1 with $k = 3$ and $L = 2$

---

**Proposition 1.** INS1 *is a $\nu^k$-preserving insertion algorithm.*

**Proposition 2.** *The asymptotic expected cost of* INS1 *is $2k$.*

*Proof.* We call only RV while processing RandomSubset$(k)$ and RandomSubset$(L)$, and each has an asymptotic expected cost of $k$ (recall $\mathbb{E}(L) = k$). $\square$

**Algorithm 2.** INS2

**Input:** a digraph $G = (V, E)$, a vertex $u \notin V$
**Output:** a digraph $G'$
1: $V' \leftarrow V + u;\ E' \leftarrow E;\ n \leftarrow |V|$
2: $L \leftarrow \mathbf{Binomial}(n, k/n)$
3: $X \leftarrow \mathbf{RandomVertex}()$
4: $W \leftarrow \emptyset$
5: **for** $1 \leq i \leq L$ **do**
6: $\quad W \leftarrow W + X$
7: $\quad D \leftarrow \mathbf{Uniform}(N_G^+(X))$
8: $\quad E' \leftarrow E' - (X, D) + (X, u)$
9: $\quad$ **if** $i < L$ **then**
10: $\quad\quad Y \leftarrow \{v \in N_G^+(X) - D \mid v \notin W\}$
11: $\quad\quad$ **if** $\mathbf{Bernoulli}(|Y|/(n-i)) = 1$ **then**
12: $\quad\quad\quad X \leftarrow \mathbf{Uniform}(Y)$
13: $\quad\quad$ **else**
14: $\quad\quad\quad X \leftarrow D$
15: $\quad\quad\quad$ **if** $X \in W$ **then**
16: $\quad\quad\quad\quad X \leftarrow \mathbf{RVAvoiding}(W \cup N_G^+(X))$
17: $\quad$ **else**
18: $\quad\quad$ **if** $\mathbf{Bernoulli}(k/n) = 1$ **then**
19: $\quad\quad\quad X \leftarrow \mathbf{Uniform}(N_G^+[X] - D)$
20: $\quad\quad$ **else**
21: $\quad\quad\quad X \leftarrow D$
22: $S \leftarrow \mathbf{RandomSubset}(k-1)$
23: **if** $X \in S$ **then**
24: $\quad S \leftarrow S - X + \mathbf{RVAvoiding}(S)$
25: $E' \leftarrow E' + (u, X) \cup \{(u, v) \mid v \in S\}$
26: $G' \leftarrow (V', E')$

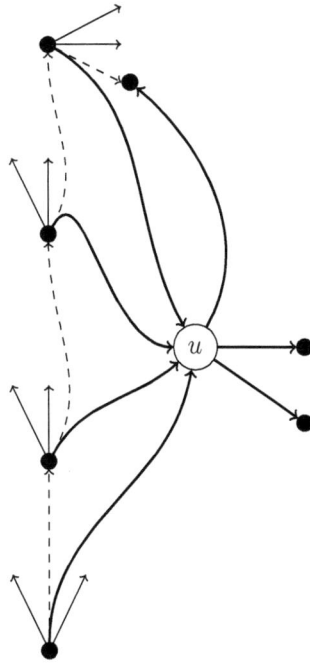

**Fig. 2.** An execution of INS2 with $k = 3$ and $L = 4$

**Insertion with Cost $k$.** We may improve the previous algorithm in order to get an expected cost of $k$, which is optimal in some sense (see Proposition 5). The idea is to reuse the neighbors of the vertices that will become the predecessors of $u$, the new inserted vertex. In the process, we have to take care not to introduce any dependencies between the outgoing neighborhoods.

**Proposition 3.** INS2 *is a $\nu^k$-preserving insertion algorithm.*

**Proposition 4.** *The asymptotic expected cost of* INS2 *is $k$.*

*Proof.* Let $\mathcal{R}$ be the number of times line 16 is processed during the execution of the algorithm. Except from this line, we call RV on line 3, while processing RandomSubset on line 22 and possibly other times if line 24 is executed. Notice this results in an expected cost of $k + \mathbb{E}(\mathcal{R})(1 + \mathcal{O}(1/n))$.

For $i \in \mathbb{N}$, let $B_i$ be a Bernoulli random variable which is 1 if line 16 is executed during the $i$th run through the loop, and 0 otherwise (or if the loop has fewer than $i$ runs); thus, $B_i = 1$ with probability bounded by $\frac{i}{n-(k-1)}\mathbb{P}(L \geq i)$. We have $\mathcal{R} = \sum_i B_i$; taking expectations,

$$\mathbb{E}(\mathcal{R}) \leq \sum_{i=1}^n \frac{i}{n-(k-1)}\mathbb{P}(L \geq i) = \frac{1}{n-(k-1)}\sum_{i=1}^n i\mathbb{P}(L \geq i).$$

Simple sum manipulations show that

$$\sum_{i=1}^n i\mathbb{P}(L \geq i) = \sum_{j=1}^n \frac{j(j+1)}{2}\mathbb{P}(L = j) = \mathbb{E}\left(\frac{L^2 + L}{2}\right).$$

Since $L$ follows a binomial distribution with expectation $k$ (hence variance less than $k$), this is bounded by $k^2/2 + k = \mathcal{O}(1)$. Thus $\mathbb{E}(\mathcal{R}) = \mathcal{O}(1/n)$. □

We conclude this section by arguing that algorithm INS2 is, in some sense, optimal.

The number of possible $k$-out graphs on a vertex set $V$ of size $n$ is $|\mathcal{G}_V^k| = \binom{n-1}{k}^n$. Taking ratios shows that there are $r_n = \binom{n}{k}(n/(n-k))^n$ times more such graphs with one more vertex. Thus, when given a uniform $k$-out graph on $n$ vertices and asked to produce a uniform $k$-out graph on $n+1$ vertices, we need at least $\log_2(r_n) = k\log_2(n) + \mathcal{O}(1)$ additional bits of information. Each call to RV gives $\log_2(n)$ bits of information, since its result is uniform on a set of size $n$.

If RV were the only source of randomness used by our algorithms, then this would prove that no algorithm can use, in expectation, fewer than $k$ such calls; but this is not the case. In fact, a single call to RV is enough to learn with high probability the whole vertex set by just exploring the connected component of a vertex, in the underlying undirected graph (which is connected with asymptotic probability 1). Subsequent calls can then be simulated once $V$ is known. This would be at the cost of a much higher time complexity, though; our algorithm, in contrast, has constant expected time complexity if one assumes the calls to random samplers take constant time. Such an assumption is valid if one uses optimal samplers as described in [13].

**Proposition 5.** *For any $\nu^k$-preserving algorithm $\mathcal{A}$ that has bounded expected time complexity and only makes calls in addition to RV to random samplers for distributions of bounded entropy, the expected number of calls to RV is, asymptotically, at least $k$.*

*Proof.* The above discussion shows that fewer than $k$ expected calls to RV, together with a $\nu^k$-distributed graph of size $n$, and a bounded number of random numbers from distributions with bounded entropy, provide strictly less binary information than a $\nu^k$-distributed graph of size $n+1$. □

Note that our algorithm INS2 does respect the conditions of the above proposition: other than calls to RV, the only additional source of randomness is in the form of uniform variables on sets of size at most $k$ (thus entropy at most $\log_2(k)$), Bernoulli variables (with entropy at most 1), and a single binomial variable with

## Algorithm 3. DEL1

**Input:** a digraph $G = (V, E)$, a vertex $u \in V$
**Output:** a digraph $G'$
1: $V' \leftarrow V - u$
2: $E' \leftarrow E|_{V'}$
3: **for all** $v \in N_G^-(u)$ **do**
4:      $X \leftarrow \mathbf{RVAvoiding}(N_G^+[v])$
5:      $E' \leftarrow E' + (v, X)$
6: $G' \leftarrow (V', E')$

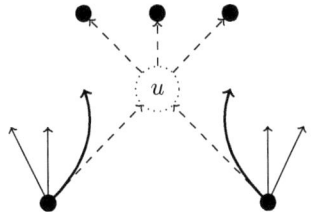

**Fig. 3.** An execution of DEL1 with $k = 3$ and $L = 2$

parameters $n$ and $k/n$, whose entropy is close to that of the limiting Poisson distribution with expectation $k$. Thus, our algorithm is, at least in this class of algorithms, optimal.

### 3.3 Deletion Algorithms

**Simple Deletion.** We first introduce the following simple deletion algorithm : if $u$ is leaving the network, this algorithm replaces any edge $(v, u)$ pointing to $u$ by an edge $(v, x)$ where $x$ is chosen uniformly at random in the new vertex set, avoiding incompatible choices for $x$.

**Proposition 6.** DEL1 *is a $\nu^k$-preserving deletion algorithm.*

**Proposition 7.** *The asymptotic expected cost of* DEL1 *is $k$.*

*Proof.* In $G$, $u$ has, in expectation, $k$ predecessors. For each of them we make a call to RVA, resulting, in expectation, in $1 + \mathcal{O}(1/n)$ calls to RV. □

**Improved Deletion.** We now describe a better deletion algorithm. The idea is to make use of the successors of the deleted vertex $u$ as suggestions in order to save calls to RV for some of its predecessors; recall that $N_G^+(u)$ is a uniform random $k$-subset of $V - u$, independent of the other outgoing neighborhoods.

RandomVector is used to correct the dependencies between the successors of $u$. For the $k$ first predecessors of $u$, we only call RVA if these suggestions cannot be accepted. In the algorithm description, we use the function Permute in order to guarantee that $N_G^+(u)$ is indeed in a random uniform order.

**Proposition 8.** DEL2 *is a $\nu^k$-preserving deletion algorithm.*

**Proposition 9.** *The asymptotic expected cost of* DEL2 *is $e^{-k}\frac{k^k}{(k-1)!}$.*

*Proof.* For $i \in \mathbb{N}$, let $B_i$ be a Bernoulli random variable which is 1 if line 10 is executed during the $i$th run through the loop, 0 otherwise (in particular, if there is no such round). Notice $B_i = 1$ with probability $\frac{k-1}{n-1}\mathbb{P}(L \geq i)$ for $i \leq k$ and with probability $\mathbb{P}(L \geq i)$ if $i > k$.

**Algorithm 4.** DEL2

**Input:** a digraph $G = (V, E)$, a vertex $u \in V$
**Output:** a digraph $G'$
1: $V' \leftarrow V - u;\ n \leftarrow |V|$
2: $E' \leftarrow E|_{V'}$
3: $u_1, \ldots, u_L \leftarrow N_G^-(u)$
4: $v_1, \ldots, v_k \leftarrow \mathbf{Permute}(N_G^+(u))$
5: $V_1, \ldots, V_k \leftarrow \mathbf{RandomVector}(v_1, \ldots, v_k, n-1)$
6: **for** $1 \leq i \leq L$ **do**
7:    **if** $i \leq k$ and $V_i \notin N_G^+[u_i]$ **then**
8:       $X \leftarrow V_i$
9:    **else**
10:       $X \leftarrow \mathbf{RVAvoiding}(N_G^+[u_i])$
11:    $E' \leftarrow E' + (u_i, X)$
12: $G' \leftarrow (V', E')$

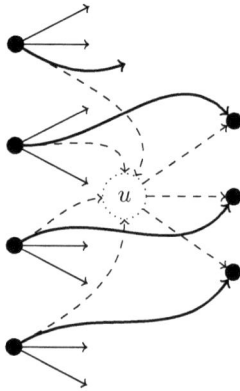

**Fig. 4.** An execution of DEL2 with $k = 3$ and $L = 4$

Let $\mathcal{R} = \sum_{i \geq 1} B_i$ and notice the expected cost of DEL2 is simply $\mathbb{E}(\mathcal{R})(1 + \mathcal{O}(1/n))$, where

$$\mathbb{E}(\mathcal{R}) = \sum_{i=1}^{n-1} \mathbb{E}(B_i) = \sum_{i=1}^{n-1} \left( \mathbb{1}_{i \leq k} \frac{k-1}{n-1} \mathbb{P}(L \geq i) + \mathbb{1}_{i > k} \mathbb{P}(L \geq i) \right)$$

$$= \frac{k-1}{n-1} \sum_{i=1}^{k} \mathbb{P}(L \geq i) + \sum_{i=k+1}^{n-1} \mathbb{P}(L \geq i).$$

The first term is $\mathcal{O}(1/n)$ since $\sum_{i=1}^{k} \mathbb{P}(L \geq i) \leq k$. Standard sum and binomial probabilities manipulations show that the second term may be rewritten as $k \cdot \mathbb{P}(L = k) + \mathcal{O}(1/n)$.

The result follows directly since $\mathbb{P}(L = k) = e^{-k} \frac{k^k}{k!} + \mathcal{O}(1/n)$. □

*Remark 3.* The Stirling approximation formula implies that the asymptotic expected cost is bounded by, and close to, $\sqrt{\frac{k}{2\pi}}$.

**Optimal Deletion.** In our previous algorithm, we improved on the natural deletion algorithm by recycling the successors of the deleted vertex $u$ instead of "wasting" this information. Our last deletion algorithm, which has expected cost $o(1)$, uses $u$'s predecessors instead of successors as suggestions to save calls to RV, and only one successor of $u$. Surprisingly, this can be done while preserving independence.

**Proposition 10.** DEL3 *is a $\nu^k$-preserving deletion algorithm.*

**Proposition 11.** *The expected cost of* DEL3 *is $o(1)$.*

## Algorithm 5. DEL3

**Input:** a digraph $G = (V, E)$, a vertex $u \in V$
**Output:** a digraph $G'$
1: $V' \leftarrow V - u$; $n \leftarrow |V|$; $E' \leftarrow E|_{V'}$
2: $u_1, ..., u_L \leftarrow$ **Permute**$(N_G^-(u))$
3: **for** $1 \leq i \leq L$ **do**
4:    **if** $i = L$ **then**
5:       $X \leftarrow$ **Uniform**$(N_G^+(u))$
6:       **if** $X \in N_G^+[u_L]$ **then**
7:          $X \leftarrow$ **RVAvoiding**$(N_G^+[u_L])$
8:    **else**
9:       $C \leftarrow \{u_j \mid j < i \text{ and } u_j \notin N_G^+(u_i)\}$
10:      **if** **Bernoulli**$(|C|/(n-1-k))$ **then**
11:         $X \leftarrow$ **Uniform**$(C)$
12:      **else if** $(u_i, u_{i+1}) \notin E'$ **then**
13:         $X \leftarrow u_{i+1}$
14:      **else**
15:         $X \leftarrow$ **RVAvoiding**$(N_G^+[u_i] \cup C)$
16:    $E' \leftarrow E' + (u_i, X)$
17: $G' \leftarrow (V', E')$

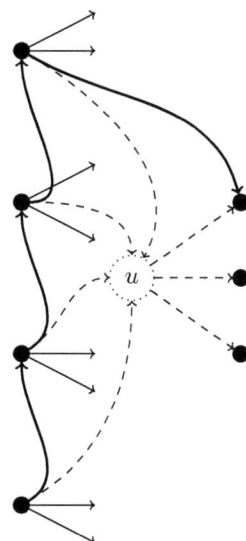

**Fig. 5.** An execution of DEL3 with $k = 3$ and $L = 4$

*Proof.* We condition on the value $\ell$ of $L$, the number of predecessors of $u$ in $G$. Notice that the probability that $L$ is larger than, say, $n/2 - k$, is exponentially small, and, conditionally on such a large value of $L$, the expected number of calls to RV from all calls to RVA is still at most $n^2$ (at most $n$ calls to RVA, each calling RV at most $n$ times in expectation); thus, the contribution to the expected cost of such abnormally large values of $L$ is exponentially small.

We now assume that $L < n/2 - k$. There are $L$ possible calls to RVA. For $i < L$, the $i$th possible call does not occur if $u_{i+1} \notin N_G^+(u_i)$, i.e., the probability that it occurs is less than $(k-1)/(n-2)$. Since the forbidden set is always of size at most $\ell + k$, the $i$th potential call contributes at most $\frac{k-1}{n-2} \cdot \frac{n}{n-\ell-k} \leq 2\frac{k-1}{n-2}$ to the expected cost.

Similarly, the $L$th possible call occurs with probability $k/(n-1)$ and has a forbidden set of size $k+1$, resulting in an expected number of calls to RV equal to $\frac{k}{n-1} \cdot \frac{n}{n-k-1} = \mathcal{O}(1/n)$.

Thus, taking expectations for $L$ – and keeping in mind that $\mathbb{E}(L|L < n/2) \leq \mathbb{E}(L) = k$, the total expected number of calls to RV is bounded above by

$$2\frac{(k-1)^2}{n-2} + \frac{nk}{(n-1)(n-k-1)},$$

which, for any fixed $k$, is $\mathcal{O}(1/n)$. □

## 4 Conclusion and Future Research

We have defined a notion of distribution preservation for update algorithms on dynamic random graphs, and have illustrated it by giving examples of such distribution preserving insertion and deletion algorithms for uniform $k$-out graphs. Our algorithms are naturally suited to implementation in a distributed setting, once the problem of getting a random vertex in the graph is solved.

A natural direction for further research is to devise similar algorithms for other, more complex distributions on random graphs. Obtaining algorithms with similar properties for uniform undirected $k$-regular graphs (for even $k$, since for odd $k$ such regular graphs do not exist for odd size) would be quite a feat, since known algorithms for sampling from this distribution tend to have poor time complexity for even moderate $k$. In the present work, the fact that the graphs are directed introduces a lot of independence that our algorithms rely upon.

Our $k$-out graphs are "simple" in that renaming the vertices in any (deterministic) way leaves the $\nu^k$ distribution unchanged. It would be interesting and challenging to apply the same ideas to random graph models where the "identities" of vertices actually have some influence over the distribution of graphs, such as any graph model that relies on vertices being placed in some geometric space.

# References

1. Frankel, J., Pepper, T.: The gnutella project, http://www.gnutelliums.com/
2. Bennett, K., Stef, T., Grothoff, C., Horozov, T., Patrascu, I.: The gnet whitepaper. Technical report, Purdue University (2002)
3. Clarke, I., Sandberg, O., Wiley, B., Hong, T.W.: Freenet: A distributed anonymous information storage and retrieval system. In: Federrath, H. (ed.) Designing Privacy Enhancing Technologies. LNCS, vol. 2009, pp. 46–66. Springer, Heidelberg (2001)
4. Liben-nowell, D., Balakrishnan, H., Karger, D.: Analysis of the evolution of peer-to-peer systems. In: ACM Symposium on Principles of Distributed Computing, pp. 233–242 (2002)
5. Pandurangan, G., Raghavan, P., Upfal, E.: Building low-diameter p2p networks. In: FOCS, pp. 492–499 (2001)
6. Cooper, C., Dyer, M., Greenhill, C.: Sampling regular graphs and a peer-to-peer network. Comb. Probab. Comput. 16(4), 557–593 (2007)
7. Bourassa, V., Holt, F.B.: Swan: Small-world wide area networks. In: International Conference on Advances in Infracstructure, SSGRR-2003s (2003)
8. Seidel, R., Aragon, C.R.: Randomized search trees. Algorithmica 16(4/5), 464–497 (1996)
9. Martínez, C., Roura, S.: Randomized binary search trees. J. ACM 45(2), 288–323 (1998)
10. Pugh, W.: Skip lists: A probabilistic alternative to balanced trees. Commun. ACM 33(6), 668–676 (1990)
11. Stoica, I., Morris, R., Karger, D., Kaashoek, M.F., Balakrishnan, H.: Chord: A scalable peer-to-peer lookup service for internet applications. SIGCOMM Comput. Commun. Rev. 31(4), 149–160 (2001)
12. Cooper, C., Klasing, R., Radzik, T.: A randomized algorithm for the joining protocol in dynamic distributed networks. Theor. Comput. Sci. 406(3), 248–262 (2008)
13. Knuth, D.E., Yao, A.C.: The complexity of nonuniform random number generation. In: Proceedings of Symposium on Algorithms and Complexity, pp. 357–428. Academic Press, New York (1976)

# Odd Graphs Are Prism-Hamiltonian and Have a Long Cycle[*]

Felipe De Campos Mesquita, Letícia Rodrigues Bueno,
and Rodrigo De Alencar Hausen

Universidade Federal do ABC (UFABC)
Santo André, Brazil
felipe.mesquita@aluno.ufabc.edu.br
leticia.bueno@ufabc.edu.br
hausen@compscinet.org

**Abstract.** The odd graph $O_k$ is the graph whose vertices are all subsets with $k$ elements of a set $\{1,\ldots,2k+1\}$, and two vertices are joined by an edge if the corresponding pair of $k$-subsets is disjoint. A conjecture due to Biggs claims that $O_k$ is hamiltonian for $k \geq 3$ and a conjecture due to Lovász implies that $O_k$ has a hamiltonian path for $k \geq 1$. In this paper, we show that the prism over $O_k$ is hamiltonian and that $O_k$ has a cycle with $.625|V(O_k)|$ vertices at least.

**Keywords:** hamiltonian cycle, prism over a graph, odd graph.

## 1 Introduction

Lovász [14] conjectured that every connected undirected vertex-transitive graph has a hamiltonian path. The odd graphs form a well-studied family of connected vertex-transitive graphs. Later, Biggs [2] hypothesized that the odd graphs are hamiltonian for all $k > 2$. Still, a related conjecture by Havel [7] claims that the bipartite double graph of the odd graph is hamiltonian.

The vertices of the *Kneser graph* $K(n,k)$ are the $k$-subsets of $\{1,2,\ldots,n\}$ and two vertices are adjacent if the corresponding $k$-subsets are disjoint. For $n = 2k+1$, the Kneser graph $K(2k+1,k)$ is called the *odd graph* and it is denoted by $O_k$ (Figures 1a and 1b).

The bipartite double graph of the Kneser graph $K(n,k)$ is known as the *bipartite Kneser graph* $B(n,k)$. The vertices are the $k$-subsets and $(n-k)$-subsets of $\{1,2,\ldots,n\}$ and the edges represent the inclusion between two such subsets. The vertex set of $B(n,k)$ can be seen as two (symmetric) layers of the $n$-dimensional cube. For $n = 2k+1$, the graph $B(2k+1,k)$ is called the *middle-layers graph* and it is denoted by $B_k$ (Figure 1(c)).

To date, Biggs' conjecture has been verified for $3 \leq k \leq 13$ [17] and there is a hamiltonian path in $O_k$ for $k \leq 19$ [3,19]. Havel's conjecture is true for $k \leq 19$ [18,19]. Also, apart from the Petersen graph, $K(n,k)$ and $B(n,k)$ are hamiltonian if $n \geq (3k+1+\sqrt{5k^2-2k+1})/2$, as reported in [5].

---

[*] Research partially supported by Brazilian agency CNPq.

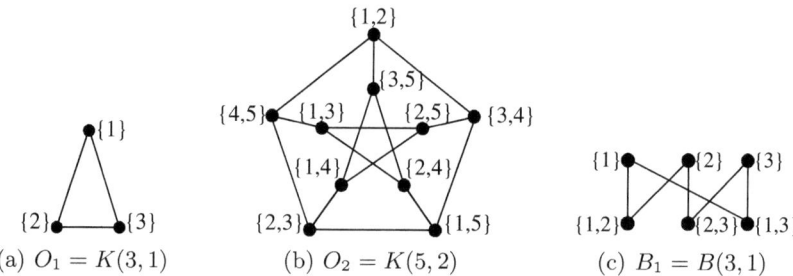

**Fig. 1.** The odd graph $O_k$ for $k = 1, 2$ and the middle-levels graph $B_k$ for $k = 1$

Since the decision problem of the hamiltonian cycle problem is NP-Complete [13], one recent trend is to search for long cycles. Johnson [11] provided a lower bound showing that $O_k$ and $B_k$ contain a cycle of length $(1 - o(1))|V(O_k)|$ and $(1 - o(1))|V(B_k)|$, respectively, where the error term $o(1)$ is of the form $\frac{c}{\sqrt{k}}$ for some constant $c$. The author does not estimate $c$ but, in other words, this means that the graphs $O_k$ and $B_k$ are asymptotically hamiltonian because, as $k$ increases, the length of the cycle increases as well. Savage and Winkler [16] showed that if $B_k$ has a hamiltonian cycle for $k \leq h$, then $B_k$ has a cycle containing a fraction $1 - \varepsilon$ of the graph vertices for all $k$, where $\varepsilon$ is a function of $h$. For example, since $B_k$ has a hamiltonian cycle for $1 \leq k \leq 19$, for $k \geq 20$ the graph $B_k$ has a cycle containing at least 87.46% of the graph vertices. For the graph $O_k$, the best lower bound currently known on the length of the longest cycle is $\sqrt{3|V(O_k)|}$, given by Babai [1] for vertex-transitive graphs in general, which is less than 3% for $O_{10}$, and asymptotically approaches zero as $k$ increases.

Another trend is to search for related structures. In this aspect, having a hamiltonian prism in a graph has been shown to be an interesting relaxation of being hamiltonian. A closed spanning walk where each vertex is traversed at most $q$ times is a $q$-*walk* and a spanning tree of maximum degree $q$ is a $q$-*tree*. A hamiltonian cycle is then a 1-walk, and a hamiltonian path is a 2-tree. It is proven [9] that a graph with a $q$-tree has a $q$-walk, and that a $q$-walk implies the existence of a $(q + 1)$-tree, resulting that

1-walk (hamiltonian cycle) $\Longrightarrow$ 2-tree (hamiltonian path) $\Longrightarrow$ 2-walk $\Longrightarrow$ 3-tree $\Longrightarrow$ 3-walk $\Longrightarrow$ ...

Recently, this hierarchy of graphs was improved by [12]. The *prism over a graph* $G$ is the Cartesian product $G \square K_2$ of $G$ with the complete graph on two vertices (Figure 2). If the prism over $G$ is hamiltonian, we say $G$ is *prism-hamiltonian*. It was shown in [12] that the property of having a hamiltonian prism is "sandwiched" between the existence of a 2-tree and the existence of a 2-walk which results in

2-tree $\Longrightarrow$ hamiltonian prism $\Longrightarrow$ 2-walk.

This means that graphs having a hamiltonian prism are "close" to being hamiltonian, even closer than graphs having a 2-walk. Previously, it was established that the prism over $B_k$ is hamiltonian [8]. Later, the counterpart of this result has been proven for $O_k$ only for $k$ even [4]. In this paper, we demonstrate that the prism over $O_k$ is hamiltonian for all $k$. Moreover, we improve the lower bound on the length of the longest cycle of $O_k$ by providing a cycle with $.625|V(O_k)|$ vertices at least.

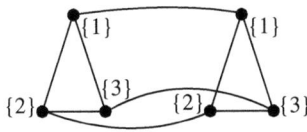

**Fig. 2.** The prism over the graph $O_1$

In Section 2, we give some auxiliary proofs to prove our main results in Section 3. In Section 4 we provide an alternative proof for $B_k$ to be prism-hamiltonian, using the proof presented in Section 3 and, therefore, relating the prism-hamiltonicity of $O_k$ and $B_k$.

## 2 Preliminaries

In this section, we introduce some notations, notions and results from [6,8,10,20] which will be frequently used throughout this paper. We need two definitions from [20]: a *spanning cactus* in a graph $G$ is a spanning connected subgraph $H$ of maximum degree 3 consisting of vertex-disjoint cycles and vertex-disjoint paths such that, if every cycle is replaced by a single vertex connected to the paths incident with it, the resulting graph will be a tree. The cactus is said to be *even* if all cycles are even.

The main idea of our results is to show that $O_k$ contains a spanning even cactus consisting of an even cycle and paths of size 1 and/or 2. Additionally, we show that such an even cycle in $O_k$ has $.625|V(O_k)|$ vertices at least. We use a tool provided in [20]:

**Proposition 1 ([20]).** *If $G$ contains a spanning even cactus, then it is prism-hamiltonian.*

There is a correspondence between the $k$-subsets and the $(n-k)$-subsets of $\{1, 2, \ldots, n = 2k+1\}$ with a set of binary strings of $n$ bits with exactly $k$ 1's and $(n-k)$ 0's. The correspondence $b_n b_{n-1} \ldots b_1 \rightarrow \{i | b_i = 1\}$ is a bijection of binary strings of $n$ bits into the subsets of $n$. The *complement* $\bar{x}$ of a binary digit $x$ is 1 if $x = 0$ and 0 if $x = 1$. The complement of a binary string extends this definition by bitwise complement. Throughout this paper we will consider the vertices of $O_k$ and $B_k$ represented by binary strings, except where noted otherwise.

We will now develop some useful tools that will be used to prove our main results in Section 3.

**Definition 1.** *Let $C = (v_1, v_2, v_3, v_4, \ldots, v_q)$ be a cycle in $O_{k-1}$. We define the sequences $C_1$ and $C_2$ as follows:*

*(i) If $q$ is even, $C_1 = (0v_11, 1v_20, 0v_31, 1v_40, \ldots, 1v_q0)$ and $C_2 = (1v_10, 0v_21, 1v_30, 0v_41, \ldots, 0v_q1)$;*
*(ii) If $q$ is odd, $C_1 = (0v_11, 1v_20, 0v_31, 1v_40, \ldots, 0v_q1)$ and $C_2 = (1v_10, 0v_21, 1v_30, 0v_41, \ldots, 1v_q0)$.*

Denote by $\overleftarrow{Q}$, a path $Q$ traversed from the last to the first vertex. Given two disjoint paths $Q_1$ and $Q_2$ such that the last vertex of $Q_1$ is adjacent to the first vertex of $Q_2$, we denote by $Q_1 \circ Q_2$ the path obtained by first traversing the vertices of $Q_1$, and then the vertices of $Q_2$.

**Lemma 1.** *Let $C$ be a cycle with $q$ vertices in $O_{k-1}$.*

*(i) If $q$ is odd then there is a cycle with $2q$ vertices in $O_k$;*
*(ii) If $q$ is even then there are two disjoint cycles each with $q$ vertices in $O_k$.*

*Proof.* By adding 1 and 0 to a vertex of $C$ we have, by definition, a vertex of $O_k$. Therefore, we construct the paths $C_1$ and $C_2$ in $O_k$ according to Definition 1. Notice, by construction, that $C_1$ and $C_2$ are paths in $O_k$ and, if $q$ is even, both are cycles as well. If $q$ is odd, since $C$ is a cycle, then there are edges $\{0v_11, 1v_q0\}$ and $\{1v_10, 0v_q1\}$ in $E(O_k)$, which implies $C_1 \circ C_2$ is a cycle in $O_k$. □

**Definition 2.** *Let $S_k$, $T_k$ and $R_k$ be three disjoint subsets of the vertices of $O_k$ such that:*

*(i) $S_k$ is the set of $k$-subsets which have element 1 or $n$, but not both;*
*(ii) $T_k$ is the set of $k$-subsets which neither has element 1 nor element $n$;*
*(iii) $R_k$ is the set of $k$-subsets which have both elements 1 and $n$.*

Notice that $V(O_k) = S_k \cup T_k \cup R_k$.

**Lemma 2.** *Each vertex $v \in T_k$ has exactly two edges to vertices of $S_k$ and $(k-1)$ edges to vertices of $R_k$.*

*Proof.* Since a vertex of $T_k$ does not contain element 1 and $n$, there are $(2k-1)$ elements to choose $k$. Therefore, there are no edges between vertices of $T_k$. However, we can choose a $k$-subset $u$ and $(k-1)$-subset $v$ in $(2k-1)$ elements. Notice that $u \in T_k$ is adjacent to $v \cup \{1\} \in S_k$ and to $v \cup \{n\} \in S_k$. □

By definition, the vertices of $S_k$ have no edges to vertices of $R_k$ and, by Lemma 2, they have two edges to vertices of $T_k$ and $(k-1)$ edges to vertices of $S_k$. The vertices of $R_k$ are adjacent only to vertices of $T_k$, therefore $O_k$ has a bipartite subgraph with bipartition $(T_k, R_k)$ such that the partition $T_k$ has degree $(k-1)$ and the partition $R_k$ has degree $(k+1)$.

**Lemma 3.** It holds that $|T_k| > |R_k|$.

*Proof.* Since $\binom{2k-1}{k} = \binom{2k-1}{k-1}$, and this value is strictly greater than any other value in the Pascal triangle for combinations of $(2k-1)$ elements, we have that $|T_k| = \binom{2k-1}{k} > \binom{2k-1}{k-2} = |R_k|$. □

**Lemma 4.** If $C_h = (v_1, v_2, v_3, v_4, \ldots, v_q)$ is a hamiltonian cycle in $O_{k-1}$ and $C_1$ and $C_2$ are constructed according to Definition 1, then

(i) $S_k = V(C_1) \cup V(C_2)$;
(ii) $|S_k| = |C_1| + |C_2| = 2|V(O_{k-1})| > 0.5|V(O_k)|$.

*Proof.* First, we show that $S_k = V(C_1) \cup V(C_2)$ and $V(C_1) \cap V(C_2) = \emptyset$. Without loss of generality, let $v \in S_k$ be such that $v = 0v'1$ for $v' \in V(O_{k-1})$. Then $v'$ has to contain $(k-1)$ 1's, that can be placed in $\binom{2k-1}{k-1}$ possible positions in the string that represents $v'$. Those combinations correspond exactly to the vertices of $O_{k-1}$. The proof is analogous for $v = 1v'0$. It follows that, by construction of $C_1$ and $C_2$, and because $C_h$ is a hamiltonian cycle, each vertex of $S_k$ is either in $C_1$ or in $C_2$. Next, we show that $|C_1| + |C_2| > 0.5|V(O_k)|$:

$$2|V(O_{k-1})| > 0.5|V(O_k)|$$

$$|V(O_k)| < 4|V(O_{k-1})|$$

$$\frac{|V(O_k)|}{|V(O_{k-1})|} < 4 \quad (I)$$

We have $|V(O_{k-1})| = \binom{2k-1}{k-1} = \frac{(2k-1)!}{k!(k-1)!}$ and

$$|V(O_k)| = \binom{2k+1}{k} = \frac{(2k+1)!}{(k+1)!k!} = \frac{(2k+1)(2k)(2k-1)!}{(k+1)(k)(k-1)!k!}.$$

It follows that $|V(O_k)| = \frac{(2k+1)(2k)}{(k+1)(k)}|V(O_{k-1})|$ and, therefore,

$$\frac{|V(O_k)|}{|V(O_{k-1})|} = \frac{(2k+1)(2k)}{(k+1)(k)} = \frac{(4k+2)}{(k+1)} \quad (II)$$

Finally, $\lim_{k\to\infty} \frac{|V(O_k)|}{|V(O_{k-1})|} = \lim_{k\to\infty} \frac{(4k+2)}{(k+1)} = \lim_{k\to\infty} (4 - \frac{2}{k+1}) = 4$.

From (I) and (II): $\frac{(4k+2)}{(k+1)} \leq 4$

$$\frac{4(k+1) - 2}{k+1} = 4 - \frac{2}{k+1} \leq 4$$

$$\frac{-2}{(k+1)} \leq 0.$$

For $k > 0$, we have $2|V(O_{k-1})| > 0.5|V(O_k)|$. □

**Lemma 5.** It holds that $|T_k| = |V(O_{k-1})|$.

*Proof.* Since $|T_k| = \binom{2k-1}{k} = \binom{2k-1}{(2k-1)-k}$ and $\binom{2k-1}{k} = \binom{2k-1}{k-1}$, we have that $|T_k| = |V(O_{k-1})|$. □

**Theorem 1.** *If there exists a hamiltonian cycle (respectively, path) $C_h = (v_1, v_2, \ldots, v_q)$ in $O_{k-1}$, then $O_k$ has a cycle (respectively, path) $C'$ such that $|C'| > 0.75|V(O_k)|$.*

*Proof.* Construct $C_1$ and $C_2$ according to Definition 1. First, assume that $C_h$ is a hamiltonian cycle. Notice that there are $q$ vertices $0\overline{v_j}0$ connecting $0v_j1$ to $1v_j0$, where $0v_j1 \in S_k$, $1v_j0 \in S_k$ and $v_j \in C_h$ for $1 \leq j \leq q$, since the complement $\overline{v_j}$ of a vertex $v_j \in C_h$ has $k$ 1's and $(k-1)$ 0's. Therefore, $0\overline{v_j}0 \in V(O_k)$. Construct $q$ paths with 3 vertices by combining the vertices of $C_1$, $C_2$ and $T_k$:

$$Q_1 = 0v_11, 0\overline{v_1}0, 1v_10$$

$$Q_2 = 1v_20, 0\overline{v_2}0, 0v_21$$

$$\vdots$$

$$Q_q = 1v_q0, 0\overline{v_q}0, 0v_q1, \text{ if } q \text{ is even or}$$

$$Q_q = 0v_q1, 0\overline{v_q}0, 1v_q0, \text{ if } q \text{ is odd.}$$

Notice that, for $Q_j$, $1 \leq j \leq q$, the first vertex of $Q_j$ is in $C_1$, the second one is in $T_k$ and the third one is in $C_2$. The $q$ vertices $0\overline{v_j}0$, for $v_j \in C_h$, are distinct, because $C_h$ is a hamiltonian cycle in $O_{k-1}$ and, therefore, the complement of the vertices of $C_h$ are distinct as well and, by Lemmas 5, consist of all vertices of $T_k$.

Concatenate the $q$ paths $Q_j$, $1 \leq j \leq q$, as follows:
$C' = Q_1 \circ \overleftarrow{Q_2} \circ Q_3 \circ \overleftarrow{Q_4} \circ \ldots \circ \overleftarrow{Q_q}$, if $q$ is even and
$C' = Q_1 \circ \overleftarrow{Q_2} \circ Q_3 \circ \overleftarrow{Q_4} \circ \ldots \circ Q_q$, if $q$ is odd.

Figure 3 illustrates the construction of $C'$ for $q$ even and $q$ odd. Notice that, in both cases, the last vertex of $C'$ is $1v_q0$, which is adjacent to the first vertex of $C'$: $0v_11$. Therefore, $C'$ is a cycle.

Assume now that $C_h$ is a hamiltonian path. Then $C'$ is a path as well, because the vertex $v_1$ is not adjacent to $v_q$ in $C_h$. By Lemmas 4 and 5, $|C'| = |C_1| + |C_2| + |T_k| = 3|V(O_{k-1})| > 0.75|V(O_k)|$. □

Since $O_k$ is hamiltonian for $3 \leq k \leq 13$ [17] and has a hamiltonian path for $k \leq 19$ [3,19], Theorem 1 give a cycle in $O_{14}$ and a path in $O_{20}$, both with at least 75% of the vertices of the graph.

Lemmas 4, 5 and Theorem 1 imply that $C'$ has all vertices of $S_k$ and $T_k$. Therefore, to make $C'$ a hamiltonian cycle or path of $O_k$, it remains to add the vertices of $R_k$ to $C'$.

## 2.1 Modular Matchings

*Modular matchings* were proposed in [6] for the middle-levels graph $B_k$. Let $A$ be a $k$-subset of $\{1, \ldots, 2k+1\}$. In a matching $m_i$, for $i = 1, \ldots, k+1$, we have that $A$ is adjacent to the set $A \cup \{\bar{a}_j\}$, where

$$j \equiv i + \sum_{a \in A} a \pmod{k+1},$$

and $\bar{a}_j$ is the $j$-th largest element in $\bar{A}$.

When working with elements of $\{1, \ldots, 2k+1\}$, we will assume a cyclical order for them, in which 1 is the successor of $2k+1$, and that every operation is done modulo $k+1$, with remainders $\{1, 2, \ldots, k+1\}$. We will use a lemma from [6], as formulated in [8].

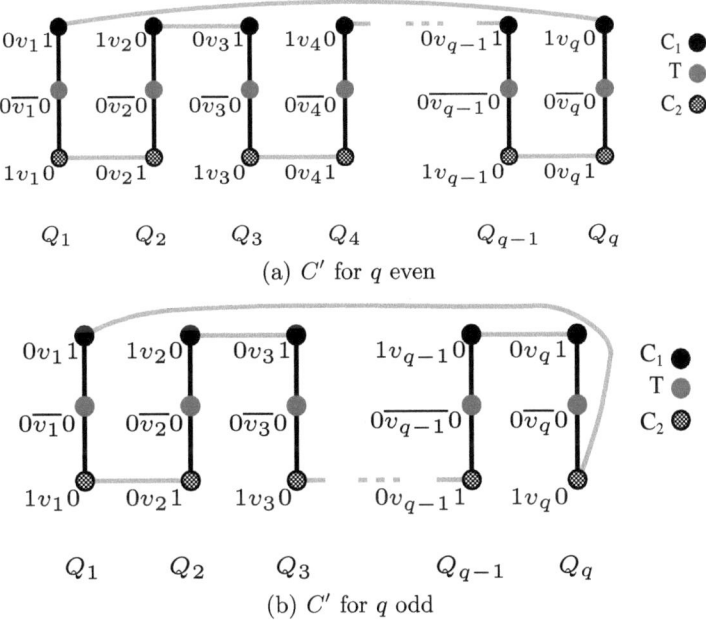

**Fig. 3.** Theorem 1: the $q$ paths with 3 vertices (bold edges) are concatenated by the edges of $C_1$ and $C_2$ (gray edges)

**Lemma 6 ([6,8]).** *For $i = 1, \ldots, k+1$, $m_i$ is a matching in $B_k$ and the set $\{m_1, \ldots, m_{k+1}\}$ is a 1-factorization of $B_k$.*

A modular matching $m_i$ in $B_k$ can be projected onto the graph $O_k$ by replacing each $(k+1)$-set $A$ by its complement $\bar{A} = \{1, \ldots, 2k+1\} \setminus A$. Notice that $\bar{A}$ is a $k$-set and, therefore, a vertex of $O_k$. Given a modular matching $m_i$ in $B_k$, denote by $\rho(m_i)$, the projection of $m_i$ in $O_k$ where, for every vertex $A$ in $V(B_k)$, $\rho(A) = A$ if $A$ is a $k$-subset and $\rho(A) = \overline{A}$ if $A$ is a $(k+1)$-subset. It has been proven [10] that $\rho(m_i)$ is a 2-factor or, if $i = \frac{k+2}{2}$, a perfect matching of $O_k$.

## 3 Proof of the Main Results

This section shows how a spanning even cactus in $O_k$ is obtained from a spanning even cactus in $O_{k-1}$. Since $O_k$ has a hamiltonian cycle for $3 \leq k \leq 13$, it suffices to prove the statement for $k \geq 14$.

**Definition 3.** *A* peyote *is a spanning even cactus of $O_k$ such that all vertices of $S_k$ and $T_k$ form an even cycle and each vertex of $R_k$ is connected to that cycle by an edge (Figure 4).*

**Lemma 7.** *If there exists a hamiltonian cycle $C_h = (v_1, v_2, \ldots, v_q)$ in $O_{k-1}$ such that $|C_h|$ is even, then $O_k$ has a peyote.*

*Proof.* Construct a cycle $C'$ as given in the proof for Theorem 1. Since $C'$ is formed by the concatenation of $q$ paths, each with three vertices, and $|C_h|$ is even, then $|C'|$ is even as well. It remains to connect the vertices of $R_k$ to $C'$. Notice that any modular matching applied to the vertices of $R_k$ gives a matching $M$ in $B_k$ which saturates each vertex of $R_k$. Since the vertices of $R_k$ have edges only to vertices of $T_k$, the projection $\rho(M)$ is in the bipartite subgraph with bipartition $(T_k, R_k)$. The matching $\rho(M)$ along with the cycle $C'$ are a peyote in $O_k$. □

**Definition 4.** *A* cactoid *is a spanning even cactus of $O_k$ such that all vertices of $S_k$ and $X$, where $X \subseteq T_k$, form an even cycle, all vertices of $T_k \setminus X$ are connected to that cycle by an edge, and each vertex of $R_k$ is connected by an edge to some vertex of $T_k$ (Figure 4).*

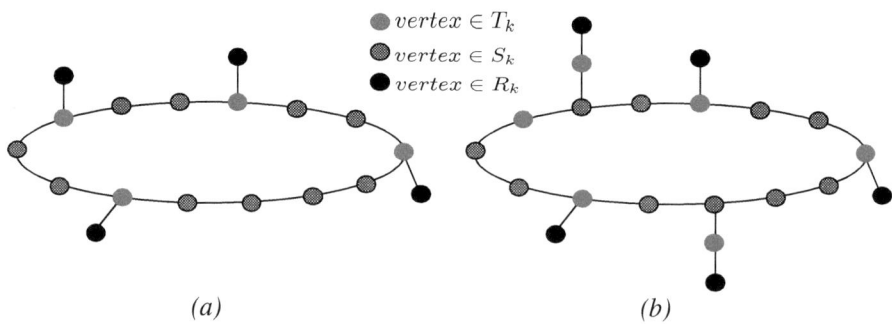

**Fig. 4.** Illustrations of (a) a peyote and (b) a cactoid where $X \subsetneq T_k$

Notice that every peyote is a cactoid where $X = T_k$. Besides, since the even cycle in a cactoid only has vertices of $S_k$ and $T_k$, each vertex of $T_k \setminus X$ is connected to the cycle by an edge to some vertex of $S_k$. Finally, the vertices of $R_k$ are connected to the vertices of $X$ or $T_k \setminus X$. In the last case, instead an edge connected to the even cycle, we have a path of length 2.

**Theorem 2.** *If there exists a cactoid in $O_{k-1}$, then $O_k$ has a cactoid.*

*Proof.* First consider that $O_{k-1}$ has a peyote and let $C' = (v_1, v_2, v_3, v_4, \ldots, v_q)$ be the even cycle of such peyote. As before, we construct $C'_1$ and $C'_2$ from $C'$ according to Definition 1: $C'_1 = (0v_11, 1v_20, 0v_31, 1v_40, \ldots, 1v_q0)$ and $C'_2 = (1v_10, 0v_21, 1v_30, 0v_41, \ldots, 0v_q1)$. For a vertex $v_i$ in $C'$, $1 \le i \le q$, that is adjacent to a vertex $w$ of $R_{k-1}$ in the peyote, we add properly to $C'_1$ and $C'_2$, the edges $\{1w0, 0v_i1\}$ and $\{0w1, 1v_i0\}$. Notice that $0w1$ and $1w0$ are vertices of $S_k$. As in Theorem 1, construct $q$ paths by combining the vertices of $C'_1$, $C'_2$ e $T_k$:

(i) If the vertex $v_i$ is not adjacent in $C'$ to a vertex of $R_{k-1}$, then construct a path with three vertices: $(0v_i1, 0\overline{v_i}0, 1v_i0)$;
(ii) If the vertex $v_i$ is adjacent in $C'$ to a vertex $w$ of $R_{k-1}$, then construct a path with five vertices: $(0v_i1, 1w0, 0\overline{w}0, 0w1, 1v_i0)$.

The concatenation of the $q$ paths is the same given in the proof for Theorem 1. Denote it by $C''$. Since each path has either three or five vertices and $q$ is even, $|C''|$ is even as well.

Clearly, all vertices of $S_k$ are in $C''$. The other vertices of $T_k$ can be added to $C''$ by joining them to one of the two adjacent vertices in $S_k$. All vertices of $T_k$ in $C''$ have degree 2 (if they are in the cycle) or 1. Choose any modular matching applied to the vertices of $R_k$ obtaining a matching $M$ in $B_k$. The matching $\rho(M)$ along with the cycle $C''$ are a cactoid in $O_k$.

In the general case where $O_{k-1}$ has a cactoid, not necessarily a peyote, proceed as in the previous case, with the only difference that the path between $0v_i1$ and $1v_i0$ can be a path with seven vertices: $(0v_i1, 1u0, 0w1, 0\overline{w}0, 1w0, 0u1, 1v_i0)$, where $u \in T_{k-1}$, $w \in R_{k-1}$ and $v_i \in S_{k-1}$. □

**Corollary 1.** *The prism over the odd graph $O_k$, $k \ge 14$, is hamiltonian.*

*Proof.* Since $O_{13}$ is hamiltonian and $|V(O_{13})|$ is even, by Lemma 7 and Theorem 2, for $k \ge 14$, the odd graph $O_k$ has a cactoid. Therefore, by Proposition 1, $O_k$ is prism-hamiltonian. □

**Theorem 3.** *The odd graph $O_k$, $k \ge 14$, has a cycle with at least $.625|V(O_k)|$ vertices.*

*Proof.* The number of vertices in the cycle in a cactoid in $O_k$, as constructed in the proof for Theorem 2, can be expressed by

$$2\sum_{i=14}^{k-1} \binom{2i+1}{i} + 3\binom{27}{13} \tag{1}$$

where $\binom{27}{13}$ is the number of vertices of $O_{13}$. For $k = 14$, the number of vertices in a cactoid is roughly 77% of $|V(O_k)|$. This fraction decreases as $k$ increases, but it is possible to show that, when divided by the number of vertices of $O_k$, Expression (1) is lower bounded by $\frac{1}{2}\sum_{i=0}^{k-14}\frac{1}{4^i}$, which is never less than 62.5% for $k \geq 15$, and approaches $\frac{2}{3}$ as $k$ increases. □

## 4 The Bipartite Kneser Graph $B_k$ Is Prism-Hamiltonian

Horák et al. [8] proved that the graph $B_k$ is prism-hamiltonian by determining a spanning 3-connected 3-regular subgraph in $B_k$, using Paulraja's proof [15] that graphs with such a spanning subgraph are prism-hamiltonian. We provide an alternative proof for the same result, which relates a cactoid in $O_k$ with a spanning even cactus in $B_k$.

**Proposition 2 ([11]).** *Let $C$ be a cycle with $q$ vertices in $O_k$.*

*(i) If $q$ is odd, then there is a cycle in $B_k$ with with $2q$ vertices;*
*(ii) If $q$ is even, then there are two disjoint cycles in $B_k$, each with $q$ vertices.*

Notice that $|V(B_k)| = 2|V(O_k)|$ and that, if $C_h$ is a hamiltonian cycle in $O_k$ such that $q = |C_h|$ is odd, Proposition 2 gives a hamiltonian cycle in $B_k$. Otherwise, if $q$ is even, then the two disjoint cycles contain all vertices of $B_k$.

**Lemma 8 ([3]).** *Consider the $k$-subset $\sigma = \{2, 4, 6, \ldots, n-1\} \in V(O_k)$. Let $\sigma + \delta$ denote the set $\{a + \delta : a \in \sigma\}$, with arithmetic modulo $n$. The subgraph of $O_k$ induced by $\sigma + \delta$ is a cycle $C_\sigma = (\sigma, \sigma+1, \sigma+2, \ldots, \sigma+(n-1))$.*

**Lemma 9.** *The cycle $C_\sigma$ has $(n-1)$ vertices of $S_k$ and one vertex of $T_k$.*

*Proof.* Let $\sigma'$ be the binary representation of the vertex $\sigma$ and $C'_\sigma$ be the cycle $C_\sigma$ with all vertices as binary strings. Notice that the vertices of $C'_\sigma$ are translations of all elements of $\sigma'$ by one to the left, repeatedly. Denote such translation of the bits by $sh(\sigma')$. Then $\sigma$ is $sh^0(\sigma')$, $\sigma+1$ is $sh^1(\sigma')$, $\sigma+2$ is $sh^2(\sigma')$ and so on. Now, notice that there are no two consecutive 1's in $\sigma'$. Therefore, $C'_\sigma$ does not have a subset with both elements 1 and $n$, i.e., a vertex of $R_k$. Since there is only a pair of consecutive zeros, $C'_\sigma$ has only one vertex without both elements 1 and $n$, i.e., a vertex of $T_k$. □

**Theorem 4.** *The prism over the middle-levels graph $B_k$, $k \geq 1$, is hamiltonian.*

*Proof.* Let $H$ be a cactoid in $O_k$. Let $W = (v_1, v_2, v_3, v_4, \ldots, v_q)$ be the even cycle contained in $H$. Let $Z$ be the set of vertices in $H \setminus W$. Recall that $q$ is even. According to Proposition 2, construct two cycles from $W$ by alternately complementing the vertices in $W$: $W_1 = (v_1, \overline{v_2}, v_3, \overline{v_4}, \ldots, v_{q-1}, \overline{v_q})$ and $W_2 = (\overline{v_1}, v_2, \overline{v_3}, v_4, \ldots, \overline{v_{q-1}}, v_q)$. Notice that, by construction, if a vertex $v \in W_1$, then $\overline{v} \in W_2$.

Let $W'_1$ and $W'_2$ be the cycles $W_1$ and $W_2$ to which the following edges and vertices have been appended:

(i) Let $\{u,v\} \in E(H)$ such that $u \in Z$ and $v \in W$. For $v \in W'_i$, add $\{v, \overline{u}\}$ and, for $\overline{v} \in W'_i$, add $\{\overline{v}, u\}$;
(ii) Let $\{u,w\} \in E(H)$ such that $u, w \in Z$. For $u \in W'_i$ (added in the previous case), add $\{u, \overline{w}\}$. For $\overline{u} \in W'_i$, add $\{\overline{u}, w\}$.

Notice that all vertices of $B_k$ are in $W'_1$ and $W'_2$. It remains to show how properly to connect $W'_1$ and $W'_2$ by an edge in order to obtain a spanning even cactus. Since the graph $B_k$ is connected, there exists an edge between some pair of vertices $v \in W'_1$ and $u \in W'_2$.

Let $v$ be a vertex in $C'_\sigma$. Notice that $|C'_\sigma| = n$ is odd. By Proposition 2, from $C'_\sigma$, there is a cycle with $2n$ vertices in $B_k$, which contains both $v$ and $\overline{v}$, for all $v \in C'_\sigma$. This cycle provides two paths between $v$ and $\overline{v}$. Therefore, there exists an edge $e = \{u,v\}$ in $C'_\sigma$ such that, in the cycle with $2n$ vertices in $B_k$ obtained from $C'_\sigma$, there is either an edge $\{u, \overline{v}\}$ or $\{\overline{u}, v\}$ connecting $W'_1$ to $W'_2$.

Without loss of generality, suppose $\{u, \overline{v}\}$ connects $W'_1$ to $W'_2$ for $u \in W'_1$ and $\overline{v} \in W'_2$. By Lemma 9, $(n-1)$ vertices of $C'_\sigma$ are in $S_k$ and a vertex in $T_k$, therefore:

(a) Let $u, v \in S_k$. If $u$ and $v$ have degree 2, then $W'_1$ and $W'_2$ are connected by $\{u, \overline{v}\}$ and the proof is finished. Otherwise, if $u$ (or $v$) has degree 3, there is a vertex $w$ of $T_k$ connected to that vertex. By Lemma 2, the vertex $w$ is adjacent to exactly two vertices $u$ and $u'$ of $S_k$ and Theorem 2 allows to replace the edge $\{u,w\}$ by the edge $\{u',w\}$ in $W$. Therefore, $\{u, \overline{v}\}$ can connect $W'_1$ and $W'_2$ in a spanning even cactus.

(b) Let $u \in S$ and $v \in T$. If $u$ has degree 3, we proceed as in the first case. If $v$ has degree 2 and $v \in Z$, there is a vertex of $R_k$ connected to $v$. By Lemma 3, $|T_k| > |R_k|$ for every odd graph $O_k$, which implies there exists at least a vertex $p \in T_k$ such that either $p$ has degree 2 and $p \in W$ or $p \notin W$ but it has degree 1. Since the graph is vertex-transitive and edge-transitive, from the cycle $C'_\sigma$, we obtain another cycle including $p$.

For two $k$-subsets of $T_k$, $A = \{a_1, a_2, \ldots, a_k\}$ and $B = \{b_1, b_2, \ldots, b_k\}$, not represented as binary strings, let $\Pi$ be a bijective function defined by the elements of $A$ and $B$ as follows: $\Pi(A)$ is the set $\{\Pi(a_i) | a_i \in A\}$ such that $\Pi(1) = 1$, $\Pi(n) = n$ and $\Pi(a_i) = b_i$, for $1 \leq i \leq k$. By Lemmas 8 and 9, the remaining elements of $\{1, 2, \ldots, n\}$ can be mapped such that $\Pi$ be a bijective function. Now, let $\Pi$ be defined by the vertices $v$ and $p$ and $\Pi(C'_\sigma)$ be the function $\Pi$ applied to each vertex of $C'_\sigma$. Notice that $\Pi(C'_\sigma)$ is a cycle which contains $n-1$ subsets of $S_k$ and one subset of $T_k$ as well. Since $\Pi(C'_\sigma)$ contains the vertex $p \in T_k$, $W'_1$ can be connected to $W'_2$. □

## References

1. Babai, L.: Long cycles in vertex-transitive graphs. Journal of Graph Theory 3(3), 301–304 (1979)
2. Biggs, N.: Some odd graph theory. Annals of the New York Academy of Sciences 319, 71–81 (1979)

3. Bueno, L.R., Faria, L., Figueiredo, C.M.H., Fonseca, G.D.: Hamiltonian paths in odd graphs. Applicable Analysis and Discrete Mathematics 3(2), 386–394 (2009)
4. Bueno, L.R., Horák, P.: On hamiltonian cycles in the prism over the odd graphs. Journal of Graph Theory 68(3), 177–188 (2011)
5. Chen, Y.C.: Triangle-free hamiltonian Kneser graphs. Journal of Combinatorial Theory Series B 89, 1–16 (2003)
6. Duffus, D.A., Kierstead, H.A., Snevily, H.S.: An explicit 1-factorization in the middle of the boolean lattice. Journal of Combinatorial Theory, Series A 65, 334–342 (1994)
7. Havel, I.: Semipaths in directed cubes. In: Fiedler, M. (ed.) Graphs and other Combinatorial Topics, pp. 101–108. Teubner-Texte Math., Teubner (1983)
8. Horák, P., Kaiser, T., Rosenfeld, M., Ryjáček, Z.: The prism over the middle-levels graph is hamiltonian. Order 22(1), 73–81 (2005)
9. Jackson, B., Wormald, N.C.: $k$-walks of graphs. Australasian Journal of Combinatorics 2, 135–146 (1990)
10. Johnson, J.R., Kierstead, H.A.: Explicit 2-factorisations of the odd graph. Order 21, 19–27 (2004)
11. Johnson, J.R.: Long cycles in the middle two layers of the discrete cube. Journal of Combinatorial Theory Series A 105(2), 255–271 (2004)
12. Kaiser, T., Ryjáček, Z., Král, D., Rosenfeld, M., Voss, H.-J.: Hamilton cycles in prisms. Journal of Graph Theory 56, 249–269 (2007)
13. Karp, R.M.: Reducibility among combinatorial problems. In: Miller, R.E., Thatcher, J.W. (eds.) Complexity of Computer Computations, pp. 85–103. Plenum Press, New York (1972)
14. Lovász, L.: Problem 11. In: Combinatorial Structures and their Applications. Gordon and Breach (1970)
15. Paulraja, P.: A characterization of hamiltonian prisms. Journal of Graph Theory 17, 161–171 (1993)
16. Savage, C.D., Winkler, P.: Monotone gray codes and the middle levels problem. J. Combin. Theory Ser. A 70(2), 230–248 (1995)
17. Shields, I., Savage, C.D.: A note on hamilton cycles in Kneser graphs. Bulletin of the Institute for Combinatorics and Its Applications 40, 13–22 (2004)
18. Shields, I., Shields, B.J., Savage, C.D.: An update on the middle levels problem. Discrete Mathematics 309(17), 5271–5277 (2009)
19. Shimada, M., Amano, K.: A note on the middle levels conjecture. CoRR abs/0912.4564 (2011)
20. Čada, R., Kaiser, T., Rosenfeld, M., Ryjáček, Z.: Hamiltonian decompositions of prisms over cubic graphs. Discrete Mathematics 286, 45–56 (2004)

# Relatively Bridge-Addable Classes of Graphs

Colin McDiarmid[1] and Kerstin Weller[2]

[1] Department of Statistics, University of Oxford, United Kingdom
cmcd@stats.ox.ac.uk
[2] Institut für Theoretische Informatik, ETH Zürich, Switzerland
kerstin.weller@inf.ethz.ch

**Abstract.** In recent years there has been a growing interest in random graphs sampled uniformly from a suitable structured class of (labelled) graphs, such as planar graphs. In particular, bridge-addable classes have received considerable attention. A class of graphs is called *bridge-addable* if for each graph in the class and each pair $u$ and $v$ of vertices in different components, the graph obtained by adding an edge joining $u$ and $v$ must also be in the class. The concept was introduced in 2005 by McDiarmid, Steger and Welsh, who showed that, for a random graph sampled uniformly from such a class, the probability that it is connected is at least $1/e$.

In this extended abstract, we generalise this result to relatively bridge-addable classes of graphs, which are classes of graphs where some but not necessarily all of the possible bridges are allowed to be introduced. We also give a bound on the expected number of vertices not in the largest component. These results are related to the theory of expander graphs. Furthermore, we investigate whether these bounds are tight, and in particular give detailed results about random forests in the bipartite graph $K_{n/2,n/2}$.

**Keywords:** random graphs, labelled graphs, bridge-addable, expander-graphs, forests in $K_{n,n}$.

## 1 Introduction

A set $\mathcal{A}$ of graphs is called *bridge-addable* if for any graph $G$ in $\mathcal{A}$ and any pair of vertices $u, v$ in different components the graph $G \cup \{uv\}$ obtained by adding the edge $uv$ to $G$ is also in $\mathcal{A}$. The concept of bridge-addability was introduced by McDiarmid et al.[9] in 2005 in the course of studying random planar graphs, and examples include forests, series-parallel graphs, planar graphs, and more generally graphs embeddable on any fixed surface, and minor-closed classes of graphs where the forbidden minors are all 2-connected.

If $\mathcal{S}$ is a finite non-empty set we write $R \in_u \mathcal{S}$ to mean that $R$ is drawn uniformly at random from $\mathcal{S}$. When we use such notation we assume implicitly that $\mathcal{S}$ is non-empty. For a collection $\mathcal{A}$ of graphs we write $R_n \in_u \mathcal{A}$ to mean that $R_n$ is drawn uniformly at random from $\mathcal{A}_n$, the set of all graphs in $\mathcal{A}$ on vertex set $[n] := \{1, ..., n\}$. By Theorem 2.2 in [9], for every finite bridge-addable set $\mathcal{A}$ of graphs and for $R \in_u \mathcal{A}$

$$\mathbb{P}(R \text{ is connected}) \geq e^{-1}; \tag{1}$$

and indeed the random number $\kappa(R)$ of components is stochastically at most $1 + \text{Po}(1)$, that is

$$\kappa(R) \leq_s 1 + \text{Po}(1). \tag{2}$$

Here $\text{Po}(1)$ denotes a random variable which has the Poisson distribution with mean 1; and for random variables $X$ and $Y$, we say that $X$ is *stochastically at most* $Y$ and write $X \leq_s Y$ if $\mathbb{P}(X \leq t) \geq \mathbb{P}(Y \leq t)$ for each $t$.

Let $\mathcal{F}$ be the class of forests (acyclic graphs). By a result of Rényi [12], for $F_n \in_u \mathcal{F}$ we have

$$\mathbb{P}(F_n \text{ is connected}) \to e^{-1/2} \quad \text{as } n \to \infty,$$

and indeed

$$\kappa(F_n) \xrightarrow{d} 1 + \text{Po}\left(\frac{1}{2}\right) \quad \text{as } n \to \infty.$$

Since forests are plausibly the 'least connected' bridge-addable class of graphs, it was natural to think that at least asymptotically (1) is not tight, and the following conjecture was made:

**Conjecture 1.1 ([8]).** *If $\mathcal{A}$ is bridge-addable and $R_n \in_u \mathcal{A}$ then*

$$\liminf_{n \to \infty} \mathbb{P}(R_n \text{ is connected}) \geq e^{-1/2}.$$

It was proved independently in [1] and [5] that under a further assumption on the class $\mathcal{A}$ – the class $\mathcal{A}$ has to be *bridge-alterable* – Conjecture 1.1 holds. A class of graphs $\mathcal{A}$ is *bridge-alterable* if it is closed under adding and deleting bridges. Thus $\mathcal{A}$ is bridge-alterable exactly when it satisfies the condition that for every graph $H$ and every bridge $e$ in $H$ the graph $H$ belongs to $\mathcal{A}$ if and only if $H \setminus e$ (the graph obtained by deleting $e$) belongs to $\mathcal{A}$. For $\mathcal{A}$ bridge-alterable and $R_n \in \mathcal{A}$ we have

$$\liminf_{n \in \mathbb{N}} \mathbb{P}(R_n \text{ is connected}) \geq e^{-1/2}. \tag{3}$$

Clearly, the class of forests is bridge-alterable and so the result in (3) is best possible. Also all the examples of graph classes mentioned previously (that is, minor-closed classes of graphs defined by forbidding 2-connected minors and graphs embeddable on a fixed surface) are bridge-alterable. For *bridge-addable* classes of graphs, Conjecture 1.1 is still open. The best known bound is due to Balister, Bollobás and Gerke [3] where it was proved that

$$\liminf_{n \to \infty} \mathbb{P}(R_n \text{ is connected}) \geq e^{-0.7983}. \tag{4}$$

Given a sequence of non-negative integer-valued random variables $X_1, X_2, \ldots$ and $Y$ we say that $X_n$ is *stochastically at most $Y$ asymptotically* and write $X_n \lesssim_s Y$, if for each fixed $t \geq 0$

$$\limsup_{n \to \infty} \mathbb{P}(X_n \geq t) \leq \mathbb{P}(Y \geq t).$$

In [6, Theorem 2.4] it was shown that the result (3) also yields a better bound on $\kappa(R_n)$: for any bridge-alterable class $\mathcal{A}$ of graphs and corresponding random graph $R_n \in_u \mathcal{A}$

$$\kappa(R_n) \lesssim_s 1 + \text{Po}\left(\frac{1}{2}\right).$$

Similarly, the proof of (4) together with [6, Lemma 3.3] give that for any bridge-addable class $\kappa(R_n) \lesssim_s 1 + \text{Po}\,(0.7983)$.

Let the *fragment* $\text{Frag}(G)$ of a graph $G$ be the subgraph induced on the vertices outside the largest component (with ties broken arbitrarily); and let the *fragment size* $\text{frag}(G)$ be the number of vertices in $\text{Frag}(G)$. In [6] (see inequality (7), which is an immediate consequence of Theorem 2.2) it was shown that for each bridge-addable class $\mathcal{A}$ and each $n$, for $R_n \in_u \mathcal{A}$

$$\mathbb{E}[\text{frag}(R_n)] < 2, \qquad (5)$$

generalising and improving Lemma 2.6 in [7]. For $F_n \in_u \mathcal{F}$, where $\mathcal{F}$ is the class of forests, we know that $\mathbb{E}[\text{frag}(F_n)] \to 1$ (see for example [7]) which leads us to the next conjecture, extending Conjecture 1.1:

**Conjecture 1.2 ([6]).** *Let $\mathcal{A}$ be bridge-addable and $R_n \in_u \mathcal{A}$. Then*

1. $\kappa(R_n) \lesssim_s 1 + \text{Po}\left(\frac{1}{2}\right)$ *and*
2. $\limsup_{n \to \infty} \mathbb{E}[\text{frag}(R_n)] \leq 1$.

In the following we will generalise the definition of bridge-addable and show that the results (1) and (5) are in fact special cases of a more general picture.

Fix a graph $G$ and call it the *host graph*. Let $\mathcal{A}$ be a collection of spanning subgraphs of $G$ (that is, subgraphs with the same vertex set as $G$). We call $\mathcal{A}$ *relatively bridge-addable (with respect to $G$)* if for every $H \in \mathcal{A}$ and every pair of vertices $u, v$ which are adjacent in $G$ and lie in different components of $H$ the graph $H \cup uv$ is also in $\mathcal{A}$. An illustration can be found in Figure 1. Observe that for $G = K_n$ *relatively bridge-addable* is equivalent to *bridge-addable* in the usual sense of [9]. An example of a relatively bridge-addable class of graphs for any host graph $G$ is the set of acyclic subgraphs of $G$.

For disjoint sets $B$ and $C$ of vertices in $G$ we let $e(B, C)$ denote the number of edges between $B$ and $C$. If $e(B, C) \geq \alpha |B||C|$ for each partition $B \cup C$ of $V(G)$ we say that $G$ is an $\alpha$-*edge-expander*. Thus the complete graph $K_n$ is a 1-edge-expander. Another natural example of an expander is a $d$-regular graph on $n$ vertices. In this case we may take $\alpha = \frac{d-\lambda}{n}$, where $\lambda$ is the second largest eigenvalue of the adjacency matrix of $G$, see [2, Theorem 9.1.2.]. Hence (or otherwise) $K_{d,d}$ is a $\frac{1}{2}$-edge-expander. Another example is the classical random graph $G_{n,p}$: with high probability $G_{n,p}$ is a $(1-\epsilon)p$-edge-expander for $p \geq C \log n/n$. To see this observe that for any fixed partition with $|B| = i$ and $|C| = n-i$, the probability that $e(B,C) < (1-\epsilon)p\,i(n-i)$ is at most $e^{-\Theta(\epsilon^2 i(n-i)p)}$ by standard Chernoff bounds, see for example [4]. Hence, for $p \geq C \log n/n$, with $C = C(\epsilon)$ large enough, we see that the probability that $G_{n,p}$ is not an $(1-\epsilon)p$-edge-expander is at most $\sum_{1 \leq i \leq n/2} n^i n^{-2i} = o(1)$.

The paper is organised as follows. Section 2 contains the generalisation of (1) and (5) to relatively bridge-addable classes of graphs. Complete proofs of both results are given. In Section 3 we look at one particular natural example: we consider random forests in the bipartite host graph $K_{n/2,n/2}$. In this section we do not give detailed proofs. Full proofs and further related results will be given in [10].

## 2 Connectivity for Edge-Expander Host Graphs

The following result generalises inequality (1), which is Theorem 2.2 of [9].

**Theorem 2.1.** *Let $0 < \alpha \leq 1$, let $G$ be an $\alpha$-edge-expander, let the set $\mathcal{A}$ of subgraphs of $G$ be relatively-bridge-addable, and let $R \in_u \mathcal{A}$. Then*

$$\kappa(R) \leq_s 1 + \mathrm{Po}(1/\alpha),$$

*and in particular*

$$\mathbb{P}(R \text{ is connected}) \geq e^{-1/\alpha}.$$

For the proof we are going to use the following definitions. For a subgraph $H$ of $G$, let $\mathrm{Bridge}(H)$ denote the set of bridges of $H$, with $|\mathrm{Bridge}(H)| = \mathrm{bridge}(H)$; and let $\mathrm{Cross}(H)$ denote the set of edges of $G$ between components of $H$, with $\mathrm{cross}(H) = |\mathrm{Cross}(H)|$.

*Proof.* Let $\mathcal{A}^k$ denote the set of graphs in $\mathcal{A}$ with $k$ components. Recall that all graphs in $\mathcal{A}$ are labelled. Let $G$ have $n$ vertices. For $k = 1, \ldots, n$ define

$$a_k = \min_{H \in \mathcal{A}^k} \mathrm{cross}(H)$$

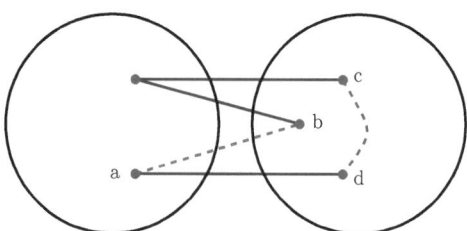

**Fig. 1.** $K_{n/2,n/2}$ as a host graph $G$. Suppose the graph with the blue edges is in our bridge-addable class. Then the graph consisting of the blue edges and the green edge also has to be in the class but not necessarily the graph with the blue edges and the red edge.

and
$$b_k = \max_{H \in \mathcal{A}^k} \text{bridge}(H).$$

Then for each $k = 1, \ldots, n-1$

$$a_{k+1}|\mathcal{A}^{k+1}| \leq \sum_{H \in \mathcal{A}^{k+1}} \text{cross}(H) \leq \sum_{H \in \mathcal{A}^k} \text{bridge}(H) \leq b_k |\mathcal{A}^k|.$$

We shall see that $b_k \leq n-k$ and $a_{k+1} \geq \alpha k(n-k)$. Then

$$|\mathcal{A}^{k+1}| \leq \frac{b_k}{a_{k+1}} |\mathcal{A}^k| \leq \frac{1}{\alpha k} |\mathcal{A}^k|$$

and so

$$\mathbb{P}(\kappa(R) = k+1) \leq \frac{1}{\alpha k} \mathbb{P}(\kappa(R) = k). \tag{6}$$

In [6, Lemma 3.3] it was shown that if for an integer-valued random variable $X$ we have $\mathbb{P}(X = k+1) \leq \frac{\beta}{k+1}\mathbb{P}(X = k)$ for each $k = 0, 1, 2, \ldots$, then $X \leq_s Y$ where $Y \sim \text{Po}(\beta)$. This result with $X = \kappa(R) - 1$ together with (6) then gives the theorem.

It remains to bound $a_{k+1}$ and $b_k$. But observe that $b_k \leq n-k$ since the number of edges in a spanning forest of a graph on $n$ vertices with $k$ components is $n-k$; so it remains now to bound $a_{k+1}$.

We want to find a graph $H$ with $k+1$ components which minimises $\text{cross}(H)$. Take an arbitrary graph $H \in \mathcal{A}$ with $k+1$ components, where the components have $n_1, \ldots, n_{k+1}$ vertices. As $G$ is an $\alpha$-edge-expander, we get

$$\text{cross}(H) \geq \frac{\alpha}{2} \sum_{i=1}^{k+1} n_i(n - n_i) = \frac{\alpha}{2}\left(n^2 - \sum_{i=1}^{k+1} n_i^2\right).$$

Thus we want to maximise $\sum_{i=1}^{k+1} n_i^2$ subject to $\sum_{i=1}^{k+1} n_i = n$ and $n_i \geq 1$ for all $i$. The maximum is attained at $n_1 = n-k$ and $n_2 = \ldots = n_{k+1} = 1$ with value $(n-k)^2 + k = n^2 - 2kn + k^2 + k$. Hence

$$\text{cross}(H) \geq \frac{\alpha}{2}\left(2nk - k^2 - k\right) = \frac{\alpha}{2}\left(2k(n-k) + k^2 - k\right) \geq \alpha k(n-k)$$

and so $a_{k+1} \geq \alpha k(n-k)$, as required. □

Recall that $\text{frag}(G)$ is the number of vertices which are not in the largest component. The following result on its expected value generalises inequality (5) above (from [6]).

**Theorem 2.2.** *Let $G$ be an $\alpha$-edge-expander, let the set $\mathcal{A}$ of subgraphs of $G$ be relatively bridge-addable, and let $R \in_u \mathcal{A}$. Then*

$$\mathbb{E}[\text{frag}(R)] < 2/\alpha.$$

*Proof.* Let $G$ have $n$ vertices. Let us prove the following claim first:

$$\mathrm{cross}(H) \geq \alpha \, \frac{n}{2} \, \mathrm{frag}(H) \tag{7}$$

for each graph $H$ in $\mathcal{A}$.

Observe that it is trivially true for $\mathrm{frag}(H) \leq \frac{n}{2}$ as $G$ is an $\alpha$-edge-expander. For the general case, let $H \in \mathcal{A}$ have $k$ components, with $n_1 \geq \cdots \geq n_k$ vertices. As $G$ is an $\alpha$-edge-expander, as earlier

$$\mathrm{cross}(H) \geq \frac{\alpha}{2} \sum_{i=1}^{k} n_i(n - n_i) = \frac{\alpha}{2} \left( n^2 - \sum_{i=1}^{k} n_i^2 \right).$$

But

$$\sum_{i=1}^{k} n_i^2 \leq \sum_{i=1}^{k} n_1 n_i = n_1 n.$$

Thus since $\mathrm{frag}(H) = n - n_1$ we get

$$\mathrm{cross}(H) \geq \frac{\alpha}{2} n(n - n_1) = \frac{\alpha \, n}{2} \, \mathrm{frag}(H),$$

which establishes the claim (7).

By (7) we have

$$\frac{\alpha \, n}{2} \sum_{H \in \mathcal{A}} \mathrm{frag}(H) \leq \sum_{H \in \mathcal{A}} \mathrm{cross}(H) \leq \sum_{H \in \mathcal{A}} \mathrm{bridge}(H).$$

Hence

$$\mathbb{E}[\mathrm{frag}(R)] \leq \frac{1}{|\mathcal{A}|} \cdot \frac{2}{\alpha \, n} \sum_{H \in \mathcal{A}} \mathrm{bridge}(H) < \frac{2}{\alpha},$$

completing the proof of the theorem. □

Are the bounds of Theorem 2.1 and Theorem 2.2 tight? In the bridge-addable case they are conjectured (Conjectures 1.1 and 1.2) not to be tight: for $G = K_n$ the 'correct' bounds are believed to be $1 + \mathrm{Po}\left(\frac{1}{2}\right)$ and $\mathbb{E}[\mathrm{frag}(R_n)] < 1$, which would correspond to the class of forests.

## 3 Forests in $K_{n/2, n/2}$

What happens for particular other $\alpha$-edge-expanders? Are the bounds of Theorem 2.1 and Theorem 2.2 tight? In the following, we will have a closer look at the complete bipartite graph $K_{n/2, n/2}$, which is a $\frac{1}{2}$-edge-expander. Here, $K_{n/2, n/2}$ really means $K_{\lfloor n/2 \rfloor, \lceil n/2 \rceil}$ but we write $K_{n/2, n/2}$ to simplify notation.

**Theorem 3.1.** *Consider $K_{n/2, n/2}$, say with parts $\{1, \ldots, \lfloor n/2 \rfloor\}$ and $\{\lceil n/2 \rceil + 1, \ldots, n\}$. Let the random graph $R_n$ be sampled uniformly at random from the forests in this graph. Then as $n \to \infty$*

(a)
$$\kappa(R_n) \to_d 1 + \text{Po}(1)$$
and so in particular $\mathbb{P}(R_n \text{ is connected}) \to e^{-1}$;
(b)
$$\mathbb{E}[\text{frag}(R_n)] \to 2;$$
and
(c) the unlabelled trees $T$ appear in $\text{Frag}(R_n)$ asymptotically independently, with distribution $\text{Po}(\lambda(T))$ where $\lambda(T) = 2/(e^{v(T)}\text{aut}(T))$.

For comparison, let $F_n$ be sampled uniformly from the $n$-vertex forests in $K_n$. Then $\mathbb{P}(F_n \text{ is connected}) \to e^{-\frac{1}{2}}$ as $n \to \infty$, and indeed $\kappa(F_n)$ converges in distribution to $1 + \text{Po}(\frac{1}{2})$, as mentioned earlier. Furthermore, as $n \to \infty$, $\mathbb{E}[\text{frag}(F_n)] \to 1$, and the unlabelled trees $T$ appear in $\text{Frag}(F_n)$ asymptotically independently, with distribution $\text{Po}(\mu(T))$ where $\mu(T) = 1/(e^{v(T)}\text{aut}(T))$ (see [9]).

Thus not only is it true asymptotically that $\mathbb{E}[\kappa(\text{Frag}(R_n))] \sim 2\,\mathbb{E}[\kappa(\text{Frag}(F_n))]$ and $\mathbb{E}[\text{frag}(R_n)] \sim 2\,\mathbb{E}[\text{frag}(F_n)]$ but it is even true for each unlabelled tree $T$ that the expected number of appearances of $T$ in $\text{Frag}(R_n)$ is twice that in $\text{Frag}(F_n)$.

We defer the (lengthy) proof of Theorem 3.1 to [10]. A proof can be based on the following exact counting result from 1962 [13]: the number of spanning trees of $K_{r,s}$ (where we are given a fixed partition of $[r+s]$ into parts of size $r$ and $s$) is
$$r^{s-1} \cdot s^{r-1}.$$

The counting result as well as further counting results on two-coloured forests in bipartite graphs can also be found in [11]. A referee observed that we may think for example of $R_n$ in Theorem 3.1 as being sampled uniformly at random from the independent sets in the cycle matroid of the graph $K_{n/2,n/2}$, and this may lead to related general questions concerning matroids.

Theorem 2.1 says that, given any relatively bridge-addable class $\mathcal{A}$ of graphs in $K_{n/2,n/2}$, for $R_n \in_u \mathcal{A}$
$$\mathbb{P}(R_n \in_u \mathcal{A}) \geq e^{-2}.$$

However, Theorem 3.1 suggests that this bound may not be tight (if relatively bridge-addable classes of graphs behave at all similarly to bridge-addable classes), and we could replace $e^{-2}$ by $e^{-1+o(1)}$. Also $\mathbb{E}(\text{frag}(R_n)) < 4$ by Theorem 2.2, but Theorem 3.1 suggests that we could replace 4 by $2 + o(1)$.

We think that indeed, analogously to the fully bridge-addable case, Theorem 2.1 is not best possible and forests are the relatively bridge-addable class of graphs that are the least likely to be connected.

**Conjecture 3.2.** *Let $G$ be a finite connected graph, let $\mathcal{A}$ be relatively bridge-addable with respect to $G$, and let $\mathcal{F}$ be the class of forests in $G$. For $R \in_u \mathcal{A}$ and $F \in_u \mathcal{F}$*
$$\mathbb{P}(R \text{ is connected}) \geq \mathbb{P}(F \text{ is connected})$$

*and indeed*
$$\kappa(R) \leq_s \kappa(F).$$

This is a strong form of the conjecture: an asymptotic form would also be very interesting.

**Acknowledgements.** The authors would like to thank Mireille Bousquet-Mélou for helpful discussions, and to thank three careful referees.

# References

1. Addario-Berry, L., McDiarmid, C., Reed, B.: Connectivity for Bridge-Addable Monotone Graph Classes. Combinatorics, Probability and Computing 21, 803–815 (2012)
2. Alon, N., Spencer, J.H.: The Probabilistic Method. John Wiley & Sons Inc. (2008)
3. Balister, P., Bollobás, B., Gerke, S.: Connectivity of addable graph classes. Journal of Combinatorial Theory, Series B 98, 577–584 (2008)
4. Bollobás, B.: Random Graphs. Cambridge University Press (2001)
5. Kang, M., Panagiotou, K.: On the connectivity of random graphs from addable classes. Journal of Combinatorial Theory, Series B 103, 306–312 (2013)
6. McDiarmid, C.: Connectivity for random graphs from a weighted bridge-addable class. Electronic Journal of Combinatorics 19, Paper 53, 20 (2012)
7. McDiarmid, C.: Random graphs from a minor-closed class. Combinatorics, Probability and Computing 18, 583–599 (2009)
8. McDiarmid, C., Steger, A., Welsh, D.J.A.: Random Graphs from Planar and Other Addable Classes. In: Topics in Discrete Mathematics. Algorithms and Combinatorics, vol. 26, pp. 231–246. Springer, Heidelberg (2006)
9. McDiarmid, C., Steger, A., Welsh, D.J.A.: Random planar graphs. Journal of Combinatorial Theory, Series B 93, 187–205 (2005)
10. McDiarmid, C., Weller, K.: Connectivity for relatively bridge-addable classes of graphs (in preparation)
11. Moon, J.W.: Counting Labelled Trees, Canadian Mathematical Congress, Montreal, Quebec (1970)
12. Rényi, A.: Some remarks on the theory of trees. Magyar Tud. Akad. Mat. Kutató Int. Közl 4, 73–85 (1959)
13. Scoins, H.I.: The number of trees with nodes of alternate parity. In: Proc. Cambridge Philos. Soc., vol. 58, pp. 12–16 (1962)

# $O(n)$ Time Algorithms for Dominating Induced Matching Problems

Min Chih Lin[1,*], Michel J. Mizrahi[1,*], and Jayme L. Szwarcfiter[2,**]

[1] CONICET, Instituto de Cálculo and Departamento de Computación
Universidad de Buenos Aires, Buenos Aires, Argentina
[2] I. Mat., COPPE and NCE, Universidade Federal do Rio de Janeiro, and
Instituto Nacional de Metrologia, Qualidade e Tecnologia,
Rio de Janeiro, Brazil
oscarlin@dc.uba.ar, michel.mizrahi@gmail.com, jayme@nce.ufrj.br

**Abstract.** We describe $O(n)$ time algorithms for finding the minimum weighted dominating induced matching of chordal, dually chordal, biconvex, and claw-free graphs. For the first three classes, we prove tight $O(n)$ bounds on the maximum number of edges that a graph having a dominating induced matching may contain. By applying these bounds, countings and employing existing $O(n+m)$ time algorithms we show that they can be reduced to $O(n)$ time. For claw-free graphs, we describe an algorithm based on that by Cardoso, Korpelainen and Lozin [4], for solving the unweighted version of the problem, which decreases its complexity from $O(n^2)$ to $O(n)$, while additionally solving the weighted version.

**Keywords:** algorithms, dominating induced matchings, graph theory.

## 1 Introduction

We consider undirected graphs $G$, denoting by $V(G)$ and $E(G)$, respectively, the sets of vertices and edges of $G$, $n = |V(G)|$ and $m = |E(G)|$. For $v \in V(G)$, $N(v)$ represents the set of neighbors of $v \in V(G)$, while $N[v] = N(v) \cup \{v\}$. For $S \subseteq V(G)$, $N(S) = \cup_{v \in S} N(v)$. We say a vertex $v \in V(G)$ such that $N[v] = V(G)$ is *universal*. Denote by $G[S]$ the subgraph of $G$ induced by the vertices of $S$. If $G[S]$ is a 0-regular graph then $S$ is an *independent set*, if it is a 1-regular graph then $S$ is the set of vertices of an *edge independent set*. An edge independent set is also known as an *induced matching*. For convenience, we may write induced matching to refer either to the set of edges or to its corresponding vertex set. Finally, we also employ the notation *matching* with its usual meaning of a set of pairwise non-adjacent edges.

Say that an edge $e \in E(G)$ *dominates* itself and every other edge adjacent to it. An *edge dominating set* of $G$ is a set of edges $E' \subseteq E(G)$, such that every

---

* Partially supported by UBACyT Grants 20020100100754 and 20020120100058, PICT ANPCyT Grant 1970 and PIP CONICET Grant 11220100100310.
** Partially supported by CNPq, CAPES and FAPERJ, research agencies.

$e \in E(G)$ is dominated by some edge of $E'$. If each $e \in E(G)$ is dominated by exactly one edge of $E'$ then $E'$ is an *efficient dominating set*. In the latter situation, $E'$ defines an induced matching, while the set of vertices not incident to $E'$ form an independent set. For this reason, an efficient edge dominating set is also called *dominating induced matching (DIM)*. Not every graph admits a DIM. The induced cycle of 4 vertices is an example of a graph with no DIM. The *DIM problem* is to determine whether a graph has such a matching, and is known to be NP-complete [6]. We will consider graphs $G$ with a weighting $\Omega$, that assigns to each edge $vw \in E(G)$ a non-negative finite weight $\omega(vw)$. The aim is to find the minimum weight of a dominating induced matching of $G$, if any. We name this problem as $DIM_\Omega(G)$.

Several polynomial-time algorithms were developed for the $DIM_\Omega(G)$ problem on restricted graph classes such as chordal graphs [9], generalized series-parallel graphs [9], bipartite permutations graphs [10], and $P_7$-free graphs [3]. On the other hand, the problem is NP-Complete even if $G$ is restricted to planar bipartite graphs of maximum degree 3 [1], or $k - regular$ graphs, for $k \geq 3$ [4]. There are also non-trivial exponential time algorithms for the $DIM_\Omega(G)$ problem for general graphs [7,8].

Since the number of edges of any DIM of $G$, if existing, is invariant, it is straightforward to generalize the problem for edges with negative weights too.

Following the definition, the DIM problem can be viewed as to decide whether there is a partition of the vertices into two sets (say a *coloring* of the vertices in *white* and *black*) such that the white set is an independent set while the black one induces a 1-regular graph. Moreover, the black set defines a DIM of the graph [4]. A coloring is *partial* if only part of the vertices of G have been assigned colors, otherwise it is *total*. A black vertex is *single* if it has no black neighbor, and is paired if it has exactly one black neighbor. Each coloring, partial or total, can be *valid* or *invalid*.

A partial coloring is valid whenever any two white vertices are non-adjacent and each black vertex is either paired, or is single having some uncolored neighbor. A total coloring is valid whenever any two white vertices are non-adjacent and each black vertex is paired.

A valid partial coloring $\Gamma$ might possibly extend into a coloring $\Gamma' \supseteq \Gamma$ by iteratively applying a set of coloring rules, compatible with $\Gamma$. In general, such rules would color some uncolored vertex $v$, whose color is uniquely determined by the colors of $\Gamma$. For instance, any uncolored neighbor of a white vertex must be colored as black, otherwise the coloring would be invalid. See [4,8] for a set of such rules. We refer to this process as *propagation*.

We prove that any chordal graph containing a DIM has at most $2n-3$ edges. Counting the edges and applying the $O(n+m)$ time algorithm by Lu, Ko and Tang [9] lead to an $O(n)$ time algorithm. For dually chordal graphs, by employing the similarity result *chordal - dually chordal* for DIMs by Brandstädt, Leitert and Rautenbach [2] also leads to solving the DIM problem in $O(n)$ time. For biconvex graphs, we prove that any $K_{3,3}$–free convex graph contains at most $2n-4$ edges. Additionally, that any biconvex graph containing a DIM is $K_{3,3}$–free. Using these two results, counting the number of edges of the given graph and employing the $O(n+m)$ time algorithm by Brandstädt, Hundt and Nevries [1] leads to solving the DIM problem for biconvex graphs in $O(n)$ time. Finally, for claw-free graphs, we describe a variation of the algorithm by Cardoso, Korpelainen and Lozin [4]. The latter solves the DIM problem, without weights, in $O(n^2)$ time, while the presently proposed algorithm requires $O(n)$ time for solving $DIM_\Omega(G)$.

## 2  Chordal, Dually Chordal and Biconvex Graphs

In this section, we remark that computing $DIM_\Omega(G)$ for any graph $G$ which is chordal, dually chordal or biconvex requires no more than $O(n)$ time.

**Lemma 1.** *[1] If $G$ contains a $K_4$ then $G$ has no DIMs.*

**Lemma 2.** *Every $K_4$-free chordal graph $G$ with at least 2 vertices has at most $2n - 3$ edges. The bound is tight even if $G$ is an interval graph.*

*Proof.* By induction on the number of vertices of the graph. For $n = 2$, the result follows since such a graph has at most one edge. Suppose the bound is valid for any graph with $n - 1$ vertices, $n \geq 3$. Let $G$ be an $n$-vertex chordal graph and $v$ a simplicial vertex of it. Since $|E(G)| = |E(G \setminus \{v\})| + d(v)$, by the induction hypothesis, the number of edges of $G \setminus \{v\}$ is bounded by $2(n-1) - 3 = 2n - 5$. Since $G$ is $K_4$-free, $d(v) \leq 2$, therefore $|E(G)| \leq 2n - 5 + 2 = 2n - 3$.

An interval graph having two universal vertices and the remaining ones having degree 2 has no $K_4$ and contains $2n - 3$ edges, meaning that the bound is tight for interval graphs. □

**Corollary 1.** *The $DIM_\Omega(G)$ problem can be solved in $O(n)$ time for (dually) chordal graphs.*

*Proof.* Let $G$ be a given chordal graph. First, we count the number of edges of $G$, up to a limit of $2n - 3$. If the bound has been exceeded then stop answering that $G$ has no DIMs. Otherwise, apply the algorithm [9] which will solve $DIM_\Omega(G)$ in $O(n)$ time. Finally, if a graph has a DIM then it is chordal if and only if it is dually chordal [2]. Consequently, $DIM_\Omega(G)$ can also be solved in $O(n)$ time for dually chordal graphs. □

Next, consider solving $DIM_\Omega(G)$ for biconvex graphs.

An ordering $<$ of $X$ in a bipartite graph $G = (X, Y, E)$ has the *interval property* if for every vertex $y \in Y$, the vertices of $N(y)$ are consecutive in the ordering $<$ of $X$. A bipartite graph $(X, Y, E)$ is *convex* if there is an ordering of $X$ or $Y$ that fulfills the interval property. Furthermore if there are orderings for both $X$ and $Y$ which fulfill the interval property the graph is *biconvex*.

**Lemma 3.** *Let $G$ be a convex bipartite graph having no subgraph isomorphic to $K_{3,3}$. Then $G$ contains at most $2n - 4$ edges, for $n \geq 3$.*

*Proof.* The proof is by induction on $n$. If $n = 3$ then it is trivial to verify that $G$ satisfies the bound since it has at most 2 edges. Let $G$ be an arbitrary $K_{3,3}$–free convex graph, $v$ its minimum degree vertex and $G'$ the graph obtained from $G$ by removing $v$.

- $d(v) \leq 2$: Clearly, $G'$ is also $K_{3,3}$–free. By inductive hypothesis, $G'$ has at most $2(n-1)-4 = 2n-6$ edges. Consequently, $G$ has at most $2n-6+d(v) \leq 2n-4$ edges.
- $d(v) > 2$: Every vertex in $G$ has degree at least 3. Let $G = (X, Y, E)$ where $X$ has the interval property. Thus for each vertex $y \in Y$, $N(y)$ consists of vertices that are consecutive. Let $\{x_1, \ldots, x_k\}$ be the ordering $<$ of $X$ and w.l.o.g. let $\{y_1, y_2, y_3\} \subseteq N(x_1)$. Since $y_1, y_2, y_3$ have at least 3 neighbors and $X$ has the interval property, it follows that $\{x_2, x_3\} \subseteq N(y_1) \cap N(y_2) \cap N(y_3)$. Therefore $\{x_1, x_2, x_3, y_1, y_2, y_3\}$ induces a $K_{3,3}$, which is a contradiction.

Hence, $G$ contains indeed at most $2n - 4$ edges. This bound is tight, $K_{2,n-2}$ is an example. □

We remark that bipartite graphs, not necessarily convex, which do not contain $K_{3,3}$ as a minor also have at most $2n - 4$ edges [5]. However, this bound does not apply to general bipartite graphs not containing $K_{3,3}$ as an induced subgraph, as shown by the example below described.

Let $G = (X, Y, E)$ be a bipartite graph where $X = \{x_0, x_1, \ldots, x_{15}\}$ and $Y = \{y_0, y_1, \ldots, y_7\}$. Add the edge $x_i y_{2j}$, if the binary representation of $i$ has the digit 0 at position $j$, while if such a binary representation contains the digit 1 at $j$ then add the edge $x_i y_{2j+1}$. It is easy to see that one of the edges $x_i y_{2j}, x_i y_{2j+1}$ will exist for all $i, j : 0 \leq i \leq 15, 0 \leq j \leq 3$. We show that $G$ is $K_{3,3}$-free: Suppose this is not true, and let $\{x_i, x_j, x_k, y_p, y_q, y_r\}$ be the vertices of an induced $K_{3,3}$. By the construction of $G$ the binary representations of $i, j, k$ have the same value for positions $\lfloor \frac{p}{2} \rfloor, \lfloor \frac{q}{2} \rfloor, \lfloor \frac{r}{2} \rfloor$. But $i, j, k$ are distinct integers $0 \leq i, j, k \leq 15$, which leads to a contradiction, since there are no three integers smaller than 16 with the above property. Consequently, $G$ is $K_{3,3}$-free. To complete the example, note that $G$ has 24 vertices and more than 44 edges.

Give $k$ copies of the graph defined above. Say $x_j^i$ is the $x_j$ vertex from the $i$-th copy. Add edges $y_7^i x_0^{(i+1)}, 0 \leq i < k$. The number of vertices is $24k$ while $m = 64k + k - 1$. This result graph is $K_{3,3}$-free bipartite and has all vertices of degree at least 4. The bound $65k - 1 \leq 48k - 4$ is not satisfied for any $k$.

**Lemma 4.** *Let $G = (X, Y, E)$ be a biconvex graph which has a DIM. Then $G$ is $K_{3,3}$–free.*

*Proof.* Suppose $G$ contains a $K_{3,3}$ given by $X' = \{x_1, x_2, x_3\} \subseteq X$ and $Y' = \{y_1, y_2, y_3\} \subseteq Y$. Consider an arbitrary DIM of the graph and its corresponding black-white coloring of the vertices as described in Section 1. Then the vertices of $X'$ and $Y'$ must have distinct colors. Suppose w.l.o.g. that the vertices $X'$ are black and those of $Y'$ are white. Let $y_1^*, y_2^*, y_3^*$ be the black neighbors of $x_1, x_2, x_3$, respectively. It follows that the graph induced by the nine vertices of $X' \cup Y' \cup \{y_1^*, y_2^*, y_3^*\}$ is not biconvex, a contradiction. □

**Corollary 2.** *The DIM problem for biconvex graphs can be solved in $O(n)$ time.*

*Proof.* Let $G$ be a biconvex graph. If $G$ contains a DIM, by Lemma 4, $G$ is $K_{3,3}$–free. Therefore $G$ has at most $2n - 4$ edges, by Lemma 3. Consequently, given an arbitrary biconvex graph, count its number of edges, up to $2n - 4$. If this number is exceeded, the graph does not contains any DIM, otherwise apply the algorithm [1], which solves the DIM problem in $O(n + m)$ time, for chordal bipartite graphs. Since convex graphs are contained in chordal bipartite, we can solve the DIM problem for biconvex graphs in $O(n)$ time. □

We remark that there are convex graphs having a quadratic number of edges that admit DIMs. For instance, $V(G) = V_1 \cup V_2 \cup V_3$, where $|V(G)| = n$, $|V_1| = |V_2| = |V_3| = \frac{n}{3}$. Let $V_i$ be an independent set for $1 \leq i \leq 3$, and let $V_1 \cup V_2$ induce a complete bipartite graph, $V_1 \cup V_3$ be an induced matching, and $V_2 \cup V_3$ be an independent set. Such a graph is bipartite, with bipartition $(V_1, V_2 \cup V_3)$, moreover it is convex bipartite since it admits a interval ordering. Also, it contains a quadratic number of edges. On the other hand, $V_1 \cup V_3$ is a DIM of it.

## 3 Claw-Free Graphs

The problem of finding a DIM of a claw-free graph, if existing, has been solved in [4] by an $O(n^2)$ time algorithm. We review the ideas of this paper and propose an improvement of it.

We assume that the given graph $G = (V(G), E(G))$ is connected, and is neither an induced cycle nor an induced path. Clearly, if $G$ is disconnected we can reduce the problem to its connected components, while if $G$ is a cycle or a path the solution is trivial.

By [4], if a claw-free graph $G$ has a DIM then each vertex $v$ of $G$ is one of the following six types:

**(1)** degree 1
**(2)** degree 2 with two non-adjacent neighbors
**(3)** degree 2 with two adjacent neighbors
**(4)** degree 3 with $G[N(v)]$ inducing a $K_1 + K_2$
**(5)** degree 3 with a $G[N(v)]$ inducing a $P_3$
**(6)** degree 4 with $G[N(v)]$ inducing a $2K_2$

Thus, we assume that each vertex of $G$ falls into one of the above types. This implies $m \leq 2n$, i.e. $m = O(n)$.

In particular, the two edges incident to a Type 4 vertex $v$, which are contained in a triangle of $G(N[v])$, are called *heavy*, while the third edge incident to $v$ is a *light*. The algorithm [4] can be viewed as a sequence of the following phases:

1. Handling three consecutive vertices of Type 2
2. Handling vertices of Type 1 which are at distance at least 3 of some Type 4 vertex

3. Coloring all vertices of Types 1,2,5 and 6
4. Coloring the remaining vertices, of Types 3 and 4

Our proposed algorithm describes new formulations for Phases 1,2 and 4, while maintaining the original Phase 3 of the algorithm [4]. We proceed by describing each of the parts.

### 3.1 Phase 1

The purpose is to eliminate the occurrence of three consecutive Type 2 vertices $v_1, v_2, v_3$, such that $N(v_2) = \{v_1, v_3\}$, $N(v_1) = \{v_2, w_1\}$ and $N(v_3) = \{v_2, w_3\}$. Consider the following alternatives:

- $w_1 = w_3$: In this case if $d(w_1) = 2$ then $G = C_4$, which contradicts $G$ not to be a cycle. Hence $d(w_1) \geq 3$, but then $G[N[w_1]]$ contains a claw, a contradiction. Thus this case does not occur.
- $w_1 w_3 \in E(G)$: If $d(w_1) = d(w_3) = 2$ then $G = C_5$ again a contradiction. Hence we may suppose $\exists u \in N(w_1) \setminus \{v_1, w_3\}$. We know that $u \notin N(v_1)$, thus in order to avoid a claw in $G[N[w_1]]$ we must assume $u \in N(w_3)$. The latter implies that no more vertices can belong to the neighborhoods of $w_1$ and $w_3$, otherwise $G$ would contain vertices outside the above six types, a contradiction.

  Any DIM of $G$ must have exactly one edge of the triangle $\{w_1, u, w_3\}$. The edge $w_1 w_3$ does not lead to a valid DIM since it forces $v_2$ to be a single black vertex without black neighbor. It is easy to verify that the possibilities are either: $\{w_1 u, v_2 v_3\}$ or $\{w_3 u, v_1 v_2\}$.

  Therefore we can eliminate vertices $v_1, v_2, v_3$ and sum the weight of edge $v_1 v_2$ to that of $w_3 u$, and sum the weight of $v_2 v_3$ to that of $w_1 u$. To guarantee that the edge $w_1 w_3$ is not chosen to enter the DIM, we assign infinite weight to it.
- $w_1 \neq w_3$ and $w_1 w_3 \notin E(G)$: In this case we use the original procedure of [4], which consists of replacing vertices $v_1, v_2, v_3$ for the edge $w_1 w_3$. However, the algorithm [4] solves the DIM problem without weights, thus, in order to guarantee the correct solution for the new weighted graph, we need to consider the following additional possibilities:
  • $w_1, w_3$ are black: Then $v_1, v_3$ are black and $v_2$ is white. The weights of edges $v_1 w_1$ and $v_3 w_3$ must be added to the weight of $w_1 w_3$
  • $w_1$ is black and $w_3$ is white: In this case, $v_2$ and $v_3$ are black while $v_1$ is white. Hence the weight of edge $v_2 v_3$ must be added to the weight of each edge of the set of edges $w_1 z$, where $z \neq v_1$
  • $w_3$ is black and $w_1$ is white: This case is symmetric to the previous one. The weight of edge $v_1 v_2$ must be added to the weight of each edge of the set $w_3 z$, where $z \neq v_3$.

These modifications to the original graph $G$ are repeated until no three consecutive vertices of Type 2 remains in the graph, leaving a new reduced graph $G' = (V(G'), E(G'))$. This can be achieved in $O(n)$ time. The algorithm now proceeds on $G'$.

## 3.2 Phase 2

In this phase, we eliminate the occurrence of Type 1 vertices, lying at distance at least 3 from some Type 4 vertex. Let $v \in V(G')$ such that $d(v) = 1$ and let $w \in V(G')$ be the vertex such that $d(w) \geq 3$ and the distance to $v$ is minimum. Note that if there is no such $w$ then $G'$ is a path, a contradiction. Therefore there is a path $v - w$ where all vertices, except $v, w$ are of Type 2. Since there are at most two consecutive vertices of Type 2, the distance between $v$ and $w$ is at most 3. It is easy to see that $w$ is of Type 4, otherwise $G'$ is not claw-free. Let $v, u_1, u_2, w$ be any path of length 3 from a vertex $v \in V(G')$ to a vertex $w \in V(G')$, with $d(v) = 1$ and $d(w) = 3$. Let $\{z_1, z_2\}$ be the $K_2$ induced by $N(w)$, and $G^*$ be the graph after deletion of vertices $\{v, u_1, u_2\}$. It is clear that any DIM $M^*$ of $G^*$ contains exactly one edge of the triangle $\{w, z_1, z_2\}$. In case $M^*$ contains the edge $z_1 z_2$, we add the edge $u_1 u_2$ to $M^*$ in order to obtain a DIM of $G$, hence to generate a DIM with the same weight in $G^*$ we set $w(z_1 z_2) = w(z_1 z_2) + w(u_1 u_2)$. In case that $M^*$ contains $wz_1$ or $wz_2$ the edge $vu_1$ is added to $M^*$. In the latter situation, we set $w(wz_1) = w(wz_1) + w(vu_1)$ and $w(wz_2) = w(wz_2) + w(vu_1)$. We repeat this process for each vertex $v \in V(G')$ such that $d(v) = 1$. Finally, we assert that every vertex of Type 1 is at distance 1 or 2 from some vertex of Type 4. These computations can be completed in $O(n)$ time.

## 3.3 Phase 3

By applying convenient propagation rules, the algorithm [4] colors a subset of vertices of the graph, including all vertices of Types 1,2,5, and 6. Let $\Gamma$ be the final coloring so obtained in the algorithm. First, check its validity. If $\Gamma$ is not valid, then $G$ has no DIMs and the algorithm terminates. If $C$ is valid and total, also terminate the algorithm, since the unique DIM of $G$ has been found. Otherwise, proceed to Phase 4.

All the above operations can be completed in $O(n)$ time. At the end of this phase, the only possibly uncolored vertices are of Types 3 and 4. Observe that the obtained coloring is unique, so there is no choice to be made concerning weights, so far.

## 3.4 Phase 4

In this phase, we extend the coloring $\Gamma$, obtained by the previous phase, into a total valid coloring. It is assumed that $\Gamma$ is a valid not total coloring, which cannot be extended by propagation. Let $U$ be the set of uncolored vertices and $S$ the set of single black vertices of the coloring $\Gamma$. Note that extending $\Gamma$ is equivalent to extending the coloring $\Gamma'$ of $G^*[U \cup S]$ (in $\Gamma'$, only vertices of $S$ are colored with black color). It can be verified that in any valid coloring, the following holds: $\forall s \in S, N[s]$ induces in $G^*[U \cup S]$ a $K_3 = \{u, v, s\}$ where $u, v \in U$.

Since vertices of $S$ and Type 3 vertices are simplicial in $G^*[U \cup S]$, any central vertex of an induced $P_3$ in $G^*[U \cup S]$ must be an uncolored Type 4 vertex. Particularly, the vertices of a cycle $C_{k \geq 4}$ are central vertices of induced $P_3$'s. Moreover, an edge of induced $P_3$ must be heavy and the other one must be light. It's easy to see that vertices of a light edge must have different colors. The following lemma is helpful to extend $\Gamma'$.

**Lemma 5.** *Let $\Gamma''$ any total valid coloring extensible from $\Gamma'$ and $P = (v_1, \ldots, v_t)$ be an induced path of $G^*[U \cup S]$ such that $v_1, v_t$ are Type 4 vertices, $v_1 v_2$ is a light edge and $v_1$ is a black vertex. Then (i) $v_i v_{i+1}$ is a light (heavy) edge if $i$ is odd (even); (ii) $v_i$ is black (white) if $i$ is odd (even).*

*Proof.* Since $P$ is an induced path, $v_2, \ldots, v_{t-1}$ are central vertices of induced $P_3$'s, they are also Type 4 vertices and the edges of $P$ are light and heavy alternately. Then (i) holds because the first edge is white. On the other hand, vertices of light edges must have different colors, while the same occurs for heavy edges if one vertex is white. Since $v_1$ is black and $v_1 v_2$ is a light edge, then $v_2$ is white and $v_3$ is black. Again, we can check that $v_3$ satisfies the same properties as $v_1$ and $v_4$ will satisfy the same properties as $v_2$. Therefore there is a unique valid coloring for vertices of $P$ which consists of alternating the colors of the vertices, where (ii) $v_i$ is black if and only if $i$ is odd. □

We proceed by finding a minimum weight DIM on each connected component $G_i$ of $G^*[U \cup S]$:

**$G_i$ Is a Chordal Graph.** In this case, for each single black vertex $s$ in $G_i$, its neighbors $u$ and $v$ form an edge and we set $\omega(uv) = \infty$. In this way any finite weight DIM of $G_i$ will not contain this edge and $s$ will be a black vertex as in $\Gamma'$. Apply the algorithm described in the previous section that computes $DIM_\Omega(G_i)$, if existing.

**$G_i$ Has an Induced Cycle $C_k, k \geq 4$.** As it was mentioned before, $C_k$ is formed by light and heavy edges, where each light edge is adjacent to heavy edges and viceversa.

**Lemma 6.** *[4]: Let $G^*$ be the resulting claw–free graph and $\Gamma$ the partial valid coloring obtained after Phase 3. If the subgraph of $G^*$ induced by uncolored vertices contains an induced cycle $C_{k \geq 4}$, then $k$ is even. Moreover, if $G^*$ admits a black-white partition, then the vertices of $C_k$ are colored alternately black and white along the cycle, and furthermore, by switching the colors of vertices of $C_k$ we again obtain a valid black-white partition of $G^*$.*

**Lemma 7.** *$G_i$ admits exactly two DIMs or none.*

*Proof.* We extend the initial coloring choosing any alternate coloring for $C_k$ and applying propagation rules. Let $\Gamma_i$ be the final coloring, so obtained. We will prove that $\Gamma_i$ is invalid or is a total valid coloring. Clearly, if $\Gamma_i$ is invalid then

$G_i$ has no DIMs by Lemma 6. If $\Gamma_i$ is a total valid coloring, then switching the colors of vertices of $C_k$, we obtain another total valid coloring of $G_i$. We show that these are the unique total valid colorings. Suppose that $\Gamma_i$ is valid but there is some uncolored vertex $u$ in $G_i$. Let $P = (v_0, \ldots, v_t = u)$ be the shortest path from a vertex $v_0$ in $C_k$. Without lost of generality we can assume that $v_0, \ldots, v_{t-1}$ are colored vertices. Clearly, $P$ is an induced path. On the other hand, $v_0 v_1$ is a heavy edge because $v_1$ must be adjacent to two consecutive vertices of $C_k$ by the claw-freeness. Hence, $v_1$ must be a black vertex and $t \geq 2$. Then $v_1, v_{t-1}$ are central vertices of induced $P_3$'s which implies that $v_1, v_{t-1}$ are Type 4 vertices and $v_1 v_2$ is a light edge. Clearly, $v_t = u$ must be Type 3 vertex because otherwise it would be colored applying Lemma 5. Hence, $v_{t-1} v_t$ is a heavy edge and $v_{t-2} v_{t-1}$ is a light edge which means that $t$ is odd and $v_{t-1}$ is a white vertex. Therefore, $v_t$ must be a black vertex which is a contradiction. Consequently, $\Gamma_i$ is a total valid coloring. □

Using these two lemmas, we can determine in linear time all DIMs of $G_i$ and return one of minimum weight, if existing.

As for the complexity of the last phase of the algorithm, observe that a cycle $C_{k \geq 4}$ of a non chordal graph can be obtained in linear time in the size of $G$, that is, $O(n)$ time. All the remaining steps can be completed in $O(n)$ time. It should be noted that the corresponding phase of the algorithm [4] requires $O(n^2)$ time. The main difference is that in the presently proposed algorithm, it is sufficient to find just one induced cycle of length $\geq 4$, and propagate the coloring to its connected component, whereas the algorithm [4] requires the computation of $O(n)$ such cycles, in subgraphs not necessarily disjoint. Since each of them needs $O(n)$ time, the overall complexity of the latter algorithm is $O(n^2)$.

Our proposed formulation computes $DIM_\Omega(G)$ in $O(n)$ time. Observe that through the algorithm the input graph is modified. However the changes do not alter the value of the $DIM_\Omega(G)$ solution. As for the actual minimizing DIM, itself, there is no difficulty retrieving it in $O(n)$ time, by backwards computation. Note that the algorithm can be easily adapted to be robust in the sense of [4], then it does not require the input graph $G$ to be claw-free.

# References

1. Brandstädt, A., Hundt, C., Nevries, R.: Efficient edge domination on hole-free graphs in polynomial time. In: López-Ortiz, A. (ed.) LATIN 2010. LNCS, vol. 6034, pp. 650–661. Springer, Heidelberg (2010)
2. Brandstädt, A., Leitert, A., Rautenbach, D.: Efficient Dominating and Edge Dominating Sets for Graphs and Hypergraphs. In: Chao, K.-M., Hsu, T.-S., Lee, D.-T. (eds.) ISAAC 2012. LNCS, vol. 7676, pp. 267–277. Springer, Heidelberg (2012)
3. Brandstädt, A., Mosca, R.: Dominating Induced Matchings for $P_7$-free Graphs in Linear Time. In: Asano, T., Nakano, S.-I., Okamoto, Y., Watanabe, O. (eds.) ISAAC 2011. LNCS, vol. 7074, pp. 100–109. Springer, Heidelberg (2011)
4. Cardoso, D.M., Korpelainen, N., Lozin, V.V.: On the complexity of the dominating induced matching problem in hereditary classes of graphs. Discrete Applied Mathematics 159, 21–531 (2011)

5. Chen, Z.-Z., Zhang, S.: Tight upper bound on the number of edges in a bipartite $K_{3,3}$-free or $K_5$-free graph with an application. Information Processing Letters 84, 141–145 (2002)
6. Grinstead, D.L., Slater, P.J., Sherwani, N.A., Holmes, N.D.: Efficient edge domination problems in graphs. Information Processing Letters 48, 221–228 (1993)
7. Lin, M.C., Mizrahi, M.J., Szwarcfiter, J.L.: Exact algorithms for dominating induced matchings. CoRR, abs/1301.7602 (2013)
8. Lin, M.C., Mizrahi, M.J., Szwarcfiter, J.L.: An $O^*(1.1939^n)$ time algorithm for minimum weighted dominating induced matching. In: Cai, L., Cheng, S.-W., Lam, T.-W. (eds.) Algorithms and Computation. LNCS, vol. 8283, pp. 558–567. Springer, Heidelberg (2013)
9. Lu, C.L., Ko, M.-T., Tang, C.Y.: Perfect edge domination and efficient edge domination in graphs. Discrete Applied Mathematics 119, 227–250 (2002)
10. Lu, C.L., Tang, C.Y.: Solving the weighted efficient edge domination problem on bipartite permutation graphs. Discrete Applied Mathematics 87, 203–211 (1998)

# Coloring Graph Powers: Graph Product Bounds and Hardness of Approximation

Parinya Chalermsook[1,*], Bundit Laekhanukit[2,**],
and Danupon Nanongkai[3,***]

[1] Max-Planck-Institut für Informatik, Saarbrücken, Germany
[2] School of Computer Science, McGill University, Canada
[3] ICERM, Brown University, USA

**Abstract.** We consider the question of computing the strong edge coloring, square graph coloring, and their generalization to coloring the $k^{th}$ power of graphs. These problems have long been studied in discrete mathematics, and their "chaotic" behavior makes them interesting from an approximation algorithm perspective: For $k = 1$, it is well-known that vertex coloring is "hard" and edge coloring is "easy" in the sense that the former has an $n^{1-\epsilon}$ hardness while the latter admits a $(1 + 1/\Delta)$-approximation algorithm, where $\Delta$ is the maximum degree of a graph. However, vertex coloring becomes easier (can be $O(\sqrt{n})$-approximated) for $k = 2$ while edge coloring seems to become much harder (no known $O(n^{1-\epsilon})$-approximation algorithm) for $k \geq 2$.

In this paper, we make a progress towards closing the gap for the edge coloring problems in the power of graphs. First, we confirm that edge coloring indeed becomes computationally harder when $k > 1$: we prove a hardness of $n^{1/3-\epsilon}$ for $k \in \{2,3\}$ and $n^{1/2-\epsilon}$ for $k \geq 4$ (previously, only NP-hardness for $k = 2$ is known). Our techniques allow us to derive an alternate proof of vertex coloring hardnesses as well as the hardness of maximum clique and stable set (a.k.a. independent set) problems on graph powers. These results rely on a common simple technique of proving bounds via *fractional coloring*, which allows us to prove some new bounds on graph products. Finally, we finish by presenting the proof of Erdös and Nešetřil conjecture on cographs, which uses a technique different from other results.

## 1 Introduction

Consider the following *broadcast scheduling* problem commonly studied in the networking research area (e.g. [31,5,22]). In this problem, we are given a graph

---

\* Work partially done while at IDSIA, Switzerland. Supported by the Swiss National Science Foundation project 200020_144491/1.
\*\* Supported by the Natural Sciences and Engineering Research Council of Canada (NSERC) grant no. 429598 and by Dr&Mrs M.Leong fellowship.
\*\*\* Work done while at Nanyang Technological University, Singapore. Supported in part by Nanyang Technological University grant M58110000, Singapore Ministry of Education (MOE) Academic Research Fund (AcRF) Tier 2 grant MOE2010-T2-2-082, and Singapore MOE AcRF Tier 1 grant MOE2012-T1-001-094.

$G$ representing a network of transceivers (represented by nodes). Each transceiver can send a message to its neighbors by broadcasting through some communication channel with the constraint that two transceivers sharing a neighbor *cannot* broadcast on the same channel since their signals interfere with each other. The objective is to minimize the number of channels we have to assign the transceivers to (while avoiding interference). More formally, given any graph $G = (V, E)$, we are interested in computing a minimum value of $C$ together with a function $c: V \to \{1, \ldots, C\}$ such that, for any pair $(u, v)$ of vertices of distance at most two, we have $c(u) \neq c(v)$.

In theoretical computer science and discrete mathematics, this problem is known as *distance-2 vertex coloring* or *coloring square graphs* (e.g., [24,28,2,3,21]) – simply view the channel assignment as a color assignment. Its natural extension, *distance-k coloring* or *coloring the $k^{th}$ power graphs*, in which our goal is to ensure that the colors of any two vertices within distance $k$ are different, is also intensively studied (e.g., [3,28,2,21]). Formally, denote by $G^k$ the $k^{th}$ power of a graph $G$ and $\chi(G)$ its chromatic number. Then we are interested in computing $\chi(G^k)$. (See Section 2 for definitions.) This problem captures a realistic situation in network where transceivers' signals interfere each other and has applications in, e.g., approximating the Hessian matrices of certain non-linear functions [28]. Strong lower bounds on the approximability of distance-$k$ vertex coloring were proved in [1,17].

The edge coloring version of the above problem, called *distance-k edge-coloring*, has received even more attention (e.g., [27,26,30,11,4,20,19]). In this case, we say that two edges are within distance $k$ if they are connected by a path on $k$ vertices (e.g., edges sharing an end-vertex are of distance one and edges connected by another edge are of distance two). If we denote by $\mathcal{L}(G)$ the *line graph* of $G$, then the distance-$k$ edge-coloring problem indeed asks for the value of $\chi(\mathcal{L}^k(G))$ (where $\mathcal{L}^k(G)$ is the $k^{th}$ power of $\mathcal{L}(G)$). A special case where $k = 2$ is known as *link scheduling* in network community and *strong edge coloring* in discrete mathematics community; for this case, $\chi'_S(G)$ is typically used to refer to $\chi((\mathcal{L}^2(G)))$. The discrete mathematics community has paid a particularly high attention to problems centering around the conjecture of Erdös and Nešetřil which states that any graph of maximum degree at most $\Delta$ needs at most $(5/4)\Delta^2$ colors for strong edge coloring, and the conjecture was later strengthened to $\Delta^2$ (see [12,13]). Recently, Laekhanukit also observed an application of strong edge coloring in proving hardness of approximation [23].

From a computational point of view, the approximability of distance-$k$ vertex and edge coloring has a chaotic behavior: While the (distance-1) vertex coloring was shown to be $n^{1-\epsilon}$ hard to approximate [16,34], there is an $O(\sqrt{n})$-approximation for distance-2 vertex coloring, suggesting that vertex coloring could be *easier* when $k$ becomes larger. However, (distance-1) edge coloring is known to admit a $(1+1/\Delta)$-approximation[1], where $\Delta$ is the maximum degree of a graph, but distance-2 edge coloring (a.k.a. strong edge coloring) does not seem to admit any $o(n)$-approximation ratio, suggesting that larger $k$ would increase

---

[1] This is an application of Vizing's theorem; see [29].

the complexity of the problem. There has also not been much progress on the lower bounds for distance-$k$ coloring: The first NP-hardness result for strong edge coloring (i.e., distance-2 edge coloring) appeared only in 2000 [27,26] (see also [11,18,2]). In sum, prior to our results, no strong approximation hardness was known for any distance-$k$ edge coloring problem when $k \geq 2$.

In this paper, we clarify the current state of the art by proving the polynomial factor hardness for the distance-$k$ edge coloring problem for $k \geq 2$. We also study, using similar techniques, other important combinatorial optimization problems in the power of graphs: approximating the *stability* (a.k.a, *independent set*) and *clique* numbers of a graph $G$, denoted by $\alpha(G)$ and $\omega(G)$, respectively[2]. (We refer to Section 2 for definitions.) Our results are summarized in Table 1.

**Table 1.** Summary of Results. The upper bounds of $n$ are trivial. All lower bounds, except for vertex coloring, are new (previously, only NP-hardness is known for **Edge Coloring** when $k = 2$). Other upper bounds follow from modifying known algorithms.

|  | Vertex Coloring | | Edge Coloring | | Stable Set | | Clique | |
|---|---|---|---|---|---|---|---|---|
|  | $k$ odd | $k$ even | $k \in \{2,3\}$ | $k \geq 4$ | $k$ odd | $k$ even | $k$ odd | $k$ even |
| Upper | $n^{2/3}$ | $n^{1/2}$ | $n$ | $n$ | $n$ | $n^{1/2}$ | $n^{2/3}$ | $n^{1/2}$ |
| Lower | $n^{1/2-\epsilon}$ [1,17] | $n^{1/2-\epsilon}$ [1,17] | $n^{1/3-\epsilon}$ | $n^{1/2-\epsilon}$ | $n^{1-\epsilon}$ | $n^{1/2-\epsilon}$ | $n^{1/2-\epsilon}$ | $n^{1/2-\epsilon}$ |

**Theorem 1.** *For any $\epsilon > 0$ and constant $k \geq 2$, given an $n$-vertex input graph $G$, there is no polynomial-time algorithm to do the followings unless* P = NP.

- DISTANCE-$k$ EDGE COLORING: *Approximate $\chi(\mathcal{L}^k(G))$ to within a factor of $n^{1/3-\epsilon}$ for $k \in \{2,3\}$ (thus the hardness of* STRONG EDGE COLORING*) and $n^{1/2-\epsilon}$ for $k \geq 4$.*
- MAXIMUM CLIQUE: *Approximate $\omega(G^k)$ to within a factor of $n^{1/2-\epsilon}$ (This is tight for all even $k$.)*
- MAXIMUM STABLE SET: *Approximate $\alpha(G^k)$ to within a factor of $n^{1/2-\epsilon}$ when $k$ is even and $n^{1-\epsilon}$ when $k$ is odd.*

Lastly, we further prove some new bounds related to the strong edge coloring of graphs, which might be of independent interests; most of the results are applications of the **fractional coloring argument** used in our hardness proofs.

- New bounds on the strong edge coloring of the *rooted, lexicographic* and *disjunctive products* of graphs: This fills in the missing pieces of Togni [32] who proved similar bounds for Cartesian, Kronecker, and strong products.

---

[2] We note that while the problems of computing $\alpha(G)$ and $\omega(G)$ are equivalent, computing $\alpha(G^k)$ and $\omega(G^k)$ are not; this is simply because the complement graph $\overline{G^k}$ of $G^k$ may not be the $k$th power of any graph.

- The conjecture of Erdös and Nešetřil for the class of *cographs*, graphs having no path on four vertices as an induced subgraph. (This is the only result proved by a different technique – a charging scheme.)
- Lower bound on the strong edge coloring number of a graph in terms of its chromatic number.

**Overview of Techniques:** We sketch the proof of the $n^{1/3-\epsilon}$ hardness of the strong edge coloring problem. The key idea is to employ the technique similar to the one used in our previous paper [6]. In short, the hardness results in [6] were obtained by, given any graph $G$, constructing a new graph $G'$ via a certain graph product, and then showing the relationship between graph measures of interest. For instance, to prove the hardness of the maximum bipartite induced matching problem, we construct the graph $G' = G^k \times K_2$ and prove that $im(G^k \times K_2) \approx \alpha(G^k)$ for sufficiently large $k$; note that $im(H)$ denotes the size of the maximum induced matching of $H$, $G^k$ is a $k$-fold disjunctive product of $G$, and $\times$ is a strong product. Since $\alpha(G)$ is hard to approximate to within a factor of $n^{1-\epsilon}$ [16], we immediately obtain an $n^{1-\epsilon}$ hardness result for approximating $im(G')$. Unfortunately, the technique in [6] involves only disjunctive and lexicographic product, which are unlikely to yield any result here because, roughly speaking, these products blow up the strong chromatic index of graphs. A new graph product and relationship between $\chi'_S(G)$ and some hard-to-approximate graph measure are needed.

Our main result in the simplified form is as below.

$$\chi'_S(G \circ S_\ell) \approx \chi(G)\ell + \chi'_S(G) \tag{1}$$

where $G \circ H$ is a rooted product between $G$ and $H$, and $S_\ell$ is a star with $\ell$ leaves. By choosing $\ell$ to be $|V(G)|^2$, the term $\chi(G)\ell$ becomes $\chi(G)|V(G)|^2$, while $\chi'_S(G) \leq |V(G)|^2$. Since the first term dominates the second one, we have $\chi'_S(G \circ S_\ell) \approx \chi(G)|V(G)|^2$. Now the hardness of $\chi(G)$ [14] immediately implies the $n^{1/3-\epsilon}$ hardness of approximating $\chi'_S(G \circ S_\ell)$. The proof of Equation 1 utilizes the "fractional coloring argument", which allows us to derive many other results.

**Related Works:** A notion closely related to the strong chromatic index of a graph is the *induced matching number* – the maximum cardinality of a set of edges $M \subseteq E(G)$ such that $M$ induces a matching in $G$ (i.e., no two edges of $M$ share an endpoint or have endpoints that are joined by some edge in $G$). Notice that each color class of a proper strong edge coloring forms an induced matching. The complexity of computing the induced matching number of a graph is better understood. It was shown to be NP-complete in [33,10] and was shown to be $n^{1-\epsilon}$-hard to approximate (implicitly) in [8]. The $n^{1-\epsilon}$-hardness holds even in bipartite graphs (see [6]), and a $\Delta^{1-\epsilon}$-hardness holds for $\Delta$-degree-bounded graphs [7].

**Organization:** The hardness of distance-$k$ edge coloring when $k = 2$ is proved in Section 3 (the proof for general $k$ uses similar ideas). We discuss other optimization problems on power graphs in Section 4.

## 2 Preliminaries

We use standard graph terminologies as in [9]. Let $G = (V, E)$ be a graph on $n$ vertices. The maximum and minimum degree of $G$ is denoted by $\Delta(G)$ and $\delta(G)$, respectively. For any vertex $v \in V(G)$, denote by $\Gamma_G(v)$ the set of neighbors of $v$ in $G$. The *k-th power of $G$*, denoted by $G^k = (V, E^k)$, is a graph with the same vertex set as $G$ such that $G^k$ has an edge $uv$ if and only if $u$ and $v$ are within distance at most $k$ in $G$. The graph $G^2$ is called the *square* of $G$.

**Edge Coloring.** The *edge-coloring* of $G$ is defined similarly as a coloring on edges such that any two edges sharing an endpoint receive different colors. The *chromatic index* (a.k.a, the edge-coloring number) of $G$, denoted by $\chi'(G)$, is the minimum number of colors needed to color edges of $G$. The *line graph* of $G$, denoted by $\mathcal{L}(G)$, is the graph whose vertex set is $E(G)$, and there is an edge $ef$ in $\mathcal{L}(G)$ joining two vertices $e, f \in V(\mathcal{L}(G))$ if and only if edges $e$ and $f$ share an endpoint in $G$ (we use $e$ and $f$ to denote both vertices in $\mathcal{L}(G)$ and edges in $G$). We say that a matching $M \subseteq E(G)$ is an *induced matching* in $G$ if and only if it is a matching such that any pair of edges $e$ and $f$ in $M$ has no edge joining their endpoints. In other words, a subgraph induced by vertices in $M$ is $M$ itself. The *strong edge coloring* of $G$ is a coloring of edges such that each color class form an *induced matching* in $G$, which is equivalent to a vertex-coloring of $\mathcal{L}^2(G)$. The *strong chromatic index* (or *strong edge coloring number*) of $G$, denoted by $\chi'_S(G)$, is the chromatic number of the square of $\mathcal{L}(G)$, i.e., $\chi'_S(G) = \chi(\mathcal{L}^2(G))$.

**Graph Products.** Given graphs $G$ and $H$ with a root $r \in V(H)$, the *rooted product* of $G$ and $H$, denoted by $G \circ H$ is defined as a graph obtained by making $|V(G)|$ copies of $H$ and unifying each vertex $u \in V(G)$ with the root $r$ of the $i$-th copy of $H$ for every $i = 1, 2, \ldots, |V(G)|$. More formally, given two graphs $G, H$ and a root $r \in V(H)$, the rooted product $G \circ H$ has the set of vertices $V(G) \times V(H)$. The set of edges is $E(G \circ H) = \{(v, r)(v', r) : vv' \in E(G)\} \cup \bigcup_{v \in V(G)} \{(v, a)(v, b) : ab \in E(H)\}$. The *disjunctive product* of $G$ and $H$, denoted by $G \vee H$, is the graph with a vertex set $V(G) \times V(H)$ and an edge set $E(G \bullet H) = \{(u, v)(u', v') : uu' \in E(G) \text{ or } vv' \in E(H)\}$. The *lexicographic product* of $G$ and $H$, denoted by $G \bullet H$, is a graph with a vertex set $V(G) \times V(H)$ and an edge set $E(G \bullet H) = \{(u, v)(u', v') : (uu' \in E(G)) or (u = u' \text{ and } vv' \in E(H)\}$.

**Our Problems.** The problems considered in this paper are defined as follows. We are given a graph $G = (V, E)$ on $n$ vertices and $m$ edges. In the strong edge coloring problem, the goal is to compute $\chi'_S(G) = \chi(\mathcal{L}^2(G))$ (and its corresponding coloring). The generalization of this problem is the distance-$k$ edge-coloring problem where we are given an additional constant $k$, and the goal is to compute $\chi(\mathcal{L}^k(G))$. In distance-$k$ vertex-coloring, maximum clique and maximum stable set problems, our goals are to compute $\chi(G^k)$, $\omega(G^k)$ and $\alpha(G^k)$, respectively.

Our results on the coloring problems are deduced from the hardness of the graph coloring problem [14], as stated in the following theorem.

**Theorem 2 ([14,16]+[34]).** *For any $\epsilon > 0$, unless* P $=$ NP, *it is hard to distinguish between the following two cases of a graph $G$ on $n$ vertices: (1)* YES-INSTANCE: $\chi(G) \leq n^\epsilon$ *and (2)* NO-INSTANCE: $\chi(G) \geq n^{1-\epsilon}$. *In particular, it is*

NP-hard to approximate the chromatic number of a graph to within a factor of $n^{1-\epsilon}$, for all $\epsilon > 0$.

For the hardness of the maximum stable set and clique problems on $G^k$, the following hardness of approximation result is needed.

**Theorem 3 ([16]+[34]).** *For any $\epsilon > 0$, unless $P = NP$, it is hard to distinguish between the following two cases of a graph $G$ on $n$ vertices: (1) Yes-Instance: $\alpha(G) \leq n^\epsilon$ (resp., $\omega(G) \geq n^{1-\epsilon}$) and (2) No-Instance: $\alpha(G) \geq n^{1-\epsilon}$ (resp., $\omega(G) \leq n^\epsilon$). In particular, it is NP-hard to approximate the stable (resp., clique) number of a graph to within a factor of $n^{1-\epsilon}$, for all $\epsilon > 0$.*

## 3  Strong Edge Coloring

In this section, we prove the approximation hardness of the the strong edge coloring problem, i.e., computing $\chi(\mathcal{L}^2(G))$. We remark that a trivial algorithm gives a $\Delta(G)$-approximation: The graph $\mathcal{L}^2(G)$ contains a clique of size $\Delta(G)$ and has maximum degree at most $2\Delta(G)^2$ (thus, $(2\Delta^2 + 1)$-colorable), which implies an $O(n)$-approximation. We now show the $n^{1/3-\epsilon}$ hardness, for any $\epsilon > 0$.

The key step lies in proving a new bound on the strong chromatic index of the rooted product of graphs. Roughly speaking, we show that

$$\chi'_S(G \circ H) = \tilde{\Theta}\left(\chi'_S(G) + \chi'_S(H - r) + \deg_H(r) \cdot \chi(G)\right)$$

where $\tilde{\Theta}(x)$ hides a polylog$(x)$ factor. More precisely, we prove the following theorem. (Note that $\chi_f(G) \leq \chi(G) \leq O(\chi_f(G) \log_2 |V(G)|)$; see, e.g., [25].)

**Theorem 4 (Rooted Product).** *Consider graphs $G$ and $H$ with a root vertex $r \in V(H)$. The following holds:*

$$\max\{\chi'_S(G), \chi'_S(H - r), \deg_H(r) \cdot \chi_f(G)\}$$
$$\leq \chi'_S(G \circ H) \leq \chi'_S(G) + \chi'_S(H - r) + \deg_H(r) \cdot \chi(G)$$

*Proof.* We first give a high level overview of the proof. First, recall that the rooted product $G \circ H$ consists of one copy of $G$ and $|V(G)|$ copies of $H$, where each copy is associated with one vertex $v \in V(G)$. The key idea in our proof is to partition edges of the graph $G \circ H$ into three parts $E_1, E_2$ and $E_3$, where $E_1$ is a copy of the edges of $G$, $E_2$ consists of disjoint copies of the edges of $H - r$, and $E_3$ consists of other edges. It is easy to color edges of $E_1$ using a strong edge coloring of $G$. Also, edges in different copies of $H - r$ are "far" from each other, so one can color edges of $E_2$, which consists of copies of $H - r$, using a strong edge coloring of $H - r$. Conversely, we can color edges in $G$ and $H - r$ using strong edge colorings of $E_1$ and $E_2$, respectively.

It remains to color the edges in $E_3$. We show that $\tilde{\Theta}(\deg_H(r) \cdot \chi(G))$ colors are needed. One direction, i.e., showing that $E_3$ needs $O(\deg_H(r) \cdot \chi(G))$ colors, is straightforward: If a vertex $v$ in $G$ has a color $c$, then we can assign colors $(c, 1), \ldots, (c, \deg_H(r))$ to edges in $E_3$ incident to $v$. Showing the converse that we

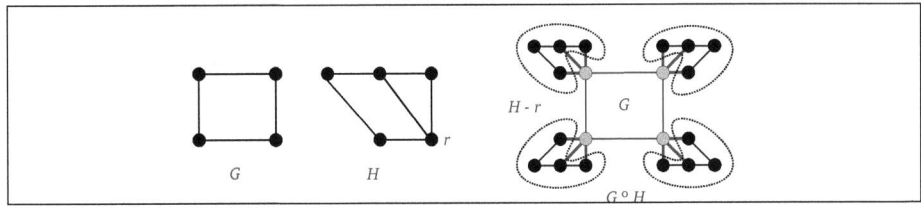

**Fig. 1.** The figure shows an example of the rooted product of $G$ and $H$ with a root $r$ and illustrates the proof of Theorem 4. The green edges are edges in $E_1$. The blue edges (in the cycles) are edges in $E_2$. The red thick edges are edges in $E_3$. The graph $G \circ H$ consists of one copy of $G$ and disjoint copies of $H$. Each copy of $H - r$ are "far" enough that it needs a path of length at least 2 to traverse to another copy.

need $\tilde{\Omega}(\deg_H(r) \cdot \chi(G))$ colors requires a more sophisticated idea – the *fractional coloring argument*.

Now, we give a formal proof. We first prove the right-hand-side: $\chi'_S(G \circ H) \leq \deg_H(r) \cdot \chi(G) + \chi'_S(G) + \chi'_S(H - r)$. The last two terms come from the fact that we can use strong edge colorings of $G$ and $H - r$ to color almost every edge of $G \circ H$. To see this, we partition edges of $G \circ H$ into three parts, i.e., $E(G \circ H) = E_1 \cup E_2 \cup E_3$, where $E_1 = \{(v,r)(w,r) : vw \in E(G)\}$, $E_2 = \bigcup_{v \in V(G)} \{(v,a)(v,b) : a,b \neq r, ab \in E(H)\}$ and $E_3 = E(G \circ H) - (E_1 \cup E_2)$. Then we show how to color $E_1, E_2,$ and $E_3$ using $\chi'_S(G), \chi'_S(H - r)$ and $\deg_H(r)\chi(G)$ colors, respectively. First, take a strong edge coloring $\sigma_1 : E(G) \to [\chi'_S(G)]$ of $G$. For each edge $(u,r)(v,r) \in E_1$, we assign to $(u,r)(v,r)$ a color $\sigma((u,r)(v,r)) := \sigma_1(uv)$. This must be a proper (partial) coloring because no edges in $E_2 \cup E_3$ join endpoints of any two edges in $E_1$. So, we finish coloring $E_1$. Now, take a strong edge coloring $\sigma_2 : E(H - r) \to [\chi'_S(H - r)]$ of $H - r$. We assign to each edge $(v,a)(v,b) \in E_2$ a color $\sigma((v,a)(v,b)) := \sigma_2(ab) + \chi'_S(G)$. (We shift the color by $\chi'_S(G)$ to avoid using the same color as $E_1$.) Again, this is a proper (partial) coloring because no edges of $E_1 \cup E_3$ join endpoints of any two edges in $E_2$. So far, we have used $\chi'_S(G) + \chi'_S(H - r)$ colors for $E_1$ and $E_2$. Finally, we color $E_3$. Take a "vertex-coloring" $\sigma_3 : V(G) \to [\chi(G)]$ of $G$. We define $\deg_H(r) \cdot \chi(G)$ new colors from $\sigma_3$, denoted by $(i, a)$ for $i \in [\chi(G)]$ and $a : ra \in E(H)$. Then, for each edge $(v,r)(v,a) \in E_3$, we assign to $(v,r)(v,a)$ a color $\sigma((v,r)(v,a)) := (\sigma_3(v), a)$. So, the total number of colors we use in this step is $\chi(G) \deg_H(r)$, thus summing up to $\chi(G) \deg_H(r) + \chi'_S(G) + \chi'_S(H - r)$ colors as desired. It only remains to show that $\sigma$ is a proper strong edge coloring of $G \circ H$.

To see this, consider a pair of edges $(v,r)(v,a), (w,r)(w,b) \in E_3$. We will check for any possible violation. If $v = w$, then $(v,r) = (w,r)$, which means that the two edges receive different colors by construction, and we are done. Hence, we may assume that $v \neq w$. Now, $(v,r)(v,a), (w,r)(w,b)$ share no endpoints. So, a possible violation is that some edge in $E(G \circ H)$ joins their endpoints. If $(v,r)(v,a)$ and $(w,r)(w,b)$ are joined by an edge $(v,r)(w,r) \in E_1$, then we must have an edge $vw \in E(G)$. So, $\sigma_3(v) \neq \sigma_3(w)$, thus implying that $\sigma((v,r)(v,a)) \neq \sigma((w,r)(w,b))$. Otherwise, $(v,r)(v,a)$ and $(w,r)(w,b)$ are joined by an edge in

$E_2 \cup E_3$. But, this is not possible since $v \neq w$ whereas edges in $E_2 \cup E_3$ are of the form $(u,a)(u,b)$ where $ab \in E(H)$. (Edges in $E_2 \cup E_3$ only join vertices in the same copy of $H$.)

Now, we prove the left-hand-side: $\max\{\chi'_S(G), \chi'_S(H-r), \deg_H(r) \cdot \chi_f(G)\} \leq \chi'_S(G \circ H)$. Clearly, $\max\{\chi'_S(G), \chi'_S(H-r)\}$ is the minimum number of colors that we need to strongly color edges of $G \circ H$ since $G \circ H$ has both $G$ and $H$ as subgraphs. So, it suffices to show that $G \circ H$ requires at least $\deg_H(r) \cdot \chi_f(G)$ colors. To prove this, we map a strong edge coloring of $G \circ H$ to a fractional vertex-coloring of $G$. Let $C_1, C_2, \ldots, C_M$ be color classes of a minimum strong edge coloring in $G \circ H$. We will show that $\chi_f(G) \leq M/\deg_H(r)$.

We define the fractional color classes of $G$ by $D_i = \{v \in V(G) : (v,r)(v,a) \in C_i \text{ for some } a \in V(H)\}$ for $i = 1, 2, \ldots, M$. Then we assign a fractional value of $1/\deg_H(r)$ to each color class $D_i$. Let us check that these color classes form a proper fractional vertex-coloring of $G$, i.e., (1) each $D_i$ is a stable set in $G$ and (2) each vertex belongs to at least $\deg_H(r)$ color classes.

Suppose the first condition does not hold. Then there is an edge $vw \in E(G)$ joining vertices $u, v$ from the same color class $D_i$. But, then there are edges $(v,r)(v,a)$ and $(w,r)(w,b)$ in the same (strong edge) color class $C_i$ that are joined by an edge $(v,r)(w,r)$, contradicting the fact that $C_i$ forms an induced matching in $G$. Next, for the second condition, we know that each vertex $(v,r) \in V(G \circ H)$ has $\deg_H(r)$ neighbors of the form $(v,a)$ where $ra \in E(H)$. So, we have $\deg_H(r)$ edges in $G \circ H$ of the form $(v,r)(v,a)$ where $ra \in E(H)$, and these edges could not have the same colors (since they share $(v,a)$ as an endpoint). It follows that $v$ belongs to $\deg_H(r)$ color classes. This completes the proof. □

The statement of the above theorem can be simplified. Using the fact that $\chi(G) \leq \chi_f(G) \log |V(G)|$ and choosing $H$ as a star:

**Corollary 1.** *For any graph $G$ and a star $S_\ell$ with a root vertex $r$ and $\ell$ leaves,*

$$\chi'_S(G \circ S_\ell) = \tilde{\Theta}(\chi'_S(G) + \ell \cdot \chi(G)).$$

### 3.1 Hardness of Approximation

Our construction is simple. We take a graph $G = (V, E)$ on $N$ vertices given by Theorem 2, and we output a graph $\widehat{G} = G \circ H$ where $H$ is a star with $\ell = N^2$ leaves. See Figure 2. Now, we invoke Corollary 1 to analyze the hardness gap. In Yes-Instance, $\chi(G) \leq N^\epsilon$ implies $\chi'_S(\widehat{G}) \leq N^2 N^\epsilon + N^2 \leq N^{2+2\epsilon}$. In No-Instance, $\chi(G) \geq N^{1-\epsilon}$ implies $\chi'_S(\widehat{G}) \geq \frac{N^2 N^{1-\epsilon}}{\log N} \geq N^{3-2\epsilon}$. So, the gap is $N^{1-\epsilon} = |V(\widehat{G})|^{1/3-\epsilon}$, thus proving Theorem 1.

### 3.2 Distance-$k$ Edge-Coloring

Here we sketch the ideas of generalizing our hardness to arbitrary $k$. For $k = 3$, we take an instance $\widehat{G} = G \circ H$ as in the previous section. Then we subdivide each edge $(v,r)(w,r)$ of $G \circ H$ by a path of length 2, namely, $((v,r), x_{vw}, (w,r))$.

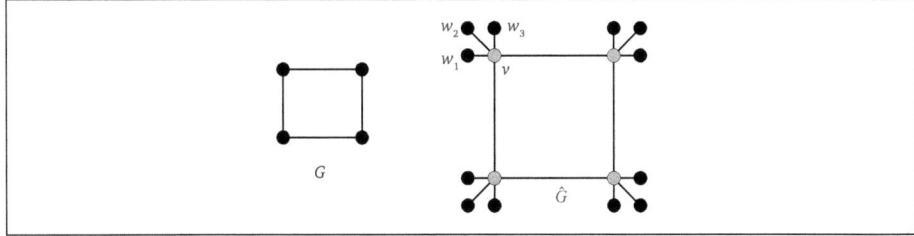

**Fig. 2.** The figure shows an example of a reduction from an instance of the graph coloring problem to the strong edge coloring problem with $\ell = 3$

The similar analysis gives a hardness of $n^{1/3-\epsilon}$ for this case. For $k = 4$, we change the choice of a graph $H$. Instead of using a star with $N^2$ leaves, we choose $H$ as a clique on $N$ vertices. The proof of the following theorem is omitted.

**Theorem 5.** *For any graph $G$ and a clique $K_\ell$, the following holds:*

$$\tilde{\Omega}\left(\frac{\ell^2 \cdot \chi(G)}{|V(G)|}\right) \leq \chi(\mathcal{L}^4(G \circ K_\ell)) \leq O(\ell^2 \cdot \chi(G) + |E(G)| + \ell|V(G)|).$$

To prove Theorem 5, we apply the same analysis as before. This approach allows us to prove a hardness of $n^{1/2-\epsilon}$ for the distance-$k$ edge coloring. For $k > 4$, we modify the construction for $k = 4$ by replacing each edge $uv$ of $G$ by a path of length $k - 3$. Detail is omitted.

### 3.3 Strong Edge Coloring of Other Graph Products

In this section, we state new bounds on the strong chromatic index of lexicographic ($\bullet$) and disjunctive ($\vee$) products of graphs.

**Theorem 6 (Lexicographic Product).** *For any graphs $G$ and $H$,*
$\max\left\{\frac{|V(H)|^2\chi'_S(G)}{\log|V(H)|}, \frac{\chi(G)\chi'_S(H)}{\log|V(G)|}\right\} \leq \chi'_S(G \bullet H) \leq |V(H)|^2\chi'_S(G) + \chi(G)\chi'_S(H).$

**Theorem 7 (Disjunctive Product).** *For any graphs $G$ and $H$,*
$\max\left\{\frac{|V(H)|^2\chi'_S(G)}{\log|V(H)|}, \frac{|V(G)|^2\chi'_S(H)}{\log|V(G)|}\right\} \leq \chi'_S(G \vee H) \leq |V(H)|^2\chi'_S(G) + |V(G)|^2\chi'_S(H).$

The proofs of these theorems use techniques from Theorem 4 and are omitted.

## 4 Other Problems on $G^k$

### 4.1 Maximum Clique

We sketch a construction for the case $k = 2$. The full proof is omitted due to the limitation of space. Given a graph $G' = (V', E')$, we construct $G$ by replacing each vertex $v$ of degree $d$ by a *star* $S_{d+1}$ and then rewiring each edge to a different vertex of $S_{d+1}$. Note that $S_{d+1}$ is a tree on $d+1$ vertices consisting of a root vertex $r$ and $d$ leaves. So, the root vertex $r$ of $S_{d+1}$ corresponds to a vertex $v$ of $G'$. We call the root vertex of each star a *canonical* vertex. Moreover, two distinct canonical vertices $r$ and $r'$ are within distance 2 of each other in $G$ if and only if their corresponding vertices $v$ and $v'$ of $G'$ are adjacent. Thus, the clique number restricted to canonical vertices corresponds to the clique number of $G'$. However, this number can be smaller than that restricted to non-canonical vertices. Hence, we make $|V(G')|$ copies of each canonical vertex to ensure that the former term (i.e., the clique number restricted to canonical vertices) is larger than the latter one (i.e., the clique number restricted to non-canonical vertices). So, we can derive the hardness of $|V(G')|^{1-\epsilon}$ from the maximum clique problem in $G'$. Since the output graph $G$ has $|V(G)| = |V(G')|^2$ vertices, we have the hardness of $n^{1/2-\epsilon}$-hardness, for any $\epsilon > 0$. For $k > 2$, we modify the construction by subdividing each edge $uv$ corresponding to an edge of $G$ by a path of length $k-1$. So, the maximum clique problem in $G^k$ admits no $n^{1/2-\epsilon}$-approximation, for any $\epsilon > 0$, unless P = NP.

### 4.2 Maximum Stable Set

We sketch tight hardness construction for the maximum stable set problem on $G^k$. For different $k$, the approximability of the problem varies, depending on the parity of $k$. Here we only discuss the case $k = 2$ and $k = 3$. The case of all odd and even $k$ can be obtained by simply modifying these two constructions. The proof is omitted.

First, consider the case when $k = 2$. Let $G$ be an input graph. The key idea is to subdivide each edge in $E(G)$ so that each pair of non-adjacent vertices are in distance more than 2 of each other. Thus, any stable set in the graph has a corresponding stable set in its square. More formally, we construct a graph $H$ by first subdividing each edge $e \in E(G)$ by a *special vertex* $x(e)$.

To make sure that these special vertices would not form a stable set in $H^2$, we add edges $x(e)x(e')$ for any pair of special vertices $x(e), x(e')$. Thus, $\alpha(G) \leq \alpha(H) \leq \alpha(G) + 1$. As $|V(H)| = |E(G)| + |V(G)| \leq 2|V(G)|^2$, we have an $n^{1/2-\epsilon}$-hardness for computing $\alpha(G^2)$. The hardness is tight because it matches the upper bounds of $O(\sqrt{n})$-approximation provided by the algorithm of Halldórsson et al. [15].

For $k = 3$, the key idea of the construction is simply ensuring that any pair of vertices are not within distance 3 of each other. Given a graph $G = (V, E)$ on $n$ vertices, we construct a graph $H$ by attaching a new vertex $v^*$ to each vertex $v \in V(G)$; we call $v$ a *white* vertex and call $v^*$ a *black* vertex. Now, we have a graph $H$ on $2|V(G)|$ vertices. Observe that a pair of black vertices $v^*$ and $w^*$ are at distance 3 of each other in $H$ if and only if their corresponding white vertices $v$ and $w$ are adjacent in $G$. Thus, a set of black vertices $S^*$ is stable in $H^3$ if and only if the set of white vertices $S = \{v : v^* \in S^*\}$ is stable in $G$. So, $\alpha(G) \leq \alpha(H^3) \leq 2\alpha(G)$. The $|V(G)|^{1-\epsilon}$-hardness follows immediately.

## 5 Conclusion and Open Problems

This paper shows the hardness of the distance-$k$ edge coloring problem and illustrates applications of the *fractional coloring argument* in proving hardness of approximation. For some of the problem, the gaps between the lower and upper bounds on approximation thresholds remain open. In particular, for the strong edge coloring problem, our result implies that the right approximation threshold lies between $n^{1/3-\epsilon}$ and $n$, but the exact value is still unknown. Proving the tight approximation hardness might require a totally new idea and thus is very challenging. The problem on bounded degree and regular graphs are also open and might be interesting as well.

**Acknowledgement.** We thank Ross Kang for pointing out interesting references and also thank anonymous reviewers for their helpful comments.

## References

1. Agnarsson, G., Greenlaw, R., Halldrsson, M.M.: On powers of chordal graphs and their colorings. Congr. Numer. 144, 41–65 (2000)
2. Agnarsson, G., Halldórsson, M.M.: Coloring powers of planar graphs. SIAM J. Discrete Math. 16(4), 651–662 (2003), also in SODA 2000
3. Alon, N., Mohar, B.: The chromatic number of graph powers. Combinatorics, Probability & Computing 11(1), 1–10 (2002)
4. Barrett, C.L., Istrate, G., Vilikanti, A.K., Marathe, M., Thite, S.V.: Approximation algorithms for distance-2 edge coloring. Tech. rep., Los Alamos National Lab., NM, US (2002)
5. Barrett, C.L., Kumar, V.S.A., Marathe, M.V., Thite, S., Istrate, G.: Strong edge coloring for channel assignment in wireless radio networks. In: PerCom Workshops, pp. 106–110 (2006)
6. Chalermsook, P., Laekhanukit, B., Nanongkai, D.: Graph products revisited: Tight approximation hardness of induced matching, poset dimension and more. In: SODA, pp. 1557–1576 (2013)
7. Chalermsook, P., Laekhanukit, B., Nanongkai, D.: Independent set, induced matching, and pricing: Connections and tight (subexponential time) approximation hardnesses. In: FOCS (2013)
8. Chlebík, M., Chlebíková, J.: Complexity of approximating bounded variants of optimization problems. Theor. Comput. Sci. 354(3), 320–338 (2006), preliminary version in FCT 2003
9. Diestel, R.: Graph Theory, 4th edn. Graduate Texts in Mathematics, vol. 173. Springer, Heidelberg (2010), http://diestel-graph-theory.com/
10. Duckworth, W., Manlove, D., Zito, M.: On the approximability of the maximum induced matching problem. J. Discrete Algorithms 3(1), 79–91 (2005)
11. Erickson, J., Thite, S., Bunde, D.P.: Distance-2 edge coloring is NP-complete. CoRR abs/cs/0509100 (2005)
12. Faudree, R.J., Gyárfás, A., Schelp, R.H., Tuza, Z.: Induced matchings in bipartite graphs. Discrete Math. 78(1-2), 83–87 (1989)
13. Faudree, R.J., Gyárfás, A., Schelp, R.H., Tuza, Z.: The strong chromatic index of graphs. Ars. Combin. 29B, 205–2011 (1990)

14. Feige, U., Kilian, J.: Zero knowledge and the chromatic number. In: CCC, pp. 278–287 (1996)
15. Halldórsson, M.M., Kratochvíl, J., Telle, J.A.: Independent sets with domination constraints. Discrete Appl. Math. 99(1-3), 39–54 (2000)
16. Håstad, J.: Clique is hard to approximate within $n^{1-\epsilon}$. In: FOCS, pp. 627–636 (1996)
17. Hell, P., Raspaud, A., Stacho, J.: On injective colourings of chordal graphs. In: Laber, E.S., Bornstein, C., Nogueira, L.T., Faria, L. (eds.) LATIN 2008. LNCS, vol. 4957, pp. 520–530. Springer, Heidelberg (2008)
18. Hocquard, H., Valicov, P.: Strong edge colouring of subcubic graphs. Discrete Appl. Math. 159(15), 1650–1657 (2011)
19. Kaiser, T., Kang, R.J.: The distance-t chromatic index of graphs. Combinatorics, Probability and Computing 23, 90–101 (2014),
http://journals.cambridge.org/article_S0963548313000473
20. Kang, R.J., Manggala, P.: Distance edge-colourings and matchings. Discrete Applied Mathematics 160(16-17), 2435–2439 (2012)
21. Král, D.: Coloring powers of chordal graphs. SIAM J. Discrete Math. 18(3), 451–461 (2004)
22. Krumke, S.O., Marathe, M.V., Ravi, S.S.: Models and approximation algorithms for channel assignment in radio network. Wireless Networks 7(6), 575–584 (2001)
23. Laekhanukit, B.: Parameters of two-prover-one-round game and the hardness of connectivity problems. To appear in SODA 2014 (2014)
24. Lloyd, E.L., Ramanathan, S.: On the complexity of distance-2 coloring. In: ICCI, pp. 71–74 (1992)
25. Lovász, L.: On the ratio of optimal integral and fractional covers. Discrete Math. 13(4), 383 (1975),
http://www.sciencedirect.com/science/article/pii/0012365X75900588
26. Mahdian, M.: The strong chromatic index of graphs. Master's thesis, University of Toronto (2000)
27. Mahdian, M.: On the computational complexity of strong edge coloring. Discrete Appl. Math. 118(3), 239–248 (2002)
28. McCormick, S.: Optimal approximation of sparse hessians and its equivalence to a graph coloring problem. Math. Program. 26, 153–171 (1983),
http://dx.doi.org/10.1007/BF02592052, doi:10.1007/BF02592052
29. Misra, J., Gries, D.: A constructive proof of vizing's theorem. Information Processing Letters 41(3), 131–133 (1992),
http://www.sciencedirect.com/science/article/pii/002001909290041S
30. Molloy, M.S.O., Reed, B.A.: A bound on the strong chromatic index of a graph. J. Comb. Theory, Ser. B 69(2), 103–109 (1997)
31. Ramanathan, S., Lloyd, E.L.: Scheduling algorithms for multihop radio networks. IEEE/ACM Trans. Netw. 1(2), 166–177 (1993), also in SODA 2000
32. Togni, O.: Strong chromatic index of products of graphs. Discrete Math. Theor. Comput. Sci. 9(1) (2007)
33. Zito, M.: Induced matchings in regular graphs and trees. In: Widmayer, P., Neyer, G., Eidenbenz, S. (eds.) WG 1999. LNCS, vol. 1665, pp. 89–100. Springer, Heidelberg (1999)
34. Zuckerman, D.: Linear degree extractors and the inapproximability of max clique and chromatic number. Theory of Computing 3(1), 103–128 (2007)

# Convexity in Partial Cubes: The Hull Number

Marie Albenque[1] and Kolja Knauer[2]

[1] LIX UMR 7161, École Polytechnique, CNRS, France
[2] I3M, Université Montpellier 2, France

**Abstract.** We prove that the combinatorial optimization problem of determining the hull number of a partial cube is NP-complete. This makes partial cubes the minimal graph class for which NP-completeness of this problem is known and improves some earlier results in the literature.

On the other hand we provide a polynomial-time algorithm to determine the hull number of planar partial cube quadrangulations.

Instances of the hull number problem for partial cubes described include poset dimension and hitting sets for interiors of curves in the plane.

To obtain the above results, we investigate convexity in partial cubes and characterize these graphs in terms of their lattice of convex subgraphs, improving a theorem of Handa. Furthermore we provide a topological representation theorem for planar partial cubes, generalizing a result of Fukuda and Handa about rank 3 oriented matroids.

## 1 Introduction

The object of this paper is the study of convexity and particularly of the hull number problem on different classes of partial cubes. Our contribution is twofold. First, we establish that the hull number problem is NP-complete for partial cubes, second, we emphasize reformulations of the hull number problem for certain classes of partial cubes leading to interesting problems in geometry, poset theory and plane topology.

Denote by $Q^d$ the hypercube graph of dimension $d$. A graph $G$ is called a *partial cube* if there is an injective mapping $\phi : V(G) \to V(Q^d)$ such that $d_G(v,w) = d_{Q^d}(\phi(v), \phi(w))$ for all $v, w \in V(G)$, where, $d_G$ and $d_{Q^d}$ denote the graph distance in $G$ and $Q^d$, respectively. It implies in particular that for each pair of vertices of $G$, at least one shortest path between them in $Q^d$ belongs also to $G$. In other words $\phi(G)$, seen as an induced subgraph of $Q^d$, is an *isometric embedding* of $G$ in $Q^d$.

Partial cubes were introduced by Graham and Pollak in [24] in the study of interconnection networks and continue to find strong applications; they form for instance the central graph class in media theory (see the recent book [18]) and frequently appear in chemical graph theory e.g. [17]. Partial cubes form a generalization of several important graph classes, thus have also many applications in different fields of mathematics. Indeed, they "present one of the central and most studied classes of graphs in all of the metric graph theory", citing [30].

This article discusses some examples of such families of graphs including Hasse diagrams of upper locally distributive lattices or equivalently antimatroids [20]

(Section 2.2), region graphs of halfspaces and hyperplanes (Section 3), and tope graphs of oriented matroids [11] (Section 5). These families contain many graphs defined on sets of combinatorial objects: flip-graphs of strongly connected and acyclic orientations of digraphs [12], linear extension graphs of posets [32], integer tensions of digraphs [20], configurations of chip-firing games [20], to name a few.

Convexity for graphs is the natural counterpart of Euclidean convexity and is defined as follows; a subgraph $G'$ of $G$ is said to be *convex* if all shortest paths in $G$ between vertices of $G'$ actually belong to $G'$. The *convex hull* of a subset $V'$ of vertices – denoted $\mathrm{conv}(V')$ – is defined as the smallest convex subgraph containing $V'$. Since the intersection of convex subgraphs is clearly convex, the convex hull of $V'$ is the intersection of all the convex subgraphs that contain $V'$.

A subset of vertices $V'$ of $G$ is a *hull set* if and only if $\mathrm{conv}(V') = G$. The *hull number* or *geodesic hull number* of $G$, denoted by $hn(G)$, is the size of a smallest hull set. It was introduced in [19], and since then has been the object of numerous papers. Most of the results on the hull number are about computing good bounds for specific graph classes, see e.g. [9, 28, 7, 6, 16, 8]. Only recently, in [15] the focus was set on computational aspects of the hull number and it was proved that determining the hull number of a graph is NP-complete. This was strengthened to bipartite graphs in [1]. On the other hand, polynomial-time algorithms have been obtained for unit-interval graphs, cographs and split graphs [15], cactus graphs and $P_4$-sparse graphs [1], distance hereditary graphs and chordal graphs [29]. Moreover, in [2], a fixed parameter tractable algorithm to compute the hull number of any graph class was obtained. Here, the parameter is the size of a vertex cover.

Let us end this introduction with an overview of the results and the organization of this paper. Section 2 is devoted to properties of convexity in partial cubes and besides providing tools for the other sections, its purpose is to convince the reader that convex subgraphs of partial cubes behave nicely. First a characterization of partial cubes in terms of their convex subgraphs is given. In particular, convex subgraphs of partial cubes behave somewhat like polytopes in Euclidean space. Namely, they satisfy an analogue of the Representation Theorem of Polytopes [35]. We then prove that for any vertex $v$ in a partial cube $G$, the set of convex subgraphs of $G$ containing $v$ ordered by inclusion forms an upper locally distributive lattice. This property leads to a new characterization of partial cube, strengthening a theorem of Handa [26].

In Section 3 the problem of determining the hull number of a partial cube is proved to be NP-complete, improving earlier results of [15] and [1]. Our proof implies a stronger result by showing that determining the hull number of a region graph of an arrangement of halfspaces and hyperplanes is also NP-complete.

In Section 4 the relation between the hull number problem for linear extension graphs and the dimension problem of posets is discussed. We present a quasi-polynomial-time algorithm to compute the dimension of a poset given its linear extension graph and conjecture that the problem is polynomial-time solvable.

Section 5 is devoted to planar partial cubes. We provide a new characterization, which is a topological representation theorem generalizing work of Fukuda and Handa on rank 3 oriented matroids [22]. This characterization is then exploited to obtain a polynomial-time algorithm that computes the hull number of planar partial cube quadrangulations.

## 2 Convexity in Partial Cubes

### 2.1 Partial Cubes and Cut-Partitions

All graphs studied in this article are supposed to be connected, simple and undirected. We use the classic graph terminology of [5]. Given a graph $G$ a *cut* $C \subseteq E$ is an inclusion-minimal set of edges whose removal disconnects $G$. The removal of a cut $C$ leaves exactly two connected components called its *sides*, denoted by $C^+$ and $C^-$. For $V' \subset V$, a cut $C$ *separates* $V'$ if both $C^+ \cap V'$ and $C^- \cap V'$ are not empty. A *cut-partition* of $G$ is a set $\mathcal{C}$ of cuts partitioning $E$. For a cut $C \in \mathcal{C}$ and $V' \subseteq V$ define $C(V')$ as $G$ if $C$ separates $V'$ and otherwise as the side of $C$ containing $V'$.

*Observation 1.* A graph $G$ is bipartite if and only if $G$ has a cut-partition.

The equivalence classes of the *Djoković-Winkler relation* of a partial cube [14, 33] can be interpreted as the cuts of a cut-partition. Reformulating some properties of these equivalence classes as well as some results from [10, 3] the following new characterization of partial cubes in terms of cut partitions can be obtained.

**Theorem 2.** *A connected graph $G$ is a partial cube if and only if $G$ admits a cut-partition $\mathcal{C}$ satisfying one of the following equivalent conditions:*

*(i) for all $u, v \in V$, there is a shortest path between them using no $C \in \mathcal{C}$ twice*
*(ii) no shortest path in $G$ uses any $C \in \mathcal{C}$ twice*
*(iii) for all $V' \subseteq V$ : $\mathrm{conv}(V') = \bigcap_{C \in \mathcal{C}} C(V')$*
*(iv) for all $v, w \in V$ : $\mathrm{conv}(v, w) = \bigcap_{C \in \mathcal{C}} C(v, w)$*

Note that (iii) resembles the *Representation Theorem for Polytopes*, see [35]; where the role of points is taken by vertices and the halfspaces are mimicked by the sides of the cuts in the cut-partition. Thanks to (iii), the hull number problem has now a very useful interpretation as a hitting set problem:

**Corollary 3.** *Let $\mathcal{C}$ be a cut-partition that satisfies Theorem 2 then $V'$ is a hull set if and only if on both sides of $C$ there is a vertex of $V'$, for all $C \in \mathcal{C}$.*

### 2.2 Partial Cubes and Upper Locally Distributive Lattices

In this subsection we present another indication for how nice partial cubes behave with respect to convexity. Generalizing a theorem of Handa [26] we characterize partial cubes in terms of their lattice of convex subgraphs, see Fig.1.

A partially ordered set or poset $\mathcal{L} = (X, \leq)$ is a *lattice*, if each pair of elements $x, y \in \mathcal{L}$ admits both a unique largest element smaller than both of them called their *meet* and denoted $x \wedge y$, and a unique smallest element larger than both of them called their *join* and denoted $x \vee y$. Since both these operations are associative, we can define $\bigvee M := x_1 \vee \ldots \vee x_k$ and $\bigwedge M := x_1 \wedge \ldots \wedge x_k$ for $M = \{x_1, \ldots, x_k\} \subseteq \mathcal{L}$. Furthermore define $\bigvee \emptyset$ and $\bigwedge \emptyset$ as respectively the minimal and maximal element of $\mathcal{L}$.

An element is called *meet-reducible* if it can be written as the meet of elements all different from itself and is called *meet-irreducible* otherwise. For $\mathcal{L} = (X, \leq)$ and $x, y \in X$, one says that $y$ covers $x$ and writes $x \prec y$ if and only if $x < y$ and there is no $z \in X$ such that $x < z < y$. The *Hasse diagram* of $\mathcal{L}$ is then the directed graph on the elements of $X$ with an arc $(x, y)$ if $x \prec y$. The classical convention is to represent a Hasse diagram as undirected graph but with a drawing in the plane such that the orientation of edges can be recovered by directing them in upward direction. It is easy to see that an element $x$ is a meet-irreducible if and only if there is exactly one edge in the Hasse diagram leaving $x$ in upward direction. (Note that the maximum of $\mathcal{L}$ is indeed meet-reducible since it can be written $\bigwedge \emptyset$.)

A lattice is called *upper locally distributive* or ULD if each of its elements admits a unique minimal representation as the meet of meet irreducibles. In other words, for every $x \in \mathcal{L}$ there is a unique inclusion-minimal set $\{m_1, \ldots, m_k\} \subseteq \mathcal{L}$ of meet-irreducibles such that $x = m_1 \wedge \ldots \wedge m_k$.

ULDs were first defined by Dilworth [13] and have thereafter often reappeared, see [31] for an overview until the mid 80s. In particular, the Hasse diagram of a ULD is a partial cube, see e.g. [20]. The following theorem sheds light on the special role played by ULDs among partial cubes with respect to convexity.

**Theorem 4.** *A graph $G$ is a partial cube if and only if for every vertex $v$ the inclusion order of convex subgraphs containing $v$ forms a ULD whose Hasse diagram contains $G$ as an isometric subgraph.*

## 3 NP-Completeness of Hull Number in Partial Cubes

The section is devoted to the proof of the following result:

**Theorem 5.** *Given a partial cube $G$ and an integer $k$ it is NP-complete to decide whether $\mathrm{hn}(G) \leq k$.*

*Proof.* Observe first that by Corollary 3, computing the convex hull of a set of vertices in a partial cube is doable in polynomial-time. It is also doable in polynomial-time in general graphs, see e.g. [15]. To prove the NP-completeness, we exhibit a reduction from the following problem, known to be NP-complete [23]:

**SAT-AM3:**
**Instance:** A formula $F$ in Conjunctive Normal Form on $m$ clauses $D_1, \ldots, D_m$, each consisting of at most three literals on variables $x_1, \ldots, x_n$. Each variable *appears in at most three clauses*.
**Question:** Is $F$ satisfiable?

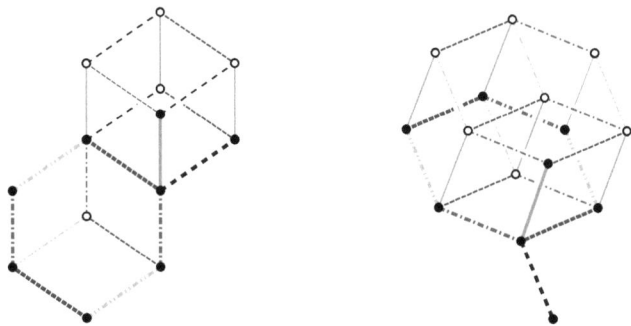

**Fig. 1.** Two ULDs obtained from the same partial cube (thick edges) by fixing a different vertex

Given an instance $F$ of SAT-AM3, we construct a partial cube $G_F$ such that $F$ is satisfiable if and only if $\text{hn}(G_F) \leq n+1$.

Given $F$ we start with two vertices $u$ and $u'$ connected by an edge. For each $1 \leq i \leq m$, introduce a vertex $d_i$ and link it to $u$. If two clauses, say $D_i$ and $D_j$, share a literal, add a new vertex $d_{i,j}$ and connect it to both $d_i$ and $d_j$.

Now for each variable $x$, introduce a copy $G_x$ of the subgraph induced by $u$ and the vertices corresponding to clauses that contain $x$ (including vertices of the form $d_{\{i,j\}}$ in case $x$ appears in the same literal in $D_i$ and $D_j$). Connect $G_x$ to the rest of the graph by introducing a matching $M_x$ connecting each original vertex with its copy in $G_x$ and call $G_F$ the graph obtained. Assume without loss of generality that each Boolean variable $x$ used in $F$ appears at least once non-negated (denoted by $x$ with a slight abuse of notations) and once negated (denoted by $\bar{x}$). Then, each literal appears at most twice in $F$ and the two possible options for $G_x$ are displayed on Fig.2. Label the vertices of $G_x$ according to that figure.

Observe first that $G_F$ is a partial cube. Define a cut partition of $G_F$ into $n+m+1$ cuts as follows. One cut consists of the edge $(u, u')$. The cut associated to a clause $D_i$ contains the edge $\{u, d_i\}$, any edge of the form $\{d_{\{i,j\}}, d_j\}$ and all the copies of such edges that belong to one of the $G_x$. Let us call this cut $C_i$. Finally, the cut associated to a variable $x$ is equal to $M_x$. This cut partition satisfies Theorem 2(i). Indeed, for a cut $C$ denote respectively $\partial^+ C$ and $\partial^- C$ the vertices in $C^+$ and $C^-$ incident to edges of $C$. Theorem 2(i) is in fact equivalent to say that, for each cut $C \in \mathcal{C}$, between any pair of vertices of $\partial^+ C$ or $\partial^- C$, there exists a shortest path that contains no edge of $C$. A case by case analysis of the different cuts in $G_F$ concludes the proof.

Assume $F$ is satisfiable and let $S$ be a satisfying assignment of variables. Let $H$ be the union of $\{u'\}$ and the subset of vertices of $G_F$ corresponding to $S$. More formally, for each variable $x$, $H$ contains the vertex $v_x$ if $x$ is set to true in $S$ or the vertex $v_{\bar{x}}$ otherwise. Let us prove that $H$ is a hull set. Since $u$ belongs

(a) The variable $y$ appears twice in $F$,   (b) the variable $x$ three times.

**Fig. 2.** General structure of the graph $G_F$, with the two possible examples of gadgets associated to a variable. Red edges correspond to the cut $M_y$ on (a) and $M_x$ on (b).

to any path between $u'$ and any other vertex, $u$ belongs to conv($H$). Moreover, for each variable $x$, the vertex $u_x$ lies on a shortest path both between $v_x$ and $u'$ and between $v_{\bar{x}}$ and $u'$, hence all the vertices $u_x$ belong to conv($H$). Next, for each literal $\ell$ and for each clause $D_i$ that contains $\ell$, there exists a shortest path between $u'$ and $v_\ell$ that contains $d_i$. Then, since $S$ is a satisfying assignment of $F$, each clause vertex belongs to conv($H$). It follows that conv($H$) also contains all vertices $d_{i,j}$.

To conclude, it is now enough to prove that for all $\ell \notin S$, the vertex $v_\ell$ also belongs to conv($H$). In the case where $\ell$ appears in only one clause $D_i$, then $v_\ell$ belongs to a shortest path between $d_i$ and $u_\ell$. In the other case, $v_\ell$ belongs to a shortest $(u_\ell, d_{i,j})$-path. Thus, conv($H$) = $G$.

Assume now that there exists a hull set $H$, with $|H| \leq n+1$. By Corollary 3, the set $H$ necessarily contains $u'$ and at least one vertex of $G_x$ for each variable $x$. This implies that $|H| = n+1$ and therefore for all variables $x$, $H$ contains exactly one vertex $h_x$ in $G_x$. Since any vertex of $G_x$ lies either on a shortest $(u', v_x)$-path or $(u', v_{\bar{x}})$-path, we can assume without loss of generality that $h_x$ is either equal to $v_x$ or to $v_{\bar{x}}$. Hence, $H$ defines a truth assignment $S$ for $F$. Now let $C_i$ be the cut associated to the clause $D_i$ and let $C_i^+$ be the side of $C_i$ that contains $d_i$. Observe that if $v_x$ belongs to $C_i^+$, then $x$ appears in $D_i$. By Corollary 3, $H$ intersects $C_i^+$, hence there exists a literal $\ell$ such that $v_\ell$ belongs to $H$. Thus, $H$ encodes a satisfying truth-assignment of $F$. □

The gadget in the proof of Theorem 5 is a relatively special partial cube and the statement can thus be strengthened. For a polyhedron $P$ and a set $\mathcal{H}$ of hyperplanes in $\mathbb{R}^d$, the *region graph* of $P \setminus \mathcal{H}$ is the graph whose vertices are the connected components of $P \setminus \mathcal{H}$ and where two vertices are joined by an edge if their respective components are separated by exactly one hyperplane of $\mathcal{H}$. The proof of Theorem 5 can be adapted to obtain:

**Corollary 6.** *Let $P \subset \mathbb{R}^d$ be a polyhedron and $\mathcal{H}$ a set of hyperplanes. It is NP-complete to compute the hull number of the region graph of $P \setminus \mathcal{H}$.*

## 4 The Hull Number of a Linear Extension Graph

Given a poset $(P, \leq_P)$, a *linear extension* $L$ of $P$ is a total order $\leq_L$ on the elements of $P$ compatible with $\leq_P$, i.e., $x \leq_P y$ implies $x \leq_L y$. The set of vertices of the *linear extension graph* $G_L(P)$ of $P$ is the set of all linear extensions of $P$ and there is an edge between $L$ and $L'$ if and only if $L$ and $L'$ differ by a neighboring transposition, i.e., by reversing the order of two consecutive elements.

Let us see that property $(i)$ of Theorem 2 holds for $G_L(P)$. Each incomparable pair $x \parallel y$ of $(P, \leq_P)$ corresponds to a cut of $G_L(P)$ consisting of the edges where $x$ and $y$ are reversed. The set of these cuts is clearly a cut-partition of $G_L(P)$. Observe then that the distance between two linear extensions $L$ and $L'$ in $G_L(P)$ is equal to the number of pairs that are ordered differently in $L$ and $L'$, i.e., no pair $x \parallel y$ is reversed twice on a shortest path. Hence $G_L(P)$ is a partial cube.

A *realizer* of a poset is a set $S$ of linear extensions such that their intersection is $P$. In other words, for every incomparable pair $x \parallel y$ in $P$, there exist $L, L' \in S$ such that $x <_L y$ and $x >_{L'} y$. It is equivalent to say that, for each cut $C$ of the cut-partition of $G_L(P)$, the sets $C^+ \cap S$ and $C^- \cap S$ are not empty. By Corollary 3, it yields a one-to-one correspondence between realizers of $P$ and hull sets of $G_L(P)$. In particular the size of a minimum realizer – called the *dimension* of the poset and denoted $\dim(P)$ – is equal to the hull number of $G_L(P)$. The dimension is a fundamental parameter in poset combinatorics, see e.g. [32]. In particular, for every *fixed* $k \geq 3$, it is NP-complete to decide if a given poset has dimension at least $k$, see [34]. But if instead of the poset its linear extension graph is considered to be the input of the problem, then we get:

**Proposition 7.** *The hull number of a linear extension graph (of size $n$) can be determined in time $O(n^{c \log n})$, i.e., the dimension of a poset $P$ can be computed in quasi-polynomial-time in $G_L(P)$.*

*Proof.* An antichain in a poset is a set of mutually incomparable elements of $P$ and the width $\omega(P)$ of $P$ is the size of the largest antichain of $P$, see [32]. It is a classic result that $\dim(P) \leq \omega(P)$. Since any permutation of an antichain appears in at least one linear extension, $\omega(P)! \leq n$ and therefore $\dim(P) \leq \log(n)$. Thus, to determine the hull-number of $G_L(P)$ it suffices to compute the convex hull of all subsets of at most $\log(n)$ vertices. Since the convex hull can be computed in polynomial-time, we get the claimed upper bound. □

In fact, since the number of linear extensions of a poset is generally exponential in the size of the poset, it seems reasonable to conjecture:

*Conjecture 8.* The dimension of a poset given its linear extension graph can be determined in polynomial-time.

## 5 Planar Partial Cubes and Rank 3 Oriented Matroids

Oriented matroids have many equivalent definitions. We refer to [4] for a thorough introduction to oriented matroids and their plenty applications. Here, we will not state a formal definition. It suffices to know that the topes of an oriented matroid $\mathcal{M}$ on $n$ elements are a subset of $\{1, -1\}^n$ satisfying several axioms. Moreover, the topes determine $\mathcal{M}$ uniquely. Joining two topes by an edge if they differ by the sign of exactly one entry yields the *tope graph* of $\mathcal{M}$.

From the axioms of oriented matroids it follows that the tope graph $G$ of an oriented matroid is an *antipodal partial cube*, i.e., $G$ is a partial cube such that for every $u \in G$ there is a $\bar{u} \in G$ with $\mathrm{conv}(u, \bar{u}) = G$, see [4]. In particular we have $hn(G) = 2$. But, not all antipodal partial cubes can be represented as the tope graphs of oriented matroids, see [26] and finding a general graph theoretical characterization is still a big problem in oriented matroid theory. The exception is for tope graphs of oriented matroids of rank at most 3 which admit a characterization as planar antipodal partial cubes, see [22]. We need a few definitions to state this characterization precisely.

A *Jordan curve* is a simple closed curve in the plane. For an arrangement $\mathcal{S}$ of Jordan curves and $S \in \mathcal{S}$, $\mathbb{R}^2 \setminus S$, the complement of $S$ has two components: one is bounded and is called its *interior*, the other one, unbounded, is called its *exterior*. The closure of the interior of the exterior of $S$ are denoted respectively $S^+$ and $S^-$. The *region graph* of an arrangement $\mathcal{S}$ of Jordan curves is the graph whose vertices are the connected components of the complement of $\mathcal{S}$ in the plane and where two vertices are neighbors if their corresponding components are separated by exactly one element of $\mathcal{S}$. Using the Topological Representation Theorem for Oriented Matroids [4] the characterization of tope graphs of oriented matroids of rank at most 3 may be rephrased as:

**Theorem 9** ([22]). *A graph $G$ is an antipodal planar partial cube if and only if $G$ is the region graph of an arrangement $\mathcal{S}$ of Jordan curves such that for every $S, S' \in \mathcal{S}$ we have $|S \cap S'| = 2$ and for $S, S', S'' \in \mathcal{S}$ either $|S \cap S' \cap S''| = 2$ or $|S^+ \cap S' \cap S''| = |S^- \cap S' \cap S''| = 1$.*

Given a Jordan curve $S$ and a point $p \in \mathbb{R}^2 \setminus S$ denote by $S(p)$ the closure of the side of $S$ not containing $p$. An arrangement $\mathcal{S}$ of Jordan curves is called *non-separating*, if for any region $p \in \mathbb{R}^2 \setminus \mathcal{S}$ and subset $\mathcal{S}' \subseteq \mathcal{S}$ the set $\mathbb{R}^2 \setminus \bigcup_{S \in \mathcal{S}'} S(p)$ is connected. Two important properties of non-separating arrangements are summarized in the following:

*Observation 10.* Let $\mathcal{S}$ a non-separating arrangement. Then the interiors of $\mathcal{S}$ form a family of *pseudo-discs*, i.e., different curves $S, S' \in \mathcal{S}$ intersect in at most two points. Moreover, $\mathcal{S}$ has the *topological Helly property*, i.e., if the interiors of $S_1, S_2, S_3 \in \mathcal{S}$ mutually intersect, then $S_1^+ \cap S_2^+ \cap S_3^+ \neq \emptyset$.

In Fig. 3 we show how violating the pseudo-disc or the topological Helly property violates the property of being non-separating.

Non-separating arrangements of Jordan curves yield a generalization of Theorem 9. The construction of the proof is exemplified in Fig. 4.

**Theorem 11.** *A graph $G$ is a planar partial cube if and only if $G$ is the region graph of a non-separating arrangement $\mathcal{S}$ of Jordan curves.*

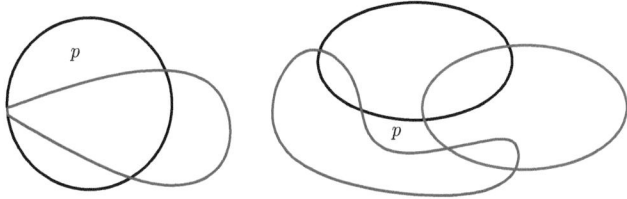

**Fig. 3.** Illustration of Observation 10. Left: Two curves intersecting in more than 2 points. Right: Three interiors of curves intersecting mutually but not having a common point. In both cases $p$ is a point proving that the arrangement is not non-separating.

*Proof.* Let $G$ be a planar partial cube with cut-partition $\mathcal{C}$. We consider $G$ with a fixed embedding and denote by $G^*$ the planar dual. By planar duality each cut $C \in \mathcal{C}$ yields a simple cycle $S_C$ in $G^*$. The set of these cycles, seen as Jordan curves defines $\mathcal{S}$. Since $(G^*)^* = G$ the region graph of $\mathcal{S}$ is isomorphic to $G$. Note that picking $p \in \mathbb{R}^2 \setminus \mathcal{S}$ and looking at all the $S(p)$ is a special choice of $\sigma \in \{+1, -1\}^{\mathcal{S}}$ and looking at all $S^{\sigma(S)}$. But for every $\mathcal{S}' \subseteq \mathcal{S}$ and $\sigma \in \{+1, -1\}^{\mathcal{S}'}$ the set $\mathbb{R}^2 \setminus \bigcup_{S \in \mathcal{S}'} S^{\sigma(S)}$ hosts a convex subgraph of $G$ namely $\bigcap_{S \in \mathcal{S}'} C_S^{-\sigma(S)}$. In particular, the region graph of $\mathcal{S}$ induced on $\mathbb{R}^2 \setminus \bigcup_{S \in \mathcal{S}'} S^{\sigma(S)}$ is connected and therefore $\mathbb{R}^2 \setminus \bigcup_{S \in \mathcal{S}'} S^{\sigma(S)}$ is connected.

Conversely, let $\mathcal{S}$ be a non-separating set of Jordan curves and suppose its region graph $G$ is not a partial cube. In particular the cut-partition $\mathcal{C}$ of $G$ arising by dualizing $\mathcal{S}$ does not satisfy Theorem 2 (i). That means there are regions $R, T$ such that every curve $S$ contributing to the boundary of $R$ contains $R$ and $T$ on the same side, i.e., for any $p \in R \cup T$ and such $S$ we have $R, T \subseteq S(p)$. Let $\mathcal{S}'$ be the union of these curves. The union $\bigcup_{S \in \mathcal{S}'} S(p)$ separates $R$ and $T$, i.e., $\mathbb{R}^2 \setminus \bigcup_{S \in \mathcal{S}'} S(p)$ is not connected. □

A set of Jordan curves is *simple* if no point of the plane is contained in more than two curves. In the following, we always assume that a set $\mathcal{S}$ of Jordan curves is encoded by a planar embedding of its 1-skeleton.

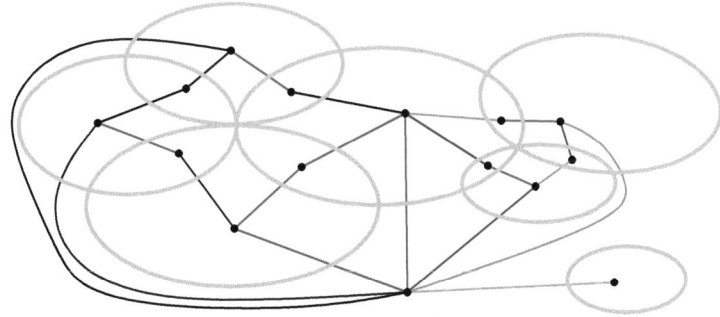

**Fig. 4.** A (non-simple) non-separating set of Jordan curves and its region graph

**Lemma 12.** *A minimal hitting set for open interiors of a non-separating simple set $S$ of Jordan curves can be computed in polynomial-time.*

*Proof.* We first prove that if all *closed* interiors of a given subset $S' \subseteq S$ intersect pairwise, then they have a non-empty common intersection. By Observation 10, the interiors of every triple $S_1, S_2, S_3 \in S'$ have a common point. We can apply the classical *Topological Helly Theorem* [27], i.e., for any family of pseudo-discs in which every triple has a point in common, all members have a point in common.

Now, since $S$ is simple the intersection of mutually intersecting interiors actually has to contain a region and not only a point.

Next, the intersection graph of open interiors of $S$ is chordal: assume that there is a chordless cycle witnessed by $S_1^+, \ldots, S_k^+$ for $k \geq 4$. Since $S$ is non-separating $S_1^+ \cap \ldots \cap S_k^+$ must be non-empty. But, if there is no edge between two vertices of the intersection graph, the corresponding open interiors must be disjoint. The set $S_1^+ \cap \ldots \cap S_k^+$ is therefore reduced to a single point, contradicting simplicity.

Chordal graphs form a subset of perfect graphs and hence by [25] their minimum clique-cover number – that is the least integer $k$ for which the graph admits a partition of its vertices into $k$ cliques – can be computed in polynomial time. Given a clique cover of the intersection graph of $S$, cliques can be assumed to be maximal. Since the intersection of several interiors actually has to contain a region, each maximal clique corresponds to one region of the region graph. Picking one point in the interior of each of those regions yields a hitting set for open interiors. □

**Theorem 13.** *A minimal hitting-set for open interiors and exteriors of a non-separating simple set $S$ of Jordan curves can be computed in polynomial-time.*

*Proof.* Viewing $S$ now as embedded on the sphere, any choice of a region $v$ as the unbounded region yields a different arrangement $S_v$ of Jordan curves in the plane. Denote the size of a minimum hitting set of the interiors of $S_v$ by $h_v$.

Let us now prove that there exist some regions $u, v$ such that $h_u < h_v$ if and only if there is a hitting set of size $h_v$ of exteriors and interiors of $S$. Let $h_u < h_v$ for some regions $u, v$. Extending the hitting set witnessing $h_u$ by the unbounded region in $S_u$ yields a hitting set of exteriors and interiors of size at most $h_v$. Conversely let $H$ be a hitting set of size $h_v$ of exteriors and interiors of $S$ and let $u \in H$. Now, because all sides hit by $u$ now are unbounded, $H \setminus u$ hits all bounded sides of $S_u$ and therefore $h_u < h_v$.

It follows that a minimum hitting set of exteriors and interiors of $S$ is of size $min_{v \in V} h_v + 1$. Since by Lemma 12 every $h_v$ can be computed in polynomial-time and $|V|$ is linear in the size of the input, we are done. □

Combining Corollary 3 and Theorems 11 and 13, we get:

**Corollary 14.** *The hull number of a plane quadrangulation that is a partial cube can be determined in polynomial-time.*

Notice that in [21], it was shown that the hitting set problem restricted to open interiors of (simple) sets of unit squares in the plane remains NP-complete and that the gadget used in that proof is indeed not non-separating.

We conclude this paper with a conjecture. Combined with Theorem 13, it would give a polynomial-time algorithm for the hull number of planar partial cubes.

*Conjecture 15.* A minimum hitting set for open interiors of a non-separating set of Jordan curves can be found in polynomial-time.

**Acknowledgments.** The authors thank Stefan Felsner, Matjaž Kovše, and Bartosz Walczak for fruitful discussions. M.A. would also like to thank Stefan Felsner for his invitation in the Discrete Maths group at the Technical University of Berlin, where this work was initiated. M.A. acknowledges the support of the ERC under the agreement "ERC StG 208471 - ExploreMap" and of the ANR under the agreement "ANR 12-JS02-001-01". K.K. was supported by TEOMATRO (ANR-10-BLAN 0207) and DFG grant FE-340/8-1 as part of ESF project GraDR EUROGIGA.

# References

[1] Araujo, J., Campos, V., Giroire, F., Nisse, N., Sampaio, L., Soares, R.: On the hull number of some graph classes. Theoret. Comput. Sci. 475, 1–12 (2013)
[2] Araujo, J., Morel, G., Sampaio, L., Soares, R., Weber, V.: Hull number: P5-free graphs and reduction rules, Tech. Report RR-8045, INRIA (2012)
[3] Bandelt, H.-J.: Graphs with intrinsic S3 convexities. J. Graph Theory 13(2), 215–228 (1989)
[4] Björner, A., Vergnas, M.L., Sturmfels, B., White, N., Ziegler, G.M.: Oriented matroids, 2nd edn. Encyclopedia of Mathematics and its Applications, vol. 46. Cambridge University Press, Cambridge (1999)
[5] Bondy, J.A., Murty, U.S.: Graph theory, vol. 244. Springer (2008)
[6] Cáceres, J., Hernando, C., Mora, M., Pelayo, I.M., Puertas, M.L.: On the geodetic and the hull numbers in strong product graphs. Comput. Math. Appl. 60(11), 3020–3031 (2010)
[7] Canoy Jr., S.R., Cagaanan, G.B., Gervacio, S.V.: Convexity, geodetic, and hull numbers of the join of graphs. Util. Math. 71, 143–159 (2006)
[8] Centeno, C.C., Penso, L.D., Rautenbach, D., Pereira de Sá, V.G.: Geodetic Number versus Hull Number in $P_3$-Convexity. SIAM J. Discrete Math. 27(2), 717–731 (2013)
[9] Chartrand, G., Harary, F., Zhang, P.: On the hull number of a graph. Ars Combin. 57, 129–138 (2000)
[10] Chepoi, V.D.: d-convex sets in graphs, Ph.D. thesis, Ph. D. dissertation, Moldova State University, Kishinev (1986) (Russian)
[11] Cordovil, R.: Sur les matroïdes orientés de rang 3 et les arrangements de pseudo-droites dans le plan projectif réel. European J. Combin. 3(4), 307–318 (1982)
[12] Cordovil, R., Forge, D.: Flipping in acyclic and strongly connected graphs (2007)
[13] Dilworth, R.P.: Lattices with unique irreducible decompositions. Ann. of Math (2) 41, 771–777 (1940)
[14] Ž Djoković, D.: Distance-preserving subgraphs of hypercubes. Journal of Combinatorial Theory, Series B 14(3), 263–267 (1973)

[15] Dourado, M.C., Gimbel, J.G., Kratochvíl, J., Protti, F., Szwarcfiter, J.L.: On the computation of the hull number of a graph. Discrete Math. 309(18), 5668–5674 (2009)
[16] Dourado, M.C., Protti, F., Rautenbach, D., Szwarcfiter, J.L.: On the hull number of triangle-free graphs. SIAM J. Discrete Math. 23, 2163–2172 (2009)
[17] Eppstein, D.: Isometric diamond subgraphs. In: Tollis, I.G., Patrignani, M. (eds.) GD 2008. LNCS, vol. 5417, pp. 384–389. Springer, Heidelberg (2009)
[18] Eppstein, D., Falmagne, J.-C., Ovchinnikov, S.: Media theory. Interdisciplinary applied mathematics, p. 328. Springer, Berlin (2008)
[19] Everett, M.G., Seidman, S.B.: The hull number of a graph. Discrete Math. 57(3), 217–223 (1985)
[20] Felsner, S., Knauer, K.: ULD-lattices and $\Delta$-bonds. Combin. Probab. Comput. 18(5), 707–724 (2009)
[21] Fowler, R.J., Paterson, M.S., Tanimoto, S.L.: Optimal packing and covering in the plane are NP-complete. Inform. Process. Lett. 12(3), 133–137 (1981)
[22] Fukuda, K., Handa, K.: Antipodal graphs and oriented matroids. Discrete Math. 111(1-3), 245–256 (1993), Graph theory and combinatorics (Luminy, 1990)
[23] Garey, M.R., Johnson, D.S.: Computers and intractability. W. H. Freeman and Co., San Francisco (1979); A guide to the theory of NP-completeness
[24] Graham, R.L., Pollak, H.O.: On the addressing problem for loop switching. Bell System Tech. J. 50, 2495–2519 (1971)
[25] Grötschel, M., Lovász, L., Schrijver, A.: Polynomial algorithms for perfect graphs. Ann. Discrete Math. 21, 325–356 (1984)
[26] Handa, K.: Topes of oriented matroids and related structures. Publ. Res. Inst. Math. Sci. 29(2), 235–266 (1993)
[27] Helly, E.: Über systeme von abgeschlossenen mengen mit gemeinschaftlichen punkten. Monatshefte für Mathematik 37(1), 281–302 (1930)
[28] Hernando, C., Jiang, T., Mora, M., Pelayo, I.M., Seara, C.: On the Steiner, geodetic and hull numbers of graphs. Discrete Math. 293(1-3), 139–154 (2005)
[29] Kanté, M.M., Nourine, L.: Polynomial time algorithms for computing a minimum hull set in distance-hereditary and chordal graphs. In: van Emde Boas, P., Groen, F.C.A., Italiano, G.F., Nawrocki, J., Sack, H. (eds.) SOFSEM 2013. LNCS, vol. 7741, pp. 268–279. Springer, Heidelberg (2013)
[30] Klavžar, S., Shpectorov, S.: Convex excess in partial cubes. J. Graph Theory 69(4), 356–369 (2012)
[31] Monjardet, B.: A use for frequently rediscovering a concept. Order 1(4), 415–417 (1985)
[32] Trotter, W.T.: Combinatorics and partially ordered sets. Johns Hopkins University Press, Baltimore (1992), Dimension theory
[33] Winkler, P.M.: Isometric embedding in products of complete graphs. Discrete Appl. Math. 7(2), 221–225 (1984)
[34] Yannakakis, M.: The complexity of the partial order dimension problem. SIAM J. Algebraic Discrete Methods 3(3), 351–358 (1982)
[35] Ziegler, G.M.: Lectures on polytopes. Graduate Texts in Mathematics, vol. 152. Springer, New York (1995)

# Connected Greedy Colourings*

Fabrício Benevides[1], Victor Campos[1], Mitre Dourado[2], Simon Griffiths[3],
Robert Morris[3], Leonardo Sampaio[4], and Ana Silva[1]

[1] Universidade Federal do Ceará (UFC) - Fortaleza, CE, Brazil
{anasilva,fabricio}@mat.ufc.br, campos@lia.ufc.br
[2] Universidade Federal do Rio de Janeiro (UFRJ) - Rio de Janeiro, RJ, Brazil
mitre@dcc.ufrj.br
[3] Instituto Nacional de Matemática Pura e Aplicada (IMPA) - Rio de Janeiro, RJ, Brazil
{rob,sgriff}@impa.br
[4] Universidade Estadual do Ceará (UECE) - Fortaleza, CE, Brazil
leonardo.sampaio@uece.br

**Abstract.** A *connected vertex ordering* of a graph $G$ is an ordering $v_1 < v_2 < \cdots < v_n$ of $V(G)$ such that $v_i$ has at least one neighbour in $\{v_1, \ldots, v_{i-1}\}$, for every $i \in \{2, \ldots, n\}$. A *connected greedy colouring* is a colouring obtained by the greedy algorithm applied to a connected vertex ordering. In this paper we study the parameter $\Gamma_c(G)$, which is the maximum $k$ such that $G$ admits a connected greedy $k$-colouring, and $\chi_c(G)$, which is the minimum $k$ such that a connected greedy $k$-colouring of $G$ exists. We prove that computing $\Gamma_c(G)$ is NP-hard for chordal graphs and complements of bipartite graphs. We also prove that if $G$ is bipartite, $\Gamma_c(G) = 2$. Concerning $\chi_c(G)$, we first show that there is a $k$-chromatic graph $G_k$ with $\chi_c(G_k) > \chi(G_k)$, for every $k \geq 3$. We then prove that for every graph $G$, $\chi_c(G) \leq \chi(G) + 1$. Finally, we prove that deciding if $\chi_c(G) = \chi(G)$, given a graph $G$, is a NP-hard problem.

**Keywords:** Vertex colouring, Greedy colouring, Connected greedy colouring.

## 1 Introduction

A $k$-*colouring* of a graph $G = (V, E)$ is a surjective mapping $\psi : V \to \{1, 2, \ldots, k\}$ such that $\psi(u) \neq \psi(v)$ for any edge $uv \in E$. A $k$-*colouring* may also be seen as a partition of the vertex set of $G$ into $k$ disjoint *stable sets* $S_i = \{v \mid \psi(v) = i\}$, $1 \leq i \leq k$. The elements of $\{1, \ldots, k\}$ are called *colours*, and the set of vertices with a given colour is a *colour class*. A graph is $k$-*colourable* if it admits a $k$-colouring. The minimum number of colours in a colouring of a graph $G$ is its *chromatic number*, defined as $\chi(G) = min\{k \mid G \text{ is } k\text{-colourable}\}$. We say that $G$ is $k$-chromatic if $\chi(G) = k$.

Graph colourings are a natural model for problems in which a set of objects is to be partitioned according to some prescribed rules. For example, problems of *scheduling* [11], *frequency assignment* [5], *register allocation* [2,3], and the *finite element*

---

* Work partially supported by CAPES/Brazil, CNPq/Brazil, FAPERJ/Brazil and FUNCAP/Brazil.

*method* [9], are naturally modelled by colourings. While it is easy to find a colouring when no bound is imposed on the number of colour classes, for most of these applications the challenge consists in finding one that minimizes the number of colours.

To decide if a graph admits a colouring with $k$ colours is an NP-complete problem, even if $k$ is not part of the input [7]. Moreover, the chromatic number is hard to approximate: for all $\epsilon > 0$, there is no algorithm that approximates the chromatic number within a factor of $n^{1-\epsilon}$ unless P = NP [8,13].

**Greedy Colourings and Its Best and Worst Case Behaviour.** The most basic and widespread algorithm producing colourings is the *greedy algorithm* or *first-fit algorithm*. Given a vertex ordering $\sigma = v_1 < v_2 < \cdots < v_n$ of $V(G)$, the greedy algorithm colours the vertices in the order $\sigma$ assigning to $v_i$ the smallest positive integer not already used in its lower-indexed neighbours. A *greedy colouring* is a colouring obtained from the greedy algorithm.

A remarkable property of the greedy algorithm is that it is always possible to find an optimal colouring by using it. That is, given any graph $G$, there exists an ordering of $V(G)$ such that the greedy algorithm produces a greedy colouring with $\chi(G)$ colours. To see that this is true, consider a colouring $S_1, S_2, \ldots, S_k$ and any vertex ordering in which the vertices of $S_i$ precede those of $S_{i+1}$, for $1 \leq i \leq k-1$. The greedy algorithm applied to any such ordering produces a greedy colouring with at most $k$ colours. When choosing $k = \chi(G)$, we get a greedy colouring with $\chi(G)$ colours.

Although greedy colourings with an optimal number of colours exist, this property is not achieved by any vertex ordering. Consider for example the path on four vertices $P_4$. Any ordering of the vertices of $P_4$ in which the vertices of degree one precede the vertices of degree two produces a greedy colouring with three colours. The worst-case behaviour of the greedy algorithm on a graph $G$ is measured by the *Grundy number* $\Gamma(G)$, which is the largest $k$ such that $G$ has a greedy $k$-colouring. It's known that $\chi(G) \leq \Gamma(G) \leq \Delta(G) + 1$. Unfortunately, colourings obtained by the greedy algorithm can be arbitrarily far from an optimal colouring. The difference $\Gamma(G) - \chi(G)$ can be arbitrarily large, even for trees. This can be seen with the *$k$-binomial-tree* $\mathcal{B}_k$, first defined in [1]. The tree $\mathcal{B}_1$ is the tree on one vertex. The tree $\mathcal{B}_k$, for $k \geq 2$, is built from a copy of $\mathcal{B}_{k-1}$ by adding $|V(\mathcal{B}_{k-1})|$ new vertices and matching them with the vertices from the copy of $\mathcal{B}_{k-1}$. A simple induction can be used to show that $\Gamma(\mathcal{B}_k) = k$, while in fact $\chi(\mathcal{B}_k) = 2$, since $\mathcal{B}_k$ is a tree.

While the Grundy number can be computed in polynomial time for trees [1] and partial $k$-trees [10], the corresponding optimization problem is NP-hard for general graphs. It remains NP-hard for complements of bipartite graphs [12], bipartite graphs and chordal graphs [6].

**Connected Greedy Colourings.** A *connected greedy colouring* of a connected graph $G = (V, E)$ is a greedy colouring obtained from a *connected ordering* $\sigma = v_1 < v_2 < \cdots < v_n$ of $V(G)$, that is, an ordering of the vertices with the property that $v_i$ has at least one neighbour in $\{v_1, \ldots, v_{i-1}\}$ for every $i \in \{2, \ldots, n\}$. In other words, if $V_i = \{v_1, \ldots, v_i\}$, then $G[V_i]$ is connected for every $i \in \{2, \ldots, n\}$. In this paper we study connected greedy colourings with a focus on upper and lower bounds for the number of colours used.

The paper is organized as follows. In Section 2 we consider the worst-case behaviour of connected greedy colourings. In order to do so we define the connected Grundy number $\Gamma_c(G)$, which is the maximum $k$ such that $G$ admits a connected greedy colouring with $k$ colours. We prove that, for a bipartite graph, the connected Grundy number is always equal to the chromatic number of the graph. We show that the difference $\Gamma_c(G) - \chi(G)$ can be arbitrarily large for chordal planar graphs. We also show that determining the connected Grundy number is NP-hard on chordal graphs and complements of bipartite graphs. In Section 3 we prove that, in contrast to what happens with greedy colourings, there are graphs $G$ for which there is no connected greedy colourings with $\chi(G)$ colours. Motivated by this fact, we define $\chi_c(G)$ as the smallest $k$ such that the graph admits a connected greedy colouring with $k$ colours. We prove that $\chi_c(G) \leq \chi(G) + 1$, for any graph $G$. We then show that, given a graph $G$, deciding if $\chi_c(G) = \chi(G)$ is a NP-hard problem.

## 2 The Worst-Case Behaviour

In order to analyse the worst-case behaviour of connected greedy colourings, we define an analogue of the Grundy number. The *connected Grundy number* of a graph $G$, denoted $\Gamma_c(G)$, is the maximum $k$ such that $G$ admits a connected greedy $k$-colouring. Clearly, $\Gamma_c(G) \leq \Gamma(G)$. The connected greedy algorithm can therefore be seen as an improved version of the greedy algorithm. Indeed, in contrast to what happens with the Grundy number, the connected greedy algorithm always finds an optimal colouring if the input graph is bipartite.

**Lemma 1.** *Let $G = (A \cup B, E)$ be a connected bipartite graph with at least one edge. Then, $\Gamma_c(G) = 2$.*

*Proof.* Let $v_1 < v_2 < \cdots < v_n$ be a connected ordering and $\psi$ be the corresponding greedy colouring. Without loss of generality, suppose $v_1 \in A$. We prove by induction on the number of coloured vertices that all coloured vertices in $A$ are coloured 1 and in $B$ are coloured 2. This is true if no vertices are coloured. Now we consider what happens when colouring $v_i$, for $1 \leq i \leq k$. If $v_i \in A$, then any coloured neighbour of $v_i$ is in $B$ and coloured 2. Therefore, $\psi(v_i) = 1$. If $v_i \in B$, then $i \neq 1$ and any coloured neighbour of $v_i$ is in $A$ and coloured 1. Furthermore, since the ordering is connected, at least one of its neighbours is already coloured so $\psi(v_i) = 2$. □

On the other hand, planar graphs and chordal graphs are examples of graph classes which have connected greedy colourings arbitrarily far from optimal. Before we prove these results, we need the following auxiliary results. If graphs $G$ and $H$ are vertex disjoint, let the *join* $G \vee H$ of $G$ and $H$ be the graph obtained from a copy of $G$, a copy of $H$ and adding all possible edges with one endpoint in $G$ and another in $H$. Say that a graph $G$ is *null* if $V(G) = \emptyset$ and *non-null* otherwise.

**Lemma 2.** *Let $G$ and $H$ be vertex disjoint non-null graphs. Also let $\psi_G$ and $\psi_H$ be greedy colourings of $G$ and $H$ with $k_G$ and $k_H$ colours, respectively. Then there is a connected greedy colouring of the join graph $G \vee H$ that uses $k = k_G + k_H$ colours.*

*Proof.* Let $\sigma_G = v_1 < v_2 < \cdots < v_p$ and $\sigma_H = u_1 < u_2 < \cdots < u_q$ be the orderings of $V(G)$ and $V(H)$ such that the greedy algorithm produces the colourings $\psi_G$ and $\psi_H$, respectively. Moreover, let $\sigma_{G \vee H} = v_1 < u_1 < u_2 < \cdots < u_q < v_2 < \cdots < v_p$ be a vertex ordering of $V(G \vee H)$. Since all vertices in $G$ are adjacent to all vertices in $H$ in the graph $G \vee H$, $\sigma_{G \vee H}$ is a connected order. The greedy algorithm applied to $\sigma_{G \vee H}$ first colours $v_1$ with colour 1, and then colours the vertices from $V(H)$ with colours $\{2, \ldots, k_H + 1\}$, assigning to $u \in V(H)$ the colour $\psi_H(u) + 1$. Now the vertices $v_2, v_3, \ldots, v_q$ all have neighbours with colours from $\{2, \ldots, k_H + 1\}$. Therefore, any vertex $v \in V(G)$ will be coloured 1 if $\psi_G(v) = 1$ and coloured $k_H + \psi_G(v)$ otherwise. □

**Corollary 1.** *If $G$ and $H$ are disjoint non-null graphs, then $\Gamma_c(G \vee H) = \Gamma(G) + \Gamma(H)$.*

*Proof.* First note that $\Gamma_c(G \vee H) \leq \Gamma(G \vee H) = \Gamma(G) + \Gamma(H)$. Now, given greedy colourings of $G$ and $H$ with $\Gamma(G)$ and $\Gamma(H)$ colours, respectively, Lemma 2 states that there is a connected greedy colouring of $G \vee H$ with $\Gamma(G) + \Gamma(H)$ colours. Therefore, $\Gamma_c(G \vee H) \geq \Gamma(G) + \Gamma(H)$ which completes the result. □

Let $K_n$ denote the complete graph on $n$ vertices.

**Proposition 1.** *For every $M \geq 0$, there is a chordal planar graph $G$ such that $\Gamma_c(G) - \chi(G) = M$.*

*Proof.* Consider a copy $H$ of the binomial tree $\mathcal{B}_{M+2}$. Clearly, $H$ is planar, as it is a tree. Every tree is *outerplanar*, meaning it admits a drawing in which every vertex is in the outer face. Therefore the graph $H' = H \vee K_1$ is also planar. Furthermore, any cycle in $H'$ must use the unique vertex $v$ in $K_1$. Therefore, $H'$ is also chordal since $v$ is adjacent to all vertices in $H$. Moreover, we have $\chi(H') = 3$. Now, Corollary 1 tells us that $\Gamma_c(H') = M + 3$. □

Now we consider the computational complexity of determining the connected Grundy number of a graph. We say that a family of graphs $\mathcal{G}$ is *closed under universal vertices* if, given a graph $G \in \mathcal{G}$, the graph $G' = G \vee K_1$ also belongs to $\mathcal{G}$.

**Proposition 2.** *Let $\mathcal{G}$ be a family of graphs closed under universal vertices such that, given $G \in \mathcal{G}$ and an integer $k$, the problem of deciding if $\Gamma(G) \geq k$ is NP-complete. Then the problem of deciding if $\Gamma_c(G) \geq k$, given $G \in \mathcal{G}$ and an integer $k$, is also NP-complete.*

*Proof.* Let $G \in \mathcal{G}$ and $k \in \mathbb{N}$. Let $G'$ be the graph $G \vee K_1$. From Corollary 1, we have that $\Gamma_c(G') = \Gamma(G) + 1$. Therefore, $\Gamma(G) \geq k$ if and only if $\Gamma_c(G') \geq k + 1$. □

Since chordal graphs and complements of bipartite graphs are graph classes that are closed under universal vertices, and because of the NP-completeness results that were mentioned before, the following result is immediate.

**Theorem 1.** *Given a graph $G$ and an integer $k$, deciding if $\Gamma_c(G) \geq k$ is a NP-complete problem. The problem remains NP-complete even if the graph $G$ is restricted to chordal graphs or complements of bipartite graphs.*

## 3 The Best-Case Behaviour

As previously mentioned, for every graph $G$ there is a greedy colouring of $G$ using $\chi(G)$ colours. In this section, we prove that the same is not true when considering connected greedy colourings. More precisely, we prove the following theorem.

**Theorem 2.** *For every $k \geq 3$, there is a $k$-chromatic graph $H_k$ with no connected greedy colouring with $k$ colours.*

Thus, it makes sense to define the minimum number of colours $\chi_c(G)$ in a connected greedy colouring of $G$. A natural question would be to ask if $\chi_c(G)$ is bounded by a function of $\chi(G)$. We prove such a function exists and that, in fact, $\chi_c(G)$ is bounded by $\chi(G) + 1$.

**Theorem 3.** *For any connected graph $G$, we have $\chi_c(G) \leq \chi(G) + 1$.*

Consider the graph $G_k$, $k \geq 3$, depicted in Figure 3.

Let $X_k$ and $Z_k$ denote the vertex sets of the copies of $K_{k-1}$ adjacent to only $\{a, b\}$ and to only $\{c, d\}$ respectively, and let $Y_k$ denote the vertex set of the remaining copy of $K_{k-1}$, which is adjacent to all four vertices.

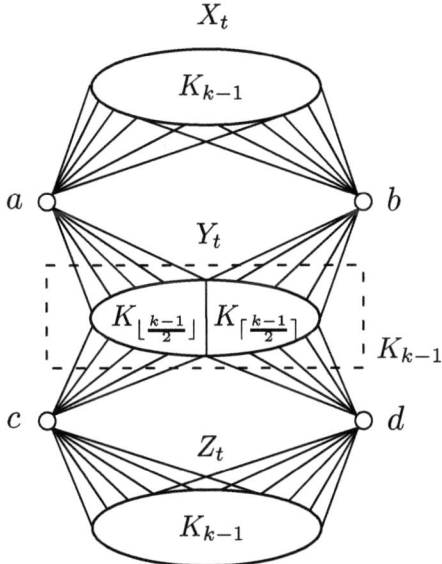

**Fig. 1.** The graph $G_k$

Since in $G_k$, $X_k \cup \{a\}$ is a clique on $k$ vertices, we have $\chi(G_k) \geq k$. To see that $\chi(G_k) = k$, consider the following $k$-colouring of $G_k$. Arbitrarily colour the vertices in $X_k$, $Y_k$ and $Z_k$ with colours in the set $\{1, \ldots, k-1\}$ and colour $a$, $b$, $c$ and $d$ with colour $k$. Indeed, in the following result we prove that $a$, $b$, $c$ and $d$ must always have the same colour in any $k$-colouring of $G_k$.

**Lemma 3.** *Let $\psi$ be a $k$-colouring of $G_k$. Then, $\psi(a) = \psi(b) = \psi(c) = \psi(d)$.*

*Proof.* In any $k$-colouring $\psi$ of $G_k$, the vertices of $X_k$ have $k-1$ distinct colours. Therefore $\psi(a) = \psi(b)$, for otherwise we would need more than $k$ colours. With a similar argument on $Z_k$ we have $\psi(c) = \psi(d)$. Since $Y_k$ induces a clique on $k-1$ vertices, it should be coloured with $k-1$ distinct colours. They should be different from $\psi(a)$, since all vertices are adjacent either to $a$ or $b$, and $\psi(a) = \psi(b)$. Therefore, since the vertices in $Y_k$ are also adjacent either to $c$ or $d$, we get that $\psi(a) = \psi(b) = \psi(c) = \psi(d)$. □

Let $v$ be a vertex of $G$ and $\alpha$ be a colour. A $(v,\alpha)$-*connected greedy colouring* of $G$ is a colouring obtained from a connected ordering that starts from $v$ by colouring $v$ with colour $\alpha$ and then colouring the remaining vertices with the greedy algorithm.

**Lemma 4.** *Let $v \in V(G_k)$ and $\alpha$ be a colour in $\{1,\ldots,k\}$. In any $(v,\alpha)$-connected greedy $k$-colouring of $G_k$, the vertices $a, b, c$ and $d$ have a colour at most $\lceil \frac{k-1}{2} \rceil + 1$.*

*Proof.* Consider a $(v,\alpha)$-connected greedy $k$-colouring of $G$ and, by symmetry, say that $v \in X_k \cup Y_k \cup \{a, b\}$. Since we follow a connected ordering and $\{c, d\}$ is a vertex cut, no vertex from $Z_k$ is coloured before at least one of $\{c, d\}$ is coloured. Let $z$ be the first vertex in the set $\{c, d\}$ that is coloured. Then the only neighbours of $z$ that may have already been coloured are the ones from $Y_k$, and therefore $z$ has at most $\lceil \frac{k-1}{2} \rceil$ coloured neighbours. Therefore, the colour of $z$ is at most $\lceil \frac{k-1}{2} \rceil + 1$, and from Lemma 3 we get that $a, b, c$ and $d$ get the same colour which is at most $\lceil \frac{k-1}{2} \rceil + 1$. □

In particular, Lemma 4 implies that a colouring of $G_k$ obtained by giving vertex $a$ (alt. $b$, $c$ or $d$) a colour greater than $\lceil \frac{k-1}{2} \rceil + 1$ and extending that colouring in a connected greedy way will always use more than $k$ colours.

We are ready to prove Theorem 2.

*Proof (of Theorem 2).* Let $H_k$ be the graph obtained as follows. Take $\lceil \frac{k-1}{2} \rceil + 2$ copies of $G_k$ and add edges joining all copies of $a$, thus forming a $(\lceil \frac{k-1}{2} \rceil + 2)$-clique $K$ with these vertices. Since all copies of $a$ are cut vertices separating the copies of $G_k$ from the clique $K$, we can paste colourings of $G_k$ with each copy of $a$ receiving a different colour to colour $H_k$ with $k$ colours. Furthermore, since $H_k$ contains at least one copy of $G_k$ we have $\chi(H_k) = k$.

Assume to the contrary that $\sigma$ is a connected ordering of $V(H_k)$ such that the greedy algorithm gives a $k$-colouring $\psi$ of $H_k$. Since $K$ forms a clique, there is at least one copy of vertex $a$ with a colour $\alpha$ at least $\lceil \frac{k-1}{2} \rceil + 2$ and call this vertex $w$. Let $v$ be the first vertex in the connected order $\sigma$ and let $W$ be the set of vertices corresponding to the copy of $G_k$ to which $w$ belongs. Furthermore, let $\sigma_W$ be $\sigma$ restricted to the vertices in $W$ and $\psi_W$ be the colouring of $G_k$ obtained from the vertices in $W$. Since $w$ is a cut vertex in $H_k$, then $\sigma_W$ is connected. Therefore, if $v \in W$, then $\psi_W$ is a $(v, 1)$-connected $k$-colouring of $G_k$ which colours $w$ with colour $\alpha$. If $v \notin W$, then $\psi_W$ is a $(w,\alpha)$-connected $k$-colouring of $G_k$. In either case, we get a contradiction to Lemma 4. □

We now show that $\chi_c(G)$ is never greater than $\chi(G) + 1$, for any graph $G$.

**Lemma 5.** *Let $G$ be a connected graph and $v$ a vertex such that $G - v$ is $k$-colourable. For any positive integer $\alpha$, there is a $(v, \alpha)$-connected greedy colouring such that no vertex in $G - v$ get a colour larger than $k + 1$.*

*Proof.* Since $G - v$ is $k$-colourable, let $S_1, \ldots, S_k$ be a partition of $V(G - v)$ into $k$ stable sets. By induction on $k$, we prove the stronger assumption that there is a $(v, \alpha)$-connected greedy colouring such that no vertex in $S_i$ gets a colour larger than $i + 1$, for $1 \leq i \leq k$. The result is valid when $k = 0$ as $G - v$ is null.

Assume $k \geq 1$. We give an algorithm to obtain the desired colouring. To do so, let $H = G - v - S_k$ and let $\mathcal{C}$ be the set of connected components of $H$. Start by colouring $v$ with $\alpha$. We then break the colouring procedure into three phases. In the first phase, we only colour vertices if $\alpha = k + 1$. In the second phase, we colour any uncoloured component of $\mathcal{C}$ that contains a neighbour of $v$. In the third phase, we colour the remaining vertices.

**Phase 1.** If $\alpha = k + 1$, then let $W = N_G(v) \cap S_k$. If $W$ is not empty, we proceed as follows. Start by colouring all vertices in $W$. Since $\alpha \geq 2$, then all vertices in $W$ are coloured 1. Let $G'$ be the graph obtained from $H$ by adding a vertex $w$ adjacent to any vertex adjacent to $W$ in $G - v$, i.e, $N_{G'}(w) = N_{G-v}(W)$. Let $C$ be the component of $G'$ that contains $w$ and note that $(S_1 \cap V(C)), \ldots, (S_{k-1} \cap V(C))$ is a partition of $C - w$ into $k - 1$ stable sets. By the induction hypothesis, there is a $(w, 1)$-connected greedy colouring $\psi_C$ of $C$ such that no vertex in $S_i$ gets a colour larger than $i + 1$, for $1 \leq i \leq k - 1$. We claim that colouring the vertices in $V(C) - \{w\}$ according to $\psi_C$ is a $(v, \alpha)$-connected colouring of $G[(V(C) - \{w\}) \cup W \cup \{v\}]$. Indeed, any vertex in $C$ is adjacent to $w$ if, and only if, it is also adjacent to a vertex of $W$ in $G$. Furthermore, any vertex $z$ in $C$ is coloured with a colour no larger than $k$ and, therefore, $z$ is coloured with colour $\psi_C(z)$ by the greedy algorithm even if $z$ is adjacent to $v$, as $v$ is coloured $\alpha = k + 1$.

At the end of Phase 1 we have the following property which is maintained until we end our colouring algorithm: any uncoloured vertex in $S_k$ has no neighbour coloured $k + 1$. Indeed, all neighbours of $v$ have been coloured if $\alpha = k + 1$, no two vertices in $S_k$ are adjacent as $S_k$ is stable and vertices in $S_i$, for $i < k$, get colours at most $k$. Also note that any component in $\mathcal{C}$ is either fully coloured or contains no coloured vertices. Furthermore, any component of $\mathcal{C}$ that is uncoloured has no neighbour in $W$.

**Phase 2.** Let $\mathcal{C}_v \subseteq \mathcal{C}$ be the set of uncoloured components of $\mathcal{C}$ that contain a neighbour of $v$. By the properties obtained at the end of Phase 1, these components have no coloured vertices and no coloured neighbour in $G$ other than $v$. Let $V_{\mathcal{C}_v} = \bigcup_{C \in \mathcal{C}_v} V(C)$. Let $\hat{G}$ be the subgraph of $G$ induced by $V_{\mathcal{C}_v} \cup \{v\}$ and note that $(S_1 \cap V_{\mathcal{C}_v}), \ldots, (S_{k-1} \cap V_{\mathcal{C}_v})$ is a partition of $\hat{G} - v$ into $k - 1$ stable sets. By the induction hypothesis, there is a $(v, \alpha)$-connected greedy colouring $\psi_{\mathcal{C}_v}$ of $\hat{G}$ such that no vertex in $S_i$ gets a colour larger than $i + 1$, for $1 \leq i \leq k - 1$. We colour the vertices in $V_{\mathcal{C}_v}$ according to $\psi_{\mathcal{C}_v}$.

At the end of Phase 2, we maintain the property that any component in $\mathcal{C}$ is either fully coloured or has no coloured vertex. Furthermore, no vertex in any uncoloured component of $\mathcal{C}$ has any coloured neighbour.

**Phase 3.** In Phase 3, if $G$ has any uncoloured vertices, then we colour the remaining vertices in a sequence of steps. At each step, we colour one vertex $w$ in $S_k$ and all components of $\mathcal{C}$ that contain at least one neighbour of $w$. At the end of each step, we maintain the property that each component in $\mathcal{C}$ is either fully coloured or contains no coloured vertex. Furthermore, no vertex in any uncoloured component of $\mathcal{C}$ has any coloured neighbour. Note that this is true initially as observed at the end of Phase 2.

The structure of each step is as follows. If $G$ has any uncoloured vertex, then there exists an uncoloured vertex $w \in S_k$ adjacent to at least one coloured vertex. Indeed this must be the case as $G$ is connected and no uncoloured component of $\mathcal{C}$ contains any coloured neighbour. Colour $w$ greedily and let this colour be $\beta$. Since $w$ has no neighbour with colour $k+1$, then $\beta \leq k+1$. From here, we follow a structure similar to what was done in Phase 2. Let $\mathcal{C}_w \subseteq \mathcal{C}$ be the set of uncoloured components of $\mathcal{C}$ that contain a neighbour of $w$. By the properties obtained at the end of Phase 2 and between steps, these components have no coloured vertices and no coloured neighbour in $G$ other than $w$. Let $V_{\mathcal{C}_w} = \bigcup_{C \in \mathcal{C}_w} V(C)$. Let $G''$ be the subgraph of $G$ induced by $V_{\mathcal{C}_w} \cup \{w\}$ and note that $(S_1 \cap V_{\mathcal{C}_w}), \ldots, (S_{k-1} \cap V_{\mathcal{C}_w})$ is a partition of $G'' - w$ into $k-1$ stable sets. By the induction hypothesis, there is a $(w, \beta)$-connected greedy colouring $\psi_{\mathcal{C}_w}$ of $G''$ such that no vertex in $S_i$ gets a colour larger than $i+1$, for $1 \leq i \leq k-1$. We colour the vertices in $V_{\mathcal{C}_w}$ according to $\psi_{\mathcal{C}_w}$.

Since we coloured all vertices in components of $\mathcal{C}$ that contain a neighbour of $w$, the desired property between steps is maintained. Therefore, this algorithm continues until all vertices in $G$ are coloured obtaining the desired colouring. □

With Lemma 5, proving Theorem 3 is simple.

*Proof (of Theorem 3).* Let $G$ be any connected $k$-chromatic graph and let $v$ be any vertex of $G$. Since $G - v$ is $k$-colourable, we can apply Lemma 5 to obtain a $(v, 1)$-connected greedy colouring of $G$ such that no vertex of $G - v$ gets a colour larger than $k+1$. Since this colouring starts by colouring $v$ with colour 1, this is a connected greedy colouring of $G$ and no vertex has colour larger than $k+1$. □

A natural question that arises is the computational complexity of deciding if $\chi_c(G) = \chi(G)$, given a connected graph $G$.

**Theorem 4.** *Let $G$ be a connected graph. To decide if $\chi_c(G) = \chi(G)$ is a NP-hard problem.*

*Proof.* Consider the $k$-COLOURABILITY problem, in which the input is a graph $G$ and the question is whether $\chi(G) \leq k$. 3-COLOURABILITY restricted to 4-regular graphs is NP-hard [4]. To see that it is also NP-hard for $k > 3$, observe that if $v$ is a universal vertex, $\chi(G) = \chi(G - v) + 1$, and therefore an instance of $(k-1)$-COLOURABILITY can be reduced to one of $k$-COLOURABILITY by adding a universal vertex. As a consequence of this reduction, $k$-COLOURABILITY is NP-hard for $k \geq 4$, even if the input graph $G$ has a universal vertex and $\chi(G) < 2k$. Let $G$ be a graph with these properties. Let $H$ be the graph obtained from $G$ as follows. For every $v \in V(G)$, add a copy $G^v$ of $G_{2k-1}$ and identify $v$ with the copy of vertex $a$. Then, since $\chi(G_{2k-1}) = 2k - 1$ and $\chi(G) < 2k$, we get that $\chi(H) = 2k - 1$. We now prove that $\chi_c(H) = \chi(H)$ if and only if $\chi(G) \leq k$.

Suppose $\chi_c(H) = \chi(H)$ and let $c$ be a greedy connected colouring of $H$ with $\chi(H)$ colours. Any vertex $v \in V(G)$ is coloured at most $k$, since otherwise there is a connected greedy colouring of $G_{2k-1}$ in which vertex $a$ has a colour in $\{k+1, \ldots, 2k-1\}$, contradicting Lemma 4. The restriction of $c$ to the copy of $G$ in $H$ is a colouring with at most $k$ colours, implying $\chi(G) \leq k$.

Suppose now that $\chi(G) \leq k$. In this case, there is a greedy colouring of $G$ that uses at most $k$ colours. Since $G$ has a universal vertex, this greedy colouring can be made a connected colouring, by rearranging the colour classes so that the universal vertex receives colour 1. Let $c$ be the partial colouring of $H$ in which the vertices from $G$ are coloured according to the previous colouring. For any vertex $v \in V(G)$, since its colour is at most $k$, we may colour the vertices in $G^v$ using only colours smaller than $2k - 1$ and while keeping the colouring connected. In this way we obtain a greedy connected colouring of $H$ that uses no colour larger than $2k - 1$. Since $\chi(G_{2k-1}) = 2k - 1$, we have that $\chi(H) \geq 2k - 1$, and therefore $\chi_c(H) = \chi(H) = 2k - 1$. □

## References

1. Beyer, T., Hedetniemi, S.M., Hedetniemi, S.T.: A linear algorithm for the grundy number of a tree. In: Proceedings of the Thirteenth Southeastern Conference on Combinatorics, Graph Theory and Computing. Utilitas Mathematica, pp. 351–363 (1982)
2. Chow, F., Hennessy, J.: Register allocation by priority-based coloring. ACM SIGPLAN Notices 19, 222–232 (1984)
3. Chow, F., Hennessy, J.: The priority-based coloring approach to register allocation. ACM Transactions on Programming Languages and Systems 12, 501–536 (1990)
4. Dailey, D.P.: Uniqueness of colorability and colorability of planar 4-regular graphs are np-complete. Discrete Mathematics 30(3), 289–293 (1980)
5. Gamst, A.: Some lower bounds for the class of frequency assignment problems. IEEE Transactions on Vehicular Technology 35(8–14) (1986)
6. Havet, F., Sampaio, L.: On the grundy and b-chromatic numbers of a graph. Algorithmica 65(4), 885–899 (2013)
7. Holyer, I.: The NP-completeness of edge-coloring. SIAM Journal on Computing 10(4), 718–720 (1981)
8. Håstad, J.: Clique is hard to approximate within $n^{1-\epsilon}$. In: Acta Mathematica, pp. 627–636 (1996)
9. Saad, Y.: Iterative Methods for Sparse Linear Systems. PWS Publishing Company, Boston (1996)
10. Telle, J.A., Proskurowski, A.: Algorithms for vertex partitioning problems on partial k-trees. SIAM Journal on Discrete Mathematics 10, 529–550 (1997)
11. Werra, D.: An introduction to timetabling. European Journal of Operations Research 19, 151–161 (1985)
12. Zaker, M.: The grundy chromatic number of the complement of bipartite graphs. Australasian Journal of Combinatorics 31, 325–329 (2005)
13. Zuckerman, D.: Linear degree extractors and the inapproximability of max clique and chromatic number. Theory of Computing 3(6) (2007)

# On the Number of Prefix and Border Tables

Julien Clément and Laura Giambruno

GREYC, CNRS-UMR 6072, Université de Caen, ENSICAEN, 14032 Caen, France
{julien.clement,laura.giambruno}@unicaen.fr

**Abstract.** For some text algorithms, the real measure for the complexity analysis is not the string itself but its structure stored in its prefix table (or border table, as border and prefix tables can be proved to be equivalent). We give a new upper bound on the number of prefix tables for strings of length $n$ (on any alphabet) which is of order $(1+\varphi)^n$ (with $\varphi = \frac{1+\sqrt{5}}{2}$ the golden mean) and present also a lower bound.

## 1 Introduction

The prefix table of a string $w$ reports for each position $i$ the length of the longest substring of $w$ that begins at $i$ and matches a prefix of $w$. This table stores the same information as the border table of the string, which memorizes for each position the maximal length of prefixes of the string $w$ ending at that position. Indeed two strings have the same border table if and only if they have the same prefix table.

Both tables are useful in several algorithms on strings. They are used to design efficient string-matching algorithms and are essential for this type of applications (see for example [8] or [3]). It has been noted that for some text algorithms (like the Knuth-Morris-Pratt pattern matching algorithm), the string itself is not considered but rather its structure meaning that two strings with the same prefix or border table are treated in the same manner. For instance, strings `abbbbb`, `baaaaa` and `abcdef` are the same in this aspect.

The study of these tables has become topical. In fact several recent articles in literature (cf. [7,4,2,5]) focus on the problem of validating prefix and border tables, that is the problem of checking if an integer array is either the prefix or the border table of at least one string. In a previous paper [10] Moore et al. represented distinct border tables by canonical strings and gave results on generation and enumeration of these string for bounded and unbounded alphabets. Some of these results were reformulated in [5] using automata-theoretic methods. Note that different words on a binary alphabet have distinct prefix/border tables. This gives us a trivial lower bound in $2^{n-1}$ (since exchanging the two letters of the alphabet does not change tables). This is no longer true as soon as the alphabet has cardinality strictly greater than 2: for instance, words `abb` and `abc` admit the same prefix table $[3, 0, 0]$.

In this paper we are interested in giving better estimates on the number of prefix/border tables $p_n$ of words of a given length $n$, that those known in literature.

For this purpose, we define the combinatorial class of *p-lists*, where a p-list $L = [\ell_1, \ldots, \ell_k]$ is a finite sequence of non negative integers. We constructively define an injection $\psi$ from the set of prefix tables to the set of p-lists which are easier to count. In particular we furnish an algorithm associating a prefix table with a p-list. We define *prefix lists* as p-lists that are images of prefix tables under $\psi$. We moreover describe an "inverse" algorithm that associates a prefix list $L = \psi(P)$ with a word whose prefix table is $P$. This result confirms the idea that prefix lists represent a more concise representation for prefix tables.

We then deduce a *new upper bound and a new lower bound* on the number $p_n$ of prefix tables (see Table 1 for first numerical values) for strings of length $n$ or, equivalently, on the number of border tables of length $n$.

Let $\varphi = \frac{1}{2}(1 + \sqrt{5}) \approx 1.618$, the golden mean, we have:

**Proposition 1 (Upper bound).** *The number of valid prefix tables $p_n$ is asymptotically upper bounded by the quantity $\frac{1}{2}\left(1 + \frac{\sqrt{5}}{5}\right)(1+\varphi)^n + o(1)$.*

**Proposition 2 (Lower bound).** *For any $\varepsilon > 0$ there exists a family of prefix tables $(\mathcal{L}_n)_{n \geq 0}$ such that $\mathrm{Card}(\mathcal{L}_n) = \Omega((1 + \varphi - \varepsilon)^n)$.*

## 2 Preliminaries

**Notations and Definitions.** Let $A$ be an ordered alphabet. A word $w$ of length $|w| = n$ is a finite sequence $w[0]w[1]\ldots w[n-1] = w[0\mathinner{.\,.}n-1]$ of letters of $A$. The language of all words is $A^*$, and $A^+$ is the set of nonempty words. The prefix (resp. suffix) of length $\ell$, $0 \leq \ell \leq n$, of $w$ is[1] the word $u = w[0\mathinner{.\,.}\ell-1]$ (resp. $u = w[n-\ell\mathinner{.\,.}n-1]$). A border $u$ of $w$ is a word that is both a prefix and a suffix of $w$ and distinct from $w$ itself. We define $\mathrm{bord}(w)$ as the set of all proper borders of $w$.

**Definition 1 (Prefix table).** *The prefix table $\mathrm{Pref}_w$ of a word $w \in A^+$ of length $n$, is the table of size $n$ defined, for $0 \leq i < n$, by*

$$\mathrm{Pref}_w[i] = \mathrm{lcp}(w, w[i\mathinner{.\,.}n-1]),$$

*where lcp denotes the maximal length of common prefixes of the two words.*

Another well-known structure used to represent the correlation structure of a string is the border table of a word.

**Definition 2 (Border table).** *The border table $\mathrm{Border}_w$ of a word $w \in A^+$ of length $n$, is the table of size $n$ defined, for $0 \leq i < n$, by*

$$\mathrm{Border}_w[i] = \max\{|u| \mid u \text{ is a border of } w[0\mathinner{.\,.}i]\},$$

---

[1] With the convention that $w[0\mathinner{.\,.}-1] = w[n\mathinner{.\,.}n-1] = \varepsilon$ is the empty word (whenever $\ell = 0$).

*Example.* Let $w$ be the word abaababa. We have the following representations for the prefix and border tables of $w$ (see also Table 2).

| $i$ | 0 | 1 | 2 | 3 | 4 | 5 | 6 | 7 |
|---|---|---|---|---|---|---|---|---|
| $w[i]$ | a | b | a | a | b | a | b | a |
| $\text{Pref}_w[i]$ | 8 | 0 | 1 | 3 | 0 | 3 | 0 | 1 |
| $\text{Border}_w[i]$ | 0 | 0 | 1 | 1 | 2 | 3 | 2 | 3 |

These structures (border and prefix tables) are in fact equivalent; actually the following proposition states a fact discussed in [3] and recently deepened in [1], where linear time conversion algorithms are given. In the following we furnish a proof of this equivalence: elements of the proof will be used in the next section.

**Proposition 3.** *Two strings have the same border table if and only if they have the same prefix table.*

*Proof (sketch).* Let $w$ be a word in $A^+$ of length $|w| = n > 0$. We can relate the border table $\text{Border}_w$ to the prefix table $\text{Pref}_w$.
For a position $j$ in $w$ of length $n$, let

$$I(j) = \{i \mid 0 < i \leq j \text{ and } i + \text{Pref}_w[i] - 1 \geq j\}.$$

The elements in $I(j)$ represent the positions $i \leq j$ for which the longest common prefixes between $w$ and $w[i \mathinner{\ldotp\ldotp} n-1]$ overlap position $j$ in $w$. Then we have

$$\text{Border}_w[j] = \begin{cases} 0 & \text{if } I(j) = \emptyset, \\ j - \min I(j) + 1 & \text{otherwise.} \end{cases} \qquad (1)$$

Conversely, given the border table $\text{Border}_w$ of $w$, we define the prefix table $\text{Pref}_w$ in the following way. First we set $\text{Pref}_w[0] = |w|$. Then let $j > 0$ be a position in $w$ and let $I'(j) = \{i \mid j \leq i < |w| \text{ and } w[j \mathinner{\ldotp\ldotp} i] \in \text{bord}(w[0 \mathinner{\ldotp\ldotp} i])\}$. We have

$$\text{Pref}_w[j] = \begin{cases} 0 & \text{if } I'(j) = \emptyset, \\ \max(I'(j)) - j + 1 & \text{otherwise.} \end{cases} \qquad (2)$$

With (1) and (2), one proves that two words have the same border table if and only if they have the same prefix table. □

Recent literature focuses on the problem of validating prefix and border tables and, in case of a valid table, providing the (canonical) word associated with it, that is the smallest in lexicographic order (cf. [7,2]).

**Previous Work.** Previous work in [10] focused on counting distinct strings of length $n$ with respect to their prefix/border tables: an upper bound is given in the form

$$b_n = \sum_{k=1}^{k^*} \left\{ {n - 2^{k-1} + k \atop k} \right\}, \qquad (3)$$

where $\left\{ {m \atop j} \right\}$ denotes the Stirling number of second kind (the number of partitions of $m$ in $j$ non empty parts), and $k^* = \lceil \log_2(n+1) \rceil$. The quantity $k^*$ is the minimal number of distinct letters to obtain all possible prefix tables of size $n$.

Numerically it is clear that $b_n$ is far from being a tight approximation of the number $p_n$ of prefix tables of size $n$. One can indeed prove that $b_n \gg p_n$. The following combinatorial lemma helps us to formalize this fact.

**Table 1.** First values: $p_n$ is the total number of prefix tables for strings of size $n$, $p_{n,k}$ is the number of prefix tables for strings of size $n$ with an alphabet of size $k$ which cannot be obtained using a smaller alphabet

| $n$ | $p_{n,1}$ | $p_{n,2}$ | $p_{n,3}$ | $p_{n,4}$ | $p_{n,5}$ | $p_{n,6}$ | $p_n$ |
|---|---|---|---|---|---|---|---|
| 1 | 1 | | | | | | 1 |
| 2 | 1 | 1 | | | | | 2 |
| 3 | 1 | 3 | | | | | 4 |
| 4 | 1 | 7 | 1 | | | | 9 |
| 5 | 1 | 15 | 4 | | | | 20 |
| 6 | 1 | 31 | 15 | | | | 47 |
| 7 | 1 | 63 | 46 | | | | 110 |
| 8 | 1 | 127 | 134 | 1 | | | 263 |
| 9 | 1 | 255 | 370 | 4 | | | 630 |
| 10 | 1 | 511 | 997 | 16 | | | 1525 |
| 11 | 1 | 1023 | 2625 | 52 | | | 3701 |
| 12 | 1 | 2047 | 6824 | 162 | | | 9034 |
| 13 | 1 | 4095 | 17,544 | 500 | | | 22,140 |
| 14 | 1 | 8191 | 44,801 | 1467 | | | 54,460 |
| 15 | 1 | 16,383 | 113,775 | 4180 | | | 134,339 |
| 16 | 1 | 32,767 | 287,928 | 11,742 | 1 | | 332,439 |
| 17 | 1 | 65,535 | 726,729 | 32,466 | 4 | | 824,735 |
| 18 | 1 | 131,071 | 1,831,335 | 88,884 | 16 | | 2,051,307 |
| 19 | 1 | 262,144 | 4,610,078 | 241,023 | 52 | | 5,113,298 |
| 20 | 1 | 524,287 | 11,599,589 | 649,022 | 168 | | 12,773,067 |
| 21 | 1 | 1,048,575 | 29,182,347 | 1,736,614 | 504 | | 31,968,041 |
| 22 | 1 | 2,097,151 | 73,430,919 | 4,623,344 | 1486 | | 80,152,901 |
| 23 | 1 | 4,194,303 | 184,845,142 | 12,253,644 | 4248 | | 201,297,338 |
| 24 | 1 | 8,388,607 | 465,567,693 | 32,356,073 | 11,983 | | 506,324,357 |
| 25 | 1 | 16,777,215 | 1,173,418,456 | 85,156,997 | 33,242 | | 1,275,385,911 |
| 26 | 1 | 33,554,431 | 2,959,762,252 | 223,493,213 | 91,297 | | 3,216,901,194 |
| 27 | 1 | 67,108,863 | 7,471,688,677 | 585,104,586 | 248,196 | | 8,124,150,323 |
| 28 | 1 | 134,217,727 | 18,877,965,663 | 1,528,508,811 | 669,799 | | 20,541,362,001 |
| 29 | 1 | 268,435,455 | 47,739,117,581 | 3,985,452,962 | 1,795,120 | | 51,994,801,119 |
| 30 | 1 | 536,870,911 | 120,831,350,575 | 10,374,418,698 | 4,784,707 | | 131,747,424,892 |
| 31 | 1 | 1,073,741,823 | 306,104,380,017 | 26,965,612,590 | 12,689,612 | | 334,156,424,043 |
| 32 | 1 | 2,147,483,647 | 776,139,381,391 | 69,999,199,986 | 33,513,035 | 1 | 848,319,578,061 |
| 33 | 1 | 4,294,967,295 | 1,969,623,334,609 | 181,500,343,408 | 88,172,789 | 4 | 2,155,506,818,106 |

**Table 2.** The nine distinct prefix/border tables for words of length 4 (as counted in Table 1) are listed together with the minimal corresponding word for lexicographical order (named canonical words in the literature)

| Prefix tables | Border tables | Canonical words |
|---|---|---|
| [4, 3, 2, 1] | [0, 1, 2, 3] | aaaa |
| [4, 2, 1, 0] | [0, 1, 2, 0] | aaab |
| [4, 1, 0, 1] | [0, 1, 0, 1] | aaba |
| [4, 1, 0, 0] | [0, 1, 0, 0] | aabb |
| [4, 0, 1, 1] | [0, 0, 1, 1] | abaa |
| [4, 0, 2, 0] | [0, 0, 1, 2] | abab |
| [4, 0, 1, 0] | [0, 0, 1, 0] | abac |
| [4, 0, 0, 1] | [0, 0, 0, 1] | abba |
| [4, 0, 0, 0] | [0, 0, 0, 0] | abbb |

**Lemma 1.** Let $\alpha_n \to \infty$ with $\alpha_n = O(n^c)$ for some $c \in ]0,1[$, one has

$$\left\{{n \atop \alpha_n}\right\} \sim \frac{(\alpha_n)^n}{\alpha_n!} \sim \sqrt{2\pi}(\alpha_n)^{n-\alpha_n+1/2}e^{\alpha_n}.$$

The proof (not detailed here) of this lemma relies on the fact that applications from $\{1,\ldots,m\}$ to $\{1,\ldots,\alpha_n\}$ (related to Stirling number of second kind) are, if $\alpha_n$ is small enough, almost always surjective.

Since $b_n$ in Equation (3) (the bound from [10]) is at least of order $\{{cn \atop d\log n}\}$ for some positive constants $c$ and $d$ (considering for instance the Stirling number for $k = k^* - 1$ in (3)), the lemma suffices to prove that $\log b_n$ is at least of order $n \log \log n$. Hence we have:

**Corollary 1.** We have $\frac{1}{n} \log b_n = \Omega(\log \log n)$.

In Section 4 we improve the bound in (3), yielding the result of Proposition 1.

## 3 Prefix Lists

The information in a valid prefix table is somewhat redundant since we do not need to use all values in the table to build a corresponding word. We introduce prefix lists which are more concise and sufficient to reconstruct such a word. We first define the combinatorial class of p-lists as it follows:

**Definition 3.** We define a p-list $L = [\ell_1, \ldots, \ell_k]$ as a finite sequence of positive integers together with a size defined for a list as $\|L\| = \sum_{i=1}^{k} \|\ell_i\|$, where the size $\|i\|$ is $i$ if $i > 0$ and $1$ if $i = 0$.

Let $\mathcal{P}$ denote the set of prefix tables and $\mathcal{L}$ the set of p-lists. In this section we define an injection $\psi : \mathcal{P} \longrightarrow \mathcal{L}$ in a constructive manner. We define prefix lists as:

**Definition 4.** Let $L$ be a p-list. We say that $L$ is a prefix list if $L = \psi(P)$ for a prefix table $P \in \mathcal{P}$.

### 3.1 Algorithms

**From Prefix Tables to Prefix Lists.** We define constructively an injection $\psi$ from $\mathcal{P}$ to $\mathcal{L}$ by defining an algorithm in a "right-to-left manner". Intuitively, the following algorithm scans the prefix table from right to left, starts with the last position $i = n-1$ and gets from the prefix table the length $\ell$ of the leftmost longest common (proper) prefix which overlaps the current position $i$, or sets $\ell = 0$ if there is no such prefix. This length is inserted at the beginning of the list and the position $i$ is updated to the position immediately before the prefix (if it exists) or just one position before (if it is not the case). The algorithm stops when the first position $i = 0$ is attained.

**Algorithm 1.** PrefixToList($P[0..n-1]$)

$L \leftarrow [\,]$
$i \leftarrow n-1$
**while** $i > 0$ **do**
$\quad I \leftarrow \{j \mid 0 < j \leq i \text{ and } j + P[j] - 1 \geq i\}$
$\quad$ **if** $I = \emptyset$ **then**
$\quad\quad (\ell, i) \leftarrow (0, i-1)$
$\quad$ **else**
$\quad\quad (\ell, i) \leftarrow (i - \min(I) + 1, \min(I) - 1)$
$\quad L \leftarrow [\ell] \cdot L$
**return** $L$

For each position $i$ in $P$, the elements in $I$ represent, as in the proof of Proposition 3, the positions less or equal to $i$, such that the longest common prefix with $w$ starting at these positions overlap position $i$.

**Definition 5.** *For a given prefix table $P$ in $\mathcal{P}$, we define $\psi(P)$ as the prefix list obtained by executing the algorithm* PrefixToList *on $P$.*

*Example 1.* Let $w$ be the word abaababa. We have the following representation for the prefix table of $w$.

| $i$ | 0 | 1 | 2 | 3 | 4 | 5 | 6 | 7 |
|---|---|---|---|---|---|---|---|---|
| $w[i]$ | a | b | a | a | b | a | b | a |
| $\text{Pref}_w[i]$ | 8 | 0 | 1 | 3 | 0 | 3 | 0 | 1 |

For this table we get the associated list $L = [0, 1, 2, 3]$. In fact, executing the algorithm PrefixToList to $\text{Pref}_w$, we start with $i = 7$ and we get that the set of starting indexes of prefixes overlapping $i$ is $I = \{5, 7\}$. Thus $\ell = i - \min(I) + 1 = 3$, the length of the overlapping prefix until $i$, is appended to $L = [\,]$. Now $i$ is initialized to 4 the position before $\min(I) = 5$. Next we have $I = \{3\}$ and so $\ell = 2$ is prefixed to $L = [3]$ and $i := 2$. Again $I = \{2\}$, $\ell = 1$ and $i := 1$. Now $I = \emptyset$, thus $\ell = 0$, $i := 0$ and the algorithm stops.

*Remark.* At first view, it would be more intuitive to define prefix lists with an algorithm visiting the prefix table from left to right. For instance let $P$ be a prefix table and $L = [\,]$ an empty list. A greedy algorithm for this construction starts with the second position $i = 1$ and appends at the beginning of $L$, the length $\ell = P[i]$ of the longest common prefix starting there. Then $i$ is updated to $i + \ell$, if $\ell > 0$ and to $i + 1$ if $\ell = 0$. Again the algorithm appends $\ell = P[i]$ and so on until position $n - 1$ is attained. However this construction of "prefix list" from left to right fails to define an injection from prefix tables to prefix lists (which is our goal for finding an upper bound). In fact let $P = [8, 0, 1, 3, 0, 3, 0, 1]$ be a valid prefix table, as in Example 1, and $P' = [8, 0, 1, 3, 0, 1, 0, 1]$ be a valid prefix table associated with $w'$, then one has

| $i$ | 0 | 1 | 2 | 3 | 4 | 5 | 6 | 7 |
|---|---|---|---|---|---|---|---|---|
| $w'[i]$ | a | b | a | a | b | a | c | a |
| $\text{Pref}_{w'}[i]$ | 8 | 0 | 1 | 3 | 0 | 1 | 0 | 1 |

Since the same list $L = [0, 1, 3, 0, 1]$ is associated with $P$ and $P'$ then the correspondence between prefix tables and these lists cannot be injective.

**From border tables to prefix lists.** In order to prove the injection we define the function $\psi$ in term of border tables: we define another function $\psi'$ from the set of border tables to the set of prefix lists. First consider the algorithm:

---
**Algorithm 2.** BorderToList($B[0..n-1]$)

$L \leftarrow []$
$i \leftarrow n - 1$
**while** $i > 0$ **do**
$\quad \ell \leftarrow B[i]$
$\quad$ **if** $B[i] = 0$ **then**
$\quad\quad i \leftarrow i - 1$
$\quad$ **else**
$\quad\quad i \leftarrow i - B[i]$
$\quad L \leftarrow [\ell] \cdot L$
Return $L$

---

From the algorithm follows the definition of $\psi'$ in an inductive way:

**Definition 6.** *For a border table $B$ of length $n$, let $\ell = B[n-1]$. We define $\psi'(B)$ as:*

$$\psi'(B) = \begin{cases} \psi'(B[0..n-1-\ell]) \cdot [\ell], & \text{if } \ell > 0; \\ \psi'(B[0..n-2]) \cdot [\ell], & \text{if } \ell = 0. \end{cases}$$

The functions $\psi$ and $\psi'$ applied on equivalent border and prefix tables give rise to the same prefix lists:

**Proposition 4.** *Let $B$ be a border table of a word $w$ and $P$ be the prefix table of $w$. Then we have that $\psi(P) = \psi'(B)$.*

*Proof.* For a given position $i$ in $w$, let $I = \{j \mid 0 \leq j \leq i \text{ et } j + P[j] - 1 \geq i\}$ as defined in the proof of Proposition 3 and in the algorithm PrefixToList for the computation of $\psi(P)$. By the conversion rules from prefix table to border table (see proof of Proposition 3), we have that

$$B[i] = \begin{cases} 0, & \text{if } I = \emptyset; \\ i - \min(I) + 1, & \text{if } I \neq \emptyset. \end{cases}$$

Thus if $I = \emptyset$ then the value $\ell = 0 = B[i]$ is inserted at the beginning of $L$ for both algorithms PrefixToList and BorderToList. If $I \neq \emptyset$ then $\ell = i - \min(I) + 1 = B[i]$ is inserted at the beginning of the list $L$ for both algorithms. Then $i$ is decremented in the same way for both the algorithms. □

Thus, for a given border table $B$, there exist $0 \leq i_1 \leq \cdots \leq i_r = n - 1$, such that $\psi'(B) = L = [B[i_1], \ldots, B[i_r]]$ and $i_j = i_{j+1} - B[i_{j+1}]$.

*Example 2.* Let $w$ be the word **abaababa**. The following table shows its border table $\text{Border}_w$ for all values of $i$.

| $i$ | 0 | 1 | 2 | 3 | 4 | 5 | 6 | 7 |
|---|---|---|---|---|---|---|---|---|
| $w[i]$ | a | b | a | a | b | a | b | a |
| $\text{Border}_w[i]$ | 0 | 0 | 1 | 1 | 2 | 3 | 2 | 3 |

The associated prefix list is $L = [0, 1, 2, 3] = [B[1], B[2], B[4], B[7]]$.

**From p-lists to words.** We now describe an "inverse" algorithm that associates a prefix list $L = \psi(P)$ with a word $w$ whose prefix table is $P$. Let $L = [\ell_1, \ldots, \ell_m]$ and $n = \|L\|$, for the length $\|\cdot\|$ defined for p-lists, then the string $w[0..n]$ is computed in the following way.

---
**Algorithm 3.** `ListToWord`$(L = [\ell_1, \ldots, \ell_m])$

$w[0] \leftarrow$ new letter
pos $\leftarrow 1$
for $i \leftarrow 1$ to $m$ do
  if $\ell_i > 0$ then
    for $j \leftarrow 0$ to $\ell_i - 1$ do
      $\lfloor\ w[\text{pos} + j] \leftarrow w[j]$
    pos $\leftarrow$ pos $+ \ell_i$
  else
    $w[\text{pos}] \leftarrow$ new letter
    pos $\leftarrow$ pos $+ 1$
$n \leftarrow$ pos
return $w[0..n]$

---

The **new letter** function returns a new letter not used so far. Informally the algorithm proceeds from left to right on the p-list input $[\ell_1, \ldots, \ell_m]$. It starts with a word reduced to one letter. Then iteratively for $i \in [1..m]$, if $\ell_i > 0$ the algorithm copies $\ell_i$ symbols, from the previous constructed word $u$, at the end of $u$, otherwise the algorithm introduces a new symbol in $w$. Note that overlapping is allowed since we are building the word from left to right.

*Example 3.* Let $w =$ **abaababa** as in Example 1, whose associated prefix list is $L = \psi(\text{Pref}_w) = [0, 1, 2, 3]$. Choosing arbitrarily the first letter to be a, one can build $w = $ a $\cdot$ b $\cdot$ a $\cdot$ ab $\cdot$ aba. A value 0 in the prefix list implies we can choose a new letter (here b at the second position).

One key property is that the word $w$ obtained by this algorithm performed on a prefix list $\psi(P)$ for a prefix table $P$ is such that $\text{Pref}_w = P$. This means that prefix lists and prefix tables are equivalent and represent the same information.

**Proposition 5.** *Given the prefix list $L = \psi(P)$ associated with a prefix table $P$ the word $w$ build by the algorithm* `ListToWord` *is such that* $\text{Pref}_w = P$.

*Proof.* We prove the proposition on border tables: let $L = \psi(P) = \psi'(B)$. We prove the result by induction on $\|L\|$. If $\|L\| = 0$, that is $L = [\,]$, then $w = a$ and $\text{Border}_w = [0] = B$.

Let $\|L\| > 0$ and $L = L' \cdot [\ell]$. We denote by $w$ and $w'$ the words built by the algorithm ListToWord on input $L$ and $L'$ respectively. By construction $w = w' \cdot v$, where $v$ consists necessarily of the first $\ell$ symbols (considered eventual overlapping) of $w'$. By the inductive definition of prefix lists there exists a decomposition $B = B' \cdot B''$ such that $L = \psi(B) = \psi(B') \cdot [\ell]$, $B''[\ell-1] = \ell$ and the length of $B''$ is equal to $\ell$. Thus $L' = \psi(B')$ and, by the inductive hypothesis, $\text{Border}_{w'} = B'$.

In general (see [5], [10]), given a border table $B = H \cdot T$, every word with border table $H$ is prolonging to a word with border table $B$. In our case $B = B' \cdot B''$ and since $B[n-1] = \ell$, the word $u$ prolonging $w'$ consists necessarily of the first $\ell$ symbols (considering possible self overlap) of $w'$ and is equal necessarily to $v$. Thus $\text{Border}_w = B$. □

### 3.2 Injectivity

**Proposition 6.** *The application $\psi$ is injective.*

*Proof.* Let us consider two prefix tables $P \neq P'$ and suppose that $\psi(P) = \psi(P') = L$. By Proposition 5 the algorithm performed on $L$ gives a word $w$ such that $\text{Pref}_w = P = P'$. Hence we must have $\psi(P) \neq \psi(P')$.

Let us remark that the application $\psi$ is not surjective. To a list $[0, 2, 2]$, we can associate a word $w = \text{a} \cdot \text{b} \cdot \text{ab} \cdot \text{ab} = \text{ababab}$ with the prefix table $\text{Pref}_w = [6, 0, 4, 0, 2, 0]$, but we have $\psi(\text{Pref}_w) = [0, 4]$.

## 4 Upper Bound

*p-lists.* We define the set of p-lists as a combinatorial class $\mathcal{L}$ of lists of positive integers
$$\mathcal{L} = \text{Seq}(\{0, 1, 2, 3, \dots\}), \tag{4}$$
together with a *special size measure* which can be defined for a p-list $L = [\ell_1, \dots, \ell_k]$ as $\|L\| = \sum_{i=1}^{k} \|\ell_i\|$, where the size $\|i\|$ is $i$ if $i > 0$ and $1$ if $i = 0$. It just means that $\|L\| = \sum_{i=1}^{k} \ell_i + \text{Card}\{i \mid \ell_i = 0\}$. The Seq operator applied to a combinatorial class $\mathcal{A}$ corresponds to all finite sequences of elements from $\mathcal{A}$, i.e., $\text{Seq}(\mathcal{A}) = \cup_{i=0}^{\infty} \mathcal{A}^i$ (reminiscent of the Kleene star operation for regular languages). By convention $\mathcal{A}^0 = \{\varepsilon\}$.

*Combinatorial specifications and generating functions.* In order to study a sequence $(a_n)_{n \in N}$, it is now usual [6] to consider its generating function $A(z)$, that is the formal power series defined by $A(z) = \sum_{n \geq 0} a_n z^n = \sum_{\alpha \in \mathcal{A}} z^{\|\alpha\|}$.

In our case, given the combinatorial specification of $\mathcal{L}$, it is easy [6] to compute the generating function $L(z) = \sum_{n \geq 0} \ell_n z^n$ where $\ell_n$ denotes the numbers of p-lists of size $n$. This is true when specification are unambiguous (in the same way as unambiguity is considered in regular expressions or formal grammars).

Indeed, the general idea is the following: here we first consider a set of atoms $\mathbb{N}$. We need a size $\|\cdot\|$ compatible with the cartesian product and disjoint union, i.e., here for $i \in \mathbb{N}$ the size of atom $i$ is $\|i\| = i$ if $i > 0$ and $\|0\| = 1$. Let us define an empty element $\varepsilon$ (the only one with size 0). Then we have the following dictionary for translating directly from combinatorial constructions to generating functions.

| Empty element: $\varepsilon \mapsto 1$ | Symbols: $\alpha \in \mathbb{N} \mapsto z^{\|\alpha\|}$, |
|---|---|
| Disjoint Union: $\mathcal{A} \cup \mathcal{B} \mapsto A(z) + B(z)$ | Sequence product: $\mathrm{SEQ}(\mathcal{A}) \mapsto \frac{1}{1-A(z)}$ |
| Cartesian product: $\mathcal{A} \times \mathcal{B} \mapsto A(z) \times B(z)$ | |

Let $\varphi = \frac{1}{2}(1+\sqrt{5}) \approx 1.618$. With this dictionary and the combinatorial description (4), we get the following result for $\ell_n = [z^n]L(z)$ the number of p-lists of size $n$.

**Proposition 7.** *The number of p-lists of size $n$ is given by*

$$\ell_n = \frac{1}{2}\left(1+\frac{\sqrt{5}}{5}\right)\varphi^n + \frac{1}{2}\left(1-\frac{\sqrt{5}}{5}\right)\varphi^{-n} = \frac{1}{2}\left(1+\frac{\sqrt{5}}{5}\right)\varphi^n + o(1).$$

*Proof.* Let $\mathcal{I} = \{0\} \cup \{1,2,3,\ldots\}$ then by definition $\mathcal{L} = \mathrm{SEQ}(\mathcal{I})$. The generating function associated with $\mathcal{I}$ is $I(z) = 2z + z^2 + z^3 + \ldots = z + z\sum_{n\geq 0} z^n = z + \frac{z}{1-z}$. With this dictionary and the combinatorial description we get

$$L(z) = \frac{1}{1-\left(z+\frac{z}{1-z}\right)} = \frac{1-z}{1-3z+z^2}.$$

This is a rational function whose denominator $1 - 3z + z^2$ has two simple roots $1+\varphi$ and $(1+\varphi)^{-1}$ with $\varphi = \frac{1}{2}(1+\sqrt{5})$. Decomposing in simple elements we can easily extract coefficients and we get

$$\ell_n = [z^n]L(z) = \frac{1}{2}\left(1-\frac{\sqrt{5}}{5}\right)(1+\varphi)^{-n} + \frac{1}{2}\left(1+\frac{\sqrt{5}}{5}\right)(1+\varphi)^n,$$

and the desired result follows. □

The main result on the upper bound (see Proposition 1) is a reformulation of the following corollary, which is a consequence of Proposition 6.

**Corollary 2.** *The number $p_n$ of prefix tables of size $n$ is upper bounded by the number $\ell_{n-1}$ of p-lists of size $n-1$.*

## 5 Lower Bound

For the lower bound, we exhibit some sets of valid prefix lists such that we are able to count them. We wish these sets to be as large as possible. In this paper, as a first step, our goal is to evaluate the exponential order growth given in Proposition 2 rather than to give a precise estimate.

The idea for proving Proposition 2 is to exhibit a language which maps bijectively to a set of prefix lists, hence maps bijectively to a set of prefix tables. Let us consider, for a fixed $k$, $\mathcal{L}_k = ab^k\left(ab^{<k}(\varepsilon + cb^*)\right)^*$.

**Proposition 8.** *Two distinct words in $\mathcal{L}_k$ admit distinct prefix tables.*

*Proof.* We prove that the set $\mathcal{L}_k$ is in bijection with a subset of prefix lists. Then, since a prefix table is associated with a unique prefix list, the desired result immediately follows.

First we prove that prefix lists associated with words in $\mathcal{L}_k$ are concatenations of non negative integers $\ell < k$. Indeed by construction, for any word $u \in \mathcal{L}_k$ we have that the longest border of a prefix of $u$ is of length strictly less than $k+1$. Let us note by $L(u)$ the prefix list associated with $u$: $L(u) = \psi(\text{Pref}_u)$. Since the elements in $L(u)$ are borders of prefixes of $u$ we get the result.

Let us prove the main statement by contradiction. Let us consider $u, v$ in $\mathcal{L}_k$ such that $u \neq v$ and $L(u) = L(v) = L = [\ell_1, \ldots, \ell_r]$ (for some $r > 0$). The prefix list $L$ induces the same factorization in $u$ and $v$. Let $i$ be the smallest position in the words such that $u[i] \neq v[i]$ and let $\ell_j$ such that $\sum_{k=1}^{j-1} \|\ell_k\| < i < \sum_{k=1}^{j} \|\ell_k\|$, where $\|\cdot\|$ is defined as in Definition 3. Let $i_1 = \sum_{k=1}^{j-1} \|\ell_k\|$ and $i_2 = \sum_{k=1}^{j} \|\ell_k\|$. If $\ell_j \neq 0$ then $u[i_1 + 1 .. i_2] = ab^{\ell_j - 1} = v[i_1 + 1 .. i_2]$ since $1 \leq \ell_j < k$, that is a contradiction since $u[i] \neq v[i]$. If $\ell_j = 0$ then the longest border of $u[0..i]$ is the empty word. Then $u[i]$ can be equal either to $b$ or to $c$. If $u[i] = b$ then, by definition of $\mathcal{L}_k$, $u[i]$ must be preceded by $cb^t$ for some $t \geq 0$. The element $v[i]$, by definition, must be preceded by $ab^s$ for some $s \geq 0$, that is a contradiction since $u[0..i-1] = v[0..i-1]$. $\square$

We are now ready to give a sketch of Proposition 2, that is: for any $\varepsilon > 0$ there exists a family of prefix tables $(\mathcal{L}_n)_{n \geq 0}$ such that $\text{Card}(\mathcal{L}_n) = \Omega((1 + \varphi - \varepsilon)^n)$.

*Proof (Sketch of the proof of Proposition 2).* For a given $k$, by using analytic combinatorics for regular expressions and since the regular expression $\mathcal{L}_k$ is unambiguous, one compute easily the generating function $L_k(z)$ for $\mathcal{L}_k$

$$L_k(z) = z^{k+1} \frac{1}{1 - \left(z \frac{1-z^k}{1-z}(1 + \frac{1}{1-z})\right)} = \frac{z^{k+1}(z-1)^2}{1 - 3z + z^2 + z^{k+1}}.$$

By general principles [6] we have that the number of words of length $n$ in $\mathcal{L}_k$ is

$$\ell_{n,k} := [z^n] L_k(z) \sim C_k \rho_k^{-n},$$

where $C_k$ is a constant and $\rho_k$ is the smallest real (simple) root of $1 - 3z + z^2 + z^{k+1}$. We write $\rho_k = (1 + \varphi - \varepsilon_k)^{-1}$, and thus are considering $\rho_k$ as a perturbation of the root $\rho = (1 + \varphi)^{-1}$ of $1 - 3z + z^2 = 0$. Solving approximately the perturbed equation when $k$ tends to $\infty$, we get

$$\varepsilon_k = \tfrac{1}{2}(1 + 3\tfrac{\sqrt{5}}{5}) \frac{1}{(1+\varphi)^k}(1 + o(1)).$$

This process of reinjecting an approximate solution in order to get better and better approximations is the essence of the so-called *bootstrapping method* (as in [9]). Hence we get that, for any $\varepsilon > 0$, one can fix $k$ such that $\ell_{n,k} = \Omega((\varphi + 1 - \varepsilon)^n)$ yielding the result of Proposition 2. $\square$

This result gives only rough information on the asymptotics of $\ell_{n,k}$. A more thorough study is in order to get better estimates. However this hints at the following conjecture.

*Conjecture 1.* There exists a constant $c > 0$ such that the number $p_n$ of prefix tables of size $n$ is asymptotically equivalent to $c(1+\varphi)^n$.

## 6 Conclusion

In this paper we have provided some bounds for the number of prefix (or border) tables. The problem of finding an asymptotic equivalent for the number of prefix tables is however still open, and would require a very fine understanding of the autocorrelation structure of words. For this purpose it would be interesting to find characterizations on prefix lists in order to get better bounds. It would be also interesting to study other families of words in bijection with prefix tables to get better lower bounds.

**Acknowledgments.** We would like to thank Maxime Crochemore, Cyril Nicaud and Giuseppina Rindone for helpful discussions, and also anonymous referees for useful remarks.

## References

1. Bland, W., Kucherov, G., Smyth, W.F.: Prefix Table Construction and Conversion. In: Lecroq, T., Mouchard, L. (eds.) IWOCA 2013. LNCS, vol. 8288, pp. 41–53. Springer, Heidelberg (2013)
2. Clement, J., Crochemore, M., Rindone, G.: Reverse engineering prefix tables. In: Albers, S., Marion, J.-Y. (eds.) 26th International Symposium on Theoretical Aspects of Computer Science (STACS 2009). Leibniz International Proceedings in Informatics (LIPIcs), vol. 3, pp. 289–300. Schloss Dagstuhl–Leibniz-Zentrum fuer Informatik (2009)
3. Crochemore, M., Hancart, C., Lecroq, T.: Algorithms on strings. Cambridge University Press, Cambridge (2007)
4. Duval, J.-P., Lecroq, T., Lefebvre, A.: Border array on bounded alphabet. Journal of Automata, Languages and Combinatorics 10(1), 51–60 (2005)
5. Duval, J.-P., Lecroq, T., Lefebvre, A.: Efficient validation and construction of border arrays and validation of string matching automata. RAIRO-Theoretical Informatics and Applications 43(2), 281–297 (2009)
6. Flajolet, P., Sedgewick, R.: Analytic Combinatorics. Cambridge University Press, Cambridge (2009)
7. Franek, F., Gao, S., Lu, W., Ryan, P.J., Smyth, W.F., Sun, Y., Yang, L.: Verifying a border array in linear time. Journal on Combinatorial Mathematics and Combinatorial Computing 42, 223–236 (2002)
8. Gusfield, D.: Algorithms on strings, trees and sequences: computer science and computational biology. Cambridge University Press, Cambridge (1997)
9. Knuth, D.: The average time for carry propagation. Indagationes Mathematicae 40, 238–242 (1978)
10. Moore, D., Smyth, W.F., Miller, D.: Counting distinct strings. Algorithmica 23(1), 1–13 (1999)

# Probabilities of 2-Xor Functions

Élie de Panafieu[1,*], Danièle Gardy[2,**], Bernhard Gittenberger[3,***],
and Markus Kuba[3,†]

[1] Univ. Paris Diderot, Sorbonne Paris Cité, LIAFA, UMR 7089, 75013, Paris, France
[2] PRISM, Univ. of Versailles, France
[3] Institute of Discrete Mathematics and Geometry, TU Wien, Austria

**Abstract.** The problem 2-Xor-Sat asks for the probability that a random expression, built as a conjunction of clauses $x \oplus y$, is satisfiable. We consider here a refinement of this question, namely the probability that a random expression computes a specific Boolean function. The answer involves a description of 2-Xor expressions as multigraphs, and uses classical methods of analytic combinatorics.

**Keywords:** multigraphs, probability of Boolean functions, 2-Xor expressions.

## 1 Introduction

In constraint satisfaction problems we ask for the probability that a random expression, built on a finite set of Boolean variables according to some rules ($k$-Sat, $k$-Xor-Sat, NAE, ...), is (un)satisfiable. The behaviour of this probability, when the number $n$ of Boolean variables and the length $m$ of the expression (usually defined as the number of clauses) tend to infinity, most specially the existence and location of a threshold from satisfiability to unsatisfiability as the ratio $m/n$ grows, has given rise to numerous studies. The literature in this direction is vast, for Xor-functions see e.g. [1,2,3,4,5].

Defining a probability distribution on Boolean functions through a distribution on Boolean expressions is *a priori* a different question. Quantitative logic aims at answering such a question, and many results have been obtained when the Boolean expression, or equivalently the random tree that models it, is a variation of well-known combinatorial or probabilistic tree models (Galton-Watson and Pólya trees, binary search trees, etc).

So we have two frameworks: on the one hand we try to determine the probability that an expression is satisfiable; on the other hand we try to identify probability distributions on Boolean functions. It is only natural that we should wish to merge these

---

[*] Supported by the ANR projects BOOLE (2009-13) and MAGNUM (2010-14).
[**] Part of the work of this author was done during a long-term visit at the Institute of Discrete Mathematics and Geometry, TU Wien. Supported by the P.H.C. Amadeus project *Probabilities and tree representations for Boolean functions* and by the ANR project BOOLE (2009-13).
[***] Supported by the FWF (Austrian Science Foundation), Special Research Program F50, grant F5003-N15, and by the ÖAD, grant Amadée F03/2013.
[†] Supported by the ÖAD, grant Amadée F03/2013.

two approaches: what if we set satisfiability problems into the framework of quantitative logic (this only requires to choose a suitable model of expressions), and ask for the probability of FALSE – this is the classical satisfiability problem – *and* of the other Boolean functions? This amounts to refining the satisfiable case, which gathers together all the functions differing from FALSE, into subcases according to the exact (class of) Boolean function(s) that is computed.

Within this unified framework one could, e.g., ask for the probability that a random expression computes a function that is satisfied by a specific number of assigments. Although this may turn out to be out of our reach for most classical satisfiability problems, there are some problems for which we may still have hope to obtain a (partial) description of the probability distribution on Boolean functions. The case of 2-Xor expressions is such a problem, and this paper is devoted to presenting our results in this domain.

The 2-Xor-Sat satisfiability problem has been studied by Creignou and Daudé [1] who established the existence of a threshold for $m = \frac{n}{2}$, then proved in [4] that this threshold is coarse. Further work by Daudé and Ravelomanana [6] and by Pittel and Yeum [7] led to a precise understanding of the transition in a window of size $n^{2/3}$.

The paper is organized as follows. We present in the next section the 2-Xor problem and the set of Boolean functions that can be attained by such expressions, then give a modelization in terms of multigraphs, before considering in Section 3 how enumeration results on classes of multigraphs allow us to compute probabilities of Boolean functions. We then give explicit results for several classes of functions in Section 4, and conclude with a discussion on the relevance and of possible extensions of our work.

## 2 Boolean Expressions and Functions, and Multigraphs

### 2.1 2-Xor Expressions and Boolean Functions

Starting from an infinite set $\{x_1, x_2, \ldots\}$ of Boolean variables, we define a 2-Xor expression as a finite conjunction of clauses $l \oplus l'$, where $l$ and $l'$ are literals, i.e. either some $x_i$ or $\bar{x}_i$. We shall denote by $m$ the number of clauses of an expression. Now each 2-Xor expression defines a Boolean function on a finite number of variables, but not all Boolean functions on a finite number of variables can be obtained from a 2-Xor expression. We define $\mathcal{X}$ as the set of functions from $\{0,1\}^{\mathbb{N}}$ to $\{0,1\}$, such that there exists at least one 2-Xor expression representing them. We also define, for each $n \geq 1$, the set $\mathcal{X}_n$ of functions in $\mathcal{X}$ such that there exists an expression representing the function, that does not use any of the variables $x_{n+1}, x_{n+2}, \ldots$ This implies that $\mathcal{X}_{n_1} \subset \mathcal{X}_{n_2}$ for $n_1 \leq n_2$, and that $\mathcal{X} = \cup_{n \geq 1} \mathcal{X}_n$.[1]

Consider now the expressions that can represent a function of $\mathcal{X}_n$. The literals in a clause are ordered (the clauses $x \oplus y$ and $y \oplus x$ are distinct), hence there are $4n^2$ distinct clauses. We assume that the $m$ clauses are drawn with a uniform probability and with replacement (i.e., a clause can appear several times), and are unordered, i.e. we are dealing with a *set* of clauses. This framework allows us to define a probability distribution on the set $\mathcal{X}_n$ : $\Pr_{[m,n]}(f) = \frac{N_{[m,n]}(f)}{N_{[m,n]}}$, with $N_{[m,n]}$ the total number of

---

[1] For brevity's sake, "(the set of) Boolean functions" in the sequel is to be understood as either the set $\mathcal{X}_n$ or the set $\mathcal{X}$, according to the context.

expressions with $m$ clauses on the variables $x_1, \ldots, x_n$, and $N_{[m,n]}(f)$ the number of these expressions that compute $f$.

## 2.2 The Sets $\mathcal{X}_n$

Rewriting a clause $l_1 \oplus l_2$ as $l_1 \sim \bar{l}_2$ (i.e., $l_1$ and $l_2$ must take opposite values for the clause to evaluate to TRUE), we see that the functions we obtain can be written as a conjunction of equivalence relations on literals: $(l_1 \sim \cdots \sim l_p) \wedge (l_{p+1} \sim \cdots \sim l_q) \wedge \cdots \wedge (l_{r+1} \sim \cdots \sim l_s)$. E.g., for $n = 7$ the expression $(x_1 \oplus x_3) \wedge (\bar{x}_6 \oplus x_5) \wedge (x_7 \oplus \bar{x}_7) \wedge (x_2 \oplus \bar{x}_3)$ computes a Boolean function $f$ that we can write as $(x_1 \sim \bar{x}_2 \sim \bar{x}_3) \wedge (x_5 \sim x_6)$, and this function partitions the Boolean variables into the subsets $\{x_1, x_2, x_3\}, \{x_4\}, \{x_7\}$ and $\{x_5, x_6\}$.

If a relation $l \sim \bar{l}$ appears in at least one of the equivalence relations, the expression simply computes FALSE. In other words: *For any $n \geq 1$, the set $\mathcal{X}_n$ comprises exactly the function FALSE and those functions that partition the set of the $n$ Boolean variables into subsets, as follows: the variables (or their negations) in a given part are equivalent; a variable which appears in no clause of an expression computing the function, or only as $l \sim l$, is put in a singleton.*

We now define the following equivalence relation on $\mathcal{X}_n$. Two Boolean functions $f$ and $g$ on $n$ variables are equivalent, if $g$ can be obtained from $f$ by permuting the variables and flipping some of them. We denote by $\mathcal{C}(f)$ the equivalence class of a function $f$. All the Boolean functions in $\mathcal{C}(f)$ share the same probability $\Pr_{[m,n]}(f)$.

Let $f \in \mathcal{X}$; we say that a Boolean variable $x$ is an *essential* variable w.r.t. $f$ iff $f|_{x=1} \neq f|_{x=0}$. Let $f \notin \{\text{TRUE}, \text{FALSE}\}$ and $e(f) \leq n$ be the number of its essential variables[2]; then $f \notin \mathcal{X}_{e(f)-1}$ but $f \in \mathcal{X}_{e(f)}$. In our example, $e(f) = 5$.

It is not hard to see that, with the exception again of FALSE that is in a class by itself, the classes we have thus defined on $\mathcal{X}_n$ are in bijection with partitions of the integer $n$; in our example the function $f$ partitions the integer 7 as $1 + 1 + 2 + 3$.

Let $\mathbf{i} = (i_\ell)_{\ell \geq 1}$ be an integer partition of $n$, written in its part-count representation. Hence $i_\ell \geq 0$ for all $\ell$ and $s(\mathbf{i}) := \sum_\ell \ell i_\ell = n$; the total number of parts (or *blocks*) is $\xi(\mathbf{i}) := \sum_\ell i_\ell$ and $i_\ell$ is the number of parts of size $\ell$. Partitions of the type $i_\ell = 0$ except $i_n = 1$ appear regularly in the sequel; we shall denote such a partition by $\mathbf{imax}(n)$. We write $\mathbf{i}(f)$ for the integer partition associated to a Boolean function $f$, and we extend the notation for the equivalence class into $\mathcal{C}_\mathbf{i} = \mathcal{C}(f)$ when $\mathbf{i} = \mathbf{i}(f)$.

Our running example corresponds to the integer partition $(n-5, 1, 1, 0, 0, 0)$ on $n \geq 5$ variables, which has $n - 3$ parts; the set partition it induces on the set of Boolean variables may be taken, for example, equal to $\{x_1, x_2\}, \{x_3, x_4, x_5\}$. The function TRUE corresponds to the integer partition $(n, 0, \ldots, 0)$ and is computed by the expressions that have only clauses of the type $l \oplus \bar{l}$.

**Proposition 1.** *Set $p(n)$ as the number of partitions of $n$; the number of classes of computable Boolean functions is then $p(n) + 1$. The class associated to a partition $\mathbf{i} = (i_\ell)$ has cardinality $\frac{2^{n-\xi(\mathbf{i})} n!}{\prod_{\ell \geq 1} i_\ell! (\ell!)^{i_\ell}}$. The number of assignments satisfying a function $f \in \mathcal{C}_\mathbf{i}$ is $2^{\xi(\mathbf{i})}$.*

---

[2] Although the constant functions can only be written as 2-Xor expressions involving one or more variables, they have no essential variable: $e(\text{TRUE}) = e(\text{FALSE}) = 0$.

## 2.3 2-Xor Expressions as Colored Multigraphs

Consider multigraphs, i.e. graphs where we allow loops and multiple edges. Set $M_{m,n}$ as the number of multigraphs on $n$ vertices[3] and $m$ edges or loops, each multigraph being weighted as follows: every loop contributes a multiplicative factor $1/2$ to the weight, each $k$-fold edge a factor $1/k!$. The generating function for weighted multigraphs is (see Janson, Knuth, Luczak and Pittel [8])

$$M(z,v) = \sum_{m,n} M_{m,n} z^m \frac{v^n}{n!} = \sum_{n \geq 0} e^{\frac{n^2}{2} z} \cdot \frac{v^n}{n!}.$$

A multigraph being a set of connected components, the g.f. for *connected* multigraphs is

$$C(z,v) = \log M(z,v) = \sum_{r \geq -1} z^r C_r(zv), \qquad (1)$$

where we have set $r = m-n$, the *excess* of the multigraph, and where $C_r(z)$ enumerates the *connected* multigraphs of fixed excess $r$.

We are now ready to define a bijection between Boolean expressions of $m$ clauses on $n$ variables, and *colored* multigraphs on $n$ vertices and with $m$ edges, i.e. multigraphs with different types (colors) of edges between any two vertices, as follows.

- Each Boolean variable $x_\ell$ corresponds to a vertex, and each 2-Xor clause to an edge between two distinct vertices, or to a loop on one vertex; each loop or edge can be repeated.
- A loop on vertex $x$ can appear in four colors : $x \oplus x$, $x \oplus \bar{x}$, $\bar{x} \oplus x$ or $\bar{x} \oplus \bar{x}$.
- An edge between two distinct vertices $x_i$ and $x_j$ can appear in eight colors: $l_i \oplus l_j$ or $l_j \oplus l_i$, where $l_i$ and $l_j$ are respectively equal to $x_i$ or its negation, and $x_j$ or its negation.

**Proposition 2.** *There is a bijection between 2-Xor expressions, and multigraphs where loops are 4-colored and other edges are 8-colored. Hence the generating function for 2-Xor expressions is $M(8z,v)$.*

*Let $f \in \mathcal{X}_n$; then $\xi(\mathbf{i}(f))$ is the number of connected components of the associated multigraph.*

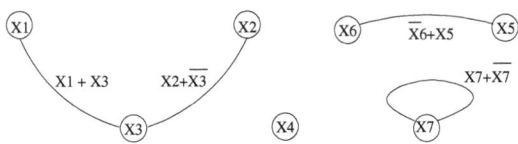

**Fig. 1.** The colored multigraph for our running example

---

[3] As is usual when enumerating such structures, we consider labels on the vertices, say $1, \ldots, n$.

## 2.4 The Different Ranges

We shall consider in the sequel the range where $m$ and $n$ are related, and set $m \sim \alpha n$ ($\alpha$ is usually assumed to be a constant). It is well known ([6]) that the probability that a random expression is satisfiable decreases from 1 to 0 when $\alpha$ increases, with a (coarse) threshold at $\frac{1}{2}$. However, *a Boolean function corresponding to a partition of the $n$ Boolean variables into $p$ blocks cannot appear before at least $n - p$ clauses have been drawn, i.e. before $m \geq n - p$.* E.g., the function $x_1 \sim \cdots \sim x_n$ cannot appear for $m < n - 1$, which means that it has a non-zero probability only for $\alpha \geq 1$, much later than the threshold – and at this point the probability of FALSE is $1 - o(1)$. This leads us to define regions according to the value of $\alpha$ when $n, m \to +\infty$:

- $\alpha < 1/2$. Here the probability of satisfiability is non-zero, but the attainable functions cannot have more than $n(1 - \alpha)$ blocks.
- $1/2 < \alpha < 1$. Some Boolean functions still have probability zero, but now the probability of satisfiability is $o(1)$ and the probability of FALSE is $1 - o(1)$. Thus any other attainable Boolean function has a vanishing probability $o(1)$.
- $1 \leq \alpha$. At this point all the attainable Boolean functions have non-zero probability, but again the probability of FALSE is tending to 1.

## 3 Probabilities on Boolean Functions

We consider here how we can obtain the probability of satisfiability (or equivalently of FALSE), or of any function in $\mathcal{X}_n$. The reader should recall that the probabilities given in the sequel are actually distributions on $\mathcal{X}_n$, i.e. they depend on $n$ and $m$. Letting $n$ and $m = m(n)$ grow to infinity amounts to specializing the probability distribution $\Pr_{[m,n]}(f)$ (defined in Section 2.1 for $f \in \mathcal{X}_n$) into $\Pr_{[m(n),n]}(f)$. We shall be interested in its limit when $n \to +\infty$ and $f$ is a function of $\mathcal{X}$. We begin with the case $f = $ FALSE (which is the usual satisfiability problem) and derive anew the probability of satisfiability in the critical window, before turning to general Boolean functions.

### 3.1 Probability of Satisfiability

**Theorem 1.** *The probability that a random expression is satisfiable is*

$$\Pr_{[m,n]}(Sat) = \frac{[z^m v^n]\sqrt{M(4z, 2v)}}{[z^m v^n]M(8z, v)}.$$

*Its asymptotic value for $n \to +\infty$ and $m = \frac{n}{2}(1 + \mu n^{-1/3})$ is*

$$n^{-1/12}\sqrt{2\pi} \sum_{r \geq 0} \frac{e_r^{(1/2)}}{2^r} A(3r + 1/4, \mu),$$

where

$$e_r^{(\sigma)} = [z^{2r}] \left( \sum_{k \geq 0} \frac{(6k!)z^{2k}}{2^{5k} 3^{2k} (2k)! (3k)!} \right)^\sigma,$$

$$A(y, \mu) = \frac{e^{-\mu^3/6}}{3^{(y+1)/3}} \sum_{k \geq 0} \frac{(3^{2/3} \mu/2)^k}{k! \Gamma\left(\frac{y+1-2k}{3}\right)}.$$

*Proof.* To obtain the g.f. for satisfiable expressions, we shall count the number of pairs {satisfiable expression, satisfying assignment}, then get rid of the number of satisfying assignments. We can assign TRUE or FALSE to each variable, and one of eight colors to an edge, hence $M(8z, 2v)$ counts all pairs {expression, assignment}.

Once we have chosen an assignment of variables, for an expression to be satisfiable we have to restrict the edges we allow. Say that $x$ and $y$ are assigned the same value; then the edges colored by $x \oplus y$, $y \oplus x$, $\bar{x} \oplus \bar{y}$ or $\bar{y} \oplus \bar{x}$ cannot appear in a satisfiable expression. For a similar reason, the only loops allowed are $x \oplus \bar{x}$ or $\bar{x} \oplus x$. We thus count multigraphs with 2 colors of loops and 4 colors of edges, which gives a g.f. equal to $M(4z, 2v)$.

Now consider the generating function $S(z, v)$ for satisfiable expressions: we claim that it is equal to $\sqrt{M(4z, 2v)}$. To see this, choose an expression computing a Boolean function $f$, and consider how many assignments satisfy it: we have seen (cf. Proposition 1) that their number is equal to $2^{\xi(f)}$, with $\xi(f)$ the number of connected components (once we have chosen the value of a single variable in a block, all other variables in that block have received their values if the expression is to be satisfiable). This means that, writing $S(z, v) = \exp \log S(z, v)$ with $\log S(z, v)$ the function for connected components, the g.f. enumerating the pairs {expression, satisfiable assignment} is equal to $\exp(2 \log S(z, v)) = S(z, v)^2$. As we have just shown that it is also equal to $M(4z, 2v)$, the value of $\Pr_{[m,n]}(Sat)$ follows.

To obtain the asymptotics in the critical window $m = n/2 + \mathcal{O}(n^{2/3})$, we use Lemma 1 below, which is an easy variation of [8, Lemma 3]. The function $A(y, \mu)$ is a variation of the classical Airy function; see for example [8, Lemma 3], [9, Theorem 11] or [10, Theorem IX.16].

**Lemma 1.** *Let us consider a positive real value $\sigma$, a bounded parameter $\mu$ and $m = \frac{n}{2}(1 + \mu n^{-1/3})$. Then, with the notations of Theorem 1,*

$$n! [z^m v^n] M(z, v)^\sigma \sim \frac{n^{2m}}{2^m m!} \sigma^{n-m} n^{(\sigma-1)/6} \sqrt{2\pi} \sum_r \sigma^r e_r^{(\sigma)} A(3r + \sigma/2, \mu).$$

**Theorem 2.** *The probability for a random satisfiable expression with $n$ variables and $m$ clauses to be satisfied by a random input, in the range $m = \frac{n}{2}(1 + \mu n^{-1/3})$, is*

$$\Pr_{[m,n]}(Sat) = \frac{[z^m v^n] M(4z, 2v)}{2^n [z^m v^n] \sqrt{M(4z, 2v)}} \sim \frac{n^{1/12}}{2^m} \frac{\sum_r e_r^{(1)} A(3r + 1/2, \mu)}{\sum_r 2^{-r} e_r^{(2)} A(3r + 1/4, \mu)}.$$

### 3.2 Probability of a Given 2-Xor Function

We now refine the probability of satisfiability, by computing the probability of a specific Boolean function $\neq$ FALSE. We first give in Proposition 3 the generating functions for all Boolean functions (except again FALSE), then use it to provide in Theorem 3 a general expression for the probability of a Boolean function, or rather of all the functions of an equivalence class $C_{\mathbf{i}}$. This theorem is at a level of generality that does not give readily precise probabilities, and we delay until Section 4 such examples of asymptotic probabilities.

**Proposition 3.** *For $\mathbf{i}$ an integer partition, define $\phi_{\mathbf{i}}(z)$ as the generating function for Boolean expressions that compute a specific Boolean function $f$ in the class $C_{\mathbf{i}}$: $\phi_{\mathbf{i}}(z) = \sum_m N_{[m,n]}(f)\, z^m$. When $\mathbf{i} = \mathbf{imax}(n)$, we set $\phi_n(z) := \phi_{\mathbf{imax}(n)}(z)$. Then*

$$\phi_n(z) = \left[\frac{v^n}{n!}\right] C(4z, v); \qquad \phi_{\mathbf{i}}(z) = \prod_{\ell \geq 1} (\phi_\ell(z))^{i_\ell}.$$

*Proof.* A canonical representant of the class $\mathbf{imax}(n)$ is the function $x_1 \sim \cdots \sim x_n$. Any expression that computes it corresponds to a connected multigraph, where we only allow the 2 types of loops that compute TRUE, and the 4 types of edges between $x_i$ and $x_j$ ($i \neq j$) that compute $x_i \sim x_j$; this gives readily the expression of $\phi_n(z)$.

As for functions whose associated multigraphs have several components, such multigraphs are a product of connected components; hence the global generating function is itself the product of the generating functions for each component.

**Theorem 3.** *1. The probability that a random expression of $m$ clauses on $n$ variables computes the function $x_1 \sim \cdots \sim x_n$ is*

$$\Pr{}_{[m,n]}(x_1 \sim \cdots \sim x_n) = \frac{\left[z^m \frac{v^n}{n!}\right] C(4z, v)}{\left[z^m \frac{v^n}{n!}\right] M(8z, v)} = \frac{m!}{n^{2m}} \left[\frac{v^n}{n!}\right] C_{m-n}(v).$$

*2. Let $f$ be a function of $\mathcal{X}$, with $q = \xi(\mathbf{i}(f))$, and $B_1, \ldots, B_q$ be the blocks of $\mathbf{i}(f)$, with $r_j$ ($1 \leq j \leq q$) the excess of the block $B_j$. The probability that a random expression of $m$ clauses on $n$ variables computes $f$ is*

$$\Pr{}_{[m,n]}(f) = \frac{m!}{n^{2m}} \sum_{\substack{r_1,\ldots,r_q \geq -1 \\ r_1 + \cdots + r_q = m-n}} \prod_{j=1}^{q} \left[\frac{v^{|B_j|}}{|B_j|!}\right] C_{r_j}(v).$$

*Proof.* By the correspondance between 2-Xor expressions and weighted multigraphs the probability that an expression of $m$ clauses on $n$ variables computes a function $f$ can be expressed as follows:

$$\Pr{}_{[m,n]}(f) = \frac{[z^m]\phi_{\mathbf{i}}(z)}{[z^m \frac{v^n}{n!}] M(8z, v)}.$$

Expressing $\phi_{\mathbf{i}}$ in terms of coefficients of powers of $C(4z, v)$, then substituting the expression (1) for $C$, gives the result after careful management of the coefficients.

## 4 Explicit Probability Computations

We now show on examples how Theorem 3 allows us to compute the asymptotic probability of a specific function.

We consider first a Boolean function $f$ with a fixed number $e(f)$ of essential variables, and consider how its probability varies when $n \to +\infty$ (i.e. when we add non-essential variables), then turn to functions that vary with $n$, either with a fixed number of blocks (this includes functions that are "close to" FALSE in the sense that they have few blocks, hence few satisfying assigments), or with a number of blocks that grows with $n$ (e.g., $\frac{n}{j}$ blocks of size $j$ for some $j \geq 2$).

### 4.1 Probability of a Fixed Function

We compute here the probability of any specific function, when $m$ is large enough so that it can be obtained, and see how it varies when $n, m \to +\infty$ with fixed ratio $\alpha$.

**Proposition 4.** *Let $f \in \mathcal{X}_n$, with $e(f)$ the number of its essential variables, and $\mathbf{i}(f) = (i_1, i_2, \dots)$ its associated integer partition. Assume $m = \alpha n \geq n - \xi(\mathbf{i}(f))$; then*

$$P_{[\alpha n, n]}(f) \sim \frac{e^{\alpha\, e(f)}}{(2n)^{\alpha n}} \prod_{\ell \geq 2} \left(\ell!\phi_\ell\left(\frac{\alpha}{2}\right)\right)^{i_\ell} \qquad (n \to +\infty).$$

### 4.2 Asymptotics for a Single-Block Function

We consider here the class of $x_1 \sim \cdots \sim x_n$, and the range $m \geq n - 1$. This corresponds to (a subset of) the third range of Section 2.4. From Theorem 3, we have

$$\Pr_{[m,n]}(x_1 \sim \cdots \sim x_n) = \frac{m!}{n^{2m}} \cdot \left[\frac{v^n}{n!}\right] C_{m-n}(v).$$

We now specialize this expression according to the possible values for the excess $r = m - n$. For the first three cases, we use the fact that for each fixed excess $r$, there is an explicit constant $K_r$ such that

$$\left[\frac{v^n}{n!}\right] C_r(z) \sim K_r \cdot n^{n + \frac{3r-1}{2}}.$$

We use the result of [11] and the alternative proof of [12] to derive the remaining cases.

1. For $r = -1$, we have $\Pr_{[m,n]}(x_1 \sim \cdots \sim x_n) = \frac{(n-1)!}{n^n} \sim \sqrt{\frac{2\pi}{n}}\, e^{-n}$.
2. For $r = 0$, we get $\Pr_{[m,n]}(x_1 \sim \cdots \sim x_n) \sim \frac{\pi}{2}\, e^{-n}$.
3. For $r \geq 1$ but still fixed, $\Pr_{[m,n]}(x_1 \sim \cdots \sim x_n) \sim K_r\, e^{-n} n^{r/2}$ where the constant $K_r$ can be made explicit.
4. For $r \to \infty$ and $r = o(\sqrt{n})$, $\Pr_{[m,n]}(x_1 \sim \cdots \sim x_n) \sim \sqrt{\frac{3}{2}} \frac{e^{r/2}}{(2\sqrt{3})^r} e^{-n} \left(\frac{n}{r}\right)^{r/2}$.
5. For $r = cn$ for a constant $c > 0$, $\Pr_{[m,n]}(x_1 \sim \cdots \sim x_n) \sim K \left(\frac{(1+c)^c \cosh \zeta}{(2\zeta)^c e^{1+c}}\right)^n$
   where $\zeta \coth \zeta = 1 + c$ and $K = \sqrt{1+c}\, \frac{e^{2\zeta} - 1 - 2\zeta}{\sqrt{\zeta(e^{4\zeta} - 1 - 4\zeta e^{2\zeta})}}$.

6. When $r \to +\infty$ and $2m/n - \log(n)$ is bounded - which covers the two previous cases - then $\Pr_{[m,n]}(x_1 \sim \cdots \sim x_n) \sim \frac{K}{(2\zeta)^r} \left(\frac{\sinh \zeta}{\zeta}\right)^n \frac{(1+r/n)^{n+r+1/2}}{e^{n+r}}$ where $\zeta$ is the positive solution of $\zeta \coth \zeta = 1 + r/n$ and $K = \frac{e^{2\zeta} - 1 - 2\zeta}{\sqrt{\zeta(e^{4\zeta} - 1 - 4\zeta e^{2\zeta})}}$. This formula is an adaptation for multigraphs of Theorem 3 of [12] and appeared originally (for graphs) in [13].
7. Finally, when $2m/n - \log(n) \to +\infty$ as $n \to +\infty$, $\Pr_{[m,n]}(x_1 \sim \cdots \sim x_n) \sim \frac{1}{2^m}$ because almost all multigraphs are connected.

## 4.3 Asymptotics for a Two-Blocks Function

We now consider a function in the class of $x_1 \sim \cdots \sim x_p, x_{p+1} \sim \cdots \sim x_n$ (the block sizes are $p$ and $n - p$), which has cardinality $2^{n-2} \frac{n!}{p!(n-p)!}$. We are again in the third range: $m \geq n - 2$, i.e. $r \geq -2$. Theorem 3 gives the generating function as

$$\phi_j(z) = \left[\frac{v^p}{p!}\right] C(4z, v) \cdot \left[\frac{w^{n-p}}{(n-p)!}\right] C(4z, w),$$

from which we readily obtain that

$$\Pr_{[m,n]}(f) = \frac{m!}{n^{2m}} \sum_{d=-1}^{r+1} \left[\frac{v^p}{p!}\right] C_d(v) \cdot \left[\frac{w^{n-p}}{(n-p)!}\right] C_{r-d}(w).$$

We now consider several cases. For simplicity we assume that, when $r$ is large, it is equal to $cn$ for a fixed positive value $c$.

1. Fixed excess $r$, and a single large part. In the range we are working in, $p$ and $d$ belong to a fixed, finite set. For some explicitly computable constant $k_f$,

$$\Pr_{[m,n]}(f) \sim k_f \cdot n^{\frac{r+3}{2} - p} e^{-n}.$$

2. Fixed excess $r$, and two proportional large parts. By symmetry, we can assume that $p \leq n - p$. We have that

$$\Pr_{[m,n]}(f) \sim \frac{2\pi}{e^n n^{2n+2r}} (n-p)^{2n+\frac{3r}{2}} \left(\frac{p}{n-p}\right)^{2p} \sum_{d=-1}^{r+1} K_d K_{r-d} \left(\frac{p}{n-p}\right)^{\frac{3d}{2}}. \quad (2)$$

Now assume for simplicity that $p = \gamma n$, then

$$\Pr_{[m,n]}(f) \sim k_f \, n^{-\frac{r+1}{2}} \beta^{2n} e^{-n} \quad \text{with} \quad \beta = (1-\gamma)^{1-\gamma} \gamma^\gamma.$$

3. Fixed excess $r$, and two non-proportional large parts. In this case, the expression (2) is still valid, but now $p/(n-p) = o(1)$ i.e. $p = \varepsilon n$ with $\varepsilon = o(1)$. Then

$$\Pr_{[m,n]}(f) \sim k_f \, e^{-n} \, n^{\frac{r-1}{2}} \, \varepsilon^{n\varepsilon - 1} (1-\varepsilon)^{(1-\varepsilon)n}.$$

A more precise evaluation of probabilities requires to know the order of growth of $p$ w.r.t. $n$. E.g.,

(a) $p = \sqrt{n}$: then $\varepsilon = n^{-1/2}$ and the probability of the function is of order $n^{-\frac{r}{2}+\frac{3}{4}}e^{-n-2\sqrt{n}}n^{-\sqrt{n}}$.

(b) $p = \log n$: now $\varepsilon = \frac{\log n}{n}$, the probability is of order $\left(\frac{\log n}{n}\right)^{\log n - 1} n^{\frac{r+1}{2}} e^{-n}$.

4. **Large excess $r$ and a single large part.** If $r = cn$ and $p$ is fixed, we obtain for some explicitly computable constant $k_f$

$$\Pr{}_{[m,n]}(f) \sim \frac{k_f}{n^{p-1}} \left(\frac{(1+c)^c \cosh(\zeta)}{e^{1+c}(2\zeta)^c}\right)^n$$

where $\zeta \coth \zeta = 1 + \frac{cn+1}{n-p}$.

5. **Large excess $r$ and two proportional large parts.** If $r = cn$ and $p = \gamma n$,

$$\Pr{}_{[m,n]}(f) \sim \frac{k_f}{n} \left(\frac{\gamma^\gamma (1-\gamma)^{1-\gamma}(1+c)^{1+c}}{2^c e^{1+c}} g(a_0)\right)^n$$

where $k_f$ is a computable constant, and $g(a_0)$ is the unique maximum in $[0;1]$ of the function

$$g(a) = \left(\frac{\cosh(\zeta_1(a))}{1+\frac{ac}{\gamma}}\right)^\gamma \left(\frac{\cosh(\zeta_2(a))}{1+\frac{(1-a)c}{1-\gamma}}\right)^{1-\gamma} \left(\frac{\gamma}{\zeta_1(a)}\right)^{ac} \left(\frac{1-\gamma}{\zeta_2(a)}\right)^{(1-a)c}$$

where the functions $\zeta_1$ and $\zeta_2$ are implicitly defined by $\zeta_1(a) \coth \zeta_1(a) = 1 + \frac{ac}{\gamma}$ and $\zeta_2(a) \coth \zeta_2(a) = 1 + \frac{(1-a)c}{1-\gamma}$.

### 4.4 Number of Blocks Proportional to $n$

A general approach via Theorem 3 seems difficult, so we assume a certain regularity: Let $f$ denote a boolean function whose associated integer partition representation has the form $\mathbf{i}(f) = (0,\ldots,0,n/g,0,\ldots)$, with $g \geq 2$. Note that the corresponding multigraph has to have at least $m = (g-1) \cdot \frac{n}{g}$ edges. Thus, in contrary to the previously discussed cases, the excess is no more bounded from below, as $n \to \infty$. Such functions may now appear even close to the threshold $1/2$. In Proposition 5, we derive an exact result for those functions, and an asymptotic result in Proposition 6.

Using directly the definition $C(z,v) = \log M(z,v)$ we can show:

**Proposition 5.** *The number of expressions $N_{[m,n]}(f)$ with $n$ variables and $m$ clauses computing a function $f$ with associated integer partition representation of the form $\mathbf{i}(f) = (0,\ldots,0,n/g,0,\ldots)$, i.e. $n/g$ blocks of size $g$, is given by*

$$N_{[m,n]}(f) = 4^m (g!)^{\frac{n}{g}} [z^m] \left(\sum_{j=1}^{g} \frac{(-1)^{j-1}}{j} e_{j,g-j}(z)\right)^{\frac{n}{g}} \tag{3}$$

with

$$e_{j,n}(z) = \sum_{\substack{\sum_{\ell=1}^{j} k_\ell = n \\ k_\ell \geq 0}} \binom{n}{k_1,\ldots,k_j} \frac{\exp\left(\sum_{\ell=1}^{j} \frac{(k_\ell+1)^2 z}{2}\right)}{\prod_{r=1}^{j}(k_\ell+1)!}.$$

For example, in the case $g = 2$ we get

$$\Pr_{[m,n]}(f) = \frac{1}{n^{2m}} \sum_{\ell=0}^{\frac{n}{2}} \binom{\frac{n}{2}}{\ell} (n+\ell)^m (-1)^{\frac{n}{2}-\ell},$$

and for $g = 3$

$$\Pr_{[m,n]}(f) = \frac{1}{n^{2m}} \sum_{\ell=0}^{\frac{n}{3}} \sum_{j=0}^{\ell} \binom{\frac{n}{3}}{\ell} \binom{\ell}{j} (\frac{n}{2}+\ell+2j)^m (-3)^{\ell-j} 2^{\frac{n}{3}-\ell}.$$

It turns out that there is no qualitative difference between constant and large excess. The relevant quantity here is the distance from the minimal possible excess. Thus we start with small $g = 2, 3, \ldots$ and assume that $m = \frac{g-1}{g} \cdot n + \kappa_n$, with $\kappa_n \geq 0$. According to [4] the interesting range is $\frac{n}{2} + \Theta(n^{2/3})$. Hence we also assume $\kappa_n = O(n^{2/3})$.

The above expression for $N_{[m,n]}(f)$, Eq. (3), is a fixed function $G(z)$ raised to a large power. Moreover, it is not hard to show that $G(z) = \sum_{\ell \geq g-1} a_\ell z^\ell$. Thus in case of constant $\kappa_n$ of equation (3) becomes a finite sum which can be computed explicitly (at least in principle).

For $\kappa_n \to \infty$ the saddle point method applies and we can compute $N_{[m,n]}(f)$ asymptotically, though the expressions quickly become messy as $g$ grows. For $g = 2$ we obtain

**Proposition 6.** *The number of expressions $N_{[m,n]}(f)$ with $n$ variables and $m$ clauses computing a function $f$ with associated integer partition representation of the form $i(f) = (0, n/2, 0, 0, \ldots)$, i.e. $n/2$ blocks of size 2, is given by*

$$N_{[m,n]}(f) = \frac{2}{\sqrt{6\pi}} 4^{m+\frac{n}{4}} r_n^{-m+\frac{n}{2}} \exp\left(\frac{3nr_n}{4} + \frac{1}{48} nr_n^2 + O(nr_n^4)\right).$$

*where $r_n$ is the unique positive solution of $\frac{z(2e^z-1)}{e^z-1} = 1 + \frac{2\kappa_n}{n}$, and satisfies*

$$r_n = \frac{4}{3} \cdot \frac{\kappa_n}{n} + O\left(\frac{\kappa_n^2}{n^2}\right).$$

## 5 Discussion

We have analysed the probability of Boolean functions generated by random 2-Xor expressions. This is strongly related to the 2-Xor-SAT problem. For people working in SAT-solver design the structure of solutions of satisfiable expressions, which corresponds to the component structure of the associated multigraphs, is also important.

We derived expressions in terms of coefficients of generating functions for the probability of satisfiability in the critical region ($m \sim \frac{n}{2} + \Theta(n^{2/3})$) as well as a general expression for the probability of any function (Theorem 3). Unfortunately, this expression is too complicated to be used for an asymptotic analysis of general functions. So, we discussed several particular classes of functions: Single block functions are completely analyzed. The asymptotic probability very much depends on the range of the excess. For two block functions, the only missing case is that of two large components which are not proportional in size. All those functions are rather close to FALSE. Finally, functions on the other edge (close to TRUE) were studied and, under some regularity conditions on the block sizes, we were able to get the asymptotic probability.

What is missing is an asymptotic analysis of functions on the boundaries TRUE and FALSE having a more irregular component structure as well as the study of functions in the intermediate range.

**Acknowledgments.** We thank Hervé Daudé and Vlady Ravelomanana for fruitful discussions.

# References

1. Creignou, N., Daudé, H.: Satisfiability threshold for random XOR-CNF formulas. Discrete Applied Mathematics 96-97, 41–53 (1999)
2. Creignou, N., Daudé, H.: Smooth and sharp thresholds for random $k$-XOR-CNF satisfiability. Theor. Inform. Appl. 37(2), 127–147 (2003)
3. Creignou, N., Daudé, H., Dubois, O.: Approximating the satisfiability threshold for random $k$-XOR-formulas. Combin. Probab. Comput. 12(2), 113–126 (2003)
4. Creignou, N., Daudé, H.: Coarse and sharp transitions for random generalized satisfiability problems. In: Mathematics and Computer Science III. Trends Math, pp. 507–516. Birkhäuser, Basel (2004)
5. Creignou, N., Daudé, H., Egly, U.: Phase transition for random quantified XOR-formulas. J. Artif. Intell. Res. 29, 1–18 (2007)
6. Daudé, H., Ravelomanana, V.: Random 2XorSat phase transition. Algorithmica 59(1), 48–65 (2011)
7. Pittel, B., Yeum, J.A.: How frequently is a system of 2-linear equations solvable? Electronic Journal of Combinatorics 17 (2010)
8. Janson, S., Knuth, D., Luczak, T., Pittel, B.: The birth of the giant component. Random Structures and Algorithms 4(3), 233–358 (1993)
9. Banderier, C., Flajolet, P., Schaeffer, G., Soria, M.: Random maps, coalescing saddles, singularity analysis, and Airy phenomena. Random Struct. Algorithms 19(3-4), 194–246 (2001)
10. Flajolet, P., Sedgewick, R.: Analytic combinatorics. Cambridge University Press, Cambridge (2009)
11. Bender, E.A., Canfield, E.R., McKay, B.D.: The asymptotic number of labeled connected graphs with a given number of vertices and edges. Random Structures and Algorithm 1, 129–169 (1990)
12. Pittel, B., Wormald, N.C.: Counting connected graphs inside-out. J. Comb. Theory, Ser. B 93(2), 127–172 (2005)
13. Bender, E.A., Canfield, E.R., McKay, B.D.: The asymptotic number of labeled connected graphs with a given number of vertices and edges. Random Struct. Algorithms 1(2), 127–170 (1990)

# Equivalence Classes of Random Boolean Trees and Application to the Catalan Satisfiability Problem[*]

Antoine Genitrini[1,2] and Cécile Mailler[3]

[1] Sorbonne Universités, UPMC Univ Paris 06, UMR 7606,
LIP6, F-75005, Paris, France
[2] CNRS, UMR 7606, LIP6, F-75005, Paris, France
Antoine.Genitrini@lip6.fr
[3] Laboratoire de Mathématiques de Versailles;
CNRS UMR 8100 and Université de Versailles Saint-Quentin-en-Yvelines,
45 avenue des États-Unis, 78035 Versailles, France
Cecile.Mailler@uvsq.fr

**Abstract.** An and/or tree is a binary plane tree, with internal nodes labelled by connectives, and with leaves labelled by literals chosen in a fixed set of $k$ variables and their negations. We introduce the first model of such Catalan trees, whose number of variables $k_n$ is a function of $n$, its number of leaves. We describe the whole range of the probability distributions depending on the functions $k_n$, as soon as it tends jointly with $n$ to infinity. As a by-product we obtain a study of the satisfiability problem in the context of Catalan trees.

Our study is mainly based on analytic combinatorics and extends the Kozik's *pattern theory*, first developed for the fixed-$k$ Catalan tree model.

**Keywords:** Random Boolean expressions, Boolean formulas, Boolean function, Probability distribution, Satisfiability, Analytic combinatorics.

## 1 Introduction

Since years, many scientists of different areas, e.g. computer scientists, mathematicians or statistical physicists, are studying satisfiability problems (like $k$–SAT problems) and some questions that arise around them: for example, phase transitions between satisfiable and unsatisfiable expressions or constraints satisfaction problems. The classical 3–SAT problem takes into consideration expressions of a specific form: conjunction of clauses that are themselves disjunctions of three literals. The literals are chosen among a set whose size is linked to the size of the expression. Then one question consists of deciding if a large random expression is satisfiable or not. Actually we know among other things, see [1] for example, that satisfiability is related to the ratio between the size of the expression and the number of allowed literals. There is a phase transition such that, if the ratio is smaller than a critical

---

[*] Partially supported by the A.N.R. project *BOOLE*, 09BLAN0011.

value, the random expression is satisfiable with probability tending to 1, when the size of the expression tends to infinity. Otherwise, when the ratio is larger than the critical value, the probability tends to 0.

An interesting paper [3] about Boolean satisfiability problems deals with random 2–XORSAT expressions. Using generating functions, in the context of analytic combinatorics, the authors describe precisely the phase transition between satisfiable and unsatisfiable expressions.

Still dealing with Boolean expressions, but in a completely different direction, researchers have studied the complete probability distribution on Boolean functions induced by random Boolean expressions. The first approach, by Lefmann and Savický [12], consists in fixing a finite set of $k$ variables, allowing the two logical connectives **and** and **or** and choosing uniformly at random a Boolean expression of *size* $n$ in this logical system. Their model is usually called the *Catalan model*. Lefmann and Savický first proved the existence of a limiting probability distribution on Boolean functions when the size of the random Boolean expressions tends to infinity. Since the seminal paper by Chauvin et al. [2], almost all quantitative studies of such Boolean distributions are deeply related to analytic combinatorics: a survey by Gardy [6] provides a wide range of models with various numerical results. Later, Kozik [11] proved a strong relation between the limiting probability of a given function and its *complexity* (i.e. the minimal *size* of an expression representing the function). His approach lies in two separate steps: (i) first let the size of the Boolean expressions taken into consideration tend to infinity, and then (ii) let the number of variables used to label the expressions tend to infinity. His powerful machinery, the *pattern theory*, easily classifies and counts large expressions according to structural constraints. The main objection to this model is about the two consecutive limits that cannot be interchanged: in order to obtain quantitative results, a function must be fixed and thus we cannot consider functions whose complexity depends on $n$. Genitrini and Kozik have proposed another model [10,9] that builds random Boolean expressions over an infinite set of variables. This approach avoids the bias induced by both successive limits. According to our knowledge, the single paper that relates the number of variables to the size is [7]: it finds a large family of functions of small complexity. However from this results we cannot derive any quantitative results of the probability of a small family of functions whose complexities depends on $n$. Moreover looking at satisfiability problems in this context seems to be very exciting.

Our paper extends the Catalan model in order to fit in the satisfiability context. By using an equivalence relation on Boolean expressions, we manage to let both the number of variables and the size of expressions tend jointly to infinity. The number of variables is a function of the size of the expressions and thus we deal with satisfiability in the context of Catalan expressions. Furthermore by extending the techniques of Kozik, we describe in details the probability distribution on functions and exhibit some threshold for the latter distribution: as soon as the number of variables is *large enough* compared to the size of the expressions, the general behaviour of the induced probability on Boolean functions does not change anymore by adding more variables.

The paper is organized as follows. Section 2 introduces our new model based on an equivalence relation of Boolean expressions. Then, Section 3 states our three main results: (1) the satisfiability question for random Catalan expressions; (2) the link between the probability of a class of functions and their common complexity; (3) the behaviour of the probability related to the dynamic between the number of variables and the size of the expressions. Section 4 is devoted to the technical core of the paper. Finally Section 5 applies our approach to and/or trees and proves the main results.

## 2 Probability Distributions on Equivalence Classes of Boolean Functions

### 2.1 Contextual Definitions

A Boolean function is an mapping from $\{0,1\}^\mathbb{N}$ into $\{0,1\}$. The two constant functions $(x_i)_{i\geq 1} \mapsto 1$ and $(x_i)_{i\geq 1} \mapsto 0$ are respectively called true and false.

An and/or **tree** is a binary plane tree whose leaves are labelled by literals, i.e. by elements of $\{x_i, \bar{x}_i\}_{i \in \mathbb{N}}$, and whose internal nodes are labelled by the connective and or the connective or, respectively denoted by $\wedge$ and $\vee$. We will say that $x_i$ and $\bar{x}_i$ are two different literals but they are respectively the positive and the negative version of the same variable $x_i$. Every and/or tree is equivalent to a Boolean expression and thus represents a Boolean function: for example, the tree in Figure 1 is equivalent to the expression $([x_1 \vee (\bar{x}_1 \vee x_2)] \vee x_3) \vee (x_4 \wedge x_1)$ and represents the function $f$ such that, for all $(x_i) \in \{0,1\}^\mathbb{N}$, $f((x_i)_{i\geq 1}) = ([x_1 \vee (\neg x_1 \vee x_2)] \vee x_3) \vee (x_4 \wedge x_1) = 1$, where $\neg x = 1 - x$ for all $x \in \{0,1\}$.

The **size** of an and/or tree is its number of leaves: remark that, for all $n \geq 1$, there is infinitely many and/or trees of size $n$.

The **complexity** of a non constant Boolean function $f$, denoted by $L(f)$, is defined as the size of its **minimal trees**, i.e. the smallest trees computing $f$. The complexity of true and false is 0. Although a Boolean function is defined on an infinite set of variables, it may actually depend only on a finite subset of *essential variables*: given a Boolean function $f$, we say that the variable $x$ is **essential** for $f$, if and only if $f_{|x \leftarrow 0} \neq f_{|x \leftarrow 1}$ (where $f_{|x \leftarrow \alpha}$ is the restriction of $f$ to the subspace where $x = \alpha$). We denote by $E(f)$ the number of essential variables of $f$. Remark that the complexity and the number of essential variables of a Boolean function are related by the following inequalities: $E(f) \leq L(f) \leq 2^{E(f)+2}$ (see e.g. [4, p. 77–78] for the second inequality).

### 2.2 Equivalence Relations

Analytic combinatorics (cf. [4]) is based on the notion of combinatorial classes. A *combinatorial class* is a denumerable (or finite) set of objects on which a size notion is defined such that each object has a non-negative size and the set of objects of any given size is finite. Thus our class of and/or trees is not a combinatorial class since there is infinitely many trees of a given size. To use analytic combinatorics, we define an equivalence relation on Boolean trees. In the rest of the paper, we define a **tree-structure** to be an and/or tree whose leaf-labels have been removed (but internal nodes remain labelled).

Informally two trees are equivalent if the leaves of first one can be relabelled (and negated) without collision in order to obtain the second tree.

**Definition 1.** *Let $A$ and $B$ be two and/or trees. Trees $A$ and $B$ are **equivalent** if (1) their tree-structures are identical, if (2) two leaves are labelled by the same variable in $A$ if and only of they are labelled by the same variable in $B$, and if (3) two leaves are labelled by the same literal in $A$ if and only of they are labelled by the same literal in $B$.*

This equivalence relation on Boolean trees induces straightforwardly an equivalence relation on Boolean functions. Note that all functions of an equivalence class have the same complexity and the same number of essential variables. In the following, we will denote by $\langle f \rangle$ the equivalence class of the function $f$.

### 2.3 Probability Distribution

Let $(k_n)_{n \geq 1}$ **be an increasing sequence that tends to infinity** when $n$ tends to infinity. In the following, we only consider trees such that: for all $n \geq 1$, **the set of variables that appear as leaf-labels** (negated or not) **of a tree of size $n$ has cardinality at most $k_n$**. Note that if $k_n \geq n$ for all $n \geq 1$, this hypothesis is not a restriction. Therefore, we will assume that $k_n \leq n$.

**Definition 2.** *We denote by $t_n$ the number of equivalence classes of trees of size $n$ in which at most $k_n$ different variables appear as leaf-labels. We define the ordinary generating function $T(z)$ as $T(z) = \sum_n t_n z^n$.*

**Proposition 1.** *The number of classes of trees of size $n$ satisfies:*

$$t_n = C_n \cdot \sum_{p=1}^{k_n} \left\{ {n \atop p} \right\} 2^{2n-1-p},$$

*where $C_n$ is the number of unlabelled binary trees[1] of size $n$ and $\left\{ {n \atop p} \right\}$ is the Stirling number of the second kind[2].*

*Proof.* Once the tree-structure of the binary tree is chosen (factor $2^{n-1}C_n$), we partition the set of leaves into $p$ parts such that two leaves that belong to the same part are labelled by the same variable. It gives the contribution $\left\{ {n \atop p} \right\}$. Then, we choose to label each leaf by a positive or negative literal: contribution $2^n$. The equivalence relation states that a tree and the one obtained from it by replacing the positive literals corresponding to a fixed variable by its negation (and conversely) are equivalent. Thus, for each class we multi-count the number of trees: correction $2^{-p}$. □

---
[1] In Proposition 1, $C_n$ is the $(n-1)$st Catalan number (see e.g. [4, p. 6–7]).
[2] In Proposition 1, $\left\{ {n \atop p} \right\}$ is the number of partitions of $n$ objects in $p$ non-empty subsets (see e.g. [4, p. 735–737]).

Given a set $\mathcal{S}$ of equivalence classes of trees and $S_n$ the number of elements of $\mathcal{S}$ of size $n$, we define the **ratio** of $\mathcal{S}$ by $\mu_n(\mathcal{S}) = S_n/t_n$. For a given Boolean function $f$, we denote by $T_n\langle f\rangle$ the number of equivalence classes of trees of size $n$ that compute a function of $\langle f\rangle$, and we define the **probability** of $\langle f\rangle$ as the ratio of $T_n\langle f\rangle$:

$$\mathbb{P}_n\langle f\rangle = \frac{T_n\langle f\rangle}{t_n}.$$

One goal of this paper consists in studying the behaviour of the probabilities $(\mathbb{P}_n\langle f\rangle)_{f\in\mathcal{F}}$ when the size $n$ of the trees tends to infinity.

## 3 Results

We state here our main result: the behaviour of $\mathbb{P}_n\langle f\rangle$ for all fixed functions $f \in \mathcal{F}$ in the framework of and/or trees. Note that $f$ is a single function, not a family of functions. Neither $f$ nor $\langle f\rangle$ depend on $n$, but we look at their representations with trees of size $n$ over at most $k_n$ variables.

The main idea of this part is that *a typical tree computing a Boolean function $f$ is a minimal tree of $f$ into which a large tree has been plugged, that does not change the function computed by the minimal tree.*

This informal but fundamental idea is central in the recent results about quantitative logics (e.g. logic of implication [5]). The proofs in the distinct studies are different, only because of technical incidences induced by the connectives under study. Thus, we are convinced that our results hold in other logics.

**Definition 3.** *Let $f$ be a Boolean functions. We denote by $L\langle f\rangle = L(f)$ (resp. $E\langle f\rangle = E(f)$) the complexity (resp. number of essential variables) of class $\langle f\rangle$. The **multiplicity** of the class $\langle f\rangle$, denoted by $R\langle f\rangle$, is the result $L\langle f\rangle - E\langle f\rangle$: it corresponds to the number of repetitions of variables in a minimal tree of a function from $\langle f\rangle$.*

We recall that a Boolean expression is said *satisfiable* if does not represent the constant function false.

**Theorem 1.** *Let $(k_n)_{n\geq 1}$ be an increasing sequence of integers tending to $+\infty$ when $n$ tends to $+\infty$. A random Catalan expression is satisfiable with probability tending to 1, when the size of the expression tends to infinity.*

**Theorem 2.** *Let $(k_n)_{n\geq 1}$ be an increasing sequence of integers tending to $+\infty$ when $n$ tends to $+\infty$. There exists a sequence $(M_n)_{n\geq 1}$ such that $M_n \sim \frac{n}{\ln n}$ (when $n$ tends to $+\infty$) and such that, for all fixed equivalence classes of Boolean functions $\langle f\rangle$, there exists a positive constant $\lambda_{\langle f\rangle}$ satisfying*

*(i) if, for all sufficiently large $n$, $k_n \leq M_n$, then, asymptotically when $n$ tends to $+\infty$,*
$$\mathbb{P}_n\langle f\rangle \sim \lambda_{\langle f\rangle} \cdot \left(\frac{1}{k_{n+1}}\right)^{R\langle f\rangle+1};$$

*(ii) if, for all sufficiently large $n$, $k_n \geq M_n$, then, asymptotically when $n$ tends to $+\infty$,*
$$\mathbb{P}_n\langle f\rangle \sim \lambda_{\langle f\rangle} \cdot \left(\frac{\ln n}{n}\right)^{R\langle f\rangle+1}.$$

Informally, $\lambda_{\langle f \rangle}$ is related to the number of places where some large trees can be plugged in minimal trees. By taking the complexity of both extremal constant functions true and false to 0, the theorem fits to their equivalence classes too.

In the *finite* context [2,11], each Boolean function is studied separately instead of being considered among its equivalence class. We can translate the result obtained by Kozik in terms of equivalence classes by summing over all Boolean functions belonging to a given equivalence class: note that there are $\binom{k}{E(f)} 2^{E(f)}$ functions in the equivalence class of $f$, therefore, the result of Kozik is equivalent to: for all classes $\langle f \rangle$, asymptotically when $k$ tends to infinity,

$$\lim_{n \to +\infty} \mathbb{P}_{n,k}\langle f \rangle = \Theta_{k \to \infty}\left(\frac{1}{k^{L(f)-E(f)+1}}\right) = \Theta_{k \to \infty}\left(\frac{1}{k^{R(f)+1}}\right).$$

Of course, interchanging the two limits is not a priori possible. However, the *finite* context looks like a degenerate case of our model where there exists an fixed integer $k$ such that $k_n = k$ for all $n \geq 1$. But let us remark that we assume in the present paper that $k_n$ tends to $+\infty$ when $n$ tend to infinity: the case $k_n = k$ is thus not a particular case of our results.

Concerning the infinite context [10,9] $k_n = +\infty$, we already noticed that the cases such that $k_n$ is larger than $n$ are equivalent to the model $k_n = n$, even if $k_n = +\infty$. Therefore, this infinite context is actually the extreme case $k_n = n$ of our model, and is thus fully treated in the present paper. In this specific setting, the Stirling numbers introduced in Proposition 1 induce Bell numbers, that naturally appear in [10,9].

## 4 Technical Key Points

In this section, we state the technical core of our results, and we demonstrate how a threshold does appear according to the dependence $k_n$ in $n$.

### 4.1 Threshold Induced by $k_n$'s Behaviour

**Definition 4.** *Let us define the following quantity: $B_{n,k_n} = \sum_{p=1}^{k_n} \{{n \atop p}\} 2^{-p}$. The number $B_{n,k_n}$ quantitatively represents the labelling constraints of leaf-labelling by variables (cf. Proposition 1).*

The following proposition, which can be seen as some particular case of Bonferroni inequalities allows to exhibit bounds on $B_{n,k_n}$.

**Proposition 2 (for example [13]).** *For all $n \geq 1$, for all $p \in \{1, \ldots, n\}$,*

$$\frac{p^n}{p!} - \frac{(p-1)^n}{(p-1)!} \leq \left\{{n \atop p}\right\} \leq \frac{p^n}{p!}.$$

In view of these inequalities and of the expression of $B_{n,k_n}$ (cf. Definition 4), both following sequences naturally appear:

**Lemma 1.** *Let $n$ be a positive integer.*

(i) *The sequence $(a_p^{(n)})_{p \in \{1,\ldots,n\}} = \left(\frac{p^n}{p!} 2^{-p}\right)_p$ is unimodal, i.e. there exists an integer $M_n$ such that $(a_p)_p$ is strictly increasing on $\{1, \ldots, M_n\}$ and strictly decreasing on $\{M_n + 1, \ldots, n\}$.*

(ii) *Moreover, the sequence $(M_n)_n$ is increasing and asymptotically satisfies: $M_n \sim n/\ln n$.*

We are now ready, to understand the asymptotic behaviour of $B_{n,k_n}$: roughly speaking, asymptotically, $B_{n,k_n}$ does essentially only depend on the terms around $M_n$ in the sum.

**Lemma 2.** *Let $(u_n)_{n \geq 1}$ be an increasing sequence such that $u_n \leq n$ for all integer $n \geq 1$ and $u_n$ tends to $+\infty$ when $n$ tends to $+\infty$.*

(i) *If, for all large enough $n$, $u_n \leq M_n$, then, for all sequences $(\delta_n)_{n \geq 1}$ such that $\delta_n = o(u_n)$ and $\frac{u_n \sqrt{\ln u_n}}{n} = o(\delta_n)$, we have, asymptotically when $n$ tends to $+\infty$,*

$$B_{n,u_n} = \Theta\left(\sum_{p=u_n-\delta_n}^{u_n} \frac{p^n}{p!} 2^{-p}\right). \quad (1)$$

(ii) *If, for large enough $n$, $u_n \geq M_n$, then, for all sequences $(\delta_n)_{n \geq 1}$ such that $\delta_n = o(u_n)$ and $\frac{u_n \sqrt{\ln u_n}}{n} = o(\delta_n)$, for all sequences $(\eta_n)_{n \geq 1}$ such that $\eta_n = o(M_n)$, $\lim_{n \to +\infty} \frac{\eta_n^2}{M_n} = +\infty$ and $\sqrt{M_n \ln(u_n - M_n)} = o(\eta_n)$, we have, asymptotically when $n$ tends to $+\infty$,*

$$B_{n,u_n} = \Theta\left(\sum_{p=M_n-\delta_n}^{\min\{M_n+\eta_n, u_n\}} \frac{p^n}{p!} 2^{-p}\right). \quad (2)$$

**Definition 5.** *Let the fraction $\mathsf{rat}_n$ be the quantitative evolution of the leaf-labelling constraints from trees of size $n-1$ to size $n$: $\mathsf{rat}_n = B_{n-1,k_n}/B_{n,k_n}$. Its asymptotic behaviour is quantified by the two following Lemmas 3 and 4.*

Let us now deduce the following results on the behaviour of $B_{n,k_n}$, when $n$ tends to infinity.

**Lemma 3.** *Let $(k_n)_{n \geq 1}$ be a sequence of integers that tends to $+\infty$ when $n$ tends to $+\infty$. Let us assume that $k_n \leq M_n$ for large enough $n$, then, asymptotically when $n$ tends to infinity,*

$$\frac{B_{n,k_n+1}}{B_{n+1,k_n+1}} = \Theta\left(\frac{1}{k_n+1}\right).$$

**Lemma 4.** *Let $(k_n)_{n \geq 1}$ be a sequence of integers that tends to $+\infty$ when $n$ tends to $+\infty$. Let us assume that $k_n \geq M_n$ for large enough $n$, then, asymptotically when $n$ tends to infinity,*

$$\frac{B_{n,k_n+1}}{B_{n+1,k_n+1}} = \Theta\left(\frac{\ln n}{n}\right).$$

## 4.2 Adjustment of Kozik's Pattern Language Theory

In 2008, Kozik [11] introduced a quite effective way to study Boolean trees: he defined a notion of pattern that permits to easily classify and count large trees according to some constraints on their structures. Kozik applied this pattern theory to study and/or trees with a finite number of variables.

We recall the definitions of patterns, illustrate them on an example and then extend Kozik's paper results to our new model.

**Definition 6.** *(i) A **pattern** is a binary tree with internal nodes labelled by $\wedge$ or $\vee$ and with external nodes labelled by $\bullet$ or $\square$. Leaves labelled by $\bullet$ are called **pattern leaves** and leaves labelled by $\square$ are called **placeholders**. A **pattern language** is a set of patterns*
*(ii) Given a pattern language $L$ and a family of trees $\mathcal{M}$, we denote by $L[\mathcal{M}]$ the family of all trees obtained by replacing every placeholder in an element from $L$ by a tree from $\mathcal{M}$.*
*(iii) We say that $L$ is **unambiguous** if, and only if, for any family $\mathcal{M}$ of trees, any tree of $L[\mathcal{M}]$ can be built from a unique pattern from $L$ into which trees from $\mathcal{M}$ have been plugged.*

The generating function of a pattern language $L$ is $\ell(x,y) = \sum_{d,p} L(d,p) x^d y^p$, where $L(d,p)$ is the number of elements of $L$ with $d$ pattern leaves and $p$ placeholders.

**Definition 7.** *We define the composition of two pattern languages $L[P]$ as the pattern language of trees which are obtained by replacing every placeholder of a tree from $L$ by a tree from $P$.*

**Definition 8.** *A pattern language $L$ is **sub-critical** for a family $\mathcal{M}$ if the generating function $m(z)$ of $\mathcal{M}$ has a square-root singularity $\tau$, and if $\ell(x,y)$ is analytic in some set $\{(x,y) : |x| \leq \tau + \varepsilon, |y| \leq m(\tau) + \varepsilon\}$ for some positive $\varepsilon$.*

**Definition 9.** *Let $L$ be a pattern language, $\mathcal{M}$ be a family of trees and $\Gamma$ a subset of $\{x_i\}_{i \geq 1}$, whose cardinality does not depend on $n$. Given an element of $L[\mathcal{M}]$,*

*(i) the number of its $L$-**repetitions** is the number of its $L$-pattern leaves minus the number of different variables that appear in the labelling of its $L$-pattern leaves.*
*(ii) the number of its $(L, \Gamma)$-**restrictions** is the number of its $L$-pattern leaves that are labelled by variables from $\Gamma$, plus the number of its $L$-repetitions.*

**Definition 10.** *Let $\mathcal{I}$ be the family of the trees with internal nodes labelled by a connective and leaves without labelling, i.e. the family of tree-structures.*

The generating function of $\mathcal{I}$ satisfies $I(z) = z + 2I(z)^2$, that implies $I(z) = (1 - \sqrt{1-8z})/4$ and thus its dominant singularity is $1/8$.

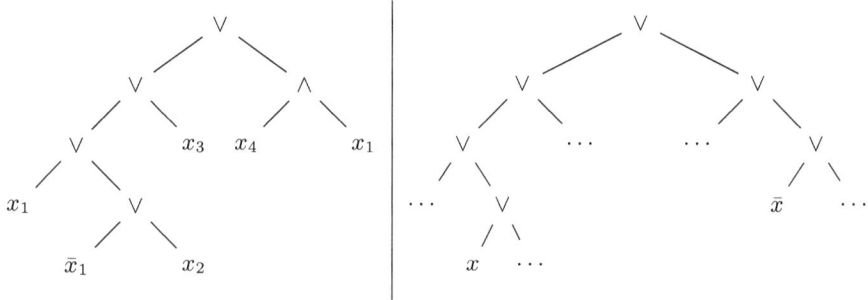

**Fig. 1.** Left: a tree that computes the function true. Right: a simple tautology.

We can, for example, define the unambiguous pattern language $N$ by induction as follows: $N = \bullet | N \vee N | N \wedge \square$, meaning that a pattern from $N$ is either a single pattern leaf, or a tree rooted by $\vee$ whose two subtrees are patterns from $N$, or a tree rooted by $\wedge$ whose left subtree is a pattern from $N$ and whose right subtree is a placeholder. Its generating function verifies, $n(x,y) = x + n(x,y)^2 + yn(x,y)$ and is equal to $n(x,y) = \frac{1}{2}(1 - y - \sqrt{(1-y)^2 - 4x})$. It is thus subcritical for $\mathcal{I}$.

On the left-hand side of Fig. 1, we have depicted a Boolean tree that computes the constant function true. It has 5 $N$-pattern leaves, 1 $N$-repetition and 2 $(N, \{x_2\})$-restrictions.

The following key lemma is a generalization of the corresponding lemma of Kozik [11, Lemma 3.8].

**Lemma 5.** *Let $L$ be an unambiguous pattern, and $\mathcal{T}$ the families of and/or trees. Let $r$ be a fixed positive integer. Let $T_n^{[r]}$ (resp. $T_n^{[\geq r]}$) be the number of labelled (with at most $k_n$ variables) trees of $L[\mathcal{T}]$ of size $n$ and with $r$ $L$-repetitions (resp. at least $r$ $L$-repetitions). We assume that $L$ is sub-critical for the family $\mathcal{I}$ of the unlabelled-leaves trees. Then, asymptotically when $n$ tends to infinity,*

$$\frac{T_n^{[r]}}{t_n} = \mathcal{O}\left(\mathrm{rat}_n^r\right) \quad \text{and} \quad \frac{T_n^{[\geq r]}}{t_n} = \mathcal{O}\left(\mathrm{rat}_n^r\right).$$

*Proof.* The number of labelled trees of $L[\mathcal{T}]$ of size $n$ and with at least $r$ $L$-repetitions is given by:

$$t_n^{[\geq r]} = \sum_{d=r+1}^{n} I_n(d) \mathrm{Lab}(n, k_n, d, r),$$

where $I_n(d)$ is the number of tree-structures with $d$ $L$-pattern leaves and the number $\mathrm{Lab}(n, k_n, d, r)$ corresponds to the number of leaf-labellings of these trees giving at least $r$ $L$-repetitions. The following enumeration contains some multi-counting and we therefore get an upper bound:

$$\mathrm{Lab}(n, k_n, d, r) \leq 2^n \cdot \sum_{j=1}^{r} \binom{d}{r+j} \begin{Bmatrix} r+j \\ j \end{Bmatrix} B_{n-r-j+1, k_n}.$$

The factor $2^n$ corresponds to the polarity of each leaf (the variable labelling it is either negated or not); the index $j$ stands for the number of different variables involved in the $r$ repetitions; the binomial factor corresponds to the choices of the pattern leaves that are involved in the $r$ repetitions; the Stirling number corresponds to the partition of $r + j$ leaves into $j$ parts; finally, the factor $B_{n-r-j+1,k_n}$ corresponds to the choices of the variable assigned to each class of leaves. Therefore,

$$t_n^{[\geq r]} \leq 2^n \cdot B_{n-r,k_n} \sum_{j=1}^{r} \begin{Bmatrix} r+j \\ j \end{Bmatrix} \sum_{d=r+j}^{n} I_n(d) \binom{d}{r+j}.$$

Let $\ell(x, y)$ be the generating function of the pattern $L$. Then, for all $p \geq 0$,

$$\frac{z^p}{p!} \frac{\partial^p \ell}{\partial x^p}(z, I(z)) = \sum_{n=1}^{\infty} \sum_{d=1}^{\infty} I_n(d) \binom{d}{p} z^n.$$

Thus,
$$\frac{t_n^{[\geq r]}}{t_{n,k_n}} \leq \frac{B_{n-r,k_n}}{B_{n,k_n}} \sum_{j=1}^{r} \begin{Bmatrix} r+j \\ j \end{Bmatrix} \frac{[z^n] z^{r+j} \frac{\partial^{r+j} \ell}{\partial x^{r+j}}(z, I(z))}{[z^n] I(z)}.$$

Since $z^{r+j} \frac{\partial^{r+j} \ell}{\partial x^{r+j}}(z, I(z))$ and $I(z)$ have the same singularity because of the subcriticality of the pattern $L$ according to $\mathcal{I}$, the previous sum tends to a constant (because $r$ is fixed) when $n$ tends to infinity and so we conclude:

$$\frac{t_n^{[r]}}{t_n} \leq \frac{t_n^{[\geq r]}}{t_n} = \mathcal{O}\left(\frac{B_{n-r,k_n}}{B_{n,k_n}}\right) = \mathcal{O}\left(\mathsf{rat}_n^r\right).$$

$\square$

## 5 Behaviour of the Probability Distribution

Once we have adapted the pattern theory to our model and proved the central Lemma 5, we are ready to quantitatively study our model. A first step consists to understand the asymptotic behaviour of $\mathbb{P}_n \langle \text{true} \rangle$. It is indeed natural to focus on this "simple" function before considering a general class $\langle f \rangle$; and moreover, it happens to be essential for the continuation of the study. In addition, the methods used to study tautologies (mainly pattern theory) will also be the core of the proof for a general equivalence class. We prove in this section the main Theorem 2 for both classes $\langle \text{true} \rangle$ and $\langle \text{false} \rangle$ of complexity zero, using the duality of both connectives $\wedge$ and $\vee$ and both positive and negative literals. Theorem 1 is then obtained as a by-product of Theorem 2. The main ideas of the proof for a general equivalence class are given in Section 5.2.

### 5.1 Tautologies

Recall that a **tautology** is a tree that represents the Boolean function true. Let $\mathcal{A}$ be the family of tautologies. In this part, we prove that the probability of $\langle \text{true} \rangle$ is asymptotically equal to the ratio of a simple subset of tautologies.

**Definition 11 (cf. right-hand side of Fig. 1).** *A* **simple tautology** *is an and/or tree that contains two leaves labelled by a variable $x$ and its negation $\bar{x}$ and such that all internal nodes from the root to both leaves are labelled by $\vee$-connectives. We denote by ST the family of simple tautologies.*

**Proposition 3.** *The ratio of simple tautologies verifies*

$$\mu_n(ST) = \frac{ST_n}{t_n} \sim \frac{3}{4}\mathsf{rat}_n, \text{ when } n \text{ tends to infinity.}$$

*Moreover, asymptotically when $n$ tends to infinity, almost all tautologies are simple tautologies.*

The latter proposition gives us for free the proof of Theorem 1. In fact, both dualities between the two connectives and positive and negative literals transform expressions computing true to expressions computing false, which implies $\mathbb{P}_n\langle\mathsf{false}\rangle = 3/4 \cdot \mathsf{rat}_n$. Moreover, the only expressions that are not satisfiable compute the function false and $\mathbb{P}_n\langle\mathsf{false}\rangle = 3/4 \cdot \mathsf{rat}_n$ tends to 0 as $n$ tends to infinity, which proves Theorem 1.

### 5.2 Probability of a General Class of Functions

With similar arguments than those used for tautologies, it is possible to prove that the probability of the class of projections (i.e. $(x_i)_{i\geq 1} \mapsto x_j$) is equivalent to $5/8 \cdot \mathsf{rat}_n$, when $n$ tends to $+\infty$.

This last section is devoted to the general result, i.e. to the study of the behaviour of $\mathbb{P}_n\langle f\rangle$ for all fixed $f \in \mathcal{F}$. The main idea of this part is that, roughly speaking, *a typical tree computing a Boolean function in $\langle f\rangle$ is a minimal tree of $\langle f\rangle$ into which a single large tree has been plugged.*

*Proof (sketch).* Our aim is to describe the asymptotic behaviour of $\mathbb{P}_n\langle f\rangle$, for a given class of Boolean functions $\langle f\rangle$.

– We first define several notions of *expansions* of a tree: the idea is to replace in a tree, a subtree $S$ by $T \wedge S$, where $T$ is chosen such that the expanded tree still computes the same function.
– The ratio of minimal trees of $\langle f\rangle$ expanded once is of the order of $\mathsf{rat}_n^{R(f)+1}$.
– The ratio of trees computing a function from $\langle f\rangle$ is asymptotically equal to the ratio of minimal trees expanded once.

The most technical part of the proof is the last one, because we need a precise upper bound of $\mathbb{P}_n\langle f\rangle$. But the ideas are more or less the same as those developed for the class $\langle\mathsf{true}\rangle$. □

## 6 Conclusion

We focussed on the logical context of and/or connectives because of the richness of this logical system: normal forms, functional completeness. However the implicational logical system (e.g. [5,9]) could also be studied in this new context and we deeply believe the general behaviour to be identical. Indeed, the key idea is that *each repetition induces a factor* $\mathsf{rat}_n$, and this remains true in all those models – although pattern theory does not adapt to every model, e.g. models with *implication*. Extending our results to these models would give nice unifications of the known results of the literature: papers [11,5,9] and [8].

The numerous results of the last decade in quantitative logics are now linked, through this new model, to satisfiability problems. Our Catalan model of expressions behaves differently than $k$–SAT or 2–XORSAT problems: asymptotically, almost all expressions are satisfiable, regardless of the ratio between the number of variables and the size of expressions. We can thus conclude that the behaviour of a SAT problem heavily depends on the considered subfamily of Boolean expressions.

**Acknoledgements.** We are grateful to Pierre Lescanne, whose remark at CLA'12 has allowed us to go beyond our initial idea and consider the more general framework presented here. We also want to thank Brigitte Chauvin and Danièle Gardy who proof-read a previous version of this paper and gave us precious advises to improve it.

# References

1. Achlioptas, D., Moore, C.: Random k-SAT: Two moments suffice to cross a sharp threshold. SIAM Journal of Computing 36(3), 740–762 (2006)
2. Chauvin, B., Flajolet, P., Gardy, D., Gittenberger, B.: And/Or trees revisited. Combinatorics, Probability and Computing 13(4-5), 475–497 (2004)
3. Daudé, H., Ravelomanana, V.: Random 2-XORSAT phase transition. Algorithmica 59(1), 48–65 (2011)
4. Flajolet, P., Sedgewick, R.: Analytic Combinatorics. Cambridge U.P. (2009)
5. Fournier, H., Gardy, D., Genitrini, A., Gittenberger, B.: The fraction of large random trees representing a given boolean function in implicational logic. Random Structures and Algorithms 40(3), 317–349 (2012)
6. Gardy, D.: Random Boolean expressions. In: Colloquium on Computational Logic and Applications, vol. AF, pp. 1–36. DMTCS (2006)
7. Genitrini, A., Gittenberger, B.: No Shannon effect on probability distributions on Boolean functions induced by random expressions. In: 21st Meeting Analysis of Algorithms, pp. 303–316 (2010)
8. Genitrini, A., Gittenberger, B., Kraus, V., Mailler, C.: Probabilities of Boolean functions given by random implicational formulas. Electronic Journal of Combinatorics 19(2), P37, 20 pages (electronic) (2012)
9. Genitrini, A., Kozik, J.: In the full propositional logic, 5/8 of classical tautologies are intuitionistically valid. Ann. of Pure and Applied Logic 163(7), 875–887 (2012)
10. Genitrini, A., Kozik, J., Zaionc, M.: Intuitionistic vs. Classical tautologies, quantitative comparison. In: Miculan, M., Scagnetto, I., Honsell, F. (eds.) TYPES 2007. LNCS, vol. 4941, pp. 100–109. Springer, Heidelberg (2008)
11. Kozik, J.: Subcritical pattern languages for And/Or trees. In: Fifth Colloquium on Mathematics and Computer Science. DMTCS Proceedings (2008)
12. Lefmann, H., Savický, P.: Some typical properties of large And/Or Boolean formulas. Random Structures and Algorithms 10, 337–351 (1997)
13. Sibuya, M.: Log-concavity of Stirling numbers and unimodality of Stirling distributions. Ann. of the Institute of Statistical Mathematics 40(4), 693–714 (1988)

# The Flip Diameter of Rectangulations and Convex Subdivisions*

Eyal Ackerman[1], Michelle M. Allen[2], Gill Barequet[3], Maarten Löffler[4]
Joshua Mermelstein[2], Diane L. Souvaine[2], and Csaba D. Tóth[5]

[1] Dept. Math., Physics, and Comp. Sci., University of Haifa at Oranim, Tivon, Israel
ackerman@sci.haifa.ac.il
[2] Department of Computer Science, Tufts University, Medford, MA, USA
{michelle.allen,joshua.mermelstein,diane.souvaine}@tufts.edu
[3] Department of Computer Science, Technion, Haifa, Israel
barequet@cs.technion.ac.il
[4] Department of Computing and Information Sciences, Utrecht University, The Netherlands
m.loffler@uu.nl
[5] Department of Mathematics, California State Univ. Northridge, Los Angeles, CA, USA
csaba.toth@csun.edu

**Abstract.** We study the configuration space of rectangulations and convex subdivisions of $n$ points in the plane. It is shown that a sequence of $O(n \log n)$ elementary *flip* and *rotate* operations can transform any rectangulation to any other rectangulation on the same set of $n$ points. This bound is the best possible for some point sets, while $\Theta(n)$ operations are sufficient and necessary for others. Some of our bounds generalize to convex subdivisions of $n$ points in the plane.

## 1 Introduction

The study of rectangular subdivisions of rectangles is motivated by VLSI floorplan design [25] and cartographic visualization [13,28]. The rich combinatorial structure of rectangular floorplans has also attracted theoretical research [6,15]. Combinatorial properties lead to efficient algorithms for the recognition and reconstruction of the *rectangular graphs* induced by the corners of the rectangles in a floorplan [19,27], the *contact graphs* of the rectangles [22,35], and the contact graphs of the horizontal and vertical *line segments* that separate the rectangles [10]. The number of combinatorially different floorplans with $n$ rectangles is known to be $B(n) = \Theta(8^n/n^4)$, the $n$th Baxter number [36].

Rectangular subdivisions in the presence of points have also been studied in the literature. Given a set $P$ of points in an axis-aligned bounding box $R$, a *rectangulation* of $(R, P)$ is a subdivision of $R$ into rectangles by pairwise noncrossing axis-parallel line segments such that every point in $P$ lies in the relative interior of a segment (Fig. 1).

Finding a rectangulation of minimum total edge length has attracted considerable attention [7,8,12,17,18,24,26] due to its applications in VLSI design and stock cutting

---
* Löffler is partially supported by the NWO (639.021.123). Allen, Mermelstein, Souvaine, and Tóth are supported in part by the NSF (CCF-0830734).

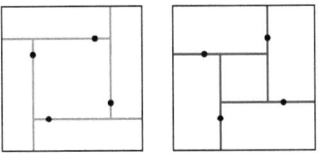

**Fig. 1.** Two different rectangulations of a set of four points

in the presence of material defects. This problem is known to be NP-hard [26], however, its complexity is unknown when the points in $P$ are in general position in the sense that they have distinct $x$- and $y$-coordinates, that is, the points are *noncorectilinear*. It is not hard to see that in this case the minimum edge-length rectangulation must consist of exactly $n$ line segments [7]. For the rest of this paper, we consider only such rectangulations.

The space of all the rectangulations of a point set $P$ in a rectangle $R$ can be explored using the following two elementary operators introduced in [2] (refer to Fig. 2).

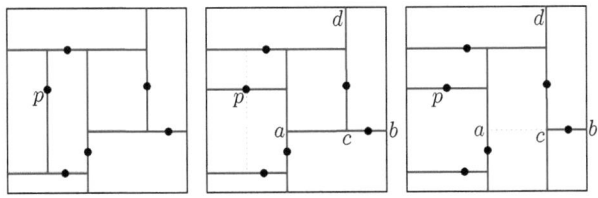

**Fig. 2.** A rectangulation $r_1$ of a set of 6 points, $r_2 = \text{FLIP}(r_1, p)$, and $r_3 = \text{ROTATE}(r_2, c)$

**Definition 1 (Flip).** *Let $r$ be a rectangulation of $P$ and let $p \in P$ be a point such that the segment $s$ that contains $p$ does not contain any endpoints of other segments. The operation $\text{FLIP}(r, p)$ changes the orientation of $s$ from vertical to horizontal or vice-versa.*

**Definition 2 (Rotate).** *Let $r$ be a rectangulation of $P$. Let $s_1 = ab$ be a segment that contains $p \in P$. Let $s_2 = cd$ be a segment such that $c$ lies in $ap \subset s_1$ and $ac$ does not contain any endpoints of other segments. The operation $\text{ROTATE}(r, c)$ shortens $s_1$ to $cb$, and extends $s_2$ beyond $c$ until it reaches another segment or the boundary of $R$.*

For a finite set of noncorectilinear points $P$, we denote by $G(P) = (V, E)$ the *graph of rectangulations* of $P$, where the vertex set is $V = \{r : r \text{ is a rectangulation of } P\}$ and the edge set is $E = \{(r_1, r_2) : \text{a single flip or rotate operation on } r_1 \text{ produces } r_2\}$. Since both operations are reversible, $G(P)$ is an undirected graph. It is not hard to show that $G(P)$ is connected [2], and there is a sequence of $O(n^2)$ flip and rotate opreations between any two rectangulations in $G(P)$ when $P$ is a set of $n$ points in $R$. It is natural to ask for the diameter of $G(P)$, which we call for short, the *flip diameter* of $P$.

**Results.** In this paper, we show that the flip diameter of $P$ is $O(n \log n)$ for every $n$-element point set $P$ (Section 2), and it is $\Omega(n \log n)$ for some $n$-element point sets

(Section 3). However, there are $n$-elements point sets with $\Theta(n)$ flip diameter (Section 4). That is, the flip diameter is always between $O(n \log n)$ and $\Omega(n)$, depending on the point configuration, and both bounds are the best possible.

We extend the flip and rotate operations and the notion of flip diameter to convex subdivisions (Section 5). A *convex subdivision* of a set $P \subset \mathbb{R}^2$ of points is a subdivision of the plane into convex faces by pairwise noncrossing line segments, halflines, and lines, each of which contains exactly one point of $P$. We show that the flip diameter for the convex subdivisions of $n$ points is always $O(n \log n)$ and sometimes $\Theta(n)$.

**Related Work.** Determining the exact number of rectangulations on $n$ noncorectilinear points remains an elusive open problem in enumerative combinatorics [2,3]. Recently, Felsner [14] proved that every combinatorial floorplan can be embedded into every set of points, hence every set of $n$ points has at least $B(n)$, i.e., $\Omega(8^n/n^4)$ rectangulations.

The currently best known upper bound, $O(18^n/n^4)$ by Ackerman [1], uses the so-called "cross-graph" charging scheme [30,31], originally developed for counting the number of (geometric) triangulations on $n$ points in the plane. This method is based on elementary "flip" operations that transform one triangulation into another. Lawson [23] proved that every triangulation on $n$ points in the plane can be transformed into the Delaunay triangulation with $O(n^2)$ flips, and this bound is the best possible by a construction due to Hurtado et al. [21]. However, for $n$ points in convex position, $2n - 10$ flips are sufficient, due to a bijection with binary trees with $n - 2$ internal nodes [32]. Hence the flip-diameter of the triangulations on $n$ points in the plane is always between $\Theta(n)$ and $\Theta(n^2)$ depending on the point configuration. Eppstein et al. [13] and Buchin et al. [5] define two elementary flip operations on floorplans, in terms of the directed dual graph, and solve optimization problems on floorplans by traversing the flip graph.

## 2 An Upper Bound on the Flip Diameter of Rectangulations

In this section, we show that for every set $P$ of $n$ noncorectilinear points in a rectangle $R$, the diameter of $G(P)$ is $O(n \log n)$.

**Theorem 1.** *For every $n$-element point set $P$, the diameter of $G(P)$ is $O(n \log n)$.*

Given a rectangulation $r$ of $P$, we construct a sequence of $O(n \log n)$ operations that transforms $r$ into a rectangulation with all vertical segments (a *canonical* rectangulation). Our method relies on the concept of "independent" points, defined in terms of the bar visibility graph. Let $r$ be a rectangulation $r$ of $P$. The bar visibility graph [11,33] on the horizontal segments of $r$ is defined as a graph $H(r)$, where the vertices correspond to the horizontal segments in $r$; and two horizontal segments $s_1$ and $s_2$ are adjacent in $H(r)$ if and only if there are points $a \in s_1$ and $b \in s_2$ such that $ab$ is a vertical segment (not necessarily in $r$) that does not intersect any other horizontal segment in $r$. It is clear that the bar visibility graph is planar.

Observe that we can always change the orientation of any line segment $s$ with $O(n)$ operations: simply shorten $s$ using rotate operations until $s$ contains no other segment endpoints, and then flip $s$. This simple procedure is formulated in the following subroutine.

Shorten&Flip$(r, s)$. Let $s$ be a segment in a rectangulation $r$. Assume $s = ab$ and $p \in P$ is in the relative interior of $s$. While $s$ contains the endpoint of some other segment, let $c_1 \in s$ and $c_2 \in s$ be the endpoints of some other segments closest to $a$ and $b$, respectively (possibly $c_1 = c_2$). If $p \notin ac_1$, then apply ROTATE$(r, c_1)$ to shorten $s = ab$ to $c_1 b$. Else, apply ROTATE$(r, c_2)$ to shorten $s$ to $ac_2$. When $s$ does not contain the endpoint of any other segment, apply FLIP$(r, p)$.

The proof of Theorem 1 follows from a repeated invocation of the following lemma.

**Lemma 1.** *Let $r$ be a rectangulation of a set of $n$ pairwise noncorectilinear points in a rectangle $R$. There is a sequence of $O(n)$ flip and rotate operations that turns at least one quarter of the horizontal segments into vertical segments, and keeps vertical segments vertical.*

*Proof.* By the four color theorem [29], $H(r)$ has an independent set $I$ that contains at least one quarter of the horizontal segments in $r$. The total number of endpoints of vertical segments that lie on some horizontal segment in $I$ is $O(n)$. Successively call the subroutine Shorten&Flip$(r, s)$ for all horizontal segment $s \in I$.

The horizontal segments in $I$ are shortened and flipped into vertical orientation. All operations maintain the invariants that (1) the segments in $I$ are pairwise disjoint, and (2) the remaining horizontal segments in $I$ form an independent set in the bar visibility graph (of all horizontal segments in the current rectangulation). It follows that each operation either decreases the number of horizontal segments in $I$ (flip), or decreases the number of segment endpoints that lie in the relative interior of a segment in $I$ (rotate). After $O(n)$ operations, all segments in $I$ become vertical. Since only the segments in $I$ are flipped (once each), all vertical segments in $r$ remain vertical, as required.

*Proof (of Theorem 1).* Let $P$ be a set of $n$ pairwise noncorectilinear points in a rectangle $R$. Denote by $r_0$ the rectangulation that consists of $n$ vertical line segments, one passing through each point in $P$.

We show that every rectangulation $r_1$ of $P$ can be transformed into $r_0$ by a sequence of $O(n \log n)$ flip and rotate operations. By Lemma 1, a sequence of $O(n)$ operations can decrease the number of horizontal segments by a factor of at least $4/3$. After at most $\log n / \log(4/3)$ invocations of Lemma 1, the number of horizontal segments drops below 1, that is, all segments become vertical and we obtain $r_0$.

## 3 A Lower Bound on the Flip Diameter of Rectangulations

We show that the diameter of the graph $G(P)$ is $\Omega(n \log n)$ when $P$ is an $n$-element *bit-reversal point set* (alternatively, Halton-Hammersley point set) [9, Section 2.2]. For every integer $k \geq 0$, we define a point set $P_k$ of size $n = 2^k$ with integer coordinates lying in the square $R = [-1, n]^2$. For an integer $m$, $0 \leq m < 2^k$, with binary representation $m = \sum_{i=1}^{k} b_i 2^{i-1}$, the bit-reversal gives $y(m) = \sum_{i=1}^{k} b_i 2^{k-i}$. The bit-reversal point set of size $n = 2^k$ is $P_k = \{(m, y(m)) : m = 0, 1, \ldots, n-1\}$. By construction, no two points in $P_k$ are corectilinear.

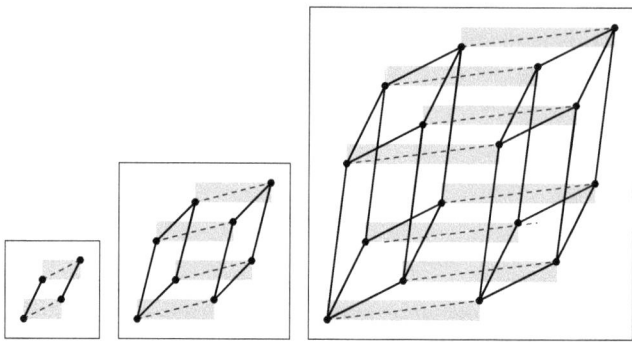

**Fig. 3.** The sets $P_2$, $P_3$, and $P_4$. The edges connect points whose binary representations differ in a single bit, showing that $P_k$ is a projection of a $k$-dimensional hypercube. The grey rectangles are spanned by point pairs whose binary representations differ exactly in the last coordinate.

We establish a lower bound of $k2^{k-3}$ for the diameter of $G(P_k)$ using a charging scheme. We define $k2^{k-1}$ empty rectangles (called *boxes*) spanned by $P_k$, and charge one unit for "saturating" a box with vertical segments (as defined below). We show that when a rectangulation with all horizontal segments is transformed into one with all vertical segments, each box becomes saturated. We also show that each rotate (resp., flip) operation contributes a total of at most 2 (resp., 4) units to the saturation of various boxes in our set. It follows that at least $(k2^{k-1})/4 = k2^{k-3} = n \log n /8$ operations are required to saturate all $k2^{k-1}$ boxes.

Consider the point set $P_k$ for some $k \in \mathbb{N}$. We say that a rectangle $B \subset [-1, n]^2$ is *spanned* by $P_k$ if two opposite corners of $B$ are in $P_k$; and $B$ is *empty* if its interior is disjoint from $P_k$.

Let $\mathcal{B}$ be the set of closed rectangular boxes spanned by point pairs in $P_k$ whose corresponding binary representation $(b_1, \ldots, b_k)$ differ in exactly one bit. See Fig. 3 for examples. Each point in $P_k$ is incident to $k$ boxes in $\mathcal{B}$, since there are $k$ bits. Every box in $\mathcal{B}$ is spanned by two points of $P_k$, thus $|\mathcal{B}| = k \cdot |P_k|/2 = k2^{k-1}$. Each point is incident to $k$ boxes of sizes $2^{i-1} \times 2^{k-i}$ for $i = 0, \ldots, k-1$, since changing the $i$th bit $b_i$ incurs an $2^{i-1}$ change in the $x$-coordinate and an $2^{k-i}$ change in the $y$-coordinate. It follows that every box in $\mathcal{B}$ is empty, and the boxes of the same size are pairwise disjoint.

We now define the "saturation" of each box $B \in \mathcal{B}$ with respect to a rectangulation of $P_k$. Let $B \in \mathcal{B}$ and let $r$ be a rectangulation of $P_k$. The vertical *extent* of $B$ is the orthogonal projection of $B$ into the $y$-axis. Consider the vertical segments of $r$ clipped in $B$ (i.e., the segments $s \cap B$ for all vertical segments $s$ in $r$). The *saturation of $B$* with respect to $r$ is the percentage of the vertical extent of $B$ covered by projections of vertical segments of $r$ clipped in $B$. See Fig. 4 for examples. By definition, the saturation of $B$ is a real number in $[0, 1]$. For every $B \in \mathcal{B}$, we have that the saturation of $B$ is 0 when $r$ is a rectangulation with all horizontal segments, and it is 1 when $r$

consists of all vertical segments. If we transform an all-horizontal rectangulation into an all-vertical one by a sequence of operations, the total saturation of all $k2^{k-1}$ boxes in $\mathcal{B}$ increases from 0 to $k2^{k-1}$. The key observation is that a single operation increases the total saturation of all boxes in $\mathcal{B}$ by at most a constant.

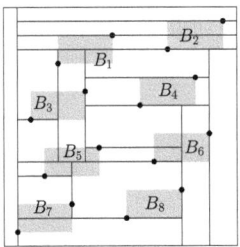

**Fig. 4.** A rectangulation of $P_4$. The saturation of box $B_1, \ldots, B_8$ is $\frac{1}{2}$, 0, 1, 0, 1, 1, 1, and 1, respectively.

It remains to bound the impact of a single operation on the saturation of a box in $\mathcal{B}$. Consider first an operation ROTATE$(r, c)$ that increases the saturation of some box $B \in \mathcal{B}$. A rotate operation shortens a segment $s_1$ and extends an orthogonal segment $s_2$. The saturation of a box $B$ can increase only if a vertical segment grows, so we may assume that $s_1$ is horizontal and $s_2$ is vertical. Denote by $s$ the newly inserted portion of $s_2$. Note that $s$ lies in a single face of the rectangulation $r$. Similarly, if an operation FLIP$(r, p)$ increases the saturation of a box in $B$, then it replaces a horizontal segment by a vertical segment passing through $p$. The new vertical segment lies in *two* adjacent faces of $r$, separated by the original horizontal segment through $p$. We represent the new vertical segment as the union of two collinear vertical segments $s \cup s'$ that meet at point $p$. In summary, an operation ROTATE$(r, c)$ inserts one vertical segment $s$ that lies in the interior of a face of $r$, and an operation FLIP$(r, p)$ inserts two such segments. We show now that if such a new vertical segment $s$ increases the saturation of some box in $B \in \mathcal{B}$, then $s$ must lie in $B$.

**Lemma 2.** *Suppose that an operation inserts a vertical segment $s$ that lies in a face $f$ of the rectangulation $r$. If the insertion of $s$ increases the saturation of a box $B \in \mathcal{B}$, then $s \subset B$.*

*Proof.* Suppose, to the contrary, that $s \not\subset B$. Let $p, q \in P_k$ denote the two opposite corners of points that span $B$, such that $p$ is the upper left or upper right corner of $B$, and $q$ is the opposite corner of $B$. Since $s$ increases the saturation of $B$, it must intersect $B$. Hence $f \cap B \neq \emptyset$. Since $s \not\subset B$, at least one of the endpoints of $s$ lies in the exterior of $B$. Assume, without loss of generality, that the upper endpoint of $s$ lies outside $B$, and $p$ is the upper left corner of $B$. Then, the top side of $f$ is strictly above the top side of $B$. Since point $p$ cannot be in the interior of $f$, the left side of the face $f$ intersects the top side of $B$. Note that $s$ and the left side of $f$ have the same orthogonal projection on the $y$-axis. Therefore, the insertion of $s$ cannot increase the saturation of $B$, contradicting our assumption. We conclude that both endpoints of $s$ lie in $B$, and $s \subset B$.

**Lemma 3.** *A rotate (resp., flip) operation increases the total saturation of all boxes in $\mathcal{B}$ by at most 2 (resp., 4).*

*Proof.* Suppose that an operation ROTATE($r, c$) inserts a vertical segment $s$, or an operation FLIP($r, p$) inserts two collinear vertical segments $s \cup s'$ that meet at $p$. By Lemma 2, the insertion of $s$ increases the saturation of a box $B \in \mathcal{B}$ of height $h$ by $|s|/h$ if $h \geq |s|$, and does not affect the saturation of boxes of height $h < |s|$. Recall that the boxes in $\mathcal{B}$ have only $k$ different sizes, $2^{i-1} \times 2^{k-i}$ for $i = 1, \ldots, k$, and the boxes of the same size are pairwise disjoint. Let $j \in \{1, 2, \ldots, k\}$ be the largest index such that $|s| \leq 2^{k-j}$. For $i = 1, \ldots, j$, segment $s$ increases the saturation of at most one box of height $h = 2^{k-i}$, and the increase is at most $|s|/h = |s| \cdot 2^{i-k}$. So $s$ increases the total saturation of all boxes in $\mathcal{B}$ by at most $\sum_{i=1}^{j} |s| 2^{i-k} \leq \sum_{i=1}^{j} 2^{i-j} < 2$, as required.

**Theorem 2.** *For every $n \in \mathbb{N}$, there is an $n$-element point set $P \subset [-1, n]^2$ such that the diameter of $G(P)$ is $\Omega(n \log n)$.*

*Proof.* First assume that $n = 2^k$ for some $k \in \mathbb{N}_0$. We have defined a set $P_k$ of $n = 2^k$ points and a set $\mathcal{B}$ of $k 2^{k-1} = n \log n / 2$ boxes spanned by $P_k$. The total saturation of all boxes in $\mathcal{B}$ is 0 in the rectangulation with horizontal segments, and $|\mathcal{B}| = n \log n / 2$ in the one with all vertical segments. By Lemma 3, a single flip or rotate operation increases the total saturation by at most 4. Therefore, at least $n \log n / 8$ operations are required to transform the horizontal rectangulation to the vertical one, and the diameter of $G(P_k)$ is at least $n \log n / 8$.

If $n$ is not a power of two, then put $k = \lfloor \log_2 n \rfloor$ and let $P \subset [-1, n]^2$ be the union of $P_k$ and $n - 2^k$ arbitrary (noncorectilinear) points in $[2^k, n]^2$. All axis-parallel segments containing the points in $P \setminus P_k$ are in the exterior of $[-1, 2^k]^2$. Therefore $k 2^{k-3} = \Omega(n \log n)$ operations are required when all segments containing the points in $P_k \subset P$ change from horizontal to vertical.

## 4 The Flip Diameter for Diagonal Point Sets

We say that a point set $P$ is *diagonal* if all points in $P$ lie on the graph of a strictly increasing function (e.g., $f(x) = x$). In this section we show that the flip diameter is $O(n)$ for any $n$-element diagonal set.

**Theorem 3.** *For every $n \in \mathbb{N}$, the diameter of $G(P)$ is at most $12n$ when $P$ is a diagonal set of $n$ points.*

We present an outline of the proof. The detailed proof is in the full version of the paper.

*Proof.* (outline) Without loss of generality, we may assume that the diagonal set is $P = \{p_i : i = 1, \ldots, n\}$, where $p_i = (i, i)$. Given a rectangulation $r$ for $P$, we construct a sequence of at most $6n$ flip and rotate operations that transforms $r$ into a rectangulation that consists of vertical segments.

Our algorithm consists of four phases (shown in Fig. 5): Phase 1 ensures that no three consecutive points lie on parallel segments, by successively calling subroutine Shorten&Flip for the middle segment of a consecutive parallel triple. Intuitively, the

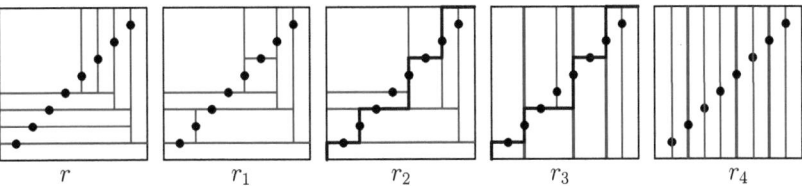

**Fig. 5.** A rectangulation $r$ of a diagonal point set. The rectangulation $r_i$, for $i = 1, 2, 3, 4$, is the output of phase $i$ of our algorithm.

middle segment is "protected" by its neighbors—the number of operations in phase 1 is bounded by $3n$. Phase 2 uses up to $n$ rotations to produce a rectangulation that contains a *staircase*, a monotone increasing path along the segments from the lower left to the upper right corner of $R$ that does not skip two or more consecutive points. Phase 3 extends the vertical segments of the staircase to maximal length, sweeping each side of the staircase independently. The final phase flips the horizontal segments of the staircase independently. The last two phases jointly use at most $2n$ operations, each of which increases the total number of segment endpoints on the boundary of $R$.

## 5 Generalization to Convex Subdivisions

Given a set $P$ of $n$ points in the plane $\mathbb{R}^2$, a *convex subdivision* for $P$ is a subdivision of the plane into convex cells by $n$ pairwise noncrossing line segments (possibly lines or half-lines) such that each segment contains exactly one point of $P$, and no three segments have a point in common.

The flip and rotate operations can be interpreted for convex subdivisions of a point set $P$, as well. The definition of the operation ROTATE$(r, c)$ is identical to the rectilinear version. The operation FLIP$(r, p)$ requires more attention, since a segment may have infinitely many possible orientations.

**Definition 3 (Flip).** *Let $r$ be a convex subdivision of $P$, let $p \in P$ be a point such that the segment $s$ containing $p$ does not contain any endpoints of other segments, and let $\sigma \in \mathbb{S}^1$ be a unit vector. The operation* FLIP$(r, p, \sigma)$ *replaces $s$ by a segment of direction $\sigma$ containing $p$.*

Similarly to the graph of rectangulations $G(P)$, we define the *graph of convex subdivisions of $P$*, $\widehat{G}(P) = (V, E)$, where the vertex set is $V = \{r : r \text{ is a convex subdivision of } P\}$ and the edge set is $E = \{(r_1, r_2) : \text{a single flip or rotate operation on } r_1 \text{ produces } r_2\}$. Our main result in this section is that even though $\widehat{G}(P)$ is an infinite graph, its diameter is $O(n \log n)$, where $n = |V|$.

**Theorem 4.** *For set $P$ of $n$ points, the graph $\widehat{G}(P)$ is connected and its diameter is $O(n \log n)$.*

We show that any convex subdivision can be transformed into a subdivision with all vertical segments through a sequence of $O(n \log n)$ operations. The subroutine

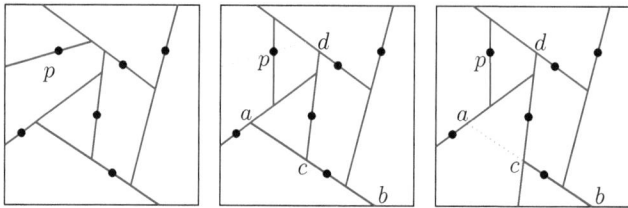

**Fig. 6.** A convex subdivision $r_1$ of 6 points, $r_2 = \text{FLIP}(r_1, p, \sigma)$, and $r_3 = \text{ROTATE}(r_2, c)$.

Shorten&Flip$(r, s)$ from Section 2 can be adapted almost verbatim: for a unit vector $\sigma \in \mathbb{S}^1$, subroutine Shorten&Flip$(r, s, \sigma)$ shortens segment $s$ maximally by rotate operations, and then flips it to direction $\sigma$.

**Lemma 4.** *Let $r$ be a convex subdivision of a set of $n$ points in the plane with distinct $x$-coordinates. There is a sequence of $O(n)$ flip and rotate operations that turns at least $\frac{1}{36}$ fraction of the nonvertical segments vertical, and keeps all vertical segments vertical.*

Before proving Lemma 4, we need to introduce a few technical terms. Consider a convex subdivision $r$ of a set of $n$ points with distinct $x$-coordinates. We say that a segment $s_1$ *hits* another segment $s_2$ if an endpoint of $s_1$ lies in the relative interior of $s_2$. An *extension of $s_1$ beyond $s_2$ hits $s_3$* if $s_1$ hits $s_2$ and $s_1$ is contained in a segment $s_1'$ such that $s_1'$ hits $s_3$ and $s_1'$ crosses at most one segment (namely $s_2$). We define the *extension visibility digraph* $\widehat{H}(r)$ on all segments in $r$, where the vertices correspond to the segments in $r$, and we have a directed edge $(s_2, s_3)$ if $s_2$ hits $s_3$ or there is a segment $s_1$ such that an extension of $s_1$ beyond $s_2$ hits $s_3$. The graph $\widehat{H}(r)$ is not necessarily planar: it is not difficult to construct a convex subdivision $r$ of a set of $\binom{t}{2}$ points where $\widehat{H}(r)$ is isomorphic to the complete graph $K_t$ (Fig. 7). Note that the number of edges in $\widehat{H}(r)$ is at most $4n$, since each segment hits at most two other segments, but some segments extend to infinity; and the extension of each segment beyond each of its endpoints hits at most one other segment. If $P$ contains $n$ points, the average degree in $\widehat{H}(r)$ is less than 8. Therefore $\widehat{H}(r)$ has an independent set of size at least $n/9$ (obtained by successively choosing minimum-degree vertices [20,34]).

*Proof (of Lemma 4).* Let $r$ be a convex subdivision of a set of $n$ points with distinct $x$-coordinates. Let $I_0$ be an independent set in the extension visibility graph $\widehat{H}(r)$ induced by all nonvertical segments. As noted above, $I_0$ contains at least $1/9$ of the nonvertical segments in $r$. Let $I_1 \subseteq I_0$ be an independent set in the bar visibility graph of the segments in $I_0$ (two nonvertical segments in $I_0$ are mutually visible if there is a vertical segment between them that does not cross any segment of the subdivision). Since the bar visibility graph is planar, we have $|I_1| \geq |I_0|/4$, and so $I_1$ contains at least $1/36$ fraction of the nonvertical segments in $r$. The total number of segment endpoints that lie in the relative interior of segments in $I_1$ is $O(n)$. A subroutine Shorten&Flip$(r, s, \sigma)$ for each segment $s \in I_1$ changes their orientation to vertical.

The operations maintain the invariants that (1) the segments in $I_1$ are pairwise disjoint, and (2) the nonvertical segments in $I_1$ form an independent set in both $\widehat{H}$ and

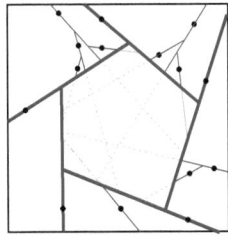

**Fig. 7.** A convex subdivision $r$ of a set of 15 points. The five bold segments induce $K_5$ in the extension visibility graph $\widehat{H}(r)$.

the bar visibility graph of all segments in $I_0$. It follows now that each operation either decreases the number of nonvertical segments in $I_1$ (flip), or decreases the number of segment endpoints that lie in the relative interior of a nonvertical segment in $I_1$ (rotate). After $O(n)$ operations, all segments in $I_1$ become vertical. Since only the segments in $I_1$ change orientation (each of them is flipped to become vertical), all vertical segments in $r$ remain vertical, as required.

*Proof (of Theorem 4).* Let $P$ be a set of $n$ points in a bounding box. We may assume, by rotating the point set if necessary, that the points in $P$ have distinct $x$-coordinates. Denote by $r_0$ the convex subdivision given by $n$ vertical line segments, one passing through each point in $P$.

Consider a convex subdivision $r_1$ of $P$. By Lemma 1, $O(n)$ operations can decrease the number of nonhorizontal segments by a factor of at least $36/35$. After at most $\log n/\log(36/35)$ invocations of Lemma 1, the number of nonvertical segments drops below 1, that is, all segments become vertical and we obtain $r_0$, as claimed.

*Linear Upper Bound for Collinear Points.* We show that the upper bound $O(n \log n)$ on the diameter of the flip graph $\widehat{G}(P)$ from Theorem 4 can be improved to $O(n)$ for some simple point configurations.

**Theorem 5.** *For every $n \in \mathbb{N}$, the diameter of $\widehat{G}(P)$ is $O(n)$ when $P$ is a set of $n$ collinear points.*

*Proof.* (outline) We may assume that $P = \{p_i : i = 1, \ldots, n\}$ where $p_i = (i, 0)$. Let $r$ be a convex subdivision of $P$. No segment in $r$ is horizontal, since each segment contains a unique point. We show that there is a sequence of $O(n)$ operations that transforms $r$ into a convex subdivision with all vertical segment. See the full paper for further details.

## 6 Conclusion

We have shown that the diameter of the flip graph $G(P)$ is between $\Omega(n)$ and $O(n \log n)$ for every $n$-element point set $P$, and these bounds cannot be improved. The diameter is $\Theta(n)$ for diagonal point sets, and $\Theta(n \log n)$ for the bit reversal point set. The flip graph $G(P)$ of a noncorectilinear set $P$ is uniquely determined by the permutation of the

$x$- and $y$-coordinates of the points [2] (e.g., diagonal point sets correspond to the identity permutation). It is an open problem to find the average diameter of $G(P)$ over all $n$-element permutations. It would already be of interest to find broader families of point sets with linear diameter: Is the diameter of $G(P)$ linear if $P$ is in convex position or unimodal, or corresponds to a separable permutation (that are defined recursively [4])?

We have shown that the diameter of the flip graph is also bounded by $O(n \log n)$ for the convex subdivisions of $n$ points in the plane. We do not know whether this bound is tight. It is possible that the flip diameter is $\Omega(n \log n)$ for the bit reversal point set defined in Section 3, but our proof of Theorem 2 heavily relies on axis-aligned boxes, and does not seem to extend to convex subdivisions.

Given a convex subdivision $r$ of a point set $P$, the flip and rotate operations can be thought of as a continuous deformation: $\mathrm{FLIP}(r, p, \sigma)$ rotates the segment containing $p$ continuously to position $\sigma$; and $\mathrm{ROTATE}(r, c)$ rotates continuously a portion of the segment containing $c$ into the extension of the segment that currently ends at $c$. The *weight* of an operation can be defined as the number of vertices swept during this continuous deformation. By Theorem 4, a sequence of $O(n \log n)$ operations can transform any convex subdivision to any other convex subdivision on $n$ points. A single operation, however, may have $\Omega(n)$ weight. We conjecture that the weighted diameter of the graph $\widehat{G}(P)$ is also $O(n \log n)$ for every $n$-elements point set $P$.

**Acknowledgments.** Research on this paper started at the Workshop on *Counting and Enumerating of Plane Graphs* at Schloss Dagstuhl. We thank Sonia Chauhan, Michael Hoffmann, and André Schulz for insightful comments on the topics of this paper.

# References

1. Ackerman, E.: Counting Problems for Geometric Structures: Rectangulations, Floorplans, and Quasi-Planar Graphs, Ph.D. thesis, Technion—Israel Inst. of Technology (2006)
2. Ackerman, E., Barequet, G., Pinter, R.Y.: On the number of rectangulations of a planar point set, J. Combin. Theory, Ser. A. 113(6), 1072–1091 (2006)
3. Asinowski, A., Barequet, G., Bousquet-Mélou, M., Mansour, T., Pinter, R.Y.: Orders induced by segments in floorplans and (2-14-3,3-41-2)-avoiding permutations. Electr. J. Comb. 20(2), P35 (2013)
4. Bose, P., Buss, J., Lubiw, A.: Pattern matching for permutations. Inf. Proc. Lett. 65, 277–283 (1998)
5. Buchin, K., Eppstein, D., Löffler, M., Nöllenburg, M., Silveira, R.I.: Adjacency-preserving spatial treemaps. In: Dehne, F., Iacono, J., Sack, J.-R. (eds.) WADS 2011. LNCS, vol. 6844, pp. 159–170. Springer, Heidelberg (2011)
6. Buchsbaum, A.L., Gansner, E.R., Procopiuc, C.M., Venkatasubramanian, S.: Rectangular layouts and contact graphs. ACM Trans. Algorithms 4, 28 (2008)
7. Calheiros, F.C., Lucena, A., de Souza, C.C.: Optimal rectangular partitions. Networks 41(1), 51–67 (2003)
8. Cardei, M., Cheng, X., Cheng, X., Du, D.Z.: A tale on guillotine cut. In: Proc. Novel Approaches to Hard Discrete Optimization, Ontario, Canada (2001)
9. Chazelle, B.: The Discrepancy Method. Cambridge University Press (2000)
10. de Fraysseix, H., de Mendez, P.O., Pach, J.: A left-first search algorithm for planar graphs. Discrete Comput. Geom. 13(1), 459–468 (1995)

11. Duchet, P., Hamidoune, Y., Las Vergnas, M., Meyniel, H.: Representing a planar graph by vertical lines joining different levels. Discrete Math. 46, 319–321 (1983)
12. Du, D.Z., Pan, L.Q., Shing, M.T.: Minimum edge length guillotine rectangular partition, Technical Report MSRI 02418-86, University of California, Berkeley, CA (1986)
13. Eppstein, D., Mumford, E., Speckmann, B., Verbeek, K.: Area-universal and constrained rectangular layouts. SIAM J. Comput. 41(3), 537–564 (2012)
14. Felsner, S.: Exploiting air-pressure to map floorplans on point sets. In: Wismath, S., Wolff, A. (eds.) GD 2013. LNCS, vol. 8242, pp. 196–207. Springer, Heidelberg (2013)
15. Felsner, S.: Rectangle and square representations of planar graphs. In: Pach, J. (ed.) Thirty Essays in Geometric Graph Theory, pp. 213–248. Springer, New York (2013)
16. Felsner, S., Fusy, É., Noy, M., Orden, D.: Bijections for Baxter families and related objects. J. Comb. Theory, Ser. A 118(3), 993–1020 (2011)
17. Gonzalez, T.F., Zheng, S.-Q.: Improved bounds for rectangular and guillotine partitions. J. of Symbolic Computation 7, 591–610 (1989)
18. Gonzalez, T.F., Zheng, S.-Q.: Approximation algorithms for partitioning a rectangle with interior points. Algorithmica 5, 11–42 (1990)
19. Hasan, M. M., Rahman, M. S., Karim, M. R.: Box-rectangular drawings of planar graphs. In: Ghosh, S.K., Tokuyama, T. (eds.) WALCOM 2013. LNCS, vol. 7748, pp. 334–345. Springer, Heidelberg (2013)
20. Hochbaum, D.S.: Efficient bounds for the stable set, vertex cover, and set packing problems. Discrete Appl. Math. 6, 243–254 (1983)
21. Hurtado, F., Noy, M., Urrutia, J.: Flipping edges in triangulations. Discrete Comput. Geom. 22(3), 333–346 (1999)
22. Koźimiński, K., Kinnen, E.: Rectangular duals of planar graphs. Networks 15, 145–157 (1985)
23. Lawson, C.: Software for $c_1$ surface interpolation. In: Rice, J. (ed.) Mathematical Software III, pp. 161–194. Academic Press, New York (1977)
24. Levcopoulos, C.: Fast heuristics for minimum length rectangular partitions of polygons. In: Proc. 2nd ACM Symp. on Computational Geometry, Yorktown Heights, NY, pp. 100–108 (1986)
25. Liao, C.C., Lu, H.I., Yen, H.C.: Compact floor-planning via orderly spanning trees. J. Algorithms 48, 441–451 (2003)
26. Lingas, A., Pinter, R.Y., Rivest, R.L., Shamir, A.: Minimum edge length rectilinear decompositions of rectilinear figures. In: Proc. 20th Allerton Conf. on Communication, Control, and Computing, Monticello, IL, pp. 53–63 (1982)
27. Rahman, M., Nishizeki, T., Ghosh, S.: Rectangular drawings of planar graphs. J. Algorithms 50, 62–78 (2004)
28. Raisz, E.: The rectangular statistical cartogram. Geogr. Rev. 24(3), 292–296 (1934)
29. Robertson, N., Sanders, D.P., Seymour, P., Thomas, R.: The four-colour theorem. J. Combin. Theory, Ser. B 70(1), 2–44 (1997)
30. Santos, F., Seidel, R.: A better upper bound on the number of triangulations of a planar point set. J. Combin. Theory, Ser. A 102, 186–193 (2003)
31. Sharir, M., Welzl, E.: Random triangulations of planar point sets. In: Proc. 22nd ACM Symp. on Comput. Geom., pp. 273–281. ACM Press (2006)
32. Sleator, D., Tarjan, R., Thurston, W.: Rotations distance, triangulations and hyperbolic geometry. J. AMS 1, 647–682 (1988)
33. Tamassia, R., Tollis, I.G.: A unified approach to visibility representations of planar graphs. Dicrete Comput. Geom. 1, 321–341 (1986)
34. Turán, P.: On an extremal problem in graph theory. Math. Fiz. Lapok 48, 436–452 (1941) (in Hungarian)
35. Ungar, P.: On diagrams representing graphs. J. London Math. Soc. 28, 336–342 (1953)
36. Yao, B., Chen, H., Cheng, C.K., Graham, R.: Floorplan representations: Complexity and connections. ACM Trans. on Design Automation of Electronic Systems 8, 55–80 (2003)

# Weighted Staircase Tableaux, Asymmetric Exclusion Process, and Eulerian Type Recurrences

Paweł Hitczenko[1,*] and Svante Janson[2,**]

[1] Department of Mathematics, Drexel University, Philadelphia, PA 19104
phitczenko@math.drexel.edu
[2] Department of Mathematics, Uppsala University, Uppsala, Sweden
svante.janson@math.uu.se

**Abstract.** We consider a relatively new combinatorial structure called staircase tableaux. They were introduced in the context of the asymmetric exclusion process and Askey–Wilson polynomials; however, their purely combinatorial properties have gained considerable interest in the past few years.

We will be interested in a general model of staircase tableaux in which symbols that appear in staircase tableaux may have arbitrary positive weights. Under this general model we derive a number of results concerning the limiting laws for the number of appearances of symbols in a random staircase tableaux.

One advantage of our generality is that we may let the weights approach extreme values of zero or infinity, which covers further special cases appearing earlier in the literature.

One of the main tools we use are generating functions of the parameters of interests. This leads us to a two–parameter family of polynomials. Specific values of the parameters cover a number of special cases analyzed earlier in the literature including the classical Eulerian polynomials.

**Keywords:** Staircase tableau, Eulerian polynomial, Asymmetric Exclusion Process.

## 1 Introduction

This note is concerned with a combinatorial structure introduced recently by Corteel and Williams [8,9] and called *staircase tableaux*. The original motivations were in connections with the asymmetric exclusion process (ASEP) on a one-dimensional lattice with open boundaries, an important model in statistical mechanics. The generating function for staircase tableaux was also used to give a combinatorial formula for the moments of the Askey–Wilson polynomials (see [9,5] for the details). Further work includes [3], where special situations in which

---

[*] Supported in part by Simons Foundation grant no. 208766.
[**] Supported in part by Knut and Alice Wallenberg Foundation.

the generating function of staircase tableaux took a particularly simple form, were considered. Furthermore, [10] deals with the analysis of various parameters associated with appearances of the Greek letters $\alpha$, $\beta$, $\gamma$, and $\delta$ in a randomly chosen staircase tableau (see below, or e.g. [9, Section 2], for the definitions and the meaning of these symbols). Moreover, there are natural bijections (see [9, Appendix]) between a class of staircase tableaux (the $\alpha/\beta$-*staircase tableaux* defined below) and *permutation tableaux* (see e.g. [4,6,7,15] and the references therein for more information on these objects and their connection to a version of the ASEP) as well as to *alternative tableaux* [20] which, in turn, are in one-to-one correspondence with *tree-like tableaux* [1].

The purpose of this extended abstract is to describe further properties of staircase tableaux, regarding them as interesting combinatorial objects in themselves. We refer to the full paper [16] for more details and proofs.

We recall the definition of a staircase tableau introduced in [8,9]:

**Definition 1.** *A staircase tableau of size $n$ is a Young diagram of shape $(n, n-1, \ldots, 2, 1)$ whose boxes are filled according to the following rules:*

(Si) *each box is either empty or contains one of the letters $\alpha$, $\beta$, $\delta$, or $\gamma$;*
(Sii) *no box on the diagonal is empty;*
(Siii) *all boxes in the same row and to the left of a $\beta$ or a $\delta$ are empty;*
(Siv) *all boxes in the same column and above an $\alpha$ or a $\gamma$ are empty.*

An example of a staircase tableau is given in Fig. 1(a).

The set of all staircase tableaux of size $n$ will be denoted by $\mathcal{S}_n$. There are several proofs of the fact that the number of staircase tableaux $|\mathcal{S}_n| = 4^n n!$, see e.g. [5,3,10] for some of them.

## 2 Staircase Tableaux and ASEP

As mentioned in the introduction, staircase tableaux were introduced in [8,9] in connection with the *asymmetric exclusion process (ASEP)*; as a background, we give some details here. In a discrete version, the ASEP is a Markov chain describing a system of particles on a line with $n$ sites $1, \ldots, n$; each site may contain at most one particle. Particles jump one step to the right with probability $u$ and to the left with probability $q$, provided the move is to a site that is empty; moreover, new particles enter site 1 with probability $\alpha$ and site $n$ with probability $\delta$, provided these sites are empty, and particles at site 1 and $n$ leave the system with probabilities $\gamma$ and $\beta$, respectively. See further [9], which also contains references and information on applications and connections to other branches of science.

Explicit expressions for the steady state probabilities of the ASEP were first given in [11]. Corteel and Williams [9] gave an expression for the steady state probabilities using staircase tableaux, their weight, and generating function for them. To describe it we first fill the tableau $S$ by labelling the empty boxes of $S$ with $u$'s and $q$'s as follows: first, we fill all the boxes to the left of a $\beta$ with

$u$'s, and all the boxes to the left of a $\delta$ with $q$'s. Then, we fill the remaining boxes above an $\alpha$ or a $\delta$ with $u$'s, and the remaining boxes above a $\beta$ or a $\gamma$ with $q$'s. When the tableau is filled, we let $N_\alpha, N_\beta, N_\gamma, N_\delta, N_u, N_q$ be the numbers of symbols $\alpha, \beta, \gamma, \delta, u, q$ in $S$. We then define its *weight* to be

$$\operatorname{wt}(S) := \alpha^{N_\alpha} \beta^{N_\beta} \gamma^{N_\gamma} \delta^{N_\delta} u^{N_u} q^{N_q}, \tag{1}$$

i.e., the product of all symbols in $S$; this is thus a monomial of degree $n(n+1)/2$ in $\alpha$, $\beta$, $\gamma$, $\delta$, $u$ and $q$. Figure 1(b) shows the tableau in Fig. 1(a) filled with $u$'s and $q$'s; its weight is $\alpha^5 \beta^2 \delta^3 \gamma^3 u^{13} q^{10}$.

Further, we let $Z_n(\alpha, \beta, \gamma, \delta, q, u)$ be the total weight of all filled staircase tableaux of size $n$, i.e.

$$Z_n(\alpha, \beta, \gamma, \delta, q, u) = \sum_{S \in \mathcal{S}_n} \operatorname{wt}(S).$$

Obviously, $Z_n$ is a homogeneous polynomial of degree $n(n+1)/2$.

To describe the connection to ASEP, define the *type* of a staircase tableau $S$ of size $n$ to be a word of the same size on the alphabet $\{\circ, \bullet\}$ obtained by reading the diagonal boxes from northeast (NE) to southwest (SW) and writing $\bullet$ for each $\alpha$ or $\delta$, and $\circ$ for each $\beta$ or $\gamma$. (Thus a type of a tableau is a possible state for the ASEP.) Figure 1(a) shows a tableau and its type.

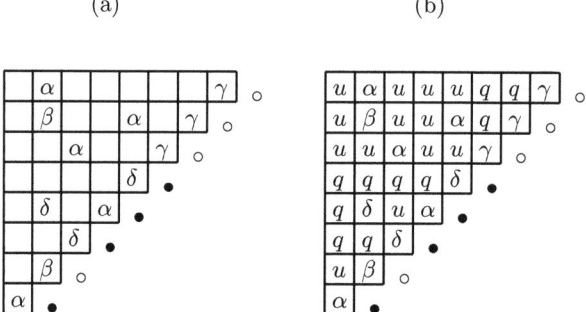

**Fig. 1.** (a) A staircase tableau of size 8; its type is $\circ \circ \circ \bullet \bullet \bullet \circ \bullet$ (b) the same tableau filled with $u$'s and $q$'s, its weight is $\alpha^5 \beta^2 \gamma^3 \delta^3 u^{13} q^{10}$

As Corteel and Williams [9,8] have shown that the steady state probability that the ASEP is in state $\sigma$ is

$$\frac{Z_\sigma(\alpha, \beta, \gamma, \delta, q, u)}{Z_n(\alpha, \beta, \gamma, \delta, q, u)},$$

where $Z_\sigma(\alpha, \beta, \gamma, \delta, q, u) = \sum_{S \text{ of type } \sigma} \operatorname{wt}(S).$

## 3  Generating Function of the Total Weight

The generating function of the total weight of tableaux in $\mathcal{S}_n$

$$Z_n(\alpha, \beta, \gamma, \delta) := Z_n(\alpha, \beta, \gamma, \delta, 1, 1) \tag{2}$$

has a particularly simple form, viz., see [5,3],

$$Z_n(\alpha, \beta, \gamma, \delta) = \prod_{i=0}^{n-1} \Big(\alpha + \beta + \gamma + \delta + i(\alpha + \gamma)(\beta + \delta)\Big). \tag{3}$$

(A proof is included in [5].) In particular, the number of staircase tableaux of size $n$ is $Z_n(1,1,1,1) = \prod_{i=0}^{n-1}(4+4i) = 4^n n!$.

Note that the symbols $\alpha$ and $\gamma$ have exactly the same role in the definition above of staircase tableaux, and so do $\beta$ and $\delta$. (This is no longer true in the connection to the ASEP, which is the reason for using four different symbols in the definition.) We say that a staircase tableau using only the symbols $\alpha$ and $\beta$ is an $\alpha/\beta$-staircase tableau, and we let $\overline{\mathcal{S}}_n \subset \mathcal{S}_n$ be the set of all $\alpha/\beta$-staircase tableaux of size $n$. We thus see that any staircase tableau can be obtained from an $\alpha/\beta$-staircase tableau by replacing some (or no) $\alpha$ by $\gamma$ and some (or no) $\beta$ by $\delta$; conversely, any staircase tableau can be reduced to an $\alpha/\beta$-staircase tableau by replacing every $\gamma$ by $\alpha$ and every $\delta$ by $\beta$.

We define the generating function of the total weight of $\alpha/\beta$-staircase tableaux by

$$Z_n(\alpha, \beta) = \sum_{S \in \overline{\mathcal{S}}_n} \mathrm{wt}(S) = Z_n(\alpha, \beta, 0, 0),$$

and note that the relabelling argument just given implies

$$Z_n(\alpha, \beta, \gamma, \delta) = Z_n(\alpha + \gamma, \beta + \delta).$$

We let $x^{\overline{n}}$ denote the rising factorial defined by

$$x^{\overline{n}} = x(x+1)\ldots(x+n-1) = \Gamma(x+n)/\Gamma(x),$$

and note that by (3),

$$Z_n(\alpha, \beta) = Z_n(\alpha, \beta, 0, 0) = \prod_{i=0}^{n-1}(\alpha + \beta + i\alpha\beta) = \alpha^n \beta^n (\alpha^{-1} + \beta^{-1})^{\overline{n}} \tag{4}$$

$$= \alpha^n \beta^n \frac{\Gamma(n + \alpha^{-1} + \beta^{-1})}{\Gamma(\alpha^{-1} + \beta^{-1})}. \tag{5}$$

In particular, as noted in [3] and [5], the number of $\alpha/\beta$-staircase tableaux is $Z_n(1,1) = 2^{\overline{n}} = (n+1)!$.

## 4  Main Result: Symbols on the Diagonal

Because of connections with ASEP, the diagonal of the staircase tableau is of natural interest. Dasse–Hartaut and Hitczenko [10] studied random staircase tableaux and in particular the symbols on the diagonal of a tableau obtained by picking a staircase tableau in $\mathcal{S}_n$ uniformly at random. Our purpose here is to consider $\alpha/\beta$-staircase tableaux for arbitrary parameters $\alpha, \beta \geq 0$ and generalize several results from [10] to this case. This generality is also useful in studying the structure of random staircase tableaux. See items (ii) and (iii) in Section 7 below for further comments and [16] for more details.

We consider the following probability measure on the set of staircase tableaux of size $n$ with weights $\alpha$, $\beta$.

**Definition 2.** *Let $n \geq 1$ and let $\alpha, \beta \in [0, \infty)$ with $(\alpha, \beta) \neq (0, 0)$. Then $S_{n,\alpha,\beta}$ is the random $\alpha/\beta$-staircase tableau in $\overline{\mathcal{S}}_n$ with the distribution*

$$\mathbb{P}_{\alpha,\beta}(S_{n,\alpha,\beta} = S) = \frac{\mathrm{wt}(S)}{Z_n(\alpha,\beta)} = \frac{\alpha^{N_\alpha(S)} \beta^{N_\beta(S)}}{Z_n(\alpha,\beta)}, \qquad S \in \overline{\mathcal{S}}_n. \qquad (6)$$

*We also allow the parameters $\alpha = \infty$ or $\beta = \infty$; in this case (6) is interpreted as the limit when $\alpha \to \infty$ or $\beta \to \infty$, with the other parameter fixed. Similarly, we allow $\alpha = \beta = \infty$; in this case (6) is interpreted as the limit when $\alpha = \beta \to \infty$. (In the case $\alpha = \beta = \infty$, we tacitly assume $n \geq 2$ or sometimes even $n \geq 3$ to avoid trivial complications.)*

*Remark 1.* There is a symmetry (involution) $S \mapsto S^\dagger$ of staircase tableaux defined by reflection in the NW–SE diagonal, thus interchanging rows and columns, together with an exchange of the symbols by $\alpha \leftrightarrow \beta$ and $\gamma \leftrightarrow \delta$, see further [3]. This maps $\overline{\mathcal{S}}_n$ onto itself, and maps the random $\alpha/\beta$-staircase tableau $S_{n,\alpha,\beta}$ to $S_{n,\beta,\alpha}$; the parameters $\alpha$ and $\beta$ thus play symmetric roles.

*Remark 2.* We can similarly define a random staircase tableaux $S_{n,\alpha,\beta,\gamma,\delta}$, with four parameters $\alpha, \beta, \gamma, \delta \geq 0$, by picking a staircase tableau $S \in \mathcal{S}_n$ with probability $\mathrm{wt}(S)/Z_n(\alpha,\beta,\gamma,\delta)$. This is the same as taking a random $S_{n,\alpha+\gamma,\beta+\delta}$ and randomly replacing each symbol $\alpha$ by $\gamma$ with probability $\gamma/(\alpha+\gamma)$, and each $\beta$ by $\delta$ with probability $\delta/(\beta+\delta)$. Our results can thus be translated to results for $S_{n,\alpha,\beta,\gamma,\delta}$. In particular, the case $\alpha = \beta = \gamma = \delta = 1$ considered in [10] corresponds to picking an $\alpha/\beta$-staircase tableau in $\overline{\mathcal{S}}_n$ at random with probability proportional to $2^{N_\alpha+N_\beta}$ and then randomly replacing some symbols; each $\alpha$ is replaced by $\gamma$ with probability $1/2$, and each $\beta$ by $\delta$ with probability $1/2$, with all replacements independent. Note that the weight $2^{N_\alpha+N_\beta}$ is the weight (1) if we choose the parameters $\alpha = \beta = 2$.

We are interested in the distribution of the symbols on the diagonal of $S_{n,\alpha,\beta}$. We define $A(S)$ and $B(S)$ as the numbers of $\alpha$ and $\beta$, respectively, on the diagonal of an $\alpha/\beta$-staircase tableau $S$, and consider the random variables $A_{n,\alpha,\beta} := A(S_{n,\alpha,\beta})$ and $B_{n,\alpha,\beta} := B(S_{n,\alpha,\beta})$; note that $A_{n,\alpha,\beta} + B_{n,\alpha,\beta} = n$ by (Sii), so it suffices to consider one of these. Moreover, by (1), $B_{n,\alpha,\beta} \stackrel{d}{=} A_{n,\beta,\alpha}$.

In order to describe the distribution of $A_{n,\alpha,\beta}$ we need some further notation. Define the numbers $v_{a,b}(n,k)$, for $a,b \in \mathbb{R}$, $k \in \mathbb{Z}$ and $n = 0,1,\ldots$, by the recursion

$$v_{a,b}(n,k) = (k+a)v_{a,b}(n-1,k) + (n-k+b)v_{a,b}(n-1,k-1), \qquad n \geq 1, \quad (7)$$

with $v_{a,b}(0,0) = 1$ and $v_{a,b}(0,k) = 0$ for $k \neq 0$ and $v_{a,b}(n,k) = 0$ for $k < 0$ and $k > n$, for all $n \geq 0$. These numbers were defined and studied by Carlitz and Scoville [2]. (Their notation is $A(n-k, k \mid a, b)$.) We give some additional properties below. Furthermore, define polynomials

$$P_{n,a,b}(x) := \sum_{k=0}^{n} v_{a,b}(n,k)x^k = \sum_{k=-\infty}^{\infty} v_{a,b}(n,k)x^k.$$

Thus, $P_{0,a,b}(x) = 1$.

In the case $a = b = 0$, we trivially have $v_{0,0}(n,k) = 0$ and $P_{n,0,0} = 0$ for all $n \geq 1$; in this case we define the substitutes, for $n \geq 2$,

$$\tilde{v}_{0,0}(n,k) := v_{1,1}(n-2, k-1) \qquad (8)$$

and

$$\tilde{P}_{n,0,0}(x) := \sum_{k=0}^{n} \tilde{v}_{0,0}(n,k)x^k = xP_{n-2,1,1}(x). \qquad (9)$$

We assume the following relation throughout the rest of this abstract: $a = \alpha^{-1}$ and $b = \beta^{-1}$. Our main result is as follows.

**Theorem 1.** *Let $\alpha, \beta \in (0, \infty]$. If $(\alpha, \beta) \neq (\infty, \infty)$, then the probability generating function $g_A(x)$ of the random variable $A_{n,\alpha,\beta}$ is given by*

$$g_A(x) := \mathbb{E}x^{A_{n,\alpha,\beta}} = \sum_{k=0}^{n} \mathbb{P}(A_{n,\alpha,\beta} = k)x^k = \frac{P_{n,a,b}(x)}{P_{n,a,b}(1)} = \frac{P_{n,a,b}(x)}{(a+b)^{\overline{n}}}$$

$$= \frac{\Gamma(a+b)}{\Gamma(n+a+b)}P_{n,a,b}(x).$$

*Equivalently,*

$$\mathbb{P}(A_{n,\alpha,\beta} = k) = \frac{v_{a,b}(n,k)}{P_{n,a,b}(1)} = \frac{v_{a,b}(n,k)}{(a+b)^{\overline{n}}} = \frac{\Gamma(a+b)}{\Gamma(n+a+b)}v_{a,b}(n,k).$$

*In the case $\alpha = \beta = \infty$, and $n \geq 2$, we have instead*

$$g_A(x) := \sum_{k=0}^{n} \mathbb{P}(A_{n,\alpha,\beta} = k)x^k = \frac{\tilde{P}_{n,0,0}(x)}{\tilde{P}_{n,0,0}(1)} = \frac{\tilde{P}_{n,0,0}(x)}{(n-1)!},$$

$$\mathbb{P}(A_{n,\alpha,\beta} = k) = \frac{\tilde{v}_{0,0}(n,k)}{\tilde{P}_{n,0,0}(1)} = \frac{\tilde{v}_{0,0}(n,k)}{(n-1)!}.$$

This result has a number of consequences; some of them we describe below. But first, because of their role in Theorem 1 and connections to other parts of mathematics we briefly discuss the polynomials $P_{n,a,b}$ and their coefficients $v_{a,b}(n,k)$.

## 5 The Polynomials $P_{n,a,b}$

For $a = 1$, $b = 0$, the recursion (7) is the standard recursion for *Eulerian numbers* $\left\langle {n \atop k} \right\rangle$, see e.g. [14, Section 6.2], [21, §26.14], [22, A008292]; thus

$$v_{1,0}(n,k) = \left\langle {n \atop k} \right\rangle.$$

These are often defined as the number of permutations of $n$ elements with $k$ descents (or ascents). See e.g. [24, Section 1.3], where also other relations to permutations are given. The corresponding polynomials

$$P_{n,1,0}(x) = \sum_{k=0}^{n} \left\langle {n \atop k} \right\rangle x^k$$

are known as *Eulerian polynomials*. We can thus see $v_{a,b}(n,k)$ and $P_{n,a,b}(x)$ as generalizations of Eulerian numbers and polynomials.

Furthermore, the cases $(a,b) = (0,1)$ and $(1,1)$ also lead to Eulerian numbers, with different indexing:

$$v_{0,1}(n,k) = v_{1,0}(n, n-k) = \left\langle {n \atop n-k} \right\rangle = \left\langle {n \atop k-1} \right\rangle, \qquad n \geq 1$$

(which is non-zero for $1 \leq k \leq n$). Similarly, by (7) and induction,

$$v_{1,1}(n,k) = v_{1,0}(n+1, k) = \left\langle {n+1 \atop k} \right\rangle, \qquad n \geq 0. \qquad (10)$$

Equivalently,

$$P_{n,0,1}(x) = x P_{n,1,0}(x), \qquad P_{n,1,1}(x) = P_{n+1,1,0}(x). \qquad (11)$$

Similarly, by the definition (8) and (10),

$$\widetilde{v}_{0,0}(n,k) = \left\langle {n-1 \atop k-1} \right\rangle, \qquad n \geq 2,$$

and by (9) and (11),

$$\widetilde{P}_{n,0,0}(x) = P_{n-1,0,1}(x) = x P_{n-1,1,0}(x).$$

As mentioned above, in the case $a = b = 0$ we trivially have

$$v_{0,0}(n,k) = 0 \quad \text{and} \quad P_{n,0,0}(x) = 0 \quad \text{for all } n \geq 1.$$

In the case when $a = 0$ or $b = 0$ we have the following simple relations, generalizing the results for Eulerian numbers and polynomials (11).

**Proposition 1.** For all $n \geq 1$,

$$v_{a,0}(n,k) = av_{a,1}(n-1,k),$$
$$v_{0,b}(n,k) = bv_{1,b}(n-1,k-1), \text{ and, equivalently,}$$
$$P_{n,a,0}(x) = aP_{n-1,a,1}(x),$$
$$P_{n,0,b}(x) = bxP_{n-1,1,b}(x).$$

We collect some further properties in the following theorems.

**Theorem 2.** For all $a,b$ and $n \geq 0$,

$$P_{n,a,b}(1) = \sum_{k=0}^{n} v_{a,b}(n,k) = \overline{(a+b)^n} = \frac{\Gamma(n+a+b)}{\Gamma(a+b)}.$$

$$P'_{n,a,b}(1) = \sum_{k=0}^{n} kv_{a,b}(n,k) = \frac{n(n+2b-1)}{2}\overline{(a+b)^{n-1}}$$

$$P''_{n,a,b}(1) = \sum_{k=0}^{n} k(k-1)v_{a,b}(n,k)$$
$$= \frac{n(n-1)(3n^2 + (12b-11)n + 12b^2 - 24b + 10)}{12}\overline{(a+b)^{n-2}}.$$

Furthermore, we have the symmetry

$$v_{a,b}(n,k) = v_{b,a}(n,n-k) \tag{12}$$

and thus

$$P_{n,a,b}(x) = x^n P_{n,b,a}(1/x). \tag{13}$$

**Remark 3.** The symmetries (12)–(13) between $a$ and $b$ are more evident if we define the homogeneous two-variable polynomials

$$\widehat{P}_{n,a,b}(x,y) := \sum_{k=0}^{n} v_{a,b}(n,k) x^k y^{n-k},$$

which satisfy the recursion

$$\widehat{P}_{n,a,b}(x,y) = \left(bx + ay + xy\frac{\partial}{\partial x} + xy\frac{\partial}{\partial y}\right)\widehat{P}_{n-1,a,b}(x,y), \qquad n \geq 1$$

and the symmetry $\widehat{P}_{n,a,b}(x,y) = \widehat{P}_{n,b,a}(y,x)$. (Note that $\widehat{P}_{n,a,b}(x,y) = y^n P_{n,a,b}(x/y)$ and $P_{n,a,b}(x) = \widehat{P}_{n,a,b}(x,1)$.)

The following theorem has important consequences for us.

**Theorem 3.** (i) If $a,b > 0$, then $v_{a,b}(n,k) > 0$ for $0 \leq k \leq n$, and $P_{n,a,b}(x)$ is a polynomial of degree $n$ with $n$ simple negative roots.

(ii) If $a > b = 0$, then $v_{a,b}(n,k) > 0$ for $0 \leq k < n$, and $P_{n,a,b}(x)$ is a polynomial of degree $n-1$ with $n-1$ simple negative roots.

(iii) If $a = 0 < b$, then $v_{a,b}(n,k) > 0$ for $1 \leq k \leq n$, and $P_{n,a,b}(x)$ is a polynomial of degree $n$ with $n$ simple roots in $(-\infty, 0]$; one of the roots is 0, provided $n > 0$.

(iv) If $a = b = 0$, then $\widetilde{v}_{0,0}(n,k) > 0$ for $1 \leq k \leq n-1$, and $\widetilde{P}_{n,0,0}(x)$ is a polynomial of degree $n-1$ with $n-1$ simple roots in $(-\infty, 0]$; one of the roots is 0, provided $n \geq 2$.

The proof that roots are distinct and negative uses an argument of Frobenius [13] for the classical Eulerian polynomials and is based on the recursion

$$P_{n,a,b}(x) = \big((n-1+b)x + a\big)P_{n-1,a,b}(x) + x(1-x)P'_{n-1,a,b}(x), \qquad n \geq 1$$

(which is easily seen to be equivalent to the recursion (7)). The proof also shows that the roots of $P_{n-1,a,b}$ and $P_{n,a,b}$ are interlaced (except that 0 is a common root when $a = 0$). More general results of this kind, can be found in e.g. [25] and [18, Proposition 3.5].

*Remark 4.* The case $a = b = 1/2$ appeared in [10]. In this case, it is more convenient to study the numbers $B(n,k) := 2^n v_{1/2,1/2}(n,k)$ which are integers and satisfy the recursion

$$B(n,k) = (2k+1)B(n-1,k) + (2n-2k+1)B(n-1,k-1), \qquad n \geq 1; \quad (14)$$

these are called *Eulerian numbers of type B* [22, A060187]. The numbers $B(n,k)$ seem to have been introduced by MacMahon [19] in number theory. They also have combinatorial interpretations, for example as the number of descents in signed permutations. The generating function (in a general symmetric case $a = b$) was found by Franssens [12, Proposition 3.1] who studied numbers $B_{n,k}(c)$ (and the resulting polynomials) given by $B_{n,k}(c) = 2^n v_{c/2,c/2}(n,k)$.

Furthermore, the case $a + b = 1$ yields polynomials $P_{n,a,1-a}(x)$ generalizing the Eulerian polynomials (the case $a = 1$, or $a = 0$); they are sometimes called (generalized) *Euler–Frobenius polynomials* and appear e.g. in spline theory; we refer to [17] for more information and references.

## 6 Consequences

**Theorem 4.** *The p.g.f. $g_A(x)$ of the random variable $A_{n,\alpha,\beta}$ has only simple roots and they are on the negative halfline $(-\infty, 0]$. As a consequence, for any given $n, \alpha, \beta$ there exist $p_1, \ldots, p_n \in (0,1)$ such that*

$$A_{n,\alpha,\beta} \stackrel{d}{=} \sum_i \mathrm{Be}(p_i), \qquad (15)$$

*where $\mathrm{Be}(p_i)$ is a Bernoulli random variable with parameter $p_i$ and the summands are independent. It follows that the distribution of $A_{n,\alpha,\beta}$ and the sequence $v_{a,b}(n,k)$, $k \in \mathbb{Z}$, are unimodal and log-concave.*

Because of the representation (15), the $A_{n,\alpha,\beta}$ will follow the central (and local) limit theorem as long as the variance $\mathrm{Var}(A_{n,\alpha,\beta}) \to \infty$ (see, e.g. [23]). But from Theorem 2 we see that

$$\mathbb{E} A_{n,\alpha,\beta} = \frac{n(n+2b-1)}{2(n+a+b-1)}$$

and

$$\mathrm{Var}(A_{n,\alpha,\beta}) = n \frac{(n-1)(n-2)(n+4a+4b-1) + 6(n-1)(a+b)^2 + 12ab(a+b-1)}{12(n+a+b-1)^2(n+a+b-2)}.$$

*Remark 5.* In the symmetric case $\alpha = \beta$ we thus obtain $\mathbb{E}(A_{n,\alpha,\alpha}) = n/2$; this is also obvious by symmetry, since $A_{n,\alpha,\alpha} \stackrel{d}{=} B_{n,\alpha,\alpha}$ by Remark 1. Regardless of the values of $\alpha$, $\beta$ we have $\mathbb{E}(A_{n,\alpha,\beta}) \sim n/2$. Thus, the effects of changing the parameters $\alpha$ and $\beta$ are surprisingly small. Typically, probability weights of the type (1) (which are common in statistical physics) shift the distributions of the random variables considerably, but here the effects are only second-order.

For the variance we similarly have

$$\mathrm{Var}(A_{n,\alpha,\beta}) \sim \frac{n}{12}.$$

This leads to the following central limit theorem:

**Theorem 5.** *Let $\alpha, \beta \in (0, \infty]$ be fixed and let $n \to \infty$. Then $A_{n,\alpha,\beta}$ is asymptotically normal:*

$$\frac{A_{n,\alpha,\beta} - n/2}{\sqrt{n}} \xrightarrow{d} N(0, 1/12).$$

*Moreover, a corresponding local limit theorem holds:*

$$\mathbb{P}(A_{n,\alpha,\beta} = k) = \sqrt{\frac{6}{\pi n}} \left( e^{-6(k-n/2)^2/n} + o(1) \right),$$

*as $n \to \infty$, uniformly in $k \in \mathbb{Z}$.*

*Remark 6.* The proof shows that the (suitably modified to take into account the asymptotics of the expected value and the variance) central limit theorem holds also if $\alpha$ and $\beta$ are allowed to depend on $n$, provided only that $\mathrm{Var}(A_{n,\alpha,\beta}) \to \infty$, which by the expression for the variance holds as soon as $n^2/(a+b) \to \infty$ or $nab/(a+b)^2 \to \infty$; hence this holds except when $a$ or $b$ is $\infty$ or tends to $\infty$ rapidly, i.e., unless $\alpha$ or $\beta$ is 0 or tends to 0 rapidly. It should be noted, however, the asymptotic normality may fail in extreme cases.

## 7 Further Remarks

(i) We concentrated here on the diagonal of a staircase tableau because of the connections to the ASEP. We can also study the total numbers $N_\alpha$ and $N_\beta$ of symbols $\alpha$ and $\beta$ in a random $S_{n,\alpha,\beta}$. This is actually simpler; we refer to [16] for the details. Similarly we can study the joint distribution of $N_\alpha$ and $N_\beta$ and the joint distribution of $N_\alpha$ and, say, $A_{n,\alpha,\beta}$.

(ii) The notion of weights brings forth the possibility of studying the distribution of the symbols in $S_{n,\alpha,\beta}$. We note that when $\alpha = \beta = \infty$ (i.e. when $\alpha = \beta \to \infty$ in (6)) then the probability measure $\mathbb{P}_{\alpha,\beta}$ is concentrated on the tableaux with the maximal total degree in $Z_n(\alpha, \beta)$, i.e. with the maximal number of symbols. As $\alpha = \beta \to \infty$ we have

$$Z_n(\alpha, \beta) \sim (\alpha + \beta) \prod_{i=1}^{n-1} (i\alpha\beta) = (n-1)! \left( \alpha^n \beta^{n-1} + \alpha^{n-1} \beta^n \right).$$

Hence there are $(n-1)!$ tableaux with $n$ $\alpha$'s and $n-1$ $\beta$'s, and $(n-1)!$ with $n-1$ $\alpha$'s and $n$ $\beta$'s for the total of $2(n-1)!$ $\alpha/\beta$-tableaux with the maximal number of symbols, $2n-1$ (similarly, the corresponding number of staircase tableaux with $2n-1$ symbols $\alpha, \beta, \gamma, \delta$ is $2^{2n}(n-1)!$, see [3]).

(iii) It follows from the previous comment that there are only at most $n-1$ symbols in the $n(n-1)/2$ off–diagonal boxes. So, it is natural to ask where they are. Here is a step towards answering this question; we believe this is the first result in this direction: for a given box of a staircase tableau we give the probability that it contains a given symbol. Let $S_{n,\alpha,\beta}(i,j)$ be a content of the $(i,j)$th box (enumerated as in a matrix). For the off-diagonal boxes we have

$$P(S_{n,\alpha,\beta}(i,j) = \alpha) = \frac{j-1+b}{(i+j+a+b-1)(i+j+a+b-2)},$$

$$P(S_{n,\alpha,\beta}(i,j) = \beta) = \frac{i-1+a}{(i+j+a+b-1)(i+j+a+b-2)},$$

$$P(S_{n,\alpha,\beta}(i,j) \neq \emptyset) = \frac{1}{i+j+a+b-1}.$$

For the diagonal boxes we can give a complete description of the distribution of the symbols. To simplify the notation let $S_n(j) := S_{n,\alpha,\beta}(n+1-j, j)$ be the symbol on the diagonal in the $j$th column and let $1 \leq j_1 < \ldots < j_\ell \leq n$. Then

$$\mathbb{P}\big(S_n(j_1) = \ldots = S_n(j_\ell) = \alpha\big) = \prod_{k=1}^{\ell} \frac{j_k - k + b}{n - k + a + b}.$$

## References

1. Aval, J.-C., Boussicault, A., Nadeau, P.: Tree-like tableaux. In: 23rd International Conference on Formal Power Series and Algebraic Combinatorics (FPSAC 2011). Discrete Math. Theor. Comput. Sci. Proc., AO, pp. 63–74 (2011)
2. Carlitz, L., Scoville, R.: Generalized Eulerian numbers: combinatorial applications. J. Reine Angew. Math. 265, 110–137 (1974)
3. Corteel, S., Dasse-Hartaut, S.: Statistics on staircase tableaux, Eulerian and Mahonian statistics. In: 23rd International Conference on Formal Power Series and Algebraic Combinatorics (FPSAC 2011), Discrete Math. Theor. Comput. Sci. Proc., AO, pp. 245–255 (2011)

4. Corteel, S., Hitczenko, P.: Expected values of statistics on permutation tableaux. In: 2007 Conference on Analysis of Algorithms, AofA 2007, Discrete Math. Theor. Comput. Sci. Proc., AH, pp. 325–339 (2007)
5. Corteel, S., Stanley, R., Stanton, D., Williams, L.: Formulae for Askey–Wilson moments and enumeration of staircase tableaux. Trans. Amer. Math. Soc. 364(11), 6009–6037 (2012)
6. Corteel, S., Williams, L.K.: A Markov chain on permutations which projects to the PASEP. Int. Math. Res. Notes, Article 17:rnm055, 27pp (2007)
7. Corteel, S., Williams, L.K.: Tableaux combinatorics for the asymmetric exclusion process. Adv. Appl. Math. 39, 293–310 (2007)
8. Corteel, S., Williams, L.K.: Staircase tableaux, the asymmetric exclusion process, and Askey–Wilson polynomials. Proc. Natl. Acad. Sci. 107(15), 6726–6730 (2010)
9. Corteel, S., Williams, L.K.: Tableaux combinatorics for the asymmetric exclusion process and Askey–Wilson polynomials. Duke Math. J. 159, 385–415 (2011)
10. Dasse-Hartaut, S., Hitczenko, P.: Greek letters in random staircase tableaux. Random Struct. Algorithms 42, 73–96 (2013)
11. Derrida, B., Evans, M.R., Hakim, V., Pasquier, V.: Exact solution of a 1D asymmetric exclusion model using a matrix formulation. J. Phys. A 26(7), 1493–1517 (1993)
12. Franssens, G.R.: On a number pyramid related to the binomial, Deleham, Eulerian, MacMahon and Stirling number triangles. J. Integer Seq. 9(4):Article 06.4.1, 34 (2006)
13. Frobenius, G.: Über die Bernoullischen Zahlen und die Eulerschen Polynome. Sitzungsberichte der Königlich Preussischen Akademie der Wissenschaften, Berlin, pp. 809–847 (1910)
14. Graham, R.L., Knuth, D.E., Patashnik, O.: Concrete Mathematics, 2nd edn. Addison-Wesley, Reading (1994)
15. Hitczenko, P., Janson, S.: Asymptotic normality of statistics on permutation tableaux. Contemporary Math. 520, 83–104 (2010)
16. Hitczenko, P., Janson, S.: Weighted random staircase tableaux. To appear in Combin. Probab. Comput., arxiv.org/abs/1212.5498
17. Janson, S.: Euler–Frobenius numbers and rounding. arxiv.org/abs/1305.3512
18. Liu, L.L., Wang, Y.: A unified approach to polynomial sequences with only real zeros. Adv. Appl. Math. 38(4), 542–560 (2007)
19. MacMahon, P.A.: The divisors of numbers. Proc. London Math. Soc. Ser. 2 19(1), 305–340 (1920)
20. Nadeau, P.: The structure of alternative tableaux. J. Combin. Theory Ser. A 118(5), 1638–1660 (2011)
21. NIST Digital Library of Mathematical Functions, http://dlmf.nist.gov/
22. The On-Line Encyclopedia of Integer Sequences, http://oeis.org
23. Petrov, V.V.: Sums of Independent Random Variables. Springer, Berlin (1975)
24. Stanley, R.P.: Enumerative Combinatorics, vol. I. Cambridge Univ. Press, Cambridge (1997)
25. Wang, Y., Yeh, Y.-N.: Polynomials with real zeros and Pólya frequency sequences. J. Combin. Theory Ser. A 109(1), 63–74 (2005)

# Counting and Generating Permutations Using Timed Languages[*],[**]

Nicolas Basset

Department of Computer Science, University of Oxford, United Kingdom
basset@cs.ox.ac.uk

**Abstract.** The signature of a permutation $\sigma$ is a word $\mathbf{sg}(\sigma) \subseteq \{\mathbf{a},\mathbf{d}\}^*$ whose $i^{th}$ letter is $\mathbf{d}$ when $\sigma$ has a descent (i.e. $\sigma(i) > \sigma(i+1)$) and is $\mathbf{a}$ when $\sigma$ has an ascent (i.e. $\sigma(i) < \sigma(i+1)$). Combinatorics of permutations with a prescribed signature is quite well explored. Here we state and address the two problems of counting and randomly generating in the set $\mathbf{sg}^{-1}(L)$ of permutations with signature in a given regular language $L \subseteq \{\mathbf{a},\mathbf{d}\}^*$. First we give an algorithm that computes a closed form formula for the exponential generating function of $\mathbf{sg}^{-1}(L)$. Then we give an algorithm that generates randomly the $n$-length permutations of $\mathbf{sg}^{-1}(L)$ in a uniform manner, that is all the permutations of a given length with signature in $L$ are equally probable to be returned. Both contributions are based on a geometric interpretation of a subclass of regular timed languages.

Generating all the permutations with a prescribed signature (described in the abstract) or simply counting them are two classical combinatorial topics (see [17] and reference therein). The random generation of permutations with a prescribed signature has been addressed very recently by Philippe Marchal [13].

A very well studied example of permutations given by their signatures are the so-called alternating (or zig-zag, or down-up) permutations (see [16] for a survey). Their signatures belong to the language expressed by the regular expression $(\mathbf{da})^*(\mathbf{d}+\varepsilon)$ (in other words they satisfy $\sigma_1 > \sigma_2 < \sigma_3 > \sigma_4...$).

To a language $L \subseteq \{\mathbf{a},\mathbf{d}\}^*$, we associate the class $\mathbf{sg}^{-1}(L)$ of permutations whose signature is in $L$. Many classes of permutations can be expressed in that way (e.g. alternating permutations, those with an even number of descents).

We state and address the two problems of counting and randomly generating in $\mathbf{sg}^{-1}(L)$ when the language of signatures $L$ is regular. We propose Algorithm 1 that returns a closed form formula for the exponential generating function (EGF) of $\mathbf{sg}^{-1}(L)$. That is a formal power series $\sum a_n \frac{z^n}{n!}$ where the $n^{th}$ coefficient $a_n$ counts the permutations of length $n$ with signature in $L$. With such an EGF, it is easy to recover the number $a_n$ and some estimation of the growth rate of $a_n$ (see [9] for an overview of analytic combinatorics). The random generation is done by

---

[*] This research is supported in part by ERC Advanced Grant VERIWARE and was also supported by the ANR project EQINOCS (ANR-11-BS02-004).
[**] Omitted proofs and detailed examples can be found in Chapter 8 of [4].

an algorithm described in Theorem 3. The regular language $L$ together with $n$ the size of permutation to generate are the inputs while the outputs are $n$-length random permutations with signatures in $L$ equally probable to be returned.

Timed automata were introduced in [1] to model and verify properties of real-time systems. Our theory is based on a geometric interpretation of timed languages recognized by timed automata initiated in [3]. In that paper the authors introduced the concept of volume and entropy of timed languages. With these authors we defined and characterized volume generating function of timed language in [2]. In this latter paper a link between enumerative combinatorics and timed languages was foreseen. Here we establish such a link. The passage from a class of permutations to a timed language is in two steps. First we associate order and chain polytopes to signatures which are particular cases of Stanley's poset polytopes [15]. Then we interpret the chain polytopes of a signature $w$ as the set of delays which together with $w$ forms a timed word of a well chosen timed language.

**Related Works.** Particular regular languages of signatures are considered in [7] under the name of consecutive descent pattern avoidance. Numerous other works treat more general cases of (consecutive) pattern avoidance (see [8], [12]) and are quite incomparable to our work. Indeed, certain classes of permutations avoiding a finite set of patterns cannot be described as a language of signatures while some classes of permutations involving regular languages cannot be described by finite pattern avoidance (e.g. the permutations with an even number of descents).

The random sampler of timed words (Algorithm 2) is an adaptation to the timed case of the so-called recursive method of [14] developed by [10]. It has been improved for the particular case of generation of words in regular languages [5].

Further connections to related works are considered at the end of section 4.

## 1 Two Problem Statements

All along the paper we use the two letter alphabet $\{\mathbf{a}, \mathbf{d}\}$ whose elements must be read as "ascent" and "descent". Words of $\{\mathbf{a}, \mathbf{d}\}^*$ are called *signatures*. For $n \in \mathbb{N}$, $[n]$ denotes $\{1, \ldots, n\}$ and $\mathfrak{S}_n$ the set of permutations of $[n]$. We use the one line notation, for instance $\sigma = 231$ means that $\sigma(1) = 2$, $\sigma(2) = 3$, $\sigma(3) = 1$.

Let $n$ be a positive integer. The *signature* of a permutation $\sigma = \sigma_1 \cdots \sigma_n$ is the word $u = u_1 \cdots u_{n-1} \in \{\mathbf{a}, \mathbf{d}\}^{n-1}$ denoted by $\mathsf{sg}(\sigma)$ such that for $i \in [n]$, $\sigma_i < \sigma_{i+1}$ iff $u_i = \mathbf{a}$ (we speak of an "ascent") and $\sigma_i > \sigma_{i+1}$ iff $u_i = \mathbf{d}$ (we speak of a "descent"), for instance $\mathsf{sg}(21354) = \mathsf{sg}(32451) = \mathbf{daad}$.

This notion appears in the literature under several different names and forms such as descent word, descent set, ribbon diagram, etc. We are interested in $\mathsf{sg}^{-1}(L) = \{\sigma \mid \mathsf{sg}(\sigma) \in L\}$: the class of permutations with signature in $L \subseteq \{\mathbf{a}, \mathbf{d}\}^*$. Given a language $L$ we denote by $L_n$ the sub-language of $L$ restricted to its $n$-length words. The exponential generating function of $\mathsf{sg}^{-1}(L)$ is

$$F_L(z) =_{\text{def}} \sum_{\sigma \in \mathsf{sg}^{-1}(L)} \frac{z^{|\sigma|}}{|\sigma|!} = \sum_{n \geq 1} |\mathsf{sg}^{-1}(L_{n-1})| \frac{z^n}{n!}.$$

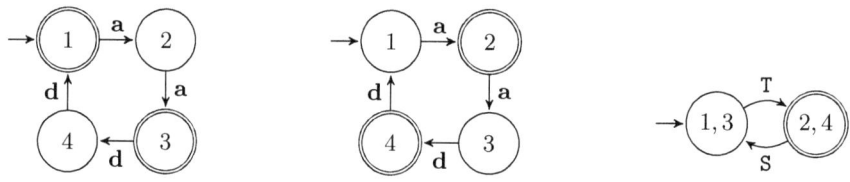

**Fig. 1.** From left to right: automata for $L^{ex}$, $L^{ex'}$ and $\mathtt{std}(L^{ex'})$

*Example 1.* Consider as a running example the class of "up-up-down-down" permutations with signature in the language[1] $L^{ex} = (\mathbf{aadd})^*(\mathbf{aa} + \varepsilon)$ recognized by the automaton depicted in the left of Figure 1. The theory developed in the paper permits to find the exponential generating function of $\mathbf{sg}^{-1}(L^{ex})$:

$$F_{L^{ex}}(z) = \frac{\sinh(z) - \sin(z) + \sin(z)\cosh(z) + \sinh(z)\cos(z)}{1 + \cos(z)\cosh(z)}.$$

Its Taylor expansion is

$$z + \frac{z^3}{3!} + 6\frac{z^5}{5!} + 71\frac{z^7}{7!} + 1456\frac{z^9}{9!} + 45541\frac{z^{11}}{11!} + 2020656\frac{z^{13}}{13!} + \cdots.$$

For instance, there are 1456 up-up-down-down permutations of length 9.

Now we state the two problems solved in this paper.

*Problem 1.* Design an algorithm which takes as input a regular language $L \subseteq \{\mathbf{a},\mathbf{d}\}^*$ and returns a closed form formula for $F_L(z)$.

*Problem 2.* Design an algorithm which takes as input a regular language $L \subseteq \{\mathbf{a},\mathbf{d}\}^*$ and $n \geq 1$ and returns a random permutation $\sigma$ uniformly in $\mathbf{sg}^{-1}(L_{n-1})$, that is the probability for each $\sigma \in \mathbf{sg}^{-1}(L_{n-1})$ to be returned is $1/|\mathbf{sg}^{-1}(L_{n-1})|$.

## 2 A Timed and Geometric Approach

In section 2.1 we recall definition of order and chain polytopes associated to signatures. We introduce a sequence of sets $\mathcal{O}_n(L) \subseteq [0,1]^n$ and see how the two problems posed can be reformulated as computing the volume generating function of this sequence and generating points uniformly in $\mathcal{O}_n(L)$. Then we define a timed language $\mathbb{L}'$ associated to $L$ as well as its volume sequence (section 2.2) and describe a volume preserving transformation between $\mathcal{O}_n(L)$ and $\mathbb{L}'_n$.

### 2.1 Order and Chain Polytopes of Signatures

We say that a collection of polytopes $(S_1, \cdots, S_n)$ is an *almost disjoint partition* of a set $A$ if it is the union of $S_i$ and they have pairwise a null volume intersection. In this case we write $S = \bigsqcup_{i=1}^n S_i$.

---

[1] We identify regular expressions with the regular languages they express.

The set $\{(\nu_1,\ldots,\nu_n) \in [0,1]^n \mid 0 \leq \nu_{\sigma_1^{-1}} \leq \ldots \leq \nu_{\sigma_n^{-1}} \leq 1\}$ is called the *order simplex*[2] of $\sigma$ and denoted by $\mathcal{O}(\sigma)$. For instance $\nu = (0.3, 0.2, 0.4, 0.5, 0.1)$ belongs to $\mathcal{O}(32451)$ since $\nu_5 \leq \nu_2 \leq \nu_1 \leq \nu_3 \leq \nu_4$ and $(32451)^{-1} = 52134$. The set $O(\sigma)$ for $\sigma \in \mathfrak{S}_n$ forms an almost disjoint partition of $[0,1]^n$. By symmetry all the order simplices of permutations have the same volume which is $1/n!$.

If $\nu$ is uniformly sampled in $[0,1]^n$ then it falls in any $O(\sigma)$ with probability $1/n!$. To retrieve $\sigma$ from $\nu$ it suffices to use a sorting algorithm. We denote by $\Pi(\nu)$ the permutation $\sigma$ returned by the sorting algorithm on $\nu$, that is such that $0 \leq \nu_{\sigma_1^{-1}} \leq \ldots \leq \nu_{\sigma_n^{-1}} \leq 1$. Moreover with probability 1, $\nu$ has pairwise distinct[3]. coordinates and one can define its *signature* $\mathsf{sg}(\nu) = u_1 \ldots u_{n-1}$ by $u_i = \mathbf{a}$ if $\nu_i < \nu_{i+1}$ and $u_i = \mathbf{d}$ if $\nu_i > \nu_{i+1}$. For instance $\mathsf{sg}(0.3, 0.2, 0.4, 0.5, 0.1) = \mathbf{daad}$.

The *order polytope* $\mathcal{O}(u)$ [15] of a signature $u \in \{\mathbf{a,d}\}^{n-1}$ is the set of vectors $\nu$ such that for all $i \leq n-1$, if $u_i = \mathbf{a}$ then $\nu_i \leq \nu_{i+1}$ and $\nu_i \geq \nu_{i+1}$ otherwise. That is the topological closure of $\{\nu \in [0,1]^n \mid \mathsf{sg}(\nu) = u\}$. It is clear that the collection of order simplices $\mathcal{O}(\sigma)$ with all $\sigma$ having the same signature $u$ form an almost disjoint partition of the order polytope $\mathcal{O}(u)$: $\mathcal{O}(u) = \bigsqcup_{\sigma \in \mathsf{sg}^{-1}(u)} \mathcal{O}(\sigma)$, for instance $\mathcal{O}(\mathbf{daa}) = \mathcal{O}(2134) \sqcup \mathcal{O}(3124) \sqcup \mathcal{O}(4123)$. Passing to volume we get:

$$\mathtt{Vol}(\mathcal{O}(u)) = \sum_{\sigma \in \mathsf{sg}^{-1}(u)} \mathtt{Vol}(\mathcal{O}(\sigma)) = \frac{|\mathsf{sg}^{-1}(u)|}{n!}. \tag{1}$$

Let $L$ be a language of signatures and $n \geq 1$, then the family $(\mathcal{O}(u))_{u \in L_{n-1}}$ forms an almost disjoint partition of a subset of $[0,1]^n$ called the $n^{th}$ *order set* of $L$ and denoted by $\mathcal{O}_n(L)$:

$$\mathcal{O}_n(L) = \bigsqcup_{u \in L_{n-1}} \mathcal{O}(u) = \bigsqcup_{\sigma \in \mathsf{sg}^{-1}(L_{n-1})} \mathcal{O}(\sigma) = \overline{\{\nu \in [0,1]^n \mid \mathsf{sg}(\nu) \in L_{n-1}\}}. \tag{2}$$

For volumes we get:

$$\mathtt{Vol}(\mathcal{O}_n(L)) = \sum_{u \in L_{n-1}} \mathtt{Vol}(\mathcal{O}(u)) = \sum_{\sigma \in \mathsf{sg}^{-1}(L_{n-1})} \mathtt{Vol}(\mathcal{O}(\sigma)) = \frac{|\mathsf{sg}^{-1}(L_{n-1})|}{n!} \tag{3}$$

The *chain polytope* [15] of a signature $u$ is the set $\mathcal{C}(u)$ of vectors $t \in [0,1]^n$ such that for all $i < j \leq n$ and $l \in \{\mathbf{a,d}\}$, $w_i \cdots w_{j-1} = l^{j-i} \Rightarrow t_i + \ldots + t_j \leq 1$.

*Example 2.* A vector $(t_1, t_2, t_3, t_4, t_5) \in [0,1]^5$ belongs to $\mathcal{C}(\mathbf{daad})$ iff $t_1 + t_2 \leq 1, t_2 + t_3 + t_4 \leq 1, t_4 + t_5 \leq 1$ iff $1 - t_1 \geq t_2 \leq t_2 + t_3 \leq 1 - t_4 \geq t_5$ iff $(1 - t_1, t_2, t_2 + t_3, 1 - t_4, t_5) \in \mathcal{O}(\mathbf{daad})$.

More generally, for $w = ul$ with $u \in \{\mathbf{a,d}\}^*$, $l \in \{\mathbf{a,d}\}$ and $n = |w|$, there is a volume preserving transformation $(t_1, \cdots, t_n) \mapsto (\nu_1, \cdots, \nu_n)$ from the chain polytope $\mathcal{C}(u)$ to the order polytope $\mathcal{O}(u)$ defined as follows.

---

[2] Order simplices, order and chain polytopes of signatures defined here are particular cases of Stanley's order and chain polytopes of posets [15].
[3] Alternatively $\mathsf{sg}(\nu) =_{\mathrm{def}} \mathsf{sg}(\Pi(\nu))$ (defined also when some coordinates are equal).

Let $j \in [n]$ and $i$ be the index such that $w_i \cdots w_{j-1}$ is a maximal ascending or descending block, that is $i$ is minimal such that $w_i \cdots w_{j-1} = l^{j-i}$ with $l \in \{\mathbf{a}, \mathbf{d}\}^*$. If $w_j = \mathbf{d}$ we define $\nu_j = 1 - \sum_{k=i}^{j} t_k$ and $\nu_j = \sum_{k=i}^{j} t_k$ otherwise.

**Proposition 1 (simple case of Theorem 2.1 of [11]).** *The mapping $\phi_{ul}$ : $(t_1, \cdots, t_n) \mapsto (\nu_1, \cdots, \nu_n)$ is a volume preserving transformation from $\mathcal{C}(u)$ to $\mathcal{O}(u)$. It can be computed in linear time using the following recursive definition:*

$$\begin{vmatrix} \nu_1 = t_1 & \text{if } w_1 = \mathbf{a} \\ \nu_1 = 1 - t_1 & \text{if } w_1 = \mathbf{d} \end{vmatrix} \text{ and for } i \geq 2: \begin{vmatrix} \nu_i = \nu_{i-1} + t_i & \text{if } w_{i-1} w_i = \mathbf{aa}; \\ \nu_i = t_i & \text{if } w_{i-1} w_i = \mathbf{da}; \\ \nu_i = 1 - t_i & \text{if } w_{i-1} w_i = \mathbf{ad}; \\ \nu_i = \nu_{i-1} - t_i & \text{if } w_{i-1} w_i = \mathbf{dd}. \end{vmatrix}$$

As a corollary of (3) and Propostion 1 the first problem can be reformulated in geometric terms as follows.

**Corollary 1.** *For every $L \in \{\mathbf{a}, \mathbf{d}\}^*$ the following equalities hold:*

$$F_L(z) = \sum_{n \geq 1} \mathrm{Vol}(\mathcal{O}_n(L)) z^n = \sum_{u \in L} \mathrm{Vol}(\mathcal{O}(u)) z^{|u|-1} = \sum_{u \in L} \mathrm{Vol}(\mathcal{C}(u)) z^{|u|-1}.$$

For the second problem, it suffices to generate uniformly a vector $\nu \in \mathcal{O}_n(L)$ and then sort it to get a permutation $\sigma = \Pi(\nu)$. As the simplices $\mathcal{O}(\sigma)$ for $\sigma \in \mathrm{sg}^{-1}(L_n)$ form an almost disjoint partition of $\mathcal{O}_n(L)$ and all these simplices have the same volume $1/n!$, they are equally probable to receive the random vector $\nu$. Hence all $\sigma \in \mathrm{sg}^{-1}(L_n)$ have the same probability to be chosen.

In fact, it is not clear how to fit the sequence of order sets (when $n$ varies) with the dynamics of the language $L$. We prefer to use a timed language for which we can write recursive equations on volumes (inspired by [3,2]). The reduction from the sequence of order sets to the timed language is mainly given by Proposition 1 since this latter language is a formal union of chain polytopes (Proposition 2).

## 2.2 Timed Semantics of a Language of Signatures: $(\mathbb{L}'_n)_{n \in \mathbb{N}}$

This section is inspired by timed automata theory and designed for non experts. We adopt a non standard[4] and self-contained approach based on the notion of clock languages introduced by [6] and used in our previous work [2].

**Timed Languages, Their Volumes and Their Generating Functions.** An alphabet of *timed events* is the product $\mathbb{R}^+ \times \Sigma$ where $\Sigma$ is a finite alphabet. The meaning of a timed event $(t_i, w_i)$ is that $t_i$ is the *time delay* before the *event* $w_i$. A *timed word* is just a word of timed events and a *timed language* a set of timed words. Adopting a geometric point of view, a timed word is a vector of delays $\boldsymbol{t} = (t_1, \ldots, t_n) \in \mathbb{R}^n$ together with a word of events $w = w_1 \cdots w_n \in \Sigma^n$. That is why we sometimes write such a timed word $(\boldsymbol{t}, w)$ instead of $(t_1, w_1) \cdots (t_n, w_n)$. With this convention, given a timed language $\mathbb{L}' \subseteq (\mathbb{R}^+ \times \Sigma)^*$, its restriction

---

[4] We refer the reader to [1] for a standard approach of timed automata theory.

to $n$-length words $\mathbb{L}'_n$ can be seen as a formal union of sets $\biguplus_{w \in \Sigma^n} \mathbb{L}'_w \times \{w\}$ where $\mathbb{L}'_w = \{\boldsymbol{t} \in \mathbb{R}^n \mid (\boldsymbol{t}, w) \in \mathbb{L}'\}$ is the set of delay vectors that together with $w$ form a timed word of $\mathbb{L}'$. In the sequel we will only consider languages $\mathbb{L}'$ for which every $\mathbb{L}'_w$ is volume measurable. To such $\mathbb{L}'_n$ one can associate a sequence of volumes and a *volume generating function* as follows:

$$\text{Vol}(\mathbb{L}'_n) = \sum_{w \in \Sigma^n} \text{Vol}(\mathbb{L}'_w); \quad VGF(\mathbb{L}')(z) = \sum_{w \in \Sigma^*} \text{Vol}(\mathbb{L}'_w) z^{|w|} = \sum_{n \in \mathbb{N}} \text{Vol}(\mathbb{L}'_n) z^n$$

**The Clock Semantics of a Signature.** A *clock* is a non-negative real variable. Here we only consider two clocks bounded by 1 and denoted by $x^{\mathbf{a}}$ and $x^{\mathbf{d}}$. A *clock word* is a tuple whose component are a starting clock vector $(x_0^{\mathbf{a}}, x_0^{\mathbf{d}}) \in [0,1]^2$, a timed word $(t_1, a_1) \cdots (t_n, a_n) \in ([0,1] \times \{\mathbf{a}, \mathbf{d}\})^*$ and an ending clock vector $(x_n^{\mathbf{a}}, x_n^{\mathbf{d}}) \in [0,1]^2$. It is denoted by $(x_0^{\mathbf{a}}, x_0^{\mathbf{d}}) \xrightarrow{(t_1, a_1) \cdots (t_n, a_n)} (x_n^{\mathbf{a}}, x_n^{\mathbf{d}})$. Two clock words $\boldsymbol{x}_0 \xrightarrow{w} \boldsymbol{x}_1$ and $\boldsymbol{x}_2 \xrightarrow{w'} \boldsymbol{x}_3$ are said to be compatible if $\boldsymbol{x}_2 = \boldsymbol{x}_1$, in this case their product is $(\boldsymbol{x}_0 \xrightarrow{w} \boldsymbol{x}_1) \cdot (\boldsymbol{x}_2 \xrightarrow{w'} \boldsymbol{x}_3) = \boldsymbol{x}_0 \xrightarrow{ww'} \boldsymbol{x}_3$. A *clock language* is a set of clock words. The product of two clock languages $\mathcal{L}$ and $\mathcal{L}'$ is

$$\mathcal{L} \cdot \mathcal{L}' = \{c \cdot c' \mid c \in \mathcal{L},\ c' \in \mathcal{L}',\ c \text{ and } c' \text{ compatible}\}. \tag{4}$$

The clock language[5] $\mathcal{L}(\mathbf{a})$ (resp. $\mathcal{L}(\mathbf{d})$) of an ascent (resp. a descent) is the set of clock words of the form $(x^{\mathbf{a}}, x^{\mathbf{d}}) \xrightarrow{(t, \mathbf{a})} (x^{\mathbf{a}} + t, 0)$ (resp. $(x^{\mathbf{a}}, x^{\mathbf{d}}) \xrightarrow{(t, \mathbf{d})} (0, x^{\mathbf{d}} + t)$) and such that $x^{\mathbf{a}} + t \in [0,1]$ and $x^{\mathbf{d}} + t \in [0,1]$ (and by definition of clocks and delays $x^{\mathbf{a}} \geq 0$, $x^{\mathbf{d}} \geq 0$, $t \geq 0$). These definitions extend inductively to all signatures: $\mathcal{L}(u_1 \cdots u_n) = \mathcal{L}(u_1) \cdots \mathcal{L}(u_n)$ (with product (4)).

*Example 3.* $(0,0) \xrightarrow{(0.7, \mathbf{d})(0.2, \mathbf{a})(0.2, \mathbf{a})(0.5, \mathbf{d})} (0, 0.5) \in \mathcal{L}(\mathbf{daad})$ since
$(0,0) \xrightarrow{(0.7, \mathbf{d})} (0, 0.7) \in \mathcal{L}(\mathbf{d}); \quad (0, 0.7) \xrightarrow{(0.2, \mathbf{a})} (0.2, 0) \in \mathcal{L}(\mathbf{a});$
$(0.2, 0) \xrightarrow{(0.2, \mathbf{a})} (0.4, 0) \in \mathcal{L}(\mathbf{a}); \quad (0.4, 0) \xrightarrow{(0.5, \mathbf{a})} (0, 0.5) \in \mathcal{L}(\mathbf{d}).$

**The Timed Semantics of a Language of Signatures.** The *timed polytope* associated to a signature $w \in \{\mathbf{a}, \mathbf{d}\}^*$ is

$$P_w =_{\text{def}} \{\boldsymbol{t} \mid (0,0) \xrightarrow{(t, w)} \boldsymbol{y} \in \mathcal{L}(w) \text{ for some } \boldsymbol{y} \in [0,1]^2\}.$$

For instance $(0.7, 0.2, 0.2, 0.5, 0.1) \in P_{\mathbf{daada}}$. The timed semantics of a language of signatures $L'$ is

$$\mathbb{L}' = \{(\boldsymbol{t}, w) \mid \boldsymbol{t} \in P_w \text{ and } w \in L'\} = \cup_{w \in L'} P_w \times \{w\}.$$

This language restricted to words of length $n$ is $\mathbb{L}'_n = \cup_{w \in L'_n} P_w \times \{w\}$, its volume is $\text{Vol}(\mathbb{L}'_n) = \sum_{w \in L'} \text{Vol}(P_w)$.

---

[5] A reader acquainted with timed automata would have noticed that the clock language $\mathcal{L}(\mathbf{a})$ (resp. $\mathcal{L}(\mathbf{d})$) corresponds to a transition of a timed automaton where the guards $x^{\mathbf{a}} \leq 1$ and $x^{\mathbf{d}} \leq 1$ are satisfied and where $x^{\mathbf{d}}$ (resp. $x^{\mathbf{a}}$) is reset.

**The Link with Order and Chain Polytopes of Signatures.** We first state the link between timed polytopes and chain polytopes.

**Proposition 2.** *Given a word $u \in \{\mathtt{a},\mathtt{d}\}^*$ and $l \in \{\mathtt{a},\mathtt{d}\}$, the timed polytope of $ul$ is the chain polytope of $u$: $P_{ul} = \mathcal{C}(u)$.*

Hence Proposition 1 links the timed polytope $P_{ul} = \mathcal{C}(u)$ of a signature of length $n+1$ and the order polytopes $\mathcal{O}(u)$ of a signature of length $n$. We correct the mismatch of length using prolongation of languages. A language $L'$ is called a *prolongation* of a language $L$ whenever the truncation of the last letter $w_1 \ldots w_n \mapsto w_1 \ldots w_{n-1}$ is a bijection between $L'$ and $L$. Every language $L$ has prolongations, for instance $L' = Ll$ for $l \in \{\mathtt{a},\mathtt{d}\}$. A prolongation of $L^{ex}$ is $L^{ex'} = (\mathtt{aadd})^*(\mathtt{aad}+\mathtt{a})$ recognized by the automaton depicted in the middle of Figure 1. Proposition 1 can be extended to language of signatures as follows.

**Corollary 2.** *Let $L \subseteq \{\mathtt{a},\mathtt{d}\}^*$ and $\mathbb{L}'$ be the timed semantics of a prolongation of $L$ then for all $n \in \mathbb{N}$, the following function is a volume preserving transformation between $\mathbb{L}'_n$ and $\mathcal{O}_n(L)$. Moreover it is computable in linear time.*

$$\phi: \mathbb{L}'_n \to \mathcal{O}_n(L) \\ (\boldsymbol{t},w) \mapsto \phi_w(\boldsymbol{t}) \qquad (5)$$

As a consequence, the two problems can be solved if we know how to compute the VGF of a timed language $\mathbb{L}'$ and how to generate timed vector uniformly in $\mathbb{L}'_n$. A characterization of the VGF of a timed language as a solution of a system of differential equations is done in [2]. Nevertheless the equations of this article are quite uneasy to handle and don't give a closed form formula for the VGF. To get simpler equations than in [2] we work with a novel class of timed languages involving two kinds of transitions S and T.

### 2.3 The S-T (Timed) Language Encoding

**The S-T-Encoding.** We consider the finite alphabet $\{\mathtt{S},\mathtt{T}\}$ whose elements must be respectively read as *straight* and *turn*. The S-T-encoding of type $l \in \{\mathtt{a},\mathtt{d}\}$ of a word $w \in \{\mathtt{a},\mathtt{d}\}^*$ is a word $w' \in \{\mathtt{S},\mathtt{T}\}^*$ denoted by $\mathtt{st}_l(w)$ and defined recursively as follows: for every $i \in [n]$, $w'_i = \mathtt{S}$ if $w_i = w_{i-1}$ and $w'_i = \mathtt{T}$ otherwise, with the convention that $w_0 = l$. The mapping $\mathtt{st}_l$ is invertible and can also be defined recursively. Indeed $w = \mathtt{st}_l^{-1}(w')$ iff for every $i \in [n]$, $w_i = w_{i-1}$ if $w'_i = \mathtt{S}$ and $w_i \neq w_{i-1}$ otherwise, with convention that $w_0 = l$. Notion of S-T-encoding can be extended naturally to languages. For the running example: $\mathtt{st}_\mathtt{d}(L^{ex'}) = (\mathtt{TS})^*\mathtt{T}$. We call an S-T-automaton, a deterministic finite state automaton with transition alphabet $\{\mathtt{S},\mathtt{T}\}$ (see Figure 1 for an S-T-automaton recognizing $\mathtt{st}_\mathtt{d}(L^{ex'})$).

**Timed Semantics and S-T-Encoding.** In the following we define clock and timed languages similarly to what we have done in section 2.2. Here we need only one clock $x$ that remains bounded by 1. We define the clock language associated to S by $\mathcal{L}(\mathtt{S}) = \{x \xrightarrow{(t,\mathtt{S})} x+t \mid x \in [0,1], t \in [0,1-x]\}$ and the clock language associated to T by $\mathcal{L}(\mathtt{T}) = \{x \xrightarrow{(t,\mathtt{T})} t \mid x \in [0,1], t \in [0,1-x]\}$. Let $L'' \subseteq \{\mathtt{S},\mathtt{T}\}^*$ we denote by $L''(x)$ the timed language starting from $x$: $L''(x) = \{(t,w) \mid \exists y \in [0,1], x \xrightarrow{(t,w)} y \in \mathcal{L}(w), w \in L''\}$. The *timed semantics* of $L'' \subseteq \{\mathtt{S},\mathtt{T}\}^*$ is $L''(0)$.

The S-T-encodings yields a natural volume preserving transformation between timed languages:

**Proposition 3.** *Let $L' \subseteq \{\mathtt{a},\mathtt{d}\}^*$, $l \in \{\mathtt{a},\mathtt{d}\}$, $\mathbb{L}'$ be the timed semantics of $L'$ and $\mathbb{L}''$ be the timed semantics of $st_l(L')$ then the function $(t,w) \mapsto (t, st_l^{-1}(w))$ is a volume preserving transformation from $\mathbb{L}''_n$ to $\mathbb{L}'_n$.*

Using notation and results of Corollary 2 and Proposition 3 we get a volume preserving transformation from $\mathbb{L}''_n$ to $\mathcal{O}_n(L)$.

**Theorem 1.** *The function $(t,w) \mapsto \phi_{st_l^{-1}(w)}(t)$ is a volume preserving transformation from $\mathbb{L}''_n$ to $\mathcal{O}_n(L)$ computable in linear time. In particular*

$$\mathtt{Vol}(\mathbb{L}''_n) = \frac{|\mathtt{sg}^{-1}(L_{n-1})|}{n!} \text{ for } n \geq 1 \text{ and } VGF(\mathbb{L}'')(z) = F_L(z).$$

Thus to solve Problem 1 it suffices to characterize the VGF of an S-T-automaton.

## 3 Solving the Two Problems

### 3.1 Characterization of the VGF of an S-T-Automaton

In this section we characterize precisely the VGF of the timed language recognized by an S-T-automaton. This solves Problem 1.

We have defined just above timed language $L''(x)$ parametrized by an initial clock $x$. Given an S-T-automaton, we can also consider the intial state $p$ as a parameter and write Kleene like systems of equations on parametric language $L_p(x)$ (similarly to [2]). More precisely, let $\mathcal{A} = (\{\mathtt{S},\mathtt{T}\}, Q, q_0, F, \delta)$ be an S-T-automaton with states $Q$, initial state $q_0 \in Q$, final states $F \subseteq Q$ and transition function $\delta : Q \times \{\mathtt{S},\mathtt{T}\} \to Q$. To every state $p \in Q$ we denote by $L_p \subseteq \{\mathtt{S},\mathtt{T}\}^*$ the language starting from $p$ that is recognized by $\mathcal{A}_p =_{\mathrm{def}} \{\{\mathtt{S},\mathtt{T}\}, Q, p, F, \delta\}$. Then for every $p \in Q$, we have a parametric language equation:

$$L_p(x) = \left[\cup_{t \leq 1-x}(t,\mathtt{S})L_{\delta(p,\mathtt{S})}(x+t)\right] \cup \left[\cup_{t \leq 1-x}(t,\mathtt{T})L_{\delta(p,\mathtt{T})}(t)\right] \cup (\varepsilon \text{ if } p \in F). \quad (6)$$

We denote by $f_p(x,z)$ and $V_p(z)$ the volume generating function of $L_p(x)$ and $L_p$ respectively and are interested in $V_{q_0}(z)$. As in [2], we pass from equation on languages (6) to equation on generating functions:

$$f_p(x,z) = z \int_x^1 f_{\delta(p,\mathtt{S})}(s,z)ds + z \int_0^{1-x} f_{\delta(p,\mathtt{T})}(t,z)dt + (1 \text{ if } p \in F) \quad (7)$$

In matrix notation:

$$\boldsymbol{f}(x,z) = zM_\mathtt{S} \int_x^1 \boldsymbol{f}(s,z)ds + zM_\mathtt{T} \int_0^{1-x} \boldsymbol{f}(t,z)dt + \boldsymbol{F} \quad (8)$$

where $\boldsymbol{f}(x,z)$, $\int_x^1 \boldsymbol{f}(s,z)ds$ and $\int_0^{1-x} \boldsymbol{f}(t,z)dt$ are the column vectors whose coordinates are respectively the $f_p(x,z)$, $\int_x^1 f_p(s,z)ds$ and $\int_0^{1-x} f_p(t,z)dt$ for $p \in Q$. The $p^{th}$ coordinate of the column vector $\boldsymbol{F}$ is 1 if $p \in F$ and 0 otherwise. The $Q \times Q$-matrices $M_{\mathtt{S}}$ and $M_{\mathtt{T}}$ are the adjacency matrices corresponding to letter S and T that is for $l \in \{\mathtt{S},\mathtt{T}\}$, $M_l(p,q) = 1$ if $\delta(p,l) = q$ and 0 otherwise.

The equation (8) is equivalent to the differential equation:

$$\frac{\partial}{\partial x} \boldsymbol{f}(x,z) = -z M_{\mathtt{S}} \boldsymbol{f}(x,z) - z M_{\mathtt{T}} \boldsymbol{f}(1-x,z) \tag{9}$$

with boundary condition

$$\boldsymbol{f}(1,z) = \boldsymbol{F}. \tag{10}$$

The equation (9) is equivalent to the following linear homogeneous system of ordinary differential equations with constant coefficients:

$$\frac{\partial}{\partial x} \begin{pmatrix} \boldsymbol{f}(x,z) \\ \boldsymbol{f}(1-x,z) \end{pmatrix} = z \begin{pmatrix} -M_{\mathtt{S}} & -M_{\mathtt{T}} \\ M_{\mathtt{T}} & M_{\mathtt{S}} \end{pmatrix} \begin{pmatrix} \boldsymbol{f}(x,z) \\ \boldsymbol{f}(1-x,z) \end{pmatrix} \tag{11}$$

whose solution is of the form

$$\begin{pmatrix} \boldsymbol{f}(x,z) \\ \boldsymbol{f}(1-x,z) \end{pmatrix} = \exp\left[xz \begin{pmatrix} -M_{\mathtt{S}} & -M_{\mathtt{T}} \\ M_{\mathtt{T}} & M_{\mathtt{S}} \end{pmatrix}\right] \begin{pmatrix} \boldsymbol{f}(0,z) \\ \boldsymbol{f}(1,z) \end{pmatrix}. \tag{12}$$

Taking $x = 1$ in (12) and using the boundary condition (10) we obtain:

$$\begin{pmatrix} \boldsymbol{F} \\ \boldsymbol{V}(z) \end{pmatrix} = \exp\left[z \begin{pmatrix} -M_{\mathtt{S}} & -M_{\mathtt{T}} \\ M_{\mathtt{T}} & M_{\mathtt{S}} \end{pmatrix}\right] \begin{pmatrix} \boldsymbol{V}(z) \\ \boldsymbol{F} \end{pmatrix} \tag{13}$$

where $\boldsymbol{V}(z)$ is the vector whose coordinates are $V_q(z)$ for $q \in Q$. Hence,

$$\boldsymbol{F} = A_1(z)\boldsymbol{V}(z) + A_2(z)\boldsymbol{F}; \quad \boldsymbol{V}(z) = A_3(z)\boldsymbol{V}(z) + A_4(z)\boldsymbol{F} \tag{14}$$

where $\begin{pmatrix} A_1(z) & A_2(z) \\ A_3(z) & A_4(z) \end{pmatrix} = \exp\left[z \begin{pmatrix} -M_{\mathtt{S}} & -M_{\mathtt{T}} \\ M_{\mathtt{T}} & M_{\mathtt{S}} \end{pmatrix}\right]$. In particular when $z = 0$, $A_1(0) = I - A_3(0) = I$ and thus the two continuous functions $z \mapsto \det A_1(z)$ and $z \mapsto \det(I - A_3(z))$ are positive in a neighbourhood of 0. We deduce that the inverses of the matrices $A_1(z)$ and $I - A_3(z)$ are well defined in a neighbourhood of 0 and thus both equations of (14) permit to express $\boldsymbol{V}(z)$ wrt. $\boldsymbol{F}$:

$$\boldsymbol{V}(z) = [A_1(z)]^{-1}[I - A_2(z)]\boldsymbol{F}; \quad \boldsymbol{V}(z) = [I - A_3(z)]^{-1} A_4(z)\boldsymbol{F} \tag{15}$$

To sum up, we address Problem 1 with the following theorem.

**Theorem 2.** *Given a regular language $L \subseteq \{\mathtt{a},\mathtt{d}\}^*$, one can compute the exponential generating function $F_L(z)$ using Algorithm 1.*

**Some Comments about the Algorithm.** In line 1, several choices are left to the user: the prolongation $L'$ of the language $L$, the type of the S-T-encoding and the automaton that realizes the S-T-encoding. These choices should be made such that the output automaton has a minimal number of states or more generally such that the matrices $M_{\mathtt{T}}$ and $M_{\mathtt{S}}$ are the simplest possible. Exponentiation of matrices is implemented in most of computer algebra systems.

**Algorithm 1.** Computation of the generating function

1: Compute an S-T-automaton $\mathcal{A}$ for an extension of $L$ and its corresponding adjacency matrices $M_\text{T}$ and $M_\text{S}$;
2: Compute $\begin{pmatrix} A_1(z) & A_2(z) \\ A_3(z) & A_4(z) \end{pmatrix} =_{\text{def}} \exp\left[z \begin{pmatrix} -M_\text{S} & -M_\text{T} \\ M_\text{T} & M_\text{S} \end{pmatrix}\right]$;
3: Compute $\boldsymbol{V}(z) = [A_1(z)]^{-1}[I - A_2(z)]\boldsymbol{F}$ (or $\boldsymbol{V}(z) = [I - A_3(z)]^{-1}A_4(z)\boldsymbol{F}$);
4: **return** $V_{q_0}(z)$ the component of $\boldsymbol{V}(z)$ corresponding to the initial state of $\mathcal{A}$.

## 3.2 An Algorithm for Problem 2

Now we can solve Problem 2 using a uniform sampler of timed words (Algorithm 2), the volume preserving transformation of Theorem 1 and a sorting algorithm.

**Theorem 3.** *Let $L \subseteq \{\mathtt{a},\mathtt{d}\}^*$ and $\mathbb{L}''$ be the timed semantics of a S-T-encoding of type $l$ (for some $l \in \{\mathtt{a},\mathtt{d}\}$) of a prolongation of $L$. The following algorithm permits to achieve a uniform sampling of permutation in $\mathtt{sg}^{-1}(L_{n-1})$.*

1. *Choose uniformly an $n$-length timed word $(\boldsymbol{t}, w) \in \mathbb{L}''_n$ using Algorithm 2;*
2. *Return $\Pi(\phi_{\mathtt{st}_l^{-1}(w)}(\boldsymbol{t}))$.*

**Uniform Sampling of Timed Words.** Recursive formulae (16) and (17) below are freely inspired by those of [3] and of our previous work [2]. They are the key tools to design a uniform sampler of timed word. This algorithm is a lifting from the discrete case of the so-called recursive method (see [5,10]). For all $q \in Q$, $n \in \mathbb{N}$ and $x \in [0,1]$ we denote by $L_{q,n}(x)$ the language $L_q(x)$ restricted to $n$-length timed words. The languages $L_{q,n}(x)$ can be recursively defined as follows: $L_{q,0}(x) = \varepsilon$ if $q \in F$ and $L_{q,0} = \emptyset$ otherwise;

$$L_{q,n+1}(x) = \left[\cup_{t \leq 1-x}(t, \mathtt{S})L_{\delta(q,\mathtt{S}),n}(x+t)\right] \cup \left[\cup_{t \leq 1-x}(t, \mathtt{T})L_{\delta(q,\mathtt{T}),n}(t)\right]. \quad (16)$$

For $q \in Q$ and $n \geq 0$, we denote by $v_{q,n}$ the function $x \mapsto \mathtt{Vol}[L_{q,n}(x)]$ from $[0,1]$ to $\mathbb{R}^+$. Each $v_{q,n}$ is a polynomial of a degree less or equal to $n$ that can be computed recursively using the recurrent formula: $v_{q,0}(x) = 1_{q \in F}$ and

$$v_{q,n+1}(x) = \int_x^1 v_{\delta(q,\mathtt{S}),n}(y)dy + \int_0^{1-x} v_{\delta(q,\mathtt{T}),n}(y)dy. \quad (17)$$

The polynomials $v_{q,n}(x)$ play a key role for the uniform sampler. They permit also to retrieve directly the terms of the wanted VGF: $\mathtt{Vol}(\mathbb{L}''_n) = v_{q_0,n}(0)$ where $q_0$ is the initial state of the S-T automaton.

**Theorem 4.** *Algorithm 2 is a uniform sampler of timed words of $\mathbb{L}''_n$, that is for every volume measurable subset $A \subseteq \mathbb{L}''_n$, the probability that the returned timed word belongs to $A$ is $\mathtt{Vol}(A)/\mathtt{Vol}(\mathbb{L}''_n)$.*

**Algorithm 2.** Recursive uniform sampler of timed words
1: $x_0 \leftarrow 0$; $q_0 \leftarrow$ initial state;
2: **for** $k = 1$ to $n$ **do**
3:     Compute $m_k = v_{q_{k-1}, n-(k-1)}(x_{k-1})$ and $p_\mathsf{S} = \int_{x_{k-1}}^1 v_{\delta(q_{k-1}, \mathsf{S}), n-k}(y) dy / m_k$;
4:     $b \leftarrow \texttt{BERNOULLI}(p_\mathsf{S})$;   (return 1 with probability $p_\mathsf{S}$ and 0 otherwise)
5:     **if** $b = 1$ **then**
6:         $w_k \leftarrow \mathsf{S}$; $q_k \leftarrow \delta(q_{k-1}, \mathsf{S})$;
7:         $r \leftarrow \texttt{RAND}([0,1])$;   (return a number uniformly sampled in $[0,1]$)
8:         $t_k \leftarrow$ the unique solution in $[0, 1-x_{k-1}]$ of $\frac{1}{m_k p_\mathsf{S}} \int_{x_{k-1}}^{x_{k-1}+t_k} v_{q_k, n-k}(y) dy - r = 0$;
9:         $x_k \leftarrow x_{k-1} + t_k$;
10:    **else**
11:        $w_k \leftarrow \mathsf{T}$; $q_k \leftarrow \delta(q_{k-1}, \mathsf{T})$;
12:        $r \leftarrow \texttt{RAND}([0,1])$;   (return a number uniformly sampled in $[0,1]$)
13:        $t_k \leftarrow$ the unique solution in $[0, 1-x_{k-1}]$ of $\frac{1}{m_k(1-p_\mathsf{S})} \int_0^{t_k} v_{q_k, n-k}(y) dy - r = 0$;
14:        $x_k \leftarrow t_k$;
15:    **end if**
16: **end for**
17: **return** $(t_1, w_1)(t_2, w_2) \ldots (t_n, w_n)$

**Some Comments about the Algorithm.** Algorithm 2 requires a precomputation of all functions $v_{q,k}$ for $q \in Q$ and $k \leq n$. They can be computed in polynomial time by a dynamic programming method using (17). The expressions in lines 8 and 13 are polynomial functions increasing on $[x, 1]$ (the derivative is the integrand which is positive on $(x, 1)$). Finding the root of such a polynomial can be done numerically and efficiently with a controlled error using a numerical scheme such as the Newton's method. A toy implementation of Algorithm 2 as well as that sketched in Theorem 3 is available on-line http://www.liafa.univ-paris-diderot.fr/~nbasset/sage/sage.htm.

## 4 Discussion, Perspectives and Related Works

We have stated and solved the problems of counting and uniform sampling of permutations with signature in a given regular language of signatures. The timed semantics of such a language is a particular case of regular timed languages (i.e. recognized by timed automata [1]). However, with the approach used, timed languages can be defined from any kind of languages of signatures. A challenging task for us is to treat the case of context free languages. For this we should use as in [2] volume of languages parametrized both by starting and ending states.

Our work can also benefit timed automata research. Indeed, we have proposed a uniform sampler for a particular class of timed languages. An ongoing work is to adapt this algorithm to all deterministic timed automata with bounded clocks using recursive equations of [3].

There is no mention of the parameter $x$ in Algorithm 1. It could be interesting to find a direct explanation of this algorithm (without using parameters). In any case, the parametric approach was crucial in the solution of the second problem.

Parametric approaches similar to ours are used in [7,13,17]. In particular recursive equations involving integrals are also described there. Technically, these approaches are based on order polytopes and yield integral operators of the form $\int_0^x$ and $\int_x^1$ while ours is based on chain polytopes and yields integral operators $\int_0^{1-x}$ and $\int_x^1$. The fact that these two operators are both null in $x = 1$ was very useful here. The main novelty is our use of Kleene like equations for regular timed languages and their volume functions (inspired by [2,3]) that allowed us to address the two problems for all regular languages of signatures.

## References

1. Alur, R., Dill, D.L.: A theory of timed automata. Theor. Comput. Sci. 126(2), 183–235 (1994)
2. Asarin, E., Basset, N., Degorre, A., Perrin, D.: Generating functions of timed languages. In: Rovan, B., Sassone, V., Widmayer, P. (eds.) MFCS 2012. LNCS, vol. 7464, pp. 124–135. Springer, Heidelberg (2012)
3. Asarin, E., Degorre, A.: Volume and entropy of regular timed languages: Analytic approach. In: Ouaknine, J., Vaandrager, F.W. (eds.) FORMATS 2009. LNCS, vol. 5813, pp. 13–27. Springer, Heidelberg (2009)
4. Basset, N.: Volumetry of timed languages and applications. PhD thesis, Université Paris-Est (2013)
5. Bernardi, O., Giménez, O.: A linear algorithm for the random sampling from regular languages. Algorithmica 62(1-2), 130–145 (2012)
6. Bouyer, P., Petit, A.: A Kleene/Büchi-like theorem for clock languages. Journal of Automata, Languages and Combinatorics 7(2), 167–186 (2002)
7. Ehrenborg, R., Jung, J.: Descent pattern avoidance. In: Advances in Applied Mathematics (2012)
8. Elizalde, S., Noy, M.: Consecutive patterns in permutations. Advances in Applied Mathematics 30(1), 110–125 (2003)
9. Flajolet, P., Sedgewick, R.: Analytic combinatorics. Camb. Univ. press (2009)
10. Flajolet, P., Zimmerman, P., Van Cutsem, B.: A calculus for the random generation of labelled combinatorial structures. Theoretical Computer Science 132(1), 1–35 (1994)
11. Hibi, T., Li, N.: Unimodular equivalence of order and chain polytopes. arXiv preprint arXiv:1208.4029 (2012)
12. Kitaev, S.: Patterns in permutations and words. Springer (2011)
13. Marchal, P.: Generating random permutations with a prescribed descent set. Presentation at Permutation Patterns (2013)
14. Nijenhuis, A., Wilf, H.S.: Combinatorial algorithms for computers and calculators. In: Computer Science and Applied Mathematics, 2nd edn., p. 1. Academic Press, New York (1978)
15. Stanley, R.P.: Two poset polytopes. Discrete & Computational Geometry 1(1), 9–23 (1986)
16. Stanley, R.P.: A survey of alternating permutations. In: Combinatorics and graphs. Contemp. Math., vol. 531, pp. 165–196. Amer. Math. Soc., Providence (2010)
17. Szpiro, G.G.: The number of permutations with a given signature, and the expectations of their elements. Discrete Mathematics 226(1), 423–430 (2001)

# Semantic Word Cloud Representations: Hardness and Approximation Algorithms

Lukas Barth[1], Sara Irina Fabrikant[2], Stephen G. Kobourov[3,*], Anna Lubiw[4], Martin Nöllenburg[1], Yoshio Okamoto[5,**], Sergey Pupyrev[3,9,*], Claudio Squarcella[6], Torsten Ueckerdt[7], and Alexander Wolff[8,***]

[1] Institute of Theoretical Informatics, Karlsruhe Institute of Technology
[2] Department of Geography, University of Zurich
[3] Department of Computer Science, University of Arizona
[4] David R. Cheriton School of Computer Science, University of Waterloo
[5] Dept. Comm. Engineering and Informatics, University of Electro-Communications
[6] Dipartimento di Ingegneria, Roma Tre University
[7] Department of Mathematics, Karlsruhe Institute of Technology
[8] Lehrstuhl für Informatik I, Universität Würzburg
[9] Institute of Mathematics and Computer Science, Ural Federal University

**Abstract.** We study a geometric representation problem, where we are given a set $\mathcal{B}$ of axis-aligned rectangles (boxes) with fixed dimensions and a graph with vertex set $\mathcal{B}$. The task is to place the rectangles without overlap such that two rectangles touch if the graph contains an edge between them. We call this problem CONTACT REPRESENTATION OF WORD NETWORKS (CROWN). It formalizes the geometric problem behind drawing word clouds in which semantically related words are close to each other. Here, we represent words by rectangles and semantic relationships by edges.

We show that CROWN is strongly NP-hard even if restricted to trees and weakly NP-hard if restricted to stars. We also consider the optimization problem MAX-CROWN where each adjacency induces a certain profit and the task is to maximize the sum of the profits. For this problem, we present constant-factor approximations for several graph classes, namely stars, trees, planar graphs, and graphs of bounded degree. Finally, we evaluate the algorithms experimentally and show that our best method improves upon the best existing heuristic by 45%.

## 1 Introduction

Word clouds and tag clouds are popular ways to visualize text. They provide an appealing way to summarize the content of a webpage, a research paper, or a political speech. Often such visualizations are used to contrast two documents; for example, word cloud visualizations of the speeches given by the candidates in the 2008 US Presidential elections were used to draw sharp contrast between them in the popular media.

---

* Supported in part by NSF grants CCF-1115971 and DEB 1053573.
** Supported by Grant-in-Aid for Scientific Research from Ministry of Education, Science and Culture, Japan, and Japan Society for the Promotion of Science (JSPS).
*** Supported by the ESF EuroGIGA project GraDR (DFG grant Wo 758/5-1).

A practical tool, Wordle [23], which is available on-line, offers high-quality design, graphics, style and functionality, but ignores relationships between words in the input. While some of the more recent word cloud visualization tools aim to incorporate semantics in the layout, none provides any guarantees about the quality of the layout in terms of semantics. We propose a mathematical model of the problem via a simple edge-weighted graph. The vertices in the graph are the words in the document. The edges in the graph correspond to semantic relatedness, with weights corresponding to the strength of the relation. Each vertex must be drawn as an axis-aligned rectangle (*box*, for short) with fixed dimensions. Usually, the dimensions will be determined by the size of the word in a certain font, and the font size will be related to the importance of the word. The goal is to "realize" as many edges as possible by contacts between their corresponding rectangles; see Fig. 1.

*Related Work.* Hierarchically clustered document collections are often visualized with self-organizing maps [15] and Voronoi treemaps [18]. The early word-cloud approaches did not explicitly use semantic information, such as word relatedness, when placing the words in the cloud. More recent approaches attempt to do so, as in ManiWordle [14] and in parallel tag clouds [4]. The most relevant approaches rely on force-directed graph visualization methods [5] and a seam-carving image processing method together with a force-directed heuristic [24]. The semantics-preserving word cloud problem is related to classic graph layout problems, where the goal is to draw graphs so that vertex labels are readable and Euclidean distances between pairs of vertices are proportional to the underlying graph distance between them. Typically, however, vertices are treated as points and label overlap removal is a post-processing step [7,11].

In *rectangle representations* of graphs, vertices are axis-aligned rectangles with non-intersecting interiors and edges correspond to rectangles with non-zero length common boundary. Every graph that can be represented this way is planar and every triangle in such a graph is a facial triangle; these two conditions are also sufficient to guarantee a rectangle representation [22]. In a recent survey, Felsner [9] reviews many rectangulation variants, including squarings. Algorithms for area-preserving rectangular cartograms are also related [21]. Area-universal rectangular representations where vertex weights are represented by area have been characterized [8] and edge-universal representations, where edge weights are represented by length of contacts have been studied [19]. Unlike cartograms, in our setting there is no inherent geography, and hence, words can be positioned anywhere. Moreover, each word has fixed dimensions enforced by its frequency in the input text, rather than just fixed area.

*Our Contribution.* The input to the problem variants we consider is a (multi)set $\mathcal{B}$ of axis-aligned boxes $B_1, \ldots, B_n$ with fixed positive dimensions and an edge-weighted undirected graph $G = (\mathcal{B}, E)$, called the *profit graph*. Box $B_i$ has an associated *size* $(w_i, h_i)$, where $w_i$ and $h_i$ are its width and height. For some of our results, boxes may be rotated by $90°$, which means exchanging $w_i$ and $h_i$. By scaling appropriately, we may always assume that all heights and widths are positive integers. The vertex set of $G$ is $\mathcal{B}$. Every edge $(B_i, B_j) \in E$ has a positive weight $p_{ij}$, called its *profit*, representing the gain for making boxes $B_i$ and $B_j$ touch. A *representation* of $\mathcal{B}$ is a map that associates with each box a position in the plane so that no two boxes overlap. A *contact* between

**Fig. 1.** A hierarchical word cloud for complexity classes. A class is above another class when the former contains the latter. The font size is the square root of millions of Google hits for the corresponding word. This is an instance of the problem variant HIER-CROWN.

two boxes is a maximal line segment *of positive length* in the boundary of both. If two boxes are in contact, we say that they *touch*. If two boxes touch and one lies above the other, we call this a *vertical contact*. We define *horizontal contact* symmetrically. We say that a representation *realizes* an edge $(B_i, B_j) \in E$ if $B_i$ and $B_j$ touch. Finally, we define the *total profit* of a representation to be the sum of profits over all edges of $G$ that the representation realizes. Our problems and results are as follows.

**Contact Representation of Word Networks (CROWN)** is to decide whether there exists a representation of the given boxes that realizes all edges of the profit graph. This is equivalent to deciding whether there is a representation whose contact graph contains the profit graph as a subgraph. If such a representation exists, we say that it *realizes the profit graph* and that the given instance of CROWN is *realizable*.

We show that CROWN is strongly NP-hard even if restricted to trees and weakly NP-hard if restricted to stars; see Theorem 1. We also consider two variants of the problem that can be solved efficiently. First we present a linear-time algorithm for CROWN on so-called irreducible triangulations; see Section 2.1. Then we turn to the problem variant **HIER-CROWN**, where the profit graph is a single-source directed acyclic graph with fixed plane embedding, and the task is to find a representation in which each edge corresponds to a vertical contact directed upwards; see Fig. 1. We solve this variant efficiently; see Section 2.2.

**MAX-CROWN** is the optimization version of CROWN where the task is to find a box representation maximizing the total profit. We present constant-factor approximation algorithms for stars, trees, and planar graphs, and a $2/\lfloor \Delta+1 \rfloor$-approximation for graphs of maximum degree $\Delta$; see Section 3. We have implemented two approximation algorithms and evaluated them experimentally in comparison to three existing algorithms (two of which are semantics-aware). Based on a dataset of 120 Wikipedia documents our best method outperforms the best previous method by more than 45%; see Section 5. We also consider an extremal version of the MAX-CROWN problem and show that the complete graph $K_n$ ($n \geq 7$) with unit profits can always be realized with total profit $2n - 2$, which is sometimes the best possible; see Section 3.2.

**AREA-CROWN** is as follows: Given a realizable instance of CROWN, find a representation that realizes the profit graph and minimizes the area of the representation's bounding box. We show that the problem is NP-hard even if restricted to paths; see Section 4.

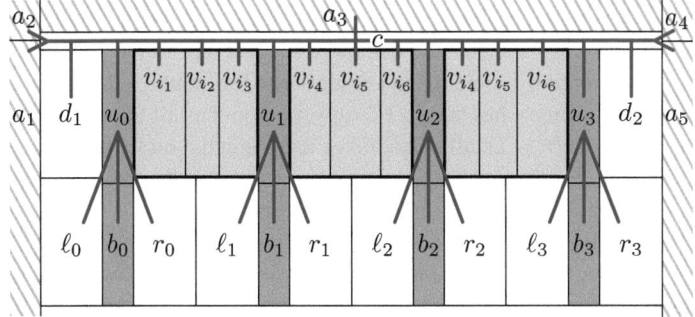

**Fig. 2.** Given an instance $S$ of 3-PARTITION, we construct a tree $T_S$ (thick red line segments) and define boxes such that $T_S$ has a realization if and only if $S$ is feasible

## 2 The CROWN Problem

In this section, we investigate the complexity of CROWN for several graph classes.

**Theorem 1.** CROWN *is (strongly) NP-hard. The problem remains strongly NP-hard even if restricted to trees and is weakly NP-hard if restricted to stars.*

*Proof.* To show that CROWN on stars is weakly NP-hard, we reduce from the weakly NP-hard problem PARTITION, which asks whether a given multiset of $n$ positive integers $a_1, \ldots, a_n$ that sum to $B$ can be partitioned into two subsets, each of sum $B/2$. We construct a star graph whose central vertex corresponds to a $(B/2, \delta)$-box (for some $0 < \delta < \min_i a_i$). We add four leaves corresponding to $(B, B)$-squares and, for $i = 1, \ldots, n$, a leaf corresponding to an $(a_i, a_i)$-square. It is easy to verify that there is a realization for this instance of CROWN if and only if the set can be partitioned.

To show that CROWN is (strongly) NP-hard, we reduce from 3-PARTITION: Given a multiset $S$ of $n = 3m$ integers with $\sum S = mB$, is there a partition of $S$ into $m$ subsets $S_1, \ldots, S_m$ such that $\sum S_i = B$ for each $i$? It is known that 3-PARTITION is NP-hard even if, for every $s \in S$, we have $B/4 < s < B/2$, which implies that each of the subsets $S_1, \ldots, S_m$ must contain exactly three elements [12].

Given an instance $S = \{s_1, s_2, \ldots, s_n\}$ of 3-PARTITION as described above, we define a tree $T_S$ on $n + 4(m-1) + 7$ vertices as in Fig. 2 (for $n = 9$ and $m = 3$). Let $K = (m+1)B + m + 1$. We make a vertex $c$ of size $(K, 1/2)$. For each $i = 1, \ldots, n$, we make a vertex $v_i$ of size $(s_i, B)$. Let $\varepsilon \in (0, B/2)$. For each $j = 0, \ldots, m$, we make a vertex $u_j$ of size $(1, B + \varepsilon)$, a vertex $b_j$ of size $(1, B - \varepsilon)$, and vertices $\ell_j$ and $r_j$ of size $(B/2, B)$. Finally, we make vertices $a_1, \ldots, a_5$ of size $(K, K)$, and vertices $d_1$ and $d_2$ of size $(B/2, B)$. The tree $T_S$ is as shown by the thick lines in Fig. 2: vertex $c$ is adjacent to all the $v_i$'s, $u_j$'s, $a$'s, and $d$'s; and each vertex $u_j$ is adjacent to $b_j$, $\ell_j$, and $r_j$.

We claim that an instance $S$ of 3-PARTITION is feasible if and only if $T_S$ can be realized with the given box sizes. It is easy to see that $T_S$ can be realized if $S$ is feasible: we simply partition vertices $v_1, \ldots, v_n$ into groups of three (by vertices $u_0, \ldots, u_m$) in the same way as their widths $s_1, \ldots, s_n$ are partitioned in groups of three; see Fig. 2.

For the other direction, consider any realization of $T_S$. Let us refer to the box of some vertex $v$ also as $v$. Since $c$ touches the five large squares $a_1, \ldots, a_5$, at least three sides of $c$ are partially covered by some $a_k$ and at least one horizontal side of $c$ is completely covered by some $a_k$. Since $c$ has height 1/2 only, but touches all the $v_i$'s and $u_j$'s and $d_1$ and $d_2$ (each of height $B > 1$), all these boxes must touch $c$ on its free horizontal side, say, the bottom side. Furthermore, the sum of the widths of the boxes exactly matches the width of $c$; so they must pack side by side in some order.

This means that the only free boundary of $u_j$ is at the bottom, and $u_j$ must make contact there with $b_j$, $\ell_j$, and $r_j$. This is only possible if $b_j$ is placed directly beneath $u_j$, and $\ell_j$ and $r_j$ make contact with the bottom corners of $u_j$. (They need not appear to the left and right as shown in Fig. 2.) Because the sum of the widths of the $b_j$'s, $\ell_j$'s, and $r_j$'s exactly matches the width of $c$, they must pack side by side, and therefore the $u_j$'s are spaced distance $B$ apart. There is a gap of width $B/2$ before the first $u_j$ and after the last $u_j$. These gaps are too wide for one box in $v_1, \ldots, v_n$ and too small for two of them since their widths are contained in the *open* interval $(B/4, B/2)$. Therefore, the boxes $d_1$ and $d_2$ must occupy these gaps, and the boxes $v_1, \ldots, v_n$ are packed into $m$ groups each of width $B$, as required. □

In case rectangles may be rotated, both proofs still hold: the weak NP-hardness proof for stars still works because all boxes are squares–except the central one. The strong NP-hardness for trees also still holds, basically because the boxes $a_1, \ldots, a_5$ are squares and because all boxes (except $c$) are at least as high as wide (and at least as wide as $c$ is high), so there is no advantage to rotating any box.

Although CROWN is NP-hard in general, on some graph classes the problem can be solved efficiently. In the remainder of this section, we investigate such a class: irreducible triangulations. We also consider a restricted variant of CROWN: HIER-CROWN.

### 2.1 The CROWN Problem on Irreducible Triangulations

A box representation is called a *rectangular dual* if the union of all rectangles is again a rectangle whose boundary is formed by exactly four rectangles. A graph $G$ admits a rectangular dual if and only if $G$ is planar, internally triangulated, has a quadrangular outer face and does not contain separating triangles [2]. Such graphs are known as *irreducible triangulations*. The four outer vertices of an irreducible triangulation are denoted by $v_N, v_E, v_S, v_W$ in clockwise order around the outer quadrangle. An irreducible triangulation $G$ may have exponentially many rectangular duals. Any rectangular dual of $G$, however, can be built up by placing one rectangle at a time, always keeping the union of the placed rectangles in staircase shape.

**Theorem 2.** CROWN *on irreducible triangulations can be solved in linear time.*

*Proof (sketch).* The algorithm greedily builds up the box representation, similarly to an algorithm for edge-proportional rectangular duals [19]. We define a *concavity* as a point on the boundary of the so-far constructed representation, which is a bottom-right or top-left corner of some rectangle. Start with a vertical and a horizontal ray emerging from the same point $p$, as placeholders for the right side of $v_W$ and the top side of $v_S$, respectively. Then at each step consider a concavity, with $p$ as the initial one.

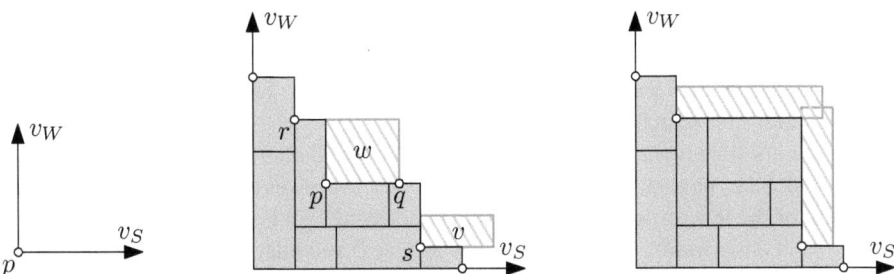

**Fig. 3.** Left: starting configuration with rays $v_S$ and $v_W$. Center: representation at an intermediate step: vertex $w$ fits into concavity $p$ and results in a staircase, vertex $v$ fits into concavity $s$ but does not result in a staircase. Adding box $w$ to the representation introduces a new concavity $q$ and allows wider boxes to be placed at $r$. Right: no box can be placed, so the algorithm terminates.

Since each concavity $p$ is contained in exactly two rectangles, there exists a unique rectangle $R_p$ that is yet to be placed and has to touch both these rectangles. If by adding $R_p$ we still have a staircase shape representation, then we do so. If no such rectangle can be added, we conclude that $G$ is not realizable; see Fig. 3. The complete proof is in the full version [1]. □

### 2.2 The HIER-CROWN Problem

The HIER-CROWN problem is a restricted variant of the CROWN problem that can be used to create word clouds with a hierarchical structure; see Fig. 1. The input is a directed acyclic graph $G$ with only one sink and with a plane embedding. The task is to find a representation that *hierarchically realizes* $G$, meaning that for each directed edge $(v, u)$ in $G$ the top of the box $v$ is in contact with the bottom of the box $u$.

If the embedding of $G$ is not fixed, the problem is NP-hard even for a tree, by an easy adaptation of the proof of Theorem 1. (Remove the vertices $a_2, a_3, a_4$, and orient the remaining edges of $T_S$ upward according to the representation shown in Fig. 2.) However, if we fix the embedding of the profit graph $G$, then HIER-CROWN can be solved efficiently.

**Theorem 3.** HIER-CROWN *can be solved in polynomial time.*

*Proof.* Let $G$ be the given profit graph, with vertex set $\mathcal{B} = \{B_1, \ldots, B_n\}$, where $B_i$ has height $h_i$ and width $w_i$, and $B_1$ is the unique sink. We first check that the orientation and embedding of $G$ are compatible, that is, that incoming edges and outgoing edges are consecutive in the cyclic order around each vertex.

The main idea is to set up a system of linear equations for the $x$- and $y$-coordinates of the sides of the boxes. Let variables $t_i$ and $b_i$ represent the $y$-coordinates of the top and bottom of $B_i$ respectively, and variables $\ell_i$ and $r_i$ represent the $x$-coordinates of the left and right of $B_i$, respectively. For each $i = 1, \ldots, n$, impose the linear constraints $t_i = b_i + h_i$ and $r_i = \ell_i + w_i$. For each directed edge $(B_i, B_j)$, impose the constraints

$t_i = b_j, r_i \geq \ell_j + 1/2$, and $r_j \geq \ell_i + 1/2$. The last two constraints force $B_i$ and $B_j$ to share some $x$-range of positive length in which they touch. Initialize $t_1 = 0$.

With these equations, variables $t_i$ and $b_i$ are completely determined since every box $B_i$ has a directed path to $B_1$. Furthermore, the values for $t_i$ and $b_i$ can be found using a depth-first-search of $G$ starting from $B_1$.

The $x$-coordinates are not yet determined and depend on the horizontal order of the boxes, which can be established as follows. We scan the boxes from top to bottom, keeping track of the left-to-right order of boxes intersected by a horizontal line that sweeps from $y = 0$ downwards. Initially the line is at $y = 0$ and intersects only $B_1$. When the line reaches the bottom of a box $B$, we replace $B$ in the left-to-right order by all its predecessors in $G$, using the order given by the plane embedding. In case multiple boxes end at the same $y$-coordinate, we make the update for all of them. Whenever boxes $B_a$ and $B_b$ appear consecutively in the left-to-right order, we impose the constraint $r_a \leq \ell_b$.

The scan can be performed in $O(n \log n)$ time using a priority queue to determine which boxes in the current left-to-right order have maximum $b_i$ value. The resulting system of equations has size $O(n)$ (because the constraints correspond to edges of a planar graph). It is straightforward to verify that the system of equations has a solution if and only if there is a representation of the boxes that hierarchically realizes $G$. The constraints define a linear program (LP) that can be solved efficiently. (A feasible solution can be found faster than with an LP, but we omit the details in this paper.) □

We can show that HIER-CROWN becomes weakly NP-complete if rectangles may be rotated, by a simple reduction from SUBSET SUM (for details, see the full version [1]).

## 3 The MAX-CROWN Problem

In this section, we study approximation algorithms for MAX-CROWN and consider an extremal variant of the problem.

### 3.1 Approximation Algorithms

We present approximation algorithms for MAX-CROWN restricted to certain graph classes. Our basic building blocks are an approximation algorithm for stars and an exact algorithm for cycles. Our general technique is to find a collection of disjoint stars or cycles in a graph. We begin with stars, using a reduction to the MAXIMUM GENERALIZED ASSIGNMENT PROBLEM (GAP) defined as follows: Given a set of bins with capacity constraints and a set of items that may have different sizes and values in each bin, pack a maximum-value subset of items into the bins. It is known that the problem is NP-hard (KNAPSACK and BIN PACKING are special cases of GAP), and there exists a $(1 - 1/e)$-approximation algorithm [10]. In the remainder, we assume that there is an $\alpha$-approximation algorithm for GAP, setting $\alpha = 1 - 1/e > 0.632$.

**Theorem 4.** *There exists an $\alpha$-approximation algorithm for* MAX-CROWN *on stars.*

*Proof.* We can solve instances with $n < 5$ by brute force exactly, so let's assume that $n \geq 5$. Let $B_1$ denote the box corresponding to the center of the star. Given an optimal

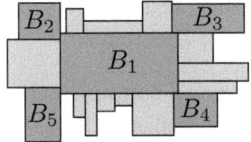

**Fig. 4.** A solution to an instance of MAX-CROWN whose profit graph is a star with center $B_1$

solution to MAX-CROWN, we can modify it (by sliding boxes along the sides of $B_1$) such that there are four boxes $B_2, B_3, B_4, B_5$ whose contact with $B_1$ has length exactly $1/2$. In particular, each of these boxes touches exactly one corner of $B_1$. The problem reduces to choosing four corner boxes and the way they touch $B_1$, and assigning the remaining boxes to one of the sides of $B_1$; see Fig. 4.

Each corner of $B_1$ can be touched in two different ways, via its incident horizontal or vertical sides. Depending on the way the corners of $B_1$ are touched, we create the following instance of GAP. We introduce eight bins, one bin for each side of $B_1$ with appropriately adjusted sizes and one bin of size 1 for each corner. For $i = 2, \ldots, n$, the value of item $B_i$ is the profit of the edge $(B_1, B_i)$. The size of $B_i$ is $w_i$ for each horizontal bin, $h_i$ for each vertical bin, and 1 for each corner bin. For each of the 16 ways the corners of $B_1$ can be touched, we apply an $\alpha$-approximation algorithm for GAP [10]. Hence, we obtain an $\alpha$-approximation for MAX-CROWN. □

In the case where rectangles may be rotated by $90°$, the MAX-CROWN problem on a star reduces to an easier problem, the MULTIPLE KNAPSACK PROBLEM, where every item has the same size and value no matter which bin it is placed in. This is because, for non-corner bins, we will always attach a rectangle $B$ to the central rectangle of the star using the smaller dimension of $B$. For the corner bins, we can try all possible choices of which box to put into which bin. There is a PTAS for MULTIPLE KNAPSACK [3]. Therefore, there is a PTAS for MAX-CROWN on stars if we may rotate rectangles.

A *star forest* is a disjoint union of stars. Theorem 4 applies to a star forest since we can combine the solutions for the disjoint stars.

**Theorem 5.** MAX-CROWN *on the class of graphs that can be partitioned in polynomial time into $k$ star forests admits an $\alpha/k$-approximation algorithm.*

*Proof.* We partition the edges of the profit graph into $k$ star forests, apply the approximation algorithm of Theorem 4 to each of them, and take the best of the $k$ solutions. We claim that this (polynomial-time) method yields the desired approximation factor.

Consider an optimum solution, and let $W_{\text{opt}}$ be its profit. By the pigeon-hole principle, our partition of the profit graph contains a star forest $F$ that realizes a profit of at least $W_{\text{opt}}/k$ in the optimum solution. Hence, on $F$, the approximation algorithm of Theorem 4 achieves a profit of at least $\alpha W_{\text{opt}}/k$. □

**Corollary 1.** MAX-CROWN *admits an $\alpha/2$-approximation algorithm on trees and an $\alpha/5$-approximation algorithm on planar graphs.*

*Proof.* It is easy to partition any tree into two star forests in linear time. Moreover, it is known that every planar graph has star arboricity at most 5, that is, it can be partitioned into at most five star forests, and such a partition can be found in polynomial time [13]. The results now follow directly from Theorem 5. □

Our algorithms involve approximating a number of GAP instances, using the LP-based algorithm of Fleischer et al. [10]. Because of this, the runtime of our approximation algorithms is dominated by the runtime of solving linear programs.

Our star forest partition method is possibly not optimal. Nguyen et al. [17] show how to find a star forest of an arbitrary weighted graph carrying at least half of the profits of an optimal star forest in polynomial time. We cannot, however, guarantee that the approximation of the optimal star forest carries a positive fraction of the total profit in an optimal solution to MAX-CROWN. Hence, approximating MAX-CROWN for general graphs remains an open problem. As a first step into this direction, we present a constant-factor approximation for profit graphs with bounded maximum degree. First we need the following lemma.

**Lemma 1.** *Given $n \geq 3$ boxes and an n-cycle defined on them, we can find a representation realizing the n-cycle in linear time.*

*Proof.* Let $C = (B_1, \ldots, B_n)$ be the given cycle, let $W$ be the sum of all the widths, that is, $W = \sum_i w_i$, and let $t$ be the maximum index such that $\sum_{i \leq t} w_i < W/2$. We place $B_1, \ldots, B_t$ in this order side by side from left to right with their bottom sides on a horizontal line $h$; see Fig. 5. We call this the *top channel*. Starting at the same point on $h$, we place $B_n, B_{n-1}, \ldots, B_{t+2}$ in this order side by side from left to right with their top sides on $h$. We call this the *bottom channel*. Note that $B_1$ and $B_n$ are in contact. It remains to place $B_{t+1}$ in contact with $B_t$ and $B_{t+2}$. It is easy to see that the following works: add $B_{t+1}$ to the channel of minimum width or, in case of a tie, place $B_t$ straddling the line $h$; see Fig. 5. □

Following the idea of Theorem 5, we can approximate MAX-CROWN by applying Lemma 1 to a partition of the profit graph into sets of disjoint cycles.

**Theorem 6.** MAX-CROWN *on the class of graphs that can be partitioned into $k$ sets of disjoint cycles (in polynomial time) admits a (polynomial-time) algorithm that achieves total profit at least $\frac{1}{k} \sum_{i \neq j} p_{ij}$. In particular, there is a $1/k$-approximation algorithm for* MAX-CROWN *on this graph class.*

**Corollary 2.** MAX-CROWN *on graphs of maximum degree $\Delta$ admits a $2/\lfloor \Delta + 1 \rfloor$-approximation.*

*Proof.* As Peterson [20] shows, the edges of any graph of maximum degree $\Delta$ can be covered by $\lceil \Delta/2 \rceil$ sets of cycles and paths, and such sets can be found in polynomial time. The result now follows from Theorem 6. □

### 3.2 An Extremal MAX-CROWN Problem

In the following, we bound the maximum number of contacts that can be made when placing $n$ boxes. It is easy to see that for $n = 2, 3, 4$ any set of boxes allows $2n - 3$ contacts. For larger $n$ we have:

**Theorem 7.** *For $n \geq 7$ and any set of $n$ boxes, the boxes can be placed in the plane to realize $2n - 2$ contacts. For some sets of boxes this is the best possible.*

*Proof.* Let $B_1, \ldots, B_n$ be any set of boxes. First we place $k \in \{5, 6, 7\}$ boxes to make $2(k - 1)$ contacts, and then place the remaining boxes to make 2 contacts each for a total of $2(k - 1) + 2(n - k) = 2n - 2$ contacts. Let $B_1$ and $B_2$ be the boxes with largest height, and $B_3$ and $B_4$ be the boxes with largest width. Let $B_5$ be any further box. Place the five boxes as in Fig. 6. This realizes 8 contacts, unless one or two of $B_1, B_2$ has the same height as $B_5$, in which case we consider one or two further boxes $B_6, B_7$ and represent in total 10 or 12 contacts as in Fig. 6.

Place the remaining boxes one by one as in the proof of Lemma 1 along the horizontal line between $B_2$ and $B_3$. Then each remaining box makes two new contacts.

Next we describe a set of $n$ boxes for which the maximum number of contacts is $2n - 2$. Let $B_i$ be a square box of side length $2^i$. Consider any placement of the boxes and partition the contacts into the set of horizontal contacts and the set of vertical contacts. From the side lengths of the boxes, it follows that neither set of contacts contains a cycle. Thus each set of contacts has size at most $n - 1$ for a total of $2n - 2$. □

## 4 The AREA-CROWN Problem

The same profit graph can often be realized by different box representations, not all of which are equally useful or visually appealing when viewed as word clouds. In this section we consider the AREA-CROWN problem and show that finding a "compact" representation that fits into a small bounding box is another NP-hard problem.

The reduction is from the (strongly) NP-hard problem 2D BOX PACKING: The input is a set $\mathcal{R}$ of $n$ rectangles with width and height functions $w \colon \mathcal{R} \to \mathbb{N}$ and $h \colon \mathcal{R} \to \mathbb{N}$, and a box of width $W$ and height $H$. All the input numbers are bounded by some polynomial in $n$. The task is to pack the given rectangles into the box. The problem is known to be NP-complete even if the box is a square, that is, if $W = H$ [16].

The BOX PACKING problem is equivalent to AREA-CROWN when the profit graph has no edges. Edges in the profit graph, however, impose additional constraints on the representation, which may make AREA-CROWN easier for certain (simple) profit graph classes. In the full version [1], we show that this is not the case.

**Theorem 8.** AREA-CROWN *is (strongly) NP-hard even on paths.*

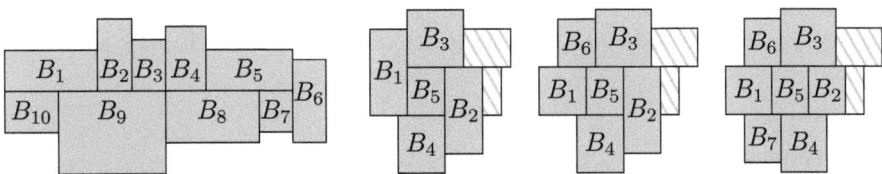

**Fig. 5.** Example for Lemma 1: Realizing the 10-cycle $(B_1, \ldots, B_{10})$

**Fig. 6.** Examples for Theorem 7: 8, 9, or 10 adjacencies with 5, 6, or 7 boxes, respectively

## 5 Experimental Results

We implemented two new methods for constructing word clouds: the STAR FOREST algorithm based on extracting star forests (Corollary 1) and the CYCLE COVER algorithm based on decomposing edges of a graph into cycle covers (Theorem 6). We compared the two algorithms to the following existing methods: WORDLE [23], CPDWCV [5], and SEAM CARVING [24]. Our dataset consists of 120 Wikipedia documents, each with 400 words or more. Frome these, we removed stop words (e.g., "the") and constructed profit graphs $G_{50}$ and $G_{100}$ for the 50 and 100 most frequent words, respectively. We set profits using the so-called *Latent Semantic Analysis* [6] based on the co-occurrence of these words within the same sentence. For details, see the full version [1].

We compare the percentage of realized profit in the box representations. Since STAR FOREST handles planar profit graphs, we first extracted maximal planar subgraphs of the profit graphs and then applied the algorithm the the planar subgraphs. The percentage of realized profit is presented in the table below. Our results indicate that, in terms of the realized profit, CYCLE COVER outperforms existing approaches, realizing more than 17% (13%) of the total profit of graphs with 50 (100) vertices, that is, 45% (55%) more than the second best known heuristic, CPDWCV. On the other hand, existing algorithms may perform better in terms of compactness, aspect ratio, and other aesthetic criteria; we leave a deeper comparison of word cloud algorithms to the future.

| Algorithm | Realized Profit of $G_{50}$ | Realized Profit of $G_{100}$ |
|---|---|---|
| WORDLE [23] | 3.4% | 2.2% |
| CPDWCV [5] | 12.2% | 8.9% |
| SEAM CARVING [24] | 7.4% | 5.2% |
| STAR FOREST | 11.4% | 8.2% |
| CYCLE COVER | 17.8% | 13.8% |

## 6 Conclusions and Future Work

We formulated the Contact Representation of Word Networks (CROWN) problem, motivated by the desire to provide theoretical guarantees for semantics-preserving word cloud visualization. We showed that some variants of CROWN are NP-hard, gave efficient algorithms for others, and presented approximation algorithms. A natural open problem is to find an approximation algorithm for general graphs with arbitrary profits.

**Acknowledgments.** This work began at Dagstuhl Seminar 12261. We thank organizers and participants (in particular Therese Biedl), as well as Steve Chaplick and Günter Rote.

## References

1. Barth, L., Fabrikant, S.I., Kobourov, S., Lubiw, A., Nöllenburg, M., Okamoto, Y., Pupyrev, S., Squarcella, C., Ueckerdt, T., Wolff, A.: Semantic word cloud representations: Hardness and approximation algorithms. Arxiv report arxiv.org/abs/1311.4778 (2013)

2. Buchsbaum, A.L., Gansner, E.R., Procopiuc, C.M., Venkatasubramanian, S.: Rectangular layouts and contact graphs. ACM Trans. Algorithms 4(1) (2008)
3. Chekuri, C., Khanna, S.: A polynomial time approximation scheme for the multiple knapsack problem. SIAM J. Comput. 35(3), 713–728 (2005)
4. Collins, C., Viégas, F.B., Wattenberg, M.: Parallel tag clouds to explore and analyze faceted text corpora. In: Proc. IEEE Symp. Vis. Analytics Sci. Tech., pp. 91–98 (2009)
5. Cui, W., Wu, Y., Liu, S., Wei, F., Zhou, M., Qu, H.: Context-preserving dynamic word cloud visualization. IEEE Comput. Graphics Appl. 30(6), 42–53 (2010)
6. Dumais, S.T.: Latent semantic analysis. Annu. Rev. Inform. Sci. Tech. 38(1), 188–230 (2004)
7. Dwyer, T., Marriott, K., Stuckey, P.J.: Fast node overlap removal. In: Healy, P., Nikolov, N.S. (eds.) GD 2005. LNCS, vol. 3843, pp. 153–164. Springer, Heidelberg (2006)
8. Eppstein, D., Mumford, E., Speckmann, B., Verbeek, K.: Area-universal and constrained rectangular layouts. SIAM J. Comput. 41(3), 537–564 (2012)
9. Felsner, S.: Rectangle and square representations of planar graphs. In: Pach, J. (ed.) Thirty Essays on Geometric Graph Theory, pp. 213–248. Springer, Heidelberg (2013)
10. Fleischer, L., Goemans, M.X., Mirrokni, V.S., Sviridenko, M.: Tight approximation algorithms for maximum separable assignment problems. Math. Oper. Res. 36(3), 416–431 (2011)
11. Gansner, E.R., Hu, Y.: Efficient, proximity-preserving node overlap removal. J. Graph Algortihms Appl. 14(1), 53–74 (2010)
12. Garey, M.R., Johnson, D.S.: Computers and Intractability: A Guide to the Theory of NP-Completeness. W. H. Freeman & Co., New York (1979)
13. Hakimi, S.L., Mitchem, J., Schmeichel, E.F.: Star arboricity of graphs. Discrete Math. 149(1-3), 93–98 (1996)
14. Koh, K., Lee, B., Kim, B.H., Seo, J.: Maniwordle: Providing flexible control over Wordle. IEEE Trans. Vis. Comput. Graph. 16(6), 1190–1197 (2010)
15. Lagus, K., Honkela, T., Kaski, S., Kohonen, T.: Self-organizing maps of document collections: A new approach to interactive exploration. In: Simoudis, E., Han, J., Fayyad, U.M. (eds.) KDD 1996, pp. 238–243. AAAI Press (1996)
16. Leung, J.Y.T., Tam, T.W., Wong, C., Young, G.H., Chin, F.Y.: Packing squares into a square. J. Parallel Distrib. Comput. 10(3), 271–275 (1990)
17. Nguyen, C.T., Shen, J., Hou, M., Sheng, L., Miller, W., Zhang, L.: Approximating the spanning star forest problem and its application to genomic sequence alignment. SIAM J. Comput. 38(3), 946–962 (2008)
18. Nocaj, A., Brandes, U.: Organizing search results with a reference map. IEEE Trans. Vis. Comput. Graphics 18(12), 2546–2555 (2012)
19. Nöllenburg, M., Prutkin, R., Rutter, I.: Edge-weighted contact representations of planar graphs. J. Graph Algorithms Appl. 17(4), 441–473 (2013)
20. Petersen, J.: Die Theorie der regulären Graphen. Acta Mathematica 15(1), 193–220 (1891)
21. Raisz, E.: The rectangular statistical cartogram. Geogr. Review 24(3), 292–296 (1934)
22. Thomassen, C.: Interval representations of planar graphs. J. Combin. Theory, Ser. B 40(1), 9–20 (1986)
23. Viégas, F.B., Wattenberg, M., Feinberg, J.: Participatory visualization with Wordle. IEEE Trans. Vis. Comput. Graphics 15(6), 1137–1144 (2009)
24. Wu, Y., Provan, T., Wei, F., Liu, S., Ma, K.L.: Semantic-preserving word clouds by seam carving. Comput. Graphics Forum 30(3), 741–750 (2011)

# The Complexity of Homomorphisms of Signed Graphs and Signed Constraint Satisfaction

Florent Foucaud[1] and Reza Naserasr[2]

[1] Universitat Politècnica de Catalunya, Barcelona, Spain
University of Johannesburg, Auckland Park, South Africa
PSL, Université Paris-Dauphine, LAMSADE - CNRS UMR 7243, France
florent.foucaud@gmail.com
[2] CNRS, Université Paris-Sud 11, LRI - CNRS UMR 8623, Orsay, France
reza@lri.fr

**Abstract.** A signed graph $(G, \Sigma)$ is an undirected graph $G$ together with an assignment of signs (positive or negative) to all its edges, where $\Sigma$ denotes the set of negative edges. Two signatures are said to be equivalent if one can be obtained from the other by a sequence of resignings (i.e. switching the sign of all edges incident to a given vertex). Extending the notion of usual graph homomorphisms, homomorphisms of signed graphs were introduced, and have lead to some extensions and strengthenings in the theory of graph colorings and homomorphisms. We study the complexity of deciding whether a given signed graph admits a homomorphism to a fixed target signed graph $[H, \Sigma]$, i.e. the $(H, \Sigma)$-COLORING problem. We prove a dichotomy result for the class of all $(C_k, \Sigma)$-COLORING problems (where $C_k$ is a cycle of length $k \geq 3$): $(C_k, \Sigma)$-COLORING is NP-complete, unless both $k$ and the size of $\Sigma$ are even. We conjecture that this dichotomy can be extended to all signed graphs in a natural way. We also introduce the more general concept of signed constraint satisfaction problems and show that a dichotomy for such problems is equivalent to the statement of the Feder-Vardi Dichotomy Conjecture.

## 1 Introduction

The Four Color Theorem (4CT), stating that every planar graph is 4-colorable, is considered to be one of the central theorems in graph theory and, considering its simple statement in the form of a map coloring theorem, attracts a wide audience. One can reason the hidden beauty of this theorem in scientific ways based on the following classic theorems:

**Theorem 1.** *Deciding if a given graph is 4-colorable is NP-complete.*

**Theorem 2.** *Deciding if a given planar graph is 3-colorable is NP-complete.*

The latter indicates that the class of planar graphs (though recognizable in linear time) is a rich class of graphs, but the 4CT shows that 4-colorability for this rich class of graphs can be easily decided (simply answer YES all the time). This is in contrast with the former theorem.

Despite being such a powerful theorem, the 4CT witnesses a special weakness. While it is very easily decidable if a graph is 2-colorable (i.e., bipartite), the 4CT proves no bound on the chromatic number of such a graph. A more complicated case is when an edge of a planar graph is replaced with a large (complete) bipartite graph. Such an operation does not change the chromatic number of the graph, but makes it far from being planar. Attempts to strengthen the 4CT so that it provides some bound in such cases has developed the theory of *signed graphs*. Coloring of graphs with signed graphs as forbidden minors have been studied, see for example Odd Hadwiger's conjecture (we refer to [8] for some recent developments), an extension of the well-known Hadwiger conjecture. Only recently, the development of the theory of homomorphisms of signed graphs has begun, see [10,14]. This paper is the first study of the *complexity* of signed graph homomorphisms. This work is also strongly related to the celebrated Dichotomy Conjecture of Feder and Vardi [7]. We proceed with some notation.

Given a graph $G$, a *signature* is an assignment of signs $+$ and $-$ to the edges of $G$. It is normally denoted by the set $\Sigma$ of negative edges (the others being positive). Given a graph together with a signature, a *resigning* at a vertex $v$ is to change the sign of all edges incident to $v$. Two signatures $\Sigma_1$ and $\Sigma_2$ are said to be equivalent if one can be obtained from the other by a sequence of resignings — equivalently, by changing the signs at the edges of an edge-cut of $G$. This is an equivalence relation on the class of all signatures of a graph. A *signed graph* is defined to be a graph together with a class of equivalent signatures. It will normally be denoted by $[G, \Sigma]$ where $\Sigma$ is any member of the equivalence class of signatures. When we want to emphasize on a specific signature, say $\Sigma_1$, then we will write $(G, \Sigma_1)$. Note that one can easily check in polynomial time whether two signatures are equivalent using a reduction to 2-SAT or using Theorem 4:

**Proposition 3.** *Let $G$ be a graph, and let $\Sigma$ and $\Sigma'$ be two signatures of $S$. One can decide in polynomial time whether $\Sigma \equiv \Sigma'$.*

An important notion here is the one of *balance* of a cycle. A cycle with even number of negative edges is called *balanced cycle* and the ones with odd number of negative edges are *unbalanced cycles*. The set of balanced or unbalanced cycles of a signed graph uniquely determines the equivalent class of signatures by the following theorem of Zaslavsky.

**Theorem 4 (Zaslavsky [15]).** *Given two signatures $\Sigma$ and $\Sigma'$ on a graphs we have $\Sigma \equiv \Sigma'$ if and only if the set of balanced (unbalanced) cycles are the same.*

A *minor* of signed graph $(G, \Sigma)$ is a signed graph $(H, \Sigma')$ which is obtained from $(G, \Sigma)$ by a sequence of the following operations: *i.* deleting vertices or edges, *ii.* contracting a positive edge (that is to identify two end vertices and delete loops) and *iii.* resigning. The last operation implies that notion of minor for $(G, \Sigma)$ is the same as that of $[G, \Sigma]$. Using this notion, a strengthening of the 4CT (which corresponds to one of the first cases of Odd Hadwiger's conjecture) was announced by Guenin in 2005 [9]: If $(G, E(G))$ has no $(K_5, E(K_5))$-minor, then $G$ is 4-colorable. Moreover, it follows from a recent work [6] that deciding if $(G, E(G))$ has a $(K_5, E(K_5))$-minor is polynomial-time solvable.

A classic way of extending the theory of graph colorings is through graph homomorphisms. The extension to signed graphs, introduced in [10] is given below. Given two signed graphs $[G, \Sigma]$ and $[H, \Sigma_1]$ we say there is a *homomorphism* of $[G, \Sigma]$ to $[H, \Sigma_1]$, and write $[G, \Sigma] \to [H, \Sigma_1]$, if there is a mapping $\phi$ of $V(G)$ to $V(H)$ such that *i.* $\phi$ preserves the adjacency (i.e., $xy \in E(G) \Rightarrow \phi(x)\phi(y) \in E(H)$, and *ii.* with respect to some choice of signature $\Sigma' \equiv \Sigma$, $\phi$ also preserves the signs. Since the existence of a homomorphism does not depend on the signature of the target graph, we may write it as homomorphism to $(H, \Sigma_1)$.

By considering signed graphs where all edges are of the same sign, we observe that graph homomorphisms are a special case of signed graph homomorphisms. From a complexity point of view, the following is then the first natural question to ask in the theory of signed graph homomorphisms.

$(H, \Sigma_1)$-COLORING
INSTANCE: A signed graph $[G, \Sigma]$.
QUESTION: Does $[G, \Sigma] \to (H, \Sigma_1)$?

The celebrated dichotomy result by Hell and Nešetřil [11] states that for any (non-signed) graph $H$, $H$-COLORING is polynomial-time if $H$ is bipartite (in which case it becomes equivalent to checking 2-colorability), and NP-complete otherwise. As an extension of this result, we believe that there is a also dichotomy in the signed case, i.e., that for any given $(H, \Sigma_1)$, either $(H, \Sigma_1)$-COLORING is polynomial-time solvable, or it is NP-complete. In fact, we believe that the problem is NP-complete unless $H$ is bipartite and $(H, \Sigma_1)$ has no unbalanced cycle (in which case $\Sigma_1 \equiv \emptyset$ and thus the problem becomes again equivalent to checking 2-colorability). As we mentioned earlier, when $\Sigma \equiv \emptyset$ or when $\Sigma \equiv E(G)$, the problem is reduced to simple graph homomorphisms and the dichotomy holds.

We point out that the ability of resigning gives the signed homomorphism problem a different flavor than classical homomorphism problems. When we do not allow resigning, we would get the concept of two-edge-colored graph homomorphisms [2], whose complexity was studied in [3]. Already for two-edge-colored cycles, a complexity classification is difficult to obtain, but the problem is significantly different; for example, the case of a 4-cycle with three blue edges and one red edge is polynomial-time solvable, see [5]. This is in contrast with the signed graph case, as we will see in Section 2.

**Our Results and Structure of the Paper.** We begin by proving a dichotomy for the set of $(C_k, \Sigma)$-COLORING problems in Section 2, where $C_k$ denotes a cycle on $k$ vertices. In Section 3, we discuss the case where the target is a signed *bipartite* graph. Indeed it is known that this case already captures all usual graph homomorphism problems, making it a good candidate for an interesting subclass to study. We also give a few more examples of signed graphs for which the corresponding homomorphism problem is NP-complete. In Section 4, we define a natural extension of signed graph homomorphisms to *signed relational structures*. We then prove that a complexity dichotomy for the class of signed constraint satisfaction problems exists if and only if Feder-Vardi's celebrated dichotomy conjecture holds. The paper is concluded in Section 5.

## 2 Mapping to Signed Cycles

In this section, we determine the complexity of $(H, \Sigma)$-COLORING when $H$ is a fixed cycle on $k$ vertices, $C_k$. Observe that there are only two signed graphs based on a cycle $C_k$: a balanced cycle, denoted $BC_k$ and which is equivalent to $(C_k, \emptyset)$, and the unbalanced cycle, denoted $UC_k$, which is equivalent to $(C_k, \{e\})$ where $e$ is any edge of $C_k$. Furthermore, for odd values of $k$, the unbalanced cycle $UC_k$ is also equivalent to $(C_k, E(C_k))$. Thus for odd values of $k$, mapping signed graphs to $BC_k$ or $UC_k$ is equivalent to mapping graphs to the odd cycle $C_k$, hence by Hell-Nešetřil's theorem [11] it is an NP-complete problem. For $BC_k$ with even values of $k$, the problem is equivalent to 2-coloring of graphs, thus it is polynomial-time solvable. The case that remains to study is $UC_k$ with even values of $k$. In this section, we prove that this is an NP-hard problem (even if the underlying graph of the input signed graph is of maximum degree 6).

**Theorem 5.** $UC_{2k}$-COLORING *is NP-complete for any* $k \geq 2$, *even when restricted to signed (bipartite) graphs of maximum degree 6.*

To prove Theorem 5, we give a reduction from MONOTONE NOT-ALL-EQUAL-3SAT,[1] which is NP-complete [13]:

MONOTONE NOT-ALL-EQUAL-3SAT
INSTANCE: A set $\mathcal{C}$ of monotone size-3-clauses from a set $X$ of boolean variables.
QUESTION: Is there a boolean assignment of the variables of $X$ such that each clause contains at least one false and one true variable?

The main idea of our proof is that it uses the resigning of specific vertices as indication for the truth assignment of the corresponding variables.

*Proof.* Without loss of generality, when mapping to $UC_{2k}$, we assume $V(UC_{2k}) = \{1, \ldots, 2k\}$ and $UC_{2k}$ has only edge $\{12\}$ in its signature (see Figure 1).

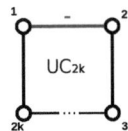

**Fig. 1.** The target graph $UC_{2k}$

Given a formula $F = \{C_1, \ldots, C_m\}$ over variable set $X$, we construct the signed graph $(G_F, \Sigma_F)$ as follows. For each clause $C = \{x_1, x_2, x_3\}$ of $F$ we

---
[1] *Monotone*: there are no negated variables.

construct a clause gadget $(G_C, \Sigma_C)$: $G_C$ has a *central vertex*, $c$; it contains three edge-disjoint copies $U_1, U_2, U_3$ of $UC_{2k}$, meeting at vertex $c$ only. For each copy $U_i$, the unique vertex that is at distance $k$ of $c$ is denoted $x_i$, and corresponds to variable $x_i$ in clause $C$. For each $x_i$, we have a distinct path $P_i$ of length $k-1$ whose first end is identified with $x_i$, its other end being denoted $y_i$. Finally, let $U_4$ be a new cycle as follows: if $k=2$, $U_4$ is a copy of $UC_6$. Otherwise, $U_4$ is a copy of $UC_{2k}$. We place $U_4$ such that it goes through $y_1, y_2$ and $y_3$ in such a way that the distances on $U_4$ between each pair of vertices $y_i, y_j$ are even, and two of them are equal to $\lfloor \frac{2k}{6} \rfloor$ (i.e., the three distances are $\{2\ell, 2\ell, 2\ell\}$ if $U_4$ has length $6\ell$, $\{2\ell, 2\ell, 2\ell+2\}$ if $U_4$ has length $6\ell+2$, and $\{2\ell, 2\ell, 2\ell+4\}$ if $U_4$ has length $6\ell+4$). Finally, we assign the following signature $\Sigma_C$ to $G_C$: $U_1, U_2, U_3$ contain exactly one negative edge each (this edge being incident to $c$); each path $P_i$ has exactly one negative edge (the one incident to $y_i$), and there is another negative edge incident to $y_i$ (the one of $U_4$ that lies on the path from $y_i$ to $y_{(i \bmod 3)+1}$). For $k=2$, the gadget is depicted in Figure 2; otherwise, see Figure 3(a).

Now, to build $(G_F, \Sigma_F)$, we consider all clause gadgets corresponding to distinct clauses and identify all vertices of type $c$ with each other. Vertices representing the same variable are identified with each other as well. $\Sigma_F$ is the union of all signatures $\Sigma_C$.

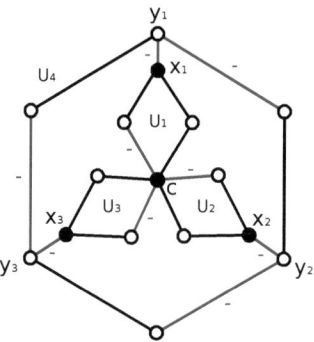

**Fig. 2.** Clause gadget for $UC_4$

We now show that $(G_F, \Sigma_F)$ maps to $UC_{2k}$ if and only $F$ is satisfiable. In the first part of the proof, no restriction on the maximum degree is shown; we explain afterward how to prove that part of the claim.

For the first part, assume that $F$ is satisfiable. We give a mapping $f$ from $(G_F, \Sigma_F)$ to $UC_{2k}$ with the corresponding signature $\Sigma$ with $\Sigma_F \equiv \Sigma$.

Consider a truth assignment $A$ of $F$. We resign each vertex $x_i$ of $(G_F, \Sigma_F)$ such that the corresponding variable $x_i$ is true in $A$, and do not resign the vertices corresponding to a false variable. We also do not resign vertex $c$. Now, since $A$ is a satisfying truth assignment, either one or two variables are true in each clause. One can see that in any clause, it is possible to resign the remaining vertices as to obtain (up to symmetry) the signature with exactly four negative edges: one

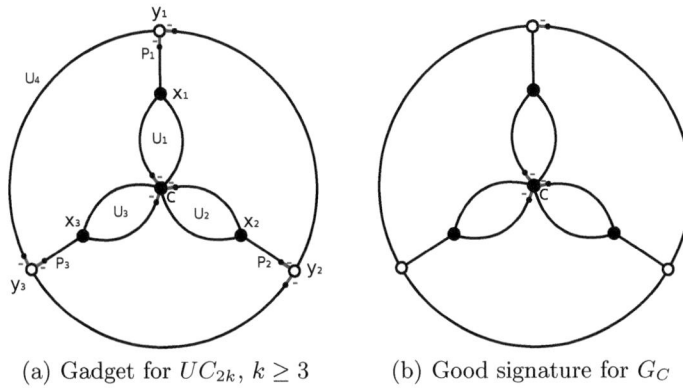

(a) Gadget for $UC_{2k}$, $k \geq 3$  (b) Good signature for $G_C$

**Fig. 3.** Clause gadget $G_C$ for larger cycles

in each cycle $U_i$ ($1 \leq i \leq 3$) being incident to $c$, and one in $U_4$ incident to some $y_j$ ($1 \leq j \leq 3$), see Figure 3(b). Let $\Sigma$ be the union of all these clause gadget signatures, and let us now construct the mapping $f$.

We map vertex $c$ to vertex 1 in $UC_{2k}$, and each vertex $x_i$ is mapped to vertex $k+1$. Observe that in $\Sigma$ and in each clause gadget, exactly one vertex among $y_1, y_2, y_3$ has an incident negative edge. Without loss of generality, we assume it to be $y_1$, the other cases follow by symmetry.

If $k = 2$, we map $y_1$ to vertex 2, whereas $y_2, y_3$ are both mapped to vertex 4. It is now easy to extend the mapping.

If $k \geq 3$, we map $y_1$ to vertex 2 (the $k$ vertices $x_1, \ldots, y_1$ of path $P_1$ are mapped to $k+1, k, \ldots, 2$). Let $\ell = \lfloor 2k/6 \rfloor$. We now distinguish two cases.

On the one hand, if vertices $y_1, y_2$ and $y_3$ are pairwise at distance $2\ell$ on the cycle $U_4$ (i.e. $U_4$ has length $2k = 6\ell$), we map vertex $y_2$ to vertex $2\ell + 2$, and vertex $y_3$ to vertex $4\ell + 2$. Note that the parity of the length of each $P_i$ ($k-1$) is the same as the parity of the distance $d$ between vertices $k+1$ and $2\ell + 2$ or $4\ell + 2$ in $UC_{2k}$, with $d < k$. It is now easy to complete the mapping.

On the other hand, if $y_1$ is at distance $2\ell + 2$ of say $y_2$, $2k = 6\ell + 2$ and $y_1$ is at distance $2\ell$ of $y_3$ (respectively, $2k = 6\ell + 4$ and $y_1$ is at distance $2\ell + 4$ of $y_2$), we map $y_2$ to vertex $2\ell + 4$ and $y_3$ to vertex $4\ell + 4$ (resp. $2\ell + 6$ and $4\ell + 6$). Again it is now easy to complete the mapping.

For the other part, suppose that $(G_F, \Sigma_F)$ maps to $UC_{2k}$. Let $f$ be the mapping, and $\Sigma$ the signature that corresponds to $f$. Without loss of generality, we can assume that vertex $c$ of $G_F$ maps to vertex 1 of $UC_{2k}$ (if not, since $C_{2k}$ is vertex-transitive, it is easily seen that we could resign $(G_F, \Sigma)$ in an appropriate manner so that this would hold for some other signature).

We claim that when obtaining $\Sigma$ from $\Sigma_F$, for each clause $C = \{x_1, x_2, x_3\}$, either one or two of the vertices $x_1, x_2, x_3$ of $G_C$ have to be resigned. In this case, setting to TRUE each variable $x_i$ such that the corresponding vertex $x_i$ has been resigned, and to FALSE otherwise, would yield a truth assignment satisfying $F$.

Observe first that the three cycles starting at vertex $c$ are unbalanced and of length $2k$. Hence, they have to map to $UC_{2k}$ in a surjective way, and each vertex $x_i$ maps to vertex $k+1$ of $UC_{2k}$. Hence, the path joining vertex $x_i$ to vertex $y_i$ has to map to a path of $UC_{2k}$ having only positive edges, because the distance between $x_i$ and $y_i$ is exactly $k-1$. Therefore, vertex $x_i$ is resigned if and only if vertex $y_i$ is not resigned. Indeed, if $y_i$ is not resigned, the edge incident to $y_i$ on the path from $y_i$ to $x_i$ remains negative. Since each edge of this path maps to a positive edge, all vertices of the path, including $x_i$, must be resigned. The other side follows from the same argument applied to the other end of the path.

We now claim that either one or two of the vertices $y_1, y_2, y_3$ of $G_C$ have to be resigned, which will complete the proof of this part.

If $k=2$, $y_1$, $y_2$ and $y_3$ lie on an unbalanced 6-cycle which hence has to map to $UC_4$ in a surjective way. The only way that this is possible is to map a path of length 3 of the 6-cycle to an edge of $UC_4$. Assume, by contradiction, that all three vertices $y_1, y_2, y_3$ have been resigned. Then, no matter how the resigning is done on the other three vertices, it is not possible to proceed to the mapping of any such path of length 3 to an edge, since the three edges of this path should all have the same sign. The case where none of them is resigned follows by symmetry because the resulting signatures on the 6-cycle are symmetric.

Now, if $k \geq 3$, $U_4$ has to map in a surjective way to $UC_{2k}$, and in the final signature $\Sigma$, exactly one of its edges must be negative. Assume that either none, or all three vertices $y_1, y_2, y_3$ have been resigned. Then the signature along $U_4$ does not change (up to symmetry): each of the paths $y_1, \ldots, y_2$, $y_2, \ldots, y_3$, $y_1, \ldots, y_3$ contains exactly one negative edge. But now, any resigning of the remaining vertices will lead to at least one negative edge on each of these three paths, a contradiction.

It now remains to prove how to restrict the maximum degree of our construction. Observe that in the above reduction, the reason for having a high maximum degree is that we identify all vertices of type $c$ and all vertices $x_i$ with each other. Instead of doing so, we can use a *replicator gadget of length* $\ell$, consisting of a sequence of $i$ unbalanced $2k$-cycles $V_1, \ldots, V_i$, where each cycle $V_i$ has vertex set $\{v_i^1, \ldots, v_i^{2k}\}$ and an edge between two consecutive vertices on this cyclic order. Each edge $\{v_i^1, v_i^2\}$ is negative. Moreover, for each $1 \leq i \leq \ell-1$, $V_i$ and $V_{i+1}$ share their edge $\{1, 2k\}$ when $i$ is odd, and the edge $\{k, k+1\}$ otherwise. An illustration is given in Figure 4.

Now, observe that in order to map a replicator gadget of length $\ell$ to $UC_{2k}$ with the signature of Figure 1, for each fixed $j$ ($1 \leq j \leq 2k$), all vertices $v_i^j$, $1 \leq i \leq \ell$ have to be identified with each other. Moreover, it can be easily checked that either all vertices of the gadget have to be resigned, or none. Now, consider the construction of $G_F$ described in the first part of the proof. Let $x_i \in X$, and let $\ell_i$ be the number of occurrence of variable $x_i$ in $F$. Instead of identifying all vertices $x_i$ with each other, we take a copy $R_i$ of the replicator gadget of length $2\ell_i$. Now, for the $j$'th clause $C$ containing $x_i$, we identify vertex $x_i$ of $G_C$ with vertex $v_{2j-1}^{k+1}$ of $R_i$, as indicated in Figure 1. Moreover, we take an additional copy $R_c$ of the replicator gadget of length $6|F|$, where $|F|$ is the number of clauses in

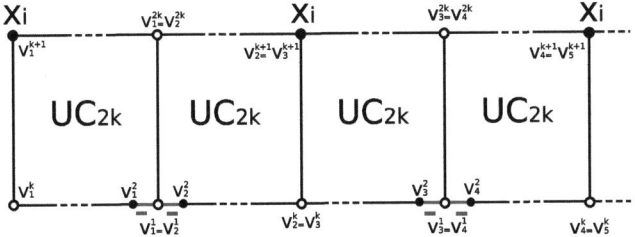

**Fig. 4.** Replicator gadget

$F$. For each clause $C_j$, we split vertex $c$ of $G_{C_j}$ into three non-adjacent vertices $c_1, c_2, c_3$, each one being part of one of the cycles $U_1, U_2, U_3$, and we identify vertex $c_i$ ($1 \leq i \leq 3$) with vertex $v^1_{2(j+i-1)}$. Observe that the created graph has maximum degree 6 (vertices $x_i$ possibly having six neighbours: three in the clause gadget, and three in the replicator gadget).

By the properties of the replicator gadget, in a mapping from $G_F$ to $UC_{2k}$, all vertices $c_i$ will be mapped to vertex 1 of $UC_{2k}$, and vertices $x_i$, to vertex $k+1$, as in the original construction. Moreover, every vertex $x_i$ is resigned if and only if every other vertex $x_i$ is resigned, and the same holds for all vertices of type $c$. Hence the same proof as earlier applies. □

## 3 Further Cases and Signed Bipartite Graphs

There are two special classes of signed graphs: signed graphs where, in some representation of signature, all edges are negative and signed bipartite graphs. These two are exactly the class of signed graphs in which all balanced cycles are even and all unbalanced cycles have a same parity. A homomorphism problem to a signed graph of former type is simply a graph homomorphism problem as all edges must be negative and then resigning does not play a role. In contrast, for the latter family, normally it is the choice of right signature that is the most difficult. However this case this case is already more difficult than graph homomorphism and graph coloring problems. It is shown in [14] that the concept of homomorphisms of signed *bipartite* graphs captures both the notion of homomorphisms of graphs and the concept of the chromatic number using the following construction. These theorems are stated based on the following construction of signed graphs from graphs: given a graph $G$, the signed bipartite graph $S(G)$ is obtained by replacing each edge $uv$ of $G$ by an unbalanced 4-cycle on four vertices $ux_{uv}vy_{uv}$, where $x_{uv}$ and $y_{uv}$ are new and distinct vertices. The following two theorems are then proved.

**Theorem 6 (Naserasr, Rollová, Sopena [14]).** *For any graph $G$, $\chi(G) \leq k$ if and only if $S(G) \to (K_{k,k}, M)$, where $M$ is a perfect matching of $K_{k,k}$.*

**Theorem 7 (Naserasr, Rollová, Sopena [14]).** *For every pair $G, H$ of graphs, $G \to H$ if and only if $S(G) \to S(H)$.*

Note that if $[G, \Sigma] \to (H, \Sigma')$, then, in particular, $G \to H$. Therefore, if $H$ is a bipartite graph, this mapping would imply that $G$ is also a bipartite graph. Hence for bipartite signed graphs, the complexity of $([H, \Sigma]$-COLORING is determined by its complexity when reduced to signed bipartite input graphs. The above mentioned theorems then imply that $(H, \Sigma)$-COLORING is NP-complete whenever $(H, \Sigma)$ is $(K_{k,k}, M)$ for $k \geq 3$, or if $(H, \Sigma)$ is equivalent to $S(G)$ for any non-bipartite graph $G$.

Moreover, using Theorem 5, one can build more examples of signed graphs for which the homomorphism problem is NP-hard:

**Theorem 8.** $(K_4, \{e\})$-COLORING *is NP-complete, where e is any edge of* $K_4$.

*Proof.* Let $x, y, z, t$ be the four vertices of $K_4$ and assume $e = xy$. Let $UC_4$ be a signed cycle on $x, y, z$ and $t$ where $xy$ is a negative edge and $yz$, $zt$, $tx$ are positive edges (thus an unbalanced 4-cycle). We claim that a signed bipartite graph $[G, \Sigma]$ maps to $(K_4, \{e\})$ if and only if it maps to $UC_4$. Since $UC_4$ is a subgraph of $[K_4, \{e\}]$, one direction is trivial. For the other direction, let $A, B$ be the bipartition of $G$ and let $\phi$ be the mapping of $[G, \Sigma]$ to $(K_4, \{e\})$ and suppose the mapping preserves the signs with respect to $\Sigma$. We define a new mapping $\phi'$ which will be a homomorphism of $[G, \Sigma]$ to $UC_4$. For each vertex $u$ in $A$, if $\phi$ maps it to $x$ or $y$, then $\phi'$ maps it to $x$, and if $\phi$ maps it to $z$ or $t$, then $\phi'$ maps it to $z$. Similarly, for each vertex $v$ in $B$, if $\phi$ maps it to $x$ or $y$, then $\phi'$ maps it to $y$, and if $\phi$ maps it to $z$ or $t$, then $\phi'$ maps it to $t$. It can now be easily checked that $\phi'$ is a homomorphism of $[G, \Sigma]$ to $UC_4$ with respect to $\Sigma$. □

## 4 Signed Constraint Satisfaction Problems

A (finite) *relational structure* $T$ is a domain of elements, denoted $V(T)$, together with a finite set of relations $R_1, \ldots, R_k$, each relation $R_i$ ($1 \leq i \leq k$) having arity $a_i$ (that is, $R_i \subseteq V(T)^{a_i}$). An element of a relation $R_i$ is called a *tuple*. This is an extension of the notion of graphs and digraphs, as a graph is a relational structure with one binary and symmetric relation. When the symmetry is not forced we have the notion of digraphs.

The notion of homomorphisms of graphs and digraphs can then be generalized to relational structures as follows: given two relational structures $S$ and $T$ over the same number of relations $R_1, \ldots, R_k$ and the same (ordered) set of arities $a_1, \ldots, a_k$ a *homomorphism* of $S$ to $T$ is a mapping $\phi : V(S) \to V(T)$ such that if $X \in V(S)^{a_i}$ belongs to $R_i$ in $S$, then the ordered set $\phi(X) = \{\phi(x), x \in X\}$ belongs to $R_i$ in $T$. We will write $S \to T$ whenever there exists a homomorphism of $S$ to $T$.

For every fixed relational structure $T$, we have the following associated decision problem, called $T$-CSP.

$T$-CSP
INSTANCE: A relational structure $S$.
QUESTION: Does $S$ admit a homomorphism to $T$?

It is a folklore fact that this notion captures some well-known problems such as various versions of SAT-problems. The class of all constraint satisfaction problems is denoted by *CSP*. The complexity of CSPs has been extensively studied, see e.g. the book [1]. One of the major open problems in the area is commonly known as the Dichotomy Conjecture, and has motivated many works such as e.g. [4,12]:

*Conjecture 9 (Dichotomy Conjecture, Feder and Vardi [7]).* For every fixed relational structure $T$, $T$-CSP is either polynomial or NP-complete.

Our aim in this section is to introduce the extended notion of signed relational structures and the related decision problem. We then show that a dichotomy for this class of problems is equivalent to the Dichotomy Conjecture.

A *signed relational structure* $(S, \Sigma)$ is a relational structure $S$ with a subset $\Sigma$ of the set of all tuples in $S$ (regardless of which relation the tuple belongs to). We say that the tuples in $\Sigma$ are *negative*, and the others are *positive*. Given an element $x$ in $S$, the resigning operation at $x$ switches the signs of all tuples in $S$ containing $x$. As for signed graphs, resigning defines an equivalence relation $\equiv$ between all signatures of $S$: two signatures are equivalent if and only if they can be obtained from the other by a sequence of resignings. Relational structure with a class of equivalent signatures is then denoted by $[S, \Sigma]$.

We say that there is an homomorphism of $[S, \Sigma_1]$ to $[T, \Sigma]$ (or, equivalently, to $(T, \Sigma)$) if there is a signature $\Sigma'_1 \equiv \Sigma_1$ and a homomorphism $f : S \to T$ which preserves the signs of tuples according to $\Sigma'_1$ and $\Sigma$.

Given a signed relational structure $(T, \Sigma)$, we define $(T, \Sigma)$-CSP analogously to $(H, \Sigma)$-COLORING:

$(T, \Sigma)$-CSP
INSTANCE: A signed relational structure $[S, \Sigma_1]$.
QUESTION: Is there a homomorphism of $[S, \Sigma_1]$ to $(T, \Sigma)$?

We call the class of all signed constraint satisfaction problems *S-CSP*. We note that, by considering $\Sigma = \emptyset$ the class of signed constraint satisfaction problems contains the class of usual constraint satisfaction problems. Thus, a dichotomy for S-CSP would imply a dichotomy for CSP, i.e., the Dichotomy Conjecture. Our aim here is to show that the inverse is also true. This follows from another result of Feder and Vardi explained after the following definitions.

The class MMSNP, short for Monotone Monadic Strict NP, is the class of decision problems whose set of positive instances can be described in existential second-order logic with a universal first-order part (that is, having no existential quantifier). In other words, they can be described as the set of instances satisfying a formula of the form $F(S) := \exists S', \forall X, \Phi(X, S, S')$, where $S$ is the instance relational structure, $S'$ is a relational structure $S'$ with $V(S) = V(S')$ (intuitively, $S'$ is the "proof" for $F(S)$ to be true), $X$ is a subset of elements in $V(S)$, and $\Phi$ is a first-order formula that only uses negations, disjunctions, conjunctions, relations of $S$ and $S'$, and the equality operator. Moreover:

- each relation of $S$ must appear in an odd number of negated subformulas in $\Phi$ (monotonicity);
- the relations in $S'$ are only defined over sets of variables of $S$, not over relations of $S$ (monadicity);
- negation cannot be applied to the equality operator (no inequality).

It can be verified that each problem in CSP is also in MMSNP. Feder and Vardi, after introducing their Dichotomy Conjecture, proved that a dichotomy for CSP implies a dichotomy or MMSNP.

**Theorem 10 (Feder and Vardi [7]).** *These three statements are equivalent:*
*(i) MMSNP has a dichotomy;*
*(ii) CSP has a dichotomy;*
*(iii) the set of digraph homomorphism problems has a dichotomy.*

Here we show that, furthermore, each S-CSP problem is also in MMSNP. Thus, while S-CSP includes CSP, it is included in MMSNP and, therefore:

**Theorem 11.** *CSP has a dichotomy if and only if S-CSP has a dichotomy.*

*Proof.* As we mentioned, it is enough to prove that each problem in S-CSP belongs to the class MMSNP.

First, consider a usual $T$-CSP problem, when there is no signature. For each instance $S$, the problem of deciding whether $S \to T$ can be express by a formula $F(S)$ in MMSNP: see [7] for details.

Now consider $(T, \Sigma)$-CSP and let $(S, \Sigma_1)$ be an input signed structure. To express $(S, \Sigma_1) \to (T, \Sigma)$ with a similar formula $F(S, \Sigma_1) := \exists S', \forall X, \Phi(X, S, S')$, one has to add that there is an assignment $s : V(S) \to \{0, 1\}$ (which encodes the set of resigned elements, and can be expressed as a unary relation in $S'$). Moreover, for each subset $X = \{x_1, \ldots, x_k\}$ of variables of $S$, not only $R_i(X)$ implies $R_i(f(X))$, but now also $(f(X), i) \in \Sigma$ implies that either $(X, i) \in \Sigma_1$ and an even number of elements in $X$ have been resigned, or $(X, i) \notin \Sigma_1$ and an odd number of elements in $X$ have been resigned. We give an example when $S, T$ have a unique binary relation $R$. The "proof" structure $S'$ is a relational structure over $V(S)$ with a unary relation $A_x$ for every element $x$ in $V(T)$ (it encodes the assignment $V(S) \to V(T)$), and a unary relation $R_S$ (which encodes the set of resigned elements in $V(S)$).

$F(S, \Sigma_1) := \exists S', \forall (x_1, x_2) \in R,$

$((R(x_1, x_2) \wedge (x_1, x_2) \in \Sigma_1 \wedge R_S(x_1) \wedge R_S(x_2)) \Rightarrow ((A_x(x_1) \wedge A_y(x_2)) \vee *))$

[*: enumerate all assignments allowed by $T$ and belonging to $\Sigma$]

$\bigwedge \ldots$ [repeat for each possibility for $(x_1, x_2) \in \Sigma_1$ and resigning $x_1$ and $x_2$]

This formula is indeed monotone, monadic and without inequality. In fact, the only difference with a usual CSP is the case distinction according to which of $x_1, x_2$ have been resigned; hence the signature can simply be treated as additional constraints. □

## 5 Conclusion

As a first study of the complexity of signed graph homomorphisms, we have proved a dichotomy for signed cycles, by showing that $(C_k, \Sigma)$-COLORING is NP-complete if and only if $(C_k, \Sigma)$ is both even and balanced. We have discussed some further cases, in particular, the case of signed *bipartite* graphs. As a natural generalization, the notion of signed constraint satisfaction problems and the corresponding class S-CSP were introduced, with S-CSP lying in between of the classes CSP and MMSNP. While it is a difficult problem to prove a dichotomy for S-CSP (equivalently, for CSP or for digraph homomorphism problems), it will be of interest to prove a dichotomy for signed graph homomorphism problems or other special cases of signed CSPs. By extending the classes in CSP/MMSNP for which a dichotomy is known, this would bring some new insight to the Dichotomy Conjecture, and therefore it is a promising direction of research.

## References

1. Rossi, F., van Beek, P., Walsh, T. (eds.): Handbook of Constraint Programming. Elsevier (2006)
2. Alon, N., Marshall, T.H.: Homomorphisms of edge-colored graphs and Coxeter groups. J. Algebr. Combin. 8(1), 5–13 (1998)
3. Brewster, R.C., Hell, P.: On homomorphisms to edge-coloured cycles. Electr. Notes Discrete Math. 5, 46–49 (2000)
4. Bulatov, A.A.: A dichotomy constraint on a three-element set. In: Proc. 43rd IEEE Symposium on Theory of Computing, pp. 649–658 (2002)
5. Charpentier, C., Naserasr, R., Sopena, E.: Analogue of Jeager-Zhang conjecture for signed bipartite graphs (manuscript)
6. Demaine, E., Hajiaghayi, M., Kawarabayashi, K.-I.: Decomposition, Approximation, and Coloring of Odd-Minor-Free Graphs. In: Proc. SODA 2010, pp. 329–344 (2010)
7. Feder, T., Vardi, M.Y.: The Computational structure of monotone monadic SNP and constraint catisfaction: a study through datalog and group theory. SIAM J. Comput. 28(1), 57–104 (1998)
8. Geelen, J., Gerards, B., Reed, B., Seymour, P., Vetta, A.: On the odd-minor variant of Hadwigers conjecture. J. Combin. Theor. Series B 99, 20–29 (2009)
9. Guenin, B.: Graphs without odd-K5 minors are 4-colourable. Talk at Oberwolfach Seminar on Graph Theory (January 2005)
10. Guenin, B.: Packing odd circuit covers: a conjecture (2005) (unpublished manuscript)
11. Hell, P., Nešetřil, J.: On the complexity of H-coloring. J. Combin. Theor. Series B 48(1), 92–110 (1990)
12. Hell, P., Nešetřil, J., Zhu, X.: Complexity of tree homomorphisms. Discrete Appl. Math. 70, 23–36 (1996)
13. Moret, B.M.E.: The Theory of Computation, ch. 7, Problem 7.1, Part 2. Addison-Wesley (1998)
14. Naserasr, R., Rollová, E., Sopena, E.: Homomorphisms of signed graphs (submitted manuscript)
15. Zaslavsky, T.: Signed graphs. Discrete Appl. Math. 4(1), 47–74 (1982)

# Complexity of Coloring Graphs without Paths and Cycles

Pavol Hell and Shenwei Huang

School of Computing Science
Simon Fraser University, Burnaby B.C., V5A 1S6, Canada
{pavol,shenweih}@sfu.ca

**Abstract.** Let $P_t$ and $C_\ell$ denote a path on $t$ vertices and a cycle on $\ell$ vertices, respectively. In this paper we study the $k$-COLORING problem for $(P_t, C_\ell)$-free graphs. It has been shown by Golovach, Paulusma, and Song that when $\ell = 4$ all these problems can be solved in polynomial time. By contrast, we show that in most other cases the $k$-COLORING problem for $(P_t, C_\ell)$-free graphs is NP-complete. Specifically, for $\ell = 5$ we show that $k$-COLORING is NP-complete for $(P_t, C_5)$-free graphs when $k \geq 4$ and $t \geq 7$; for $\ell \geq 6$ we show that $k$-COLORING is NP-complete for $(P_t, C_\ell)$-free graphs when $k \geq 5$, $t \geq 6$; and additionally, we prove that 4-COLORING is NP-complete for $(P_t, C_\ell)$-free graphs when $t \geq 7$ and $\ell \geq 6$ with $\ell \neq 7$, and that 4-COLORING is NP-complete for $(P_t, C_\ell)$-free graphs when $t \geq 9$ and $\ell \geq 6$ with $\ell \neq 9$. It is known that, generally speaking, for large $k$ the $k$-COLORING problem tends to remain NP-complete when one forbids an induced path $P_t$ with large $t$. Our findings mean that forbidding an additional induced cycle $C_\ell$ (with $\ell > 4$) does not help. We also revisit the problem of $k$-COLORING $(P_t, C_4)$-free graphs, in the case $t = 6$. (For $t = 5$ the $k$-COLORING problem is known to be polynomial even on just $P_5$-free graphs, for every $k$.) The algorithms of Golovach, Paulusma, and Song are not practical as they depend on Ramsey-type results, and end up using tree-decompositions with very high widths. We develop more practical algorithms for 3-COLORING and 4-COLORING on $(P_6, C_4)$-free graphs. Our algorithms run in linear time if a clique cutset decomposition of the input graph is given. Moreover, our algorithms are certifying algorithms. We provide a finite list of all minimal non-$k$-colorable $(P_6, C_4)$-free graphs, for $k = 3$ and $k = 4$. Our algorithms output one of these minimal obstructions whenever a $k$-coloring is not found. In fact, we prove that there are only finitely many minimal non-$k$-colorable $(P_6, C_4)$-free graphs for any fixed $k$; however, we do not have the explicit lists for higher $k$, and thus no certifying algorithms. (We note there are infinitely many non-$k$-colorable $P_5$-free, and hence $P_6$-free, graphs for any given $k \geq 4$, according to a result of Hoàng, Moore, Recoskie, Sawada, and Vatshelle.)

## 1 Introduction

We say that $G$ is $\mathcal{H}$-free if it does not contain, as an induced subgraph, any graph $H \in \mathcal{H}$. If $\mathcal{H} = \{H\}$ or $\mathcal{H} = \{H_1, H_2\}$, we say that $G$ is $H$-free or $(H_1, H_2)$-free.

Given any positive integer $t$, let $P_t$ and $C_t$ be the path and cycle on $t$ vertices, respectively. The *neighborhood* of $x \in V$ is denoted by $N(x)$. Given a set $S$ of vertices of $G$, we let $N_S(x) = N(x) \cap S$. The *degree* of $x$ is denoted by $d(x)$ and we denote $\delta(G)$ by the minimum degree of $G$. For two disjoint vertex subsets $X$ and $Y$ we say that $X$ is *complete*, respectively *anti-complete*, to $Y$ if every vertex in $X$ is adjacent, respectively non-adjacent, to every vertex in $Y$.

The problem $k$-COLORING asks, for an input graph $G$, whether $G$ admits a $k$-coloring. A graph $G$ is called a *minimal obstruction* for $k$-COLORING if $G$ is not $k$-colorable but any proper induced subgraph of $G$ is $k$-colorable. We also call $G$ a *minimal non-$k$-colorable* graph. A minimal non-$(k-1)$-colorable graph is also called a *$k$-critical* graph. A graph is *critical* if it is $k$-critical for some $k$. We shall use $n$ and $m$ to denote the number of vertices and edges of $G$, respectively. Since $k$-COLORING is known to be NP-complete for any fixed $k \geq 3$, there has been considerable interest in studying the complexity of $k$-COLORING restricted to various graph classes. It is well known for instance that $k$-COLORING is polynomially solvable for perfect graphs [13]. More information on this classical result and related work on coloring problems restricted to graph classes can be found in several surveys, e.g, [22,25].

One type of graph classes that has been given wide attention in recent years is the class of $H$-free graphs, for various graphs $H$ [3,4,11,14,21,26]. For example, if $H$ contains a cycle, then $k$-COLORING is NP-complete for $H$-free graphs. This follows from the fact, proved by Kamiński and Lozin [17] and independently Král et al [18], that, for any fixed $k \geq 3$ and $g \geq 3$, $k$-COLORING is NP-complete for the class of graphs of girth at least $g$. Similarly, if $H$ is a forest with a vertex of degree at least 3, then $k$-COLORING is NP-complete for $H$-free graphs; this follows from [15] and [20]. Combining these results we conclude that $k$-COLORING is NP-complete for $H$-free graphs, as long as $H$ is not a linear forest, i.e., a union of disjoint paths. This focused attention on the case when $H$ is a path. Woeginger and Sgall [26] proved that 4-COLORING is NP-complete for $P_{12}$-free graphs, and that 5-COLORING is NP-complete for $P_8$-free graphs. Later on, these results were improved by various groups of researchers [3,4,11,19]. The strongest results so far are due to Huang [16] who proved that 4-COLORING is NP-complete for $P_7$-free graphs, and that 5-COLORING is NP-complete for $P_6$-free graphs. On the positive side, Hoàng et al. [14] developed an elegant recursive algorithm showing that $k$-COLORING can be solved in polynomial time for $P_5$-free graphs for any fixed $k$. These results give a complete classification of the complexity of $k$-COLORING $P_t$-free graphs for any fixed $k \geq 5$, and leave only 4-COLORING $P_6$-free graphs open for $k = 4$. It should be noted that deciding the complexity of 3-COLORING for $P_t$-free graphs seems difficult. It is not even known that whether or not there exists any $t$ such that 3-COLORING is NP-complete on $P_t$-free graphs. Randerath and Schiermeyer [21] gave the first polynomial time algorithm for 3-COLORING $P_6$-free graphs. As far as we know, this result has been extended to $P_7$-free graphs by Chudnovsky et al. [6,7].

In this paper we undertake a systematic examination of the complexity of $k$-COLORING with inputs restricted to $(P_t, C_\ell)$-free graphs. Since $k$-COLORING is NP-complete for $P_t$-free graphs for most values of $k$ and $t$, we are asking whether or not forbidding the additional cycle makes the problem easier.

**Known Facts:** Some facts can be derived from known results. For instance, in all the above polynomial cases for $k$-COLORING $P_t$-free graphs, we have the same for $(P_t, C_\ell)$-free graphs, for all $\ell$. This including the cases with $t \leq 5$ or $k = 3, t \leq 7$. When $\ell = 3$, each $k$-COLORING is polynomial for $t \leq 6$, as $(P_6, C_3)$-free graphs have bounded cliquewidth; on the other hand, for $t \geq 164$, 4-COLORING is NP-complete for $(P_t, C_3)$-free graphs [11]. When $\ell = 4$, each $k$-COLORING for is polynomial for $(P_t, C_4)$-free graphs [11]. When $\ell \geq 5$, 4-COLORING is NP-complete for $(P_t, C_\ell)$-free graphs as long as $t$ is large enough with respect to $\ell$ [11]. (For $\ell = 5$, the bound on $t$ is $t \geq 21$.)

**Our Contributions:** We prove (in Section 2) that $k$-COLORING is NP-complete for $(P_t, C_5)$-free graphs when $k \geq 4$ and $t \geq 7$, and that $k$-COLORING is NP-complete for $(P_t, C_\ell)$-free graphs when $\ell \geq 6$ and $k \geq 5, t \geq 6$. Moreover, we show that $k$-COLORING is also NP-complete for $(P_t, C_7)$-free graphs if $k = 4$ and $t \geq 9$. The first and last of these results is proved by extending a framework one of us introduced in [16]. That framework, however, does not apply to $C_5$-free graphs, and we give a new type of reduction to derive the second result. This almost completely settles the complexity of $k$-COLORING for $(P_t, C_\ell)$-free graphs when $\ell \geq 4, k \geq 4$. The few remaining open problems are listed in the last section. We also propose better algorithms for $k$-COLORING $(P_6, C_4)$-free graphs. The algorithms from [11] are linear time in theory, but they rely on several Ramsey-type results and use a tree decomposition with very high width, so they are not practical. We show that for $t = 6$ and any $k$ there are only finitely many minimal non-$k$-colorable $(P_6, C_4)$-free graphs. Thus there will be polynomial time certifying algorithms for each $k$-COLORING problem restricted to the class of $(P_6, C_4)$-free graphs. We explicitly describe all the minimal non-$k$-colorable $(P_6, C_4)$-free graphs for $k = 3, 4$, and construct corresponding certifying algorithms for 3-COLORING and 4-COLORING restricted to $(P_6, C_4)$-free graphs. Our algorithms make use of the clique cutset decomposition algorithm of Tarjan [24]. This is the most time intensive task, and once a clique decomposition of the input graph $G$ is given, the running time is $O(m + n)$. In any event, we believe our algorithms are more practical than those of [11].

## 2 NP-Completeness

Recently, Huang [16] proved that 5-COLORING is NP-complete for $P_6$-free graphs and 4-COLORING is NP-complete for $P_7$-free graphs. The proof used a novel general framework. In fact, the framework can be used to prove stronger results on $(P_t, C_\ell)$-free graphs. We recall the framework below. We call a $k$-critical graph *nice* if $G$ contains three independent vertices $\{c_1, c_2, c_3\}$ such that the clique number $\omega(G - \{c_1, c_2, c_3\}) = \omega(G) = k - 1$. For example, any odd cycle of length at least 7 is a nice 3-critical graph.

To prove the desired NP-completeness results Huang [16] gave a reduction from 3-SAT. Let $I$ be any 3-SAT instance with variables $X = \{x_1, x_2, \ldots, x_n\}$ and clauses $\mathcal{C} = \{C_1, C_2, \ldots, C_m\}$, and let $H$ be a nice $k$-critical graph with three specified independent vertices $\{c_1, c_2, c_3\}$. Huang [16] constructed the graph $G_I$ as follows.

- Introduce for each variable $x_i$ a *variable component* $T_i$ which is isomorphic to $K_2$, labeled by $x_i \bar{x}_i$. Call these vertices $X$-type.
- Introduce for each variable $x_i$ a vertex $d_i$. Call these vertices $D$-type.
- Introduce for each clause $C_j = y_{i_1} \vee y_{i_2} \vee y_{i_3}$ a *clause component* $H_j$ which is isomorphic to $H$, where $y_{i_t}$ is either $x_{i_t}$ or $\bar{x}_{i_t}$. Denote three specified independent vertices in $H_j$ by $c_{i_t j}$ for $t = 1, 2, 3$. Call $c_{i_t j}$ $C$-type and all remaining vertices $U$-type.

For any $C$-type vertex $c_{ij}$ we call $x_i$ or $\bar{x}_i$ its *corresponding literal vertex*, depending on whether $x_i \in C_j$ or $\bar{x}_i \in C_j$.

- Connect each $U$-type vertex to each $D$-type and $X$-type vertices.
- Connect each $C$-type vertex $c_{ij}$ to $d_i$ and its corresponding literal vertex.

We refer to [16] for the proof of the following two lemmas.

**Lemma 1.** *Let $H$ be a nice $k$-critical graph. Suppose $G_I$ is the graph constructed from $H$ and a 3-SAT instance $I$. Then $I$ is satisfiable if and only if $G_I$ is $(k+1)$-colorable.*

**Lemma 2.** *Let $H$ be a nice $k$-critical graph. Suppose $G_I$ is the graph constructed from $H$ and a 3-SAT instance $I$. If $H$ is $P_t$-free where $t \geq 6$, then $G_I$ is $P_t$-free as well.*

We explain how this framework can be used to prove stronger NP-completeness results. To obtain NP-completeness results for $(P_t, C_\ell)$-free graphs, we need one more lemma.

**Lemma 3.** *Let $\ell \geq 6$. If $H$ is $C_\ell$-free, then $G_I$ is $C_\ell$-free.*

**Proof.** Let $Q = v_1 \ldots v_\ell$ be an induced $C_\ell$ in $G_I$. Let $C_i$ (resp. $\bar{C}_i$) be the set of $C$-type vertices that connect to $x_i$ (resp. $\bar{x}_i$). Let $G_i = G[\{T_i \cup \{d_i\} \cup C_i \cup \bar{C}_i\}]$. Note that $G - U$ is disjoint union of $G_i$, $i = 1, 2, \ldots, n$. If $Q \cap U = \emptyset$, then $Q \subseteq G_i$ for some $i$. It is easy to see that $G_i$ is $C_\ell$-free as $\ell \geq 6$. Thus, $Q \cap U \neq \emptyset$. Without loss of generality, we assume that $v_1$ is a $U$-type vertex where $v_1$ is in the $j$th clause component $H_j$. If $v_2$ and $v_\ell$ are both in $H_j$, then $Q \subseteq H_j$, which contradicts our assumption that $H_j = H$ is $C_\ell$-free. If $v_2$ and $v_\ell$ are both in $X \cup D$, then as $U$-type vertices are complete to $X$-type and $D$-type vertices, all other vertices on $Q$ are of $C$-type. This is impossible since $C$ is independent. The last case is $v_\ell$ is in $H_j$ and $v_2$ is in $X \cup D$. Similar to the second case, we have $v_4, v_5, \ldots v_{l-1}$ are $C$-type vertices. This contradicts that $v_4 v_5$ is an edge. □

The following theorem follows now directly from the above lemmas.

**Theorem 1.** *Let $\ell \geq 6$. Then $k$-COLORING is NP-complete for $(P_t, C_\ell)$-free graphs whenever there exists a $(P_t, C_\ell)$-free nice $(k-1)$-critical graph.*

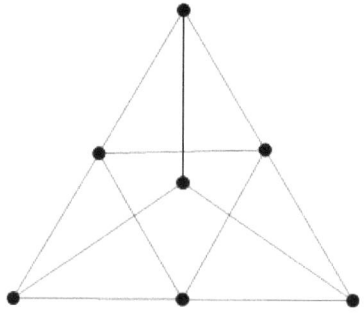

**Fig. 1.** $G_1$

We apply Theorem 1 to derive a series of hardness results on $(P_t, C_\ell)$-free graphs for various values of $k$ and $t$.

**Theorem 2.** *Let $k \geq 5$, $t \geq 6$ and $\ell \geq 6$ be fixed integers. $k$-COLORING is NP-complete for $(P_t, C_\ell)$-free graphs.*

**Proof.** It is easy to check that the graph $G_1$ shown in Figure 1 is a nice 4-critical $(P_6, C_\ell)$-free graph for any fixed $\ell \geq 6$. Applying Theorem 1 with $G_1$ will complete our proof. □

**Theorem 3.** *4-COLORING is NP-complete for $(P_t, C_\ell)$-free graphs when $t \geq 7$ and $\ell \geq 6$ with $\ell \neq 7$; and 4-COLORING is NP-complete for $(P_t, C_\ell)$-free graphs when $t \geq 9$ and $\ell \geq 6$ with $\ell \neq 9$.*

**Proof.** It is easy to check that $C_7$ is a nice 3-critical $(P_t, C_\ell)$-free graph for any $t \geq 7$ and $\ell \geq 6$ except $\ell = 7$, and that $C_9$ is a nice 3-critical $(P_t, C_\ell)$-free graph for any $t \geq 9$ and $\ell \geq 6$ except $\ell = 9$. Applying Theorem 1 with $C_7$ and $C_9$ will complete the proof. □

Theorem 1 is not sufficient to prove NP-completeness result for $(P_t, C_5)$-free graphs as the graph $G_I$ in the above reduction contains an induced $C_5$, regardless of the choice of the clause component. We shall use a different reduction to prove the next result.

**Theorem 4.** *4-COLORING is NP-complete for $(P_7, C_5)$-free graphs.*

**Proof.** We reduce NOT-ALL-EQUAL 3-SATISFIABILITY with positive literals only (NAE 3-SAT PL for short) to our problem. The NAE 3-SAT PL is NP-complete [23] and is defined as follows. Given a set $X = \{x_1, x_2, \ldots, x_n\}$ of logical variables, and a set $\mathcal{C} = \{C_1, C_2, \ldots, C_m\}$ of three-literal clauses over $X$ in which all literals are positive, does there exist a truth assignment for $X$ such that each clause contains at least one true literal and at least one false literal? Given an instance $I$ of NAE 3-SAT PL we construct a graph $G_I$ as follows.

- For each variable $x_i$ we introduce a single vertex named as $x_i$. Call these vertices $X$-type.
- For each variable $x_i$ we introduce a "truth assignment" component $F_i$ where $F_i$ is isomorphic to $P_4$ whose vertices are labeled by $d_i e'_i e_i d'_i$.
- For each clause $C_j = x_{i_1} \vee x_{i_2} \vee x_{i_3}$ we introduce two copies of $C_7$ denoted by $H_j$ and $H'_j$. Choose three independent vertices of $H_j$ and name them as $c_{i_1 j}$, $c_{i_2 j}$ and $c_{i_3 j}$. Choose three independent vertices of $H'_j$ and name them as $c'_{i_1 j}$, $c'_{i_2 j}$ and $c'_{i_3 j}$. Call these vertices $C$-type and $C'$-type, respectively. The remaining vertices in clause components are said to be of $U$-type.
- Connect each $U$-type vertex to each $X$-type vertex and each vertex in $F_i$ for $1 \leq i \leq n$.
- Connect each $C$-type vertex $c_{ij}$ to $x_i$ and $d_i$ and connect each $C'$-type vertex $c'_{ij}$ to $x_i$ and $d'_i$.

This completes the construction of $G_I$. It is easy to see that $d_i$ and $d'_i$ have no common neighbor in $G - U$ and same for $e_i$ and $e'_i$.

**Claim 1.** *The instance $I$ is satisfiable if and only if $G_I$ is 4-colorable.*

**Proof.** Suppose first that $G_I$ is 4-colorable and $\phi$ is a 4-coloring of $G_I$. Without loss of generality, we may assume that the two adjacent $U$-type vertices in $H_1$ receive color 1 and 2, respectively. Now as $U$ is complete to $X \cup F$, it follows that each $x_i$ and each vertex in $F_i$ receives color 3 or 4. Further, $\phi(d_i) \neq \phi(d'_i)$ for each $i$. We define a truth assignment as follows.

- We set $x_i$ to be TRUE if $\phi(x_i) = \phi(d_i)$ and to be FALSE if $\phi(x_i) \neq \phi(d_i)$.

We show that every clause $C_j$ contains at least one ture literal and one false literal. Suppose $x_{i_1}$, $x_{i_2}$, and $x_{i_3}$ are all TRUE. Then it implies that $\phi(d'_{i_j}) \neq \phi(x_{i_j})$ for all $j = 1, 2, 3$. As a result, $c'_{i_j}$ must be colored with color 1 or 2 under $\phi$. Moreover, all $U$-type vertices in $H'_j$ are colored with 1 or 2 under $\phi$. This contradictions the fact that $H'_j = C_7$ is not 2-colorable. If $x_{i_1}$, $x_{i_2}$, and $x_{i_3}$ are all FALSE we would reach a similar contradiction. Conversely, suppose that every clasue $C_j$ contains at least one ture literal and one false literal. We define a 4-coloring $\phi$ as follows.

- Set $\phi(x_i) = 3$ if $x_i$ is TRUE and $\phi(x_i) = 4$ if $x_i$ is FALSE.
- We color vertices in $F_i$ alternately with color 3 and 4 starting from setting $\phi(d_i) = 3$.
- Let $C_j = x_{i_1} \vee x_{i_2} \vee x_{i_3}$ be a clause. Without loss of generality, we may assume that $x_{i_1}$ is TRUE and $x_{i_2}$ is FALSE. It follows from the definiton of $\phi$ that $\phi(x_{i_1}) = \phi(d_{i_1}) = 3$. Hence, we can color $c_{i_1 j}$ with color 4, so that $H_j - c_{i_1 j}$ can be colored with colors 1 and 2. Similarly, we can 4-color $H'_j$. □

**Claim 2.** *$G_I$ is $C_5$-free.*

**Claim 3.** *$G_I$ is $P_7$-free.*

We omit proofs for Claims 2 and 3. □

The following result is a direct corollary of Theorem 4.

**Theorem 5.** *Let $k \geq 4$ and $t \geq 7$. Then $k$-COLORING is NP-complete for $(P_t, C_5)$-free graphs.*

## 3 Certifying Algorithms

We have shown that in general $k$-COLORING $P_t$-free graphs remains hard even if we forbid some induced $C_\ell$ where $\ell \geq 5$. In a sharp contrast, forbidding $C_4$ does make the problem easier. Golovach et al. [11] have proved that $k$-COLORING can be solved in linear time for $(P_t, K_{r,s})$-free graphs for any fixed $k, r, s, t$. As $C_4 = K_{2,2}$, their result implies that $k$-COLORING becomes polynomial solvable for $P_t$-free graphs when we also forbid an induced $C_4$. In this section we shall present polynomial time certifying algorithms for $k$-COLORING $(P_6, C_4)$-free graphs in the special case when $k = 3$ and $k = 4$. The algorithms of Golovach et al. [11] depends on Ramsey-type results. This means that they are not certifying. Moreover, even though they run in linear time using tree-decompositions, the multiplicative constants in the algorithms depend on the treewidth, which is quite high because of a double use of Ramsey-type results. Our algorithms use a clique cutset decomposition, and the overall running time is $O(mn)$. However, if a clique cutset decomposition is given, then our algorithms run in linear time. Even in the case the clique cutset decomposition has to be found, we believe our $O(mn)$ algorithm has much better practical performance on realistic-sized graphs than the algorithms of Golovach et al. [11].

In Section 3.1 we develop some preliminary results on $(P_6, C_4)$-free graphs that will be used in our algorithms. Then we give certifying algorithms for 3-COLORING and 4-COLORING in Section 3.2. Our algorithms will output a $k$-coloring if $G$ has one, or a subgraph of $G$ that certifies that $G$ is not $k$-colorable.

### 3.1 Imperfect $(P_6, C_4)$-Free Graphs

Let $G$ be a connected imperfect $(P_6, C_4)$-free graph. By the Strong Perfect Graph Theorem [8], $G$ must contain an induced $C = C_5 = v_0 \ldots v_4$. We call a vertex $v \in V \setminus C$ a *p-vertex* with respect to $C$ if $v$ has exactly $p$ neighbors on $C$, i.e., $|N_C(v)| = p$. We denote by $S_p$ the set of $p$-vertices for $0 \leq p \leq 5$. In the following all indices are modulo 5. Let $S_1(v_i)$ be the subset of $S_1$ containing all 1-vertices that have $v_i$ as their neighbor on $C$. Let $S_3(v_i)$ be the subset of $S_3$ containing all 3-vertices that have $v_{i-1}, v_i$ and $v_{i+1}$ as their neighbors on $C$. Let $S_2(v_i)$ be the subset of $S_2$ containing all 2-vertices that have $v_{i-2}$ and $v_{i+2}$ as their neighbors on $C$. Alternatively, we also denote $S_2(v_i, v_{i+1})$ by the set of 2-vertices that have $v_i$ and $v_{i+1}$ as their neighbors on $C$. Clearly, $S_2(v_i) = S_2(v_{i-2}, v_{i+2})$. We shall use either of notation whichever is convenient. $S_p = \bigcup_{i=0}^{4} S_p(v_i)$ for $p = 1, 2, 3$. It follows easily from the $(P_6, C_4)$-freeness of $G$ that $S_5$ and each $S_3(v_i)$ are cliques and $S_4 = \emptyset$.

**Observation 1.** *(1) $S_1(v_i)$ is complete to $S_1(v_{i+2})$ and anti-complete to $S_1(v_{i+1})$; if both $S_1(v_i)$ and $S_1(v_{i+2})$ are nonempty, both sets are cliques. (2) $S_2(v_i)$ is complete to $S_2(v_{i+1})$ and anti-complete to $S_2(v_{i+2})$. (3) $S_3(v_i)$ is anti-complete to $S_3(v_{i+2})$.*

**Observation 2.** *(1) $S_1(v_i)$ is anti-complete to $S_2(v_j)$ if $i \neq j$. Further, if $y \in S_2(v_i)$ is not anti-complete to $S_1(v_i)$, then $y$ is an universal vertex in $S_2(v_i)$. (2) $S_1(v_i)$ is anti-complete to $S_3(v_{i+2})$. (3) $S_2(v_i)$ is anti-complete to $S_3(v_i)$.*

**Observation 3.** *(1) One of $S_1(v_i)$ and $S_2(v_{i+1})$ is empty; (2) One of $S_2(v_i)$, $S_2(v_{i-2})$ and $S_2(v_{i+2})$ is empty; (3) If both $S_1(v_{i-1})$ and $S_1(v_{i+1})$ are nonempty, then $S_2 = \emptyset$. If both $S_1(v_i)$ and $S_1(v_{i+1})$ are nonempty, then $S_2 = S_2(v_i, v_{i+1})$; (4) Let $x \in S_3(v_i)$. If both $S_2(v_{i-1})$ and $S_2(v_{i+1})$ are nonempty, then $x$ is either complete or anti-complete to $S_2(v_{i-1})$ and $S_2(v_{i+1})$. In the case of complete, $S_2(v_{i-1})$ and $S_2(v_{i+1})$ are cliques. Moreover, if $S_2(v_i)$ is also nonempty, then $x$ is anti-complete to $S_2(v_{i-1})$ and $S_2(v_{i+1})$.*

The proofs for Observations 1-3 follow readily from the fact that $G$ is $(P_6, C_4)$-free. Brandstädt and Hoàng [2] proved the following important property on $(P_6, C_4)$-free graphs. A subset $S \subseteq V$ is *dominating* if every vertex not in $S$ has a neighbor in $S$.

**Lemma 4.** *([2]) Let $G$ be a $(P_6, C_4)$-free graph without clique cutset. Then every induced $C_5$ is dominating.*

The following Observation is based on Lemma 4.

**Observation 4.** *(1) $S_1(v_i)$ is complete to $S_3(v_i)$. (2) If $S_1(v_i)$ is not anti-complete to $S_2(v_i)$ then $S_1 = S_1(v_i)$.*

### 3.2   3-COLORING and 4-COLORING $(P_6, C_4)$-Free Graphs

In this section we shall develop certifying algorithms for 3-COLORING and 4-COLORING $(P_6, C_4)$-free graphs by describing all minimal non-3-colorable and non-4-colorable $(P_6, C_4)$-free graphs. The following lemma is folklore.

**Lemma 5.** *A minimal obstruction $G$ for k-COLORING has no clique cutset and $\delta(G) \geq k$.*

In 2010 Bruce et al. [5] successfully characterized all minimal non-3-colorable $P_5$-free graphs. Here we characterize all minimal non-3-colorable $(P_6, C_4)$-free graphs.

**Theorem 6.** *There are exactly four minimal non-3-colorable minimal non-4-colorable $(P_6, C_4)$-free graphs, depicted in Figure 2.*

**Proof.** Let $G$ be a minimal $(P_6, C_4)$-free obstruction for 3-COLORING. $G$ contains no clique cutset and $\delta(G) \geq 3$ by Lemma 5. If $G$ is perfect, then $G = K_4$. Therefore, we may assume that $G$ is not perfect and $K_4$-free. By Strong Perfect Graph Theorem, $G$ must contain an induced $C = C_5 = v_0 v_1 \ldots v_4$. We define

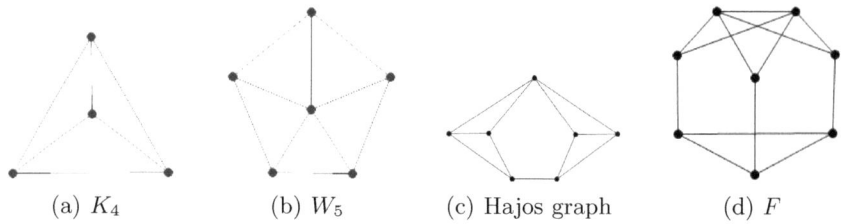

**Fig. 2.** All minimal non-3-colorable $(P_6, C_4)$-free graphs

$S_p$, $S_1(v_i)$, $S_2(v_i)$ and $S_3(v_i)$ in the same way we defined in the beginning of this section. By Lemma 4, $S_0 = \emptyset$. It is easy to see that $|S_5| \leq 1$. If $|S_5| = 1$, then $G = W_5$. So we may assume that $S_5 = \emptyset$. If there exists an index $i$ such that $S_3(v_i) \neq \emptyset$ and $S_3(v_{i+2}) \neq \emptyset$, then $G$ is Hajos graph. Hence, at most two $S_3(v_i)$'s are nonempty. Further, each $S_3(v_i)$ is clique and contains at most one vertex since $G$ is $(C_4, K_4)$-free. Therefore, $|S_3| \leq 2$. We distinguish three cases.

**Case 1.** $|S_3| = 2$. Without loss of generality, assume that $S_3(v_0) = \{x\}$ and $S_3(v_1) = \{y\}$. $xy \notin E$ as $G$ is $K_4$-free. $S_1(v_3) = \emptyset$ otherwise let $t \in S_1(v_3)$ and then $tv_3v_2yv_0x = P_6$. Moreover, $x$ (resp. $y$) is complete to $S_2(v_3, v_4)$ (resp. $S_2(v_2, v_3)$). Otherwise there exists some vertex $z \in S_2(v_3, v_4)$ with $xz \notin E$. Then $zv_3v_2yv_0x = P_6$. Hence, $S_2(v_3, v_4)$ and $S_2(v_3, v_2)$ are cliques and each of them contains at most one vertex. As $d(v_3) \geq 3$ and $S_1(v_3) = \emptyset$, at least one of them is nonempty. Suppose first that $p \in S_2(v_3, v_4)$ and $q \in S_2(v_2, v_3)$. $xp \in E$ and $yq \in E$. It follows from $S_1(v_3) = \emptyset$ and Observation 3 (1) that $S_1 = \emptyset$. Further, $S_2(v_1, v_2) = S_2(v_0, v_4) = \emptyset$ by Observation 3 (4). Hence, $S_2 = \{p, q\}$ by Observation 3 (2) and $N(x) = \{v_4, v_1, v_0, p\}$. Since $G$ is a minimal obstruction, there exists a 3-coloring $\phi$ of $G-x$. Note that we must have $\phi(v_4) = \phi(q) = \phi(v_1)$ and $\phi(p) = \phi(v_2) = \phi(v_0)$. Consequently, we can extend $\phi$ to $G$ be setting $\phi(x) = \{1, 2, 3\} \setminus \{\phi(v_0), \phi(v_1)\}$. This contradicts that $G$ is not 3-colorable. Therefore, exactly one of $S_2(v_3, v_4)$ and $S_2(v_3, v_2)$ is empty. Without loss of generality, assume that $S_2(v_3, v_4) = \emptyset$ and let $z \in S_2(v_2, v_3)$. Note that $N(v_3) = \{v_4, v_2, z\}$. Let $\phi$ be a 3-coloring of $G - v_3$, and we must have $\phi(v_4) = \phi(v_1) = \phi(z)$. Thus we can extend $\phi$ to $G$. This is a contradiction.

**Case 2.** $|S_3| = 1$.

**Case 3.** $|S_3| = 0$.

We omit the lengthy analysis of the last two cases in this version. □

The above proof can be easily turned into a linear time algorithm for 3-COLORING. We first test if $G$ is chordal. If so, we can tell whether or not $G$ is 3-colorable. Otherwise we have an induced $C = C_\ell$ for some $\ell \geq 4$. Up to this point every step can be done in linear time (see, e.g., [12]). If $\ell = 4$ or $\ell \geq 7$ then $G$ is not $(P_6, C_4)$-free. If $\ell = 5$ we follow the above proof, and it can be readily checked that every step can be done in linear time. The remaining case is $\ell = 6$, and we can now assume $G$ is also $C_5$-free. Brandstädt and Hoàng [2] stated that either $C$ is dominating or $G$ belongs to a specific graph class for which $k$-COLORING can be solved in linear time. Therefore, we assume that $C$ is dominating. We define $p$-vertices and $S_p$ with respect to $C$. We either find that $G$ is not $(P_6, C_4)$-free or $V = C \cup S_6 \cup S_3$. Finally, in linear time we either find a $K_4$ or conclude that $|G| \leq 13$ in which case a 3-coloring of $G$ can be obtained.

Using similar techniques, with a more sophisticated analysis, we are able to describe all minimal non-4-colorable $(P_6, C_4)$-free graphs. It is clear that every minimal non-3-colorable $(P_6, C_4)$-free graph with one more dominating vertex is a minimal non-4-colorable $(P_6, C_4)$-free graph. Moreover, there are nine additional graphs. The proof can be transformed into a certifying 4-COLORING algorithm for $(P_6, C_4)$-free graphs, analogously to the application of Theorem 6.

**Theorem 7.** *There are exactly* 13 *minimal non-4-colorable* $(P_6, C_4)$-*free graphs.*

For larger values of $k$, we prove the following theorem.

**Theorem 8.** *For any fixed $k$, there are only finitely many minimal non-$k$-colorable $(P_6, C_4)$-free graphs.*

We note however, that even though the number of minimal obstructions to $k$-COLORING $(P_6, C_4)$-free graphs is finite, this fact alone does not yield an efficient, nor a certifying, algorithm, since the proof of the finiteness does not explicitly describe these obstructions, and such a task seems hopeless for large values of $k$.

## 4 Conclusions

We have undertaken a systematic study of complexity of $k$-COLORING for graphs forbidding an induced cycle $C_\ell$ and an induced path $P_t$. For most values of $k$, $t$ and $\ell$ we have obtained hardness results. We have also given certifying algorithms for $k$-COLORING $(P_6, C_4)$-free graphs when $k = 3$ and $k = 4$. Our algorithms make use of the clique cutset decomposition of Tarjan [24] and run in $O(n+m)$ time given a clique cutset decomposition of the input graph. We expect our algorithms to have a good performance in practice. We have proved that for any $k$ there are only finitely many minimal non-$k$-colorable $(P_6, C_4)$-free graphs, and have described all of them for $k = 3$ and $k = 4$. The proofs allowed us to give polynomial certifying algorithms for 3-coloring and 4-coloring $(P_6, C_4)$-free graphs. However, for larger $k$, we do not know certifying $k$-COLORING algorithms for $(P_6, C_4)$-free graphs. Our hardness results come close to classifying the complexity all cases of $k$-COLORING for $(P_t, C_\ell)$-free graphs. There seem to be two stubborn cases about which not much can be said with the current tools, when $k = 3$ or $\ell = 3$. (But note [6,7].) Beyond these cases, our results leave only the following remaining open problems.

*Problem 1.* What is the complexity of $k$-COLORING $(P_6, C_5)$-free graphs for $k \geq 4$?

*Problem 2.* What is the complexity of 4-COLORING $(P_6, C_6)$-free graphs?

*Problem 3.* What is the complexity of 4-COLORING $(P_t, C_7)$-free graphs for $t = 7$ and $t = 8$?

In [16] Huang conjectured that 4-COLORING is polynomial time solvable for $P_6$-free graphs. If the problems in Problem 1 for $k = 4$ or Problem 2 are polynomial, this would add evidence to the conjecture.

## References

1. Bondy, J.A., Murty, U.S.R.: Graph Theory. Springer Graduate Texts in Mathematics, vol. 244 (2008)
2. Brandstädt, A., Hoàng, C.T.: On clique separators, nearly chordal graphs, and the Maximum Weight Stable Set Problem. Theoretical Computer Science 389, 295–306 (2007)
3. Broersma, H.J., Fomin, F.V., Golovach, P.A., Paulusma, D.: Three complexity results on coloring $P_k$-free graphs. European Journal of Combinatorics (2012) (in press)
4. Broersma, H.J., Golovach, P.A., Paulusma, D., Song, J.: Updating the complexity status of coloring graphs without a fixed induced learn forest. Theoret. Comput. Sci. 414, 9–19 (2012)
5. Bruce, D., Hoàng, C.T., Sawada, J.: A certifying algorithm for 3-colorability of $P_5$-free graphs. In: Dong, Y., Du, D.-Z., Ibarra, O. (eds.) ISAAC 2009. LNCS, vol. 5878, pp. 594–604. Springer, Heidelberg (2009)
6. Chudnovsky, M., Maceli, P., Zhong, M.: Three-coloring graphs with no induced six-edge path I: the triangle-free case (in preparation)
7. Chudnovsky, M., Maceli, P., Zhong, M.: Three-coloring graphs with no induced six-edge path II: using a triangle (in preparation)
8. Chudnovsky, M., Robertson, N., Seymour, P., Thomas, R.: The strong perfect graph theorem. Annals of Mathematics 64, 51–229 (2006)
9. Dabrowski, K., Golovach, P., Paulusma, D.: Colouring of graphs with Ramsey-type forbidden subgraphs (submitted)
10. Garey, M.R., Johnson, D.S.: Computers and Intractability: A Guide to the Theory of NP-Completeness. Freeman San Faranciso (1979)
11. Golovach, P.A., Paulusma, D., Song, J.: Coloring graphs without short cycles and long induced paths (2013), http://www.dur.ac.uk/daniel.paulusma/Papers/Submitted/girth.pdf
12. Golumbic, M.C.: Algorithmic graph theory and perfect graphs, San Diego (1980)
13. Grötschel, M., Lovász, L., Schrijver, A.: Polynomial algorithms for perfect graphs. Ann. Discrete Math. 21, 325–356 (1984), Topics on Perfect Graphs
14. Hoàng, C.T., Kamiński, M., Lozin, V.V., Sawada, J., Shu, X.: Deciding $k$-colorability of $P_5$-free graphs in polynomial time. Algorithmica 57, 74–81 (2010)
15. Holyer, I.: The NP-completeness of edge coloring. SIAM J. Comput. 10, 718–720 (1981)
16. Huang, S.: Improved complexity results on $k$-coloring $P_t$-free graphs. In: Chatterjee, K., Sgall, J. (eds.) MFCS 2013. LNCS, vol. 8087, pp. 551–558. Springer, Heidelberg (2013)
17. Kamiński, M., Lozin, V.V.: Coloring edges and vertices of graphs without short or long cycles. Contrib. Discrete. Mah. 2, 61–66 (2007)

18. Král', D., Kratochvíl, J., Tuza, Z., Woeginger, G.J.: Complexity of coloring graphs without forbidden induced subgraphs. In: Brandstädt, A., Le, V.B. (eds.) WG 2001. LNCS, vol. 2204, pp. 254–262. Springer, Heidelberg (2001)
19. Le, V.B., Randerath, B., Schiermeyer, I.: On the complexity of 4-coloring graphs without long induced paths. Theoret. Comput. Sci. 389, 330–335 (2007)
20. Leven, D., Galil, Z.: NP-completeness of finding the chromatic index of regular graphs. J. Algorithm 4, 35–44 (1983)
21. Randerath, B., Schiermeyer, I.: 3-Colorability $\in$ P for $P_6$-free graphs. Discrete Appl. Math. 136, 299–313 (2004)
22. Randerath, B., Schiermeyer, I.: Vertex colouring and fibidden subgraphs-a survey. Graphs Combin. 20, 1–40 (2004)
23. Schaefer, T.J.: The complexity of satisfiability problems. In: Proc. STOC 1978, pp. 216–226 (1978)
24. Tarjan, R.E.: Decomposition by clique separators. Discrete Mathematics 55, 221–232 (1985)
25. Tuza, Z.: Graph colorings with local restrictions-a survey. Discuss. Math. Graph Theory 17, 161–228 (1997)
26. Woeginger, G.J., Sgall, J.: The complexity of coloring graphs without long induced paths. Acta Cybernet. 15, 107–117 (2001)

# Approximating Real-Time Scheduling on Identical Machines*

Nikhil Bansal[1], Cyriel Rutten[2], Suzanne van der Ster[3], Tjark Vredeveld[2], and Ruben van der Zwaan[1]

[1] Eindhoven University of Technology
{n.bansal,g.r.j.v.d.zwaan}@tue.nl
[2] Maastricht University
cyrielrutten@gmail.com, t.vredeveld@maastrichtuniversity.nl
[3] Vrije Universiteit
suzanne.vander.ster@vu.nl

**Abstract.** We study the problem of assigning $n$ tasks to $m$ identical parallel machines in the real-time scheduling setting, where each task recurrently releases jobs that must be completed by their deadlines. The goal is to find a partition of the task set over the machines such that each job that is released by a task can meet its deadline. Since this problem is co-NP-hard, the focus is on finding $\alpha$-approximation algorithms in the resource augmentation setting, i.e., finding a feasible partition on machines running at speed $\alpha \geq 1$, if some feasible partition exists on unit-speed machines.

Recently, Chen and Chakraborty gave a polynomial-time approximation scheme if the ratio of the largest to the smallest relative deadline of the tasks, $\lambda$, is bounded by a constant. However, their algorithm has a super-exponential dependence on $\lambda$ and hence does not extend to larger values of $\lambda$. Our main contribution is to design an approximation scheme with a substantially improved running-time dependence on $\lambda$. In particular, our algorithm depends exponentially on $\log \lambda$ and hence has quasi-polynomial running time even if $\lambda$ is polynomially bounded. This improvement is based on exploiting various structural properties of approximate demand bound functions in different ways, which might be of independent interest.

## 1 Introduction

The *sporadic task system* is one of the most widely adopted models for infinitely recurring executions in real-time systems. Specifically, each sporadic task $\tau = (c_\tau, d_\tau, p_\tau)$ is specified by the amount of processing needed by its jobs $c_\tau$, a deadline $d_\tau$ by which a job must be completed, relative to its arrival time, and a minimum interarrival time $p_\tau$ between two consecutive jobs, which is called the period of the task. Such a sporadic task releases a possibly infinite sequence of jobs. A sporadic task system $\mathcal{T}$ consists of $n$ sporadic tasks. A task system

---
* Supported by the NWO VIDI grant 639.022.211.

is said to be *feasible* on a computing platform if for any job sequence that can be possibly generated by the system, there exists a schedule for the task system, such that all jobs from all tasks meet their deadlines. In this paper, we consider the feasibility question of scheduling a set of sporadic tasks to multiple identical machines (processors). This problem and related problems in real-time scheduling have received great attention in the last years; see for example [1,4,6] and the references therein.

*Single-processor case:* Determining the feasibility of a task system on a single (preemptive[1]) processor is quite well-understood. It is well-known that the hardest case for feasibility is when the first jobs of all tasks arrive simultaneously and all subsequent jobs arrive as rapidly as legally possible [5]. That is, we can assume that for each task $\tau$ in the task system, the jobs of $\tau$ arrive at times $0, p_\tau, 2p_\tau, \ldots$. This sequence of job-arrivals is called the *synchronous arrival sequence*. Another well-known fact [11] is that the Earliest Deadline First (EDF) algorithm, that schedules at any time the job with the earliest absolute deadline, will always produce a valid schedule for any sequence of jobs that is feasible.

Although one can validate whether a task system is feasible by running EDF, this does not provide an efficient polynomial-time feasibility test. The problem is that the periodic nature of jobs together with their relative deadlines can introduce complicated long-range dependencies. In particular, the infeasibility may occur only at a very late time in the schedule, say close to the hyperperiod (which is the least common multiple of the periods of the tasks). In fact, no polynomial-time feasibility test on a uniprocessor is likely to exist, unless P=co-NP [10]. For more results on scheduling sporadic task systems on a single processor, we refer to Baruah and Goosens [4].

*Multiprocessor case:* For multiprocessor systems, there are two main paradigms for scheduling: *global* vs. *partitioned* scheduling. In partitioned scheduling each task is assigned to one of the machines and all jobs corresponding to this task must be scheduled on that machine. In global scheduling, tasks can use all machines and jobs can even be migrated. Partitioned scheduling is used much more than global scheduling as it is easier to implement and has no communication overhead, which is required if a single task is split between multiple processors. The communication may also lead to security issues. In this paper, we only consider partitioned scheduling.

Observe that in this setting, given a partition of the tasks over the machines, determining its feasibility simply reduces to several independent uniprocessor feasibility problems - one for each machine. Together with the facts for uniprocessor feasibility, the problem we study can be viewed as follows: Find a partition of tasks among machines, such that for each machine, the synchronous arrival sequence for tasks assigned to that machine is feasible for EDF. Clearly, this problem is also co-NP-hard and, as we shall see, it is also NP-hard.

---

[1] That is, any job can be interrupted arbitrarily during its execution and resumed later from the point of interruption without any penalty. Throughout the paper we consider the preemptive setting.

*Resource Augmentation and $\alpha$-feasibility:* The hardness of the problem leads us to finding a good approximation algorithm. As usual, we consider the resource augmentation setting, where our algorithm is allowed some additional speedup per machine.

Given a parameter $\alpha \geq 1$, we call an algorithm an *$\alpha$-feasibility test* if it

1. either returns a partition of the tasks into sets $\{\mathcal{T}_i\}_{i \in [m]}$ such that each $\mathcal{T}_i$ can be feasibly scheduled on a machine that runs at speed $\alpha$; or,
2. returns 'infeasible' if no feasible partition of tasks exists which can be scheduled on $m$ machines running at unit speed.

Alternatively, the algorithm always finds a partition $\mathcal{T}_1, \ldots, \mathcal{T}_m$ of $\mathcal{T}$ such that each $\mathcal{T}_i$ can be feasibly scheduled on a speed-$\alpha$ machine, provided there exists some feasible partition on unit-speed machines.

Let us call a family of feasibility tests a polynomial-time approximation scheme (PTAS), if for any arbitrarily small constant $\epsilon > 0$, there exists a $(1+\epsilon)$-feasibility test in this family with running time polynomial in $n$ and $m$. Note that the running time dependence on $\epsilon$ can be any arbitrary function. If the running time dependence on $1/\epsilon$ is also polynomial, we call the test a fully polynomial-time approximation scheme (FPTAS).

## 1.1 Related Previous Results

In the single processor case, an FPTAS feasibility test is known [7]. We will describe a related test later, as its structure will play a key role in our algorithm (see Observation 2 and Theorem 3 below). In particular in this test, one only needs to check the feasibility of the EDF schedule for the job sequence at about $(1/\epsilon) \log(d_{\max}/d_{\min})$ time steps, where $d_{\max}$ and $d_{\min}$ are the maximum and minimum task deadlines in the instance.

For partitioned scheduling on multiple machines, Chen and Chakraborty [9] gave a PTAS, generalizing a previous result of [3], if the maximum to minimum deadline ratio is bounded by a constant. Let us call this ratio $\lambda$. The idea of [9] is to view the problem as a vector scheduling problem in (roughly) $\ell = (1/\epsilon) \log \lambda$ dimensions. That is, each task is viewed as a $\ell$-dimensional vector, and the tasks can be feasibly scheduled on machine, if the corresponding vectors can be feasibly packed in a unit $\ell$-dimensional bin. This connection essentially follows from the property for the single-processor test mentioned above. Then the known PTAS for vector scheduling [8] is used in a black-box manner to obtain a $(1+\epsilon)$-approximate feasibility test that runs in time roughly[2] $n^{O(\exp((\frac{1}{\epsilon}) \log \lambda))}$. Note that this running time is doubly exponential in $\log \lambda$, and while this is polynomial time for constant $\lambda$, it is super-polynomial if $\lambda$ is super-constant.

If $m = O(1)$, Marchetti-Spaccamela et al. [12] design a PTAS, even for the case that the execution time of a task is machine-dependent.

---
[2] For clarity, we suppress some dependence on terms involving $\log \log \lambda$.

## 1.2 Our Contribution

We provide a $(1+\epsilon)$-feasibility test which substantially improves upon the result of Chen and Chakraborty [9]. In particular we show the following result.

**Theorem 1.** *Given $\epsilon > 0$, a task set $\mathcal{T}$ consisting of $n$ tasks and $m$ parallel identical processors, there is a $(1+\epsilon)$-feasibility test in the partitioned scheduling setting, with running time $O\left(m^{O(f(\epsilon)\log(\lambda))}\right)$. Here $\lambda = d_{\max}/d_{\min}$ and $f(\epsilon) := \exp(O(\frac{1}{\epsilon}\log(\frac{1}{\epsilon})))$ is a function depending solely on $\epsilon$.*

Note that the running time of our algorithm only has a singly exponential dependence on $\log \lambda$, and hence gives an exponential improvement over the result of Chen and Chakraborty [9]. Thus our algorithm can run over a substantially wider range of input instances, beyond just the ones with $\lambda = O(1)$. For example, even if $\lambda$ is polynomially large in $n$, our algorithm runs in time $n^{O(\log n)} = 2^{O(\log^2 n)}$ and hence yields a quasi-polynomial-time approximation scheme, as opposed to exponential time by the algorithm in [9].

## 1.3 High-Level Idea

As in Chen and Chakraborty [9], our result is also based on reducing the feasibility problem to vector scheduling in roughly $\ell = (1/\epsilon)\log \lambda$ dimensions. However, we crucially exploit the special structure of the vectors that arise in this transformation and give a faster vector scheduling algorithm for such instances. In fact, as we show in a companion paper [2], exploiting this structure is necessary to obtain any major improvement. In particular, in [2] we show that any PTAS for a general $\ell$-dimensional vector scheduling must incur a running time of $\exp((1/\epsilon)^{\Omega(\ell)})$ (under suitable complexity assumptions), and hence the running time in [9] is essentially the best one can hope for if one uses vector scheduling as a black-box.

The starting observation is that even though the vectors corresponding to tasks have $(1/\epsilon)\log \lambda$ coordinates, the number of relevant coordinates are essentially $1/\epsilon$. In particular, only $1/\epsilon$ consecutive coordinates of a vector can have "arbitrary" values, and all subsequent coordinates have an identical value (see (2) and Lemma 2 for the precise statement). This follows from the slack provided by the $1 + \epsilon$ speedup, as the demand bound function[3] of a task can be approximated so that a task $\tau$ has no "complicating" influence at time points $\geq d_\tau/\epsilon$.

To exploit this structure of vectors, we design a sliding-window based algorithm for vector scheduling, where we carefully build a schedule by considering the coordinates in left to right order, and only keeping track of the relevant short-range information in the dynamic program. The main technical difficulty is to combine the sliding-window approach with the exhaustive enumeration techniques of [8] for vector scheduling. In particular, to ensure that the sliding

---

[3] The reader unfamiliar with basic concepts of real-time scheduling such as demand bound functions, may wish to first look at Sections 2 and 3 before reading this part.

window does not build up too much error as it moves over the various coordinates, we keep track of different coordinates for a task with different accuracy. Moreover, to keep the running time low, we need more refined enumeration techniques to handle and combine small and large vectors.

## 2 Preliminaries

The input consists of a task system $\mathcal{T}$ consisting of tasks $\tau_1, \ldots, \tau_n$ and a set of $m$ identical processors. Each task releases a sequence of jobs throughout time. Each task $\tau$ is characterized by three parameters; the worst-case processing time $c_\tau$, the period $p_\tau$ and the relative deadline $d_\tau$. For notational convenience, we write $d_j$, $c_j$ and $p_j$ instead of $d_{\tau_j}$, $c_{\tau_j}$ and $p_{\tau_j}$. Without loss of generality, we assume that all these parameters are integers. The synchronous arrival sequence for $\mathcal{T}$ is defined to be the collection of job arrivals in which each task in $\mathcal{T}$ generates a job at time instant zero and subsequent jobs arrive as soon as permitted by the period parameters, i.e., task $\tau$ releases jobs at times $0, p_\tau, 2p_\tau, 3p_\tau, \ldots$ with deadlines $d_\tau, p_\tau + d_\tau, 2p_\tau + d_\tau, 3p_\tau + d_\tau, \ldots$. The utilization of a task $\tau$ is $u_\tau := \frac{c_\tau}{p_\tau} \leq 1$, and indicates the share of the processor used by this task in the long run. Without loss of generality, we assume that tasks are ordered such that $d_1 \leq \ldots \leq d_n$. We use $[n] := \{1, 2, \ldots, n\}$ to denote the set of integers from 1 up to $n$.

In the partitioned scheduling paradigm, we want to find a partition $\mathcal{T}_1, \mathcal{T}_2, \ldots, \mathcal{T}_m$ of $\mathcal{T}$ such that all jobs generated by the tasks in $\mathcal{T}_i$ can be feasibly scheduled on machine $i$, for all $i \in [m]$. Furthermore, task set $\mathcal{T}_i$ can be feasibly scheduled on machine $i$ if the synchronous arrival sequence for tasks in $\mathcal{T}_i$ can be scheduled feasibly by the Earliest Deadline First (EDF) algorithm. This implies that a task set $\mathcal{T}_i$ can be feasibly scheduled on the machine if and only if the total workload of the jobs generated by tasks in $\mathcal{T}_i$ that need to be finished by time $t$ is not more than the amount of work machine $i$ can do up to time $t$.

*Demand bound function:* It is known [5] that task system $\mathcal{T}_i$ is feasible upon a preemptive uniprocessor if and only if

$$\sum_{\tau \in \mathcal{T}_i} \max\left\{0, \left\lfloor \frac{t + p_\tau - d_\tau}{p_\tau} \right\rfloor c_\tau\right\} \leq t, \qquad \forall t > 0. \tag{1}$$

The term $\max\left\{0, \left\lfloor \frac{t+p_\tau-d_\tau}{p_\tau} \right\rfloor c_\tau\right\}$ is known as the *demand bound function* (dbf) of task $\tau$ at time point $t$ and is denoted by $\mathrm{dbf}_\tau(t)$. It expresses the amount of processing task $\tau$ needs up to time $t$. The left-hand side of (1) is called the dbf for the task set $\mathcal{T}_i$ and denoted by $\mathrm{dbf}_{\mathcal{T}_i}(t)$.

Condition (1) can be weakened slightly, and it is easy to see that it suffices to check (1) only for times $t$ that are deadlines of some job and $t \leq p_{lcm}$ where $p_{lcm}$ denotes the least common multiple of the tasks' periods. However, given that the feasibility testing problem is co-NP-hard, it is unlikely that the number of points where (1) must be tested can be reduced substantially.

As mentioned above, our goal is to develop a $(1+\epsilon)$-feasibility test for any $\epsilon > 0$. As we shall see soon, if we only care about $(1+\epsilon)$-feasibility, it suffices to check condition (1) at only $\log_{(1+\epsilon)}(d_n/(\epsilon^2 d_1)) \approx O(\log(d_n/d_1)/\epsilon)$ time points. This allows us to transform the feasibility problem into the so-called vector scheduling problem, which is defined as follows.

**Definition 1 (Vector Scheduling).** *We are given a set $A$ consisting of $n$ $d$-dimensional vectors with each coordinate in the range $[0,1]$ (i.e., vectors in $[0,1]^d$), and a positive integer $m$. The goal is to determine whether there is a partition of $A$ into $m$ sets $A_1, \ldots, A_m$ such that for each set $A_i$, the sum of vectors in that set does not exceed 1 in any coordinate. That is, $\max_{1 \leq i \leq m} \left\| \sum_{a \in A_i} a \right\|_\infty \leq 1$.*

Chekuri and Khanna [8] showed the following result for vector scheduling.

**Theorem 2 ([8]).** *Given any $\epsilon > 0$, there is a $(1+\epsilon)$-approximation algorithm, i.e., an algorithm that finds a partition with $\max_{1 \leq i \leq m} \left\| \sum_{a \in A_i} a \right\|_\infty \leq 1 + \epsilon$, for the vector scheduling problem that runs in time $n^{O(s)}$, where $s = (\frac{\ln d}{\epsilon})^d$.*

In the following section, we show how $(1+\epsilon)$-feasibility reduces to vector scheduling with $d = O((1/\epsilon) \log(d_n/d_1))$. While similar results have been used before (e.g., [9,12]), we will repeat the proof here, as we explicitly need the structure of the vectors in the resulting vector scheduling instance, which our algorithm will crucially exploit later.

## 3 From Sporadic Task System to Vector Scheduling

We begin with the notion of *approximate* demand bound functions. Observe that over the long run, a task $\tau$ uses $c_\tau$ units of time every $p_\tau$ units of time, but the relative deadlines, that may be different from the periods, complicate the demand bound function. The demand bound function has sharp jumps at the (absolute) deadlines $d_\tau, p_\tau + d_\tau, 2p_\tau + d_\tau, \ldots$, but the effects of these jumps become milder as time progresses. A machine that is $1+\epsilon$ times faster gives $\epsilon t$ units of extra processing time up to time $t$, which lets us ignore these sharp jumps after a certain point in time and instead it is sufficient to use the *utilization* (the average processing requirement).

The next lemma shows that we only need to check the demand bound function at time points which are a factor $1+\epsilon$ apart.

**Lemma 1.** *For any task $\tau$, if $\mathrm{dbf}_\tau((1+\epsilon)^k d_1) \leq (1+\epsilon)^k d_1 \alpha$ for all $k \in \mathbb{N}_{\geq 0}$, then $\mathrm{dbf}_\tau(t) \leq (1+\epsilon)\alpha t$ for all $t \geq 0$.*

*Proof.* For any $t$, define integer $k_t$ such that $(1+\epsilon)^{k_t-1} d_1 < t \leq (1+\epsilon)^{k_t} d_1$. Then
$$\mathrm{dbf}_\tau(t) \leq \mathrm{dbf}_\tau((1+\epsilon)^{k_t} d_1) \leq (1+\epsilon)^{k_t} d_1 \alpha < (1+\epsilon)\alpha t,$$
where the first inequality follows from the demand bound function being non-decreasing. □

We consider an approximate demand bound function $\mathrm{dbf}^*_\tau(t)$ used by Marchetti-Spaccamela et al. [12]. Let $L$ be the smallest integer such that $1 \leq (1+\epsilon)^{L-1}\epsilon^2$. Note that $L \leq 2 + \log_{(1+\epsilon)}(1/\epsilon^2)$. Let

$$\mathrm{dbf}^*_\tau(t) = \begin{cases} \left\lfloor \frac{t+p_\tau - d_\tau}{p_\tau} \right\rfloor c_\tau & \text{if } t < (1+\epsilon)^L d_\tau, \\ u_\tau t & \text{otherwise.} \end{cases} \quad (2)$$

Note that $\mathrm{dbf}^*$ differs from $\mathrm{dbf}$ only when $t \geq d_\tau(1+\epsilon)/\epsilon^2$, and is proportional to the utilization of $\tau$ in that case. The following lemma shows that it is a good approximation to $\mathrm{dbf}$.

**Lemma 2 ([12]).** *For every task $\tau$ and every time $t \geq 0$,*

$$\frac{1}{(1+\epsilon)} \mathrm{dbf}_\tau(t) \leq \mathrm{dbf}^*_\tau(t) \leq (1+\epsilon) \mathrm{dbf}_\tau(t).$$

Another obvious property of $\mathrm{dbf}$ and $\mathrm{dbf}^*$ is the following, which allows us to start our feasibility analysis at the first deadline only.

**Observation 1.** *For all tasks $\tau$, for all $t < d_\tau$, we have that*

$$\mathrm{dbf}_\tau(t) = \mathrm{dbf}^*_\tau(t) = \max\{0, \lfloor (t+p_\tau - d_\tau)/p_\tau \rfloor\} = 0.$$

*In particular, $\mathrm{dbf}_\tau(t) = \mathrm{dbf}^*_\tau(t) = 0$ for all $t < d_1$ and all tasks $\tau \in \mathcal{T}$.*

Using Lemma 1, Lemma 2 and Observation 1, we can encode our approximate demand bound function $\mathrm{dbf}^*_\tau$ into a vector $w^\tau$. More precisely, we will use a *normalized* demand bound function which is $\mathrm{dbf}^*_\tau(t)/t$. Let $t_{end} = (1+\epsilon)^L d_n$, and let $K := \lceil \log_{(1+\epsilon)} t_{end}/d_1 \rceil$. For each task $\tau$ we define the vector $w^\tau$, with coordinates $w^\tau_k$ as follows:

$$w^\tau_k := \begin{cases} \frac{\mathrm{dbf}^*_\tau((1+\epsilon)^{k-1} d_1)}{(1+\epsilon)^{k-1} d_1} & \text{if } k = 1, \ldots, K-1, \\ u_\tau & \text{if } k = K. \end{cases}$$

Note that the first $K-1$ coordinates of these vectors consider times that are powers of $(1+\epsilon)$ and lie between $d_1$ and $(1+\epsilon)^L d_n$. Recall that for $t \geq t_{end}$, it holds that $\mathrm{dbf}^*_\tau(t) = u_\tau t$ for each task $\tau_1, \ldots, \tau_n$. Thus, there is no need to consider additional coordinates. The coordinate $K$ is equal to the utilization and will play a special role in our algorithm.

We note the following structural property of the vectors $w_\tau$.

**Observation 2.** *A task $\tau$ is associated to a vector $w^\tau$ from $[0,1]^K$ with $K := 1 + \lceil \log_{(1+\epsilon)} \frac{d_n}{\epsilon^2 d_1} \rceil$:*

$$w^\tau_k := \begin{cases} 0 & \text{if } k \leq k_\tau, \\ \frac{\mathrm{dbf}^*_\tau(t_k)}{t_k} & \text{if } k = k_\tau + 1, \ldots, k_\tau + L, \\ u_\tau & \text{otherwise,} \end{cases} \quad (3)$$

*where $t_k = (1+\epsilon)^k d_1$ and $k_\tau = \lceil \log_{(1+\epsilon)}(d_\tau/d_1) \rceil - 1$.*

*In particular, each vector has initial coordinates zero, followed by at most $L$ entries of arbitrary value, followed by all entries equal to $u_\tau$.*

The following theorem connects the vector scheduling problem formally to the sporadic task system scheduling, and follows directly from Lemmas 1 and 2, (1) and Observation 1.

**Theorem 3.** *Define the vectors $w^\tau$ as in (3). Given is a partition of vectors $w^\tau$ into $m$ sets $W_1, \ldots, W_m$ and the corresponding partition of tasks $\tau \in \mathcal{T}$ into $m$ sets $\mathcal{T}_1, \ldots, \mathcal{T}_m$. Then, for all machines $i$,*

*(i) if $\left\|\sum_{w^\tau \in W_i} w^\tau\right\|_\infty \leq \alpha$, then $\mathrm{dbf}_{\mathcal{T}_i}(t) \leq (1+\epsilon)^2 \alpha t$ for all $t \geq 0$;*
*(ii) if $\mathrm{dbf}_{\mathcal{T}_i}(t) \leq \alpha t$ for all $t \geq 0$, then $\left\|\sum_{w^\tau \in W_i} w^\tau\right\|_\infty \leq (1+\epsilon)\alpha.$*

This theorem tells us that if we can partition the set of vectors $w^\tau \in W$ into sets $W_1, \ldots W_m$ such that $\|\sum_{w^\tau \in W_i} w^\tau\|_\infty \leq 1 + \epsilon$, for all $i \in [m]$, then we can feasibly schedule the corresponding tasks in set $\mathcal{T}_i$ on machine $i$ if this machine receives a speedup factor $(1+\epsilon)^3$. Moreover, if $\mathcal{T}$ can be partitioned into sets $\mathcal{T}_i$ such that each of these can be scheduled on a unit-speed machine, then the corresponding sets of vectors $W_i$ satisfy $\|\sum_{w^\tau \in W_i} w^\tau\|_\infty \leq 1 + \epsilon$, for all $i \in [m]$.

Thus, a $(1+\epsilon)$-approximation for vector scheduling implies a $(1+\epsilon)^2(1+\epsilon) = 1 + O(\epsilon)$-feasibility test for partitioned scheduling. The result of Chen and Chakraborty [9] follows directly from this connection, and applying Theorem 2. In the next section we show how the running time can be improved for vector scheduling by exploiting the special structure of the vectors $w^\tau$ as described in Observation 2.

## 4 Solving the Special Case Vector Scheduling Problem

In this section we develop a substantially faster $(1+\epsilon)$-approximation algorithm for vector scheduling, which is specifically tailored towards vectors described in Observation 2. We combine several techniques from bin packing and vector scheduling and design a "sliding window" dynamic programming approach. The time complexity of our algorithm is given in the following theorem. Note that this theorem and Theorem 3 of the previous section suffice to prove Theorem 5.

**Theorem 4.** *Given $\epsilon > 0$, let $C = \left(\left\lceil\frac{8L+19}{\epsilon}\right\rceil\right)^L \left\lceil\frac{K(8L+19)}{\epsilon}\right\rceil$ where $L = 1 + \lceil\log_{(1+\epsilon)}(1/\epsilon^2)\rceil$ and $K = 1 + \lceil\log_{(1+\epsilon)}(d_n/\epsilon^2 d_1)\rceil$. Then, given a set of vectors $W$ from $[0,1]^K$ as defined in Observation 2, Algorithm 1 determines in $O(m^{O(C)})$ time whether the set of vectors $W$ can be scheduled on $m$ machines such that in every coordinate the load is at most $1+\epsilon$, or whether no feasible assignment exists.*

*Proof.* The theorem follows easily from Lemma 3 which will follow in Section 4.3. Setting $\eta := \epsilon/(8L+19)$ in Lemma 3 leads to a schedule with height at most $1 + (8L+19)\eta = 1 + \epsilon$. □

The main idea of the algorithm is, after some rounding of the vectors, to first classify the vectors, then determine how the vectors of one class can possibly be scheduled and finally to combine the schedule of the classes into one overall schedule. To give a high-level overview of our algorithm in Section 4.2 and some of the details in the subsequent subsection, we first need to introduce some notation and concepts in the following subsection.

### 4.1 Notation and Definitions

Given $\epsilon > 0$, let $L$ and $K$ be defined as above. Let $0 < \eta < 1$ be a small constant and define $\delta = \eta/K$.

We associate each task $\tau$ to a vector $w^\tau$ from $[0,1]^K$ as defined in (3). We classify these vectors into several classes depending on the index of the first non-zero coordinate. Hereto, we say that a vector is a $t$-**vector** if its first non-zero coordinate is coordinate $t$.

A $t$-**configuration** is an $(L+1)$-tuple $(f_1, \ldots, f_L, f_u)$ with, for all $k \in [L]$, $f_k \in \{0, \eta, 2\eta, \ldots, \eta\lfloor 1/\eta \rfloor, 1\}$ and $f_u \in \{0, \delta, 2\delta, \ldots, \delta\lfloor 1/\delta \rfloor, 1\}$. We say that (a set of vectors assigned to) a machine $i$ **conforms to** a $t$-configuration $f = (f_1, \ldots, f_L, f_u)$ if the contribution to coordinate $t - 1 + k$ is at most $f_k$, for all $k \in [L]$, and if the contribution to all coordinates $k \geq t + L$ is at most $f_u$. As the first $L$ elements in a $t$-configuration can attain one of $\lceil 1/\eta \rceil$ different values and the last element can attain one of $\lceil 1/\delta \rceil$ different values, the number of different $t$-configurations, denoted by $C$, is $C := \left(\left\lceil \frac{1}{\eta} \right\rceil\right)^L \left\lceil \frac{1}{\delta} \right\rceil$.

A $t$-**profile** $Q$ defines a $t$-configuration for each machine. Therefore, it can be represented by an $m$-tuple $Q = (q_1, \ldots, q_m)$ where $q_i$ denotes the $t$-configuration corresponding to machine $i$. On the other hand, as the number of $t$-configurations is bounded by $C$ and the machines are identical, a $t$-profile can also be represented by a $C$-tuple $Q = \langle n_1, \ldots, n_C \rangle$ where $n_f$ denotes the number of machines that conform to configuration $f$. As the numbers $n_f$ sum up to $m$, we find that the number of different $t$-profiles is at most $m^C$.

Finally, we define the addition of a $t$-profile $Q$ and a vector $e = (e_1, \ldots, e_L, e_u) \in [0,1]^{L+1}$, $Q + e = Q' = (q'_1, \ldots, q'_m)$, as the pointwise addition of the vector $e$ to each configuration $q_i \in Q$, i.e., $q'_i = q_i + e$ for all $i \in [m]$.

### 4.2 Overview of the Algorithm

Our algorithm, given in Algorithm 1, determines whether we can feasibly schedule all vectors with a load of at most $1 + \epsilon$ in every coordinate on each machine. It first applies two rounding steps (see Step 2 and 3), to limit the number of different vectors.

In Step 4 of the algorithm, we determine for each $t = 1, \ldots, K$ and $t$-profile $R$ whether all $t$-vectors can be scheduled to conform to $R$. Due to lack of space the proof of this is omitted.

Once we know, for every $t$, conforming to which $t$-profiles the set of all $t$-vectors can be scheduled, we can determine conforming to which $t$-profiles all vectors together can be scheduled. Hereto, we design a *sliding window DP* to determine whether all $k$-vectors ($k < t$) can be combined with all $t$-vectors to conform to a given $t$-profile $Q$ (Section 4.3). The final result can then easily be obtained by taking $t = K$ and $Q$ equal to the all-1 profile, i.e., $q_i = 1$ for all $i$. When $T[K, 1]$ returns true, Algorithm 1 also can be used to find the corresponding solution.

Both Step 4 and Step 5 of Algorithm 1 need to be able to determine whether a $t$-profile $R$ and a $(t-1)$-profile (or $t$-profile) $S$ can be combined into a $t$-profile $Q$. This can be determined in advance in $O(m^{O(C)})$ time, but the proof is omitted.

---

**Algorithm 1.** Vector Scheduling algorithm
---
**Require:** Input: a set $W$ of vectors $w^\tau$ as defined in Section 3, and $\eta > 0$.
1: Define $\delta := \eta/K$.
2: For each vector $w^\tau$ round each component $w_k^\tau$ down to the nearest power of $\frac{1}{1+\eta}$.
3: Modify each vector
$$z_k^\tau := \begin{cases} 0 & \text{if } w_k^\tau \leq \delta \, \|w^\tau\|_\infty, \\ w_k^\tau & \text{otherwise.} \end{cases}$$
4: Determine whether all $t$-vectors can be scheduled conforming to $t$-profile $R$, for all possible $t$-profiles $R$ and all $t$.
5: Let $T[t, Q]$ be true if all $k$-vectors with $k \leq t$ can be scheduled conforming to $t$-profile $Q$, and false otherwise. Determine $T[t, Q]$ for all possible $t$-profiles $Q$ and all $t$.
6: Return $T[K, 1]$.

---

### 4.3 The Sliding Window Dynamic Program

In this subsection, we introduce a dynamic program to determine whether all $k$-vectors with $k \leq t$ can be scheduled conforming to $t$-profile $Q$. To be precise, we compute the values $T[t, Q]$, which essentially evaluates to true if all $k$-vectors with $k \leq t$ can be scheduled conforming to $t$-profile $Q$. The dynamic program works in $K$ phases as it moves from the first coordinate to coordinate $K$. While scheduling all $t$-vectors in a certain phase $t$, the DP also looks ahead to the next $L - 1$ coordinates and the last utilization coordinate to ensure no conflicts arise in these coordinates. That is, we slide a window covering $L$ coordinates from coordinate 1 to coordinate $K$ in as many phases.

Intuitively, phase $t$ corresponds to scheduling the $t$-vectors, given a partial schedule for all $k$-vectors with $k < t$. To determine the value of $T[t, Q]$, we split the $t$-profile $Q$ into a $t$-profile $R$ and a $(t-1)$-profile $S$ that capture the division of space per machine and per coordinate between the $t$-vectors and the other $k$-vectors with $k < t$.

Since the $t$-configurations are "coarse valued" (all values are multiples of either $\eta$ or $\delta$), it is unclear how to split the $t$-profile $Q$: perhaps a coordinate $f_k$ of the $t$-configuration can be split into two parts yielding a feasible $t$-profile $R$ and a $(t-1)$-profile $S$, but not in such a way that the two parts are multiples of $\eta$. In that case, the corresponding DP-cell is erroneously evaluated to false. To circumvent this issue, an additional small error in each phase of the sliding window DP is allowed. For this reason the vector $(\eta, \ldots, \eta, \delta)$ is added to the $(t-1)$-profile $S$.

The boolean value $B[t, R]$, that essentially denotes whether all $t$-vectors can be scheduled conforming to $t$-profile $R$, can be precomputed in $O(m^{O(C)})$ time (proof is omitted). Once these values are known, the recursive formula for $T$ can be easily computed by considering all possible combinations of $t$-profiles $R$ and $(t-1)$-profiles $S$ that can be combined to a $t$-profile $Q$, and determining whether or not all $t$-vectors can be scheduled conforming to $R$ and all other $k$-vectors with $k < t$ can be scheduled conforming to $S + (\eta, \ldots, \eta, \delta)$. That is, for $t > 1$,

$$T[t, Q] = \bigvee_{(R,S) \in \mathcal{W}(Q)} (B[t, R] \wedge T[t-1, S+(\eta, \ldots, \eta, \delta)]), \tag{4}$$

where $\mathcal{W}(Q)$ contains all tuples $(R, S)$ of $t$-profiles $R$ and $(t-1)$-profiles $S$ that $Q$ can be split into. The base case of the recursion is

$$T[1, Q] = B[1, Q]. \tag{5}$$

To evaluate the running time of computing $T[K, Q]$, we note that $B[t, R]$ is precomputed and can be accessed in $O(1)$ time. The proof of the following lemma is omitted due to lack of space.

**Lemma 3.** *Let $W$ be a set of vectors $w^\tau$ as defined in (3) and let $\eta > 0$ be small enough. Algorithm 1 decides in $O(Km^{O(C)})$ time whether there exists a partition of the vectors $W$ into $m$ sets $W_1, \ldots, W_m$ such that $\left\|\sum_{w^\tau \in W_i} w^\tau\right\|_\infty < 1 + (8L + 19)\eta$ for all $i$, or that there does not exist a partition with $\left\|\sum_{w^\tau \in W_i} w^\tau\right\|_\infty \leq 1$.*

Note that if we choose $\eta = \epsilon/(8L+19)$, we prove Theorem 4 and find a partition of height at most $1 + \epsilon$.

## 5 Conclusion

Combining Theorem 3 and Theorem 4 yields the desired result.

**Theorem 5.** *Given $\epsilon > 0$, a task set $\mathcal{T}$ and $m$ parallel identical processors, there is an algorithm which correctly decides in $O\left(m^{O(f(\epsilon) \log \lambda)}\right)$ time whether $\mathcal{T}$ can be feasibly partitioned with a speedup of $1 + \epsilon$, or no feasible partition exists in case the machines run at unit speed, where $\lambda = d_n/d_1$, the ratio between the largest and smallest deadline, and $f(\epsilon)$ is a function depending solely on $\epsilon$.*

*Proof.* Theorem 4 determines in $O(m^{O(C)})$ time whether a feasible solution to the vector scheduling problem exists with a speedup factor of $1 + \epsilon$, or whether no such partition of the vectors to the machines exists without speedup. Thus in light of Theorem 4, Theorem 3 implies that if there exists a feasible partition for the vector scheduling problem, then this partition is feasible for our real-time scheduling problem if the machines receive a speedup factor of $(1+\epsilon)^3$, and that if no feasible partition for the vector scheduling problem exists, then no feasible partition exists for the real-time scheduling problem in case the machines run at speed $1/(1+\epsilon)$. Rescaling $\epsilon$ appropriately yields the stated result.

# References

1. Baker, T.P., Baruah, S.K.: Schedulability analysis of multiprocessor sporadic task systems. In: Handbook of Real-Time and Embedded Systems, ch. 3. CRC Press (2007)
2. Bansal, N., Vredeveld, T., van der Zwaan, R.: Approximating vector scheduling: Almost matching upper and lower bounds. In: Pardo, A., Viola, A. (eds.) LATIN 2014. LNCS, vol. 8392, pp. 47–59. Springer, Heidelberg (2014)
3. Baruah, S., Fisher, N.: The partitioned multiprocessor scheduling of sporadic task systems. In: Proceedings of 26th IEEE Real-Time Systems Symposium, pp. 321–329. IEEE (2005)
4. Baruah, S., Goossens, J.: Scheduling real-time tasks: Algorithms and complexity. In: Leung, J.Y.-T. (ed.) Handbook of Scheduling: Algorithms, Models and Performance Evalution, ch. 28. CRC Press (2004)
5. Baruah, S., Mok, A., Rosier, L.: Preemptively scheduling hard-real-time sporadic tasks on one processor. In: Proceedings of 11th IEEE Real-Time Systems Symposium, pp. 182–190. IEEE (1990)
6. Baruah, S.K., Pruhs, K.: Open problems in real-time scheduling. Journal of Scheduling 13, 577–582 (2010)
7. Chakraborty, S., Künzli, S., Thiele, L.: Approximate schedulability analysis. In: Proceedings of 23rd IEEE Real-Time Systems Symposium, pp. 159–168. IEEE (2002)
8. Chekuri, C., Khanna, S.: On multi-dimensional packing problems. SIAM Journal on Computing 33(4), 837–851 (2004)
9. Chen, J.-J., Chakraborty, S.: Partitioned packing and scheduling for sporadic real-time tasks in identical multiprocessor systems. In: Proceedings of 24th Euromicro Conference on Real-Time Systems, pp. 24–33 (2012)
10. Eisenbrand, F., Rothvoß, T.: EDF-schedulability of synchronous periodic task systems is coNP-hard. In: Proceedings of 21st Symposium on Discrete Algorithms, pp. 1029–1034 (2010)
11. Liu, C., Layland, J.: Scheduling algorithms for multiprogramming in a hard real-time environment. Journal of the ACM 20, 46–61 (1973)
12. Marchetti-Spaccamela, A., Rutten, C., van der Ster, S., Wiese, A.: Assigning sporadic tasks to unrelated parallel machines. In: Czumaj, A., Mehlhorn, K., Pitts, A., Wattenhofer, R. (eds.) ICALP 2012, Part I. LNCS, vol. 7391, pp. 665–676. Springer, Heidelberg (2012)

# Integrated Supply Chain Management via Randomized Rounding

Lehilton L.C. Pedrosa[1,*] and Maxim Sviridenko[2,**]

[1] Institute of Computing, University of Campinas, Brazil
[2] Department of Computer Science, University of Warwick, UK

**Abstract.** We consider the supply chain problem of minimizing ordering, distribution and inventory holding costs of a supply chain formed by a set of warehouses and retailers over a finite time horizon, that we call *Production and Distribution Problem* (PDP). This is a common generalization of the classical Metric Facility Location Problem and Joint Replenishment Problem, that coordinates the network design and inventory management decisions in an integrated manner. This coordination can represent significant economy for many applications, where network design and operational costs are normally considered separately. This problem is considered when the instances satisfy assumptions such as metric space of warehouse and retailer locations, and monotonic increasing inventory holding costs. In this work, we give a 2.77-approximation based on the randomized rounding of the natural mixed integer programming relaxation. Also, we give a 5-approximation for the case that objective function includes retailer ordering costs.

## 1 Introduction

We consider a supply chain management problem of minimizing ordering, distribution and inventory holding costs of a supply chain formed by a set of warehouses and retailers. Each retailer can face a demand in a given time of a discretized planning horizon. This demand should be satisfied by the items currently held in the inventory, that were previously ordered and transported from any of the warehouses. The objective is to determine an inventory replenishment policy for each retailer, minimizing the overall cost of inventory holding, and transportation. These costs are balanced by a fixed ordering setup cost that depends on the warehouse, but is independent of the number of items produced.

Traditionally, network design, distribution and inventory replenishment decisions are made separately. In the location theory, the literature focuses on the strategic decisions of network design, such as where to place facilities and how to assign one facility to each client. On the other hand, in the inventory theory, a static network design is usually assumed. Such a static network design

---

* Research supported by FAPESP (grant number 2012/17634-0). This work was done while the author was visiting the Dep. of Computer Science, University of Warwick.
** Research supported by EPSRC grant EP/J021814/1, EP/D063191/1, FP7 Marie Curie Career Integration Grant and Royal Society Wolfson Research Merit Award.

has defined assignments between facilities and clients, and the decisions are concentrated on determining replenishment policies for the inventory. However, the lack of coordination between inventory and shipment costs when determining the network design leads to sub-optimality [12].

*Problem definition.* We consider a generalization of the Facility Location Problem (FLP), that we call *Production and Distribution Problem* (PDP). We are given a set of warehouses $P$, and a set of retailers $Q$. Each retailer $q$ may face a demand for $d_{qt} \in \mathbb{Z}_+$ units of item in each time $t$ of a discretized planning horizon with steps $\{1, \ldots, T\}$. Each demand can only be satisfied with items that are currently held in the stock of the corresponding retailer, *i.e.*, we have a make-to-stock scenario. The stock is initially empty, and may be replenished by placing orders to any warehouse. There is no stock at warehouse facilities, so every time $s$ that a warehouse $p$ receives an order, the demanded items are produced at a setup cost $k_p$, that is independent of the number of items produced, or the number of retailers participating in the order. Once the items have been produced, each unit is transported to the requesting retailer $q$ at a cost $c_{pq}$. We assume that the transportation time is negligible, so each unit of item is held in stock of retailer $q$ from the time $s$ it is was produced until the time $t$ it was delivered. The holding cost incurred for this item is $h_{qst}$. The objective is to minimize the overall sum of ordering, distribution, and holding costs. An equivalent mixed integer programming (MIP) formulation is given in Subsection 2.2.

*Related works.* The literature on integrated supply chain problems have considered several design and inventory problems, with different network structures, objective functions, and constraints. For reviews, see [12,10]. In different models, each retailer must be assigned to one supplier [16,12,8], or can be assigned to different suppliers in different times [11,1]. Several approaches have been used to deal with these problems, such as Lagrangean relaxation [5], column generation [13,16], and meta-heuristics [3,1]. Pochet and Wosley have discussed valid inequalities for the mixed integer programming formulation of a generalized PDP [11, Section 13.4] with warehouse storage and production capacity. Approximation algorithms have been proposed for few problems, such as the Warehouse-Retailer Network Design Problem (WRND) [16,8], and the Stochastic Transportation-Inventory Network Design Problem (STIND) [14,8]. Both admit 3-approximation factors via primal-dual [8]. These problems have infinite time horizons, with constant (WRND), or stochastic demand rates (STIND).

*Our results and techniques.* We study the PDP under a natural assumption that the transportation and holding cost functions satisfy a generalization of the triangle inequality. The intuition for this assumption is that in many applications it is cheaper to transport one item from the warehouse to the retailer of destination directly, rather than using other retailers as storage midpoints. The main contribution of this work is a 2.77-approximation for the PDP. Our algorithm is based on the randomized rounding of the natural LP relaxation, and uses clustering of demand points, in the spirit of FLP algorithms [15,4,2], but has to carefully select the order to place for each cluster, due to the additional temporal restriction. This extra step leads to a worse service approximation factor, when compared to the standard FLP. To balance the ratios for ordering and service costs, we use two different approaches. In the first approach, we place two orders for each cluster. This results in high ordering cost, but reduces significantly the expected service cost. In the second approach, we use the filtering technique [9] parameterized by some $\alpha$ to obtain the opposite imbalance on the approximation guarantee. Combining the two approaches is done by the use of a probability distribution over the choice of parameter $\alpha$, or the use of the first approach. We also consider the PDP when there is a positive setup cost $k_q$ for each time a retailer $q$ places an order. For this variant, we give a 5-approximation.

In Section 2, we describe the assumptions on the service cost structure, give a mixed integer linear program formulation for the PDP, and review the filtering technique. In Section 3, we present the approximation for the PDP. Finally, in Section 4, we extend this algorithm to obtain an approximation for the PDP with retailer ordering costs.

## 2 Assumptions and Basic Techniques

### 2.1 Holding and Transportation Costs Model

The *service cost* of a demand point is the sum of distribution and inventory holding costs. We describe assumptions on the cost functions in the following.

For the most inventory problems considered in the literature, the holding cost is modeled on a *per unit* and *per time step* basis, that is, in traditional inventory models a non-negative cost $h_t^q$ is incurred to hold one unit of item from time step $t$ to time step $t+1$ in the stock of retailer $q$. For the PDP, the holding cost is modeled by the more general function $h_{qst}$. We assume that this holding cost function is monotonic, that is, the holding cost can only decrease if the period that an item is kept in the stock is shortened.

**Assumption 1.** *Fix a retailer $q$. Let $s, s', t, t'$ be time steps such that $s \leq t$ and $s' \leq t'$. If $[s, t] \subseteq [s', t']$, then $h_{qst} \leq h_{qs't'}$.*

For network design problems, such as the Facility Location Problem and $k$-medians, it is common to make the assumption that facilities and clients are in a metric space, that is, the distance between facilities and clients is a symmetric function that satisfies the triangle inequality. Indeed, if no restrictions on the distance function are made, then Facility Location Problem is hard to approximate by a factor better than $O(\log n)$. For many distribution problems, however, the assumption of triangle inequality can be made without loss of generality. The reason is that one can create a modified instance where the new distance function is defined the by the lengths of the shortest paths in the original graph. This new instance can be solved assuming the triangle inequality, and a solution of non-greater cost for the original problem can be obtained by rerouting direct routes by the shortest paths of the graph. We define an analogous notion for the PDP.

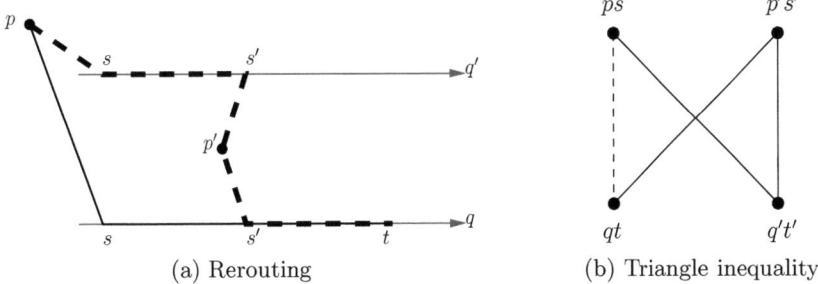

**Fig. 1.** Mixed holding and transportation costs metric

**Assumption 2.** *For all retailers $q$ and $q'$, warehouses $p$ and $p'$, and time steps $s, s'$ and $t$, with $s \leq s' \leq t$, it holds: $c_{pq} + h_{qst} \leq c_{pq'} + h_{q'ss'} + c_{p'q'} + c_{p'q} + h_{qs't}$.*

This assumption states that the cost to transport one item from the warehouse $p$ directly to the retailer $q$, and holding it in $q$ until the delivery time $t$ is cheaper than the following alternative route: transporting the item from the warehouse $p$ to the retailer retailer $q'$, holding it from time $s$ to time $s'$, then transporting it again from retailer $q'$ to retailer $q$ through warehouse $p'$, and holding it until time $t$ (see Figure 1a). This inequality can also be interpreted in the following way: the cost to serve demand point $(q,t)$ directly by order $(p,s)$ is not greater then the overall cost to serve demand $(q,t)$ by order $(p',s')$ and some demand $(q',t')$ by orders $(p,s)$ and $(p',s')$ (Figure 1b). Since this inequality resembles the triangle inequality, we say that a pair of holding and transportation costs that satisfy Assumption 2 forms a metric service cost for the PDP.

## 2.2 A Linear Programming Relaxation

The PDP has a natural formulation as a MIP problem. This is analogous to that of facility location problems, where a client is a demand point, and a facility is a warehouse order. For PDP, however, a demand can only be served by orders placed before the arrival time. In the following, let $\mathcal{D} \subseteq Q \times [T]$ be the set of all positive demand points. Also, let $\mathcal{P} = P \times [T]$ be the set of all potential warehouse orders, and for each $t \in [T]$, let $\mathcal{P}_t = P \times [t]$ be the set of all warehouse orders that can serve a demand point at time $t$. Variable $x_{ps}^{qt}$ means that demand $(q,t)$ is served by order $(p,s)$, and $y_{ps}$ means means that order $(p,s)$ is placed.

$$
\begin{aligned}
\text{minimize} \quad & \sum_{(p,s) \in \mathcal{P}} y_{ps} k_p + \sum_{(q,t) \in \mathcal{D}} \sum_{(p,s) \in \mathcal{P}_t} x_{ps}^{qt} d_{qt}(h_{qst} + c_{pq}) \\
\text{subject to} \quad & \sum_{(p,s) \in \mathcal{P}_t} x_{ps}^{qt} \geq 1 & (q,t) \in \mathcal{D} \\
& x_{ps}^{qt} \leq y_{ps} & (q,t) \in \mathcal{D}, (p,s) \in \mathcal{P}_t \\
& x_{ps}^{qt} \geq 0 & (q,t) \in \mathcal{D}, (p,s) \in \mathcal{P}_t \\
& y_{ps} \in \{0,1\} & (p,s) \in \mathcal{P}
\end{aligned} \quad (1)
$$

A linear relaxation can be obtained by replacing the integrality constraints by constraints $y_{ps} \geq 0$ for all $(p, s) \in \mathcal{P}$. The dual program of the relaxation is

$$\text{maximize} \sum_{(q,t) \in \mathcal{D}} b_{qt}$$

$$\begin{aligned}
\text{subject to } & b_{qt} \leq d_{qt}(h_{qst} + c_{pq}) + z_{ps}^{qt} & (q,t) \in \mathcal{D}, (p,s) \in \mathcal{P}_t \\
& \sum_{(q,t) \in \mathcal{D}: t \geq s} z_{ps}^{qt} \leq k_p & (p,s) \in \mathcal{P} \\
& z_{ps}^{qt}, b_{qt} \geq 0 & (q,t) \in \mathcal{D}, (p,s) \in \mathcal{P}
\end{aligned} \quad (2)$$

Given any feasible solution $(x, y)$, we define for each demand $j = (q, t)$, the set $\mathcal{S}_j$ of all orders $(p, s) \in \mathcal{P}_t$ such that $x_{ps}^{qt} > 0$. This set is called the *service set* of $j$, and the interval $[\min_{(p,s) \in \mathcal{S}_j} s, t]$ is its *service window*. The fractional service cost of demand $j$ is $S_j = C_j + H_j$, where $C_j = \sum_{(p,s) \in \mathcal{S}_j} x_{ps}^{qt} d_{qt} c_{pq}$ is the fractional distribution cost of $j$, and $H_j = \sum_{(p,s) \in \mathcal{S}_j} x_{ps}^{qt} d_{qt} h_{qst}$ is the fractional holding cost of $j$.

Let $(x^*, y^*)$ be an optimal solution for the LP relaxation of a given PDP instance. The cost is divided in ordering cost $K^* = \sum_{(p,s) \in \mathcal{P}} y_{ps}^* k_p$, and service cost $S^* = C^* + H^*$, where $C^* = \sum_{(q,t) \in \mathcal{D}, (p,s) \in \mathcal{P}_t} x_{ps}^{*qt} d_{qt} c_{pq}$ is the distribution cost, and $H^* = \sum_{(q,t) \in \mathcal{D}, (p,s) \in \mathcal{P}_t} x_{ps}^{*qt} d_{qt} h_{qst}$ is the holding cost.

### 2.3 Complete Solutions and Filtering

**Complete Solutions.** In a complete solution, each demand point is fully served by the most economical orders. This is formalized as following: for a fixed demand point $j = (q, t)$, consider a permutation $\pi_j$ of warehouse orders $\mathcal{P}_t$ such that the elements are listed in non-decreasing order of service cost, that is, if $\pi_j = ((p_1, s_1), \ldots, (p_k, s_k))$, where $k = |\mathcal{P}_t|$, then $c_{p_1 q} + h_{q s_1 t} \leq \cdots \leq c_{p_k q} + h_{q s_k t}$. Assume that for each demand point $j$, permutation $\pi_j$ is unique, by breaking ties arbitrarily, but in a fixed way (two elements appear in the same order in all permutations in which they tie). A solution $(x, y)$ of the LP is said to be *complete* if for every demand point $(q, t)$ there is an index $l$, such that $x_{p_i s_i}^{qt} = y_{p_i s_i}$ if $i \leq l$, and $x_{p_i s_i}^{qt} = 0$ if $i > l$.

Any feasible solution can be transformed into a complete solution of no greater cost. Indeed, let $(x, y)$ be a feasible solution of the LP. We create a new solution $(\bar{x}, \bar{y})$, such $\bar{y} = y$, and $\bar{x}$ is given by serving each demand point $j = (q, t)$ greedily by the orders in the permutation $\pi_j$. More precisely, for each $(q, t) \in \mathcal{D}$, let $l$ be minimum such that $\sum_{i=1}^{l} y_{p_i s_i} \geq 1$. Since $(x, y)$ is feasible, we know that there is such an $l$. Now, we define $\bar{x}_{p_i s_i}^{qt} = y_{p_i s_i}$ for each $i \leq l$, and $\bar{x}_{p_i s_i}^{qt} = 0$ for each $i > l$. We assume without loss of generality that $(\bar{x}, \bar{y})$ is complete, that is, $\sum_{i=1}^{l} y_{p_i s_i} = 1$. In the case that the solution is not complete, we can always replace $p_l$ by two warehouses $p'_l$ and $p''_l$ at the same location, and split its fractional ordering $y_{p_l s_l}$ between $p'_l$ and $p''_l$, such that $y_{p'_l s_l} = 1 - \sum_{i=1}^{l-1} y_{p_i s_i}$, and $y_{p''_l s_l} = y_{p_l s_l} - y_{p'_l s_l}$. Repeating this for each demand point, we obtain an equivalent instance with corresponding complete solution (the arguments are completely analogous to Lemma 1, in [15]).

**Filtering.** This technique was introduced in by Lin and Vitter [9], and was used by many algorithms for the FLP [15,2]. The idea is that if $(x, y)$ is a complete solution, then for each demand $j = (q, t)$ we can consider only the subset of orders in $\mathcal{S}_j$ that is the most economical. This subset is formed by the orders in the minimal prefix of permutation $\pi_j$ that serves an $\alpha$ fraction of the demand. Formally, given a parameter $\alpha \in (0, 1]$, let $l$ be the minimal index such that $\sum_{i=1}^{l} y_{p_i s_i} \geq \alpha$. The $\alpha$-*neighborhood* of a demand point $j$ is the set $\mathcal{N}_j(\alpha) = \{(p_i, s_i) : i \leq l\}$. The radius $R_j(\alpha)$ of this neighborhood is the maximum cost paid to serve $j$ by an order in $\mathcal{N}_j(\alpha)$, that is, $R_j(\alpha) = \max_{(p,s) \in \mathcal{N}_j(\alpha)} d_{qt}(c_{pq} + h_{qst})$.

Intuitively, if we increase the amount of fractional orders of an LP solution, then the average service cost of a given demand should decrease. Indeed, given a solution $(x, y)$, the filtering technique consists of scaling up the $y$ variables by $1/\alpha$, then defining $x$ such that each demand is fully satisfied by orders in its $\alpha$-neighborhood. We will obtain a complete solution $(\bar{x}, \bar{y})$ (by splitting warehouse fractional ordering if necessary). For a demand $j = (q, t)$, we denote the average service cost by $W_j(\alpha) = \sum_{(\hat{p}, \hat{s}) \in \mathcal{N}_j(\alpha)} \bar{x}_{\hat{p}\hat{s}}^{qt} d_{qt}(c_{\hat{p}q} + h_{q\hat{s}t})$.

## 3 Approximation for the Metric PDP

### 3.1 Clustering

Many algorithms for the metric FLP are based on a clustering technique. In such algorithms, we are given an optimal solution for the LP relaxation, and construct the support graph corresponding to this solution, that is, the bipartite graph that contains an edge for each pair of client and facility such that the client is fractionally served by the facility in the LP solution. In the support graph, two clients are called *neighbors* if they are adjacent to a common facility. A partition of the clients is then obtained, such that any client in a given cluster is neighbor to a leading client, that is called the *cluster center*. It is required that no two cluster centers are neighbors. The algorithms for the FLP use the following greedy procedure: while not all clients are clustered, choose a cluster center with a certain greedy criterion, and create a new cluster with this center and all its neighbors. Different greedy criteria lead to different algorithms and analyses.

We use a clustering algorithm for the PDP. However, in the PDP, we are not aiming to locate facilities to be opened, rather, the warehouses are already established, and we want to select the set of time steps at which we place orders for each warehouse. Therefore, we can think of an order formed by a pair of warehouse and time step as a single facility. Analogously, each demand can be though of as a single client, that is willing to be connected to this "facility". We can then construct the corresponding support graph, and proceed to the clustering algorithm, in a way similar to facility location algorithms.

Formally, the support graph $G$ is the bipartite graph such that the vertices are formed by the disjoint union $\mathcal{P} \cup \mathcal{D}$, and there is an edge between order $(p, s) \in \mathcal{P}$

and demand $(q,t) \in \mathcal{D}$ if $(p,s) \in \mathcal{S}_j$. Notice that, contrary to the case of the facility location, when a non-center client could always be indirectly connected to any facility that served the cluster center, for the PDP, it can happen that a non-center demand $(q,t)$ cannot be served by an order adjacent to its cluster center $(q',t')$. This happens when demand $(q,t)$ arrives before $(q',t')$, that is $t < t'$, and cluster center $(q',t')$ is adjacent to some order $(p,s)$ with $s > t$. To guarantee that every demand in a cluster is served, we place orders at the beginning of the cluster center's service window.

**Algorithm 1** (Clustering algorithm)
We are given a complete solution $(x,y)$ for the LP relaxation, and an ordered list of demand points $L$. The algorithm returns a set $F'$ of placed orders, and a clustering $\mathcal{C}$ of demand points.
1. Construct the support graph $G$.
2. While there are unclustered demands points:
   (a) Create cluster $D$ with the next unclustered element $j'$ in $L$ as center.
   (b) Add all unclustered demand points that are neighbors of $j'$ to $D$.
   (c) Add $D$ to clustering $\mathcal{C}$.
3. For each cluster $D$ with center $j'$:
   (a) Choose one order $(\bar{p}, \bar{s}) \in \mathcal{S}_{j'}$ with probability $y_{\bar{p}\bar{s}}$.
   (b) Let $s' = \min_{(\hat{p},\hat{s}) \in \mathcal{S}_{j'}} \hat{s}$, and $p' = \bar{p}$.
   (c) Add $(p', s')$ to set $F'$.

Different choices of the list $L$ lead to algorithms with different approximation guarantees. In Subsection 3.2, list $L$ will be the set of demands in increasing order of $(C^*_{j'} + 2b^*_{j'})/d_{j'}$, and in Subsection 3.3 the demands will be chosen by order of $(W_{j'}(\alpha) + 2R_{j'}(\alpha))/d_{j'}$, for some parameter $\alpha$.

Suppose that we run Algorithm 1 for an optimal LP solution $(x^*, y^*)$, and some arbitrary list $L$. Let $K_{F'} = \sum_{(p,s) \in F'} k_p$ be the total cost of the orders in the set $F'$. The next lemmas calculates the expected value of $K_{F'}$, and the expected cost to serve one unit of a demand point.

**Lemma 1.** *Let $K_{F'}$ be the ordering cost Algorithm 1, then $E[K_{F'}] \leq K^*$.*

**Lemma 2.** *Let $j = (q,t)$ be a demand point, and $j' = (q', t')$ be the corresponding cluster center. Then, $E\left[\min_{(p,s) \in F'}(c_{pq} + h_{qst})\right] \leq (C^*_{j'} + 2b^*_{j'})/d_{j'} + b^*_j/d_j$.*

*Proof.* Let $(p', s')$ be the order placed by Algorithm 1 at step 3(c) corresponding to cluster center $j'$, and $(\bar{p}, \bar{s})$ be the order drawn in step 3(a). It is enough to bound the expected cost to serve one unit of $j$ by order $(p', s')$.

Since $j'$ is the cluster center corresponding to $j$, we know that there is an order $(p,s) \in \mathcal{S}_j \cap \mathcal{S}_{j'}$. We obtain $s' \leq s$, since $s$ is in the service window of $j'$ and $s'$ is the minimum time step in this service window. Also, we get $s \leq t$, because demand $j$ is fractionally served by $(p,s)$. Similarly, demand $j'$ is fractionally served by $(p,s)$, so $s \leq t'$ (see Figure 2). Using Assumption 2, we obtain

$$c_{p'q} + h_{qs't} \leq c_{p'q'} + h_{q's's} + c_{pq'} + c_{pq} + h_{qst}$$
$$\leq c_{p'q'} + h_{q's't'} + c_{pq'} + c_{pq} + h_{qst},$$

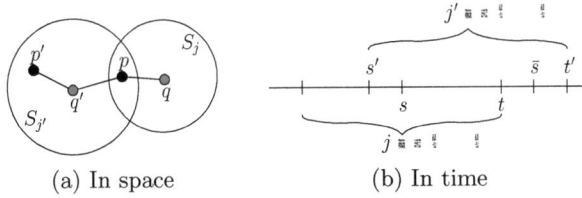

**Fig. 2.** Possible configuration

where the last inequality follows from Assumption 1 and the fact that $s \le t'$. Since $(p, s) \in S_j$, we obtain $x_{ps}^{qt} > 0$. By complementary slackness, it follows that $b_j^* = d_j(h_{qst} + c_{pq}) + z_{ps}^{qt}$, and thus $h_{qst} + c_{pq} \le b_j^*/d_j$. Similarly, we also get $h_{q's't'} \le b_{j'}^*/d_{j'}$, and $c_{pq'} \le b_{j'}^*/d_{j'}$. Finally, the expected value of $c_{p'q'}$ is

$$E[c_{p'q'}] = E[c_{\bar{p}q'}] = \sum_{(\hat{p},\hat{s}) \in S_{j'}} y_{\hat{p}\hat{s}}^* c_{\hat{p}q'} = \sum_{(\hat{p},\hat{s}) \in S_{j'}} x_{\hat{p}\hat{s}}^{*q't'} c_{\hat{p}q'} = C_{j'}^*/d_{j'},$$

where the third equality holds because the solution is complete. Adding up all terms, we obtain the desired statement.

### 3.2 Balancing Using Extra Orders

Lemmas 1 and 2 give bounds to the ordering and service costs of the solution yielded by Algorithm 1. One can observe the imbalance between the low value of ordering cost and the high value of the service cost. Indeed, Lemma 2 bounds the demand service cost in the worst case, when a demand point is served through its cluster center. For the FLP, the algorithms [15,4,2] would first try to connect a client to its close facilities. In fact, they would open the unclustered facilities after the clustering phase, so that facilities in service sets are opened independently, or negatively correlated, with probability equal to the fractional opening. This implies that there is, with constant probability, an open close facility.

For the PDP, however, this approach cannot be applied directly. The reason is that the orders placed for each cluster are moved to earlier times, and thus serving demand points by the corresponding orders would incur an extra holding cost, that is potentially unbounded. Instead, for the PDP, we place an extra set of orders, in their original time positions. This increases the total ordering cost, but such increase is compensated by the decrease of the service cost.

**Algorithm 2** (Balancing algorithm)
We are given an instance of the PDP. The algorithm returns a solution for this instance, formed by a set of orders $F' \cup F''$.
1. Solve the LP relaxation, and obtain solution $(x^*, y^*)$ and $b^*$.
2. Make the solution complete, splitting fractional ordering if necessary.
3. Run Algorithm 1 using $(x^*, y^*)$, and $L$ as the set of demand points $j'$ in increasing order of $(C_{j'}^* + 2b_{j'}^*)/d_{j'}$. Obtain obtain set of orders $F'$.
4. For each $(p, s) \in \mathcal{P}$, add order $(p, s)$ to set $F''$ with probability $y_{ps}^*$.

5. Place an order for each element of $F' \cup F''$.
6. Serve each demand point $(q,t)$ by the order $(p,s)$ that minimizes $c_{pq} + h_{qst}$.

The following lemma bounds the expected service cost of the cheapest placed order in a subset $A$ of $\mathcal{P}$, conditioned to the event that there is one placed order in $A$. Since versions of this lemma have appeared in several LP rounding algorithms for the FLP (for instance, see Lemma 4.2 in [2]), we omit the proof.

**Lemma 3.** $E\left[\min_{(p,s)\in A\cap F''}(c_{pq} + h_{qst})\middle| A\cap F'' \neq \emptyset\right] \leq \sum_{(p,s)\in A} y^*_{ps}(c_{pq} + h_{qst}).$

**Theorem 1.** *The balancing algorithm produces a solution for the PDP with expected cost at most $(2 + \frac{3}{e})K^* + (1 + \frac{3}{e})C^* + (1 + \frac{2}{e})H^*$.*

*Proof.* Let $K_{F''}$ be the expected cost of orders in $F''$. We obtain $E[K_{F''}] = E[\sum_{(p,s)\in F''} k_p] = \sum_{(p,s)\in \mathcal{P}} y^*_{ps} k_p = K^*$.

Consider a demand point $j = (q,t)$. We calculate the service cost $c_j$ to serve $j$ by a placed order in $F' \cup F''$ if we used the following, suboptimal, algorithm: if $S_j \cap F''$ is not empty, then we serve $j$ by the closest order in $S_j \cap F''$, otherwise we serve it indirectly through its cluster center $j'$. Let $p_c$ be the probability that $S_j \cap F'' \neq \emptyset$. We have $p_c = 1 - \prod_{(p,s)\in S_j}(1 - y^*_{ps}) \geq 1 - e^{-\sum_{(p,s)\in S_j} y^*_{ps}} = 1 - e^{-1}$. Now, combining with Lemmas 2 and 3, we obtain

$$E[c_j] \leq p_c \sum_{(p,s)\in S_j} y^*_{ps} d_j(c_{pq} + h_{qst}) + (1 - p_c) d_j((C^*_{j'} + 2b^*_{j'})/d_{j'} + b^*_j/d_j)$$
$$\leq p_c (C^*_j + H^*_j) + (1 - p_c) d_j((C^*_j + 2b^*_j)/d_j + b^*_j/d_j)$$
$$\leq (1 - e^{-1})(C^*_j + H^*_j) + e^{-1}(C^*_j + 3b^*_j),$$

where the second inequality follows since $j'$ was chosen as cluster center, thus $(C^*_{j'} + 2b^*_{j'})/d_{j'} \leq (C^*_j + 2b^*_j)/d_j$, and the last inequality follows since, by complementary slackness, $C^*_j + H^*_j \leq C^*_j + 3b^*_j$, and $p_c \geq 1 - e^{-1}$. Finally, we get

$$E[K_{F'}] + E[K_{F''}] + \sum_{j\in D} E[c_j]$$
$$\leq K^* + K^* + \sum_{j\in D}\left((1 - e^{-1})(C^*_j + H^*_j) + e^{-1}(C^*_j + 3b^*_j)\right)$$
$$= K^* + K^* + (1 - e^{-1})(C^* + H^*) + e^{-1}(C^* + 3(K^* + C^* + H^*))$$
$$= (2 + \frac{3}{e})K^* + (1 + \frac{3}{e})C^* + (1 + \frac{2}{e})H^*.$$

### 3.3 Balancing Using Filtering

Other way to fix the imbalance between the production and the service cost is using the filtering technique: the fractional ordering $y_{ps}$ is scaled by a factor $1/\alpha$, for some $\alpha \in (0,1]$. Notice that $y_{ps}$ can become larger than 1; in this case, a copy of the warehouse $p$ is made, and the fractional ordering $y_{ps}$ is split in the filtering step. Intuitively, placing "more times" each warehouse order should increase the probability of a client being served by a cheap order.

**Algorithm 3** (Filtering algorithm)
Given an instance of the PDP and a parameter $\alpha$, the algorithm returns a solution for this instance, formed by warehouses orders $F'$.
1. Solve the LP relaxation and obtain solution $(x^*, y^*)$.
2. Scale up the ordering variables $y^*$ by $1/\alpha$. Change variables $x^*$, and obtain a complete solution $(\bar{x}, \bar{y})$, splitting warehouses orders if necessary.
3. Run Algorithm 1 with solution $(\bar{x}, \bar{y})$, passing as list $L$ the set of demand points $j'$ in increasing order of $(W_{j'}(\alpha) + 2R_{j'}(\alpha))/d_j$, and obtain set $F'$.
4. Serve each demand point $(q, t)$ by the order $(p, s)$ that minimizes $h_{qst} + c_{pq}$.

The following lemmas are similar to Lemmas 1 and 2. Recall that $R_j(\alpha)$ is the maximum service cost in the $\alpha$-neighborhood of demand $j$, an $W_j(\alpha)$ is the average service cost in this neighborhood.

**Lemma 4.** *Let $K_{F'}$ be the ordering cost of Algorithm 3, then $E[K_{F'}] \leq 1/\alpha K^*$.*

**Lemma 5.** *If $j = (q, t)$, then $E\left[\min_{(p,s) \in F'} d_j(c_{pq} + h_{qst})\right] \leq W_j(\alpha) + 3R_j(\alpha)$.*

Lemma 4 shows that the approximation factor of the filtering algorithm for the ordering cost depends only on the parameter $\alpha$. On the other hand, as it can be seen in Lemma 5, the service cost depends on the neighborhood radius function of each demand point $j$. Therefore, the input instance can be characterized by summation of such radius functions, as in the following definition.

**Definition 1.** *Given an instance of the PDP, and an optimal fractional solution $(x^*, y^*)$, the characteristic function $r : [0, 1] \to \mathbb{R}^+$ is $r(\alpha) = \sum_{j \in \mathcal{D}} R_j(\alpha)/S^*$.*

*Remark 1.* The characteristic function $r(\alpha)$ satisfies to: $\int_0^1 r(t)dt = 1$.

**Corollary 1.** *The total expected service cost of filtering algorithm is bounded by $\left((1/\alpha)\int_0^\alpha r(t)dt + 3r(\alpha)\right)S^*$.*

### 3.4 Combining Different Algorithms

Algorithm 2 is a bifactor approximation algorithm, that achieves factor $2 + 3/e$ for the ordering cost, and factor $1 + 3/e$ for the service cost (*i.e.*, holding and transportation costs). Similarly, for each value of parameter $\alpha$, Algorithm 3 is a bifactor approximation with factors $1/\alpha$ and $(1/\alpha)\int_0^\alpha r(t)dt + 3r(\alpha)$.

To combine the two algorithms, we use the following strategy: with a given probability $\delta$, we run Algorithm 2, and with probability $1 - \delta$ we run Algorithm 3 with parameter $\alpha$ drawn from a probability density function $f : (0, 1] \to \mathbb{R}^+$. A similar approach has been done in the algorithm by Li [7], for the FLP. Let $SOL$ be the cost corresponding to this algorithm, thus

$$E[SOL] \leq (A_2)\,\delta + \left(\int_0^1 A_3(\alpha)f(\alpha)d\alpha\right)(1-\delta), \tag{3}$$

where $A_2 = (2 + \frac{3}{e})K^* + (1 + \frac{3}{e})S^*$ is the expected cost of Algorithm 2, and $A_3(\alpha) = 1/\alpha\, K^* + \left((1/\alpha)\int_0^\alpha r(t)dt + 3r(\alpha)\right)S^*$ is the expected cost of Algorithm 3 with parameter $\alpha$.

For simplicity, we let $g(\alpha) = (1-\delta)f(\alpha)$, so that $\delta + \int_0^1 g(\alpha)d\alpha = 1$. We may rewrite (3) as $E[SOL] \leq \beta(\delta, g)K^* + \gamma(\delta, g, r)S^*$, where $\beta(\delta, g)$ and $\gamma(\delta, g, r)$ are the obtained approximation factors for ordering and service costs of the combining algorithm, respectively. We obtain

$$\beta(\delta, g) = (2 + \frac{3}{e})\delta + \int_0^1 \frac{1}{\alpha} g(\alpha)d\alpha, \tag{4}$$

$$\gamma(\delta, g, r) = (1 + \frac{3}{e})\delta + \int_0^1 \left((1/\alpha)\int_0^\alpha r(t)dt + 3r(\alpha)\right)g(\alpha)d\alpha. \tag{5}$$

We define $g(\alpha)$ and $\delta$ as follows, where $\alpha_0 \in (0, 1]$, and $c > 0$ are constants to be defined later:

$$g(\alpha) = \begin{cases} 0 & \alpha < \alpha_0 \\ c\alpha^{\frac{1}{3}} & \alpha \geq \alpha_0 \end{cases} \quad \text{and} \quad \delta = 1 - \int_0^1 g(\alpha)d\alpha. \tag{6}$$

Substituting $g(\alpha)$, when $\alpha \in [\alpha_0, 1]$, we can simplify the indefinite integral in (5)

$$\int \left((1/\alpha)\int_0^\alpha r(t)dt + 3r(\alpha)\right)g(\alpha)d\alpha = \int\int_0^\alpha r(t)dt\, c\alpha^{-\frac{2}{3}}d\alpha + \int 3r(\alpha)c\alpha^{\frac{1}{3}}d\alpha$$

$$= \int_0^\alpha r(t)dt \cdot 3c\alpha^{\frac{1}{3}} - \int 3c\alpha^{\frac{1}{3}} r(\alpha)d\alpha + \int 3r(\alpha)c\alpha^{\frac{1}{3}}d\alpha = 3c\int_0^\alpha r(t)dt \cdot \alpha^{\frac{1}{3}}d\alpha,$$

that used integration by parts in the second equality. Now, we obtain calculating the definite integrals that

$$\beta(\delta, g) = (2 + \frac{3}{e})\delta + 3c(1^{\frac{1}{3}} - \alpha_0^{\frac{1}{3}}), \tag{7}$$

$$\gamma(\delta, g, r) = (1 + \frac{3}{e})\delta + 3c\left[\int_0^1 r(t)dt \cdot 1^{\frac{1}{3}} - \int_0^{\alpha_0} r(t)dt \cdot \alpha_0^{\frac{1}{3}}\right]$$

$$\leq (1 + \frac{3}{e})\delta + 3c, \tag{8}$$

where the inequality comes from Remark 1, and the fact that $r(t)$ is non-negative. Notice that the last expression bounds $\gamma(\delta, g, r)$, and does not depend on $r$, so it is independent of the input instance. By appropriately choosing $c$ and $\alpha_0$, we are ready to give an approximation factor for the combining algorithm.

**Theorem 2.** *There exists a pair $(\delta, g)$ such that, for any characteristic function $r$, we obtain that $\max\{\beta(\delta, g), \gamma(\delta, g, r)\} \leq 2.77$.*

## 4 PDP with Retailer Ordering Costs

For this PDP variant, the objective function also includes a cost $k_q$ for each time a retailer $q$ places an order, independent of the number of demanded items. In addition to Assumption 2, we also assume traditional holding costs.

**Assumption 3.** *For each retailer $q$, the are non-negative numbers $h_i^q$, for $i = 1, \ldots, T$, such that for each $s, t$, it holds $h_{qst} = \sum_{i=s}^{t} h_i^q$.*

Combining the shift procedure used for the joint replenishment problem [6], and a modification of Algorithm 3, we can obtain the following approximation.

**Theorem 3.** *There is a 5-approximation for the PDP with retailer order costs.*

# References

1. Boudia, M., Prins, C.: A memetic algorithm with dynamic population management for an integrated production–distribution problem. European J. Oper. Research 195(3), 703–715 (2009)
2. Byrka, J., Aardal, K.: An Optimal Bifactor Approximation Algorithm for the Metric Uncapacitated Facility Location Problem. SIAM J. on Comp. 39, 2212–2231 (2010)
3. Chan, F.T.S., Chung, S.H., Wadhwa, S.: A hybrid genetic algorithm for production and distribution. Omega 33(4), 345–355 (2005)
4. Chudak, F.A., Shmoys, D.B.: Improved Approximation Algorithms for the Uncapacitated Facility Location Problem. SIAM J. on Comp. 33(1), 1–25 (2004)
5. Daskin, M.S., Coullard, C.R., Shen, Z.-J.: An Inventory-Location Model: Formulation, Solution Algorithm and Computational Results. Annals of Oper. Research 110(1-4), 83–106 (2002)
6. Levi, R., Roundy, R., Shmoys, D.B., Sviridenko, M.: A Constant Approximation Algorithm for the One-Warehouse Multiretailer Problem. Management Science 54(4), 763–776 (2008)
7. Li, S.: A 1.488 Approximation Algorithm for the Uncapacitated Facility Location Problem. In: Aceto, L., Henzinger, M., Sgall, J. (eds.) ICALP 2011, Part II. LNCS, vol. 6756, pp. 77–88. Springer, Heidelberg (2011)
8. Li, Y., Shu, J., Wang, X., Xiu, N., Xu, D., Zhang, J.: Approximation Algorithms for Integrated Distribution Network Design Problems. In: INFORMS J. Comp. (2012)
9. Lin, J.-H., Vitter, J.S.: $\varepsilon$-approximations with minimum packing constraint violation (extended abstract). In: Proc. of the Twenty-Fourth Annual ACM Symposium on Theory of Computing, pp. 771–782 (1992)
10. Melo, M.T., Nickel, S., Saldanha-da Gama, F.: Facility location and supply chain management – A review. European J. Oper. Research 196(2), 401–412 (2009)
11. Pochet, Y., Wolsey, L.A.: Production planning by mixed integer programming. Springer series in operations research and financial engineering. Springer, New York (2006)
12. Shen, Z.-J.: Integrated Stochastic Supply-Chain Design Models. Computing in Science Engineering 9(2), 50–59 (2007)
13. Shen, Z.-J., Coullard, C., Daskin, M.S.: A Joint Location-Inventory Model. Transportation Science 37(1), 40–55 (2003)
14. Shu, J., Teo, C.-P., Shen, Z.-J.: Stochastic Transportation-Inventory Network Design Problem. Oper. Research 53(1), 48–60 (2005)
15. Sviridenko, M.: An Improved Approximation Algorithm for the Metric Uncapacitated Facility Location Problem. In: Cook, W.J., Schulz, A.S. (eds.) IPCO 2002. LNCS, vol. 2337, pp. 240–257. Springer, Heidelberg (2002)
16. Teo, C.-P., Shu, J.: Warehouse-Retailer Network Design Problem. Oper. Research 52(3), 396–408 (2004)

# The Online Connected Facility Location Problem

Mário César San Felice[1,*], David P. Williamson[2], and Orlando Lee[1,**]

[1] Unicamp, Institute of Computing, Campinas SP 13083-852, Brazil
{felice,lee}@ic.unicamp.br
[2] Cornell University, School of Operations Research and Information Engineering,
Ithaca NY 14853-3801, USA
dpw@cs.cornell.edu

**Abstract.** In this paper we propose the Online Connected Facility Location problem (OCFL), which is an online version of the Connected Facility Location problem (CFL). The CFL is a combination of the Uncapacitated Facility Location problem (FL) and the Steiner Tree problem (ST). We give a randomized $O(\log^2 n)$-competitive algorithm for the OCFL via the sample-and-augment framework of Gupta, Kumar, Pál, and Roughgarden and previous algorithms for Online Facility Location (OFL) and Online Steiner Tree (OST). Also, we show that the same algorithm is a deterministic $O(\log n)$-competitive algorithm for the special case of the OCFL with $M = 1$, where $M$ is a scale factor for the edge costs.

**Keywords:** Online Algorithms, Competitive Analysis, Connected Facility Location, Steiner Tree, Approximation Algorithms, Randomized Algorithms.

## 1 Introduction

We start by presenting several problems that are relevant to this work.

In the Facility Location (FL) problem, we have a set of clients and a set of facilities in a metric space. Each facility has a cost associated with opening the facility. The cost of assigning a client to a facility is the distance between the two points. The goal of the problem is to select a set of facilities to open and to assign clients to open facilities so that the total cost of opening the facilities plus the cost of connecting clients to their assigned facilities is minimized. FL is an NP-complete problem that has been well-studied; several constant ratio approximation algorithms are known for it [1–4]. It is particularly interesting that several different design techniques, such as LP rounding, primal-dual and local search, are successful at achieving good approximation ratios for this problem.

The online version of FL is the Online Facility Location problem (OFL), in which the clients are revealed one at a time and each one needs to be connected to an open facility before the next one arrives. As time progresses, no connection

---

[*] Grant No. 2012/06728-3, São Paulo Research Foundation (FAPESP).
[**] Supported by Bolsa de Produtividade do CNPq Proc. 303947/2008-0. Edital Universal CNPq 477692/2012-5.

can be changed or opened facility can be closed. Algorithms for online problems are analyzed via competitive analysis [5]. An $\alpha$-competitive algorithm returns a solution whose cost is within a factor of $\alpha$ of the cost of an optimal solution to the corresponding offline problem; $\alpha$ is called the competitive ratio of the algorithm. There are randomized and deterministic $O(\log n)$-competitive algorithms known for the OFL [6–10], where $n$ is the number of clients. Also, the best lower bound for the competitive ratio of an algorithm for OFL is $\Omega\left(\frac{\log n}{\log \log n}\right)$ [7].

The Steiner Tree problem (ST) is a network design problem defined in a graph with edge costs. Its input is a set of terminals that need to be connected to each other. A solution for ST is a tree that contains all terminal nodes and that can contain other nodes, called Steiner nodes. The goal is to minimize the total cost of edges in the tree. The ST is also a well-studied NP-complete problem for which several different constant ratio approximation algorithms are known [11, 12], such as greedy, primal-dual, and randomized rounding algorithms.

The online version of ST is the Online Steiner Tree problem (OST), in which the terminals are revealed one at a time and each one needs to be connected to the current tree before the next one arrives. Also, no edge in the tree can be removed in the future. There are $O(\log n)$-competitive algorithms known for OST [13, 14], where $n$ is the number of terminals. Also, the best lower bound for the competitive ratio of an algorithm for OST is $\Omega(\log n)$ [13].

The Connected Facility Location problem (CFL) is a network design problem with two layers; it is motivated by the necessity of building networks in which the end users are connected to servers, with less expensive lower bandwidth connections, and the servers are connected to each other, through more expensive higher bandwidth connections. The input to the CFL is the same as the FL, except that there is a facility that is designated the root that represents the connection of the network to the outside world, and a parameter $M \geq 1$, which is a cost scaling factor. A solution for CFL is a set of open facilities (including the root), an assignment of clients to open facilities, and a tree spanning the open facilities. The goal of the problem is to minimize the total cost of opening facilities plus the total cost of connecting clients to their assigned facilities plus $M$ times the cost of the edges in the tree spanning the open facilities. The CFL is an NP-complete problem; it has randomized and deterministic constant ratio approximation algorithms [15–20] that use techniques such as sample-and-augment, LP rounding and primal-dual. The CFL can be seen as a combination of FL with ST, using the cost scaling factor $M$.

*Our Contributions.* In this paper we propose the Online Connected Facility Location problem (OCFL) that is the online version of CFL. In the OCFL the clients are revealed one at a time and each one needs to be connected with a facility before the next one arrives. If a new facility is opened, it needs to be connected to the tree spanning the other opened facilities immediately. Also, no connection can be changed, opened facility can be closed or edge used in the tree can be removed in the future. We can also view the OCFL as the combination of OFL and OST, using a cost scaling factor $M$. Since the OST can be reduced to the OCFL, there is a lower bound of $\Omega(\log n)$ on the competitive ratio of any algorithm for the OCFL.

Our main result is a randomized $O(\log^2 n)$-competitive algorithm for the OCFL that uses the sample-and-augment technique of Gupta, Kumar, Pál, and Roughgarden [15], where $n$ is the number of clients. We also show that the same algorithm is a deterministic $O(\log n)$-competitive algorithm for the special case of the OCFL with $M = 1$.

The sample-and-augment technique was developed by Gupta et al. [15] for a number of different problems; we illustrate it here with the single-source rent-or-buy problem. In this problem, we are given as input a graph with edge costs, a set of terminals, a root node, and a cost scaling factor $M \geq 1$. We must connect each terminal to the root. We can either buy an edge (at $M$ times its cost), or rent an edge (but then each terminal using the edge must pay its cost). The sample-and-augment technique marks each terminal with probability $\frac{1}{M}$; it then uses an approximation algorithm for ST on the marked terminals and buys the edges in the resulting tree; then it rents edges as needed to connect the other terminals to the tree of bought edges. The key idea of this technique is that by sampling with probability $\frac{1}{M}$ the algorithm balances the costs of the edges that should be bought and the cost of the edges that should be rented. We will use it in a similar way.

## 2 Problem Definitions

In this section we formally define the Online Connected Facility Location problem (OCFL). First we define its offline version, the Connected Facility Location problem (CFL), and then we describe what changes in the online version.

As mentioned previously the CFL combines the Facility Location problem (FL) with the Steiner Tree problem (ST). The input to the problem is a complete graph $G = (V, E)$, distances $d : E \to R^+$ that respects the triangle inequality, possible facilities $F \subseteq V$, facility opening costs $f : F \to R^+$, clients $D \subseteq V$, a root $r \in V$ and a parameter $M \geq 0$, that we call the cost scaling factor.

The goal is to serve the clients with the minimum cost. To serve the clients one must open a subset of facilities $F' \subseteq F$, connect each client in $D$ to an opened facility in $F'$ or to the root $r$, and give a tree $T$ connecting the facilities in $F'$ and $r$ to each other. The cost to be minimized is the sum of the cost of the open facilities, the distance of each client to its assigned open facility, and $M$ times the cost of the tree $T$ that connects $\{r\} \cup F'$. Namely:

$$\sum_{i \in F'} f(i) + \sum_{j \in D} d(j, a(j)) + M \sum_{e=(i,j) \in T} d(i,j) ,$$

where $a : D \to V$ is a function that assigns each client in $D$ to the root $r$ or to a facility in $F'$.

The OCFL is the online version of CFL, so it combines the Online Facility Location problem (OFL) with the Online Steiner Tree problem (OST), just as CFL does with FL and ST. In the OCFL the clients in $D$ arrive one at a time and the one that just arrived must be served before the next one arrives; in particular, it must be assigned to an open facility. The algorithm can open a facility for this purpose, but then the opened facility must be connected to the other opened facilities in the tree $T$. All decisions of the algorithm are irrevocable. In this case, this means that the algorithm cannot decide to remove from the current solution any facility previously opened, change to which facility a client is connected, even if a closer facility was opened, or remove an edge from the tree $T$.

## 3 Notation and Definitions

In what follows,

- $n = |D|$ is the number of clients. Notice that for CFL and OCFL the number of open facilities is upper bounded by $n$,
- compFL is the $c_{OFL}$-competitive primal-dual algorithm for the OFL from Fotakis [8] and Nagarajan and Williamson [9] papers, with $c_{OFL} \leq 4 \log n$,
- compST is the $c_{OST}$-competitive greedy algorithm for the OST from Imase and Waxman [13] paper, with $c_{OST} \leq \log n$,
- $i = a(j)$ means that $j$ is connected to $i$ by the online algorithm we are analyzing,
- $i = a^*(j)$ means that $j$ is connected to $i$ in the offline optimal solution with which we are comparing,
- $path(j, S)$ is the shortest path connecting $j$ to $v$, being $v$ the closest node to $j$ in $S$,
- $ALG_{OCFL}(D) = O + C + S$ is the cost of the Online CFL algorithm when serving the clients $D$, where $O$ is the facility opening cost, $C$ is the client connection cost and $S$ is the Steiner tree cost,
- $OPT_{CFL}(D) = O^* + C^* + S^*$ is the cost of the CFL offline optimal solution with which we compare the cost of the online algorithm. It is also divided in facility opening, client connection and Steiner tree cost.

## 4 The Online CFL Algorithm

In this section we present a sample-and-augment algorithm for the OCFL that is based on the algorithm for the CFL presented in Eisenbrand et al. [20] paper. Our algorithm uses the algorithm compFL as a subroutine when deciding which facilities to open and how to connect the clients. Also, it simulates the behavior of the compST algorithm when creating the tree that connects the open facilities.

The algorithm keeps a virtual solution that is competitive for the OFL. This solution serves all the clients that arrive and may have more open facilities, called virtual facilities, than the algorithm's actual solution. When a client $j$ arrives it is served by compFL and connected to a virtual facility $i$. Also, the client $j$ is marked with probability $\frac{1}{M}$. A virtual facility is actually opened by the algorithm only when a client that was connected to it is marked. If the client

$j$ is marked then $i$ is opened and $j$ is connected to it. Otherwise $j$ is connected to the closest actually opened facility. Notice that $i$ could already be open due to some previously marked client.

The algorithm builds the tree $T$ that connects the facilities as follows. When a facility $i$ is actually opened, due to a client $j$ that was marked, the algorithm connects the client $j$ to the tree $T$, using the shortest path. This is the behaviour of the compST algorithm. Then, it augments $T$ to connect $i$ to $j$. Although connecting the facility $i$ directly to the tree $T$ seems more intuitive, this behavior of the algorithm is useful during the analysis.

**Input:** $G = (V, E)$, $d$, $f$, $F$, root $r$ and $M$
$D \leftarrow \emptyset$; $D' \leftarrow \emptyset$; $F' \leftarrow \emptyset$; $T \leftarrow \emptyset$;
make $f(r) \leftarrow 0$ and initialize compFL with $G$, $d$, $f$, $F$;
send $r$ to compFL as its first client;
$F' \leftarrow F' \cup \{r\}$; $V(T) \leftarrow V(T) \cup \{r\}$;
**while** *a new client $j$ arrives* **do**
  send $j$ to compFL;
  include $j$ in $D'$ with probability $\frac{1}{M}$;
  **if** $j \in D'$ *and is connected to a virtual facility $i$ not opened* **then**
    $F' \leftarrow F' \cup \{i\}$;
    $T \leftarrow T \cup \{(i,j)\} \cup \{path(j, V(T))\}$;
  **end**
  let $i$ be the closest open facility to $j$;
  $D \leftarrow D \cup \{j\}$; $a(j) \leftarrow i$;
**end**
**return** $(F' \setminus \{r\}, T, a)$;

**Algorithm 1.** The Online CFL algorithm

### 4.1 Analysis of the Online CFL Algorithm

During this analysis we let $D' \subseteq D$ denote the clients marked by the Online CFL algorithm. Note that $D'$ is a random set.

First we bound the facility opening cost of the algorithm.

**Lemma 1.** $O \leq c_{\text{OFL}}(O^* + C^*)$.

*Proof.* Let $O_{\text{compFL}}$ be the facility opening cost paid by compFL to serve $\{r\} \cup D$ and $\text{OPT}_{\text{FL}}$ be an offline optimal solution for the FL. Once our algorithm opens a subset of the facilities opened by compFL to serve $\{r\} \cup D$ we have that:

$$O \leq O_{\text{compFL}} \leq c_{\text{OFL}} \text{OPT}_{\text{FL}}(\{r\} \cup D) \leq c_{\text{OFL}}(O^* + C^*) \; , \qquad (1)$$

where the last inequality follows since part of an optimal solution for CFL is a feasible solution for the OFL. □

Now we bound the expected cost of the Steiner tree $T$ that connects the root and the opened facilities to each other.

**Lemma 2.** $E[S] \leq c_{\text{OST}}(S^* + C^*) + c_{\text{OFL}}(O^* + C^*)$.

*Proof.* Define $D''$ as the set of marked clients that were responsible for opening the facilities opened by the algorithm and $\text{OPT}_{\text{ST}}$ as an offline optimal solution for the ST. The Online CFL algorithm builds a tree $T$ by connecting the root $r$ to each client in $D''$ simulating the compST algorithm. Then it augments $T$ connecting each client $j \in D''$ to the facility $i$ that was opened by it. So, we have that:

$$S \leq M\text{compST}(\{r\} \cup D'') + M \sum_{j \in \{r\} \cup D''} d(j, a(j))$$

$$\leq Mc_{\text{OST}}\text{OPT}_{\text{ST}}(\{r\} \cup D'') + M \sum_{j \in D''} d(j, a(j))$$

$$\leq Mc_{\text{OST}}\text{OPT}_{\text{ST}}(\{r\} \cup D') + M \sum_{j \in D'} d(j, a(j)) , \qquad (2)$$

where the second inequality follows because $d(r, a(r)) = 0$ and the last inequality follows because $D'' \subseteq D'$.

We bound the expected cost of $\text{OPT}_{\text{ST}}$ when serving $\{r\} \cup D'$ as follows:

$$E[\text{OPT}_{\text{ST}}(\{r\} \cup D')] \leq E\left[\frac{S^*}{M}\right] + E\left[\sum_{j \in D'} d(j, a^*(j))\right]$$

$$\leq \frac{S^*}{M} + \sum_{j \in D} \frac{1}{M} d(j, a^*(j)) \leq \frac{S^*}{M} + \frac{C^*}{M} , \qquad (3)$$

where the first inequality follows because the union of the optimal Steiner tree for CFL, $T^*$, and a connection from each client in $D'$ to its facility in $T^*$ contains a tree that spans $\{r\} \cup D'$, since $r \in V(T^*)$; the second inequality follows because the probability that a client is marked is $\frac{1}{M}$.

Let $a_{\text{compFL}}(j)$ be the facility to which compFL connected $j$ and $C_{\text{compFL}}$ be the client connection cost of compFL when serving $\{r\} \cup D$. We bound the expected cost of $\sum_{j \in D'} d(j, a(j))$ as follows:

$$E\left[\sum_{j \in D'} d(j, a(j))\right] = E\left[\sum_{j \in D'} d(j, a_{\text{compFL}}(j))\right]$$

$$= \sum_{j \in D} \frac{1}{M} d(j, a_{\text{compFL}}(j))$$

$$= \frac{C_{\text{compFL}}}{M}$$

$$\leq \frac{c_{\text{OFL}}}{M} \text{OPT}_{\text{FL}}(\{r\} \cup D) \leq \frac{c_{\text{OFL}}}{M}(O^* + C^*) , \qquad (4)$$

where the first equality follows because $a(j) = a_{\text{compFL}}(j)$ for $j \in D'$ and the last equality follows because $d(r, a_{\text{compFL}}(r)) = 0$. The last inequality follows since part of an optimal solution for CFL is a feasible solution for the OFL.

Using the last three inequalities we have:

$$E[S] \leq E\left[M c_{\text{OST}} \text{OPT}_{\text{ST}}(\{r\} \cup D')\right] + E\left[M \sum_{j \in D'} d(j, a(j))\right]$$

$$\leq M c_{\text{OST}}\left(\frac{S^*}{M} + \frac{C^*}{M}\right) + M\left(\frac{c_{\text{OFL}}}{M}(O^* + C^*)\right)$$

$$\leq c_{\text{OST}}(S^* + C^*) + c_{\text{OFL}}(O^* + C^*) \; , \tag{5}$$

which concludes the lemma. □

Using the two previous lemmas we can bound the expectation of the facility opening cost $O$ and of the Steiner tree cost $S$ of the Online CFL algorithm. Now we will bound the client connection cost.

**Lemma 3.** $E[C] \leq c_{\text{OFL}}(O^* + C^*) + c_{\text{OST}}(S^* + C^* + c_{\text{OFL}}(O^* + C^*))$.

*Proof.* First we define cost shares that divide the expected cost of compFL($\{r\} \cup D'$) and compST($F'$) among the clients. For each client $j$ we call its cost share the buying cost $b_j$. We also divide the client connection cost $C$ among the clients and, for each client $j$ we call its share the renting cost $r_j$.

Then we analyze the algorithms compFL and compST to show that, when each client $j$ arrives, the expected renting cost of $j$ is at most its expected buying cost. Finally, using the linearity of expectation and summing over all the clients in $D$, we conclude the lemma.

Remember that $D'$ is the set of marked clients, $a(j)$ is the open facility to which the Online CFL algorithm connected $j$ and $a_{\text{compFL}}(j)$ is the open facility to which compFL connected $j$. Also, let $n(j)$ be the position of client $j$ in the sequence of clients and $F'_{n(j)}$ be the set of facilities that were opened after the first $n(j)$ clients were served by the algorithm. Let $F'_{n(j)-1}$ be the facilities opened in the time step prior to the arrival of $j$.

Now we define the buying cost of a client $j$ as:

$$b_j = \begin{cases} Md(j, a_{\text{compFL}}(j)) + Md(a_{\text{compFL}}(j), F'_{n(j)-1}) & \text{if } j \in D' \; , \\ 0 & \text{if } j \in D \setminus D' \; . \end{cases}$$

Summing over all the clients in $D$ we have:

$$\sum_{j \in D} b_j = \sum_{j \in D'} b_j$$

$$= \sum_{j \in D'} Md(j, a(j)) + \sum_{j \in D'} M(d(a(j), F'_{n(j)-1}))$$

$$= M \sum_{j \in D'} d(j, a(j)) + M\text{compST}(\{r\} \cup F') \; . \tag{6}$$

Let $C_{\text{compFL}}$ be the client connection cost of compFL when serving $\{r\} \cup D$. We have that:

$$E\left[\sum_{j \in D'} d(j, a(j))\right] = \sum_{j \in D} \frac{1}{M} d(j, a_{\text{compFL}}(j)) = \frac{C_{\text{compFL}}}{M} . \qquad (7)$$

Similarly:

$$E\left[\sum_{j \in D'} d(j, a^*(j))\right] = \sum_{j \in D} \frac{1}{M} d(j, a^*(j)) = \frac{C^*}{M} . \qquad (8)$$

Bounding the expected cost of compST when it is serving $\{r\} \cup F'$ we have:

$$E[\text{compST}(\{r\} \cup F')] \le c_{\text{OST}} E[\text{OPT}_{\text{ST}}(\{r\} \cup F')]$$

$$\le c_{\text{OST}} E\left[\frac{S^*}{M} + \sum_{j \in D'} d(j, a^*(j)) + \sum_{j \in D'} d(j, a(j))\right]$$

$$\le c_{\text{OST}} \left(\frac{S^*}{M} + \frac{C^*}{M} + \frac{C_{\text{compFL}}}{M}\right) , \qquad (9)$$

where the second inequality follows because the union of an optimal Steiner tree for CFL, $T^*$, along with each client in $D'$ connected to its facility in $T^*$ and to its facility in the online solution, contains a tree that spans $\{r\} \cup F'$, since $r \in V(T^*)$.

Using the previous inequalities we bound the expected value of the total buying cost by:

$$E\left[\sum_{j \in D} b_j\right] \le M \cdot E\left[\sum_{j \in D'} d(j, a(j))\right] + M \cdot E[\text{compST}(\{r\} \cup F')]$$

$$= C_{\text{compFL}} + c_{\text{OST}}(S^* + C^* + C_{\text{compFL}})$$

$$\le c_{\text{OFL}} \text{OPT}_{\text{FL}}(\{r\} \cup D) + c_{\text{OST}}(S^* + C^* + c_{\text{OFL}} \text{OPT}_{\text{FL}}(\{r\} \cup D))$$

$$\le c_{\text{OFL}}(O^* + C^*) + c_{\text{OST}}(S^* + C^* + c_{\text{OFL}}(O^* + C^*)) . \qquad (10)$$

We define the renting cost of a client $j$ as follows:

$$r_j = d(j, F'_{n(j)}) .$$

Summing over all the clients in $D$ we have:

$$\sum_{j \in D} r_j = \sum_{j \in D} d(j, F'_{n(j)}) = C , \qquad (11)$$

where the last equality holds because the Online CFL algorithm connects each new client to the closest facility that was open at that moment.

Now we upper bound the expected renting cost of a client $j$ using the expected buying cost of $j$. First we analyze the expectation of buying and of renting conditioned on the result of the first $n(j) - 1$ coin tosses. We denote these respectively as $E[b_j | n(j) - 1]$ and $E[r_j | n(j) - 1]$.

Recalling that a client $j$ has probability $\frac{1}{M}$ of being maked and probability $\frac{M-1}{M}$ of not, we have:

$$\begin{aligned}
E[r_j | n(j) - 1] &= \frac{M-1}{M} d(j, F'_{n(j)-1}) + \frac{1}{M} d(j, F'_{n(j)}) \\
&\leq d(j, F'_{n(j)-1}) \\
&\leq \frac{1}{M} \left( M d(j, a_{\text{compFL}}(j)) + M d(a_{\text{compFL}}(j), F'_{n(j)-1}) \right) \\
&= E[b_j | n(j) - 1] ,
\end{aligned} \tag{12}$$

where the first equality holds because $F'_{n(j)} = F'_{n(j)-1}$ when the client $j$ is not maked and because $F'_{n(j)-1} \subseteq F'_n(j)$; the second inequality follows by the triangle inequality and the fact that the distance from $j$ to the closest facility to $a_{\text{compFL}}(j)$ in $F'_{n(j)-1}$ has to be at least the distance from $j$ to its closest facility in $F'_{n(j)-1}$.

Since this is true for all possible outcomes of the $n(j) - 1$ first coin tosses, the result is true unconditionally, i.e.:

$$E[r_j] \leq E[b_j] . \tag{13}$$

We conclude the lemma using (10), (11) and (13), relying on the linearity of expectation as follows:

$$\begin{aligned}
E[C] = \sum_{j \in D} E[r_j] &\leq \sum_{j \in D} E[b_j] \\
&\leq c_{\text{OFL}}(O^* + C^*) + c_{\text{OST}}(S^* + C^* + c_{\text{OFL}}(O^* + C^*)) .
\end{aligned} \tag{14}$$

□

Using the three previous lemmas and that the competitive ratio of compFL and compST is $O(\log n)$, we prove our main result in the next theorem.

**Theorem 1.** $E[\text{ALG}_{\text{OCFL}}(D)] \in O(\log n^2) \text{OPT}_{\text{CFL}}(D)$.

*Proof.*

$$\begin{aligned}
E[\text{ALG}_{\text{OCFL}}(D)] &= E[O + S + C] \\
&\leq c_{\text{OFL}}(O^* + C^*) + (c_{\text{OST}}(S^* + C^*) + c_{\text{OFL}}(O^* + C^*)) \\
&\quad + (c_{\text{OFL}}(O^* + C^*) + c_{\text{OST}}(S^* + C^* + c_{\text{OFL}}(O^* + C^*))) \\
&\leq 14 \log n \text{OPT}_{\text{CFL}}(D) + 4 \log^2 n \text{OPT}_{\text{CFL}}(D) \\
&= O(\log^2 n) \text{OPT}_{\text{CFL}}(D) ,
\end{aligned} \tag{15}$$

where the last inequality follows because $c_{\text{OFL}} \leq 4 \log n$ and $c_{\text{OST}} \leq \log n$. □

## 4.2 Analysis of the Special Case of the Online CFL Problem with $M = 1$

Here we analyze the algorithm Online CFL on instances of the Online Connected Facilty Location problem for which $M = 1$.

In this analysis we let $D' \subseteq D$ denote the clients maked by the Online CFL algorithm. In this special case, since $M = 1$, all clients are marked by the Online CFL algorithm, so that $D' = D$. In fact, this means that the Online CFL algorithm is deterministic in this case.

First we bound the facility opening cost $O$ and the client connection cost $C$.

**Lemma 4.** $O + C \leq c_{\text{OFL}}(O^* + C^*)$.

*Proof.* Let $O_{\text{compFL}}$ be the facility opening cost paid by compFL to serve $\{r\} \cup D$, and let $C_{\text{compFL}}$ be the client connection cost of compFL when serving $\{r\} \cup D$. Since the OCFL algorithm opens exactly the facilities opened by compFL to serve $\{r\} \cup D$ and connects each client to the closest facility opened when that client was served, we have that:

$$O + C = O_{\text{compFL}} + C_{\text{compFL}}$$
$$\leq c_{\text{OFL}}\text{OPT}_{\text{FL}}(\{r\} \cup D) \leq c_{\text{OFL}}(O^* + C^*) , \qquad (16)$$

where the last inequality follows since part of an optimal solution for CFL is a feasible solution for the OFL. □

Now we bound the cost of the Steiner tree that connects the root and the opened facilities.

**Lemma 5.** $S \leq c_{\text{OST}}(S^* + C^*) + c_{\text{OFL}}(O^* + C^*)$.

*Proof.* Define the set $D''$ to be the set of marked clients that were responsible for opening a facility. The Online CFL algorithm builds a tree $T$ connecting the root $r$ to each client in $D''$ simulating the compST algorithm, and then connects each client $j \in D''$ with the facility $i$ that was opened by it. So we have that:

$$S \leq M \text{compST}(\{r\} \cup D'') + M \sum_{j \in \{r\} \cup D''} d(j, a(j))$$
$$\leq c_{\text{OST}}\text{OPT}_{\text{ST}}(\{r\} \cup D) + \sum_{j \in D} d(j, a(j))$$
$$\leq c_{\text{OST}}\text{OPT}_{\text{ST}}(\{r\} \cup D) + C , \qquad (17)$$

where the second inequality follows because $M = 1$, $D'' \subseteq D$ and $d(r, a(r)) = 0$.
We bound the cost of $\text{OPT}_{\text{ST}}$ when serving $\{r\} \cup D$ as follows:

$$\text{OPT}_{\text{ST}}(\{r\} \cup D) \leq S^* + C^* , \qquad (18)$$

where the inequality follows because the union of the optimal Steiner tree for CFL, $T^*$, together with a connection from each client in $D$ to its optimal facility, contains a tree that spans $\{r\} \cup D$, since $r \in V(T^*)$.

The cost $C$ can be bounded using the previous lemma:

$$C \leq c_{\text{OFL}}(O^* + C^*) . \tag{19}$$

Using the last three inequalities we have:

$$\begin{aligned} S &\leq c_{\text{OST}} \text{OPT}_{\text{ST}}(\{r\} \cup D) + C \\ &\leq c_{\text{OST}}(S^* + C^*) + c_{\text{OFL}}(O^* + C^*) , \end{aligned} \tag{20}$$

what concludes the lemma. □

Using the two previous lemmas and the fact that the competitive ratio of compFL and compST is $O(\log n)$, we prove the next theorem.

**Theorem 2.** *When $M = 1$ we have that* $\text{ALG}_{\text{OCFL}}(D) \in O(\log n \text{OPT}_{\text{CFL}}(D))$.

*Proof.*

$$\begin{aligned} \text{ALG}_{\text{OCFL}}(D) &= O + C + S \\ &\leq c_{\text{OFL}}(O^* + C^*) + (c_{\text{OST}}(S^* + C^*) + c_{\text{OFL}}(O^* + C^*)) \\ &\leq 9 \log n \text{OPT}_{\text{CFL}}(D) \\ &= O(\log n) \text{OPT}_{\text{CFL}}(D) , \end{aligned} \tag{21}$$

where the last inequality follows because $c_{\text{OFL}} \leq 4 \log n$ and $c_{\text{OST}} \leq \log n$. □

It is worth noticing that an algorithm that does not mark the clients, but instead solves the OFL part of the problem and connects each open facility in an online Steiner tree, can be shown to be $O(M \log n)$-competitive by an analysis very similar to the previous one.

## 5 Conclusion and Future Work

In this paper we proposed the Online Connected Facility Location problem and presented a randomized $O(\log^2 n)$-competitive algorithm for it. The algorithm uses the sample-and-augment technique. We also showed that this algorithm is a deterministic $O(\log n)$-competitive algorithm for the special case of the OCFL with $M = 1$, which is the best possible.

Two natural questions arise: first, is our algorithm $O(\log n)$-competitive for the general case of the OCFL? Second, is the best possible lower bound to the competitive ratio of OCFL algorithms $\omega(\log n)$?

Also, some directions for future work are to analyze our algorithm using the known distribution model from stochastic analysis, instead of the worst case model from competitive analysis, and to find a deterministic algorithm for the OCFL problem.

# References

1. Shmoys, D.B.: Approximation algorithms for facility location problems. In: Jansen, K., Khuller, S. (eds.) APPROX 2000. LNCS, vol. 1913, pp. 27–32. Springer, Heidelberg (2000)
2. Mahdian, M., Ye, Y., Zhang, J.: Approximation algorithms for metric facility location problems. SIAM Journal on Computing 36, 411–432 (2006)
3. Byrka, J., Aardal, K.: An optimal bifactor approximation algorithm for the metric facility location problem. SIAM Journal on Computing 39, 2212–2231 (2010)
4. Li, S.: A 1.488 approximation algorithm for the uncapacitated facility location problem. Information and Computation 222, 45–58 (2013)
5. Borodin, A., El-Yaniv, R.: Online Computation and Competitive Analysis. Press Syndicate of the University of Cambridge (1998)
6. Meyerson, A.: Online facility location. In: Proceedings of the 42nd IEEE Symposium on Foundations of Computer Science, pp. 426–431 (2001)
7. Fotakis, D.: On the competitive ratio for online facility location. Algorithmica 50, 1–57 (2008)
8. Fotakis, D.: A primal-dual algorithm for online non-uniform facility location. Journal of Discrete Algorithms 5(1), 141–148 (2007)
9. Nagarajan, C., Williamson, D.P.: Offline and online facility leasing. In: Lodi, A., Panconesi, A., Rinaldi, G. (eds.) IPCO 2008. LNCS, vol. 5035, pp. 303–315. Springer, Heidelberg (2008)
10. Fotakis, D.: Online and incremental algorithms for facility location. SIGACT News 42(1), 97–131 (2011)
11. Vazirani, V.: Approximation Algorithms. Springer (2003)
12. Williamson, D.P., Shmoys, D.B.: The Design of Approximation Algorithms. Cambridge University Press (2011)
13. Imase, M., Waxman, B.M.: Dynamic Steiner tree problem. SIAM Journal on Discrete Mathematics 4(3), 369–384 (1991)
14. Buchbinder, N., Naor, J.S.: The design of competitive online algorithms via a primal-dual approach. Foundations and Trends in Theoretical Computer Science 3, 93–263 (2009)
15. Gupta, A., Kumar, A., Pál, M., Roughgarden, T.: Approximation via cost sharing: Simpler and better approximation algorithms for network design. Journal of the ACM 54(3), Article 11 (2007)
16. Gupta, A., Srinivasan, A., Tardos, É.: Cost-sharing mechanisms for network design. Algorithmica 50, 98–119 (2008)
17. Swamy, C., Kumar, A.: Primal-dual algorithms for connected facility location problems. Algorithmica 40(4), 245–269 (2004)
18. Hasan, M.K., Jung, H., Chwa, K.Y.: Approximation algorithms for connected facility location. Journal of Combinatorial Optimization 16, 155–172 (2008)
19. Jung, H., Hasan, M.K., Chwa, K.Y.: A 6.55 factor primal-dual approximation algorithm for the connected facility location problem. Journal of Combinatorial Optimization 18, 258–271 (2009)
20. Eisenbrand, F., Grandoni, F., Rothvoß, T., Schäfer, G.: Connected facility location via random facility sampling and core detouring. Journal of Computer and System Sciences 76(8), 709–726 (2010)

# Multiply Balanced $k-$Partitioning

Amihood Amir[1,2,*], Jessica Ficler[1], Robert Krauthgamer[3,**], Liam Roditty[1], and Oren Sar Shalom[1]

[1] Department of Computer Science, Bar-Ilan University, Ramat-Gan 52900, Israel
{amir,liamr}@cs.biu.ac.il, {jessica.ficler,oren.sarshalom}@gmail.com
[2] Department of Computer Science, Johns Hopkins University, Baltimore, MD 21218
[3] Weizmann Institute of Science, Rehovot, Israel
robert.krauthgamer@weizmann.ac.il

**Abstract.** The problem of partitioning an edge-capacitated graph on $n$ vertices into $k$ balanced parts has been amply researched. Motivated by applications such as load balancing in distributed systems and market segmentation in social networks, we propose a new variant of the problem, called Multiply Balanced $k$ Partitioning, where the vertex-partition must be balanced under $d$ vertex-weight functions simultaneously.

We design bicriteria approximation algorithms for this problem, i.e., they partition the vertices into up to $k$ parts that are nearly balanced simultaneously for all weight functions, and their approximation factor for the capacity of cut edges matches the bounds known for a single weight function times $d$. For the case where $d = 2$, for vertex weights that are integers bounded by a polynomial in $n$ and any fixed $\epsilon > 0$, we obtain a $(2 + \epsilon, O(\sqrt{\log n \log k}))$-bicriteria approximation, namely, we partition the graph into parts whose weight is at most $2+\epsilon$ times that of a perfectly balanced part (simultaneously for both weight functions), and whose cut capacity is $O(\sqrt{\log n \log k}) \cdot \mathrm{OPT}$. For unbounded (exponential) vertex weights, we achieve approximation $(3, O(\log n))$.

Our algorithm generalizes to $d$ weight functions as follows: For vertex weights that are integers bounded by a polynomial in $n$ and any fixed $\epsilon > 0$, we obtain a $(2d + \epsilon, O(d\sqrt{\log n \log k}))$-bicriteria approximation. For unbounded (exponential) vertex weights, we achieve approximation $(2d + 1, O(d \log n))$.

## 1 Introduction

In the $k$-BALANCED PARTITIONING problem (aka MINIMUM $k$-PARTITIONING) the input is an edge-capacitated graph and an integer $k$, and the goal is to partition the graph vertices into $k$ parts of equal size, so as to minimize the total capacity of the cut edges (edges connecting vertices in different parts). The problem has many applications, ranging from parallel computing and VLSI design to social

---

[*] Partly supported by NSF grant CCR-09-04581, ISF grant 347/09, and BSF grant 2008217.
[**] Work supported in part by a US-Israel BSF grant #2010418, ISF grant 897/13, and by the Citi Foundation.

networks, as we discuss further below. The above problem is known to be NP-hard. Even the special case where $k = 2$ (called minimum bisection) is already NP-hard [7] and several approximation algorithms were designed [13,5,2,17]. For constant $k$ the polynomial-time algorithm of MacGregor [14] can solve the problem on trees. However, if $k$ is not constant the problem is hard to approximate within any finite factor [1]. Several heuristics were proposed, see e.g. [10,16,15,9] but they do not guarantee any upper bounds on the cut capacity. It is therefore common to consider a bicriteria approximation, which relaxes the balance constraint.

Formally, let $G = (V, E)$ be a graph of $n$ vertices. In a $(k, \nu)$-*balanced partition*, the vertex set $V$ is partitioned into at most $k$ parts, each of size at most $\nu n/k$, and the cut capacity is compared against an optimal (minimum cut capacity) perfectly balanced $k$-partition [12,18,4,1,11,6]. In the *weighted* version, every vertex $v \in V$ has a weight $w(v) \geq 0$, and now in a $(k, \nu)$-balanced partitioning, there are at most $k$ parts, and the total weight of every part is at most $\nu W/k$, where $W$ is the total weight of all the vertices. Let us emphasize that we always consider graphs with edge capacities; the terms weighted or unweighted graphs refer only to vertex weights. Throughout, we assume that there exists a perfectly balanced $k$-partition, e.g., in the unweighted version this means that $k$ divides $n$.

**Definition 1.** *An algorithm for $k$-BALANCED PARTITIONING is said to give a $(\nu, \alpha)$-bicriteria approximation if it finds a $(k, \nu)$-balanced partition whose cut capacity is at most $\alpha$ OPT, where OPT is the cut capacity of an optimal perfectly balanced $k$-partition.*

The $k$-BALANCED PARTITIONING problem has numerous applications. Specifically, in parallel computing, each vertex typically represents a task, its weight represents the amount of processing time needed for that task, and edges represent the communication costs. In this example, $k$ is the number of available processors. However, this formulation does not support the case where we want to distribute the load by two parameters, for example processing time and memory. A similar unsolved problem arises in social networks and marketing: vertices represent people, edges are the strength of the relationship between two people, and each person has a value (potential revenue) for a marketing campaign. The goal is to partition the people into $k$ groups, such that there will be the least connection between the groups, and the groups are balanced both by their size and by their total marketing value.

**Definition 2.** *In the DOUBLY BALANCED $k$-PARTITIONING problem, the input is a graph $G = (V, E, w_1, w_2, c)$ and an integer $k$, where $w_1, w_2 : V \to \mathbb{R}_{\geq 0}$ are the vertex-weight functions and $c : E \to \mathbb{R}_{\geq 0}$ is the edge capacity function. The goal is to find a partition of the graph into at most $k$ parts that are balanced by both weight functions, so as to minimize the total capacity of the cut between the different parts.*

We emphasize that $k$-BALANCED PARTITIONING refers to the case where there is a single vertex-weight function, in contrast to DOUBLY BALANCED $k$-PARTITIONING.

This is true even when one of the two vertex-weight functions above is constant (aka uniform weights), which means balancing with respect to the sizes (cardinalities) of the parts.

The problem can be generalized to $d$ vertex-weight functions:

**Definition 3.** *In the* MULTIPLY BALANCED $k$ PARTITIONING *problem, the input is a graph $G = (V, E, w_1, w_2, ..., w_d, c)$ and an integer $k$, where $w_1, ..., w_d : V \to \mathbb{R}_{\geq 0}$ are the vertex-weight functions and $c : E \to \mathbb{R}_{\geq 0}$ is the edge capacity function. The goal is to find a partition of the graph into at most $k$ parts that are balanced by all $d$ weight functions, so as to minimize the total capacity of the cut between the different parts.*

## 2 Bicriteria Approximations and Our Results

The DOUBLY BALANCED $k$-PARTITIONING problem is hard to approximate within any finite factor, simply because setting $w_2(v) = 0$ (or $w_2(v) = w_1(v)$) for all $v \in V$ yields the $k$-BALANCED PARTITIONING problem as a special case. We therefore aim at a bicriteria approximation for the problem. Throughout, for $S \subseteq V$, $1 \leq j \leq 2$, and a vertex-weight function $w$, we define $w(S) := \sum_{v \in S} w_j(v)$, and let $W := w(V)$ denote the total weight of all the vertices.

**Definition 4.** *A partition $\{P_i\}$ of $V$ is called $(k, \nu)$-doubly balanced if it has at most $k$ parts, and for each part $P_i$ and each $j = 1, 2$ it hold that $w_j(P_i) \leq \nu\, w_j(V)/k$.*

Before defining the precise guarantees of our algorithm, we need to understand the criteria that we are trying to approximate. In the unweighted version, balanced partition asserts that every part is of size at most $\lceil \frac{n}{k} \rceil$, which guarantees that there always exists a perfectly balanced partition. In the weighted version, this might not be possible at all. For example, if the graph has a single vertex whose weight exceeds that of all other vertices together, then obviously there is no perfectly balanced partition. Therefore, in all existing algorithms there is an implicit assumption that there exists a perfectly balanced partition (which we are trying to approximate).

In DOUBLY BALANCED $k$-PARTITIONING, we could assume the existence of a perfectly doubly balanced partition as well. However, this might be an unreasonable assumption in many applications, and thus we weaken the requirement — we only assume that there is a perfectly balanced partition for each weight separately, but not necessarily together.

**Definition 5.** *A $(\nu, \alpha)$-bicriteria approximation for the* DOUBLY BALANCED $k$-PARTITIONING *problem finds a $(k, \nu)$-doubly balanced partition, whose cut capacity is at most $\alpha\, \mathrm{OPT}$, where $\mathrm{OPT}$ is the maximum of the two optimal $(k, 1)$-balanced partitions.*

Throughout, the term OPT in the context of Doubly Balanced $k$-Partitioning refers to the above value. Notice that the cut capacity of a perfectly $k$-doubly balanced partition (if it exists) might be substantially larger than each of the $k$-balanced partitions. Nevertheless, because we relax the balance constraints, we require our algorithm to return a near $k$-doubly balanced partition whose cut capacity is comparable to the larger of the two different partitions.

**Definition 6.** *A partition $\{P_i\}$ of the vertices is called $(k^+, \nu)$-balanced if $w(P_i) \leq \nu W/k$ for every part $P_i$.*

Notice that in a $(k^+, \nu)$-balanced partition, unlike a $(k, \nu)$-balanced partition, there can be more than $k$ parts.

**Definition 7.** *An algorithm for $k$-BALANCED PARTITIONING is said to give a $(\nu, \alpha)^+$-bicriteria approximation if it finds a $(k^+, \nu)$-balanced partition whose cut capacity is at most $\alpha$ OPT, where OPT is the cut capacity of an optimal perfectly balanced $k$-partition.*

Notice that because every $(k, \nu)$-balanced partition is also a $(k^+, \nu)$-balanced partition, then every $(\nu, \alpha)$-bicriteria approximation algorithm is also a $(\nu, \alpha)^+$-bicriteria approximation.

### 2.1 Our Results

**Theorem 1.** *The Multiply Balanced $k$ Partitioning problem admits a polynomial-time $(\nu, \alpha)$-bicriteria approximation, according to the following table:*

| vertex-weight functions | $\nu$ | $\alpha$ |
|---|---|---|
| polynomial | $2d + \epsilon$ | $O(d\sqrt{\log n \log k})$ |
| arbitrary (exponential) | $2d + 1$ | $O(d \log n)$ |

In particular, for the Doubly Balanced $k$-Partitioning problem, where $d = 2$ we have:

| vertex-weight functions | $\nu$ | $\alpha$ |
|---|---|---|
| polynomial | $2 + \epsilon$ | $O(\sqrt{\log n \log k})$ |
| arbitrary (exponential) | $3$ | $O(\log n)$ |

When we say that the weights are polynomial we mean that the length necessary to encode each weight is poly-logarithmic in $n$.

We now provide a high-level overview of the algorithm for the Doubly Balanced $k$-Partitioning problem. The full details are presented in Section 3. Let $A$ be a $(\nu, \alpha)^+$-bicriteria approximation algorithm for the $k$-BALANCED PARTITIONING problem.

1. *Partition Stage*: Divide the vertices into some number of parts $t$, with cut capacity at most $\alpha$ OPT, and the respective weight of each part is bounded by $\nu W_1/k$ and $\nu W_2/k$ (simultaneously). If $t \leq k$ then the balance requirements are met. Otherwise proceed to stage 2.

2. *Union Stage*: Combine these $t$ parts into $k$ parts carefully, so that each part $S$ has weights $w_1(S) \leq (1+\nu)W_1/k$ and $w_2(S) \leq (1+\nu)W_2/k$. This new partition meets the same approximation factor for the cut capacity, because combining parts can only decrease the capacity of the cut.

We present two different algorithms, each based on a different $k$-balanced partition approximation algorithm, to achieve the two bounds stated in Theorem 1. We first present the special case $d = 2$ in Section 3, and then prove its generalization to $d$ weight functions in Section 4.

### 2.2 Polynomial Weights

We now show how it is possible to extend the approximation ratio for $k$-BALANCED PARTITIONING to hold also for weighted graphs.

Andreev and Räcke [1] showed a $(1+\epsilon, \log^{1.5} n)$ bicriteria approximation for any constant $\epsilon > 0$. Their work balances the graph with respect to the sizes of the parts, but can be extended in a straightforward manner to the case where the vertices of the graph have polynomial weights and the goal is to balance the weight among the parts.

**Theorem 2.** *Every $(\nu, \alpha)$-bicriteria approximation algorithm $\mathcal{A}$ for the $k$-balanced partitioning problem in unweighted graphs can be used also in (polynomially) weighted graphs with the same approximation factors.*

*Proof.* Will appear at the full version.

If the running time of the unweighted version algorithm is $f(n)$, where $n$ is the number of vertices, then the modified running time would be $f(W)$, where $W$ is the total weight. If the (integer) weights of the vertices are polynomial in $n$, then the algorithm runs in polynomial time as well. Since the length necessary to encode each weight is polylogarithmic in $n$, then it guarantees that the total weight is polynomial.

For any fixed $0 < \epsilon < 1$, Feldman and Foschini [6] presented a $(1+\epsilon, O(\log n))$ bicriteria approximation for unweighted graphs. Krauthgamer, Naor and Schwartz [11] presented a $(2, O(\sqrt{\log n \log k}))$ bicriteria approximation algorithm. Their algorithm can be considered as a $(1+\epsilon, O(\sqrt{\log n \log k}))^+$ bicriteria approximation algorithm, since during its main procedure it finds a $(k^+, 1+\epsilon)$-balanced partition. As explained above, we can modify this algorithm to support weighted graphs.

### 2.3 Unrestricted Weights

To our knowledge, the only algorithm that achieves a bicriteria approximation for graphs with exponential weight function is that of Even, Naor, Rao and Schieber [4]. Their algorithm uses an algorithm for the $\rho$-separator problem in order to achieve a $(2, \log n)$ bicriteria approximation with an exponential weight function.

## 3 Bicriteria Approximation Algorithm for $d = 2$

Let $\mathcal{A}$ be a $(\nu, \alpha)^+$-bicriteria approximation algorithm for the $k$-balanced partitioning problem. For convenience sake, we will normalize the weights such that for every $v \in V$ we have $w_j(v) \leftarrow w_j(v) \cdot \frac{k}{w_j(V)}$, where $j = 1, 2$. From the definition of $(k, \nu)$-doubly balanced partition each part is of weight at most $\frac{\nu w_j(V)}{k}$, thus after the normalization each part is of weight at most $\frac{\nu w_j(V)}{k} \cdot \frac{k}{w_j(V)} = \nu$. Moreover, after the normalization $w_j(V) = k$.

The algorithm works as follows. First, partition $G$ using algorithm $\mathcal{A}$ with respect to weight function $w_1$. Let $\mathcal{P} = \{P_1, P_2, \ldots, P_{\ell_1}\}$ be the resulting partition. It holds for every $P \in \mathcal{P}$, that $w_1(P) \leq \nu$. Let $\mathcal{P}^{>} = \{P \mid P \in \mathcal{P}, w_2(P) > \nu\}$. In case that $\mathcal{P}^{>} = \emptyset$, then if $\ell_1 \leq k$ then we have at most $k$ parts and each part satisfies the balance condition with respect to both $w_1$ and $w_2$. Let $OPT_j$ be the cut capacity of an optimal perfectly balanced $k$-partition with respect to weight function $w_j$, $j = 1, 2$. The cut capacity of partition $\mathcal{P} \leq \alpha \, OPT_1 \leq \alpha \, OPT$.

In case that $\mathcal{P}^{>} \neq \emptyset$ we partition $G$ using algorithm $\mathcal{A}$ with respect to weight function $w_2$. Let $\mathcal{Q} = \{Q_1, Q_2, \ldots, Q_{\ell_2}\}$ be the resulting partition.

Fix a part $P \in \mathcal{P}^{>}$. Let $R_i(P) = Q_i \cap P$, where $Q_i \in \mathcal{Q}$ and $1 \leq i \leq \ell_2$. As $w_2(Q_i) \leq \nu$ it follows that $w_2(R_i(P)) \leq \nu$ for every $1 \leq i \leq \ell_2$.

Consider now the partition $\mathcal{R}$ that is composed of all the parts that are in $\mathcal{P} \setminus \mathcal{P}^{>}$ and the parts $R_i(P) = Q_i \cap P$, where $1 \leq i \leq \ell_2$, for every $P \in \mathcal{P}^{>}$. Each part of this partition has weight at most $\nu$ with respect to $w_1$ and to $w_2$. The cut capacity of this partition is at most $\alpha \, OPT_1 + \alpha \, OPT_2 \leq O(\alpha) \cdot OPT$.

The only problem with the partition $\mathcal{R}$ is that the number of its parts might be as large at $\ell_1 \cdot \ell_2$ and this may be larger than $k$.

In subsection 3.1 we describe a process that takes as an input this partition and combines parts of it until it reaches a final partition with at most $k$ parts each of weight at most $1 + \nu$. As the the final partition is obtained only by combining parts of the input partition, its cut capacity cannot exceed the cut capacity of the input partition. In subsection 3.2 we show lower bounds for the method of subsection 3.1.

### 3.1 Combining Partitions via Bounded Pair Scheduling

Let $\mathcal{R} = \{R_1, R_2, \ldots, R_{\ell_3}\}$. Each $R_i \in \mathcal{R}$ is represented by a pair of coordinates $(x_i, y_i)$, where $x_i = w_1(R_i)$, $y_i = w_2(R_i)$ and $1 \leq i \leq \ell_3$. Moreover, $0 \leq x_i, y_i < \nu$ and $\sum_{i=1}^{\ell_3} x_i = \sum_{i=1}^{\ell_3} y_i = k$. In case that $\ell_3 \leq k$ then the partition has all the desired properties. Hence, we assume that $\ell_3 > k$. This problem resembles a known NP-hard problem, called VECTOR SCHEDULING with 2 dimensions. Formally:

**Definition 8.** (VECTOR SCHEDULING) *We are given a set $J$ of $n$ rational $d$-dimensional vectors $p_1, \ldots, p_n$ from $[0, \infty)^d$ and a number $m$. A valid solution is a partition of $J$ into $m$ sets $A_1, \ldots, A_m$. The objective is to minimize $\max_{1 \leq i \leq m} \|\bar{A}_i\|_\infty$ where $\bar{A}_i = \sum_{j \in A_i} p_j$ is the sum of the vectors in $A_i$.*

When $d$ is constant, [8] shows a $(d+1)$ approximation, and a later work [3] gives a PTAS for the problem.

Our problem is a special case of the VS, namely with $d = 2$. However, the existing algorithms approximate the objective with respect to the optimal solution that can be achieved for a specific instance. We need to design an approximation algorithm that bounds the maximal objective for a *family* of instances, and not just for a specific instance. The family of input instances are the vectors $p_i$ such that $\|p_i\|_\infty < \nu$ and for all $1 \leq j \leq d$, $\sum_{i=1}^n p_i^j = k$, where $p^j$ is the $j$'th element of vector $p$.

Formally, we need to solve the following problem:

**Definition 9.** (BOUNDED PAIR SCHEDULING)

INPUT: A number $k$ and a set $\mathcal{R}$ of $n > k$ elements, such that each element is a pair $(x, y)$ that holds $0 \leq x, y < \nu$, and $\sum_{i=1}^n x_i = \sum_{i=1}^n y_i = k$.

OUTPUT: A partition of $\mathcal{R}$ into a set of $k$ elements $R_1, \ldots, R_k$, such that for all $i = 1, \ldots, k$ and $R_i = (x^{(i)}, y^{(i)})$, it holds that $0 \leq x^{(i)} \leq 1 + \nu$ and $0 \leq y^{(i)} \leq 1 + \nu$.

The algorithm below solves the bounded pair scheduling problem.

Consider now the following sets of elements:

1. $S = \{(x, y) \mid x < 1, y < 1\}$
2. $A = \{(x, y) \mid 1 \leq x < 1 + \nu,\ 1 \leq y < 1 + \nu\}$
3. $B_x = \{(x, y) \mid 1 \leq x < \nu,\ y < 1\}$
4. $B_y = \{(x, y) \mid x < 1,\ 1 \leq y < \nu\}$
5. $C_x = \{(x, y) \mid \nu \leq x < 1 + \nu,\ y < 1\}$
6. $C_y = \{(x, y) \mid x < 1,\ \nu \leq y < 1 + \nu\}$

Elements in $A$ are balanced, and the minimum weight in each coordinate exceeds 1, therefore, if all our elements were of type $A$ we would be done - there are no more than $k$ balanced elements.

The $B$ elements are "almost" balanced and the union of every element in $B_x$ with a element in $B_y$ is a element in $A$.

The elements in $C$ are not balanced and can not be trivially combined with any other elements. The main effort of our algorithm is dealing with these elements.

The $S$ elements are the ones which present a difficulty since both their coordinates are not bounded below. Thus there may be a very large number of them. However, they do give us the necessary maneuverability in the combining process.

The auxiliary sets span the input set $\mathcal{R}$, and because their criteria are exclusive, every element in $\mathcal{R}$ fits to exactly one of these sets. We begin by dividing the input set to the appropriate auxiliary sets. Clearly, the $C$ sets remain empty at this stage. As we show next, the algorithm iteratively combines elements. The meaning of combining elements $R_i$ and $R_j$ is creating a new element $(x_i + x_j, y_i + y_j)$ instead of them and assigning it to the appropriate set.

As long as there are two elements $R_i, R_j \in S$ such that $x_i + x_j < 1$ and $y_i + y_j < 1$ we pick such two elements $R_i$ and $R_j$ and combine them. At the end of this stage it is guaranteed for every $R_i, R_j \in S$ that either $x_i + x_j \geq 1$ or $y_i + y_j \geq 1$.

Next, as long as there is a pair $R_i$ and $R_j \in S$ such that $x_i + x_j \geq 1$ and $y_i + y_j \geq 1$ then it also holds for such a pair that $x_i + x_j \leq 2 < 1 + \nu$ because $x_i < 1$ and $x_j < 1$. Similarly, $y_i + y_j \leq 2 < 1 + \nu$. We combine such a pair to $R_{ij}$, and add it to $A$.

At the end of this stage it is guaranteed that for every $R_i, R_j \in S$ either $x_i + x_j \geq 1$ and $y_i + y_j < 1$ or $y_i + y_j \geq 1$ and $x_i + x_j < 1$.

**Lemma 1.** *At this stage of the algorithm, for every pair $R_i, R_j \in S$ it holds that $(x_i + x_j, y_i + y_j)$ fits to one of the elements $B_x, B_y, C_x$ and $C_y$.*

*Proof.* Will appear at the full version.

The algorithm proceeds as follows. We iteratively choose a pair $R_i, R_j \in S$ that minimizes $max\{x_i + x_j, y_i + y_j\}$. Since we choose the pair that minimizes the maximum of the two coordinates it is guaranteed that all the pairs that their combination is either in $B_x$ or in $B_y$ will be chosen before all the pairs that their combination is either in $C_x$ or in $C_y$. As long as there is a pair whose combination belongs to $B_x$ (or $B_y$), we combine it.

If we reach to a point that $B_x$ and $B_y$ are not empty and there is no longer a pair of elements that its combination belong to either $B_x$ or $B_y$ we do the following. As long as both $B_x$ and $B_y$ are not empty we combine an arbitrary pair $R_i \in B_x$ and $R_j \in B_y$. Notice that the combined element belongs to $A$ because $1 \leq x_i + x_j < 1 + \nu$ as $1 \leq x_i < \nu$ and $x_j < 1$, and similarly, $1 \leq y_i + y_j < 1 + \nu$. After that, at most one of $B_x$ and $B_y$ is not empty. Assume that one of them is not empty, and wlog let it be $B_x$.

Consider the following state of the algorithm: the sets $B_y$, $C_x$ and $C_y$ are empty and the sets $B_x$ and $S$ are not empty. We now distinguish between two cases. The case that there is at most one $R \in S$ such that $w_1(R) < w_2(R)$ and the case that there is more than one such element in $S$. For the first case we prove:

**Lemma 2.** *If $B_y$ and $C_y$ are empty elements, and there is at most one $R \in S$ such that $w_1(R) < w_2(R)$, then there is a way to combine the elements of $S$ so that the total number of different elements is at most $k$ and every element is of weight at most $1 + \nu$.*

*Proof.* Will appear at the full version.

It stems from the lemma above that if we are in the case that there is at most one $R \in S$ such that $w_1(R) < w_2(R)$ then we can reach the desired partition. Thus, we assume now that there are at least two elements $R_j, R_q \in S$ such that $w_1(R_j) < w_2(R_j)$ and $w_1(R_q) < w_2(R_q)$. We choose an arbitrary element $R_i \in B_x$. We know that $x_i < \nu$, hence $x_i + x_j + x_q < 1 + \nu$. So even if we combine $R_i$ with $R_j$ and $R_q$ the $x$-coordinate is in the right range for $A$. The only question is if the $y$-coordinate fits. If $y_i + y_j \geq 1$, then we combine $R_i$ and $R_j$ as both $y_i$ and $y_j$ are less than 1 we can remove them and add their combination to $A$. If $y_i + y_j < 1$ then combine the elements $R_i, R_j, R_q$ and add them to $A$, because $y_i + y_j < 1$ and $y_q < 1$ then $y_i + y_j + y_q < 2 < 1 + \nu$, and because $y_j + y_q \geq 1$ by our assumption. We continue with this process until $B_x$ gets empty.

Now both $B_x$ and $B_y$ are empty, and the next pair $R_i, R_j \in S$ that minimizes $max\{x_i + x_j, y_i + y_j\}$ fits into $C_x$ or $C_y$. Assume, wlog it belongs to $C_x$. There are two possible cases:

There is a third element $R_q \in S$ such that $y_i + y_q \geq 1$ or $y_j + y_q \geq 1$. Assume, wlog, that $y_i + y_q \geq 1$, which leads to $x_i + x_q < 1$ and therefore $x_i + x_q + x_j < 2 < 1 + \nu$. Additionally, $y_i + y_j < 1$, $y_i + y_q \geq 1$ therefore, $1 \leq y_i + y_j + y_q < 2 < 1 + \nu$. We can remove these three elements from $S$ and add their combination to $A$.

If there is no such third element it holds for each element $R_q \in S$ that $y_i + y_q < 1$ and $y_j + y_q < 1$. We show that in such a case there is at most one element $R_q \in S$ such that $x_q < y_q$. Assume for the sake of contradiction that there are two elements $R_q, R_t \in S$ such that $x_q < y_q$ and $x_t < y_t$. We know that either $x_i + x_q \geq x_i + x_j$ or $y_i + y_q \geq x_i + x_j$, because otherwise a different pair of elements would have obtain the minimum of the maximum. Because $y_i + y_q < 1$ then $x_i + x_q > x_i + x_j \geq \nu$. Also, $x_q \geq x_i$ because otherwise a different pair of elements would have obtain the minimum of the maximum. From the same considerations $x_q \geq x_j$. Recall that $x_i + x_j \geq \nu$, thus, $x_q \geq \frac{\nu}{2}$. Recall that by our assumption $y_q > x_q$, thus $y_q > \frac{\nu}{2}$. In the same way we can show that $x_t \geq \frac{\nu}{2}, y_t > \frac{\nu}{2}$, and therefore the combination of $R_q$ and $R_t$ belongs to $A$, which contradicts the fact that no combination of any pairs in $S$ belongs to $A$. Therefore, the conditions of the Lemma 2 are satisfied and we can apply it. After each time the algorithm combines two elements, it checks whether we are left with exactly $k$ pairs, and if so it stops and outputs the result. Therefore we do not explicitly mention this check in the algorithm itself.

### 3.2 Lower Bounds

This section considers the tightness of the bounds of the union stage. For a given $\nu$, the partition stage produces many parts such that each part $R$ has weights $w_1(R) \leq (1+\nu)W_1/k$ and $w_2(R) \leq (1+\nu)W_2/k$. It was shown in [4] that any $(k,\nu)$-balanced partitioning problem with $\nu > 2$ can be reduced to a $(k',\nu')$-balanced partitioning problem with $k' \leq k$ and $\nu \leq 2$, i.e., that it is only necessary to analyze the problem for values of $\nu$ at most 2. Therefore we can express $\nu$ as $1 + \epsilon$ when $0 < \epsilon \leq 1$.

**Lemma 3.** *There exist an input to the Bounded Pair Scheduling problem that can not be combined to $k$ parts without exceeding $2 + \frac{2\epsilon}{2+\epsilon}$.*

*Proof.* Our example consists of two types of elements: type $A = (1+\epsilon, 0)$ and type $B$, which will be defined later. First we set up an input with $s$ elements whose total weight is $(s, s)$. We use only a single element of type $A$, so we are left with $s-1$ type $B$ elements. The total value of all of the $y$'s is $s$ and only the $s-1$ type $B$ elements contribute to this value. Therefore the $y$ value of each such element is $\frac{s}{s-1}$. The total value of all of the $x$'s is $s$, the type $A$ element contributes $1+\epsilon$ to this sum, so the $y$ value of each of the type $B$ elements is $\frac{s-(1+\epsilon)}{s-1}$. We call this the *basic structure*. The basic structure will be replicated many times, as we'll show later.

Since type $A$ elements get the maximal value $(1+\epsilon)$, combining two such elements yields a high value. Therefore, we need to balance between the combined value of $A$ and $B$ compared to the combined value of two $B$'s. Combining a type $A$ with a type $B$ pair yields $x$ value $1+\epsilon+\frac{s-(1+\epsilon)}{s-1}$ and $y$ value $\frac{s}{s-1}$, i.e. the $x$ value is greater.

Combining two type $B$ pairs yields $x$ value $2(\frac{s-(1+\epsilon)}{s-1})$ and $y$ value $2\frac{s}{s-1}$, i.e. the $y$ value is greater.

If we want to balance the $x$ and $y$ values, we would need to have $1+\epsilon+\frac{s-(1+\epsilon)}{s-1}$ equals to $\frac{2s}{s-1}$. For this to happen compute $s$ as a function of $\epsilon$:

$1+\epsilon+\frac{s-(1+\epsilon)}{s-1} = \frac{2s}{s-1}$

$s = 2+\frac{2}{\epsilon}$.

We can assume that $s$ is an integer. Otherwise we can represent $s$ as a ratio of two integers $s = \frac{n}{d}$, and replicate each of the elements $d$ times.

The $y$ value of type $B$ elements $= \frac{s}{s-1} = \frac{2+\frac{2}{\epsilon}}{2+\frac{2}{\epsilon}-1} = \frac{2+2\epsilon}{2+\epsilon} = 1+\frac{\epsilon}{2+\epsilon} < 1+\epsilon$.

The $x$ value of type $B$ elements $= \frac{s-(1+\epsilon)}{s-1} = 1-\frac{\epsilon}{s-1} = 1-\frac{\epsilon}{1+\frac{2}{\epsilon}} = 1-\frac{\epsilon^2}{2+\epsilon} < 1+\epsilon$. Therefore type $B$ elements do not reach the $1+\epsilon$ threshold and are valid elements.

The above scenario is not interesting since the total value of $W_1$ and $W_2$ equals to the number of parts, so no parts should be combined. Therefore we tweak the example. Modify the basic structure as follows: For each basic structure subtract an infinitesimally small value $\delta \ll \epsilon$ from each of the coordinates of type $B$ elements. This decreases the total value of each coordinate by $(s-1)\delta$. Now we will replicate the whole set $\frac{s-(s-1)\delta}{(s-1)\delta} = \frac{s}{(s-1)\delta}-1$ times. This leaves enough free space for an additional basic structure.

At this point we have to combine at least two elements. If we combine two type $A$ elements, their $x$ value will be $2+2\epsilon$. If we combine two type $B$ elements, their $y$ value will be $2+\frac{2\epsilon}{2+\epsilon}-2\delta \approx 2+\frac{2\epsilon}{2+\epsilon}$. The last possible combination is combining a type $A$ element with a type $B$. The $x$ value will be $(1+\epsilon)+(1-\frac{\epsilon^2}{2+\epsilon})-\delta = 2+\frac{2\epsilon+\epsilon^2-\epsilon^2}{2+\epsilon}-\delta = 2+\frac{2\epsilon}{2+\epsilon}-\delta \approx 2+\frac{2\epsilon}{2+\epsilon}$. Therefore no matter which pair we decide to combine, we get a value, as $k$ goes to infinity of $2+\frac{2\epsilon}{2+\epsilon}$. □

**Conclusion.** Our algorithm is within $1+\frac{\epsilon^2}{4(1+\epsilon)}$ of the optimal.

*Proof.* Our algorithm achieves $1+\nu = 2+\epsilon$, which is within $\frac{2+\epsilon}{2+\frac{2\epsilon}{2+\epsilon}} = \frac{2+\epsilon}{\frac{4+2\epsilon+2\epsilon}{2+\epsilon}} = \frac{4+4\epsilon+\epsilon^2}{4+4\epsilon} = 1+\frac{\epsilon^2}{4(1+\epsilon)}$ of the example. □

## 4 Generalization to $d$ Weight Functions

The important observation is that the algorithm of Subsection 3.1 can be viewed as a subroutine whose input is a partition of the vertices into $k$ subsets, each having weight bounded by $\nu$, and another partition into $k$ subsets, each having weight bounded by $\nu$. The result of the subroutine is a partition into $k$ subsets, each having weight bounded by $\nu + 1$. Call this subroutine COMBINE. As presented, the sum of the weights in each of the two coordinates is bounded by $k$. We need to use the subroutine in a more general fashion, where the sum of the weights of each coordinate is bounded by $mk$, for a parameter $m$. The necessary change to COMBINE is in the definitions of the $A, B,$ and $C$ sets. It now becomes:

1. $S = \{(x, y) \mid x < m, y < m\}$
2. $A = \{(x, y) \mid m \leq x < m + \nu,\ m \leq y < m + \nu\}$
3. $B_x = \{(x, y) \mid m \leq x < \nu,\ y < m\}$
4. $B_y = \{(x, y) \mid x < m,\ m \leq y < \nu\}$
5. $C_x = \{(x, y) \mid \nu \leq x < m + \nu,\ y < m\}$
6. $C_y = \{(x, y) \mid x < m,\ \nu \leq y < m + \nu\}$

The result of the subroutine is a partition of the vertices into $k$ subsets, each having weight bounded by $\nu + m$. Observe also that the cut capacity of the partition after subroutine COMBINE is bounded by the sum of the two initial cut capacities.

Assume now that we have $d$ weight functions. Assume also that $d$ is a power of 2. Using COMBINE we can construct $\frac{d}{2}$ partitions of the graph vertices, each into $k$ subsets, and each subset having weight bounded by $1 + \nu$, where in the $i$-th partition the weights considered are $w_{2i-1}$ and $w_{2i}$. We prepare these $\frac{d}{2}$ partitions for the next iteration, by considering partition $i$ as having weight function $w_{1,i} = \max(w_{2i-1}, w_{2i})$.

We can now do the same process, but we now have only $\frac{d}{2}$ partitions. Use COMBINE to produce $\frac{d}{4}$ partitions of the graph vertices, each into $k$ subsets, and each subset having weight bounded by $2 + \nu$. Again, we prepare these $\frac{d}{4}$ partitions for the next iteration, by considering partition $i$ as having weight function $w_{2,i} = \max(w_{1,2i-1}, w_{1,2i})$.

After $\lceil \log d \rceil$ iterations, we have a partition into $k$ subsets, each having weight bounded by $2^{\lceil \log d \rceil} - 1 + \nu \leq 2d - 1 + \nu$ in every weight function.

Since we employ subroutine COMBINE $\lceil \log d \rceil$ times, the final cut capacity as a result of our algorithm is the cut capacity resulting from a single partitioning multiplied by $O(d)$.

## References

1. Andreev, K., Racke, H.: Balanced graph partitioning. Theory of Computing Systems 39(6), 929–939 (2006)
2. Arora, S., Rao, S., Vazirani, U.: Expander flows, geometric embeddings, and graph partitionings. In: 36th Annual Symposium on the Theory of Computing, pp. 222–231 (May 2004)

3. Chekuri, C., Khanna, S.: On multidimensional packing problems. SIAM Journal on Computing 33(4), 837–851 (2004)
4. Even, G., Naor, J., Rao, S., Schieber, B.: Fast approximate graph partitioning algorithms. In: Proceedings of the 8th Annual ACM-SIAM Symposium on Discrete Algorithms, pp. 639–648. ACM, New York (1997)
5. Feige, U., Krauthgamer, R.: A polylogarithmic approximation of the minimum bisection. SIAM J. Comput. 31(4), 1090–1118 (2002)
6. Feldmann, A.E., Foschini, L.: Balanced partitions of trees and applications. In: 29th International Symposium on Theoretical Aspects of Computer Science (STACS 2012), vol. 14, pp. 100–111. Schloss Dagstuhl–Leibniz-Zentrum fuer Informatik, Dagstuhl (2012)
7. Garey, M.R., Johnson, D.S., Stockmeyer, L.: Some simplified NP-complete graph problems. Theoret. Comput. Sci. 1(3), 237–267 (1976)
8. Garofalakis, M.N., Ioannidis, Y.E.: Parallel query scheduling and optimization with time-and space-shared resources. SORT 1(T2), T3 (1997)
9. Hendrickson, B., Leland, R.W.: A multi-level algorithm for partitioning graphs. SC 95, 28 (1995)
10. Karypis, G., Kumar, V.: A fast and high quality multilevel scheme for partitioning irregular graphs. SIAM J. Sci. Comput. 20(1), 359–392 (1998)
11. Krauthgamer, R., Naor, J.S., Schwartz, R.: Partitioning graphs into balanced components. In: 20th Annual ACM-SIAM Symposium on Discrete Algorithms, pp. 942–949. SIAM (2009)
12. Leighton, F., Makedon, F., Tragoudas, S.: Approximation algorithms for VLSI partition problems. In: Proceedings of the IEEE International Symposium on Circuits and Systems, pp. 2865–2868. IEEE Computer Society Press (1990)
13. Leighton, T., Rao, S.: Multicommodity max-flow min-cut theorems and their use in designing approximation algorithms. J. ACM 46(6), 787–832 (1999)
14. MacGregor, R.: On Partitioning a Graph: A Theoretical and Empirical Study. Memorandum UCB/ERL-M. University of California, Berkeley (1978)
15. Patkar, S.B., Narayanan, H.: An efficient practical heuristic for good ratio-cut partitioning. In: 16th International Conference on VLSI Design, pp. 64–69. IEEE (2003)
16. Portugal, D., Rocha, R.: Partitioning generic graphs into $k$ regions. In: 6th Iberian Congress on Numerical Methods in Engineering (CMNE 2011), Coimbra, Portugal (June 2011)
17. Räcke, H.: Optimal hierarchical decompositions for congestion minimization in networks. In: 40th Annual ACM Symposium on Theory of Computing, pp. 255–264. ACM (2008)
18. Simon, H.D., Teng, S.: How good is recursive bisection? SIAM J. Sci. Comput. 18(5), 1436–1445 (1997)

# On Some Recent Approximation Algorithms for MAX SAT

Matthias Poloczek[1,*], David P. Williamson[1,**], and Anke van Zuylen[3,***]

[1] School of Operations Research and Information Engineering, Cornell University, Ithaca, NY, USA
poloczek@orie.cornell.edu, dpw@cs.cornell.edu
[2] Department of Mathematics, College of William and Mary, Williamsburg, VA, USA
anke@wm.edu

**Abstract.** Recently a number of randomized $\frac{3}{4}$-approximation algorithms for MAX SAT have been proposed that all work in the same way: given a fixed ordering of the variables, the algorithm makes a random assignment to each variable in sequence, in which the probability of assigning each variable true or false depends on the current set of satisfied (or unsatisfied) clauses. To our knowledge, the first such algorithm was proposed by Poloczek and Schnitger [7]; Van Zuylen [8] subsequently gave an algorithm that set the probabilities differently and had a simpler analysis. Buchbinder, Feldman, Naor, and Schwartz [1], as a special case of their work on maximizing submodular functions, also give a randomized $\frac{3}{4}$-approximation algorithm for MAX SAT with the same structure as these previous algorithms. In this note we give a gloss on the Buchbinder et al. algorithm that makes it even simpler, and show that in fact it is equivalent to the previous algorithm of Van Zuylen. We also show how it extends to a deterministic LP rounding algorithm, and we show that this same algorithm was also given by Van Zuylen [8]. Finally, we describe a data structure for implementing these algorithms in linear time and space.

## 1 Introduction

The maximum satisfiability problem (MAX SAT) is a fundamental problem in discrete optimization. In the problem we are given $n$ boolean variables $x_1, \ldots, x_n$ and $m$ clauses that are conjunctions of the variables or their negations. With each clause $C_j$, there is an associated weight $w_j \geq 0$. We say a clause is *satisfied* if one of its positive variables is set to true or if one of its negated variables is set to false. The goal of the problem is to find an assignment of truth values to the variables so as to maximize the total weight of the satisfied clauses. The problem is NP-hard via a trivial reduction from satisfiability.

---

[*] Supported by the Alexander von Humboldt Foundation within the Feodor Lynen program, and in part by NSF grant CCF-1115256.
[**] Supported in part by NSF grant CCF-1115256.
[***] Supported in part by a Suzann Wilson Matthews Summer Research Award.

We say we have an α-approximation algorithm for MAX SAT if we have a polynomial-time algorithm that computes an assignment whose total weight of satisfied clauses is at least α times that of an optimal solution; we call α the performance guarantee of the algorithm. A randomized α-approximation algorithm is a randomized polynomial-time algorithm such that the expected weight of the satisfied clauses is at least α times that of an optimal solution. The 1974 paper of Johnson [5], which introduced the notion of an approximation algorithm, also gave a $\frac{1}{2}$-approximation algorithm for MAX SAT. This algorithm was later shown to be a $\frac{2}{3}$-approximation algorithm by Chen, Friesen, and Zheng [2] (see also the simpler analysis of Engebretsen [3]). Yannakakis [10] gave the first $\frac{3}{4}$-approximation algorithm for MAX SAT; it uses network flow and linear programming computation as subroutines. Goemans and Williamson [4] subsequently showed how to use randomized rounding of a linear program to obtain a $\frac{3}{4}$-approximation algorithm for MAX SAT. Subsequent approximation algorithms which use semidefinite programming have led to still better performance guarantees.

In 1998, Williamson [9, p. 45] posed the question of whether it is possible to obtain a $\frac{3}{4}$-approximation algorithm for MAX SAT without solving a linear program. This question was answered positively in 2011 by Poloczek and Schnitger [7]. They give a randomized algorithm with the following particularly simple structure: given a fixed ordering of the variables, the algorithm makes a random assignment to each variable in sequence, in which the probability of assigning each variable true or false depends on the current set of satisfied (or unsatisfied) clauses. Subsequently, Van Zuylen [8] gave an algorithm with the same structure that set the probabilities differently and had a simpler analysis. In 2012, Buchbinder, Feldman, Naor, and Schwartz [1], as a special case of their work on maximizing submodular functions, also gave a randomized $\frac{3}{4}$-approximation algorithm for MAX SAT with the same structure as these previous algorithms[1]. Poloczek [6] gives evidence that the randomization is necessary for this style of algorithm by showing that a deterministic algorithm that sets the variables in order (where the next variable to set is chosen adaptively) and uses a particular set of information about the clauses cannot achieve performance guarantee better than $\frac{\sqrt{33}+3}{12} \approx .729$. However, Van Zuylen [8] shows that it is possible to give a deterministic $\frac{3}{4}$-approximation algorithm with the same structure given a solution to a linear programming relaxation.

The goal of this paper is to give an interpretation of the Buchbinder et al. MAX SAT algorithm that we believe is conceptually simpler than the one given there. We also restate the proof in terms of our interpretation. We further show that the Buchbinder et al. algorithm is in fact equivalent to the previous algorithm of Van Zuylen. We extend the algorithm and analysis to a deterministic LP rounding algorithm. We conclude with a description of a data structure that allows to implement greedy algorithms as well as LP rounding algorithms, such

---

[1] In the extended abstract of [1], the authors claim the MAX SAT result and omit a part of the proof, but it is not difficult to reconstruct the proof from the rest of the paper.

as the ones above, using linear time and space. Buchbinder et al. also state (without proof) that there is a linear-time implementation of the algorithm.

Here we give the main idea of our perspective on the algorithm. Consider greedy algorithms that set the variables $x_i$ in sequence. A natural greedy algorithm sets $x_i$ to true or false depending on which increases the total weight of the satisfied clauses by the most. An alternative to this algorithm would be to set each $x_i$ so as to increase the total weight of the clauses that are not yet *unsatisfied* given the setting of the variable (a clause is unsatisfied if all the variables of the clause have been set and their assignment does not satisfy the clause). The algorithm is in a sense a randomized balancing of these two algorithms. It maintains a bound that is the average of two numbers, the total weight of the clauses satisfied thus far, and the total weight of the clauses that are not yet unsatisfied. For each variable $x_i$, it computes the amount by which the bound will increase if $x_i$ is set true or false; one can show that the sum of these two quantities is always nonnegative. If one assignment causes the bound to decrease, the variable is given the other assignment (e.g. if assigning $x_i$ true decreases the bound, then it is assigned false). Otherwise, the variable is set randomly with a bias towards the larger increase.

This paper is structured as follows. Section 2 sets up some notation we will need. Section 3 gives the randomized $\frac{3}{4}$-approximation algorithm and its analysis. Section 4 extends these to a deterministic LP rounding algorithm. Section 5 explains how the algorithm is equivalent to the previous algorithm of Van Zuylen. Section 6 addresses how to implement the algorithms in linear time. We conclude with some open questions in Section 7.

## 2 Notation

We assume a fixed ordering of the variables, which for simplicity will be given as $x_1, x_2, \ldots, x_n$. As the algorithm proceeds, it will sequentially set the variables; let $S_i$ denote some setting of the first $i$ variables. Let $W = \sum_{j=1}^{m} w_j$ be the total weight of all the clauses. Let $\text{SAT}_i$ be the total weight of clauses satisfied by $S_i$, and let $\text{UNSAT}_i$ be the total weight of clauses that are unsatisfied by $S_i$; that is, clauses that only have variables from $x_1, \ldots, x_i$ and are not satisfied by $S_i$. Note that $\text{SAT}_i$ is a lower bound on the total weight of clauses satisfied by our final assignment $S_n$ (once we have set all the variables); furthermore, note that $W - \text{UNSAT}_i$ is an upper bound on the total weight of clauses satisfied by our final assignment $S_n$. We let $B_i = \frac{1}{2}(\text{SAT}_i + (W - \text{UNSAT}_i))$ be the midpoint between these two bounds; we refer to it simply as the bound on our partial assignment $S_i$. For any assignment $S$ to all of the variables, let $w(S)$ represent the total weight of the satisfied clauses. Then we observe that for the assignment $S_n$, $w(S_n) = \text{SAT}_n = W - \text{UNSAT}_n$, so that $w(S_n) = B_n$. Furthermore, $\text{SAT}_0 = 0$ and $\text{UNSAT}_0 = 0$, so that $B_0 = \frac{1}{2}W$.

Note that our algorithm will be randomized, so that $S_i$, $\text{SAT}_i$, $\text{UNSAT}_i$, and $B_i$ are all random variables.

## 3  The Algorithm and Its Analysis

The goal of the algorithm is at each step to try to increase the bound; that is, we would like to set $x_i$ randomly so as to increase $E[B_i - B_{i-1}]$. We let $t_i$ be the value of $B_i - B_{i-1}$ in which we set $x_i$ true, and $f_i$ the value of $B_i - B_{i-1}$ in which we set $x_i$ false. Note that the expectation is conditioned on our previous setting of the variables $x_1, \ldots, x_{i-1}$, but we omit the conditioning for simplicity of notation. We will show momentarily that $t_i + f_i \geq 0$. Then the algorithm is as follows. If $f_i \leq 0$, we set $x_i$ true; that is, if setting $x_i$ false would not increase the bound, we set it true. Similarly, if $t_i \leq 0$ (setting $x_i$ true would not increase the bound) we set $x_i$ false. Otherwise, if either setting $x_i$ true or false would increase the bound, we set $x_i$ true with probability $\frac{t_i}{t_i+f_i}$.

**Lemma 1.** *For $i = 1, \ldots, n$,*
$$t_i + f_i \geq 0.$$

*Proof.* We note that any clause that becomes unsatisfied by $S_{i-1}$ and setting $x_i$ true must be then be satisfied by setting $x_i$ false, and similarly any clause that becomes unsatisfied by $S_{i-1}$ and setting $x_i$ false must then be satisfied by setting $x_i$ true. Let $\text{SAT}_{i,t}$ be the clauses that are satisfied by setting $x_i$ true given the partial assignment $S_{i-1}$, and $\text{SAT}_{i,f}$ be the clauses satisfied by setting $x_i$ false given the partial assignment $S_{i-1}$. We define $\text{UNSAT}_{i,t}$ ($\text{UNSAT}_{i,f}$) to be the clauses unsatisfied by $S_{i-1}$ and $x_i$ set true (respectively false). Our observation above implies that $\text{SAT}_{i,f} - \text{SAT}_{i-1} \geq \text{UNSAT}_{i,t} - \text{UNSAT}_{i-1}$ and $\text{SAT}_{i,t} - \text{SAT}_{i-1} \geq \text{UNSAT}_{i,f} - \text{UNSAT}_{i-1}$.

Let $B_{i,t} = \frac{1}{2}(\text{SAT}_{i,t} + (W - \text{UNSAT}_{i,t}))$ and $B_{i,f} = \frac{1}{2}(\text{SAT}_{i,f} + (W - \text{UNSAT}_{i,f}))$. Then $t_i = B_{i,t} - B_{i-1}$ and $f_i = B_{i,f} - B_{i-1}$; our goal is to show that $t_i + f_i \geq 0$, or

$$\frac{1}{2}(\text{SAT}_{i,t} + (W - \text{UNSAT}_{i,t})) + \frac{1}{2}(\text{SAT}_{i,f} + (W - \text{UNSAT}_{i,f}))$$
$$- \text{SAT}_{i-1} - (W - \text{UNSAT}_{i-1}) \geq 0.$$

Rewriting, we want to show that

$$\frac{1}{2}(\text{SAT}_{i,t} - \text{SAT}_{i-1}) + \frac{1}{2}(\text{SAT}_{i,f} - \text{SAT}_{i-1})$$
$$\geq \frac{1}{2}(\text{UNSAT}_{i,f} - \text{UNSAT}_{i-1}) + \frac{1}{2}(\text{UNSAT}_{i,t} - \text{UNSAT}_{i-1}),$$

and this follows from the inequalities of the previous paragraph.  □

Let $x^*$ be a fixed optimal solution. Following both Poloczek and Schnitger, and Buchbinder et al., given a partial assignment $S_i$, let $\text{OPT}_i$ be the assignment in which variables $x_1, \ldots, x_i$ are set as in $S_i$, and $x_{i+1}, \ldots, x_n$ are set as in $x^*$. Thus if OPT is the value of an optimal solution, $w(\text{OPT}_0) = \text{OPT}$, while $w(\text{OPT}_n) = w(S_n)$.

The following lemma is at the heart of both analyses (see Section 2.2 in Poloczek and Schnitger [7] and Lemma III.1 of Buchbinder et al. [1]).

**Lemma 2.** *For $i = 1, \ldots, n$, the following holds:*

$$E[w(\text{OPT}_{i-1}) - w(\text{OPT}_i)] \leq E[B_i - B_{i-1}].$$

Before we prove the lemma, we show that it leads straightforwardly to the desired approximation bound.

**Theorem 1.**
$$E[w(S_n)] \geq \frac{3}{4}\text{OPT}.$$

*Proof.* We sum together the inequalities from the lemma, so that

$$\sum_{i=1}^{n} E[w(\text{OPT}_{i-1}) - w(\text{OPT}_i)] \leq \sum_{i=1}^{n} E[B_i - B_{i-1}].$$

Using the linearity of expectation and telescoping the sums, we get

$$E[w(\text{OPT}_0) - w(\text{OPT}_n)] \leq E[B_n] - E[B_0].$$

Thus
$$\text{OPT} - E[w(S_n)] \leq E[w(S_n)] - \frac{1}{2}W,$$

or
$$\text{OPT} + \frac{1}{2}W \leq 2E[w(S_n)],$$

or
$$\frac{3}{4}\text{OPT} \leq E[w(S_n)]$$

as desired, since $\text{OPT} \leq W$. □

The following lemma is the key insight of proving the main lemma, and is the randomized balancing of the two greedy algorithms mentioned in the introduction. The bound holds whether $x_i^*$ is true or false.

**Lemma 3.**
$$E[w(\text{OPT}_{i-1}) - w(\text{OPT}_i)] \leq \max\left(0, \frac{2 t_i f_i}{t_i + f_i}\right).$$

*Proof.* Assume for the moment that $x_i^*$ is set false; the proof is analogous if $x_i^*$ is true. We claim that if $x_i$ is set true while $x_i^*$ is false, then $w(\text{OPT}_{i-1}) - w(\text{OPT}_i) \leq 2f_i$. If $f_i \leq 0$, then we set $x_i$ true and the lemma statement holds given the claim. If $t_i \leq 0$, we set $x_i$ false; then the assignment $\text{OPT}_i$ is the same as $\text{OPT}_{i-1}$ so that $w(\text{OPT}_i) - w(\text{OPT}_{i-1}) = 0$ and the lemma statement again holds. Now assume both $f_i > 0$ and $t_i > 0$. We set $x_i$ false with probability $f_i/(t_i + f_i)$, so that again $w(\text{OPT}_i) - w(\text{OPT}_{i-1}) = 0$. We set $x_i$ true with probability $t_i/(t_i + f_i)$. If the claim holds then the lemma is shown, since

$$E[w(\text{OPT}_{i-1}) - w(\text{OPT}_i)] \leq \frac{f_i}{t_i + f_i} \cdot 0 + \frac{t_i}{t_i + f_i} \cdot 2f_i = \frac{2t_i f_i}{t_i + f_i}.$$

If $x_i$ is set true while $x_i^*$ is false, $\text{OPT}_{i-1}$ differs from $\text{OPT}_i$ precisely by having the $i$th variable set false. Hence, $w(\text{OPT}_{i-1}) - w(\text{OPT}_i)$ is the difference in the weight of the satisfied clauses made by flipping the $i$th variable from true to false in $\text{OPT}_i$. Since both assignments have the first $i-1$ variables set as in $S_{i-1}$, they both satisfy at least $\text{SAT}_{i-1}$ total weight, so the increase of flipping the $i$th variable from true to false is at most $\text{SAT}_{i,f} - \text{SAT}_{i-1}$. Additionally, both assignments leave at least $\text{UNSAT}_{i-1}$ total weight of clauses unsatisfied, so that flipping the $i$th variable from true to false leaves at least $\text{UNSAT}_{i,f} - \text{UNSAT}_{i-1}$ additional weight unsatisfied. In particular, flipping the $i$th variable will unsatisfy an additional $\text{UNSAT}_{i,f} - \text{UNSAT}_{i-1}$ weight of the clauses that only have variables from $x_1, \ldots, x_i$ and may unsatisfy additional clauses as well. Thus, if $x_i$ is set true and $x_i^*$ is false,

$$w(\text{OPT}_{i-1}) - w(\text{OPT}_i) \leq (\text{SAT}_{i,f} - \text{SAT}_{i-1}) - (\text{UNSAT}_{i,f} - \text{UNSAT}_{i-1})$$
$$= (\text{SAT}_{i,f} + (W - \text{UNSAT}_{i,f}))$$
$$\quad - (\text{SAT}_{i-1} + (W - \text{UNSAT}_{i-1}))$$
$$= 2(B_{i,f} - B_{i-1}) = 2f_i.$$

□

Now we can prove the main lemma.

*Proof of Lemma 2.* If either $t_i \leq 0$ or $f_i \leq 0$, then by Lemma 3, we set $x_i$ deterministically so that the bound does not decrease and $B_i - B_{i-1} \geq 0$. Since then $t_i f_i \leq 0$, by Lemma 3

$$E[w(\text{OPT}_{i-1}) - w(\text{OPT}_i)] \leq \max(0, 2t_i f_i / (t_i + f_i)) \leq 0,$$

and the inequality holds.

If both $t_i, f_i > 0$, then

$$E[B_i - B_{i-1}] = \frac{t_i}{t_i + f_i}[B_{i,t} - B_{i-1}] + \frac{f_i}{t_i + f_i}[B_{i,f} - B_{i-1}]$$
$$= \frac{t_i^2 + f_i^2}{t_i + f_i},$$

while by Lemma 3

$$E[w(\text{OPT}_{i-1}) - w(\text{OPT}_i)] \leq \frac{2t_i f_i}{t_i + f_i}.$$

Therefore in order to verify the inequality, we need to show that when $t_i, f_i > 0$,

$$\frac{2t_i f_i}{t_i + f_i} \leq \frac{t_i^2 + f_i^2}{t_i + f_i},$$

which follows since $t_i^2 + f_i^2 - 2t_i f_i = (t_i - f_i)^2 \geq 0$. □

## 4  A Deterministic LP Rounding Algorithm

We can now take essentially the same algorithm and analysis, and use it to obtain a deterministic LP rounding algorithm.

We first give the standard LP relaxation of MAX SAT. It uses decision variables $y_i \in \{0,1\}$, where $y_i = 1$ corresponds to $x_i$ being set true, and $z_j \in \{0,1\}$, where $z_j = 1$ corresponds to clause $C_j$ being satisfied. Let $P_j$ be the set of variables that occur positively in clause $C_j$ and $N_j$ be the set of variables that occur negatively. Then the LP relaxation is:

$$\text{maximize} \quad \sum_{j=1}^{m} w_j z_j$$

$$\text{subject to} \quad \sum_{i \in P_j} y_i + \sum_{i \in N_j} (1 - y_i) \geq z_j, \quad \forall C_j = \bigvee_{i \in P_j} x_i \vee \bigvee_{i \in N_j} \bar{x}_i,$$

$$0 \leq y_i \leq 1, \quad i = 1, \ldots, n,$$
$$0 \leq z_j \leq 1, \quad j = 1, \ldots, m.$$

Note that given a setting of $y$, we can easily find the value of $z$ that maximizes the LP objective, by setting $z_j = \min(1, \sum_{i \in P_j} y_i + \sum_{i \in N_j}(1 - y_i))$; thus we can determine the best possible value of the LP for a given $y$. Let $\text{OPT}_{\text{LP}}$ be the optimal value of the LP.

Let $y^*$ be an optimal solution to the LP relaxation. As before, our algorithm will sequence through the variables $x_i$, deciding at each step whether to set $x_i$ to true or false; now the decision will be made deterministically. Let $B_i$ be the same bound as before, and as before let $t_i$ be the increase in the bound if $x_i$ is set true, and $f_i$ the increase if $x_i$ is set false. The concept corresponding to $\text{OPT}_i$ in the previous algorithm is $\text{LP}_i$, the value of the LP for a vector $\hat{y}$ in which the first $i$ elements are 0s and 1s corresponding to our assignment $S_i$, while the remaining entries are the values of the optimal LP solution $y^*_{i+1}, \ldots, y^*_n$. Thus $\text{LP}_0 = \text{OPT}_{\text{LP}}$ and $\text{LP}_n = w(S_n)$, the weight of our assignment. We further introduce the notation $\text{LP}_{i,t}$ ($\text{LP}_{i,f}$), which correspond to the value of the LP for the vector $\hat{y}$ in which the first $i-1$ elements are 0s and 1s corresponding to our assignment $S_{i-1}$, the entries for $i+1$ to $n$ are the values of the optimal LP solution $y^*_{i+1}, \ldots, y^*_n$, and the $i$th entry is 1 (0, respectively). Note that after we decide whether to set $x_i$ true or false, either $\text{LP}_i = \text{LP}_{i,t}$ (if we set $x_i$ true) or $\text{LP}_i = \text{LP}_{i,f}$ (if we set $x_i$ false).

The following lemma is the key to the algorithm and the analysis; we defer the proof.

**Lemma 4.** *For each $i$, $i = 1, \ldots, n$, both of the following two inequalities are true:*

$$\text{LP}_{i-1} - \text{LP}_{i,t} \leq 2(1 - y^*_i) f_i \text{ and } \text{LP}_{i-1} - \text{LP}_{i,f} \leq 2 y^*_i t_i.$$

We remark that the first inequality in the lemma is a "fractional version" of the inequality from the proof of Lemma 3: if $x_i$ is set to true, then $w(\text{OPT}_{i-1}) - w(\text{OPT}_i) \leq 2 f_i$. The second inequality is a fractional version of the analogous inequality $w(\text{OPT}_{i-1}) - w(\text{OPT}_i) \leq 2t_i$ which holds if $x_i$ is set to false.

The algorithm is then as follows: when we consider variable $x_i$, we check whether $2y_i^* t_i \leq f_i$: if the inequality holds, we set $x_i$ false (and thus $\text{LP}_i = \text{LP}_{i,f}$), otherwise we set $x_i$ true (and thus $\text{LP}_i = \text{LP}_{i,t}$).

The following lemma, which is analogous to Lemma 2, now follows easily.

**Lemma 5.** *For $i = 1, \ldots, n$,*

$$\text{LP}_{i-1} - \text{LP}_i \leq B_i - B_{i-1}.$$

*Proof.* If $2y_i^* t_i \leq f_i$, we set $x_i$ to false, so that $\text{LP}_{i-1} - \text{LP}_i = \text{LP}_{i-1} - \text{LP}_{i,f}$ and $B_i - B_{i-1} = f_i$. By the second inequality of Lemma 4, $\text{LP}_{i-1} - \text{LP}_{i,f} \leq 2y_i^* t_i$, and by the condition for setting $x_i$ to false, this is at most $f_i$.

If $2y_i^* t_i > f_i$, we set $x_i$ to true, so that $\text{LP}_{i-1} - \text{LP}_i = \text{LP}_{i-1} - \text{LP}_{i,t}$ and $B_i - B_{i-1} = t_i$. Using the first inequality of Lemma 4, $\text{LP}_{i-1} - \text{LP}_{i,t} \leq 2(1 - y_i^*) f_i$. Since $f_i < 2y_i^* t_i$, this is less than $4(1 - y_i^*) y_i^* t_i \leq t_i$, where the final inequality uses the fact that $0 \leq y_i^* \leq 1$. □

Given the lemma, we can prove the following.

**Theorem 2.** *For the assignment $S_n$ computed by the algorithm,*

$$w(S_n) \geq \frac{3}{4}\text{OPT}.$$

*Proof.* As in the proof of Theorem 1, we sum together the inequalities given by Lemma 5, so that

$$\sum_{i=1}^n (\text{LP}_{i-1} - \text{LP}_i) \leq \sum_{i=1}^n (B_i - B_{i-1}).$$

Telescoping the sums, we get

$$\text{LP}_0 - \text{LP}_n \leq B_n - B_0, \quad \text{or} \quad \text{OPT}_{\text{LP}} - w(S_n) \leq w(S_n) - \frac{W}{2}.$$

Rearranging terms, we have

$$w(S_n) \geq \frac{1}{2}\text{OPT}_{\text{LP}} + \frac{W}{4} \geq \frac{3}{4}\text{OPT},$$

since both $\text{OPT}_{\text{LP}} \geq \text{OPT}$ (since the LP is a relaxation) and $W \geq \text{OPT}$. □

Now to prove Lemma 4.

*Proof of Lemma 4.* We observe that $\text{LP}_{i-1} - \text{LP}_{i,t}$ is equal to the outcome of changing an LP solution $y$ from $y_i = 1$ to $y_i = y_i^*$; all other entries in the $y$ vector remain the same. Both $\text{LP}_{i-1}$ and $\text{LP}_{i,t}$ satisfy at least $\text{SAT}_{i-1}$ weight of clauses;

any increase in weight of satisfied clauses due to reducing $y_i$ from 1 to $y_i^*$ must be due to clauses in which $x_i$ occurs negatively; if we reduce $y_i$ from 1 to 0, we would get a increase of the objective function of at most $\text{SAT}_{i,f} - \text{SAT}_{i-1}$, but since we reduce $y_i$ from 1 to $y_i^*$, we get $(1-y_i^*)(\text{SAT}_{i,f} - \text{SAT}_{i-1})$. Additionally both $\text{LP}_{i-1}$ and $\text{LP}_{i,t}$ have at least $\text{UNSAT}_{i-1}$ total weight of unsatisfied clauses; any increase in the weight of unsatisfied clauses due to reducing $y_i$ from 1 to $y_i^*$ must be due to clauses in which $x_i$ occurs positively; if we reduce $y_i$ from 1 to 0, we would get a decrease in the objective function of at least $\text{UNSAT}_{i,f} - \text{UNSAT}_{i-1}$, but since we reduce $y_i$ from 1 to $y_i^*$ we get $(1-y_i^*)(\text{UNSAT}_{i,f} - \text{UNSAT}_{i-1})$. Then we have

$$\begin{aligned}\text{LP}_{i-1} - \text{LP}_{i,t} &\leq (1-y_i^*)(\text{SAT}_{i,f} - \text{SAT}_{i-1} - (\text{UNSAT}_{i,f} - \text{UNSAT}_{i-1}))\\ &\leq 2(1-y_i^*)f_i,\end{aligned}$$

The proof that $\text{LP}_{i-1} - \text{LP}_{i,f} \leq 2y_i^* t_i$ is analogous. □

## 5 The Algorithms of Van Zuylen

In this section, we show that Van Zuylen's randomized algorithm is equivalent to the algorithm of Section 3, and Van Zuylen's deterministic LP rounding algorithm is equivalent to the algorithm of Section 4. Van Zuylen's randomized algorithm [8] uses the following quantities to decide how to set each variable $x_i$. Let $W_i$ be the weight of the clauses that become satisfied by setting $x_i$ true and unsatisfied by setting $x_i$ false, and let $\bar{W}_i$ be the weight of the clauses satisfied by setting $x_i$ false and unsatisfied by setting $x_i$ true. Let $F_i$ be the weight of the clauses that are satisfied by setting $x_i$ true, and are neither satisfied nor unsatisfied by setting $x_i$ false; similarly, $\bar{F}_i$ is the weight of the clauses that are satisfied by setting $x_i$ false, and are neither satisfied nor unsatisfied by setting $x_i$ true. Then Van Zuylen calculates a quantity $\alpha$ as follows:

$$\alpha = \frac{W_i + F_i - \bar{W}_i}{F_i + \bar{F}_i}.$$

Note that the case $F_i + \bar{F}_i = 0$ is trivial, hence we may assume that $F_i + \bar{F}_i > 0$ and $\alpha$ is well defined. Van Zuylen's algorithm sets $x_i$ false if $\alpha \leq 0$, true if $\alpha \geq 1$, and sets $x_i$ true with probability $\alpha$ if $0 < \alpha < 1$.

We observe that in terms of our prior quantities, $W_i + F_i = \text{SAT}_{i,t} - \text{SAT}_{i-1}$, while $\bar{W}_i = \text{UNSAT}_{i,f} - \text{UNSAT}_{i-1}$. Then

$$\begin{aligned}W_i + F_i - \bar{W}_i &= \text{SAT}_{i,t} - \text{SAT}_{i-1} + (W - \text{UNSAT}_{i,t}) - (W - \text{UNSAT}_{i-1})\\ &= 2(B_{i,t} - B_{i-1}) = 2t_i.\end{aligned}$$

Similarly, $\bar{W}_i + \bar{F}_i - W_i = 2f_i$. Furthermore,

$$\begin{aligned}
F_i + \bar{F}_i &= (F_i + W_i) + (\bar{F}_i + \bar{W}_i) - W_i - \bar{W}_i \\
&= (\text{SAT}_{i,t} - \text{SAT}_{i-1}) + (\text{SAT}_{i,f} - \text{SAT}_{i-1}) \\
&\quad - (\text{UNSAT}_{i,t} - \text{UNSAT}_{i-1}) - (\text{UNSAT}_{i,f} - \text{UNSAT}_{i-1}) \\
&= [(\text{SAT}_{i,t} + (W - \text{UNSAT}_{i,t}) - (\text{SAT}_{i-1} + (W - \text{UNSAT}_{i-1})] \\
&\quad + [(\text{SAT}_{i,f} + (W - \text{UNSAT}_{i,f}) - (\text{SAT}_{i-1} + (W - \text{UNSAT}_{i-1})] \\
&= 2(B_{i,t} - B_{i-1}) + 2(B_{i,f} - B_{i-1}) \\
&= 2(t_i + f_i).
\end{aligned}$$

We remark that the above gives an alternative proof for Lemma 1, and that our assumption that $F_i + \bar{F}_i > 0$ implies that $t_i + f_i > 0$.

Now, $\alpha \leq 0$ if and only if $t_i \leq 0$, and in this case $x_i$ is set false. Also, $\alpha \geq 1$ if and only if

$$W_i + F_i - \bar{W}_i \geq F_i + \bar{F}_i, \quad \text{or} \quad \bar{W}_i + \bar{F}_i - W_i \leq 0,$$

which is equivalent to $f_i \leq 0$, and in this case $x_i$ is set true. Finally, $0 < \alpha < 1$ if and only if $t_i > 0$ and $f_i > 0$ and in this case, $x_i$ is set true with probability $\alpha = \frac{2t_i}{2(t_i+f_i)} = \frac{t_i}{t_i+f_i}$. Thus Van Zuylen's algorithm and the algorithm of Section 3 are equivalent.

We now turn to Van Zuylen's deterministic LP rounding algorithm. It also makes use of the quantity $\alpha$ defined above. If $y^*$ is an optimal LP solution, then it sets $x_i$ to false if $\alpha \leq 0$, or if $\alpha \in (0,1)$ and $y_i^* \leq \frac{1-\alpha}{2\alpha}$ and to true otherwise. We show that this is equivalent to the algorithm from Section 4, which sets $x_i$ to false if $2y_i^* t_i \leq f_i$ and to true otherwise, except for the choice it makes in the "degenerate" case when $t_i = f_i = 0$ and $\alpha$ is not defined. Recall that we have assumed $t_i + f_i > 0$.

First, consider the case $\alpha \in (0,1)$. We have shown above that $\alpha = \frac{t_i}{t_i+f_i}$, which implies that $\frac{1-\alpha}{2\alpha} = \frac{f_i}{2t_i}$ and $t_i > 0$ since $\alpha > 0$. Thus, $y_i^* \leq \frac{1-\alpha}{2\alpha}$ is equivalent to $2y_i^* t_i \leq f_i$, and hence Van Zuylen's algorithm employs the same rounding as our algorithm from Section 4 in the case when $\alpha \in (0,1)$.

If $\alpha \leq 0$, then the fact that $t_i + f_i > 0$ implies that $t_i \leq 0$ and $f_i > 0$. Hence, $2y_i^* t_i < f_i$ holds, and indeed Van Zuylen's algorithm sets $x_i$ to false in this case. Finally, if $\alpha \geq 1$, Van Zuylen's algorithm sets $x_i$ to true. Note that $\alpha \geq 1$ implies that $t_i \geq t_i + f_i$, and hence $t_i > 0$ and $f_i \leq 0$. Hence, it must be that case that $2y_i^* t_i > f_i$.

## 6  A Linear Time Implementation

We describe a linear time and linear space implementation of greedy algorithms for MAX SAT. The data structure we propose covers the randomized greedy algorithm studied in Section 3, its variants in [7,6], both LP rounding algorithms, and Johnson's algorithm equipped with either an online or a random variable ordering [5,2,7].

There are two main problems to address: When given a variable $x_i$, one needs to access all clauses that contain $x_i$ and are not satisfied by $S_{i-1}$, the partial assignment to the first $i-1$ variables. In particular, the algorithms we consider require the weight and the number of unfixed literals for any such clause in order to decide $x_i$. Then, after fixing $x_i$, one has to "update" the clause set, i.e. remove satisfied and unsatisfied clauses to avoid the computational overhead of processing a clause that has been decided.

We assume that the formula is given as a collection of $m$ clauses, where each clause $C_j$, $j \in [m]$, consists of a list of literals and has a nonnegative weight $w_j$. The size of the formula, i.e. its encoding length, is denoted by $F$. In particular, $F$ is proportional to the total number of literals in the formula, counting multiple occurrences. In the following presentation we omit the space required to represent the clause weights, as they can be assumed to be in binary both in the input and in the data structure. Buchbinder et al. state that their version of the randomized greedy algorithm can be implemented in linear time (Theorem IV.2 in [1]), but defer the proof to a full version of their paper. We propose a data structure that yields a time and space complexity of $O(F)$.

Assume that the number of variables is $n$ and their indices are $\{1, 2, \ldots, n\}$. For each literal $x$ we create a doubly linked list $L_x$ that will provide access to the clauses that are still undecided and contain $x$. Thus, there are two lists $L_{x_i}$ and $L_{\bar{x}_i}$ for each variable $x_i$ that can be accessed via an auxiliary array using the variable index. Now we perform a single run through the clause set: For clause $C_j$ we create a clause object $O_j$ that stores the weight $w_j$ and the number of unfixed literals in $C_j$ with respect to the current partial assignment. Further, for each literal $x \in C_j$ we append a new element $E_j$ to $L_x$ that contains a pointer to the clause object $O_j$, and store a pointer to $E_j$ in $O_j$. The pointers in $O_j$ can be kept in a simple linked list. We may charge the cost of these operations to the literals involved: The occurrence of literal $x$ in clause $C_j$ is charged for element $E_j$ in $L_x$ as well as the two pointers, and the first literal of $C_j$ is charged for $O_j$. Thus, the data structure can be constructed in time and space $O(F)$.

We run the algorithm as follows. The algorithm specifies an ordering on the variables, for example the ordering given by the variable indices. If variable $x_i$ is to be decided next, the algorithm cycles once through $L_{x_i}$ and $L_{\bar{x}_i}$ to collect the information used to decide $x_i$. In case of the algorithms presented in paper this means computing $t_i$ and $f_i$. Assume that the algorithm sets $x_i$ to true (the case of false will be handled analogously). Then we remove the satisfied clauses from the formula by cycling through $L_{x_i}$: for each element $E_j \in L_{x_i}$ we access the corresponding clause object $O_j$ and remove the clause from the $L$-lists of those literals that appear in $C_j$, using the pointers stored in the clause object $O_j$. Here we utilize that the $L$-lists are doubly linked and hence the corresponding elements can be removed from the respective list in constant time. Then we delete the clause object $O_j$ itself. Finally, we pass through $L_{\bar{x}_i}$ and decrement the number of unfixed literals for the corresponding clauses. If the counter drops to zero for some clause, we also remove it from the data structure by the procedure described above. Thus, at any time the data structure contains only undecided clauses. Observe that each literal $x$ is fixed (at most) once by the algorithm, thus every occurrence of $x$ is charged with constant cost.

## 7 Conclusions

A natural question is whether there exists a simple deterministic $\frac{3}{4}$-approximation algorithm for MAX SAT that does not require the use of linear programming. The paper of Poloczek [6] rules out certain types of algorithms, but other types might still be possible. Another question is whether randomization is inherently necessary for the $\frac{1}{2}$-approximation algorithm for (nonmonotone) submodular function maximization of Buchbinder et al. [1]; that is, can one achieve a deterministic $\frac{1}{2}$-approximation algorithm? It might be possible to show that given a fixed order of items and a restriction on the algorithm that it must make an irrevocable decision on whether to include an item in the solution set or not, a deterministic $\frac{1}{2}$-approximation algorithm is not possible. However, it seems that there must be some reasonable restriction on the number of queries made to the submodular function oracle.

## References

1. Buchbinder, N., Feldman, M., Naor, J.S., Schwartz, R.: A tight linear time (1/2)-approximation for unconstrained submodular maximization. In: Proceedings of the 53rd Annual IEEE Symposium on the Foundations of Computer Science, pp. 649–658 (2012)
2. Chen, J., Friesen, D.K., Zheng, H.: Tight bound on Johnson's algorithm for maximum satisfiability. J. Comput. Syst. Sci. 58, 622–640 (1999)
3. Engebretsen, L.: Simplified tight analysis of Johnson's algorithm. Inf. Process. Lett. 92, 207–210 (2004)
4. Goemans, M.X., Williamson, D.P.: New 3/4-approximation algorithms for the maximum satisfiability problem. SIAM Journal on Discrete Mathematics 7, 656–666 (1994)
5. Johnson, D.S.: Approximation algorithms for combinatorial problems. J. Comput. Syst. Sci. 9, 256–278 (1974)
6. Poloczek, M.: Bounds on greedy algorithms for MAX SAT. In: Demetrescu, C., Halldórsson, M.M. (eds.) ESA 2011. LNCS, vol. 6942, pp. 37–48. Springer, Heidelberg (2011)
7. Poloczek, M., Schnitger, G.: Randomized variants of Johnson's algorithm for MAX SAT. In: Proceedings of the 22nd Annual ACM-SIAM Symposium on Discrete Algorithms, pp. 656–663 (2011)
8. van Zuylen, A.: Simpler 3/4-approximation algorithms for MAX SAT. In: Solis-Oba, R., Persiano, G. (eds.) WAOA 2011. LNCS, vol. 7164, pp. 188–197. Springer, Heidelberg (2012)
9. Williamson, D.P.: Lecture notes in approximation algorithms, Fall 1998. IBM Research Report RC 21409, IBM Research (1999)
10. Yannakakis, M.: On the approximation of maximum satisfiability. Journal of Algorithms 17, 475–502 (1994)

# Packet Forwarding Algorithms in a Line Network

Antonios Antoniadis[1,*], Neal Barcelo[1], Daniel Cole[1], Kyle Fox[2,**],
Benjamin Moseley[3], Michael Nugent[1], and Kirk Pruhs[1,***]

[1] University of Pittsburgh, Pittsburgh PA 15260, USA
antoniosantoniadis@gmail.com,
{ncb30,dcc20,mnugent,kirk}@cs.pitt.edu
[2] University of Illinois at Urbana-Champaign, Urbana IL 61801, USA
kylefox2@illinois.edu
[3] Toyota Technological Institute at Chicago, Chicago IL 60637, USA
moseley@ttic.edu

**Abstract.** We initiate a competitive analysis of packet forwarding policies for maximum and average flow in a line network. We show that the policies Earliest Arrival and Furthest-To-Go are scalable, but not constant competitive, for maximum flow. We show that there is no constant competitive algorithm for average flow.

## 1 Introduction

The Internet Protocol (IP) layer of the TCP/IP suite is responsible for transporting (essentially fixed-sized) datagrams/packets from a source host, through intermediate routers, to a destination host specified by an IP address. The utilization of any imaginable economically-sustainable network will be sufficiently high so that routers will usually have a backlog of packets waiting to be forwarded. Thus routers need a policy that specifies which packets to forward first in the event of a backlog. Ideally the goal of this forwarding policy should be to provide the best possible quality of service (QoS) to the application layer clients, although this is a problematic goal as one consequence of the layering/encapsulation principle of the protocol suite design is that the overlying applications are generally hidden from the IP layer. Thus, a reasonable fallback goal for this forwarding policy would be to provide good QoS to the packets.

The most natural QoS measure for an individual packet $j$ is the response/flow time $C_j - r_j$, the duration of time between the time $r_j$ when the packet $j$ is

---

[*] With support by a fellowship within the Postdoc-Programme of the German Academic Exchange Service (DAAD).
[**] Research by this author is supported in part by the Department of Energy Office of Science Graduate Fellowship Program (DOE SCGF), made possible in part by the American Recovery and Reinvestment Act of 2009, administered by ORISE-ORAU under contract no. DE-AC05-06OR23100.
[***] Supported in part by NSF grants CCF-1115575, CNS-1253218, and an IBM Faculty Award.

injected into the IP layer at the source host, until the time $C_j$ when the packet $j$ is ejected from the IP layer at the destination host. The most natural QoS measure for a collection of packets is then to take a $p$-norm, for $p \in [1, \infty]$, of the flow time of individual packets. The $\infty$-norm, or maximum flow time, is usually the most mathematically tractable norm, and is the second most commonly considered norm in the systems literature. The 1-norm, or average flow time, is usually the second most mathematically tractable norm, and is the most commonly considered norm in the systems literature.

The goal of the research we report on in this paper is to initiate a competitive analysis of packet forwarding policies for these natural QoS measures. In this paper, we generally assume that the network topology is a line. Even for this simplest of topologies, we find the subtlety of the algorithm analysis and design process to be surprising.

There are three natural packet forwarding policies that play a central role in our findings:

**Furthest-To-Go (FTG):** The FTG policy always forwards a packet with the most hops left to go. It is not too difficult to see that FTG minimizes the makespan, the time that the last packet is delivered. We actually show in Section 5 the stronger statement that for every router $i$ and for every time $t$, FTG has forwarded the maximum number of packets possible over router $i$ by time $t$. Intuitively, this implies that FTG maximizes the amount of parallel processing possible in the network.

**Earliest Arrival (EA):** The EA policy always forwards the packet that was first injected into the network layer. For a one-edge network, it is well-known and obvious that the policy EA is optimal for maximum flow. For general line networks, it is obvious that EA is not optimal for maximum flow because there are situations where the optimal algorithm needs to forward a younger further-to-go packet. Earliest Arrival is also known in the literature as Longest-In-System.

**Shortest-To-Go (STG):** Shortest-To-Go always forwards the packet that is the fewest number of hops from its destination router. We show in Lemma 11 that STG achieves optimal average flow time if all packets are injected into the system at the same time. We recently learned that the same result was proven independently by Kowalski et al. and will appear in [1]. Shortest-To-Go is also known in the literature as Nearest-To-Go.

In this paper, we report on the progress that we made beyond these initial observations. Namely, we show that:

**Maximum Flow on a Line:** Our initial conjecture was that EA is $O(1)$-competitive.
- In Section 3 we show that in fact EA is not $O(1)$-competitive. The lower bound instance results from a rather intricate recursive construction that increases the age of a packet by a fixed amount on each recursion. Intuitively, this shows that a competitive algorithm must take into account the path lengths of the packets.

- In Section 4 we show that EA is however scalable, that is, $O(1)$-competitive with arbitrarily small speed augmentation. Intuitively, this shows that EA should be reasonable until the network utilization is near the capacity of the network.
- In Section 5 we show that FTG is also scalable.

**Average Flow on a Line:** Initially we had two competing intuitions as to the "right" policy for average flow time. One might reasonably think that as Shortest Remaining Processing Time is the optimal scheduling policy for average flow for arbitrary sized jobs on a single processor, that analogously the policy STG, which forwards a packets with the fewest hops to go, should be a good policy for average flow. Alternatively, one might reasonably think that FTG should be a good policy for average flow because it maximizes parallelism.

- In Section 6 we show that there is no $O(1)$-competitive online algorithm. Intuitively in our lower bound construction, if the online algorithm initially forwards packets using STG, then there is a future in which FTG was the right initial policy, and if instead the online algorithm initially forwards packets using FTG, there is another future in which STG was the right initial policy (and there is no intermediate policy that is good in both futures). So in some sense, this shows that there is no possible resolution to the conflicting intuitions favoring STG and FTG.

**Maximum Flow on a Tree:** In Section 7 we show that there is no $O(1)$-speed $O(1)$-competitive deterministic local online algorithm. This shows that generalizing the problem to the second most simple network already makes the problem harder.

We then conclude by stating the two open problems "discovered" by this research that we find appealing.

## 1.1 Related Work

Previous work on routing algorithms under the adversarial model has, to the best of our knowledge, revolved around two distinct models. In the first one, stability is studied, i.e., whether the number of packets in the system will remain bounded as the system runs for an arbitrarily long period of time. In general this depends on the protocol studied, on the size and topology of the network and on the maximum rate at which the adversary is allowed to inject packets into the network. We refer to [2–13] for some representative papers under this model. In the second model, a subset of the packets has to be dropped, due to some restriction. For example, there might be a limit in the size of the buffer, a maximum delay (per packet) allowed by the system, or packets may come with a deadline. The objective is to maximize a function of the transmitted packets, for instance their number, size, or (weighted) value. Work in this model has mostly employed competitive analysis, see [14–25]. The problem has also been studied when the routers have shared but limited memory [26]. Our work significantly differs from both these models, since (1) instead of considering the stability of a network, we use competitive analysis with respect to specific objective functions, and (2) in our model every single packet has to be transmitted to its destination.

We would like to point out that [22] also studies the Furthest-To-Go algorithm on line networks, but with fixed-size buffers under the objective of maximizing the throughput. They show that for this model on a line of length $k$ every greedy algorithm (including Furthest-To-Go) has a competitive ratio of $O(k)$, and also give a matching lower bound of $\Omega(k)$ on the competitive factor of Furthest-To-Go. Also, Angelov et al. [19] as well as Azar and Zachut [18], give centralized online algorithms with a polylogarithmic competitive ratio for the problem of maximizing throughput on the line. The special case of information gathering, where all packets have a common destination, was studied on the line by Rosén and Scalosub [25].

## 2 Preliminaries and Notation

We begin by introducing a formal model for the problems we consider and some notation. In the problems we consider there are $k$ routers labeled 1 through $k$ on a line network. The routers are ordered in increasing order on the line from left to right. Over time $n$ packets arrive. We say that a packet $j$ arrives at time $r_j$. We assume arrival times are integral. Each packet is associated with a path $P_j$. The path $P_j$ consists of a set of routers that packet $j$ must be processed on to be completed. This corresponds to sending a packet from its source to its destination. Since we are considering a line network, $P_j$ will consist of a set of adjacent routers on the line. A router can process one packet at each time unit, which corresponds to sending this packet to the router to the right of this router on the line. A packet can only be processed if it is on a router and a packet $j$ starts at the leftmost router in $P_j$. Note that a packet must be processed by every router in $P_j$ and therefore $|P_j|$ is a lower bound on the amount of time a packet requires to be sent to its destination.

We will consider two different objective functions, namely total (average) flow time and maximum flow time. For some schedule, let $C_j$ denote the time packet $j$ is finished being processed. The flow time of $j$ is $C_j - r_j$. For total flow time we are interested in minimizing $\sum_{j \in [n]} (C_j - r_j)$ and for maximum flow time we are interested in minimizing $\max_{j \in [n]} \{C_j - r_j\}$. We will be considering algorithms that possibly use resource augmentation. If an algorithm is given $s + 1/c$ speed the algorithm is allowed to send $s$ packets every time step at a particular router and an additional packet every $c$ time steps. Here $s$ and $c$ will be assumed to be integral. Note that we assume packets are sent only at discrete time steps. Therefore, a packet can only move one router in a time step.

We will compare our algorithms against a fixed optimal solution for a given objective and problem instance. We denote the optimal solution as OPT. For an algorithm $A$, we let $Q^A(t)$ be the packets alive at time $t$ and $Q_i^A(t)$ as the packets available for processing on router $i$ at time $t$. We let $p_i^A(t)$ denote the number of packets processed on router $i$ by time $t$ for $A$ and, likewise, $p_i^O(t)$ for OPT. For a packet $j$ and a fixed algorithm, which will be clear by the context,

$P_j(t)$ is the remaining routers $j$ needs to use to be completed at time $t$ and $d_j(i,t)$ is the distance of packet $j$ to router $i$. Note that $P_j(r_j) = P_j$. The value of $n_i^A(t',t)$ denotes the number of packets released by time $t'$ that still need to use router $i$ at time $t$ for $A$ and $n_i^A(t')$ is short for $n_i^A(t',t')$. Note that the packets that contribute to $n_i^A(t',t)$ do not necessarily have to be in $Q_i^A(t)$. Let $A_j$ be the total flow time for packet $j$ for $A$ and $\text{OPT}_j$ be total flow time for packet $j$ in the optimal schedule.

For an input $I$, let $A(I)$ and $\text{OPT}(I)$ denote the final objective value for running $I$ on $A$ and OPT respectively. We may use $A$ and OPT to denote the objective when $I$ is clear from context. Finally, we say an algorithm $A$ is $s$-speed $c$-competitive if $\frac{A(I)}{\text{OPT}(I)} \leq c$ for any input $I$ when $A$ runs at $s$ speed and OPT runs at unit speed.

Some proofs are ommitted due to space constraints.

## 3  Lower Bound for Earliest Arrival for Maximum Flow

We show that the EA policy is not constant competitive for maximum flow.

**Theorem 1.** *There exists an $n_0$ so that for each integer $L > n_0$, there exists an instance $I$ with $OPT(I) = \Theta(L)$ and $\frac{EA(I)}{OPT(I)} \geq OPT(I)$. Furthermore, there exist instances $I_{n,k}$ with $n$ packets and $k$ routers so that $\frac{EA(I_{n,k})}{OPT(I_{n,k})} \geq n^{1/3}$, and $\frac{EA(I_{n,k})}{OPT(I_{n,k})} \geq k^{1/2}$.*

Let $K$ and $C$ be sufficiently large even integers such that $C < K/2 - 3$. Set an input with $C \cdot \frac{K}{2} - C + 3$ routers. We designate two sets of routers, the *stream-routers* and the *gap-routers*. The routers are defined as follows:

- There is a stream-router with index $R^S(0,0) = 1$. At time $T^S(0,0) = 0$ there are $K/2$ *short stream-packets* released to $R^S(0,0)$ with destination 1 and $K/2$ *long stream-packets* released to $R^S(0,0)$ with destination $K/2$.
- For each $p$ and $q$ with $1 \leq p \leq C$ and $0 \leq q \leq p-1$ there is a stream-router with index $R^S(p,q) = p \cdot \frac{K}{2} - 2p + 2 + q$. At time $T^S(p,q) = p(K-1) + \frac{K}{2}\left(\frac{p(p+1)}{2} - 1\right) - (p-1)\left(\frac{K}{2}+1\right) + q\left(\frac{K}{2}+1\right)$ there are $K/2$ short stream-packets released to $R^S(p,q)$ with destination $p \cdot \frac{K}{2} - 2p + 2 + q$ and $K/2$ long stream-packets released to $R^S(p,q)$ with destination $(p+1) \cdot \frac{K}{2} - p + 1$.
- For each $p$ with $1 \leq p \leq C$ there is a gap-router with index $R^G(p,0) = p \cdot \frac{K}{2} - p + 2$. At time $T^G(p,0) = p(K-1) + \frac{K}{2}\left(\frac{p(p+1)}{2}\right) + 1$ there are $K/2$ short gap-packets released to $R^G(p,0)$ with destination $p \cdot \frac{K}{2} - p + 2$ and $K/2$ long gap-packets released to $R^G(p,0)$ with destination $(p+1) \cdot \frac{K}{2} - p + 1$.

Note that the instance created above consists of $k = \Theta(K^2)$ routers assuming $C = \Omega(K)$. Further, $n = \Omega(K^3)$. We have the following observation.

*Note 2.* The set of stream-routers and the set of gap-routers are disjoint. Further, for any two stream- or gap-routers $i_1, i_2$ with $i_1 < i_2$, the packets on router $i_1$ are released earlier than the packets for $i_2$.

We now compare the performance of the optimal schedule to EA for minimizing maximum flow time.

**Lemma 3.** *The maximum flow time for the optimal schedule is at most $K + C$ for the given instance.*

**Lemma 4.** *EA has total flow time of at least $(C+1) \cdot \frac{K}{2}$ on the given instance.*

We can finally prove Theorem 1.

*Proof.* By setting $C = \frac{K}{2} - 4$, we have that on the above instance, $OPT = \Theta(K), EA = \Theta(K^2)$, and again, the number of routers is $k = \Theta(K^2)$ and the number of packets $n = \Omega(k^3)$. This proves the theorem. □

## 4 Analysis of EA for Maximum Flow

We show that EA is scalable for maximum flow.

**Theorem 5.** *EA is $(1+\varepsilon)$-speed $4/\varepsilon$-competitive for any $\varepsilon > 0$.*

We first prove a useful fact for the algorithm Earliest Arrival First. This fact is useful because it will give us an upper bound on how long it takes at time $t$ for a packet $j$ to be completed assuming no more packets arrive. We know that $|P_j(t)|$ is the remaining path length for packet $j$ and we know packet $j$ will need to wait at least this long to be completed. Further, $n_i^A(r_j, t)$ is the total number of packets with strictly higher priority than $j$ that need to use router $i$. Thus, $j$ may have to wait on all these packets and therefore $j$ may need to wait $\max_{i \in P_j(t)}\{n_i^A(r_j, t)\}$ time. Intuitively, we would like to show that in fact $j$ waits at most $|P_j(t)| + \max_{i \in P_j(t)}\{n_i^A(r_j, t)\}$ time to be completed assuming no more packets arrive. To do this, we would like to show that $|P_j(t)| + \max_{i \in P_j(t)}\{n_i^A(r_j, t)\}$ decreases each time step. Knowing that if this expression reaches 0 then $j$ has reached its destination ($|P_j(t)| = 0$), this would show that $j$ waits at most this much time. We will not be able to show this directly, but rather will show a slightly more involved expression decreases in a similar manner. Fix an input $I$.

**Lemma 6.** *Let $A$ be any algorithm (possibly with speedup). Let $j$ be any packet alive at time $t$ and suppose $A$ processes the $\min\{s, Q_i^A(t)\}$ packets with earliest release time on each router $i$ at time $t$. Then for any constant $c \geq s$*

$$\max_{i \in P_j(t+1)}\{n_i^A(r_j, t+1) - c \cdot d_j(i, t+1)\} + c|P_j(t+1)|$$
$$\leq \max_{i \in P_j(t)}\{n_i^A(r_j, t) - c \cdot d_j(i, t)\} + c|P_j(t)| - s.$$

We can now prove Theorem 5.

*Proof.* Assume the theorem is false for a contradiction. Let $t$ be the earliest time such that there is some packet $j$ with flow time greater than $\frac{4}{\varepsilon}\text{OPT}$ so that $t - r_j > \frac{4}{\varepsilon}\text{OPT}$. By Lemma 6, $\max_{i \in P_j(t)}\{n_i^A(r_j, t) - 2d_j(i, t)\} + 2|P_j(t)|$ decreases by 1 every time step except for every $1/\varepsilon$ time steps where it decreases by 2. Further, it does not reach 0 until $j$'s completion. Therefore,

$$\max_{i \in P_j} n_i^A(r_j) + 2|P_j| \geq \max_{i \in P_j(t)}\{n_i^A(r_j, r_j) - 2d_j(i, r_j)\} + 2|P_j(r_j)|$$
$$> \frac{4}{\varepsilon}\text{OPT} + 4\text{OPT} - 1.$$

Let $i$ be a router in $P_j$ that maximizes the value $n_i^A(r_j)$. As $|P_j| \leq \text{OPT}$ and $\text{OPT} \geq 1$, we have

$$n_i^A(r_j) > \left(\frac{4}{\varepsilon} + 1\right)\text{OPT}. \tag{1}$$

Packet $j$ is the first packet with flow greater than $\frac{4}{\varepsilon}\text{OPT}$, so any packets contributing to $n_i^A(r_j)$ have age at most $\frac{4}{\varepsilon}\text{OPT}$ at time $r_j$. These packets have arrival time between $r_j - \frac{4}{\varepsilon}\text{OPT}$ and $r_j$. Further, the optimal schedule must complete them by time $r_j + \text{OPT}$, so the total amount of time the optimal schedule can process them is $r_j + \text{OPT} - (r_j - \frac{4}{\varepsilon}\text{OPT}) = (\frac{4}{\varepsilon} + 1)\text{OPT}$. The optimal schedule can only process one of these packets at a time on router $i$, so we observe $n_i^A(r_j) \leq (\frac{4}{\varepsilon} + 1)\text{OPT}$. Finally, we combine the previous expression with (1) to yield

$$\left(\frac{4}{\varepsilon} + 1\right)\text{OPT} > \left(\frac{4}{\varepsilon} + 1\right)\text{OPT},$$

a contradiction based on our assumption that the theorem is false. □

## 5 Analysis of FTG for Maximum Flow

We show that FTG is scalable for maximum flow.

**Theorem 7.** *FTG is $(1 + \varepsilon)$-speed $3/\varepsilon$-competitive for any $\varepsilon > 0$.*

We begin by proving a fact for the algorithm FTG that will prove useful later and is interesting in its own right. Fix an input $I$. The following lemma essentially states that this algorithm gets the maximum amount of 'parallelism' possible by showing that for this algorithm at any point in time each router will have processed the most number of packets possible.

**Lemma 8.** *Let $A$ be any algorithm (possibly with speedup) for which each router $i$ processes a packet with furthest final destination at least once every time step if any are available. We have $p_i^A(t) \geq p_i^O(t)$ at all routers $i$ and times $t$.*

In order to prove Theorem 7, we need to formalize how efficiently routers are processing packets with extra speed. We say router $i$ *fully processes* a set $J$ of packets at time $t$ if router $i$ processes as many packets from $J$ at time $t$ as the speedup allows. We have the following lemma.

**Lemma 9.** *Run FTG with speed $1+\varepsilon$ on input $I$. Let $j$ be a packet and let $i \in P_j$. Let $t$ be any time step strictly before router $i$ processes packet $j$. If $t \geq r_j + d_j(i, r_j) + 1$, then router $i$ processes at least one packet at time $t$ with destination at least as far as packet $j$'s. Further, if $t \geq r_j + (d_j(i, r_j) + 1)/\varepsilon$, then router $i$ fully processes packets with destination at least as far as $j$'s at time $t$.*

We may now prove Theorem 7.

*Proof.* Assume the theorem is false for a contradiction. Let OPT be the maximum flow in the optimal schedule. Run FTG with speed $1 + \varepsilon$ and let $t$ be the earliest time such that there is some packet $j$ with flow time greater than $\frac{3}{\varepsilon}$OPT. Let $i$ be the router upon which $j$ is queued (or being processed) at time $t$. Let $t_0$ be the earliest time such that FTG always processes at least one packet every step of the interval $[t_0, t)$. Let $J$ be the set of packets processed by $i$ during the interval $[t_0, t)$.

Each packet arrives at most distance OPT $-1$ from its destination. Therefore, every packet in $J$ arrives no earlier than $t_0 - $ OPT. Otherwise, Lemma 9 implies router $i$ processes a packet at time $t_0 - 1$. Further, every packet in $J$ is released strictly before time $t$ and completed by the optimal algorithm by time $t+$ OPT. Therefore, the optimal algorithm spends strictly less than $t - t_0 + 2$OPT time steps processing all packets in $J$ on router $i$.

By assumption, $t > r_j + \frac{3}{\varepsilon}$OPT $= r_j + \frac{\text{OPT}}{\varepsilon} + \frac{2}{\varepsilon}$OPT. Therefore, router $i$ has fully processed packets over $\frac{2}{\varepsilon}$OPT consecutive time steps by Lemma 9. Over this time period, there are at least 2OPT time steps where $i$ processes 2 packets instead of 1. As $i$ processes at least one packet every time step since $t_0$, we have $|J| \geq t - t_0 + 2$OPT. The optimal algorithm spends strictly less than $|J|$ time steps to process every packet in $J$ on router $i$, a contradiction. □

## 6 Average Flow on a Line

We now show that there is no constant competitive algorithm for average flow on a line.

**Theorem 10.** *Any randomized algorithm for packet routing on a line is $\Omega(k)$ competitive for average flow in the oblivious adversary model.*

*Proof.* We prove the theorem for any deterministic algorithm. The proof can be easily generalized to randomized algorithms. Let $A$ be any deterministic algorithm for packet routing on a line. Consider the following input. For each $i \in \{1, \ldots, k/2\}$, we have $k$ *short* packets arrive at time 0 with source $i$ and destination $i$. In addition, $k$ *long* packets arrive at time 0 with source 1 and destination $k$. The rest of the input is determined by $A$'s behavior.

Suppose $A$ processes fewer than $k/4$ packets on router $k/2$ at time $k$. Then there are at least $3k/4$ long packets that still need processing on router $k/2$. Assume they all wait at router $k/2$. At each time step from $2k - 1$ to some sufficiently large $T$, a packet arrives with source $k$ and destination $k$. Algorithm $A$ will still have $k/4$ long packets remaining at time $2k-1$, so it will always have $k/4$

packets alive with destination $k$ and total flow time $\Omega(kT)$. However, consider FTG which finishes all long packets by time $2k-1$ and has only one packet pending at each time step after $3k-1$. The optimal total flow time is $O(T)$, and the competitive ratio of $A$ is $\Omega(k)$.

Now, suppose $A$ processes at least $k/4$ packets on router $k/2$ at time $k$. Then each router $i \in \{1,\ldots,k/2\}$ has at least $k/4$ short packets remaining, for a total of $k^2/8$ short packets remaining. For each time step from $k$ to some sufficiently large $T$ and for each router $i \in \{1,\ldots,k\}$, a packet with source $i$ and destination $i$ arrives (so $k$ packets in total arrive at each time step). Algorithm $A$ has at least $k^2/8 + k$ packets alive at each time step, so it has total flow time $\Omega(k^2 T)$. However, consider the algorithm that always schedules a packet with nearest destination. This algorithm finishes all the short packets by time $k$, so it has $2k$ packets remaining at each time step after $k$. The optimal flow time is $O(kT)$, and the competitive ratio of $A$ is $\Omega(k)$. □

We show that STG achieves optimal average flow time if all packets are injected into the system at the same time.

**Lemma 11.** *STG is optimal for the objective of average flow if all packets are released at time 0.*

## 7 Maximum Flow on Trees

We briefly explore extending our ideas to work with networks of routers that do not necessarily lie on a directed line. It turns out the problem of minimizing maximum flow time becomes much more difficult in this setting, unless we allow routers some way to communicate with one another.

To make the idea of communication concrete, define a *local* algorithm to be one where each router $i$ processes packets based only on the existence of the packets currently queued at router $i$, their arrival times, and their distance to their destination. Define a *router network* as a directed graph $G = (\{1,\ldots,m\}, E)$ with arbitrary edge set. An input for a router network $G$ can only move a packet from router $i_1$ to router $i_2$ in one processing step if $i_1 i_2$ is an edge in $G$. A *tree network* is a router network where the undirected edges form a spanning tree. We have the following theorem.

**Theorem 12.** *No deterministic local algorithm for packet routing on a tree network is $s$-speed $c$-competitive for any constants $s$ and $c$.*

*Proof.* Fix a deterministic local algorithm $A$ and let $k$ be a sufficiently large integer multiple of $s$. We define a tree network and associated input for every $L \in \{0,\ldots,k-1\}$ recursively as follows. The construction is based on identifying routers of one or more paths of length $k$, where the routers within a single path are indexed 1 through $k$. We maintain the invariant that algorithm $A$ queues a packet on the $L+1$st router of some path $P$ in network $L$ at time $kL/s$ when $L < k$, and $A$ completes a packet at time $kL/s$ when $L = k$. If $L < k$, then we refer to the router mentioned in the invariant as a *shared router*.

For $L = 0$, we use one path $P$ of length $k$. At time 0, we have 1 packet arrive with a source of the first router in $P$ and a destination of the $k$th router in $P$. The invariant trivially holds.

For $L > 0$, we create $k$ copies of network $L-1$ and its associated input. As $A$ is deterministic and local, we know which path $P$ maintains the invariant for $L-1$ in each of these copies. Identify the $k$ shared routers that are guaranteed by the invariant. It takes $k/s$ time steps for $A$ to process the $k$ packets on the now common shared router, so either the $L+1$st router in some path receives a packet at time $kL/s$ or $A$ completes a packet at time $kL/s$.

We see immediately that the maximum flow time for $A$ in network $k$ is $k^2/s$. However, an optimal schedule will always give precedence to the one packet at each router that will later need processing on a shared router as described above. There are never more than $k$ packets that need to use a single router, and all but one will will not go to another shared router, so that optimal maximum flow time is $2k-1$ for a competitive ratio of $\Omega(\frac{k^2}{ks})$. Setting $k$ sufficiently high proves the lemma. □

## 8 Conclusions

We initially found it surprising that there was no prior literature on the packet forwarding problems considered in this paper as they seem quite natural. Part of the explanation may be that even for a line network, the problems are surprisingly challenging. The two most natural open problems "discovered" by our research are:

- Is there an $O(1)$-competitive algorithm for maximum flow on a line? The authors are in disagreement amongst themselves about which answer is most likely, and there is a modest wager on the outcome. As evidence that there might be an $\omega(1)$ lower bound, even finding a reasonable candidate policy seems challenging. We are able to show that all of the following policies are not $O(1)$-competitive:
  - $c$-EA/FTG - Forward using EA every $c$ steps, and FTG the rest of the time, where $c > 0$ is any constant.
  - $c$-OPT/FTG Threshold - If we know OPT, the value of the optimal solution, forward the furthest-to-go packet that has age at least $c \cdot$ OPT, otherwise send the furthest-to-go packet, where $c > 0$ is any constant.
  - $\frac{1}{c}$-Local [Global] FTG Threshold - Forward the furthest-to-go packet with age at least $1/c$ of the maximum age of any packet at the current router [or in the network], where $c > 0$ is any constant.
- Is there an $O(1)$-speed $O(1)$-competitive (or even scalable) policy for average flow on a line? In this case, the authors all conjecture that the answer is yes, for the traditional reason that we were not able to prove otherwise. Here candidate policies are abundant, but it is not clear how to do the analysis. STG is a good algorithm when all release times are the same, which suggests that it is a good candidate for an $O(1)$-speed $O(1)$-competitive algorithm,

as processor-sharing is for one processor [27]. Also running STG and FTG simultaneously is another obvious candidate algorithm. The main issue with the analysis is that neither the standard potential function approach [28] nor the standard application of linear programming duality seem to work.

# References

1. Kowalski, D., Nussbaum, E., Segal, M., Milyeykovsky, V.: Scheduling problems in transportation networks of line topology. Optimization Letters (2013) (to appear)
2. Aiello, W., Kushilevitz, E., Ostrovsky, R., Rosén, A.: Adaptive packet routing for bursty adversarial traffic. J. Comput. Syst. Sci. 60(3), 482–509 (2000)
3. Andrews, M., Awerbuch, B., Fernández, A., Leighton, F.T., Liu, Z., Kleinberg, J.M.: Universal-stability results and performance bounds for greedy contention-resolution protocols. J. ACM 48(1), 39–69 (2001)
4. Andrews, M.: Instability of FIFO in the permanent sessions model at arbitrarily small network loads. ACM Transactions on Algorithms 5(3) (2009)
5. Andrews, M., Zhang, L.: The effects of temporary sessions on network performance. SIAM J. Comput. 33(3), 659–673 (2004)
6. Borodin, A., Kleinberg, J.M., Raghavan, P., Sudan, M., Williamson, D.P.: Adversarial queuing theory. J. ACM 48(1), 13–38 (2001)
7. Broder, A.Z., Frieze, A.M., Upfal, E.: A general approach to dynamic packet routing with bounded buffers. J. ACM 48(2), 324–349 (2001)
8. Leighton, F.T., Maggs, B.M., Rao, S.: Packet routing and job-shop scheduling in $O$(Congestion + Dilation) steps. Combinatorica 14(2), 167–186 (1994)
9. Ostrovsky, R., Rabani, Y.: Universal $O$(Congestion + Dilation + $\log^{1+epsilon} N$) local control packet switching algorithms. In: STOC, pp. 644–653 (1997)
10. Rabani, Y., Tardos, É.: Distributed packet switching in arbitrary networks. In: STOC, pp. 366–375 (1996)
11. Chlebus, B.S., Kowalski, D.R., Rokicki, M.A.: Adversarial queuing on the multiple access channel. ACM Transactions on Algorithms 8(1), 5 (2012)
12. Gamarnik, D.: Stability of adaptive and nonadaptive packet routing policies in adversarial queueing networks. SIAM J. Comput. 32(2), 371–385 (2003)
13. Díaz, J., Koukopoulos, D., Nikoletseas, S.E., Serna, M.J., Spirakis, P.G., Thilikos, D.M.: Stability and non-stability of the FIFO protocol. In: SPAA, pp. 48–52 (2001)
14. Srinivasan, A., Teo, C.P.: A constant-factor approximation algorithm for packet routing and balancing local vs. global criteria. SIAM J. Comput. 30(6), 2051–2068 (2000)
15. Awerbuch, B., Azar, Y., Plotkin, S.A.: Throughput-competitive on-line routing. In: FOCS, pp. 32–40 (1993)
16. Kesselman, A., Mansour, Y., van Stee, R.: Improved competitive guarantees for QoS buffering. Algorithmica 43(1-2), 63–80 (2005)
17. Andelman, N., Mansour, Y., Zhu, A.: Competitive queueing policies for QoS switches. In: SODA, pp. 761–770 (2003)
18. Azar, Y., Zachut, R.: Packet routing and information gathering in lines, rings and trees. In: Brodal, G.S., Leonardi, S. (eds.) ESA 2005. LNCS, vol. 3669, pp. 484–495. Springer, Heidelberg (2005)
19. Angelov, S., Khanna, S., Kunal, K.: The network as a storage device: Dynamic routing with bounded buffers. Algorithmica 55(1), 71–94 (2009)

20. Adler, M., Rosenberg, A.L., Sitaraman, R.K., Unger, W.: Scheduling time-constrained communication in linear networks. Theory Comput. Syst. 35(6), 599–623 (2002)
21. Gordon, E., Rosén, A.: Competitive weighted throughput analysis of greedy protocols on DAGs. ACM Transactions on Algorithms 6(3) (2010)
22. Aiello, W., Ostrovsky, R., Kushilevitz, E., Rosén, A.: Dynamic routing on networks with fixed-size buffers. In: SODA, pp. 771–780 (2003)
23. Chin, F.Y.L., Chrobak, M., Fung, S.P.Y., Jawor, W., Sgall, J., Tichý, T.: Online competitive algorithms for maximizing weighted throughput of unit jobs. J. Discrete Algorithms 4(2), 255–276 (2006)
24. Kesselman, A., Lotker, Z., Mansour, Y., Patt-Shamir, B., Schieber, B., Sviridenko, M.: Buffer overflow management in QoS switches. SIAM J. Comput. 33(3), 563–583 (2004)
25. Rosén, A., Scalosub, G.: Rate vs. buffer size-greedy information gathering on the line. ACM Transactions on Algorithms 7(3), 32 (2011)
26. Kesselman, A., Mansour, Y.: Harmonic buffer management policy for shared memory switches. Theor. Comput. Sci. 324(2-3), 161–182 (2004)
27. Edmonds, J., Pruhs, K.: Scalably scheduling processes with arbitrary speedup curves. ACM Transactions on Algorithms 8(3), 28 (2012)
28. Im, S., Moseley, B., Pruhs, K.: A tutorial on amortized local competitiveness in online scheduling. SIGACT News 42(2), 83–97 (2011)

# Survivability of Swarms of Bouncing Robots

Jurek Czyzowicz[1], Stefan Dobrev[2], Evangelos Kranakis[3], and Eduardo Pacheco[3]

[1] Université du Québec en Outaouais, Gatineau, Québec J8X 3X7, Canada
[2] Slovak Academy of Sciences, 840 00 Bratislava, Slovak Republic
[3] Carleton University, Ottawa, Ontario K1S 5B6, Canada

**Abstract.** Bouncing robots are mobile agents with limited sensing capabilities adjusting their movements upon collisions either with other robots or obstacles in the environment. They behave like elastic particles sliding on a cycle or a segment. When two of them collide, they instantaneously update their velocities according to the laws of classical mechanics for elastic collisions. They have no control on their movements which are determined only by their masses, velocities, and upcoming sequence of collisions.

We suppose that a robot arriving for the second time to its initial position dies instantaneously. We study the survivability of collections of swarms of bouncing robots. More exactly, we are looking for subsets of swarms such that after some initial bounces which may result in some robots dying, the surviving subset of the swarm continues its bouncing movement, with no robot reaching its initial position.

For the case of robots of equal masses and speeds we prove that all robots bouncing in the segment must always die while there are configurations of robots on the cycle with surviving subsets. We show the smallest such configuration containing four robots with two survivors. We show that any collection of less than four robots must always die. On the other hand, we show that $|\mathcal{S}^+ - \mathcal{S}^-|$ robots always die where $\mathcal{S}^+$ (and $\mathcal{S}^-$) is the number of robots starting their movements in clockwise (respectively counterclockwise) direction in swarm $\mathcal{S}$.

When robots bouncing on a cycle or a segment have arbitrary masses we show that at least one robot must always die. Further, we show that in either environment it is possible to construct swarms with $n-1$ survivors. We prove, however, that the survivors in the segment must remain static (i.e, immobile) indefinitely, while in the case of the cycle it is possible to have surviving collections with strictly positive kinetic energy.

Our proofs use results on dynamics of colliding particles. As far as we know, this is the first time that these particular techniques have been used in order to analyze the behavior of mobile robots from a theoretical perspective.

**Keywords and Phrases:** Mobile robots, elastic collisions, weak robots, bouncing, survivability, synchronous, salmon problem.

## 1 Introduction

Mobile robots have been widely used to perform tasks that, otherwise carried out by humans, would be dangerous, less efficient, and expensive, for instance, environment exploration, perimeter patrolling, mapping, pattern formation and localization.

Mobile robots are entities that have the ability to move within their environment, to interact with other robots, to perceive the information of the environment, and to process this information. In order to reduce power consumption and to prevent scalability problems it helps to design autonomous robots that have limited capabilities. Distributed applications are often concerned about the use of entities of very limited communication and sensing capabilities, mainly due to the limited production cost, size, and battery power. Consequently, mobile robots fit perfectly for designing distributed applications.

We are interested in studying systems of *weak* mobile robots, i.e., systems of robots with very limited capabilities. More precisely, we investigate the model of *bouncing robots* studied in [1–8]. In this model, mobile robots are deployed either on a cycle or a segment. Each robot is given an initial direction (clockwise or counterclockwise in the cycle and left-to-right or right-to-left on the segment) at which it starts its movement. The interaction among robots is limited to colliding elastically among themselves (or to the endpoints, in case of segment). Each robot possesses a clock to measure the times of its collisions. Moreover, bouncing robots have available the measured times of their collisions and the knowledge of their own velocities.

Large collections of weak mobile robots, like bouncing robots, are frequently called *swarms*. Despite their simplicity, they are used to perform complicated tasks like surveillance and monitoring in hazardous or hard to access environments. In most situations, involving such robots of extremely limited capabilities, the fundamental research question concerns the feasibility of solving a given task (cf. [1, 2, 9, 10]). Due to the nature of the environments on which mobile robots are frequently deployed, they may get trapped or *die* while performing some hazardous task. For instance, while a robot is exploring a terrain, it can be destroyed by enemy forces or by stepping on a mine. Understanding their survivability will help us understand which measures could be taken in order to ensure that they fulfill their purpose.

We study the survivability of swarms of bouncing robots. To do so, we mark the starting position of every robot as *deadly* in the sense that if a robot ever returns to its starting position it dies. A robot *survives* if it never returns to its starting position. We investigate the necessary conditions for swarms of bouncing robots to have surviving subsets in the cycle and in the segment. Since bouncing robots do not have any control over their movements, it might seem that in most configurations all the robots must die. We prove that this is not always the case, thus answering an open question first posed in [8] for the case of robots of equal masses and speeds.

### 1.1 Related Work

The dynamics of bouncing robots is similar to the one observed in some systems of particles. The study of the dynamics of particles sliding in a surface that collide among themselves has been of great interest in physics for a long time.

Much of this work has been motivated by the need of understanding the dynamic properties of gas particles [6, 11–13].

The simplest models of such particle systems assume either a line or a cycle as the environment in which particles move. For instance, Jepsen [7] considers particles of equal masses and arbitrary velocities moving in a cycle. He assumes the conservation of momentum and conservation of energy principles, such that when two particles collide they simply exchange velocities. Jepsen studies the probabilistic velocity distribution of particles because of its importance for understanding some gas equilibrium properties.

The dynamical system that emerges from a set of particles colliding while moving either on a line or a cycle is quite rich and not well understood yet [4]. However, there are some results regarding the total number of collisions if the particles move within an infinite line. Sevryuk [6] proved that the total number of collisions between particles that collide elastically on an infinite line is finite. He proved that the number of collisions is bounded by $2\left(8n^2(n-1)\frac{m_{max}}{m_{min}}\right)^{n-2}$, where $n$ is the total number of particles and $m_{max}$ and $m_{min}$ are the largest and smallest masses of the particles, respectively. Other results concerning the number of collisions for higher dimensions can be found in [14]. For the case of particles colliding in a cycle, some interesting results have been found, for instance, Mittag [15] proved that the dynamical system of three particles of different masses and arbitrary velocities elastically colliding in a cycle is equivalent to a standard billiard flow.

On the other hand, the dynamics of these systems of particles have motivated the design of algorithms for mobile robots. Susca et al. in [16] designed a system of mobile robots that imitates the impact behavior of elastic particles sliding in a cycle. They do so in order to carry out perimeter surveillance. Synchronization of mobile robots in the line is another example [17].

[1–3] studied the feasibility of the *localization* task in the cycle and in the segment by bouncing robots. The problem of localization consists of each robot finding the starting position (relative to its own position) of all the other robots in finite time. E.g. in [1], authors proved that if robots have arbitrary initial velocities but equal masses the task of localization can be performed if and only if none of the velocities is equal to the average of all the velocities.

This type of mobility of bouncing robots is also found in other models like those in [18, 19], where the so-called *population protocols* were studied. The agents of population protocols follow mobility patterns totally out of their control. This is called *passive mobility*. Passive mobility is intended to model, e.g., some unstable environment, like a flow of water, chemical solution, human blood, wind or unpredictable mobility of agents' carriers (e.g. vehicles or flocks of birds).

Other works have studied the destruction of mobile agents while visiting some specific location of their environment. Dobrev et al. in [20] introduced the *black hole search task*. They consider a set of mobile agents moving in a ring searching for a highly harmful item called *black hole*. A black hole is a stationary process that destroys any visiting agent upon its arrival without leaving any trace of it.

This task requires that at least one robot survives in order to report the location of the black hole.

Part of our motivation to study the survivability of bouncing robots was due to the equivalence between the dynamics of some systems of bouncing robots and the dynamics of salmon fries of the *salmon problem*, introduced by Moshe Rosenfeld in [8]. The salmon problem is inspired by the life cycle of salmons: a salmon after being hatched lives in the ocean for a period of several years, then it returns to its place of birth to spawn and die. The salmon problem is stated in [8] as follows: Consider $n$ *salmon fries* distributed on a ring, each fry moving with constant speed either clockwise or counterclockwise. When two fries collide they reverse direction and when a fry returns to its initial position, it dies. Death has priority over collisions. Is it possible that some fries live forever? Is there an efficient algorithm to decide whether all fries will die?

Rosenfeld gives an example of a configuration of five salmons of which one dies and the remaining four live forever. However, Rosenfeld's example has a flaw. We show what the problem is with his example and give the first correct example of a swarm with survivors. We also consider the salmon problem for mobile robots in a more general setting as well as different environments of deployment.

## 1.2 Results

In Section 2, we prove some general properties regarding the dynamics of bouncing robots that are crucial for understanding their survivability. In Section 3, we study the survivability of swarms of robots of equal masses and speeds. We prove that if the robots are deployed in a segment they all must die. In contrast, in the cycle we show the existence of swarms with two survivors for a swarm of $n \geq 4$ robots and we prove that the smallest swarm with survivors has four robots with exactly two survivors. We also prove a lower bound on the number of robots that die if not all robots have the same initial direction. For the case of robots of arbitrary masses and velocities, in Section 4, we prove that, if robots are deployed in a segment, the survivors, if any, must become indefinitely immobile. However, this is not the case when the robots are deployed in a cycle. We show that in the cycle at least one robot dies and that the maximum number of robots that can survive either in the cycle or in the segment is $n - 1$. We conclude by listing some open problems in Section 5. Omitted proofs can be found in the complete version of the paper.

## 1.3 Preliminaries

Let $\mathcal{S}_n = (\mathcal{M}, \mathcal{H}, \mathcal{U})$ be a swarm of $n$ bouncing robots $r_0, r_1, \ldots, r_n$ with masses $\mathcal{M} = (m_1, m_1, \ldots, m_n)$, starting positions or home bases $\mathcal{H} = (h_1, h_1, \ldots, h_n)$, and non-zero initial velocities $\mathcal{U} = (u_1, u_2, \ldots, u_n)$, respectively. We call the *size* of a swarm the total number of robots.

We study the question concerning robots moving in a segment and in a cycle. We assume that each robot starts moving at the same time with constant speed until a collision takes place. When two robots collide, they update their velocities following the conservation of momentum and conservation of energy principles. For the case of the segment, its end points model walls in which robots can collide with. When a robot collides with a wall it reverses direction but keeps moving with the same speed. Throughout this paper, we assume that collisions are elastic and that in any collision no more than two robots participate. Thus, if two robots of equal masses collide, they simply exchange velocities. If robots $r_1$ and $r_2$ of masses $m_1$ and $m_2$, and velocities $u_1$ and $u_2$ respectively, collide, after their collision they get new velocities $v_1$ and $v_2$, respectively, where:

$$v_1 = \frac{m_1 - m_2}{m_1 + m_2} u_1 + \frac{2m_2}{m_1 + m_2} u_2, \quad v_2 = \frac{2m_1}{m_1 + m_2} u_1 + \frac{m_2 - m_1}{m_1 + m_2} u_2. \quad (1)$$

A cycle of perimeter $l$ is modeled by the real interval $[0, l]$, with 0 and $l$ corresponding to the same point. By $r_i(t) \in [0, l]$ we denote the position of robot $r_i$ at time $t$. We suppose that originally each robot $r_i$ occupies the position $r_i(0) = h_i$ of its environment and that $0 \leq r_1(0) < r_2(0) < \cdots < r_n(0)$. Each robot is given an initial direction, clockwise (CW) or counterclockwise (CCW) in the cycle and left-to-right or right-to-left on the segment, according to which it starts its movement. By $dir_i$ we denote the starting direction of robot $r_i$ and we set $dir_i = 1$ if $r_i$ starts its movement in the counterclockwise direction around the ring or the left-to-right direction along the segment. By $dir_i = -1$ we denote the clockwise starting direction (on the ring) or right-to-left (on the segment).

By $\mathcal{S}^+$ we denote the number of robots in the swarm $\mathcal{S}$ whose initial direction is counterclockwise and by $\mathcal{S}$ the number of robots in the swarm $\mathcal{S}^-$ whose initial direction is clockwise. We identify the counterclockwise and left-to-right directions as positive.

A robot dies if at some time it returns to its home base. If a robot $r_a$ dies, it disappears from the environment and does not interact with any other robot any more. On the other hand, we say that a robot survives if it does not die. If $D$ is a subset of robots of a swarm $\mathcal{S}_n$ that die at some time, $\mathcal{S}_n \setminus D$ denotes the resulting swarm of survivors after the death of the robots in $D$.

The death of a robot takes priority over collisions, i.e, if two robots collide at the home base of one of them, the death of the robot returning to its home base takes place first and the other robot keeps moving without updating any of its parameters.

If the collisions of a swarm are not concurrent (no two different collisions take place at the same time), we can enumerate the collisions. Notice that this is not possible in most cases since it is plausible that two different pairs of robots may collide at the same time. If collisions can be enumerated, we denote the $i$-th collision of a given swarm by $C_i = (a^{(dir)}, b^{(dir')})$, where $a$ and $b$ are the robots involved in the $i$-th collision, and $dir, dir' \in \{+, -\}$, denote the directions that the robots had before colliding. We use $(+)$ to denote the CCW direction and $(-)$ to denote the CW direction. We use the notation $u_i^{(j)}$ to indicate the velocity of robot $r_i$ resulting after the $j$-th collision of the swarm, where $u_i^{(0)} = u_i$ denotes the initial velocity of robot $r_i$.

Since we use some properties of classical mechanics to prove our results, we define some useful concepts that extend from the properties of systems of particles. Suppose we have a swarm $\mathcal{S}_n$ of bouncing robots $r_1, r_2, \ldots, r_n$, of initial velocities $u_1, u_2, \ldots, u_n$ and masses $m_1, m_2, \ldots, m_n$, respectively. Analogously to systems of particles we define the momentum of the swarm as $P(\mathcal{S}_n) = \sum_{i=1}^{n} m_i u_i$, where $m_i u_i$ is the linear momentum of robot $r_i$. The velocity of the swarm is defined as $U(\mathcal{S}_n) = \frac{P(\mathcal{S}_n)}{M}$, where $M = \sum_{i}^{n} m_i$. Finally, the kinetic energy of $\mathcal{S}_n$ is given by $KE(\mathcal{S}_n) = \sum_{i=1}^{n} m_i u_i^2$.

## 2  General Behavior of Swarms of Bouncing Robots

In this section we focus on studying the dynamics of bouncing robots as they are deployed either in a segment or a cycle. In all these results, we do not assume that robots die since we focus only on studying their dynamics. We denote the infinite line by $\mathcal{L} = (-\infty, \infty)$, and the half positive semi-line by $\mathcal{L}^+ = (0, \infty)$, where 0 represents the wall on which the leftmost robot may collide. We say that a swarm $\mathcal{S}_n$ deployed on $\mathcal{L}$, *expands to the right* (respectively to the left) if and only if:

1. there exists $t_0 \geq 0$, such that for every $t \geq t_0$, no more collision takes place, and
2. for any $b > 0$, there exists some robot $r_m$ and time $t_m$ such that $r_m(t_m) > b$ (respectively $r_m(t_m) < a$, for arbitrary $a < 0$).

We say that $\mathcal{S}_n$ expands in both directions, if and only if $\mathcal{S}_n$ expands to the right and to the left.

**Theorem 1.** *Let $\mathcal{S}_n$ be any swarm of bouncing robots deployed on $\mathcal{L}$, such that $KE(\mathcal{S}_n) \neq 0$. Then, for any finite segment $[a,b] \subset \mathcal{L}$, there exists a time $t^\star$, such that for any $t' > t^\star$ some robot $r_m(t') \notin [a,b]$. Moreover, if the swarm has either positive or negative or zero momentum, the swarm expands to the right or to the left or in both directions, respectively.*

*Proof.* Let $\mathcal{S}_n$ be a swarm of robots, such that $KE(\mathcal{S}_n) \neq 0$, and let $[a,b] \subset \mathcal{L}$ be any finite segment. We assume that $h_i \in [a,b]$, for $i \geq 1$. Since the number of elastic collisions in $\mathcal{L}$ for any system of particles is finite [6], and the robots of $\mathcal{S}_n$ behave exactly as particles, the number of collisions in $\mathcal{S}_n$ is also finite. More precisely, there exists some $t_0 \geq 0$, such that for any $t \geq t_0$ no more collisions take place among the robots of the swarm. Because of the principle of conservation of kinetic energy, $KE(\mathcal{S}_n) \neq 0$ at any time, meaning that at any time there exists one robot $r_m$ that is still moving on the line. Let us assume that $r_m(t_0) \in [a,b]$ and that $r_m$ is moving to the right with velocity $v$, then $r_m(t') > b$ for any time $t' > t^\star$, where $t^\star = \frac{|b - r_m(t_0)|}{|v|}$, for arbitrary $b \in \mathcal{L}$. Analogously, if $r_m$ is moving to the left, $t^\star = \frac{|a - r_m(t_0)|}{|v|}$. This finishes the proof of the first part of the theorem. Notice that since we are assuming the principle of conservation of momentum, at any time bigger than $t_0$ the momentum of the system remains the same. Thus, if $P(\mathcal{S}_n) < 0$, not all robots might have positive velocities. Thus, $\mathcal{S}_n$ expands to the left. Analogously, if $P(\mathcal{S}_n) > 0$, there must exist some robot moving with positive velocity. If $P(\mathcal{S}_n) = 0$, the swarm must expand in both directions. ∎

In the next corollary, we assume that robots are deployed on the half infinite line. The origin models a wall, if the leftmost robot collides with the wall, it bounces back, i.e, the robot reverses direction but keeps moving with the same speed.

**Lemma 1.** *Let $\mathcal{S}_n$ be any swarm of bouncing robots deployed on $\mathcal{L}^+$. There exists a swarm $\mathcal{S}_{2n}$ of bouncing robots deployed on $\mathcal{L}$, such that $P(\mathcal{S}_{2n}) = 0$ and half of its robots mimic the dynamics of the robots in $\mathcal{S}_n$.*

The following corollary follows from the Theorem 1 and Lemma 1.

**Corollary 1.** *Any swarm $\mathcal{S}_n$ of bouncing robots of arbitrary masses and velocities that are deployed on $\mathcal{L}^+$, such that $KE(\mathcal{S}_n) \neq 0$, expands to the right.*

*Proof.* Take swarm $\mathcal{S}_{2n}$ of Lemma 1. Since $P(\mathcal{S}_{2n}) = 0$ and $KE(\mathcal{S}_{2n}) \neq 0$, Theorem 1 implies that $\mathcal{S}_{2n}$ expands in both directions. It follows that $\mathcal{S}_n$ expands to the right. ∎

We now consider swarms deployed on a cycle. Let $D_i^{(+)}(t)$ denote the total distance that robot $r_i$ traveled until time $t$ in the CCW direction, and $D_i^{(-)}(t)$ - the total distance traveled by $r_i$ in the CW direction. Denote $D_i(t) = D_i^{(+)}(t) - D_i^{(-)}(t)$. The next theorem establishes a relationship between the momentum of a swarm and the total distance that any robot traverses.

**Theorem 2.** *If $\mathcal{S}_n$ is a swarm of bouncing robots of same masses but arbitrary speeds, such that $P(\mathcal{S}_n) \neq 0$, then all robots eventually complete a full cycle.*

*Proof.* Without loss of generality assume $P(\mathcal{S}_n) > 0$, then we have that $U(\mathcal{S}_n) > 0$ which implies that $t \cdot (\sum_{i=1}^{n} v_i) > 0$, for any $t > 0$. Therefore, there exists a big enough $t^\star > 0$ such that $D_i(t^\star) \geq 1$ for any $i \geq 1$. ∎

## 3 Robots of Equal Masses and Equal Speeds

In this section, we study the survivability of bouncing robots of equal masses and speeds that are deployed either in a segment or a cycle. The following result shows that there are no swarms of equal masses and speeds in the segment that contain surviving robots.

**Theorem 3.** *All bouncing robots die of any swarm deployed in the segment.*

*Proof.* Let $\mathcal{S}_n$ be a swarm of bouncing robots of same speeds and masses. Notice that in $\mathcal{S}_n$ during a collision either with a robot or with a wall, robots simply reverse direction, so if $KE(\mathcal{S}_n) \neq 0$, no robot can remain static at any time even in the presence of the death of a robot. We prove this theorem by induction on the size of the swarm, so we assume that in any swarm of size $n-1$ all robots die. Let $\mathcal{S}_n$ be a swarm of size $n$. It is enough to prove that one of the extreme robots of $\mathcal{S}_n$ dies. Let $r_1$ and $r_n$ be the leftmost robot and the rightmost robot, respectively. Notice that if $dir_1 = 1$, $r_1$ will die after reversing direction. Without loss of generality, let us assume that $dir_1 = -1$. If no robot dies, robot $r_1$ has to be bumping against the wall and its neighbor $r_2$ so that it never returns to its home base, for the same reason $r_2$ is bumping against $r_1$ and $r_3$ without reaching its home base and so on. This can only happen if robots $r_1, \ldots, r_{n-1}$ are at the left of their home bases indefinitely, however this can not be true for $r_n$ which after colliding with $r_{n-1}$ reverses direction and dies. The remaining system has $n-1$ robots and because of the induction hypothesis all of them die. ∎

**Observation 1.** *An interested reader may notice that Theorem 3 also holds for robots of equal masses but arbitrary non-zero speeds.*

**Theorem 4.** *There exist swarms of size four in a cycle containing two survivors.*

*Proof.* Let $\mathcal{S}_4$ be a swarm of size four deployed in a cycle of perimeter 80. Let $h_1 = 10, h_2 = 25, h_3 = 30, h_4 = 80$ and $u_1 = u_2 = u_3 = 1, u_4 = -1$ be the corresponding home bases and initial velocities of the robots, respectively. We assume that all the robots have unitary masses. It is easy to check that robots $r_1$ and $r_2$ survive while the other two robots die. ∎

The configuration from Theorem 4 is the first correct example of surviving subset of robots. We argue below that the example from [8], represented in Table 1, which supposedly contains a swarm of five robots with four survivors is not correct. The table describes the positions (in degrees) of the robots (in the cycle) at the given time, where + and − indicate the current directions of the robots (+ for CCW and − for CW). The time is given in seconds and it takes 128 seconds for a robot to complete a full cycle. For instance, $r_1$ has starting position at 120 degrees and CW initial direction and it dies after 110 seconds while the remaining robots live eternally.

**Table 1.** Example given in [8]

| time | $r_1$ | $r_2$ | $r_3$ | $r_4$ | $r_5$ |
|------|-------|-------|-------|-------|-------|
| 0    | −120  | −92   | +55   | +41   | +51   |
| 110  | die   | −102  | +100  | +0    | −74   |
| 123  |       | +115  | −87   | −115  | +87   |

Although it is correct that the first robot to die is $r_1$, it is easy to check that robot $r_3$ or robot $r_2$ must die as well. This is because both of them have to reverse direction and inevitably one of them must return to its home base. The following theorem states that there are no swarms of smaller size than the swarm of Theorem 4 with survivors.

**Theorem 5.** *In the cycle any swarm of size less than four has no survivors.*

For two points $p, q$ in the ring, by $d^{(+)}(p,q)$ we denote the counterclockwise distance from $p$ to $q$ in the cycle, i.e. the distance which needs to be traveled in the counterclockwise direction in order to arrive at $q$ starting from $p$. We denote the time of the $j$-th collision of robot $r_i$ by $t_i^{(j)}$. For simplicity, we scale the perimeter of the cycle of Theorem 4 so as to have a unitary cycle.

**Observation 2.** *The following observations are crucial to understand the survivability of the survivors of the swarm of Theorem 4.*

1. For $i = 1, 2, 3$, we have that:
   (a) $t_i^{(1)} = \frac{d^{(+)}(h_i, h_4)}{2}$. Thus for $r_i$ not to die after reversing direction in its first collision, it should get its second hit before time $2t_i^{(1)}$. This can only happen if, there exists some robot $r_c$ such that $dir_c = 1$ and $d^{(+)}(h_c, h_i) < 2t_i^{(1)}$.
   (b) robot $r_i$ after its first collision every time that it moves in CW direction it does so by exactly $\frac{1}{2}$ time.
2. For $r_1$ and $r_2$ to survive, the second hit of $r_1$ must take place within the interval $(h_1, h_2)$. To guarantee so, for each robot $r_l$, such that $dir_l = 1$, $2\left(t_1^{(1)} - d^{(+)}(h_1, h_2)\right) < d^{(+)}(h_l, h_2)$, and
3. If robot $r_3$ participates in a fifth collision, its total distance traversed would be $1 + t_3^{(1)} - \frac{d^{(+)}(h_2, h_3) + d^{(+)}(h_1, h_2)}{2}$. So it must hold that $d^{(+)}(h_3, h_4) - \left(d^{(+)}(h_2, h_3) + d^{(+)}(h_1, h_2)\right) > 0$, since robot $r_3$ dies after its fourth collision.

The next theorem is an extension of Theorem 4.

**Theorem 6.** *There exists a swarm in the cycle of size n of two survivors, for any $n \geq 4$.*

*Proof.* The idea is to simply extend the swarm of Theorem 4 by inserting an arbitrary number of robots $r_a$ that copy the behavior of $r_3$, i.e, that die after their fourth collision without disturbing the survivability of $r_1$ and $r_2$. Notice that if we add a new robot $r_a$ between robots $r_2$ and $r_3$, we have that its first collision takes place at time $t_a^{(1)} = \frac{d^{(+)}(h_a, h_4)}{2} > t_3^{(1)}$. Further, because of Observation 2, it holds that $t_a^{(1)} - \left(d^{(+)}(r_2, r_a) + d^{(+)}(r_1, r_2)\right) > 0$. Thus, at time $t_a^{(5)}$, the total distance covered by $r_a$ would be $1 + t_a^{(1)} - \frac{d^{(+)}(h_2, h_a) + d^{(+)}(h_1, h_2)}{2}$ which is also bigger than one and therefore $r_a$ would die after its fourth collision. Moreover, after inserting $r_a$, robot $r_3$ still dies since the total distance covered in ccw direction until its new fourth collision would be $d^{(+)}(h_a, h_3) + d^{(+)}(h_2, h_a) < d^{(+)}(h_2, h_3)$. Finally, notice that the addition of $r_a$ would not make the second hit of $r_1$ to happen outside of $\frac{d^{(+)}(h_a, h_3)}{2}$ since Observation 2 holds for $r_3$. It is easy to check that Observation 1 holds after the insertion of $r_a$. We can repeat this procedure as many times as we want by adding new robots and still having $r_1$ and $r_2$ as survivors. Therefore, the theorem holds. ■

The following corollary is an immediate consequence of Theorem 2.

**Corollary 2.** *For any swarm $\mathcal{S}$ in the cycle at least $|\mathcal{S}^+ - \mathcal{S}^-|$ robots die.*

*Proof.* Let $\mathcal{S}$ be a swarm of bouncing robots of equal masses and speeds, we have that $P(\mathcal{S}) = ms(\mathcal{S}^+ - \mathcal{S}^-)$, where $s$ and $m$ denote the values of the common speed and mass of the robots, respectively. Theorem 2 guarantees that unless $\mathcal{S}^+ - \mathcal{S}^- = 0$, the swarm moves forward in the direction of the majority and thus there exists some robot that returns to its home base and dies. It follows that in $\mathcal{S}$ at least $|\mathcal{S}^+ - \mathcal{S}^-|$ robots die. ∎

## 4 Robots of Arbitrary Masses and Velocities

Recall that if the collisions of a swarm are not concurrent, $C_i = (a^{(dir)}, b^{(dir')})$ denotes the $i$-th collision in the swarm, where $a$ and $b$ are the robots involved in such collision and $dir, dir' \in \{+, -\}$, denote the directions that the robots had before they collide.

**Theorem 7.** *For any $n \geq 2$ there exists a swarm $\mathcal{S}_n$ of bouncing robots of size $n$ in the segment, such that:*

1. *$u_1 > 0$ and $u_i < 0$, for all $2 \leq i \leq n$.*
2. *$C_i = (r_i^{(+)}, r_{i+1}^{(-)})$, for all $1 \leq i \leq n-1$, such that:*
   *(a) $u_i^{(i)} = 0$, robot $r_i$ stops.*
   *(b) $u_{i+1}^{(i)} > 0$, robot $r_{i+1}$ reverses direction.*
3. *only $r_n$ dies.*
4. *$u_{i+1} = \frac{m_{i+1} - m_i}{2m_{i+1}} u_1^0 \prod_{j=2}^{i} \left( \frac{2m_{j-1}}{m_{j-1}+m_j} + \frac{(m_j - m_{j-1})^2}{(m_{j-1}+m_j)2m_j} \right)$.*

Notice that Theorem 7 can be extended to the cycle since the construction is independent of the environment.

**Corollary 3.** *For any $n \geq 2$ there exists a swarm $\mathcal{S}_n$ of bouncing robots of size $n$ in the cycle in which exactly $n-1$ robots survive, such that properties 1 – 4 in Theorem 7 are valid.*

Theorem 3 shows that in the segment all robots die if they have all the same masses and speeds. This is because, the laws of classical mechanics for particles of equal masses and speeds force each robot to have at any time a fraction of the total kinetic energy of the swarm so they can never be static. However, if robots have arbitrary masses and velocities, it is possible that some robot carrying all the kinetic energy of the swarm dies, then all the remaining static robots would survive.

**Theorem 8.** *Let $\mathcal{S}_n$ be any swarm of bouncing robots of different masses and arbitrary velocities deployed in the segment and let $D$ be the subset of robots of $\mathcal{S}_n$ that do not survive. Therefore, the robots of the resulting swarm after the death of the robots in $D$ must be static, i.e, $KE(\mathcal{S}_n \setminus D) = 0$.*

In contrast to the case of the segment in which all survivors must be static, in the cycle there are swarms of non-static survivors.

**Theorem 9.** *In the cycle there exist a swarm $\mathcal{S}_3$ of size three in which only one robot dies and the kinetic energy of the survivors is different from zero.*

*Proof.* Let $h_a = 0, h_b = 20, h_c = 16$ be the home bases of robots $r_a, r_b$ and $r_c$, respectively, and let $m_a = 1, m_b = 3, m_c = 3$ and $u_a = 3, u_b = 1, u_c = -1$ be the values of their respective masses and velocities. The resulting swarm satisfies the properties of the theorem. ∎

The next theorem states that the number of survivors can never be larger than $n - 1$ in any swarm of arbitrary masses and speeds deployed in the cycle.

**Theorem 10.** *Any swarm $\mathcal{S}_n$ of bouncing robots of arbitrary masses and speeds in the segment or a cycle has some robot that dies.*

*Proof.* Consider first a segment environment. As the initial kinetic energy of the swarm is positive and it stays the same after any bounce it must stay positive until some robot dies. By Theorem 8, at least one robot has to die. Consider now a cycle environment. Take the the segment representation of the cycle. It is enough to prove that one of the two robots at the end points of the segment dies. Theorem 1 guarantees that if no robot dies, all the robots can not remain colliding among themselves in the interval $(h_1, h_n)$ for an indefinite period of time but that there exists a time $t^*$ in which no more collisions take place and that some robot would eventually leave the interval $[h_1, h_n]$. Notice that the first robots that can leave the interval $[h_1, h_n]$ can only be $r_1$ or $r_n$. Thus, whatever the direction of the expansion of the swarm is, robot $r_1$ or robot $r_n$ will die. ∎

## 5 Conclusions

One open problem is to study the survivability of robots when the deadly zones are not just points but rather regions of the environment. It also remains open to investigate the largest number of survivors for robots of equal masses and speeds. Towards this goal, we wrote a program in Java to simulate the survivability of bouncing robots in the cycle. Our program randomly generated positions and initial directions for $n$ robots of unitary masses and speeds and tested their survivability. Our program could only find examples with two survivors for $1 \leq n \leq 10$. It also remains open to figure out the largest number of non-static survivors for swarms of different masses and velocities. We only considered elastic collisions between robots, so we think that it would be interesting to study the case of inelastic collisions and different physical interactions among the robots.

## References

1. Czyzowicz, J., Kranakis, E., Pacheco, E.: Localization for a system of colliding robots. In: Fomin, F.V., Freivalds, R., Kwiatkowska, M., Peleg, D. (eds.) ICALP 2013, Part II. LNCS, vol. 7966, pp. 508–519. Springer, Heidelberg (2013)

2. Czyzowicz, J., Gąsieniec, L., Kosowski, A., Kranakis, E., Ponce, O.M., Pacheco, E.: Position discovery for a system of bouncing robots. In: Aguilera, M.K. (ed.) DISC 2012. LNCS, vol. 7611, pp. 341–355. Springer, Heidelberg (2012)
3. Friedetzky, T., Gąsieniec, L., Gorry, T., Martin, R.: Observe and remain silent (Communication-less agent location discovery). In: Rovan, B., Sassone, V., Widmayer, P. (eds.) MFCS 2012. LNCS, vol. 7464, pp. 407–418. Springer, Heidelberg (2012)
4. Cooley, B., Newton, P.: Iterated impact dynamics of $n$-beads on a ring. SIAM Rev. 47(2), 273–300 (2005)
5. Cooley, B., Newton, P.: Random number generation from chaotic impact collisions. Regular and Chaotic Dynamics 9(3), 199–212 (2004)
6. Sevryuk, M.: Estimate of the number of collisions of $n$ elastic particles on a line. Theoretical and Mathematical Physics 96(1), 818–826 (1993)
7. Jepsen, D.: Dynamics of a simple many-body system of hard rods. Journal of Mathematical Physics 6, 405 (1965)
8. Rosenfeld, M.: Some of my favorite "lesser known" problems. Ars Mathematica Contemporanea 1(2), 137–143 (2008)
9. Das, S., Flocchini, P., Santoro, N., Yamashita, M.: On the computational power of oblivious robots: forming a series of geometric patterns. In: PODC, pp. 267–276 (2010)
10. Suzuki, I., Yamashita, M.: Distributed anonymous mobile robots: Formation of geometric patterns. SIAM J. Comput. 28(4), 1347–1363 (1999)
11. Murphy, T.: Dynamics of hard rods in one dimension. Journal of Statistical Physics 74(3), 889–901 (1994)
12. Tonks, L.: The complete equation of state of one, two and three-dimensional gases of hard elastic spheres. Physical Review 50(10), 955 (1936)
13. Wylie, J., Yang, R., Zhang, Q.: Periodic orbits of inelastic particles on a ring. Physical Review E 86(2), 026601 (2012)
14. Murphy, T., Cohen, E.: Maximum number of collisions among identical hard spheres. Journal of Statistical Physics 71(5-6), 1063–1080 (1993)
15. Glashow, S.L., Mittag, L.: Three rods on a ring and the triangular billiard. Journal of Statistical Physics 87(3-4), 937–941 (1997)
16. Susca, S., Bullo, F.: Synchronization of beads on a ring. In: 46th IEEE Conference on Decision and Control, pp. 4845–4850 (2007)
17. Wang, H., Guo, Y.: Synchronization on a segment without localization: algorithm and applications. In: International Conference on Intelligent Robots and Systems, IROS, pp. 3441–3446 (2009)
18. Angluin, D., Aspnes, J., Diamadi, Z., Fischer, M.J., Peralta, R.: Computation in networks of passively mobile finite-state sensors. Distributed Computing 18(4), 235–253 (2006)
19. Angluin, D., Aspnes, J., Eisenstat, D.: Stably computable predicates are semilinear. In: PODC, pp. 292–299 (2006)
20. Dobrev, S., Flocchini, P., Prencipe, G., Santoro, N.: Mobile search for a black hole in an anonymous ring. Algorithmica 48(1), 67–90 (2007)

# Emergence of Wave Patterns on Kadanoff Sandpiles*

Kévin Perrot[1,2] and Éric Rémila[3]

[1] Université de Lyon - LIP (UMR 5668 - CNRS - ENS de Lyon - Université Lyon 1)
46 allée d'Italie 69364 Lyon Cedex 7, France
[2] Universidad de Chile - Departamento de Ingeniería Matemática - CMM
(UMI 2807 - CNRS) Blanco Encalda 2120, Santiago, Chile
[3] Université de Lyon - Groupe d'Analyse de la Théorie Economique Lyon
Saint-Etienne - (UMR 5824 - CNRS - Université Lyon 2) - Site stéphanois,
6 rue Basse des Rives, 42 023 Saint-Etienne Cedex 2, France
kevin.perrot@ens-lyon.fr, eric.remila@univ-st-etienne.fr

**Abstract.** Emergence is a concept that is easy to exhibit, but very hard to formally handle. This paper is about cubic sand grains moving around on nicely packed columns in one dimension (the physical sandpile is two dimensional, but the support of sand columns is one dimensional). The Kadanoff Sandpile Model is a discrete dynamical system describing the evolution of a finite number of stacked grains —as they would fall from an hourglass— to a stable configuration (fixed point). Grains move according to the repeated application of a simple local rule until reaching a fixed point. The main interest of the model relies in the difficulty of understanding its behavior, despite the simplicity of the rule. In this paper we prove the emergence of wave patterns periodically repeated on fixed points. Remarkably, those regular patterns do not cover the entire fixed point, but eventually emerge from a seemingly highly disordered segment. The proof technique we set up associated arguments of linear algebra and combinatorics, which interestingly allow to formally state the emergence of regular patterns without requiring a precise understanding of the chaotic initial segment's dynamic.

**Keywords:** sandpile model, discrete dynamical system, emergence, fixed point.

## 1 Introduction

Understanding and proving properties on discrete dynamical systems (DDS) is challenging, and demonstrating the global behavior of a DDS defined with local rules is at the heart of our comprehension of natural phenomena [28, 15]. Sandpile models are a class of DDS defined by local rules describing how grains move

---
* Partially supported by IXXI (Complex System Institute, Lyon), ANR projects Subtile, Dynamite and QuasiCool (ANR-12-JS02-011-01), and Modmad Federation of U. St-Etienne.

in discrete space and time. We start from a finite number of stacked grains —in analogy with an hourglass—, and try to predict the asymptotic shape of stable configurations.

Bak, Tang and Wiesenfeld introduced sandpile models as systems presenting *self-organized criticality* (SOC), a property of dynamical systems having critical points as attractors [1]. Informally, they considered the repeated addition of sand grains on a discretized flat surface. Each addition possibly triggers an avalanche, consisting of grains falling from column to column according to simple local rules, and after a while a heap of sand has formed. SOC is related to the fact that a single grain addition on a stabilized sandpile has a hardly predictable consequence on the system, on which fractal structures may emerge [2]. This model can be naturally extended to any number of dimensions.

## 1.1 Kadanoff Sandpile Model (KSPM)

A one-dimensional sandpile configuration can be represented as a sequence $(h_i)_{i \in \mathbb{N}}$ of non-negative integers, $h_i$ being the number of sand grains stacked on column $i$. The evolution starts from the initial configuration $h$ where $h_0 = N$ and $h_i = 0$ for $i > 0$, and in the classical sandpile model a grain falls from column $i$ to column $i + 1$ if and only if the height difference $h_i - h_{i+1} > 1$. One-dimensional sandpile models were well studied in recent years [11, 4, 12, 5, 27, 22, 6].

Kadanoff *et al.* proposed a generalization of classical models in which a fixed parameter $p$ denotes the number of grains falling at each step [17]. Starting from the initial configuration composed of $N$ stacked grains on column 0, we iterate the following rule: if the difference of height (the slope) between column $i$ and $i + 1$ is greater than $p$, then $p$ grains can fall from column $i$, and one grain reaches each of the $p$ columns $i + 1, i + 2, \ldots, i + p$ (Figure 1). The rule is applied once (non-deterministically) during each time step.

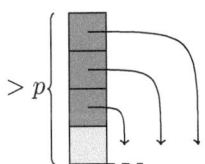

**Fig. 1.** KSPM($p$) rule. When $p$ grains leave column $i$, the slope $b_{i-1}$ is increased by $p$, $b_i$ is decreased by $p+1$ and $b_{i+p}$ is increased by 1. The slope of other columns are not affected.

Formally, this rule is defined on the space of ultimately null decreasing integer sequences where each integer represents a column of stacked sand grains. Let $h = (h_i)_{i \in \mathbb{N}}$ denote a *configuration* of the model, $h_i$ is the number of grains on column $i$. The words *column* and *index* are synonyms. In order to consider only the relative height between columns, we represent configurations as sequences of *slopes* $b = (b_i)_{i \in \mathbb{N}}$, where for all $i \geqslant 0$, $b_i = h_i - h_{i+1}$. This latter is the main representation we are using (also the one employed in the definition of the model), within the space of ultimately null non-negative integer sequences. We denote by $0^\omega$ the infinite sequence of zeros that is necessary to explicitly write the value of a configuration.

**Definition 1.** *KSPM with parameter $p > 0$, KSPM(p), is defined by two sets:*

- Configurations. *Ultimately null non-negative integer sequences.*
- Transition rules. *There is a possible transition from a configuration $b$ to a configuration $b'$ on column $i$, and we note $b \xrightarrow{i} b'$ when:*

  - $b'_{i-1} = b_{i-1} + p$ *(for $i \neq 0$)*
  - $b'_i = b_i - (p+1)$
  - $b'_{i+p} = b_{i+p} + 1$
  - $b'_j = b_j$ *for $j \notin \{i-1, i, i+p\}$.*

In this case we say that $i$ is *fired*. For the sake of imagery, we always consider indices to be increasing on the right (Figure 1). Remark that according to the definition of the transition rules, $i$ may be fired if and only if $b_i > p$, otherwise $b'_i$ is negative. We note $b \to b'$ when there exists an integer $i$ such that $b \xrightarrow{i} b'$. The transitive closure of $\to$ is denoted by $\xrightarrow{*}$, and we say that $b'$ is *reachable* from $b$ when $b \xrightarrow{*} b'$. A basic property of the KSPM model is the *diamond property*: if there exists $i$ and $j$ such that $b \xrightarrow{i} b'$ and $b \xrightarrow{j} b''$, then there exists a configuration $b'''$ such that $b' \xrightarrow{j} b'''$ and $b'' \xrightarrow{i} b'''$.

We say that a configuration $b$ is *stable*, or a *fixed point*, if no transition is possible from $b$. As a consequence of the diamond property, one can easily check that, for each configuration $b$, there exists a unique stable configuration, denoted by $\pi(b)$, such that $b \xrightarrow{*} \pi(b)$. Moreover, for any configuration $b'$ such that $b \xrightarrow{*} b'$, we have $\pi(b') = \pi(b)$ (see [14] for details). For convenience, we denote by $N$ the initial configuration $(N, 0^\omega)$, such that $\pi(N)$ is the sequence of slopes of the fixed point associated to the initial configuration composed of $N$ stacked grains. This paper is devoted to the study of $\pi(N)$ according to $N$. An example of evolution is pictured on figure 2.

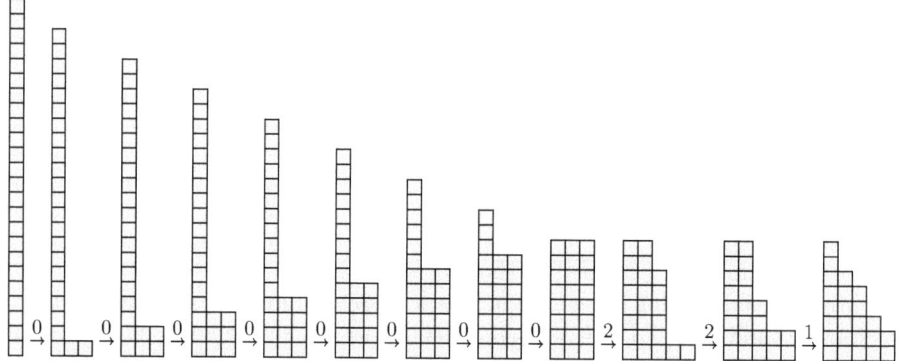

**Fig. 2.** A possible evolution in KSPM(2) from the initial configuration for $N = 24$ to $\pi(24)$. $\pi(24) = (2, 1, 2, 1, 2, 0^\omega)$ and its shot vector (definition in Subsection 2.1) is $(8, 1, 2, 0^\omega)$.

## 1.2 Our Result

For a configuration $b$ we denote by $b_{[n,\infty[}$ the infinite subsequence of $b$ starting from index $n$, and * is the Kleene star denoting finite repetitions of a regular expression (see for example [16] for details). In this paper we prove the following precise asymptotic form of fixed points, presenting an emergent regular structure stemming from a seemingly complex initial segment (note that the support of $\pi(N)$ is in $\Theta(\sqrt{N})$, because fixed points are non-degenerated rectangular triangles or area $N$):

**Theorem 1.** *There exists an $n$ in $\mathcal{O}(\log N)$ such that*

$$\pi(N)_{[n,\infty[} \in (p \cdot \ldots \cdot 2 \cdot 1)^* \, 0 \, (p \cdot \ldots \cdot 2 \cdot 1)^* \, 0^\omega.$$

Omitted proofs and illustrations of the result can be found in [26].

The result above presents an interesting feature: we asymptotically completely describe the form of stable configurations, though there is a part of asymptotically null relative size which remains mysterious. Furthermore, proven regularities are directly stemming from this messy part. Informally, it means that we prove the emergence of a very regular behavior, after a short transitional and complex phase. Most interestingly, the proof technic we develop does not require to understand precisely this complex initial segment.

In some previous works ([23–25]), we obtained a similar result for the smallest parameter $p = 2$ (the case $p = 1$ is the well known Sandpile Model) using arguments of combinatorics, but, for the general case, we have to introduce a completely different approach. The main ideas are the following: we first relate different representations of a sandpile configuration (Subsection 2.1), which leads to the construction of a DDS on $\mathbb{Z}^{p+1}$ such that the orbit of a well chosen point (according to the number of grains $N$) describes the fixed point configuration we want to characterize. This system is quasi-linear in the sense that the image of a point is obtained by a linear contracting transformation followed by a rounding (in order to remain in $\mathbb{Z}^{p+1}$) which we do not precisely predict. We want to prove that this system converges rapidly, but the unknown rounding makes the analysis of the system very difficult (except for $p = 2$). The key point (Subsection 2.2) is the reduction of this system to another quasi-linear system in $\mathbb{Z}^p$, for which we have a clear intuition (Subsection 2.3), and which allows to conclude the convergence of the system to points involving wavy patterns on fixed points (Subsections 2.4 and 2.5). The reader can refer to Figure 3 (at the end of the paper) for an illustration of the various representations that will be used along these developments.

## 1.3 The Context

The problem of describing and proving regularity properties suggested by numerical simulations, for models issued from basic dynamics is a present challenge for physicists, mathematicians, and computer scientists. There exist a lot of conjectures on discrete dynamical systems with simple local rules (sandpile model [3] or chip firing games, but also rotor router [19], the famous Langton's ant [8, 9]...) but very few results have actually been proved. Regarding KSPM(1), the *prediction problem* (namely, the problem of computing the fixed point $\pi(k)$)

has been proven in [21] to be in $\mathbf{NC}^2 \subseteq \mathbf{AC}^2$ for the one dimensional case, the model of our purpose (improved to $\mathbf{LOGCFL} \subseteq \mathbf{AC}^1$ in [20]), and **P**-complete when the dimension is $\geq 3$. A recent study [13] showed that in the two dimensional case the avalanche problem (given a configuration $\sigma$ and two columns $i$ and $j$, does adding one grain on column $i$ have an influence on columnn $j$?) is **P**-complete for KSPM($p$) with $p > 1$, which points out an inherently sequential behavior. The two dimensional case for $p = 1$ is still open, though we know from [7] that wires cannot cross.

## 2 Analysis

We consider the parameter $p$ to be fixed. We study the "*internal dynamic*" of fixed points, via the construction of a DDS in $\mathbb{Z}^{p+1}$, such that the orbit of a well chosen point according to the number of grains $N$ describes $\pi(N)$. The aim is then to prove the convergence of this orbit in $\mathcal{O}(\log N)$ steps, such that the values it takes involve the form described in Theorem 1.

### 2.1 Internal Dynamic of Fixed Points

A useful representation of a configuration reachable from $(N, 0^\omega)$ is its *shot vector* $(a_i)_{i \in \mathbb{N}}$, where $a_i$ is the number of times that the rule has been applied on column $i$ from the initial configuration (see figure 2 for an example). A fixed point $\pi(N)$ can also be represented as a sequence of slopes $(b_i)_{i \in \mathbb{N}}$ (*i.e.*, $b_i = \pi(N)_i$ for all $i$), and those two representations are obviously linked in various ways. In particular for any $i$ we can compute the slope at index $i$ provided the number of firings at $i - p$, $i$ and $i + 1$, because $b_i$ is initially equal to 0 (the case $i = 0$ is discussed below), and: a firing at $i - p$ increases $b_i$ by 1; a firing at $i$ decreases $b_i$ by $p+1$; a firing at $i+1$ increases $b_i$ by $p$; and any other firing has no consequence on the slope $b_i$. Therefore, $b_i = a_{i-p} - (p+1)\, a_i + p\, a_{i+1}$, with $0 \leq b_i \leq p$ since $\pi(N)$ is a fixed point, and thus

$$a_{i+1} = -\frac{1}{p} a_{i-p} + \frac{p+1}{p} a_i + \frac{1}{p} b_i$$

This equation expresses the value of the shot vector at position $i+1$ according to its values at positions $i-p$ and $i$, and a bounded perturbation $0 \leq \frac{b_i}{p} \leq 1$. As an initial condition, we consider a virtual column of index $-p$ that has been fired $N$ times: $a_{-p} = N$ and $a_i = 0$ for $-p < i < 0$, representing the fact that column 0 is the only one receiving $N$ times 1 unit of slope.

*Remark 1.* Note that $a_{i+1} \in \mathbb{N}$, thus $-a_{i-p} + (p+1)\, a_i + b_i \equiv 0 \mod p$. As a consequence, the value of $b_i$ is *nearly determined*: given $a_{i-p}$ and $a_i$, there is only one possible value of $b_i$, except when $-a_{i-p} + (p+1)\, a_i \equiv 0 \mod p$ in which case $b_i$ equals $0$ or $p$.

For example, consider $\pi(2000)$ for $p = 4$ (see Figure 3b). We have $a_8 = 120$ and $a_4 = 189$, so $-a_4 + 5\, a_8 = 411 \equiv 3 \mod p$. From this knowledge, $b_8$ is determined to be equal to 1, so that $a_9 = -\frac{1}{4} a_4 + \frac{5}{4} a_8 + \frac{1}{4} b_8 = 103$ is an integer.

We rewrite this relation as a linear system we can manipulate easily. $a_{i+1}$ is expressed in terms of $a_{i-p}$ and $a_i$, so we choose to construct a sequence of vectors $(X_i)_{i \in \mathbb{N}}$ with $X_i \in \mathbb{N}^{p+1}$ and such that $X_i = {}^t(a_{i-p}, a_{i-p+1}, \ldots, a_i)$ where ${}^t v$ stands for the transpose of $v$. Note that we consider only finite configurations, so there always exists an integer $i_0$ such that $X_i = \mathbb{0}$ for $i_0 \leqslant i$, with $\mathbb{0} = {}^t(0, \ldots, 0)$.

Given $X_i$ and $b_i$ we can compute $X_{i+1}$ with the relation

$$X_{i+1} = A X_i + \frac{b_i}{p} J \quad \text{with} \quad A = \begin{pmatrix} 0 & 1 & 0 & 0 \\ & \ddots & & \\ 0 & 0 & 1 & 0 \\ 0 & 0 & 0 & 1 \\ -\frac{1}{p} & 0 & 0 & \frac{p+1}{p} \end{pmatrix} \quad J = \begin{pmatrix} 0 \\ \vdots \\ 0 \\ 0 \\ 1 \end{pmatrix}$$

in the canonical base $B = (e_0, e_1, \ldots, e_p)$, with $A$ a square matrix[1] of size $(p+1) \times (p+1)$.

This system expresses the shot vector around position $i + 1$ (via $X_{i+1}$) in terms of the shot vector around position $i$ (via $X_i$) and the slope at $i$ (via $b_i$). Thus the orbit of the point $X_0 = (N, 0, \ldots, 0, a_0)$ in $\mathbb{N}^{p+1}$ describes the shot vector of the fixed point composed of $N$ grains.

Note that it may look odd to study the sequence $(b_i)_{i \in \mathbb{N}}$ using a DDS whose iterations presuppose the knowledge of $(b_i)_{i \in \mathbb{N}}$. It is actually helpful because of the underlined fact that values $b_i$ are *nearly determined* (Remark 1): in a first phase we will make no assumption on the sequence $(b_i)_{i \in \mathbb{N}}$ (except that $b_i \leqslant p$ for all $i$) and prove that the system converges exponentially quickly in $N$; and in a second phase we will see that from an $n$ in $\mathcal{O}(\log N)$ such that the system has converged, the sequence $(b_i)_{i \geqslant n}$ is *determined* to have a regular wavy shape.

The system we get is a linear map plus a perturbation induced by the discreteness of values of the slope. Though the perturbation is bounded by a global constant at each step ($b_i \leqslant p$ for all $i$ since $\pi(N)$ is a fixed point), it seems that the non-linearity prevents classical methods to be powerful enough to decide the convergence of this model.

We denote by $\phi$ the corresponding transformation from $\mathbb{Z}^{p+1}$ to $\mathbb{Z}^{p+1}$, which is composed of two parts: a matrix and a perturbation. Let $R(x) = x^{p-1} + \frac{p-1}{p} x^{p-2} + \cdots + \frac{2}{p} x + \frac{1}{p}$, the characteristic polynomial of $A$ is $(1-x)^2 R(x)$. We can first notice that 1 is a double eigenvalue. A second remark, which helps to get a clear picture of the system, is that all the other eigenvalues are distinct and less than 1 (from a bound by Enerström and Kakeya [10]). We will especially use these remarks in Subsection 2.3. Therefore there exists a basis such that the matrix of $\phi$ is in Jordan normal form with a Jordan block of size 2. Then, we could project on the $p - 1$ other components to get a diagonal matrix for the transformation, hopefully exhibiting an understandably contracting behavior.

We tried to express the transformation $\phi$ in a basis such that its matrix is in Jordan normal form, but we did not manage to handle the effect of the

---

[1] As a convention, blank spaces are 0s and dotted spaces are filled by the sequence induced by its endpoints.

perturbation expressed in such a basis. Therefore, we rather express $\phi$ in a basis such that the matrix and the perturbation act harmoniously. The proof of the Theorem 1 is done in three steps:

1. the construction of a new dynamical system: we first express $\phi$ is a new basis $B'$, and then project along one component (Subsection 2.2);
2. the behavior of this new dynamical system is easily tractable, and we will see that it converges exponentially quickly (in $\mathcal{O}(\log N)$ steps) to a uniform vector (Subsection 2.3);
3. finally, we prove that as soon as the vector is uniform, then the wavy shape of Theorem 1 takes place (Subsections 2.4 and 2.5).

### 2.2 Making the Matrix and the Perturbation Act Harmoniously

From the dynamical system $X_{i+1} = A X_i + \frac{b_i}{p} J$ in the canonical basis $B$, we construct a new dynamical system for $\phi$ in two steps: first we change the basis of $\mathbb{Z}^{p+1}$ in which we express $\phi$, from the canonical one $B$, to a well chosen $B'$; then we project the transformation along the first component of $B'$. The resulting system on $\mathbb{Z}^p$, called *averaging system*, is very easily understandable, very intuitive, and the proof of its convergence to a uniform vector can then be completed straightforwardly.

Let $B' = \begin{pmatrix} 1 & 0 & & 0 \\ 1 & 1 & & 0 \\ \vdots & \vdots & \ddots & \\ 1 & 1 & \cdots & 1 \end{pmatrix}$ and $B'^{-1} = \begin{pmatrix} 1 & & 0 & 0 \\ -1 & \ddots & 0 & 0 \\ & \ddots & \ddots & \\ 0 & & -1 & 1 \end{pmatrix}$ be square matrices of size $p+1$.

$B' = (e'_0, \ldots, e'_p)$ (with $e'_i$ the $(i+1)^{th}$ column of the matrix $B'$) is a basis of $\mathbb{Z}^{p+1}$, and we have

$$B'^{-1} X_{i+1} = B'^{-1} A B' B'^{-1} X_i + \frac{b_i}{p} B'^{-1} J$$
$$\iff X'_{i+1} = A' X'_i + \frac{b_i}{p} J'$$

with

$$X'_i = B'^{-1} X_i \qquad A' = B'^{-1} A B' = \begin{pmatrix} 1 & 1 & & 0 \\ & \ddots & & \\ & & 0 & 0 & 1 \\ & & 0 & \frac{1}{p} & \cdots & \frac{1}{p} \end{pmatrix} \qquad J' = B'^{-1} J = \begin{pmatrix} 0 \\ \vdots \\ 0 \\ 1 \end{pmatrix}$$

We now proceed to the second step by projecting along $e'_0$. Let $P$ denote the projection in $\mathbb{Z}^{p+1}$ along $e'_0$ onto $\{0\} \times \mathbb{Z}^p$. We can notice that $e'_0$ is an eigenvector of $A'$, hence projecting along $e'_0$ simply corresponds to erasing the first coordinate of $X'_i$. For convenience, we do not write the zero component of objects in $\{0\} \times \mathbb{Z}^p$.

The new DDS we now have to study, which we call *averaging system*, is

$$Y_{i+1} = M Y_i + \frac{b_i}{p} K \qquad (1)$$

with the following elements in $\mathbb{Z}^p$ (in $\{0\} \times \mathbb{Z}^p$):

$$Y_i = PX'_i \qquad M = PA' = \begin{pmatrix} 0 & 1 & & 0 \\ & & \ddots & \\ 0 & 0 & & 1 \\ \frac{1}{p} & \frac{1}{p} & \cdots & \frac{1}{p} \end{pmatrix} \qquad K = PJ' = \begin{pmatrix} 0 \\ \vdots \\ 0 \\ 1 \end{pmatrix}$$

Let us look in more details at $Y_i$ and what it represents concerning the shot vector. We have $X_i = {}^t(a_{i-p}, a_{i-p+1}, \ldots, a_i)$, thus

$$Y_i = PB'^{-1}X_i = \begin{pmatrix} a_{i-p+1} - a_{i-p} \\ \vdots \\ a_{i-1} - a_{i-2} \\ a_i - a_{i-1} \end{pmatrix} \text{ and for initialization } Y_0 = \begin{pmatrix} -N \\ 0 \\ \vdots \\ 0 \\ a_0 \end{pmatrix}.$$

$Y_i$ represents differences of the shot vector, which may of course be negative. In Subsection 2.3 we will see that the averaging system is easily tractable and converges exponentially to a uniform vector. Subsection 2.4 concentrates on the implications following this uniform vector, *i.e.*, the emergence of a wavy shape.

## 2.3 Convergence of the Averaging System

The averaging system is understandable in simple terms. From $Y_i$ in $\mathbb{Z}^p$, we obtain $Y_{i+1}$ by:

1. shifting all the values one row upward;
2. for the bottom component, computing the mean of values of $Y_i$, and adding a small perturbation (a multiple of $\frac{1}{p}$ between 0 and 1) to it.

Let $y_i$ be the first component of $Y_i$, we therefore have $Y_i = {}^t(y_i, \ldots, y_{i+p-1})$.

*Remark 2.* $(Y_i)_{i \in \mathbb{N}}$ are still integer vectors, hence the perturbation added to the last component is again nearly determined: let $m_i$ denote the mean of value of $Y_i$, we have $(m_i + \frac{b_i}{p}) \in \mathbb{Z}$ and $0 \leq \frac{b_i}{p} \leq 1$. Consequently, if $m_i$ is not an integer then $b_i$ is determined and equals $p(\lceil m_i \rceil - m_i)$, otherwise $b_i$ equals 0 or $p$.

For example, consider $\pi(2000)$ for $p = 4$ (see Figure 3a, be careful that it pictures $a_i - a_{i+1}$ at position $i$). We have $Y_{13} = {}^t(-3, -5, -7, -7)$, then $y_{13} = -\frac{11}{2}$ and $b_{13}$ is forced to be equal to 2 so that $Y_{14} = {}^t(-5, -7, -7, -5)$ is an integer vector.

We can foresee what happens as we iterate this dynamical system and new values are computed: on a *large scale* —when values are large compared to $p$— the system evolves roughly toward the mean of values of the initial vector $Y_0$, and on a *small scale* —when values are small compared to $p$— the perturbation lets the vector wander a little around. Previous developments where intending to allow a simple argument to prove that those wanderings do not prevent the exponential convergence towards a uniform vector.

The study of the convergence of the averaging system works in three steps:

(i) state a linear convergence of the whole system; then express $Y_n$ in terms of $Y_0$ and $(b_i)_{0 \leq i \leq n}$;
(ii) isolate the perturbations induced by $(b_i)_{0 \leq i \leq n}$ and bound them;
(iii) prove that the other part (corresponding to the linear map $M$) converges exponentially quickly.

From (ii) and (iii), a point converges exponentially quickly into a ball of constant radius, then from (i) this point needs a constant number of extra iterations in order to reach the center of the ball, that is, a uniform vector.

**Proposition 1.** *There exists an $n$ in $\mathcal{O}(\log N)$ such that $Y_n$ is a uniform vector.*

*Proof.* Let $m_i$ (respectively $\overline{m}_i$, $\underline{m}_i$) denote the mean (respectively maximal, minimal) of values of $Y_i$. We will prove that $\overline{m}_i - \underline{m}_i$ converges exponentially quickly to 0, which proves the result.

We start with $Y_0 = {}^t(-N, 0, \ldots, 0, a_0)$, thus $\overline{m}_0 - \underline{m}_0 = N + a_0 \leq \frac{p+1}{p} N$ since $a_0 \leq \frac{N}{p}$ (recall that $a_0$ is the number of times column 0 has been fired).

This proof is composed of two parts. Firstly, the system converges exponentially quickly on a large scale. Intuitively, when $\overline{m}_i - \underline{m}_i$ is large compared to $p$, the perturbation is negligible.

**Lemma 1.** *There exists a constant $\alpha$ and $n_0$ in $\mathcal{O}(\log N)$ s.t. $\overline{m}_{n_0} - \underline{m}_{n_0} < \alpha$.*

*Proof (Proof sketch).* Complete proof in [26]. Let $M_i = (m_i, \ldots, m_i)$ in $\mathbb{Z}^p$. Since $Y_i$ converges roughly towards the mean of its values, we consider the evolution of $Z_i = Y_i - M_i$. We easily establish a relation of the form $Z_{i+1} = O Z_i + C_i$ with $C_i$ a bounded perturbation vector and $O$ a linear transformation. It follows that $Z_n = O^n Z_0 + \sum_{i=0}^{n-1} O^{n-1-i} C_i$.

Moreover, the characteristic polynomial of $O$ is $R(x)$ (proof omitted). One proves that $R(x)$ has $p-1$ distinct roots $\lambda_1, \ldots, \lambda_{p-1}$ (using coprimality of $R(x)$ and $R'(x)$) and for all $k$, $|\lambda_k| \leq \frac{p-1}{p} < 1$ (using a bound by Eneström and Kakeya, see for example [10]).

Consequently, $O$ is a contraction operator, and $\sum_{i=0}^{n-1} O^{n-1-i} C_i$ can be upper bounded by a constant independent of $n$ and the number of grains $N$. We can therefore conclude that there exists an $n_0$ in $\mathcal{O}(\log N)$ such that $Z_{n_0} = O^{n_0} Z_0 + \sum_{i=0}^{n_0-1} O^{n_0-1-i} C_i$ is upper bounded.

Secondly, on a small scale, the system converges linearly.

**Lemma 2.** *The value of $\overline{m}_i - \underline{m}_i$ decreases linearly: if $\underline{m}_i \neq \overline{m}_i$, then there is an integer $c$, with $0 \leq c \leq p$ such that $\overline{m}_{i+c} - \underline{m}_{i+c} < \overline{m}_i - \underline{m}_i$.*

*Proof.* If $\underline{m}_i \neq \overline{m}_i$, that is, if the vector $Y_i$ is not uniform, the mean value is strictly between the greatest and smallest values: $\underline{m}_i < m_i < \overline{m}_i$. Consequently $\underline{m}_i < y_{i+p} = m_i + \frac{b_i}{p} \leq \overline{m}_i$ (since the perturbation added is at most one and the resulting number is an integer, we cannot reach a greater integer). Therefore, we get $\underline{m}_i \leq \underline{m}_{i+1} \leq \overline{m}_{i+1} \leq \overline{m}_i$.

This reasoning applies while $\underline{m}_{i+j} \neq \overline{m}_{i+j}$, from which we get $\underline{m}_{i+j} < y_{i+p+j} \leq \overline{m}_{i+j}$ and $\underline{m}_{i+j} \leq \underline{m}_{i+j+1} \leq \overline{m}_{i+j+1} \leq \overline{m}_{i+j}$.

If there exists $c \leq p$ such that $\underline{m}_{i+c} = \overline{m}_{i+c}$, then, we are done. Otherwise, for $0 \leq j < p$, we have, $\underline{m}_i \leq \underline{m}_{i+j} < y_{i+j+p} \leq \overline{m}_{i+j} \leq \overline{m}_i$), thus $\underline{m}_i < \underline{m}_{i+p} \leq \overline{m}_{i+p} \leq \overline{m}_i$.

To conclude, we start with $\overline{m}_0 - \underline{m}_0$ in $\mathcal{O}(N)$, we have a constant $\alpha$ and a $n_0$ in $\mathcal{O}(\log N)$ such that $\overline{m}_{n_0} - \underline{m}_{n_0} < \alpha$ thanks to the exponential decrease on a large scale (Lemma 1). Then after $p$ iterations the value of $\overline{m}_{n_0+p} - \underline{m}_{n_0+p}$ is decreased by at least 1 (Lemma 2), hence there exists $\beta$ with $\beta \leq p\alpha$ such that after $\beta$ extra iterations we have $\overline{m}_{n_0+\beta} - \underline{m}_{n_0+\beta} = 0$. Thus $Y_{n_0+\beta}$ is a uniform vector, and $n_0 + \beta$ is in $\mathcal{O}(\log N)$.

In this proof, neither the discrete nor the continuous studies is conclusive by itself. On one hand, the discrete study gives a linear convergence but not an exponential convergence. On the other hand, the continuous study gives an exponential convergence towards a uniform vector, but in itself the continuous part never reaches a uniform vector but tends asymptotically towards it. It is the simultaneous study of those modalities (discrete and continuous) that allows to reach the conclusion.

*Remark 3.* Note that for $p = 1$, the averaging system has a trivial dynamics. For $p = 2$, the behavior is a bit more complex, but major simplifications are found: the computed value is equal to the mean of two values, hence in this case the difference $\overline{m}_i - \underline{m}_i$ decreases by a factor of two *at each time step*.

## 2.4 Emergence of a Loosely Wavy Shape

We call *wave* the pattern $p \cdot \ldots \cdot 2 \cdot 1$ in the sequence of slopes. Lemma 1 shows that there exists an $n = \mathcal{O}(\log N)$ such that $Y_n$ is a uniform vector. In this subsection, we prove that if $Y_n$ is a uniform vector, then the shape of the sandpile configuration is exclusively composed of waves and 0s from the index $n$.

**Lemma 3.** $Y_n$ is a uniform vector of $\mathbb{Z}^p$ implies

$$\pi(N)_{[n,\infty[} \in \Big(0 + (p \cdot p{-}1 \cdot \ldots \cdot 1)\Big)^* 0^\omega$$

*Proof (Proof sketch).* Complete proof in [26]. We straightforwardly apply Remark 2. If $Y_n$ is a uniform vector, we notice that the value of $b_n$ is 0 or $p$. If it is 0 then $Y_{n+1}$ is still a uniform vector; if it is $p$, then the sequence $(b_i)_{n \leq i < n+p}$ is determined to be equal to $p \cdot p - 1 \cdot \ldots \cdot 1$, and $Y_{n+p}$ is a uniform vector. The following diagram illustrates those observations: the grey node represents a uniform vector, and arrows are labelled by values of the sequence $b_i$. If we start on the grey node, any path's labels verify the statement of the Lemma.

*Remark 4.* Composing Proposition 1 and Lemma 3 allows us to prove the emergence, from a logarithmic column, of the shape given in Lemma 3.

## 2.5 Avalanches to Complete the Proof

In order to prove Theorem 1, we refine Remark 4 to show that there is at most one set of two non empty and consecutive columns of equal height, called *plateau* of size two, and corresponding to a slope equal to 0. It seems necessary to overcome the "static" study —for a given fixed point— presented above, and consider the dynamic of sand grains from $\pi(0)$ to $\pi(N)$. The fixed point $\pi(N)$ can be computed inductively, using the relation

$$\text{for all } k > 0, \ \pi(\pi(k-1)^{\downarrow 0}) = \pi(k)$$

where $\sigma^{\downarrow 0}$ denotes the configuration obtained by adding one grain on column 0 of $\sigma$ (see [26] for details). We start from $\pi(0)$ and inductively compute $\pi(1), \pi(2), \ldots, \pi(N-1)$ and $\pi(N)$ by repeating the addition of one grain on column 0 and reaching a stable configuration. The sequence of firings from $\pi(k-1)^{\downarrow 0}$ to $\pi(k)$ is called the $k^{th}$ *avalanche*.

We studied the structure of avalanches in [23–25], and first proved that each column is fired at most once in an avalanche $s^k$. Secondly, we showed that as soon as $p$ consecutive columns are fired, then the avalanche fires a set of consecutive columns —without any *hole* $i$ such that $i \notin s^k$ and $(i+1) \in s^k$—, and saw that this property leads to important regularities in successive fixed points. The following proof uses those observations: the structure of an avalanche on a wave pattern is very constrained, and as soon as an avalanche goes beyond a wave, it necessarily fires every columns of that wave, thus it fires a set of $p$ consecutive columns as mentioned above. We detail in the following proof why this property of avalanches on wave patterns ensures that if there is at most one plateau of size two (one slope equal to 0) in-between wave patterns of a fixed point $\pi(N-1)$, then there remains at most one plateau of size two on the wave patterns of the fixed point $\pi(N)$.

*Proof (Proof of Theorem 1).* We prove the result by induction on $N$. From Proposition 1 and Lemma 3, there is an index $n$ (resp. $n'$) in $\mathcal{O}(\log N)$ from which $\pi(N)$ (resp. $\pi(N-1)$) is described by the expression given in Lemma 3. Moreover, there is an index $l$ in $\mathcal{O}(\log N)$ such that the $N^{th}$ avalanche fires a set of consecutive columns on the right of $l$ (we omit the proof of this observation, see [26] for details). Without loss of generality, we consider that $l \geq n, n'$, and will prove that if $\pi(N-1)_{[l,\infty[}$ has at most one value 0, then so has $\pi(N)_{[l,\infty[}$.

If the avalanche ends before column $l - p$ (if $\max s^N < l - p$), the result holds. In the other case, we simply notice by contradiction that as soon as the avalanche reaches the wave patterns, it necessarily ends on the first value 0 it encounters, otherwise the resulting configuration is not stable (we recall that from $l$ a set of consecutive columns is fired exactly once). The consequence is that the 0 "climbs" one wave to the left, preserving the invariant of having at most one value 0 in-between wave patterns.

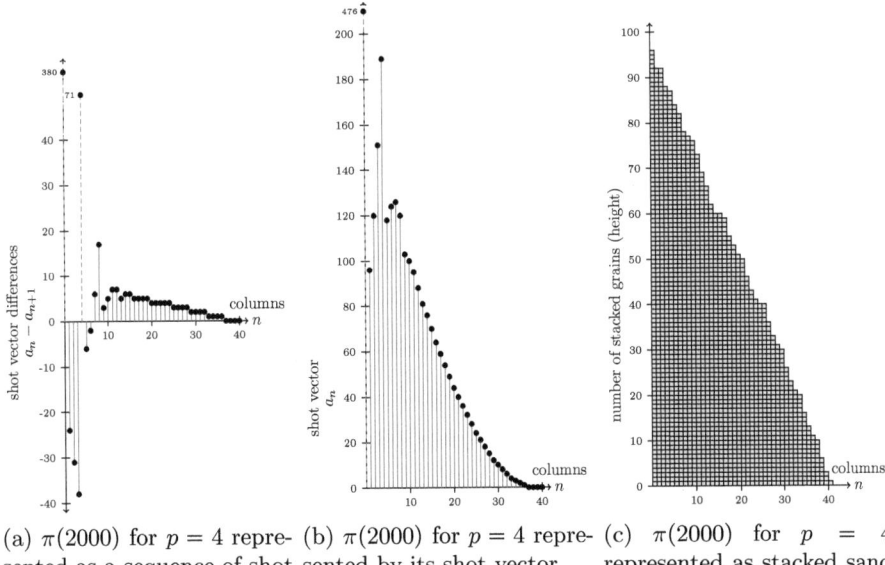

**Fig. 3.** Representations of $\pi(2000)$ for $p = 4$ used in the developments of the paper: differences of the shot vector (Figure 3a), shot vector (Figure 3b) and height (Figure 3c). $\pi(2000) = (4,0,4,1,3,2,4,1,1,3,4,3,4,2,0,1,4,2,2,1,4,3,2,1,0,4,3,2,1,4,3,2,1,4,3,2,1,4,3,2, 1,0^\omega)$ We can notice on Figure 3a that the shot vector differences contract towards some "*steps*" of length $p$, which corresponds to the statement of Lemma 1 that the vector $Y_n$ becomes uniform exponentially quickly (note that this graphic plots the opposite of the values of the components of $Y_n$). The shot vector representation on Figure 3b corresponds to the values of the components of $X_n$, which we did not manage to tackle with classical methods. Figure 3c pictures the sandpile configuration on which the wavy shape appears starting from column 20.

## 3  Concluding Discussion

The proof technic we set up in this paper allowed us to prove the emergence of regular patterns periodically repeated on fixed points, without requiring a precise understanding of the initial segment's dynamic. Arguments of linear algebra allowed to prove a rough convergence of the system (when the dynamic is not precisely known but coarsely bounded), completed with arguments of combinatorics, using the discreteness of the model, to prove the emergence of precise and regular wave patterns.

This result stresses the fact that sandpile models are on the edge between discrete and continuous systems. Indeed, when there are very few sand grains, each one seems to contribute greatly to the global shape of the configuration. However, when the number of grains is very large, a particular sand grain seems to have no importance to describe the shape of a configuration. The result also suggests a separation of the discrete and continuous parts of the system. On one hand, the seemingly unordered initial segment, interpreted as reflecting the discrete behavior, prevents regularities from emerging. On the other hand, the asymptotic and ordered part, interpreted as reflecting the continuous behavior, lets a regular and smooth pattern come into view.

Nevertheless, the separation between discrete and continuous behaviors may be challenged because the continuous part emerges from the discrete part. We have two remarks about this latter fact. Firstly, the consequence seems to be a slight bias appearing on the continuous part: it is not fully homogeneous —that is, with exactly the same slope at each index— which would have been expected for a continuous system, but a —very small— pattern is repeated. It looks like this bias comes from the gap between the unicity of the *border* column on the left side at index $-1$ compared to the rule which has a parameter $p$, because we still observe the appearance of wave patterns starting from variations of the initial configuration (for example starting from $p$ consecutive columns of height $N$, thus $pN$ grains). Secondly, if we consider the asymptotic form of a fixed point, the relative size of the discrete part is null. This, regarding the intuition described above that when the number of grains is very large then a particular grain has no importance, is satisfying.

Finally, the emergence of regularities in this system hints at a clear qualitative distinction between some sand grains and a heap of sand. Let us save the last words to a distracting application to the famous *sorites paradox*. Someone who has a very little amount of money is called *poor*. Someone *poor* who receives one cent remains *poor*. Nonetheless, if the increase by 1 cent is repeated a great number of times then the person becomes *rich*. The question is: when exactly does the person becomes *rich*? An answer may be that *richness* appears when money creates waves...

# References

1. Bak, P., Tang, K., Wiesenfeld, C.: Self-organized criticality. Phys. Rev. A 38(1), 364–374 (1988)
2. Creutz, M.: Cellular automata and self organized criticality. In: Some New Directions in Science on Computers (1996)
3. Dartois, A., Magnien, C.: Results and conjectures on the sandpile identity on a lattice. In: Discrete Model for C.S. Discrete Math. and T.C.S., pp. 89–102 (2003)
4. Durand-Lose, J.O.: Parallel transient time of one-dimensional sand pile. Theor. Comput. Sci. 205(1-2), 183–193 (1998)
5. Formenti, E., Masson, B., Pisokas, T.: Advances in symmetric sandpiles. Fundam. Inform. 76(1-2), 91–112 (2007)
6. Formenti, E., Van Pham, T., Phan, T.H.D., Tran, T.T.H.: Fixed point forms of the parallel symmetric sandpile model. CoRR, abs/1109.0825 (2011)
7. Gajardo, A., Goles, E.: Crossing information in two-dimensional sandpiles. Theor. Comput. Sci. 369(1-3), 463–469 (2006)
8. Gajardo, A., Moreira, A., Goles, E.: Complexity of Langton's ant. Discrete Applied Mathematics 117(1-3), 41–50 (2002)
9. Gale, D., Propp, J., Sutherland, S., Troubetzkoy, S.: Further travels with my ant. Mathematical Entertainments Column, Mathematical Intelligencer 17, 48–56 (1995)

10. Gardner, R.B., Govil, N.K.: Some generalizations of the Eneström-Kakeya theorem. Acta Math. Hungar. 74(1-2), 125–134 (1997)
11. Goles, E., Kiwi, M.A.: Games on line graphs and sand piles. Theor. Comput. Sci. 115, 321–349 (1993)
12. Goles, E., Latapy, M., Magnien, C., Morvan, M., Phan, T.H.D.: Sandpile models and lattices: a comprehensive survey. Theor. Comput. Sci. 322(2), 383–407 (2004)
13. Goles, E., Martin, B.: Computational Complexity of Avalanches in the Kadanoff Two-dimensional Sandpile Model. In: Proc. of JAC 2010, pp. 121–132 (December 2010)
14. Goles, E., Morvan, M., Phan, T.H.D.: The structure of a linear chip firing game and related models. Theor. Comput. Sci. 270(1-2), 827–841 (2002)
15. Grauwin, S., Bertin, É., Lemoy, R., Jensen, P.: Competition between collective and individual dynamics. Nat. Ac. of Sciences USA 106(49), 20622–20626 (2009)
16. Hopcroft, J.E., Motwani, R., Ullman, J.D.: Introduction to automata theory, languages, and computation - international edition, 2nd edn. Addison-Wesley (2003)
17. Kadanoff, L.P., Nagel, S.R., Wu, L., Zhou, S.: Scaling and universality in avalanches. Phys. Rev. A 39(12), 6524–6537 (1989)
18. Katok, A., Hasselblatt, B.: Introduction to the Modern Theory of Dynamical Systems. Encycl. of Math. and App. Cambridge University Press (1996)
19. Levine, L., Peres, Y.: Spherical asymptotics for the rotor-router model in z d. Indiana Univ. Math. J., 431–450 (2008)
20. Miltersen, P.B.: The computational complexity of one-dimensional sandpiles. Theory Comput. Syst. 41(1), 119–125 (2007)
21. Moore, C., Nilsson, M.: The computational complexity of sandpiles. Journal of Statistical Physics 96, 205–224 (1999), doi:10.1023/A:1004524500416
22. Perrot, K., Phan, T.H.D., Van Pham, T.: On the set of Fixed Points of the Parallel Symmetric Sand Pile Model. In: AUTOMATA 2011 (November 2011)
23. Perrot, K., Rémila, É.: Avalanche Structure in the Kadanoff Sand Pile Model. In: Dediu, A.-H., Inenaga, S., Martín-Vide, C. (eds.) LATA 2011. Perrot, K., Rémila, É, vol. 6638, pp. 427–439. Springer, Heidelberg (2011)
24. Perrot, K., Rémila, É.: Transduction on Kadanoff Sand Pile Model Avalanches, Application to Wave Pattern Emergence. In: Murlak, F., Sankowski, P. (eds.) MFCS 2011. LNCS, vol. 6907, pp. 508–519. Springer, Heidelberg (2011)
25. Perrot, K., Rémila, É.: Kadanoff sand pile model. Avalanche structure and wave shape. Theor. Comput. Sci. 504, 52–72 (2013)
26. Perrot, K., Rémila, É.: Kadanoff Sand Piles, following the snowball (January 2013), Research Report available on arXiv at http://arxiv.org/abs/1301.0997
27. Phan, T.H.D.: Two sided sand piles model and unimodal sequences. ITA 42(3), 631–646 (2008)
28. Weaver, W.: Science and Complexity. American Scientist 36(536) (1948)

# A Divide and Conquer Method to Compute Binomial Ideals

Deepanjan Kesh[1,*] and Shashank K. Mehta[2]

[1] Indian Institute of Technology Guwahati
Guwahati, India
deepkesh@iitg.ernet.in
[2] Indian Institute of Technology Kanpur
Kanpur, India
skmehta@iitk.ac.in

**Abstract.** A binomial is a polynomial with at most two terms. In this paper, we give a *divide-and-conquer* strategy to compute binomial ideals. This work is a generalization of the work done by the authors in [12,13] and is motivated by the fact that any algorithm to compute binomial ideals spends a significant amount of time computing Gröbner basis and that Gröbner basis computation is very sensitive to the number of variables in the ring. The divide and conquer strategy breaks the problem into subproblems in rings of lesser number of variables than the original ring. We apply the framework on five problems – radical, saturation, cellular decomposition, minimal primes of binomial ideals, and computing a generating set of a toric ideal.

## 1 Introduction

Consider the polynomial ring $k[x_1, \ldots, x_n]$. A **binomial** in such a ring is a polynomial of the form $c \cdot \mathbf{x}^\alpha + d \cdot \mathbf{x}^\beta$, where $c, d \in k$ and $\alpha, \beta \in \mathbb{N}^n$. An ideal in the polynomial ring which has a generating set comprising only of binomials is called a **binomial ideal**. In this paper, we will be concerned with computing various binomial ideals.

Binomial ideals, unlike general polynomial ideals, possess rich combinatorial structure which can be exploited while computing various structures derived from them, for example Gröbner bases, primary decomposition, and associated primes [17,10]. Pure difference binomials are binomials of the form $\mathbf{x}^\alpha - \mathbf{x}^\beta$. The varieties of pure difference prime binomial ideals are exactly the toric varieties. Hence, such ideals are also known as toric ideals [7,6]. There is a large literature studying applications and computations of toric ideals [14,1]. Moreover, quotients of polynomial rings by pure difference binomial ideals form commutative semigroup rings [9].

Apart from a purely academic interest in the subject of binomial ideals, their study is also motivated by the fact that they are often encountered in interesting

---
* Support from IMPECS is acknowledged.

problems in diverse fields. These include solving integer programs [11,3,18,16], computing primitive partition identities [14, Chapters 6,7], and solving scheduling problems [15]. In algebraic statistics, closures of discrete exponential families have been identified with nonnegative toric varieties [8].

The theory of binomial ideals was developed in a seminal paper by Eisenbud and Sturmfels [6]. Their paper not only showed various properties of binomial ideals – for example, the radicals and associated primes of binomial ideals are themselves binomial ideals – but they also show how to compute these structures.

In [12], we had dealt with the computation of toric ideals. In [13], we had extended our approach to compute the saturation of binomial ideals. In this paper, we present a general framework to compute several of such binomial ideals, namely radical, saturation, minimal primes and cellular decompositions. This work is motivated by two crucial observations –

1. Most of these computations involve computing a Gröbner basis of certain ideals, and
2. Buchberger's algorithm [2] to compute Gröbner basis is very sensitive to the number of variables in the underlying polynomial ring.

In the light of these observations, we propose a *divide-and-conquer* technique to solve the aforementioned problems, with the hope that this strategy can also be applied to a host of other problems related to binomial ideals, like computing associated primes, primary decomposition, primary component, and so on. The essence of the strategy is the following. Consider the polynomial ring $k[x_1, \ldots, x_n]$, and a binomial ideal $I \subseteq k[x_1, \ldots, x_n]$. We compute the image of $I$ under the natural homomorphism in the derived rings $k[x_2, \ldots, x_n]$ and $k[x_1^{\pm}, x_2, \ldots, x_n]$ and perform the same computation on these ideals(Intuitively, $\mathbf{x}_1^{\pm}$ implies that we allow both positive and negative integers as exponents for $x_1$). Then we "lift" the results in the original ring and combine them to compute a solution of the original problem. Both these rings are isomorphic to polynomial rings with one less variable [13], hence Gröbner basis (actually such basis does not exist in some of these new rings but we use a variant for the computations) can be computed more efficiently.

The paper has been arranged as follows. In the next section, we briefly present some background required for the paper and define some notations. Section 3 defines two maps from ideals of a Laurent ring to certain derived rings, and state some useful properties of these maps. These two maps form the basis of the reduction of the problem into the subproblems, discussed earlier. Section 4 contains the main contribution of the paper – discussion of the proposed divide-and-conquer framework. In Section 5, we use this framework to compute radical, saturation, cellular decomposition, minimal primes of binomial ideals, and a generating set of a toric ideal.

## 2 Background

A detailed treatment of the background required for the paper, like the notions of localization, Laurent polynomial rings, or of various kinds of ideals like radical, prime, saturation, etc., was not included here due to constraint of page limit, but the reader can refer to [4,5].

We will just state a few notations used in the paper. For a ring $R$, if $r_1, \ldots, r_s \in R$, then $\langle r_1, \ldots, r_n \rangle$ will denote the ideal generated by $r_1, \ldots, r_n$. For an ideal $I \subseteq R$, $\sqrt{I} = \{\, r \mid r^m \in I,\ m \geq 0 \,\}$ is the radical of $I$. $I : r^\infty = \{\, s \mid sr^j \in I,\ \text{for some } j \geq 0 \,\}$ will denote the saturation of $I$ w.r.t. $r$.

For a field $k$, we will use the standard notation of $k[x_1, \ldots, x_n]$ to denote the polynomial ring in $n$ variables. If $U \subseteq R$ is a multiplicatively closed set of $R$, then $R[U^{-1}]$ is the localization of $R$ w.r.t. $U$. If the ring $k[x_1, \ldots, x_n]$ is localized w.r.t. $x_1, \ldots, x_i$, then the partial Laurent polynomial ring $k[x_1^\pm, \ldots, x_i^\pm, x_{i+1}, \ldots, x_n]$ will be denoted by the tuple $(k, X, L)$, where $k$ is the field, $X$ is the set of variables, and $L$ is the set of variables w.r.t. which $k[X]$ has been localized.

## 3 Two Ring Homomorphisms

### 3.1 Modulo Map

Let $r$ be an element of a Noetherian ring $R$. Then $\theta : R \to R/\langle r \rangle$ denotes the natural homomorphism $\theta(a) = [a] = a + \langle r \rangle$, $\forall a \in R$. Here, $[a]$ or $a + \langle r \rangle$ denotes the coset of $a$ in $R/\langle r \rangle$. This induces a map $\Theta$ from the ideals in $R$ containing $r$ and the ideals of $R/\langle r \rangle$ as follows –

$$\Theta(I) = \{\, [a] \mid a \in I \,\},$$

where $I \subseteq R$ is an ideal containing $r$. Similarly, we define a map $\Theta^{-1}$ from the ideals of $R/\langle r \rangle$ to the ideals of $R$ containing $r$ as follows –

$$\Theta^{-1}(J) = \{\, x \mid [x] \in J \,\},$$

where $J \subseteq R/\langle r \rangle$ is an ideal. We state, without proof, some basic properties of $\Theta$.

**Lemma 1.** *The maps $\Theta$ and $\Theta^{-1}$ have the following properties –*

*(i) $\Theta$ and $\Theta^{-1}$ preserve set inclusion.*
*(ii) $\Theta$ is a bijection.*
*(iii) $\Theta$ and $\Theta^{-1}$ map primes to primes.*
*(iv) $\Theta$ and $\Theta^{-1}$ distribute over finite intersections.*
*(v) In a Noetherian ring, $\Theta(\sqrt{I}) = \sqrt{\Theta(I)}$*
*(vi) $\Theta^{-1}(\langle [f_1], \ldots, [f_n] \rangle) = \langle f_1, \ldots, f_n \rangle + \langle r \rangle$*

### 3.2 Localization Map

Let $r$ be a nonzero-divisor of a Noetherian ring $R$. Let $U$ denote the set of all powers of $r$, $U = \{\, r^i \mid i \geq 0 \,\}$. Since $r$ is not nilpotent, $U$ does not contain zero. $U$ is also multiplicatively closed. Therefore $R[U^{-1}]$ is well defined.

Let $\phi : R \to R[U^{-1}]$ be the natural homomorphism given by $\phi(a) = a/1$, $\forall a \in R$. We define a map, $\Phi$, induced by $\phi$, from the ideals in $R$ saturated w.r.t. $r$ to the ideals of $R[U^{-1}]$ as follows –

$$\Phi(I) = \langle\, \{\, a/1 \mid a \in I\, \}\, \rangle,$$

where $I \subseteq R$ is an ideal saturated w.r.t. $r$. Similarly, we will define a map, $\Phi^{-1}$, from the ideals in $R[U^{-1}]$ to the ideals in $R$ which are saturated with respect to $r$ as follows –

$$\Phi^{-1}(J) = \{\, a \mid a/r^k \in J,\ k \geq 0\, \}.$$

We present some straight forward properties of $\Phi$ and $\Phi^{-1}$ without proof.

**Lemma 2.** *The maps $\Phi$ and $\Phi^{-1}$ have the following properties –*

*(i) $\Phi$ and $\Phi^{-1}$ preserve set inclusion.*
*(ii) $\Phi$ is a bijection.*
*(iii) $\Phi$ and $\Phi^{-1}$ map primes to primes.*
*(iv) $\Phi$ and $\Phi^{-1}$ distribute over finite intersections.*
*(v) For $x \in R, \Phi(I : x^\infty) = \Phi(I) : x^\infty$.*
*(vi) In a Noetherian ring $\Phi(\sqrt{I}) = \sqrt{\Phi(I)}$*
*(vii) $\Phi^{-1}(\langle\, f_1/r^{a_1}, \ldots, f_n/r^{a_n}\, \rangle) = \langle\, f_1, \ldots, f_n\, \rangle : r^\infty$.*

## 4  A Divide-and-Conquer Method

In this section, we focus on the main objective of this paper. We present a general algorithm (Algorithm 1) based on *divide-and-conquer* technique which is useful in computing several binomial ideals associated with a given binomial ideal. The algorithm takes as input the following 3 objects (i) A ring $(k, X, L)$, (ii) A set of binomials, $S$, generating an ideal $I$, and (iii) A set of variables $V \subseteq X \setminus L$ called *forbidden* set. The objective of the algorithm is to compute $\mathbb{A}(\langle\, S\, \rangle)$, where $\mathbb{A}$ is some object associated with the binomial ideal $I$. In Section 5 we demonstrate how Algorithm 1 computes (i) Radical of $I$, (ii) Saturation of $I$ w.r.t. all the variables in the ring, (iii) Generating basis of a toric ideal from $I$, (iv) Minimal Primes of $I$, and (v) Cellular decomposition of $I$.

We restate, from the introduction, the two crucial observations behind this algorithm –

1. Most computations involving binomial ideals compute Gröbner basis of certain ideals, and
2. Buchberger's algorithm [2] to compute Gröbner basis is very sensitive to the number of variables in the underlying polynomial ring.

The motivation behind the algorithm is to divide the problem suitably into smaller subproblems, solve these subproblems in rings with less variables than the original ring, and combine these results to solve the original problem.

Let $x \in (X \setminus L) \setminus V$, and consider the maps (i) $\Theta : (k, X, L) \to (k, X \setminus \{x\}, L)$, (ii) $\Phi : (k, X, L) \to (k, X, L \cup \{x\})$, and (iii) $f : (k, X, L) \to (k, X, L)$

**Algorithm 1.** A framework for computing binomials ideals – A

**Data**: A ring $(k, X, L)$, where $k$ is algebraically closed, and $\mathsf{char}(k) = 0$; forbidden set $V \subseteq X \setminus L$; a binomial generating set $S$ of an ideal in the ring.
**Result**: $\mathsf{A}(\langle\, S\, \rangle)$

1 **if** $X = \phi$ **then**                                // The ring is a field
2   | Nothing to do ;
3 **else if** $X = L$ **then**                          // Laurent polynomial ring
4   | Compute $\mathsf{A}(\langle\, S\, \rangle)$ and **return** ;
5 **else if** $V = X \setminus L$ **then**           // No more reductions
6   | Compute $\mathsf{A}(\langle\, S\, \rangle)$ and **return** ;
7 **end**
8 Let $x \in (X \setminus L) \setminus V$ ;
  /* computing $\mathsf{A}(\Theta(\langle\, S\, \rangle + \langle\, x\, \rangle))$ and lift                             */
9 Call A with ideal $\Theta(\langle\, S\, \rangle + \langle\, x\, \rangle)$, ring $(k, X \setminus \{x\}, L)$ and forbidden set $V$ ;
10 Compute $\Theta^{-1}(\mathsf{A}(\Theta(\langle\, S\, \rangle + \langle\, x\, \rangle)))$ ;
  /* computing $\mathsf{A}(\Phi(\langle\, S\, \rangle : x^\infty))$ and lift                                     */
11 Call A with ideal $\Phi(\langle\, S\, \rangle : x^\infty)$, ring $(k, X, L \cup \{x\})$ and forbidden set $V$ ;
12 Compute $\Phi^{-1}(\mathsf{A}(\Phi(\langle\, S\, \rangle : x^\infty)))$ ;
  /* computing $\mathsf{A}(f(\langle\, S\, \rangle : x^\infty))$                                                    */
13 Call A with ideal $f(\langle\, S\, \rangle)$, ring $(k, X, L)$ and forbidden set $V \cup \{x\}$ ;
  /* Computing $\mathsf{A}(\langle\, S\, \rangle)$                                                                         */
14 Combine $\Theta^{-1}(\mathsf{A}(\Theta(\langle\, S\, \rangle + \langle\, x\, \rangle)))$, $\Phi^{-1}(\mathsf{A}(\Phi(\langle\, S\, \rangle : x^\infty)))$ and $\mathsf{A}(f(\langle\, S\, \rangle))$ to get $\mathsf{A}(\langle\, S\, \rangle)$ ;
  /* Return                                                                                                                        */
15 **return** $\mathsf{A}(\langle\, S\, \rangle)$ ;

which depends on the problem $\mathsf{A}()$. The reduction step involves solution of the subproblems (i) $\mathsf{A}(\Theta(I + \langle\, x\, \rangle))$, in ring $(k, X \setminus \{x\}, L)$ and forbidden set $V$(step 9), (ii) $\mathsf{A}(\Phi(I : x^\infty))$, in ring $(k, X, L \cup \{x\})$ and forbidden set $V$(step 11), (iii) $\mathsf{A}(f(I))$ in ring $(k, X, L)$ and forbidden set $V \cup \{x\}$(step 13). The first subproblem is in a ring with one less variable compared to the original ring. In the case of the second subproblem, Gröbner bases are not defined in the context of partial Laurent polynomial rings $(k, X, L)$. But pseudo-Gröbner bases [13], briefly discussed later in this section, can effectively substitute Gröbner bases for binomial ideal computations. The time complexity of the algorithm to compute pseudo Gröbner basis was shown in that paper to be dependent on the number of variables in $X \setminus L$. Hence, this subproblem is also justifiably "smaller".

The role of the forbidden set of variables is that reduction must not be done with respect to these variables. Thus, if $V = X \setminus L$, then the computation $A(I)$ must be easy to perform without further reduction. In addition, the third subproblem should be such that it does not require the computation of a Gröbner basis since in this case the ring is same as in the original problem and involves no reduction in ring size. Here is a motivating example to justify the use of forbidden set. Suppose we want to compute the saturation, $I : (x_1 \cdots x_n)^\infty$, while $I$ is already saturated w.r.t. $x_1, x_2$. Then reduction with these variables is futile. Hence we can put these variables in the forbidden set.

Next, the algorithm computes the inverse images of $\mathsf{A}(\Theta(I + \langle\, x\, \rangle))$ (step 10) and $\mathsf{A}(\Phi(I : x^\infty))$ (step 12) in the original ring $(k, X, L)$. In the applications discussed in the next section, $\mathsf{A}(I)$ is either an ideal (as in the case of radical of $I$) or a set of ideals (as in the case of minimal primes of $I$). Hence these images are well defined. Abusing notation, we denote these inverse images respectively by $\Theta^{-1}(\mathsf{A}(\Theta(I + \langle\, x\, \rangle)))$ and $\Phi^{-1}(\mathsf{A}(\Phi(I : x^\infty)))$.

Finally in step 14, $\mathsf{A}(I)$ is to be constructed from these images and $\mathsf{A}(f(I))$. One can easily observe that the algorithm terminates, as in each step either cardinality of $X$ decreases, or that of $L$ or $V$ increases. This algorithm is a general method and can be tuned to a particular problem by specifying the following three steps in the context of that problem –

**(steps 4, 6)** $V = X \setminus L$: Give the method to compute $A(I)$ in these base cases.

**(step 13)** : Specify function $f$.

**(step 14)** : Show how to combine the results of the subproblems.

In the next few subsections we show how to compute $\Theta$, $\Phi$, and their inverses using a generating set of the input ideal.

### 4.1 Computing Modulo

Let $L = \{y_1, \ldots, y_k\}$ and $X = \{x_1, \ldots, x_l\} \cup \{z\} \cup L$. Maps $\theta$ and $\Theta$ from $(k, X, L) \to (k, X \setminus \{z\}, L)$ are computed as follows. Consider an arbitrary polynomial in $(k, X, L)$, $f = \sum_i \mathbf{x}^{\alpha_i} \mathbf{y}^{\beta_i} + \sum_j \mathbf{x}^{\alpha_j} \mathbf{y}^{\beta_j} z^{c_j}$. Then, $\theta(f) = \sum_i \mathbf{x}^{\alpha_i} \mathbf{y}^{\beta_i}$. Further, suppose $S \subset (k, X, L)$ is a set of binomials. Then, $\Theta(\langle\, S\, \rangle) = \langle\, \theta(f)\, |\, f \in S\, \rangle$. Conversely, if $S' \subset (k, X \setminus \{z\}, L)$, then $\Theta^{-1}(\langle\, S'\, \rangle) = \langle\, S' \cup \{z\}\, \rangle$, from Lemma 1.

### 4.2 Computing Localization

Consider the ring $(k, X, L)$ as defined in the previous subsection. If $g \in (k, X, L)$, then $\phi(g) = g/1$.

Computing $\Phi$ and $\Phi^{-1}$ is also easy. For any $S \subset (k, X, L)$, $\Phi(\langle\, S\, \rangle) = \langle\, \{\, g/1\, |\, g \in S\, \}\, \rangle$. In the reverse direction, for any $S' \subset (k, X, L \cup \{z\})$, we define $\Phi^{-1}(\langle\, S'\, \rangle)$ as follows. Let $S' = \{g_1/z^{a_1}, \ldots, g_k/z^{a_k}\}$. Then $\Phi^{-1}(\langle\, S'\, \rangle) = \langle\, g_1, \ldots, g_k\, \rangle : z^\infty$. The correctness follows from Lemma 2.

To see how we can compute saturation with respect to $z$ in a partial Laurent polynomial ring, we briefly revisit the results on *pseudo-Gröbner basis* in [13].

### 4.3 Pseudo-Gröbner Basis

Gröbner bases are defined for ideals in rings $k[x_1, \ldots, x_n]$ ([4, Chapter 2]). This notion has been generalized for binomial ideals in partial Laurent polynomial rings, called pseudo-Gröbner bases in [13, Section 5]. Here we reproduce some relevant results.

**Definition 1.** *A binomial $a\mathbf{x}^\alpha + b\mathbf{x}^\beta \in (k, X, L)$ is said to be **balanced** if $x_i \in X \setminus L$ implies $\alpha_i = \beta_i$.*

**Definition 2.** *For every finite binomial set $G$, $G_1$ and $G_2$ will denote its partition, where the former will represent the set of non-balanced binomials and the latter will represent the set of balanced binomials of $G$.*

**Definition 3.** *A binomial basis $G = (G_1, G_2)$ of a binomial ideal $I$ will be called a pseudo Gröbner basis with respect to a given term-order, if $G_1$ reduces every binomial of $I$ to $0(\text{mod}(G_2))$.*

**Theorem 1.** *[13, Theorem 3] Every binomial ideal in $(k, X, L)$ has a pseudo-Gröbner basis with respect to any term-order.*

The Buchberger's algorithm to compute Gröbner basis has been adopted to compute pseudo-Gröbner basis in [13, Algorithm 4]. Finally, the following theorem shows that saturation can be computed in similar way as in $k[x_1, \ldots, x_n]$.

**Theorem 2.** *[13, Theorem 3] Let $(G_1, G_2)$ be a pseudo Gröbner basis of a homogeneous binomial ideal in $(k, X, L)$ with respect to a graded reverse lexicographic term order with the variable $x_i \notin L$ being the least. Then $(G'_1 = G_1 \div x_i^\infty, G'_2 = G_2 \div x_i^\infty)$ is a pseudo Gröbner basis of $I : x_i^\infty$.*

Here $S \div x^\infty$ is the result of the division of each polynomial in $S$ by the largest possible power of $x$.

## 5 Computing $\mathbf{A}(I)$

As mentioned in the previous section, we will describe the steps 4, 6, 13 and 14 of the algorithm in context of five problems – (i) radical of a binomial ideal, (ii) the saturation of a binomial ideal with respect to all variables in the ring, (iii) computing toric ideal, (iv) the minimal prime ideals of a binomial ideal, and (v) cellular decomposition of a binomial ideal.

### 5.1 Radical Ideal: A = Radical

**Theorem 3.** *Let $R$ be a Noetherian ring, $r \in R$ a non-zero-divisor, and $I \subseteq R$ be an ideal. Then, $\sqrt{I + \langle\, r\, \rangle} \cap \sqrt{I : r^\infty} = \sqrt{I}$, for some $r \in R$.*

*Proof.* We know that every radical ideal in a Noetherian ring has a prime decomposition. Let the prime decomposition of $\sqrt{I}$ be $\sqrt{I} = P_1 \cap P_2 \cap \ldots \cap P_n$. Let the collection of the primes in the decomposition be denoted by $\mathcal{P}$. Define two ideals $\mathcal{P}_r = (\cap_{r \in P \in \mathcal{P}} P)$, and $\overline{\mathcal{P}_r} = (\cap_{r \notin P \in \mathcal{P}} P)$. It is easy to see that $I + \langle\, r\, \rangle \subseteq \mathcal{P}_r$. Hence, $\sqrt{I + \langle\, r\, \rangle} \subseteq \mathcal{P}_r$. Next, we want to show that $\sqrt{I : r^\infty} \subseteq \overline{\mathcal{P}_r}$.

Let $f \in I : r^\infty$. Then, $r^n f \in I$ for some $n \geq 0$. This implies that for all $P \in \mathcal{P}, r^n f \in P$. In particular, if $r \notin P$, then $f \in P$. We deduce that $I : r^\infty \subseteq \overline{\mathcal{P}_r}$, and hence $\sqrt{I : r^\infty} \subseteq \overline{\mathcal{P}_r}$. Putting the two observation together we have $\sqrt{I + \langle\, r\, \rangle} \cap \sqrt{I : r^\infty} \subseteq \mathcal{P}_r \cap \overline{\mathcal{P}_r} = \sqrt{I}$

The converse containment $\sqrt{I} \subseteq \sqrt{I + \langle\, r\, \rangle} \cap \sqrt{I : r^\infty}$ is obvious. □

The following theorem will help us in the formulation of step 14.

**Theorem 4.** *Let $R$ be a Noetherian ring, $r \in R$ a non-zero-divisor, and $I \subseteq R$ be an ideal. Then, $\sqrt{I} = \Theta^{-1}(\sqrt{\Theta(I + \langle\, r\, \rangle)}) \cap \Phi^{-1}(\sqrt{\Phi(I : r^\infty)})$.*

*Proof.* We will continue to use the notations defined in the previous theorem. From the proof of Theorem 3, we have $I + \langle r \rangle \subseteq \mathcal{P}_r$. From the containment preserving property and the commutation with intersection property of $\Theta$, we have $\Theta(I + \langle r \rangle) \subseteq \Theta(\cap_{r \in P \in \mathcal{P}} P) = \cap_{r \in P \in \mathcal{P}} \Theta(P)$. Similarly $\sqrt{\Theta(I + \langle r \rangle)} \subseteq \sqrt{\cap_{r \in P \in \mathcal{P}} \Theta(P)} = \cap_{r \in P \in \mathcal{P}} \sqrt{\Theta(P)}$. The last equality is due to the fact that intersection of radicals is equal to the radical of intersections.

As the $P$s are primes, from Lemma 1 we know that the $\Theta(P)$s are primes and since prime ideals are radical, we have $\sqrt{\Theta(I + \langle r \rangle)} \subseteq (\cap_{r \in P \in \mathcal{P}} \Theta(P))$. Hence $\Theta^{-1}(\sqrt{\Theta(I + \langle r \rangle)}) \subseteq \mathcal{P}_r$.

Similarly, starting from the following relation given in the proof of theorem 3 $I : r^\infty \subseteq \overline{\mathcal{P}_r}$ we can deduce that $\Phi^{-1}(\sqrt{\Phi(I : r^\infty)}) \subseteq \overline{\mathcal{P}_r}$. Combining the two results gives $\Theta^{-1}(\sqrt{\Theta(I + \langle r \rangle)}) \cap \Phi^{-1}(\sqrt{\Phi(I : r^\infty)}) \subseteq \sqrt{I}$.

To prove the converse, from Lemmas 1 and 2 we have $\sqrt{I} \subseteq \Theta^{-1}(\sqrt{\Theta(I + \langle\, r\, \rangle)} \cap \Phi^{-1}(\sqrt{\Phi(I : r^\infty)})$. □

We will not use the $\mathsf{A}(f(I))$ branch of the reduction for this problem. Thus, Theorem 3 shows that the *combine* step (step 14) is intersection. Also, we will have $V = \emptyset$. The base case computation in step 4 of the algorithm is trivial because all binomial ideals in a Laurent polynomial ring are already radical as shown below.

**Theorem 5 (Corollary 2.2, [6]).** *Let $J$ be a binomial ideal in the ring $(k, X, \phi)$. Then, if $k$ is algebraically closed and $\mathsf{char}(k) = 0$, then $J : (\Pi_{x \in X} x)^\infty$ is radical.*

**Corollary 1.** *Let $k$ be an algebraically closed field, with $\mathsf{char}(k) = 0$. Then, all binomial ideals in $(k, X, X)$ are radical.*

*Proof.* Let $J$ be a binomial ideal in the ring $(k, X, X)$, where $X = \{x_1, \ldots, x_n\}$. Consider the ideal localization map, $\Phi_n$, from $(k, X, X \setminus \{x_n\})$ to $(k, X, X)$. Under this map, we know that $\Phi_n^{-1}(J)$ is saturated w.r.t $x_n$. Similarly, if we consider the map $\Phi_{n-1}$ from $(k, X, X \setminus \{x_{n-1}, x_n\})$ to $(k, X, X \setminus \{x_n\})$, then the ideal $\Phi_{n-1}^{-1}(\Phi_n^{-1}(J))$ is saturated w.r.t. $x_{n-1}$. So we have $\Phi_n^{-1}(J) = \Phi_n^{-1}(J) : x_n^\infty$. Hence, $\Phi_{n-1}^{-1}(\Phi_n^{-1}(J)) = \Phi_{n-1}^{-1}(\Phi_n^{-1}(J) : x_n^\infty) = \Phi_{n-1}^{-1}(\Phi_n^{-1}(J)) : x_n^\infty$ (Lemma 2) Thus, $\Phi_{n-1}^{-1}(\Phi_n^{-1}(J))$ is saturated w.r.t. $\{x_{n-1}, x_n\}$. Continuing this argument we see that $\Phi_1^{-1}(\cdots (\Phi_n^{-1}(J)) \cdots)$, in the ring $(k, X, \phi)$, is saturated w.r.t. $\{x_1, \ldots, x_n\}$. From the previous theorem $\Phi_1^{-1}(\cdots (\Phi_n^{-1}(J)))$ is radical. Now, by repeated application of Lemma 2 we deduce that $J$ is radical too. □

**Analysis:** The proposed algorithm uses two out of the three branches of the *Divide-and-Conquer* strategy (Algorithm 1), so if $n$ is the number of variables in the input ideal, this algorithm requires $2^n$ Gröbner basis computations. Compare this with $n!$ computations in [6, Algorithm 9.1].

## 5.2 Saturation : A = Saturation

Suppose $I$ is saturated with respect to $\{x_{i_1}, \ldots, x_{i_j}\}$ then we begin the computation with $V = \{x_{i_1}, \ldots, x_{i_j}\}$. For this problem, we only use the $\mathsf{A}(I : x^\infty)$ branch of the reduction. The base case for this algorithm occurs when $X \setminus L = V$ (step 6). As $\Phi$ preserves saturation (Lemma 2), the ideal is already saturated in this case. Since the algorithm uses only one branch of the reduction, step 14 is redundant.

**Analysis:** In this proposed algorithm, the number of variables in the image space is 1 in the first iteration, 2 in the second iteration, and so on. Symbolically, if $\mathcal{G}(k)$ denotes the time complexity of Buchberger's algorithm in a $k$ variable ideal, then the cost of the proposed algorithm is $\sum_{k=1}^{n} \mathcal{G}(k)$, where $n$ is the number of variables in input ideal. On the other hand, the cost of the Sturmfels' algorithm [14, Lemma 12.1] is $n\mathcal{G}(n)$.

## 5.3 Toric Ideals: A = Toric

Pure difference prime binomial ideals are called toric ideals. So, they are a special class of general binomial ideals and, as pointed out in Section 1, perhaps the most useful of all binomial ideals from an application perspective. The goal in this case is also to saturate a given binomial ideal, but we are guaranteed that the saturated ideal will be a toric ideal. The solution of Section 5.2 applies to toric ideals as well and our proposed algorithm do not exploit the fact that the solution is known to be a toric ideal. But there are algorithms that do, namely the *project and lift* algorithm due to Hemmecke and Malkin [10], and it is much faster than the Sturmfels' Algorithm alluded to in the previous section.

**Analysis:** Using the notation $\mathcal{G}(k)$ introduced in the previous section, the cost of *project and lift* algorithm is $\sum_{i=k}^{n} \mathcal{G}(i) + k\mathcal{G}(k)$, where $k$ is dependant on the input. $n$, as in the previous cases, denote the number of variables in the input ideal. We note that the cost of the proposed algorithm is $\sum_{i=1}^{n} \mathcal{G}(i)$. Thus, the proposed algorithm matches *project and lift* in the worst case, and does better in all other cases.

## 5.4 Prime Decomposition: A = Prime

In this case, as in the computation of a radical, the $\mathsf{A}(f(I))$ branch will not be used. We will first handle the base case, i.e. how to compute the minimal primes of a binomial ideal in a Laurent polynomial ring (step 4). To do this, we will mention (without proof) a set of results from [6].

**Definition 4.** *A **partial character** on $\mathbb{Z}^n$ is a homomorphism $\rho$ from a sublattice $L_\rho$ of $\mathbb{Z}^n$ to the multiplicative group $k^*(= k \setminus \{0\})$. A partial character will always refer to the tuple $(\rho, L_\rho)$.*

For a binomial ideal $I$ in $(k, X, X)$, let $L(I) = \{\, \alpha \mid \mathbf{x}^\alpha - c \in I \,\}$. It is easy to verify that $L(I)$ is a lattice. We define a function $\rho$ as $\rho(\alpha) = c$, where $\mathbf{x}^\alpha - c \in I$. Thus, $(\rho, L(I))$ is a partial character. Conversely, given a partial character $(\rho, L)$, we will define a binomial ideal as $I(\rho) = \langle \mathbf{x}^\alpha - c | \alpha \in L, \rho(\alpha) = c \rangle$.

**Theorem 6.** *For any proper binomial ideal in $(k, X, X)$, there is a unique partial character $\rho$ on $\mathbb{Z}^n$ such that $I = I(\rho)$.*

**Definition 5.** *If $L$ is a sublattice of $\mathbb{Z}^n$, then the* **saturation** *of $L$ is the lattice* $\mathsf{Sat}(L) = \{ m \in \mathbb{Z}^n \mid dm \in L \text{ for some } d \in \mathbb{Z} \}$.

We can compute $\mathsf{Sat}(L)$ for any lattice $L$ by a change of variables in $(k, X, X)$.

**Definition 6.** *If $(\rho, L_\rho)$ is a partial character, any partial character $(\rho', \mathsf{Sat}(L_\rho))$ is called a* **saturation** *of $(\rho, L_\rho)$ if $\rho'$ coincides with $\rho$ when restricted to $L_\rho$.*

**Theorem 7.** *If $g$ is the order of the group $\mathsf{Sat}(L_\rho)/L_\rho$, then there are $g$ distinct saturations of $\rho$: $\rho_1, \ldots, \rho_g$. Also $I(\rho) = \cap_{j=1}^{g} I(\rho_j)$.*

**Theorem 8.** *The radical of a cellular ideal is of the form $I(\rho) + M(\mathscr{E})^{(d)}$ ($d$ is vector with all 1s), and its minimal primes are the lattice ideals with the saturations of $\rho$.*

So in a Laurent polynomial ring, to determine the set of minimal primes of a binomial ideal $I = I(\rho)$, all we need to do is to compute the saturations of $\rho$. The lattice ideals corresponding to these saturations are the associated primes of $I(\rho)$. The minimal of these ideals constitute the prime decomposition.

Now, let us discuss how we can combine the results from the modulo and the localization branch (step 14). From the recursive calls of the algorithm we have computed the minimal primes of $\Theta(I + \langle r \rangle)$ and $\Phi(I : r^\infty)$. Let the set of minimal primes be denoted by $\mathcal{P}_\Theta$ and $\mathcal{P}_\Phi$, respectively. So, we have $\sqrt{\Theta(I + \langle r \rangle)} = \cap_{P \in \mathcal{P}_\Theta} P$, $\sqrt{\Phi(I : r^\infty)} = \cap_{P \in \mathcal{P}_\Phi} P$. From Theorem 4, we have $\sqrt{I} = \Theta^{-1}(\sqrt{\Theta(I + \langle r \rangle)}) \cap \Phi^{-1}(\sqrt{\Phi(I + \langle r \rangle)})$. Thus $I = (\cap_{P \in \mathcal{P}_\Theta} \Theta^{-1}(P)) \cap (\cap_{P \in \mathcal{P}_\Phi} \Phi^{-1}(P))$. We know that $\Theta$ and $\Phi$ map primes to primes (Lemmas 1 and 2). The desired set of prime ideals is $\{ \Theta^{-1}(P) \mid P \in \mathcal{P}_\Theta \} \cup \{\Phi^{-1}(P) \mid P \in \mathcal{P}_\Phi\}$. We just need to remove the redundant ones.

**Analysis:** In this case we have only used the modulo and localization branches. So, the cost of the algorithm is $2^n$ Gröbner basis computations for an input ideal containing $n$ variables. The cost of algorithm 9.2 proposed in [6] is $2^n$ iterations, where in each iteration a Gröbner basis and a cellular decomposition is computed. Our proposed solution, thus, has gotten rid of the necessity of computing cellular decomposition in each of the $2^n$ iterations.

## 5.5 Cellular Decomposition: A = Cellular

In this section we will generalize the notion of **cellular ideals** to partial Laurent polynomial rings, establish that every ideal has a cellular decomposition, and use our framework to compute such a decomposition.

Let $(k, X, L)$ be the underlying partial Laurent polynomial ring. For a given set of variables $\mathscr{E} \subseteq (X \setminus L)$ and an integer vector $d = (d_i)_{i \in (X \setminus L) \setminus \mathscr{E}}$, the ideal $M(\mathscr{E})^{(d)}$ is defined as $\langle \{ x_i^{d_i} \mid i \in (X \setminus L) \setminus \mathscr{E} \} \rangle$.

**Definition 7.** *An ideal $I$ of $(k, X, L)$ is **cellular** if for some $\mathscr{E} \subseteq (X \setminus L)$, we have $I = I : (\prod_{i \in \mathscr{E}} x_i)^\infty$ and $I$ contains $M(\mathscr{E})^{(d)}$ for some vector $d$.*

**Observation 1.** *An ideal $I$ is cellular iff $\exists \mathscr{E} \subseteq (X \setminus L)$ and an integer vector $d = (d_i)_{i \in (X \setminus L) \setminus \mathscr{E}}$, such that $I = (I + M(\mathscr{E})^d) : (\prod_{i \in \mathscr{E}} x_i)^\infty$. It is denoted by $I_{\mathscr{E}}^{(d)}$.*

**Lemma 3.** *$\Phi^{-1}$ preserves cellular ideals.*

*Proof.* Let $\Phi^{-1}$ be a map from $(k, X, L)$ to $(k, X, L \setminus \{x\})$, where $x \in L$, and consider the cellular ideal $I = I_{\mathscr{E}}^{(d)}$ in $(k, X, L)$. As $\Phi^{-1}(I)$ is saturated w.r.t. $x$, it corresponds to the cellular ideal $\Phi^{-1}(I)_{\mathscr{E} \cup \{x\}}^{(d')}$, where $d'$ is the same vector as $d$, except that it does not contain the component corresponding to $x$. □

**Lemma 4.** *Let $s \in \mathbb{N}$ be such that $I : r^s = I : r^\infty$ in some Noetherian ring $R$. Then, $I = (I + \langle r^s \rangle) \cap (I : r^s)$.*

*Proof.* Let $g \in (I + \langle r^s \rangle) \cap (I : r^s)$. Then $g = i + hr^s \in I : r^s$ for some $i \in I, h \in R \implies gr^s = ir^s + hr^{2s} \in I$. This, coupled with the fact that $I : r^{2s} = I : r^s$, we have $g \in I$. □

Now we state how to compute a cellular decomposition of $I$. The computation will not use $\mathsf{A}(\Theta(I))$ branch of the reduction. $f(I)$ is defined as $I + \langle x^s \rangle$, where $s \in \mathbb{N}$ is such that $I : x^s = I : x^\infty$. By using Lemma 3, we see that cellular decomposition of $\Phi(I : x^\infty)$ gives us a cellular decomposition of $I : x^s$. To combine the decompositions of $\mathsf{A}(I : x^s)$ and $\mathsf{A}(f(I))$, we use Lemma 4.

Ideals in the base cases (i.e., $X = L \cup V$) are already cellular because variables in $V = X \setminus L$ are nilpotents of the ideals. Hence, there is no computation required in steps 4 and 6.

**Analysis:** As our algorithm uses only two branches, the cost of our algorithm is $2^n$ Gröbner basis computations for an input ideal containing $n$ variables. Algorithm 9.3 of [6] also needs to perform the same number of Gröbner basis computations. So, in this case, we do not see an improvement in the performance of our algorithm over existing algorithms. Advantage, if any, is the proposed generalized and unified approach which, according to the authors, is much simpler and cleaner.

# References

1. Bigatti, A.M., Scala, R., Robbiano, L.: Computing toric ideals. J. Symb. Comput. 27(4), 351–365 (1999)
2. Buchberger, B.: A theoretical basis for the reduction of polynomials to canonical forms. SIGSAM Bull. 10(3), 19–29 (1976)
3. Conti, P., Traverso, C.: Buchberger algorithm and integer programming. In: Mattson, H.F., Mora, T., Rao, T.R.N. (eds.) Applied Algebra, Algebraic Algorithms and Error-Correcting Codes. LNCS, vol. 539, pp. 130–139. Springer, Heidelberg (1991)

4. Cox, D.A., Little, J., O'Shea, D.: Ideals, Varieties, and Algorithms: An Introduction to Computational Algebraic Geometry and Commutative Algebra, 3/e (Undergraduate Texts in Mathematics). Springer-Verlag New York, Inc., Secaucus (2007)
5. Eisenbud, D.: Commutative Algebra with a View toward Algebraic Geometry. Springer, New York (1995)
6. Eisenbud, D., Sturmfels, B.: Binomial ideals. Duke Mathematical Journal 84(1), 1–45 (1996)
7. Fulton, W.: Introduction to toric varieties. Annals of Mathematics Studies, vol. 131. Princeton University Press, Princeton (1993)
8. Geiger, D., Meek, C., Sturmfels, B.: On the toric algebra of graphical models. The Annals of Statistics 34(3), 1463–1492 (2006)
9. Gilmer, R.: Commutative semigroup rings. University of Chicago Press, Chicago (1984)
10. Hemmecke, R., Malkin, P.N.: Computing generating sets of lattice ideals and Markov bases of lattices. Journal of Symbolic Computation 44(10), 1463–1476 (2009)
11. Hosten, S., Sturmfels, B.: Grin: An implementation of Gröbner bases for integer programming. In: Balas, E., Clausen, J. (eds.) IPCO 1995. LNCS, vol. 920, Springer, Heidelberg (1995)
12. Kesh, D., Mehta, S.K.: Generalized Reduction to Compute Toric Ideals. In: Dong, Y., Du, D.-Z., Ibarra, O. (eds.) ISAAC 2009. LNCS, vol. 5878, pp. 483–492. Springer, Heidelberg (2009)
13. Kesh, D., Mehta, S.K.: A Saturation Algorithm for Homogeneous Binomial Ideals. In: Wang, W., Zhu, X., Du, D.-Z. (eds.) COCOA 2011. LNCS, vol. 6831, pp. 357–371. Springer, Heidelberg (2011)
14. Sturmfels, B.: Gröbner Bases and Convex Polytopes. University Lecture Series, vol. 8. American Mathematical Society (December 1995)
15. Tayur, S.R., Thomas, R.R., Natraj, N.R.: An algebraic geometry algorithm for scheduling in the presence of setups and correlated demands. Mathematical Programming 69(3), 369–401 (1995),
citeseer.ist.psu.edu/tayur94algebraic.html
16. Thomas, R., Weismantel, R.: Truncated Gröbner bases for integer programming. Applicable Algebra in Engineering, Communication and Computing 8(4), 241–256 (1997), dx.doi.org/10.1007/s002000050062
17. Thomas, R.R.: A Geometric Buchberger Algorithm for Integer Programming. Mathematics of Operations Research 20, 864–884 (1995)
18. Urbaniak, R., Weismantel, R., Ziegler, G.M.: A variant of the Buchberger algorithm for integer programming. SIAM J. Discret. Math. 10(1), 96–108 (1997)

# How Fast Can We Multiply Large Integers on an Actual Computer?

Martin Fürer*

Department of Computer Science and Engineering
Pennsylvania State University
University Park, PA 16802, USA
furer@cse.psu.edu
http://cse.psu.edu/~furer

**Abstract.** We provide two complexity measures that can be used to measure the running time of algorithms to compute multiplications of long integers. The random access machine with unit or logarithmic cost is not adequate for measuring the complexity of a task like multiplication of long integers. The Turing machine is more useful here, but fails to take into account the multiplication instruction for short integers, which is available on physical computing devices.

An interesting outcome is that the proposed refined complexity measures do not rank the well known multiplication algorithms the same way as the Turing machine model.

**Keywords:** Integer multiplication, RAM models, FFT.

## 1 Introduction

The use of asymptotic time to measure the complexity of algorithms has been enormously successful in driving the search for new algorithmic ideas and more efficient algorithms. Nevertheless, it has some shortcomings.

One example is the complexity of the fastest known integer multiplication algorithm in the original version [6] as well as the modular version [5]. We call these algorithms F-R and DKSS-R respectively, as both operate over a ring. Obviously, these algorithms running in time $n \log n \, 2^{O(\log^* n)}$ are asymptotically faster than the previously fastest algorithm [11] with a running time of $O(n \log n \log \log n)$.

Nevertheless, the Wikipedia entry "Multiplication algorithm" [14] says, "However, these latter algorithms are only faster than Schönhage-Strassen for impractically large inputs." And this judgment is not uncommon. It seems to be implied by the following argument. For any practical large length $n$ (from $2^{16}+1$ to well beyond astronomical in length), $\log^* n$ is 5, resulting in $2^{\log^* n} = 32$, whereas for practical values of $n$, we have $\log \log n \leq 6$. (All logarithms are to the base 2 in this paper.)

In this special case, this reasoning is particularly faulty, because the exponent $\log^* n$ is used as an upper bound for the number of nested recursive calls, which

---

* Research supported in part by NSF Grant CCF-0964655 and CCF-1320814.

should really be max$\{0, \log^* n - 4\}$. The reason is that for small $n$, a different algorithm would be used in practice. Indeed, for any practical length $n$, the number of nested recursive calls and thus the exponent should be at most 1. The practical performance of the newer algorithm would have to be determined by an implementation, which might well be competitive, even though not by a large factor.

Thus we have these two very different methods of evaluating a multiplication algorithm, the theoretical asymptotic time bound and the practical implementation. The purpose of this paper is to build a bridge between the two, i.e., to propose models of computation, that are theoretically rigorous, yet can better predict the practical running time.

We have to stress, that we are focussing here on unbounded integer operations. The models developed here are valid for similar tasks. This is in contrast to many areas, like graph algorithms, where the unit cost RAM provides a perfectly good complexity measure, because most natural algorithms involve only numbers of length $O(\log n)$ which can be implemented to fit into a computer word.

Another example shows our concern more clearly. Schönhage and Strassen [11] have designed two fast integer multiplication algorithms. The first one (SS-C) is based on numerical approximations of complex roots of unity. It runs in time $O(n \log^2 n)$ for integers of length $n$ if school multiplication is used for all recursive calls. The second one (SS-F) is a discrete algorithm based on integers modulo Fermat numbers. It runs in time $O(n \log n \log \log n)$. Yet for a long time, the first algorithm has been routinely used for the extensive integer multiplications needed to find large prime numbers.

The first algorithm, SS-C, is asymptotically slower and somewhat unnatural for a discrete problem. It requires the tedious task of controlling rounding errors. It seems justified to ask whether the greater simplicity of SS-C is a sufficient reason to select it. After all, implementers don't stick with the simplicity, but add many clever ideas to speed up an implementation.

The running times cited above are bounds that hold simultaneously for the Turing machine time, as well as for the Boolean circuit size. These computation models and their time complexity are very natural and have facilitated many new algorithmic ideas in a vast number of areas including fast integer multiplication.

Sometimes, Turing machines are viewed as being impractical, because they function quite differently from physical computing machines. But when there is strong locality of memory access and no need for random access, as is the case for the Fast Fourier Transforms (FFTs) used in these fast integer multiplication algorithms, a Turing machine actually works just fine. Indeed, the first implementation [10] of a fast integer multiplication algorithm was on a Turing machine (a versatile multi-tape Turing machine with alphabet size equal to $2^w$, where $w$ is the word-length of the machine).

So in this particular application, the disadvantage of the Turing machine is not the lack of random access. What is missing in the Turing machine model are the built-in arithmetic operations of an actual computing device, in particular the multiplication instruction. Thus, we are aiming at a version of a Random Access Machine (RAM) [12,4] with multiplication.

The unit cost RAM with multiplication is not a viable model, because it can quickly build up large numbers, and its operations are then far too powerful compared to a real world computing device. We want a version of a RAM that is both theoretically appealing and closely modeling the capabilities of actual computers.

The basic model proposed here is a log-RAM. It can do arithmetic operations of length $O(\log n)$ in constant time. While this model might be too restricted for very short inputs $n$, it is very realistic for input lengths from thousands to trillions and beyond.

A more refined model aims to be even more realistic, adjusting for the varying costs of different operations, and accounting for the benefits of temporal and spatial locality.

## 2 The Basic Model

Our log-RAM model is a random access machine [12,4] augmented with arithmetic, Boolean, shift, input and output operations on nonnegative integers stored in binary. Furthermore, it has conditional and unconditional jump operations. It can do direct and indirect addressing. It has register $R_i$ for every $i \in \mathbb{N}$, as well as an input and output register.

To be specific, we give now a precise definition of the instruction set of a log-RAM. Nevertheless, the important part of a log-RAM is not the instruction set, but the time complexity.

**Definition 1.** *A* log-RAM *has the the following instructions set*

- $R_i = A \text{ op } A'$.
  This is an assignment. Here, and in the following, the arguments $A$ and $A'$ can be of one of 3 possible forms: nonnegative integer constant, $R_j$, or $R_{R_j}$. The operation is one of $\{\vee, \wedge, \oplus, \neg, +, -, *, \div\}$, where the Boolean operations (or, and, sum modulo 2, and negation) are vector operations, and div is integer division $(A \div A' = \lfloor A/A' \rfloor)$. The special case $R_i = R_j + 0$ copies registers.
  Registers and arguments are non-negative integers stored in binary. For Boolean operations, they are interpreted as bit vectors.
- $R_i = R_j$ cyclic shift $\pm A$ of length $A'$.
  Here the $A'$ right-most bits are cyclicly shifted by $A$ positions to the left $(+A)$ or to the right $(-A)$.
- Input to $R_i$ from $A$ to $A'$.
  This instruction has the effect of reading into $R_i$ the $A' - A + 1$ bits from position $A$ to position $A'$ of the input register.
- Output $R_j$ of length $A$.
  This instruction has the effect of concatenating the rightmost $A$ bits of $R_j$ to the output on the right hand side.

– *Jump if $A = A'$.*
  *This instruction allows a jump conditioned on two registers being equal, but also a jump when 0 ($R_i = 0$) and an unconditional jump (0 = 0).*

**Definition 2.** *We assume the machine knows the length $n$ of the input, e.g., it is written in register $R_0$ before a computation starts.*
*All registers, except the input and output register, are only allowed to be assigned bit strings of length $O(\log n)$ encoding nonnegative integers up to $n^{O(1)}$. Arguments denoting positions and lengths in the input, output, and shift instructions are only allowed to have values $O(\log n)$.*
*The time for any ($O(\log n)$ long) operation is $O(1)$ including for multiplication and division. We therefore might refer to the machine model as a* unit cost log-RAM.

As the log-RAM realistically mimics physical computing machines, it allows for a speed-up of $\Theta(\log n)$ for additions and shifts of length $\Theta(\log n)$ and a speed-up of $\Theta(M(\log n))$ for multiplications of binary integers compared to Turing machine time, where $M(n)$ is the multiplication time of a Turing machine.

An alternative definition, equivalent for our purposes, would be to allow arbitrary long registers and arguments, but to charge $(\lceil \ell / \log n \rceil)^2$ for multiplication and division instructions, 1 for jump instructions, and $\lceil \ell / \log n \rceil$ for all other instructions. Here, $\ell = \max\{\ell', 2^{\ell''}\}$, where $\ell'$ is the maximal length of any operand, and $\ell''$ the maximal length of a position or length operand.

Naturally, the equivalence only holds for somewhat efficient computations. The alternative definition of the log-RAM would define a universal computing device, while by our definition, a log-RAM can only define polynomial space functions.

**Proposition 1.** *A partial function is computable by a log-RAM if and only if it is computable by a Turing machine in polynomial space.*

*Proof.* Only if part: Note that the the log-RAM can only access the first $n^{O(1)}$ registers, which each are allowed to hold bit strings of length $O(\log n)$. The log-RAM has no way to address other registers. Each operation can easily be simulated by a Turing machine.

If part: The log-RAM can simulate a polynomially space bounded Turing machine by storing the contents of tape cell $i$ in the register $R_{i+2}$ and storing the head position in register $R_2$. □

## 3 Differences to the Traditional RAM

Naturally, the definition of the log-RAM is very similar to the traditional definition of a RAM [12,4] (see [1, pp. 5 ff.]). Nevertheless, there are important differences.

The unit cost RAM provides an excellent cost measure for well behaved algorithms, e.g., for many graph algorithms. It gets useless if a multiplication operation is allowed, as one could quickly produce integers of huge lengths, resulting (for some simple instruction sets) in the power of unbounded parallel machines, which can handle PSPACE in polynomial time [7]. Even without multiplication, length $T(n)$ integers can be produced in time $T(n)$, resulting in unrealistically cheap additions.

These drastic problems are avoided by the traditional RAM with logarithmic cost, which is a cost proportional to the length of a binary integer. Nevertheless, this type of RAM is still not suitable to provide a practical cost measure for a task like the multiplication of long integers. Obviously an instruction multiplying in one step with a cost of $O(n)$ would make the task trivial. Without a multiplication instruction, the logarithmic cost RAM still has the random access advantage over Turing machines (which we don't need here). But it does not have the practical advantage of real computers, that can do operations like additions of reasonable numbers almost as fast as a bit operation.

Our new log-RAM provides a complexity measure that is much closer to the computation time of a real computer. It allows a theoretical investigation that can better predict the practicality of an algorithm in a domain like large integer multiplication.

One could even define a log-RAM with a more explicit cost function. If all registers have lengths bounded by $k \log n$, then the cost of a multiplication could be defined as $k^2$ and the cost of any other operation could be defined as $k$. This time measure would avoid any large hidden constant factors.

## 4 Performance of the log-RAM on Multiplication Algorithms

First let us review the most important multiplication algorithms.

### 4.1 The Traditional Multiplication Algorithms

The first multiplication algorithm with a non-trivial asymptotic running time is due to Karatsuba [8]. It multiplies $a_1 2^{n/2} + a_0$ with $b_1 2^{n/2} + b_0$ recursively by computing the 3 products $a_1 b_1$, $a_0 b_0$, and $(a_1 + a_0)(b_1 + b_0)$, to obtain the product with a few additions and subtractions. The running time is $O(n^{\log 3})$.

It is straightforward to see that working with numbers of length $O(\log n)$ allows us to use the full computational power of the log-RAM. Thus, the time for school multiplication is $O(n^2 / \log^2 n)$, while the time for Karatsuba's algorithm is $O(n^{\log 3} / \log^2 n)$. Karatsuba's algorithm can be viewed as multiplying 2 linear polynomials by evaluating them at 0, 1, and $\infty$, followed by multiplying the values and interpolating. Toom's algorithm [13] (analyzed and implemented on the Turing machine by Cook [3]) instead uses higher degree polynomials. Degree 2 (with 5 coefficients in the product polynomial) is often used for moderately large numbers. Clearly, on a log-RAM, we get the same factor $\Theta(\log^2 n)$ speed-up.

The first Schönhage-Strassen integer multiplication algorithm, SS-C [11] partitions the factors into pieces of length $\Theta(\log n)$, to be used as coefficients of two polynomials over the complex numbers $\mathbb{C}$. The polynomials are evaluated at all powers of a primitive root of unity by a Fast Fourier Transform (FFT). After the multiplications of corresponding values, interpolation is done by an inverse FFT. $O(\log n)$ accuracy of these numerical computations is sufficient to recover the precise result by rounding. The running time is $O(n \log^2 n)$ on the Turing machine.

The second Schönhage-Strassen integer multiplication algorithm, SS-F [11] partitions the factors into pieces of length $\Theta(\sqrt{n})$, to be used as coefficients of two polynomials over the ring of integers modulo $2^{\sqrt{n}} + 1$, where $\sqrt{n}$ is rounded to a power of 2. The polynomials are evaluated at all powers of a principal root of unity in this ring by an FFT. The multiplications of values is done recursively. It is followed by interpolation with an inverse FFT. This faster method requires a depth $O(\log \log n)$ of nested recursions, and runs in time $O(n \log n \log \log n)$ on the Turing machine.

It is important to notice that the slower SS-C algorithm does small multiplications on every level of the FFT. Thus it can really profit from a built-in multiplication instruction for short integers. The faster SS-F algorithm does relatively simple shift operations at all levels of the FFT. This makes it fast for Turing machines. On the other hand, it cannot benefit from a built-in integer multiplication for small integers, except at the bottom of the recursion.

**Theorem 1.** *(a) The running time of the first Schönhage-Strassen integer multiplication algorithm, SS-C, is $O(n)$ on the log-RAM.*
*(b) The running time of the second Schönhage-Strassen integer multiplication algorithm, SS-F, is $O(n \log \log n)$ on the log-RAM.*

*Proof.* (a) The analysis of the Turing machine algorithm accounts for $O(n \log n)$ simple bit operations to do shifts, copies and additions for the FFT and its inverse. Furthermore, it accounts for $O(n/\log n)$ multiplications of length $O(\log n)$ at each of the $O(\log n)$ levels of the FFT. As all coefficients have lengths $\Omega(\log n)$, all simple operations, including input and output, allow for a speed-up by a factor of $\Theta(\log n)$ on the log-RAM compared to the Turing machine. Furthermore, all the $O(n)$ small integer multiplications are done in constant time each on the log-RAM, resulting in an overall linear time.

(b) When coefficients reach a length of $O(\log n)$, recursion is no longer required. Then multiplications can be done directly on the log-RAM. This reduces the recursion depth of the FFT from $\log \log n$ to $\log \log n - \log \log \log n$, representing no asymptotic speed-up. The Turing machine algorithm spends time $O(n \log n)$ at each recursion level. With the shortcut just described, all numbers involved have lengths $\Omega(\log n)$, resulting in a speed-up by a factor of $\Theta(\log n)$ on the log-RAM for the additions and shifts. Therefore, the time for all additions and shifts is $O(n \log \log n)$. There are $O(n 2^{\log \log n - \log \log \log n} / \log n) = O(n/\log \log n)$ short multiplications. Thus the total cost of $O(n \log \log n)$ is determined by the additions and shifts. □

## 4.2 The Newest Multiplication Algorithms

The asymptotically fastest multiplication algorithm F-R [6] does the FFT over a ring of polynomials $\mathcal{R} = \mathbb{C}[x]/(x^P + 1)$ with both, the value of $P$ and the length of the coefficients being of order $\log n$. Of the $O(\log n)$ levels of the FFT, only every $\log \log n$-th level requires expensive multiplications in the ring $\mathcal{R}$, while the multiplications at the other levels are done by a version of cyclic shifts.

Multiplication in the ring $\mathcal{R}$ itself is done by an FFT with $O(\log \log n)$ levels of cheap multiplications by cyclic shifts. For the asymptotic analysis of the Turing machine algorithm, the multiplication of values is done recursively. This introduces the factor of $2^{O(\log^* n)}$ for the $\log^* n$ depth of recursive calls with geometrically increasing cost from one recursion depth to the next. A practical implementation would not do such recursive calls, but instead use the built-in multiplication instruction of the actual machine. Similarly, the log-RAM does each such multiplication with 1 machine operation (or $O(1)$ operations with a small hidden factor).

**Theorem 2.** *The running time of the F-R integer multiplication algorithm is $O(n)$ on the log-RAM.*

*Proof.* A multiplication in the ring $\mathcal{R}$ requires $O(\log \log n)$ levels of $O(\log n)$ easy operations of length $O(\log n)$ and direct multiplication of $O(\log n)$ values of length $O(\log n)$ each. The resulting cost for one multiplication in $\mathcal{R}$ is $O(\log n \log \log n)$ for the easy operations and $O(\log n)$ for the direct multiplications. Thus the total time is $O(\log n \log \log n)$.

The FFT and its inverse have $O(\log n)$ levels involving shifts and additions of integers of length $\Omega(\log n)$ and a total length of $O(n)$ on each level. This requires time $O(n)$ on the log-RAM.

Furthermore, there are also $O(\log n / \log \log n)$ levels, with each requiring $O(n/\log^2 n)$ multiplications in $\mathcal{R}$ at a cost of $O(\log n \log \log n)$ each. Thus this part also requires time $O(n)$. □

The discrete variant DKSS-R [5] of the F-R algorithm operating over a ring approximating $p$-adic numbers has the same time analysis for the log-RAM as the original F-R algorithm.

## 4.3 Comparisons of the log-RAM Algorithms

In summary, the algorithms SS-C, SS-F, F-R, and DKSS-R still have very similar running times in the log-RAM model. It is interesting to see that the ranking changes. The previously slowest (SS-C) and the most advanced (F-R and DKSS-R) now have the same linear running time. The elegant SS-F algorithm, that achieves multiplication mainly by shifts is now less competitive, because it does not make much use of the power of the built-in multiplication instruction available on the log-RAM and in practical computers.

Still it is hard to judge these algorithms just based on the log-RAM complexity. The algorithms SS-C and F-R have the significant drawback of using non-discrete numerical computations requiring sufficient precision and rounding.

## 5 Related Tasks on the log-RAM and the Storage Modification Machine

**Division and Elementary Functions.** The results about multiplication have immediate corollaries about division and other tasks based on multiplication. Division and $n$-bit approximations of algebraic numbers can be computed with the help of these algorithms with the same asymptotic running time, while $n$-bit approximate evaluations of elementary functions or constants like $e$ and $\pi$ can be computed with only an additional factor of $\log n$ time [2].

**Storage Modification Machine.** It is worth noticing that there is another theoretically interesting modification of the random access machine (RAM), namely the storage modification machine. Here the modification goes in the opposite direction. Instead of allowing the more powerful multiplication instruction, even the weaker addition instruction is not allowed. Instead, there is just a successor and a predecessor instruction. As numbers never get too long, the unit cost measure makes sense here.

Schönhage [9] has obtained a very powerful result about multiplication on a storage modification machine. It can be done in linear time. Naturally, this is a far more difficult result than our linear time results on the log-RAM. On the other hand, his algorithm is very sophisticated with sorting and table lookup for the mass production of short products. This is seemingly an impractical algorithm. In the linear time log-RAM algorithms, on the other hand, we can use natural practical procedures and take advantage of easily available hardware.

**Addition versus Multiplication.** An added bonus of the log-RAM model is also the distinction between the easy task of addition and the complicated task of multiplication. On the storage modification machine both tasks take linear time, just with a huge difference of the constant factors involved. The log-RAM model makes the distinction clear. Using the built-in addition instruction, long additions of length $n$ take time $O(n/\log n)$, which is significantly better than the $O(n)$ time for multiplication.

**Open Problem.** For the Turing machine model, there is no super-linear lower bound known for integer multiplication. For the log-RAM model the input only provides a trivial $\Omega(n/\log n)$ lower bound. Naturally, we conjecture an $\Omega(n)$ lower bound for the log-RAM.

This conjectured $\Omega(n)$ lower bound for multiplication on the log-RAM might be independent of the well known $\Omega(n \log n)$ lower bound conjectured for multiplication on the Turing machine. The log-RAM can only benefit strongly from a Turing machine computation if it operates on $\Theta(\log n)$ long chunks of data, while the Turing machine can simulate a log-RAM efficiently only if the log-RAM does not jump around much.

## 6 The Refined Model log-RAM with Depth-Cost

The proposed log-RAM model realistically models the advantage provided by the availability of instructions (including multiplication) operating on words in real machines. Still, it does not account for the higher cost of a multiplication instruction over a simple Boolean vector operation on a word. More importantly, the log-RAM model does not model the speed-up provided by local access patterns on real machines.

For this purpose, we propose the *log-RAM with Depth-Cost* as a refined model. We basically keep the instruction set of the log-RAM, but define a different cost measure. The cost is modified only slightly, as the log-RAM already quite well approximates the cost on an actual machine.

**Definition 3.** *A log-RAM with Depth-Cost has all the instructions of the (unit-cost) log-RAM plus an additional vector copying instruction. The vector copying instruction puts a copy of the vector of the memory cells from $R_A$ to $R_{A+A''-1}$ into the memory cells from $R_{A'}$ to $R_{A'+A''-1}$. The cost of this operation is $\max\{\log A, \log A', A''\}$. For the other operations, there is a cost associated with accessing a register and cost associated with the operation itself.*
*(a) The cost of an operation is the order of the parallel time to do this operation efficiently on operands of length $O(\log n)$ by a bounded fan-in Boolean circuit. In particular, the cost of Boolean operations is $O(1)$, while the cost of arithmetic operations and shifts is $O(\log \log n)$.*
*(b) The cost of accessing register $R_i$ is $O(\max\{1, \log i\})$.*

The cost of accessing $R_i$ reflects the idea that registers with low index $i$ are in a faster cache and therefore less costly to access. The cost of the vector copying operation reflects locality of access. Access in a single memory cell is expensive. Accessing a sequence of adjacent memory cells with the vector copying instructions is cheaper.

Due to the locality of access for doing FFTs, the cost of memory access can be bounded by the cost of arithmetic operations, if one uses a cache hierarchy, with geometrically increasing chunks of data being brought in or moved out. A higher cache level just means shorter addresses, i.e., being closer to the top of the memory.

## 7 Multiplication on the log-RAM with Depth-Cost

Trivial tasks, like input, output and addition of length $n$ numbers now take $O(n/\log n)$ steps costing $O(\log \log n)$ each, resulting in time $O(n \log \log n / \log n)$ per task.

The simple multiplication algorithms (school, Karatsuba, Toom) gain a factor of $\Omega(\log n / \log \log n)$ compared to Turing machine cost, because they operate with chunks of length $\Theta(\log n)$ stored in a register costing $O(\log \log n)$ per operation.

Accessing far away registers is efficient for these and all the other multiplication algorithms studied here. They would just be brought to the front in vectors of $O(\log n)$ length at a cost of $O(1)$ per register.

**Theorem 3.** *The first Schönhage-Strassen SS-C algorithm has a running time of $O(n \log \log n)$ on the log-RAM with Depth-Cost.*

*Proof.* All operations now cost $O(\log \log n)$ instead of $O(1)$ in the unit cost log-RAM model. □

**Theorem 4.** *The second Schönhage-Strassen SS-F algorithm has a running time of $O(n (\log \log n)^2)$ on the log-RAM with Depth-Cost.*

*Proof.* A direct implementation (still with bringing vectors of length $O(\log n)$ to the front at once) costs an additional factor of $O(\log \log n)$ compared to the unit cost log-RAM model because of the cyclic shifts and additions. □

The cost of $O(\log \log n)$ for an addition could be reduced to $O(1)$ by storing numbers in a redundant form. Every number would be represented as the sum of 2 registers. An actual sum would require the replacement of 4 summands by 2, which can be accomplished by $O(1)$ Boolean and shift operations, as there are no long carries to handle. But these cost savings are useless, because the shift operations still cost $O(\log \log n)$ each.

**Theorem 5.** *The F-R algorithm has a running time of $O(n \log \log n)$ on the log-RAM with Depth-Cost.*

*Proof.* In the F-R algorithm, most levels of the FFT do cheap operations with cyclic shifts of coefficients within the ring $\mathcal{R}$. The coefficients themselves are not subjected to shifts, they are just cyclicly interchanged. These shifts can be done by $O(1)$ vector copy operations at a cost of $O(\log n)$ per operation involving a vector of length $O(\log n)$. The coefficients are subject to the addition operation. But this time, as there are no cyclic shifts of the summands, redundant additions are sufficient at a cost of $O(1)$ per operation.

Only every $O(\log \log n)$-th level, expensive operations (arbitrary multiplications in $\mathcal{R}$) have to be done. In each of the $O(\log n / \log \log n)$ expensive levels, $O(n/\log^2 n)$ multiplications in $\mathcal{R}$ have to be done at a cost of $O(\log n (\log \log n)^2)$ each. This results in a total cost of $O(n \log \log n)$. □

## 8  Conclusions

We have investigated the cost of integer multiplication and noticed that neither the Turing machine nor the standard random access machine (RAM) models provide good complexity measures. The Turing machine, as well as the RAM without multiplication instruction cannot model the advantage of physical machines with a multiplication instruction on words.

If the unit cost RAM had a multiplication instruction, then it would not be a good computation model, because it has the power of parallel machines by creating huge numbers. The logarithmic cost RAM could have a multiplication instruction. But it would show too low a cost for long integers and too high a cost for short integers. It would not be a reasonable model for measuring the complexity of long integer multiplication.

We have proposed two RAM variants as better complexity measures for long multiplication and similar tasks. The measures reflect the fact that addition is faster than multiplication by known algorithms. They provide interesting practical complexity results for the various known algorithms. It may be surprising that the order of the standard multiplication algorithms by their time complexity in the new measures is different from the corresponding order in the Turing machine measure. The often used SS-C algorithm, based on numerical approximation in $\mathbb{C}$, is actually faster in the new measures than the discrete SS-F algorithm, which is faster in the Turing machine model.

## References

1. Aho, A.V., Hopcroft, J.E., Ullman, J.D.: The Design and Analysis of Computer Algorithms. Addison-Wesley, Reading (1974)
2. Brent, R.P.: Fast multiple-precision evaluation of elementary functions. J. Assoc. Comput. Mach. 23, 242–251 (1976)
3. Cook, S.A.: On the minimum computation time of functions. PhD thesis, Harvard University (1966)
4. Cook, S.A., Reckhow, R.A.: Time bounded random access machines. Journal of Computer and System Sciences 7(4), 354–375 (1973)
5. De, A., Kurur, P., Saha, C., Saptharishi, R.: Fast integer multiplication using modular arithmetic. In: STOC 2008: Proceedings of the 40th Annual ACM Symposium on Theory of Computing, pp. 499–506. ACM, New York (2008)
6. Fürer, M.: Faster integer multiplication. SIAM Journal on Computing 39(3), 979–1005 (2009)
7. Hartmanis, J., Simon, J.: On the power of multiplication in random access machines. IEEE Annual Symposium on Foundations of Computer Science, 13–23 (1974)
8. Karatsuba, A., Ofman, Y.: Multiplication of multidigit numbers on automata. Doklady Akademii Nauk SSSR 145(2), 293–294 (1962) (in Russian), English translation in Soviet Physics-Doklady 7, 595–596 (1963)
9. Schönhage, A.: Storage modification machines. SIAM J. Comput. 9(3), 490–508 (1980)
10. Schönhage, A., Grotefeld, A.F.W., Vetter, E.: Fast algorithms: A Turing machine implementation. B.I. Wissenschaftsverlag, Mannheim-Leipzig-Wien-Zürich (1994)
11. Schönhage, A., Strassen, V.: Schnelle Multiplikation grosser Zahlen. Computing 7, 281–292 (1971)
12. Shepherdson, J.C., Sturgis, H.E.: Computability of recursive functions. Journal of the ACM 10(2), 217–255 (1963)
13. Toom, A.L.: The complexity of a scheme of functional elements simulating the multiplication of integers. Dokl. Akad. Nauk SSSR 150, 496–498 (1963) (in Russian), English translation in Soviet Mathematics 3, 714–716 (1963)
14. Wikipedia, the free encyclopedia. Multiplication algorithm (October 2013), http://en.wikipedia.org/wiki/Multiplication_algorithm

# Sorting Permutations by Prefix and Suffix Versions of Reversals and Transpositions

Carla Negri Lintzmayer and Zanoni Dias

Institute of Computing, University of Campinas (Unicamp), Brazil
{carlanl,zanoni}@ic.unicamp.br

**Abstract.** Reversals and transpositions are the most common kinds of genome rearrangements, which allow us to establish the divergence between individuals along evolution. When the rearrangements affect segments from the beginning or from the end of the genome, we say they are prefix or suffix rearrangements, respectively. This paper presents the first approximation algorithms for the problems of Sorting by Prefix Reversals and Suffix Reversals, Sorting by Prefix Transpositions and Suffix Transpositions and Sorting by Prefix Reversals, Prefix Transpositions, Suffix Reversals and Suffix Transpositions, all of them with factor 2. We also present the intermediary algorithms that lead us to the main results.

## 1 Introduction

We assume that the evolution distance between two individuals is given by the minimum number of rearrangements needed to transform one genome into another. If we represent them as permutations and assume that one is the identity, the problem is to find the minimum number of operations that sort the other.

The problems of Sorting by Reversals and Sorting by Transpositions (the most common rearrangements) are well studied, so that their best-known algorithms have approximation factor 1.375 [3,8]. In addition, both are NP-hard [6,5].

When rearrangements affect segments from the beginning of the genome they are prefix rearrangements. For Sorting by Prefix Reversals and for Sorting by Prefix Transpositions, the best-known algorithms have approximation factor of 2 [9,7]. The former was proved to be NP-hard [4] while the latter remains an open problem. Sharmin et al. [11] considered a variation in which prefix reversals and prefix transpositions were allowed and gave a 3-approximation algorithm.

In addition to rearrangements restricted to the prefix of a permutation, it is also possible to consider their suffix version. It is reasonable to believe that it is easier to break a genome at one point than at two or more. Besides, if this happens, either the first or the second part could be reversed; thus, characterizing the prefix/suffix reversals. The same analogy can be used for the prefix/suffix transpositions since it would require breaking a genome at two points, which can be more difficult, but it is still easier than at three points, as a transposition would. However, notice that if a problem involves only prefix rearrangements, there is no need to study a problem that allows only the suffix versions of the same rearrangements, since they are equivalent. Hence, this paper will study problems of sorting permutations by reversals and transpositions involving both prefix and suffix versions of them. Note that there are no records of a similar study in the literature.

The paper is divided as follows: Section 2 presents important definitions related to our problems. Section 3 describes the algorithms developed. Section 4 shows the results. Finally, Section 5 concludes and suggests future work.

## 2 Definitions

Given a permutation $\pi = (\pi_1 \pi_2 \ldots \pi_n)$, the identity permutation $\iota = (1\,2\,\ldots\,n)$ and the reverse permutation $\eta = (n\,\ldots\,2\,1)$, we introduce now some concepts important to this paper and to the Genome Rearrangements area.

A **composition** between two permutations $\pi$ and $\sigma$ is the operation "·" in which $\pi \cdot \sigma = (\pi_{\sigma_1} \pi_{\sigma_2} \ldots \pi_{\sigma_n})$. The **inverse** permutation of $\pi$ is $\pi^{-1}$, in which $\pi^{-1}_{\pi_i} = i$, for $1 \leq i \leq n$, and it satisfies $\pi \cdot \pi^{-1} = \iota$.

We extend $\pi$ by setting $\pi_0 = 0$ and $\pi_{n+1} = n+1$. We can now define some important concepts related to permutations.

A **reversal** $\rho(i,j)$, $1 \leq i < j \leq n$, is a rearrangement that transforms $\pi$ into $\pi \cdot \rho(i,j) = (\pi_1 \ldots \pi_{i-1}\ \underline{\pi_j\ \pi_{j-1}\ \ldots\ \pi_{i+1}\ \pi_i}\ \pi_{j+1}\ \ldots\ \pi_n)$. A **prefix reversal** $\rho_p(j)$ is a reversal $\rho(1,j)$, $1 < j \leq n$, while a **suffix reversal** $\rho_s(i)$ is $\rho(i,n)$, $1 \leq i < n$.

A **transposition** $\tau(i,j,k)$, $1 \leq i < j < k \leq n+1$, is a rearrangement that transforms $\pi$ into $\pi \cdot \tau(i,j,k) = (\pi_1 \ldots \pi_{i-1}\ \underline{\pi_j\ \pi_{j+1}\ \ldots\ \pi_{k-1}}\ \underline{\pi_i\ \pi_{i+1}\ \ldots\ \pi_{j-1}}\ \pi_k \ldots \pi_n)$. A **prefix transposition** $\tau_p(j,k)$ is a transposition $\tau(1,j,k)$, $2 \leq j < k \leq n+1$, while a **suffix transposition** $\tau_s(i,j)$ is a transposition $\tau(i,j,n+1)$, $1 \leq i < j \leq n$.

If a problem involves some kind of reversal, a breakpoint exists between a pair of consecutive elements $\pi_i$ and $\pi_{i+1}$ if $|\pi_{i+1} - \pi_i| \neq 1$. For both Sorting by Prefix Reversals (SBPR) and Sorting by Prefix Reversals and Prefix Transpositions (SBPRPT), $1 \leq i \leq n$ and $\pi_0$ and $\pi_1$ never form a breakpoint. For both Sorting by Suffix Reversals (SBSR) and Sorting by Suffix Reversals and Suffix Transpositions (SBSRST), $0 \leq i \leq n-1$ and $\pi_n$ and $\pi_{n+1}$ never form a breakpoint. For them, $\iota$ is the unique permutation without breakpoints. For both Sorting by Prefix Reversals and Suffix Reversals (SBPRSR) and Sorting by Prefix Reversals, Prefix Transpositions, Suffix Reversals and Suffix Transpositions (SBPRPTSRST), $1 \leq i \leq n-1$ and neither $\pi_0$ and $\pi_1$, nor $\pi_n$ and $\pi_{n+1}$ form breakpoints. For them, $\iota$ and $\eta$ are the unique permutation without breakpoints.

If a problem involves only some kind(s) of transposition(s), then a breakpoint exists between a pair of consecutive elements $\pi_i$ and $\pi_{i+1}$ if $\pi_{i+1} - \pi_i \neq 1$. For Sorting by Prefix Transpositions (SBPT), $1 \leq i \leq n$ and $\pi_0$ and $\pi_1$ never form a breakpoint. For Sorting by Suffix Transpositions (SBST), $0 \leq i \leq n-1$ and $\pi_n$ and $\pi_{n+1}$ never form a breakpoint. For Sorting by Prefix Transpositions and Suffix Transpositions (SBPTST), $1 \leq i \leq n-1$ and neither $\pi_0$ and $\pi_1$, nor $\pi_n$ and $\pi_{n+1}$ form breakpoints. For them, $\iota$ is the unique permutation without breakpoints.

Given a set $\beta$ of rearrangements allowed in a sorting problem, we denote the number of breakpoints of a permutation by $b_\beta(\pi)$. If there is no breakpoint between two elements, we say there is an **adjacency** between them. The **sorting distance** of a permutation $\pi$, denoted by $d_\beta(\pi)$, is defined as the minimum number of operations in $\beta$ needed to transform $\pi$ into $\iota$. Since the identity has the smallest number of breakpoints and it is (usually) the only one with this feature, we can say that sorting $\pi$ is equivalent to reducing its number of breakpoints. This allows us to establish lower bounds for rearrangement distances.

For SBR, it was proved [1] that $d_\rho(\pi) \geq \lceil b_\rho(\pi)/2 \rceil$. For SBPR, $d_{\rho_p}(\pi) \geq b_{\rho_p}(\pi)$, as demonstrated by Fischer and Ginzinger [9]. For SBT [2], $d_\tau(\pi) \geq \lceil b_\tau(\pi)/3 \rceil$. For SBPT, Dias and Meidanis [7] showed that $d_{\tau_p}(\pi) \geq \lceil b_{\tau_p}(\pi)/2 \rceil$. For SBPRPT, Sharmin et al. showed that $d_{\rho_p \tau_p}(\pi) \geq \lceil b_{\rho_p \tau_p}(\pi)/2 \rceil$.

Since the other problems were yet to be considered in the literature, we now define their trivial lower bounds. Because of the equivalences, for SBSR, $d_{\rho_s}(\pi) \geq b_{\rho_s}(\pi)$, for SBST, $d_{\tau_s}(\pi) \geq \lceil b_{\tau_s}(\pi)/2 \rceil$, and for SBSRST, $d_{\rho_s \tau_s}(\pi) \geq \lceil b_{\rho_s \tau_s}(\pi)/2 \rceil$.

**Theorem 1.** *For an arbitrary permutation $\pi$, $d_{\rho_p \rho_s}(\pi) \geq b_{\rho_p \rho_s}(\pi)$, $d_{\tau_p \tau_s}(\pi) \geq \lceil b_{\tau_p \tau_s}(\pi)/2 \rceil$, and $d_{\rho_p \tau_p \rho_s \tau_s}(\pi) \geq \lceil b_{\rho_p \tau_p \rho_s \tau_s}(\pi)/2 \rceil$.*

A **strip** is a subsequence $\pi_i, \ldots, \pi_j$ of $\pi$, with $1 \leq i \leq j \leq n$, such that (i) either $i = 1$ or $\pi_{i-1}$ and $\pi_i$ form a breakpoint; (ii) either $j = n$ or $\pi_j$ and $\pi_{j+1}$ form a breakpoint; and (iii) the other elements of the subsequence form adjacencies. A strip of length greater or equal to two is *ascending* if $\pi_k = \pi_{k+1} - 1$ for all $i \leq k < j$. It is *descending* if $\pi_k = \pi_{k+1} + 1$. Otherwise, it is a *singleton*.

A **breakpoint graph** [1] of a permutation $\pi$ is a graph $G(\pi) = (V, E)$ in which $V = \{\pi_0, \pi_1, \ldots, \pi_{n+1}\}$ and $E$ contains *black edges* and *gray edges*. A black edge $e$ exists if and only if (i) $e = (\pi_i, \pi_{i+1})$ and $\pi_i$ and $\pi_{i+1}$ form a breakpoint, $0 \leq i \leq n$; (ii) $e = (\pi_0, \pi_1)$ and prefix operations are involved; and (iii) $e = (\pi_n, \pi_{n+1})$ and suffix operations are involved. A gray edge $e$ exists if and only if $e = (\pi_i, \pi_j)$ for some $0 \leq i < j \leq n+1$ with $\pi_j = \pi_i \pm 1$ and $j \neq i+1$. The convention is to draw black edges as straight lines and gray edges as dashed arcs.

Let $(\pi_i, \pi_j)$ be a gray edge. Since $\pi_j = \pi_i \pm 1$, at least one black edge either begins or ends at $\pi_i$ as well as at least one black edge either begins or ends at $\pi_j$. Hence, we can classify such edge into at least one of the four types in Fig. 1.

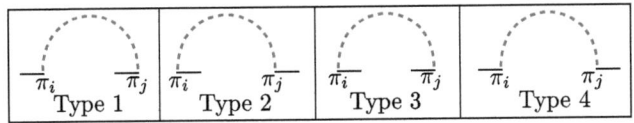

**Fig. 1.** Classification of gray edges

## 3 Algorithms

The following subsections describe the algorithms that we have developed, in addition to some of the existing algorithms related to ours.

## 3.1 Algorithms for Sbprsr

Fischer and Ginzinger [9] published the first 2-approximation algorithm (and the best so far) for SBPR, which we will call 2-PR. They used the breakpoint graph and defined requirements for each type of gray edge. Based on that, it is possible to establish what to do when considering not only prefix reversals, but also suffix reversals. Lemma 1 shows what we called *good edges*: edges for which is possible to remove one breakpoint with one or two reversals. 2-PR only deals with *good prefix edges*, that is, items 1, 3, 5, and 7 of Lemma 1 without the constraints over $j$.

**Lemma 1.** *Let $\pi$ be an arbitrary permutation. There is a sequence of at most two prefix reversals or suffix reversals that removes one breakpoint if $G(\pi)$ contains at least one gray edge $(\pi_i, \pi_j)$: (1) of type 1 with $i=1$ and $j \leq n$; (2) of type 2 with $j=n$ and $i \geq 1$; (3) of type 3 with $i=1$ and $j \leq n$; (4) of type 3 with $j=n$ and $i \geq 1$; (5) of type 2 with $i \neq 0$ and $j \leq n$; (6) of type 1 with $j \neq n+1$ and $i \geq 1$; (7) of type 3 with $i > 1$ and $j \leq n$; (8) of type 3 with $j < n$ and $i \geq 1$.*

*Proof.* If $(\pi_i, \pi_j)$ is a gray edge, then $\pi_j = \pi_i \pm 1$. To create an adjacency between $\pi_i$ and $\pi_j$ without creating new breakpoints one must perform, for each type of edge: (1) one prefix reversal $\rho_p(j-1)$; (2) one suffix reversal $\rho_s(i+1)$; (3) one prefix reversal $\rho_p(j-1)$; (4) one suffix reversal $\rho_s(i+1)$; (5) two prefix reversals $\rho_p(j)$ and $\rho_p(j-i)$; (6) two suffix reversals $\rho_s(i)$ and $\rho_s(n+1-(j-i))$; (7) two prefix reversals $\rho_p(i)$ and $\rho_p(j-1)$; (8) two suffix reversals $\rho_s(j)$ and $\rho_s(i+1)$. □

If a permutation $\pi$ does not contain good prefix edges, then it is of the form $\pi = (\underbrace{p_1 \ldots 1}_{\ell_1} \underbrace{p_2 \ldots p_1+1}_{\ell_2} \ldots \ldots t \underbrace{\ldots p_{b_{\rho_p}(\pi)-1}+1}_{\ell_{b_{\rho_p}(\pi)}} t+1\ t+2\ \ldots\ n)$ with $t \leq n$, that is, $\pi$ consists in $b_{\rho_p}(\pi) \geq 2$ decreasing strips of size $\ell_i \geq 2$, $1 \leq i \leq b_{\rho_p}(\pi)$ [9]. The following $2b_{\rho_p}(\pi)$ prefix reversals transforms $\pi$ into $\iota$: $\rho_p(t) \cdot \rho_p(t-\ell_1) \cdot \rho_p(t) \cdot \rho_p(t-\ell_2) \cdot \ldots \cdot \rho_p(t) \cdot \rho_p(t-\ell_{b_{\rho_p}(\pi)})$.

2-PR scans $\pi$ from left to right trying to find a good prefix edge in $G(\pi)$ in the order that the four appear in Lemma 1. If a good prefix edge does not exist, the algorithm applies the sequence given above, guaranteeing that $d_{\rho_p}(\pi) \leq 2b_{\rho_p}(\pi)$. Using the lower bound, we can see that it is indeed a 2-approximation algorithm.

By searching for a good prefix edge from right to left on a permutation, we created a new algorithm, 2-PRg. Other than that, it works exactly as 2-PR. Then we simply modified 2-PR and 2-PRg for SBSR and called the new algorithms 2-SR and 2-SRg, respectively. They only deal with the gray edges presented in items 2, 4, 6, and 8 of Lemma 1 without the constraints over $i$, called *good suffix edges*.

Finally, we created two algorithms for SBPRSR, which will be called 2-PRSR and 2-PRSRg, respectively. Both of them search for any of the eight good edges given by Lemma 1, in that order. The only difference between them is how to scan the permutation: 2-PRSR searches for good prefix edges from right to left and for good suffix edges from left to right and 2-PRSRg does the opposite.

When a permutation does not contain a good edge, it is of one of the forms shown by Lemma 2. Now, we can transform it into $\iota$ with at most $b_{\rho_p \rho_s}(\pi) + 2$

reversals, as Lemma 3 shows. Despite this, 2-PRSR and 2-PRSRg still use 2 operations to eliminate one breakpoint sometimes, leading to either the identity or the reverse permutation. Therefore, $d_{\rho_p\rho_s}(\pi) \leq 2b_{\rho_p\rho_s}(\pi)+1$ and both algorithms have asymptotic approximation factor of 2.

**Lemma 2.** *If a permutation $\pi$ does not contain a good edge, then it is of one of the three forms: (1) $\eta$; or (2) $\sigma^1 = (\underbrace{p_1 \ldots 1}_{\ell_1} \underbrace{p_2 \ldots p_1+1}_{\ell_2} \ldots\ldots \underbrace{n \ldots p_b+1}_{\ell_{b+1}})$; or (3) $\sigma^2 = (\underbrace{p_b+1 \ldots n}_{\ell_1} \ldots\ldots \underbrace{p_1+1 \ldots p_2}_{\ell_b} \underbrace{1 \ldots p_1}_{\ell_{b+1}})$ where $b = b_{\rho_p\rho_s}(\pi)$ and $\ell_i \geq 2$ for all $1 \leq i \leq b+1$.*

*Proof.* Omitted due to space restrictions. □

**Lemma 3.** *Let $\pi$ be one of the three permutations shown by Lemma 2. If $\pi = \eta$, one reversal $\rho_p(n)$ sorts $\pi$. Otherwise, at most $b_{\rho_p\rho_s}(\pi)+2$ reversals sort $\pi$.*

*Proof.* Let $b = b_{\rho_p\rho_s}(\pi)$. If $\pi = \sigma^1$ and $b$ is an odd number, then the $b+1$ reversals $\rho_s(\ell_1+1) \cdot \rho_p(n-\ell_2) \cdot \rho_s(\ell_3+1) \cdot \rho_p(n-\ell_4) \cdot \ldots \cdot \rho_s(\ell_b+1) \cdot \rho_p(n-\ell_{b+1})$ transform $\pi$ into $\iota$, as we show next. Let $\pi^k$, $1 \leq k \leq \frac{b-1}{2}$, be the permutation we obtain after applying the first $2k$ prefix reversals of the sequence given above: $\rho_s(\ell_1+1)\cdot\rho_p(n-\ell_2)\cdot\ldots\cdot\rho_s(\ell_{2k-1}+1)\cdot\rho_p(n-\ell_{2k})$. We will show by induction on $k$ that $\pi^k$ is of the form $(\underbrace{p_{2k+1}\ldots p_{2k}+1}_{\ell_{2k+1}} \underbrace{p_{2k+2}\ldots p_{2k+1}+1}_{\ell_{2k+2}}$
$\ldots\ldots \underbrace{n\ldots p_b+1}_{\ell_{b+1}} \underbrace{1\,2\,\ldots\, p_{2k-2}+1\ldots p_{2k-1}\, p_{2k-1}+1\ldots p_{2k}}_{\ell_1+\ell_2+\ldots\ell_{2k}})$.

It is easy to see that it holds for $k = 1$. Now, assume that $\pi^{k-1}$ is of the form given above. Since $\pi^k = \pi^{k-1} \cdot \rho_s(\ell_{2k-1}+1) \cdot \rho_p(n-\ell_{2k})$, the result follows. At the end, $\iota = \pi^{(b-1)/2} \cdot \rho_s(\ell_b+1) \cdot \rho_p(n-\ell_{b+1})$.

If $\pi = \sigma^2$ and $b$ is odd, one must apply $\rho_p(n)$ to transform it into $\sigma^1$ and then apply the $b+1$ reversals given above.

If $\pi = \sigma^2$ and $b$ is an even number, then the $b+1$ reversals $\rho_p(n-\ell_{b+1}) \cdot \rho_s(\ell_b+1) \cdot \rho_p(n-\ell_{b-1}) \cdot \rho_s(\ell_{b-2}+1) \cdot \ldots \cdot \rho_p(n-\ell_3) \cdot \rho_s(\ell_2+1) \cdot \rho_p(n-\ell_1)$ sort $\pi$. This also can be shown by a similar induction as the one above. If $\pi = \sigma^1$ and $b$ is even, one must apply $\rho_p(n)$ to transform it into $\sigma^2$ and then apply the reversals given. □

### 3.2 Algorithms for Sbptst

Dias and Meidanis [7] presented a 2-approximation algorithm for SBPT, here called 2-PT, which always removes one breakpoint with one prefix transposition. They also demonstrated that there is at most one prefix transposition that removes two breakpoints at once, which leads to a greedy 2-approximation algorithm for the problem, called 2-PTg [10]: first it tries to remove two breakpoints and if this is not possible, it removes only one, as 2-PT does. Therefore, both guarantee that $d_{\tau_p}(\pi) \leq b_{\tau_p}(\pi)-1$, since the last transposition always removes two breakpoints [7], and both are 2-approximation algorithms.

It is simple to make suffix versions of both 2-PT and 2-PTg, which we will call 2-ST and 2-STg, respectively. 2-STg also tries to remove two breakpoints at once. If this is not possible, then it removes only one, as 2-ST does. So, both also guarantee that $d_{\tau_s}(\pi) \leq b_{\tau_s}(\pi) - 1$.

We created two algorithms for SBPTST, called 2-PTST and 2-PTSTg. The former always removes one breakpoint at a time, randomly choosing between a prefix or a suffix transposition to do so. To remove one breakpoint with one prefix transposition $\tau_p(i+1,j)$, let $\pi_i$ be the last element of the first strip of an arbitrary permutation $\pi$. If $\pi_i = n$, then choose $j = \pi^{-1}_{\pi_1-1}+1$. Otherwise, choose $j = \pi^{-1}_{\pi_i+1}$. To remove one breakpoint with one suffix transposition $\tau_s(i+1,j)$, let $\pi_j$ be the first element of the last strip of $\pi$. If $\pi_j = 1$, then choose $i = \pi^{-1}_{\pi_n+1}-1$. Otherwise, choose $i = \pi^{-1}_{\pi_j-1}$. The basic idea is to increase the first or last strip with either their previous or their next element.

2-PTSTg is more interesting, since it tries to remove two breakpoints using either prefix or suffix. A prefix transposition $\tau_p(i,j)$ removes two breakpoints from $\pi$ if $j = \pi^{-1}_{\pi_1-1}+1$, $i = \pi^{-1}_{\pi_j-1}+1$ and $2 \leq i < j \leq n$. It is easy to see that $\pi_1$ determines uniquely $j$ and $j$ determines uniquely $i$. A suffix transposition $\tau_s(i,j)$ removes two breakpoints from $\pi$ if $i = \pi^{-1}_{\pi_n+1}$, $j = \pi^{-1}_{\pi_{i-1}+1}$ and $2 \leq i < j \leq n$. Again, $\pi_n$ determines uniquely $i$ and $i$ determines uniquely $j$. If this removal is not possible, then we have to choose how to remove only one breakpoint as described above, which is always possible. Therefore, $d_{\tau_p\tau_s}(\pi) \leq b_{\tau_p\tau_s}(\pi)$. Also, note that the last transposition removes only one breakpoint. Hence, 2-PTST and 2-PTSTg are 2-approximation algorithms.

### 3.3 Algorithms for Sbprptsrst

Sharmin et al. [11] presented the Sorting by Prefix Reversals and Prefix Transpositions problem and provided a 3-approximation algorithm, called here 3-PRPT. It also uses the breakpoint graph to decide which operation to perform and it is similar to 2-PR; however, the use of a second operation allows the four types of gray edges to be considered *good edges*. In addition, they gave an important concept for their algorithm and for ours, presented in Lemma 4. Now, based on their work we can define what to do with each type of gray edge while also considering suffix reversals and suffix transpositions, as Lemma 5 shows.

**Lemma 4.** *[11] Let $(\pi_i, \pi_j)$ be a gray edge of type 1. Then there is at least one black edge $(\pi_{k-1}, \pi_k)$ for some $i < k < j$, that is called a **trapped black edge**.*

**Lemma 5.** *Let $\pi$ be an arbitrary permutation. There is a sequence of at most three prefix reversals, prefix transpositions, suffix reversals or suffix transpositions that removes at least one breakpoint if $G(\pi)$ contains at least one gray edge $(\pi_i, \pi_j)$: (1) of type 4 with $\pi_1 \neq 1$, $i = 1$, and $j \leq n$; (2) of type 4 with $\pi_n \neq n$, $j = n$, and $i \geq 1$; (3) of type 1 with $\pi_1 \neq 1$, $i = 1$, and $j \leq n$; (4) of type 2 with $\pi_n \neq n$, $j = n$, and $i \geq 1$; (5) of type 3 with $\pi_1 = 1$, $i \geq 1$, and $j \leq n$;*

(6) of type 3 with $\pi_n = n$, $i \geq 1$, and $j \leq n$; (7) of type 2 with $\pi_1 = 1$, $i \geq 1$, and $j \leq n$, where $\pi_i$ is the last element of the first strip of $\pi$; (8) of type 1 with $\pi_n = n$, $i \geq 1$, and $j \leq n$, where $\pi_j$ is the first element of the last strip of $\pi$.

*Proof.* If there is a gray edge $(\pi_i, \pi_j)$ then $\pi_j = \pi_i \pm 1$. To create an adjacency between $\pi_i$ and $\pi_j$ without creating new breakpoints one must perform, respectively, for each type of edge: (1) one prefix transposition $\tau_p(k, j+1)$ where $(\pi_{k-1}, \pi_k)$ is a trapped black edge, $i < k < j$; (2) one suffix transposition $\tau_s(k, j+1)$ where $(\pi_{k-1}, \pi_k)$ is a trapped black edge, $i < k < j$; (3) one prefix reversal $\rho_p(j-1)$; (4) one suffix reversal $\rho_s(i+1)$; (5) one prefix transposition $\tau_p(i+1, j)$; (6) one suffix transposition $\tau_s(i+1, j)$; (7) one prefix reversal $\rho_p(j)$, followed by one prefix reversal $\rho_p(j-i)$, and by one operation to handle an edge of type 4 or 1; (8) one suffix reversal $\rho_s(i)$, followed by one suffix reversal $\rho_s(n+1-(j-i))$, and by one operation to handle an edge of type 4 or 2. □

3-PRPT only deals with the gray edges shown in items 1, 3, 5, and 7 of Lemma 5 without the constraints over $j$, also called *good prefix edges*. It scans $G(\pi)$ from left to right to find its first good prefix edge, it decides its type (in the order that the four appear in the lemma), in addition to performing the required operation(s). Thus, it guarantees that $d_{\rho_p \tau_p}(\pi) \leq 3b_{\rho_p \tau_p}(\pi)/2$, which, using the lower bound, proves that it has an approximation factor of 3.

As explained, if a good prefix edge is of types 3 or 4, the algorithm applies one prefix transposition. However, since a prefix transposition can remove at most 2 breakpoints, we developed a greedy version, which we will call 3-PRPTg, whose features are: (i) it scans the permutation from right to left; (ii) it tries to find gray edges in a different order, namely items 1, 5, 3, and 7 of Lemma 5; (iii) when there is an edge $(\pi_i, \pi_j)$ of type 4, it tries to find the best trapped black edge $(\pi_{k-1}, \pi_k)$, $i < k < j$, such that $\pi_{j+1} = \pi_k \pm 1$ and $j \leq n-1$; (iv) when it is trying to find an edge of type 3, it searches for a $\pi_j$ such that $\pi_1 = \pi_{j-1} \pm 1$.

The suffix versions of both 3-PRPT and 3-PRPTg will be called 3-SRST and 3-SRSTg, respectively. Of course, they only deal with gray edges given by items 2, 4, 6, and 8 of Lemma 5 without the constraints over $i$, which are *good suffix edges*. They work similarly to their prefix versions, but 3-SRST scans the permutation from right to left and 3-SRSTg scans from left to right. Besides, 3-SRSTg (i) searches for gray edges in the order of the items 2, 6, 4, and 8 of Lemma 5; (ii) when there is an edge $(\pi_i, \pi_j)$ of type 4, it tries to find the best trapped black edge $(\pi_{k-1}, \pi_k)$, $i < k < j$, such that $\pi_{j+1} = \pi_k \pm 1$ and $i \geq 2$; and (iii) tries to find a $\pi_i$ such that $\pi_n = \pi_{i+1} \pm 1$ when it is searching for a type 3 edge.

Finally, we created 2-PRPTSRST and 2-PRPTSRSTg, algorithms for SBPRPT-SRST. They can handle all the good edges described in Lemma 5, but they do not consider the edges described in items 7 and 8. When the other six edges does not exist, $\pi$ is of one of the forms shown by Lemma 6 and the algorithms perform either a prefix reversal $\rho_p(n)$ or a prefix transposition to concatenate the first strip with the last one. Because of this, they can never separate the elements $n$ and 1, unless the black edge between them is the last one (disregard the edges $(\pi_0, \pi_1)$ and $(\pi_n, \pi_{n+1})$). This will guarantee that $d_{\rho_p \tau_p \rho_s \tau_s}(\pi) \leq b_{\rho_p \tau_p \rho_s \tau_s}(\pi) + 2$,

---
**Algorithm 1.** Good edges of type 4
---
PRPTSRST_EDGE_TYPE_4($\pi$, n)
1  **if** $\pi_1 \neq 1$ **and** $G(\pi)$ has a GPE $(1, \pi_{\mathrm{jp}})$ of type 4 **and** jp $\leq$ n **then**
2      (kp − 1, kp) ← trapped black edge;
3      **if** $\pi_\mathrm{n} \neq$ n **and** $G(\pi)$ has a GSE $(\pi_{\mathrm{is}}, \mathrm{n})$ of type 4 **and** is $\geq 1$ **then**
4          (ks − 1, ks) ← trapped black edge;
5          **if** $\pi_{\mathrm{kp}} - 1 = \pi_{\mathrm{jp}+1} \pm 1$ **then** $\quad \pi \leftarrow \pi \cdot \tau_p(\mathrm{kp}, \mathrm{jp} + 1)$
6          **else if** $\pi_{\mathrm{ks}} - 1 = \pi_{\mathrm{js}+1} \pm 1$ **then** $\quad \pi \leftarrow \pi \cdot \tau_s(\mathrm{ks}, \mathrm{js} + 1)$
7          **else if** jp < n − is **then** $\quad \pi \leftarrow \pi \cdot \tau_p(\mathrm{kp}, \mathrm{jp} + 1)$
8          **else** $\quad \pi \leftarrow \pi \cdot \tau_s(\mathrm{ks}, \mathrm{js} + 1)$
9      **else** $\quad \pi \leftarrow \pi \cdot \tau_p(\mathrm{kp}, \mathrm{jp} + 1)$
10 **else if** $\pi_\mathrm{n} \neq$ n **and** $G(\pi)$ has a GSE $(\pi_{\mathrm{is}}, \mathrm{n})$ of type 4 **and** is $\geq 1$ **then**
11     (ks-1, ks) ← trapped black edge;
12     $\pi \leftarrow \pi \cdot \tau_s(\mathrm{ks}, \mathrm{js} + 1)$
13 **return** $\pi$
---

---
**Algorithm 2.** Good prefix edges of type 1 and good suffix edges of type 2
---
PRPTSRST_EDGE_TYPE_1_2($\pi$, n)
1  **if** $\pi_1 \neq 1$ **and** $G(\pi)$ has a GPE $(1, \pi_{\mathrm{jp}})$ of type 1 **and** jp $\leq$ n **then**
2      **if** $\pi_\mathrm{n} \neq$ n **and** $G(\pi)$ has a GSE $(\pi_{\mathrm{is}}, \mathrm{n})$ of type 2 **and** is $\geq 1$ **then**
3          **if** jp < n − is **then** $\quad \pi \leftarrow \pi \cdot \rho_p(\mathrm{jp} - 1)$
4          **else** $\quad \pi \leftarrow \pi \cdot \rho_s(\mathrm{is} + 1)$
5      **else** $\quad \pi \leftarrow \pi \cdot \rho_p(\mathrm{jp} - 1)$
6  **else if** $\pi_\mathrm{n} \neq$ n **and** $G(\pi)$ has a GSE $(\pi_{\mathrm{is}}, \mathrm{n})$ of type 2 **and** jp $\leq$ n **then**
7      $\pi \leftarrow \pi \cdot \rho_s(\mathrm{is} + 1)$
8  **return** $\pi$
---

as Theorem 2 shows. With the lower bound, we can prove that the asymptotic approximation factor of both algorithms is 2.

The difference between 2-PRPTSRST and 2-PRPTSRSTg is that the former searches for good prefix edges from left to right, searches for good suffix edges from right to left and follows the order given by the lemma. The latter searches for good prefix edges from right to left, searches for good suffix edges from left to right, follows the order of items 1, 2, 5, 6, 3, and 4 of Lemma 5, and tries to find edges of types 3 and 4 that allow the removal of 2 breakpoints at once. Algs. 4 and 5 present them, respectively. In the algorithms, GPE stands for good prefix edge while GSE stands for good suffix edge.

**Lemma 6.** *Let $\pi \neq \iota$ be a permutation without the first six edges of Lemma 5. Then $\pi$ is either $\eta$, or of the form $\pi = (1\ 2\ \ldots\ k\ \ldots\ldots\ k{+}i\ \ldots\ k{+}2\ k{+}1$ $\ldots\ldots\ j{-}1\ j{-}2\ \ldots\ j{-}\ell\ \ldots\ldots\ j\ j{+}1\ \ldots\ n)$ with $i \geq 2$ and $\ell \geq 2$, or of the form $\pi = (n\ n{-}1\ \ldots\ j\ \pi_{n-j+2}\ \ldots\ldots\ \pi_{n-k}\ k\ k{-}1\ \ldots\ 1)$ with $\pi_{n-j+2} \neq j-1$ and $\pi_{n-k} \neq k+1$.*

**Lemma 7.** *Let $\pi \neq \eta$ be of one of the two other permutations given in Lemma 6. One transposition $\tau_p(i+1, n+1)$, where $\pi_i$ is the last element of the first strip, transforms $\pi$ into either $\pi \cdot \tau_p = (\ldots\ldots\ j\ j{+}1\ \ldots\ n{-}1\ n\ 1\ 2\ \ldots\ k{-}1\ k)$ or $\pi \cdot \tau_p$*

**Algorithm 3.** Good edges of type 3

PRPTSRST_EDGE_TYPE_3($\pi$, n)
1  **if**  $\pi_1 = 1$ **and** $G(\pi)$ has a GPE $(\pi_{\text{ip}}, \pi_{\text{jp}})$ of type 3 **then**
2      **if** $\pi_n = n$ **and** $G(\pi)$ has a GSE $(\pi_{\text{is}}, \pi_{\text{js}})$ of type 3 **then**
3          **if** $\pi_1 = \pi_{\text{jp}-1} \pm 1$ **then**      $\pi \leftarrow \pi \cdot \tau_p(\text{ip}+1, \text{jp})$
4          **else if** $\pi_n = \pi_{\text{is}+1} \pm 1$ **then**  $\pi \leftarrow \pi \cdot \tau_s(\text{is}+1, \text{js})$
5          **else if** jp $<$ n $-$ is **then**     $\pi \leftarrow \pi \cdot \tau_p(\text{ip}+1, \text{jp})$
6          **else**   $\pi \leftarrow \pi \cdot \tau_s(\text{is}+1, \text{js})$
7      **else**  $\pi \leftarrow \pi \cdot \tau_p(\text{ip}+1, \text{jp})$
8  **else if** $\pi_n = n$ **and** $G(\pi)$ has a GPE $(\pi_{\text{is}}, \pi_{\text{js}})$ of type 3 **then**
9      $\pi \leftarrow \pi \cdot \tau_s(\text{is}+1, \text{js})$
10 **return** $\pi$

**Algorithm 4.** A 2-approximation algorithm for SBPRPTSRST

2-PRPTSRST($\pi$, n)
1  **while** $\pi \neq \iota$ **do**
2      **if** $\pi_1 \neq 1$ **and** $G(\pi)$ has a GPE of type 4 **or**
       $\pi_n \neq n$ **and** $G(\pi)$ has a GSE of type 4 **then**
3          $\pi \leftarrow$ PRPTSRST_EDGE_TYPE_4($\pi$, n)
4      **else if** $\pi_1 \neq 1$ **and** $G(\pi)$ has a GPE of type 1 **or**
       $\pi_n \neq n$ **and** $G(\pi)$ has a GSE of type 2 **then**
5          $\pi \leftarrow$ PRPTSRST_EDGE_TYPE_1_2($\pi$, n)
6      **else if** $\pi_1 = 1$ **and** $G(\pi)$ has a GPE of type 3 **or**
       $\pi_n = n$ **and** $G(\pi)$ has a GSE of type 3 **then**
7          $\pi \leftarrow$ PRPTSRST_EDGE_TYPE_3($\pi$, n)
8      **else if** $\pi = \eta$ **then**
9          $\pi \leftarrow \pi \cdot \rho_p(n)$
10     **else**
11         Let k be the position of the last element of the first strip of $\pi$
12         $\pi \leftarrow \pi \cdot \tau_p(k+1, n+1)$

$= (\ldots\ldots\ k\ k{-}1\ \ldots\ 2\ 1\ n\ n{-}1\ \ldots\ j{+}1\ j)$ *without changing the number of breakpoints. After that, it is always possible to keep removing at least one breakpoint with one operation, if the algorithms never separate the elements* 1 *and* n.

**Lemma 8.** *Let* $(\pi_i, \pi_j)$ *be a gray edge of type 4 of either prefix or suffix of an arbitrary permutation* $\pi \neq \iota$. *If the edge between the elements* 1 *and* n *is the only trapped black edge between* $\pi_i$ *and* $\pi_j$, *then actually* $i = 1, j = n$ *and the permutation is either of the form* $\pi' = (k\ k{-}1\ k{-}2\ \ldots\ 2\ 1\ n\ n{-}1\ \ldots\ k{+}2\ k{+}1)$ *or of the form* $\pi'' = (k{+}1\ k{+}2\ \ldots\ n{-}1\ n\ 1\ 2\ \ldots\ k{-}2\ k{-}1\ k)$.

*Besides, by acting on such edge,* 2-PRPTSRST *and* 2-PRPTSRSTg *will be performing either their last or their last but one operation.*

**Lemma 9.** *The operation explained at Lemma 7 is performed at most once by* 2-PRPTSRST *and* 2-PRPTSRSTg, *if both never separate the elements* 1 *and* n.

**Theorem 2.** *Both algorithms* 2-PRPTSRST *and* 2-PRPTSRSTg *sort any permutation* $\pi \neq \iota$ *using at most* $b_{\rho_p \tau_p \rho_s \tau_s}(\pi) + 2$ *operations.*

**Algorithm 5.** A 2-approximation algorithm for SBPRPTSRST, greedy version

2-PRPTSRSTG($\pi$, n)
1   **while** $\pi \neq \iota$ **do**
2     **if** $\pi_1 \neq 1$ **and** $G(\pi)$ has a GPE of type 4 **or**
      $\pi_n \neq n$ **and** $G(\pi)$ has a GSE of type 4 **then**
3       $\pi \leftarrow$ PRPTSRST_EDGE_TYPE_4($\pi$, n)
4     **else if** $\pi_1 = 1$ **and** $G(\pi)$ has a GPE of type 3 **or**
      $\pi_n = n$ **and** $G(\pi)$ has a GSE of type 3 **then**
5       $\pi \leftarrow$ PRPTSRST_EDGE_TYPE_3($\pi$, n)
6     **else if** $\pi_1 \neq 1$ **and** $G(\pi)$ has a GPE of type 1 **or**
      $\pi_n \neq n$ **and** $G(\pi)$ has a GSE of type 2 **then**
7       $\pi \leftarrow$ PRPTSRST_EDGE_TYPE_1_2($\pi$, n)
8     **else if** $\pi = \eta$ **then**
9       $\pi \leftarrow \pi \cdot \rho_p(\mathbf{n})$
10     **else**
11       Let k be the position of the last element of the first strip of $\pi$
12       $\pi \leftarrow \pi \cdot \tau_p(\mathbf{k}+1, \mathbf{n}+1)$

*Proof.* Directly from Lemmas 6, 7, 8, and 9, whose proofs were omitted due to space restrictions.

## 4 Results

All the algorithms have complexity $O(n^2)$, since the distance is $O(n)$ and they spent linear time to choose and to apply an operation at each step. They were implemented in C language and executed in a Intel Core 2 of 2.13 GHz, 4GB RAM running Ubuntu 12.04.2 LTS under the same set of 190000 arbitrary permutations, being 10000 of each size $n$, for $n$ varying between 10 and 1000 in intervals of 5. Figure 2 shows the results. The $x$-axis presents a value of $n$ and the $y$-axis presents the average of the approximation factors of the permutations of that size, calculated using the theoretical lower bound of the distance.

We can see that the simple change of scanning the permutation at a different order had better results. This means that bigger operations (specially reversals) are preferable. As expected, problems which involve only prefix rearrangements are equivalent to their suffix versions. Besides, it was expected that problems with both prefix and suffix versions of a rearrangement would obtain better results than those that allow only the prefix version. It is interesting to notice that this did not happen for 2-PTST. Finally, for $n \geq 100$, the average approximation factor of 2-PRSRg is below 1.131, of 2-PTSTg is below 1.314 and of 2-PRPTSRSTg is below 1.382, which are the best algorithms for the three new problems we presented. Besides, the maximum factor of this three problems over all permutations tested is below 1.342, 1.596 and 1.600, respectively, when $n \geq 100$.

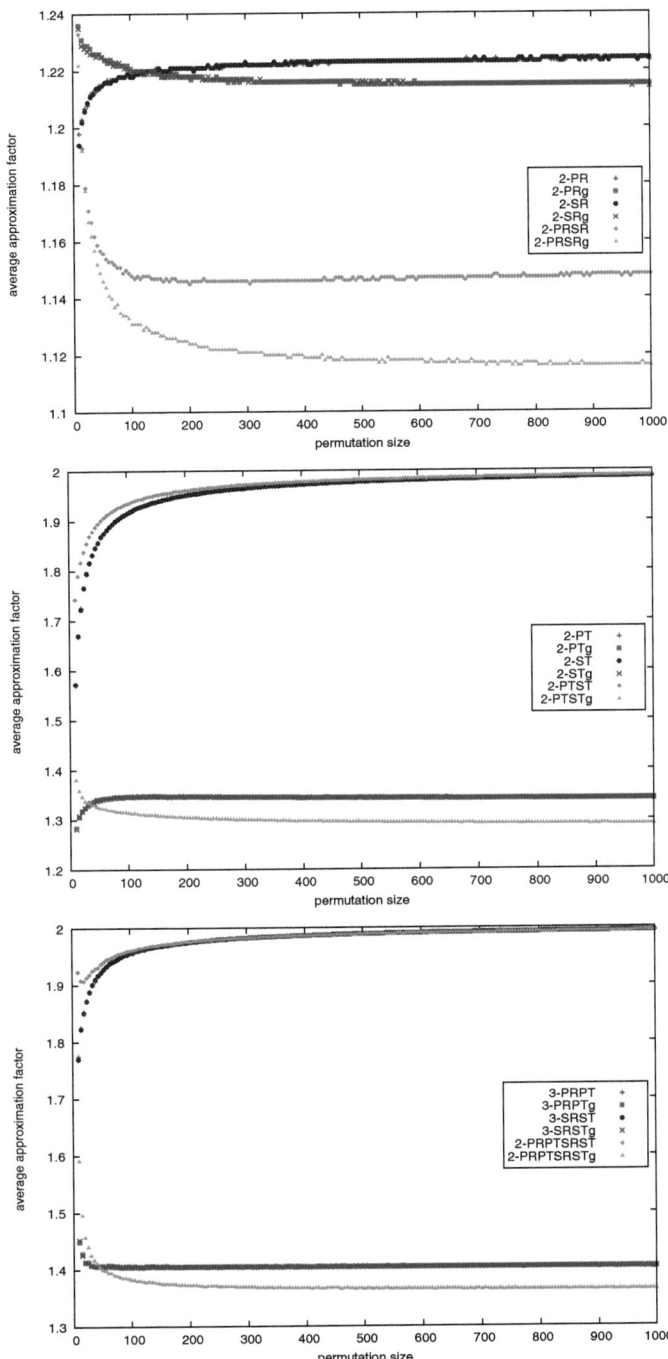

**Fig. 2.** Average approximation factor of all implemented algorithms when the permutation size grows

## 5 Conclusion

We introduced the study of suffix rearrangements along with prefix rearrangements. We showed lower bounds for the distances and described approximation algorithms of factor 2 to three new problems, considering some existing algorithms. Simple considerations, such as bigger operations and greedy choices, proved to be better options and improved the first versions of the algorithms. Future work will be directed not only to create new algorithms, but also to find results related to both distance and diameter of the problems.

**Acknowledgements.** This work was partially supported by São Paulo Research Foundation - FAPESP (grants 2013/01172-0 and 2013/08293-7) and National Counsel of Technological and Scientific Development - CNPq (grants 477692/2012-5 and 483370/2013-4). We thank Espaço da Escrita - Coordenadoria Geral da Universidade - UNICAMP - for the language services provided.

## References

1. Bafna, V., Pevzner, P.A.: Genome Rearrangements and Sorting by Reversals. In: Proceedings of the 34th Annual Symposium on Foundations of Computer Science (FOCS 1993), pp. 148–157 (1993)
2. Bafna, V., Pevzner, P.A.: Sorting by Transpositions. SIAM Journal on Discrete Mathematics 11(2), 224–240 (1998)
3. Berman, P., Hannenhalli, S., Karpinski, M.: 1.375-Approximation Algorithm for Sorting by Reversals. In: Möhring, R.H., Raman, R. (eds.) ESA 2002. LNCS, vol. 2461, pp. 200–210. Springer, Heidelberg (2002)
4. Bulteau, L., Fertin, G., Rusu, I.: Pancake flipping is hard. In: Rovan, B., Sassone, V., Widmayer, P. (eds.) MFCS 2012. LNCS, vol. 7464, pp. 247–258. Springer, Heidelberg (2012)
5. Bulteau, L., Fertin, G., Rusu, I.: Sorting by Transpositions Is Difficult. SIAM Journal on Computing 26(3), 1148–1180 (2012)
6. Caprara, A.: Sorting Permutations by Reversals and Eulerian Cycle Decompositions. SIAM Journal on Discrete Mathematics 12(1), 91–110 (1999)
7. Dias, Z., Meidanis, J.: Sorting by Prefix Transpositions. In: Laender, A.H.F., Oliveira, A.L. (eds.) SPIRE 2002. LNCS, vol. 2476, pp. 65–76. Springer, Heidelberg (2002)
8. Elias, I., Hartman, T.: A 1.375-Approximation Algorithm for Sorting by Transpositions. IEEE/ACM Transactions on Computational Biology and Bioinformatics 3(4), 369–379 (2006)
9. Fischer, J., Ginzinger, S.W.: A 2-Approximation Algorithm for Sorting by Prefix Reversals. In: Brodal, G.S., Leonardi, S. (eds.) ESA 2005. LNCS, vol. 3669, pp. 415–425. Springer, Heidelberg (2005)
10. Galvão, G.R., Dias, Z.: On the performance of sorting permutations by prefix operations. In: Proceedings of the 4th International Conference on Bioinformatics and Computational Biology (BICoB 2012), Las Vegas, Nevada, USA, pp. 102–107 (2012)
11. Sharmin, M., Yeasmin, R., Hasan, M., Rahman, A., Rahman, M.S.: Pancake Flipping with Two Spatulas. Electronic Notes in Discrete Mathematics 36, 231–238 (2010), International Symposium on Combinatorial Optimization (ISCO 2010)

# Algorithmic and Hardness Results for the Colorful Components Problems

Anna Adamaszek[1,*] and Alexandru Popa[2]

[1] Max-Planck-Institut für Informatik, Saarbrücken, Germany
anna@mpi-inf.mpg.de
[2] Faculty of Informatics, Masaryk University, Brno, Czech Republic
popa@fi.muni.cz

**Abstract.** In this paper we investigate the *colorful components* framework, motivated by applications emerging from comparative genomics. The general goal is to remove a collection of edges from an undirected vertex-colored graph $G$ such that in the resulting graph $G'$ all the connected components are *colorful* (i.e., any two vertices of the same color belong to different connected components). We want $G'$ to optimize an objective function, the selection of this function being specific to each problem in the framework.

We analyze three objective functions, and thus, three different problems, which are believed to be relevant for the biological applications: minimizing the number of singleton vertices, maximizing the number of edges in the transitive closure, and minimizing the number of connected components.

Our main result is a polynomial-time algorithm for the first problem. This result disproves the conjecture of Zheng et al. that the problem is $NP$-hard (assuming $P \neq NP$). Then, we show that the second problem is $APX$-hard, thus proving and strengthening the conjecture of Zheng et al. that the problem is $NP$-hard. Finally, we show that the third problem does not admit polynomial-time approximation within a factor of $|V|^{1/14-\epsilon}$ for any $\epsilon > 0$, assuming $P \neq NP$ (or within a factor of $|V|^{1/2-\epsilon}$, assuming $ZPP \neq NP$).

## 1 Introduction

In this paper we consider the following framework.

COLORFUL COMPONENTS FRAMEWORK: Given a simple, undirected graph $G = (V, E)$ and a coloring $\sigma : V \to C$ of the vertices with colors from a given set $C$, remove a collection of edges $E' \subseteq E$ from the graph such that each connected component in $G' = (V, E \backslash E')$ is a *colorful component* (i.e., it does not contain two identically colored vertices). We want the resulting graph $G'$ to be optimal according to some fixed *optimization measure*.

---
* Supported by the Alexander von Humboldt Foundation.

We consider three optimization measures and, respectively, three different problems: *Minimum Singleton Vertices (MSV)*, *Maximum Edges in Transitive Closure (MEC)*, and *Minimum Colorful Components (MCC)*. We now introduce the optimization measures for all these problems.

*Problem 1 (Minimum Singleton Vertices).* The goal is to minimize the number of connected components of $G'$ that consist of one vertex.

*Problem 2 (Maximum Edges in Transitive Closure).* The goal is to maximize the number of edges in the transitive closure of $G'$.

If a graph consists of $k$ connected components, each containing respectively $a_1, a_2, \ldots, a_k$ vertices, the number of edges in the transitive closure equals

$$\sum_{i=1}^{k} \frac{a_i \cdot (a_i - 1)}{2} \ .$$

*Problem 3 (Minimum Colorful Components).* The goal is to minimize the number of connected components in $G'$.

The first two problems have been introduced in [12], while the third one is newly introduced in this paper.

*Motivation.* The colorful components framework is motivated by applications originating from comparative genomics [10,12], which is a fundamental branch of bioinformatics that studies the relationship of the genome structure between different biological species. Research performed in this field can help scientists to improve the understanding of the structure and the functions of human genes and, consequently, find treatments for many diseases [8].

As pointed out in [10,12], one of the key problems in this area, the multiple alignment of gene orders, can be captured as a graph theoretical problem, using the colorful components framework. We refer the reader to [12] for an overview of the connection between the multiple alignment of gene orders and the graph theoretic framework considered, and for a discussion about the biological motivation of two particular problems we consider, MSV and MEC.

*Related work.* We now discuss the collection of known problems which fit into the colorful components framework.

We start with a problem named either *Colorful Components* [5,4] or *Minimum Orthogonal Partition* [7,12], since this problem has received the most attention so far. In this problem the objective function is to minimize the number of edges removed from $G$ to obtain the graph $G'$ in which all the components are colorful. Bruckner et al. show [5] that the problem is $NP$-hard for three or more colors and they study fixed-parameter algorithms for the problem. Their $NP$-hardness reduction can be modified slightly (starting the reduction from a version of 3SAT when each variable occurs only $O(1)$ times, instead of from the general 3SAT) to show the APX-hardness of the problem. Zheng et al. [12] and

Bruckner et al. [4] study heuristic approaches for the problem, and He et al. [7] present an approximation algorithm for some special case of the problem. As the general problem is a special case of the Minimum Multi-Multiway Cut, it admits a $O(\log |C|)$ approximation algorithm [2].

Other objective functions have been proposed, with the hope that some of them are both tractable and biologically meaningful. The MSV and the MEC problems have been introduced by Zheng et al. [12], who presented heuristic algorithms for the problems, without giving any worst-case approximation guarantee. They also conjectured both problems to be NP-hard.

Tremblay-Savard and Swenson [11] consider a Maximum Orthogonal Edge Cover Problem (MAX-OREC), which is a dual problem to MSV. There, the goal is to cover a maximum number of vertices of a graph using vertex-disjoint, non-singleton connected colorful subgraphs. In [11], a 2/3-approximation algorithm for MAX-OREC is presented. Although an approximation algorithm for MAX-OREC does not yield an approximation algorithm for MSV, an optimal solution for MSV gives also an optimal solution for MAX-OREC.

We are not aware of any other results concerning the MSV and MEC problems, or of any previous research on the MCC problem.

*Our results.* Our main result is a polynomial-time *exact* algorithm for the MSV problem, presented in Section 2. This disproves the conjecture of Zheng et al. [12] that the problem is $NP$-hard (assuming $P \neq NP$). Our algorithm maintains a feasible solution $G' = (V, E')$ for the MSV problem, starting with an edgeless graph $G' = (V, \emptyset)$. Then, in each step $G'$ is modified by applying to it a carefully chosen alternating path $p$, starting at a singleton vertex. The alternating path consists of the edges of $G$, and its every second edge is in $G'$. Applying $p$ to $G'$ means that the edges from $p$ which are not in $G'$ are added to $G'$, and at the same time the edges of $p$ which are in $G'$ are removed from $G'$. The algorithm ensures that at each step $G'$ is a feasible solution to the problem, and satisfies an invariant that all connected components in $G'$ are either singletons, edges or stars. In the analysis we show that when the algorithm does not find any new alternating path, the number of singleton components in $G'$ matches the lower bound presented in Section 2.1.

In Section 3 we study the MEC problem and we show that the problem is $NP$-hard and $APX$-hard when the number of colors in the graph is at least 4. This proves the conjecture of Zheng et al [12]. We show the result via a reduction from the version of the MAX-3SAT problem where each variable appears at most some constant number of times in the formula (see [1], Section 8.4).

Finally, in Section 4 we consider the MCC problem, which is introduced for the first time in this paper. We prove that MCC does not admit polynomial-time approximation within a factor of $|V|^{1/14-\epsilon}$, for any $\epsilon > 0$, unless $P = NP$ (or within a factor of $|V|^{1/2-\epsilon}$, unless $ZPP = NP$), even if each vertex color appears at most two times. We show the inapproximability result via a reduction from Minimum Clique Partition which is equivalent to Minimum Graph Coloring [9].

Due to space constraints some proofs have been omitted and will only appear in the full version of the paper.

## 2 A Polynomial-Time Exact Algorithm for MSV

In this section we present a polynomial-time algorithm MSVEXACT which finds an optimal solution for the MSV problem. First, in Section 2.1 we show a lower bound on the number of singleton vertices in any feasible solution for the problem. Then, in Section 2.2 we describe the algorithm, with its key procedure presented in Section 2.3. The analysis of the algorithm is made in Section 2.4.

### 2.1 Lower Bound

Let a graph $G = (V, E)$, together with a coloring $\sigma : V \to C$, be an instance of the MSV problem. For any color $c \in C$ let $V_c \subseteq V$ denote the set of vertices of color $c$. For any set of vertices $V' \subseteq V$ we denote by $N(V')$ the set of neighbors of $V'$ in $G$, i.e. $N(V') = \{v \in V \setminus V' : \exists v' \in V' \text{ s.t. } (v', v) \in E\}$. For any set of colors $C' \subseteq C$ and set of vertices $V' \subseteq V$ we denote by $N_{C'}(V')$ the set of neighbors of $V'$ in $G$ which have colors in $C'$, i.e. $N_{C'}(V') = \{v \in N(V') : \sigma(v) \in C'\}$.

**Lemma 1.** *For any color $c \in C$ let*

$$s_c = \max_{V' \subseteq V_c} \left( |V'| - |N_{C \setminus \{c\}}(V')| \right).$$

*Then in any feasible solution for MSV there are at least $s_c$ singletons of color $c$.*

*Proof.* Let $G' = (V, E')$, where $E' \subseteq E$, be a feasible solution for $G$. Fix a color $c$ for which $s_c > 0$ and let $V' \subseteq V_c$ be the subset maximizing the value of $s_c$. (Notice that $s_c$ depends only on the graph $G$, and not on $G'$.) For each vertex $v' \in V'$ which is not a singleton in $G'$ we pick an arbitrary neighbor $n(v')$ in $G'$. We have $n(v') \in N_{C \setminus \{c\}}(V')$. As any two vertices from $V'$ belong to different connected components in $G'$, the vertices $n(v')$ are pairwise different. The number of vertices of $V'$ which are not singletons in $G'$ is therefore at most $|N_{C \setminus \{c\}}(V')|$. The number of singletons amongst vertices from $V'$, and also the number of singletons of color $c$, is at least $|V'| - |N_{C \setminus \{c\}}(V')| = s_c$. □

**Corollary 1.** *Any feasible solution for MSV has at least $\sum_{c \in C} s_c$ singletons.*

### 2.2 Idea of the Algorithm

We now present an algorithm MSVEXACT which finds an optimal solution for MSV. The input consists of a simple, undirected graph $G = (V, E)$, together with a coloring $\sigma : V \to C$. The algorithm maintains a feasible solution $G' = (V, E')$ (i.e., $G'$ is a subgraph of the input graph $G$, and every connected component of $G'$ is a colorful component), starting with an edgeless graph $G' = (V, \emptyset)$. In each step the graph $G'$ is modified by applying to it a carefully chosen alternating path $p$. The alternating path consists of the edges of $G$, and its every second edge is in $G'$. Applying $p$ to $G'$ means that the edges from $p$ which are not in $G'$ are added to $G'$, and at the same time the edges of $p$ which are in $G'$ are removed from $G'$. See Algorithm 1 for the formal description of the algorithm.

```
Input: A simple, undirected graph G = (V, E), a coloring σ : V → C
Output: A subgraph of G minimizing the number of connected components,
        and in which each connected component is colorful
1  G' := (V, ∅)
2  foreach c ∈ C do
3  |   while p=ALTERNATING_PATH(G, σ, G', c) is a path do
4  |   |   apply p to G'
5  |   end
6  end
```

**Algorithm 1.** MSVEXACT$(G, \sigma)$

The path $p$ is chosen in such a way, that applying it to $G'$ decreases the number of singleton vertices of color $c$, without increasing the number of singleton vertices of other colors. Additionally, at each step of the algorithm $G'$ satisfies the invariant that each connected component of $G'$ is a singleton, an edge, or a star (where a star is a tree of diameter 2, in particular it has at least 3 vertices).

We will show that when the algorithm stops, i.e., when it does not find any alternating path $p$ which can be applied to $G'$ to decrease the number of singletons of any color, the number of singleton vertices in $G'$ matches the lower bound from Corollary 1.

### 2.3 Finding an Alternating Path

Let $G' = (V, E')$ be a feasible solution for an instance $(G = (V, E), \sigma)$ of MSV, such that each connected component of $G'$ is a singleton vertex, an edge, or a star. Let $c \in C$ be an arbitrary color, and let $S_c \subseteq V$ be the set of all singletons of color $c$ in $G'$. We describe a procedure ALTERNATING_PATH$(G, \sigma, G', c)$ which outputs an alternating path $p$ for $G'$ in $G$. In the following section we prove that $p$ satisfies all properties outlined in Section 2.2, and that when no path is found, the number of singletons of color $c$ in $G'$ matches the lower bound from Lemma 1.

The idea behind the path construction is as follows. We want to find a path starting in some singleton vertex of color $c$, connecting each vertex of color $c$ with a vertex of color different than $c$ using an edge $e \in E \setminus E'$; and each vertex of color different than $c$ with an vertex of color $c$ using an edge $e \in E'$. We end the construction of the path when the current endpoint $v \notin V_c$ of the path belongs to a connected component of $G'$ to which we can attach an additional vertex of color $c$ (possibly while splitting the component into two parts). Such a case occurs when $v$ is a leaf of a star (which will result in removing $v$ from the star-component and connecting it with the vertex of color $c$), or when the connected component of $v$ does not contain color $c$. Then applying the alternating path to the graph $G'$ results in "switching" vertices of color $c$ between different connected components of $G'$, and removing one singleton of color $c$, as the start point of the path will not be a singleton in the new graph. The algorithm performs a BFS search for the path satisfying the required conditions, starting with the collection of all singleton vertices of color $c$. See Procedure 2 for a formal description.

```
    Input: A simple, undirected graph G = (V, E), a coloring σ : V → C, a feasible
           subgraph G' = (V, E') of G, and a color c ∈ C
    Output: A path p or NO_PATH_FOUND
 1  V' := S_c
 2  N' := N_{C\{c}}(V')                                           // Neighbors in G
 3  ∀v ∈ N' pred(v) := any v' ∈ S_c s.t. (v, v') ∈ E
 4  while |N'| > 0 do
 5      if ∃v ∈ N' : v is a leaf of a star in G' then
 6          p := PATH_FROM(v)
 7          return p ∪ {(v, v')} s.t. (v, v') ∈ E'
 8      end
 9      if ∃v ∈ N' : the connected component of v in G' has no color c then
10          p := PATH_FROM(v)
11          return p
12      end
13      V'' := {v'' ∈ V_c : ∃v ∈ N' s.t. (v, v'') ∈ E'}
14      ∀v'' ∈ V'' pred(v'') := any v ∈ N' s.t. (v, v'') ∈ E'
15      V' := V' ∪ V''
16      N' := N_{C\{c}}(V') \ N_{C\{c}}(V' \ V'')
17      ∀v ∈ N' pred(v) := any v' ∈ V'' s.t. (v, v') ∈ E
18  end
19  return NO_PATH_FOUND
```

**Procedure 2.** ALTERNATING_PATH$(G, \sigma, G', c)$

```
   Input: A vertex v ∈ V
   Output: A path starting in S_c and ending in v
 1 if pred(v) ∈ S_c then
 2     return (pred(v),v)
 3 end
 4 return PATH_FROM(pred(v)) ∪ {(pred(v),v)}
```

**Procedure 3.** PATH_FROM$(v)$

Procedure ALTERNATING_PATH constructs the path $p$ as follows. It keeps a set of vertices $V'$ of color $c$, initially setting $V' := S_c$ (line 1). For each element $v \notin S_c$ considered by the procedure, its *predecessor* pred($v$) is fixed (line 3, 14, 17). Intuitively pred($v$) is an element such that (pred($v$), $v$) $\in E$, and processing pred($v$) by the procedure resulted in adding $v$ to one of the sets $V', N'$. Procedure PATH_FROM($v$), invoked in lines 6 and 10, can then reconstruct the whole path, starting from the final vertex $v$ and finding the predecessors until it reaches a vertex from $S_c$ (see Procedure 3 for a formal description).

Each loop of the algorithm (lines 4 – 18) considers the set $N'$ of new neighbors of the vertices from $V'$ (i.e., the neighbors of $V'$ which have not been considered in the previous loops), see lines 2 and 16, in search for vertices which can yield an end of the path (see lines 5, 9). If no such vertex is found, the set $V'$ will be further increased to include the neighbors of $N'$ of color $c$ (line 13, 15). The process continues until an appropriate vertex $v$ is found in $N'$ (lines 5, 9), and then the algorithm returns the path reconstructed from $v$, or the set $N'$ becomes empty, in which case the answer NO_PATH_FOUND is returned (line 19).

## 2.4 Analysis

**Lemma 2.** *When the procedure* ALTERNATING_PATH$(G, \sigma, G', c)$ *invoked for a graph $G'$ which is a feasible solution for MSV, and s.t. each connected component of $G'$ is a singleton, an edge or a star returns* NO_PATH_FOUND, *then* $|S_c| = s_c$.

*Proof.* If the procedure ALTERNATING_PATH returns NO_PATH_FOUND, then it returns in line 19, i.e., after checking the condition "$|N'| > 0$" (line 4) failed. We show that just before the procedure ends, the following inequality holds:

$$|V'| - |N_{C\setminus\{c\}}(V')| \geq |S_c| .$$

If the loop in line 4 has never been entered, we have $V' = S_c$, $N_{C\setminus\{c\}}(V') = N' = \emptyset$, and therefore $|V'| - |N_{C\setminus\{c\}}(V')| = |S_c|$.

Each vertex $v \in N_{C\setminus\{c\}}(V')$ has been inserted into $N'$ at some step of the procedure (line 2 or 16), and subsequently processed in line 5 and 9. As that did not cause the algorithm to return in line 7 or 11, we must have:

- $v$ is not a leaf of a star in $G'$, and
- the connected component containing $v$ contains a vertex colored with $c$.

As each connected component in $G'$ is a singleton, an edge or a star, and the color of $v$ is different from $c$, we have two possibilities:

- the connected component of $G'$ containing $v$ is an edge, and the other endpoint of the edge has color $c$, or
- the connected component of $G'$ containing $v$ is a star containing a vertex of color $c$, and $v$ is the center of the star.

As any connected component of $G'$ has at most one vertex $v$ satisfying one of the above conditions, any two elements of $N_{C\setminus\{c\}}(V')$ are in different connected components of $G'$. From the conditions above we also know that each vertex $v \in N_{C\setminus\{c\}}(V')$ has some neighbor $n(v)$ of color $c$ in $G'$. Each vertex $n(v)$ has been added to the set $V'$ when the element $v$ has been processed by the procedure (line 13, 15). As any two elements $v_1, v_2 \in N_{C\setminus\{c\}}(V')$ are in different connected components of $G'$, any two vertices $n(v_1), n(v_2)$ are different. As the elements from $S_c$ are singletons in $G'$, and therefore cannot be equal to any $n(v)$, and as $S_c \subseteq V'$, we get $|V'| \geq |S_c| + |N_{C\setminus\{c\}}(V')|$. We obtain the desired inequality.

We have shown that for the set of vertices $V'$ we have $|V'| - |N_{C\setminus\{c\}}(V')| \geq |S_c|$. As $V' \subseteq V_c$, we get $|S_c| \leq \max_{V'' \subseteq V_c}(|V''| - |N_{C\setminus\{c\}}(V'')|) = s_c$. As $s_c$ is a lower bound on $|S_c|$ (see Lemma 1), we get $|S_c| = s_c$. □

**Lemma 3.** *Let $G' = (V, E')$ be a feasible solution for MSV s.t. each connected component of $G'$ is a singleton, an edge or a star. Let $p$ be a path returned by* ALTERNATING_PATH$(G, \sigma, G', c)$, *and let $G''$ be the result of applying $p$ on $G'$. Then:*

a) $p$ is an alternating path for $G'$ in $G$,
b) the number of singleton vertices of color $c$ in $G''$ is smaller than in $G'$; the number of singleton vertices of any other color does not increase,
c) each connected component of $G''$ is a colorful component, and it is a singleton, an edge or a star.

Using Lemmas 2 and 3 we can show the main result of this section.

**Theorem 1.** *The algorithm* $\text{MSVEXACT}(G, \sigma)$ *finds an optimal solution for the MSV problem in time* $O(|V| \cdot |E|)$.

## 3 Hardness of MEC

In this section we prove the $NP$-hardness and the $APX$-hardness of the MEC problem, for $|C| \geq 4$. We show our result via a reduction from MAX-3SAT($\beta$), a version of the MAX-3SAT problem where each variable appears at most $\beta$ times in the formula. For $\beta = 3$ the problem is $APX$-hard (see [1], Section 8.4).

### 3.1 Reduction from MAX-3SAT($\beta$)

Given an instance of the MAX-3SAT($\beta$) problem, i.e., a 3-CNF formula $\phi$ with $m$ clauses and $n$ variables, where each variable appears at most $\beta$ times, we construct an instance of the MEC problem. Our instance is a vertex colored graph $G = (V, E)$, where the vertices are colored with colors from a four-element set $\{a, b, c, v\}$. An example of the reduction is illustrated in Figure 1.

First we describe the set of vertices $V$.

1. We add to $V$ a set of vertices $c_1, \ldots, c_m$, each colored with color $c$, where vertex $c_i$ corresponds to the $i$-th clause of the formula.
2. For a variable $x$, let $n_x$ be the number of occurrences of the literals $x$ and $\neg x$ in $\phi$. For each variable $x$, we add to $V$: $n_x$ vertices of color $a$ (denoted by $a_1^x, a_2^x, \ldots, a_{n_x}^x$), $n_x$ vertices of color $b$ (denoted by $b_1^x, b_2^x, \ldots, b_{n_x}^x$), and $2n_x$ vertices of color $v$ (denoted by $v_1^x, v_2^x, \ldots, v_{n_x}^x$ and $w_1^x, w_2^x, \ldots, w_{n_x}^x$). Intuitively, the vertices $v_i^x$ and $w_i^x$ are associated with $x$ and $\neg x$, respectively.

We now show how to construct the set of edges $E$.

1. For each variable $x$, we construct a cycle of length $4n_x$ by adding to $E$ the collection of edges $(a_i^x, v_i^x)$, $(v_i^x, b_i^x)$, $(b_i^x, w_i^x)$ and $(w_i^x, a_{(i \bmod n_x)+1}^x)$ for $i = 1, .., n_x$.
2. For each clause we add to $E$ three edges, where each edge connects the vertex $c_i$ representing the clause with a vertex representing one literal of $c_i$. More formally, if a literal $x$ ($\neg x$) occurs in the $i$-th clause, we add to $E$ an edge connecting $c_i$ with some vertex $v_j^x$ ($w_j^x$, respectively). We do this operation in such a way, that each vertex $v_j^x$ and $w_j^x$ representing a literal is incident with at most one clause-vertex $c_i$. Notice that since we have more vertices $v_j^x$ and $w_j^x$ than actual literals, some of the vertices $v_j^x$ and $w_j^x$ will not be connected with any clause-vertex $c_i$.

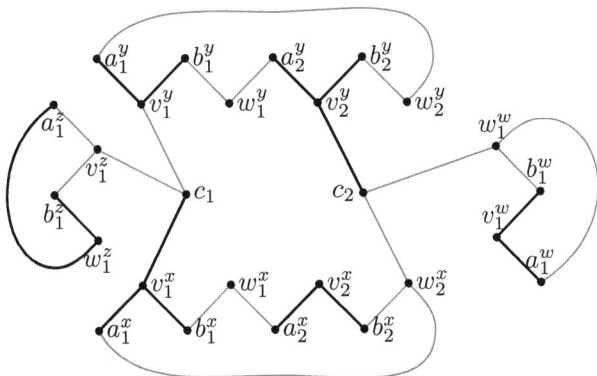

**Fig. 1.** An instance $G$ of the MEC problem corresponding to the 3SAT formula $(x \vee y \vee z) \wedge (\neg x \vee y \vee \neg w)$ (both black and gray edges). The subgraph $G''$ consisting of all vertices and only black edges represents a solution for $G$ corresponding to the following assignment: $f(x) = f(y) = f(w) = \text{TRUE}$, $f(z) = \text{FALSE}$.

### 3.2 Analysis of the Reduction

Let $\phi$ be a MAX-3SAT$(\beta)$ formula on $m$ clauses, and $G = (V, E)$ a vertex-colored graph obtained from $\phi$ by our reduction. Let $G' = (V, E')$ be a subgraph of $G$ which is an optimal solution for the MEC problem on $G$.

**Lemma 4.** *If the formula $\phi$ is satisfiable, then the transitive closure of $G'$ has at least $12m$ edges.*[1]

*Proof.* We construct a graph $G'' = (V, E'')$ which is a subgraph of $G$ in the following way (see Figure 1). Fix a satisfying assignment $f$ for $\phi$. For each clause, represented by a vertex $c_i$, we choose arbitrarily a literal $x$ ($\neg x$) which is satisfied by the assignment $f$. Let $v_j^x$ ($w_j^x$, respectively) be the vertex corresponding to the chosen literal which is incident with $c_i$ in $G$. We add the edge $(c_i, v_j^x)$ $((c_i, w_j^x)$, respectively) to $G''$. Additionally, each vertex $v_j^x$ and $w_j^x$ associated with a literal satisfied by $f$ is connected in $G''$ with the neighboring vertices of color $a$ and $b$.

It is straightforward to check that $G''$ is a feasible solution for the MEC problem (i.e., each connected component of $G''$ is colorful), and that $G''$ has $m$ connected components containing 4 vertices, $2m$ connected components containing 3 vertices, and $3m$ singletons. The transitive closure of $G''$ has $6 \cdot m + 3 \cdot 2m = 12m$ edges. As $G'$ is an optimal solution for the MEC problem in $G$, the transitive closure of $G'$ has at least as many edges as the transitive closure of $G''$. □

**Lemma 5.** *If any assignment can satisfy at most a $(1 - \epsilon)$ fraction of the $m$ clauses of the formula $\phi$, then the transitive closure of $G'$ has at most $12m - \Theta(\epsilon)m$ edges.*

---

[1] It can be proven that in this case the transitive closure of $G'$ has exactly $12m$ edges, but that is not needed in the later part of the reasoning.

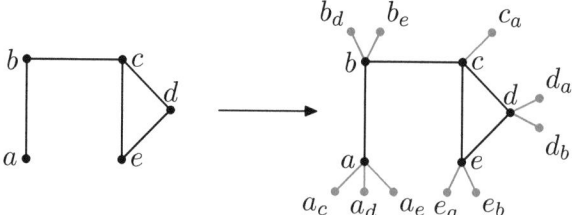

**Fig. 2.** Creating an instance of the MCC problem (right) from an instance of the Minimum Clique Partition (left). Base vertices and edges are drawn in black, and the additional ones in gray. An optimal solution for both problems is obtained by removing an edge $(b, c)$.

**Theorem 2.** *The Maximum Edges in Transitive Closure (MEC) problem is APX-hard, even for graphs with only four colors.*

## 4 Hardness of MCC

In this section we prove that the MCC problem does not admit polynomial-time approximation within a factor of $|V|^{1/14-\epsilon}$, for any $\epsilon > 0$, unless $P = NP$, or within a factor of $|V|^{1/2-\epsilon}$, unless $ZPP = NP$. The results hold even if each vertex color appears at most two times in the input graph. We prove our results via a reduction from the Minimum Clique Partition problem.

**Minimum Clique Partition:** Given a simple, undirected graph $G = (V, E)$, find a partition of $V$ into a minimum number of subsets $V_1, \ldots, V_k$ such that the subgraph of $G$ induced by each set of vertices $V_i$ is a complete graph.

The Minimum Clique Partition problem is equivalent to Minimum Graph Coloring [9], and therefore it cannot be approximated in polynomial time within a factor of $|V|^{1/7-\epsilon}$ for any $\epsilon > 0$ [3], unless $P = NP$, or within a factor of $|V|^{1-\epsilon}$, unless $ZPP = NP$ [6].

### 4.1 Reduction from Minimum Clique Partition

Let $G = (V, E)$ be an instance of the Minimum Clique Partition problem. We create an instance of the MCC problem, i.e., a vertex colored graph $G' = (V', E')$, as follows. The reduction is illustrated in Figure 2.

1. The vertex set $V' = V'_b \cup V'_a$ consists of two parts. The set $V'_b = V$ is the set of all vertices in $G$, each colored with a distinct color. We term these vertices *base vertices*. The set $V'_a$ has two vertices, $u_v$ and $v_u$, for each pair of vertices $u, v \in V$ such that $(u, v) \notin E$. Both vertices $u_v$ and $v_u$ have the same color, which is different from other colors in the graph. We refer to the vertices from $V'_a$ as *additional vertices*. We emphasize that each color appears *at most* two times in $G'$.

2. The set of edges $E' = E'_b \cup E'_a$ consists of two parts. First, $E'_b = E$ is the set of edges in $G$, which we term *base edges*. The set $E'_a$ has two edges, $(u_v, u)$ and $(v_u, v)$, for each pair of vertices $u, v \in V$ such that $(u, v) \notin E$ (i.e., each additional vertex $u_v$ is connected with a base vertex $u$). We refer to the edges from $E'_a$ as *additional edges*.

### 4.2 Analysis of the Reduction

Let $G = (V, E)$ be an instance of the Minimum Clique Partition problem, and $G' = (V', E')$ the corresponding instance of MCC, obtained by our reduction. We first compare the costs of the optimal solution for both problem instances, which leads to the main theorem of this section.

**Lemma 6.** *If there is a partition of $G$ into $k$ cliques, then the optimal solution for the MCC problem for $G'$ has cost at most $k$.*

*Proof.* Let $G$ be a graph which can be partitioned into $k$ cliques. We have to show that there is a collection of edges $E'' \subseteq E'$ in $G'$, such that after removing $E''$ from $G'$ we obtain a graph consisting of at most $k$ colorful components. The set of edges $E''$ is exactly the set of base edges that have been removed from $G$ to obtain the collection of $k$ cliques.

As we do not remove any additional edges of $G'$ (i.e., the edges from the set $V'_a$), the resulting graph consists of $k$ connected components. The only pairs of vertices sharing the same color are pairs $u_v, v_u$ such that $u, v \in V$ and $(u, v) \notin E$. Then $u$ and $v$ must be in different connected components of the clique partition, and so $u$ and $v$ (and therefore also $u_v$ and $v_u$) are in different connected components of the constructed graph. Each connected component of the constructed graph is colorful. □

**Lemma 7.** *If the optimal solution for the MCC problem for $G'$ has cost $k$, then there exists a partition of $G$ into $k$ cliques.*

**Theorem 3.** *The Minimum Colorful Components (MCC) problem does not admit polynomial-time approximation within a factor of $n^{1/14-\epsilon}$, for any $\epsilon > 0$, unless $P = NP$, or within a factor of $n^{1/2-\epsilon}$, for any $\epsilon > 0$, unless $ZPP = NP$, where $n$ is the number of vertices in the input graph.*

## 5 Open Problems

The $APX$-hardness result for the MEC problem requires that the input graphs are colored with at least four colors. A natural question is, thus, to settle the complexity of the problem for three colors (as for the case of two colors MEC is easily solvable in polynomial time, using a maximum matching algorithm). Another open question is to design approximation algorithms for the MEC problem or to strengthen the hardness of approximation result.

From the biological perspective it is interesting to analyze how our MSV algorithm behaves on real data. Finally, we mention that an intriguing and challenging task is to find problems in this framework that admit practical algorithms and are also meaningful for the biological applications.

**Acknowledgements.** We would like to thank Guillaume Blin for introducing us to the problem and for useful discussions.

# References

1. Ausiello, G., Protasi, M., Marchetti-Spaccamela, A., Gambosi, G., Crescenzi, P., Kann, V.: Complexity and Approximation: Combinatorial Optimization Problems and Their Approximability Properties, 1st edn. Springer-Verlag New York, Inc., Secaucus (1999)
2. Avidor, A., Langberg, M.: The multi-multiway cut problem. Theoretical Computer Science 377(1-3), 35–42 (2007)
3. Bellare, M., Goldreich, O., Sudan, M.: Free bits, PCPs, and nonapproximability - towards tight results. SIAM Journal on Computing 27(3), 804–915 (1998)
4. Bruckner, S., Hüffner, F., Komusiewicz, C., Niedermeier, R.: Evaluation of ILP-based approaches for partitioning into colorful components. In: Bonifaci, V., Demetrescu, C., Marchetti-Spaccamela, A. (eds.) SEA 2013. LNCS, vol. 7933, pp. 176–187. Springer, Heidelberg (2013)
5. Bruckner, S., Hüffner, F., Komusiewicz, C., Niedermeier, R., Thiel, S., Uhlmann, J.: Partitioning into colorful components by minimum edge deletions. In: Kärkkäinen, J., Stoye, J. (eds.) CPM 2012. LNCS, vol. 7354, pp. 56–69. Springer, Heidelberg (2012)
6. Feige, U., Kilian, J.: Zero knowledge and the chromatic number. Journal of Computer and System Sciences 57(2), 187–199 (1998)
7. He, G., Liu, J., Zhao, C.: Approximation algorithms for some graph partitioning problems. Journal of Graph Algorithms and Applications 4(2) (2000)
8. Mushegian, A.R.: Foundations of Comparative Genomics. Elsevier Science (2010)
9. Paz, A., Moran, S.: Non deterministic polynomial optimization problems and their approximations. Theoretical Computer Science 15(3), 251–277 (1981)
10. Sankoff, D.: OMG! Orthologs for multiple genomes - competing formulations - (keynote talk). In: Chen, J., Wang, J., Zelikovsky, A. (eds.) ISBRA 2011. LNCS (LNBI), vol. 6674, pp. 2–3. Springer, Heidelberg (2011)
11. Savard, O.T., Swenson, K.M.: A graph-theoretic approach for inparalog detection. BMC Bioinformatics 13(S-19), S16 (2012)
12. Zheng, C., Swenson, K., Lyons, E., Sankoff, D.: OMG! Orthologs in multiple genomes - competing graph-theoretical formulations. In: Przytycka, T.M., Sagot, M.-F. (eds.) WABI 2011. LNCS, vol. 6833, pp. 364–375. Springer, Heidelberg (2011)

# On the Stability of Generalized Second Price Auctions with Budgets

Josep Díaz[1,*], Ioannis Giotis[1,2,*], Lefteris Kirousis[3], Evangelos Markakis[2], and Maria Serna[1,*]

[1] Departament de Llenguatges i Sistemes Informatics
Universitat Politecnica de Catalunya, Barcelona
[2] Department of Informatics
Athens University of Economics and Business, Greece
[3] Department of Mathematics, National & Kapodistrian University of Athens,
Greece and Computer Technology Institute & Press "Diophantus"
{diaz,igiotis,mjserna}@lsi.upc.edu,
lkirousis@math.uoa.gr, markakis@gmail.com

**Abstract.** The Generalized Second Price (GSP) auction used typically to model sponsored search auctions does not include the notion of budget constraints, which is present in practice. Motivated by this, we introduce the different variants of GSP auctions that take budgets into account in natural ways. We examine their stability by focusing on the existence of Nash equilibria and envy-free assignments. We highlight the differences between these mechanisms and find that only some of them exhibit both notions of stability. This shows the importance of carefully picking the right mechanism to ensure stable outcomes in the presence of budgets.

## 1 Introduction

Advertising on Internet search engines has evolved into a phenomenal driving force both for the search engines and the advertising businesses. It is a modern and rapidly growing method that is now being implemented in various other popular sites beyond search engines, such as blogs, and social networking sites. Although some rightful concern has been raised regarding privacy issues and distinguishability from non-sponsored results, there are clear advantages to the advertisers who can efficiently reach their target audiences and observe the results of their ad campaign within days or even hours. At the same time, online ads account for a large share of the profits for search engines and other participating web-sites. Even the web-user experience can be enhanced, by the delivery of additional information relevant to their queries.

---

* Josep Díaz, Maria J. Serna and Ioannis Giotis supported by the CICYT project TIN-2007-66523 (FORMALISM). This research has also been co-financed by the European Union (European Social Fund ESF) and Greek national funds through the Operational Program "Education and Lifelong Learning" of the National Strategic Reference Framework (NSRF) - Research Funding Program: Thales. Investing in knowledge society through the European Social Fund.

In a typical instance, a user queries a search engine for a particular keyword of commercial interest, and the search engine determines the ads to be displayed by means of an auction. The prevailing system uses a pay-per-click policy, i.e., it only charges an advertiser when the user clicks on the corresponding link and is diverted to the advertiser' s web-site.

The mechanism used can be viewed as an auction for multiple homogenous indivisible items, the advertisement areas available, with single-demand buyers, since it is not desirable for the same advertisement to appear more than once. Such auctions can find applications in a wide variety of scenarios besides Internet advertising where buyers have a valuation per unit of a particular good but the setting is restricted to selling only single fixed-sized bundles. For example, consider selling different fixed-sized shipments of a food product when the seller cannot send more than one shipment to the same destination or frequency spectrum auctions of different fixed-sized bandwidths where regulations do not permit buying more than one continuous bandwidth. Noting that the mechanisms can be applied in much different scenarios, we will work in the context of Internet search advertising both because of its wide-spread application today and also because of the significant focus it receives in related literature.

In the early history of sponsored search auctions, the allocation of slots to advertisers was determined by a *first-price auction*, as in the systems originally used by Overture. Later on, Google was the first to switch to a *second-price auction*, an approach which demonstrated superior characteristics and was quickly adopted by the rest of the major search engines. The main idea is that the advertisers declare how much they are willing to pay for a click to their ad but they are charged instead a lesser amount equal to the next lower competing bid. Apart from its elegant simplicity, this scheme has been quite successful in terms of its generated revenue as well. In the literature, this system is commonly known as the *Generalized Second-Price* (GSP) auction and, by now, a large volume of work has emerged on the study of GSP auctions and related mechanisms, see e.g., the surveys [15] and [17].

However, an aspect that has been often ignored, especially in the early literature, is the presence of a budget constraint, requested from the advertisers to limit their exposure and expenditure. We believe this is a key parameter, essential in accurately understanding and evaluating the systems used in practice.

**Our Contribution.** Our main conceptual contribution is a study of Generalized Second-Price auctions under the presence of budgets. First we showcase that ignoring these constraints might lead to unstable outcomes. We then introduce three simple and natural extensions of the GSP mechanism, that take budgets into account. As it is not straight-forward to define a single natural mechanism, we define these variants motivated by the key desirable properties of second-price auctions. For all mechanisms, we investigate the existence of Nash equilibria and envy-free assignments, which are the main notions of stability that have been considered in the literature. For the first mechanism, we show that a Nash equilibrium might not always exist, yet an envy-free assignment is always achievable. For the other two mechanisms, we show that they always possess envy-free Nash equilibria, in fact we show that any envy-free assignment can be realized as an equilibrium of the mechanisms under consideration. In our model, we consider the budget as part of a bidder's strategy, i.e., we have *private* budgets. An interesting and surprising outcome of our study is that in the case of public budgets Nash equilibria do not always exist, despite the existence of envy-free assignments. In contrast to mechanism design problems, where having public budgets usually eases the design of a truthful algorithm, here we realized that having public budgets may eliminate the existence of stable profiles. Of separate interest might be an algorithmic process than can construct an envy-free assignment in our model.

## 1.1 Related Work

Varian [19] and Edelman et al. [11] have been the two seminal works on equilibrium analysis of GSP auctions without budgets. They established the existence of a Nash equilibrium which also satisfies other desirable stability properties such as being welfare-maximizing and envy-free. A further analysis of envy-free Nash equilibria, by taking into account the quality factor of the advertisers was also provided in [16].

The notion of budget constraints has been introduced in various models and objectives, such as, among others, in [4], [6], [12], [7] and [13]. Recent work on truthful mechanism design, mainly inspired by the clinching auction of Ausubel [3], has led to the introduction of truthful Pareto-optimal mechanisms in the presence of budgets, see e.g. [10,14,9]. Ashlagi et al.introduced the model we'll be using in [2]. However, all these mechanisms employ techniques that are very different from the GSP scheme in order to achieve truthfulness. As a result, they lack the simplicity of second price auctions at the expense of achieving better properties.

The work of Arnon and Mansour [1] is conceptually closer to our approach. They studied second-price auctions with budget constraints but their model simplifies the items for sale to clicks, as opposed to the slots, allowing a player to potentially receive more or less clicks than a single slot could offer. This deviates from the one player per slot paradigm used in practice.

Finally, a different direction that has been pursued recently is the performance of mechanisms in terms of the generated social welfare. Price of Anarchy analysis for auctions was initiated in [8], see also [5], for sponsored search auctions without budgets. For certain settings with budget constraints, some results have been recently obtained in [18] (which however do not have any implications for our proposed mechanisms). Our work does not focus on Price of Anarchy, which we leave for future research, but on existence of stability concepts.

## 2 Preliminaries

### 2.1 Model

Our model is the same as in Ashlagi et al. [2], a natural extension to budget limited players of the model introduced by Varian [19] which is widely adopted in related literature.

We assume we have $k$ slots, each with a fixed, distinct[1] and publicly known click-through rate (CTR), $\theta_j$ for slot $j$, representing the number of clicks received in a fixed time period (typically a day), independently of the advertisement displayed. Let us order the slots such that $\theta_1 > \theta_2 > \ldots > \theta_k$. Even though the click-through rates are probabilistic in nature, we will make the typical assumption that they are *deterministically* realized for simplification purposes; $\theta_i$ will really correspond to the expected click-through rate of slot $i$. Contrary to the numerical ordering, we will typically use the terminology "higher" and "lower" slots referring to slots of higher and lower CTR. Finally, for ease of illustration, we will ignore the bidder-dependent quality factor that is usually incorporated in calculating click-through rates in the separable model.

We have $n \geq k$ players (advertisers). Each player $i$ has a private *valuation* $v_i$ representing the perceived value per click. Each player also has a *budget constraint* $B_i$, indicating the total amount he is willing to spend in a fixed time period, not on a per click basis. We will also assume that these budget values are pairwise distinct. This assumption has been necessary in other works as well [2], and affects many properties in related mechanisms [10,2]. In fact, as we will exhibit later on, envy-free assignments, which is one of the stability concepts we are interested in, are not guaranteed to exist when budgets are not distinct. Hence, similarly to previous works, we also choose to adopt the distinctness of budgets.

Each player $i$ is interested in maximizing $\theta_{s(i)}(v_i - p(i))$, where $s(i)$ is the slot assigned to $i$ and $p(i)$ the accompanying price per click requested by the respective mechanism. At the same time, he must also satisfy the budget constraint, $\theta_{s(i)}p(i) \leq B_i$. If this condition holds, we say the player can afford slot $s(i)$. We wish to enforce strict budget constraints so we define the utility of the players whose budget constraints are violated to be minus infinity as is typically done in the literature; any other negative value would also serve our purpose (i.e., budget violations are less desirable than not getting a slot). More formally,

$$u_i = \begin{cases} 0, & \text{if } i \text{ was not awarded a slot,} \\ \theta_{s(i)}(v_i - p(i)), & \text{if } \theta_{s(i)}p(i) \leq B_i, \\ -\infty, & \text{otherwise.} \end{cases}$$

### 2.2 Second-Price Auctions under Budgets

The players submit *value-bids* $b_i$, representing the maximum amount they are willing to pay per click. These bids do not necessarily form a truthful declaration of the players' values to the mechanism. Similarly, the players also submit a *budget-bid* $g_i$ to declare their budget. We will use the term bid to refer to the combination of these two types or to one particular type when clear from context. In the case of ties, we assume there exists a fixed a priori defined ordering of the players based on which tie-breaking is resolved.

It is not trivial to introduce mechanisms that take budgets into account in a straightforward and natural way. To address this, we first ponder what con-

---
[1] This assumption is derived from the distinct space these slots occupy on a web-page.

stitutes a second-price mechanism by noting some key properties of generalized second-price auctions:

- The slot allocation should be performed by a simple and efficient process.
- The allocation should be in accordance with the bid ordering. If a player raises his bid he should be getting at least the slot he was getting before and should he lower his bid he should be getting at most the previous slot.
- Furthermore, if a bidder raises his bid, this should cause his total payment to potentially rise and respectively lowering his bid potentially lowers his payment.
- Finally, the price per click for each slot should be determined by either the next lower bid, the bid of the player awarded the next slot or the minimum bid required to obtain the slot. While these three concepts coincide in the regular GSP mechanism, this is not the case when one introduces budgets.

We first consider the GSP mechanism without budgets in our context, only to highlight that ignoring budgets can lead to unstable outcomes.

**Definition 1 (Budget-Oblivious).** *The* budget-oblivious *second-price auction, in short,* BOSP, *orders the value-bids in decreasing order and then assigns the slots in that order,* ignoring *the budget constraints. Naturally, the price for each slot is determined by the immediately lower value-bid.*

Since BOSP, as is shown in Section 3, does not have good stability properties, we turn our attention on mechanisms that respect the budget constraints by not assigning slots/prices to players that can afford them as declared by their budget-bids. The first interpretation of second-price pricing, charging the next lower bid, leads us to the following mechanism.

**Definition 2 (Budget-Conscious by Price).** *We define the* budget-conscious by price *second-price auction, in short,* BCSP(PRICE), *as the mechanism which first orders the value-bids in decreasing order and assigns a price per click for each player equal to the immediately lower value-bid in the bid ordering. Then,* BCSP-(PRICE) *assigns the players to slots in order of decreasing value-bids, respecting the budget constraints of each player as declared by their budget-bids, by assigning each player to the highest unassigned slot he can afford with his assigned price. If the player cannot afford any slot, he is left unassigned and he is not charged anything.*

Note that under this mechanism a player assigned to a slot might end up paying more per click than a player in a higher slot but we are guaranteed that all budget constraints of assigned players are satisfied. Also note that some slots might end up unassigned if no player can afford to occupy them.

The mechanism above is a natural way to guarantee budget compliance but raises a fairness issue, as players might be declaring value-bids as the maximum amount they are willing to pay and getting a slot that they cannot afford to pay if they were to pay their own bid. As will be evident in the later sections, this can lead to players intentionally raising their bid to just below their competitor's bid. The following mechanism addresses this.

**Definition 3 (Budget-Conscious by Bid).** *We define the* budget-conscious by bid *second-price auction, in short,* BCSP(BID), *similarly to* BCSP(PRICE) *except the mechanism now requires the players to be able to afford their slot if they were to pay a price per click equal to their own value-bid. The players are ordered in decreasing order of value-bids and the price of each player is set to the next lower bid. Then, the players are assigned from the highest bidder to the lower, one by one, to the highest available unassigned slot that they can afford should they were to pay their own bid.*

The second way of interpreting second-price prices, charging prices equal to the value-bid of the player ending up occupying the next slot, first, has definitional issues since we cannot know if our player can afford a slot without knowing who gets the next one and secondly, might charge a player more than his bid. For these reasons, we do not investigate mechanisms of this type in this work.

Finally, we introduce a mechanism that considers what the players are willing and afford to pay for a slot by considering the minimum as implied from their value and budget bids. This mechanism essentially captures pricing by charging the minimum amount required to obtain a slot.

**Definition 4 (Best Offer Budget-Conscious).** *We define the* best offer budget-conscious *second-price auction, in short,* BCSP(BEST OFFER), *as the mechanism which intuitively awards each slot to the player that can offer the most "money" but charges them the next lower amount offered. More formally, each slot s, one by one from higher to lower, is awarded to the unassigned player with the largest* $\min\{b_i, g_i/\theta_s\}$ *and he is charged a price per click equal to the second largest such value among unassigned players. We note that under this mechanism, the price charged for each slot is the minimum bid required to secure the slot. Alternatively, one can think of the slot rewarded to the player with the highest* $\min\{\theta_s b_i, g_i\}$, *and paying the second highest such amount, representing the total offer of the player and the total price charged.*

It should be pointed out that a slot in BCSP(BEST OFFER), in turn of decreasing CTR, is offered to the player who is willing to pay the most, given that does not violate his declared budget; he is then charged what the second such player would pay (not counting players who already got a slot). Whereas in the previous two budget-conscious mechanisms, each player, in turn of its bid, chooses the best object he can afford and pays the bid of the next player in line. The above two approaches are obviously equivalent in any mechanism that does not refuse giving a slot to a player who cannot afford it. However, once we introduce into the mechanism the additional requirement of refusing to give objects to anybody who cannot afford it, then the above distinction becomes necessary.

In all mechanisms, we assume that players not awarded a slot are not charged a payment and that the price of the lowest bidding player is zero, should he be awarded a slot. Finally, given a finite set of players, we note that the allocation of slots and pricing can be determined efficiently in all defined mechanisms.

## 2.3 Stable Assignments

It is easy to see that none of the mechanisms defined above are *incentive-compatible*. There are cases where a player might receive a higher utility by "lying" about his value and getting a lower slot at a beneficial price, even if other players are truthfully bidding their values. Naturally, we turn our attention to notions of stability, a requirement to analyze significant properties of these auctions and in general a desired property for the advertisers as well. As usual, we will focus on the notion of *Nash equilibrium*.

**Definition 5 (Nash Equilibrium).** *A profile of bids, $\langle b_i, g_i \rangle$ for each player $i$, forms a Nash equilibrium if no player has an incentive to deviate to a different strategy $\langle b'_i, g'_i \rangle$, for any $\langle b'_i, g'_i \rangle$.*

In related work [19,11], the notion of *symmetric* or *envy-free* equilibrium was defined. Under the generalized second-price auction without budget constraints, this class of envy-free equilibria is a subset of Nash equilibria.

**Definition 6 (Envy-Free Assignment).** *We define an envy-free assignment as a slot allocation $s(\cdot)$, where no slot is left unassigned, along with a set of prices per click $p(\cdot)$, assigning slot $s(i)$ to player $i$ and charging him $p(i)$ per click, such that for all players $i$ we have*

$$\forall i' \text{ with } 1 \leq s(i') \leq k, u_i \geq \begin{cases} \max\{\theta_{s(i')}(v_i - p(i')), 0\}, & \text{if } \theta_{s(i')} p(i') \leq B_i, \\ 0, & \text{otherwise,} \end{cases}$$

*where $u_i$ is the utility of player $i$ as defined earlier.*

Note that an envy-free assignment also guarantees *rationality*: $p(i) \leq v_i$ for all players $i$. We say that an envy-free assignment is *realizable* under a certain mechanism, if a set of bids exists such that the allocation and pricing generated by the mechanism under this set of bids matches the allocation and pricing of the envy-free assignment.

Under BOSP, where slot allocation depends only on the value-bids and not on the budgets or budget-bids, the constraints on the bids that realize an envy-free assignment are stricter than those of a Nash equilibrium. The same holds for BCSP(BID), as all players can pay their own bid for their slot and intuitively cannot be forced out of position by someone else's bid[2]. Similarly, under BCSP(BEST OFFER), a player cannot get a higher slot without paying more than the current player occupying the slot; again a realizable envy-free assignment effectively produces a Nash equilibrium. Hence, under the mechanisms BOSP, BCSP(BID), BCSP(BEST OFFER), the realizable envy-free assignments form a subset of the set of Nash equilibria.

Under BCSP(PRICE) however, the slot allocation is dependent on budgets and intuitively, one could alter the allocation to his benefit by forcing other

---

[2] In more detail, the instability arises when some player can alter his bid to raise someone else's price, forcing the mechanism to evict him from his slot based on budget constraints and subsequently benefiting the first player.

players out of budget, hence there might exist bids that realize an envy-free assignment but do not form a Nash equilibrium.

For the other direction, it is trivial to find an example where the outcome of a Nash equilibrium is not an envy-free assignment in all mechanisms building on the intuition that someone might be envious of someone else' s higher slot but they are not able to get it at that price.

## 3  The Budget-Oblivious Second-Price Auction

BOSP lacks the notions of stability defined earlier.

**Theorem 1.** *There are settings where no Nash equilibrium exists under* BOSP.

A simple counterexample is a game with two slots with rates 1 and 0.4 and three players with value/budget 50/50, 16/5 and 8/2 and the proof is presented in the full version.

Since under BOSP, realizable envy-free assignments are a subset of Nash equilibria it follows that:

**Corollary 1.** *There are settings where no realizable envy-free assignment exists under* BOSP.

Let us point out here that in Lemma 3, stated below, we show that envy-free assignments always exist (under the assumption of distinctness of budgets). By the above corollary, such assignments are not realizable under BOSP.

## 4  The Budget-Conscious by Price Second-Price Auction

We now turn our attention to BCSP(PRICE). We first show

**Theorem 2.** *There are settings in which no Nash equilibrium exists for* BCSP-(PRICE).

To prove Theorem 2, we will first show that Nash equilibria do not always exist in the special case where players are budget-bidding their true constraints and then extend the result to the general case.

**Lemma 1.** *There are settings where players are budget-bidding their true constraints, and no Nash equilibrium exists under* BCSP(PRICE).

*Proof.* Consider two slots with $\theta_1 = 1, \theta_2 = 0.4$ and 3 players with attributes as shown in Figure 1. We assume that the tie-breaking ordering favors player 3 and then player 2. In order to show that a Nash equilibrium does not exist we have to consider all orderings of bids and for each such case all possible slot assignments. Intuitively, we will showcase two types of instability. If a bid is low so that the player paying it has the budget constraint satisfied then it can be easily overbid or if a bid is high then underbidding below it will force that player out of budget for the slot.

|         | Player 1 | Player 2 | Player 3 |
|---------|----------|----------|----------|
| Value   | 50       | 16       | 8        |
| Budget  | 50       | 5        | 2        |

**Fig. 1.** Example of non-existence of Nash equilibrium under BCSP(PRICE), having two slots with $\theta_1 = 1, \theta_2 = 0.4$

We start by considering the case where $b_1 > b_2 > b_3$ and slot 1 is assigned to player 1 and slot 2 to player 2. This means that these players can afford these slots, therefore we must have $b_2 \leq 50$ and $b_3 \leq 12.5$. If $b_2 > 5$ then player 1 can bid $b_2 - \varepsilon > b_3$, for some small $\varepsilon$, and still get slot 1 at a lower price, since player 2 cannot afford it. If $b_2 \leq 5$ then player 3 can bid $b_2 + \varepsilon < b_1$ and gain strictly positive utility.

If player 1 is assigned to slot 2 because he cannot afford it and player 2 gets slot 1, we must have $b_2 > 50$ and $b_3 \leq 5$. If player 1 bids $b_2 - \varepsilon > 5 > b_3$, then player 2's bid will be the highest bid but he will not be able to afford slot 1 anymore, which will end up at player 1 for a lower price per click than before.

The rest of the cases follow similarly and we present them in the full version of this paper. □

**Lemma 2.** *Under* BCSP(PRICE), *if a Nash equilibrium exists then a Nash equilibrium also exists where the players are budget-bidding their true constraints.*

To prove Lemma 2, we show that when a player changes his budget-bid to his true budget the same slot allocation can be achieved. This is combined with an adequate shifting of value-bids that maintains the same prices and that concludes the proof of Lemma 2 and of Theorem 2.

Despite the non-existence of Nash equilibria, all envy-free assignments are realizable under BCSP(PRICE) (and there exists at least one such assignment).

**Theorem 3.** *There exists an envy-free assignment which is realizable under* BCSP(PRICE).

The proof of the Theorem is based on Lemma 3 and Lemma 4 below.

**Lemma 3.** *Under budget constraints, there always exists an envy-free assignment.*

*Proof outline.* We present a proof outline. We first note that an envy-free assignment can always be obtained from the mechanism in [2]. We present here an alternate way to obtain an envy-free assignment. Our procedure produces a different assignment than that of [2] in some settings, has a shorter proof and we believe it has some value on its own.

We run the following procedure that completes with all slots assigned in an envy-free assignment. The slots are initially free and have an assigned infinite price. We pick a free slot and lower its price until some player can obtain non-negative utility, at which point we award him the slot at that price and stop lowering it.

We repeat similarly with other slots but allow players that are assigned to request the particular slot if its price makes it more beneficial for them. This leads to a reassignment of the player. We then effectively maintain the envy-free conditions during the whole process.

We have to be more careful when a player can receive equal utility from his assigned slot and the slot whose price is being lowered as we cannot further lower either price without potentially inducing envy or infinite switch loops among the players. We deal with this by simultaneously lowering both prices in a uniform manner and by a careful analysis on potential outcomes and progress measures to guarantee the termination and maintenance of envy-free conditions. For more details, we refer the reader to the full version of the paper. □

**Lemma 4.** *Any envy-free assignment is realizable under* BCSP(PRICE).

We define value bids such that the mechanism produces the same allocation and prices as in the envy-free assignment. The budget-bids can be set to match the true budgets of the players.

## 5 Budget-Conscious by Bid and Best Offer Second-Price Auctions

Recall that under BCSP(BID) and BCSP(BEST OFFER), realizable envy-free assignments are a subset of Nash equilibria. We are going to show that for these two mechanisms, Nash equilibria exist, by establishing that envy-free assignments are realizable in both BCSP(BID) and BCSP(BEST OFFER). Given an envy-free assignment, we pick the bids such that the mechanism assigns the players in the desired slots and achieves the same prices. The proofs of the following theorems are presented in the full version.

**Theorem 4.** *Under* BCSP(BID), *there is always a Nash equilibrium that produces an envy-free assignment.*

**Theorem 5.** *Under* BCSP(BEST OFFER), *there is always a Nash equilibrium that produces an envy-free assignment.*

Note that in both theorems we need to set the budget-bids of the players different from their true budgets. It turns out this is necessary, a perhaps surprising result as having budgets publicly known was necessary for truthful mechanisms in related work [10,2].

**Theorem 6.** *There are settings with public budgets, where a Nash equilibrium does not exist under both* BCSP(BID) *and* BCSP(BEST OFFER).

As envy-free assignments realizable under BCSP(BID) or BCSP(BEST OFFER) form a subset of Nash equilibira of these two mechanisms, respectively, the Theorem above implies that there are settings where neither of these mechanisms can realize any envy-free assignment (guaranteed to exist by Lemma 3), unless players are allowed to bid non-true budgets.

## 6 Discussion

The consideration of several variants of mechanisms, introduced not out of idle curiosity, but as representations of all the natural answers to natural questions raised by the introduction of budgets, and the examination of their properties and differences is what we consider as our primary contribution in this work. We believe our work can serve as a starting point for studying further the properties of GSP auctions under budget constraints. Although we studied these auctions in the context of sponsored search, second-price auctions are widely used in many different settings both in off-line and on-line scenarios. As such, our results are applicable in a much wider context.

In the area of sponsored search, a further step is required towards the more accurate modeling of the deployed systems. In practice, a player can transition between slots during a time period, as players are moderated according to their budget depletion rate. Similarly to the majority of related work on keyword auctions with budgets, we chose to study the static setting first both as a stepping stone and in its own interest for settings outside of keyword auctions. An analysis on the effects of budget constraints in a dynamic setting that would extend upon our results, would contribute towards a more accurate modeling of sponsored search auctions. Another interesting direction for future research is to evaluate the performance of these mechanisms in terms of the generated welfare. This type of Price of Anarchy analysis for non-truthful auctions was initiated in [8], and for certain settings with budget constraints, some results have been recently obtained in [18].

**Acknowledgements.** We would like to thank Konstantinos Gavriil for pointing out to us the counterexample that without distinct budgets, envy-free assignments may fail to exist. We also want to thank Giorgos Birbas for valuable discussions during the preparation of this work. Finally, we thank the anonymous reviewers for their very helpful comments.

## References

1. Arnon, A., Mansour, Y.: Repeated budgeted second price ad auction. In: Proceedings of the 4th Symposium on Algorithmic Game Theory. pp. 7–18. Springer-Verlag (2011)
2. Ashlagi, I., Braverman, M., Hassidim, A., Lavi, R., Tennenholtz, M.: Position auctions with budgets: existence and uniqueness. B.E. journal of Theoretical Economics Advances (to appear, 2013)
3. Ausubel, L.M.: An efficient ascending-bid auction for multiple objects. The American Economic Review 94(5), 1452–1475 (2004)
4. Borgs, C., Chayes, J., Immorlica, N., Mahdian, M., Saberi, A.: Multi-unit auctions with budget-constrained bidders. In: ACM Conference on Electronic Commerce (EC). pp. 44–51 (2005)
5. Caragiannis, I., Kaklamanis, K., Kanellopoulos, P., Kyropoulou, M., Lucier, B., Paes Leme, R., Tardos, E.: On the efficiency of equilibria in generalized second price auctions. arxiv:1201.6429 (2012)

6. Chakrabarty, D., Zhou, Y., Lukose, R.: Budget constrained bidding in keyword auctions and online knapsack problems. In: Workshop on Internet and Network Economics (WINE). pp. 566–576 (2008)
7. Charles, D., Chakrabarty, D., Chickering, M., Devanur, N.R., Wang, L.: Budget smoothing for internet ad auctions: a game theoretic approach. In: Proceedings of the fourteenth ACM conference on Electronic commerce. pp. 163–180. EC '13, ACM, New York, NY, USA (2013), http://doi.acm.org/10.1145/2482540.2482583
8. Christodoulou, G., Kovács, A., Schapira, M.: Bayesian combinatorial auctions. In: ICALP (1). pp. 820–832 (2008)
9. Colini-Baldeschi, R., Henzinger, M., Leonardi, S., Starnberger, M.: On multiple keyword sponsored search auctions with budgets. In: Czumaj, A., Mehlhorn, K., Pitts, A.M., Wattenhofer, R. (eds.) Automata, Languages, and Programming - 39th International Colloquium, ICALP 2012. Lecture Notes in Computer Science, vol. 7392, pp. 1–12. Springer (2012)
10. Dobzinski, S., Lavi, R., Nisan, N.: Multi-unit auctions with budget limits. In: Proceedings of the 2008 49th Annual IEEE Symposium on Foundations of Computer Science. pp. 260–269. FOCS '08, IEEE Computer Society, Washington, DC, USA (2008), http://dx.doi.org/10.1109/FOCS.2008.39
11. Edelman, B., Ostrovsky, M., Schwarz, M.: Internet advertising and the generalized second-price auction: Selling billions of dollars worth of keywords. The American Economic Review 97(1), 242–259 (2007)
12. Feldman, J., Muthukrishnan, S., Pal, M., Stein, C.: Budget optimization in search-based advertising auctions. In: ACM Conference on Electronic Commerce (EC). pp. 40–49 (2007)
13. Fiat, A., Leonardi, S., Saia, J., Sankowski, P.: Single valued combinatorial auctions with budgets. In: Proceedings of the 12th ACM conference on Electronic commerce. pp. 223–232. EC '11, ACM, New York, NY, USA (2011), http://doi.acm.org/10.1145/1993574.1993609
14. Goel, G., Mirrokni, V.S., Leme, R.P.: Polyhedral clinching auctions and the adwords polytope. In: ACM Symposium on Theory of Computing (STOC). pp. 107–122 (2012)
15. Lahaie, S., Pennock, D., Saberi, A., Vohra, R.: Sponsored search auctions. In: Nisan, N., Roughgarden, T., Tardos, E., Vazirani, V. (eds.) Algorithmic Game Theory, chap. 28, pp. 699–716. Cambridge University Press (2007)
16. Lahaie, S., Pennock, D.: Revenue analysis of a family of ranking rules for keyword auctions. In: Proc. ACM Conference on Electronic Commerce (EC). pp. 50–56. San Diego, California, USA (June 2007)
17. Maillé, P., Markakis, E., Naldi, M., Stamoulis, G.D., Tuffin, B.: Sponsored search auctions: an overview of research with emphasis on game theoretic aspects. Electronic Commerce Research 12(3), 265–300 (2012)
18. Syrgkanis, V., Tardos, É.: Composable and efficient mechanisms. In: ACM Symposium on Theory of Computing (STOC 2013). pp. 211–220 (2013)
19. Varian, H.: Position auctions. International Journal of Industrial Organization 25, 1163–1178 (2005)

# Approximation Algorithms for the Max-Buying Problem with Limited Supply*

Cristina G. Fernandes and Rafael C.S. Schouery

Department of Computer Science, University of São Paulo, Brazil
{cris,schouery}@ime.usp.br

**Abstract.** We consider the Max-Buying Problem with Limited Supply, in which there are $n$ items, with $C_i$ copies of each item $i$, and $m$ bidders such that every bidder $b$ has valuation $v_{ib}$ for item $i$. The goal is to find a pricing $p$ and an allocation of items to bidders that maximize the profit, where every item is allocated to at most $C_i$ bidders, every bidder receives at most one item and if a bidder $b$ receives item $i$ then $p_i \leq v_{ib}$. Briest and Krysta presented a 2-approximation for this problem and Aggarwal et al. presented a 4-approximation for the Price Ladder variant where the pricing must be non-increasing (that is, $p_1 \geq p_2 \geq \cdots \geq p_n$). We present a randomized $e/(e-1)$-approximation for the Max-Buying Problem with Limited Supply and, for every $\varepsilon > 0$, a $(2+\varepsilon)$-approximation for the Price Ladder variant.

**Keywords:** pricing problem, unit-demand auctions, approximation algorithms.

## 1 Introduction

One interesting economic problem faced by companies that sell products or provide services to consumers is to choose the price of products or services in order to maximize profit. If prices are high then some consumers will not want (or will not be able) to buy the product and if the prices are low, the company might obtain a low profit. This is a vastly studied problem, with different models for different situations and a great diversity of approaches [13,14,16,17].

One way to address this problem is through the nonparametric approach [15], where the company collects the preferences of consumers groups (for example, using a website) and optimizes according to some assumptions on the consumer behavior.

In this scenario, we have $n$ products or services, that we will call items, and there are $m$ consumers, that we will call bidders, in the market. At first, we consider that there is an unlimited supply of every item. The auctioneer (the price setter) wants to assign a price $p_i$ for every item $i$ with the objective of maximizing his profit (the sum of the prices of sold items considering multiplicities). For this, the auctioneer gathers information about the valuations of the

---

* Research partially supported by CNPq (Proc. 309657/2009-1 and 475064/2010-0) and FAPESP (Proc. 2009/00387-7 and 2013/03447-6).

bidders, that is, the largest amount that a bidder $b$ is willing to pay for item $i$, denoted by $v_{ib}$.

First of all, we consider that our market is unit-demand, that is, every bidder desires to buy exactly one item. But a bidder $b$ will only buy an item if it is not too expensive (that is, if $p_i \leq v_{ib}$). We call such items feasible for $b$. If there is no feasible item for $b$, then $b$ does not buy any item.

Three models were introduced by Rusmevichientong et al. [15]. In the **Min-Buying Problem**, every bidder buys one of the least expensive items that are feasible (for him). In the **Rank-Buying Problem**, one is also given the preference order among the items for every bidder, and a bidder buys the most preferred feasible item. Finally, in the **Max-Buying Problem**, every bidder buys one of the most expensive items that are feasible.

There are also some known variants of the problems mentioned above. In one of the variants, called **limited supply**, every item has a maximum number of copies that can be sold. Observe that this is a generalization of the problem, as one can take the maximum number of copies of each item to be the number of bidders, and this corresponds to the original problem. The second variant is a restriction on the original problem. Sometimes a company knows (or desires) an ordering in the prices of its products, so one can require a **price ladder**, that is, one is interested only in pricings $p_1 \geq p_2 \geq \cdots \geq p_n$. Finally, one can also focus on **uniform budgets**, where every bidder $b$ has a set of items $I_b$ and a value $V_b$ such that, for every item $i$, $v_{ib} = V_b$ if $i \in I_b$ and $v_{ib} = 0$ otherwise.

Rusmevichientong et al. [15] showed that, if we impose a price ladder, then one can solve the Min-Buying Problem with Uniform Budgets in polynomial time. Later on, Aggarwal et al. [1] proved several results for these models considering non-uniform budgets. They presented a polynomial-time approximation scheme for the Max-Buying Problem with a Price Ladder and showed how to reduce the Rank-Buying Problem with a Price Ladder to the Max-Buying Problem with a Price Ladder. They also presented a 4-approximation algorithm for the Max-Buying Problem with a Price Ladder and Limited Supply. For the case without the price ladder requirement, Aggarwal et al. presented an $e/(e-1)$-approximation algorithm for the Max-Buying Problem along with a lower bound of 16/15, a $\log(m)$-approximation that can be used for the three models and a $1+\varepsilon$ lower bound for the Min-Buying Problem, for some constant $\varepsilon$.

Another variant considered by Aggarwal et al. [1] is the online version of the Max-Buying Problem with Limited Supply, where we know the valuation matrix $v$ in advance but we do not know the arrival order of the bidders and we have to choose a pricing. When a bidder arrives, he buys the most expensive feasible item that still has an unsold copy. They proved that, for any fixed pricing, the revenue obtained by any ordering of the bidders is at least $1/2$ of the revenue obtained by an optimal ordering of the bidders. From this, it follows that any $\alpha$-approximation for the Max-Buying Problem with Limited Supply (with or without a price ladder) is also $2\alpha$-competitive for the online version.

Briest and Krysta [3] showed that the Min-Buying Problem (with or without a price ladder) is not approximable within $O(\log^\varepsilon m)$ for some positive constant $\varepsilon$, unless NP $\subseteq$ DTIME($n^{O(\log \log n)}$), within $O(\ell^\varepsilon)$ where $\ell$ is an upper bound on the number of non-zero valuations per bidder, and within $O(n^\varepsilon)$ unless NP $\subseteq$ DTIME($2^{O(n^\delta)}$) for every $\delta > 0$. They also showed that the Max-Buying Problem with a Price Ladder is strongly NP-hard and they presented a 2-approximation algorithm for the Max-Buying Problem with Limited Supply (without a price ladder).

For the Min-Buying Problem with Uniform Budgets, Briest [2] showed that the problem cannot be approximated within $O(\log^\varepsilon |B|)$ for some $\varepsilon > 0$ if we assume some specific hardness of refuting random 3SAT-instances or approximating the balanced bipartite independent set problem in constant degree graphs. Later on, Chalermsook et al. [5] presented lower bounds of $\Omega(\ell^{1/2})$ (unless P = NP) and $\Omega(\log^{1-\varepsilon}(m+n))$ for any positive constant $\varepsilon$ (unless NP $\subseteq$ DTIME($n^{O(\log^\delta n)}$), where $\delta$ is a constant depending on $\varepsilon$). Finally, Chalermsook et al. [6] presented a lower bound of $\Omega(\min(\ell^{1-\varepsilon}, n^{1/2-\varepsilon}))$ (using the Exponential Time Hypothesis) for any positive constant $\varepsilon$ for this problem.

Guruswami et al. [9] studied the Envy-Free Pricing Problem, where a bidder $b$ must receive an item in the set $D_b = \{i \in I : p_i < v_{ib}\}$ that maximizes $v_{ib} - p_i$. If such set is empty, then $b$ must either receive no item or receive an item such that $p_i = v_{ib}$. The problem was considered in a more general setting where bidders have valuations for bundles of items (like in a combinatorial auction), but their work focus on unit-demand auctions and also on single-minded bidders. Guruswami et al. proved that the Envy-Free Pricing Problem for the unit-demand case is APX-hard and provided an $O(\log n)$-approximation for it. Also, as the Envy-Free Pricing Problem with Uniform Budgets for the unit-demand case is the same problem as the Min-Buying Problem with Uniform Budgets, the lower bounds from Briest [2], Chalermsook et al. [5], and Chalermsook et al. [6] also hold for the Envy-Free Pricing Problem.

## 1.1 Our Results

We present two new approximation algorithms for the Max-Buying Problem with Limited Supply, one for the general case and another for the case where a price ladder is required. Both algorithms improve the previously best know approximation ratio for these problems. (A version of our with all the proofs can be found in the arXiv.)

For the Max-Buying Problem with Limited Supply (without the price ladder requirement), we present an $e/(e-1)$-approximation improving the previous upper bound of 2 by Briest and Krysta [3]. (Recall that $e/(e-1) < 1.582$.) Also, this algorithm has the same approximation ratio as the algorithm for the Max-Buying Problem (with unlimited supply) presented by Aggarwal et al. [1]. Recall that unlimited supply is a particular case of our problem where the number of copies of every item is the number of bidders. We believe that the algorithm is interesting by itself: it uses an integer programming formulation with an exponential number of variables to do a probabilistic rounding and also it explores some structure of the problem that could be useful when developing approximations for the other problems previously described.

For the Max-Buying Problem with Limited Supply and a Price Ladder, we present a family of algorithms parametrized by a positive rational $\varepsilon$ such that the algorithm is polynomial for constant $\varepsilon$ and provides an approximation ratio of $2 + \varepsilon$.

Notice that, using the result presented by Aggarwal et al. [1], our first algorithm is $2e/(e-1)$-competitive for the online version of the Max-Buying Problem with Limited Supply and our second algorithm is $(4 + \varepsilon)$-competitive for the online version of the Max-Buying Problem with Limited Supply and a Price Ladder.

This paper is organized as follows. In the next section, we present some notation and describe formally the problem that we address. In Sect. 3 we present the randomized $e/(e-1)$-approximation for the Max-Buying Problem with Limited Supply (that can be derandomized) and in Sect. 4 we present, for a given positive rational $\varepsilon$, a $(2 + \varepsilon)$-approximation for the Max-Buying Problem with Limited Supply and a Price Ladder. Finally, in Sect. 5 we present our final remarks.

## 2 Model and Notation

We denote by $B$ the set of bidders and by $I$ the set of items.

**Definition 1.** *A* valuation *matrix is a non-negative integer matrix $v$ indexed by $I \times B$. The number $v_{ib}$ is the value of item $i$ to bidder $b$.*

**Definition 2.** *A* pricing *is a non-negative rational vector indexed by $I$.*

**Definition 3.** *An* allocation *is a vector $x$ indexed by $B$ where $x_b$ is the item allocated to bidder $b$. If a bidder $b$ does not receive an item, then $x_b = \emptyset$.*

Note that an allocation is not necessarily a matching, as the same item can be assigned to more than one bidder (each one receives a copy of the item).

**Definition 4.** *Given a valuation $v$ and a pricing $p$, an item $i$ is* feasible *for a bidder $b$ if $v_{ib} \geq p_i$.*

**Definition 5.** *The* Max-Buying-Limited Problem *consists of, given a valuation $v$ and a integer positive vector $C$ indexed by $I$, finding a pricing $p$ and an allocation $x$ that maximize the auctioneer's profit, and such that every item is allocated to at most $C_i$ bidders and every bidder either receives no item or receives a feasible item.*

Notice that, in spite of the name of the problem, we do not demand that a bidder $b$ receives a feasible item $i$ that maximizes $p_i$. That is, we allow bidder $b$ to receive another item (or none at all) if the more expensive items that are feasible to $b$ are sold out (that is, all copies are allocated to other bidders).

Next we formalize the variant of the problem with the price ladder requirement.

**Definition 6.** *The* Max-Buying-PL-Limited Problem *is the variant of the Max-Buying-Limited Problem in which the prices must be non-increasing, that is, $p_1 \geq \cdots \geq p_n$, where $n$ is the number of items.*

## 3  An Algorithm for Limited Supply

Next, we present a new approximation algorithm for the Max-Buying-Limited Problem with a better ratio than the approximation presented by Briest and Krysta [3]. This approximation applies also to the Max-Buying Problem (with unlimited supply) achieving the same ratio of the best known approximation for this problem [1].

First, consider a solution $(x, p)$ of the Max-Buying-Limited Problem, an item $i$, and the set $S$ of bidders that bought item $i$ according to $x$. Note that, because $i$ is feasible for every bidder in $S$, we have that $p_i \leq \min\{v_{ib} : b \in S\}$. If $S \neq \emptyset$ and $p_i < \min\{v_{ib} : b \in S\}$, then $(x, p)$ cannot be an optimal solution because one could increase the price of $i$ to obtain a strictly better solution. So we may assume, w.l.o.g., that $p_i = \min\{v_{ib} : b \in S\}$ for every item $i$ that is bought by a set $S$ of bidders.

We will present an IP formulation for the Max-Buying-Limited Problem that is heavily based on the observation above and on the next definitions. From this formulation, we will design a randomized rounding approximation algorithm.

**Definition 7.** *For an item $i$, let $\mathcal{S}(i) = \{(i, S) : S \subseteq B, |S| \leq C_i\}$. We call $(i, S) \in \mathcal{S}(i)$ a star of $i$ and we denote by $\mathcal{S}$ the set of all stars, that is, $\mathcal{S} = \bigcup_{i \in I} \mathcal{S}(i)$.*

For $S \subseteq B$ and an item $i$, note that $S \cup \{i\}$ induces a star in the complete bipartite graph where the parts of the bipartition are $I$ and $B$.

**Definition 8.** *For a star $(i, S)$, let $P_{(i,S)} = \min\{v_{ib} : b \in S\}$, that is, $P_{(i,S)}$ is the price of item $i$ when sold to the set $S$ of bidders.*

Notice that a feasible solution of the Max-Buying-Limited Problem can be seen as a collection of stars, one for each item, with every bidder in at most one star, and the price of an item $i$ being $P_{(i,S)}$, where $(i, S)$ is the star of item $i$ in the collection.

Next, we present our formulation, called (SF) for star formulation, in which we have a vector $x$ of binary variables, with $|\mathcal{S}|$ positions, where $x_{(i,S)}$ is equal to 1 if and only if the set of bidders that receive item $i$ is precisely $S$. The goal is to determine $x$ that

$$\begin{aligned}
\text{(SF)} \quad \text{maximizes} \quad & \sum_{(i,S) \in \mathcal{S}} |S| P_{(i,S)} x_{(i,S)} \\
\text{subject to} \quad & \sum_{(i,S) \in \mathcal{S}(i)} x_{(i,S)} = 1, & \forall i \in I \\
& \sum_{(i,S) \in \mathcal{S} : b \in S} x_{(i,S)} \leq 1, & \forall b \in B \\
& x_{(i,S)} \in \{0, 1\}, \quad \forall (i, S) \in \mathcal{S}.
\end{aligned}$$

The (SF) formulation can be seen as a reduction of our problem to the Set Packing Problem [8]. A similar idea was used by Hochbaum [10] to obtain an $O(\log n)$-approximation for the Metric Uncapacitated Facility Location

Problem by reducing it to the Set Cover Problem [8]. In our case, we can use the weight structure of the sets to obtain a constant factor approximation for our problem.

This formulation can have $\Omega(|I|2^{|B|})$ variables. But, fortunately, it is possible to solve its linear relaxation in polynomial time using a procedure similar to the one used by Karmarkar and Karp [11]. First, we solve the dual in polynomial time (using a polynomial-time separation algorithm) and then solve (in polynomial time) the primal restricted to the columns corresponding to the constraints used in the resolution of the dual.

**Lemma 9.** *The linear relaxation of the* (SF) *formulation can be solved in polynomial time.* □

We will use this IP formulation to design an approximation algorithm for our problem using probabilistic rounding. Next we present our algorithm.

STARROUNDING$(I, B, v, C)$
1  Let $x$ be an optimal solution of the linear relaxation of (SF) for $(I, B, v, C)$
2  **for** every item $i \in I$
3      Choose a star $S_i \in \mathcal{S}(i)$ with probability $\mathbb{P}(S_i = (i, S)) = x_{(i,S)}$
4      Set the price of $i$ as $P_{S_i}$
5  **for** every bidder $b \in B$
6      Let $i$ be an item such that $S_i = (i, S)$ with $b \in S$ and maximum $P_{S_i}$
7      **if** there is no such item
8          bidder $b$ does not receive an item
9      **else** sell item $i$ to bidder $b$

**Theorem 10.** STARROUNDING *is a randomized $\frac{e}{e-1}$-approximation for the Max-Buying-Limited Problem.*

*Proof.* First, notice that the objective function of (SF) can be rewritten as $\sum_{b \in B} \sum_{(i,S) \in \mathcal{S}: b \in S} P_{(i,S)} x_{(i,S)}$ and that this value for the $x$ in Line 1 is an upper bound on the value of an optimal solution of our problem. We will prove that the expected price paid by bidder $b$ in the solution produced by the algorithm is at least $\frac{e-1}{e} \sum_{(i,S) \in \mathcal{S}: b \in S} P_{(i,S)} x_{(i,S)}$, from where the result will follow.

Consider a bidder $b$ and a non-increasing (in $P_{(i,S)}$) ordering of the stars $(i, S)$ with $b \in S$ and $x_{(i,S)} > 0$. Let $k$ be the number of such stars. If $k = 0$, then the result trivially holds. From now on, we assume that $k > 0$.

We will denote the $\ell$-th star in this ordering simply by $\ell$, its price by $P_\ell$, and its primal variable by $x_\ell$. If the star $\ell$ is $(i, S)$, we define $c(\ell) = i$, that is, $c(\ell)$ is the item of star $\ell$. Also, we define $y_\ell = \sum \{x_{\ell'} : \ell' < \ell \text{ and } c(\ell') = c(\ell)\}$. Finally, we denote by $E_\ell$ the event in which star $\ell$ was chosen by STARROUNDING.

Let profit$(b)$ be the profit that we obtain from bidder $b$. For $1 \le \ell \le k$, note that $\mathbb{E}[\text{profit}(b)|\overline{E}_1, \ldots, \overline{E}_{\ell-1}, E_\ell] = P_\ell$ because STARROUNDING will allocate $c(\ell)$ (or another item with the same price) to bidder $b$ as $c(\ell)$ is one of the most expensive items that has $b$ in the chosen star. Also, notice that $\mathbb{P}(E_\ell|\overline{E}_1, \overline{E}_2, \ldots, \overline{E}_{\ell-1}) = \frac{x_\ell}{1-y_\ell}$, because we choose the star of an item independently of the star chosen for other items.

Let $f(z) = \frac{1-e^{-z}}{z}$. For $1 \leq \ell \leq k$, we denote $f(\sum_{i=\ell}^k x_i)$ simply by $f_\ell$. Observe that $f_\ell \leq 1$ for $1 \leq \ell \leq k$, because $1 - z \leq e^{-z}$ for every $z$.

Using the observations above, we will prove that, for every $1 \leq \ell \leq k$, $\mathbb{E}[\text{profit}(b)|\overline{E}_1, \overline{E}_2, \ldots, \overline{E}_{\ell-1}] \geq f_\ell \sum_{i=\ell}^k P_i x_i$. Observe that the theorem follows from this statement, as it reduces to $\mathbb{E}[\text{profit}(b)] \geq f_1 \sum_{i=1}^k P_i x_i$ for $\ell = 1$, and $f_1 \geq \frac{e-1}{e}$ because $f(z)$ is decreasing.

So we proceed with the proof of the statement, by induction on $k - \ell$. Note that $y_k = 1 - x_k$, so $\mathbb{E}[\text{profit}(b)|\overline{E}_1, \ldots, \overline{E}_{k-1}] = P_k x_k/(1 - y_k) = P_k \geq f_k P_k x_k$, thus the statement is valid for $\ell = k$, the base case. Now, for $\ell < k$, assume the statement is valid for $\ell + 1$. We have that

$$\mathbb{E}[\text{profit}(b)|\overline{E}_1, \ldots, \overline{E}_{\ell-1}] = \mathbb{E}[\text{profit}(b)|\overline{E}_1, \ldots, \overline{E}_{\ell-1}, E_\ell] \frac{x_\ell}{1-y_\ell}$$
$$+ \mathbb{E}[\text{profit}(b)|\overline{E}_1, \ldots, \overline{E}_{\ell-1}, \overline{E}_\ell] \left(1 - \frac{x_\ell}{1-y_\ell}\right)$$
$$= P_\ell \frac{x_\ell}{1-y_\ell} + \left(1 - \frac{x_\ell}{1-y_\ell}\right) \mathbb{E}[\text{profit}(b)|\overline{E}_1, \ldots, \overline{E}_\ell].$$

Using the induction hypothesis, we deduce that

$$\mathbb{E}[\text{profit}(b)|\overline{E}_1, \ldots, \overline{E}_{\ell-1}] \geq P_\ell \frac{x_\ell}{1-y_\ell} + \left(1 - \frac{x_\ell}{1-y_\ell}\right) f_{\ell+1} \sum_{i=\ell+1}^k P_i x_i$$
$$= \frac{x_\ell}{1-y_\ell} \left(P_\ell - f_{\ell+1} \sum_{i=\ell+1}^k P_i x_i\right) + f_{\ell+1} \sum_{i=\ell+1}^k P_i x_i.$$

As $\frac{x_\ell}{1-y_\ell} \geq x_\ell$, $1 - z \leq e^{-z}$ for every $z$, $P_\ell \geq P_\ell \sum_{i=\ell}^k x_i \geq \sum_{i=\ell}^k P_i x_i$ (because $P_\ell \geq P_i$ for every $i \geq \ell$) and $f_{\ell+1} \leq 1$, it follows that

$$\mathbb{E}[\text{profit}(b)|\overline{E}_1, \ldots, \overline{E}_{\ell-1}] \geq x_\ell \left(P_\ell - f_{\ell+1} \sum_{i=\ell+1}^k P_i x_i\right) + f_{\ell+1} \sum_{i=\ell+1}^k P_i x_i$$
$$\geq (1 - e^{-x_\ell}) \left(P_\ell - f_{\ell+1} \sum_{i=\ell+1}^k P_i x_i\right) + f_{\ell+1} \sum_{i=\ell+1}^k P_i x_i$$
$$= (1 - e^{-x_\ell}) P_\ell + e^{-x_\ell} f_{\ell+1} \sum_{i=\ell+1}^k P_i x_i$$
$$= (1 - e^{-x_\ell} - e^{-x_\ell} f_{\ell+1} x_\ell) P_\ell + e^{-x_\ell} f_{\ell+1} \sum_{i=\ell}^k P_i x_i.$$

In order to proceed, let us first show that $1 - e^{-x_\ell} - e^{-x_\ell} f_{\ell+1} x_\ell \geq 0$. For that, let $h(z) = 1 - e^{-z} - e^{-z} t z$, for some $0 < t \leq 1$. Notice that $h(0) = 0$

and that $h'(z) = e^{-z} + e^{-z}tz - e^{-z}t \geq e^{-z}tz$. Therefore, $h(z)$ is non-decreasing for non-negative $z$, from which we conclude that $h(z) \geq 0$ for every non-negative $z$. So, in particular, $1 - e^{-x_\ell} - e^{-x_\ell} f_{\ell+1} x_\ell \geq 0$, as we wished. Also, recall that $P_\ell \sum_{i=\ell}^{k} x_i \geq \sum_{i=\ell}^{k} P_i x_i$. Combining this with the previous, we deduce that

$$(1 - e^{-x_\ell} - e^{-x_\ell} f_{\ell+1} x_\ell) P_\ell + e^{-x_\ell} f_{\ell+1} \sum_{i=\ell}^{k} P_i x_i$$

$$\geq \left(1 - e^{-x_\ell} - e^{-x_\ell} f_{\ell+1} x_\ell\right) \frac{\sum_{i=\ell}^{k} P_i x_i}{\sum_{i=\ell}^{k} x_i} + e^{-x_\ell} f_{\ell+1} \sum_{i=\ell}^{k} P_i x_i$$

$$= \left(1 - e^{-x_\ell} - e^{-x_\ell} f_{\ell+1} x_\ell + e^{-x_\ell} f_{\ell+1} \sum_{i=\ell}^{k} x_i\right) \frac{\sum_{i=\ell}^{k} P_i x_i}{\sum_{i=\ell}^{k} x_i}$$

$$= \left(1 - e^{-x_\ell} + e^{-x_\ell} f_{\ell+1} \sum_{i=\ell+1}^{k} x_i\right) \frac{\sum_{i=\ell}^{k} P_i x_i}{\sum_{i=\ell}^{k} x_i}.$$

To finish the induction, recall that $f_{\ell+1} \sum_{i=\ell+1}^{k} x_i = 1 - e^{-\sum_{i=\ell+1}^{k} x_i}$, and conclude that

$$\mathbb{E}[\text{profit}(b) | \overline{E}_1, \ldots, \overline{E}_{\ell-1}] \geq \left(1 - e^{-x_\ell} + e^{-x_\ell}(1 - e^{-\sum_{i=\ell+1}^{k} x_i})\right) \frac{\sum_{i=\ell}^{k} P_i x_i}{\sum_{i=\ell}^{k} x_i}$$

$$= \left(1 - e^{-x_\ell} + e^{-x_\ell}(1 - e^{-\sum_{i=\ell+1}^{k} x_i})\right) = f_\ell \sum_{i=\ell}^{k} P_i x_i,$$

which ends the induction and the proof of the theorem. □

It is also possible to obtain a deterministic version of STARROUNDING using the method of conditional expectations [7,18]. Moreover, it is easy to prove that the analysis is tight, as we do in the next lemma.

**Lemma 11.** *For every $\varepsilon > 0$, there is an instance where the expected profit of a solution found by* STARROUNDING *is smaller than $((e-1)/e + \varepsilon)$OPT, where OPT is the value of an optimal solution for this instance.*

*Proof.* Consider this simple instance: we have a set of items $I$ and only one bidder $b$ such that $v_{ib} = 1$, for every $i \in I$. It is easy to see that an optimal solution for this instance has value 1. It is also clear that an optimal solution of the linear relaxation has value 1. One of such optimal solutions is $x$ such that $x_{(i,\{b\})} = 1/|I|$ and $x_{(i,\emptyset)} = 1 - 1/|I|$, for every item $i \in I$. Notice that bidder $b$ pays 1 if any of the stars $(i, \{b\})$ is chosen and pays 0 (because is unallocated) otherwise. Thus we have $\mathbb{E}[\text{profit}(b)] = 1 - (1 - 1/|I|)^{|I|} \xrightarrow{|I| \to \infty} 1 - \frac{1}{e}$ and the result follows. □

## 4 An Algorithm for Limited Supply with a Price Ladder

In this section we present, for every $\varepsilon > 0$, a $(2+\varepsilon)$-approximation for the Max-Buying-PL-Limited Problem. We use some ideas from the 4-approximation algorithm for Max-Buying-PL-Limited Problem developed by Aggarwal et al. [1], but in a different way, in order to obtain a better approximation ratio.

**Definition 12.** *Let $\alpha > 1$ be a rational, $t$ be a positive integer, $v$ be a valuation matrix indexed by $I \times B$ and, for a non-negative integer $k$, let $d_k = \max\{v_{ib} : i \in I, b \in B\}/\alpha^k$. The Max-Buying-PL-Limited-$(\alpha, t)$ Problem is a variant of the Max-Buying-PL-Limited Problem where every price is in the set $\{d_0, d_1, \dots\}$ and every bidder receives, for each non-negative $r$, at most one item with price in $\{d_{rt}, d_{rt+1}, \dots, d_{(r+1)t-1}\}$.*

Notice that, in the Max-Buying-PL-Limited-$(\alpha, t)$ Problem, a bidder can receive more than one item but, for every multiple $s$ of $t$, he can receive at most one item with price in $\{d_s, d_{s+1}, \dots, d_{s+t-1}\}$.

The next result guarantees that one needs to consider only a finite number of possible prices.

**Lemma 13.** *For every valuation matrix $v$ indexed by $I \times B$ and positive integer vector $C$ indexed by $I$, there is an optimal solution for the instance $(I, B, v, C)$ of the Max-Buying-PL-Limited-$(\alpha, t)$ Problem whose smallest price of an item is $d_\ell$, with $\ell = \lceil \log_\alpha V \rceil$, for $V = \max\{v_{ib} : i \in I, b \in B\}$.* □

We will prove that this problem can be solved in polynomial time.

**Theorem 14.** *The Max-Buying-PL-Limited-$(\alpha, t)$ Problem can be solved in polynomial time for fixed $\alpha$ and $t$.*

*Proof.* For non-negative integers $r$ and $j$, let $F(j, r)$ be the maximum profit achievable for the instance of Max-Buying-PL-Limited-$(\alpha, t)$ with only the first $j$ items, allowing only prices in $\{d_0, \dots, d_{(r+1)t-1}\}$. Also, for the variant of Max-Buying-PL-Limited-$(\alpha, t)$ where each bidder is restricted to buy at most one item, for $i \leq j$, let $P(i, j, r)$ be the maximum profit achievable, considering only items $i, i+1, \dots, j$ and prices in $\{d_{rt}, \dots, d_{(r+1)t-1}\}$. We have the following recurrence:

$$F(j,r) = \begin{cases} 0, & \text{if } j = 0, \\ P(1,j,r), & \text{if } j > 0 \text{ and } r = 0, \\ \max_{0 \leq i \leq j} \{F(i, r-1) + P(i+1, j, r)\}, & \text{otherwise.} \end{cases}$$

Let $\ell$ be as in Lemma 13 and notice that $F(|I|, \lceil \ell/t \rceil)$ is the value of an optimal solution for the Max-Buying-PL-Limited-$(\alpha, t)$ Problem (because the last price considered is $d_{(\lceil \ell/t \rceil + 1)t - 1} \leq d_\ell$). So, for fixed $\alpha$ and $t$, if we can compute $P(i, j, r)$ in polynomial time, then this recurrence can be solved in polynomial time.

In order to compute $P(i, j, r)$ in polynomial time, we can enumerate all possible valid pricings (there are at most $|I|^t$ such pricings) and construct a bipartite graph $G$ with bipartition sides $\{i,\ldots,j\}$ and $B$, where, for every item $k$ in $\{i,\ldots,j\}$ and every bidder $b \in B$, we have an edge $\{k,b\} \in E(G)$ of weight $p_k$ if and only if $v_{kb} \geq p_k$. Then it remains to find a maximum weighted $C^+$-matching [4,12] on such graph, where $C_k^+ = C_k$ for every item $k$, and $C_b^+ = 1$ for every bidder $b$ (that is, every bidder is matched to at most one item and every item $k$ is matched to at most $C_k$ bidders.) □

We will now establish some relations involving the value of an optimal solution for the Max-Buying-PL-Limited Problem and the value of an optimal solution for the Max-Buying-PL-Limited-$(\alpha,t)$ Problem.

**Lemma 15.** *Let* OPT *be the value of an optimal solution for the Max-Buying-PL-Limited Problem and* OPT′ *be the value of an optimal solution for the Max-Buying-PL-Limited-$(\alpha,t)$ Problem. We have that* OPT′ $\geq$ OPT$/\alpha$. □

**Lemma 16.** *Given a solution of the Max-Buying-PL-Limited-$(\alpha,t)$ Problem of value $w$, a solution of the Max-Buying-PL-Limited Problem of value at least $\frac{\alpha^t-1}{\alpha^t-1+\alpha^{t-1}} w$ can be computed in polynomial time.*

*Proof.* Consider a solution of the Max-Buying-PL-Limited-$(\alpha,t)$ Problem. We will construct a solution for the Max-Buying-PL-Limited Problem by assigning to every bidder $b$ the most expensive item bought by $b$ (which can be done in polynomial time).

For a bidder $b$, let $K$ be the set of integers such that $k \in K$ if and only if $b$ bought an item of price $d_k$ and let $d_i$ denote the most expensive item bought by $b$. Remember that, for every $r$, a bidder can buy at most one item of price $d_{rt+k}$ for $0 \leq k < t$. From this we conclude that $\sum_{k \in K} d_k \leq d_i + \sum_{k \in K \setminus \{i\}} d_{\lfloor k/t \rfloor t} \leq d_i + \sum_{r \geq 0} d_{i+rt+1}$.

Now, notice that $d_{i+rt+1} = d_i/\alpha^{rt+1}$, from which we conclude that

$$\sum_{k \in K} d_k \leq d_i + \sum_{r \geq 0} d_{i+rt+1} = d_i \left(1 + \frac{1}{\alpha} \sum_{r \geq 0} \left(\frac{1}{\alpha^t}\right)^r\right) \leq d_i \left(1 + \frac{1}{\alpha}\left(\frac{1}{1-\frac{1}{\alpha^t}}\right)\right).$$

Notice that the profit obtained from $b$ in the solution of the Max-Buying-PL-Limited-$(\alpha,t)$ is exactly $\sum_{k \in K} d_k$ and the profit obtained from $b$ in the solution found for the Max-Buying-PL-Limited Problem is exactly $d_i$. But we concluded that $d_i \geq \frac{\alpha^t-1}{\alpha^t-1+\alpha^{t-1}} \sum_{k \in K} d_k$, so the result follows. □

Combining the results of Lemmas 15 and 16, we can derive an approximation for the Max-Buying-PL-Limited Problem.

**Corollary 17.** *For every positive integer $t$ and rational $\alpha > 1$, there is an $\frac{\alpha(\alpha^t-1+\alpha^{t-1})}{\alpha^t-1}$-approximation for the Max-Buying-PL-Limited Problem.*

*Proof.* The algorithm is very simple: find an optimal solution $(x,p)$ of the Max-Buying-PL-Limited-$(\alpha,t)$ Problem using the algorithm described in Theorem 14 and return $(\tilde{x},p)$, where the item allocated to a bidder $b$ in $\tilde{x}$ is the most expensive item allocated to $b$ in $(x,p)$.

Let SOL denote the value of the solution found, OPT denote the value of an optimal solution of the Max-Buying-PL-Limited Problem and OPT' denote the value of an optimal solution of the Max-Buying-PL-Limited-$(\alpha,t)$ Problem. By Lemma 15, we have that OPT' $\geq$ OPT$/\alpha$ and, by Lemma 16, we have that SOL $\geq \frac{\alpha^t-1}{\alpha^t-1+\alpha^{t-1}}$ OPT'. Hence, we conclude that SOL $\geq \frac{\alpha^t-1}{\alpha(\alpha^t-1+\alpha^{t-1})}$ OPT and we obtain the desired approximation ratio. □

**Corollary 18.** *For every $0 < \varepsilon < 1$, there is a $(2+\varepsilon)$-approximation of the Max-Buying-PL-Limited Problem.*

*Proof.* Let $\alpha = 1 + \frac{\varepsilon}{2}$ and $t = \lceil \log_\alpha(\frac{2}{\varepsilon}+1) \rceil$ (notice that $t$ is a positive integer, because $\varepsilon < 1$). We have that

$$\frac{\alpha(\alpha^t - 1 + \alpha^{t-1})}{\alpha^t - 1} = 1 + \alpha + \frac{1}{\alpha^t - 1} \leq 1 + \left(1 + \frac{\varepsilon}{2}\right) + \frac{\varepsilon}{2} = 2 + \varepsilon.$$

That is, by using $\alpha$ and $t$ as above, the algorithm from Corollary 17 achieves the desired ratio. □

## 5 Final Remarks

In this paper we focused on the Max-Buying Problem when we have limited supply, considering the case with the price ladder restriction and without this restriction.

Our results improve the previously best known approximation ratios for both problems (with and without the price ladder restriction). The technique of enumerating all possible allocations of items to bidders, used in the price ladder case, might help in other pricing problems.

We believe that pricing problems with limited supply are very interesting because this is a realistic restriction and also a hard one to be considered from the approximation algorithm perspective. Even though in general the Max-Buying Problem seems to be simpler than the Min-Buying Problem, the Rank-Buying Problem, and the Envy-Free Pricing Problem, it is not trivial to develop good approximations for it when we have limited supply.

There are also some open problems. It is interesting to notice that our algorithm, when applied for the Max-Buying Problem, has the same ratio as the algorithm presented by Aggarwal et al. [1]. It would be nice to develop an approximation with ratio better than $e/(e-1)$ for the Max-Buying Problem (if possible, for limited supply) or to prove that this value is a lower bound on the approximation ratio of every algorithm for these problems.

In the case of the Max-Buying-PL-Limited Problem, it would be interesting to design a PTAS (since there is a PTAS for the unlimited supply case [1]) or to prove that the problem is APX-hard. Also, notice that the price ladder requirement is not helping to achieve a better approximation ratio as it happens for the unlimited supply version. This is somehow against our intuition that knowing the prices order would make it easier to find good pricings. We do not know if this is something intrinsic to this problem or if there are other ways to exploit the price ladder in order to obtain better approximations.

# References

1. Aggarwal, G., Feder, T., Motwani, R., Zhu, A.: Algorithms for multi-product pricing. In: Díaz, J., Karhumäki, J., Lepistö, A., Sannella, D. (eds.) ICALP 2004. LNCS, vol. 3142, pp. 72–83. Springer, Heidelberg (2004)
2. Briest, P.: Uniform budgets and the envy-free pricing problem. In: Aceto, L., Damgård, I., Goldberg, L.A., Halldórsson, M.M., Ingólfsdóttir, A., Walukiewicz, I. (eds.) ICALP 2008, Part I. LNCS, vol. 5125, pp. 808–819. Springer, Heidelberg (2008)
3. Briest, P., Krysta, P.: Buying cheap is expensive: hardness of non-parametric multi-product pricing. In: Proceedings of the 18th Annual ACM-SIAM Symposium on Discrete Algorithms (2007)
4. Burkard, R.E., Dell'Amico, M., Martello, S.: Assignment problems. SIAM (2009)
5. Chalermsook, P., Chuzhoy, J., Kannan, S., Khanna, S.: Improved hardness results for profit maximization pricing problems with unlimited supply. In: Gupta, A., Jansen, K., Rolim, J., Servedio, R. (eds.) APPROX 2012 and RANDOM 2012. LNCS, vol. 7408, pp. 73–84. Springer, Heidelberg (2012)
6. Chalermsook, P., Laekhanukit, B., Nanongkai, D.: Independent set, induced matching, and pricing: Connections and tight (subexponential time) approximation hardnesses. In: 54th Annual IEEE Symposium on Foundations of Computer Science (2013)
7. Erdös, P., Selfridge, J.: On a combinatorial game. Journal of Combinatorial Theory, Series A 14(3), 298–301 (1973)
8. Garey, M.R., Johnson, D.S.: Computers and intractibility. W. H. Freeman (1979)
9. Guruswami, V., Hartline, J.D., Karlin, A.R., Kempe, D., Kenyon, C., McSherry, F.: On profit-maximizing envy-free pricing. In: Proceedings of the 16th Annual ACM-SIAM Symposium on Discrete Algorithms, pp. 1164–1173 (2005)
10. Hochbaum, D.: Heuristics for the fixed cost median problem. Mathematical Programming 22(1), 148–162 (1982)
11. Karmarkar, N., Karp, R.: An efficient approximation scheme for the one-dimensional bin-packing problem. In: Proceedings of the 23rd Annual Symposium on Foundations of Computer Science, pp. 312–320 (1982)
12. Kuhn, H.W.: The Hungarian Method for the assignment problem. Naval Research Logistics Quarterly 2(1-2), 83–97 (1955)
13. Oren, S., Smith, S., Wilson, R.: Product line pricing. Journal of Business 57(1), S73–S79 (1984)
14. Oren, S.S., Smith, S.A., Wilson, R.B.: Multi-product pricing for electric power. Energy Economics 9(2), 104–114 (1987)
15. Rusmevichientong, P., Roy, B.V., Glynn, P.W.: A nonparametric approach to multiproduct pricing. Operations Research 54(1), 82–98 (2006)
16. Sen, S.: Issues in optimal product design. In: Analytic Approaches to Product and Marketing Planning: The Second Conference, pp. 265–274 (1982)
17. Smith, S.A.: New product pricing in quality sensitive markets. Marketing Science 5(1), 70–87 (1986)
18. Spencer, J.: Ten lectures on the probabilistic method. Society for Industrial and Applied Mathematics (1987)

# Budget Feasible Mechanisms for Experimental Design

Thibaut Horel[1], Stratis Ioannidis[2], and S. Muthukrishnan[3]

[1] École Normale Supérieure
thibaut.horel@normalesup.org
[2] Technicolor
stratis.ioannidis@technicolor.com
[3] Rutgers University
muthu@cs.rutgers.edu

**Abstract.** We present a deterministic, polynomial time, budget feasible mechanism scheme, that is approximately truthful and yields a constant ($\approx$ 12.98) factor approximation for the *Experimental Design Problem* (EDP). By applying previous work on budget feasible mechanisms with a submodular objective, one could *only* have derived either an exponential time deterministic mechanism or a randomized polynomial time mechanism. We also establish that no truthful, budget-feasible mechanism is possible within a factor 2 approximation, and show how to generalize our approach to a wide class of learning problems, beyond linear regression.

## 1 Introduction

In the classic setting of experimental design [25, 3], an *experimenter* E has access to a population of $n$ potential experiment subjects. Each subject $i \in \{1, \ldots, n\}$ is associated with a set of parameters (or features) $x_i \in \mathbb{R}^d$, known to the experimenter. E wishes to measure a certain inherent property of the subjects by performing an experiment: the outcome $y_i$ of the experiment on a subject $i$ is unknown to E before the experiment is performed.

Typically, E has a hypothesis on the relationship between $x_i$'s and $y_i$'s. Due to its simplicity, as well as its ubiquity in statistical analysis, a large body of work has focused on linear hypotheses: *i.e.*, it is assumed that there exists a $\beta \in \mathbb{R}^d$ such that $y_i = \beta^T x_i + \varepsilon_i$, for all $i \in \{1, \ldots, n\}$, where $\varepsilon_i$ are zero-mean, i.i.d. random variables. Conducting the experiments and obtaining the measurements $y_i$ lets E estimate $\beta$, *e.g.*, through linear regression.

The above experimental design scenario has many applications. Regression over personal data collected through surveys or experimentation is the cornerstone of marketing research, as well as research in a variety of experimental sciences such as medicine and sociology. Crucially, statistical analysis of user data is also a widely spread practice among Internet companies, which routinely use machine learning techniques over vast records of user data to perform inference and classification tasks integral to their daily operations. Beyond linear regression, there is a rich literature about estimation procedures, as well as about means of quantifying the quality of the produced estimate [25]. There is also an extensive theory on how to select subjects if E can conduct only a limited number of experiments, so the estimation process returns a $\beta$ that approximates the true parameter of the underlying population [15, 20, 9, 6].

We depart from this classical setup by viewing experimental design in a strategic setting, and by studying budgeted mechanism design issues. In our setting, experiments cannot be manipulated and hence measurements are reliable. E has a total budget of $B$ to conduct all the experiments. There is a cost $c_i$ associated with experimenting on subject $i$ which varies from subject to subject. This cost $c_i$ is determined by the subject $i$ and reported to E; subjects are strategic and may misreport these costs. Intuitively, $c_i$ may be viewed as the cost $i$ incurs when tested and for which she needs to be reimbursed; or, it might be viewed as the incentive for $i$ to participate in the experiment; or, it might be the intrinsic worth of the data to the subject. The economic aspect of paying subjects has always been inherent in experimental design: experimenters often work within strict budgets and design creative incentives. Subjects often negotiate better incentives or higher payments. However, we are not aware of a principled study of this setting from a strategic point of view, when subjects declare their costs and therefore determine their payment. Such a setting is increasingly realistic, given the growth of these experiments over the Internet.

Our contributions are as follows. *First*, we initiate the study of experimental design in the presence of a budget and strategic subjects. In particular, we formulate the *Experimental Design Problem* (EDP) as follows: the experimenter E wishes to find a set $S$ of subjects to maximize

$$V(S) = \log \det \left( I_d + \sum_{i \in S} x_i x_i^T \right) \quad (1)$$

subject to a budget constraint $\sum_{i \in S} c_i \leq B$, where $B$ is E's budget. When subjects are strategic, the above problem can be naturally approached as a *budget feasible mechanism design* problem, as introduced by Singer [26]. The objective function, which is the key, is formally obtained by optimizing the information gain in $\beta$ when the latter is learned through ridge regression, and is related to the so-called *D-optimality* criterion [25, 3]. *Second*, we present a polynomial time mechanism scheme for EDP that is approximately truthful and yields a constant factor ($\approx 12.98$) approximation to the optimal value of (1). In contrast to this, we show that no truthful, budget-feasible mechanisms are possible for EDP within a factor 2 approximation.

We note that the objective (1) is submodular. Using this fact, applying previous results on budget feasible mechanism design under general submodular objectives [26, 10] would yield either a deterministic, truthful, constant-approximation mechanism that requires exponential time, or a non-deterministic, (universally) truthful, poly-time mechanism that yields a constant approximation ratio only *in expectation* (*i.e.*, its approximation guarantee for a given instance may in fact be unbounded).

From a technical perspective, we propose a convex optimization problem and establish that its optimal value is within a constant factor from the optimal value of EDP. In particular, we show our relaxed objective is within a constant factor from the so-called multi-linear extension of (1), which in turn can be related to (1) through pipage rounding. We establish the constant factor to the multi-linear extension by bounding the partial derivatives of these two functions; we achieve the latter by exploiting convexity properties of matrix functions over the convex cone of positive semidefinite matrices.

Our convex relaxation of EDP involves maximizing a self-concordant function subject to linear constraints. Its optimal value can be computed with arbitrary accuracy in polynomial time using the so-called barrier method. However, the outcome of this computation may not be monotone, a property needed in designing a truthful mechanism. Nevertheless, we construct an algorithm that solves the above convex relaxation and is "almost" monotone; we achieve this by applying the barrier method on a set perturbed constraints, over which our objective is "sufficiently" concave. In turn, we show how to employ this algorithm to design a poly-time, $\delta$-truthful, constant-approximation mechanism for EDP.

In what follows, we describe related work in Section 2. We briefly review experimental design and budget feasible mechanisms in Section 3 and define EDP formally. We present our convex relaxation to EDP in Section 4 and use it to construct our mechanism in Section 5. We conclude in Section 6. All proofs of our technical results are provided in the full version of this paper [16].

## 2 Related Work

*General Submodular Functions.* Singer [26] considers the problem of maximizing an arbitrary submodular function subject to a budget constraint in the *value query* model. He shows that there exists a randomized, 112-approximation mechanism that is *universally truthful* (*i.e.*, it is a randomized mechanism sampled from a distribution over truthful mechanisms). Chen et al. [10] improve this result to a 7.91-approximate mechanism, and show a corresponding lower bound of 2 among universally truthful randomized mechanisms. The above approximation guarantees hold for the expected value of the randomized mechanism: for a given instance, the approximation ratio provided by the mechanism may be unbounded. No deterministic, truthful, constant approximation mechanism that runs in polynomial time is presently known for submodular maximization. Assuming access to an oracle providing the optimum in the full-information setup, Chen et al. propose a truthful, 8.34-approximate mechanism; in cases for which the full-information problem is NP-hard, as EDP, this mechanism is not poly-time, unless P=NP. Chen et al. also prove a $1 + \sqrt{2}$ lower bound for truthful deterministic mechanisms, improving upon an earlier bound of 2 by Singer [26].

*Specific Problems.* Improved uper and lower bounds *and* deterministic polynomial mechanisms are known for specific submodular objectives [26, 10, 27]. The deterministic mechanisms for KNAPSACK [10] and COVERAGE [27] follow the same general framework, which we also employ. In these two cases, the framework approximates the optimal solution to the underlying combinatorial problem by a linear program (LP) relaxation [1]. No such relaxation exists for EDP, which is unlikely to be approximable through an LP due to its logarithmic objective. We develop instead a convex relaxation to EDP; though, contrary to the above LP relaxations, this cannot be solved exactly, we show how to incorporate it in the framework of [10, 27] to yield a $\delta$-truthful mechanism for EDP.

*Beyond Submodular Objectives.* Beyond submodular objectives, it is known that no truthful mechanism with approximation ratio smaller than $n^{1/2-\varepsilon}$ exists for maximizing fractionally subadditive functions (a class that includes submodular functions) assuming access to a value query oracle [26]. Assuming access to a stronger oracle (the *demand* oracle), there exists a truthful, $O(\log^3 n)$-approximate mechanism [11] as well as a universally truthful, $O(\frac{\log n}{\log \log n})$-approximate mechanism for subadditive maximization [5]. Moreover, in a Bayesian setup, assuming a prior distribution among the agent's costs, there exists a truthful mechanism with a 768/512-approximation ratio [5]. Posted price, rather than direct revelation mechanisms, are also studied in [4].

*Monotone Approximations in Combinatorial Auctions.* Relaxations of combinatorial problems are prevalent in *combinatorial auctions*, in which an auctioneer aims at maximizing social welfare. As noted by Archer et al. [2], approximations to this maximization must preserve incentive compatibility. Most approximation algorithms do not preserve this property, hence specific relaxations, and corresponding roundings to an integral solution, must be constructed [2, 19, 12, 7]. Because of the specificity of our relaxation, and because we seek a determinist mechanism and $\delta$-truthfulness, not truthfulness-in-expectation, none of the techniques present in these works apply to our setting.

*$\delta$-Truthfulness and Differential Privacy.* The notion of $\delta$-truthfulness has attracted considerable attention recently in the context of differential privacy (see, *e.g.*, the survey by Pai and Roth [24]). McSherry and Talwar [21] were the first to observe that any $\varepsilon$-differentially private mechanism must also be $\delta$-truthful in expectation, for $\delta = 2\varepsilon$. This property was used to construct $\delta$-truthful (in expectation) mechanisms for a digital goods auction [21] and for $\alpha$-approximate equilibrium selection [17]. Nissim et al. [23] propose a framework for converting a differentially private mechanism to a truthful-in-expectation mechanism by randomly selecting between a differentially private mechanism with good approximation guarantees, and a truthful mechanism. They apply their framework to the FACILITYLOCATION problem. We depart from the above works in seeking a deterministic mechanism for EDP, and using a stronger notion of $\delta$-truthfulness.

## 3 Preliminaries

### 3.1 Linear Regression and Experimental Design

The theory of experimental design [25, 3, 9] considers the following formal setting. Suppose that an experimenter E wishes to conduct $k$ among $n$ possible experiments. Each experiment $i \in \mathcal{N} \equiv \{1, \ldots, n\}$ is associated with a set of parameters (or features) $x_i \in \mathbb{R}^d$, normalized so that $b \leq \|x_i\|_2^2 \leq 1$, for some

$b > 0$. Denote by $S \subseteq \mathcal{N}$, where $|S| = k$, the set of experiments selected; upon its execution, experiment $i \in S$ reveals an output variable (the "measurement") $y_i$, related to the experiment features $x_i$ through a linear function, *i.e.*, $y_i = \beta^T x_i + \varepsilon_i$ where $\beta$ is a vector in $\mathbb{R}^d$, commonly referred to as the *model*, and $\varepsilon_i$ (the *measurement noise*) are independent, normally distributed random variables with mean 0 and variance $\sigma^2$.

For example, each $i$ may correspond to a human subject; the feature vector $x_i$ may correspond to a normalized vector of her age, weight, gender, income, *etc.*, and the measurement $y_i$ may capture some biometric information (*e.g.*, her red cell blood count, a genetic marker, etc.). The magnitude of the coefficient $\beta_i$ captures the effect that feature $i$ has on the measured variable, and its sign captures whether the correlation is positive or negative.

The purpose of these experiments is to allow E to estimate the model $\beta$. In particular, assume that the experimenter E has a *prior* distribution on $\beta$, *i.e.*, $\beta$ has a multivariate normal prior with zero mean and covariance $\sigma^2 R^{-1} \in \mathbb{R}^{d^2}$ (where $\sigma^2$ is the noise variance). Then, E estimates $\beta$ through *maximum a posteriori estimation*: *i.e.*, finding the parameter which maximizes the posterior distribution of $\beta$ given the observations $y_S$. Under the linearity assumption and the Gaussian prior on $\beta$, maximum a posteriori estimation leads to [14]:

$$\hat{\beta} = \arg\max_{\beta \in \mathbb{R}^d} \Pr(\beta \mid y_S) = \arg\min_{\beta \in \mathbb{R}^d} \left( \sum_{i \in S}(y_i - \beta^T x_i)^2 + \beta^T R \beta \right) \qquad (2)$$
$$= (R + X_S^T X_S)^{-1} X_S^T y_S$$

where the last equality is obtained by setting $\nabla_\beta \Pr(\beta \mid y_S)$ to zero and solving the resulting linear system; in (2), $X_S \equiv [x_i]_{i \in S} \in \mathbb{R}^{|S| \times d}$ is the matrix of experiment features and $y_S \equiv [y_i]_{i \in S} \in \mathbb{R}^{|S|}$ are the observed measurements. This optimization, commonly known as *ridge regression*, includes an additional quadratic penalty term $\beta^T R \beta$ compared to the standard least squares estimation.

Let $V : 2^\mathcal{N} \to \mathbb{R}$ be a *value function*, quantifying how informative a set of experiments $S$ is in estimating $\beta$. The classical experimental design problem amounts to finding a set $S$ that maximizes $V(S)$ subject to the constraint $|S| \le k$. A variety of different value functions are used in literature [25, 6]; one that has natural advantages is the *information gain*, $V(S) = I(\beta; y_S) = H(\beta) - H(\beta \mid y_S)$ which is the entropy reduction on $\beta$ after the revelation of $y_S$ (also known as the mutual information between $y_S$ and $\beta$). Hence, selecting a set of experiments $S$ that maximizes $V(S)$ is equivalent to finding the set of experiments that minimizes the uncertainty on $\beta$, as captured by the entropy reduction of its estimator. Under the linear model, and the Gaussian prior, the information gain takes the following form (see, *e.g.*, [9]):

$$I(\beta; y_S) = \frac{1}{2} \log \det(R + X_S^T X_S) - \frac{1}{2} \log \det R \qquad (3)$$

Maximizing $I(\beta; y_S)$ is therefore equivalent to maximizing $\log \det(R + X_S^T X_S)$, which is known in literature as the Bayes *D*-optimality criterion [25, 3, 9].

Our analysis will focus on the case of a *homotropic* prior, in which the prior covariance is the identity matrix, *i.e.*, $R = I_d \in \mathbb{R}^{d \times d}$. Intuitively, this corresponds to the simplest prior, in which no direction of $\mathbb{R}^d$ is a priori favored;

equivalently, it also corresponds to the case where ridge regression estimation (2) performed by E has a penalty term $\|\beta\|_2^2$. A generalization of our results to arbitrary covariance matrices $R$ can be found in [16].

### 3.2 Budget-Feasible Experimental Design: Full Information Case

Instead of the cardinality constraint in classical experimental design discussed above, we consider a budget-constrained version. Each experiment is associated with a cost $c_i \in \mathbb{R}_+$. The cost $c_i$ can capture, e.g., the amount the subject $i$ deems sufficient to incentivize her participation in the experiment. The experimenter E is limited by a budget $B \in \mathbb{R}_+$. In the full-information case, experiment costs are common knowledge; as such, the experimenter wishes to solve:

EXPERIMENTALDESIGNPROBLEM (EDP)

$$\text{Maximize} \quad V(S) = \log\det(I_d + X_S^T X_S) \tag{4a}$$
$$\text{subject to} \quad \sum_{i \in S} c_i \leq B \tag{4b}$$

W.l.o.g., we assume that $c_i \in [0, B]$ for all $i \in \mathcal{N}$, as no $i$ with $c_i > B$ can be in an $S$ satisfying (4b). Denote by

$$OPT = \max_{S \subseteq \mathcal{N}} \left\{ V(S) \,\Big|\, \sum_{i \in S} c_i \leq B \right\} \tag{5}$$

the optimal value achievable in the full-information case. EDP, as defined above, is NP-hard; to see this, note that KNAPSACK reduces to EDP with $d = 1$ by mapping the weight of each item, say, $w_i$, to an experiment with $x_i^2 = w_i$.

The value function (4a) has the following properties, which are proved in [16]. First, it is non-negative, i.e., $V(S) \geq 0$ for all $S \subseteq \mathcal{N}$. Second, it is also monotone, i.e., $V(S) \leq V(T)$ for all $S \subseteq T$, with $V(\emptyset) = 0$. Finally, it is submodular, i.e., $V(S \cup \{i\}) - V(S) \geq V(T \cup \{i\}) - V(T)$ for all $S \subseteq T \subseteq \mathcal{N}$ and $i \in \mathcal{N}$. The above imply that a greedy algorithm yields a constant approximation ratio to EDP. In particular, consider the greedy algorithm in which, for $S \subseteq \mathcal{N}$ the set constructed thus far, the next element $i$ included is the one which maximizes the *marginal-value-per-cost*, i.e., $i = \arg\max_{j \in \mathcal{N} \setminus S} (V(S \cup \{i\}) - V(S))/c_i$. This is repeated until adding an element in $S$ exceeds the budget $B$. Denote by $S_G$ the set constructed by this heuristic and let $i^* = \arg\max_{i \in \mathcal{N}} V(\{i\})$ be the element of maximum singleton value. Then, the following algorithm:

$$\text{if } V(\{i^*\}) \geq V(S_G) \text{ return } \{i^*\} \text{ else return } S_G \tag{6}$$

yields an approximation ratio of $\frac{5e}{e-1}$ [26]; this can be further improved to $\frac{e}{e-1}$ using more complicated greedy set constructions [18, 28].

### 3.3 Budget-Feasible Experimental Design: Strategic Case

We study the following *strategic* setting, in which the costs $c_i$ are *not* common knowledge and their reporting can be manipulated by the experiment subjects.

The latter are strategic and wish to maximize their utility, which is the difference of the payment they receive and their true cost. Note that, though subjects may misreport $c_i$, they cannot lie about $x_i$ (*i.e.*, all public features are verifiable prior to the experiment) nor $y_i$ (*i.e.*, the subject cannot falsify her measurement). Experimental design thus reduces to a *budget feasible reverse auction*, as introduced by Singer [26]; we review the formal definition in [16]. In short, given a budget $B$ and a value function $V : 2^\mathcal{N} \to \mathbb{R}_+$, a *reverse auction mechanism* $\mathcal{M} = (S, p)$ comprises (a) an *allocation function* $S : \mathbb{R}_+^n \to 2^\mathcal{N}$, determining the set of experiments to be purchased, and (b) a *payment function* $p : \mathbb{R}_+^n \to \mathbb{R}_+^n$, determining the payments $[p_i(c)]_{i \in \mathcal{N}}$ received by experiment subjects.

We seek mechanisms that are *normalized* (unallocated experiments receive zero payments), *individually rational* (payments for allocated experiments exceed costs), have *no positive transfers* (payments are non-negative), and are *budget feasible* (the sum of payments does not exceed the budget $B$). We relax the notion of truthfulness to *δ-truthfulness*, requiring that reporting one's true cost is an *almost-dominant* strategy: no subject increases their utility by reporting a cost that differs more than $\delta > 0$ (*e.g.*, a tenth of a cent) from their true cost. Under this definition, a mechanism is truthful if $\delta = 0$. In addition, we would like the allocation $S(c)$ to be of maximal value; however, δ-truthfulness, as well as the hardness of EDP, preclude achieving this goal. Hence, we seek mechanisms with that are $(\alpha, \beta)$-*approximate*, *i.e.*, there exist $\alpha \geq 1$ and $\beta > 0$ s.t. $OPT \leq \alpha V(S(c)) + \beta$, and are *computationally efficient*, in that $S$ and $p$ can be computed in polynomial time.

We note that the constant approximation algorithm (6) breaks truthfulness. Though this is not true for all submodular functions (see, *e.g.*, [26]), it is true for the objective of EDP: we show this in [16], motivating our study of more complex mechanisms.

## 4 Approximation Results

Previous approaches towards designing truthful, budget feasible mechanisms for KNAPSACK [10] and COVERAGE [27] build upon polynomial-time algorithms that compute an approximation of $OPT$, the optimal value in the full information case. Crucially, to be used in designing a truthful mechanism, such algorithms need also to be *monotone*, in the sense that decreasing any cost $c_i$ leads to an increase in the estimation of $OPT$; the monotonicity property precludes using traditional approximation algorithms.

In the first part of this section, we address this issue by designing a convex relaxation of EDP, and showing that its solution can be used to approximate $OPT$. The objective of this relaxation is concave and self-concordant [6] and, as such, there exists an algorithm that solves this relaxed problem with arbitrary accuracy in polynomial time. Unfortunately, the output of this algorithm may not necessarily be monotone. Nevertheless, in the second part of this section, we show that a solver of the relaxed problem can be used to construct a solver that is "almost" monotone. In Section 5, we show that this algorithm can be used to design a δ-truthful mechanism for EDP.

## 4.1 A Convex Relaxation of EDP

A classical way of relaxing combinatorial optimization problems is *relaxing by expectation*, using the so-called *multi-linear* extension of the objective function $V$ (see, *e.g.*, [8, 30, 13]). This is because this extension can yield approximation guarantees for a wide class of combinatorial problems through *pipage rounding*, a technique by Ageev and Sviridenko [1]. In general, such relaxations preserve monotonicity which, as discussed, is required in mechanism design.

Formally, let $P_\mathcal{N}^\lambda$ be a probability distribution over $\mathcal{N}$ parametrized by $\lambda \in [0,1]^n$, where a set $S \subseteq \mathcal{N}$ sampled from $P_\mathcal{N}^\lambda$ is constructed as follows: each $i \in \mathcal{N}$ is selected to be in $S$ independently with probability $\lambda_i$, i.e., $P_\mathcal{N}^\lambda(S) \equiv \prod_{i \in S} \lambda_i \prod_{i \in \mathcal{N} \setminus S}(1 - \lambda_i)$. Then, the *multi-linear* extension $F : [0,1]^n \to \mathbb{R}$ of $V$ is defined as the expectation of $V$ under the distribution $P_\mathcal{N}^\lambda$:

$$F(\lambda) \equiv \mathbb{E}_{S \sim P_\mathcal{N}^\lambda}[V(S)] = \mathbb{E}_{S \sim P_\mathcal{N}^\lambda}\left[\log \det\left(I_d + \sum_{i \in S} x_i x_i^T\right)\right], \quad \lambda \in [0,1]^n. \quad (7)$$

Function $F$ is an extension of $V$ to the domain $[0,1]^n$, as it equals $V$ on integer inputs: $F(\mathbb{1}_S) = V(S)$ for all $S \subseteq \mathcal{N}$, where $\mathbb{1}_S$ denotes the indicator vector of $S$. Contrary to problems such as KNAPSACK, the multi-linear extension (7) cannot be optimized in polynomial time for the value function $V$ we study here, given by (4a). Hence, we introduce an extension $L : [0,1]^n \to \mathbb{R}$ s.t.

$$L(\lambda) \equiv \log \det\left(I_d + \sum_{i \in \mathcal{N}} \lambda_i x_i x_i^T\right), \quad \lambda \in [0,1]^n. \quad (8)$$

Note that $L$ also extends $V$, and follows naturally from the multi-linear extension by swapping the expectation and log det in (7). Crucially, it is *strictly concave* on $[0,1]^n$, a fact that we exploit in the next section to maximize $L$ subject to the budget constraint in polynomial time.

Our first technical lemma relates $L$ to the multi-linear extension $F$:

**Lemma 1.** *For all $\lambda \in [0,1]^n$, $\frac{1}{2} L(\lambda) \leq F(\lambda) \leq L(\lambda)$.*

The proof of this lemma can be found in [16]. In short, exploiting the concavity of the log det function over the set of positive semi-definite matrices, we first bound the ratio of all partial derivatives of $F$ and $L$. We then show that the bound on the ratio of the derivatives also implies a bound on the ratio $F/L$.

Armed with this result, we subsequently use pipage rounding to show that any $\lambda$ that maximizes the multi-linear extension $F$ can be rounded to an "almost" integral solution. More specifically, given a set of costs $c \in \mathbb{R}_+^n$, we say that a $\lambda \in [0,1]^n$ is feasible if it belongs to the set $\mathcal{D}_c = \{\lambda \in [0,1]^n : \sum_{i \in \mathcal{N}} c_i \lambda_i \leq B\}$. Then, the following lemma holds:

**Lemma 2 (Rounding).** *For any feasible $\lambda \in \mathcal{D}_c$, there exists a feasible $\bar{\lambda} \in \mathcal{D}_c$ such that (a) $F(\lambda) \leq F(\bar{\lambda})$, and (b) at most one of the coordinates of $\bar{\lambda}$ is fractional.*

The proof, also in [16], follows the main steps of the pipage rounding method of Ageev and Sviridenko [1]. Together, Lemma 1 and Lemma 2 imply that $OPT$, the optimal value of EDP, can be approximated by solving the following convex optimization problem:

$$\text{Maximize:} \quad L(\lambda) \quad \text{subject to:} \quad \lambda \in \mathcal{D}_c \qquad (P_c)$$

In particular, for $L_c^* \equiv \max_{\lambda \in \mathcal{D}_c} L(\lambda)$, the following holds [16]:

**Proposition 1.** $OPT \leq L_c^* \leq 2OPT + 2\max_{i \in \mathcal{N}} V(i)$.

As we discuss in the next section, $L_c^*$ can be computed by a poly-time algorithm at arbitrary accuracy. However, the outcome of this computation may not necessarily be monotone; we address this by converting this poly-time estimator of $L_c^*$ to one that is "almost" monotone.

### 4.2 Polynomial-Time, Almost-Monotone Approximation

The log det objective function of $(P_c)$ is strictly concave and *self-concordant* [6]. The maximization of a concave, self-concordant function subject to a set of linear constraints can be performed through the *barrier method* (see, e.g., [6] Section 11.5.5 for general self-concordant optimization as well as [29] for a detailed treatment of the log det objective). The performance of the barrier method is summarized in our case by the following lemma:

**Lemma 3 (Boyd and Vandenberghe [6]).** *For any $\varepsilon > 0$, the barrier method computes an approximation $\hat{L}_c^*$ that is $\varepsilon$-accurate, i.e., it satisfies $|\hat{L}_c^* - L_c^*| \leq \varepsilon$, in time $O\left(\text{poly}(n, d, \log\log \varepsilon^{-1})\right)$. The same guarantees apply when maximizing $L$ subject to an arbitrary set of $O(n)$ linear constraints.*

Clearly, the optimal value $L_c^*$ of $(P_c)$ is monotone in $c$: formally, for any two $c, c' \in \mathbb{R}_+^n$ s.t. $c \leq c'$ coordinate-wise, $\mathcal{D}_{c'} \subseteq \mathcal{D}_c$ and thus $L_c^* \geq L_{c'}^*$. Hence, the map $c \mapsto L_c^*$ is non-increasing. Unfortunately, the same is not true for the output $\hat{L}_c^*$ of the barrier method: there is no guarantee that the $\varepsilon$-accurate approximation $\hat{L}_c^*$ exhibits any kind of monotonicity.

Nevertheless, we prove that it is possible to use the barrier method to construct an approximation of $L_c^*$ that is "almost" monotone. More specifically, given $\delta > 0$, we say that $f : \mathbb{R}^n \to \mathbb{R}$ is $\delta$-*decreasing* if $f(x) \geq f(x + \mu e_i)$, for all $i \in \mathcal{N}, x \in \mathbb{R}^n, \mu \geq \delta$, where $e_i$ is the $i$-th canonical basis vector of $\mathbb{R}^n$. In other words, $f$ is $\delta$-decreasing if increasing any coordinate by $\delta$ or more at input $x$ ensures that the output will be at most $f(x)$.

We achieve this by restricting the optimization over a subset of $\mathcal{D}_c$ at which the concave relaxation $L$ is "sufficiently" concave. Formally, for $\alpha \geq 0$ let

$$\mathcal{D}_{c,\alpha} \equiv \{\lambda \in [\alpha, 1]^n : \sum_{i \in \mathcal{N}} c_i \lambda_i \leq B\} \subseteq \mathcal{D}_c.$$

Note that $\mathcal{D}_c = \mathcal{D}_{c,0}$. Consider the following perturbed problem:

$$\begin{aligned}\text{Maximize:} &\quad L(\lambda)\\ \text{subject to:} &\quad \lambda \in \mathcal{D}_{c,\alpha}\end{aligned} \qquad (P_{c,\alpha})$$

Restricting the feasible set to $\mathcal{D}_{c,\alpha}$ ensures that the gradient of the optimal solution with respect to $c$ is bounded from below. This implies that an approximate solution to $P_{c,\alpha}$ given by the barrier method is $\delta$-decreasing with respect to the costs. On the other hand, by taking $\alpha$ small enough, we ensure that the approximate solution to $P_{c,\alpha}$ is still an $\varepsilon$-accurate approximation of $L_c^*$. This methodology is summarized in the following proposition, whose proof can be found in [16].

**Proposition 2.** *For any $\delta \in (0,1]$ and any $\varepsilon \in (0,1]$, using the barrier method to solve $(P_{c,\alpha})$ for $\alpha \equiv \varepsilon(\delta/B + n^2)^{-1}$ with accuracy $\frac{1}{2^{n+1}B}\alpha\delta b$ yields a $\delta$-decreasing, $\varepsilon$-accurate approximation of $L_c^*$. The running time of the algorithm is $O\bigl(\text{poly}(n, d, \log\log \frac{B}{b\varepsilon\delta})\bigr)$.*

## 5 Mechanism for EDP

The $\delta$-decreasing, $\varepsilon$-accurate algorithm solving the convex optimization problem $(P_c)$ can be used to design a mechanism for EDP. The construction follows a methodology proposed in [26] and employed by Chen et al. [10] and Singer [27] to construct mechanisms for KNAPSACK and COVERAGE respectively. We briefly outline this below (see [16] for a detailed description).

Recall from Section 3.2 that $i^* \equiv \arg\max_{i \in \mathcal{N}} V(\{i\})$ is the element of maximum value, and $S_G$ is a set constructed greedily, by selecting elements of maximum marginal value per cost. The general framework used by Chen et al. [10] and by Singer [27] for the KNAPSACK and COVERAGE value functions contructs an allocation as follows. First, a polynomial-time, monotone approximation of $OPT$ is computed over all elements excluding $i^*$. The outcome of this approximation is compared to $V(\{i^*\})$: if it exceeds $V(\{i^*\})$, then the mechanism constructs an allocation $S_G$ greedily; otherwise, the only item allocated is the singleton $\{i^*\}$. Provided that the approximation used is within a constant from $OPT$, the above allocation can be shown to also yield a constant approximation to $OPT$. Furthermore, Myerson's Theorem [22] implies that this allocation combined with *threshold payments* (see Lemma 4 below) constitute a truthful mechanism.

The approximation algorithms used in [10, 27] are LP relaxations, and thus their outputs are monotone and can be computed exactly in polynomial time. We show that the convex relaxation $(P_c)$, solved by an $\varepsilon$-accurate, $\delta$-decreasing algorithm, can be used to construct a $\delta$-truthful, constant approximation mechanism.

To obtain this result, we use the following modified version of Myerson's theorem [22], whose proof we provide in [16].

**Lemma 4.** *A normalized mechanism $\mathcal{M} = (S, p)$ for a single parameter auction is $\delta$-truthful if: (a) $S$ is $\delta$-monotone, i.e., for any agent $i$ and $c_i' \leq c_i - \delta$, for any fixed costs $c_{-i}$ of agents in $\mathcal{N} \setminus \{i\}$, $i \in S(c_i, c_{-i})$ implies $i \in S(c_i', c_{-i})$, and (b) agents are paid threshold payments, i.e., for all $i \in S(c)$, $p_i(c) = \inf\{c_i' : i \in S(c_i', c_{-i})\}$.*

Lemma 4 allows us to incorporate our relaxation in the above framework, yielding the following theorem:

**Theorem 1.** *For any $\delta \in (0,1]$, and any $\varepsilon \in (0,1]$, there exists a $\delta$-truthful, individually rational and budget feasible mechanim for EDP that runs in time $O\big(poly(n,d,\log\log\frac{B}{b\varepsilon\delta})\big)$ and allocates a set $S^*$ such that*

$$OPT \leq \frac{10e - 3 + \sqrt{64e^2 - 24e + 9}}{2(e-1)} V(S^*) + \varepsilon \simeq 12.98 V(S^*) + \varepsilon.$$

*Furthemore, there is no 2-approximate, truthful, budget feasible, individually rational mechanism for EDP.*

The detailed description of our proposed mechanism as well as the proof of the theorem can be found in [16].

## 6 Conclusions

We have proposed a convex relaxation for EDP, and showed how to use it to design a $\delta$-truthful, constant approximation mechanism that runs in polynomial time. A natural question to ask is to what extent ou results generalize to other machine learning tasks beyond linear regression. We outline a path to such a generalization in [16]: for a wide class of models in which experiment outcomes are perturbed by independent noise, the information gain exhibits submodularity. In light of this, it would be interesting to investigate whether our convex relaxation approach generalizes to other tasks in this broader class. Moreover, the literature on experimental design includes several other optimality criteria [25, 3], many of which are convex [6]. Exploiting this to design budget feasible mechanisms is an additional open problem of interest.

**Acknowledgments.** We thank Francis Bach for our helpful discussions on approximate solutions of convex optimization problems, and Yaron Singer for his comments and suggestions, and for insights into budget feasible mechanisms.

## References

[1] Ageev, A.A., Sviridenko, M.: Pipage rounding: A new method of constructing algorithms with proven performance guarantee. J. Comb. Optim. 8(3), 307–328 (2004)

[2] Archer, A., Papadimitriou, C., Talwar, K., Tardos, E.: An approximate truthful mechanism for combinatorial auctions with single parameter agents. Internet Mathematics 1(2), 129–150 (2004)

[3] Atkinson, A., Donev, A., Tobias, R.: Optimum experimental designs, with SAS. Oxford University Press, Oxford (2007)

[4] Badanidiyuru, A., Kleinberg, R., Singer, Y.: Learning on a budget: posted price mechanisms for online procurement. In: EC (2012)

5. Bei, X., Chen, N., Gravin, N., Lu, P.: Budget feasible mechanism design: from prior-free to bayesian. In: STOC (2012)
6. Boyd, S., Vandenberghe, L.: Convex Optimization. Cambridge University Press (2004)
7. Briest, P., Krysta, P., Vöcking, B.: Approximation techniques for utilitarian mechanism design. In: ACM STOC (2005)
8. Calinescu, G., Chekuri, C., Pál, M., Vondrák, J.: Maximizing a submodular set function subject to a matroid constraint (Extended abstract). In: Fischetti, M., Williamson, D.P. (eds.) IPCO 2007. LNCS, vol. 4513, pp. 182–196. Springer, Heidelberg (2007)
9. Chaloner, K., Verdinelli, I.: Bayesian experimental design: A review. Statistical Science, 273–304 (1995)
10. Chen, N., Gravin, N., Lu, P.: On the approximability of budget feasible mechanisms. In: SODA (2011)
11. Dobzinski, S., Papadimitriou, C.H., Singer, Y.: Mechanisms for complement-free procurement. In: ACM EC (2011)
12. Dughmi, S.: A truthful randomized mechanism for combinatorial public projects via convex optimization. In: EC (2011)
13. Dughmi, S., Roughgarden, T., Yan, Q.: From convex optimization to randomized mechanisms: toward optimal combinatorial auctions. In: STOC (2011)
14. Friedman, J., Hastie, T., Tibshirani, R.: The elements of statistical learning. Springer Series in Statistics, vol. 1 (2001)
15. Ginebra, J.: On the measure of the information in a statistical experiment. Bayesian Analysis 2(1), 167–211 (2007)
16. Horel, T., Ioannidis, S., Muthukrishnan, S.: Budget feasible mechanisms for experimental design (2013), http://arxiv.org/abs/1302.5724
17. Kearns, M., Pai, M.M., Roth, A., Ullman, J.: Private equilibrium release, large games, and no-regret learning (2012), http://arxiv.org/abs/1207.4084v1
18. Krause, A., Guestrin, C.: A note on the budgeted maximization of submodular functions. Tech. Rep. CMU-CALD-05-103, CMU (2005)
19. Lavi, R., Swamy, C.: Truthful and near-optimal mechanism design via linear programming. Journal of the ACM 58(6), 25 (2011)
20. Le Cam, L.: Comparison of experiments: a short review. Lecture Notes-Monograph Series, pp. 127–138 (1996)
21. McSherry, F., Talwar, K.: Mechanism design via differential privacy. In: FOCS (2007)
22. Myerson, R.: Optimal auction design. Mathematics of Operations Research 6(1), 58–73 (1981)
23. Nissim, K., Smorodinsky, R., Tennenholtz, M.: Approximately optimal mechanism design via differential privacy. In: ITCS (2012)
24. Pai, M., Roth, A.: Privacy and mechanism design. SIGecom Exchanges (2013)
25. Pukelsheim, F.: Optimal design of experiments. Society for Industrial Mathematics, vol. 50 (2006)
26. Singer, Y.: Budget feasible mechanisms. In: FOCS (2010)
27. Singer, Y.: How to win friends and influence people, truthfully: influence maximization mechanisms for social networks. In: WSDM (2012)
28. Sviridenko, M.: A note on maximizing a submodular set function subject to a knapsack constraint. Oper. Res. Lett. 32(1), 41–43 (2004)
29. Vandenberghe, L., Boyd, S., Wu, S.: Determinant maximization with linear matrix inequality constraints. SIAM Journal on Matrix Analysis and Applications 19(2), 499–533 (1998)
30. Vondrak, J.: Optimal approximation for the submodular welfare problem in the value oracle model. In: ACM STOC (2008)

# LZ77-Based Self-indexing with Faster Pattern Matching

Travis Gagie[1], Paweł Gawrychowski[2], Juha Kärkkäinen[1],
Yakov Nekrich[3], and Simon J. Puglisi[1]

[1] University of Helsinki, Finland
[2] Max Planck Institute, Germany
[3] University of Kansas, United States

**Abstract.** To store and search genomic databases efficiently, researchers have recently started building self-indexes based on LZ77. As the name suggests, a self-index for a string supports both random access and pattern matching queries. In this paper we show how, given a string $S[1..n]$ whose LZ77 parse consists of $z$ phrases, we can store a self-index for $S$ in $\mathcal{O}(z \log(n/z))$ space such that later, first, given a position $i$ and a length $\ell$, we can extract $S[i..i + \ell - 1]$ in $\mathcal{O}(\ell + \log n)$ time; second, given a pattern $P[1..m]$, we can list the $occ$ occurrences of $P$ in $S$ in $\mathcal{O}(m \log m + occ \log \log n)$ time.

## 1 Introduction

With the advance of DNA-sequencing technologies comes the problem of how to store many individuals' genomes compactly but such that we can search them quickly. Any two human genomes are almost the same but self-indexes based on compressed suffix arrays, the Burrows-Wheeler Transform or LZ78 (see [24] for a survey) do not take full advantage of this similarity [20]. Researchers have recently started building self-indexes based on context-free grammars and LZ77 [30], which better compress such highly repetitive strings.

A self-index for a string $S[1..n]$ stores $S$ in compressed form such that later, first, given a position $i$ and a length $\ell$, we can quickly extract $S[i..i + \ell - 1]$; second, given a pattern $P[1..m]$, we can quickly list the $occ$ occurrences of $P$ in $S$. In this paper we describe a self-index that takes $\mathcal{O}(z \log(n/z))$ space, where $z$ is the number of phrases in the LZ77 parse of $S$, and supports extraction in $\mathcal{O}(\ell + \log n)$ time and pattern matching in $\mathcal{O}(m \log m + occ \log \log n)$ time. Our model throughout is the word RAM with $\Theta(\log n)$-bit words and we measure space in terms of these words. We assume $S$ is over a fixed alphabet.

Several authors have designed self-indexes for repetitive data sets: e.g., Arroyuelo, Navarro and Sadakane [3]; Claude and Navarro [7]; Do et al. [8]; Huang et al. [15]; Kreft and Navarro [19]; Mäkinen et al. [20]; Maruyama et al. [21]; Russo and Oliveira![25]; Wandelt et al. [28]; Yang et al. [29]; and ourselves [13]. Most of these indexes have bounds depending on values other than $m$, $n$ and $z$, however, making them difficult to compare directly to our results in this paper.

Here we are concerned with theoretical worst-case bounds and, as far as we are aware, when any known self-index's worst-case bounds are expressed in terms only of $n$ and $z$, they are somehow worse than the bounds we give here.

In Section 2 we briefly describe LZ77 and other preliminaries: straight-line programs, bookmarking, and Karp-Rabin fingerprinting. In Section 3 we describe Kärkkäinen and Ukkonen's [16] LZ77-based index. Their index makes use of Patricia trees [22] and planar orthogonal range reporting [2,6], and we assume knowledge of these concepts here. Although not itself a self-index — because it requires $S$ to be stored separately — the basic structure of Kärkkäinen and Ukkonen's index has been copied in many of the self-indexes listed above. In a previous paper [13] we combined their index with a bookmarked straight-line program for $S$ to obtain a self-index that takes $\mathcal{O}(z \log(n/z))$ space and supports extraction in $\mathcal{O}(\ell + \log n)$ time and pattern matching in $\mathcal{O}(m^2 + (m + occ) \log \log n)$ time. In Section 4 we use Karp-Rabin fingerprinting, fat binary search and bookmarked fingerprinting to reduce pattern-matching time to $\mathcal{O}(m \log m + occ \log \log n)$, which is the contribution of this paper.

## 2 Preliminaries

### 2.1 LZ77

The LZ77 algorithm [30] compresses $S$ by parsing it into $z = \mathcal{O}(n/\log n)$ phrases such that if $S[i..j]$ is a phrase, then $S[i..j-1]$ occurs in $S[1..j-2]$ but $S[i..j]$ does not occur in $S[1..j-1]$ (unless $j = n$, in which case $S[i..j]$ may occur in $S[1..j-1]$; it is common to delimit strings with a unique end-of-file character $\$$ to avoid this). We store $S[i..j]$ as a pointer to the leftmost occurrence of $S[i..j-1]$ in $S$, followed by $j - i$ and $S[j]$. The leftmost occurrence of $S[i..j-1]$ is called the source of phrase $S[i..j]$. For example, LZ77 parses the delimited Fibonacci word $abaababaabaab\$$ into phrases $a$, $b$, $aa$, $bab$, $aabaa$ and $b\$$, which it encodes as $(null, 0, a), (null, 0, b), (1, 1, a), (2, 2, b), (3, 4, a)$ and $(2, 1, \$)$. We can compute the LZ77 parse of $S$ in $\mathcal{O}(n)$ time.

Suppose $S[i..j]$ is a phrase with source $S[i'..j']$. Farach and Thorup [12] observed that if a substring is completely contained in $S[i..j-1]$ — so that substring neither ends at nor crosses a phrase boundary — then it is equal to the corresponding substring in $S[i'..j']$. Kärkkäinen and Ukkonen [16] defined primary substrings to be those that cross or end at phrase boundaries, and secondary substrings to be those that do not. By Farach and Thorup's observation, we can compute the locations of all the secondary occurrences of $P$ in $S$ from the structure of the parse and a list of the locations of $P$'s primary occurrences: we find all the phrase sources that include the occurrences in the list; compute the positions of all the corresponding secondary occurrences in the phrases with those sources; add those secondary occurrences to the list; and recurse on them.

### 2.2 Straight-Line Programs

A straight-line program (SLP) for $S$ is a context-free grammar in Chomsky normal form that generates $S$ and only $S$. For example, Figure 1 shows an SLP for

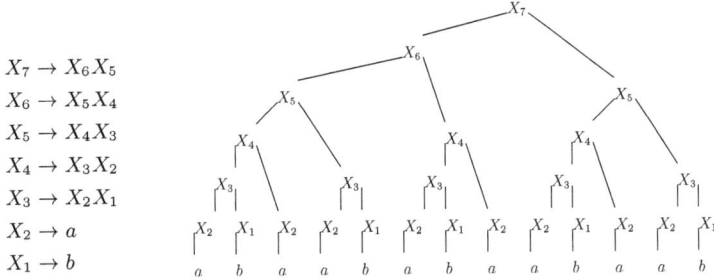

**Fig. 1.** An SLP for *abaababaabaab* (left) and the corresponding parse tree (right)

*abaababaabaab* together with the corresponding parse tree. An SLP is balanced if the height of each symbol is logarithmic in the length of its expansion. The height of a terminal is defined to be 0; if $X \to y$ then $height(X) = 1$ (because the SLP is in Chomsky normal form so $y$ is a terminal); if $X \to YZ$, then $height(X) = \max(height(Y), height(Z)) + 1$.

Rytter [26] and Charikar et al. [9] showed how we can turn the LZ77 parse of $S$ into a balanced SLP for $S$ with $\mathcal{O}(z \log(n/z))$ rules. If we store such an SLP for $S$ together with the length of each non-terminal's expansion, which takes a total of $\mathcal{O}(z \log(n/z))$ space, then we can support extraction in $\mathcal{O}(\ell + \log n)$ time, which nearly matches a lower bound by Verbin and Yu [27]. Bille et al. [10] showed how we can store even an unbalanced SLP in space proportional to the number of rules such that we can support extraction in $\mathcal{O}(\ell + \log n)$ time.

**Theorem 1 (Rytter, 2003; Charikar et al., 2005).** *Given a string $S[1..n]$ whose LZ77 parse consists of $z$ phrases, we can store $S$ in $\mathcal{O}(z \log(n/z))$ space such that later, given a position $i$ and a length $\ell$, we can extract $S[i..i + \ell - 1]$ in $\mathcal{O}(\ell + \log n)$ time.*

### 2.3 Bookmarking

In a previous paper [13] we proved the following lemma, which we used to store $S$ in $\mathcal{O}(z \log(n/z))$ space such that we can support extraction of primary substrings in $\mathcal{O}(\ell)$ time. In fact, our construction can be applied to any set of pre-specified positions to allow fast extraction of substrings that cross one or more of those positions, hence the name "bookmarking". In this paper, however, we care only about fast extraction of primary substrings.

**Lemma 1 (Gagie et al., 2012).** *Given a balanced SLP for $S$ and positions $i$ and $j$, we can find two non-terminals with height $\mathcal{O}(\log(j - i))$ such that the concatenation of their expansions includes $S[i..j]$.*

We now review how we use Lemma 1 to support bookmarked extraction. Suppose we have stored a balanced SLP for $S$ together with the length of each non-terminal's expansion, so we can support extraction in $\mathcal{O}(\ell + \log n)$ time. This is fast enough when the target primary substring has length $\Omega(\log n)$. Without loss of generality, therefore, we need only consider how, given a length $\ell = o(\log n)$ and the position $b$ of the character immediately following a phrase boundary, we can extract $S[b..b+\ell-1]$ in $\mathcal{O}(\ell)$ time. Extracting characters preceding the boundary is symmetric.

We first consider the case when $\log \log n < \ell \leq \log n$. For each phrase boundary, we find two non-terminals $Y$ and $Z$ with height $\mathcal{O}(\log \log n)$ such that their concatenation contains $S[b..b + \log n - 1]$. We store the rule $X \to YZ$, where $X$ is a new non-terminal; the length of $X$'s expansion, which is in $2^{\mathcal{O}(\log \log n)} = \log^{\mathcal{O}(1)} n$; and the offset of $S[b]$ in that expansion. This takes a total of $\mathcal{O}(z)$ space for all phrase boundaries.

We now have a balanced SLP for $X$'s expansion together with the length of each non-terminal's expansion. Therefore, for $\log \log n < \ell \leq \log n$, we can extract $S[b..b+\ell-1]$ in $\mathcal{O}(\ell + \log \log n) = \mathcal{O}(\ell)$ time. If we recurse $\log^* n$ times, we use a total of $\mathcal{O}(z \log^* n) \subset \mathcal{O}(z \log(n/z))$ extra space and can support extraction of primary substrings in $\mathcal{O}(\ell)$ time for any $\ell$.

## 2.4 Karp-Rabin Fingerprinting

A Karp-Rabin [18] fingerprinting function $\phi$ has the form

$$\phi(S[1..n]) = \sum_{i=1}^{n} (S[i] \cdot \sigma^{i-1}) \bmod p,$$

where $S[i]$ is treated as that character's lexicographic rank (counting from 0) in the alphabet, $\sigma$ is the size of the alphabet and $p$ is a prime. The value $\phi(S)$ is called $S$'s Karp-Rabin fingerprint. Two strings $S_1$ and $S_2$ of length at most $n$ have the same fingerprint if and only if $p$ divides $|S_1 - S_2|$ when $S_1$ and $S_2$ are viewed as $\sigma$-ary numbers. Since $|S_1 - S_2| < \sigma^n$, it has $\mathcal{O}(n/\log n)$ distinct prime factors when $\sigma$ is fixed. There are $\Theta(n^3/\log n)$ primes less than $n^3$ so, if we choose $p$ uniformly at random from among them, the probability $S_1$ and $S_2$ have the same fingerprint is $\mathcal{O}(1/n^2)$. By the union bound, the probability no two distinct substrings of $S$ have the same fingerprint, is $1 - \mathcal{O}(1/n)$.

A useful feature of Karp-Rabin fingerprinting functions is that, if we already have the fingerprints $\phi(S_1)$ and $\phi(S_2)$ of $S_1$ and $S_2$, then we can compute the fingerprint $\phi(S_1 S_2) = \phi(S_1) + \phi(S_2) \cdot \sigma^{|S_1|}$ of their concatenation $S_1 S_2$ in $\mathcal{O}(1)$ time. Similarly, we can compute $\phi(S_1)$ from $\phi(S_2)$ and $\phi(S_1 S_2)$, or $\phi(S_2)$ from $\phi(S_1)$ and $\phi(S_1 S_2)$, in $\mathcal{O}(1)$ time. In an unpublished extension [14] of our previous paper we observed that, if we store a balanced SLP for $S$ together with the length and Karp-Rabin fingerprint of each non-terminal's expansion, then we can compute the fingerprint of any substring $S[i..j]$ of $S$ in $\mathcal{O}(\log n)$ time. To do this, we find $\mathcal{O}(\log n)$ non-terminals such that the concatenations of their expansions is $S[i..j]$, and combine their fingerprints.

Bille et al. [11] recently and independently made the same observation, and showed that we can store even an unbalanced SLP in space proportional to the number of rules such that we can support fingerprinting substrings in $\mathcal{O}(\log n)$

time. More importantly for our results in this paper, they showed how to use Karp, Miller and Rosenberg's [17] renaming algorithm to check in $\mathcal{O}(n \log n)$ time that no two distinct substrings of $S$ whose lengths are powers of 2 have the same fingerprint.

## 3 Kärkkäinen and Ukkonen's Index

Kärkkäinen and Ukkonen's [16] index for $S$ consists of two parts: the first uses two Patricia trees, access to a plain-text representation of $S$ and 4-sided range reporting such that, given a pattern $P[1..m]$, in $\mathcal{O}(m^2 + (m + occ) \log \log z)$ time it returns a list of the primary occurrences of $P$ in $S$; the second then uses that list and 2-sided range reporting to return in $\mathcal{O}(occ \log \log n)$ time a list of the secondary occurrences of $P$ in $S$. (We assume the index uses the latest range-reporting data structures.) Since it uses a plain-text representation of $S$, their index supports extraction in $\mathcal{O}(\ell)$ time; pattern matching takes a total of $\mathcal{O}(m^2 + m \log \log z + occ \log \log n)$ time.

### 3.1 Finding Primary Occurrences

To be able to find primary occurrences quickly, we store one Patricia tree $T_1$ for the reversed LZ77 phrases and another $T_2$ for the suffixes of $S$ that start at phrase boundaries. At each node of each tree we store the left-to-right ranks of that node's leftmost and rightmost leaf descendants. We also store a plain-text representation of $S$. Finally, we store a data structure for 4-sided range reporting [6] on the $z \times z$ grid on which there is a point $(x, y)$ if the lexicographically $x$th reversed phrase is followed in $S$ by the lexicographically $y$th suffix starting at a phrase boundary; we store the position of the boundary between those phrases as satellite data. For example, the reversed phrases, suffixes starting at phrase boundaries, and grid for abaababaabaab$ are shown in Figure 2, with the end-of-file character $ considered lexicographically less than the characters in the alphabet. Altogether, these data structures take $\mathcal{O}(z \log(n/z))$ space in addition to the plain string $S$.

Recall that primary occurrences of $P[1..m]$ in $S$ are those that cross or end at phrase boundaries. Therefore, for any primary occurrence, there is a smallest value $i \leq m$ such that $P[1..i]$ lies immediately to the left of a phrase boundary and $P[i+1..m]$ lies immediately to the right of it. To find all such primary occurrences, we search in $T_1$ for a node $w_1$ such that either the reversed phrases prefixed by $(P[1..i])^{rev}$, where $rev$ indicates reversal, are precisely those identified by $w_1$'s leaf descendants, or there are no such reversed phrases. We search in $T_2$ for a node $w_2$ such that either the suffixes that start at phrase boundaries and are prefixed by $P[i+1..m]$ are precisely those identified by $w_2$'s leaf descendants, or there are no such suffixes.

We use the plain-text representation of $S$ to check that the reversed phrases and suffixes indicated by $w_1$'s and $w_2$'s leaf descendants in $T_1$ and $T_2$ really are prefixed by $(P[1..i])^{rev}$ and $P[i+1..m]$, respectively. If so, we compute the left-to-right ranks $x_1$ and $x_2$ of the leftmost and rightmost of $w_1$'s leaf descendants

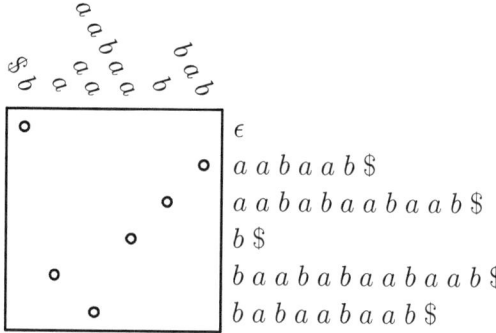

**Fig. 2.** The reversed LZ77 phrases (top; in lexicographic order from left to right), the suffixes starting at phrase boundaries (right) and the grid (center) for $abaababaabaab\$$

in $T_1$, and the left-to-right ranks $y_1$ and $y_2$ of the leftmost and rightmost of $w_2$'s leaf descendants in $T_2$. Finally, we perform a 4-sided range-reporting query for $[x_1, x_2] \times [y_1, y_2]$; each point we find corresponds to a phrase boundary that has $P[1..i]$ immediately to its left and $P[i+1..m]$ immediately to its right.

Searching in $T_1$ and $T_2$ takes $\mathcal{O}(m)$ time; finding $x_1$, $x_2$, $y_1$ and $y_2$ takes $\mathcal{O}(1)$ time; and 4-sided range reporting takes $\mathcal{O}((1+k)\log\log z)$ time, where $k$ is the number of points returned. Therefore, for $1 \leq i \leq m$, we use a total of $\mathcal{O}(m^2 + (m + occ)\log\log z)$ time.

### 3.2 Finding Secondary Occurrences

To be able to find secondary occurrences quickly, we store a data structure for 2-sided range reporting on the $n \times n$ grid on which there is a point $(x, y)$ if there is a phrase with source $S[x..y]$; we store as satellite data the locations of all the phrases with that source. Given the list of primary occurrences of $P$ in $S$, for each occurrence $S[a..b]$ in the list we use 2-sided range reporting to find each phrase source $S[x..y]$ with $x \leq a$ and $y \geq b$ (i.e., that completely contains $S[a..b]$); look up each phrase $S[i..j]$ with source $S[x..y]$; add the corresponding substring $S[i + a - x..i + b - x] = S[a..b] = P$ to the list; and recurse on the newly added occurrences. Since we perform a 2-sided range-reporting query for each occurrence of $P$ in $S$, we use a total of $\mathcal{O}(occ \log \log n)$ time.

**Theorem 2 (Kärkkäinen and Ukkonen, 1996).** *Given a string $S[1..n]$ whose LZ77 parse consists of $z$ phrases, we can store $\mathcal{O}(z)$ words such that later, given the list of primary occurrences of a pattern $P$ that occurs occ times in $S$, we can list all $P$'s occurrences in $\mathcal{O}(occ \log \log n)$ time.*

## 4 Faster Pattern Matching

If we replace the plain-text representation of $S$ in Kärkkäinen and Ukkonen's index by the data structure described in Theorem 1 — i.e., by a balanced

SLP augmented as described in Subsection 2.2 — then the index becomes an $\mathcal{O}(z\log(n/z))$-space self-index that supports extraction in $\mathcal{O}(\ell + \log n)$ and pattern matching in $\mathcal{O}(m^2 + m\log n + occ\log\log n)$ time. We showed in our previous paper [13] that, if we augment the SLP as described in Subsection 2.3 to support extraction of primary substrings in $\mathcal{O}(\ell)$ time, then we can reduce the time for pattern matching to $\mathcal{O}(m^2 + m\log\log z + occ\log\log n)$.

Next, we augment the Patricia trees to support the fat binary search method of Belazzougui, Boldi, Pagh, and Vigna [4]. We also augment the SLP to support fingerprinting substrings in $\mathcal{O}(\log n)$ time, as described in Subsection 2.4. Then, given $P$, we first compute the Karp-Rabin fingerprints $\phi(P[1]), \ldots, \phi(P[1..m])$ in $\mathcal{O}(m)$ time. After this, for $1 \leq i \leq m$, in $\mathcal{O}(\log m)$ time we can find in the Patricia tree $T_1$ for the reversed phrases a node $w_1$ such that, with high probability, either the reversed phrases prefixed by $(P[1..i])^{rev}$ are precisely those identified by $w_1$'s leaf descendants, or there are no such reversed phrases. Similarly, in $\mathcal{O}(\log m)$ time we can find in the Patricia tree $T_2$ for the suffixes starting at phrase boundaries, a node $w_1$ such that, with high probability, either the suffixes that start at phrase boundaries and are prefixed by $P[i+1..m]$ are precisely those identified by $w_2$'s leaf descendants, or there are no such suffixes.

We use the SLP to compute in $\mathcal{O}(\log n)$ time the fingerprint of the prefix of length $i$ of the reversed phrase indicated by one of $w_1$'s leaf descendants in $T_1$, and the fingerprint of the prefix of length $m-i$ of the suffix indicated by one of $w_2$'s leaf descendants in $T_2$. Comparing these two fingerprints to $\phi((P[1..i])^{rev})$ and $\phi(P[i+1..m])$ tells us, with high probability, whether the reversed phrases and suffixes indicated by the leaf descendants of $w_1$ and $w_2$ really are prefixed by $(P[1..i])^{rev}$ and $P[i+1..m]$. We still use $\mathcal{O}((m+occ)\log\log z)$ time for 4-sided range reporting.

Summing up, so far we have an LZ77-based self-index that takes $\mathcal{O}(z\log(n/z))$ space and supports extraction in $\mathcal{O}(\ell + \log n)$ time and pattern matching in $\mathcal{O}(m\log n + (m+occ)\log\log n)$ time, with a small probability of pattern-matching errors. In the rest of this section we strengthen this index in three ways: first, we reduce the $m\log n$ term to $m\log m$ using bookmarked fingerprinting; second, we eliminate the $m\log\log n$ term; and finally, we derandomize the self-index. We will continue to use Theorems 1 and 2 unchanged to support extraction and finding secondary occurrences, respectively; from now on, we are concerned only with speeding up how we find primary occurrences.

### 4.1 Bookmarked Fingerprinting

For bookmarked fingerprinting, we want to augment the SLP for $S$ such that we can compute the Karp-Rabin fingerprint of any primary substring of length $\ell$ in $\mathcal{O}(\log \ell)$ time. As with the bookmarked extraction described in Subsection 2.3, our construction here works for any set of pre-specified positions, but in this paper we care only about computing the fingerprints of primary substrings.

For the sake of the analysis, suppose we are willing to use $\mathcal{O}(f(n)\log \ell)$ time to compute the fingerprint of any primary substring of length $\ell$. We will later set

$f(n) = 2$ but, for now, it is useful to write it as a function to stop it disappearing in asymptotic notation.

With the SLP augmented with the lengths and fingerprints of non-terminals' expansions, as described in Subsection 2.4, we can compute the fingerprint of any substring in $\mathcal{O}(\log n)$ time. This is fast enough when the target primary substring has length $n^{\Omega(1/f(n))}$. Without loss of generality, therefore, we need only consider how, given a length $\ell = n^{o(1/f(n))}$ and the position $b$ of the character immediately following a phrase boundary, we can compute $\phi(S[b..b+\ell-1])$ in $\mathcal{O}(f(n) \log \ell)$ time. Computing the fingerprint of substrings preceding the boundary is symmetric.

We first consider the case when $n^{1/f(n)^2} < \ell \leq n^{1/f(n)}$. By Lemma 1, for each phrase boundary, we can find two non-terminals $Y$ and $Z$ with height $\mathcal{O}((1/f(n)) \log n)$ such that their concatenation contains $S[b..b + n^{1/f(n)} - 1]$. We store the rule $X \to YZ$, where $X$ is a new non-terminal; the length of $X$'s expansion, which is in $2^{\mathcal{O}((1/f(n))\log n)} = n^{\mathcal{O}(1/f(n))}$; the fingerprint of $X$'s expansion; and the offset of $S[b]$ in that expansion. This takes a total of $\mathcal{O}(z)$ space for all phrase boundaries.

We now have a balanced SLP for $X$'s expansion together with the length and fingerprints of each non-terminal's expansion. Therefore, for $n^{1/f(n)^2} < \ell \leq n^{1/f(n)}$, we can compute $\phi(S[b..b+\ell-1])$ in $\mathcal{O}((1/f(n))\log n) = \mathcal{O}(f(n)\log \ell)$ time. If we recurse $\log_{f(n)} \log n$ times, we use a total of $\mathcal{O}\left(z \log_{f(n)} \log n\right)$ extra space and can support fingerprinting of primary substrings in $\mathcal{O}(f(n)\log \ell)$ time. Setting $f(n) = 2$ (or any constant greater than 1) means we use $\mathcal{O}(z \log \log n)$ space, which is $\mathcal{O}(z \log(n/z))$ since $z = \mathcal{O}(n/\log n)$, and we can support fingerprinting of primary substrings in $\mathcal{O}(\log \ell)$ time.

### 4.2 Removing the $m \log \log z$ Term

Our index now supports pattern matching in $\mathcal{O}(m \log m + m \log \log z + occ \log \log n)$ time. If $m \geq \log z$ then this is $\mathcal{O}(m \log m + occ \log \log n)$, so we need consider only the case when $m < \log z$, which we split into two subcases: when $m < \log \log z$, and when $\log \log z \leq m < \log z$.

We handle the first subcase with a technique by Bille and Gørtz [5]: at each node in the top $\log \log z$ levels in the Patricia tree $T_1$ for the reversed phrases, we store a data structure for 1-dimensional range reporting on the union of the range of columns for that node in the $z \times z$ grid. Since each point on the grid appears in $\mathcal{O}(\log \log z)$ unions of ranges, all the ranges together contain a total of $\mathcal{O}(z \log \log z)$ points. Alstrup, Brodal and Rauhe [1] and Mortensen, Pagh and Pătrașcu [23] showed how we can store these data structures in a total of $\mathcal{O}(z \log \log z) \subseteq \mathcal{O}(z \log(n/z))$ space such that range reporting takes $\mathcal{O}(1+k)$ time, where $k$ is the number of points returned. When $m < \log \log z$, then our search in $T_1$ will return a node $w_1$ with depth less than $\log \log z$, so we can implement the 4-sided range reporting in $\mathcal{O}(1+k)$ time.

To deal with the second subcase, we build a Patricia tree for the set of $\mathcal{O}(z \log z)$ substrings of $S$ that cross a phrase boundary, start at most $\log z$

characters before the first phrase boundary they cross, and end exactly $\log z$ characters after it (or at $S[n]$, whichever comes first). At the leaf corresponding to each such substring, we store $\mathcal{O}(\log \log z)$ bits indicating the position in the substring where it first crosses a phrase boundary. In total this Patricia tree takes $\mathcal{O}(z \log \log z)$ space.

If $\log \log z < m < \log z$, we search for $P$ in this new Patricia tree, which takes $\mathcal{O}(m)$ time. Suppose our search ends at node $v$. If $P$ occurs in $S$, then the leaves in $v$'s subtree store the distinct positions in $P$'s primary occurrences where they cross phrase boundaries. To determine whether $P$ occurs in $S$, it suffices for us to choose any one of those positions, say $i$, and check whether there is a phrase boundary immediately preceded by $P[1..i]$ and immediately followed by $P[i+1..m]$. To do this, we search in our first two augmented Patricia trees and perform a range-emptiness query. If $m \leq \log \log z$ time then we can perform the range-emptiness query with the one-dimensional range-reporting data structures in $\mathcal{O}(1)$ time; otherwise, we perform the range-emptiness query with our data structure for 4-sided range reporting in $\mathcal{O}(\log \log z) \subseteq \mathcal{O}(m)$ time. If we learn that $P$ does not occur in $S$, then we stop here, having used a total of $\mathcal{O}(m)$ time. If we learn that $P$ does occur in $S$, then in $\mathcal{O}(occ)$ time we traverse $v$'s subtree to obtain the full list of distinct positions in $P$'s primary occurrences where they first cross phrase boundaries. For each such position, we search in our first two augmented Patricia trees and perform a 4-sided range-reporting query. This takes $\mathcal{O}(m \log m + occ \log \log z)$ time and gives us the positions of all $P$'s primary occurrences in $S$.

### 4.3 Derandomization

Using Karp-Rabin fingerprints means our index can fail in two ways: first, if two fingerprints are the same in one of our two Patricia trees, then a fat binary search can fail to return the correct node and we may miss some occurrences of $P$ in $S$ (i.e., a false-negative error); second, if the fingerprints of a prefix $P[1..i]$ and suffix $P[i+1..m]$ collide with the fingerprints of the substrings of length $i$ and $m-i$ immediately before and after a phrase boundary, then we may report occurrences where none exist (i.e., a false-positive error).

False-negative errors seem more serious, but we can prevent them by testing fingerprinting functions until we find one such that all the fingerprints in each Patricia tree are distinct. Recall from Subsection 2.4 that, if we choose the prime of the fingerprinting function uniformly at random from the primes less than $n^3$, then the probability no two distinct substrings have the same fingerprint is $1 - \mathcal{O}(1/n)$. It follows that with high probability it takes $\mathcal{O}(z)$ time to find a fingerprinting function such that the fingerprints in each Patricia tree are distinct.

Dealing with false-positive errors is more complicated because we do not have the query patterns at construction time; thus, we cannot check in advance for collisions between fingerprints of pattern substrings and and fingerprints of substrings of $S$. Of course, at query time, we could use bookmarked extraction to check each supposed primary occurrence, but this would increase our bound for pattern matching to $\mathcal{O}(m^2 + occ \log \log n)$.

Recall from Subsection 2.4 that Bille et al. [11] showed how to check in $\mathcal{O}(n \log n)$ time that no two distinct substrings of $S$ whose lengths are powers of 2 have the same fingerprint. It follows that with high probability it takes $\mathcal{O}(n \log n)$ time to find a fingerprinting function such that no two distinct substrings of $S$ whose lengths are powers of 2 have the same fingerprint. With such a fingerprinting function, we can test — with no risk of false positives or negatives — the equality of any two substrings $S[i_1..j_1]$ and $S[i_2..j_2]$ of $S$ with $j_1 - i_1 = j_2 - i_2$, by checking that $\phi\left(S\left[i_1..i_1 + 2^{\lfloor \log(j_1 - i_1 + 1)\rfloor} - 1\right]\right) = \phi\left(S\left[i_2..i_2 + 2^{\lfloor \log(j_1 - i_1 + 1)\rfloor} - 1\right]\right)$ and $\phi(S[j_1 - 2^{\lfloor \log(j_1 - i_1 + 1)\rfloor} + 1..j_1]) = \phi(S[j_2 - 2^{\lfloor \log(j_1 - i_1 + 1)\rfloor} + 1..j_2])$.

Suppose we have found a fingerprinting function such that all the fingerprints in each Patricia tree are distinct and no two distinct substrings of $S$ whose lengths are powers of 2 have the same fingerprint. To avoid the possibility of false positives, we modify slightly how we find primary occurrences. Given a pattern $P[1..m]$, we first compute the fingerprints $\phi(P[1..i]), \ldots, \phi(P[1..m])$ in $\mathcal{O}(m)$ time, as before. We then use fat binary search for each reversed prefix $(P[1..i])^{rev}$ of $P$ in the Patricia tree $T_1$ for reversed phrases, and for each suffix $P[i+1..m]$ of $P$ in the Patricia tree $T_2$ for suffixes starting at phrase boundaries. We now use bookmarked fingerprinting and a single bookmarked extraction to check all the results of the fat binary searches in $\mathcal{O}(m \log m)$ time, as we explain next. Finally, we apply 4-sided range reporting for the remaining candidate matches.

Without loss of generality, we consider only how to check the results of the fat binary searches for suffixes of $P$; checking the results of searches for reversed prefixes is symmetric. For each node $w$ returned by a search, we take the position $b$ indicated by one of $w$'s leaf descendants. After doing this, we obtain a list $i_1, \ldots, i_k$ in $P$, where $k \leq m$, and a list $b_1, \ldots, b_k$ of positions immediately after phrase boundaries in $S$. For each $j \leq k$ we want to check whether $P[i_j..m] = S[b_j..b_j + m - i_j]$. For the sake of simplicity, assume $i_1, \ldots, i_k$ are decreasing, so $P[i_j..m]$ is a suffix of $P[i_{j+1}..m]$ for $1 \leq j < m$.

Notice that, if we can ensure that $S[b_j..b_j + m - i_j]$ is a suffix of $S[b_{j+1}..b_{j+1} + m - i_{j+1}]$, for $1 \leq j < m$, then we need only extract $S[b_k..b_k + m - i_k]$ and find its longest common suffix with $P[i_k..m]$, which takes $\mathcal{O}(m)$ time. Therefore, for $1 \leq j < m$, we use bookmarked fingerprinting to check in $\mathcal{O}(\log m)$ time whether $S[b_j..b_j + m - i_j]$ is a suffix of $S[b_{j+1}..b_{j+1} + m - i_{j+1}]$. Since we are comparing substrings of $S$ for equality, this cannot lead to a false match.

If $S[b_j..b_j + m - i_j]$ is not a suffix of $S[b_{j+1}..b_{j+1} + m - i_{j+1}]$, then we compute the fingerprints of the (overlapping) prefix and suffix of $S[b_j..b_j + m - i_j]$ of length $2^{\lfloor \log(m - i_j + 1)\rfloor}$, and compare them to $\phi\left(P\left[i_j..i_j + 2^{\lfloor \log(m - i_j + 1)\rfloor} - 1\right]\right)$ and $\phi\left(P\left[m - 2^{\lfloor \log(m - i_j + 1)\rfloor} + 1..m\right]\right)$, respectively. This takes $\mathcal{O}(\log m)$ time. If these fingerprints do not match, then we know $S[b_j..b_j + m - i_j] \neq P[i_j..m]$, so we discard $i_j$ from the list and check whether $S[b_{j-1}..b_{j-1} + m - i_{j-1}]$ is a suffix of $S[b_{j+1}..b_{j+1} + m - i_{j+1}]$.

If the fingerprints do match, on the other hand, then we know the fingerprints of the corresponding substrings of $S[b_{j+1}..b_{j+1} + m - i_{j+1}]$ cannot match $\phi\left(P\left[i_j..i_j + 2^{\lfloor\log(m-i_j+1)\rfloor} - 1\right]\right)$ and $\phi\left(P\left[m - 2^{\lfloor\log(m-i_j+1)\rfloor} + 1..m\right]\right)$. (Otherwise, $S[b_j..b_j + m - i_j]$ would be a suffix of $S[b_{j+1}..b_{j+1} + m - i_{j+1}]$, contrary to assumption.) Therefore, we discard $i_{j+1}$ from the list and check that $S[b_j..b_j + m - i_j]$ is a suffix of $S[b_{j+2}..b_{j+2} + m - i_{j+2}]$.

### 4.4 Summary

Combining the results in this section, we obtain the following theorem:

**Theorem 3.** *Given a string $S[1..n]$ whose LZ77 parse consists of $z$ phrases, we can store $S$ in $\mathcal{O}(z \log(n/z))$ space such that later, given a pattern $P[1..m]$ that occurs occ times in $S$, we can list the primary occurrences of $P$ in $S$ in $\mathcal{O}(m \log m + occ \log \log n)$ time.*

Our construction is randomized and takes $\mathcal{O}(n \log n)$ time with high probability, but the resulting index does not use randomization and cannot fail. Combining Theorem 3 with Theorems 1 and 2, we obtain our desired result:

**Theorem 4.** *Given a string $S[1..n]$ whose LZ77 parse consists of $z$ phrases, we can store a self-index for $S$ in $\mathcal{O}(z \log(n/z))$ space such that later, first, given a position $i$ and a length $\ell$, we can extract $S[i..i + \ell - 1]$ in $\mathcal{O}(\ell + \log n)$ time; second, given a pattern $P[1..m]$, we can list the occ occurrences of $P$ in $S$ in $\mathcal{O}(m \log m + occ \log \log n)$ time.*

**Acknowledgments.** Many thanks to Djamal Belazzougui, Francisco Claude, Veli Mäkinen, Gonzalo Navarro and Jorma Tarhio, for helpful discussions.

## References

1. Alstrup, S., Brodal, G., Rauhe, T.: Optimal static range reporting in one dimension. In: Proc. STOC, pp. 476–482 (2001)
2. Alstrup, S., Brodal, G.S., Rauhe, T.: Pattern matching in dynamic texts. In: Proc. SODA, pp. 819–828 (2000)
3. Arroyuelo, D., Navarro, G., Sadakane, K.: Stronger Lempel-Ziv based compressed text indexing. Algorithmica 62(1-2), 54–101 (2012)
4. Belazzougui, D., Boldi, P., Pagh, R., Vigna, S.: Monotone minimal perfect hashing: searching a sorted table with $\mathcal{O}(1)$ accesses. In: Proc. SODA, pp. 785–794 (2009)
5. Bille, P., Gørtz, I.L.: Substring range reporting. In: Giancarlo, R., Manzini, G. (eds.) CPM 2011. LNCS, vol. 6661, pp. 299–308. Springer, Heidelberg (2011)
6. Chan, T.M., Larsen, K.G., Pătraşcu, M.: Orthogonal range searching on the RAM, revisited. In: Proc. SoCG, pp. 1–10 (2011)
7. Claude, F., Navarro, G.: Self-indexed grammar-based compression. Fund. Inf. 111(3), 313–337 (2011)
8. Do, H.H., Jansson, J., Sadakane, K., Sung, W.-K.: Fast relative Lempel-Ziv self-index for similar sequences. Theor. Comp. Sci. (to appear)

9. Charikar, M., et al.: The smallest grammar problem. IEEE Trans. Inf. Theory 51(7), 2554–2576 (2005)
10. Bille, P., et al.: Random access to grammar-compressed strings. In: Proc. SODA, pp. 373–389 (2011)
11. Bille, P., Cording, P.H., Gørtz, I.L., Sach, B., Vildhøj, H.W., Vind, S.: Fingerprints in compressed strings. In: Dehne, F., Solis-Oba, R., Sack, J.-R. (eds.) WADS 2013. LNCS, vol. 8037, pp. 146–157. Springer, Heidelberg (2013)
12. Farach, M., Thorup, M.: String matching in Lempel-Ziv compressed strings. In: Proc. STOC, pp. 703–712 (1995)
13. Gagie, T., Gawrychowski, P., Kärkkäinen, J., Nekrich, Y., Puglisi, S.J.: A faster grammar-based self-index. In: Dediu, A.-H., Martín-Vide, C. (eds.) LATA 2012. LNCS, vol. 7183, pp. 240–251. Springer, Heidelberg (2012)
14. Gagie, T., Gawrychowski, P., Kärkkäinen, J., Nekrich, Y., Puglisi, S.J.: A faster grammar-based self-index. Technical Report 1109.3954v6, arxiv.org (2012)
15. Huang, S., Lam, T.W., Sung, W.K., Tam, S.L., Yiu, S.M.: Indexing similar DNA sequences. In: Chen, B. (ed.) AAIM 2010. LNCS, vol. 6124, pp. 180–190. Springer, Heidelberg (2010)
16. Kärkkäinen, J., Ukkonen, E.: Lempel-Ziv parsing and sublinear-size index structures for string matching. In: Proc. WSP, pp. 141–155 (1996)
17. Karp, R.M., Miller, R.E., Rosenberg, A.L.: Rapid identification of repeated patters in strings, trees and arrays. In: Proc. STOC, pp. 125–136 (1972)
18. Karp, R.M., Rabin, M.O.: Efficient randomized pattern-matching algorithms. IBM J. Res. Dev. 31(2), 249–260 (1987)
19. Kreft, S., Navarro, G.: On compressing and indexing repetitive sequences. Theor. Comp. Sci. 483, 115–133 (2013)
20. Mäkinen, V., Navarro, G., Sirén, J., Välimäki, N.: Storage and retrieval of highly repetitive sequence collections. J. Comp. Bio. 17(3), 281–308 (2010)
21. Maruyama, S., Nakahara, M., Kishiue, N., Sakamoto, H.: ESP-index: A compressed index based on edit-sensitive parsing. J. Dis. Alg. 18, 100–112 (2013)
22. Morrison, D.R.: PATRICIA - Practical algorithm to retrieve information coded in alphanumeric. J. ACM 15(4), 514–534 (1968)
23. Mortensen, C.W., Pagh, R., Pătraşcu, M.: On dynamic range reporting in one dimension. In: Proc. STOC, pp. 104–111 (2005)
24. Navarro, G., Mäkinen, V.: Compressed full-text indexes. ACM Comp. Surv. 39(1) (2007)
25. Russo, L.M.S., Oliveira, A.L.: A compressed self-index using a Ziv-Lempel dictionary. Inf. Retr. 11(4), 359–388 (2008)
26. Rytter, W.: Application of Lempel-Ziv factorization to the approximation of grammar-based compression. Theor. Comp. Sci. 302(1-3), 211–222 (2003)
27. Verbin, E., Yu, W.: Data structure lower bounds on random access to grammar-compressed strings. In: Fischer, J., Sanders, P. (eds.) CPM 2013. LNCS, vol. 7922, pp. 247–258. Springer, Heidelberg (2013)
28. Wandelt, S., Leser, U.: QGramProjector: Q-gram projection for indexing highly-similar strings. In: Catania, B., Guerrini, G., Pokorný, J. (eds.) ADBIS 2013. LNCS, vol. 8133, pp. 260–273. Springer, Heidelberg (2013)
29. Yang, X., Wang, B., Li, C., Wang, J., Xie, X.: Efficient direct search on genomic data. In: Proc. ICDE, pp. 961–972 (2013)
30. Ziv, J., Lempel, A.: A universal algorithm for sequential data compression. IEEE Trans. Inf. Theory 23(3), 337–343 (1977)

# Quad-$K$-d Trees

Nikolett Bereczky[1], Amalia Duch[2], Krisztián Németh[1], and Salvador Roura[2]

[1] Budapest University of Technology and Economics,
Department of Telecommunications and Media Informatics, Budapest, Hungary
{nikolett.bereczky,krisztian.nemeth}@tmit.bme.hu
[2] Universitat Politècnica de Catalunya,
Departament de Llenguatges i Sistemes Informàtics, Barcelona, Catalonia, Spain
{duch,roura}@lsi.upc.edu

**Abstract.** We introduce the Quad-$K$-d tree (or simply $QK$-d tree) a hierarchical and general purpose data structure for the storage of multi-dimensional points, which is a generalization of point quad trees and $K$-d trees at once. $QK$-d trees can be tuned by means of insertion heuristics to obtain trade-offs between their costs in time and space. We propose three such heuristics and show analytically and experimentally their competitive performance. On the one hand, our analytical results back the experimental outcomes and suggest that $QK$-d trees could constitute a general framework for the study of inherent properties of trees akin to $K$-d trees and quad trees. On the other hand, our experimental results indicate that the $QK$-d tree is a flexible data structure, which can be tailored to the resource requirements of a given application.

## 1 Introduction

Associative retrieval is a fundamental computing task [18, 32]. Given a file $\mathcal{F}$, which is a collection of $n$ records, each *record* of $\mathcal{F}$ is an ordered $K$-tuple of values (the attributes or coordinates of the record) drawn from a totally ordered domain. A *query* of $\mathcal{F}$ is a retrieval of the records whose attributes satisfy certain conditions. The query is considered *associative* when its conditions deal with more than one of the attributes. Examples of associative queries are: (i) *nearest-neighbor* queries, to retrieve the record in $\mathcal{F}$ that is the closest to a given record under a given distance metric, (ii) *partial match* queries, to retrieve all the records in $\mathcal{F}$ whose attributes match the attributes of the query record that are specified, or (iii) *orthogonal range* queries, to retrieve all the records in $\mathcal{F}$ that fall inside a given hyper-rectangle whose sides are parallel to the coordinate axes.

The way of dealing with associative queries varies. In particular, when several types of associative queries are required, general purpose multidimensional data structures, such as $K$-d trees and quad trees [18, 32], are typically used.[1] Therefore, it is important to know as much as possible about their performance. Here, we are interested in the amount of memory required by the data structure to store the collection of records (its *cost in space*), and in the execution time

---

[1] Throughout this work, we will write "quad trees" to refer to point quad trees [32].

of each of the supported operations (its *cost in time*). Both measures strongly depend on the computer model under consideration and on the way to store the data structures. We will follow Samet's [32] customary representation of multidimensional trees in main memory. Other approaches such as succinct representation of trees or the consideration of memory hierarchies are beyond the scope of this paper. Nevertheless, the techniques used in those research areas are applicable to our work in the same way that apply to $K$-d trees and quad trees (see Chapters 27, 34 and 37 of [25] and [2, 6, 8, 29]).

For a set of $n$ $K$-dimensional records, both $K$-d trees and quad trees require a basic amount of $\Theta(Kn)$ memory for the records. In addition, $K$-d trees are binary trees, so each of the $n$ nodes has two pointers, to its left and right children. Note that this $\Theta(n)$ increment of space is diminishable for large $K$. By contrast, quad trees are $2^K$-ary trees, so the total amount of space is $\Theta(2^K n)$, which is considerably greater than the optimal space required by $K$-d trees. In fact, for large $K$, most of the pointers of a quad tree point to empty subtrees, incurring in an important waste of space [23].

Regarding the execution time, there are several results in the literature analyzing the cost of performing insertions, deletions, exact match, partial match, orthogonal or nearest neighbor queries in both $K$-d trees (and a wide range of its variants) and quad trees [4, 5, 7, 9–11, 13, 16, 24]. A measure closely related to the time performance of $K$-d trees and quad trees in almost any of their available operations is the average height of the tree, or equivalently, its Internal Path Length (*IPL* from now on) [20]. It turns out that the *IPL* of perfectly balanced or random quad trees is asymptotically optimal and is considerably smaller than the one of perfectly balanced or random $K$-d trees [22].

From the previous discussion, one could argue that $K$-d trees are space-optimal while quad trees are *IPL*-optimal. It is a natural question, therefore, whether it is possible to define hierarchical multidimensional data structures that combine the two kinds of optimality.

Much work has been done to optimize the time performance of ad-hoc multidimensional data structures towards some specific associative queries [3, 18, 19, 30, 34]. However, some of these techniques substantially increment the cost in space. Moreover, it is often the case that ad-hoc data structures do not perform well in a general setting.

There are also several proposals to keep general multidimensional trees balanced in order to optimize either the execution time of most of the query operations or the required space [7, 9, 14, 15, 18, 26, 27, 35, 33]. Unfortunately, in such settings the cost of update operations increases significantly.

Other works adapt the behavior of general purpose multidimensional data structures either to the distribution of the input data or to the most frequent kind of multidimensional queries that they have to support [12, 18, 28], at the cost of increments in space and in the execution time of the update operations.

To the best of our knowledge, there is no work combining simultaneously tunable space/time trade-off as well as adaptability to the distribution of the input data. This is what $QK$-d trees aim to achieve. $QK$-d trees are general purpose, multidimensional trees in which each node may have any number of discriminants, instead of a fixed number as it is the usual case for multidimensional trees. (For example, [21] uses as discriminants a variable set of the attributes, but the set is the same for all the nodes in the tree.) This flexibility allows us to play with the compromise between space and time costs. Using insertion heuristics (we propose three of them) it is then possible to produce the whole range of trade-offs that lie between $K$-d trees and quad trees.

The rest of the paper is organized as follows. In Section 2, we introduce $QK$-d trees together with the *Random-Split*, the *Distance-Dependent* and the *Probability-Dependent* insertion heuristics. In Section 3 we provide the theoretical value of the *IPL* of 3-dimensional random $QK$-d trees built using two of the proposed heuristics, while Section 4 shows the experimental analysis of $QK$-d trees built under these heuristics.

## 2 Quad-$K$-d Trees

Informally, $QK$-d trees are multidimensional trees in which each node discriminates by a number $i$ (specific for each node), $1 \leq i \leq K$, of coordinates (and thus it has $2^i$ subtrees). The idea behind them arises from the following observation. $K$-d trees consist of nodes that discriminate by exactly one of the coordinates of the $K$-dimensional keys that they store, while in quad-trees each node discriminates by all the $K$ coordinates of their stored keys. Indeed, $K$-d trees and quad trees can be seen as extreme cases on the number of discriminating coordinates that a node in the tree can have (minimal for $K$-d trees and maximal for quad trees). Thus, it is natural to envisage multidimensional trees where the number of discriminating coordinates for each node in the tree is between 1 and $K$.

**Definition 1.** *A Quad-$K$-d (or simply $QK$-d) search tree $T$ of size $n \geq 0$ stores a set of $n$ $K$-dimensional records, each holding a key $x = (x_0, \ldots, x_{K-1}) \in D$, where $D = D_0 \times \cdots \times D_{K-1}$, and each $D_j$, $0 \leq j < K$, is a totally ordered domain. The $QK$-d tree $T$ is such that*

- *either it is empty when $n = 0$, or*
- *its root stores both a record with key $x$ and a coordinate split boolean vector $s = (s_0, \ldots, s_{K-1})$ that contains exactly $i$ ones, where $1 \leq i \leq K$ is the order of vector $s$, and the node has $2^i$ subtrees that store the $n-1$ remaining points as follows: each subtree, let us call it $T_w$, is associated to a string $w = w_0 w_1 \ldots w_{K-1} \in \{0, 1, \#\}^K$, such that $\forall w \in \{0, 1, \#\}^K$, $T_w$ is a $QK$-d tree and, for any key $y \in T_w$ and $0 \leq j < K$, it holds that*
  - $w_j = \#$ *iff* $s_j = 0$
  - *if* $s_j = 1$ *and* $w_j = 0$, *then* $y_j < x_j$
  - *if* $s_j = 1$ *and* $w_j = 1$, *then* $y_j > x_j$.

We assume, without loss of generality, that $D = [0,1]^K$ and that all the key coordinates in each dimension are different.

In Figure 1, we show an example of a 3-dimensional $QK$-d tree. Inside each node we show (between brackets) the 3-dimensional key that it has associated

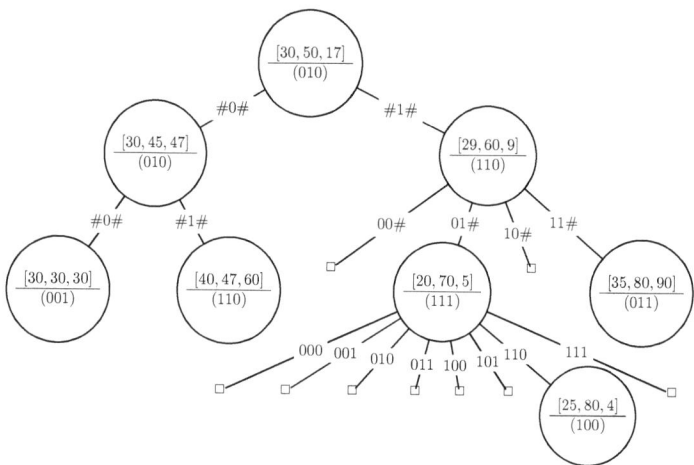

**Fig. 1.** An example of 3-dimensional $QK$-d tree

and its split boolean vector (in parenthesis). Next to every edge we show the label (the string $w$) of the subtree it points to.

As we have already mentioned, both $K$-d trees and quad trees are special cases of $QK$-d trees. In fact, for any $QK$-d tree $T$ of size $n$, if the order of the split vector $s$ associated to every node of $T$ is exactly $K$ then $T$ is a quad tree, and if this order is exactly one for every node and the splitting coordinates are used cyclically then $T$ is a $K$-d tree.

The algorithms for exact search and insertion are similar to those of $K$-d trees and quad trees, except that in this case insertions require a specific method in order to generate the split vector of each newly inserted node.

In general, if we want to search for a record in a $QK$-d tree, we have to traverse it, starting at the root, and examine the values in the split vector of the root: for every 1 value in the vector we have to compare the corresponding coordinate of the root with the one of the record we are looking for, and finally follow the subtree whose associated string matches the result of all the comparisons. The search continues recursively until we find a node containing the requested record (successful search) or an empty subtree (unsuccessful search).

Alternatively, if we want to insert a record that is not in the tree, we have to search for it as described above, until we reach an empty subtree which we replace by a new node containing the inserted key and empty subtrees. At this point, a predefined algorithm to generate the split vector of the new node is required. In the following paragraphs we propose three heuristics for this purpose: (i) *Random-Split* (*RS* for short), (ii) *Distance-Dependent* (*DD* for short) and (iii) *Probability-Dependent* (*PD* for short). As we will show experimentally, any of these heuristics provides a "smooth" transition between the space of $K$-d trees and the one of quad trees as well as between their *IPL*. Therefore, using them $QK$-d trees can be tuned for the specific application requirements.

The *Random-Split (RS)* heuristic generates randomly the members of the split vector of each new node independently of each other, using the same Bernoulli distribution, where the probability of occurrence of the value 0 is a given constant, referred to as *Prob-of-0*. Note that it can happen that every coordinate in the split vector is zero. In this case, a randomly chosen position will be set to one.

The *Distance-Dependent (DD)* heuristic decides, for a given node with key $x = (x_0, x_1, \ldots, x_{K-1})$ and bounding box[2] $[\ell_0, u_0] \times \ldots \times [\ell_{K-1}, u_{K-1}]$, whether to discriminate or not by coordinate $j$, $0 \leq j < K$, depending on the distance of $x_j$ from both $\ell_j$ and $u_j$. More precisely, given a fixed value $0 < z < 1$ (referred to as *Split Threshold*), we discriminate by coordinate $j$ if and only if $\frac{x_j - \ell_j}{u_j - \ell_j} > z$ and $\frac{u_j - x_j}{u_j - \ell_j} > z$. For example: let $v$ be the root of a 2-dimensional tree with associated key $(0.35, 0.52)$ and bounds array $[0, 1] \times [0, 1]$. If the value of $z$ is 0.3, then we discriminate by the two coordinates (and consequently we have a split vector $s = (1, 1)$) because $0.30 < 0.35 < 0.70$ and $0.30 < 0.52 < 0.70$. However, if the value of $z$ is 0.4, then the split vector is $s = (0, 1)$, and consequently the first coordinate will not be used to discriminate. Note that as the value of *Split Threshold* grows the probability of splitting decreases. The idea behind this heuristic is to split only when the given coordinate of the node is useful to discriminate, which happens when it splits more or less evenly the subsequent keys. For uniformly distributed keys a coordinate is useful to discriminate if it is close to the center of the bounding box in the given dimension: this indicates that the nodes inserted under the current one will probably be uniformly distributed in the forthcoming subtrees. Again, if every coordinate in the split vector happens to be zero, a randomly chosen one will be set to one.

The *Probability-Dependent (PD)* heuristic is an extension of the *DD* heuristic for points drawn from continuous but non-uniform distributions. In this case, given a node of the tree with key $x$, its $j$-th coordinate is used to discriminate only if both quotients $\frac{F(x_j) - F(\ell_j)}{F(u_j) - F(\ell_j)}$ and $\frac{F(u_j) - F(x_j)}{F(u_j) - F(\ell_j)}$ are greater than a given *Split Threshold* constant, where $F$ is the cumulative distribution function and $u_j$ and $\ell_j$ are the coordinate bounds. Informally, the $j$-th coordinate of a given key is going to discriminate if it is "centered" within the probability distribution of the set of keys.

To analyze the expected performance of $QK$-d trees we are going to use the same probabilistic model generally used to analyze the expected performance of $K$-d trees and quad trees, i. e., we say that a $K$-d tree or a quad tree built from a given set of $n$ random keys is *random* if it is built with identical probability from any of the $n!^K$ possible input sequences. As a consequence, the insertion of $n$ points independently drawn from a continuous distribution in $[0, 1]^K$ into an initially empty $K$-d tree or quad tree will produce random $K$-d trees, random quad trees or random $QK$-d trees.

---

[2] The *bounding box* (or *bounds array*) of a node $x$ is the region of the space corresponding to the leaf in which $x$ fell when it was inserted into the tree.

## 3 The *IPL* of *QK*-d Trees

Here we study the asymptotic average *IPL* of the random *QK*-d trees built under the *RS* and the *DD* heuristics presented above.

Let us start with the recurrence relation for the expected cost $C_n$ of a random exact search in a random binary search tree (or in a random *K*-d-tree), with $i$ keys in its left subtree

$$C_n = 1 + \sum_{i=0}^{n-1} \frac{1}{n}\left(\frac{i}{n}C_i + \frac{n-i-1}{n}C_{n-i-1}\right) = 1 + \sum_{i=0}^{n-1} \frac{2i}{n^2}C_i . \quad (1)$$

If we denote by $w_i$ the weight of $C_i$, i.e., $w_i = 2i/n^2$, we observe that those weights asymptotically fit the shape of the function $f_1(z) = 2z$ between $[0..1]$, in the sense that $w_i = f_1(i/n)/n \simeq \int_{i/n}^{(i+1)/n} f_1(z)dz$. Informally speaking, $f_1(z)$ is like a continuous probability distribution, with indeed $\int_0^1 f_1(z)dz = 1$.

As shown in [31], a recurrence like (1) has always a solution of the form $C_n \sim c \ln n$ for some constant $c$. Using this fact, we can compute $c$ by means of integrals:

$$c \ln n \sim 1 + \sum_{i=0}^{n-1} w_i c \ln i \sim 1 + \int_0^1 f_1(z) c \ln(zn) dz$$

$$= 1 + c \int_0^1 f_1(z) \ln z \, dz + c \ln n \int_0^1 f_1(z) dz .$$

Since $\int_0^1 f_1(z)dz = 1$, the terms $c \ln n$ vanish and we get

$$0 = 1 + c\left[z^2 \ln z - \frac{z^2}{2}\right]_0^1 = 1 + c\left(-\frac{1}{2}\right) ,$$

which implies $c = 2$. The expected cost is certainly $\sim 2 \ln n = 2 \ln 2 \log_2 n \simeq 1.39 \log_2 n$.

Let $f_K(z)$ be the function that describes the asymptotic shape of the weights of the recurrence for the average cost to search in a quad-tree with $K$ dimensions. For instance, we already have $f_1(z) = 2z$. To compute $f_2(z)$, note that using two dimensions to discriminate is like using each of the two dimensions one after another. The "density of probability" to reach some $z$ from 1 in two steps is therefore $f_2(z) = \int_z^1 f_1(x)f_1(z/x)/x \, dx$. (For every $x$ between $z$ and 1, is the "probability" to reach $x$ in one step—$f_1(x)$—times the "probability" to reach $z/x$ in another step, which is $f_1(z/x)$ scaled by the factor $1/x$ so that the integral of the probability distribution $f_1$ between 0 and $x$ adds up to 1.) Hence, $f_2(z) = 4z \int_z^1 1/x \, dx = -4z \ln z$.

For comparison with the traditional (discrete) approach, the recurrence for the cost of a search in a two-dimensional quad-tree can be seen to be exactly (see [17], page 102)

$$C_n = 1 + \sum_{i=0}^{n-1} \frac{4i}{n^2}(H_n - H_i)C_i ,$$

where $H_n \sim \ln n$ denotes as usual the $n$-th harmonic number $\sum_{i=1..n} 1/i$.
Note how the weights fit the function $f_2(z)$:

$$4i/n(H_n - H_i)/n \sim -4i/n \ln(i/n)/n = f_2(i/n)/n \ .$$

We have the following lemma for general $K$.

**Lemma 1.** *For* $K \geq 1$, $f_K(z) = k_K z (\ln z)^{K-1}$, *where* $k_K = -(-2)^K/(K-1)!$.

*Proof.* The proof is by induction. We already know that $f_1(z) = k_1 z (\ln z)^0$. Assuming that the lemma is true up to a certain $K-1$, and following a similar reasoning as for the computation of $f_2(z)$, we have

$$f_K(z) = \int_z^1 f_1(x) f_{d-1}(z/x)/x\,dx = \int_z^1 2k_{K-1} z/x (\ln(z/x))^{K-2} dx \ .$$

By the change of variable $y = z/x$, we get

$$f_K(z) = 2k_{K-1} z \int_z^1 (\ln y)^{K-2}/y\, dy = 2k_{K-1} z [(\ln y)^{K-1}/(K-1)]_z^1$$
$$= -2k_{K-1}/(K-1) z (\ln z)^{K-1} \ .$$

Let $f(z)$ be any continuous probability distribution for the asymptotic shape of the weights $w_i$ of a recurrence like $C_n = 1 + \sum_{i=0}^{n-1} w_i C_i$. As we have already seen, the solution is $C_n \sim c \ln n$ for some constant $c$, which can be computed by solving $c \ln n = 1 + \int_0^1 f(z) c \ln(zn) dz$. Once the terms $c \ln n$ cancel from both sides, what we have left is

$$c = \frac{-1}{\int_0^1 f(z) \ln z\, dz} \ . \tag{2}$$

The following technical lemma will be useful.

**Lemma 2.** *For* $K \geq 0$, *let* $I_K = \int_0^1 z (\ln z)^K dz$. *Then* $I_K = -K!/(-2)^{K+1}$.

*Proof.* By induction. For the base case, we have $I_0 = 1/2$. For $K \geq 1$,

$$I_K = \int_0^1 z(\ln z)^K dz = \left[\frac{z^2 (\ln z)^K}{2}\right]_0^1 - \int_0^1 \frac{K}{2} z (\ln z)^{K-1} dz = -\frac{K}{2} I_{K-1} \ .$$

By combining this lemma with Equation 2, we can easily compute the constant of searching in a quad-tree with $K$ dimensions:

$$c(K) = -1/\int_0^1 k_K z (\ln z)^K = -1/(k_K I_K) = 2/K \ ,$$

as expected.

More interesting is the analysis of some of the new variants presented in this paper. For instance, consider a 3-dimensional $QK$-d tree built using $RS$. Let $p$

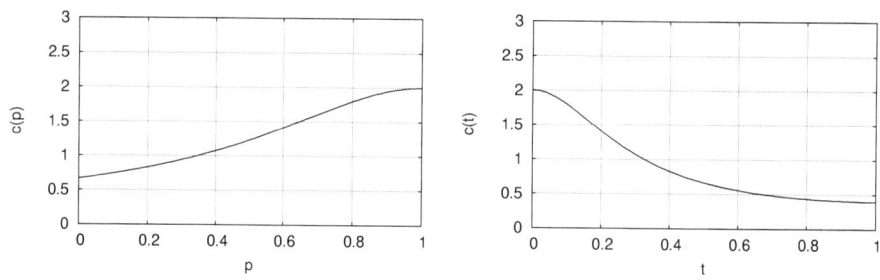

**Fig. 2.** $c(p)$ (left) and $c(t)$ (right)

be the probability of zero, and $q = 1 - p$ be the probability of one. Then, with probability $q^3$ the current node discriminates w.r.t. the three dimensions, with probability $3pq^2$ w.r.t. two dimensions, and with probability $3p^2q + p^3$ w.r.t. one dimension. Therefore, $f(z) = q^3 f_3(z) + 3pq^2 f_2(z) + (3p^2q + p^3) f_1(z)$ and, the constant of the search is thus

$$c(p) = \frac{-1}{q^3 k_3 I_3 + 3pq^2 k_2 I_2 + (3p^2q + p^3) k_1 I_1} = \frac{2}{p^3 - 3p + 3}.$$

We can see $c(p)$ at the left of Figure 2.

From this point, the analysis of the $DD$ variant under random keys is straightforward. It is enough to note that $p$ can be set in terms of $t$, that is, replace $p$ by $1 - 2t$ (and $q$ by $2t$). Therefore, as a function of $t$, we have

$$c(t) = \frac{2}{(1 - 2t)^3 + 6t} = -\frac{2}{8t^3 - 12t^2 - 1}.$$

We can see $c(t)$ at the right of Figure 2.

Some remarks are in order. On the one hand, both the $RS$ and the $DD$ heuristics can also be analyzed for dimensions other than three. On the other, note that the tools used above do not allow us to compute the constant for expected number of leaves of variants presented in this paper. The reason is that those tools produce results that rely exclusively upon asymptotic information. However, the number of leaves strongly depends on what the algorithms do when the number of remaining points is small. (For instance, if we decide to always discriminate w.r.t. one dimension when only a few points remain, then the number of leaves will significantly drop.) Therefore, the computation of such a constant would require more sophisticated techniques.

## 4 Experimental Analysis

In this section we study the performance of $QK$-d trees experimentally. Specifically, we show the experimental results that we obtained by comparing the $IPL$ (as a measure of data access time) and the number of empty subtrees (as a measure of storage space usage) of random $K$-d trees and quad trees versus $QK$-d trees built using the *RS*, *DD* and the *PD* heuristics. All measures, unless otherwise stated, were obtained by averaging the values of 100 3-dimensional trees holding each 100,000 nodes, although we have also performed similar experiments for 2, 4 and 5-dimensional trees obtaining equivalent results.

We start in a uniform experimental setting comparing randomly built $QK$-d trees (built from uniformly distributed and independently generated keys from a continuous domain) with $K$-d trees and quad trees built from the same set of keys inserted in the same order.

In Figure 3 (top) we show the performance of $QK$-d trees built using the *RS* heuristic. As expected, we can see that as the probability of discriminating by a coordinate decreases (i.e., *Prob-of-0* increases) the *IPL* of the $QK$-d tree grows and the space it needs diminishes. Indeed, the more ones we have in the split vector, the more $QK$-d trees resemble quad trees, and the less ones we have, the more $QK$-d trees resemble $K$-d trees. Note how the experimental results coincide with the theoretical values shown in Figure 2.

In the case of uniformly distributed keys, the *DD* heuristic performs well. In Figure 3 (middle) we show how the *IPL* and the number of empty subtrees of such a $QK$-d tree behave depending on the value of the *Split Threshold*. Again, note how the theoretical values match the experimental results shown in Figure 2. As it can be seen in the diagrams, $QK$-d trees are between quad trees and $K$-d trees considering both the *IPL* and the number of the empty subtrees. The curves also show that when the value of the *Split Threshold* grows, the *IPL* grows and the space requirements diminish.

Comparing $QK$-d trees built using the *RS* heuristic against $QK$-d trees built using the *DD* heuristic (see Fig. 3 (bottom)), which combines the previous diagrams), it can be seen that for randomly built trees the *IPL* of the latter is better. On the other hand, concerning the number of empty subtrees, their performance are practically identical (the curves are overlapping). These together mean that the more sophisticated *DD* heuristic outperforms the *RS* heuristic. However, for applications for which the extra *IPL* is not a problem, it might be worth using the trees built by the *RS* heuristic because of its simplicity.

Anyway, the fact that the experiments and the theoretical results for the *IPL* match so well provides us with some confidence about the soundness of our experiments for other quantities, especially those concerning the number of empty subtrees that cannot be computed using the techniques of Section 3.

We have measured also the *IPL* and number of empty subtrees while the number of the nodes in the tree grows. Our results show that these metrics are linearly proportional to the size of the tree for $K$-d trees, quad trees and $QK$-d trees built using the *DD* heuristic.

Our experiments show also that increasing the number of dimensions has considerable effect on the results. In fact, for higher dimensions, using $QK$-d trees one can avoid the waste of space of quad trees while keeping a short *IPL*.

We have carried out experiments with non-uniform settings comparing $K$-d trees and quad trees with $QK$-d trees built using the *RS*, *DD* and *PD* heuristics. For these experiments we have used trees built by independently generated

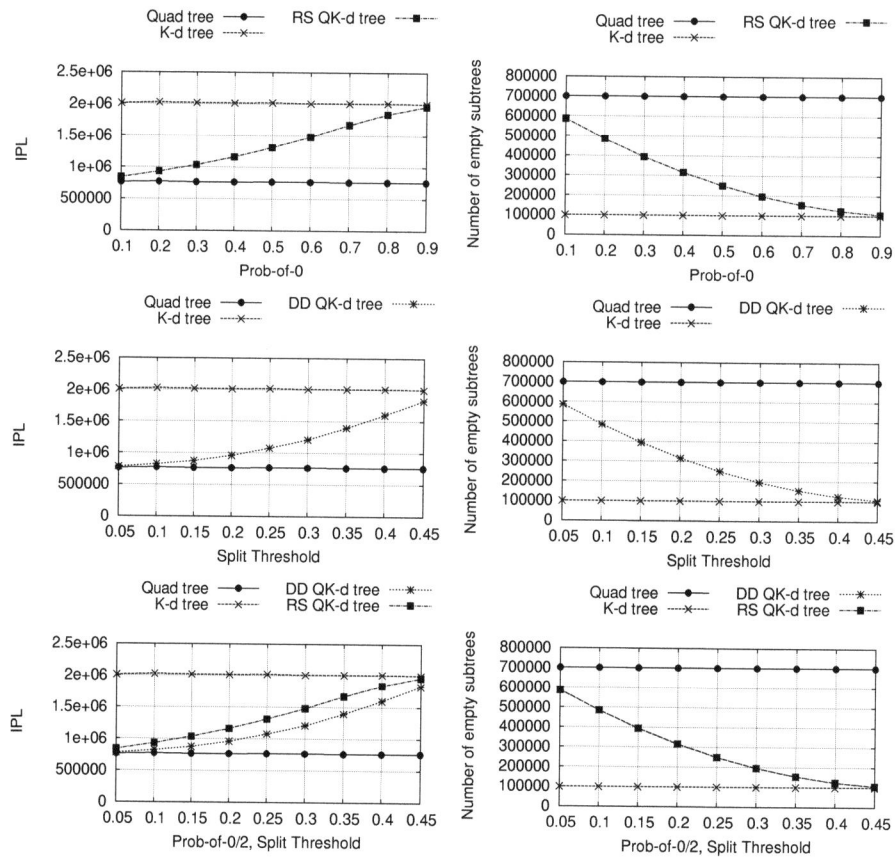

**Fig. 3.** Performance of the *RS* heuristic (top), the *DD* heuristic (middle) and their comparison (bottom)

keys from exponential and normal distributions (the latter is usually used for modeling non-uniformly distributed data in multidimensional settings [1]). The results show that $QK$-d trees built using the *PD* heuristic are very competitive, specially regarding the *IPL*.

More experimental results using as input real data are omitted for lack of space.

## 5 Conclusions

We have introduced a simple and flexible multidimensional data structure, the $QK$-d tree, which includes $K$-d trees and quad trees as particular cases. We have shown through formal analysis and experiments that, as expected, the performance of randomly built $QK$-d trees is between the one of quad trees and the one of $K$-d trees considering time and space efficiency.

We have also proposed three insertion heuristics that allow "a la carte" space and time trade-offs. Among these heuristics, *Random-Split* is the simplest, yet it is adjustable and has a stable performance in all the examined cases. On the other hand, the results for the *Distance-Dependent* heuristic are more favorable in the case of uniform data distributions. Finally, if the distribution of the input data is not uniform, but still known, then the *Probability-Dependent* heuristic is preferable. Using these heuristics, the resources used by the data structure can be tailored to the application requirements.

A first—and challenging—line of future work consists of a formal analysis approach that takes $QK$-d trees as a framework to analyze inherent properties (*IPL*, average cost of update and search operations, and so on) of the whole family of hierarchical multidimensional trees akin to $K$-d trees and quad trees. A second line of further interest addresses the possible practical applicability of $QK$-d trees.

## References

1. Ang, C.H., Samet, H.: Node Distribution in a PR Quadtree. In: Buchmann, A.P., Günther, O., Smith, T.R., Wang, Y.-F. (eds.) Design and Implementation of Large Spatial Databases. LNCS, vol. 409, pp. 233–252. Springer, Heidelberg (1989)
2. Benoit, D., Munro, R., Raman, R.: Representing trees of higher degree. Algorithmica 43 (2005)
3. Bentley, J.L., Friedman, J.H.: Data structures for range searching. ACM Computing Surveys 11(4), 397–409 (1979)
4. Chanzy, P., Devroye, L., Zamora-Cura, C.: Analysis of range search for random $k$-d trees. Acta Informatica 37, 355–383 (2001)
5. Chern, H.H., Hwang, H.K.: Partial match queries in random $k$-d trees. SIAM Journal on Computing 35(6), 1440–1466 (2006)
6. Choi, M.G., Ju, E., Chang, J.W., Lee, J., Kim, Y.J.: Linkless octree using multilevel perfect hashing. Comput. Graph. Forum 28(7), 1773–1780 (2009)
7. Cunto, W., Lau, G., Flajolet, P.: Analysis of kdt-trees: $k$d-trees improved by local reorganisations. In: Dehne, F., Santoro, N., Sack, J.-R. (eds.) WADS 1989. LNCS, vol. 382, pp. 24–38. Springer, Heidelberg (1989)
8. Darragh, J.J., Cleary, J.G., Witten, I.H.: Bonsai: A compact representation of trees. SOFTPREX: Software–Practice and Experience 23 (1993)
9. Devroye, L., Jabbour, J., Zamora-Cura, C.: Squarish $k$-d trees. SIAM Journal on Computing 30, 1678–1700 (2000)
10. Duch, A., Estivill-Castro, V., Martínez, C.: Randomized $K$-dimensional binary search trees. In: Chwa, K.-Y., Ibarra, O.H. (eds.) ISAAC 1998. LNCS, vol. 1533, pp. 199–209. Springer, Heidelberg (1998)
11. Duch, A., Martínez, C.: On the average performance of orthogonal range search in multidimensional data structures. Journal of Algorithms 44(1), 226–245 (2002)
12. Duch, A., Martínez, C.: Improving the performance of multidimensional search using fingers. ACM Journal of Experimental Algorithms - JEA 10 (2005)
13. Duch, A., Martínez, C.: Updating Relaxed K-d Trees. ACM Transactions on Algorithms 6(1) (2009)
14. Duncan, C., Goodrich, M., Kobourov, S.: Balanced Aspect Ratio Trees: Combining the Advantages of k-d Trees and Octrees. Journal of Algorithms 38, 303–333 (2001)

15. Eppstein, D., Goodrich, M., Sun, J.Z.: Skip Quadtrees: Dynamic Data Structures for Multidimensional Point Sets. Int. J. Comput. Geom. Appl. 18 (2008)
16. Flajolet, P., Puech, C.: Partial match retrieval of multidimensional data. Journal of the ACM 33(2), 371–407 (1986)
17. Flajolet, P., Gonnet, G.H., Puech, C., Robson, J.M.: The Analysis of Multidimensional Searching in Quad-Trees. In: SODA, pp. 100–109 (1991)
18. Gaede, V., Günther, O.: Multidimensional access methods. ACM Computing Surveys 30(2), 170–231 (1998)
19. Indyk, P., Motwani, R., Raghavan, P., Vempala, S.: Locality-preserving hashing in multidimensional spaces. In: STOC 1997 Proceedings of the Twenty-Ninth Annual ACM Symposium on Theory of Computing, pp. 618–625 (1997)
20. Knuth, D.E.: The Art of Computer Programming: Sorting and Searching, 2nd edn., vol. 3. Addison–Wesley (1998)
21. Lin, K.I., Jagadish, H., Faloutsos, C.: The TV-Tree: An Index Structure for High-Dimensional Data. Proc. VLDB Journal 3(4), 517–542 (1994)
22. Mahmoud, H.M.: Evolution of Random Search Trees. Wiley-Interscience series in discrete mathematics and optimization (1991)
23. Mahmoud, H.M., Pittel, B.: Analysis of the Space of Search Trees Under the Random Insertion Algorithm. Journal of Algorithms 10, 52–75 (1989)
24. Martínez, C., Panholzer, A., Prodinger, H.: Partial match queries in relaxed multidimensional search trees. Algorithmica 29(1-2), 181–204 (2001)
25. Mehta, D.P., Sahni, S.: Handbook of Data Structures and Applications. Computer & Information Science Series. Chapman & Hall/CRC (2005)
26. Mulmuley, K.: Randomized Multidimensional Search Trees: Lazy Balancing and Dynamic Shuffling. In: FOCS: 32nd Annual Symposium on Foundations of Computer Science, pp. 180–196 (1991)
27. Overmars, M.H., van Leeuwen, J.: Dynamic Multi-dimensional Data Structures Based on Quad and K-d Trees. Acta Informatica 17(3), 267–285 (1982)
28. Park, E., Mount, D.M.: A Self-Adjusting Data Structure for Multidimensional Point Sets. In: Epstein, L., Ferragina, P. (eds.) ESA 2012. LNCS, vol. 7501, pp. 778–789. Springer, Heidelberg (2012)
29. Raman, R., Raman, V., Rao, S.S.: Succinct indexable dictionaries with applications to encoding k-ary trees and multisets. In: Proceedings of the 13th Annual ACM-SIAM Symposium on Discrete Mathematics (SODA 2002), pp. 233–242. ACM Press (2002)
30. Rotem, D.: Clustered multiattribute hash files. In: PODS 1989 Proceedings of the Eighth ACM SIGACT-SIGMOD-SIGART Symposium on Principles of Database Systems, pp. 225–234 (1989)
31. Roura, S.: Improved master theorems for divide-and-conquer recurrences. J. ACM 48(2), 170–205 (2001)
32. Samet, H.: The Design and Analysis of Spatial Data Structures. Addison-Wesley (1990)
33. Sherk, M.: Self-Adjusting k-ary Search Trees. Journal of Algorithms 19(1), 25–44 (1995)
34. Tamminen, M.: The extendible cell method for closest point problems. BIT Numerical Mathematics 22, 27–41 (1982)
35. Vaishnavi, V.K.: Multidimensional Height-Balanced Trees. IEEE Transactions on Computers 33(4), 334–343 (1984)

# Biased Predecessor Search*

Prosenjit Bose[1], Rolf Fagerberg[2], John Howat[3], and Pat Morin[1]

[1] School of Computer Science
Carleton University
{jit,morin}@scs.carleton.ca
[2] Department of Mathematics and Computer Science
University of Southern Denmark
rolf@imada.sdu.dk
[3] School of Computing
Queen's University
howat@cs.queensu.ca

**Abstract.** We consider the problem of performing predecessor searches in a bounded universe while achieving query times that depend on the distribution of queries. We obtain several data structures with various properties: in particular, we give data structures that achieve expected query times logarithmic in the entropy of the distribution of queries but with space bounded in terms of universe size, as well as data structures that use only linear space but with query times that are higher (but still sublinear) functions of the entropy. For these structures, the distribution is assumed known. We also consider data structures with general weights on universe elements, as well as the case when the distribution is not known in advance.

## 1 Introduction

The notion of *biased searching* has received significant attention in the literature on ordered dictionaries. In this setting, each element $i$ of the data structure has some probability $p_i$ of being queried, and we wish for predecessor queries—that is, queries for the largest element stored in that data structure that is smaller than than a given query element—to take a time related to the inverse of the probability of that query. For example, a *biased search tree* can answer a query for item $i$ in time $O(\log 1/p_i)$ [5].[1] Recall that $\sum_i p_i \log(1/p_i)$ is the *entropy* of the distribution of queries. In terms of this quantity, we note that the *average* query time in a biased search tree is linear in the entropy of the query distribution.

Predecessor searches have also been researched extensively in the context of *bounded universes*. Let $\mathcal{U} = \{0, 1, \ldots, U-1\}$, and consider a static subset $S = \{s_1, s_2, \ldots, s_n\} \subseteq \mathcal{U}$, where $s_1 < s_2 < \cdots < s_n$. Predecessor searches in this context admit data structures that are not only a function of $n$, but also of $U$. For example, van Emde Boas trees [16] can answer predecessor queries in time $O(\log \log U)$.

---

* This research was partially supported by NSERC and MRI.
[1] In this paper, we define $\log x = \log_2(x+2)$.

A natural question—but one which has been basically unexplored—is how to combine these two areas of study to consider biased searches in bounded universes. In this setting, we have a probability distribution $D = \{p_0, p_1, \ldots, p_{U-1}\}$ over the universe $\mathcal{U}$ such that the probability of receiving $i \in \mathcal{U}$ as a query is $p_i$ and $\sum_{i=0}^{U-1} p_i = 1$. We wish to preprocess $\mathcal{U}$ and $S$, given $D$, such that the time for a query is related to $D$.

The motivation for such a goal is the following. Let $H = \sum_{i=0}^{U-1} p_i \log(1/p_i)$ be the entropy of the distribution $D$. Recall that the entropy of a $\mathcal{U}$-element distribution is between 0 and $\log U$. Therefore, if a query time of $O(\log H)$ can be achieved, this for any distribution will be at most $O(\log \log U)$, which matches the performance of van Emde Boas trees [16]. However, for lower-entropy distributions, this will be faster. In other words, such a structure will allow bias in the query sequence to be exploited for ordered dictionaries over bounded universes. Hence, perhaps the most natural way to frame the line of research in this paper is by analogy: the results here are to biased search trees as van Emde Boas trees (and similar structures) are to binary search trees.

*Our Results.* The results presented here can be divided into several categories. We give three variants of a data structure that obtains $O(\log H)$ query time but space that is bounded in terms of $U$, as well as a solution that obtains space that is linear in $n$ but has query time $O(\sqrt{H})$. We also consider the cases of general weights on universe elements and of query times related to the working-set number (which is defined as the number of distinct predecessors reported since the last time a particular predecessor was reported), so that the query distribution need not be known in advance.

*Organization.* The rest of the paper is organized in the following way. We first complete the introduction by reviewing related work. Section 2 shows how to obtain good query times at the expense of large space. Section 3 shows how to obtain good space at the expense of larger query times. We conclude in Section 4 with a summary of the results obtained and possible directions for future research.

### 1.1 Related Work

It is a classical result that predecessor searches in bounded universes can be performed in time $O(\log \log U)$. This was first achieved by van Emde Boas trees [16], and later by $y$-fast tries [17], and Mehlhorn and Näher [13]. Of these, van Emde Boas trees use $O(U)$ space, while the other two structures use $O(n)$ space. These bounds can be improved to

$$O\left(\min\left\{\frac{\log \log U}{\log \log \log U}, \sqrt{\frac{\log n}{\log \log n}}\right\}\right)$$

using $n^{O(1)}$ space [3]. By paying an additional $O(\log \log n)$ factor in the first half of this bound, the space can be improved to $O(n)$. Pătraşcu and Thorup later effectively settled this line of research with a set of time-space tradeoffs [14].

Departing the bounded universe model for a moment and considering only biased search, perhaps the earliest such data structure is the optimum binary search tree [11], which is constructed to be the best possible static binary search tree for a given distribution. Optimum binary search trees take a large amount of time to construct; in linear time, however, it is possible to construct a binary search tree that answers queries in time that is within a constant factor of optimal [12]. Even if the distribution is not known in advance, it is still possible to achieve the latter result (e.g., [2,15]).

Performing biased searches in a bounded universe is essentially unexplored, except for the case where the elements of $S$ are drawn from $D$ rather than the queries [4]. In that result, $D$ need not be known, but must satisfy certain smoothness constraints, and a data structure is given that supports $O(1)$ query time with high probability and $O\left(\sqrt{\log n/\log \log n}\right)$ worst-case query time, using $O(n^{1+\epsilon})$ bits of space, which can be reduced to $O(n)$ space at the cost of a $O(\log \log n)$ query time (with high probability). It is worth noting that this data structure is also dynamic.

A related notion is to try to support query times that are related to the distribution in a less direct way. For example, *finger searching* can be supported in time $O\left(\sqrt{\log d/\log \log d}\right)$ where $d$ is the number of keys stored between a *finger* pointing at a stored key and the query key [1]. There is also a data structure that supports such searches in expected time $O(\log \log d)$ for a wide class of input distributions [10]. Finally, a query time of $O(\log \log \Delta)$, where $\Delta$ is the difference between element queried and the element returned, can also be obtained [7]. Of these last two results, note that the former gives a larger bound in terms of rank space, while the latter gives a smaller bound in terms of the universe.

Other problems in bounded universes can also be solved in similar ways. A *priority queue* that supports insertion and deletion in time $O(\log \log D)$, where $D$ is the difference between the successor and predecessor (in times of priority) of the query is known [9], as well as a data structure for the *temporal precedence problem*, wherein the older of two query elements must be determined, that supports query time $O(\log \log \delta)$, where $\delta$ is the temporal distance between the given elements [8].

## 2 Supporting $O(\log H)$ Query Time

In this section, we describe how to achieve query time $O(\log H)$ using space that is somewhat large in terms of $U$. In bounded universe problems, it is desirable to have query time that is a function of $n$ rather than $U$.

## 2.1 Using $O(n + U^\epsilon)$ Space

Let $\epsilon > 0$. We will place all elements $i \in \mathcal{U}$ with probability $p_i \geq (1/U)^\epsilon$, along with their predecessor in $S$ (which never changes since $S$ is static) into a hash table $T$. All elements of $S$ are also placed into a $y$-fast trie over the universe $\mathcal{U}$. Since there are at most $U^\epsilon$ elements with probability greater than $(1/U)^\epsilon$, it is clear that the hash table requires $O(U^\epsilon)$ space. Since the $y$-fast trie requires $O(n)$ space, we have that the total space used by this structure is $O(n + U^\epsilon)$. To execute a search, we check the hash table first. If the query (and thus the answer) is not stored there, then a search is performed in the $y$-fast trie to answer the query.

The expected query time is thus

$$\sum_{i \in T} p_i O(1) + \sum_{i \in \mathcal{U} \setminus T} p_i O(\log \log U)$$

$$= O(1) + \sum_{i \in \mathcal{U} \setminus T} p_i O(\log \log U)$$

$$= O(1) + \sum_{i \in \mathcal{U} \setminus T} p_i O\left(\log \log \left((U^\epsilon)^{1/\epsilon}\right)\right)$$

$$= O(1) + \sum_{i \in \mathcal{U} \setminus T} p_i O((\log(1/\epsilon))(\log U^\epsilon))$$

$$= O(1) + \sum_{i \in \mathcal{U} \setminus T} p_i O(\log(1/\epsilon)) + \sum_{i \in \mathcal{U} \setminus T} p_i O(\log \log U^\epsilon)$$

$$= O(1) + O(\log(1/\epsilon)) + \sum_{i \in \mathcal{U} \setminus T} p_i O\left(\log \log \frac{1}{1/U^\epsilon}\right)$$

$$\leq O(1) + O(\log(1/\epsilon)) + \sum_{i \in \mathcal{U} \setminus T} p_i O(\log \log(1/p_i))$$

The last step here follows from the fact that, if $i \in \mathcal{U} \setminus T$, then $p_i \leq (1/U)^\epsilon$, and so $1/(1/U)^\epsilon \leq 1/p_i$. Recall Jensen's inequality, which states that for concave functions $f$, $E[f(X)] \leq f(E[X])$. Since the logarithm is a concave function, we therefore have

$$\sum_{i \in \mathcal{U} \setminus T} p_i O(\log \log(1/p_i)) \leq \log \sum_{i \in \mathcal{U} \setminus T} p_i O(\log(1/p_i)) \leq O(\log H)$$

therefore, the expected query time is $O(\log(1/\epsilon)) + O(\log H) = O(\log(H/\epsilon))$.

**Theorem 1.** *Given a probability distribution with entropy $H$ over the possible queries in a universe of size $U$, it is possible to construct a data structure that performs predecessor searches in expected time $O(\log(H/\epsilon))$ using $O(n + U^\epsilon)$ space for any $\epsilon > 0$.*

Theorem 1 is a first step towards our goal. For $\epsilon = 1/2$, for example, we achieve $O(\log H)$ query time, as desired, and our space usage is $O(n) + o(U)$. This dependency on $U$, while sublinear, is still undesirable. In the next section, we will see how to reduce this further.

## 2.2 Using $O(n + 2^{\log^\epsilon U})$ Space

To improve the space used by the data structure described in Theorem 1, one observation is that we can more carefully select the threshold for "large probabilities" that we place in the hash table. Instead of $(1/U)^\epsilon$, we can use $(1/2)^{\log^\epsilon U}$ for some $0 < \epsilon < 1$. The space used by the hash table is thus $O(2^{\log^\epsilon U})$, which is $o(U^\epsilon)$ for any $\epsilon > 0$. The analysis of the expected query times carries through as follows

$$\sum_{i \in T} p_i O(1) + \sum_{i \in \mathcal{U} \setminus T} p_i O(\log \log U) = O(1) + \sum_{i \in \mathcal{U} \setminus T} p_i O(\log \log U)$$

$$= O(1) + \sum_{i \in \mathcal{U} \setminus T} p_i \epsilon (1/\epsilon) O(\log \log U)$$

$$= O(1) + \sum_{i \in \mathcal{U} \setminus T} p_i (1/\epsilon) O(\log ((\log U)^\epsilon))$$

$$= O(1) + \sum_{i \in \mathcal{U} \setminus T} p_i (1/\epsilon) O\left(\log \log \left(2^{\log^\epsilon U}\right)\right)$$

$$\leq O(1) + \sum_{i \in \mathcal{U} \setminus T} p_i (1/\epsilon) O(\log \log(1/p_i))$$

$$\leq O(1) + (1/\epsilon) \sum_{i \in \mathcal{U} \setminus T} p_i O(\log \log(1/p_i))$$

$$\leq O((1/\epsilon) \log H)$$

**Theorem 2.** *Given a probability distribution with entropy $H$ over the possible queries in a universe of size $U$, it is possible to construct a data structure that performs predecessor searches in expected time $O((1/\epsilon) \log H)$ using $O(n + 2^{\log^\epsilon U})$ space for any $0 < \epsilon < 1$.*

### 2.3 Alternate Solution

Results similar to those of Theorem 2 can be achieved through other means, albeit with slightly higher space requirements. However, as we shall see, this variant has an interesting property that the previous data structures lack. Recall from Section 1.1 the predecessor search data structure that achieves $O(\log \log \Delta)$ expected[2] query time, where $\Delta$ is the distance between the query element and the element returned [7]. We add *guide points* to $S$, which have their predecessors (in the original set $S$) precomputed, to ensure that the distance between $i$ and its predecessor is at most $\Delta_i \leq (1/p_i)^{\log^{(1/\epsilon)-1} 1/p_i}$ for some $1/2 < \epsilon < 1$.

---

[2] In this instance, the expectation is taken over random choices made by the algorithm, not by the distribution of queries.

To perform a search, we query the data structure. If we happen to be returned a guide point, then we use its precomputed predecessor to answer the query. We then obtain the following expected running time

$$\sum_{i \in \mathcal{U}} p_i O(\log \log \Delta_i) \leq \sum_{i \in \mathcal{U}} p_i O\left(\log \log(1/p_i)^{\log^{(1/\epsilon)-1}(1/p_i)}\right)$$

$$= \sum_{i \in \mathcal{U}} p_i O\left(\log \left(\log(1/p_i) \log^{(1/\epsilon - 1)}(1/p_i)\right)\right)$$

$$= \sum_{i \in \mathcal{U}} p_i O\left(\log \log^{1/\epsilon} 1/p_i\right)$$

$$= \sum_{i \in \mathcal{U}} p_i O((1/\epsilon) \log \log 1/p_i)$$

$$= \sum_{i \in \mathcal{U}} (1/\epsilon) p_i O(\log \log 1/p_i)$$

$$= (1/\epsilon) \sum_{i \in \mathcal{U}} p_i O(\log \log 1/p_i)$$

$$\leq O((1/\epsilon) \log H)$$

It remains to determine the amount of space required for this structure. The predecessor search data structure uses space $O(n \log \log \log U)$. However, we must also store guide points, and so we now count the number of guide points we add. Let us consider how such points were added. For each element $i \in \mathcal{U}$, in order of the highest probability to lowest, we check to see if the predecessor of $i$ is within distance $(1/p_i)^{\log^{(1/\epsilon)-1} 1/p_i}$. Observe that the answer to this question is always "yes" if $(1/p_i)^{\log^{(1/\epsilon)-1} 1/p_i} \geq U$. We have

$$(1/p_i)^{\log^{(1/\epsilon)-1} 1/p_i} \geq U$$
$$\iff \log(1/p_i)^{\log^{(1/\epsilon)-1} 1/p_i} \geq \log U$$
$$\iff \left(\log^{(1/\epsilon)-1}(1/p_i)\right)(\log(1/p_i)) \geq \log U$$
$$\iff \log^{1/\epsilon}(1/p_i) \geq \log U$$
$$\iff 1/p_i \geq 2^{\log^\epsilon U}$$
$$\iff p_i \leq 1/2^{\log^\epsilon U}$$

Observe that there are at most two elements of $U$ with probability between 1 and 1/2, at most four elements with probability between 1/2 and 1/4, at most eight elements with probability between 1/4 and 1/8, and so on. In general, there are at most $2^{i+1}$ points with probability between $1/2^i$ and $1/2^{i+1}$. In the worst case, we always place a guide point. Therefore, we place at most $2^{i+1}$ guide points on elements with probabilities between $1/2^i$ and $1/2^{i+1}$, but never for elements with probability at most $1/2^{\log^\epsilon U}$. The number of guide points is thus $\sum_{j=0}^{\log^\epsilon U} 2^i = O(2^{\log^\epsilon U})$, for a total space requirement of $O((n + 2^{\log^\epsilon U}) \log \log \log U)$.

This result has an extra factor of $O(\log \log \log U)$ in the space requirements, but has one interesting property that the structure in Theorem 2 lacks. Observe that, for this data structure, an individual query for element $i$ can be executed in time $O((1/\epsilon)\log\log 1/p_i)$ time. This is in contrast to Theorem 2, where an individual query takes either $\Theta(1)$ or $\Theta(\log\log U)$ time. As a result, this technique can be used to support arbitrarily weighted elements in $\mathcal{U}$. Suppose each element $i \in \mathcal{U}$ has a real-valued weight $w_i > 0$ and let $W = \sum_{i=0}^{U-1} w_i$. By assigning each element probability $p_i = w_i/W$, we achieve an expected query time of $O((1/\epsilon)\log\log(W/w_i))$, which is analogous to the $O(\log W/w_i)$ query time of biased search trees [5].

**Theorem 3.** *Given a positive real weight $w_i$ for each element $i$ in a universe of size $U$, such that the sum of all weights is $W$, it is possible to construct a data structure that performs a predecessor search for item $i$ in expected time $O((1/\epsilon)\log\log(W/w_i))$ using $O\big((n + 2^{\log^\epsilon U})\log\log\log U\big)$ space for any $1/2 < \epsilon < 1$.*

## 3 Supporting $O(n)$ Space

In this section, we describe how to achieve space $O(n)$ by accepting a larger query time $O\big(\sqrt{H}\big)$. We begin with a brief note concerning input entropy vs. output entropy.

*Input vs. Output Distribution.* Until now, we have discussed the *input* distribution, *i.e.*, the probability that $i \in \mathcal{U}$ is the *query*. We could also discuss the *output* distribution, *i.e.*, the probability that $i \in \mathcal{U}$ is the *answer* to the query. This distribution can be defined by letting $p_i^* = 0$ if $i \notin S$ and $p_i^* = \sum_{j=s_i}^{s_{i+1}-1} p_j$ otherwise.

Suppose we can answer a predecessor query for $i$ in time $O\big(\log\log 1/p_{pred(i)}^*\big)$ where $pred(i)$ is the predecessor of $i$. Then the expected query time is

$$\sum_{i \in \mathcal{U}} p_i O\big(\log\log 1/p_{pred(i)}^*\big)$$

Since $p_i \leq p_{pred(i)}^*$ for all $i$, this is at most $\sum_{i \in \mathcal{U}} p_i \log\log 1/p_i$, *i.e.*, the entropy of the input distribution. It therefore suffices to consider the output distribution.

Our data structure will use a series of data structures for predecessor search [3] that increase doubly-exponentially in size in much the same way as the working-set structure [2]. Recall from Section 1.1 that there exists a linear space data structure that is able to execute predecessor search queries in time $O\Big(\min\Big\{\frac{\log\log n \cdot \log\log U}{\log\log\log U}, \sqrt{\frac{\log n}{\log\log n}}\Big\}\Big)$ [3]. We will maintain several such structures $D_1, D_2, \ldots$, where $D_j$ is over the entire universe $\mathcal{U}$ but contains only $2^{2^j}$ elements. In particular, $D_j$ contains the $2^{2^j}$ elements of highest probability that are not contained in any $D_k$ for $k < j$. Note that here, "highest probability" refers to the highest *output* probability.

Searches are performed by doing a predecessor search in each of $D_1, D_2, \ldots$. Along with each element we also store its successor. When we receive the predecessor of the query in $D_j$, we check its successor to see if that successor is larger than the query. If so, the predecessor in $D_j$ is the answer to the query. Otherwise, the real predecessor is somewhere between the predecessor in $D_j$ and the query, and can be found by continuing the search; this technique is already known [6].

We now consider the search time in this data structure. Suppose the correct predecessor of the query $i$ is found in $D_j$ where $j > 1$ (otherwise, the predecessor was found in $D_1$ in $O(1)$ time). All $2^{2^{j-1}}$ elements of $D_{j-1}$ have (output) probability greater than $p^*_{pred(i)}$, and so $p^*_{pred(i)} \leq 1/2^{2^{j-1}}$. Equivalently, $j$ is $O\left(\log \log 1/p^*_{pred(i)}\right)$. The total time spent searching is bounded by

$$\sum_{k=1}^{j} \sqrt{\frac{\log 2^{2^k}}{\log \log 2^{2^k}}} \leq O\left(\sqrt{\frac{2^j}{j}}\right) \leq O\left(\sqrt{\log 1/p^*_{pred(i)}}\right)$$

Therefore, since $p_i \leq p^*_{pred(i)}$ for all $i$, the expected query time is

$$\sum_{i \in \mathcal{U}} p_i \sqrt{\log 1/p^*_{pred(i)}} \leq \sum_{i \in \mathcal{U}} p_i \sqrt{\log 1/p_i} \leq \sqrt{H}$$

The final step above follows from Jensen's inequality. To determine the space used by this data structure, observe that every element stored in $S$ is stored in exactly one $D_j$. Since each $D_j$ uses space linear in the number of elements stored in it, the total space usage is $O(n)$.

**Theorem 4.** *Given a probability distribution with entropy $H$ over the possible queries in a universe of size $U$, it is possible to construct a data structure that performs predecessor searches in expected time $O\left(\sqrt{H}\right)$ using $O(n)$ space.*

Observe that we need not know the exact distribution $D$ to achieve the result of Theorem 4; it suffices to know the sorted order of the keys in terms of non-increasing probabilities.

Furthermore, since the predecessor search structure used is in fact dynamic [3], we can even obtain a bound similar to the working-set property: a predecessor search for item $i$ can be answered in time $O\left(\sqrt{\log w(i)}\right)$ where $w(i)$ is the number of distinct predecessors reported since the last time the predecessor of $i$ was reported. This can be accomplished using known techniques [2], similar to the data structure of Theorem 4, except instead of ordering the elements by their probabilities, we order them in increasing order of their working-set numbers $w(i)$. Whenever an element is reported, we place it in the first substructure and shift elements from the end to fill the space created, just as in the working-set structure [2]. This result also uses $O(n)$ space.

**Theorem 5.** *Let $w(i)$ denote the number of distinct predecessors reported since the last time the predecessor of $i$ was reported, or $n$ if the predecessor of $i$ has not yet been reported. It is possible to construct a data structure that performs a predecessor search for item $i$ in time $O\left(\sqrt{\log w(i)}\right)$ using $O(n)$ space.*

## 4 Conclusion

In this paper, we have introduced the idea of biased predecessor search in bounded universes. Two different categories of data structures were considered: one with query times that are logarithmic in the entropy of the query distribution (with space that is a function of $U$), and one with linear space (with query times larger than logarithmic in the entropy). We also considered the cases of general weights and of query times related to the working-set number.

Our results leave open several possible directions for future research:

1. Is it possible to achieve $O(\log H)$ query time and $O(n)$ space?
2. The reason for desiring a $O(\log H)$ query time comes from the fact that $H \leq \log U$ and the fact that the usual data structures for predecessor searching have query time $O(\log \log U)$. Of course, this is not optimal: other results have since improved this upper bound [3,14]. Is it possible to achieve a query time of, for example, $O(\log H / \log \log U)$?
3. What lower bounds can be stated in terms of either the input or output entropies? Clearly $O(U)$ space suffices for $O(1)$ query time, and so such lower bounds must place restrictions on space usage.

## References

1. Andersson, A., Thorup, M.: Dynamic ordered sets with exponential search trees. Journal of the ACM 54(3), Article 13 (2007)
2. Bădoiu, M., Cole, R., Demaine, E.D., Iacono, J.: A unified access bound on comparison-based dynamic dictionaries. Theoretical Computer Science 382(2), 86–96 (2007)
3. Beame, P., Fich, F.E.: Optimal bounds for the predecessor problem and related problems. Journal of Computer and System Sciences 65(1), 38–72 (2002)
4. Belazzougui, D., Kaporis, A.C., Spirakis, P.G.: Random input helps searching predecessors. arXiv:1104.4353 (2011)
5. Bent, S.W., Sleator, D.D., Tarjan, R.E.: Biased search trees. SIAM Journal on Computing 14(3), 545–568 (1985)
6. Bose, P., Howat, J., Morin, P.: A distribution-sensitive dictionary with low space overhead. In: Dehne, F., Gavrilova, M., Sack, J.-R., Tóth, C.D. (eds.) WADS 2009. LNCS, vol. 5664, pp. 110–118. Springer, Heidelberg (2009)
7. Bose, P., Douïeb, K., Dujmović, V., Howat, J., Morin, P.: Fast local searches and updates in bounded universes. In: Proceedings of the 22nd Canadian Conference on Computational Geometry (CCCG 2010), pp. 261–264 (2010)
8. Brodal, G.S., Makris, C., Sioutas, S., Tsakalidis, A., Tsichlas, K.: Optimal solutions for the temporal precedence problem. Algorithmica 33(4), 494–510 (2002)

9. Johnson, D.B.: A priority queue in which initialization and queue operations take $O(\log \log D)$ time. Theory of Computing Systems 15(1), 295–309 (1981)
10. Kaporis, A.C., Makris, C., Sioutas, S., Tsakalidis, A., Tsichlas, K., Zaroliagis, C.: Improved bounds for finger search on a RAM. In: Di Battista, G., Zwick, U. (eds.) ESA 2003. LNCS, vol. 2832, pp. 325–336. Springer, Heidelberg (2003)
11. Knuth, D.E.: Optimum binary search trees. Acta Informatica 1(1), 14–25 (1971)
12. Mehlhorn, K.: Nearly optimal binary search trees. Acta Informatica 5(4), 287–295 (1975)
13. Mehlhorn, K., Näher, S.: Bounded ordered dictionaries in $O(\log \log N)$ time and $O(n)$ space. Information Processing Letters 35(4), 183–189 (1990)
14. Pătraşcu, M., Thorup, M.: Time-space trade-offs for predecessor search. In: STOC 2006: Proceedings of the 38th Annual ACM Symposium on Theory of Computing, pp. 232–240 (2006)
15. Sleator, D.D., Tarjan, R.E.: Self-adjusting binary search trees. Journal of the ACM 32(3), 652–686 (1985)
16. van Emde Boas, P.: Preserving order in a forest in less than logarithmic time and linear space. Information Processing Letters 6(3), 80–82 (1977)
17. Willard, D.E.: Log-logarithmic worst-case range queries are possible in space $\Theta(N)$. Information Processing Letters 17(2), 81–84 (1983)

# Author Index

Ackerman, Eyal 478
Adamaszek, Anna 683
Ahn, Hee-Kap 273
Alam, Muhammad Jawaherul 144
Albenque, Marie 421
Allen, Michelle M. 478
Allen, Peter 355
Amir, Amihood 586
Antoniadis, Antonios 610
Ayala-Rincón, Mauricio 202

Bae, Sang Won 120
Bansal, Nikhil 47, 550
Barash, Mikhail 190
Barba, Luis 84
Barcelo, Neal 610
Barequet, Gill 478
Barth, Lukas 514
Basset, Nicolas 502
Bekos, Michael A. 144
Benevides, Fabrício 433
Bereczky, Nikolett 743
Bhattacharya, Binay 330
Bose, Prosenjit 84, 108, 755
Böttcher, Julia 355
Bueno, Letícia Rodrigues 379

Campos, Victor 433
Castelli Aleardi, Luca 168
Chakraborty, Sourav 306
Chalermsook, Parinya 409
Cheilaris, Panagiotis 96
Clément, Julien 442
Cole, Daniel 610
Correa, José R. 35
Czyzowicz, Jurek 342, 622

De Campos Mesquita, Felipe 379
de Panafieu, Élie 454
Dereniowski, Dariusz 342
Di Giacomo, Emilio 132
Dias, Zanoni 671
Díaz, Josep 695
Diekert, Volker 1

Dobrev, Stefan 622
Dourado, Mitre 433
Duan, Ran 285
Duch, Amalia 743
Duchon, Philippe 367
Durocher, Stephane 156, 294
Duvignau, Romaric 367

Englert, Matthias 318

Fabrikant, Sara Irina 514
Fagerberg, Rolf 755
Farach-Colton, Martín 250
Felsner, Stefan 156
Fernandes, Cristina G. 707
Feuilloley, Laurent 35
Ficler, Jessica 586
Figueiredo, Celina M.H. 13
Filho, Helio B. Macêdo 13
Filtser, Omrit 294
Foucaud, Florent 526
Fox, Kyle 610
Fraser, Robert 294
Fürer, Martin 72, 660
Fusy, Éric 168

Gagie, Travis 731
Gardy, Danièle 454
Gąsieniec, Leszek 342
Gawrychowski, Paweł 731
Genitrini, Antoine 466
Giambruno, Laura 442
Giotis, Ioannis 695
Gittenberger, Bernhard 454
Griffiths, Simon 433

Haeusler, Edward Hermann 202
Hàn, Hiệp 355
Hausen, Rodrigo De Alencar 379
Hell, Pavol 538
Herlihy, Maurice 214
Hitczenko, Paweł 490
Horel, Thibaut 719
Howat, John 755
Huang, Shenwei 538

Ioannidis, Stratis  719
Ishii, Toshimasa  238

Janson, Svante  490

Kameda, Tsunehiko  330
Kärkkäinen, Juha  731
Kaufmann, Michael  144
Kesh, Deepanjan  648
Khramtcova, Elena  96
Kim, Sang-Sub  273
Kindermann, Philipp  144
Kirousis, Lefteris  695
Klasing, Ralf  342
Klein, Rolf  261
Klimann, Ines  180
Knauer, Kolja  421
Kobourov, Stephen G.  144, 514
Kohayakawa, Yoshiharu  355
Korman, Matias  120
Kosowski, Adrian  342
Kostrygin, Anatolii  168
Kranakis, Evangelos  622
Krauthgamer, Robert  586
Kuba, Markus  454

Laekhanukit, Bundit  409
Lampis, Michael  24
Langerman, Stefan  84, 96
Lee, Orlando  574
Levcopoulos, Christos  261
Lin, Min Chih  399
Lingas, Andrzej  261
Lintzmayer, Carla Negri  671
Liotta, Giuseppe  132
Löffler, Maarten  478
Lubiw, Anna  514

Machado, Raphael C.S.  13
Mailler, Cécile  466
Markakis, Evangelos  695
Matsakis, Nicolaos  318
McDiarmid, Colin  391
Mehrabi, Ali D.  294
Mehrabi, Saeed  156, 294
Mehta, Shashank K.  648
Mermelstein, Joshua  478
Mitsou, Valia  24
Mizrahi, Michel J.  399
Mondal, Debajyoti  156
Montecchiani, Fabrizio  132

Morin, Pat  755
Morris, Robert  433
Moseley, Benjamin  610
Mucha, Marcin  318
Muthukrishnan, S.  719
Myasnikov, Alexei G.  1

Nanongkai, Danupon  409
Naserasr, Reza  526
Nekrich, Yakov  731
Németh, Krisztián  743
Nöllenburg, Martin  514
Nugent, Michael  610

Okamoto, Yoshio  120, 514
Okhotin, Alexander  190
Ono, Hirotaka  238

Pacheco, Eduardo  622
Pająk, Dominik  342
Papadopoulou, Evanthia  96
Pedrosa, Lehilton L.C.  562
Perrot, Kévin  634
Person, Yury  355
Picantin, Matthieu  180
Poloczek, Matthias  598
Popa, Alexandru  683
Pratap, Rameshwar  306
Pruhs, Kirk  610
Puglisi, Simon J.  731
Pupyrev, Sergey  514

Rajsbaum, Sergio  214
Raynal, Michel  214
Rémila, Éric  634
Rey, Anja  60
Roditty, Liam  586
Rothe, Jörg  60
Roura, Salvador  743
Roy, Sasanka  306
Rutten, Cyriel  550

Sampaio, Leonardo  433
San Felice, Mário César  574
Saraf, Shubhangi  306
Sar Shalom, Oren  586
Schouery, Rafael C.S.  707
Serna, Maria  695
Shinn, Tong-Wook  226
Silva, Ana  433
Song, Zhao  330

## Author Index

Soto, José A.    35
Souvaine, Diane L.    478
Squarcella, Claudio    514
Stainer, Julien    214
Sviridenko, Maxim    562
Szwarcfiter, Jayme L.    399

Takaoka, Tadao    226
Tóth, Csaba D.    478
Tsai, Meng-Tsung    250

Ueckerdt, Torsten    514
Uno, Yushi    238

van der Ster, Suzanne    550
van der Zwaan, Ruben    47, 550
van Renssen, André    108
van Zuylen, Anke    598
Vredeveld, Tjark    47, 550

Wang, Haitao    120
Weiß, Armin    1
Weller, Kerstin    391
Williamson, David P.    574, 598
Wolff, Alexander    144, 514

MIX
Papier aus verantwortungsvollen Quellen
Paper from responsible sources
FSC® C105338

If you have any concerns about our products,
you can contact us on
**ProductSafety@springernature.com**

In case Publisher is established outside the EU,
the EU authorized representative is:
**Springer Nature Customer Service Center GmbH
Europaplatz 3, 69115 Heidelberg, Germany**

Printed by Libri Plureos GmbH
in Hamburg, Germany